Handbook of Dynamic System Modeling

CHAPMAN & HALL/CRC
COMPUTER and INFORMATION SCIENCE SERIES

Series Editor: Sartaj Sahni

PUBLISHED TITLES

ADVERSARIAL REASONING: COMPUTATIONAL APPROACHES TO READING THE OPPONENT'S MIND
Alexander Kott and William M. McEneaney

DISTRIBUTED SENSOR NETWORKS
S. Sitharama Iyengar and Richard R. Brooks

DISTRIBUTED SYSTEMS: AN ALGORITHMIC APPROACH
Sukumar Ghosh

FUNDEMENTALS OF NATURAL COMPUTING: BASIC CONCEPTS, ALGORITHMS, AND APPLICATIONS
Leandro Nunes de Castro

HANDBOOK OF ALGORITHMS FOR WIRELESS NETWORKING AND MOBILE COMPUTING
Azzedine Boukerche

HANDBOOK OF APPROXIMATION ALGORITHMS AND METAHEURISTICS
Teofilo F. Gonzalez

HANDBOOK OF BIOINSPIRED ALGORITHMS AND APPLICATIONS
Stephan Olariu and Albert Y. Zomaya

HANDBOOK OF COMPUTATIONAL MOLECULAR BIOLOGY
Srinivas Aluru

HANDBOOK OF DATA STRUCTURES AND APPLICATIONS
Dinesh P. Mehta and Sartaj Sahni

HANDBOOK OF DYNAMIC SYSTEM MODELING
Paul A. Fishwick

HANDBOOK OF SCHEDULING: ALGORITHMS, MODELS, AND PERFORMANCE ANALYSIS
Joseph Y.-T. Leung

THE PRACTICAL HANDBOOK OF INTERNET COMPUTING
Munindar P. Singh

SCALABLE AND SECURE INTERNET SERVICES AND ARCHITECTURE
Cheng-Zhong Xu

SPECULATIVE EXECUTION IN HIGH PERFORMANCE COMPUTER ARCHITECTURES
David Kaeli and Pen-Chung Yew

Handbook of Dynamic System Modeling

Edited by
Paul A. Fishwick
University of Florida
Gainesville, U.S.A.

 CRC Press
Taylor & Francis Group
Boca Raton London New York

CRC Press is an imprint of the
Taylor & Francis Group, an **informa** business

A CHAPMAN & HALL BOOK

Chapman & Hall/CRC
Taylor & Francis Group
6000 Broken Sound Parkway NW, Suite 300
Boca Raton, FL 33487-2742

© 2007 by Taylor & Francis Group, LLC
Chapman & Hall/CRC is an imprint of Taylor & Francis Group, an Informa business

No claim to original U.S. Government works
Printed in the United States of America on acid-free paper
10 9 8 7 6 5 4 3 2 1

International Standard Book Number-10: 1-58488-565-3 (Hardcover)
International Standard Book Number-13: 978-1-58488-565-8 (Hardcover)

This book contains information obtained from authentic and highly regarded sources. Reprinted material is quoted with permission, and sources are indicated. A wide variety of references are listed. Reasonable efforts have been made to publish reliable data and information, but the author and the publisher cannot assume responsibility for the validity of all materials or for the consequences of their use.

No part of this book may be reprinted, reproduced, transmitted, or utilized in any form by any electronic, mechanical, or other means, now known or hereafter invented, including photocopying, microfilming, and recording, or in any information storage or retrieval system, without written permission from the publishers.

For permission to photocopy or use material electronically from this work, please access www.copyright.com (http://www.copyright.com/) or contact the Copyright Clearance Center, Inc. (CCC) 222 Rosewood Drive, Danvers, MA 01923, 978-750-8400. CCC is a not-for-profit organization that provides licenses and registration for a variety of users. For organizations that have been granted a photocopy license by the CCC, a separate system of payment has been arranged.

Trademark Notice: Product or corporate names may be trademarks or registered trademarks, and are used only for identification and explanation without intent to infringe.

Library of Congress Cataloging-in-Publication Data

Handbook of dynamic system modeling / editor, Paul A. Fishwick.
    p. cm. -- (Chapman & Hall/CRC computer & information science series)
  Includes bibliographical references and index.
  ISBN-13: 978-1-58488-565-8 (alk. paper)
  ISBN-10: 1-58488-565-3 (alk. paper)
  1. Computer simulation. 2. System design. I. Fishwick, Paul A. II. Title. III. Series.

QA76.9.C65H345 2007
003'.3--dc22
                                                      2007000718

Visit the Taylor & Francis Web site at
http://www.taylorandfrancis.com

and the CRC Press Web site at
http://www.crcpress.com

Contents

Abstract xiii
Intended Audience xv
Preface xvii
Editor xix
Contributors xxi

I Introduction

1 The Languages of Dynamic System Modeling *Paul A. Fishwick* 1-1
   1.1 Introduction 1-1
   1.2 Dynamic System Modeling Examples 1-2
   1.3 Taxonomic Approaches 1-3
   1.4 Language 1-3
   1.5 Syntax 1-5
   1.6 Semantics 1-7
   1.7 Pragmatics 1-10
   1.8 Summary 1-10

2 The Dynamics of the Computational Modeling of Analogy-Making
   Robert M. French 2-1
   2.1 Introduction 2-1
   2.2 Analogy-Making as *Sameness* 2-2
   2.3 Analogy-Making as a Means of "Bootstrapping" Cognition 2-3
   2.4 The Necessity of Malleable Representations 2-3
   2.5 The Dynamics of Representation-Building in Analogy-Making 2-5
   2.6 Context-Dependent Computational Temperature 2-6
   2.7 Interaction between Top-Down and Bottom-Up Processes: An Example ... 2-6
   2.8 Computational Models Implementing this Bottom-Up/Top-Down Interaction ... 2-8
   2.9 Architectural Principles 2-8
   2.10 How this Type of Program Works: The Details 2-12
   2.11 How Tabletop Finds a Reasonable Solution 2-13
   2.12 The Issue of Scaling Up 2-16
   2.13 The Potential Long-Term Impact of the Mechanisms Presented 2-16
   2.14 Conclusions 2-16

3 Impact of the Semantic Web on Modeling and Simulation
John A. Miller, Congzhou He and Julia I. Couto **3**-1
- 3.1 Introduction **3**-1
- 3.2 Semantic Web: Relevant Issues **3**-2
- 3.3 Conceptual Basis for Discrete-Event Simulation **3**-4
- 3.4 Types of Mathematical Models **3**-6
- 3.5 Adding Semantics to Simulation Models **3**-10
- 3.6 Overview of DeSO **3**-12
- 3.7 Overview of DeMO **3**-14
- 3.8 Summary **3**-18

4 Systems Engineering *Andrew P. Sage* **4**-1
- 4.1 Introduction **4**-1
- 4.2 Systems Engineering **4**-2
- 4.3 The Importance of Technical Direction and Systems Management **4**-4
- 4.4 Other Parts of the Story **4**-8
- 4.5 Summary **4**-9

5 Basic Elements of Mathematical Modeling *Clive L. Dym* **5**-1
- 5.1 Principles of Mathematical Modeling **5**-2
- 5.2 Dimensional Consistency and Dimensional Analysis **5**-3
- 5.3 Abstraction and Scale **5**-9
- 5.4 Conservation and Balance Principles **5**-17
- 5.5 The Role of Linearity **5**-19
- 5.6 Conclusions **5**-20

6 DEVS Formalism for Modeling of Discrete-Event Systems *Tag Gon Kim* **6**-1
- 6.1 Introduction **6**-1
- 6.2 System-Theoretic DES Modeling **6**-3
- 6.3 DEVS Formalism for DES Modeling **6**-3
- 6.4 DES Analysis with DEVS Model **6**-7
- 6.5 Simulation of DEVS Model **6**-10
- 6.6 Conclusion **6**-12

II Modeling Methodologies

7 Domain-Specific Modeling *Jeff Gray, Juha-Pekka Tolvanen, Steven Kelly, Aniruddha Gokhale, Sandeep Neema and Jonathan Sprinkle* **7**-1
- 7.1 Introduction **7**-1
- 7.2 Essential Components of a Domain-Specific Modeling Environment **7**-2
- 7.3 Case Studies in DSM **7**-6
- 7.4 Overview of Supporting Tools **7**-17
- 7.5 Conclusion **7**-18

8 Agent-Oriented Modeling in Simulation: Agents for Modeling, and Modeling for Agents *Adelinde M. Uhrmacher and Mathias Röhl* **8**-1
- 8.1 Introduction **8**-1
- 8.2 Agents for Modeling in Simulation **8**-3

	8.3	Modeling and Simulation for Agents	8-6
	8.4	Conclusion	8-10

9 Distributed Modeling   *Simon J. E. Taylor* ... 9-1
- 9.1 Introduction ... 9-1
- 9.2 Modeling with COTS Simulation Packages ... 9-2
- 9.3 Distributed Simulation ... 9-3
- 9.4 CSP-Based Distributed Simulation ... 9-5
- 9.5 A Standards-Based Approach ... 9-7
- 9.6 Case Study ... 9-13
- 9.7 Conclusion ... 9-16

10 Model Execution   *Kalyan S. Perumalla* ... 10-1
- 10.1 Introduction ... 10-1
- 10.2 Time-Stepped Execution ... 10-5
- 10.3 Discrete-Event Execution ... 10-7
- 10.4 Summary ... 10-13

11 Discrete-Event Simulation of Continuous Systems   *James Nutaro* ... 11-1
- 11.1 Introduction ... 11-1
- 11.2 Simulating a Single Ordinary Differential Equation ... 11-2
- 11.3 Simulating Coupled Ordinary Differential Equations ... 11-6
- 11.4 DEVS Representation of Discrete-Event Integrators ... 11-8
- 11.5 The Heat Equation ... 11-13
- 11.6 Conservation Laws ... 11-16
- 11.7 Two-Point Integration Schemes ... 11-19
- 11.8 Conclusions ... 11-21

III   Multiobject and System

12 Toward a Multimodel Hierarchy to Support Multiscale Simulation
Mark S. Shephard, E. Seegyoung Seol and Benjamin FrantzDale ... 12-1
- 12.1 Introduction ... 12-1
- 12.2 Functional and Information Hierarchies in Multiscale Simulation ... 12-4
- 12.3 Constructing a Multimodel: Design of Functional Components to Support Multiscale Simulations ... 12-10
- 12.4 Example Multimodel Simulation Procedures ... 12-14
- 12.5 Closing Remarks ... 12-15

13 Finite Elements   *Marc Hoit and Gary Consolazio* ... 13-1
- 13.1 Finite Element Theory ... 13-1
- 13.2 Membrane Elements ... 13-9
- 13.3 Flat Plate and Shell Elements ... 13-12
- 13.4 Solid Elements ... 13-15
- 13.5 Dynamics ... 13-16
- 13.6 Summary ... 13-21

14 Multimodeling *Minho Park, Paul A. Fishwick and Jinho Lee* 14-1
- 14.1 Introduction 14-1
- 14.2 Scene Construction 14-5
- 14.3 Multimodeling Exchange Language (MXL) 14-11
- 14.4 Dynamic Exchange Language (DXL) 14-13
- 14.5 A Boiling Water Example 14-20
- 14.6 Conclusion 14-27

15 Hybrid Dynamic Systems: Modeling and Execution *Pieter J. Mosterman* 15-1
- 15.1 Introduction 15-1
- 15.2 Hybrid Dynamic Systems 15-3
- 15.3 Hybrid Dynamic System Behaviors 15-9
- 15.4 An Implementation 15-12
- 15.5 Advanced Topics in Hybrid Dynamic System Simulation 15-17
- 15.6 Pathological Behavior Classes 15-22
- 15.7 Conclusions 15-23

16 Theory and Practice for Simulation Interconnection: Interoperability and Composability in Defense Simulation *Ernest H. Page* 16-1
- 16.1 Introduction 16-1
- 16.2 The Practice of Simulation Interconnection—Simulation Interoperability 16-2
- 16.3 The Theory of Simulation Interconnection—Simulation Composability 16-6
- 16.4 Conclusions 16-9

IV Model Types

17 Ordinary Differential Equations *Francisco Esquembre and Wolfgang Christian* 17-1
- 17.1 Introduction 17-1
- 17.2 Numerical Solution 17-3
- 17.3 Taylor Methods 17-5
- 17.4 Runge–Kutta Methods 17-7
- 17.5 Implementation 17-8
- 17.6 Adaptive Step 17-11
- 17.7 Implementation of Adaptive Step 17-12
- 17.8 Performance and Other Methods 17-15
- 17.9 State Events 17-19
- 17.10 The OSP Library 17-20

18 Difference Equations as Discrete Dynamical Systems *Hassan Sedaghat* 18-1
- 18.1 Introduction 18-1
- 18.2 Basic Concepts 18-2
- 18.3 First-Order Difference Equations 18-4
- 18.4 Higher Order Difference Equations 18-8

19 Process Algebra *J.C.M. Baeten, D.A. van Beek and J.E. Rooda* 19-1
- 19.1 Introduction 19-1
- 19.2 Syntax and Informal Semantics of the χ Process Algebra 19-4
- 19.3 Algebraic Reasoning and Verification 19-11
- 19.4 Conclusions 19-19

20 Temporal Logic *Antony Galton* 20-1
- 20.1 Propositional Logic 20-1
- 20.2 Introducing Temporal Logic 20-3
- 20.3 Syntax and Semantics 20-5
- 20.4 Models of Time 20-6
- 20.5 Further Extensions to the Formal Language 20-11
- 20.6 Illustrative Examples 20-11
- 20.7 Conclusion 20-14
- 20.8 Further Reading 20-14

21 Modeling Dynamic Systems with Cellular Automata
Peter M.A. Sloot and Alfons G. Hoekstra 21-1
- 21.1 Introduction 21-1
- 21.2 A Bit of History 21-2
- 21.3 Cellular Automata to Model Dynamical Systems 21-3
- 21.4 One-Dimensional CAs 21-3
- 21.5 Lattice Gas Cellular Automata Models of Fluid Dynamics 21-5

22 Spatio-Temporal Connectionist Networks *Stefan C. Kremer* 22-1
- 22.1 Introduction 22-1
- 22.2 Connectionist Networks (CNs) 22-2
- 22.3 Spatio-Temporal Connectionist Networks 22-4
- 22.4 Representational Power 22-6
- 22.5 Learning 22-6
- 22.6 Applications 22-8
- 22.7 Conclusion 22-9

23 Modeling Causality with Event Relationship Graphs *Lee Schruben* 23-1
- 23.1 Introduction 23-1
- 23.2 Background and Definitions 23-2
- 23.3 Enrichments to Event Relations Graphs 23-7
- 23.4 Relationships to Other Discrete-Event System Modeling Methods 23-10
- 23.5 Simulation of Event Relationship Graphs 23-16
- 23.6 Event Relationship Graph Analysis 23-16
- 23.7 Experimenting with ERGs 23-17

24 Petri Nets for Dynamic Event-Driven System Modeling *Jiacun Wang* 24-1
- 24.1 Introduction 24-1
- 24.2 Petri Net Definition 24-1
- 24.3 Transition Firing 24-3
- 24.4 Modeling Power 24-4
- 24.5 Petri Net Properties 24-5
- 24.6 Analysis of Petri Nets 24-7
- 24.7 Colored Petri Nets 24-10
- 24.8 Timed Petri Nets 24-12
- 24.9 Concluding Remark 24-16

25 Queueing System Models *Christos G. Cassandras* 25-1
- 25.1 Introduction 25-1
- 25.2 Specification of Queueing System Models 25-2

	25.3	Performance of a Queueing System	25-4
	25.4	Queueing System Dynamics	25-6
	25.5	Little's Law	25-7
	25.6	Simple Markovian Queueing Models	25-8
	25.7	Markovian Queueing Networks	25-11
	25.8	Non-Markovian Queueing Systems	25-17

26 Port-Based Modeling of Engineering Systems in Terms of Bond Graphs
Peter Breedveld 26-1

26.1	Introduction	26-1
26.2	Structured Systems: Physical Components and Interaction	26-4
26.3	Bond Graphs	26-5
26.4	Multiport Generalizations	26-21
26.5	Conclusion	26-28

27 System Dynamics Modeling of Environmental Systems *Andrew Ford* 27-1

27.1	Introductory Examples	27-1
27.2	Comparison of the Flowers and Sales Models	27-4
27.3	Background on Daisy World	27-6
27.4	The Daisy World Model	27-6
27.5	The Daisy World Management Flight Simulator	27-9

28 Dynamic Simulation with Energy Systems Language *Clay L. Montague* 28-1

28.1	Introduction	28-1
28.2	Reading an Energy Systems Language Diagram	28-4
28.3	Translating a Diagram to Dynamic Equations	28-8
28.4	Calibration of Model Constants	28-21
28.5	Preparation for Simulation	28-22
28.6	Dynamic Output of the Marsh Sector Model	28-27
28.7	A Brief Comparison with Forrester's Systems Dynamics Approach	28-29
28.8	Conclusions	28-31

29 Ecological Modeling and Simulation: From Historical Development to Individual-Based Modeling *David R.C. Hill and P. Coquillard* 29-1

29.1	Introduction	29-1
29.2	An Old Story?	29-2
29.3	Determinism or Probability?	29-5
29.4	Modeling Techniques	29-5
29.5	The Use of Models in Ecology	29-6
29.6	Models are Scientific Instruments	29-7
29.7	Levels of Organization and Methodological Choices	29-8
29.8	Individual-Based Models	29-9
29.9	Applications	29-12
29.10	Conclusion	29-15

30 Ontology-Based Simulation in Agriculture and Natural Resources
Howard Beck, Rohit Badal and Yunchul Jung 30-1

30.1	Introduction	30-1
30.2	Ways in Which Ontologies can be Applied to Simulation	30-2

Contents

30.3 How to Build an Ontology-Based Simulation— Bioprocessing Example **30**-6
30.4 Tools for Ontology-Based Simulation **30**-10
30.5 Conclusions **30**-12

31 Modeling Human Interaction in Organizational Systems
Stewart Robinson **31**-1
- 31.1 Introduction **31**-1
- 31.2 Systems and Human Interaction **31**-2
- 31.3 Why Model Human Interaction? **31**-3
- 31.4 Modeling Human Interaction: Research and Practice **31**-4
- 31.5 The KBI Methodology **31**-5
- 31.6 A Case Study: Modeling Human Decision Making at Ford Motor Company **31**-8
- 31.7 Conclusion **31**-12

32 Military Modeling *Roger Smith* **32**-1
- 32.1 Introduction **32**-1
- 32.2 Applications **32**-1
- 32.3 Representation **32**-2
- 32.4 Dynamics **32**-4
- 32.5 Modeling Approach **32**-8
- 32.6 Military Simulation Systems **32**-11
- 32.7 Conclusion **32**-12

33 Dynamic Modeling in Management Science *Michael Pidd* **33**-1
- 33.1 Introduction **33**-1
- 33.2 An Approach to Dynamic Systems Modeling in Management Science **33**-4
- 33.3 Discrete Event Simulation **33**-7
- 33.4 System Dynamics in Management Science **33**-16
- 33.5 Model Validation **33**-21
- 33.6 Chapter Summary **33**-22

34 Modeling and Analysis of Manufacturing Systems
E. Lefeber and J.E. Rooda **34**-1
- 34.1 Introduction **34**-1
- 34.2 Preliminaries **34**-2
- 34.3 Analytical Models for Steady-State Analysis **34**-3
- 34.4 Discrete-Event Models **34**-7
- 34.5 Effective Process Times **34**-8
- 34.6 Control of Manufacturing Systems: A Framework **34**-10
- 34.7 Standard Fluid Model and Extensions **34**-12
- 34.8 Flow Models **34**-16
- 34.9 Conclusions **34**-18

35 Sensor Network Component-Based Simulator
Boleslaw K. Szymanski and Gilbert Gang Chen **35**-1
- 35.1 The Need for a New Sensor Network Simulator **35**-1
- 35.2 Component Simulation Toolkit **35**-3
- 35.3 Wireless Sensor Network Simulation **35**-7
- 35.4 Conclusions **35**-15

V Case Studies

36 Multidomain Modeling with Modelica
Martin Otter, Hilding Elmqvist and Sven Erik Mattsson 36-1
    36.1   Modelica Overview 36-1
    36.2   Modelica Basics 36-3
    36.3   Modelica Libraries 36-17
    36.4   Symbolic Processing of Modelica Models 36-19
    36.5   Outlook 36-25

37 On Simulation of Simulink® Models for Model-Based Design
Rohit Shenoy, Brian McKay and Pieter J. Mosterman 37-1
    37.1   Introduction 37-1
    37.2   The Case Study Example 37-3
    37.3   Designing with Simulation 37-4
    37.4   Obtaining Computational Models 37-4
    37.5   The Robotic Arm Model 37-7
    37.6   Using Computational Models for Control Design 37-12
    37.7   Testing with Model-Based Design 37-16
    37.8   Conclusions 37-19

Index **I**-1

Abstract

A dynamic model is a model that describes how a system changes in time. In some cases, time is the only independent variable, and in others, there are additional variables such as those of spatial frames. Even though the concept of creating a dynamic model, capturing the dynamics of a system, is ubiquitous, the topic tends to be splintered across numerous disciplines, from mathematical modeling and computer simulation to more qualitative models for software design and science. Moreover, models have a variety of representations from the traditional notations of mathematics to diagrammatic, and even immersive representations. The purpose of this volume is to provide a text that brings together all of these expressions for dynamic models.

This book emphasizes presenting a "computer science slant" toward the **problems of model** design, representation, and analysis. As such, each chapter will be of a ***tutorial*** **or** ***survey*** nature, including mathematical descriptions, pseudocode, and diagrams wherever possible. The idea is to inform the reader as to how to use the model(s). A Web page with chapter author links will be provided on the CRC Web site.

The book is novel for the following key reasons:

- Most comprehensive books available on the topic of dynamic modeling—there have been many books on modeling—tend to be particular to a specific community, rather than cutting across interdisciplinary lines. This book spans numerous disciplines to educate the reader about the wide variety of modeling methods available for dynamic systems.
- It provides the readers with pseudocode, diagrams, and methods for representing, simulating, and analyzing models.
- It brings together theory, foundations, a catalog of model types, as well as applications that use them. In this sense, the book can be used as a reference guide or in the classroom as a primary text or as a secondary reference.

Intended Audience

This book is meant to be a comprehensive and complete reference guide for dynamic modeling. As such, we expect a very large audience from core disciplines (mathematics, computer science, and physics) to all science and engineering disciplines. The applications section will inform readers how to apply modeling methods to several areas.

Preface

Dynamic system modeling is defined as creating time-dependent models for physical systems. In particular, we are concerned with digital computer models rather than scale or analog models for dynamic phenomena. This subject matter can be placed, more broadly, within the overall topic of *modeling*. When we study a system, we build many sorts of models to describe it, including capturing salient system characteristics of geometry, information, knowledge, and dynamics. Therefore, dynamics is only one type of a model, but a vital one if we are to represent how a system behaves and changes state over time. Dynamic system modeling tends to be highly interdisciplinary. So, one is likely to see practitioners belonging to different areas from computer science and engineering to philosophy and the life sciences. The topic does not fit into a singular mold.

Another interesting aspect of dynamic modeling is the range of coverage from the very general to the very specific. One may build models that capture the dynamics over a wide set of domains. Other models are more targeted toward one specific domain. For example, a model defining the topological character of a digital circuit is classified within the domain of electrical engineering; however, a bond graph can also capture the dynamics of the digital circuit using a more general energy-based specification. The reader may wonder if it is better to be general or specific when modeling. An underlying thesis of this book is that one must maintain a balanced approach by understanding specific models on the one hand, while appreciating generalized dynamic knowledge on the other. This approach naturally results in *variety in modeling* in which handbooks excel. Model types have different presentations: some are text-based using symbols while others have associated diagrams. Also, all models have issues involving connectivity—how to connect components within models at the same level of abstraction, or how to capture multiple levels of abstraction or aggregation using varying scales.

This handbook is meant to be an aid to readers interested in dynamic models for the purpose of a quick "lookup" reference or a more in-depth study in the area of dynamic modeling as a discipline. The book can also be used within the classroom if the goal is to educate students on variety within model types and the practice of dynamic modeling. The handbook is organized through an introduction to dynamic models by way of general issues, representations, and philosophy. This introduction is followed by modeling methodologies defined as approaches to the modeling process. Modeling methodologies also contain the topic of how to execute models on a computer once those models have been designed. The "multiobject and system" section addresses the issues of scale, heterogeneity, and composition. The next section covers specific model types often characterized by specific visual or text-based grammars. The section on application domains is primarily concerned with models that tend to be tied to a specific discipline or a set of closely related subdisciplines. The handbook concludes with case studies using two well-known commercial packages that support dynamic model construction, simulation, and analysis.

I hope that the reader will obtain useful dynamic model types for an application of interest and will find that the handbook serves in the capacity of a useful reference whenever it is necessary to consider how systems function, and are represented, over time and space.

Paul A. Fishwick

Editor

Paul A. Fishwick is professor of Computer and Information Science and Engineering at the University of Florida. He received his BS in Mathematics from the Pennsylvania State University, MS in Applied Science from the College of William and Mary, and PhD in Computer and Information Science from the University of Pennsylvania in 1986. He has six years of industrial/government production and research experience working at Newport News Shipbuilding and Dry Dock Co. (doing CAD/CAM parts definition research) and at NASA Langley Research Center (studying engineering database models for structural engineering). His research interests are computer simulation modeling and analysis methods for complex systems. He is a Senior Member of the IEEE and a Fellow of the Society for Computer Simulation. He is also a member of the IEEE Society for Systems, Man and Cybernetics; ACM; and AAAI. Dr. Fishwick founded the comp.simulation Internet news group (Simulation Digest) in 1987, which has served numerous subscribers. He has chaired several workshops and conferences in the area of computer simulation, including General Chair of the 2000 Winter Simulation Conference. He was chairman of the IEEE Computer Society Technical Committee on Simulation (TCSIM) for two years (1988–1990) and has served on the editorial boards of several journals including the *ACM Transactions on Modeling and Computer Simulation*; *IEEE Transactions on Systems, Man and Cybernetics*; *Transactions of the Society for Computer Simulation*; *International Journal of Computer Simulation*; and *Journal of Systems Engineering*. He has delivered 12 international keynote addresses at major conferences relating to simulation. He has published over 180 technical papers, authored one textbook, coedited two Springer-Verlag volumes in modeling and simulation, and published 12 book chapters. His recent book *Aesthetic Computing* was published in April 2006 by MIT Press.

Contributors

Rohit Badal
Agricultural and Biological
 Engineering Department
University of Florida
Gainesville, Florida

J.C.M. Baeten
Department of Mathematics
 and Computer Science
Eindhoven University of Technology
Eindhoven, The Netherlands

Howard Beck
Agricultural and Biological
 Engineering Department
University of Florida
Gainesville, Florida

Peter Breedveld
Faculty of Electrical
 Engineering, Mathematics
 and Computer Science
University of Twente
Enschede, The Netherlands

Christos G. Cassandras
Department of Manufacturing
 Engineering and Center for
 Information and Systems
 Engineering
Boston University
Brookline, Massachusetts

Gilbert Gang Chen
Department of Computer Science
Rensselaer Polytechnic Institute
Troy, New York

Wolfgang Christian
Davidson College
Charlotte, North Carolina

Gary Consolazio
Department of Civil and
 Coastal Engineering
University of Florida
Gainesville, Florida

P. Coquillard
Gestion de la biodiversité
Université de Nice-Sophia Antipolis
Nice, France

Julia I. Couto
Computer Science Department
University of Georgia
Athens, Georgia

Clive L. Dym
Department of Engineering
Harvey Mudd College
Claremont, California

Hilding Elmqvist
Dynasim AB
Lund, Sweden

Francisco Esquembre
Universidad de Murcia
Spain

Paul A. Fishwick
Department of Computer
 and Information Sciences
 and Engineering
University of Florida
Gainesville, Florida

Andrew Ford
School of Earth and
 Environmental Sciences
Washington State University
Pullman, Washington

Benjamin FrantzDale
Scientific Computation
 Research Center
Troy, New York

Robert M. French
LEAD-CNRS UMR 5022
University of Burgundy
Dijon, France

Antony Galton
University of Exeter
Exeter, United Kingdom

Aniruddha Gokhale
Institute for Software Integrated
 Systems
Vanderbilt University
Nashville, Tennessee

Jeff Gray
Computer and Information Sciences
University of Alabama
 at Birmingham, Alabama

Congzhou He
Computer Science Department
University of Georgia
Athens, Georgia

David R.C. Hill
ISIMA/LIMOS – UMR CNRS 6158
Blaise Pascal University
Aubiere, France

Alfons G. Hoekstra
Section Computational Science
University of Amsterdam
Amsterdam, The Netherlands

Marc Hoit
Department of Civil and Coastal
 Engineering
University of Florida
Gainesville, Florida

Yunchul Jung
Agricultural and Biological
 Engineering Department
University of Florida
Gainesville, Florida

Steven Kelly
MetaCase
Jyväskylä, Finland

Tag Gon Kim
Department of Electrical
  Engineering and
  Computer Science
KAIST
Daejeon, Korea

Stefan C. Kremer
Department of Computing
  and Information Science
University of Guelph
Guelph, Ontario, Canada

Jinho Lee
Samsung Electronics
Suwon, Korea

E. Lefeber
Department of Mechanical
  Engineering
Eindhoven University of
  Technology
Eindhoven, The Netherlands

Sven Erik Mattsson
Dynasim AB
Lund, Sweden

Brian McKay
The MathWorks, Inc.
Natick, Massachusetts

John A. Miller
Computer Science Department
University of Georgia
Athens, Georgia

Clay L. Montague
University of Florida
Gainesville, Florida

Pieter J. Mosterman
The MathWorks, Inc.
Natick, Massachusetts

Sandeep Neema
Institute for Software
  Integrated Systems
Vanderbilt University
Nashville, Tennessee

James Nutaro
Oak Ridge National Laboratory
Oak Ridge, Tennessee

Martin Otter
DLR Institute of Robotics
  and Mechatronics
Germany

Ernest H. Page
The MITRE Corporation
McLean, Virginia

Minho Park
Department of Computer Science
Stephen F. Austin State University
Nacogdoches, Texas

Kalyan S. Perumalla
Oak Ridge National Laboratory
Oak Ridge, Tennessee

Michael Pidd
Department of Management
  Science
Lancaster University
  Management School
Lancaster, United Kingdom

Stewart Robinson
Warwick Business School
University of Warwick
Coventry, United Kingdom

Mathias Röhl
University of Rostock
Rostock, Germany

J.E. Rooda
Department of Mechanical
  Engineering
Eindhoven University of Technology
Eindhoven, The Netherlands

Andrew P. Sage
Department of Systems Engineering
  and Operations Research
George Mason University
Fairfax, Virginia

Lee Schruben
University of California
Berkeley, California

Hassan Sedaghat
Department of Mathematics
Virginia Commonwealth University
Richmond, Virginia

E. Seegyoung Seol
Scientific Computation
  Research Center
Troy, New York

Jonathan Sprinkle
Electrical Engineering and
  Computer Science
University of Arizona
Tucson, Arizona

Rohit Shenoy
The MathWorks, Inc.
Natick, Massachusetts

Mark S. Shephard
Scientific Computation Research
  Center
Troy, New York

Peter M.A. Sloot
Section Computational Science
University of Amsterdam
Amsterdam, The Netherlands

Roger Smith
Modelbenders LLC
Orlando, Florida

Boleslaw K. Szymanski
Department of Computer Science
Rensselaer Polytechnic Institute
Troy, New York

Simon J.E. Taylor
School of Information Systems,
  Computing and Mathematics
Brunel University
Uxbridge, United Kingdom

Juha-Pekka Tolvanen
MetaCase
Jyväskylä, Finland

Adelinde M. Uhrmacher
University of Rostock
Rostock, Germany

D.A. van Beek
Department of Mechanical
  Engineering
Eindhoven University of Technology
Eindhoven, The Netherlands

Jiacun Wang
Department of Software Engineering
Monmouth University
West Long Branch, New Jersey

Introduction

I

1
The Languages of Dynamic System Modeling

Paul A. Fishwick
University of Florida

1.1 Introduction 1-1
1.2 Dynamic System Modeling Examples 1-2
1.3 Taxonomic Approaches 1-3
1.4 Language 1-3
1.5 Syntax 1-5
1.6 Semantics 1-7
1.7 Pragmatics 1-10
1.8 Summary 1-10

1.1 Introduction

Just as we use natural language to communicate verbally or in writing, we use *models* as a form of scientific language to communicate about the world. Models are more compact than natural languages, and tend to be more structured using mathematical expressions, diagrams, or analog machinery. Morgan and Morrison (1999) underscore the relevance of models in theory construction as being vital to science: models mediate between humans and phenomena. We begin the study of dynamic models with a brief historical review of the word "model" and then we continue with an overview of "dynamic model" as a subtype. Hodges (2005) provides a concise description of how the word originated as well as different uses for the word:

> In late Latin a *modellus* was a measuring device, for example to measure water or milk. By the vagaries of language, the word generated three different words in English: mould, module, model. Often a device that measures out a quantity of a substance also imposes a form on the substance. We see this with a cheese mould, and also with the metal letters (called *moduli* in the early 17th century) that carry ink to paper in printing. So *model* comes to mean an object in hand that expresses the design of some other objects in the world: the artist's model carries the form that the artist depicts, and Christopher Wren's *module* of St Paul's Cathedral serves to guide the builders.

> Later on, models began to include mathematical forms, in addition to the relatively recent computational forms (i.e., computer-based text and graphics). All models serve as a convenient substitute for a phenomenon, but the type of substitution differs slightly depending on the context. For example, a model house captures the mould legacy since one can imagine a machine that stamps out real houses based on the model house "mould." This is an example of using *model* as an archetype or prototype: the model precedes the phenomenon. Contrast this use of "model" with the artifact that interests us: the dynamic system model. This kind of model captures only the behavioral, or dynamic, aspects of the phenomenon, which precedes

the model. It is possible to see this *difference in precedence* as the duality of synthesis (i.e., model as prototype) and analysis (i.e., model as theory). The idea of dynamic system models as theories might be debated, but they act in this way, as hypotheses that remain to be proven within the context of real-world phenomena. This process is known as *verification and validation*, with verification being assumed to test truth against requirements and validation to test system behavior against empirical observation. The mathematical topic of *model theory* adds to the literature on models, by introducing models of logical formulae.

We will introduce three distinct ways of categorizing models:

- *Synthesis*: Model X is a *model for* Y, with X being the prototype and Y being the instance model. The model house X is a model for a specific house Y. The *mapping* between X and Y is achieved through property inheritance, much as derived classes from base classes in object-oriented languages.
- *Analysis*: Model X is a *model of* Y, with X and Y being instance objects. For example, the wave tank is a model of the ocean, or the Petri net is a model of an asynchronous communication network. The *mapping* involves a transformation of objects in X to those in Y.
- *Theory*: The *mapping* is a model of X, with Y generated as a logical consequence. For example, for X being the formula $\forall x, y (P(x,y) \wedge Q(x) \wedge R(y))$, the mapping ($P \equiv bigger, Q \equiv whale, R \equiv human$) is a model for X since whales are bigger than humans. Model theory captures this approach since a model is the mapping itself, rather than an object specified as a model. The model is the interpretation that makes specific logical formulae true.

1.2 Dynamic System Modeling Examples

Within the prior discussion of model types, *dynamic models* fall under *analysis*; however, other model types (i.e., synthesis and theory) may play subsidiary roles. We may have a set of equations that dynamically models the population dynamics within an ecosystem, or a diagram that is translated into this set. Our dynamic models are also "digital" since they are executed on a digital computer. Models can take many forms as indicated in the examples shown in Figure 1.1.

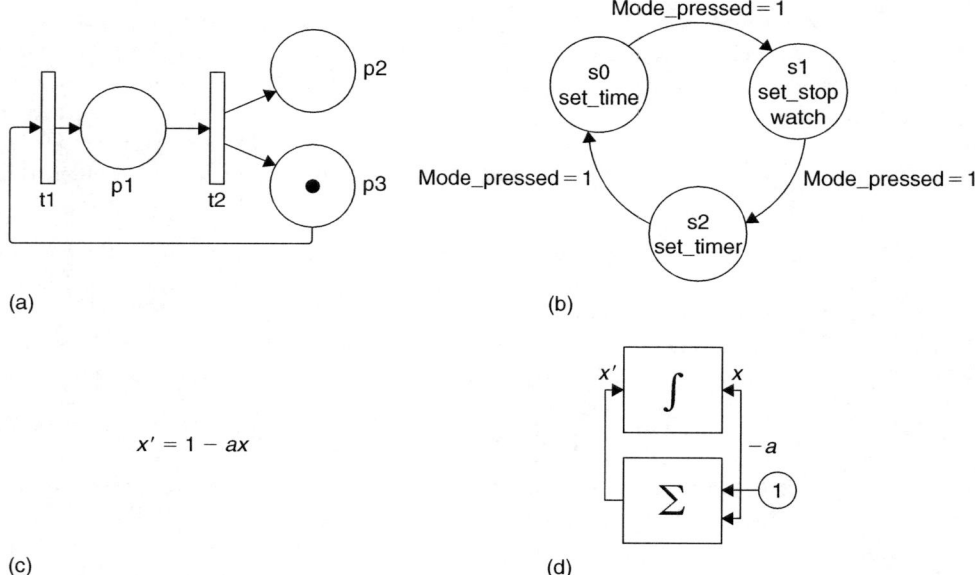

FIGURE 1.1 Four types of dynamic system models. (a) Petri net. (b) Finite state machine. (c) Ordinary differential equation. (d) Functional block model.

The Languages of Dynamic System Modeling

We observe the following from this illustration:

- Except for Figure 1.1(c), all the models are graphical, which suggests that models can take many forms. Not only can models look different but they might be made of different media, in different styles according to metaphor, and using different forms of human–computer interaction (HCI).
- Figure 1.1(c) and Figure 1.1(d) have almost equivalent semantics, although their syntax is different. This difference is surfaced through the use of different metaphors that form the structural bases.
- Some models are static in appearance, whereas others have natural dynamics during model execution as in Figure 1.1(a). This is not apparent from a static figure, but the solid circle moves through the network.
- Models operate in different scales over spacetime. So, models in Figure 1.1(a) and Figure 1.1(b) have discrete events and those in Figure 1.1(c) and Figure 1.1(d) operate over a temporal continuum offering no recognizable events separate from the initial and final conditions necessary for unique solution.

1.3 Taxonomic Approaches

The field of dynamic system modeling is vast and there has been research in creating theories and approaches for studying and classifying systems. The initial work in cybernetics (Wiener, 1948; Ashby, 1963) and, subsequently, systems theory (Kalman et al., 1962; Padulo and Arbib, 1974; Bertalanffy, 1968) provided the context for viewing dynamic systems as control systems with feedback. More recent work has identified a systems theoretical approach to specific types of systems, such as discrete event (Zeigler et al., 2000). Other taxonomic approaches are based on a systems philosophy (Klir, 1985; Ören, 1984, 1987) or a language-based approach (Fishwick, 1995, 1996; Nance, 1995). The latter approach is one that we stress in this chapter as a way to understand the nature of dynamic models as language components. By treating the models in Figure 1.1 as "statements" within a language (i.e., the type or category of the model), we are able to surface a comprehensive approach to the study of dynamic system modeling.

Taxonomies have always played a key role in the way a group perceives what they do. By thinking of a task in terms of the metaphor of "agent," one cannot but help envision agents acting in a way similar to human agents. The metaphor (Lakoff and Johnson, 2003; Lakoff and Nunez, 2000) provides an important mechanism for reasoning how the model is presented to the user as well as how it operates when executed. Here are three different ways of viewing models:

- *Models as mathematical constructs*: Models are mathematical structures, encoded in traditional textually based mathematical notation.
- *Models as physical constructs*: Models are physical objects made of organic or engineered materials.
- *Models are language constructs*: Models are formal languages with syntax, semantics, and pragmatics.

These views are not complete and views may easily be combined; however, the key observation is that the view taken by someone determines their philosophy about modeling. The adopted views also provide different emphases: a view of modeling as mathematical construct de-emphasizes human presentation and interaction, whereas a language view emphasizes presentation and semantics equally, along with how the human interacts with the symbols. Fortunately, a combination of views fosters a more pluralistic view of modeling, and allows for mathematical rigor to be combined with human interaction.

1.4 Language

Pinker (1994) begins his manuscript with "As you are reading these words, you are taking part in one of the wonders of the natural world." It is amazing that language exists and works as well as it does. Language provides us with the ability to communicate, and it is pervasive within human societies. The interesting thing is that language operates over many levels and layers. The oldest and most prominent type of language is *natural language*. However, there are regional dialects, colloquialisms, codes, signs, and formal languages that grew from mathematics. It is somewhat ironic that while we think of mathematics as rigorously defined, the semantics of mathematics are fundamentally based on natural language, which includes the gesticulations of numerous parents and teachers. For example, to know the meaning of *integral*, one must know the meaning of summation, and that is ultimately gleaned through examples, gesture, analogy, and metaphor, all of which are delivered with the assistance of natural language. Still, natural language is notoriously ambiguous and so if science is to progress, we need other more rule-based approaches to model building. The consensual agreement in a scientific community of signs and rules is termed *formalism*. Formalisms, such as logic-based systems with axioms and rules, tend to minimize the number of components in an attempt to reduce potential semantic ambiguity.

Semiotics (Noth, 1990) is the theory of language (i.e., or more atomistically of signs) first discussed in depth by de Saussure (1916). Along with de Saussure, Pierce (Hartshorne and Cambridge, 1997) is considered to be the theory's cofounder. de Saussure's definition of a "sign" is a dyadic structure containing a signifier and a signified. The signifier, for de Saussure, was primarily phonetic, although one can easily broaden this perspective to more materialistic interpretations. Pierce invented a triadic view of the sign with these components: the representamen (i.e., the sign), the object, and the interpretant. He proceeded to create a somewhat complex taxonomy, but we will focus on what Morris (1925) defined, since his taxonomy of language has since permeated computing: syntax, semantics, and pragmatics. We will define these, and place them into the context of dynamic system models:

- *Syntax*: relation of signifiers to each other—structure of models.
- *Semantics*: relation of the signifier to the signified—meaning of models.
- *Pragmatics*: relation of the signifier and signified to the human—interaction of humans with models.

In computing, one normally uses these three terms within the context of computational linguistics for natural-language processing, or programming languages for formal languages. Figure 1.2 is based on Morris' taxonomy with slight simplifications.

The "sign vehicle" represents the complete process where syntax, semantics, and pragmatics can take place. Let us consider dynamic system models from this perspective.

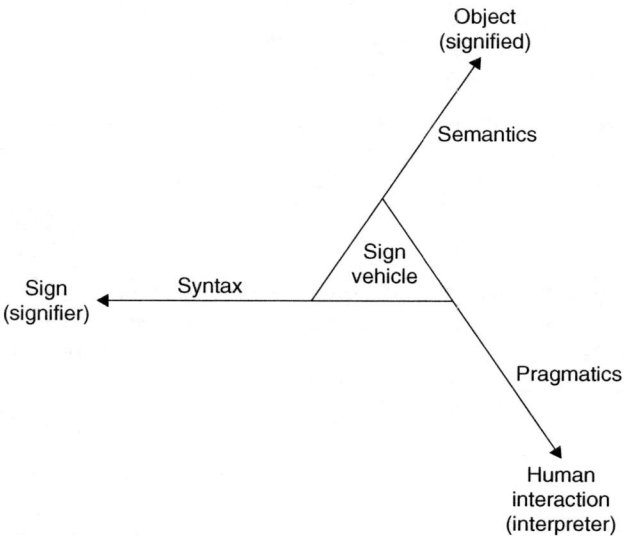

FIGURE 1.2 Triangular relationship in sign formation (After Morris, C.W. *Symbolism and Reality: A Study in the Nature of Mind*, Benjamin Pub. Co., Philadelphia, PA, 1993.)

Syntax, semantics, and pragmatics interact in several ways, and looking at these topics from a mathematical perspective is enlightening. Since the language of set theory underpins most formalisms, let us consider syntax as the mathematical *structure* defined by elements such as variables, tuples, graphs, and trees. Semantics captures the concept of the *relation*, or its restriction in the form of *function*. Pragmatics is difficult to define mathematically except to note that it involves the *human context* of doing mathematical modeling. The syntax/semantics duality exists elsewhere as in architecture: syntax is *form* and semantics is *function*. Syntax, therefore, is concerned not only with structure, but also what the structure looks and sounds like–its presentation. The essence of semantics is in the act of *representation*. In the chain of meaning, we translate from one form to another, and still yet, to another in a seemingly endless set of transformations. Meaning is generated or induced through these transformations, where an increasing knowledge level provides the human with understanding. This is true of dynamic system models as well since we translate from one model to another model, and eventually to behavior. Whether a specific model is viewed as syntax or semantics is relative to the goal of the modeling exercise. A mathematical equation may be viewed as the semantic specification for a diagrammatic model, or it may be viewed as the syntax where the semantics are defined in a programming language such as C++ or Java.

The leading edge for language development in computing has evolved from programming languages, and currently manifests itself in the area of the semantic Web (Berners Lee et al., 2001), which is an extension and evolution of the World Wide Web (WWW). The Web provides a comprehensive infrastructure for discussing the three categories of language for both natural and formal languages. The original Web focused on markup for human consumption with regard to the underlying semantic content. The semantic Web, by contrast, is concerned with the development of ontologies (Maedche, 2002), which can be defined as an evolutionary structure beginning with the concept of a glossary, proceeding to a taxonomy, and gradually maturing into a semantic network complete with logical axioms and formulae that can be used for reasoning. The semantic Web begins with the extensible markup language (XML) and contains a slew of additional languages for transformation using extensible stylesheet transformation (XSLT). XML is described as being used for communication among machines, whereas one may present XML using a variety of methods to humans using XSLT, for example. It may seem odd that XML is humanly readable for a language meant for consumption by machines; however, humans must still cooperate with each other first in standardizing the structure. Also, while the Web maintains its early vestiges of "document markup," the documents have become full-fledged objects in their own right and the metaphor of "documentation" seems outdated. For ontology specification, there is the resource description language (RDF) and the ontology Web language (OWL). The semantic Web represents a new way of thinking about system models using the three categories.

1.5 Syntax

Syntax is concerned with *notation*: the way a system model looks; however, we might expand this to the other senses. For example, parts of the model may be associated with a sound or a tactile sense. For the graph-based models in Figure 1.1, we realize that the icons are positioned in certain ways, and with regularly specified connections. This is part of the syntax of the model: how we sense and organize the signs that comprise the model structure. The semantic Web has dramatic consequences for the way in which we think about dynamic system models. In the original Web, as in most traditional media, the artifact in Figure 1.1(c) would have been considered to represent the raw mathematical expression. Within the parlance of the semantic Web (i.e., hereafter referred to as the "Web"), however, this figure is one *presentation* of the underlying *content*. There are multiple presentations for the same content. The content of Figure 1.1(c) is expressed in the mathematics markup language (MathML) defined in Figure 1.3.

From the Web's perspective, this represents the "real" mathematics, and model artifacts such as Figures 1.1(c) and Figure 1.1(d) are viewed as *alternate presentations* of this machine-readable content. There are equally as expressive XML languages for other dynamic system models (Kim et al., 2002), and

```
<math>
    <apply>
        <eq/>
        <apply>
            <diff/>
            <bvar>
                <ci>t</ci>
            </bvar>
            <ci>x</ci>
        </apply>
        <apply>
            <minus/>
            <cn>1</cn>
            <apply>
                <times/>
                <ci>a</ci>
                <ci>x</ci>
            </apply>
        </apply>
    </apply>
</math>
```

FIGURE 1.3 Content MathML for $\frac{dx}{dt} = 1 - ax$.

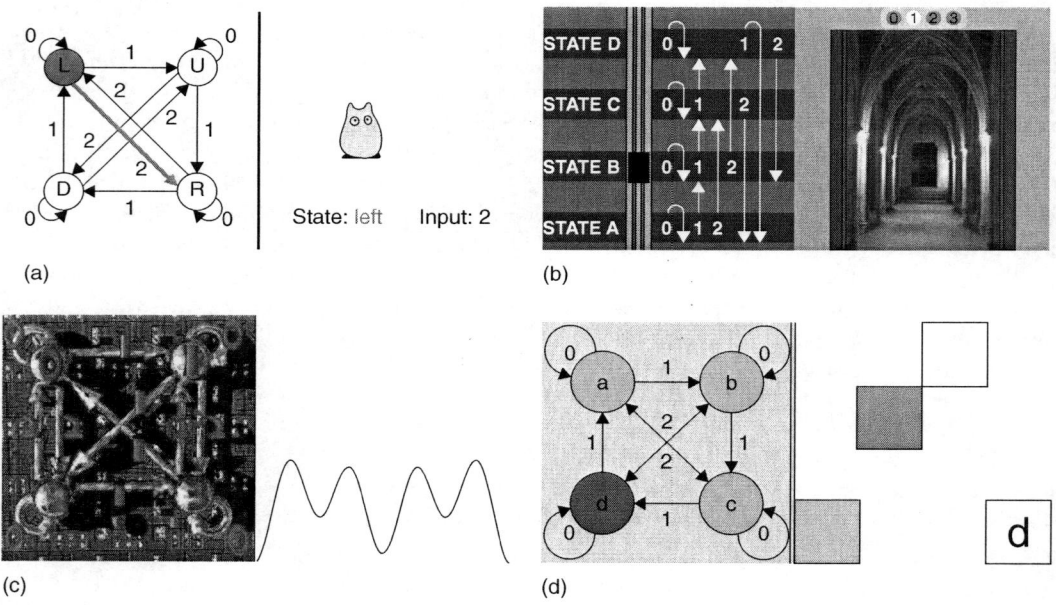

FIGURE 1.4 Four interpretations of the same state machine. (a) Christina Sirois. (b) Tim Winfree. (c) Brett Baskovich. (d) Emily Welles.

for most model types one generally finds researchers on the Web who are pioneering their own XML *applications*.

Figure 1.1 shows example presentations of dynamic models; however, we are not limited to two dimensions, the use of a single color, or a digital medium (i.e., as long as the interaction can be communicated digitally) (Fishwick, 2006). For example, Figure 1.4 demonstrates the creative application of students in a computer simulation class being given a canonical iconographic representation of a finite state machine

(FSM) and then being allowed to create different representations for both the model itself (left-hand side) and the behavior of the model (right-hand side). This concept mirrors Minsky's (1988) adage of knowing something "in more than one way." It also captures the importance of *multiple representations* discussed in mathematics education (Kaput, 1989).

In viewing these models, we are ultimately led to think about which presentations would be used under which circumstances. This modeling aspect is covered in Section 1.7. Besides capturing the syntax of signs such as $\frac{d}{dt}$, the other key area of syntax is in specifying rules for which icons can appear next to each other, and in what order, or using which types of connectors. For example, given the Petri net in Figure 1.1(a), where is the rule which states that Petri net transitions (i.e., the "t" identifiers) must be connected to places (i.e., the "p" identifiers)? For this, as in natural language, we need a *grammar*. In semiotics, the difference between the raw syntactic connections and grammar is captured, respectively, by the terms *syntagmatic* and *paradigmatic* (Chandler, 2002). For English, we have paradigms such as "noun" and "verb," and for Petri nets, we have "place" and "transition." Grammars can be expressed in XML, but simpler notations based on the Backus-Naur Form (BNF) capture the grammar. For example,

```
<eqn>     ::=   <diff> = <expr>
<num>     ::=   0 | 1 | ... | 9
<var>     ::=   x | y | z
<diff>    ::=   D( <var> )
<expr>    ::=   <expr> + <term> | <expr> - <term> | <term>
<term>    ::=   <term> * <factor> | <factor>
<factor>  ::=   ( <expr> ) | <var> | <diff> | <num>
```

represents a grammar that can parse the equation in Figure 1.1(c) along with many other equations. This is not a particularly robust grammar since it is limited to restricted equational forms; there are no functions aside from "D," and only single digit numbers. However, it serves the purpose of exemplifying the BNF. The tokens in angle brackets are *nonterminals* and the other tokens are *terminals*. Only nonterminals appear on the left-hand side of a rule. "|" is logical disjunction. The most powerful aspect of this grammar is its use of recursion in defining expressions. "D" is the differential operator, presumed to be defined relative to "t" so that $D(x) \equiv \frac{dx}{dt}$.

The study of grammars obtains significant headway in computing as a result of Chomsky's (1956) hierarchy, which defines different types of grammars according to their relative power (regular, context-free, context-sensitive, and unrestricted). The unrestricted grammars generate recursive languages. In practice, the context-free type of language is the most common and easily implemented.

Nontext models in Figure 1.1 can be formalized with *graph grammars* that follow the same approach as the expression grammar. However, instead of terminal separators such as whitespace and newline, one uses spatial relations such as "connected to." It is not particularly common for newly introduced system model structures to come with formally specified grammars. This may seem odd, although this condition may be partially due to the different historical paths taken by systems modeling and programming languages.

1.6 Semantics

If syntax is form and structure, not only of *content* and the *presentation* of the content, then semantics is about function and meaning. Semantics and syntax are intertwined in that functional semantic mappings involve taking one form of syntax and manifesting another. In this sense, semantics is *syntax that is presumed to be more uniformly understood*. To understand what X means, define X in terms of Y where Y is better understood by a given population. The essence of this translation lies with the mapping processes inherent within sign formation. Thus, the mapping of XML content to presentation is a semantic act in itself even though the source and target structures have their own syntax. The topic of *behavior* in language can be considered to be part of semantics. The argument for this view begins with the core notion of computability: input to a machine, system, or model that undergoes state and event changes resulting in output. From the standpoint of a program, it makes no difference whether the input and output are different models or signals since this is a matter of encoding and decoding. In the semiotic sense, the "box" separating input and output is the sign vehicle and so one may view the output from the box, whether structural or signal in nature, as being part of semantics.

Meaning is often associated, at least in informal languages, with two concepts: *denotation* and *connotation*. Denotation is what a sign formally represents: the relation of signifier to signified. However, connotation is a kind of *secondary meaning* attributed to signs. When one looks at a Petri net (Peterson, 1981), there may be the thought of biological cells splitting and joining. This "meaning" is secondary to the formal semantics, and yet through metaphor allows the modeler to better understand the modeling mechanism. Finer threads of meaning could result from someone seeing a blue-colored square icon and thinking of a blue sky or a familiar road intersection reminiscent of a square shape. In semiotics, these kinds of associations are discussed at length by Barthes (1977). Perhaps, a more uniform way of viewing semantics is that a model may have multiple threads or connections to other concepts, as well as models, and that some of these connections are more pronounced depending on the human interpreter. Thus, the denotation is the primary, or the most used, reference.

The way that meaning is defined varies substantially based on community and discipline. Possibilities for defining meaning include

1. *Natural language*: while informal and sometimes ambiguous, natural languages such as English are widely used in textbooks and journals where computational models are being defined.
2. *Pseudocode*: a more constrained form of natural language frequently used to define semantics.
3. *Computer language:* a language such as assembly language, Java, C++, or FORTRAN may serve well to define what an artifact means.
4. *Mathematics:* this is a textual language that, while undergoing change, tends to have standard notations (Cajori, 1993), which reduce the risk of ambiguity.

Frequently, natural language is interspersed with diagrams and textual mathematics to clarify what a modeling component "means" and how the coupled components can be legally connected together. For example, a Petri net model can be defined as a four-tuple $NET = <P, T, I, O>$, but then natural language is necessary to define how these symbols are supposed to be understood. Without the natural language, the symbols and their corresponding functions would literally make *no sense*. It is possible to have a purely mathematical explanation, and yet very little is learned unless the explanation is described or related to an example. Even though the Petri net might be seen as a purely algebraic construct without any further attributed meaning, the visual presentation of Petri nets has come to be associated with what the Petri net actually means, and as a guide to its operational characteristics. The Petri net is composed of *places* and *tokens*. The tokens move around the network. So far, this is a fairly straightforward metaphor that relies on a mathematical map involving the functions I and O. I and O are visually associated with object (i.e., token) motion. There is another biologically inspired metaphor of joining or splitting that is required to complete the modeler's understanding of what it means to execute the net. In the case of the Petri net, these metaphors are naturally associated with generating meaning.

In the fields of programming languages and software engineering, semantics are associated with *formal methods* (Winskel, 1993), where one uses languages that are based on either lambda calculus or first-order logic, with set theory underlying both. For system dynamics models, we can use systems theory which similarly relies on set theory but is more tuned to defining temporal change. System models are viewed semantically in terms of changes in states, events, and flow among interconnected functions, possibly involving feedback. Consider the formal semantics for Figure 1.1(d):

$$x(t) = \int x'(t) dt$$
$$x'(t) = 1 - ax(t)$$

It represents the original model using another syntactical form that has a more standardized, and more widely accepted, meaning and so we term it the *semantics* of Figure 1.1(d).

For dynamic models involving discrete events, one may use a variety of semantic specifications. For example, the discrete-event system specification (DEVS) (Zeigler et al., 2000) is defined as a tuple $<I, S, O, \delta_{int}, \delta_{ext}, \lambda, ta>$, where

- I and O represent the set of input and output values, respectively.
- S is the set of system states
- $\delta_{int} : S \to S$ is the *internal transition* function
- $\delta_{ext} : Q \times I \to S$ is the *external transition* function, where $Q = \{(s, e) \mid s \in S, 0 \leq e \leq ta(s)\}$ is the *total state* set, and e the *time elapsed* since the last transition.
- $\lambda : S \to O$ is the output function
- $ta : S \to \mathcal{R}^+_{0,\infty}$ defines the time advance function.

DEVS extends traditional systems theory with the semantics of *event transition*, with key differences being the addition of an internal transition function δ_{int} and events associated with elapsed time dictated by the *ta* function. Another event-based semantic language is based on the event graph model (Schruben, 1983; Buss and Sanchez, 2002) where the time advance function of DEVS, for example, is constructed at a layer beneath the definition of event causality (i.e., specific events and their times are behavioral artifacts of event causality).

A concern for formal methods is how they are actually implemented. The degree of meaning ascribed to a formalism is based on shared consensus for its notation, and even shared consensus is problematic since there are layers of machine implementation that form the "real meaning." One approach to address this is to define semantics at a lower machine level where the semantics of the formalism are one step closer to hardware implementation. For example, consider the necessary circuit for an FSM of the Moore variety depicted in Figure 1.5. This "next state" logic for this machine is termed "FSM synthesis" in digital design, and can be manually accomplished through a Karnaugh map, which leads to an expression in Boolean algebra suitable for translation to a combinational circuit. The clocked register stores the state, which is then fed back into the combinational box. Other types of dynamic system models have direct correlations at the hardware design level, facilitating a rigorous and formal understanding of their behavior. If there is a shared understanding of the nature of digital circuits, this can lead to a clearer understanding of system semantics.

The realization that model meaning and sign chains are accomplished through transformation leads us to the study of the nature of the transformations themselves, which can be seen as either intra- or intermodel. As discussed, the set-theoretic concepts of relation and function lie at the heart of the transformation process. We have the following possibilities, among others: (1) linear chains forming a graph and (2) encapsulation forming a hierarchy. Making models work at both levels is challenging (Zeigler,

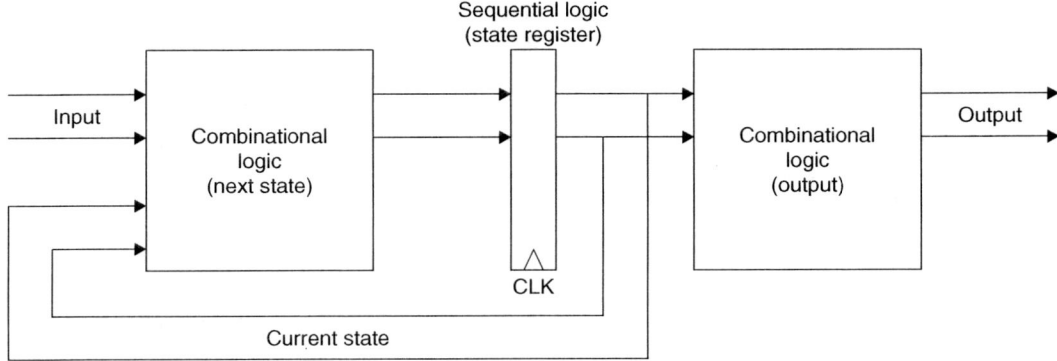

FIGURE 1.5 FSM Moore machine semantics.

1990; Mosterman and Vangheluwe, 2004; Fishwick and Zeigler, 1992) and there are several higher-level approaches to mapping. It is possible to create a mathematical morphism linking structures or use a method such as graph transformation, or a production system containing rules. The mapping itself may preserve structure, behavior, or neither. While it may be convenient to consider complete preservation when dealing with semantic mappings, each representation may have slightly different connotations, complicating the notion of identical levels of meaning. A classic example of this is in models that have a diagrammatic presentation. These models are laden with spatial metaphors that may be missing from a text-based model from which a transformation is occurring.

The *semantic Web* plays a key role in the definition of semantics and how semantics may be practically encoded. In particular, the construction of an ontology (Maedche, 2002; Fishwick and Miller, 2004; Lacy and Gerber, 2004) can span everything from a simple index or taxonomy to a semantic network that also defines a series of simple and complex logical constraints. The application of ontologies to modeling will gradually result in model taxonomies that are capable of being queried or processed. Models can then be more easily related to one another, possibly through morphism and transformation. With regard to the latter, XSLT offers one approach to generating one model from another using a pattern-based language.

1.7 Pragmatics

Pragmatics introduces the human into the mixture of model structure, transformation, and execution. Broadly speaking, pragmatics makes itself apparent in any computer-based task through topics found in areas such as HCI, pervasive computing, tangible computing, and virtual environments. By reviewing the methods in these areas of study and asking how these methods can be used in the practice of modeling, we find an abundance of approaches to how humans can interact with models.

When seeking answers as to how pragmatics affects modeling, we need to address the following:

- *Purpose*: What is the goal of creating the model? In what larger context is the model being designed, tested, and evaluated? Is the model to be used for science, engineering, education, or for entertainment? Are there multiple goals to be achieved?
- *Context*: Is the model part of a larger model that includes the human? This kind of situation is associated with experiments where the experimental apparatus can be simultaneously viewed as being yet another higher-level model or a form of pragmatics.
- *Decision making*: How does one build a model to facilitate human planning and decision making (Sage, 1977)? What support tools are available to assist the human in choosing the right model for the goals of a project?
- *Analysis*: To what extent is there an adequate set of tools and methodologies to support analyzing the model either via the *process* of verification or validation (Overstreet and Nance, 1985; Balci et al., 1990)?
- *Interaction*: What modalities are employed to allow the human to interact with the model? Is the model to be purely digital, or rather a hybrid analog/digital artifact? How will the human experience the model: can they touch it, move it around, or hear it (Singhal and Zyda, 1999; Burdea and Coiffet, 2003)? Is the model interacted with over the Web (Page et al., 2000) or using other more exotic interfaces? Is the model construction process collaborative or accomplished in phases?
- *Culture*: Are there modeling modes that encourage customization or personalization of a model structure depending on the individual or group that needs to model? Is a balance between form and function evident in the modeling practice?

1.8 Summary

We have presented the concept of dynamic system models as languages. As seen through this particular lens, models have structure (syntax), meaning and behavior (semantics), and a relationship with the human (pragmatics). As with any taxonomy, we are promoting a different way or world-view for viewing the modeling practice. As such, using language we are naturally led to linkages that other taxonomies might not stress. For example, the *language focus* suggests a concrete connection to media, communication, and human interaction. With language, there is a requirement to formalize not only the model semantics but also the textual or graphical syntax that defines legal model structures of a particular sort. One might also view the language orientation as a return of sorts to an understanding of system models that is closer to computer science or possibly to cybernetics, where the language and communication emphasis tends to be pronounced.

References

Ashby, W. R. (1963). *An Introduction to Cybernetics*. New York: Wiley.
Balci, O., R. E. Nance, E. J. Derrick, E. Page and J. L. Bishop (December 1990). Model generation issues in a simulation support environment. In *Winter Simulation Conference*, New Orleans, LA, pp. 257–263.
Barthes, R. (1977). *Image, Music, Text*. New York: Hill and Wang.
Berners Lee, T., J. Hendler and O. Lassila (May 2001). The Semantic Web. *Scientific American*.
Bertalanffy, L. V. (1968). *General System Theory*. New York: George Braziller.
Burdea, G. C. and P. Coiffet (2003). *Virtual Reality Technology* (2nd ed.). New York: Wiley.
Buss, A. H. and P. J. Sanchez (2002). Modeling very large scale systems: Building complex models with LEGOs (listener event graph objects). In *Proceedings of the Winter Simulation Conference*, San Diego, CA, pp. 732–737.
Cajori, F. (1993). *A History of Mathematical Notations*. New York: Dover.
Chandler, D. (2002). *Semiotics: The Basics*. London: Routledge.
Chomsky, N. (1956). Three models for the description of language. *IRE Transactions on Information Theory*, 2, 113–124.
de Saussure, F. (1959). *Course in General Linguistics*. New York: Philosophical Library.
Fishwick, P. (1995). *Simulation Model Design and Execution: Building Digital Worlds*. Upper Saddle River, NJ: Prentice-Hall.
Fishwick, P. (1996). A taxonomy for simulation modeling based on programming language principles. *IIE Transactions*, 30, 811–820.
Fishwick, P. (Ed.) (2006). *Aesthetic Computing*. Cambridge, MA; MIT Press.
Fishwick, P. and J. A. Miller (December 2004). Ontologies for modeling and simulation: Issues and approaches. In *Proceedings of the Winter Simulation Conference*, Washington, DC, pp. 259–264.
Fishwick, P. A. and B. P. Zeigler (1992). A multimodel methodology for qualitative model engineering. *ACM Transactions on Modeling and Computer Simulation*, 2(1), 52–81.
Hartshorne, C. and A. W. Cambridge (Eds.) (1997). *Collected Papers of Charles Pierce: Vols 1–6: 1931–58*. Cambridge, MA: Harvard University Press.
Hodges, W. (2005). Model theory (entry). http://plato.stanford.edu.
Kalman, R. E., P. L. Falb and M. A. Arbib (1962). *Topics in Mathematical Systems Theory*. New York: McGraw-Hill.
Kaput, J. J. (1989). Linking representations in the symbolic systems of algebra. In S. Wagner and C. Kiernan (Eds.), *Research Agenda for Mathematics Education: Research Issues in the Learning and Teaching of Algebra*, National Council of Teachers, pp. 167–194.
Kim, T., J. Lee and P. Fishwick (July 2002). A two-stage modeling and simulation process for web-based modeling and simulation. *ACM Transactions on Modeling and Computer Simulation*, 12(3), 230–248.
Klir, G. J. (1985). *Architecture of Systems Problem Solving*. New York: Plenum Press.
Lacy, L. and W. J. Gerber (2004). Potential modeling and simulation applications of the web ontology language-OWL. In *Proceedings of the Winter Simulation Conference*, pp. 265–270.
Lakoff, G. and M. Johnson (2003). *Metaphors We Live By* (2nd ed.). Chicago: University of Chicago Press.
Lakoff, G. and R. Nunez (2000). *Where Mathematics Comes From: How the Embodied Mind Brings Mathematics into Being*. New York: Basic Books.

Maedche, A. (2002). *Ontology Learning for the Semantic Web*. Nocwell, MA: Kluwer Academic Publishers.

Minsky, M. (1988). *Society of Mind*. New York: Simon & Schuster.

Morgan, M. S. and M. Morrison (1999). *Models as Mediators*. Cambridge: Cambridge University Press. Perspective on Natural and Social Science.

Morris, C. W. (1993). *Symbolism and Reality: A Study in the Nature of Mind*, Philadelphia, PA: Benjamin Pub. Co.

Mosterman, P. J. and H. Vangheluwe (2004). *Computer Automated Multi-Paradigm Modeling: An Introduction*, 80(9), 433–450. Special Issue: Grand Challenges for Modeling and Simulation.

Nance, R. E. (December 1995). Simulation programming languages: An abridged history. In *Winter Simulation Conference*, Washington, DC, pp. 1307–1313.

Noth, W. (1990). *Handbook of Semiotics*. Bloomington, IN: Indiana University Press.

Ören, T. I. (1984). Model-based activities: A paradigm shift. In T. I. Oren, B. P. Zeigler, and M. S. Elzas (Eds.), *Simulation and Model-Based Methodologies: An Integrative View*, pp. 3–40, Berlin: Springer.

Ören, T. I. (1987). Simulation: Taxonomy. In M. G. Singh (Ed.), *Systems and Control Encyclopedia*, pp. 4411–4414. New York: Pergamon Press.

Overstreet, C. M. and R. E. Nance (February 1985). A specification language to assist in analysis of discrete event simulation models. *Communications of the ACM*, 28(2), 190–201.

Padulo, L. and M. A. Arbib (1974). *Systems Theory: A Unified State Space Approach to Continuous and Discrete Systems*. Philadelphia, PA: W. B. Saunders.

Page, E., A. Buss, P. Fishwick, K. Healy, N. Richard and R. Paul (2000). Web-based simulation: Revolution or evolution? *ACM Transactions on Modeling and Computer Simulation*, 10(1), 3–17.

Peterson, J. L. (1981). *Petri Net Theory and the Modeling of Systems*. Englewood Cliffs, NJ: Prentice-Hall.

Pinker, S. (1994). *The Language Instinct: How the Mind Creates Language*. New York: Harper Collins Publishers.

Sage, A. P. (1977). *Methodology for Large-Scale Systems*. New York: McGraw-Hill.

Schruben, L. W. (1983). Simulation modeling with event graphs. *Communications of the ACM*, 26(11), 957–963.

Singhal, S. and M. Zyda (1999). *Networked Virtual Environments: Design and Implementation*. New York: Addison-Wesley.

Wiener, N. (1948). *Cybernetics: Or the Control of Communication in the Animal and the Machine*. Cambridge, MA: MIT Press.

Winskel, G. (1993). *The Formal Semantics of Programming Languages: An Introduction*. Cambridge, MA: MIT Press.

Zeigler, B. P. (1990). *Object-Oriented Simulation with Hierarchical, Modular Models: Intelligent Agents and Endomorphic Systems*. San Diego, CA: Academic Press.

Zeigler, B. P., T. G. Kim and H. Praehofer (2000). *Theory of Modeling and Simulation* (2nd ed.). New York: Academic Press. First edition published in 1976.

2
The Dynamics of the Computational Modeling of Analogy-Making

Robert M. French
University of Burgundy

2.1 Introduction 2-1
2.2 Analogy-Making as *Sameness* 2-2
2.3 Analogy-Making as a Means of "Bootstrapping" Cognition 2-3
2.4 The Necessity of Malleable Representations 2-3
2.5 The Dynamics of Representation-Building in Analogy-Making 2-5
2.6 Context-Dependent Computational Temperature 2-6
2.7 Interaction between Top-Down and Bottom-Up Processes: An Example 2-6
2.8 Computational Models Implementing this Bottom-Up/Top-Down Interaction 2-8
2.9 Architectural Principles 2-8
The "Slipnet," a Semantic Network • The Workspace • The Worldview • Codelets • The Coderack • Dynamic Codelet Selection via Computational Temperature • Local, Stochastic Processing • Integration of Representation-Building and Correspondence-Finding
2.10 How this Type of Program Works: The Details 2-12
Copycat • Tabletop
2.11 How Tabletop Finds a Reasonable Solution 2-13
2.12 The Issue of Scaling Up 2-16
2.13 The Potential Long-Term Impact of the Mechanisms Presented 2-16
2.14 Conclusions 2-16

2.1 Introduction

In this chapter we begin by introducing a notion of analogy-making that is considerably broader than the normal construal of this term. We argue that analogy-making, thus defined, is one of the most fundamental and powerful capacities in our cognitive arsenal. We claim that the standard separation of the representation-building and mapping phases cannot ultimately succeed as a strategy for modeling analogy-making. In short, the context-specific representations that we use in short-term memory—and that computers will someday use in their short-term memories—must arise from a continual, dynamic interaction between high-level knowledge-based processes and low-level, largely unconscious associative memory processes. We further suggest that this interactive process must be mediated by *context-dependent computational temperature*, a means by which the system dynamically monitors its own activity, ultimately allowing it to settle on the appropriate representations for a given context.

It is important to be clear about the goals of this chapter. It is not intended to be a review of computational models of analogy-making. For such a review, see, for example, Hall (1989), Gentner et al. (2001), French (2002), or Kokinov and French (2003). Rather, I will present a particular class of models developed, in the main, by Hofstadter and colleagues from the mid-1980s, in which dynamic, stochastic control mechanisms play a defining role. This, of course, is not to say that no other computer models of analogy-making incorporate dynamic control mechanisms. Certainly, for example, the settling mechanisms of Holyoak and Thagard's (1989) ACME, a constraint-satisfaction connectionist model, or the mechanisms of dynamic binding over distributed representations of Hummel and Holyoak's (1997) LISA model, are dynamic. The models by Gentner and colleagues (e.g., Gentner, 1983; Falkenhainer et al., 1989; Forbus et al., 1995) clearly have dynamic mechanisms built into them. Why, then, do I choose to discuss the Hofstadter family of models?

Several points set these models apart from all others (with the exception of a model, independently developed by Kokinov (1994) that adopted a similar design philosophy). One key principle is the eschewal of hand-coded representations. Instead, these programs rely on a dynamic feedback loop between the program's workspace and its long-term semantic memory that allows it to gradually converge on context-appropriate representations. This architecture was explicitly designed to allow scaling up without combinatorial explosion. The second key feature was the use of a context-dependent computational temperature function that mediated the degree to which the activity of the program was deterministic: the higher the temperature, the more random the program's choices became. Temperature is a measure of the overall quality of the structures perceived and as that structure becomes more and more coherent, the temperature gradually falls and the program settles into a set of coherent, stable representations. When the temperature is low enough, the program will stop.

2.2 Analogy-Making as *Sameness*

Before entering into a discussion of the dynamics of computational modeling of analogy-making, we must first make clear what we mean *analogy-making*. Frequently, what is understood by analogy-making is the classic, but more restricted, Aristotelian notion of *proportional* analogies. These take the form "A is to B as C is to D." For example, "*Left* is to *right* as *up* is to *down*" is an example of this kind of analogy. While this is certainly part of the story, I will take a broader view of analogy-making, one originally adopted, among others, by Hofstadter (1984), Mitchell and Hofstadter (1990), Chalmers et al. (1992), Mitchell (1993), French (1995), Hofstadter et al. (1995), and Kokinov (1994). In this view, analogy-making involves our ability to view a novel object, experience, or situation that belongs to one category as being *the same as* some other object, experience, or situation, generally belonging to another category. This view is summed up by French (1995, p. xv) as follows:

> If only by definition, it is impossible for two things, *any* two things, to be exactly the same. And yet, there is nothing puzzling or inappropriate about our everyday use of the word "same." We see nothing odd or wrong about ordinary utterances such as: "That's the same man I saw yesterday at lunch," or "We both wore the same outfit," or, "I felt the same way when Julie and I broke up," or, finally, "That's the same problem the Americans had in Vietnam." What makes all these uses of "the same" (and this one, too) the same?
>
> The answer is: analogy-making. .... Since no two things are ever identical, what we really mean when we say "X is the same as Y," is that, within the particular context under discussion, X is the *counterpart* of Y. In other words, X is analogous to Y.

This way of looking at analogy-making, unlike the classic view of proportional analogy, allows us to speak of a *continuum* of analogy-making. This continuum runs from simple recognition—an apple we have never seen before is recognized as being a member of the category Apple because it is "the same as" other apples we have seen before—to deep "structural" analogies where elements in one situation are mapped to completely dissimilar elements in another situation—a fellow baseball player once said of homerun king, Hank Aaron, "Trying to sneak a fastball by Hank Aaron is like trying to sneak the sun past a rooster." In this analogy, Hank Aaron is mapped, completely naturally, to a rooster, and fastballs are mapped to the sun, even though, under normal circumstances, roosters have precious little to do with Hank Aaron and fastballs have even less to do with the sun.

2.3 Analogy-Making as a Means of "Bootstrapping" Cognition

The ability to see new things as being already familiar things with a twist is, unquestionably, one of the most powerful tools in the human cognitive arsenal from early infancy to adulthood. This ability to use analogies to understand novel situations allows infants (and adults in unfamiliar settings) to "bootstrap" new knowledge based on previously learned knowledge. In short, analogy-making allows us to comprehend new situations by seeing them as being "the same" as familiar situations that we already know how to handle, even if they require a bit of behavioral fine tuning.

In its simplest form, analogy-making involves finding a set of correspondences between a "base" object (or situation, experience, etc.) and a corresponding "target." For example, you understand the statement "A credit card is like a check book," because, even though credit cards do not physically resemble check books in the least, you effortlessly extract the appropriate parts of the representations of both—in this case, attributes related to their monetary function—and bring them into correspondence.

Young children, as we have said, constantly engage in analogy-making of the most complex kind. A perfectly run-of-the-mill example was provided one day by my not-yet-3-year-old son. He lightly touched the front bumpers of two of his little toy cars together and told me, "The cars are giving each other a little kiss, Dada." What most people fail to realize, because it happens so often and seems so completely natural—"Of course little kids say things like that! It's so cute!" remarked one of my friends, thoroughly unimpressed—is that his remark constitutes an amazing cognitive leap for a 2 year old. Think of the machinery put into play for him to have made that remark: people must be mapped to cars, a car's front bumpers to lips, touching them together lightly (as opposed to slamming them together) constitutes "a little kiss," etc.

2.4 The Necessity of Malleable Representations

Well over a century ago, William James (1890) recognized just how malleable our representations of the world had to be:

> There is no property ABSOLUTELY essential to one thing. The same property which figures as the essence of a thing on one occasion becomes a very inessential feature upon another. Now that I am writing, it is essential that I conceive my paper as a surface for inscription .... But if I wished to light a fire, and no other materials were by, the essential way of conceiving the paper would be as a combustible material .... The essence of a thing is that one of its properties which is so important for my interests that in comparison with it I may neglect the rest .... The properties which are important vary from man to man and from hour to hour .... many objects of daily use—as paper, ink, butter, overcoat—have properties of such constant unwavering importance, and have such stereotyped names, that we end by believing that to conceive them in those ways is to conceive

them in the only true way. Those are no truer ways of conceiving them than any others; there are only more frequently serviceable ways to us. (James, 1890, pp. 222–224)

This point, largely overlooked throughout the heyday of traditional artificial intelligence (AI), was reiterated by Chalmers et al. (1992). They apply James' key insight to the domain of analogy-making and provide a number of examples to illustrate their point.

Consider, they say, deoxyribonucleic acid (DNA). There is an obvious—physical—analogy between DNA and a zipper. Here, we focus on two strands of paired nucleotides "unzipping" for the purposes of replication. A second analogy—focusing on information this time—involves comparing DNA to the source code of a computer program. What comes to mind now is the fact that information in the DNA gets "compiled" into enzymes, which correspond to machine code (i.e., executable code). In the latter analogy, the focus of the representation of DNA is radically different: the DNA is seen as an information-bearing entity, whose physical aspects, so important to the first analogy, are of virtually no consequence. While we obviously have a single rich representation of DNA in long-term memory, very different facets of this large, passive representational structure are selected out as being relevant, depending on the pressures of a particular analogy context. The key point is that, independent of the vast, passive content of our long-term representation of DNA, the active content that is processed at a given time is determined by a very flexible representational process.

But one might argue that examples of this kind do not really prove much of anything. One might claim that they are vanishingly small in number and, therefore, need not be taken into consideration when attempting to model the broad sweep of cognitive mechanisms, any more than one must worry about monotremes (i.e., egg-laying mammals, like the duck-billed platypus) when discussing mammals, in general.

For this reason, we consider the representation of an utterly ordinary object—in this case, a *credit card*—and demonstrate that the representational problems encountered for DNA also exist for this ordinary object, and, by extension, *any* real-world object that we try to represent. We will consider a number of simple statements involving a "credit card" that will serve to illustrate how fluidly our representational focus changes depending on the analogy we wish to make. All of the following utterances are perfectly ordinary and easy to understand (for us humans):

A credit card is like:

- money (here, the representational emphasis is on its *function*)
- a check book (again, an emphasis on its *function*)
- a playing card (emphasis on its size, shape, and relative rigidity; function becomes unimportant)
- a key (a thief can slide it between the door-stop strips and the door frame to open the door)
- a CD (bits of information are stored on its magnetic strip; bits of information are stored on the CD)
- a ruler (emphasis on its straight edges)
- Tupperware (both are made of plastic)
- a leaf (drop one from a high building and watch how it falls)
- a Braille book (emphasis on the raised letters and numbers on its surface)
- the Lacoste crocodile (emphasis on its potential snob-appeal)
- a suitcase full of clothes (advertising jingle: "You can travel light when you have a Visa card ...")
- a ball-and-chain (rack up too much debt on one and its like wearing one of these ...)
- a letter opener (emphasis on its thin, rigid not-too-sharp edge)
- a hospital, a painting, a pair of skis, a Scotsman's kilt, etc. (left to the reader).

The point should by now be clear: With a little effort, by focusing on particular aspects of our passive long-term memory representations of "credit card" and the object to be put into correspondence with it, we can invariably come up with a context for which the two objects are "the same." In short, given the right contextual pressure, we can see virtually *any* object as being like any other.

We will now describe a process, which we call *dynamic context-dependent representation-building*, by means of which the analogical alignment between two objects (or experiences or situations, etc.) takes place in a progressive fashion, involving a continual interaction between long-term (passive) memory and working (active) memory. Most importantly, we claim that this dynamic process is machine-implementable, thereby allowing a machine to develop representations without the guiding hand of a programmer who knows ahead of time the analogy that the machine is supposed to produce.

2.5 The Dynamics of Representation-Building in Analogy-Making

The representation of anything is, as I hope to have shown, highly context-dependent. Unfortunately, however, the early work of computational modeling of cognition was dominated by a far more rigid view of representations, one which Lakoff (1987) has called *objectivism*, described as follows:

> On the objectivist view, reality comes complete with a unique correct, complete structure in terms of entities, properties and relations. This structure exists, independent of any human understanding.

Were this actually the case, a universal representation language and independent representation modules would make sense. "Divide and conquer" strategies for making progress in AI would be reasonable, with some groups working on the "representation problem" and other groups independently working on how to process representations.

We argue, however, that such a division of labor is simply not possible (Chalmers et al., 1992; Mitchell, 1993; Hofstadter et al., 1995: French, 1995, 1999), at least in the broad area of computational modeling of analogy-making. While analogy-making does, indeed, consists of *representing* two situations and *mapping* the corresponding parts of each of these representations onto one another, these two operations are neither independent nor sequential (i.e., first, representation, then mapping). Representation and mapping in real analogy-making are inextricably interwoven: the only way to achieve context-appropriate representations is to continually and dynamically take into consideration the mapping process. And conversely, to produce a coherent mapping, the system must continually and dynamically take into consideration the representations being processed. In other words, representations depend on mapping and vice versa and only a continual interaction of the two will allow a gradual convergence to an appropriate analogy.

The inevitable problem with any dissociation of representation and mapping can be summed up as follows. The complete representation of any object must include every possible aspect of that object, in all possible contexts, to be able to produce the vast range of potential analogies with other objects. Now, presumably, we have stored in our long-term memory just this kind of megarepresentation of all the objects, situations, experiences, etc., with which we are familiar. The problem is, though, that, even if it was possible (or desirable) to activate in working memory the full megarepresentations for both the base and target objects, the very size of these representations would produce a combinatorial explosion of possible mappings between them. To determine precisely which parts of each megarepresentation mapped onto one another would require a highly complex process of selection, filtering, and organizing of the information available in each representation. But in that case, we are back to square one, because the very reason for separating representation from mapping was to avoid this process of mapping-dependent filtering and organizing! In contrast, if, to avoid the problem of this combinatorial explosion of mappings, you use smaller, reduced representations of each object, then the obvious question becomes, "On what basis did you reduce the original representation?" And this basis has been, almost invariably: the mapping that is to be obtained by the program!

The problem then is: How do we arrive at the "right" representations that allow us to map people to cars, famous baseball players to roosters, and credit cars to kilts? We suggest that representations must be built up gradually by means of a continual interaction between (bottom-up) perceptual processes and (top-down) mapping processes. If a particular representation seems approximately appropriate for a given mapping, then that representation will continue to be developed and the associated mappings will continue to be made. If, however, the representation seems less promising, then alternative directions will be explored by the bottom-up (perceptual) processes. These two processes of representation-building and mapping are interleaved at all stages. In this way, the system gradually converges to an appropriate analogy, based on overall representations that dovetail with the final mappings.

2.6 Context-Dependent Computational Temperature

The entire representation-building/mapping process is dynamically mediated by a feedback loop (*context-dependent computational temperature*, Hofstadter, 1984; Mitchell and Hofstadter, 1990; Mitchell, 1993; French, 1995) that indicates the quality of the overall set of representations and mappings that have been discovered. It is important to distinguish context-dependent computational temperature from the temperature function that drives simulated annealing, as described in Kirkpatrick et al. (1983). The former depends on the overall quality of the representational structures and mappings that have been built between them, whereas the latter is based on a preestablished annealing schedule.

The role of temperature is to allow the system to gradually, stochastically settle into a "high-quality" set of representations and mappings. It measures, roughly speaking, how willing the system is to take risks. At high temperature, it is very willing to do so, to abandon already created structures, to make mappings that would, under normal circumstances, be considered implausible, etc. Conversely, at low temperature, the program acts far more conservatively and takes few risks. This all-important mechanism will be discussed in greater detail later in this chapter. Suffice it to say, for the moment, that temperature is inversely proportional to the program's perception of the overall quality and coherence of the representations and mappings between them. As these representations and mappings gradually improve, the temperature falls. With each temperature, there is a probability that the program will stop and select as "the analogy" the set of mappings that it has developed up to that point. The lower the temperature, the more probable that the program will conclude that it has found a set of good representations and mappings and stop.

Note that the use of a dynamic, context-sensitive computational temperature function is not limited to the computational modeling of analogy-making. For example, in the area of human mate choice, French and Kus (2006) have built a model centered on context-sensitive computational temperature applied to individuals in a population. They likened this variable (actually, the inverse of it) to an individual's *choosiness* in picking a mate: the higher the temperature, the less choosy an individual is about his/her mate; the lower the temperature, the more choosy.

2.7 Interaction between Top-Down and Bottom-Up Processes: An Example

Let us begin by illustrating what we mean by this interaction between top-down and bottom-up processing that, we claim, is the way for a system to converge to context-appropriate representations and mappings for an analogy. The example below actually occurred. I was asked to review a paper and I accepted. But, unfortunately, I was unable to complete the review on time. Finally, the Action Editor contacted me and said, "I need your review quickly." I thought, "OK, I'll put the paper to review in plain sight next to my computer keyboard. Its continual presence there will goad me into getting it done." I wanted to describe my strategy to the editor (a friend) in a more creative way than simply saying, "The paper is right next to my computer which will constantly remind me to get it done." I chose to resort to an analogy. Below is my thought process as I recorded it immediately afterward (Figure 2.1). Notice that there is nothing out of the ordinary about this process—in fact, it is a completely everyday cognitive occurrence—but it clearly illustrates the back-and-forth switching between bottom-up, stochastic processes and top-down, knowledge-based processes.

The problem, then, is to find an analogy with the base situation: "Putting the paper in plain sight, so as to continually remind me to get the review done."

The Dynamics of the Computational Modeling of Analogy-Making

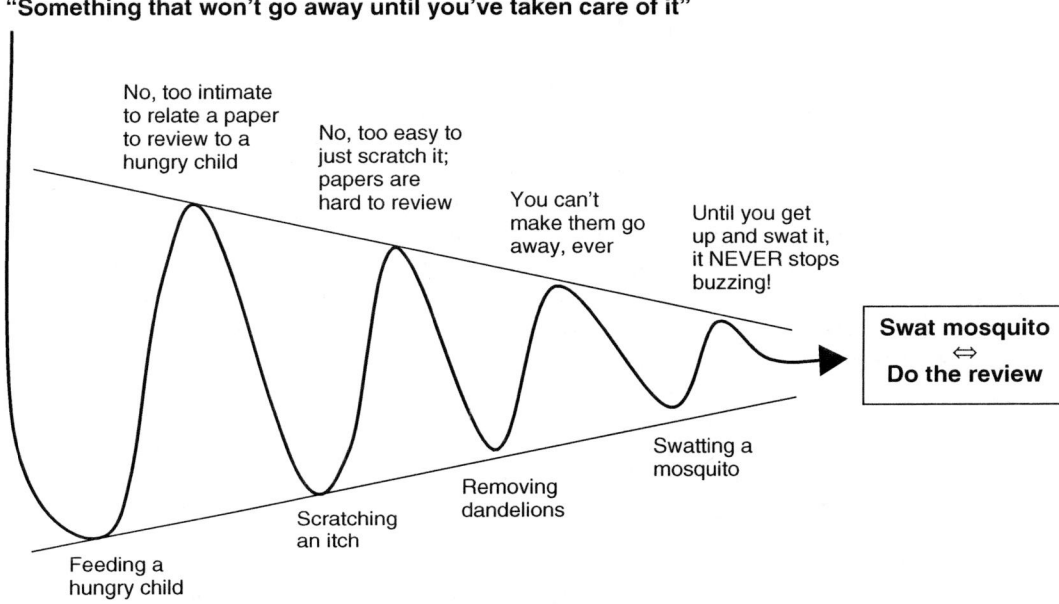

FIGURE 2.1 The gradual interaction between top-down (knowledge-based) and bottom-up (stochastic) processes that lead to context-appropriate representations.

Problem: Find an analogy for "something that won't go away until you've taken care of it."
Route to a response:

Bottom-up: "Feeding a crying child who is hungry" popped up as something that doesn't stop until you take care of it.

Top-down: Comparing a paper to review to feeding a crying, hungry child feels inappropriate because one is intimate and personal, while the other is purely professional. This then modifies the representation of the original situation: we realize that it is not merely something that won't go away until you complete it, but it is a *nonintimate* that won't go away until you do it and, thus, doesn't map well, to *intimate* situations.

Bottom-up: "Scratching an itch."

Top-down: But normal itches are *too easy* to get rid of, you simply scratch them and they disappear. Again, this redefines the representation of the original situation. It is a nonintimate something that won't go away until you've done it, but it's something that, nonetheless, takes some energy and effort to achieve.

Bottom-up: "Removing dandelions from your yard."

Top-down: No, getting rid of dandelions is *too hard*! Reviewing a paper isn't *that* hard. So, the target object has to be something that demands our attention and won't go away until we attend to it, is not too intimate, is not too easy to take care of, but not too hard, either.

Bottom-up: "Swatting a mosquito that's preventing you from sleeping." Who has never had their sleep troubled by the supremely irritating high-pitched whine of a mosquito? And, as we all know, it NEVER goes away until you get up out of bed, turn on the light, and chase it all over the room with a rolled up newspaper. And only with the Swat! that transforms the offending mosquito into a speck on the wall does the problem end. So, it fits the above criteria: it's a problem that won't go away till you take care of it, it's not too easy to do (unfortunately . . .), and it's not too hard either. That is the analogy I used.

It is crucial to note the aspects of my extensive (and passive) long-term memory representation of "mosquito," which became active in working memory, were entirely dependent on the mapping that I wanted to achieve. In other words, the working-memory representation does not include myriad other characteristics of mosquitoes, such as their disease-bearing nature, that they breed in water, that they feed on blood, that they have six legs, that they do not weigh very much, that they live from 3 to 100 days, that most of their head consists of their compound eye, etc. Further, in finding this analogy, we also modified our working-memory representation of "review a paper," in particular, focusing on reviews' strictly professional, nonintimate nature, on the fact that reviews are neither too easy nor very hard to do, of the fact that no amount of waiting will allow you to make the slightest progress on the review, etc., all aspects that were not part of my working-memory representation of "review a paper" when I began looking for an appropriate analogy.

2.8 Computational Models Implementing this Bottom-Up/Top-Down Interaction

"Hybrid" models of analogy-making (Kokinov and French, 2003) explicitly attempt to integrate mechanisms from symbolic and connectionist approaches to AI. Crucially, they rely on a distinction between long-term memory and working memory, although the implementation details of this separation may vary.

The overarching principles of the class of models that I will briefly describe were originally formulated in Douglas Hofstadter's research group from the 1980s to the present. Many people have contributed to the development and implementation of these ideas, including Hofstadter (1984), Defays (1990), Mitchell and Hofstadter (1990), Hofstadter and Mitchell (1992), Mitchell (1993), French (1995), Hofstadter et al. (1995), McGraw (1995), Bolland and Wiles (2000), Marshall (2002a, 2002b), Bolland (2005), and so on.

These models—Copycat, Numbo, Tabletop, Letter Spirit, Metacat, the Fluid Analogies Engine, etc.—are all agent-driven models that operate in microworlds that are, by design, felt to be rich enough to demonstrate the architectural principles of the models, but not so complicated that extraneous details interfere with understanding. It is beyond the scope of this chapter to present a detailed defense of microworlds. Suffice it to say that microdomains can be appropriately likened to the "ideal objects" studied in elementary physics to understand the principles of movement, gravity, force, reaction, etc. Only once these principles are mastered in the "microdomain" of these ideal objects does it make sense to be concerned with friction, air resistance, dilation due to heat, etc. The microdomains in which these analogy-making programs work are designed to help isolate the principles of analogy-making. Microdomains are to analogy-making what "ideal objects" are to physics.

2.9 Architectural Principles

The basic features of the Hofstadter family of architectures are described in the following section (cf. French, 1995, chap. 3; see also, Bolland [1997] for a particularly clear presentation of Copycat, the mainstay program of this class of programs).

2.9.1 The "Slipnet," a Semantic Network

This is a semantic network with fixed concepts, but in which the distances between concepts vary according to the situation being processed. This context-dependent malleability is a crucial feature of the architecture.

A rather vivid example (Bonnetain, 1986, personal communication) will serve to illustrate how significantly context can modify distances between concepts in a semantic network. In France, unlike in the United States, a driver's license is needed *only* when one is operating a motor vehicle and does not serve as an identity card. Further, while laws aimed at preventing under-age drinking in France do exist, they are rarely enforced. In other words, no one ever checks young people's age at the entrance to a bar in France, much less asks for a driver's license. As a result, in the mind of a 19-year-old French college student, the concepts "beer" and "driver's license" are every bit as distant as, say, "refrigerator" and "tennis ball." But were the same French college student to go to the United States for a year—where drinking age *is* enforced and a driver's license is invariably used as the means of checking ages—his/her conceptual distance between "driver's license" and "beer" would soon be radically altered. This would result in a change in the Slipnet structure—namely, the shortening of the conceptual distance between the concepts "beer" and "driver's license."

2.9.2 The Workspace

This is where all the representational structures needed to perceive the situation being examined are gradually built up. Consider the following group of letters to be parsed:

AAB

There are two obvious possible parsings of this three-letter sequence, namely

- An AA group, followed by a B.
- An A, followed by an AB group.

However, neither representation is a priori more correct than the other. So, if the AAB group is embedded in a context, such as

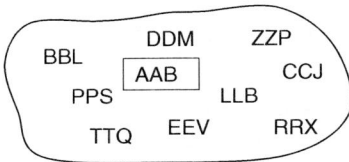

the chances are that the letter string will be parsed as a "An AA group, followed by a B."

In contrast, if the context in which this image is embedded is

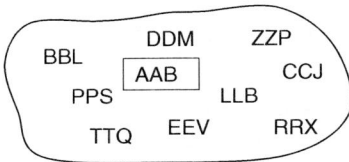

this will significantly increase the probability that it will be parsed as a "An A, followed by an AB group."

So, in a setting in which the program, to make an analogy, must parse various configurations of **AAB**, both of the following structures:

AA and **AB**

are likely to be included in the workspace. These two potential representations—which cannot coexist simultaneously in conscious perception, i.e., in the Worldview (see below)—will then compete with each other based on what else is found on the table. Ultimately, one will win out and will be used in the final representation.

2.9.3 The Worldview

These are the structures that are currently, consciously being perceived. The assumption here is that, just as in a Necker cube (or the Old Woman/Young Woman illusion, or any other ambiguous figure), one cannot perceive both of the possible interpretations simultaneously. Thus, there is one representation, based on the context as it is perceived up to that moment (say, for example "a pair of forks with a knife to the right"). Structures move in and out of the Worldview during the process of building up the final representation of the objects on the table. The longer a given representation remains in the Worldview, the lower the computational temperature of the simulation, meaning that the program believes that it has found a good, coherent way of representing the situation at hand. When the program stops, what is in the Worldview is considered to be the analogy that the program has discovered.

2.9.4 Codelets

These are the task-specific "worker ants" of the program. They carry out all of the processes necessary to produce an emerging understanding of the situation being represented and the mappings being made. None of these task-specific agents has an overview of what is going on, any more than a single ant has an overview of the anthill in which it toils. Agents are released based on numerous triggers, such as activation levels of concepts in the semantic network, computational temperature, and structures that have already been discovered. They each carry out very simple tasks. They build and break correspondences between structures, they discover relations between individual objects, etc. They perform a single task and then "die."

2.9.5 The Coderack

This is the "staging ground" for the task-specific agents. When an agent is called, it does not run immediately, but rather it is put onto the coderack. Each agent has an "urgency" value that expresses its importance and agents are probabilistically selected to run based on their urgency.

2.9.6 Dynamic Codelet Selection via Computational Temperature

This is where computational temperature plays a key role. Assume that there are N agents on the coderack waiting to run, each of which has an urgency u. The general idea of temperature-based selection is that, if the overall temperature (T) is very high ($T = 1$ is the "neutral" temperature), the probability of selection should be essentially independent of their urgency, and each agent will be as likely to be selected as any other (i.e., a uniform selection distribution over all agents, regardless of their urgency). In contrast, the lower the temperature, the more codelet selection will take place based strictly on the value of a codelet's urgency (i.e., in this case, selection is essentially deterministically based on urgency).

If there are N agents, $A_1, A_2, \ldots, A_N$, each with a raw urgency, u_i, then for a given computational temperature, T, the probability that a given codelet, A_i, will be selected is given by

$$\Pr(A_i) = \frac{u_i^\alpha}{\sum_{k=1}^{N} u_i^\alpha} \tag{2.1}$$

where $\alpha = 1/T$.

It is clear that, as $T \to \infty$ (or gets large), Eq. (2.1) reduces to

$$\Pr(A_i) = \frac{u_i^0}{\sum_{k=1}^{N} u_i^0} = \frac{1}{\sum_{k=1}^{N} 1} = \frac{1}{N} \tag{2.2}$$

which is what is required, i.e., a uniform selection distribution. In contrast, when temperature is low, α becomes large and the agent with the highest urgency will, if the temperature is low enough, be picked almost deterministically over its rivals with lower raw urgency.

An example will be useful to clarify this. Assume that there are three agents, $A_1, A_2,$ and A_3, with respective urgencies, u_i, of 2, 3, and 5. When the temperature, T, is high, $\alpha \to 0$ and we have

$$\Pr(A_1) = \frac{2^0}{2^0 + 3^0 + 5^0} = \frac{1}{3}$$

$$\Pr(A_2) = \frac{3^0}{2^0 + 3^0 + 5^0} = \frac{1}{3}$$

$$\Pr(A_3) = \frac{3^0}{2^0 + 3^0 + 5^0} = \frac{1}{3}$$

When $T = 1$, then $\alpha = 1$. This is the "neutral" selection condition, i.e., each agent has a selection probability determined by its urgency divided by the total urgency of all agents on the Coderack.

$$\Pr(A_1) = \frac{2^1}{2^1 + 3^1 + 5^1} = \frac{2}{10} = 0.2$$

$$\Pr(A_2) = \frac{3^1}{2^1 + 3^1 + 5^1} = \frac{3}{10} = 0.3$$

$$\Pr(A_3) = \frac{5^1}{2^1 + 3^1 + 5^1} = \frac{5}{10} = 0.5$$

Finally, assume the temperature falls to, say, 1/5. This means that $\alpha = 5$ and we have

$$\Pr(A_1) = \frac{2^5}{2^5 + 3^5 + 5^5} = \frac{32}{3400} \approx 0.01$$

$$\Pr(A_2) = \frac{3^5}{2^5 + 3^5 + 5^5} = \frac{243}{3400} \approx 0.07$$

$$\Pr(A_3) = \frac{5^5}{2^5 + 3^5 + 5^5} = \frac{5}{3400} \approx 0.92$$

In this case, the selection probability of agent A_3 leaps to 0.92. In other words, the fact that A_3 has a raw urgency of 5, as opposed to its rivals with urgencies of 3 and 2, means that it will be chosen to run 92 times out of 100 when $T = 1/5$.

One might wonder why this mechanism is of any importance. In the language of classic tree-searching AI, the answer is that it allows the program to occasionally (especially when the temperature is high) explore branches of the search tree that it would never have explored, thereby potentially arriving at a truly unexpected, excellent solution to a problem at the end of an exploration path that would not have been taken without this mechanism.

But, at the same time, one does not want the program to be searching these unusual paths very often. But one must not restrict the program to *never* search these paths. And, especially, in the event of hitting what appears to be a "dead end" when attempting to find the answer to a problem, computational temperature, as implemented above, allows the program to break structures it has already made, to reform structures and make mappings in a new way, and then to gradually settle using these new structures. If another dead end is encountered, the process will repeat.

One might wonder what would prevent this process from going on forever in the event that there are no good structures to be found. The answer is that at each temperature, there is a "StopWork" agent that is on the Coderack and, when it runs, all further processing comes to a halt. The structures that are currently in the Worldview, good or bad, are taken to be the best answer for that run. The "StopWork" agent will eventually run and bring to an end this cycle of building-breaking structures brought about by temperature oscillations. What frequently does, in fact, happen is that the program will find what it believes to be good structures, the temperature will therefore fall, but the overall coherence never comes and so the temperature rises. But then the program once again discovers the same structures that led it down the fruitless path, and so on. In short, the program must remember where it has been, a capability which the earlier versions of this architecture did not have. This problem has been dealt with, albeit to a limited extent, in later versions of this architecture (e.g., Marshall, 2002a, 2002b).

In conclusion, Mitchell (1993, p. 44) sums up the importance of context-dependent computational temperature as follows:

> ... [Temperature] has two roles: it measures the degree of perceptual organization in the system (its value at any moment is a function of the amount and quality of structure built so far) and it controls the degree of randomness used in making decisions .... Higher temperatures reflect the fact that there is little information on which to base decisions; lower temperatures reflect the fact that there is greater certainty about the basis for decisions.

2.9.7 Local, Stochastic Processing

It is worth reiterating here that there is nothing that resembles a global executive in these programs, any more than there is a global executive in an anthill. All results emerge from the limited actions of a large number of agents responsible for finding/breaking structures, making maps, communicating with the semantic network, etc. All decisions—whether or not to group certain elements, whether or not to put certain structures into correspondence, etc.—are made stochastically.

2.9.8 Integration of Representation-Building and Correspondence-Finding

Finally, and arguably, most importantly, there is no separation of the processes of building representations and finding correspondences between elements of different representations. A fundamental tenet of Copycat and Tabletop is that it is impossible to separate the process of representation-building from the process of correspondence-building (Chalmers et al., 1992).

2.10 How this Type of Program Works: The Details

2.10.1 Copycat

Certainly the best known implementation of this type of architecture is the Copycat program (Mitchell and Hofstadter, 1990; Mitchell, 1993). The best place to find a reasonably complete, yet succinct description of this program is in Mitchell (2001), an article that was written by the program's author. There is also an online executable version of Copycat, complete with a description of how the program works on Melanie Mitchell's website: http://web.cecs.pdx.edu/~mm/how-to-get-copycat.html. In addition, Scott Bolland at the University of Queensland wrote a Java version of Copycat, which can be downloaded from http://www2.psy.uq.edu.au/CogPsych/Copycat and http://www2.psy.uq.edu.au/CogPsych/Copycat/Tutorial/.

2.10.2 Tabletop

In what follows we will illustrate the principles discussed above by briefly looking at another of these programs, Tabletop (French, 1995), designed to work in a simulated tabletop microworld. The basic idea is that there are some ordinary objects (forks, spoons, cups, saucers, glasses, plates, salt and pepper shakers, etc.) on a small table. There are two people (Henry and Eliza) seated across from one another at the table. Henry touches one object and says to Eliza, "Do this!" (meaning, of course, "Touch the object you think is analogous to the one I just touched") (see Figure 2.2).

So, for example, if there is a cup in front of Henry, and only silverware in front of Eliza, and Henry touches his cup, Eliza will probably reach across the table and touch Henry's cup (reason: silverware is

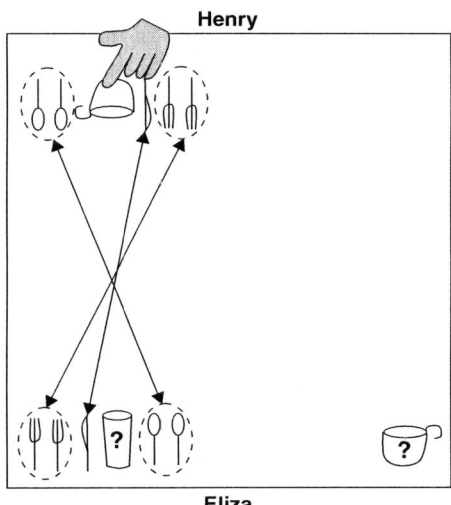

FIGURE 2.2 A table configuration where the context provided by the silverware surrounding Henry's cup and Eliza's glass influences Eliza's initial bottom-up "impulse" to respond by simply touching the cup on her side of the table.

semantically "too far" from the object that Henry touched, so she might as well touch exactly the same object he touched). Or suppose there is a cup in front of Henry and a glass and two forks in front of Eliza. Henry touches his cup. In this case, Eliza will (most likely) touch the glass in front of her, because, even though it is not the same as the cup Henry touched, it is in an analogous place on the table and the glass is semantically not too far from the cup.

The contextual factors that can be brought to bear to shift the initial bottom-up pressures to touch one or the other object are almost limitless. Suppose Henry's cup is surrounded by two forks and a knife on one side of the cup and two spoons on the other and Eliza has, on her side of the table, only an isolated cup and an isolated glass. Henry touches his cup. Eliza will, almost certainly, touch the cup on her side of the table. If, however, her glass is surrounded by the same objects as Henry's cup, i.e., by two forks and a knife on one side and two spoons on the other (Figure 2.2), she will considerably be more likely to allow this context to shift her choice toward touching her glass rather than the isolated cup. In other words, even if her first (bottom-up) response might have been to touch the cup on her side of the table, noticing the objects (and how they are grouped) around Henry's cup and "the same" objects around her glass affects her initial bottom-up impulse to touch her cup. The probability of her touching her glass is considerably increased by the "top-down" pressures induced by the discovery of the mappings between the groups of silverware surrounding the glass and the cup.

2.11 How Tabletop Finds a Reasonable Solution

How the program, Tabletop, finds a solution to this particular problem is explained in detail for the interested reader in French (1995, pp. 95–111). We will give an overview of the algorithm below.

As we described earlier, the architecture consists of a semantic network (the Slipnet), a Workspace, and a launching area for the codelets, the exploratory/structure-building agents (the Coderack). There is continual interaction between these three areas, as illustrated in Figure 2.4.

First, consider the semantic memory (called the "Slipnet"). This spreading-activation network, similar to the networks from Collins and Loftus (1975), is designed to represent the interconnections of the concepts acquired through a lifetime of interacting with the world. It represents, in some sense, our "top-down"

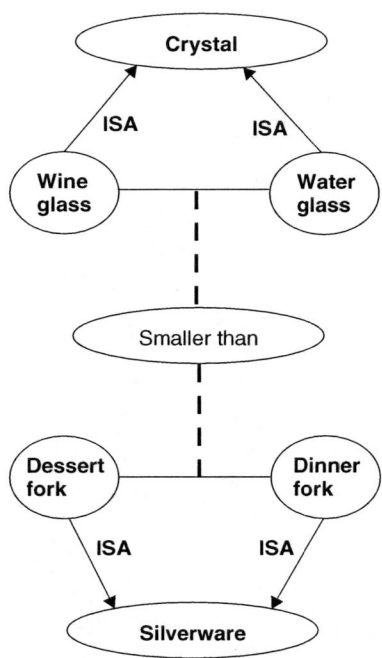

FIGURE 2.3 Part of the semantic memory (Slipnet) used in solving the Tabletop configuration shown in Figure 2.2.

knowledge of the world. The links represent connections between concepts in the world and their length corresponds to the semantic distances between them (Shepard, 1962). Most importantly, as mentioned above, it is *malleable*, i.e., it can change dynamically depending on the concepts that are activated. If, for example, the concept "smaller than" becomes active because, for example, we notice that dessert forks are smaller than dinner forks, then we become more sensitive to this relationship for all pairs of concepts for which it applies (e.g., saucers versus dinner plates, wine glasses versus water glasses, etc.) (see Figure 2.3). Among the concepts in the semantic memory is "group." Concepts in semantic memory correspond, at least approximately, to the agents that will be building structures in Working Memory.

Let us start the process by launching some agent, for example, a "Find-Group" agent. It looks for a "group" on the table. It does not see forks, knives, spoons, cups, etc. It merely looks for clusters of objects, any objects. It finds one (say, the group closest to Henry). This causes several things to happen:

- Activation is sent by the agent to the "group" node in the semantic network. The "group" node in the semantic network becomes more active.
- It spawns (i.e., places on the Coderack) a number of finer-grained agents that will, if they run, look for things that might be inside the just-discovered group. For example, a Find-Group agent that has found a group on the table, would put several Find-End-Object agents on the Coderack, each having as a parameter the Group that has just been found. When one of these Find-End-Object agents runs it will explore an extremity of the Group that was found by its parent Find-Group agent to determine what the objects at the extremity of the group are.

The added activity of the "group" node in the semantic network causes new "Find-Group" agents to be put on the Coderack, waiting there to be selected to run. (Here we see the permanent interaction between the bottom-up processes of the agents and the semantic network.) After a while the Find-Group agents will be unable to find any more groups. Agents that happen to be on the Coderack will continue to be selected for a while, but since they would not find any new groups, activation of "group" in the Slipnet will fall. This will mean that no more new Find-Group agents will be put on the Coderack. The activity of exploring the

The Dynamics of the Computational Modeling of Analogy-Making

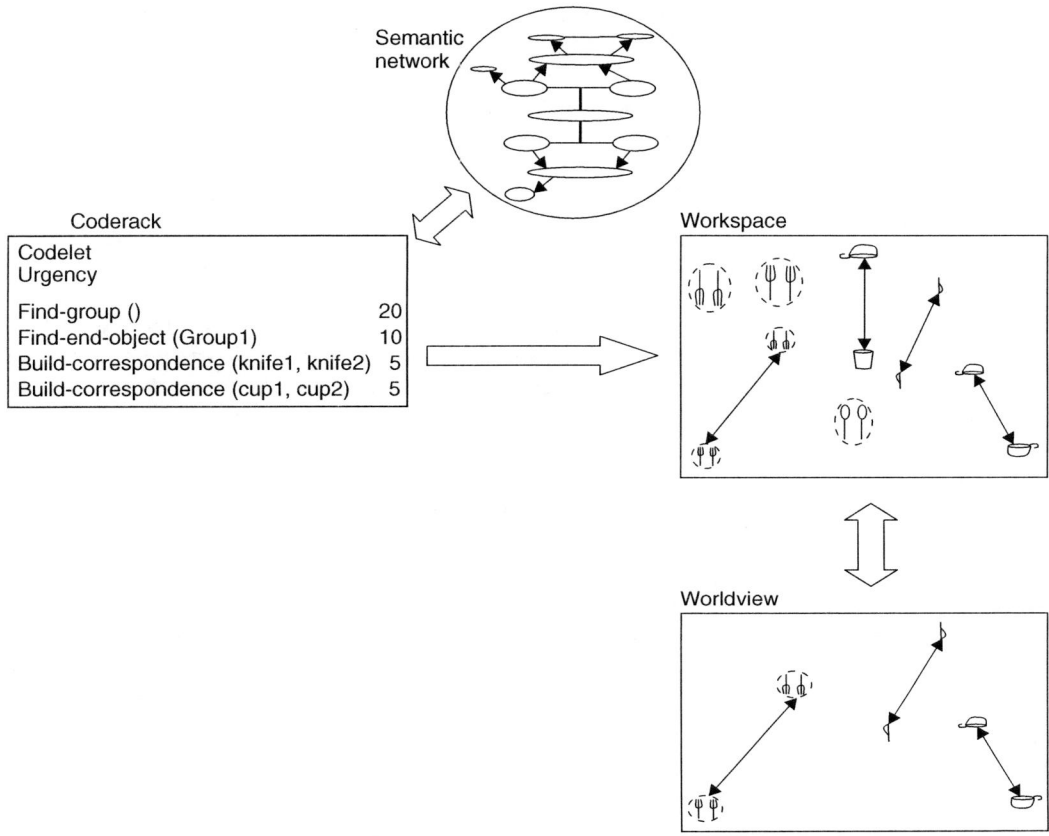

FIGURE 2.4 An illustration of Tabletop in the process of building its representations of the table configuration in Figure 2.2. Notice that at this relatively early stage of a run of the program, the Worldview contains a cup–cup correspondence, a correspondence between the two fork groups and one between the two knives. Were the program to stop now, Eliza would touch the cup on her side of the table. In general, it will end up replacing the cup–cup correspondence by the more globally coherent cup–glass correspondence, once it has found the mappings between items and subgroups of items in the two groups of silverware.

table will now be in the hands of other codelets, some of which have been spawned by the Find-Group agents.

As structures are gradually discovered—for example, the fork subgroups, the spoon subgroups, the correspondence between the two groups of two spoons (a "correspondence" being a structure, just like a "group" is a structure), etc.—they are put into the program's Workspace, where they can be accessed by the agents.

The best, most coherent structures discovered to date are put into the Worldview. But structures are continually competing to get into the Worldview from the Workspace. So, for example, early in the exploration of the table configuration in Figure 2.2, the structure mapping the two cups onto one another was part of the Worldview (see Figure 2.4). In other words, early on it was considered to be the best mapping of the object touched by Henry to the object that Eliza should touch. But as the program gradually discovered the other structures surrounding Henry's cup and Eliza's glass, the overall coherence of the mapping between Henry's cup and Eliza's glass grew stronger, until (usually, at least) it displaces the cup–cup mapping in the Worldview.

The overall coherence of structures in the Worldview determines the temperature of the program. If there is little coherence, then the temperature rises, causing structures in the Workspace to be broken. This allows new agents to try again and gives the program the possibility of discovering new structures from the "rubble" of the broken structures. However, if the structures in the Worldview are good and coherent, temperature falls. When temperature falls low enough, the program stops and picks as its answer the correspondence in the Worldview containing the touched object.

2.12 The Issue of Scaling Up

One of the important claims about this architecture is that, at least theoretically, it should be capable of scale up. This is because the program, by design, *never examines all possible structures* on its way to an answer. It starts its exploration very broadly and gradually narrows its focus as time goes on, based on what it has already found. This means that many structures, many potential mappings, many potential groupings, etc., never get examined to any depth (or, in some cases, are not seen at all) as the program moves toward an answer.

So, in the example shown in Figure 2.2, if there were, say, 30 groups of objects on the table, instead of two as there are now, the program would, almost certainly, not explore (or potentially, even find) all of them. Why? Because very soon after the discovery of the initial, most salient groups, finer-grained agents would already be exploring these groups. Most probably, if there were a reasonable answer to be found in these groups, it would be found. Only if this exploration did not lead to an answer in a reasonable time, would temperature rise and the old structures be broken, thus pushing the program to explore in new directions. In other words, the amount of exploration that is done is not proportional to the amount of potential exploration.

2.13 The Potential Long-Term Impact of the Mechanisms Presented

We have attempted to explain, by means of a number of simple everyday examples, the flexibility of human analogy-making capabilities. This argues for the importance of programs that develop representations by means of a continual interaction between a knowledge store and working memory. This crucial ability will potentially allow scaled-up versions of these programs to navigate safely between the Scylla of hand-tailored representations designed with the desired mappings in mind and the Charybdis of unfiltered megarepresentations that entail a combinatorial explosion of mappings. We believe that the mechanism of context-dependent computational temperature will be a necessary ingredient in programs that will one day be called truly creative. It allows programs to explore—infrequently, but at least occasionally—highly improbable areas of space where wonderfully creative answers to a problem could lie.

2.14 Conclusions

I would go so far as to say that analogy-making is, to steal a phrase from the literature on consciousness, The Hard Problem of AI. Not the Impossible Problem, but Very, Very Hard, all the same, and perhaps also The Central Problem. And one that will not be fully solved for a very long time. Unlike chess, unlike optical character recognition, unlike wending your way across a desert with a large database of what to expect, unlike finding someone's fingerprints or face in a huge database of fingerprints or faces, etc.—all of which, however difficult, can be precisely specified—the mechanisms of analogy-making are much harder to pin down. They involve representing situations so that they can be seen one way in one context, another way in another context. They involve enormous problems of extracting and aligning just the right parts of different representations. And what makes this so hard is that almost anything can, under the right circumstances, be "like" something else. A claw hammer is like a back scratcher. Stiletto heels are like a Bengal tiger. Thinking a high IQ is enough to succeed at research is like thinking that being tall is enough to succeed at professional basketball. And so on, ad infinitum. Analogies cover all domains of human thought. We abstract something from an event in one domain and explain in by an analogy to something in a completely different domain. And we do this constantly. It is one of our most powerful means of explaining something new or unusual. Progress in this area will, no doubt, be slow but will, without question, be one of the key means by which we will move forward in AI.

References

Bolland, S. (2005). *FAE: The Fluid Analogies Engine. A Dynamic, Hybrid model of Perception and Mental Deliberation.* Unpublished Ph.D. dissertation, Computer Science and Electrical Engineering, University of Queensland, Brisbane, Australia.

Bolland, S. and Wiles, J. (2000). Adapting Copycat to visual object recognition. *Proceedings of the Fifth Biennial Australasian Cognitive Science Conference*, Adelaide, Australia.

Bolland, S. (1997). http://www2.psy.uq.edu.au/CogPsych/Copycat/ and http://www2.psy.uq.edu.au/CogPsych/Copycat/Tutorial/

Chalmers, D. J., French, R. M., and Hofstadter, D. R. (1992). High-level perception, representation, and analogy: A critique of artificial intelligence methodology. *Journal of Experimental & Theoretical Artificial Intelligence*, 4, 185–211.

Collins, A. M., and Loftus, E. F. (1975). A spreading activation theory of semantic memory. *Psychological Review*, 82, 407–428.

Defays, D. (1990). Numbo: A study in cognition and recognition. *The Journal for the Integrated Study of Artificial Intelligence, Cognitive Science and Applied Epistemology. Issues in Connectionism: Part I*, 7(2), 217–243.

Falkenhainer, B., Forbus, K. D., and Gentner, D. (1989). The structure-mapping engine. *Artificial Intelligence*, 41(1), 1–63.

Forbus, K., Gentner, D., and Law, K. (1995). MAC/FAC: A model of similarity-based retrieval. *Cognitive Science*, 19(2), 141–205.

French, R. M. and Kus, E. (2006). KAMA: A temperature-driven model of mate-choice using dynamically evolving representations. *Adaptive Behavior* (under review).

French, R. M. (1995). *The Subtlety of Sameness*. Cambridge, MA: MIT Press.

French, R. M. (1997). When coffee cups are like old elephants or why representation modules don't make sense. In A. Riegler, M. Peschl, and A. Von Stein (Eds.), *Proceedings of the International Conference New Trends in Cognitive Science*, pp. 158–163 Austrian Society for Cognitive Science.

French, R. M. (2002). The computational modeling of analogy-making. *Trends in Cognitive Sciences*, 6(5), 200–205.

Gentner, D. (1983). Structure-mapping: A theoretical framework for analogy. *Cognitive Science*, 7(2), 155–170.

Gentner, D., Holyoak, K., and Kokinov, B. (Eds.) (2001). *The Analogical Mind: Perspectives from Cognitive Science*. Cambridge, MA: MIT Press.

Hall, R. (1989). Computational approaches to analogical reasoning: A comparative analysis. *Artificial Intelligence*, 39, 39–120.

Hofstadter, D. R., and Mitchell, M. (1992). An overview of the Copycat project. In K. J. Holyoak and J. Barnden (Eds.), *Connectionist Approaches to Analogy, Metaphor, and Case-Based Reasoning*. Norwood, NJ: Ablex.

Hofstadter, D. R. and the Fluid Analogies Research Group (1995). *Fluid Concepts and Creative Analogies*. New York: Basic Books.

Hofstadter, D. R. (1984). *The Copycat Project: An Experiment in Nondeterminism and Creative Analogies.* AI Memo No. 755, Massachusetts Institute of Technology, Cambridge, MA.

Holyoak, K., and Thagard, P. (1989). Analogical mapping by constraint satisfaction. *Cognitive Science*, 13, 295–355

Hummel, J., and Holyoak, K. (1997). Distributed representations of structure: A theory of analogical access and mapping. *Psychological Review*, 104, 427–466.

James, W. (1890). *The Principles of Psychology.* New York: Henry Holt & Co.

Kirkpatrick, S., Gelatt Jr., C. D., and Vecchi, M. P. (1983). Optimization by simulated annealing. *Science*, 220, 671–680.

Kokinov, B. (1994). A hybrid model of analogical reasoning. In K. Holyoak and J. Barnden (Eds.), *Advances in Connectionist and Neural Computation Theory, Vol. 2, Analogical Connections.* Norwood, NJ: Ablex.

Kokinov, B. and French, R. M. (2003). Computational models of analogy-making. In L. Nadel (Ed.), *Macmillan Encyclopedia of Cognitive Science*, vol. 1, pp. 113–118. London: Nature Publishing Group.

Lakoff, G. (1987). *Women, Fire, and Dangerous Things: What Categories Reveal about the Mind.* Chicago: University of Chicago Press.

Marshall, J. (2002a). METACAT: http://www.cs.pomona.edu/~marshall/metacat/

Marshall, J. (2002b). Metacat: A self-watching cognitive architecture for analogy-making. In W. D. Gray and C. D. Schunn (Eds.), *Proceedings of the 24th Annual Conference of the Cognitive Science Society*, pp. 631–636. Mahwah, NJ: Lawrence Erlbaum Associates.

McGraw, G. (1995) *Letter Spirit (Part One): Emergent High-Level Perception of Letters Using Fluid Concepts.* Unpublished Ph.D. dissertation, Indiana University, Bloomington, IN. Available at: http://www.cogsci.indiana.edu/farg/mcgrawg/thesis.html>.

Mitchell, M. and Hofstadter, D. R. (1990). The emergence of understanding in a computer model of concepts and analogy-making. *Physica D*, 42, 322–334.

Mitchell, M. (1993). *Analogy-Making as Perception: A Computer Model.* Cambridge, MA: MIT Press.

Mitchell, M. (2001). Analogy-making as a complex adaptive system. In L. Segel and I. Cohen (Ed.), *Design Principles for the Immune System and Other Distributed Autonomous Systems*, pp. 335–359. New York: Oxford University Press.

Shepard, R. N. (1962). The analysis of proximities: Multidimensional scaling with an unknown distance function. I. *Psychometrika*, 27, 125–140.

3
Impact of the Semantic Web on Modeling and Simulation

John A. Miller
University of Georgia

Congzhou He
University of Georgia

Julia I. Couto
University of Georgia

3.1 Introduction 3-1
3.2 Semantic Web: Relevant Issues 3-2
3.3 Conceptual Basis for Discrete-Event Simulation 3-4
3.4 Types of Mathematical Models 3-6
  Classification Based on State • Time-Based Classification • Causality-Based Classification • Classification Based on Determinism
3.5 Adding Semantics to Simulation Models 3-10
3.6 Overview of DeSO 3-12
3.7 Overview of DeMO 3-14
3.8 Summary 3-18

3.1 Introduction

During the mid-1990s, the World Wide Web began to substantially impact the use of computer technology. This sparked the development of the field of Web-based simulation, which is still advancing today. This chapter will examine how an ongoing major initiative involving the Web, the semantic Web, may further impact modeling and simulation (M&S).

More specifically, this chapter considers the issue of using semantics in M&S. The impetus for this is the large initiative to develop the next-generation Web, the semantic Web being developed by the artificial intelligence (AI), database and information retrieval communities. A complimentary parallel track is represented by the model-driven architecture (MDA) approach being developed by object management group (OMG) and the software engineering community. The goal of this initiative is for all software development to be model-driven.

Semantics (and the semantic Web) will likely impact the M&S community in two ways. First, the community should develop ontology to delineate, define, and relate the concepts in the field. Ontology for M&S should be logically connected to more general (or higher level) ontology, e.g., one for mathematics such as Monet (Caprotti et al., 2004) or one for general knowledge upper ontology such as the suggested upper merged ontology (SUMO) (Niles and Pease, 2001). Second, simulation models, model components, and other artifacts should be provided with richer semantic descriptions. The least disruptive way to do this is through annotation in which the artifacts refer to semantic models (e.g., a concept in ontology).

The fact that the semantic Web is being developed and simulation artifacts can be semantically annotated, begs the question of why do it. This question relates to the basic motivation for having the semantic Web and its services. For the M&S community, semantics represented in ontology provides standard terminology to the community and beyond, so that common understanding of concepts and relationships can be achieved, which, in turn, increases the potential for application interoperability and reuse of simulation artifacts. Semantic Web technology can also be used for the discovery of simulation components, composition of simulation components, implementation assistance, verification, and automated testing.

To make the discussions in this chapter more directed, we will develop as we go a small ontology for discrete-event simulation (DeSO). The purpose of this ontology is to provide a general conceptual foundation for M&S. Every effort was made to keep the ontology from becoming convoluted. If concepts were too complex to be defined in a straightforward way, they were left out. At this point, DeSO is a toy example. Later, we plan to expand and merge it with the more developed discrete-event modeling ontology (DeMO). DeMO is oriented toward discrete-event modeling techniques such as Markov chains, finite state machines, Petri nets, and event graphs. We are making DeSO more general in the following ways:

1. Include concepts related to common methodologies for creating simulation models, e.g., those built using simulation languages (or programming languages augmented with simulation libraries). DeMO is more oriented toward formal modeling techniques (of course, that was a sensible place to start since these are well defined, at least mathematically).
2. Take a first step to extend DeMO with concepts from combined continuous and discrete simulation, without obscuring the discrete event concepts.
3. Include enough concepts to allow, say, a simulation engine to interact with an animation engine. The animation engine would permit realistic (or at least interesting) rendering using 2D or 3D graphics. Either engine (simulation or animation) could include software such as a physics engine to enhance the realism of animation. This is part of the motivation for item 2, allowing, for example, smooth continuous motion governed by Newton's laws of motion.

Building such a large ontology is a daunting task, which needs guidance from well-established foundational knowledge. In this work, we use the following foundational sources: modeling, simulation, systems theory, physics, mathematics, and philosophy.

We endeavored to make our definitions as compatible as we could with existing definitions within these fields. Many sources were used for this including Wikipedia (Wikipedia, 2006a), WordNet (Miller et al., 1990), OpenMath (OpenMath, 2006), SUMO (Niles and Pease, 2001), Stanford Encyclopedia of Philosophy (Zalta, 2006), AstroOnto (Shaya et al., 2006), Simulation reference markup language (SRML) (Reichenthal, 2002), extensible modeling and simulation framework (XMSF) (Brutzman, 2004), and discrete-event systems specification (DEVS, 2005) as well as textbooks and papers in a variety of fields (see References).

Finding and defining the concepts is hard enough, but the subsequent step of determining a minimal set of useful properties is even more difficult. More important to get right are the relationships between the concepts. This is where much of the formal semantics comes in, since many of the concepts are defined in natural (not formal) languages. Indeed, certain semantically primitive concepts are not formally definable.

The rest of this chapter is organized as follows. In Section 3.2, we overview developments in the semantic Web relevant to creating and using ontology for M&S. Section 3.3 provides a conceptual framework suitable for defining the top concepts for such ontology. This is followed, in Section 3.4, by high-level classifications based on these main concepts. Techniques for adding semantics to simulation models are given in Section 3.5. A summary of DeSO is presented in Section 3.6. An overview of DeMO is given Section 3.7. Finally, Section 3.8 summarizes the chapter.

3.2 Semantic Web: Relevant Issues

Ever since the article was published by Berners-Lee et al. (2001), in the *Scientific American*, there has been a great deal of research and development on the semantic Web. Indeed, much of it is rooted in prior research in knowledge representation, distributed AI, database systems, and information retrieval. A large portion of the current Web consists of hypertext markup language (HTML) pages (either static or dynamic) intended for humans to read. To make the Web more accessible by programs (or agents), the Web content needs to be organized better, linking meaning with content. An obvious first step is to replace the formatting tags of HTML, with ones that are related to content. This is the purpose of the extensible markup language (XML) and its schema languages: data type definition (DTD) and XML schema definition (XSD). XML is good for representing nested structures in documents, but is weak regarding named relationships.

The resource description framework (RDF) is useful for indicating that certain entities of interest are discussed in a document and that these entities are related to other entities in this and other documents. In this way, it permits logical connections within and between documents. Although, one might think that hyperlinks in HTML or XLinks in XML documents play a similar role, from a program's perspective these are akin to untyped pointers. RDF provides a richer modeling language, and although RDF syntax can be represented using XML, the underlying abstract models for the two languages are fundamentally different. The abstract model for XML is tree based, while the model for RDF is graph based (Berners-Lee, 1998; Johnston, 2005).

The above additions to the Web mainly provide it with better organization, which is key in making the Web more useful to programs. The real goal of the semantic Web is to make the Web content more understandable to programs. One approach is to use natural language processing and text understanding. Long-term research efforts in these areas are beginning to bear fruit, and various algorithms have been designed to process text at morphological, syntactic, semantic, and discoursal levels with reasonable accuracy (Mitkov, 2003). However, they are not the principal focus of current semantic Web research. As already mentioned, the tags used by XML are more meaningful than the tags used by HTML (e.g., <h3> ... </h3> versus <address> ... </address>. While certainly true, this meaningfulness is mainly attributed to human understanding, but what does it mean to a program? An initial step to make documents more understandable to a program is to lessen the program's need to understand all of the documents individually. This can be done by relying on a schema that applies to several documents of the same kind. If the program knows the XSD for a group of documents, then it can more readily process the document. Furthermore, if the program knows the RDF schema (RDFS) for this group, it can process relationships between entities in this group of documents. This capability is particularly useful for semantic search (Sheth et al., 2005). Whereas, Web search engines such as Yahoo and Google use keyword search and page ranking schemes, semantic search follows meaningful links, and has the potential, in specific domains, to enhance precision and recall of documents as well as direct one to relevant portions of documents (Noronha and Silva, 2004). (Precision means the fraction of retrieved documents that are relevant; recall means the fraction of relevant documents that are retrieved.) Still, the depth of program understanding is rather shallow (useful, but shallow).

Deep understanding approaching human levels is such a long-term goal that something more intermediate is needed. For one thing, it would be better to give the tags used in XML documents more precise definitions. A key aspect of the semantic Web is to provide standard (i.e., agreed upon) definitions of terms or concepts in a variety of domains. A terminology defines a set of related terms, which may be classified to form a taxonomy. When named relationships are added, it may be referred to as ontology. Specifically, ontology concerns the classification of concepts (or classes) as well as their subclasses, properties, and relationships to other concepts. These defined concepts can also be used to annotate the content of documents. Finally, instances of these concepts can be created by extracting content from Web pages. Together the classes, properties, and instances form a knowledge base. The Web ontology language (OWL) provides this capability for the semantic Web (OWL comes in three types: OWL-Lite, OWL-DL, where DL stands for description logic, and OWL-Full). Other possible languages for modeling ontology include the entity-relationship model (Chen, 1976), unified modeling language (UML) (Rumbaugh et al., 1998), knowledge interchange format (Genesereth and Fikes, 1992), and resource description framework (Klyne and Carroll, 2004).

Having introduced the term "knowledge base," we should mention that typically they may also include rules (or something equivalent). Indeed, the latest part of the semantic Web undergoing standardization is the semantic Web rule language (SWRL). Rules allow new facts to be generated from existing facts and relevant rules, thus greatly increasing the expressivity of the knowledge base. Unfortunately, as the expressivity goes up, so does its complexity. Table 3.1 shows the current set of languages used in the semantic Web, and includes the complexity class for basic inferencing operations such as subsumption. (We also

TABLE 3.1 Semantic Web Languages.

Acronym	Language	Schema	Complexity of ⊒
XML	Extensible Markup Language	DTD, XSD, Relax NG	–
RDF	Resource Description Language	RDFS	PTIME
OWL-Lite	Web Ontology Language	OWL Schema Portion	EXPTIME-Complete
OWL-DL	Web Ontology Language	OWL Schema Portion	NEXPTIME-Complete
OWL-Full	Web Ontology Language	OWL Schema Portion	Semi-decidable
SWRL	Semantic Web Rule Language	–	Semi-decidable

note that the ontology definition metamodel (ODM) supporting UML which is under development by OMG has a proposal to have a description logic core.)

A more general and deeper discussion of semantics and ontology as well as their relationship to the semantic Web is given in the appendix. Although all the semantic Web languages are important, the OWL is the most relevant to this chapter. We may use it to define and relate terms or concepts in the fields of M&S. In the next section, we develop a conceptual foundation or basis for M&S. From this conceptualization, we create an OWL. This ontology is broad, but currently shallow. This ontology also includes a few SWRL rules. Later in this chapter, we overview DeMO, which is narrower and deeper.

3.3 Conceptual Basis for Discrete-Event Simulation

In this section, we develop a conceptual framework that is needed to clearly capture the foundational concepts of discrete-event M&S. A secondary goal is to provide a very general framework for discrete-event modeling, including combined discrete-continuous modeling. A tertiary goal is to keep this framework as simple as possible. This last goal, may allow naivety to creep in, especially with regards to continuous modeling. For example, with continuous modeling, energy-based modeling may work better than state-based modeling (Cellier, 1991). Keep in mind that the main purpose of this framework is to create basic ontological concepts for understanding the field of discrete-event simulation and modeling.

The model world begins as an empty void with space and time coordinates (Wikipedia, 2006b).

- *Time.* Let $t \in T$ indicate a point in time. Typically, T would be a subset of real numbers $\mathbb{R}$ or integers $\mathbb{Z}$.
- *Space.* Let $x \in X$ indicate a point or location in a vector space X. For example, X could be the 3D vector space $\mathbb{R}^3$ or something more abstract. Taken together, space and time may be viewed as a space-time continuum (as in relativity theory). We, however, continue to treat time as a special dimension in space-time. The void is then filled with objects, which are principally entities. Entities are the things that exist in the model world. If entities do not interact, the entities that exist at the start of simulation would simply move at a constant velocity (or remain at rest) forever. To allow entities to interact, additional entities need to be introduced into the model world. These agents may cause changes to entities such as entity creation, destruction, property updates, and acceleration. Events are used to model changes that occur instantaneously (or nearly so). Forces are used to model changes that occur smoothly over time.
- *Entity.* An entity k is an object that exists in space-time. It is also uniquely identifiable. Examples of entities include customers in banks, golf balls flying through the air, and even molecules in boiling water. As the number of entities becomes very large, modeling techniques that deal with aggregations of entities offer advantages. In many cases, the models will deal with properties of aggregations such as pressure, temperature, or weight rather than the entities (or aggregate entities) themselves.
- *Event.* An event is an object that does not exist in space-time, rather it exists only in the time dimension. It has a creation time, but the important time is its occurrence time. When the event occurs, it may affect other entities, trigger other events, or modify forces. For example, it may create

Impact of the Semantic Web on Modeling and Simulation 3-5

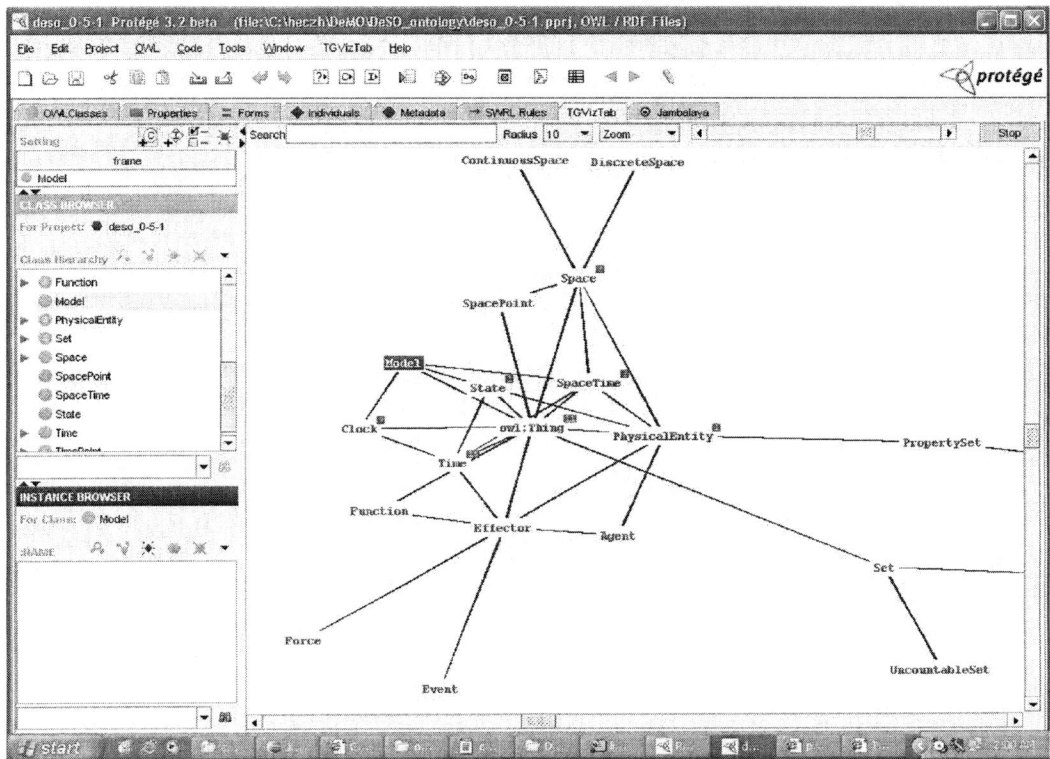

FIGURE 3.1 DeSO Visualization.

(or destroy/cancel) other events, increase (or decrease) forces, move entities, or change entity properties. An event is considered to occur instantaneously and therefore can produce discontinuities in the trajectories of entities within space-time (see below and Figure 3.1). To relate the event to space-time, we assume that it is associated with a particular main entity or agent. Finally, the event must specify what action is to be performed. The action may be specified as algebraic equations, difference equations, or in general using action logic (all of which may be implemented using a programming or simulation language). The type of action determines the type of the event (e.g., an arrival event or a departure event). The complete set of event types is denoted by the set E.

- *Force.* Complementary to events that have immediate effects, forces make changes over time. This corresponds to the worldview provided by classical physics, e.g., as exemplified by Newton's laws of motion. Force laws are typically expressed as differential equations. A common force to use is gravity, which in simulations/animations makes the motion of entities look more realistic.
- *State.* Let $\{w(t)|t \geq 0\}$ be the process (e.g., a stochastic process) representing the evolution of the model world over time. We would like to be able to stop the process at some time t', and save the minimal amount of information from the initial conditions $w(t_0)$ and the current world $w(t')$, so that the process can be resumed without affecting any future results. How this is done and how the dynamics are expressed in terms of this information largely defines the type of modeling technique that is applied. We assume that the following are defined at the beginning of the world $w(t_0)$: the range of time for the model world to exist, the space for entities in the model world to exist in, the event action logic (for discrete changes), the forces (for continuous changes), and an initial event (or events) to initiate the simulation.

- *No force case.* At first, let us keep things simple and suppose there are no forces. The dynamic state of the model world $s(t)$ is simply the aggregation of the configurations (location and, if need be, property values) of all the currently existing entities.

$$s(t) = (x_1(t), \ldots, x_k(t), \ldots) \quad \text{where } x_k(t) \in X \tag{3.1}$$

Besides the current state, we must know the future events that are ordered in the time dimension based upon their occurrence time.

$$r(t) = ((e_1, t_1), \ldots, (e_j, t_j), \ldots) \tag{3.2}$$

where $e_j \in E$ represents an event type occurring at time $t_j \geq t$.

The model world $w(t)$ at time t can be reconstructed from the initial configuration of the world, the current state of the entities and the list of future events.

$$w(t) = f(w(t_0), s(t), r(t)) \tag{3.3}$$

- *Force case.* Adding forces to the simulation can be done in many ways. Pritsker (1986) talks about three types of interactions: type 1, a discrete event makes a discrete change to a continuous variable (a discontinuity); type 2, a discrete event affects the physical laws governing the behavior of entities (equations governing continuous variable); type 3, a "state" event triggers a "time" event. A state event is said to occur when an entity or variable reaches a certain threshold, say $x_k \geq c$.

3.4 Types of Mathematical Models

Now that we have defined the fundamental concepts of time, space, entity, state, event, and force, we may classify mathematical models by differentiating them based on these fundamental concepts. Recall that the first two concepts, time T and space X, may form a time–space continuum in which entities exist. Finally, the agents of change are events E and forces. We may consider changes in time t and state $s(t)$ as specified by a clock function c and a transition function f.

$$h = c(s(t), t) \tag{3.4}$$
$$s(t+h) - s(t) = f(s(t), t)g(h) \tag{3.5}$$

where the clock function c determines the time increment h, while the transition function f and the time influence function g determine the next state. For continuous-time models, we let $g(h) = h$; otherwise, we let $g(h) = 1$. (Typically, for the discrete case, one writes $s(t+h) = f(s(t), t)$; however, we use f to indicate state difference, thus making it easier to show the relationship between the continuous and discrete cases.)

3.4.1 Classification Based on State

The state of the model world, which is a snapshot at a particular time, is based on the more primitive notion of space as well as the notion of entities that populate the space. If there is only one entity and it has no varying properties, then the two concepts space and state may be unified (i.e., $s(t) = x(t)$). Since the concepts of space and state go together, we will classify them together, in regard to whether they are discrete or continuous. The distinction between discrete and continuous simply depends on the cardinality of the state space S. The state space is discrete if its cardinality is less than or equal to the cardinality of the natural numbers, $\mathbb{N}$, denoted by $\aleph_0$; otherwise, we consider it to be continuous, i.e.,

$$|S| \leq \aleph_0 \quad \text{or} \quad |S| > \aleph_0 \tag{3.6}$$

Similarly, one could say that discrete means that the state space is finite or countably infinite (like the integers, $\mathbb{Z}$), whereas continuous means that the state space is uncountable (like the reals, $\mathbb{R}$).

3.4.2 Time-Based Classification

Our second classification is based upon time, particularly the clock function indicating how time is advanced within the model: continuous-time, discrete-event, discrete-time, or static.

3.4.2.1 Continuous-Time Models

In continuous-time models, the clock moves smoothly and continuously. The next instant of time $t + h$ is infinitesimally beyond the current time t. If we let $g(h) = h$ and consider the limit as h tends to 0, we obtain the following:

$$\lim_{h \to 0} \frac{s(t+h) - s(t)}{h} = f(s(t), t) \tag{3.7}$$

$$\frac{d}{dt} s(t) = f(s(t), t) \tag{3.8}$$

This equation is a first-order ordinary differential equation (ODE). Rather than having a function to describe the state trajectory (i.e., the values of $s(t)$ over time), the function in Eq. (3.8) describes the rate of state change. The trajectory can then be determined using some solution technique (e.g., integrating factors and Runga-Kutta). Writing the equation in terms of the derivative allows one to concisely express commonly occurring phenomena such as systems with constant growth rates and many laws of classical physics such as Newton's Second Law of Motion. For one entity (e.g., a ball) whose coordinates in space-time are given by $(x(t), t)$ and acted upon by a constant force $-g$, Newton's second law becomes the following:

$$-g = m \frac{d^2}{dt^2} x(t) \tag{3.9}$$

(For simplicity, we assume x is one-dimensional in the vertical direction.) Note that this is a second-order ODE. However, by introducing another variable, velocity v, into the state, this second-order ODE can be converted into two coupled first-order ODEs.

$$-g = m \frac{d}{dt} v(t) \tag{3.10}$$

$$v(t) = \frac{d}{dt} x(t) \tag{3.11}$$

In this and in many other cases, enlarging the state from $x(t)$ to $(x(t), v(t))$ can allow one to model phenomena using first-order ODEs. This is similar to the technique of enlarging the state space to make stochastic systems Markovian. Still, much of physics requires more than time derivates, e.g., partial derivates, leading to a partial differential equation (PDE) such as the heat equation or Schrödinger equation. We do not include PDEs, because the goal of this work is simply to generalize the DeMO papers so as to address hybrid discrete-continuous simulations as well as basic physics engines used in animations.

3.4.2.2 Discrete-Event Models

The major division of the field of M&S is between continuous-time and discrete-event models. Both have very large and long-established communities. Discrete-event models are very general and include discrete-time models. They can handle anything except an infinite number of infinitesimal changes that requires calculus. Discrete-event models are the focus of DeMO presented in Miller et al. (2004), Fishwick and Miller (2004), Miller and Baramidze (2005), and Silver et al. (2006).

In this case, model dynamics are simplified in that state changes can only occur at a countable number of points and hence a simulation may focus on these time points (also known as event occurrences). Therefore, the evolution of the model world is driven by events. Events represent things that can happen which may cause state changes (i.e., nothing else can cause the state $s(t)$ to change). Besides making state changes, events may also trigger other events to happen at the current time or in the future. The clock and transitions are given as follows: Letting $g(h) = 1$, we have

$$h = c(s(t), t) \tag{3.12}$$

$$s(t+h) - s(t) = f(s(t), t) \tag{3.13}$$

The causality owing to events, in their most general form, is embedded in the clock c and transition function f. However, it is often useful to think of the clock and transition functions as working together to advance the model forward in time to the next state, based on the type of event $e \in E$ that occurs. This is indicated by making the c and f functions parametrically dependent on e. Processing the event advances the clock to the event's occurrence time $t + h$ and transitions the state $s(t)$ to the next state $s(t + h)$.

$$t + h = c(s(t), t; e) + t \tag{3.14}$$

$$s(t+h) = f(s(t), t; e) + s(t) \tag{3.15}$$

The event type e is a point in a finite set E of event types such as {arrival, departure}. This now begs the question of how e is chosen. In general, determination of e can be complex, since events are created by other events and can indeed be canceled by other events. A future event may be created and put in, for example, a future event list (FEL). If the future event is not canceled, it will eventually come to the front of the time-ordered FEL, and become the imminent event to be processed next (i.e., used in the evaluation of c and f). Abstractly, this may be denoted by introducing activation and cancellation functions.

3.4.2.3 Discrete-Time Models

For discrete-time models, the state $s(t)$ may change only at event occurrence times which happen with regularity, i.e., every h, a fixed constant number of time units. Although, h can be any fixed constant, it can also be rescaled to one (i.e., let $g(h) = h = 1$). Then the clock function simply returns 1 every time, while the state change is as follows:

$$s(t+1) - s(t) = f(s(t), t) \tag{3.16}$$

This equation is a difference equation, which is the discrete analog to a differential equation. Here, the state s may be a discrete random variable. If we add the following restrictions:

1. let the time be discrete, for example, let $T = \mathbb{Z}$,
2. let the clock function be the successor function,
3. let the transition function be time homogeneous, that is, invariant over time, and
4. let E be a singleton set or equivalently the event is embedded into transition probabilities.

Then, the difference equation becomes the balance equation for discrete-time Markov chains.

$$P(s(t+1) = s_j) = \sum_i P(s(t+1) = s_j | s(t) = s_i) P(s(t) = s_i) \tag{3.17}$$

where $P(s(t+1) = s_j)$ is the probability that the next state is s_j. DeMO gives a step-by-step development of more and more restrictive Markov models (e.g., from generalized semi-Markov processes to semi-Markov processes to Markov chains).

3.4.2.4 Static Models

So far, we have been simplifying how we deal with time. Obviously, the simplest thing to do is freeze time (or equivalently eliminate it all together). This moves us from a dynamic worldview to a static one. Although simulation is mainly concerned with dynamic models, static models (often called Monte Carlo models) are also useful.

3.4.3 Causality-Based Classification

Causality is a well-established principle in philosophy as well as classical physics. Some modern theories such as general relativity, quantum mechanics, and string theory challenge the simple notion of causality. Yet for the simulations we are considering, we assume causality (or cause and effect). Causes or agents of change may cause changes in the current state $s(t)$ or even defer their effects to the future. The effects may introduce a gradual or sudden change. Sudden changes are captured as events that theoretically happen instantaneously. In reality they may not, but it is reasonable to represent them in this way in the model. Gradual changes happen over time by an ongoing application of force. In physics, the primary causes of changes are forces such as the four fundamental ones: gravity, electromagnetic, weak nuclear, and strong nuclear.

We may classify models based upon a characterization of the causes of change. Change may be modeled discretely or continuously. In the discrete case, changes happen discretely at specific event times. Between these times, what is happening in the simulation may be ignored. Thus, discrete-event simulations, for efficiency sake, jump discretely through time, processing event after event.

This discrete motion may present an issue for animation in the following sense: After leaving one service center, a customer goes to the next. From a simulation point of view, this may not be problematic. However, from an animation point of view, the customer should smoothly go from one center to another. This can be handled by adding animation friendly events to the simulation, or by doing event interpolation and adjustment in an animation engine.

The bottom line is that between events, the system being modeled need not be static (e.g., entities or particles may be moving), it is just that these changes are not judged to be important for the purposes of the model. Typically, in higher fidelity models which more faithfully represent the system, more events will be represented and animations should look more realistic.

In summary, changes to the model may occur discretely or continuously. Discrete changes owing to events make discontinuous changes (or jumps) to entities in the model world. Continuous changes owing to forces make infinitesimal changes in an infinitesimal amount of time to produce typically smooth changes.

Note that the character of a model's space–time does not determine whether changes occur discretely or continuously, in general, but clearly, continuous changes require space (or state) and time to be continuous. Since there are three concepts (time, state, and change) where the discrete versus continuous dichotomy applies, there are eight distinct possibilities. Of these eight possibilities only five make sense, as shown in Table 3.2.

3.4.4 Classification Based on Determinism

Determinism has been a hotly debated subject in philosophy and physics for a long time. In classical physics, mathematical models were basically deterministic. Probability was introduced to deal with lack of knowledge or just to simplify the model. Modern physics, however, postulates that probability is a fundamental part of reality. For these reasons, we classify models as either deterministic or stochastic. For a deterministic model, given the input, the output is uniquely determined. However, for a stochastic model this is not necessarily the case, since randomness is included in the model.

TABLE 3.2 Discrete (D) vs. Continuous (C).

State	Time	Change	Model Type
C	C	C	Continuous System Simulation
C	C	D	Discrete Event Simulation
D	C	D	e.g., Generalized Semi-Markov Processes
C	D	D	e.g., Time Series
D	D	D	e.g., Discrete-time Markov Chains

3.5 Adding Semantics to Simulation Models

In this section, we claim that adding semantics to simulation models is going to become more and more important in the future. A skeptic might counterclaim that semantics is not crucial for M&S, because they are general purpose techniques that achieve their usefulness through abstraction. In this way, a bank and drive-through restaurant can be modeled in similar ways using abstract queues with different parameter values for interarrival and service times. It is great to be able to abstract out the essential features and discover fundamental similarities. The M&S community has been doing this successfully for decades. In our opinion, this paradigm has three weaknesses:

1. The mapping from the real world to the abstract model is largely in the mind of the simulation analyst.
2. High fidelity, multifaceted modeling is difficult to achieve.
3. Building models out of model components is limited, in part due to lack of semantics.

As the semantic Web progresses, one might consider how it could positively impact the development and use of mathematical models in general and simulation models in particular. The purpose of a mathematical model is to create an abstract representation of a system or mini-world (be it real or artificial). As an abstract rather than concrete representation, the model could be mapped to multiple systems or mini-worlds. Being abstract, the model can be more easily analyzed and manipulated than an actual system. In this sense, abstraction is good. However, too much abstraction can result in the loss of realism and meaningfulness. One way to lessen the loss is by increasing the amount of explicit semantics given.

Previously, simulation was concerned with getting the numbers right. Have we created a simulation model (validation) and a program implementation (verification) that produce accurate estimates or predictions? Animation puts an additional constraint on this. The time evolution of the model should "look right." Still, what is the relationship between the model and the system? What are the things moving around in the model? How do they compare to similar things found in other models? How do they relate to existing knowledge such as the laws of classical physics?

A small step in this direction is to document the model (or even the program implementing the model). However, this is likely to be minimal and certainly informal. An alternative would be annotate the model as well as the model elements so that, for example, one would know what an entity looks like and what it means. Defining the meaning of something is only feasible if related terms are already defined. Then clearly these related terms only make sense if terms related to them are defined. These terms should be logically organized into a well-defined conceptualization and made readily available using the Web. This is one of the central thrusts of the semantic Web and Web-based ontology.

Since simulation is used for modeling of a vast array of fields, the above prescription is really quite challenging. First, ontology needs to be developed for simulation and modeling methodology. Then ontology from application domains (e.g., health care and transportation) needs to be utilized. Fortunately, many domains are developing ontology as shown for scientific domains in Table 3.3. There are many more ontology listings on the open biomedical ontologies (OBO) site (obo.sourceforge. net/cgi-bin/table.cgi) with 52 at last count.

The focus of this chapter, however, is more on ontology for the simulation and modeling methodology. One way to start is to try to understand what a model is and what it is used for (its purpose). The word itself has many definitions (e.g., in Merriam-Webster's dictionary and in WordNet). We are interested in its usage for abstract, conceptual, or mathematical models. From Wikipedia (2006b), we have the following definition:

> An abstract model (or conceptual model) is a theoretical construct that represents physical, biological or social processes, with a set of variables and a set of logical and quantitative relationships between them. Models in this sense are constructed to enable reasoning within an idealized logical framework about these processes.

TABLE 3.3 Ontology for Scientific Domains

Name	Title	Domain
PhysicsOnto	Ontology of Physics archive.astro.umd.edu/ont/Physics.owl	Physics
AstroOnto	Ontology of Astronomy archive.astro.umd.edu/ont/Astronomy.owl	Astronomy
ChemOnto	Chemical Ontology www.personal.leeds.ac.uk ~ scs1tvp/onto/chemonto.owl	Chemistry
GO	Gene Ontology archive.godatabase.org/latest − termdb/go_daily − termdb.owl.gz	Biology
SO	Sequence Ontology cvs.sourceforge.net/viewcvs.py/song/ontology/so.obo	Biology
MDEG	Microarray Gene Expression Data mged.sourceforge.net/ontologies/MGEDOntology.owl	Biology

The purpose of a model or modeling in general is even harder to capture. In an idealized sense, a model is the essence of science. Since the real world or real systems are so complex, models are constructed that can be manipulated logically or mathematically. The models help us dissect, understand, and make predictions about the real world. For science to be self-correcting, the models (or hypothesis or theories) must be falsifiable. In other words, tests and experiments must be developed to show that the model has deficiencies that need to be corrected either by improving the model or throwing it out completely. Besides empirical validation, models need to be consistent with other models or theories.

Let us now examine in greater detail, the problem of defining or describing a model in terms of (i) statics and (ii) dynamics. The statics of an entity define its type (types of properties) and immutable state. The statics can be described at a high level using, for example, a UML class diagram or OWL. The dynamics of an entity define its behavior. There are several ways to describe behavior in UML (e.g., sequence diagrams, collaboration diagrams, statechart diagrams, or activity diagrams). In addition, other formalisms such as process algebras, Petri nets, bond graphs, activity cycle diagrams, and event graphs may be used. The current state of affairs is that there are several competing approaches and none are as successful as the approaches used for statics. Clearly, the problem is much more difficult.

Ontology is ideal for describing things, so statics can be well handled. Dynamics or behavioral specifications are more challenging. Although knowledge representation languages in AI, such as Frames (Minsky, 1974), support the use of procedural attachments, current semantic Web initiatives are avoiding this complexity because of the need to effectively support querying and inferencing on a Web scale. Still there is, however, ongoing research work on behavioral specifications for semantic Web services. Behavioral units (such as operations) can be annotated with functional semantics, such as functional category, inputs, outputs, preconditions, and postconditions (Sivashanmugam et al., 2003). For simple cases, preconditions and postconditions (or effects) may be expressed using an ontology language like OWL, while for more complex conditions a rule language like SWRL is more suitable. One could apply such an annotation approach to simulation in several ways. For example, in the event-scheduling paradigm, the behavior can be captured in the logic of an occur operation (event routine). Similarly, in the process-interaction paradigm, the behavior can be captured in the logic of the entity's script (a network of operations).

The complexity of modeling dynamics is testified by the plethora of modeling techniques used: message charts, collaboration diagrams, state charts, activity diagrams in UML, the three simulation worldviews, event-scheduling, process-interaction, and activity scanning in simulation, as well as, event graphs, state machines, Petri nets, process models, and process algebras.

The goal is to capture what an entity does short of providing the code to implement the behavior. (The specification should provide the basis for verification of the code, and hence, cannot be the code. Yet to allow automatic verification, the specification must be machine-interpretable.) This is the essence of providing a semantic description of behavior. Entities can be coupled by (i) shared state, (ii) invocations, or (iii) events, with each being more loosely coupled than the former. The complexity of verification goes up dramatically

TABLE 3.4   Ontology for Modeling and Simulation

Name	Title	Domain
Monet	Mathematics on the Web www.cs.man.ac.uk/~dturi/ontologies/monet/allmonet.owl	Mathematics
GeomOnto	Ontology of Geometry archive.astro.umd.edu/ont/Geometry.owl	Mathematics
StatOnto	Ontology of Statistics archive.astro.umd.edu/ont/Statistics.owl	Statistics
DeMO	Discrete-event Modeling Ontology chief.cs.uga.edu/~jam/jsim/DeMO/	Simulation
DeSO	Discrete-event Simulation Ontology chief.cs.uga.edu/~jam/jsim/DeSO/	Simulation
MSOnto	Agent Ontology for Modeling and Simulation www.nd.edu/~schristl/research/ontology/agents.owl	Simulation

if the shared state space is large, and so it should be kept to a minimum. In the software engineering as well as the agent and semantic Web services communities, the semantics of invocations is often modeled using inputs, outputs, preconditions, and postconditions. Since inputs and outputs are objects, they may be described ontologically. For example, in the proposed semantic annotations for Web services description language (SAWSDL) standard (based on WSDL-S) (Akkiraju et al., 2005; Farrell and Lausen, 2006) they may be annotated with model references to OWL or UML. Pre- and postconditions (alternatively effects) may be modeled with a constraint language such as SWRL or UML's object constraint language (OCL). Although, this approach can be used to describe an invocation, it says nothing about the sequencing or ordering of invocations, leading to the need to describe interactions via a protocol specification. To more fully capture behavior, richer languages are necessary. Unfortunately, use of Turing-complete languages makes inferencing fundamentally challenging, leading to a trade-off between the ability to capture detailed behavior versus the ability to analyze it.

We conclude this section by listing ontology in Table 3.4 used for M&S as well as ontology that could provide foundations for M&S.

3.6   Overview of DeSO

DeSO is an initial attempt at providing a concise, but adequately precise ontology for the most fundamental concepts often referred to in M&S. Such an effort, however, is by no means easy, as to precisely define the basic concepts would mean to define many of the relevant concepts in mathematics, philosophy, and physics. We have no intention of making DeSO a huge, self-contained ontology, yet we have taken significant measures to make it work beyond its size.

The current DeSO includes six elementary concepts in M&S, namely, time, space, physical entity, state, effector, and model. We present an overall picture of how models are classified based on certain properties of these basic concepts as described in Section 3.4. These concepts are also complemented in DeSO by some other related concepts to provide more accurate definitions and to reflect the complicated relations between them.

The first measure we have taken to compact DeSO is to begin by importing the SUMO, as opposed to starting from scratch. The SUMO upper ontology was developed at TeKnowledge to cover around 1000 of the most general concepts intended for use by middle-level and domain ontology such as DeSO. We summarize a few of the advantages of importing an upper ontology:

- First, an obvious advantage is the reuse of established knowledge systems. DeSO, for example, uses directly many definitions and relations already in the SUMO, such as TimePoint, Set, and FiniteSet.
- Second, an important use of ontology is to facilitate information sharing through the use of common vocabulary, as it effectively reduces ambiguity in communication and facilitates machine

understanding. Since OWL does not enforce the unique name assumption, the same class name may be used to refer to different concepts in different ontological specifications. (Part of developing ontology is to handle synonyms and homonyms. Different names for the same concept are synonyms, while different concepts with the same name are homonyms.) The following two approaches will guarantee that the same name in DeSO and the SUMO refers to the same concept.

The first approach is to use the needed SUMO class directly without reproducing the same concept in DeSO. This sometimes requires that we create some new DeSO classes whose existence is dependent on the SUMO. For example, if we need to define two new classes "DeterministicFunction" and "StochasticFunction" in DeSO and we can choose to use the superclass "Function" in the SUMO directly, then the two DeSO classes would be created as the subclasses of SUMO:Function. While this approach is favorable theoretically, current ontology editors such as Protégé, do not support the notion of a package view as one would see in Javadoc, so the classes in the SUMO and DeSO are mixed up structurally, and the classes of DeSO may be buried deep inside a SUMO hierarchy. Such a mixture often deprives us of the ability to freely create new class relationships and to generate visualization, and, thus, the freedom to express what we want to say in the ontology.

Because of the limitations of the current ontology editors, common vocabulary between the SUMO and DeSO is realized through the second approach: using the equivalentClass restriction in OWL. Our way of using the SUMO classes is to generate new classes in DeSO and restrict these classes to be equivalent to the classes in the SUMO where appropriate. We choose to use the same class names as in the SUMO for easy understanding, although identical class names are not required for the equivalentClass restriction. For example, to conform to the naming system of the SUMO, the "Entity" class that we discussed in Section 3.3 is called "PhysicalEntity" in DeSO. This latter approach provides us with desirable autonomy of DeSO, the flexibility of generating class relations as we want, as well as common vocabulary between the two ontological specifications. In DeSO, for example, a new TimePoint class is defined as equivalent to the SUMO TimePoint class and is used in all the relationships involving TimePoint in DeSO.

- Third, the use of an upper ontology makes inferences across different domains easier, as relations are established across OWL files by sharing the generic concepts in the upper ontology.

The second characteristic of DeSO is that it utilizes SWRL on top of OWL so that the expressiveness of the ontology increases considerably without substantial addition in size. For instance, we would have needed 24 additional classes for all the combinations of model types based on the four classification criteria we mentioned in Section 3.4; instead, we use nine rules that express the same ideas and more. A very simple example of these SWRL rules is

$$isStochastic(?m, false) \rightarrow isDeterministic(?m, true) \tag{3.18}$$

which means "if a Model m is not stochastic, then m is deterministic." This short rule allows us to reuse the definition of stochastic model and saves us the trouble of defining deterministic model. However, rules are not always so short. More often than not, we need to write long rules to define some concepts. In DeSO, stochastic model is defined with the following rule:

$$existsIn(?m, ?st) \wedge populatedBy(?st, ?pe) \wedge effectedBy(?pe, ?ef) \wedge computes(?ef, ?sf)$$
$$\wedge StochasticFunction(?sf) \rightarrow isStochastic(?m, true) \tag{3.19}$$

To put it in plain English, the rule says "if a Model m exists in some SpaceTime st, and st is populated by some PhysicalEntity pe, and pe is effected by some Effector ef which computes a StochasticFunction sf, then m is Stochastic." SWRL rules are more than simple definitions of concepts, in that they reflect the relations between the concepts in addition to the definitions of the concepts themselves.

DeSO has been developed using Protégé with its OWL plugin. The SWRL editor (an extension to the Protégé OWL plugin) has been used to edit the SWRL rules in DeSO. Figure 3.1 is a visualization of the DeSO classes and their relationships created by OntoViz (one of the several popular visualization plugins for Protégé).

In short, DeSO is a middle-level ontology built upon SUMO that includes the most fundamental concepts in the domain of M&S. As one of the first attempts at using SWRL in an ontology, DeSO achieves maximal expressiveness within minimal volume.

3.7 Overview of DeMO

Work on the DeMO began in 2003 (Miller et al., 2004; Fishwick and Miller, 2004) to explore issues and challenges in developing ontology for simulation and modeling. As its name suggests, it is focused on discrete events models in which state changes discretely over time owing to the occurrence of events. It used the OWL language to define over 60 classes and many properties. Figures 3.2–3.5 contains visualizations created by OntoViz showing the DeMO classes and their relationships. The ontology consists of four main parts: `ModelConcept`, `DeModel`, `ModelComponent`, and `ModelMechanism`. `DeModel` is itself divided into four parts based on the three simulation worldviews plus a fourth representing state models, namely, `StateOrientedModel`, `ActivityOrientedModel`, `EventOrientedModel`, and `ProcessOrientedModel`. (Note that the images show DeMO version 1.8, which is missing the `ProcessOrientedModel` subtree, but is going into version 1.9.)

As illustrated by the OBO site, it is often better to have several (but not too many) interrelated ontology, rather than one huge monolithic ontology. Along these lines, DeMO as it is extended, could be divided into more than one ontology. In addition, DeMO ignores much of the M&S domain, including continuous models, statistical modeling, output analysis, random variates, etc. Also, DeMO at present has few instances. One could attempt to populate the ontology (or knowledgebase) with information about simulation engines, available simulation models, model components, etc. This could be done by writing extractors which scan the Web for information or by providing a mechanism for publication. Alternatively, one could simply use the ontology for annotation of simulation artifacts, as is done in the proposed WSDL-S standard (Akkiraju et al., 2005). Then special semantic search engines could precisely retrieve the information requested.

There can be several ways to approach the descriptions of different modeling formalisms (formalism specification) for the purpose of ontology engineering. One way is to consider each model separately and define it from scratch. This may be called a "problem-in-hand" approach—given a problem, define a modeling formalism that "fits" the problem well. This is a natural approach from a practical point of view: different modeling formalisms "fit" differently into different problems; some are more fitting for one purpose, some for another. Another way is to define some very general formalism and consider all other models to be some sort of subformalisms—restrictions on a general framework such as the DEVS. This view is logical and natural as well, because many of the existing modeling approaches have a formal description and it is only a question of finding a general enough framework that encompasses all the existing formalisms and from which new subformalisms can be derived. However, if this philosophy is taken to the extreme, it can lead to unnecessary complexity and awkward notations.

DeMO utilizes a middle-ground approach, where several general (upper-level) formalisms are defined independently of each other. (Of course, they do not have to be completely independent of each other and may themselves be derived from some even more general formal framework.) These upper-level formalisms can be viewed as root classes for a taxonomic tree for the discrete-event M&S domain. All other modeling formalisms are defined as restrictions on one of the root classes.

Importantly, DeMO uses a uniform approach based upon common modeling formalisms. Each DeModel is considered as having model components and model mechanisms (syntax and semantics of the model), which in turn are defined using fundamental model concepts. This approach allows for great flexibility and straightforwardness in constructing an ontology and defining new formalisms.

We close by giving a simple application for DeMO involving Petri nets. Figure 3.6 shows a screenshot from the Protégé ontology editor of the PetriNet class in DeMO. One thing to notice in this diagram is where the class fits in the class hierarchy. Another important aspect is the properties section. The former indicates how PetriNets relate to other modeling formalisms, while the latter indicates what is in a PetriNet

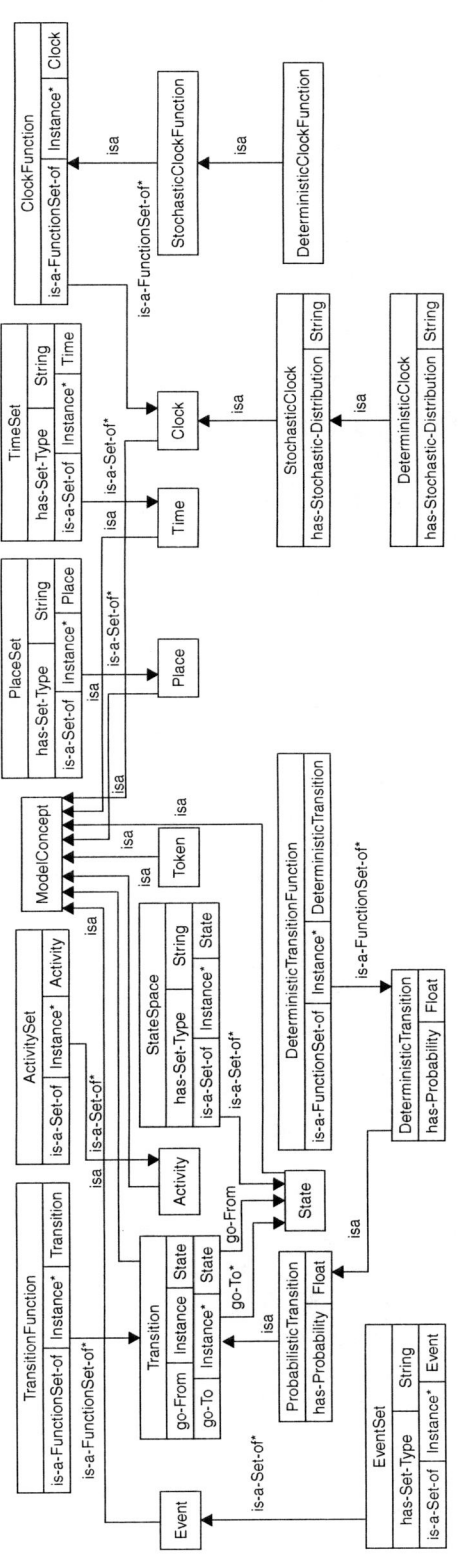

FIGURE 3.2 DeMO model concept.

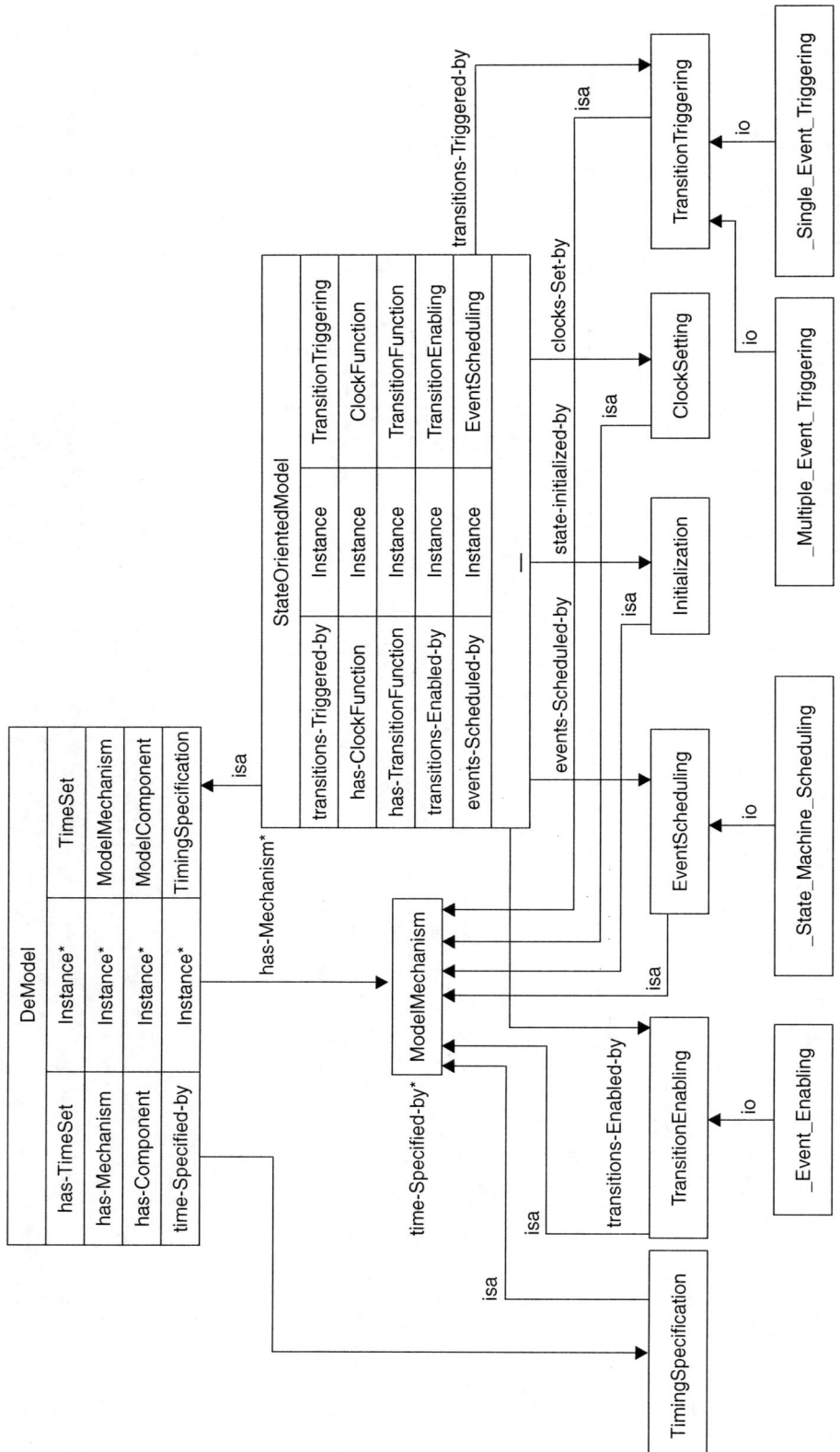

FIGURE 3.3 DeMO model component class hierarchy.

Impact of the Semantic Web on Modeling and Simulation

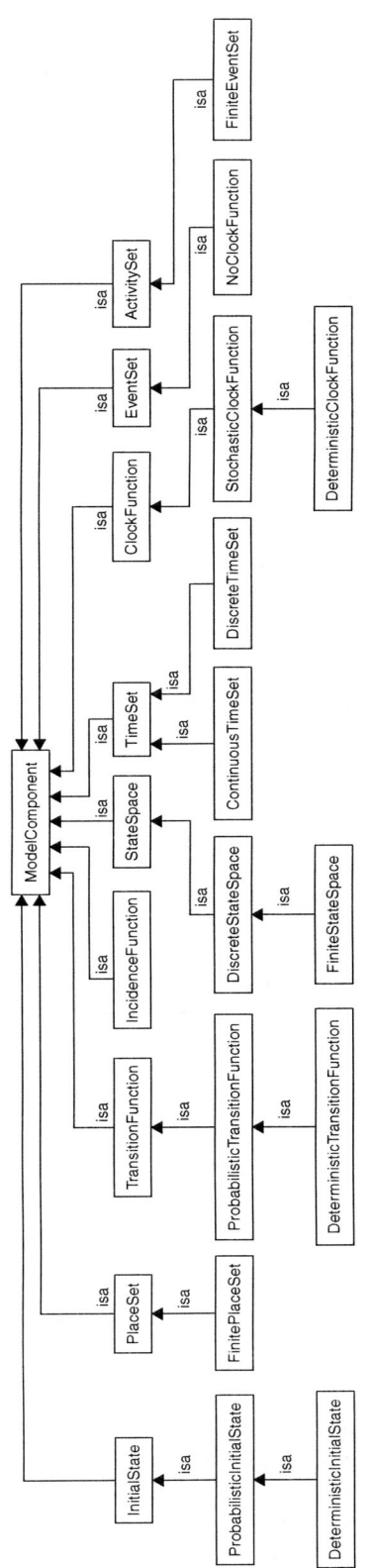

FIGURE 3.4 DeMO model mechanism.

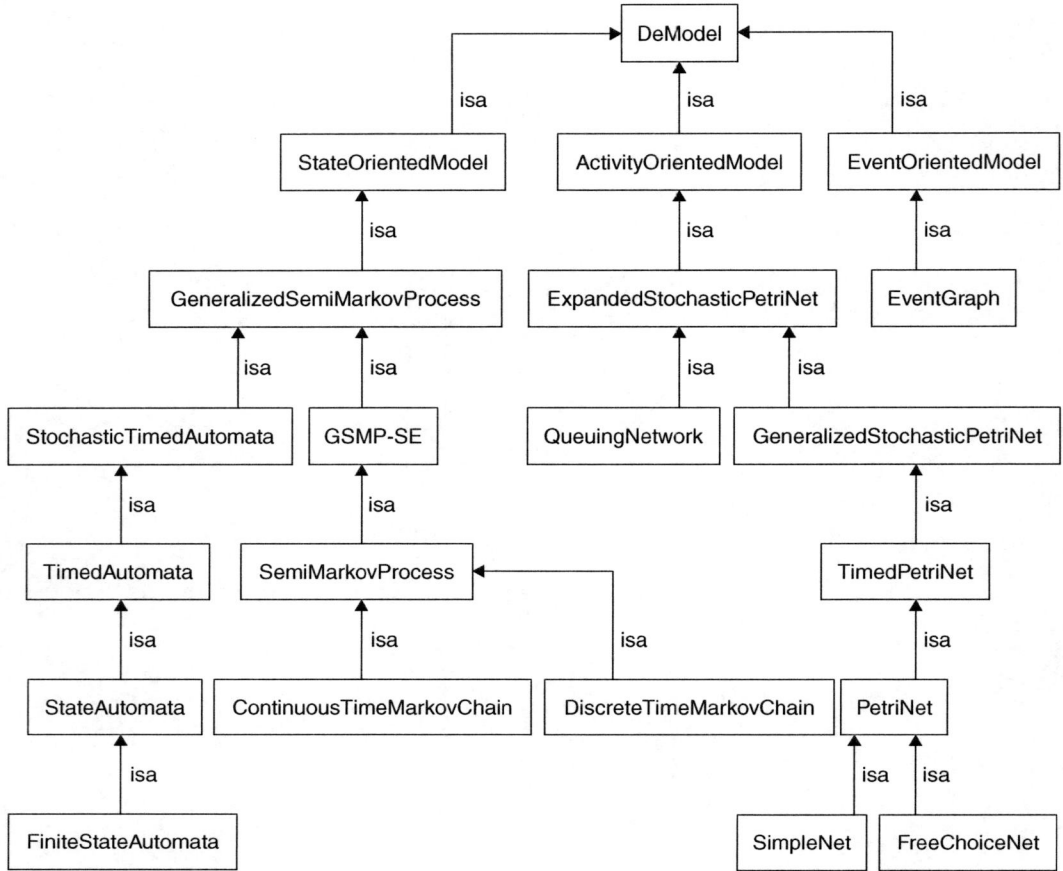

FIGURE 3.5 DeMO DeModel class hierarchy.

(for example, the following properties, has-ActivitySet, has-ArcSet, has-Component, has-Mechanism, has-PlaceSet, has-TimeSet, time-Specified-by, define the structure and mechanics of a PetriNet). Since Petri nets formalism is very popular, there are several simulators that run Petri nets. For the purposes of standardization and interoperability, the Petri net markup language (PNML; Jngel et al., 2000) has been created. Several of the simulators accept input in this format. One existing application of DeMO is the automatic generation of PNML specifications from instances stored in DeMO. Note that DeMO maintains topological information on the PetriNet, while PNML requires geometrical coordinates. Rules could be developed to select layout algorithms that will take the topological information and convert it into geometrical coordinates. This could lead to visually appealing animations of Petri net executions.

3.8 Summary

New developments in the semantic Web, especially in the ontology layer, present many opportunities for the M&S community. This chapter has highlighted several of them. We have developed a general conceptual framework for M&S as represented by DeSO and DeMO shown in Figures 3.1–3.5. The potential impact of semantic Web research on the M&S communities has been discussed. In particular, the use of OWL and SWRL have been demonstrated in the development of DeSO. Several issues in the construction and use of ontology for M&S have also been addressed in this chapter.

Impact of the Semantic Web on Modeling and Simulation 3-19

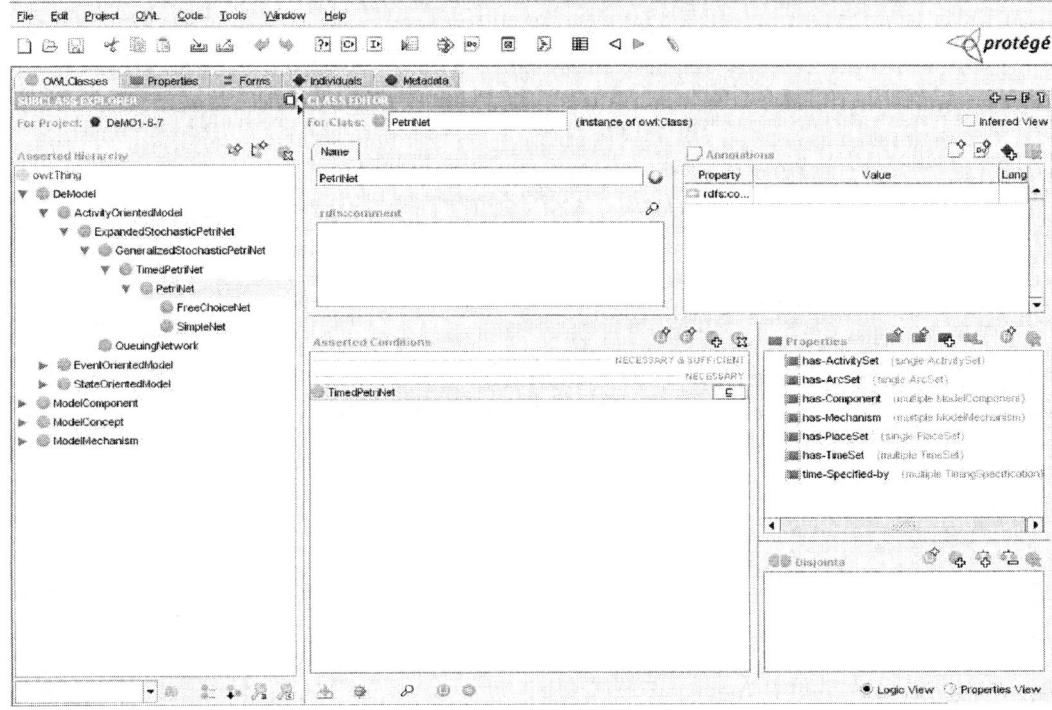

FIGURE 3.6 Protégé screenshot of DeMO.

Appendix: Semantics—Some Perspectives

Semantics has been a major topic of inquiry for a long time. As a traditional branch of linguistics, it refers to the study of the meaning of language. Deeply rooted in philosophy, semantics was first formalized in logic in the nineteenth century and was later expanded to deal with programming-language semantics. Now, the semantic Web initiative is bringing new life to this research and is attempting to make it practical and scalable at the Web level. In this appendix, we look at semantics from the standpoint of philosophy, linguistics, logic, programming languages, and the semantic Web.

Linguistics divides the world distinctly into "language" and "meaning." Words, go in a lexicon, axioms encoding meaning go in an ontology, and semantic lexicographers create bidirectional mappings between the two. A lexicon of words or symbols is mapped to concepts, listing multiple concept types for words that have more than one meaning. With many variations of notation and terminology, the basis for most systems in computational linguistics consists of the following (Sowa, 2000): (i) lexicon—a set of symbols, (ii) grammar—rules governing the ordering of symbols, and (iii) ontology—a topology of concepts. A mapping that involves all three serves as a foundation for establishing meaning.

Given a description in a formal language, a difficult question is "what does it mean." Such a description, as with natural language, will contain objects/nouns and actions/verbs. The meaning of the nouns and verbs may be refined using adjectives and adverbs, respectively. The elements of the description can therefore be naturally decomposed into the two parts: objects and actions.

The study of the nature of objects has long been pursued in the field of ontology. First, defined by Aristotle as "the science of being *qua* being," ontology studies the existence of things. It concerns the nature and meaning of existence as well as the basic categories of entities. Ontology is considered fundamental because it tells people what words refer to entities, and it provides the classification (including classes and subclasses) of entities as well as their properties and relationship to other entities. Languages for modeling ontology (or the related notion of schema) include the entity-relationship model (Chen, 1976), UML (Rumbaugh et al., 1998), knowledge interchange format (Genesereth and Fikes, 1992), RDF (Klyne and Carroll, 2004), OWL (McGuinness and Van Harmelen, 2004), and SWRL (Horrocks et al., 2003). The last four of these languages are based on logic, primarily, description logic, and first-order predicate logic.

Now that we have a way of describing entities, statically, we need to describe their dynamics. Issues of behavior and interaction come to the forefront. One might look for an analog to ontology used to describe nouns, objects, or entities that would work for verbs or actions. Unfortunately, dynamics is much more challenging that statics. The first phase of science is to describe the entities (e.g., genes and proteins), while the second phase is to describe (better yet predict) how they will behave or interact (e.g., biochemical pathway models). Dynamic models involve entities that change (appear, disappear, move, change properties, and affect others) over time.

Verbs are most naturally captured in ontology as relationships such as "student A *enrolls in* course B." However, this begs the question, what does *enrolls in* mean. The verb is not so much modeled as it is used in the model of student. Still, one could in OWL define the *enrolls in* property to be a subproperty of *takes* to claim some semantics is provided. A few comprehensive attempts at verb classification have been done, e.g., see VerbNet (Palmer, 2004).

A more complete treatment of dynamics calls for space-time models, which have a collection of interacting entities that change over time. (Note, for generality, space is often represented abstractly as a state which may include coordinates as well as other types of information.)

In formal logic, semantics provides a way to show that a statement (logical expression) is true. The most prevalent approach is model-theoretic semantics (Tarski, 1983). A logical expression consists of constants, variables, logical connectives, functions, and predicates. In first-order logic, variables can be quantified, while in second-order logic, functions and predicates may also be quantified. Unless, the expression is a tautology, a "model" is required to determine its truth value. The "model" (not to be confused with a simulation model) will indicate the domain that variables can range over, as well as, how to evaluate the functions and predicates. If the "model" relates to something meaningful (e.g., a part of the real world) then the expression can be meaningfully interpreted. There are also other alternative approaches such as proof-theoretic semantics (Gentzen, 1969).

The semantics of programming languages formally or mathematically deals with the meaning of programming languages. The symbols and the allowable orderings of these symbols are defined using the language's lexicon (what symbols) and grammar (what order). The lexicon is often described using a regular language, while the grammar is often defined using a context-free language. Together these constitute the syntax of the language. Capturing what a sequence of symbols means is not so easy. For example, what does $x + y$ mean? Does the addition operator mean integer addition, floating point addition, or string concatenation? There are three approaches to defining the meaning of programs: denotational semantics, operational semantics, and axiomatic semantics (Hoare, 1969; Scott-Strachey, 1971; Plotkin, 1981).

There is an ongoing debate about whether the semantic Web is really semantic (i.e., will it explicate the meaning of resources on the Web). This debate involves open issues in philosophy and science, which are not likely to be resolved any time soon. Hence, we simply claim that the approach makes things "more" meaningful, in the sense of being easier to find, use, and understand. Whether the machine truly understands it, is an issue for others to tackle.

References

Akkiraju, R., J. Farrell, J. Miller, M. Nagarajan, M. Schmidt, A. Sheth, and K. Verma (2005). Web service semantics—wsdl-s. http://www.w3.org/Submission/WSDL-S/.

Berners-Lee, T. (1998). Why rdf model is different from the xml model. http://www.w3.org/DesignIssues/RDF-XML.html.

Berners-Lee, T., J. Hendler, and O. Lassila (2001). The semantic web. *Scientific American 284*(5), 34–43.

Brutzman, D. (2004). Extensible modeling and simulation framework (xmsf). http://www.movesinstitute.org/xmsf.

Caprotti, O., M. Dewar, and D. Turi (2004). Mathematical service matching using description logic and owl. http://monet.nag.co.uk/cocoon/monet/publicdocs/monet_onts.pdf.

Cellier, F. E. (1991). *Continuous System Modeling*. New York: Springer.

Chen, P. P. (1976). The entity-relationship model—Toward a unified view of data. *ACM Transactions on Database Systems* 1(1), 9–36.

DEVS (2005). Devs. http://www.sce.carleton.ca/faculty/wainer/standard/.

Farrell, J. and H. Lausen (2006). Semantic annotations for wsdl. http://www.w3.org/2002/ws/sawsdl/spec/SAWSDL.html.

Fishwick, P. A. and J. A. Miller (2004). Ontologies for modeling and simulation: Issues and approaches. In *Proceedings of the 2004 Winter Simulation Conference (WSC'04)*, Washington, DC, pp. 259–264.

Genesereth, M. and R. Fikes (1992). *Knowledge Interchange Format, Version 3.0 Reference Manual*. Stanford, CA: Computer Science Department, Stanford University.

Gentzen, G. (1969). *Investigations into Logical Deduction*. Amsterdam: North-Holland.

Horrocks, I., P. F. Patel-Schneider, H. Boley, S. Tabet, B. Grosof, and M. Dean (2003). Swrl: A semantic web rule language combining owl and ruleml. http://www.daml.org/2003/11/swrl/.

Hoare, C. (1969). An axiomatic basis for computer programming. *Communications of the ACM* 12(10), 576–585.

Jngel, M., E. Kindler, and M. Weber (2000). The Petri net markup language. The Workshop AWPN, Koblenz, Germany.

Johnston, P. (2005). Xml, rdf, and dcaps. http://www.ukoln.ac.uk/metadata/dcmi/dc-elem-prop/.

Klyne, G. and J. J. Carroll (2004). Resource description framework (rdf): Concepts and abstract syntax. http://www.w3.org/TR/2004/REC-rdf-concepts-20040210/.

McGuinness, D. L. and F. Van Harmelen (2004). Xml, rdf, and dcaps. http://www.w3.org/TR/owl-features/.

Miller, G., R. Beckwith, C. Fellbaum, D. Gross, and K. Miller (1990). Introduction to WordNet: An on-line lexical database. *International Journal of Lexicography* 3(4), 235–244.

Miller, J. A. and G. Baramidze (2005). Simulation and the semantic web. In *Proceedings of the 2005 Winter Simulation Conference (WSC'05)*, Orlando, FL, pp. 2371–2377.

Miller, J. A., G. Baramidze, P. A. Fishwick, and A. P. Sheth (2004). Simulation and the semantic web. In *Proceedings of the 37th Annual Simulation Symposium (ANSS'04)*, Arlington, Virginia, pp. 55–71.

Minsky, M. (1974). A framework for representing knowledge. *MIT-AI Laboratory Memo 306*. Cambridge, MA: MIT.

Mitkov, R. (2003). *The Oxford Handbook of Computational Linguistics*. Oxford: Oxford University Press.

Niles, I. and A. Pease (2001). Towards a standard upper ontology. In C. Welty and B. Smith (Eds.), *Proceedings of the 2nd International Conference on Formal Ontology in Information Systems (FOIS-2001)*, Ogunquit, Maine.

Noronha, N. and M. J. Silva (2004). Using the semantic web for web searches. http://xldb.di.fc.ul.pt/data/Publications_attach/NormanPaperInteraccao2004.pdf.

OpenMath (2006). Openmath. http://www.openmath.org/cocoon/openmath/index.html.

Palmer, M. (2004). Verbnet. http://www.cis.upenn.edu/ mpalmer/project_pages/VerbNet.htm.

Plotkin, G. D. (1981). A structural approach to operational semantics. *Tech. Rep. DAIMI FN-19*.

Pritsker, A. A. (1986). *Introduction to Simulation and SLAM II* (3rd ed.). New York, NY: Wiley.

Reichenthal, S. (2002). Srml: A foundation for representing boms and supporting reuse. In *Proceedings of the 2002 Fall Simulation Interoperability Workshop*, Orlando, Florida.

Rumbaugh, J., I. Jacobson, and G. Booch (1998). *The Unified Modeling Language Reference Manual* (Addison-Wesley Object Technology Series). Essex, UK: Addison-Wesley Longman Ltd.

Scott, D. and C. Strachey (1971). Toward a mathematical semantics for computer languages. In *Proceedings of the Symposium on Computers and Automata*, New York, pp. 19–46.

Shaya, E., B. Thomas, P. Huang, and P. Teuben (2006). Astroonto. http://archive.astro.umd.edu/.

Sheth, A., C. Ramakrishnan, and C. Thomas (2005). Semantics for the semantic web: The implicit, the formal and the powerful. *International Journal on Semantic Web and Information Systems* 1(1), 1–18.

Silver, G., L. Lacy, and J. A. Miller (2006). Ontology based representations of simulation models following the process interaction world view. In *Proceedings of the 2006 Winter Simulation Conference*. Monterey, CA, pp. 1168–1176.

Sivashanmugam, K., K. Verma, A. P. Sheth, and J. A. Miller (2003). Adding semantics to web services standards. In *Proceedings of the 1st International Conference on Web Services (ICWS'03)*, Las Vegas, Nevada, pp. 395–401.

Sowa, J. F. (2000). *Knowledge Representation: Logical, Philosophical, and Computational Foundations*. Pacific Grove, CA: Brooks/Cole Publishing Co.

Tarski, A. (1983). *Logic, Semantics, Metamathematics* (2nd ed.). Indianapolis, IN: Hackett.

Wikipedia (2006a). Wikipedia. http://www.wikipedia.org/.

Wikipedia (2006b). Wikipedia. http://en.wikipedia.org/wiki/Model_%28abstract%29.

Zalta, E. N. (2006). Stanford encyclopedia of philosophy. http://plato.stanford.edu/contents.html.

4
Systems Engineering

Andrew P. Sage
George Mason University

4.1	Introduction	**4-1**
4.2	Systems Engineering	**4-2**
4.3	The Importance of Technical Direction and Systems Management	**4-4**
4.4	Other Parts of the Story	**4-8**
4.5	Summary	**4-9**

4.1 Introduction

There are many ways in which we can define and describe systems engineering. It can be described according to structure, function, and purpose. It may also be described in terms of efforts needed at the levels of systems management, systems methodology, and systems engineering methods and tools. We can speak of systems engineering organizations in terms of their organizational management facets, in terms of their business processes or product lines, or in terms of their products or services. We can speak of systems engineering in terms of the knowledge principles, knowledge practices, and knowledge perspectives necessary for present and future success in systems engineering. This chapter takes a multifaceted and transdisciplinary view of systems engineering. It attempts to describe systems engineering in terms of this relatively large number of trilogies. the process view of systems engineering is expanded on in some detail. Within this, a large number of necessary roles for systems engineering are described. A brief discussion of systems engineering from the perspective of each of these necessary roles concludes the chapter. We provide brief discussions on the theme of this handbook: dynamic system modeling.

The main objective of this chapter is to provide a multifaceted perspective on systems engineering and, within that, systems management. This is a major challenge for a single chapter, one that is particularly focused on the role of systems engineering in dynamic system modeling. Hopefully, this objective will be realized. We believe that some appreciation for the overall process of systems engineering will lead naturally to a discussion of the important role for systems management, and the applications of this to important areas such as how best to use dynamic system modeling in the engineering of systems of all types.

We are concerned with the engineering of large-scale systems, or systems engineering (Sage, 1992a), especially strategic-level systems engineering, or systems management (Sage, 1992b). We begin by first discussing the need for systems engineering, and then providing several definitions of systems engineering. We next present a structure describing the systems engineering process. The result of this is a life-cycle model for systems engineering processes. This is used to motivate discussion of the functional levels, or considerations, involved in systems engineering efforts: systems engineering methods and tools, systems methodology, and systems management. There is a natural hierarchical relationship among these levels and this is shown in Figure 4.1. There will be some discussions throughout this chapter on systems engineering methods. Simulation and modeling is one of the major methods of systems engineering and, of course, the theme of this work. Our primary focus here, however, is on systems engineering processes and systems management for the technical direction of efforts that are intended to ultimately result in appropriate systems, products or services. These result from an appropriate set of systems engineering methods and

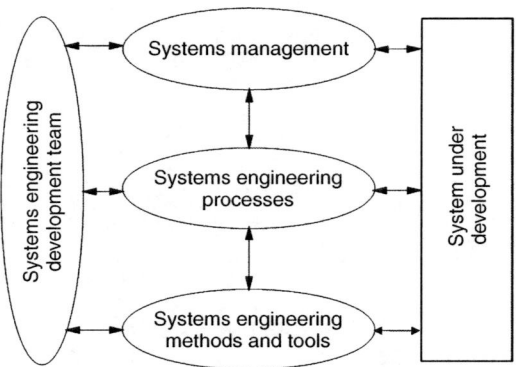

FIGURE 4.1 Conceptual illustration of the three levels for systems engineering. (From Sage, A.P., *Systems Management for Information Technology & Software Engineering*, Wiley, Hoboken, NJ, 1995.)

tools, the resulting product line or process effort, and are guided by efforts at systems management, as suggested in Figure 4.1. Considerably more details are presented in Sage (1992a, 1992b), which are the sources from which much of this chapter is derived.

4.2 Systems Engineering

Systems engineering is a transdisciplinary management technology. Technology is organization, application, and delivery of scientific knowledge for the betterment of a client group. This is a functional definition of technology as a fundamentally human activity. A technology inherently involves a purposeful human extension of one or more natural processes. For example, the stored program digital computer is a technology in that it enhances the ability of a human to perform computations and, in more advanced forms, to process information.

Management involves the interaction of the organization with the environment. The purpose of management is to enable organizations to cope better with their environments such as to achieve purposeful goals and objectives. Consequently, a management technology involves the interaction of technology, organizations concerned with both the evolution and use of technologies, and the environment.

Information and associated knowledge are the catalysts that enable these necessary interactions and allows them to be satisfactory. Information and knowledge are very important quantities that are assumed to be present in the management technology that is systems engineering. This strongly couples notions of systems engineering with those of technical direction or systems management of technological development, rather than exclusively with one or more of the methods of systems engineering, important as they may be for the ultimate success of a systems engineering effort. It suggests that systems engineering is the management technology that controls a total system life-cycle process, which involves and results in the definition, development, and deployment of a system that is of high quality, trustworthy, and cost-effective in meeting user needs. This process-oriented notion of systems engineering and systems management will be emphasized here.

As suggested in Sage (1992a, 1992b) systems engineering knowledge comprises three types of knowledge (Sage, 1987a). Knowledge principles generally represent formal problem-solving approaches to knowledge, and are employed in new situations and unstructured environments. Knowledge practices represent the accumulated wisdom and experiences that have led to the development of standard operating policies for well-structured problems. Knowledge perspectives represent the view that is held relative to future directions and realities in the technological area under consideration. Clearly, one form of knowledge leads to another. Knowledge perspectives may create the incentive for research that leads to the discovery of new knowledge principles. As knowledge principles emerge and are refined, they generally become embedded in the form of knowledge practices. Knowledge practices are generally the major influences of

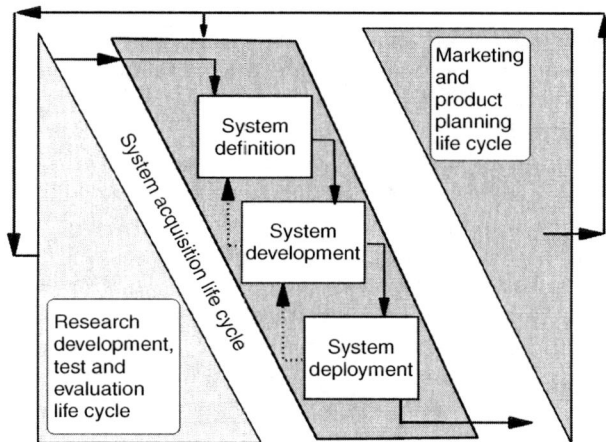

FIGURE 4.2 Three primary systems engineering life cycles. (From Sage, A.P., *Systems Management for Information Technology & Software Engineering*, Wiley, Hoboken, NJ, 1995.)

the systems that can be acquired or fielded. These knowledge types interact with each other and support one another. In a non-exclusive way, they each support one of the principal life cycles associated with systems engineering. There are a number of feedback loops that are associated with learning to enable continual improvement in performance over time. This supports our view that it is a serious mistake to consider these life cycles in isolation from one another.

It is on the basis of the appropriate use of these knowledge types that we are able to accomplish the technological system planning and development and the management system planning and development that lead to a new innovative system, product or service. All three types of knowledge are needed. We envision three different life cycles for technology evolution: system planning and marketing; research, development, test and evaluation (RDT&E); and system acquisition, production, or procurement. Each of these are generally needed, and each primarily involves the use of one of the three types of knowledge. We will discuss these briefly here, and will illustrate how and why these make major but non-exclusive use of knowledge principles, practices, and perspectives. Figure 4.2 illustrates interactions across these life cycles for one particular three-phase realization of a system acquisition life cycle.

It is important to define an area of intellectual inquiry for a better understanding. We have provided one definition of systems engineering thus far. It is primarily a structural and process-oriented definition. A related definition, in terms of purpose, is that systems engineering is a management technology to assist and support policy making, planning, decision making, and associated resource allocation or action deployment. Systems engineers accomplish this by quantitative and qualitative formulation, analysis, and interpretation of the impacts of action alternatives upon the needs perspectives, the institutional perspectives, and the value perspectives of their clients or customers. Each of these three steps is generally needed in solving systems engineering problems, and models are especially useful supports in achieving these ends. Issue formulation is an effort to identify the needs to be fulfilled and the requirements associated with these in terms of objectives to be satisfied, constraints and alterables that affect issue resolution, and generation of potential alternate courses of action. Issue analysis and assessment enables us to determine the impacts of the identified alternative courses of action, including possible refinement of these alternatives. It is in this step that model development and use are of particular value. Issue interpretation enables us to rank order the alternatives in terms of need satisfaction and to select one for implementation or additional study. This particular listing of three systems engineering steps and their descriptions is rather formal. The steps of formulation, analysis, and interpretation may also be accomplished in an "as if" basis by application of a variety of often useful heuristic approaches. These may well be quite appropriate in situations where the problem solver is experientially familiar with the task at hand, and the environment into which the task is embedded.

The key words in this definition are formulation, analysis, and interpretation. In fact, all of systems engineering can be thought of as consisting of formulation, analysis, and interpretation efforts, together with the systems management and technical direction efforts necessary to bring this about. We may exercise these in a formal sense, or in an as if or experientially based intuitive sense. These are the stepwise or microlevel components that comprise a part of the structural framework for systems engineering.

Finally, we can think of a functional definition of systems engineering. Systems engineering is the art and science of producing a product, based on phased efforts, that satisfies user needs. The system is functional, reliable, of high quality, and trustworthy; and has been developed within cost and time constraints through the use of an appropriate set of methods and tools.

In our first definition of systems engineering, we indicated that systems engineers are concerned with the appropriate definition, development, and deployment of systems. These comprise a set of phases for a systems engineering life cycle, as actually illustrated within the systems acquisition, production, or procurement life cycle of Figure 4.2. There are many ways to describe the life-cycle phases of the systems engineering process, and we have described a number of them in Sage (1992a, 1992b). Each of these basic life-cycle models, and those which are outgrowths of them, comprises these three phases. For pragmatic reasons, a typical life cycle will almost always contain more than three phases. Generally, it takes on the "waterfall" pattern illustrated in many works, although there are a number of modifications of the basic waterfall, or grand-design life cycle, to allow for incremental and evolutionary development of systems, as discussed in Sage (1992a, 1992b) and many other sources.

4.3 The Importance of Technical Direction and Systems Management

Systematic measurements are essential for appropriate practice of systems management. The use of the terms reactive measurements, interactive measurements, and proactive measurements may seem unusual. We may, however, approach measurement, and systems engineering and management in general, from at least four perspectives:

1. *Inactive*: This denotes an organization that does not use metrics, or that does not measure at all except perhaps in an intuitive and qualitative manner.
2. *Reactive*: This denotes an organization that will perform an outcome assessment and after it has detected a problem, or failure, will diagnose the cause of the problem and, often, will get rid of the symptoms that produce the problem.
3. *Interactive*: This denotes an organization that will measure an evolving product as it moves through various phases of the life-cycle process to detect problems as soon as they occur, diagnose their causes, and correct the difficulty through recycling, feedback, and retrofit to and through that portion of the life-cycle process in which the problem occurred.
4. *Proactive*: Proactive measurements are those designed to predict the potential for errors and synthesis of an appropriate life-cycle process that is sufficiently mature such that error potential is minimized.

We can also refer to the systems management style of an organization as inactive, reactive, interactive, or proactive. All of these perspectives on measurement purpose, and on systems management, are needed. Inactive and reactive measurements are associated with organizations that have a low level of process maturity. As one moves to further higher levels of process maturity, the lower level forms of measurements become less used. In part, this is so because a high level of process maturity results in such appropriate metrics for systems management that final product errors, which can be detected through a reactive measurement approach, tend to occur very infrequently. While reactive measurement approaches are used, they are not at all the dominant focus of measurement. In a very highly mature organization, they might be only needed on the rarest of occasions. In many situations, models and associated simulations of

Systems Engineering

systems are needed to obtain measurements from these, especially when we are in the preliminary design and architecting phases of a systems engineering effort and the actual system has yet to be engineered, even in a preliminary manner.

Management of the systems engineering processes, which we call systems management, is very essential for success. There are many evidences of systems engineering failures at the level of systems management. Often, one result of these failures is that the purpose, function, and structure of a new system are not identified sufficiently before the system is defined, developed, and deployed. These failures, generally, are the result of costly mistakes that could truly have been avoided. A major objective of systems engineering, at the strategic level of systems management, is to take proactive measures to avoid these difficulties.

Concerns associated with the definition, development, and deployment of tools such that they can be used efficiently and effectively have always been addressed, but often this has been on an implicit and "trial-and-error" basis. When tool designers were also tool users, which was more often than not the case for the simple tools, machines, and products of the past, the resulting designs were often good initially, or soon evolved into good designs through this trial-and-error effort. When physical tools, machines, and systems become so complex that it is no longer possible to design them by a single individual who might even also be the intended user of the tool, and a design team is necessary, then a host of new problems emerge. This is very much the condition today and it is especially the case with respect to system models and simulations. To cope with this, a number of methodologies associated with systems engineering have evolved. Through these, it has been possible to decompose large design issues into smaller component subsystem design issues, design the subsystems, and then build the complete system as a collection of these subsystems.

These phased efforts of definition, development, and deployment represent the macro structure of a systems engineering framework. Each of them need to be employed for each of the three life cycles of formulation, analysis, and interpretation. Thus, we see that our relatively simple description of systems engineering is becoming more and more complex.

Figure 4.3 illustrates how these three steps, three phases, and three life cycles comprise a more complete methodological, structural, or process-oriented view of systems engineering. Even in this relatively simple

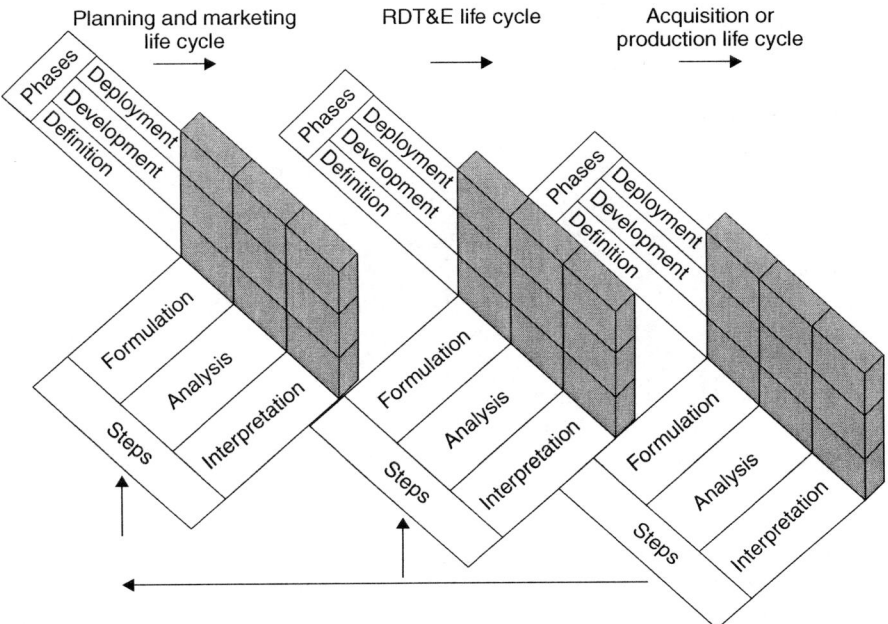

FIGURE 4.3 Three iterative systems engineering life cycles and phases and steps within each life cycle. (From Sage, A.P., *Systems Management for Information Technology & Software Engineering*, Wiley, Hoboken, NJ, 1995.)

methodological framework, which is simultaneously incomplete but relatively complex, we have a total of 27 cells of activity. In a much more realistic view of the steps and phases, as would need to be the case in actual systems development, we might well have seven phases and seven steps of effort. This yields a total of 147 cells of activity. Each of the three levels—systems engineering methods, systems engineering processes, and systems management—are necessarily associated with applicable environments to assure an appropriate systems engineering process, including the very necessary client interaction during system definition, development, and deployment. The use of appropriate systems methods and tools as well as systems methodology (Sage, 1977; Sage and Armstrong, 2000) and systems management constructs enables system design for more efficient and effective human interaction (Sage, 1987b).

System management and associated architecting and integration issues are of major importance in achieving effectiveness, efficiency, and overall functionality of systems engineering efforts. To achieve a high measure of functionality, it must be possible for a system design to be efficiently and effectively produced, used, maintained, retrofitted, and modified throughout all phases of a life cycle. This life cycle begins with need conceptualization and identification, through specification of system requirements and architectures, to ultimate system installation, operational implementation or deployment, evaluation, and maintenance throughout a productive lifetime.

In reality, there are many difficulties associated with the production of functional, reliable, and trustworthy systems of large scale and scope. These potential difficulties, when they are allowed to develop, can create many problems that are difficult to resolve. Among these are inconsistent, incomplete, and otherwise imperfect system requirement specifications; system requirements that do not provide for change as user needs evolve over time; and poorly defined management structures for product design and delivery. These lead to delivered products that are difficult to use, that do not solve the intended problem, that operate in an unreliable fashion, that are unmaintainable, and that—as a result—are not used. Sometimes these failures are so great that operational products and systems are never even fully developed, much less operationally deployed, before plans for the product or system are abruptly canceled.

These same studies generally show that the major problems associated with the engineering of trustworthy systems, or systems engineering, have a great deal more to do with the organization and management of complexity than with direct technological concerns that affect individual subsystems and specific physical science areas. Often the major concern should be more associated with the definition, development, and use of an appropriate process, or product line, for production of a product than it is with over attention to the internal design aspects of the actual product itself, in the sense that exclusive attention to the product or service without appropriate attention to the process leads to the fielding of a low-quality and expensive product or service. Models of both are needed, of course.

In our previous section, we provided structural, functional, and purposeful definitions of systems engineering. There are, of course, other definitions. Two closely related and appropriate definitions are provided by MIL-STD-499A (1974) and MIL-STD-499B (1991), which, although no longer current, have been the benchmark for many subsequent standards. According to MIL-STD-499B, systems engineering is "an interdisciplinary approach to evolve and verify an integrated and life-cycle balanced set of system product and process solutions that satisfy the customers needs. Systems engineering: encompasses the scientific and engineering efforts related to the development, manufacturing, verification, deployment, operations, support, and disposal of system products and processes; develops needed user training equipment, procedures, and data; establishes and maintains configuration management of the system; and develops work breakdown structures and statements of work, and provides information for management decision making." This definition attempts to illustrate and combine structural, functional, and purposeful views of systems engineering. There are many subsequent and reasonably comparable definitions.

We have illustrated three hierarchical levels for systems engineering in Figure 4.1. We now expand on this to indicate some of the ingredients at each of these levels. The functional definition, or lowest level, of systems engineering says that we will be concerned with the various tools and techniques and methods that enable us to design systems. Often, these will be systems science and operations research tools that enable the formal analysis of systems, including modeling. They can also include specific system design tools and components. With respect to information technology and software engineering applications, these would certainly include a variety of computer science and programming tools or methods. It should be, strictly speaking, more appropriate to refer to these as product-level methods. Then we could also refer to process methods and systems management methods. When the term "method(s)" is used alone and without a modifier, what is generally being referred to are product-level methods. The specific nature of the most useful methods and tools will naturally depend, greatly, on the particular life cycle and life-cycle phase that is being considered and the particular product, service, or system, that is ultimately to be acquired.

The functional definition of systems engineering also mentions that we will be concerned with a combination of these tools. In systems engineering, we obtain this combination as the result of using systems methodology. For our purpose, a methodology is an open set of procedures for problem solving. This brings about such important notions as appropriate development life cycles, operational quality assurance issues, and configuration management procedures, which are very important and are discussed in much more detail in Sage (1992b). Each of these reflects a structural, or methodological, perspective on systems engineering. How to best bring about these will vary from product to product and across each of the three life cycles leading to that product or system, or service.

The structural definition of systems engineering tells us that we are concerned with a framework for problem resolution that, from a formal perspective at least, consists of three fundamental steps: issue formulation, issue analysis and assessment, and issue interpretation. These are each conducted at each of the life-cycle phases that have been chosen for definition, development, and deployment. Regardless of the way in which the systems engineering life-cycle process is characterized, and regardless of the type of product or system, or service that is being designed, all characterizations of systems engineering life cycles will necessarily involve (Sage, 1992a, 1992b; Sage, 1982):

1. formulation of the issue—in which the needs and objectives of a client group are identified, and potentially acceptable alternatives, or options, are identified or generated;
2. analysis and assessment of the alternatives—in which the impacts of the identified options are identified and evaluated; and
3. interpretation and selection—in which the options, or alternative courses of action, are compared by means of an evaluation of the impacts of the alternatives and how these are valued by the client group. The needs and objectives of the client group are necessarily used as a basis for evaluation. The most acceptable alternative is selected for implementation or further study in a subsequent phase of systems engineering.

We note these three steps again, because of their great importance in systems engineering. Our model of the steps of the fine structure of the systems engineering process, is based upon this conceptualization. These three steps can be disaggregated into a number of others. Each of these steps of systems engineering is accomplished for each of the life-cycle phases. There are generally three different systems engineering life cycles. Thus we may imagine a three-dimensional model of systems engineering that comprises steps associated with each phase of a life cycle, the phases in the life cycle, and the life cycles that comprise the coarse structure of systems engineering. This is one of the many possible morphological frameworks for systems engineering (Sage, 1992a).

Without question, we have presented a formal rational model of the way in which these three systems engineering functions of formulation, analysis, and interpretation are accomplished. Even within this formal framework, there is the need for much iteration from one step back to an earlier step when it is discovered that improvements in the results of an earlier step are needed to obtain a quality result at a later step, or phase, of the systems engineering effort. Also, this description does not emphasize the key role of information and information requirement determination.

Even when these realities are associated with the morphological, or form-based, framework, it still represents an incomplete view of the way in which people do, could, or should accomplish planning, design, development, or other problem-solving activities. The most that can be argued is that this framework is correct in an "as if" manner. This introduces the notion that humans use a variety of approaches to assist them in acquisition, representation, and use of information. Although many have contributed to this area (Sage, 1990), the fundamental work of Rasmussen and his colleagues (1994) is especially significant. Three types of human information-processing activities are described: skill-, rule-, or formal-knowledge-based reasoning. The choice of which form of reasoning to employ is based primarily on the experiential familiarity of an individual with the situation at hand, the task that is in need of being performed, and the environment into which the task and situation are embedded. Additional details of this model of information processing are provided in the works of Rasmussen and his colleagues (1994) and this and other models are also discussed in Sage (1992a, 1992b). Recent application of organizational and human models are discussed in a valuable work (Sage, 1990) for those concerned with organizational models and simulations.

Systems engineering efforts are very concerned with technical direction and management of systems definition, development, and deployment, or systems management. By adopting and applying the management technology of systems engineering, we attempt to be sure that correct systems are designed, and not just that system products are correct according to some potentially ill-conceived notions of what the system should do. Appropriate metrics to enable efficient and effective error prevention and detection at the level of systems management, and at the process and product level will enhance the production of systems engineering products that are "correct" in the broadest possible meaning of this term. To ensure that correct systems are produced requires that considerable emphasis be placed on the front end of each of the systems engineering life cycles.

In particular, there needs to be considerable emphasis on the accurate definition of a system, what it should do, and how people should interact with it before one is produced and implemented. In turn, this requires emphasis upon conformance to system requirement specifications, and the development of standards to insure compatibility and integratibility of system products. Areas such as documentation and communication are important in all of this. Thus, we see the need for the technical direction and management technology efforts that comprise systems engineering, and the strong role for process and systems management-related concerns in this.

4.4 Other Parts of the Story

There are many ingredients associated with the development of trustworthy systems. From a top-down perspective, the following ingredients are surely present:

1. Systems engineering processes, including process development life cycle and process configuration management;
2. process risk, operational-level quality assurance and evaluation, and product risk and development standards;
3. metrics for quality assurance, and process and product evaluation;
4. metrics for cost estimation, and product cost and operational effectiveness evaluation;
5. strategic quality assurance and management, or total quality management;
6. organizational cultures, leadership, and process maturity;
7. reengineering at the levels of systems management, organizational processes and product lines, and product.

and related issues that concern enterprise management and systems integration, economic systems analysis, cognitive ergonomics, and system assessment and evaluation.

These are the principal issues addressed in Sage (1992a, 1992b) and we could only begin to suggest their importance here. A handbook of systems engineering and management (Sage and Rouse, 1999) provides many perspectives in these needs.

These ingredients interact with one another. One of the first efforts in systems management is to identify an appropriate process life cycle for the production of a trustworthy system. As we have already discussed, this life cycle involves a sequence of phases. These phases include identification of client requirements, translation of these requirements into (hardware and) software requirement specifications, development of system architectures, detailed design through coding, operational implementation and evaluation, and maintenance of the delivered product. The precise life cycle that is followed will depend upon the client needs. It will also depend upon environmental factors such as the presence of existing system components, or subsystems, into which a new system must be integrated, and the presence of existing software modules that may be retrofitted and reused as a part of the new system. This need for system integration brings about a host of systems management and, in many cases, legal issues (Beutel, 1991) that are much larger in scale and scope than those associated with program development only. In a similar manner, the development of appropriate system-level architectures is very important in that efficiency and effectiveness in systems architecting is very influential of the ease with which systems can be integrated and maintained and, therefore, of the extent to which an operational system is viewed as trustworthy and of high quality.

Following the identification of an appropriate systemic process-development life cycle, configuration management plans are identified. This involves using the life cycle and defining a specific development process for the set of life cycle tasks at hand. Metrics are needed to enable this to be done effectively. These metrics are the metrics of cost analysis, or cost estimation for systems engineering. They also include effectiveness analysis or estimation of system productivity indices using various metrics. This couples the notion of development of a product into notions concerning the process needs associated with developing this product. It is becoming widely recognized that these metrics must form a part of a process-management approach for process, and ultimately product.

We have set forth some of the very large number of issues associated with systems engineering as a catalyst for innovation and quality in this chapter. There are four principal messages developed that provide benchmarks for continuing efforts:

1. Much contemporary thought concerning innovation, productivity, and quality can be easily cast into a systems engineering framework.
2. This framework can be valuably applied to systems engineering in general and information technology and software engineering in particular.
3. The information technology revolution provides the necessary tool base that, together with systems engineering and systems management, provides the wherewithal to allow the needed process-level improvements for the development of systems of all types.
4. Development of appropriate system models is an essential ingredient in essentially all of this.

Further, the relatively large number of ingredients necessary to accomplish the needed change fit well within a systems engineering framework. Our discussions also illustrate that systems engineering constructs are useful not just for managing big systems engineering projects according to government and industry requirements, but for creative management of the organization itself.

4.5 Summary

The top-level objectives for systems engineering might be stated as the reduction of cost, and improvement of quality in the technical direction and management of definition, development, and deployment of modern products, systems, and services through the use of systems engineering processes. Together with the ingredients to be dealt with in a systems engineering study—humans, organizations, and technologies, and the environments that surround these—we see that systems engineering is a mixture of a management discipline and a technology discipline, and that is why we have illustrated and described systems engineering as a management technology. Finally, we indicate perhaps the major objective of systems engineering, and that is to bring together people, organizations, and technology, and all within a suitable environmental context, for creative issue resolution and innovation. This is clearly a transdisciplinary endeavor and one to which models and associated simulations can be of major use, as especially illustrated in Rouse and Boff (2005) and Sage and Rouse (1999), which are efforts discussing the substantial role of systems engineering, and modeling, in simulation-based acquisition (Sage and Olson, 2001; Olson and Sage, 2003).

References

Beutel, R. A., *Contracting for Computer Systems Integration,* Michie Co., Charlottesville, VA, 1991.
MIL-STD-499A, Engineering Management Standards, May 1974.
MIL-STD-499B, Engineering Management Standards, May 1991.
Olson, S. R. and Sage, A. P., Simulation based acquisition, in H. Booher (Ed.), *Handbook of Human Systems Integration,* Wiley, New York, 2003, Chap. 9, pp. 265–293.
Rasmussen, J., Pejtersen, A. and Goodstein, L. P. *Cognitive Systems Engineering,* Wiley, Hoboken, NJ, 1994.
Rouse, W. B. and Boff, K. R. (Eds.), *Organizational Simulation,* Wiley, Hoboken, NJ, 2005.
Sage, A. P., *Methodology for Large Scale Systems,* McGraw-Hill, New York, 1977.
Sage, A. P., Methodological considerations in the design of large scale systems engineering processes, in Y. Y. Haimes (Ed.), *Large Scale Systems,* North-Holland, Amsterdam, 1982, pp. 99–141.
Sage, A. P., Knowledge transfer: An innovative role for information engineering education, *IEEE Transactions on Systems, Man and Cybernetics,* Vol. 17, No. 5, 1987, pp. 725–728.
Sage, A. P. (Ed.), *System Design for Human Interaction,* IEEE Press, New York, 1987.
Sage, A. P. (Ed.), *Concise Encyclopedia of Information Processing in Systems and Organizations,* Pergamon Press, Oxford, UK, 1990.
Sage, A. P., *Systems Engineering,* Wiley, Hoboken, NJ, 1992.
Sage, A. P., *Systems Management for Information Technology and Software Engineering,* Wiley, Hoboken, NJ, 1992.
Sage, A. P. and Armstrong, J. E., *An Introduction to Systems Engineering,* Wiley Hoboken, NJ, 2000.
Sage, A. P. and Rouse, W. B., *Handbook of Systems Engineering and Management,* Wiley, Hoboken, NJ, 1999.
Sage, A. P. and Olson, S., Modeling and simulation in systems engineering: Whither simulation based acquisition?" *Modeling and Simulation,* Part I, May 2001, Part 2, June 2001.

Basic Elements of Mathematical Modeling

Clive L. Dym
Harvey Mudd College

5.1 Principles of Mathematical Modeling **5**-2
5.2 Dimensional Consistency and Dimensional Analysis ... **5**-3
    Dimensions and Units • Dimensional Homogeneity • The Basic Method of Dimensional Analysis • The Buckingham Pi Theorem of Dimensional Analysis
5.3 Abstraction and Scale **5**-9
    Abstraction, Scaling, and Lumped Elements • Geometric Scaling • Scale in Equations: Size and Limits • Consequences of Choosing a Scale • Scaling and Perceptions of Data Presentations
5.4 Conservation and Balance Principles **5**-17
5.5 The Role of Linearity **5**-19
    Linearity and Geometric Scaling
5.6 Conclusions **5**-20

A dictionary definition states that a *model* is "a miniature representation of something; a pattern of something to be made; an example for imitation or emulation; a description or analogy used to help visualize something (e.g., an atom) that cannot be directly observed; a system of postulates, data, and inferences presented as a mathematical description of an entity or state of affairs." This definition suggests that *modeling* is an activity, a *cognitive activity* in which one thinks about and makes models to describe how devices or objects of interest behave. Since there are many ways in which devices and behaviors can be described—words, drawings or sketches, physical models, computer programs, or mathematical formulas—it is worth refining the dictionary definition for the present purposes to define a mathematical model as a *representation in mathematical terms* of the behavior of real devices and objects.

Scientists use mathematical models to *describe* observed behavior or results; *explain why* that behavior and results occurred as they did; and to *predict* future behaviors or results that are as yet unseen or unmeasured. *Engineers* use mathematical models to describe and analyze objects and devices to predict their behavior because they are interested in *designing* devices and processes and systems. Design is a consequential activity for engineers because every new airplane or building, for example, represents a model-based prediction that the plane will fly and the building will stand without dire, unanticipated consequences. Thus, especially in engineering, it is important to ask: How are such mathematical models or representations created? How are they validated? How are they used? And, is their use limited, and how?

To answer these and related questions, this chapter first sets out some basic principles of mathematical modeling and then goes on to briefly describe:

- dimensional consistency and dimensional analysis;
- abstraction and scaling;
- conservation and balance laws; and
- the role of linearity.

5.1 Principles of Mathematical Modeling

Mathematical modeling is a principled activity that has both principles behind it and methods that can be successfully applied. The principles are overarching or metaprinciples are almost philosophical in nature, and can be both visually portrayed (see Figure 5.1) and phrased as questions (and answers) about the intentions and purposes of mathematical modeling:

- **Why?** What are we looking for? Identify the need for the model.
- **Find?** What do we want to know? List the data we are seeking.
- **Given?** What do we know? Identify the available relevant data.
- **Assume?** What can we assume? Identify the circumstances that apply.
- **How?** How should we look at this model? Identify the governing physical principles.
- **Predict?** What will our model predict? Identify the equations that will be used, the calculations that will be made, and the answers that will result.
- **Valid?** Are the predictions valid? Identify tests that can be made to validate the model, i.e., is it consistent with its principles and assumptions?
- **Verified?** Are the predictions good? Identify tests that can be made to verify the model, i.e., is it useful in terms of the initial reason it was done?

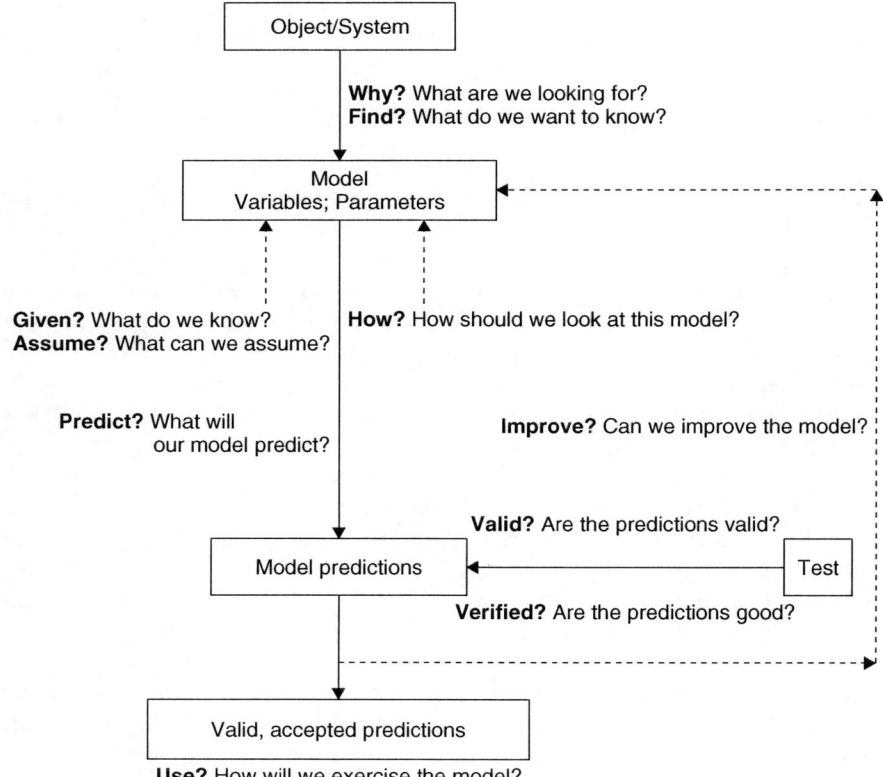

FIGURE 5.1 A first-order view of *mathematical modeling* that shows how the questions asked in a principled approach to building a model relate to the development of that model. (Inspired by Carson, E. and Cobelli, C. (Eds.), *Modelling Methodology for Physiology and Medicine*, Academic Press, San Diego, 2001.)

- **Improve?** Can we improve the model? Identify parameter values that are not adequately known, variables that should have been included, and/or assumptions/restrictions that could be lifted. Implement the iterative loop that we can call "model-validate-verify-improve-predict."
- **Use?** How will we exercise the model? What will we do with the model?

It is worth noting that the "final" principle, **Use?**, is often considered early in the modeling process, along with **Why?** and **Find?**, because the way a model is to be used is often intimately connected with the reason it is created.

Note also that this list of questions and instructions is *not* an algorithm for building a good mathematical model. However, the underlying ideas are key to mathematical modeling, as they are key to problem formulation generally. Thus, the individual questions will recur often during the modeling process, and the list should be regarded as a general approach to *ways of thinking* about mathematical modeling. In a similar vein, it is also worth remembering the associated modeling "methods" that are presented below (e.g., dimensional analysis, and abstraction and scaling) *support* rather than lead the modeling process; these modeling methods are neither algorithms themselves or susceptible to being encapsulated into some overarching "modeling algorithm."

Having a clear picture of why the model is wanted or needed is of prime importance to the model-building enterprise. For example, a first estimate of the available power generated by a dam on a large river, say The Three Gorges Dam on the Yangtze River in Hubei Province in the People's Republic of China, would not require a model of the dam's thickness or the strength of its foundation. However, its height would be essential, as would some model and estimates of river flow quantities. By contrast, a design of the actual dam would need a model that incorporates all of the dam's physical characteristics (e.g., dimensions, materials, and foundations) and relates them to the dam site and the river flow conditions. Thus, defining the task is the first essential step in model formulation.

The next step would be to list what is known, for example, river flow quantities and desired power levels, as a basis for listing variables or parameters that are not yet known. One should also list any relevant assumptions. For example, levels of desired power may be linked to demographic or economic data, so any assumptions made about population and economic growth should be spelled out. Assumptions about the consistency of river flows and the statistics of flooding should also be spelled out.

Which physical principles apply to this model? The mass of the river's water must be conserved, as must its momentum, as the river flows and energy is both dissipated and redirected as water is allowed to flow through turbines in the dam (and hopefully not spill over the top!). Mass must be conserved, within some undefined system boundary, because dams do accumulate water mass from flowing rivers. There are well-known equations that correspond to these physical principles. They could be used to develop an estimate of dam height as a function of power desired. The model can be validated by ensuring that all equations and calculated results have the proper dimensions, and the model can be exercised against data from existing hydroelectric dams to get empirical data and validation.

If the model is inadequate or that it fails in some way, an *iterative loop* is then entered in which one cycles back to an earlier stage of the model building to reexamine any assumptions, known parameter values, the principles chosen, the equations used, the means of calculation, and so on. This iterative process is essential because it is the only way that models can be improved, corrected, and validated.

5.2 Dimensional Consistency and Dimensional Analysis

There is a very powerful idea that is central to mathematical modeling: Every equation used must be *dimensionally homogeneous* or *dimensionally consistent*, that is, every term in a balance of mass should have the dimension of mass, and when forces are summed to ensure equilibrium, every term in that summation must have the physical dimension of force. Equations that are dimensionally consistent are also called *rational* equations. Ensuring the (dimensional) rationality of equations is very useful for validating newly developed mathematical models or for confirming formulas and equations before doing calculations with them. There is a technique called *dimensional analysis*, which H. L. Langhaar has defined as "a method by which we deduce information about a phenomenon from the single premise that the phenomenon can be described by a dimensionally correct equation among certain variables." Some of the available tools of dimensional analysis are now described.

5.2.1 Dimensions and Units

The physical quantities used to model objects or systems represent *concepts*, such as time, length, and mass, to which are also attached *numerical* values or measurements. If the width of a soccer field is said to be 60 m, the concept invoked is length or distance, and the numerical measure is 60 m. A numerical measure implies a comparison with a standard that enables (1) communication about and (2) comparison of objects or phenomena without their being in the same place. In other words, common measures provide a frame of reference for making comparisons.

The physical quantities used to describe or model a problem are either *fundamental* or *primary* quantities, or they are *derived* quantities. A quantity is fundamental if it can be assigned a measurement standard independent of that chosen for the other fundamental quantities. In mechanical problems, for example, mass, length, and time are generally taken as the fundamental mechanical variables, while force is derived from Newton's law of motion. For any given problem, enough fundamental quantities are required to express each derived quantity in terms of these primary quantities.

The word *dimension* is used to relate a derived quantity to the fundamental quantities selected for a particular model. If mass, length, and time are chosen as primary quantities, then the dimensions of area are (length)^2, of mass density are mass/(length)^3, and of force are (mass × length)/(time)^2. The notation of brackets [ ] is introduced to read as "the dimensions of." If M, L, and T stand for mass, length, and time, respectively, then

$$[A = \text{area}] = (L)^2, \quad [\rho = \text{density}] = M/(L)^3, \quad [F = \text{force}] = (M \times L)/(T)^2 \tag{5.1}$$

The *units* of a quantity are the numerical aspects of a quantity's dimensions expressed in terms of a given physical standard. By definition, then, a unit is an arbitrary multiple or fraction of that standard. The most widely accepted international standard for measuring length is the meter (m), but length can also be measured in units of centimeters (1 cm = 0.01 m) or of feet (0.3048 m). The magnitude or size of the attached number obviously depends on the unit chosen, and this dependence often suggests a choice of units to facilitate calculation or communication. For example, a soccer field width can be said to be 60 m, 6000 cm, or ~197 ft.

Dimensions and units are related by the fact that identifying a quantity's dimensions allows us to compute its numerical measures in different sets of units, as we just did for the soccer field width. Since the physical dimensions of a quantity are the same, there must exist numerical relationships between the different systems of units used to measure the amounts of that quantity (e.g., 1 foot [ft] $\cong$ 30.48 centimeters [cm], and 1 hour [h] = 60 minutes [min] = 3600 seconds [s]). This equality of units for a given dimension allows units to be changed or converted with a straightforward calculation, for example,

$$65\frac{\text{mi}}{\text{h}} = 65\frac{\text{mi}}{\text{h}} \times 5280\frac{\text{ft}}{\text{mi}} \times 0.3048\frac{\text{m}}{\text{ft}} \times 0.001\frac{\text{km}}{\text{m}} \cong 104.6\frac{\text{km}}{\text{h}} \tag{5.2}$$

Each of the multipliers in this conversion equation has an effective value of unity because of the equivalencies of the various units, that is, 1 mi = 5280 ft, and so on. This, in turn, follows from the fact that the numerator and denominator of each of the above multipliers have the same physical dimensions.

5.2.2 Dimensional Homogeneity

A *rational equation* is dimensionally homogeneous, which means each independent term in that equation has the same net dimensions. Simply put, length cannot be added to area in the same equation, or mass to time, or charge to stiffness—although quantities having the same dimensions but expressed in different units can be added, although with great care, e.g., length in meters and length in feet. The fact that equations must be rational in terms of their dimensions is central to modeling because it is one of the best—and easiest—checks to make to determine whether a model makes sense, has been correctly derived, or even correctly copied!

A dimensionally homogeneous equation is independent of the units of measurement being used. However, unit-dependent versions of such equations can be created for convenience in doing repeated calculations or as a memory aid. In an example familiar from mechanics, the period (or cycle time), T_0, of a pendulum undergoing small-angle oscillations can be written in terms of the pendulum's length, l, and the acceleration of gravity, g:

$$T_0 = 2\pi \sqrt{\frac{l}{g}} \tag{5.3}$$

This dimensionally homogeneous equation is independent of the system of units chosen to measure length and time. However, it may be convenient to work in the metric system, in which case $g = 9.8$ m/s², from which it follows that

$$T_0(s) = 2\pi \sqrt{\frac{l}{9.8}} \cong 2\sqrt{l} \tag{5.4}$$

Eq. (5.4) is valid *only* when the pendulum's length is measured in meters. In the so-called British system, where $g = 32.17$ ft/s²,

$$T_0(s) = 2\pi \sqrt{\frac{l}{32.17}} \cong 1.1\sqrt{l} \tag{5.5}$$

Eq. (5.4) and Eq. (5.5) are not dimensionally homogeneous. So, while these formulas may be appealing or elegant, their limited ranges of validity must be kept in mind.

5.2.3 The Basic Method of Dimensional Analysis

Dimensional analysis is the process by which dimensional consistency is ensured. First, the dimensions of all derived quantities are checked to see that they are properly represented in terms of the chosen primary quantities and their dimensions. Second, the proper *dimensionless groups* of variables—ratios and products of problem variables and parameters that are themselves dimensionless—are identified. There are two different techniques for identifying such dimensionless groups, the *basic method* and the *Buckingham Pi theorem*.

The basic method of dimensional analysis is a rather informal, unstructured approach for determining dimensional groups. It depends on being able to construct a functional equation that contains all of the relevant variables, for which we know the dimensions. The proper dimensionless groups are then identified by the thoughtful elimination of dimensions.

To illustrate the basic method, consider the mutual revolution of two bodies in space that is caused by their mutual gravitational attraction. The goal is to find a dimensionless function that relates the period of oscillation, T_R, to the two masses and the distance r between them:

$$T_R = T_R(m_1, m_2, r) \tag{5.6}$$

The dimensions for the four variables in Eq. (5.6) are

$$[m_1], [m_2] = \text{M}, \quad [T_R] = \text{T}, \quad [r] = \text{L} \tag{5.7}$$

Note that in this formulation, none of the dimensions are more than once, except for the two masses. So, while the masses can be expected to appear in a dimensionless ratio, how can the period and distance be kept in the problem? The answer is that a parameter containing the dimensions heretofore missing to the functional equation (5.6) must be added. Newton's gravitational constant, G, is such a variable, so that the functional equation (5.6) can be restated as

$$T_R = T_R(m_1, m_2, r, G) \tag{5.8}$$

where the dimensions of G are

$$[G] = L^3/MT^2 \tag{5.9}$$

The complete list of variables for this problem, consisting of Eq. (5.7) and Eq. (5.9), includes enough variables to account for all of the dimensions.

Applying the basic method to Eq. (5.8) as the assumed functional equation for two revolving bodies, the dimension of time is eliminated first. Time appears directly in the period T_R and as a reciprocal squared in the gravitational constant G. It follows dimensionally that

$$[T_R\sqrt{G}] = \sqrt{\frac{L^3}{M}} \tag{5.10}$$

where the right-hand side of Eq. (5.10) is independent of time. Thus, the corresponding revised functional equation for the period would be

$$T_R\sqrt{G} = T_{R1}(m_1, m_2, r) \tag{5.11}$$

The length dimension can be eliminated simply by noting that

$$\left[\frac{T_R\sqrt{G}}{\sqrt{r^3}}\right] = \sqrt{\frac{1}{M}} \tag{5.12}$$

which leads to a further revised functional equation

$$\frac{T_R\sqrt{G}}{\sqrt{r^3}} = T_{R2}(m_1, m_2) \tag{5.13}$$

The mass dimension can be eliminated from Eq. (5.13) by multiplying it by the square root of one of the two masses. Choosing the square root of the second mass, $\sqrt{m_2}$, suggests that

$$\left[\frac{T\sqrt{Gm_2}}{\sqrt{r^3}}\right] = 1 \tag{5.14}$$

This means that Eq. (5.13) becomes

$$\frac{T_R\sqrt{Gm_2}}{\sqrt{r^3}} = \sqrt{m_2}\,T_{R2}(m_1, m_2) \equiv T_{R3}\left(\frac{m_1}{m_2}\right) \tag{5.15}$$

where a dimensionless mass ratio has been introduced in Eq. (5.15) to recognize that this is the only way that the function T_{R3} can be both dimensionless *and* a function of the two masses. It then follows from Eq. (5.15) that

$$T_R = \sqrt{\frac{r^3}{Gm_2}}\,T_{R3}\left(\frac{m_1}{m_2}\right) \tag{5.16}$$

This example shows that it is important to start problems with complete sets of variables. Recall that the gravitational constant G was not included until it became clear that a wrong path was being followed,

Basic Elements of Mathematical Modeling

after which it was included to rectify an incomplete analysis. In hindsight, it might be argued that the attractive gravitational force must somehow be accounted for, and including G would have achieved that. This argument, however, demands insight and judgment whose origins may have little to do with the particular problem at hand.

This single application of the basic method of dimensional analysis shows that it does not have a formal algorithmic structure, it can be described as a series of steps to take:

1. List all of the variables and parameters of the problem and their dimensions.
2. Anticipate how each variable qualitatively affects quantities of interest, that is, does an increase in a variable cause an increase or a decrease?
3. Identify one variable as depending on the remaining variables and parameters.
4. Express that dependence in a functional equation (i.e., the analog of Eq. ([5.6]).
5. Eliminate one of the primary dimensions to obtain a revised functional equation.
6. Repeat step 3 until a revised, *dimensionless* functional equation is found.
7. Review the final *dimensionless* functional equation to see whether the apparent behavior accords with the behavior anticipated in step 2.

5.2.4 The Buckingham Pi Theorem of Dimensional Analysis

Buckingham's Pi theorem, fundamental to dimensional analysis, can be stated as follows: "A dimensionally homogeneous equation involving n variables in m primary or fundamental dimensions can be reduced to a single relationship among $n-m$ independent dimensionless products." A rational (or dimensionally homogeneous) equation is one in which every independent, additive term in the equation has the same dimensions. This means that any one term can be defined as a function of all of the others. If Buckingham's Π notation is introduced to represent a dimensionless term, his famous Pi theorem can be written as

$$\Pi_1 = \Phi(\Pi_2, \Pi_3, \ldots, \Pi_{n-m}) \tag{5.17a}$$

or, equivalently,

$$\Phi(\Pi_1, \Pi_2, \Pi_3, \ldots, \Pi_{n-m}) = 0 \tag{5.17b}$$

Eq. (5.17a) and Eq. (5.17b) state that a problem with n derived variables and m primary dimensions or variables requires $n-m$ dimensionless groups to correlate all of its variables.

The Pi theorem is applied by first identifying the n derived variables in a problem: $A_1, A_2, \ldots, A_n$. Then m of these derived variables are chosen such that they contain all of the m primary dimensions, say, A_1, A_2, A_3 for $m = 3$. Dimensionless groups are then formed by permuting each of the remaining $n-m$ variables ($A_4, A_5, \ldots, A_n$ for $m = 3$) in turn with those m values already chosen:

$$\Pi_1 = A_1^{a_1} A_2^{b_1} A_3^{c_1} A_4,$$
$$\Pi_2 = A_1^{a_2} A_2^{b_2} A_3^{c_2} A_5,$$
$$\vdots$$
$$\Pi_{n-m} = A_1^{a_{n-m}} A_2^{b_{n-m}} A_3^{c_{n-m}} A_n \tag{5.18}$$

The a_i, b_i, and c_i are chosen to make each of the permuted groups Π_i dimensionless.

A classical physics problem—modeling the small angle, free vibration of an ideal pendulum (viz. Figure 5.2)—will now be used to illustrate the application of Buckingham's Pi theorem. There are six variables to consider in this problem, and they are listed along with their fundamental dimensions in Table 5.1. In this case $m = 6$ and $n = 3$, so that three dimensionless groups are expected. If l, g, and m are

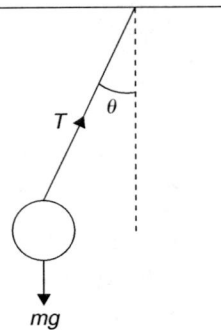

FIGURE 5.2 The classical pendulum oscillating through angle θ owing to gravitational acceleration g.

TABLE 5.5.1 The Six Derived Quantities Chosen to Model the Oscillating Pendulum

Derived Quantities	Dimensions
Length (l)	L
Gravitational acceleration (g)	L/T²
Mass (m)	M
Period (T)	T
Angle (θ)	1
String tension (T)	(M × L)/T²

chosen as the variables around which to permute the remaining three variables (T_0, θ, T) to obtain the three groups, it follows that

$$\Pi_1 = l^{a_1} g^{b_1} m^{c_1} T_0,$$
$$\Pi_2 = l^{a_2} g^{b_2} m^{c_2} \theta, \qquad (5.19)$$
$$\Pi_3 = l^{a_3} g^{b_3} m^{c_3} T$$

The Pi theorem applied here then yields three dimensionless groups:

$$\Pi_1 = \frac{T_0}{\sqrt{l/g}},$$
$$\Pi_2 = \theta, \qquad (5.20)$$
$$\Pi_3 = \frac{T}{mg}$$

These groups show how the period depends on the pendulum length l and the gravitational constant g (recall Eq. [5.3]), and the string tension T on the mass m and g. The second group also shows that the (dimensionless) angle of rotation stands alone, that is, it is apparently not related to any of the other variables. This follows from the assumption of small angles, which makes the problem linear, and makes the magnitude of the angle of free vibration a quantity that cannot be determined.

One of the "rules" of applying the Pi theorem is that the m chosen variables include all n of the fundamental dimensions, but no other restrictions are given. So, it is natural to ask how this analysis would change if one started with three different variables. For example, suppose T_0, g, and m were chosen as the variables around which to permute the remaining three variables (l, θ, T) to obtain the three groups.

Basic Elements of Mathematical Modeling

In this case

$$\Pi'_1 = T_0^{a_1} g^{b_1} m^{c_1} l,$$
$$\Pi'_2 = T_0^{a_2} g^{b_2} m^{c_2} \theta, \qquad (5.21)$$
$$\Pi'_3 = T_0^{a_3} g^{b_3} m^{c_3} T$$

Applying the Pi theorem to Eq. (5.21) can then be shown to yield the following three "new" dimensionless groups:

$$\Pi'_1 = \frac{l/g}{T_0^2} = \frac{1}{\Pi_1^2},$$
$$\Pi'_2 = \theta = \Pi_2, \qquad (5.22)$$
$$\Pi'_3 = \frac{T}{mg} = \Pi_3$$

Eq. (5.22) presents the same information as Eq. (5.20), albeit in a slightly different form. In particular, it is clear that and Π_1 and Π'_1 contain the same dimensionless group, which suggests that the number of dimensionless groups is unique, but that the precise forms that these groups may take are not. This last calculation demonstrates that the dimensionless groups determined in any one calculation are unique in one sense, but they may take on different, yet related forms when done using a slightly different calculation.

Note that these applications of the basic method and of the Buckingham Pi theorem of dimensional analysis can be cast in similar, step-like structures. However, experience and insight are key to applying both methods, even for elementary problems.

5.3 Abstraction and Scale

While still dealing with dimensions, the focus now shifts to issues of *scale*, that is, issues of *relative size*. Size, whether absolute or relative, is very important because it affects both the form and the function of those objects or systems being modeled. Scaling influences—indeed, often controls—the way objects interact with their environments, for objects in nature, the design of experiments, or the representation of data by smooth, nice-looking curves. This section briefly discusses abstraction and scale, size and shape, size and function, scaling and conditions that are imposed at an object's boundaries, and some of the consequences of choosing scales in both theory and experimental measurements.

5.3.1 Abstraction, Scaling, and Lumped Elements

An important decision in modeling is choosing an appropriate level of detail for the problem at hand, and thus knowing what level of detail is prescribed for the attendant model. This process is called *abstraction* and it typically requires a thoughtful and organized approach to identifying those phenomena that will be emphasized, that is, to answering the fundamental question about why a model is being sought or developed. Further, thinking about finding the right level of abstraction or the right level of detail often requires finding the right *scale* for the model being developed. Stated differently, thinking about *scaling* means thinking in terms of the magnitude or size of quantities measured with respect to a standard that has the same physical dimensions.

For example, a linear elastic spring can be used to model more than just the relation between force and relative extension of a simple coiled spring, as in an old-fashioned butcher's scale or an automobile spring. For example, it is possible to use $F = kx$ to describe the static load-deflection behavior of a diving board, but the spring constant k should reflect the stiffness of the diving board taken as a whole, which in turn reflects more detailed properties of the board, including the material of which it is made and its own dimensions. The validity of using a linear spring to model the board can be ascertained by measuring and plotting the deflection of the board's tip as it changes with standing divers of different weight.

The classic spring equation is also used to model the static and dynamic behavior of tall buildings as they respond to wind loading and to earthquakes. These examples suggest that a simple, highly abstracted model of a building can be developed by aggregating various details within the parameters of that model. That is, the stiffness k for a building would incorporate or lump together a great deal of information about how the building is framed, its geometry, its materials, and so on. For both a diving board and a tall building, detailed expressions of how their respective stiffnesses depended on their respective properties would be needed. It is not possible to do a detailed design of either the board or of the building without such expressions. Similarly, using springs to model atomic bonds means that their spring constants must be related to atomic interaction forces, atomic distances, subatomic particle dimensions, and so on.

Thus, the spring can be used at both much smaller, *microscales* to model atomic bonds, as well as at much larger *macroscales*, as for buildings. The notion of scaling includes several ideas, including the effects of geometry on scale, the relationship of function to scale, and the role of size in determining limits—all of which are needed to choose the right scale for a model in relation to the "reality" we want to capture.

Another facet of the abstraction process occurs whenever, for example, a statement is made that, for some well-defined purposes, a "real," three-dimensional object behaves like a simple spring. Thus, the concept of a *lumped element* model is introduced wherein the actual physical properties of some real object or device are aggregated or *lumped* into a less detailed, more abstract expression. An airplane, for example, can be modeled in very different ways, depending on the modeling goals. To lay out a flight plan or trajectory, the airplane can simply be considered as a point mass moving with respect to a spherical coordinate system. The mass of the point can simply be taken as the total mass of the plane, and the effect of the surrounding atmosphere can also be modeled by expressing the retarding drag force as acting on the mass point itself with a magnitude related to the relative speed at which the mass is moving. To model and analyze the more immediate, more local effects of the movement of air over the plane's wings, a model would be build to account for the wing's surface area and be complex enough to incorporate the aerodynamics that occur in different flight regimes. To model and design the flaps used to control the plane's ascent and descent, a model would be developed to include a system for controlling the flaps and to also account for the dynamics of the wing's strength and vibration response.

Clearly, a discussion about finding the right level of abstraction or the right level of detail is simultaneously a discussion about finding the right *scale* for the model being developed. *Scaling* or imposing a scale includes assessing the effects of geometry on scale, the relationship of function to scale, and the role of size in determining limits. All of these ideas must be addressed when the determination is made on how to scale a model in relation to the "reality" that is being captured.

The scale of things is often examined with respect to a magnitude that is set within a standard. Thus, when talking about freezing phenomena, temperatures are typically referenced to the freezing point of materials included in the model. Similarly, the models of Newtonian mechanics work extraordinarily well for virtually all earth- and space-bound applications. Why is that so? Simply because the speeds involved in all of these calculations are far smaller than c, the speed of light in a vacuum. Thus, even a rocket fired at escape speeds of 45,000 km/h seems to stand still when its speed is compared with $c \approx 300,000$ km/s $= 1.080 \times 10^9$ km/h!

These scaling ideas not only extend the ideas discussed earlier about dimensionless variables, but they also introduce the notion of *limits*. For example, in Einstein's general theory of relativity, the mass of a particle moving at speed, v, is given as a (dimensionless) fraction of the rest mass, m_0, by

$$\frac{m}{m_0} = \frac{1}{\sqrt{1 - (v/c)^2}} \qquad (5.23)$$

The scaling issue here is to find the limit that supports the customary practice of taking the masses or weights of objects to be constants in everyday life and in normal engineering applications of mechanics. A box of candy is not expected to weigh any more whether one is standing still, riding in a car at 120 km/h (75 mi/h), or flying across the country at 965 km/h (600 mi/h). This means that the square of the dimensionless speed ratio in Eq. (5.23) is much less than 1, so that $m \cong m_0$. According to Eq. (5.23), for that box of candy flying across the country at 965 km/h $= 268$ m/s, that factor in the denominator of

Basic Elements of Mathematical Modeling

the relativistic mass formula is

$$\sqrt{1 - \left(\frac{v}{c}\right)^2} = \frac{m_0}{m} = \sqrt{1 - 7.98 \times 10^{-13}} \cong 1 - 3.99 \times 10^{-13} \cong 1 \qquad (5.24)$$

Clearly, for practical day-to-day existence, such relativistic effects can be neglected. However, it remains the case that Newtonian mechanics is a good model only on a scale where all speeds are very much smaller than the speed of light. If the ratio v/c becomes sufficiently large, the mass can no longer be taken as the constant rest mass, m_0, and Newtonian mechanics must be replaced by relativistic mechanics.

5.3.2 Geometric Scaling

Consider now two cubes, one of which has sides of unit length in any system of units, that is, the cube's volume could be 1 in^3 or 1 m^3 or 1 km^3. The other cube has sides of length L in the same system of units, so its volume is either L^3 (in^3) or L^3 (m^3) or L^3 (km^3). Thus, for comparison's sake, the units in which the two cubes' sides are actually measured can be ignored. The total area and volume of the first cube are, respectively, 6 and 1, while the corresponding values for the second cube are $6L^2$ and L^3. An instance of *geometric scaling* can be immediately seen, that is, the area of the second cube changes as does L^2 and its volume scales as L^3. Thus, doubling the side of a cube increases its surface area by a factor of four and its volume by a factor of eight.

Geometric scaling has been used quite successfully in many spheres of biology, for example, to compare the effects of size and age in animals of the same species, and to compare qualities and attributes in different species of animals. As an instance of the latter, consider Figure 5.4, wherein are plotted the total weight of the flight muscles, W_{fm}, of quite a few birds against their respective body weights, W_b. How many birds are "quite a few"? The figure caption states that the underlying study actually included 29 birds, but the figure shows data only within the range 10 ≤ bird number ≤23. For the 14 birds shown in Figure 5.3 there

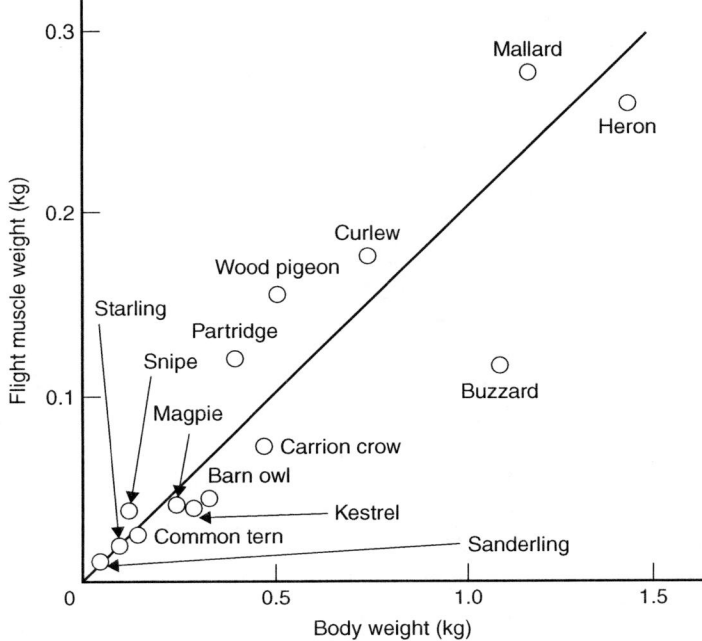

FIGURE 5.3 A simple linear fit on a plot of the total weight of the flight muscles against body weight for 14 of the 29 birds studied, including starlings, barn owls, kestrels, common terns, mallards, and herons. (From Alexander, R. M., *Size and Shape*, Edward Arnold, London, 1971.)

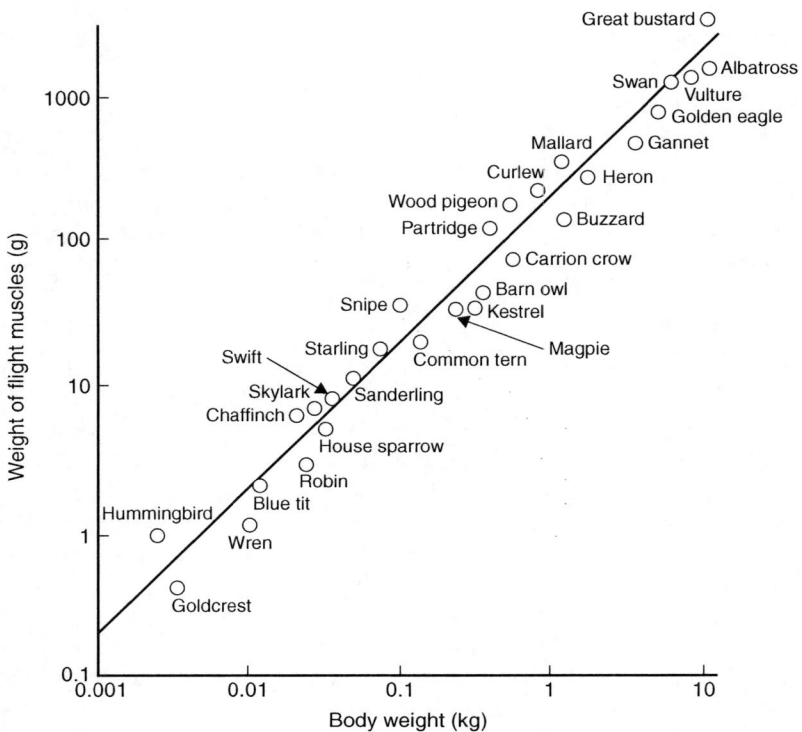

FIGURE 5.4 A "log–log" plot of the total weight of the flight muscles against body weight for 29 birds, including hummingbirds, wrens, terns, mallards, eagles, and albatrosses (From Alexander, R. M., *Size and Shape*, Edward Arnold, London, 1971.). Compare this with the linear plot of the data of Figure 5.3.

seems to be a fairly nice straight line fit for the data presented. While fitted by eye, that straight line can be determined to be

$$W_{\text{fm}} \cong 0.18 W_{\text{b}} \quad (5.25)$$

Eq. (5.25) suggests that flight muscle makes up about 18% of a bird's body weight, and that flight muscle weight *scales linearly* with—or is proportional to—body weight, a result that seems reasonable enough from our everyday observations of the birds around us.

What happened to the other 15 birds in the small scaling study just described? (Among those discriminated against in Figure 5.4 are hummingbirds, wrens, robins, skylarks, vultures, and albatrosses.) These birds were not included because the bird weights studied spanned a fairly large range, which made it hard to include the heavier birds (e.g., vultures and albatrosses) in the plot of Figure 5.4 without completely squashing the data for the very small birds (e.g., hummingbirds and goldcrests). This suggests a problem in organizing and presenting data, in itself an interesting aspect of scaling.

There is a straightforward way to include the heretofore left-out data: Construct log–log plots in which the *logarithms of the data* (normally to base 10) are graphed, as shown in Figure 5.4. In fact, the complete data set was plotted, essentially doubling the number of included data points, and a statistical regression analysis was applied to determine that the straight line shown in Figure 5.4 is given by

$$W_{\text{fm}} = 0.18 W_{\text{b}}^{0.96} \quad (5.26)$$

We could observe that Eq. (5.26) is not exactly linear because, after all, $0.96 \neq 1$. However, it is clear that Eq. (5.25) and Eq. (5.26) are sufficiently close that it is still quite reasonable to conclude that flight muscle weight scales linearly with body weight.

Basic Elements of Mathematical Modeling

The above example makes clear that large ranges of data can be handled by introducing log–log plots to extend the graphical range. Of course, with modern computational capabilities, one could skip the "old-fashioned" method of laboriously plotting data and simply enter tables of data points and let the computer spit out an equation or a curve. But thinking through such issues without a computer forces one to think about the actual magnitudes being analyzed and to develop a feel for the magnitudes of the parameters or variables being analyzed.

5.3.3 Scale in Equations: Size and Limits

As noted above, limits occur quite often in mathematical modeling, and they may control the size and shape of an object, the number, kind of variables and the range of validity of an equation, or even the application of particular physical models—or "laws," as they are often called.

Modern electronic components and computers provide ample evidence of how limits in different domains have changed the appearance, performance, and utility of a wide variety of devices. The bulky radios that were made during the 1940s, or the earliest television sets, were very large because their electronics were all done in old-fashioned circuits using vacuum tubes. These tubes were large and threw off an enormous amount of heat energy. The wiring in these circuits looked very much like that in standard electrical wiring of a house or office building. Now, of course, people carry television sets, personal digital assistants, and wireless telephones on their wrists. These new technologies have emerged because the limits on fabricated electrical circuits and devices have dramatically changed, as they have also on the design and manufacturing of small mechanical objects. And this is true beyond electronics. The scale at which surgery is done on people has changed because of new abilities to "see" inside the human body with greater resolution—with increasingly sophisticated scans and imagers, as well as with fiber-optic television cameras—and to design visual, electronic, and mechanical devices that can operate inside a human eye, and in arteries and veins. Things are being engineered at the molecular level in the emerging field of *nanotechnology*. Thus, the mathematical models will change, as will the resulting devices and "machines."

In certain situations, scaling may shift limits or points on an object's boundary where *boundary conditions* are applied. For example, to approximate the hyperbolic sine,

$$\sinh x = \tfrac{1}{2}(e^x - e^{-x}) \tag{5.27}$$

For large values of x, the term e^x will be much larger than the term e^{-x}. The approximation problem is one of defining an appropriate criterion for discarding the smaller term, e^{-x}. For dimensionless values of x greater than 3, the second term on the right-hand side of Eq. (5.27), e^{-x}, becomes very small (less than 4.98×10^{-2}) compared with e^x for $x = 3$, which is 20.09. Hence, one could generally take $\sinh x \cong 1/2 e^x$. All that must be decided is a value of x for which the approximation $e^{2x} - 1 \cong e^{2x}$ is acceptable.

This problem can be approached by introducing a *scale factor*, λ, which can be used to look for values of x for which the approximation

$$\sinh(x/\lambda) \cong \tfrac{1}{2} e^{x/\lambda} \tag{5.28}$$

can be made. Putting a scale factor, λ, in the approximation of Eq. (5.28) obviously means that it will affect the value of x for which that approximation is acceptable. Now the comparison is one which wants

$$e^{2x/\lambda} - 1 \cong e^{2x/\lambda} \tag{5.29}$$

For $\lambda = 1$, the approximation is good for $x \geq 3$, while for $\lambda = 5$ the approximation works for $x \geq 15$. Thus, by introducing the scale factor λ we can make the approximation valid for different values of x because we are now saying that $e^{-x/\lambda}$ is sufficiently small for $x/\lambda \geq 3$. Changing λ has in effect changed a boundary condition because it has changed the expression of the boundary beyond which the approximation is acceptable to $x \geq 3\lambda$.

Recall that functions such as the exponentials of Eq. (5.28) and Eq. (5.29), as well as sinusoids and logarithms, are *transcendental functions* that can always be represented as power series. For example, the power series for the exponential function is

$$e^{x/\lambda} = 1 + \frac{x}{\lambda} + \frac{1}{2!}\left(\frac{x}{\lambda}\right)^2 + \frac{1}{3!}\left(\frac{x}{\lambda}\right)^3 + \Lambda + \frac{1}{n!}\left(\frac{x}{\lambda}\right)^n + \Lambda \qquad (5.30)$$

It is clear that the argument of the exponential must be dimensionless because without this property Eq. (5.30) would not itself be a rational equation. Furthermore, one could not calculate numerical values for the exponential—or any other transcendental—function, if its argument was not dimensionless. The presence of a scale factor in Eq. (5.30) renders the exponential's argument dimensionless, and so numerical calculations can be performed.

Now, a charged capacitor draining through a resistor produces a voltage drop $V(t)$ at a rate proportional to the value of the voltage at any given instant. The mathematical model is

$$\frac{dV(t)}{dt} = -\lambda V(t) \qquad (5.31)$$

which can be rewritten as

$$\frac{dV(t)}{V(t)} = -\lambda \, dt \qquad (5.32)$$

For this rate equation to be a rational equation, the net dimensions of each side of Eq. (5.32) must be the same, which means that each side must be dimensionless. The left-hand side is clearly dimensionless because it is the ratio of a voltage change to the voltage itself. The right hand will be dimensionless *only* if the scale factor, λ, has physical dimensions such that $[\lambda] = 1/T$. Furthermore, the dimensionless product λt can be used to derive a measure of the time that it takes to discharge the capacitor being modeled. Thus, define a *decay* or *characteristic time* as the time it takes for the voltage to decrease to a specified fraction of its initial value, say 1/10. The characteristic or decay time of the charged capacitor would then be

$$V(t_{\text{decay}}) \equiv \frac{V_0}{10} \qquad (5.33)$$

The value of the characteristic time t_{decay} can be calculated from the solution to Eq. (5.32) as

$$\lambda \cong \frac{2.303}{t_{\text{decay}}} \qquad (5.34)$$

Equation (5.34) says that the scale factor λ for the discharging capacitor is inversely proportional to the characteristic (decay) time, and so the voltage in the capacitor can then be written as

$$V(t) \cong V_0 e^{-2.303(t/t_{\text{decay}})} \qquad (5.35)$$

5.3.4 Consequences of Choosing a Scale

Since all actions have consequences, it should come as no surprise that the acquisition of experimental data, its interpretation, and its perceived meaning(s) generally can be very much affected by the choice of scales for presenting and organizing data. To illustrate how scaling affects data acquisition, consider the diagnosis of a malfunctioning electronic device such as an audio amplifier. Such amplifiers are designed to reproduce their electrical input signals without any distortion. The outputs are distorted when the input signal has frequency components beyond the amplifier's range, or when the amplifier's power resources are exceeded. Distortion also occurs when an amplifier component fails, in which case the failure must be diagnosed to identify the particular failed component(s).

A common approach to doing such diagnoses is to display (on an oscilloscope) the device's output to a known input signal. If the device is working properly, a clear, smooth replication of the input would

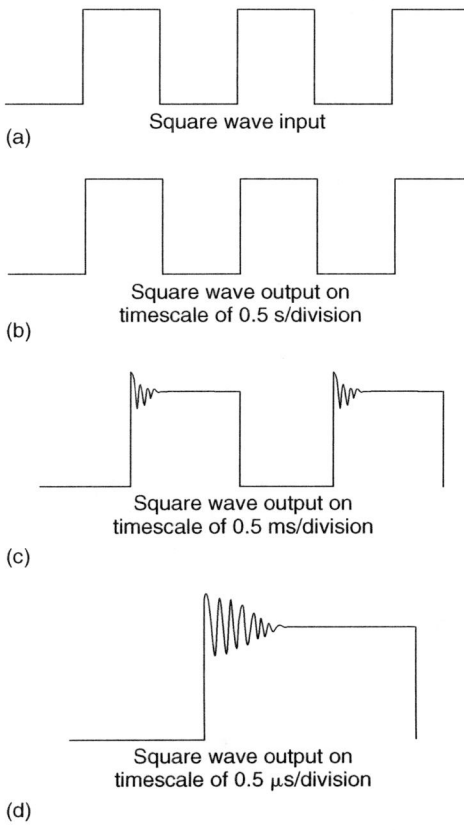

FIGURE 5.5 A square wave (a) is the input signal to a (hypothetical) malfunctioning electronic device. Traces of the output signals are shown at three different timescales (i.e., long, short, and shorter): (b) 0.5 s/division; (c) 0.5 ms/division; and (d) of 0.5 µs/division.

be expected. One standard test input is the square wave shown in Figure 5.5(a). A nice replication of that square wave is shown in Figure 5.5(b), and it seems just fine until it is noticed that the horizontal timescale is set at a fairly high value, that is, 0.5 s/division. To ensure that something that might not show up on this scale is not overlooked, the same signal we spread out on shorter timescales of 0.5 ms/division (Figure 5.5[c]) and 0.5 µs/division (Figure 5.5[d]), neither of which is a nice square wave. This suggests that the device is malfunctioning. Had the oscilloscope not been set to shorter, more appropriate timescales, an erroneous conclusion might have been reached. Thus, it is important to understand that scaling issues are central not only to displaying experimental data, but also to its measurement and interpretation.

5.3.5 Scaling and Perceptions of Data Presentations

The scales used to present modeling "results" also significantly influence how such data are perceived, no matter whether those models are analytical or experimental in nature. Indeed, individuals and institutions have been known to choose scales and portrayals to disguise or even deny the realities they purport to present. Thus, whether by accident or by intent, scales can be chosen to persuade. While this is more of a problem in politics and the media than it is in the normal practice of engineering and science, it seems useful to touch on it briefly here since the underlying issue is a consequence of scale.

Figure 5.6 and Figure 5.7 illustrate the consequences of scale in contexts somewhat beyond the normal professional concerns of engineers and scientists. Both examples are shown because they use the same

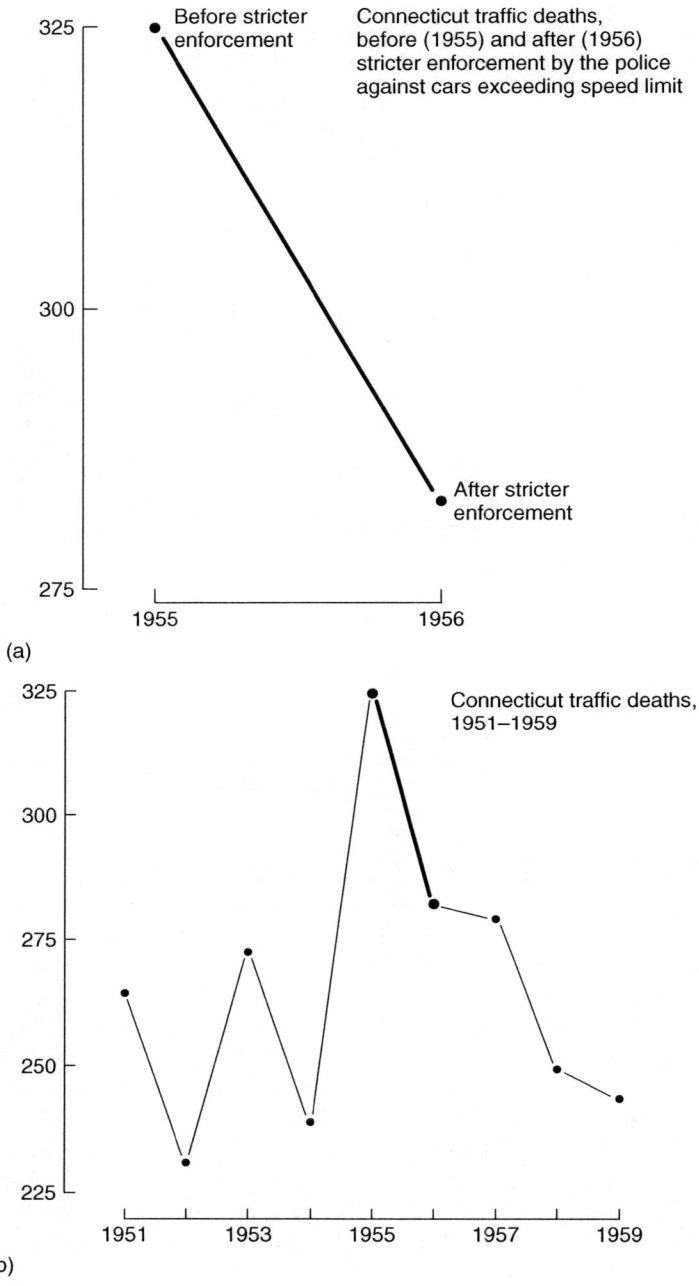

FIGURE 5.6 Plots of traffic fatalities in the state of Connecticut, showing the dangers of truncating scales and deleting comparative data: (a) Connecticut data for 1955–1956; (b) Connecticut data for 1951–1959; and (c) normalized data for Connecticut and three neighboring states for 1951–1959. (From Tufte, E. R., *The Visual Display of Quantitative Information*, Graphics Press, Cheshire, Connecticut, 1983.)

technique of carefully choosing a scale in a figure to present data out of context. Figure 5.6(a) shows a rather dated picture of traffic deaths in the state of Connecticut during the time interval 1956–1957, and a sharp drop in traffic deaths can be seen to have occurred then. But, was that drop real? And, in comparison to what? It turns out that if more data are added, as in Figure 5.6(b), the drop is seen to follow a rather

Basic Elements of Mathematical Modeling

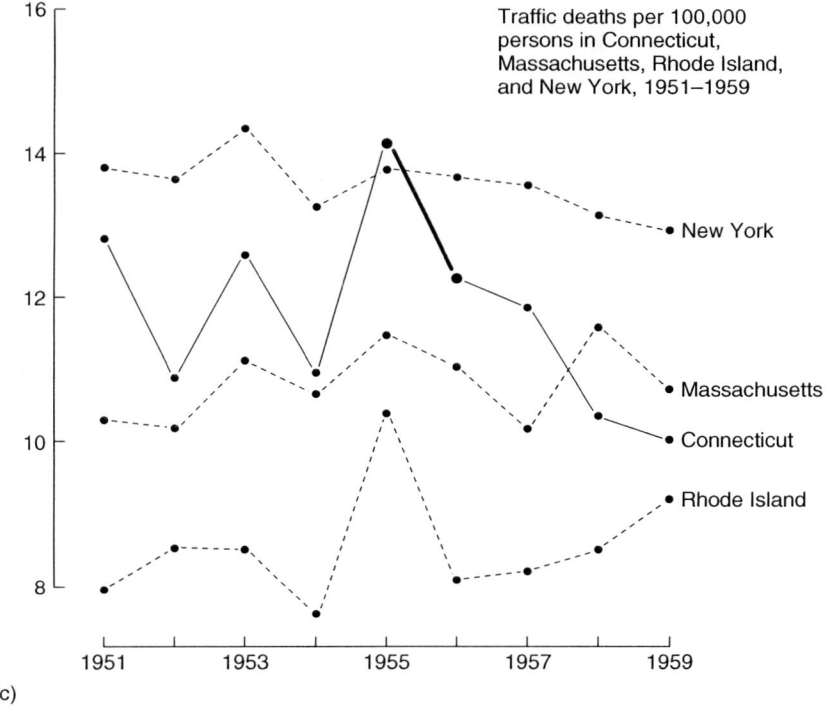

(c)

FIGURE 5.6 *(Continued)*

precipitous increase in the number of traffic fatalities. Further, data from adjacent states were added and the number of deaths was normalized against a common base, as shown in Figure 5.6(c), it would then be seen that the numbers of Connecticut's traffic fatalities was similar to those of its neighbors, although the impact of the stricter enforcement is still visible after 1955.

Similarly, one of the most often shown graphics in the financial pages of newspapers, or in their televised equivalents, are graphics such as that shown in Figure 5.7. Here, the immediate sense conveyed is that the bottom has dropped out of the market because the scale used on the ordinate (or y- or vertical axis) has been so foreshortened that it includes only one week's trading activities. Thus, a decline of a few percent in a stock market barometer such as the Dow Jones Industrial Average (DJIA) appears initially like a much more precipitous decline—especially if the curve itself is drawn in red ink!

5.4 Conservation and Balance Principles

The development of mathematical models often starts with statements that indicate that some property of an object or system is being conserved. For example, the motion of a body moving on an ideal, frictionless path could be analyzed by noting that its *energy is conserved*. Sometimes, as when modeling the population of an animal colony or the volume of a river flow, *quantities that cross a defined boundary* (whether individual animals or water volumes) *must be balanced*. Such *balance* or *conservation principles* are applied to assess the effect of maintaining or conserving levels of important physical properties. Conservation and balance equations are related—in fact, conservation laws are special cases of balance laws.

The mathematics of balance and conservation laws are straightforward at this level of abstraction. Denoting the physical property being monitored as $Q(t)$ and the independent variable time as t, a balance

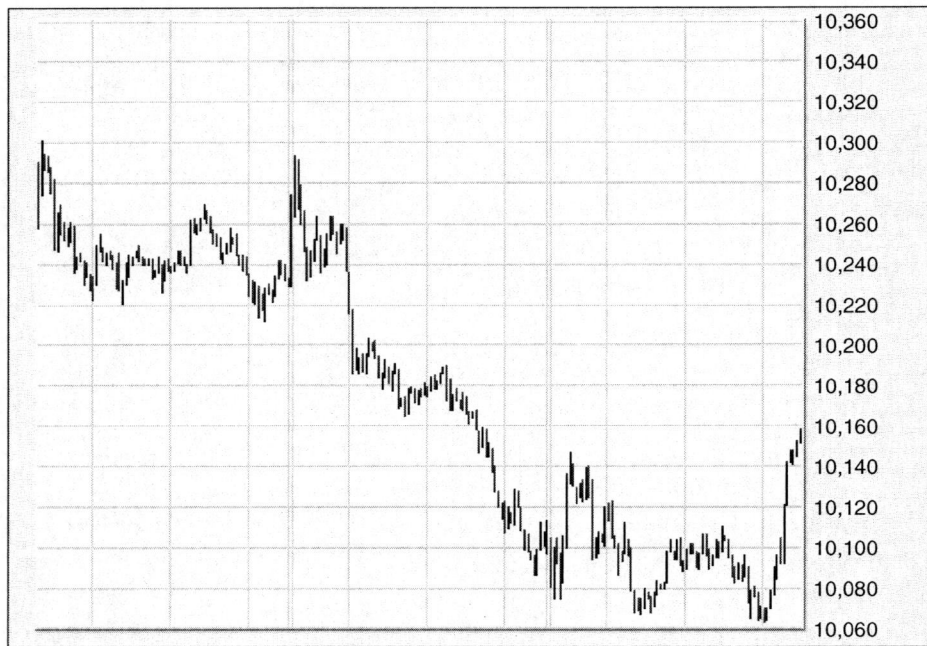

FIGURE 5.7 A plot of the performance of the New York Stock Exchange during 13–15 May 2002, as exemplified by that universally-cited barometer, the Dow Jones Industrial Average (DJIA). (From www.bigcharts.com, 2002)

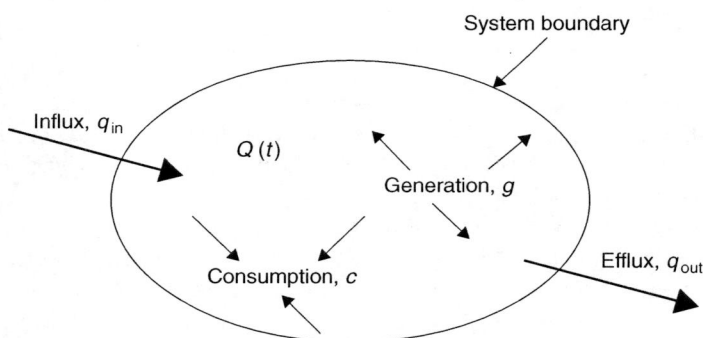

FIGURE 5.8 A system boundary surrounding the object or system being modeled. The influx $q_{\text{in}(t)}$, efflux $q_{\text{out}(t)}$, generation $g(t)$, and consumption $c(t)$ affect the rate at which the property of interest, $Q(t)$, accumulates within the boundary. (After Cha, P. D., Rosenberg, J. J., and Dym, C. L., *Fundamentals of Modeling and Analyzing Engineering Systems*, Cambridge University Press, New York, 2000.)

law for the *temporal* or time rate of change of that property within the system boundary depicted in Figure 5.8 can be written as

$$\frac{dQ(t)}{dt} = q_{\text{in}}(t) + g(t) - q_{\text{out}}(t) - c(t) \tag{5.36}$$

where q_{in} and q_{out} represent the flow rates of $Q(t)$ into (the *influx*) and out of (the *efflux*) the system boundary, $g(t)$ is the rate at which Q is generated within the boundary, and $c(t)$ the rate at which Q is consumed within that boundary. Note that Eq. (5.36) is also called a *rate equation* because each term has both the meaning and dimensions of the rate of change with time of the quantity $Q(t)$.

Basic Elements of Mathematical Modeling

In those cases where there is no generation and no consumption within the system boundary (i.e., when $g = c = 0$), the balance law in Eq. (5.36) becomes a *conservation law*:

$$\frac{dQ(t)}{dt} = q_{in}(t) - q_{out}(t) \tag{5.37}$$

Here, then, the rate at which $Q(t)$ accumulates within the boundary is equal to the difference between the influx, $q_{in}(t)$, and the efflux, $q_{out}(t)$.

5.5 The Role of Linearity

Linearity is one of the most important concepts in mathematical modeling. Models of devices or systems are said to be *linear* when their basic equations—whether algebraic, differential, or integral—are such that the magnitude of their behavior or response produced is *directly proportional* to the excitation or input that drives them. Even when devices like the classic pendulum are more fully described by nonlinear models, their behavior can often be approximated by linearized or perturbed models, in which cases the mathematics of linear systems can be successfully applied.

Linearity is applied during the modeling of the behavior of a device or system that is forced or pushed by a complex set of inputs or excitations. The response of that device or system to the sum of the individual inputs is obtained by adding or *superposing* the separate responses of the system to each individual input. This important result is called the *principle of superposition*. Engineers use this principle to predict the response of a system to a complicated input by decomposing or breaking down that input into a set of simpler inputs that produce known system responses or behaviors. However, some typical behaviors cannot be captured by linear models, in which case it is important to be careful not to oversimplify inappropriately.

5.5.1 Linearity and Geometric Scaling

The geometric scaling arguments discussed earlier can also be used to demonstrate some ideas about linearity in the context of *geometrically similar* objects, that is, objects whose basic geometry is essentially the same. Figure 5.9 shows two pairs of drinking glasses: one pair are right circular cylinders of radius r and the second pair are right circular inverted cones having a common semi-vertex angle α. If the first pair is filled to heights h_1 and h_2 respectively, the total fluid volume in the two glasses is

$$V_{cy} = \pi r^2 h_1 + \pi r^2 h_2 = \pi r^2 (h_1 + h_2) \tag{5.38}$$

Eq. (38) demonstrates that the volume is *linearly proportional* to the height of the fluid in the two cylindrical glasses. Further, since the total volume can be obtained by adding or *superposing* the two heights, the volume V_{cy} is a *linear function* of the height h. Note, however, that the volume is *not* a linear function of radius r.

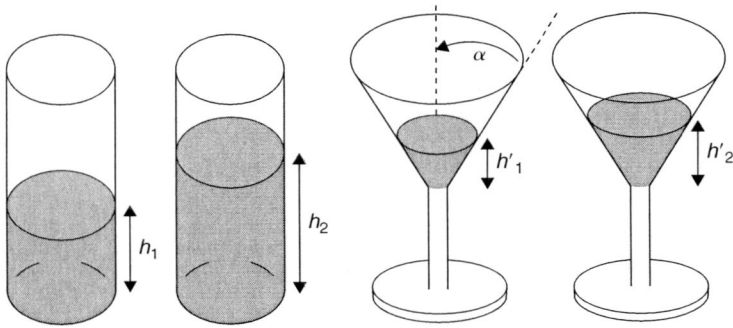

FIGURE 5.9 Two pairs of drinking glasses: one pair are cylinders of radius r, the second pair are inverted cones (sometimes referred to as martini glasses) having a common semivertex angle α.

In the two conical glasses, the radii vary with height. In fact, the volume, V_{co}, of a cone with semivertex angle, α, filled to height, h, is

$$V_{co} = \frac{\pi}{3} \frac{h^3}{\tan^2 \alpha} \tag{5.39}$$

Hence, the total volume of fluid in the two conical glasses of Figure 5.9 is

$$V_{co} = \frac{\pi}{3} \frac{h_1'^3}{\tan^2 \alpha} + \frac{\pi}{3} \frac{h_2'^3}{\tan^2 \alpha} \neq \frac{\pi}{3} \frac{(h_1' + h_2')^3}{\tan^2 \alpha} \tag{5.40}$$

That is, the relationship between volume and height is nonlinear for the conical glasses, the total volume cannot be calculated just by superposing the two fluid heights, h_1' and h_2'. Note that this result, while to a simple, even obvious case, is emblematic of what happens to superposition when a linearized model is replaced by its (originating) nonlinear version.

5.6 Conclusions

This chapter has provided a very brief summary of the most basic foundations of mathematical modeling. In this context, the discussion began with a statement of principles under which the activity of mathematical modeling could be properly performed. This was followed by a discussion of basic foundational matters, including dimensional homogeneity and dimensional analysis, abstraction and scaling, balance and conservation laws, and an introduction to the role of linearity. It is important to note that this overview emphasized brevity, dictated by chapter length limitations, and so it will hopefully serve as a stimulant to the reader's appetite for further reading and application of these basic ideas and methods.

Acknowledgment

The author is very grateful to Elsevier Academic Press for permission to summarize (or scale down!) the first three chapters of Dym, C. L., *Principles of Mathematical Modeling*, 2nd Edition, Elsevier Academic Press, Orlando, Florida, 2004, and to reprint Figures 5.1–5.9 (as numbered here) from that book.

References

Alexander, R. M., *Size and Shape*, Edward Arnold, London, 1971.
Carson, E. and Cobelli, C. (Eds.), *Modelling Methodology for Physiology and Medicine*, Academic Press, San Diego, 2001.
Cha, P. D., Rosenberg, J. J., and Dym, C. L., *Fundamentals of Modeling and Analyzing Engineering Systems*, Cambridge University Press, New York, 2000.
Dym, C. L., *Principles of Mathematical Modeling*, 2nd Edition, Elsevier Academic Press, Orlando, Florida, 2004.
Tufte, E. R., *The Visual Display of Quantitative Information*, Graphics Press, Cheshire, Connecticut, 1983.
Tufte, E. R., *Envisioning Information*, Graphics Press, Cheshire, Connecticut, 1990.

6
DEVS Formalism for Modeling of Discrete-Event Systems

Tag Gon Kim
KAIST

6.1 Introduction 6-1
6.2 System-Theoretic DES Modeling 6-3
6.3 DEVS Formalism for DES Modeling 6-3
   Atomic DEVS Model • Coupled DEVS Model • Example of DEVS Modeling: Ping-Pong Protocol • State Equation Form of Atomic DEVS
6.4 DES Analysis with DEVS Model 6-7
   Composition of Atomic DEVS Models • System Analysis by Composed DEVS Model
6.5 Simulation of DEVS Model 6-10
   DEVS Modeling Simulation Methodology and Environment • Simulation Speedup and Simulators Interoperation
6.6 Conclusion 6-12

This chapter introduces the discrete-event systems specification (DEVS) formalism for modeling of discrete-event systems (DESs). Based on set theory, the formalism specifies DESs in a hierarchical, modular manner. Models specified by the formalism can be used for analysis as well as performance simulation of DESs.

6.1 Introduction

A DES consists of a collection of components that interact with each other via events exchange to perform a given function. A component of such a system is represented by a discrete states set and operations defined on the set. The operations are mainly a set of rules for states transition, which is performed only with an occurrence of instantaneous events over time. An event in DES may occur either by an external stimulus to the component (external event) or an internal condition within the component (internal event). A message arrival and a timeout in a communication system are examples of an external and an internal event, respectively. Consider a DES of a ping-pong protocol system whose state transition diagram is shown in Figure 6.1.

The protocol system consists of two components, SENDER and RECEIVER, each of which has its own states and associated transition rules. Note that components, SENDER and RECEIVER, are coupled together via two events: *msg* and *ack*. Let us first give an informal description of each component and then the interaction between the two. SENDER has two states, "*Send*" and "*Receive*," the input event *?ack* and the output event *!msg*. Initially, SENDER stays at the "*Send*" state at which an output event *!msg* is generated with a state transition to the "*Receive*" state. It then waits for an input event

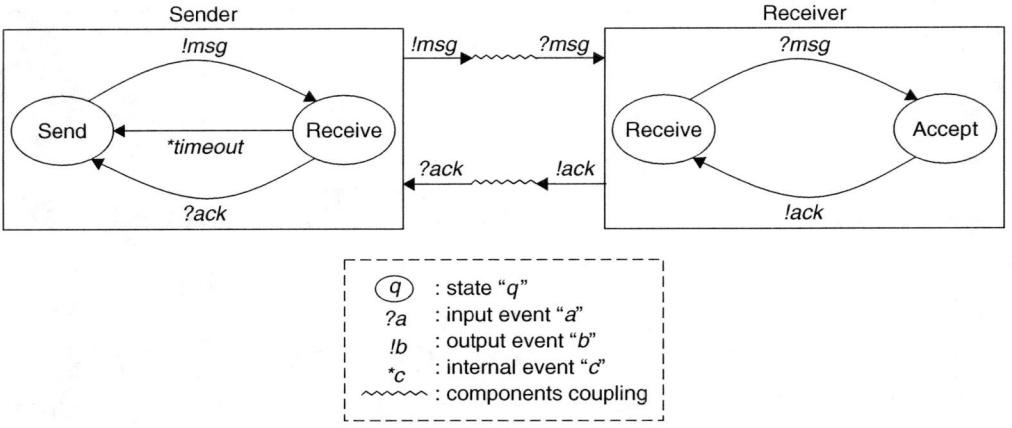

FIGURE 6.1 State transition diagram for ping-pong protocol.

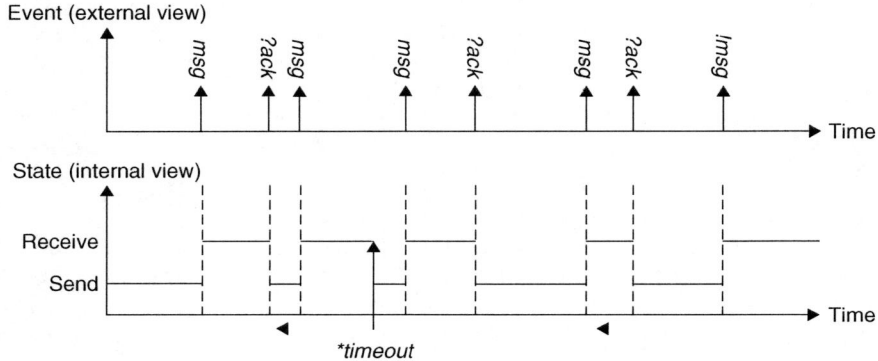

FIGURE 6.2 External and internal views for SENDER.

?ack to be arrived from RECEIVER. When SENDER receives the *?ack* event at the "*Receive*" state it returns to the "*Send*" state again. However, if the *?ack* event is not arrived before a specified period SENDER changes its state to the "*Send*" state. The state change is based on an assumption that *!msg* or *?ack* is lost during transmission through components coupling. Once SENDER comes back to the "*Send*" state it sends *!msg* again, which may be a new one or the one previously sent. Note that SENDER changes its state from "*Receive*" to "*Send*" either by an external input event *?ack* or an internal event **timeout* of the timeout condition. The operation of RECEIVER can be described similarly.

The informal description of SENDER allows one to identify an external and an internal view of a DES. The external view is a sequence of two events *!msg* and *?ack*. The order of the two depends on a rate of lost events and/or a value of the timeout. Thus, a legal events sequence may be one such as *!msg* → *?ack* → *!msg* → *!msg* → *!msg* ... → *!ack*. However, an events sequence *!msg* → *?ack* → *?ack* is not a legal one. In contrast, the internal view of SENDER is a sequence of states which is a piecewise constant function over time. For example, "*Send*" → "*Receive*" → "*Send*" → "*Receive*" ... → "*Send*" is one such states sequence. Figure 6.2 shows an external as well as an internal view of SENDER. Note that a state transition from "*Receive*" to "*Send*" has been made by an internal event **timeout*.

A DES can be viewed as a system whose abstraction level in modeling is at the discrete-event one. In the level, a modeler is interested only in what happens if an event occurs while ignoring details of system behavior between events occurrences. In this sense most systems can be modeled at the discrete-event level which we call a DES. Examples of DES include a communication protocol viewed at or above the data link layer, a computer system viewed at or above the operating system level, a military war game at the operation level, and others.

Rest of this chapter is organized as follows. Section 6.2 presents DES modeling in the system-theoretic view and Section 6.3 introduces the DEVS formalism. Analysis and simulation of DES with DEVS models are given in Sections 6.4 and 6.5, respectively. Section 6.6 concludes this chapter.

6.2 System-Theoretic DES Modeling

The system-theoretic approach for systems modeling views a system as an object in which its representation and associated operations are explicitly defined. In the view a system is represented by three sets: inputs set, outputs set, and states set; operations on the sets are defined as a collection of rules, or functions, for state transition and output generation.

In DES, both inputs and outputs are finite event sets. However, a states set is not finite. To be precise recall the definition of state at t in system theory: information required at t which uniquely determines output at $t' > t$. In fact, an output in DES is generated at a specified time when a certain condition is satisfied. Thus, a state at t is represented by a discrete state, s, and an associated elapsed time, e, which is a real number. Of course, the maximum elapsed time for each discrete state is predefined, which we call the maximum sojourn time, r, from now on. Then, a state (s,e) means that the discrete state s has been kept for the e time unit without any external input. If no input is arrived at the state before the maximum sojourn time r an output would be generated. Of course, an input event can be arrived anytime before the r time unit. In such a case, the discrete state s is changed to a new one s' that has its own maximum sojourn time r'. To be clear we call s a discrete state and $q = (s,e)$ a total state or just a state of DES. Since e is a real number q is not finite.

We are now ready to explain how system theory defines state transition functions of a dynamic system. In the theory, two state transitions are considered: one with an input and the other without an input. For example, a well-known vector differential equation for state transition of a linear continuous system is $dQ/dt = AQ + BX$, where Q is a state set, X an inputs vector, and A and B are the coefficient matrices. Note that $dQ/dt = AQ$ specifies state transition without inputs and $dQ/dt = BX$ specifies that with inputs X. As will be shown later, the same view would be applied in DES modeling. More specifically DES has a state transition either with an external input event or with a condition internal to a system, which causes an internal event. Such a condition at a state $q = (s,e)$ includes $e = r$, meaning that an elapsed time e is reached to the maximum sojourn time r at s. From now on we call state transition with an input as external state transition and that without an input as internal state transition. Similarly, an output is a function of both an input and a state. However, an output function can be represented only by state information, for the information memorizes history of inputs information.

As shown in Figure 6.1, a DES consists of components that are connected together for interaction with events exchange. Thus, specification of DES should have a means to specify the connection. In sum, modeling of DES in the system-theoretic approach requires the following expression:

- each component needs inputs set, outputs set, states set, state transition functions, both external and internal transitions, and output function;
- connection of components needs a means for coupling between events associated with components.

The DEVS formalism to be introduced in the following chapter supports such an expression in a formal manner for modeling of DES.

6.3 DEVS Formalism for DES Modeling

The DEVS formalism, developed by Zeigler, specifies a DES with the following three major features (Zeigler, 1984; Zeigler et al., 2000):

- set theory-based formalism

- system-theoretic representation
- hierarchical, modular specification.

The DEVS formalism is constructed based on set theory in which a static structure, such as states, is represented by a set and dynamic behavior, such as state transition, is specified by operations on sets. Set-theoretic modeling is known to be very intuitive in a modeling process. Moreover, with the set theory the formalism represents a DES in a system-theoretic view. With the view the formalism specifies a system as a static representation of inputs, outputs, and states sets and dynamic operations defined on sets. Finally, the formalism specifies a system in a hierarchical, modular manner. This feature allows a modeler to decompose a discrete-event model into a collection of modular submodels, each of which in turn is decomposed into modular submodels, and so on. To do so develops a discrete-event model with a powerful combination of top-down model design and bottom-up model implementation.

6.3.1 Atomic DEVS Model

The DEVS formalism defines a DES into two classes of models: atomic model and coupled model. An atomic model is a nondecomposable model in hierarchical decomposition; a coupled one is a collection of either atomic or coupled model. An atomic model specifies state transition and output generation over time. In contrast, a coupled model specifies a list of components and their coupling information. Formally, an atomic DEVS model (AM) is defined as follows:

$$AM = <X, S, Y, \delta_{ext}, \delta_{int}, \lambda, ta>$$

where
  X: a set of input events;
  S: a set of sequential states;
  Y: a set of output events;
with the following constraints:
  $ta: S \to R$ (nonnegative real number), time advance function;
  $\delta_{int}: Q \to Q$, an external transition function;
  where $Q = S \times R = \{(s, e) \mid s \in S$ and $0 \le e \le ta(s)\}$, states set of M;
  $\lambda: Q \to Y$, an output function;
  $\delta_{ext}: Q \times X \to Q$, an external transition function.

As shown in the above definition, AM has three sets and four functions which we call characteristic functions. Note that AM has two transitions, δ_{ext} and δ_{int}. δ_{ext} is state transition with an input event and δ_{int} is that without an input event but with an internal condition. The internal transition, $\delta_{int}: Q \to Q$, changes states from $q \in Q$ to $q' \in Q$, which is different from that of $\delta_{int}: S \to S$ in the original DEVS formalism (Zeigler, 2000). Note that $q = (s, e) \in Q$ represents a discrete state and an associated elapsed time at the state. Thus, the use of Q in the definition of internal transition here allows us to explicitly specify a condition for when an internal transition would have occurred. For similar reasons, external transition and output functions are defined with Q, which is slightly different from that in Zeigler (2000). As will be shown later, the transitions with Q allow us to formulate state/output equations for DEVS that would be analogous to the ones for continuous dynamic systems. Note also that S is defined as a set of discrete states although it can represent a continuous set.

Let us briefly explain the four functions. Time advance function defines the maximum sojourn time for which each discrete state can stay unless an external input is arrived before the time. Note that the sojourn time at a discrete state s, defined by $ta(s)$, should be updated whenever a discrete state is changed. The sojourn time can model such information as delay time for information transmission, work time for processing tasks, and interdeparture time for generation of events. The internal transition specifies state transition without an input but with an internal condition. The condition is whether an elapsed time at a discrete state s is reached to the maximum sojourn time $ta(s)$ at s. More specifically, assume that a discrete state of an atomic model is s and that no input is arrived until the $ta(s)$ time unit. Then, a state

$q = (s, \text{ta}(s))$ of the model is changed into another state $q' = (s', 0)$. Recall that the internal transition of $Q \to Q$, instead of $S \to S$, explicitly specifies when the transition occurs. That is, no internal transition occurs at a state $q'' = (s, e)$, where $e < \text{ta}(s)$, although discrete states for both $q = (s, \text{ta}(s))$ and $q'' = (s, e)$ are identical. Note that the elapsed times at the discrete state in q and that in q' at which an internal transition occurs are always $\text{ta}(s)$ and 0, respectively.

Definition of the external transition can be similarly explained in which an elapsed time for a new state immediately after the transition is also 0. The output function, $Q = S \times R \to Y$, specifies which discrete state in S generates what output in Y at what time in R. Recall that the output function, $S \to Y$, only specifies which discrete state in S generates what output in Y with no information about what time.

6.3.2 Coupled DEVS Model

A coupled DEVS model is a composition of DEVS models, each of which can be either atomic or coupled, thus supporting hierarchical construction of a complex model. A well-known DEVS property of closed undercoupling is a theoretical basis for such hierarchical models construction, similar to a process of assembling a complex hardware from pieces of components. A formal definition of a coupled DEVS model (CM) is as follows:

$$CM = <X, Y, D, \{M_i \mid i \in D\}, \text{EIC}, \text{EOC}, \text{IC}, \text{Select}>$$

where
  X, Y: same as in AM;
  D: a set of component names;
  M_i: a component DEVS model, atomic or coupled;
with the following constraints:
  $\text{EIC} \subseteq X \times \bigcup_{i \in D} X_i$, external input coupling relation;
  $\text{EOC} \subseteq \bigcup_{i \in D} Y_i \times Y$, external output coupling relation;
  $\text{IC} \subseteq \bigcup_{i \in D} Y_i \times \bigcup_{j \in D} X_j$, internal coupling relation;
  Select: $2^D - \emptyset \to D$, tie-breaking selector.

A coupled DEVS model has three sets and four functions. A set of components M_i is coupled to form a coupled model. The coupling specification is defined by three mathematical relations: external input, external output, and internal coupling relations. Each relation is a set of ordered pairs of events, each of which is represented by (e1, e2), indicating that an event e1 is coupled to an event e2. With the coupling in DEVS theory, all information in e1 is transmitted to e2 without any time delay. Let us look into the three relations. The external input coupling relation, EIC, specifies how an input event of CM is routed to input events of component models. The external output coupling relation, EOC, specifies how an output event of a component is connected to an output event of CM. Lastly, the internal coupling relation, IC, specifies how an output event of a component of CM is coupled to input events of other components of CM. Note that the selection function, *select*, designates a component to be selected out of many if the selection is required. The function is activated when more than one component is ready to generate their output events while events can be handled one by one.

6.3.3 Example of DEVS Modeling: Ping-Pong Protocol

Let us illustrate DEVS modeling of ping-pong protocol. We first specify atomic DEVS models for SENDER and RECEIVER, and then the overall coupled DEVS model for the protocol. We call two atomic DEVSs AMsender and AMreceiver for SENDER and RECEIVER, respectively, and the overall DEVS model CMppp. The atomic DEVS model of AMsender is defined as follows:

$$\text{AMsender} = <X, S, Y, \delta_{\text{ext}}, \delta_{\text{int}}, \lambda, ta>$$

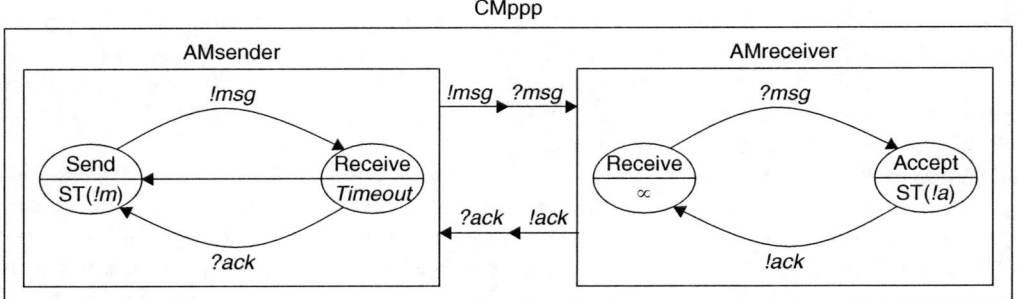

FIGURE 6.3 DEVS model for ping-pong protocol.

where
 $X = \{?ack\}; Y = \{!msg\}; S = \{\text{Send, Receive}\};$
 $ta(\text{Send}) = ST(!m)$ // Sending time of *!msg*;
 $ta(\text{Receive}) = \text{Timeout}$ // Maximum waiting time for input *?ack*;
 $\delta_{int}(\text{Send}, ST(!m)) = (\text{Receive}, 0);$
 $\delta_{int}(\text{Receive}, \text{Timeout}) = (\text{Send}, 0);$
 $\lambda(\text{Send}, ST(!m)) = !msg;$
 $\delta_{ext}((\text{Receive}, e < \text{Timeout}), ?ack) = (\text{Send}, 0).$

AMreceiver can be similarly specified. Figure 6.3 shows AMsender and AMreceiver. Note that each discrete state in Figure 6.3 has an associated sojourn time defined by the time advance function of ta, such as ST(!m), Timeout, ST(!a), and ∞. Interpretation of the time is as follows. ST(!m) associated with the discrete state "*Send*" is a time to be spent for transmission of *!msg* from SENDER to RECEIVER. It may be modeled as a fixed real number including zero or a random number depending on a modeling objective. Timeout associated with the discrete state "*Receive*" is the maximum waiting time for the input *?ack* to arrive. Thus, as shown in Figure 6.3 SENDER can change its discrete state either by the condition of Timeout or by the input event *?ack*, depending on which one occurs first between the condition and the input. Meaning of ST(!a) associated with the discrete state "*Accept*" is similar to that of ST(!m). The sojourn time of ∞ at the discrete state "*Receive*" in RECEIVER means that RECEIVER does not know how long it should wait for the input *?msg* to arrive from SENDER.

The internal transition of AMsender, $\delta_{int}(\text{Send}, ST(!m)) = (\text{Receive}, 0)$, means that if an elapsed time at the discrete state "*Send*" is reached to the maximum sojourn time ST(!m) then AMsender changes its state to the beginning of the discrete state "*Receive*." The output function, $\lambda(\text{Send}, ST(!m)) = !msg$, means that if an elapsed time at "*Send*" is reached to the maximum sojourn time ST(!m) then the output *!msg* is generated. Note that the output generation at a given state occurs at the same time as an internal transition at the state. The external transition function of $\delta_{ext}((\text{Receive}, e < \text{Timeout}), ?ack) = (\text{Send}, 0)$ specifies what to do when the input *?ack* has arrived at an elapsed time *e* before the maximum sojourn time of Timeout. In this case AMsender changes its state to (Send, 0). Note that both internal and external transition functions are piecewise constant functions over time.

Let us specify the coupled model CMppp. As shown in Figure 6.3, CMppp consists of two atomic models AMsender and AMreceiver. Thus, CMppp is defined as

$$\text{CMppp} = <X, Y, \{M_i\}, \text{EIC, EOC, IC, Select}>$$

where
 $X = Y = \{\ \}$ (no input and output to environment);
 $\{M_i\} = \{\text{AMsender, AMreceiver}\};$
 $\text{EIC} = \text{EOC} = \{\ \}$ (no interaction with external world);

IC = {(AMsender.!msg, AMreceiver.?msg),
(AMreceiver.!ack, AMsender.?ack)};
Select({AMsender, AMreceiver}) = AMsender.

Note that CMppp has no interaction with an external environment, thus $X = Y = \text{EIC} = \text{EOC} = \{\ \}$. However, if the coupled model has any such interaction specification of the sets it should reflect it. Specification of $\{M_i\}$ is self-evident. Specification of IC employs a list of ordered pairs each of which represents coupling information between two models. An ordered pair of ($M1.m1$, $M2.m2$) means that an output event $m1$ of a model $M1$ is coupled to an input event $m2$ of a model $M2$. By the coupling, all information associated with $m1$ is transferred to $m2$ with no delay. Thus, by the coupling (AMsender.!msg, AMreceiver.?msg), two models change their states simultaneously, "*Send*" to "*Receiver*" for AMsender and "*Receiver*" to "*Accept*" for AMreceiver. If a model is coupled to more than one model all such couplings should be listed in ordered pairs by following the definition of mathematical relation on two sets. Finally, the select function *Select* specifies that if AMsender and AMreceiver are ready to change its state at the same time then AMsender executes the change first, then AMreceiver. Such selection priority should be carefully specified with the deep domain knowledge of a target system to be modeled. The priority does not matter in system analysis for some cases, but it does matter for most cases.

6.3.4 State Equation Form of Atomic DEVS

As described earlier, the DEVS formalism specifies a DES in the system-theoretic viewpoint. Recall that the viewpoint considers system's dynamics both with and without inputs. Following the viewpoint the DEVS formalism has two transition functions: internal and external. Thus, a state transition in an atomic DEVS model can occur either by an internal event (i.e., timeout condition) or by an external event. No specification is given for a state transition to be performed by the two events that have occurred at the same time. A priority rule for selection of such conflict events is specified at a coupled DEVS model to which the atomic DEVS belongs as a component. Thus, a state equation of an atomic DEVS is represented as

$$q' = \delta_{\text{int}}(q) \oplus \delta_{\text{ext}}(q,x) \text{ or } (s',r') = \delta_{\text{int}}(s,r) \oplus \delta_{\text{ext}}((s, e \leq \text{ta}(s)), x) \tag{6.1}$$

$$y = \lambda(q) = \lambda(s,r) \tag{6.2}$$

where the binary operator, $\oplus$, is used to represent that a state transition can occur by either δ_{int} or δ_{ext}, but not by both at any time. Note that Eq. (6.1) is similar to the state equation $dQ/dt = AQ + BX$ for a continuous dynamic system, where Q is a states set, X an inputs vector, and A and B are the coefficient matrices. The state equation indicates that a state change (dQ/dt) is composed of a state change without input ($dQ/dt = AQ$) and a state change with input ($dQ/dt = BX$). Comparison of the two state equations show that $\delta_{\text{int}}(q)$ and $\delta_{\text{ext}}(q, x)$ in DEVS correspond to $dQ/dt = AQ$ and $dQ/dt = BX$ in a continuous system, respectively.

6.4 DES Analysis with DEVS Model

Generally, the purpose of systems modeling is twofold: verification of desired behavior and performance evaluation. A DEVS model for a DES can be used for such purposes. Verification of behavior for a DES includes properties of the system such as liveness and safeness, desired states/events sequences, and others. Safeness is a property which claims that a bad thing will never happen; liveness is another property which claims that a good thing will eventually happen. An example of safeness is deadlock-free and that of liveness is an arrival of a message *?ack* at SENDER in the ping-pong protocol introduced in Section 6.1. A desired events sequence is one that satisfies functionality of the system to be modeled. An example of a desired events sequence in the ping-pong protocol is *!msg* → *?msg* → *!ack* → *?ack*, meaning of which is self-explained.

6.4.1 Composition of Atomic DEVS Models

Verification of DES usually relies on a state space exploration approach. The approach generates a global state space of an overall system model by composition of component models. Composition of atomic DEVS models considers all atomic models on a whole as a timed state transition model. Although the composition deals with more than two components, we restrict our definition of composition on two atomic models without loss of generality. A composed model of two atomic DEVS models of AM_i and AM_j, noted by ($AM_i \parallel AM_j$), is defined as follows:

$$AM_i \parallel AM_j = <E, S, T, \text{ta}, \{AM_i, AM_j\}>$$

where
 E: events set;
 S: composed discrete states set;
 T: transition relation of composed discrete states;
 ta: time advance function;
with the following constraints:
 $E = (X_i \cup Y_i \cup \{\phi\}) \times (X_j \cup Y_j \cup \{\phi\})$, where ϕ is a null event;
 $S \subseteq S_i \times S_j$;
 $T \subseteq S \times E \times S$;
 ta: $S \rightarrow R$.

To complete the definition, we define transition relation and time advance function with the following three rules:

- Rule 1: Transit AM_i only
  — Transition relation

    If $(s_i, s_i') \in \delta_{\text{int}_i}$ and $(s_i, !a) \in \lambda_i$ and $(s_j, ?a, s_j') \notin \delta_{\text{ext}_j}$
    then $((s_i, s_j), (!a, \phi), (s_i', s_j)) \in T$ for all $s_j \in S_j$

  — Time advance

    $\text{ta}((s_i', s_j)) = \min\{\text{ta}_i(s_i'), \text{ta}_j(s_j) - \text{ta}_i(s_i)\}$

- Rule 2: Transit AM_j only
  — Transition relation

    If $(s_j, s_j') \in \delta_{\text{int}_j}$ and $(s_j, !b) \in \lambda_j$ and $(s_i, ?b, s_i') \notin \delta_{\text{ext}_i}$
    then $((s_i, s_j), (\phi, !b), (s_i, s_j')) \in T$ for all $s_i \in S_i$

  — Time advance

    $\text{ta}((s_i, s_j')) = \min\{\text{ta}_i(s_i) - \text{ta}_j(s_j), \text{ta}_j(s_j')\}$

- Rule 3: Transit both AM_i and AM_j
  — Transition relation

    If $(s_i, s_i') \in \delta_{\text{int}_i}$ and $(s_i, !c) \in \lambda_i$ and $(s_j, ?c, s_j') \in \delta_{\text{ext}_j}$
     (or [$(s_j, s_j') \in \delta_{\text{int}_j}$ and $(s_j, !c) \in \lambda_j$ and $(s_i, ?c, s_i') \in \delta_{\text{ext}_i}$])
    then $((s_i, s_j), (!c, ?c), (s_i', s_j')) \in T$
     (or [$((s_i, s_j), (?c, !c), (s_i', s_j')) \in T$]) for all $s_i \in S_i$ and $s_j \in S_j$

  — Time advance

    $\text{ta}((s_i', s_j')) = \min\{\text{ta}_i(s_i'), \text{ta}_j(s_j')\}$

Note that each rule separates specifications of transition for a composed state from time advance for the state. The separation allows us to specify a composed DEVS model as a timed state automation which is closed from an external world. Although the composed model does not have an explicit state representation, a composed state should be represented by $q = ((s_i, s_j), e) \in Q$, where e is an elapsed time at the composed discrete state (s_i, s_j).

An event $e \in E$ is an ordered pair of two events of component models. An event $(!m, ?m) \in E$ in transition relation means that an output event $!m$ of AM_i is successfully transmitted to AM_j as an input event $?m$. Likewise, an event $(?m, !m)$ in transition relation means that an output event $!m$ of AM_j is successfully transmitted to AM_i as an input event $?m$. In the above cases, both AM_i and AM_j concurrently perform state transitions of their own. However, $(\phi, !m)$ or $(!m, \phi)$ represents a failure of such concurrent state transitions; instead, it represents that either one of the two performs its own state transition.

Figure 6.4 shows a composed model of AMsender || AMreceiver by application of the above composition rules. For the composition, we assume that $!msg$ is never lost and that $!ack$ may be lost during transmission. Also assume that Timeout $> ST(!a)$. In the figure a state of AMsender || AMreceiver is represented by $q = ((s1, s2), r)$, where s1 and s2 are discrete states of AMsender and AMreceiver, respectively, and $r = ta((s1, s2))$. Initially, the states of AMsender and AMreceiver are at SD("*Send*") and RV("*Receive*"), respectively, thus being represented by (SD, RV). Time advance of (SD, RV) is the minimum of ta(SD) of AMsender and ta(RV) of AMreceiver which is $ST(!m)$. Combining the event and an associate sojourn time, a state ((SD, RV), $ST(!m)$) is represented in an oval with two partitions in Figure 6.4: the top, (SD, RV), representing the composed discrete state and the bottom, $ST(!m)$, representing the sojourn time of (SD, RV). With the event $(!msg, ?msg)$, (SD, RV) changes to (RV, AP) after $ST(!m)$ is completely elapsed. The discrete state (RV, AP) stays for $ST(!a)$ time. At the end of $ST(!a)$, (RV, AP) is changed either to (SD, RV) with $(?ack, !ack)$ or to (RV, RV) with $(\phi, !ack)$. Now note that the maximum sojourn time at (RV, RV) is min{Timeout $- ST(!a)$, ∞} = Timeout $- ST(!a)$ according to Rule 2. In fact, Timeout $- ST(!a)$ is a remaining time for which the discrete state (RV, RV) is changed to the discrete state (SD, RV) without any event.

6.4.2 System Analysis by Composed DEVS Model

We now are ready to analyze a DES whose analysis model is represented by a composed DEVS model. We first exemplify an intuitive analysis of the ping-pong protocol in Figure 6.3 using the composed DEVS model of Figure 6.4. We then briefly introduce a method for automatic verification of DESs using composed DEVS models.

As explained in Section 6.4.1, Figure 6.4 represents a global state transition with a sojourn time between each transition. Thus, an intuitive investigation of Figure 6.4 can answer questions on states sequence as

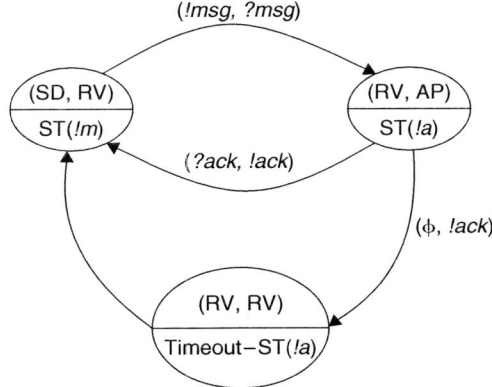

FIGURE 6.4 Composed DEVS model of ping-pong protocol (SD, send; RV, receive; AP, accept).

well as timing property of the protocol to be analyzed. The following questions–answers for ping-pong protocol can be performed using Figure 6.4:

- Question 1 on safeness: Is the protocol deadlock-free?
  Answer 1: Yes, because each state in the Figure 6.4 has a next state to move.
- Question 2 on liveness: Can 10 !*msg* be eventually transmitted to RECEIVER?
  Answer 2: Yes, because SENDER will eventually receive 10 ?*ack* if the probability of loss of !*ack* is less than 1.0.
- Question 3 on a discrete state sequence: Is (SD, RV) → (RV, AP) → (SD, AP) a legal state sequence?
  Answer 3: No, because no such sequence can be constructed in Figure 6.4.
- Question 4 on the minimum and maximum round trip times from (SD, RV) to (SD, RV)?
  Answer 4: ST(!*m*) + ST(!*a*) is minimum; ST(!*m*) + Timeout is maximum.

Although the above example shows the concepts of DES analysis using composed DEVS models, computerized automatic verification needs a systematic method for construction of a complete global state space from a composed DEVS model. Generally, the global state space is represented by an infinite number of timed events/states sequences. The sequences can be constructed by a combination of basic sequences each of which is a loop in the composed model. Once all possible events/states sequences are constructed, a complete verification of a system's property and behavior is possible.

Consider Figure 6.4, the composed DEVS model of Figure 6.3, and assume that states sequences are of interest in analysis. Intuitively, Figure 6.4 has the following two basic states sequences:

- ((SD, RV), ST(!*m*)) → ((RV, AP), ST(!*a*)) → ((SD, RV), ST(!*m*));
- ((SD, RV), ST(!*m*)) → ((RV, AP), ST(!*a*)) → ((RV, RV), Timeout − ST(!*a*)) → ((SD, RV), ST(!*m*)).

The first sequence is a sequence for a successful message transmission, and the second for a failure one. Let us call the successful one and the failure one *ss* and *fs*, respectively. Combinations of the two sequences can construct all possible states sequences of the composed DEVS model, which include *ss* → *ss* → *ss* ... → *ss*, *ss* → *fs* → *fs* → *ss* ... → *fs*, *fs* → *fs* → *ss* → *ss* → ... → *ss*, and so on. Note that each sequence may be infinite in general, but it can be finite if conditions, such as a total number of messages to be transmitted and a failure rate of each transmission are given. A method for the systematic generation of the minimum set of basic loops of states/events sequences is a research issue and one such method can be found in Hong and Kim (2005).

Once all possible sequences of a composed DEVS model are given, questions on the system can be answered. To get an answer required is an efficient search method that finds properties translated from questions in the state space. Recall that the size of a global state space of DES is infinite. Thus, a construction of the state space causes a well-known state explosion problem, a general solution of which has not been proposed so far. To solve the problem, subclasses of DEVS, such as schedule-preserved DEVS (SP-DEVS) (Hwang and Cho, 2004) and schedule-controllable DEVS (SC-DEVS) (Hwang, 2005), have been proposed in which some restrictions are applied to bound an infinite state space to a finite one.

6.5 Simulation of DEVS Model

DEVS models can be used for performance simulation of a DES. Since DEVS modeling is based on the concept of the object-oriented (OO) worldview so does simulation of such models. Recall that DEVS defines two model classes, atomic and coupled models, with which a hierarchical construction of a complex model is specified.

6.5.1 DEVS Modeling Simulation Methodology and Environment

Figure 6.5 shows a generic architecture for a DEVS-based modeling and simulation environment using a programming language L, where L can be any OO language such as C++ and Java.

Note that the DEVS modeling environment and the simulation engine are explicitly separated. Within the environment, modelers can develop DEVS models using modelers' interface which is a set of application

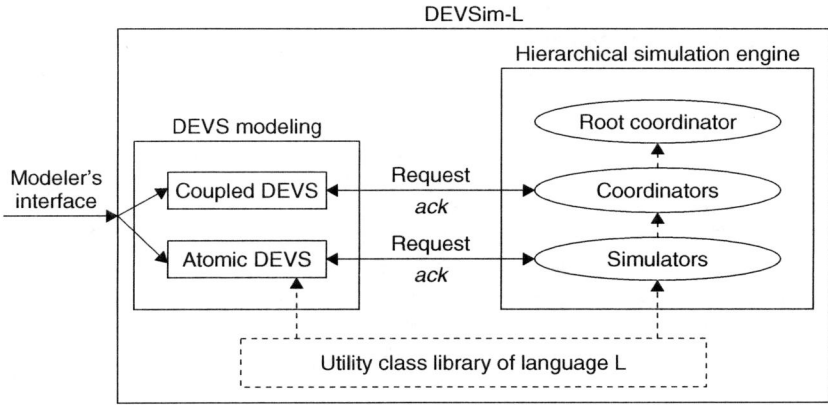

FIGURE 6.5 DEVS modeling/simulation environment.

programming interfaces (APIs) to specify DEVS models in DEVS semantics. Thus, APIs for the specification of DEVS models are defined such that there is a one-to-one correspondence between APIs in the formalism. For example, APIs for specification of atomic DEVS models in a C++ modeling/simulation environment include TimeAdvanceFn(), ExtTransFn(Message &, e), IntTransFn(), and OutputFn(Message &). APIs for the specification of coupled DEVS models are similarly defined.

The hierarchical simulation engine realizes the concepts of abstract simulators (or simulation algorithms) developed in Zeigler (1984). The abstract simulators are a set of distributed simulation algorithms which can be implemented in a sequential as well as a distributed computing environment. Also, it is natural to implement the abstract simulators algorithm in an OO language such as C++. The first implementation is DEVSim++ (Kim and Park, 1992), which is a C++ environment for DEVS modeling and simulation. Different implementations are available in public domains and efforts for standardization of DEVS modeling/simulation environments is ongoing by DEVS Standardization Organization (DEVS-STD, 2005).

The main purpose of OO implementations of DEVS modeling/simulation is to exploit the reusability of DEVS models in models development. In fact, a carefully designed environment would support two-dimensional (2D) reusability of DEVS models: one from the OO paradigm and the other from the DEVS methodology (Kim and Ahn, 1996). The former exploits inheritance mechanism in the paradigm; the latter does modular, hierarchical model construction in the methodology (Kim, 1991). Reuse metrics for DEVS models developed in the DEVSim++ environment are proposed and measure of such reusability was reported in Choi and Kim (1997).

Simulation-based performance analysis requires careful design of experimental frame, the concepts of which has been proposed in Zeigler (1984). The experimental frame is a coupled DEVS model, independent of the DEVS model of a target system to be simulated. The frame usually includes at least two models: one for the generation of events which is input scenarios to the simulated DEVS model and the other for the collection of simulation data which is output from the simulated DEVS model. Design of experimental frame is objective-driven, meaning that different design objectives require different experimental frames. More specifically, a set of simulation objectives is transformed into a set of performance indices. Then, an experimental frame is designed such that the desired performance indices are measured by simulation.

For example, assume that the ping-pong protocol described in Section 6.1 is simulated and that the simulation objective is to know how fast messages are to be transmitted with the protocol. Then, we employ the DEVS model, CMppp, in Figure 6.3 and identify that the throughput of message transmission is a performance index to be measured. The throughput is measured as a ratio of the total messages successfully transmitted to the total time spent for the transmission. To measure the ratio, an experimental frame consists of at least one atomic DEVS model which is connected to CMppp via appropriate coupling relations. Of course, ST($!m$) for AMsender and ST($!a$) for AMreceiver should be identified before simulation starts. The values may be fixed ones or random ones depending on a simulation objective. However, the values should be identical or statistically equivalent to those in a real ping-pong protocol. Data modeling of the values in terms of a distribution function with statistical parameter(s) is an important activity in a modeling process.

6.5.2 Simulation Speedup and Simulators Interoperation

Simulation for performance evaluation of practical scale DESs may spend a large amount of time. To reduce such simulation time in DEVS simulation, some attempts have been made. Among others three approaches are introduced here. The first one is distributed simulation of DEVS models in which an overall simulation is partitioned into a set of distributed computing resources (Kim et al., 1996; Seong et al., 1995). Speedup for the approach depends not only on a balance between partitioned workloads but also on a communication overhead between distributed computing resources. The second one is a hybrid simulation method in which simulation models are a combination of DEVS models and analytic models (Ahn and Kim, 1994). Speedup would be made without sacrificing accuracy if analytic models satisfy certain assumptions in the system behavior. The third one is a model composition approach in which an overall DEVS model is first composed into an atomic DEVS model and then simulate the atomic DEVS model (Lee and Kim, 2003). A tool for automatic composition of an overall DEVS model into an atomic model is developed and experimental results show about five times faster simulation than of the original DEVS model.

Recently, simulation interoperation between heterogeneous simulators is of interest first in the military domain and then in the civilian one. Accordingly, a standard for such interoperation has been proposed. High-level architecture (HLA) is a standard specification for simulators interoperation; run-time infrastructure (RTI) is implementation of HLA. HLA was first adapted as a defense modeling and simulation office (DMSO) standard in 1966 and then as an institute of electrical and electronics engineers (IEEE) standard in 2000. In conjunction with the simulators interoperation, the DEVSim++ environment has been extended such that simulators developed in DEVSim++ is HLA-compliant, meaning that the simulators can interoperate with other simulators via RTI interface. The extended environment, called DEVSimHLA (Kim, 1999; Kim and Kim, 2005) has been developed based on the concepts of a simulation bus of DEVS-BUS (Kim and Kim, 2003). The DEVSimHLA environment has been successfully employed for the development of HLA-compliant military war game simulators in Korea, which have been certified by DMSO.

6.6 Conclusion

Analysis and performance simulation of DESs should be based on a mathematical model for the systems. The DEVS formalism is one such modeling means, which supports specification of DESs in a hierarchical, modular manner. An advantage of the DEVS formalism is that it provides us with a unified model which can not only be used in analysis but also to study the performance simulation of a system. Analysis of DES with composed DEVS models requires a computerized tool for a state space construction and an associate search method for a state space exploration. Composition of more than two DEVS models can be done in an incremental manner for which only two models are composed at a time. Performance simulation of DEVS models needs a modeling/simulation environment which is usually implemented in OO programming languages such as C++ or Java.

References

M. S. Ahn and T. G. Kim, "A Framework for Hybrid Modeling/Simulation of Discrete Event Systems," *Proceedings of AIS'94*, pp. 199–205, December 1994, Gainesville, FL.

Y. Choi and T. G. Kim, "Reusability Measure of DEVS Simulation Models in DEVSim++ Environment," *Proceedings of SPIE-97*, pp. 244–255, 1997, Orlando, FL.

DEVS-STD, 2005, http://www.devs-world.org.

K. J. Hong and T. G. Kim, "Timed I/O Sequences for Discrete Event Model Verification," *LNAI 3397: Artificial Intelligence and Simulation*, pp. 275–284, Springer, Berlin, 2005.

M. H. Hwang, "Generating Finite-State Global Behavior of Reconfigurable Automation Systems: DEVS Approach," *Proceedings of 2005 IEEE-CASE*, August 1–2, 2005, Edmonton, Canada.

M. H. Hwang and S. K. Cho, "Timed Behavior Analysis of Schedule Preserved DEVS," *Proceedings of 2004 Summer Computer Simulation Conference*, pp. 173–178, July 2004, San Jose, USA.

J. H. Kim and T. G. Kim, "DEVS Framework and Toolkits for Simulators Interoperation Using HLA/RTI," *Proceedings of Asia Simulation Conference*, pp. 16–21, 2005, China.

K. H. Kim, Y. R. Seong and T. G. Kim, "Distributed Simulation of Hierarchical DEVS Models: Hierarchical Scheduling Locally and Time Warp Globally," *Transactions for SCS*, vol. 13, no. 3, pp. 135–154, 1996.

T. G. Kim, "Hierarchical Development of Model Classes in DEVS-Scheme Simulation Environment," *Expert Systems with Applications*, vol. 3, no. 3, pp. 343–351, 1991.

T. G. Kim, *DEVSimHLA User's Manual*, SMSLab, Department of EECS, KAIST, 1999, http://smslab.kaist.ac.kr.

T. G. Kim and M. S. Ahn, "Reusable Simulation Models in an Object-Oriented Framework," Chapter 4, *Object-Oriented Simulation*, IEEE Press, Piscataway, NJ, 1996.

T. G. Kim and S. B. Park, "The DEVS Formalism: Hierarchical Modular Systems Specification in C++," *Proceedings of European Simulation Multiconference*, pp. 152–156, June 1992, York, England.

Y. J. Kim and T. G. Kim, "Heterogeneous Simulation Framework Using DEVS Bus," *SIMULATION*, vol. 79, no. 1, pp. 3–18, 2003.

W. B. Lee and T. G. Kim, "Performance Evaluation of Concurrent System Using Formal Method: Simulation Speedup," *IEICE Transactions on Fundamentals of Electronics, Communications and Computer Sciences*, vol. E86-A, pp. 2755–2766, 2003.

Y. R. Seong, T. G. Kim, and K. H. Park, "Mapping Modular, Hierarchical Discrete Event Models in a Hypercube Multicomputer," *Simulation Practice and Theory*, vol. 2, no. 6, pp. 257–275, 1995.

B. P. Zeigler, *Multifacetted Modelling and Discrete-Event Simulation*, Academic Press, London, 1984.

B. P. Zeigler, H. Praehofer, and T. G. Kim, *Theory of Modelling and Simulation* (2nd ed.), Academic Press, New York, 2000.

II

Modeling Methodologies

7
Domain-Specific Modeling

Jeff Gray
University of Alabama at Birmingham

Juha-Pekka Tolvanen
MetaCase

Steven Kelly
MetaCase

Aniruddha Gokhale
Vanderbilt University

Sandeep Neema
Vanderbilt University

Jonathan Sprinkle
University of Arizona

7.1 Introduction 7-1
7.2 Essential Components of a Domain-Specific Modeling Environment 7-2
    Language Definition Formalism • Domain-Specific Modeling Environments • Model Generators • Key Application Areas of DSM
7.3 Case Studies in DSM 7-6
    A Customized Petri Net Modeling Language in the Generic Modeling Environment • Modeling and Generating Mobile Phone Applications in MetaEdit+
7.4 Overview of Supporting Tools 7-17
    A Retrospective of Metamodeling Tools • Modern Metamodeling Tools
7.5 Conclusion 7-18

7.1 Introduction

Since the inception of the software industry, modeling tools have been a core product offered by commercial vendors. In fact, the first software product sold independently of a hardware package was Autoflow, which was a flowchart modeling tool developed in 1964 by Martin Goetz of Applied Data Research (Johnson, 1998). Although software modeling tools have historical relevance in terms of offering productivity benefits, there are a few limitations that have narrowed their potential.

The primary drawback of most software and system modeling tools is that they are constrained to work with a fixed notation. That is, the tool vendor has defined a notation and environment that must be used in a prescribed way, regardless of the unique requirements of the user. Such inflexibility forces the user to adopt a language that may not be suitable in all cases for their distinct needs. Examples of such modeling tools include early flowchart tools, or more recent environments supporting object-oriented modeling, such as the unified modeling language (UML) (Booch et al., 1998). As an alternative to fixed-language modeling tools, many users desire a customized modeling environment that can be tailored to the concepts represented in the user's problem domain.

A movement within the software modeling community is advancing the concept of tailorable modeling languages (Bézivin, 2005), in opposition to a universal language that attempts to offer solutions for a broad category of users. This newer breed of tools enables domain-specific modeling (DSM), in which a metamodel is used to express the definition of a modeling language that represents the key abstractions and intentions of an expert in a particular domain (DSMFORUM; Gray et al., 2004; Pohjonen and Kelly, 2002). For example, the conference phone registration case study presented later in this chapter allows an end-user to describe the essence of a registration system at a very high-level of abstraction using concepts that are much more aligned to the problem domain, rather than the programming language solution space. Figure 7.8, shown later in the chapter, illustrates the manner in which an end-user can specify the rules of conference registration, with a complete application generated from the domain-specific model (see Figure 7.9 for the model-generated version of the phone registration application). In this case, Python code was generated from the high-level model, but the end-user is unaware of the specific technology used at the implementation level.

A contributing factor to the rising interest in DSM comes from the realization of productivity gains that have been attributed to a shift in focus toward software represented at higher levels of abstraction. In the past, abstraction was improved when programming languages evolved toward higher levels of specification. DSM takes a different approach, by raising the level of abstraction, while at the same time narrowing down the design space, often to a single range of products for a single domain. When applying DSM, the models consist of elements representing things that are part of the domain world, not the code world. The language follows the domain abstractions and semantics, allowing developers to perceive themselves as working directly with domain concepts.

In the next section of this chapter, the essential characteristics of DSM are presented, including a discussion regarding those domains that are most likely to benefit from DSM adoption. The chapter also contains a case study section where two different examples are presented in two different metamodeling tools. An overview of the history of metamodeling tools is also provided as well as concluding comments.

7.2 Essential Components of a Domain-Specific Modeling Environment

Domain-specific languages (DSLs) that are of a textual nature have been deeply investigated over the past several decades (van Deursen et al., 2000). Language tools for textual DSLs are typically tied to a grammar-based system that supports the definition of new languages (Henriques et al., 2005). A set of patterns to guide the construction of DSLs exists (Spinellis, 2001) as well as principles for general use of DSLs (Mernik et al., 2005). In comparison, this section offers a description of the essential characteristics of DSM, which is typically focused on graphical models as opposed to the textual representation of a DSL.

As illustrated in Table 7.1, there are several similarities that can be observed between DSM and other artifacts that are specified by a meta-definition (e.g., programming languages and databases). In DSM, the highest layer of the meta-stack is a meta-metamodel that defines the notation to be used to describe the modeling language of a specific domain (e.g., the metamodel). Instances of the metamodel represent a real system that can also be translated into an executable application. This four-layered meta-stack is also evident in programming language specification (where the meta-meta level is typically extended Backus–Naur form used to define a grammar) and database table definition (where the SQL data definition

TABLE 7.1 Comparison of Metamodeling to Programming Language and Database Definition

	Domain-Specific Modeling	Programming Language Definition	Database Schema Definition
Schema definition notation	Meta-metamodel (e.g., UML/OCL)	Language specification formalism (e.g., EBNF)	Database definition formalism (e.g., SQL Data Definition Language)
Schema definition	Metamodel for a specific domain (e.g., Petri net)	Grammar for a specific language (e.g., Java)	Table, constraint, and stored procedure definitions for a specific domain (e.g., University payroll database)
Schema instance	An instance of the metamodel (e.g., Petri net model of a teller machine)	A program written in a specific language	Intension of a database at a specific instance in time (e.g., the June 2006 payroll instance)
Schema execution	Executing application	Executing program	Transactions and behavior of stored procedures in an executing application

Domain-Specific Modeling

language is the meta-meta level that defines the database schema). The first to acknowledge a meta-stack for schema definition were Kotteman and Konsynski (1984). Despite these similarities, there exist core differences between metamodeling and other schema definition approaches. This section highlights some of the essential parts of a modeling environment to support the concepts of DSM.

7.2.1 Language Definition Formalism

A language, L, in its most basic form, provides a set of usable expressions as well as rules for expression composition. Well-formed composed expressions define a *program* that may be executed. We define a modeling language in Eq. [7.1], where C is the concrete syntax of the language, A the abstract syntax, S the semantics of program execution, M_s the semantic mapping (a function mapping from the abstract syntax to the semantics, as in Eq. [7.2]), and M_c the syntactic mapping (a function mapping from the concrete syntax to the abstract syntax, as in Eq. [7.3]). The composition rules are found in M_s, the well-formedness rules found in S as execution errors, and in A as a constraint layer.

$$L = <C, A, S, M_s, M_c> \tag{7.1}$$

$$M_s : A \rightarrow S \tag{7.2}$$

$$M_c : C \rightarrow A \tag{7.3}$$

The *concrete syntax* of a language defines how expressions are created, and their appearance. It is the concrete syntax that programmers see when using a language. Concrete syntax can be textual or graphical. The *abstract syntax* of a language defines the set of all possible expressions that can be created (note that it also defines possible expressions that may not be well-formed under the execution rules of S). The abstract and concrete syntax, along with the function M_c, make up the *structural* portion of a language. The semantics S makes up the *semantic domain* portion of the language, and the function M_s makes up the *semantic mapping* portion of the language.

DSM requires a language that is by definition linked to the domain over which it is valid. A domain-specific modeling language (DSML) is a language that includes domain concepts as members of the sets A or C (i.e., first-class objects of the language). The presence of other concepts that are not domain-specific affects the restrictiveness of the DSML as a language. A DSML's level of restrictiveness is only vaguely measurable, but generally is inversely proportional to the amount of freedom a developer has to address problems outside the domain. More restrictive languages reduce the modeling space, and thus reduce the possibility for errors unrelated to domain concepts (e.g., buffer overruns).

A DSML can be defined in more than one way. For instance, the DSML can be layered on top of an existing language, which is known as "piggybacking" (Mernik et al., 2005). Examples of piggybacking include programming libraries that define new classes with behaviors that reflect domain concepts. This layered style of DSML design is very unrestrictive, because it does not preclude the use of non-DSML expressions. DSMLs that use this layered style are often accompanied by a coding style guide. Implementation of a DSML via definition of a new language from scratch is also possible. Examples of this kind include VHDL (very high speed integrated circuits [VHSIC] hardware description language) for hardware description, and simulation program with integrated circuit emphasis (SPICE) for circuit design. This *language style* of DSML design is very restrictive, because the language is self-contained and more difficult to extend.

Implementation of a language coupled with its own development environment through rigorous planning and software engineering is also possible. In this case, an application with an interface for accessing the concrete syntax items of the language is the programming environment. This integrated development environment, or *IDE style*, of DSML design is also very restrictive, though it is important to note that the language definition is often obscured in the environment design, rather than decoupled from it. When a modeling environment is domain-specific, we call it a domain-specific modeling environment (DSME). The difference between a DSME and a DSML is that the DSME will provide interfaces for such activities as expression building, model execution, and well-formedness checking (among others).

The final way to define a DSML involves the co-creation and synthesis of the structural portion (i.e., C, A, and M_c) of the language (DSML), and DSME through the use of a metamodeling environment.

This *metamodeling style* of DSML design is also somewhat restrictive. This style produces similar results as the IDE style of design, though it is significantly more sophisticated because the definition of the language is used to define the DSME, rather than a design-time result of the development of the DSME.

7.2.2 Domain-Specific Modeling Environments

DSMEs provide the tools necessary for a system developer to rapidly build systems belonging to a specific domain that are *syntactically correct-by-construction*. DSMEs leverage the power of DSMLs to provide the model engineers with the building blocks necessary to develop systems rapidly and correctly. To enable syntactically correct by construction systems, a DSME must incorporate only those syntactic elements that are defined by the DSML while strictly abiding by the semantics. The modeling elements, which form the building blocks provided by the DSME, correspond to the concrete syntax defined in a DSML. The DSME must permit the composition and associations between these building blocks, which is guided by the syntax of the language. A powerful DSME provides a complete IDE and often has the following characteristics:

- Metamodeling support—a DSME must include the metamodel representing the DSML along with its syntactic elements, semantics and constraints. Only then can a DSME enable a developer to use only those artifacts that belong to the desired domain and build systems that are syntactically correct by construction.
- Separation of concerns—the DSME should enable separation of concerns, wherein it can provide multiple views corresponding to the different stakeholders and their concerns. For example, different development teams of a large project must be able to view only those artifacts that are part of their responsibility. At the same time, the DSME must maintain seamless coordination between the different views.
- Change management—a DSME must provide runtime support for issues such as change notification. For example, a DSME must be able to reflect changes made to the models in one view to appear in other views.
- Generative capabilities—a DSME must be able to provide the capabilities to transform the models into the desired artifacts. These could include code, configuration, and deployment details, or testing scripts. This feature requires that a single DSME be able to support multiple model interpreters, each of which performs a different task. Note that the modeling editor of a DSME will enable a developer to create syntactically correct systems. However, this does not ensure that the behavior and the output of a system will be correct. To validate and verify that systems perform correctly will require the generative capabilities in a DSME to transform the models into artifacts that are useful by third-party verification and validation tools.
- Model serialization—a DSME must ideally provide capabilities for serializing the models so that they can be made persistent. Model serialization provides an archival representation of a specific model such that it can be used in a later modeling session, or stored in a version control system. The process of model serialization manifests all of the model attributes and connections, while resolving all of the hierarchical relationships, in a manner that can be stored persistently in a specific file format. This capability is essential because the DSM philosophy mandates that models are the most important part of system design and implementation. Code and other artifacts, such as those related to configuration and deployment, are all generated. Thus, the models and associated generators must be maintained over time. Additional benefits of serialization are driven by the desire to share models among different tools.
- Plug-in capabilities—although not a strictly required feature, a DSME could provide the capabilities to plug-in third-party tools, such as model checkers and simulation tools.

7.2.3 Model Generators

Model generators are at the heart of model-driven development by forming the generative programming capabilities of a DSME. A fundamental benefit of generative programming is to increase the productivity, quality, and time-to-market of software by generating portions of a system from higher-level abstractions

(Czarnecki and Eisenecker, 2000). This concept is particularly applicable to the realm of product-line architectures, which are software product families that illustrate numerous commonalities in system design (Clements and Northrop, 2002). Product variants within the software family represent the parameterization points for customization. Generative programming makes it easier to manage large product-line architectures by generating product variants rapidly and correctly. This vision is being explored in further depth by the software factories movement (Greenfield et al., 2004).

Generators (in the form of model interpreters) are useful in synthesizing code artifacts or metadata used for deployment and configuration. There are numerous challenges in this space. For example, a modeled system may need to be deployed across a heterogeneous distributed system. This requires the generated code artifacts to be tailored to and optimized for the platform on which the systems will execute. Deployment and configuration metadata must address the heterogeneity in configuring and fine-tuning the platforms on which the systems will execute. The platform models typically include descriptions of the hardware, networks, operating systems, and middleware stacks. Thus, generators must incorporate optimizers and intelligent decision logic so that the generated artifacts are highly optimized for the target platforms.

Generative capabilities at the modeling level are useful in transforming models into numerous other artifacts (e.g., input to model-checking tools to verify properties like deadlock and race conditions; simulations for validating system performance and tolerance to failures; or empirical testing used for systems regression testing). These capabilities are important in the overall verification and validation of the modeled software, so that ultimately the systems developed using DSMEs and their generative capabilities are truly correct by construction.

7.2.4 Key Application Areas of DSM

As with all technologies, it is helpful to understand the situations where the technology is most likely to succeed as well as the limitations that prevent the technology from offering benefit in some scenarios. This section discusses characteristics of particular domains that suggest the applicability of DSM.

7.2.4.1 Areas Where DSM is Most Applicable

From our collective experience, DSM has been very successful in the following domains:

- Factory automation systems, where a tight coupling between the hardware configuration and software exists. As an example, the configuration of an automotive factory may be changed several times during a year to manufacture different models of a product line (Long et al., 1998). In a manual approach to software evolution, the associated software is written in an unproductive and error-prone fashion. By applying DSM, the hardware configuration can be captured in models and the associated software generated automatically from hardware configuration changes.
- Deeply embedded microcontroller systems, where the embedded system's control logic is developed using higher-level abstractions (e.g., VHDL), and low-level code (possibly assembly language) is generated and burned into microprocessor chips (Karsai et al., 2003).
- Large distributed systems, particularly those that are heterogeneous, network-centric and distributed, having stringent performance and dependability requirements, and are developed and deployed using middleware solutions (Gokhale et al., 2004).

There are general characteristics about these domains that suggest scenarios when DSM would be useful. Each of these examples represents a type of configuration problem with numerous choices (e.g., multiple "knobs" are available to configure a system). Furthermore, each of these examples is based upon an underlying execution platform that may often change. The accidental complexities associated with evolving source code in the presence of platform adaptation are very hard to accomplish using ad hoc techniques based on low-level manual coding. This makes a system brittle because of the tight coupling to the execution platform. Moreover, these systems are constantly evolving by virtue of changes in the hardware and software platform, and due to changes in requirements. Therefore, there is a need to incorporate several degrees of concern separation through higher levels of system representation.

Recently, DSM has had success in product-line modeling because the commonalities and variabilities of a software product line are best captured and represented in model forms, while the generative techniques can be used to tailor a product to a platform. The commonalities and variabilities of product lines represent the different configurations of the systems belonging to the family. Application of DSM can help decouple these systems from the specific platforms on which they are deployed, and generative techniques can then seamlessly synthesize platform-specific configurations. Other uses of DSM arise when the same high-level representation of a system can be used to accomplish a variety of other activities, such as regression testing where such code can be autogenerated, or model checking for behavioral correctness. Verifying the correctness of a system is of paramount importance particularly for large and complex mission-critical systems, such as avionics mission computing (Balasubramanian et al., 2005).

7.2.4.2 Situations Where DSM is Not Very Useful

We have found that DSM is not useful in systems that are very static and do not evolve much over time. In such systems, even though it is conceivable to have product families, the range of configurations is very limited and the choice of platforms usually does not exist. Therefore, most system development begins from scratch using low-level artifacts.

DSM can be difficult to use in autonomous systems, which entail self-healing and self-optimization. In such systems, the DSME is required to be used during system runtime where the modeling environment is driven by systemic conditions as input from which the system must infer the next course of action. Dynamic changes to models and subsequent autonomous actions are a significant area of future research.

7.2.4.3 Trade-Off Analysis for DSM Development

Not all domains are easily lumped into "easy-to-use" or "difficult-to-use" categories. One task of an expert in the domain is to perform a trade-off analysis of the effort required to produce the environment versus the effort saved by that environment. Even complex domains that are fairly established and static may benefit from a DSM effort if they will be widely used and incur frequent changes.

7.3 Case Studies in DSM

There are multiple approaches that can be adopted to achieve the goals of DSM. This section presents two separate modeling languages in two different tools to provide an overview of the different styles of metamodeling to support DSM.

7.3.1 A Customized Petri Net Modeling Language in the Generic Modeling Environment

An approach called model integrated computing (MIC) has been under development since the early 1990s at Vanderbilt University to support DSM (Sztipanovits and Karsai, 1997). A core application area of MIC is computer-based systems that have a tight integration between a hardware platform and its associated software, such that changes to the hardware configuration (e.g., an automobile assembly floor) necessitate large software adaptations. In MIC, the configuration of a system from a specific domain is modeled, resulting in an application that is generated from the model specification.

The generic modeling environment (GME) realizes the principles of MIC by creating DSMEs that are defined from a metamodel (Lédeczi et al., 2001). An overview of the process for creating a new DSME in the GME is shown in Figure 7.1, where a metamodeling interface allows a language designer to describe the essential characteristics of the language. In the GME, the metamodel definition is specified in UML and object constraint language (OCL), which is translated into a DSME that provides a model editor that permits creation and visualization of models using icons and abstractions appropriate to the domain (note that both the metamodeling interface and the subsequent DSME are hosted within the GME; i.e., the GME has a meta-metamodel available for defining metamodels). In the model interpretation process, a specific

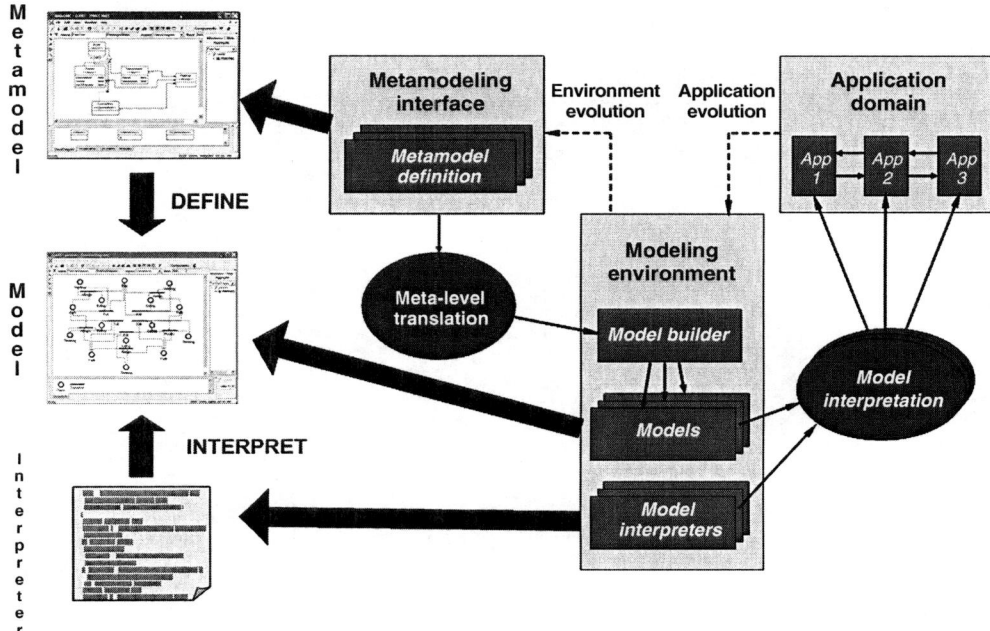

FIGURE 7.1 An overview of the metamodeling process in GME.

model interpreter is selected to traverse the model and translate it into a different representation (e.g., code or simulation scripts).

The left side of Figure 7.1 shows a metamodel for a Petri net (Peterson, 1977) language (top left), with an instance of the Petri net representing the dining philosophers (middle left). An interpreter for the Petri net language is capable of generating Java source code to allow execution of the Petri net (bottom left). The remainder of this subsection presents an overview of the Petri net modeling environment as modeled in the GME. This language is intentionally simple in nature so that the details do not overwhelm the reader in such a short overview. However, the GME has been used to create many very rich DSMEs that have several hundred modeling concepts (Balasubramanian et al., 2005).

7.3.1.1 Defining the Modeling Language

Figure 7.2 shows a screenshot of the GME to define the metamodel for the Petri net language. It should be noted that the metamodel is specified in MetaGME, which is the meta-metamodel for the GME representing a subset of UML class diagrams with OCL. In this metamodel, a `PetriNetDiagram` is defined to contain `Connections`, `Transitions`, and `Places`. The `AbstractElement` entity is a generalization of the two main diagram types that may appear in a Petri net (i.e., places and transitions). Each place has a text attribute that represents the number of tokens that exist in a particular state at a specific moment in time. Both places and transitions have names and descriptions that are inherited from their abstract parent. A `Connection` associates a `Transition` with two `Places`. Visualization attributes can also be associated with each modeling entity (e.g., the `Place` icon will be rendered from the "place.bmp" graphics file, which represents an open circle).

In addition to the class diagram from Figure 7.2, a metamodel also contains constraints that are enforced whenever a domain model is created as an instance of the metamodel. A constraint is used to specify properties of the domain that cannot be defined in a static class diagram. For example, the metamodel of Figure 7.2 would actually allow a `Place` to connect directly to another `Place`, or a `Transition` to connect directly to another `Transition`. This is not allowed in a traditional Petri net, and an OCL constraint is used to restrict such illegal connections. In the GME, constraints are specified in a different

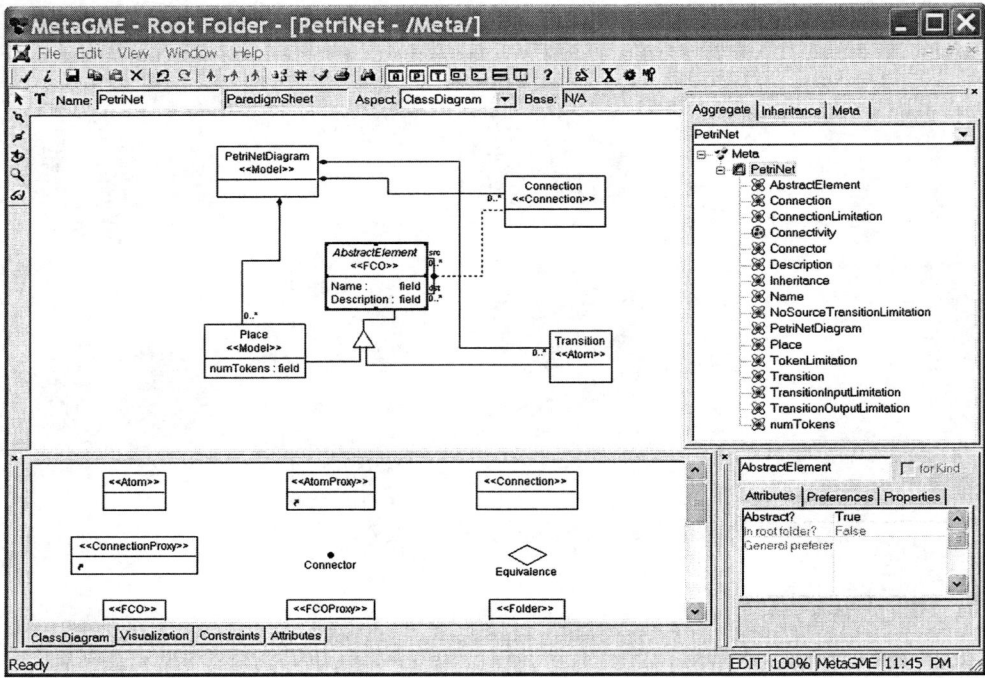

FIGURE 7.2 The Petri net metamodel represented within the GME.

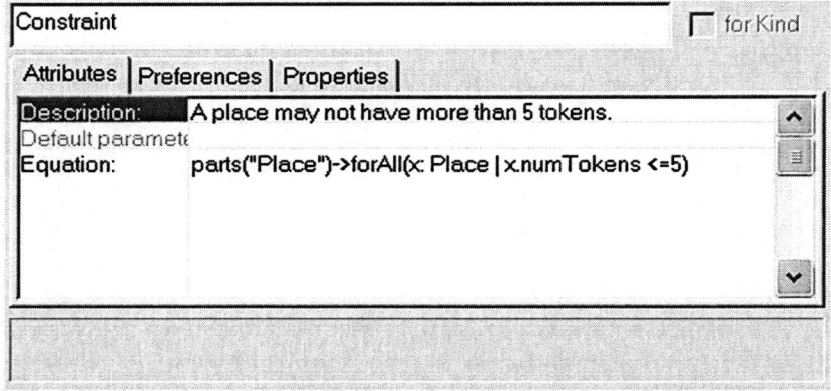

FIGURE 7.3 A constraint limiting the number of tokens for a Petri net place.

context diagram from that shown in Figure 7.2. The attribute panel shown in Figure 7.3 contains a sample constraint for the Petri net metamodel. This constraint specifies that a `Place` may not have more than five tokens. The first part of the OCL equation obtains a collection of all `Places` that appear in a model. A quantification predicate (i.e., the "`forAll`" statement) is associated with the collection to state that all such places must have its `numTokens` attribute less than or equal to five.

7.3.1.2 The Dining Philosophers in the Petri Net Language

After creating the metamodel, the Petri net language can be used to create an instance of the language, such as the dining philosophers model shown in Figure 7.4. In this model, the states (e.g., eating, thinking, and

Domain-Specific Modeling　　　　　　　　　　　　　　　　　　　　　　　　　　　　　　　　　　　　7-9

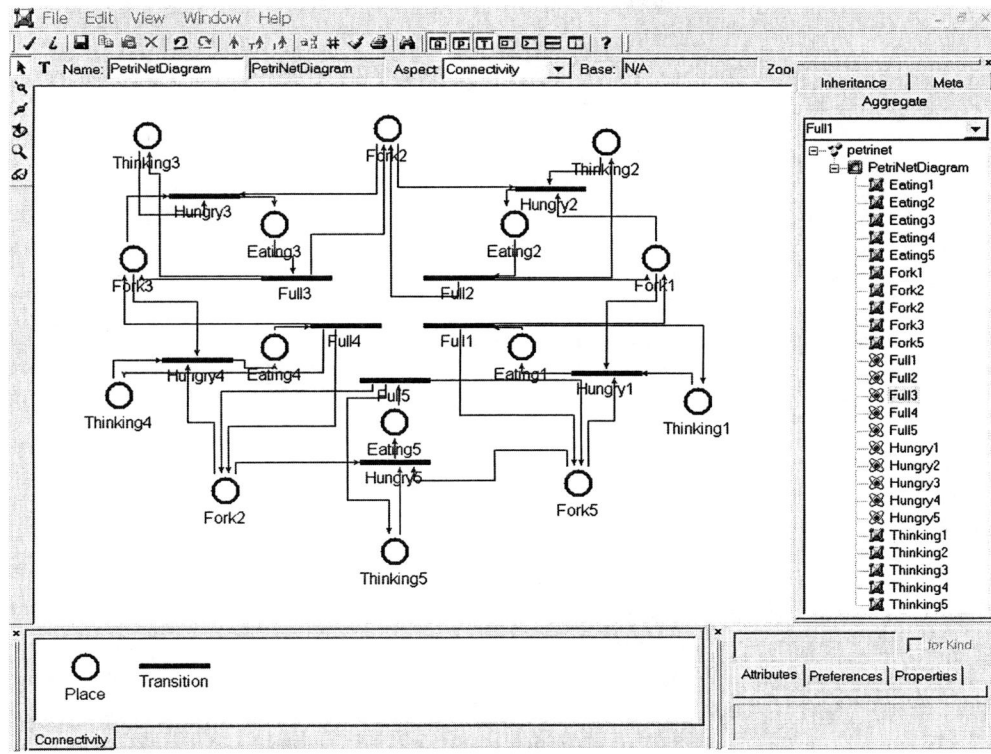

FIGURE 7.4　Dining philosophers expressed in the Petri net modeling language.

full) of five philosophers are modeled along with the representation of five forks. At this level, if the model engineer creates a Petri net that violates the metamodel in any way (e.g., connecting a `Place` directly to another `Place`, or adding more than five tokens to any `Place`), an error dialog is presented to indicate that the model is in an incorrect state.

7.3.1.3 Generating Applications

In the GME, a model interpreter is a plug-in that is associated with a particular metamodel and can be invoked from within the modeling environment. The GME provides an API for accessing the internal structure of a model, which can be navigated like an abstract syntax tree in a compiler to generate code at each node. A model interpreter is typically written in C++ and can be compiled to a Windows DLL that is registered to the GME.

An interpreter has been written to generate Java code that simulates the execution of a Petri net. The generated code will interact with a user to display a list of places and enabled transitions, and ask the user to select which transition to fire. Two different segments of the Petri net interpreter are shown in Listing 7.1. The top part of this listing contains the portion of the interpreter that generates the Java main method, which obtains all of the `Places` from a model as a collection (note that the `outf` file stream represents the .java file that is generated). The collection of `Places` is then inserted into a Java `ArrayList` for future processing. Although not shown here, a similar fragment of code is used to obtain the `Transitions` and associated connections. The `CBuilderAtomList` and `CBuilderAtom` are generic data structures within the GME that provide access to the underlying model representation. In the case of the top portion of Listing 7.1, the code fragment simply iterates over the collection of atoms that correspond to `Places` in the model. The bottom part of Listing 7.1 generates the corresponding Java code that will report to the user the names of available `Places` and the enabled `Transitions`.

```
...
outf << "public static void main(String[] args)" << endl;
outf << "{" << endl;
outf << " ArrayList places = new ArrayList();" << endl;

const CBuilderAtomList *allPlaces = petrinet->GetAtoms("Places");
pos = allNets->GetHeadPosition();

int tokens = 0;
CBuilderAtom *next;
CString strNumTokens, strPlaceName, strPlaceDescription;

while(pos)
{
   next = allPlaces->GetNext(pos);
   next->GetAttribute("numTokens", strNumTokens);
   tokens = atoi(strNumTokens);

   next->GetAttribute("Name", strPlaceName);
   next->GetAttribute("Description", strPlaceDescription);

   outf << "places.add(new Place(\"" << strPlaceName << "\", \"" <<
                    strPlaceDescription << "\", " <<
                    strNumberOfTokens << "));" << endl;
}
...
```
```
...
outf << "System.out.println(\"These places have tokens:\");" << endl;
outf << "for (int i = 0; i < places.size(); i++) {" << endl;
outf << "    Place p = (Place) places.get(i);" << endl;
outf << "    if (p.numTokens() > 0)" << endl;
outf << "        System.out.println(p.name() + \" - \" + p.description() +
                    \" - \" + p.numTokens());" << endl;
outf << "}" << endl;

outf << "System.out.println(\"The following transitions are " +
        "enabled:\");" << endl;
outf << "for (int i = 0; i < transitions.size(); i++) {" << endl;
outf << "    Transition t = (Transition) transitions.get(i);" << endl;
outf << "    if (t.isEnabled())" << endl;
outf << "        System.out.println(t.name() + \"-\" +
                    t.description());" << endl;
outf << "    }" << endl;

outf << "System.out.print(\"Select transition to fire:\");" << endl;
...
```

LISTING 7.1 Model interpreter for Petri net language.

Domain-Specific Modeling

```
private void fireTransition(String transitionName)
{
   System.out.println("Firing transition " + transitionName +
      "...\n");
   int tranIndex = 0;
   for (; tranIndex<transitionCount; ++tranIndex)
      if (transitionList[ind].equalsIgnoreCase(transitionName.trim()))
         break;

   // the transitionName was not found
   if (tranIndex == transitionCount){
      System.out.println("Transition not valid.\n");
      return;
   }

   // the associated place does not have the proper number of tokens
      // to fire
   // the transition
   for (int i = 0; i < transitionInput[tranIndex]; ++i)
      if (tokens[transitionInputIndex[tranIndex][i]] == 0){
         System.out.println("Transition not enabled.\n");
         return;
      }

   // remove the tokens from the input place and add to the output
      // place
   for (int i=0; i<transitionInput[tranIndex]; ++i)
      tokens[transitionInputIndex[tranIndex][i]]--;
   for (int i=0; i<transitionOutput[tranIndex]; ++i)
      tokens[transitionOutputIndex[tranIndex][i]]++;
}
```

LISTING 7.2 Java code generated from the dining philosophers model.

Listing 7.2 shows a small fragment of the Java code that was generated from the Petri net model interpreter. This particular piece of generated code represents the firing of a transition based on the transition name entered by the user. When executed, this code will check to see if the transition name exists and if it is enabled (i.e., the proper number of tokens are available in all of its input places). After firing, this code will decrement the tokens from the input places, and increment the tokens in the output places.

7.3.2 Modeling and Generating Mobile Phone Applications in MetaEdit+

This second example deals with modeling and generating enterprise applications for mobile phones based on Symbian/S60 (Nokia S60) and its Python framework. This framework provides a set of APIs and expects a specific programming model for the user interface (Nokia Python). To enable model-based generation, a modeling language and generator must follow the Nokia framework.

The example is implemented with MetaEdit+, a commercial tool for defining and using DSMLs and generators (MetaCase). The emphasis of MetaEdit+ is to make modeling language creation fast and easy—tool support is implemented without writing a single line of code. At any point in time, a language definition and the associated generators can be executed and tested. MetaEdit+ provides a metamodeling tool suite for entering the modeling concepts, their properties, associated rules, and symbols.

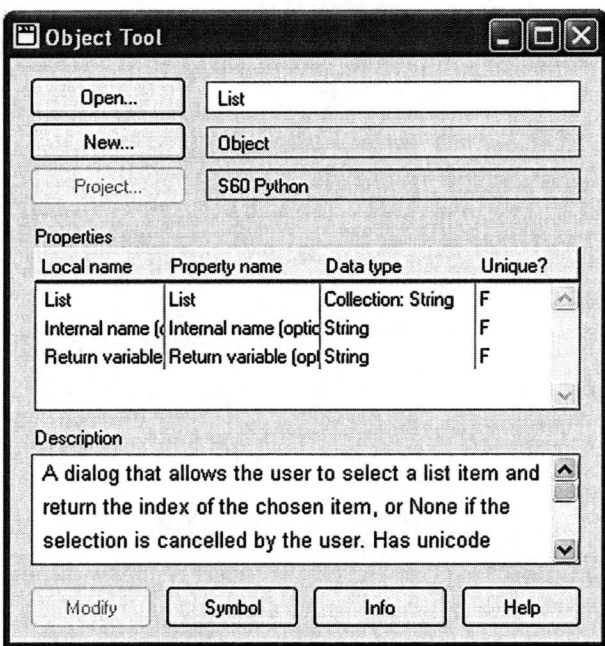

FIGURE 7.5   Defining the language concept "List."

This definition is stored as a metamodel in the MetaEdit+ repository allowing future modifications, which reflect automatically to models and generators (Kelly et al., 2005). Design data can be edited and viewed in diagram, table, matrix, or textual representations. Teamwork is supported with multiple concurrent users through a repository. Integration with other tools uses a Web Services-based API, with XML import and export also being supported. In addition to graphical metamodeling similar to GME in the previous case study, MetaEdit+ metamodels can be specified through interaction with form-based tools.

7.3.2.1 Defining the Modeling Language

The DSML in this example aims to hide the programming details by raising the abstraction level to phone concepts. This is achieved by defining modeling concepts directly based on the phone's services and user-interface (UI) widgets. These concepts include "Sending text message," "Note," "Form," and "Pop-up." Figure 7.5 shows how a language concept "List" is defined in MetaEdit+. In this figure, the concept name and its properties (e.g., collection of list items, optional internal name, and return variable for the selection) are entered into the form. Other main language constructs are defined in a similar manner.

The behavioral logic of the application is modeled using a flow model that allows user navigation to be specified in the application in a manner similar to how phone services are accessed. The navigation actions (e.g., acceptance, opening a menu, and canceling a selection) are defined with connections between the modeling concepts. The language definition also includes domain rules that follow the phone's UI programming model, supporting early error prevention, model consistency, and reuse. For example, in an S60 phone, after sending a short message server (SMS) message, only one UI element or phone service can be triggered. Accordingly, the metamodel allows only one flow from an SMS element. This rule is defined in Figure 7.6. In MetaEdit+, these rules are treated as data and can be changed at any time, even while developers are using the language. MetaEdit+ also updates the models correspondingly and delivers the domain rules automatically to the developers.

Models based on a DSML are usually represented in some format using graphical models, matrices, or tables. In MetaEdit+, the symbols are drawn or imported with a Symbol Editor tool. Figure 7.7 shows the symbol definition for the List concept. The properties of the List symbol include shape, size, and

Domain-Specific Modeling 7-13

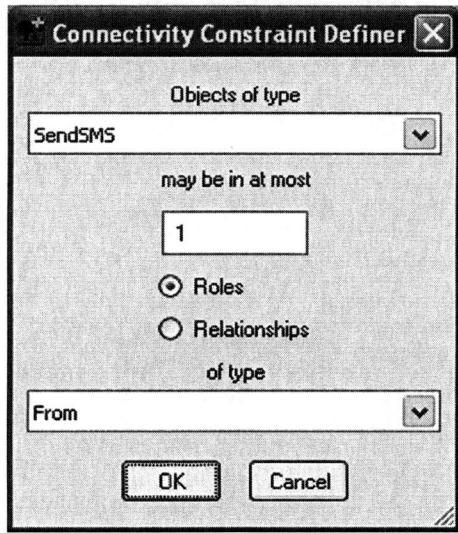

FIGURE 7.6 Choosing rules for the language constructs.

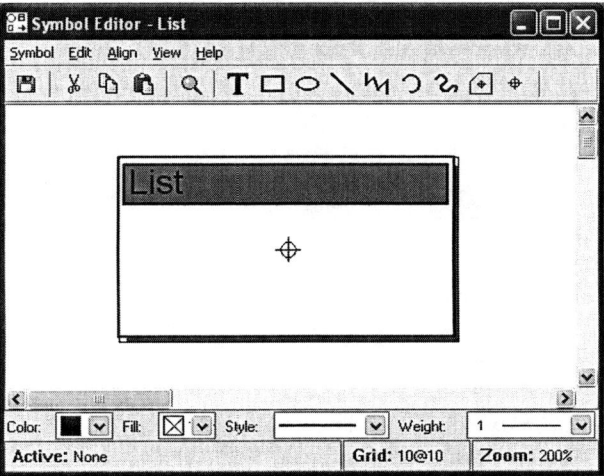

FIGURE 7.7 Drawing the notational element for language concept "List."

color. A symbol definition also declares the location for the property values to be shown in a model. The selection list is displayed in the middle of the symbol, aligned to the top left along with the font settings. This corresponds to the similar appearance of an actual list on a real S60 phone.

7.3.2.2 DSM in Use

The DSM language is illustrated in Figure 7.8 using a sample application design representing a conference registration application. This design should be intuitive to any model engineer who has experience using basic phone applications (e.g., phone book or calendar applications). A user can register for a conference using text messages, choose a payment method, view program and speaker data, browse the conference program on the Web, or cancel the registration.

As can be seen from the model, all of the implementation concepts are hidden and are not even necessary to know (i.e., the focus is on the specification of the problem in the domain of interest). The modeling

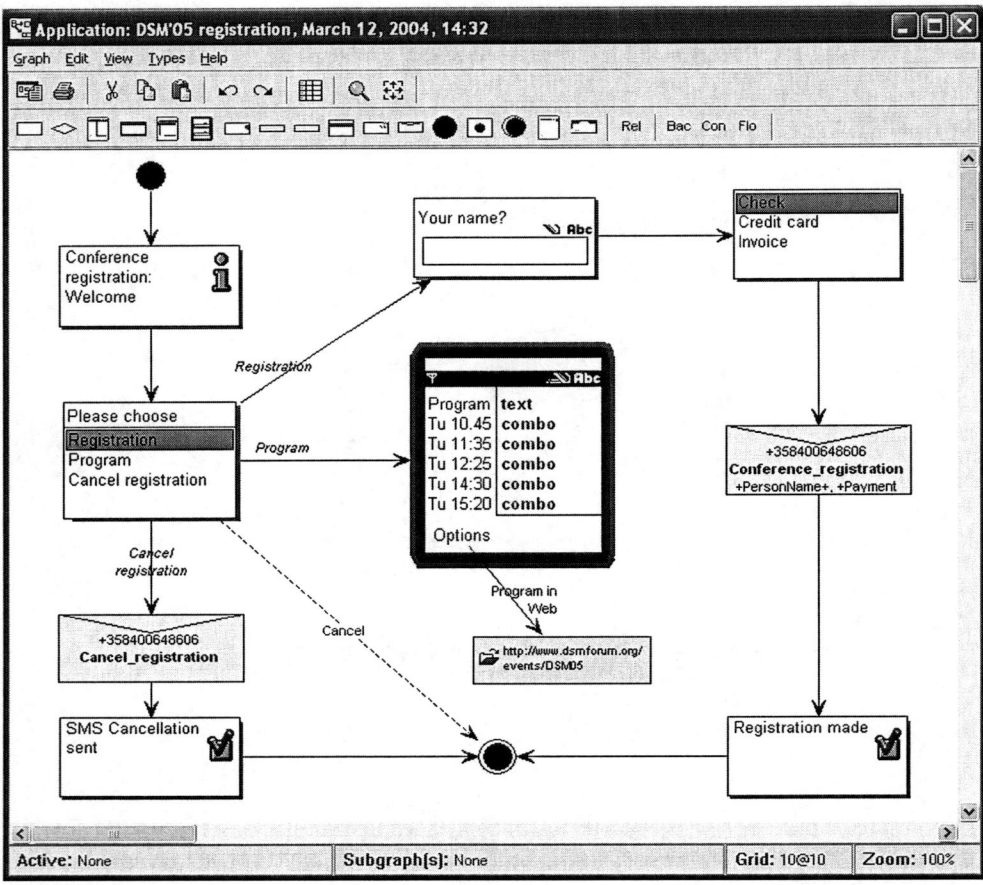

FIGURE 7.8 Design of a conference registration application for a Symbian/S60 mobile phone.

language also ensures that the architectural rules and the required programming model are followed as defined in the metamodel. As the descriptions capture all the required static and behavioral aspects of the application, it is possible to generate the application fully from the models.

7.3.2.3 Generating Applications

From the designs expressed in the model, a generator can be invoked to produce code that can be executed either in an emulator for testing purposes, or in the actual target device. The generator itself is structured into modules, often with one generator module per modeling concept (e.g., one generator module takes care of Lists, and another generator exists for SMS messages). A simple example of a generator definition for a Note dialog is presented in Listing 7.3. The Note opens a dialog with information (such as the "Conference registration: Welcome" dialog in Figure 7.8). Lines 1 and 6 are simply the structure for a generator. Line 2 creates the function definition signature and line 3 provides a comment. Function naming is based on an internal name that the generator can produce if the developer does not want to give each symbol its own function name.

Line 4 produces the call for the platform service. It uses the design data from the model (e.g., the value for the Text property of the Note element). Similarly, the developer may choose the "Note type" value in the model from a list of available notification types, such as the "info" or "confirmation" values that are used in the "Registration made" task of Figure 7.8. As several concepts require similar code to be generated, parts of the generator definitions are made into modules called by other generator modules. For example, the `_next element` generator is used by other dialogs to generate transitions. The generator

```
1 report '_Note'
2   'def '; subreport; '_Internal name'; run; '():'; newline;
3   '# Note '; :Text; newline;
4   ' appuifw.note(u"'; :Text; '", '''; :Note type; ''')'; newline;
5   subreport; '_next element'; run;
6 endreport
```

LISTING 7.3 Code generator for Note dialog.

```
01  import appuifw
02  import messaging
03
04  # This application provides conference registration by SMS.
...
33  def List3_5396():
34  # List Check Credit card Invoice
35    global Payment
36    choices3_5396 = [u"Check", u"Credit card", u"Invoice"]
37    Payment = appuifw.selection_list(choices3_5396)
38    if Payment == None:
39         return Query3_1481
40    else:
41         return SendSMS3_677
..
85  def SendSMS3_677():
86  # Sending SMS Conference_registration
87  # Use of global variables
88    global PersonName
89    global Payment
90    string = u"Conference_registration "\
91            +unicode(str(PersonName))+", "\
92            +unicode(str(Payment))
93    messaging.sms_send("4912345678", string)
94    return Note3_2227
...
101 def Stop3_983():
102 # This applications stops here
103   return appuifw.app.set_exit
...
107 f = Note3_2543
108 while True:
109   f = f()
```

LISTING 7.4 Python code generated for the conference registration model.

also includes some framework code for dispatching and for multiview management, as shown by different tabs in the pane of the UI.

A part of the generated code from the designs illustrated in Figure 7.8 is shown in Listing 7.4. The generator produces the module-importing statements (lines 1–2) based on the services used (e.g., importing the messaging module that provides SMS sending services). This is followed by documentation specified in the design. Next in the listing, each service and widget is defined as a function. Lines 33–41 describe the code for the payment method selection that uses a list widget. After defining the function name and comment, the "Payment" variable is declared and made available for the whole application. Line 36 shows the list values as Unicode in a local variable. Line 37 calls the List widget provided by the framework.

SMS sending (lines 85–94) is handled in a similar way to the List widget. Line 93 calls the imported SMS module and its sms_send function. Parameters to the function (e.g., recipient number, message keyword, and content) are taken from the model by the generator to assist in forming the correct message syntax.

The end of a function includes code for calling the next function based on user input. In SMS sending, the generator simply follows the application flow (line 94). In the case of list selection, the situation is a bit more complex. Depending on selections from the list, different alternatives can exist; for example, the cancel operation (i.e., pressing the cancel/back button on the phone) is also possible (lines 38–41). Where necessary, the generator creates the operation-cancel code to return to the previous widget. This choice minimizes the need to model this concept explicitly and guarantees that exceptions are acknowledged. As a last function, the application exit code is created based on the end-state (lines 101–103). Finally, a dispatcher starts the application by calling the first function (line 107) with tail recursion to reduce stack depth (lines 108–109).

The DSML and corresponding generators allow the application developers to focus on finding solutions using the problem domain concepts directly, while ignoring the low-level details and accidental complexities associated with coding in the S60 architecture. The cost and expertise needed to make other types of enterprise applications on Symbian/S60 phones is now greatly reduced. As shown in Figure 7.9, the generated application can be executed on a S60 simulator to observe the resulting behavior. When the applications need to be changed, it is easier to understand and make the change directly to the problem domain concepts than to the code. Additionally, if the platform changes (e.g., the API for accessing the List changes), the code generator needs to be changed in only one place, rather than manually modifying all the List usage code. Another example that examines the benefits of DSM applied to mobile devices can be found in Davis et al. (2005).

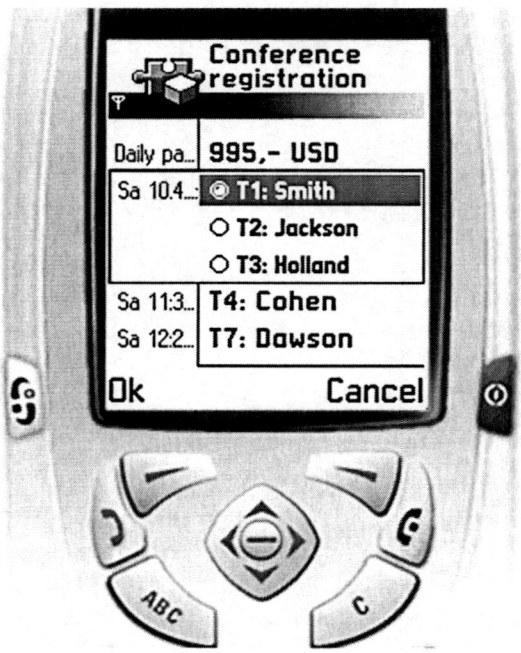

FIGURE 7.9 Generated conference registration application executing within a S60 simulator.

7.4 Overview of Supporting Tools

In addition to GME and MetaEdit+, there are several other metamodeling tools that are available, ranging from research prototypes to fully supported commercial products. From a historical context, the system encyclopedia manager (SEM) is one of the earliest metatools. SEM was developed by Dan Teichroew at the University of Michigan (Teichroew et al., 1980) and it was applied to information requirements modeling of various categories of systems. Like SEM, many of the early metatools are no longer available, but a summary of several representative examples are provided in the next subsection.

7.4.1 A Retrospective of Metamodeling Tools

The MetaPlex tool had a textual rather than graphical notation (Chen and Nunamaker, 1989). However, it is worthy of mention as one of the earliest metaCASE tools. It used a textual language to define metamodels, which were interpreted rather than compiled, and even included some rudimentary functionality to generate help text for method users from the metamodel.

The virtual software factory (VSF) used the set-theoretical and propositional calculus language CANTOR to define the conceptual data in metamodels and its constraints, and the graphical and design language (GDL) to specify graphical representations (Pocock, 1991). The latter was somewhat complicated: 15 lines of code was needed to represent a simple data flow arrow. VSF's strong point was its ability to define complex constraints. A clear weakness was the complicated nature of metamodeling: the time to construct a metamodel in VSF would be considerably longer than with today's leading tools.

The ToolBuilder metaCASE system was originally reported as a research tool in Alderson (1991) and later commercialized. It consisted of three components: the *specification component*—used to create the specification of the tool; the *generation component*—used to transform the specification into parameters for the generic tool; and the *runtime component*—the generic CASE tool itself. The first two are contained in the method specification capture component (METHS) system, and the third is called DEASEL, which provided standard CASE functionality to support multiple users on a true repository. EASEL, from "easy language," was the language generated by METHS for configuring the generic CASE tool, DEASEL. The name DEASEL came from EASEL, and as a pun related to the use of a generator: both diesel and meths are fuels used by generators. METHS captured four kinds of information: (1) the data model upon which data capture and output generation is based, (2) the frame model upon which the views are based, (3) the diagrammatic notation for each diagram frame, and (4) the textual presentation for each structured text frame. The data model of Toolbuilder was an entity-relationship (ER) model that was extended with some constraints and the ability to have attributes whose values are derived from other attributes. It allowed triggers on events applying to attributes and relationships.

7.4.2 Modern Metamodeling Tools

The GME and MetaEdit+ emerged toward the end of the first period of metaCASE tools (i.e., they each have over a decade of research and development), and are the only ones from that period that are still available. There are three other metamodeling tools that are more recent and deserve mention: AToM3, eclipse modeling framework (EMF), and the Microsoft DSL tools.

AToM3 is a research metamodeling tool that has been under development at McGill University (de Lara and Vangheluwe, 2002). A focus of AToM3 is *multiparadigm modeling*, which is a realization of the benefits of modeling a system at multiple levels of abstraction using several different formalisms (e.g., Petri nets, state machines, and differential equations). The underlying representation of an AToM3 model is represented as a graph, and the modeling environment provides a transformation system from which models can be manipulated by graph rewrite rules. A collection of preexisting metamodels is available for download, including ER diagrams and structure charts.

The EMF is relevant to this chapter because of the major influence it has made on the general modeling community (Budinsky et al., 2004). The EMF provides its own metamodel, called the ECore, which is used to describe application data models that can be fed to EMF in several formats. The native input format is an XML file, which can be produced from UML class models by tools like Rational Rose. Java source code that is annotated appropriately can also be converted into ECore models. EMF provides a code generator (called java emitter templates, or JET) that is capable of translating models that conform to the ECore into Java. Based on the input to ECore, the EMF.Codegen can generate a basic model editor following the schema. The editor uses classes from the EMF.Edit framework to provide standard table and property sheet views. Most modeling projects have extended the EMF.Edit framework with customized capabilities. The EMF itself does not produce graphical editors—other eclipse plug-ins that provide such capabilities have been integrated with EMF.

Microsoft has committed to the DSM community by initiating a metamodeling tool that is integrated within Visual Studio (Microsoft DSL) and tied to the software factory vision (Greenfield et al., 2004). At the time of this writing, this tool is still in beta mode and represents one of the newer entries in this space of representative tools. The current state of the MS DSL tools support the definition of a modeling editor as specified from a project wizard. A template-based code generator is available from which generic code can be instantiated with various placeholders representing data obtained from a model.

Given the numerous tool suites that support DSM, it is often desirable to share models among different tools. In fact, initial consideration has been given toward bridging the gap between EMF and Microsoft DSL models (Bézivin et al., 2005). The ability to exchange models among tools has several obstacles such as: (1) the *syntactic problem* of sharing the information within a model across different data formats, (2) the *semantic problem* of resolving the meaning of a model as expressed across different metamodels that may describe common properties of the same domain, and (3) the *infrastructure problem* resulting from tools that are open, but have different APIs to access the underlying representation. An approach to tool integration that addresses these problems is to consider patterns of interaction and configuration within a tool integration framework (Karsai et al., 2005).

7.5 Conclusion

This chapter presented an introduction to DSM, including an overview of general concepts as well as case studies to illustrate the potential for application. From our own collective experience, DSM offers an order of magnitude improvement in productivity in those environments that are tied to software product lines that can be configured across multiple alternative design spaces. Although the future outlook for DSM looks promising, there is still much research and development needed to improve the capabilities provided by supporting modeling tools. Issues related to group or team modeling (e.g., version control of models that are distributed among a globally dispersed team) are being explored, but additional functionality is needed to make DSM tools popular beyond specific niche domains. In addition to future technical developments, there is also a need to combat the organizational culture to promote adoption. This has been stymied by the stigma associated with the past failures of CASE tools (i.e., many who are first introduced to DSM mentally create a link back to the limitations of past CASE environments). This chapter described the flexibility and productivity that can be achieved by a modeling environment that can be tailored to a specific domain to generate applications from higher-level abstractions.

Acknowledgment

This work was supported in part by an NSF CAREER grant (CCF-0643725).

References

Alderson, A. Meta-CASE Technology. 1991. *Software Development Environments and CASE Technology, Proceedings of European Symposium*, Königswinter, Germany, pp. 81–91.

Balasubramanian, K., J. Balasubramanian, J. Parsons, A. Gokhale, and D. Schmidt. 2005. A platform-independent component modeling language for distributed real-time and embedded systems. *IEEE Real-Time and Embedded Technology and Applications Symposium*, San Francisco, California, pp. 190–199.

Bézivin, J. 2005. On the unification power of models. *Journal of Software and System Modeling*, 4(2): 171–188.

Bézivin, J., G. Hillairet, F. Jouault, I. Kurtev, and W. Piers. 2005. Bridging the gap between the MS/DSL tools and the Eclipse Modeling Framework. *OOPSLA Software Factories Workshop*, San Diego, California.

Booch, G., J. Rumbaugh, and I. Jacobson. 1998. *The Unified Modeling Language User Guide*. Redwood City: Addison-Wesley.

Budinsky, F., D. Steinberg, E. Merks, R. Ellersick, and T. Grose. 2004. *Eclipse Modeling Framework*. Redwood City: Addison-Wesley.

Chen, M. and J. Nunamaker. 1989. METAPLEX: An integrated environment for organization and information systems development. *Proceedings of the Tenth International Conference on Information Systems*, Boston, Massachusetts, pp. 141–151.

Clements, P. and L. Northrop. 2002. *Software Product Lines: Practices and Patterns*. Redwood City: Addison-Wesley.

Czarnecki, K. and U. Eisenecker. 2000. *Generative Programming: Methods, Tools, and Applications*. Redwood City: Addison-Wesley.

Davis, V., J. Gray, and J. Jones. 2005. Generative approaches for application tailoring of mobile devices. *43rd ACM Southeast Conference*, Kennesaw, Georgia, pp. 237–241.

de Lara, J. and H. Vangheluwe. 2002. Using AToM3 as a meta-Case tool. *Proceedings of the International Conference on Enterprise Information Systems*, Ciudad Real, Spain, pp. 642–649.

DSMForum. The Domain-Specific Modeling Forum, http://www.dsmforum.org/

Gokhale, A., D. Schmidt, B. Natarajan, J. Gray, and N. Wang, 2004. Model-Driven Middleware. In *Middleware for Communications*, Q. Mahmoud (Ed.), New York: Wiley, Chap. 7, pp. 163–187.

Gray, J., M. Rossi, and J.-P. Tolvanen. 2004. Preface: Special issue on domain-specific modeling. *Journal of Visual Languages and Computing*, 15(3–4): 207–209.

Greenfield, J., K. Short, S. Cook, and S. Kent. 2004. *Software Factories: Assembling Applications with Patterns, Models, Frameworks, and Tools*. New York: Wiley.

Henriques, P. R., M. J. V. Pereira, M. Mernik, M. Lenic, J. Gray, and H. Wu. 2005. Automatic generation of language-based tools using LISA, *IEE Proceedings—Software*, 152(2): 54–69.

Johnson, L. (James). 1998. A view from the 1960s: How the software industry began. *IEEE Annals of the History of Computing*, 20(1): 36–42.

Kelly, S., M. Rossi, and J.-P. Tolvanen. 2005. What is needed in a MetaCASE environment? *Journal of Enterprise Modeling and Information Systems Architectures*, 1(1): 25–35.

Karsai, G., A. Lang, and S. Neema. 2005. Design patterns for open tool integration. *Journal of Software and System Modeling*, 4(2): 157–170.

Karsai, G., J. Sztipanovits, Á. Lédeczi, and T. Bapty. 2003. Model-integrated development of embedded software. *Proceedings of the IEEE*, 91(1): 145–164.

Kotteman, J. and B. Konsynski. 1984. Information systems planning and development: Strategic postures and methodologies. *Journal of Management Information Systems*, 1(2): 45–63.

Lédeczi, Á., A. Bakay, M. Maroti, P. Volgyesi, G. Nordstrom, J. Sprinkle, and G. Karsai. 2001. Composing domain-specific design environments. *IEEE Computer*, 34(11): 44–51.

Long, E., A. Misra, and J. Sztipanovits. 1998. Increasing productivity at Saturn. *IEEE Computer*, 31(8): 35–43.

Mernik, M., J. Heering, and A. Sloane. 2005. When and how to develop domain-specific languages. *ACM Computing Surveys*, 37(4): 316–344.

MetaCase. *MetaEdit+ 4.5 User's Guide*. http://www.metacase.com

Microsoft DSL. *Visual Studio Launch: Domain-Specific Language (DSL) Tools: Visual Studio 2005 Team System*. http://msdn.microsoft.com/vstudio/dsltools/

Nokia Python. *Python for Series 60: API reference.* http://www.forum.nokia.com

Nokia S60. *S60 SDK Documentation.* http://www.forum.nokia.com

Peterson, J. 1977. Petri Nets. *Computing Surveys*, 9(3): 223–252.

Pocock, J. VSF and its relationship to open systems and standard repositories. 1991. In *Software Development Environments and CASE Technology*, A. Endres and H. Weber (Eds.), Berlin: Springer-Verlag.

Pohjonen, R., and S. Kelly. 2002. Domain-specific modeling. *Dr. Dobbs Journal*, 27(8): 26–35.

Spinellis, D. 2001. Notable design patterns for domain specific languages. *Journal of Systems and Software*, 56(1): 91–99.

Sztipanovits, J. and G. Karsai. 1997. Model-integrated computing. *IEEE Computer*, 30(4): 110–111.

Teichroew, D., P. Macasovic, E. Hershey, and Y. Yamamoto. 1980. Application of the entity-relationship approach to information processing systems modeling. In *Entity-Relationship Approach to Systems Analysis and Design*, P. P. Chen (Ed.), Amsterdam: North-Holland.

van Deursen, A., P. Klint, and J. Visser. 2000. Domain-specific languages: An annotated bibliography. *ACM SIGPLAN Notices*, 35(6): 26–36.

8
Agent-Oriented Modeling in Simulation: Agents for Modeling, and Modeling for Agents

Adelinde M. Uhrmacher
University of Rostock

Mathias Röhl
University of Rostock

8.1 Introduction 8-1
8.2 Agents for Modeling in Simulation 8-3
     Using the Agent Metaphor for Understanding in Simulation • Using the Agent Metaphor for Designing in Simulation
8.3 Modeling and Simulation for Agents 8-6
     Modeling and Simulation for Understanding Multiagent Systems • Simulation for Designing Multiagent Systems
8.4 Conclusion 8-10

8.1 Introduction

The relations between agents, modeling, and simulation are manifold (Uhrmacher et al., 2001). In relating agents and simulation research, we will concentrate on using agents as a metaphor to model the system of interest and on using virtual dynamic environments to analyze the behavior of multiagent software systems. Thereby, modeling aspects rather than the many challenges of an efficient execution shall be the focus.

Before starting our exploration, we will shortly introduce agents and discuss some characteristics that also influence the relation between agents and simulation.

A common perspective originates agents in artificial intelligence (AI) and distributed artificial intelligence (DAI) methods; even if most existing agent-oriented approaches seem to be complacent with agents that utilize only a small portion of intelligence (Wooldridge and Jennings, 1998). The goal of AI is to let machines do things that would require intelligence if done by humans: intelligence being typically interpreted as the ability to learn and understand, to solve problems, and to make decisions (Russell and Norvig, 2003). The interaction between AI software and environment was typically mediated by the user: an intelligent machine should help humans to make decisions (Negnevitsky, 2002). Real-time expert systems aimed at online diagnosis were among the first to address issues of a direct interaction with a dynamic, continuous environment including sensing and acting, timeliness of reactions, incompleteness, and uncertainty of available information. This interaction with an environment which is continuous, not entirely accessible, and dynamic has become an intrinsic part of the agent metaphor and has challenged and propelled research in knowledge representation, evaluation, planning, and learning likewise.

While AI starts from individual cognition and individual systems, DAI examines the properties of distributed, interacting AI programs. Whereas individual AI systems are increasingly designed as single agent systems working in dynamic environments, e.g., Pollack et al. (2003), DAI focuses on multiagent systems (O'Hare and Jennings, 1996): "DAI is the study, construction, and application of multiagent systems, that is, systems in which several interacting, intelligent agents pursue some set of goals or perform some set of tasks" (Weiss, 2000). One motivation for DAI was that intelligent behavior might more easily come into being by the coordination of multiple entities. Intelligent behavior as an emergent phenomenon holds some natural attraction, because at first glance it seems to release researchers from the problem to implement intelligent behavior explicitly. However, it still has to be decided how entities on microlevel should look like to enforce the desired effect on macrolevel and what the communication mechanisms and conventions of a "civilized" discourse for effective problem solving are. This discussion has been circling around reactiveness and deliberativeness and their role in realizing intelligence (Brooks, 1991; Etzioni, 1993). Whereas in the latter the entities constitute knowledgeable agents with symbolic representation and reasoning skills, based on explicit knowledge about themselves and their environment, the former do not possess such an explicit model, their behavior is determined by simply "reacting" to environmental changes. How to combine the ability to react and the deliberativeness to solve problems effectively is one of the major challenges in designing agents for dynamic, uncertain environments.

Sociology has had an impact on agent-based research from the outset, interpreting multiagent systems as societies. The functioning of societies and what role individual behavior plays is the subject of individualistic social theory. One argumentation line explains the development of societies as structured exchange relations between independently acting individuals. Social order and cooperation emerges from the unintended interaction of multiple agents in a framework of decentralized coordination. Another argumentation line emphasizes contracts, norms, and intentional cooperation in achieving desirable goals in the society and thus reflects a view of agents as being "deliberative" rather than "reactive" (Conte et al., 2001; Gilbert and Troitzsch, 2005). "Reactive" agents and the functioning of societies is also the subject of biological studies on ant and bee societies, insights which were also used in developing software agent solutions, e.g., in network routing. Thus, social theories have an impact in designing better, e.g., more robust and efficient software systems. In the opposite direction, multiagent systems as a modeling metaphor support the testing of sociological hypotheses by simulation. Thereby, the knowledge about social systems is enhanced.

Rather the efficient and reliable handling of multiple, concurrent, and interacting individuals than the achievement of the desired intelligent overall behavior motivated the involvement of the area of distributed systems in multiagent research. This becomes most obvious in the field of mobile agents, where agents are no longer dependent on one computer to be executed. Mobile code and data migrate from one computer to the next. Thereby, they cope more easily with network connections that are only temporarily available and have a limited bandwidth (Tanenbaum and van Steen, 2002). Typically, distributed systems require a balance between decentralized and central strategies, which requires answering how much a single node must know, who is in control, and what cooperation strategies to apply. This leads us back to central questions in DAI and social theories.

By illuminating the background of agents, we have already revealed their central characteristics. The least common denominator for most researchers and users is to associate agents with *autonomy*, which refers to the ability to act without permanent guidance.

Some kind of intelligence is needed if agents shall decide for themselves what they need to do to satisfy their design objectives. In this context, reactive, proactive, and social abilities are required (Wooldridge, 2000).

Since the beginning of agent-based development, the question of differentiation has loomed: is it an agent or just a program, an object, or a component for reuse? Although a general answer to this question seems elusive, compared with objects, agents embody a stronger notion of autonomy in the sense that they have control over their actions: agents can say "no" and "go" (Odell, 2002). Thus, agents can be interpreted as reflective concurrent objects and a community of agents as a special type of concurrent, distributed system. Components can be used to implement agents (Melo et al., 2004); however, with components neither an own thread of control is associated, nor a functioning in a dynamic environment (Casagni and Lyell, 2003).

Most agent definitions make use of the term "environment." For agents it is central to be able to operate robustly in rapidly changing, unpredictable, and open environments (Wooldridge, 2002). This dynamic environment, the described properties, and the different backgrounds shape the multiple facets of the relationship that has formed between simulation and agents.

A simulation is an experiment based on a model and performed on a computer, aimed at exploring the behavior of a dynamic system. If the dynamic system itself or parts of it are interpreted as agents, the modeling is shaped accordingly to distinguish between these different autonomously interacting entities. The resulting model design is necessarily nonmonolithic, and similarities between agent-based modeling and multilevel, individual-based, and/or object-oriented modeling are plentiful (Uhrmacher, 1997).

Another dimension that relates agents to simulation is the question whether agent-oriented software approaches can be utilized in realizing and implementing flexible simulation environments, e.g., to facilitate reusing models and interoperating between different simulation systems. In the latter case, agents are not subject of modeling and simulation but a metaphor used to develop flexible simulation systems.

Agent-based systems are often mission critical (or even safety critical) and, like other software systems, must be tested and evaluated before being deployed. Thus, a third dimension in relating agents and simulation is that simulation can help designing multiagent systems. It is used as an experimental technique for evaluation.

Whereas in the last case simulation helps to develop software agents, in the former two cases agents are used for modeling and designing simulation systems. Thus, we find both directions: simulation for agents and agents for simulation (Uhrmacher et al., 2001; Yilmaz and Ören, 2005). As a metaphor, agents improve understanding and designing certain systems. This also applies to simulation. Simulation is typically used in certain areas for a better understanding, in other areas to support the concrete design of systems. Furthermore, both agents and simulation are used for entertainment and training.

In the following, our exploration shall be focused on how the agent metaphor is used for modeling and thereby enhances the understanding or design of the system that is simulated, and subsequently how simulation is used as an experimental technique to improve the understanding and to support the design of software agents.

8.2 Agents for Modeling in Simulation

The multiagent metaphor, perceiving a dynamic system as a community of autonomously interacting entities, has started permeating many application areas of modeling and simulation. Modeling means structuring the knowledge about a given system. In psychology, sociology, and biology, where little is known about the system of interest, modeling and simulation are used for falsifying or supporting theories (Gilbert and Troitzsch, 2005). The more is known about a system, the more modeling and simulation are aimed at designing or manipulating systems. This design-oriented approach is traditionally found when dealing with "man-made" systems, e.g., in manufacturing and traffic systems. At first glance, agents as a metaphor seems similar to other metaphors used before. However, at second glance it shows multiple facets, as, although having often an anthropomorphic flavor, the ingredients of agents are not commonly agreed upon. On the one side, this has made deriving a uniform agent modeling formalism or reference model for agents illusive, on the other, it has propelled the propagation of the metaphor in diverse application fields.

8.2.1 Using the Agent Metaphor for Understanding in Simulation

Generally, simulation of biological, ecological, or social systems focuses on analyzing and predicting rather than on designing systems. These systems are typically not well understood, component libraries as in technical domains are not available, and thus typically modeling starts from scratch. The question when to use agents for modeling is only one facet of the central question in modeling: at which level of detail shall the system be described (Gilbert and Troitzsch, 2005)? Whereas simplicity of a model has long been interpreted as a quality in itself, this seems no longer be commonly agreed upon, neither in sociology or economy (Axtell, 2000) nor in biology (Chen et al., 2004).

If we are interested in the activities of a small number of actors with frequent stochastic fluctuations and stepwise motions, we would likely opt for a discrete, stochastic model to describe the involved actors individually. In contrast, behaviors that involve a large amount of entities are typically described at the macrolevel. Quantitative changes are represented traditionally and effectively by differential equations.

Sometimes it makes sense to aggregate and other times to explicitly and individually track entities within the simulation resulting in individual-based simulations. Individual-based simulation belongs to the class of multilevel simulation where phenomena observed at the macrolevel are produced by the entities located on microlevel (Knorr-Cetina and Cicourel, 1981; Bunge, 1979). When the individuals in the model are usefully characterized as having some sort of cognitive processes, individual-based turns into an agent-based approach (Gilbert, 2005).

Thus, agent-based models shift the focus to the individuals' intentions, desires, and beliefs (Ferber, 1999). Although intentions, like the wish to reproduce or to allay hunger and thirst, might direct the behavior of individuals, these decision processes are often not part of the model, as they are not of main interest. So, the appropriateness of the agent metaphor depends on the system to be described and on the objective of the simulation study, for example, we can treat a light switch as a (very cooperative) agent; however, "... it does not buy us anything, since we essentially understand the mechanism sufficiently to have a simpler, mechanistic description of its behaviour" (Shoham, 1993). Agent-based models are neither ipso facto preferable to other modeling approaches nor vice versa. The complexity of the model should be related to the complexity of the problem (Edmonds and Möhring, 2005).

As do multilevel simulations in general, agent-based approaches turn the attention to interacting individuals and the emerging macropatterns on the institutional level (David et al., 2003). In addition, they support the simulation of heterogeneous social communities that embrace reactive and deliberative actors likewise (Doran et al., 1996). In combination with cellular automata, agent-based models allow to capture the interplay of deliberative decisions and spatial dynamics, e.g., the growth of cities (Batty, 2005). Thereby, agent-based models move the interrelations between decision processes based on norms and preferences and the dynamics of communities into the focus of exploration. Let us illustrate this with an example.

A premodern town has been described based on agents (Ewert et al., 2003). Three actor groups, i.e., merchants, craftsmen, and laborers, form the town's population and are modeled as agents which behave according to the assumption of utility and profit maximization. They are interacting as consumers and suppliers via several markets, e.g., a grain market, a consumer good market, and a labor market. The local authorities are modeled as a planning agent which decides upon a course of interventions into market and social structures.

Disasters are induced into the model, provoking reactions of actors according to their preferences and intentions. Macrolevel effects can be observed that are not intended by the individual actors, but are characteristic for premodern mortality crises. Thus, this simulation model allows to experiment with actors' intentions and preferences, to mimic disasters leading to mortality crises, and to trace the courses of economic and demographic developments in the aftermath of such crises. The local authorities are described as a typical belief-desire-intention (BDI) agent (Rao and Georgeff, 1995), whereas the other actors are utility-based agents (Russell and Norvig, 2003). Figure 8.1 shows the beliefs of the local authorities about the situation at hand, the current desires, and the actions that are planned to take. The local authorities belief that laborers are not contented and the price for grain is high, whereas for labor it is low. To content the laborers and turn prices to normal, stored grain is supplied to the market and labor is demanded via a job program by the local authorities. Because these actions also take effect on the merchant's and craftsmen's contentedness, the local authorities have to decide on further interventions on the tax and the guild system.

The project revealed that the desires of the local authorities have a significant influence on the pace of disaster discovery—especially that short-term leveraging in favor of single actor groups, e.g., merchants, may even intensify a crisis in the long run. However, the central problems in an agent-based modeling and simulation remain: How to justify the made assumptions and validate the chosen model (Doran, 1997)? Whereas the utility-based agents could be based upon theories in microeconomy, the design of the local authorities was based on knowledge about frequent intervention schemes in premodern towns and speculation. Only a face validation of the model was possible. Still the model and simulation increased the understanding of the system and inspired some "Gedankenexperimente" that a less expressive model would not have motivated.

Thus, agent-based simulation appears what some areas have been waiting for, i.e., "... to provide the social sciences with conceptual and experimental tools, namely the capacity to model, and make up in

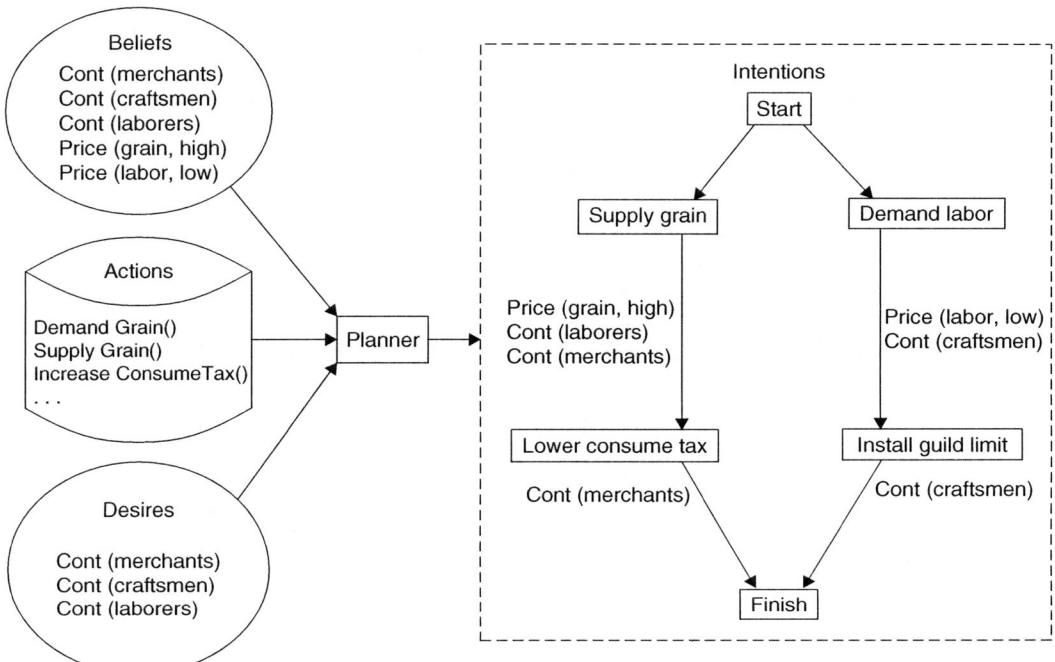

FIGURE 8.1 Sample plan generation by a deliberative agent that represents the local authorities in a premodern town.

parallel, reactive and cognitive systems, and the means to observe their interactions and emerging effects" (Conte and Castelfranchi, 1995).

8.2.2 Using the Agent Metaphor for Designing in Simulation

Simulation is also used for designing systems, i.e., to explore the implications of different design alternatives. If the performance of the system under study depends on the behavior of human individuals or institutions, the agent metaphor can be exploited. As in the previous section, the agent metaphor is used for modeling. However, it is no longer the objective of the simulation study to get insights into the behavior and motivations of human or biological communities, but to mimic the behavior sufficiently well to allow a good design of the technical system.

A human behavior representation (HBR) is a "model that mimics either the behavior of a single human or the collective action of a team of humans" (Pew and Mavor, 1998). Research on HBRs has strongly been driven by the military community, namely the U.S. Department of Defense Modeling and Simulation Office (DMSO), and lately by research on home security. Typically, an HBR model is understood primarily as a "representation of the decision making processes [...] of humans" (Wise et al., 2001), and "the details of how individual humans actually accomplish tasks—that is, the psychological theory—is [...] irrelevant as long as the behavioral output seems realistic" (Pew and Mavor, 1997). In contrast to work in related fields, such as human modeling or human factors research, HBR research does neither try to output any new evidence on human cognition nor does it need to realistically mimic physiological processes. Nevertheless, the fields are interrelated: HBR design is based on psychological and medical findings (Wise et al., 2001; Silverman, 2003).

Traffic simulation is one of the prominent areas where agents are employed to describe the behavior of individuals (Helbing and Nagel, 2004) and help exploring the effects of different traffic control strategies and thereby the design of traffic systems. To enhance the validity of traffic simulations, human behavior representation plays a central role. Typically, a physical motion model (including information about streets, intersections, surface, and speed limits) is combined with a mental model, which represents individual car drivers. Over the years models of the individuals have been gradually enriched. Classical traffic simulations use origin–destination matrices that aggregate individual mobility. Current approaches take a hierarchical approach. Overall behavior patterns are selected, appropriate locations assigned, timings determined, and finally routes calculated (Balmer et al., 2004). These are more realistic than traffic models based on averaged behavior and they are easily extensible and amenable to parallel execution. An important agent feature for traffic simulation is learning: route choice may be revised day today and within the day given the situation at hand.

Human behavior modeling in traffic systems exploits the agent metaphor to simulate realistic individual behavior and thereby stimulate emergent phenomena (e.g., traffic jams). Herein, agents work without explicit communication and use implicit coordination and social laws (Klügl and Bazzan, 2004).

Whereas to model humans seem the natural place to employ the agent metaphor, also technical systems sometimes lend themselves to be perceived as communities of interacting autonomous entities. If characteristics like distribution, heterogeneity, and self-organization match, the system to be designed can be thought of as a multiagent system. If agents and related concepts become the key abstractions for structuring these systems, they shape also modeling these systems.

The agent metaphor is especially useful if autonomous entities should be developed that are situated within an environment, e.g., to develop suitable decentral control strategies for manufacturing systems (Parunak, 1999), or if heterogeneous entities are involved and coordination and communication becomes important, e.g., to manage supply chains efficiently (Baumgaertel et al., 2003), which involve different alternative producers. Adopting the agent metaphor in modeling those systems helps reducing the complexity by decomposing the whole system into smaller pieces.

In contrast to human behavior representation, these approaches take no advantage of intensive cognition abilities of an individual agent, but use a set of more light-weight agents for distributed problem solving (Parunak, 1997). Herein, the autonomy of the agents is complemented with social abilities and communication becomes the enabling factor for coordination. For example, controlling manufacturing processes based on unmanned transportation vehicles is a nontrivial problem involving routing, collision avoidance, and deadlocks avoidance. Centralized approaches to control manufacturing processes are computationally complex and time-consuming. Weyns et al. (2005b) suggest a decentralized control based on autonomous, cooperative agents to realize flexible and open manufacturing systems. Flexibility is gained from agents that are able to locally commit to transport jobs and react to opportunities as they appear. Openness stems from the easiness of adding (or removing) vehicles to (from) the model and the facility.

So far, we have concentrated on the agent metaphor that is used in describing the system under study. However, as the last example has shown, using the metaphor in modeling for designing can imply that the system to be designed is perceived as a community of agents. If this system comprises software, obviously pieces of the software to be designed are interpreted as agents. Thus, we arrive at the point where simulation is used to test software agents. In the next section, we will no longer emphasize the view of using the agent metaphor for modeling, but approach the field from the point of view that there are software agents that shall be evaluated and tested in virtual dynamic environments.

8.3 Modeling and Simulation for Agents

Simulation provides a means for systematically analyzing the behavior of agents in dynamic virtual environments (Helleboogh et al., 2006). The development of agents faces problems associated with traditional distributed concurrent systems. Additional difficulties arise from complex interactions between autonomous problem-solving components (Jennings et al., 1998). Agent-based systems are often mission critical (or even safety critical) and, like other software systems, must be tested and evaluated before being deployed. However, their autonomy and the open heterogeneous nature of the environment in which they operate make testing and evaluation difficult. Often, existing formal notations are either too weak to express the structure and state of the agent and its environment, or if formalization is possible, the resulting formulations are intractable. Consequently, it can be difficult to verify their properties. One alternative is to test them in virtual environments. Thereby, the design of multiagent systems might become an application area for simulation tools similar to the design of manufacturing systems or network protocols.

We wish to now explore the role of simulation to help understanding and designing agents. This discussion will be based on a controversy put forward in a publication by Paul Cohen, Steve Hanks, and Martha Pollack at the beginning of the 1990s. They discussed the role of controlled experimentation in agent design in general and that of small world testing in particular (Hanks et al., 1993). In the following discussion, the focus will change from agents to their environment.

8.3.1 Modeling and Simulation for Understanding Multiagent Systems

Test beds offer sets of test scenarios are aimed at revealing prototypical problems in dynamic environments. They fill the gap between purely synthetic benchmarks and the deployment of agent systems within real environments. Since the beginning of research on agents, a series of test beds have been developed that provide dynamic environments to conduct controlled experimentation with agents in well-defined settings (Hanks et al., 1993). Test beds are simplified problem domains and facilitate testing in the small, i.e., rather isolated questions can be explored. The primary intent is not to confront the agent with a valid model of the concrete environment the agent shall dwell in. Instead, it aims at presenting a simplistic world in which properties of a concrete agent or a simplified version of an agent can be analyzed in isolation—of course, finding such "correct" simplifications is a very crucial part in the experimentation process. Test beds help to explain and understand why agents behave as they do by illuminating behavior facets of a given agent architecture.

To assess the behavior of an agent architecture, experimenting with one test scenario will hardly suffice. In general, software systems are not evaluated based on one benchmark result only. The implementation and application of dynamic scenarios requires more effort than that of conventional benchmarks. Accordingly rare are the examples where an agent architecture has been systematically tested in several dynamic scenarios.

Many scenarios have been developed together with the agent architectures to be tested. Thus, agent architecture and test environment are often closely coupled. There are only few examples where one scenario has been used to test different agent architectures. One of them is the TILEWORLD (Pollack and Ringuette, 1990) scenario that was developed to investigate commitment strategies. TILEWORLD consists of a grid of cells with tiles, obstacles, holes, and agents on it. The agents' task is to fill holes with tiles. The developed plans of how to optimally fill the different holes are subject to change as the environment changes by its own dynamics (holes and tiles appear randomly).

TILEWORLD has been picked up by several other research groups to evaluate agent architectures (Philips and Bresina, 1991; Kinny and Georgeff, 1991; Goldman and Rosenschein, 1994; Clark et al., 1997). In the course of adopting the TILEWORLD to test single- or multiagent systems, however, it has been adapted to the researchers' objectives and needs, which hampers comparing different agent architectures based on the results achieved. Therefore, it has not the role of a benchmark, but provides a type of scenario in which design alternatives can be tested. The extension of the original TILEWORLD scenario to analyze the behavior of multiagent systems (Ephrati et al., 1995) brings it close to another type of scenario which permeates the literature: variants of predator prey models explicitly designed for multiagent systems (Agre and Chapman, 1987; Doran et al., 1994; Wooldridge and Vandekerckhove, 1994; Luck et al., 1997). In this context, the interaction, coordination, and cooperation strategies are of central interest.

Cooperation, i.e., teamwork, and fast reactions are at the bottom of another test scenario for multiagent systems, which has gained popularity since the International Joint Conference on Artificial Intelligence 1997 in Japan: the soccer game. In addition to a robot league, a simulation league is held (Kitano et al., 1997; Tambe et al., 1999), in which a soccer simulator provides the common environment for evaluating different strategies of virtual robots.

In 1999, another scenario has been proposed to test agent architectures: RoboCup-Rescue (RCR) (Kitano and Tadokoro, 2001) and spurred a lot of research activities. The challenges are the number of agents (100 and more), a heterogeneous, partly hostile environment, real-time pressure, uncertain and limited

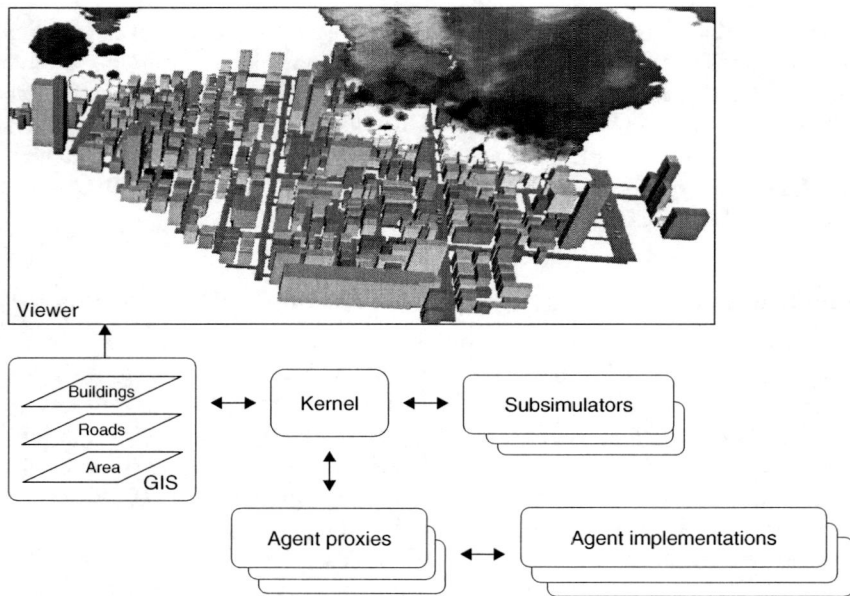

FIGURE 8.2 Architecture of a typical testbed for evaluating agents: The RoboCup Rescue simulator. (From Rescue Technical Committee, T. R. (2000). RoboCup Rescue simulator manual. Ver. 0, Rev. 4.)

information access, and distributed control. Simulators are used to describe the dynamics of typical disasters, e.g., such as the spread of fire, the collapse of buildings, and blocking of highways. Rescue complements RoboCup-Soccer with challenges in logistics, long-range planning, and collaboration among heterogeneous agents. RCR takes a special position among test beds, as it is not only a challenging artificial environment, but is *directly* aimed at socially significant problems. Besides the ability to put agent architectures to test with the Rescue simulator, RCR is aimed to make progress in emergency decision support for large-scale disaster scenarios (Kitano et al., 1999). Furthermore, the RCR-simulator is intended to function as a standard benchmark environment to compare and evaluate crisis management and provide the ground for digitally empowered rescue brigades. Figure 8.2 depicts the structure of the RCR-simulator (Rescue Technical Committee, 2000).

Each of these test beds supports testing of agents within predefined, dynamic scenarios. They may cast light on what is important and concentrate attention on key problems in a scientific area as they operationalize scientific paradigms (Sim et al., 2003). However, as Hanks points out, testing should ultimately be aimed at analyzing how agents will perform in the environment they are constructed for. With agents, AI methods have moved toward embedded applications, "the ultimate interest being not simplified systems and environments but rather real world systems deployed in complex environments" (Hanks et al., 1993, p. 18). Consequently, recent benchmarks (Kitano and Tadokoro, 2001; Weyns et al., 2005a) blur the border between artificial and real domains.

8.3.2 Simulation for Designing Multiagent Systems

Unlike testing in the small, where general strategies of agents are tested, this section discusses concrete agents designed for concrete applications. Simulation is used for testing agents in a virtual environment. Therefore, a valid model of the concrete environment the agent shall dwell in is required.

Generally, testing checks an implementation against some requirements. To this end, a set of test cases has to be generated to cover the requirements as much as possible. If testing cannot be done exhaustively, selecting suitable test cases becomes crucial. Efficient testing relies on the use of explicit models representing the conditions under which an implementation has to operate—its environment. Whereas most of the testing approaches concentrate on models of the software itself, i.e., the specification of requirements from which test cases are derived, model-based testing uses additional explicit models for test case selection (Pretschner et al., 2004). To constrain the set of test cases, model-based testing introduces a model of the test scenario to distinguish significant test cases. A random selection would reduce the number of test cases arbitrarily. In contrast, the integration of further knowledge, e.g., about critical situations or common user behavior, allows a goal-directed focusing of test cases. This is of particular importance for testing agent software, as they are usually intended to operate in rather complex settings.

The usage of a virtual environment in contrast to the real environment typically reduces costs and efforts and allows to test a system's behavior in "rare event situations." Virtual environments are easier to observe and to control, and probe effects are easier to manage. Environment models can be used to generate different test cases dynamically during simulation, including specific interaction patterns and time constraints. Thereby, the validity of the environmental models is crucial, independent of whether abstract models of agents are experimentally evaluated, single agent modules are embedded for testing, or entire agent systems are plugged into the virtual environment (Röhl and Uhrmacher, 2005).

For example, have a look at peer-to-peer (P2P) networks (Moro et al., 2005), which differ from typical client–server architectures in assigning to each node equivalent capabilities and responsibilities. P2P networks are characterized by autonomous nodes and a dynamic topology as nodes might leave or enter the network dynamically. For publishing and requesting information services in P2P networks, multiagent software systems are designed (König-Ries et al., 2004). To evaluate different protocol strategies, classical network simulators are considered as too inflexible and being on a too detailed level of abstraction (Klein, 2003). Therefore, simulators are developed from scratch, which simulate the underlying network at a rather abstract level (Minar et al., 1999; Huebsch et al., 2003), e.g., by counting hops. Often, more detail is required on the application level; e.g., instead of using simple "Churn"-rates, which describe the dynamics of the topology (Joseph, 2003), the user's individual activities are taken into account (König-Ries et al., 2004). As those users can be modeled as agents, human behavior representation becomes part of the environment software agents are tested in.

This has also been the case in developing a test environment for Autominder. Autominder is a distributed monitoring and reminder system developed at the University of Michigan (Pollack et al., 2003). It is being designed to support older adults with mild to moderate cognitive impairment in their activities of daily living. The software is installed in the client's house, together with a set of sensors and given a list of actions that must be performed during the day. At runtime, Autominder evaluates sensor echoes from the environment, reasons about ongoing user activity, and compares its findings to the given client plan to detect forgotten actions and remind the user of their execution.

Autominder complements a human in-place nurse and thus has a huge responsibility for its client's safety. Consequently, it must be tested in a variety of application scenarios before its release. Since those scenarios are not always safe for a human client and may be hard to observe during human-in-the-loop field tests, a virtual testing environment for Autominder has been developed (Gierke et al., 2006). Its layout is shown in Figure 8.3. The testing environment focuses on the model of an elderly person who acts as the Autominder user. Besides that, it includes a model of the client's living environment where Autominder is supposed to work. The Elderly model is designed as a human behavior representation that represents an elderly person in terms of actions and mental state (Gierke and Uhrmacher, 2005). Therefore, it provides two basic client functionalities: emulate suitable elderly behavior and react to incoming reminders.

Depending on its initial memory fitness and current stress level, the Elderly model may forget some of the plan actions. If the model receives an external reminder, it interrupts this default behavior as reminders should be evaluated with priority. However, there is no guarantee the Elderly model will cooperate with Autominder and execute the requested action. Depending on its initial cooperativeness and current degree of annoyance, the Elderly model may as well choose to ignore reminders.

In this scenario, we find agents as a metaphor for modeling, i.e., part of the virtual environment, the Elderly model, is described utilizing the agent metaphor HBR; and we find agents as a subject for being tested, i.e., the Autominder software. In many environments for which software agents are designed, human

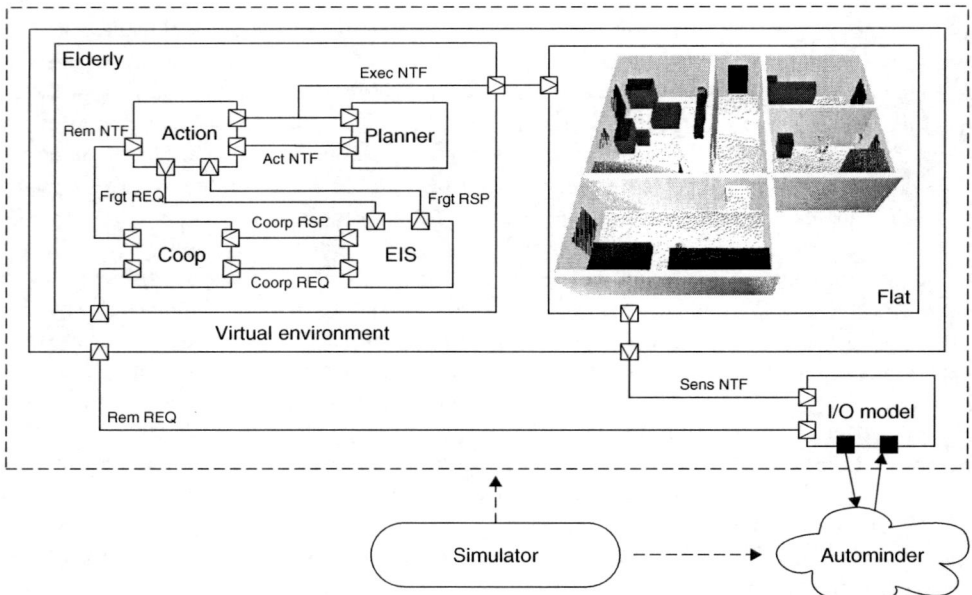

FIGURE 8.3 Testing of an agent application (Autominder) by coupling it to a simulated environment. (From Gierke, M., J. Himmelspach, M. Röhl, and A. Uhrmacher (2006). Modeling and simulation of tests for agents. In *German Conference on Multi-Agent System Technologies MATES'06*, to appear.)

beings play an important role and thus have to be taken into account when designing virtual environments for testing.

8.4 Conclusion

Agents are a metaphor used for designing software systems and for modeling dynamic systems that shall be evaluated by simulation. Multiagent systems as communities of interacting autonomous entities are supposed to function in dynamic environments. Thus, simulation is used as a means to evaluate and test agents. Accordingly, the relationship that ties agents and simulation is multifaceted.

The objective of the chapter has been to entangle the relations between agents and modeling against the background of simulation from two starting points: using the agent metaphor for modeling and modeling the virtual environment for analyzing agents. Thereby, the focus moved from the structure of single agents to their environments. These two approaches were further structured by the criteria whether analysis or design was the objective that motivated the modeling effort. As it turned out, many dependencies between these perspectives exist, which blur the distinction at some points. Figure 8.4 shows the introduced dimensions in relating agents and simulation from a modeling perspective.

For example, if we look at robots playing soccer as a possible application, a soccer simulator would help designing the software to run on the robots. However, current RoboCup simulators are not aimed toward this goal. Therefore, their models would need to include more physical and sensory details of the robots and the playing ground. Consequently, although simulators are used for designing soccer-playing robots, these are quite different from the simulators used in the RoboCup-Soccer league illustrating quite nicely that the objective influences the suitability of models. However, current approaches are aimed at enriching the RoboCup-Soccer simulator for supporting the design of soccer-playing robots, thus, bridging the gap between testing in the small and testing in the large, or simulation for understanding and simulation for design.

Agent-Oriented Modeling in Simulation

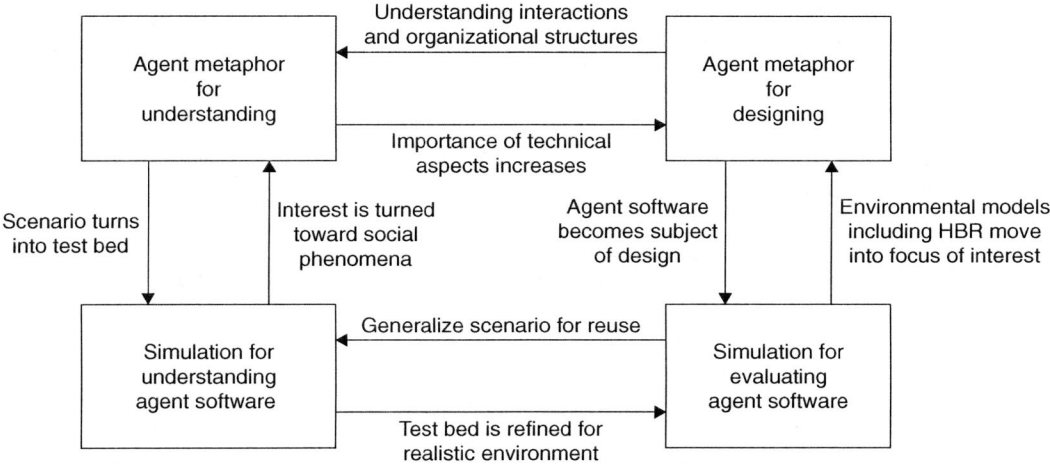

FIGURE 8.4 Interdependencies between agents for simulation and simulation for agents.

To analyze the strategies of multiagent systems, scenarios from ecology and sociology have often been used. Scenarios like RCR show that not only implemented strategies and the resulting emergent phenomena shall be analyzed by a virtual environment, but it is also hoped to learn more about decision structures in and the ability to cope with real disaster scenarios. In this case, modeling and simulation is not only used for agents, i.e., as a test bed to get a better understanding about coordination strategies when multiple heterogeneous agents are involved, but also the agent metaphor helps to understand dynamic systems, e.g., when dealing with real disaster scenarios, where the agent metaphor serves for modeling and simulating this system.

If technical systems like unguided vehicles in manufacturing systems are perceived as agents and described in the simulation based on the agent metaphor, their implementation is likely to be agent-based as well. Thus, as the implementation and design of the system progresses, we gradually turn from the view of using the agent metaphor during simulation, toward using simulation for analyzing the implemented agents. If, as in testing the agent software Autominder or information retrieval in P2P networks, human factors play a role, agent as a metaphor for modeling humans naturally become part of the virtual environment aimed at evaluating the agent software. In this context, the ideas of using agents as a metaphor for modeling and using modeling (and simulation) to evaluate agents are combined.

To conclude, let us summarize our discussion:

1. Agents are used as a *metaphor for modeling* a dynamic system.

    (a) The agent metaphor supports a better *understanding* of certain phenomena in social, ecological, or biological communities. It typically enriches traditional individual and multilevel modeling approaches by sophisticated internal decision models of the individuals, and thus helps to analyze phenomena that are supposed to depend on decision processes and preference models (e.g., urban disasters).

    (b) Whereas in sociology, simulation serves a better understanding, many technical areas depend on simulation for designing. In this area we find the agent metaphor as well. If the technical system can only be evaluated taking human behavior into account, agents are used as a metaphor to mimic the behavior of the humans, e.g., in the HBR approach. The goal is to have a realistic behavior pattern rather than to have a valid model of human decision and behavior processes, e.g., in traffic simulation. If the technical system shows characteristics of a multiagent system, i.e., a community of autonomously and concurrently interacting entities, the agent metaphor is used for describing this system, which brings it close to the relation described in Section 8.2.2 (e.g., manufacturing system).

2. The other dimension that relates agents and simulation is using *simulation for evaluating software agents*. This turns the focus of modeling from the agent toward the environment.

   (a) Small world testing presents a simplistic world in which properties of a concrete agent or a simplified version of an agent can be analyzed in isolation, a better *understanding of agents and their behaviors* being its purpose. Scenarios are often chosen by adopting social or biological systems. However, less the understanding of the social society but more the implications of the specific problem solving and cooperation strategies are the focus of interest (examples are TILEWORLD, RoboCup simulation league).

   (b) Testing in the large requires a valid model of the environment the agents are supposed to work in. The focus is on *evaluating* design alternatives of concrete *software agent applications*. If the environment comprises humans, they will likely be presented as agents. However, less with the intention to understand humans in their interaction with software, but with the intention to mimic the human behavior reasonably well to assess the performance of the software agent in interaction with humans (examples are P2P systems and Autominder).

By focusing on the modeling level, research areas like using agents for realizing simulation systems, e.g., in supporting data-driven, online simulations (Low et al., 2005) have not been considered. Similarly, the many challenges of an efficient and effective execution of multiagent systems and their environments have not been discussed, nor any of the steps toward addressing these challenges, e.g., by flexible simulation layers that support different types of executions and synchronization between agents and simulation (Himmelspach and Uhrmacher, 2004), or e.g., by an efficient handling of shared states in distributed, parallel execution (Lees et al., 2005), nor were the different simulation tools that exist, most of which originated in the realm of social simulations (Tobias and Hofmann, 2004), presented.

References

Agre, P. and D. Chapman (1987). PENGI: An implementation of a theory of activity. In *Sixth National Conference on Artificial Intelligence (AAAI-87)*, Seattle, WA, pp. 268–272.

Axtell, R. (2000). Why agents? On the varied motivations for agent computing in the social sciences. Working Paper No. 17.

Balmer, M., K. Nagel, and B. Raney (2004). Large-scale multi-agent simulations for transportation applications. *ITS Journal 8*, 1–17.

Batty, M. (2005). *Cities and Complexity: Understanding Cities with Cellular Automata, Agent-Based Models, and Fractals*. London: MIT Press.

Baumgaertel, H., S. Brueckner, V. Parunak, R. Vanderbok, and J. Wilke (2003). Agent models of supply network dynamics. In T. P. Harrison, H. L. Lee, and J. J. Neale (Eds.), *The Practice of Supply Chain Management*, International Series on Operations Research and Management Science. New York: Kluwer.

Brooks, R. A. (1991). Intelligence without reason. In J. Myopoulos and R. Reiter (Eds.), *Proceedings of the 12th International Joint Conference on Artificial Intelligence (IJCAI-91)*, Sydney, Australia, pp. 569–595. Morgan Kaufmann: San Mateo, CA, USA.

Bunge, M. (1979). *Ontology II: A World of Systems*, Volume 4 of *Treatise of Basic Philosophy*. Dordrecht: Reidel.

Casagni, M. and M. Lyell (2003). Comparison of two component frameworks: The fipa-compliant multi-agent system and the web-centre j2ee platform. In *ICSE '03: Proceedings of the 25th International Conference on Software Engineering*, Washington, DC, USA, pp. 341–351. IEEE Computer Society.

Chen, S., S. Ganguli, and C. A. Hunt (2004). An agent-based computational approach for representing aspects of in vitro multi-cellular tumor spheroid growth. In *Proceedings of the 26th Annual International Conference of the IEEE EMBS*, pp. 691–694.

Clark, M., K. Irvig, and W. Wobcke (1997). Emergent properties of teams of agents in the Tileworld. In L. Cavedon, A. Rao, and W. Wobcke (Eds.), *Intelligent Agent Systems*, Volume 1209 of Lecture Notes in Artificial Intelligence, pp. 164–176. Berlin: Springer.

Conte, R. and C. Castelfranchi (Eds.) (1995). *Cognitive and Social Action*. London: UCL Press.

Conte, R., B. Edmonds, S. Moss, and R. K. Sawyer (2001). Sociology and social theory in agent based social simulation: A symposium. *Computational & Mathematical Organization Theory 7*(3), 183–205.

David, N., J. S. Sichman, and H. Coelho (2003). Towards an emergence-driven software process for agent-based simulation. In J. Sichman, F. Bousquet, and P. Davidsson (Eds.), *MABS 2002*, Volume 2581 of Lecture Notes in Artificial Intelligence, pp. 89–104. Berlin: Springer.

Doran, J. (1997). From computer simulation to artificial societies. *SCS Transaction on Computer Simulation 14*(2), 69–78.

Doran, J., N. Gilbert, U. Müller, and K. Troitzsch (Eds.) (1996). *Object-Oriented and Agent-Oriented Simulation – Implications for Social Science Applications*, Lecture Notes in Economics and Mathematical Systems, Berlin: Springer.

Doran, J., M. Palmer, N. Gilbert, and P. Mellars (1994). The EOS project: Modeling upper paleolithic social change. In N. Gilbert and J. Doran (Eds.), *Simulating Societies*, pp. 195–223. London: UCL Press.

Edmonds, B. and M. Möhring (2005). Agent-based simulation modeling in social and organizational domains. *Simulation 81*(3), 173–174.

Ephrati, E., M. Pollack, and S. Ur (1995). Deriving multi-agent coordination through filtering strategies. In *14th International Joint Conference on Artificial Intelligence*, pp. 679–687. Morgan Kaufmann: San Mateo, CA.

Etzioni, O. (1993). Intelligence without robots: A reply to brooks. *AI Magazine 14*(4), 7–13.

Ewert, U. C., M. Röhl, and A. M. Uhrmacher (2003). What good are deliberative interventions in large scale disasters? Exploring the consequences of crisis management in pre-modern towns with agent-oriented simulation. In *Agent Based Computational Demography*. Berlin: Physica Verlag (Springer).

Ferber, J. (1999). *Multi-Agent System: An Introduction to Distributed Artificial Intelligence*. Boston, MA: Addison-Wesley.

Gierke, M., J. Himmelspach, M. Röhl, and A. Uhrmacher (2006). Modeling and simulation of tests for agents. In *German Conference on Multi-Agent System Technologies MATES'06*. LNCS Volume 4196, pp. 49–60. Heidelberg: Springer.

Gierke, M. and A. M. Uhrmacher (2005). Modeling elderly behavior for simulation-based testing on agent software. In *Proceedings of the Conceptual Modeling and Simulation Conference (CMS 2005)*. San Diego, CA: SCS.

Gilbert, N. (2005). When does social simulation need cognitive models? In R. Sun (Ed.), *Cognition and Multi-Agent Interaction: From Cognitive Modeling to Social Simulation*. Cambridge: Cambridge University Press.

Gilbert, N. and K. G. Troitzsch (2005). *Simulation for the Social Scientist* (2nd ed.). Maidenhead: Open University Press.

Goldman, G. and J. Rosenschein (1994). Emergent coordination through the use of cooperative state-changing rules. In *National Conference on Artificial Intelligence*, Seattle, WA, pp. 432–437. Menlo Park, CA: American Association for Artificial Intelligence.

Hanks, S., M. E. Pollack, and P. R. Cohen (1993). Benchmarks, test beds, controlled experimentation and the design of agent architectures. *AAAI*(Winter), 17–42.

Helbing, D. and K. Nagel (2004). The physics of traffic and regional development. *Contemporary Physics 45*(5), 405–426.

Helleboogh, A., G. Vizzari, A. Uhrmacher, and F. Michel (2007). Modeling dynamic environments in multi-agent simulation. *Journal of Autonomous Agents and Multi-Agent Systems 14*(1), 87–116.

Himmelspach, J. and A. M. Uhrmacher (2004). A component-based simulation layer for James. In *PADS '04: Proceedings of the Eighteenth Workshop on Parallel and Distributed Simulation*, Kufstein, Austria, pp. 115–122. New York: IEEE Computer Society Press.

Huebsch, R., J. M. Hellerstein, N. Lanham, B. T. Loo, S. Shenker, and I. Stoica (2003). Querying the internet with pier. In *VLDB*, pp. 321–333.

Jennings, N. R., K. Sycara, and M. Wooldridge (1998). A roadmap of agent resarch and development. *Autonomous Agents and Multi-Agent Systems 1*(1), 275–306.

Joseph, S. (2003). An extendible open source P2P simulator, *P2P Journal*, pp. 1–15.

Kinny, D. and M. Georgeff (1991). Commitment and effectiveness of situated agents. In *Proceedings of the International Joint Conference*, pp. 82–88. IJCAI, Morgan Kaufmann: San Mateo, CA.

Kitano, H. and S. Tadokoro (2001). RoboCup rescue: A grand challenge for multiagent and intelligent systems. *AI Magazine 22*(1), 39–52.

Kitano, H., S. Tadokoro, H. Noda, I. Matsubara, T. Takhasi, A. Shinjou, and S. Shimada (1999). Robocup-rescue: Search and rescue for large scale disasters as a domain for multi-agent research. In *Proceedings of the IEEE Conference on Systems, Men, and Cybernetics*.

Kitano, H., M. Tambe, P. Stone, M. Veloso, S. Coradeschi, E. Osawa, H. Matsubara, I. Noda, and M. Asada (1997). The RoboCup synthetic agent challenge 1997. In *International Joint Conference on Artificial Intelligence IJCAI'97*. Morgan Kaufmann: San Mateo, CA.

Klein, M. (2003). DIANEmu: A java based generic simulation environment for distributed protocols. Technical Report 2003–7, University of Karlsruhe.

Klügl, F. and A. L. C. Bazzan (2004). Route decision behaviour in a commuting scenario: Simple heuristics adaptation and effect of traffic forecast. *JASSS 7*(1).

Knorr-Cetina, K. and A. Cicourel (Eds.) (1981). *Advances in Social Theory and Methodology—Toward an Integration of Micro and Macro Sociologies*. Boston: Routledge and Kegan Paul.

König-Ries, B., M. Klein, and T. Breyer (2004). Activity-based user modeling in wireless networks. *Mobile Networks and Applications*. Special Issue on Internet Wireless Access: 802.11 and Beyond.

Lees, M., B. Logan, R. Minson, T. Oguara, and G. Theodoropoulos (2005). Distributed simulation of MAS. In P. Davidson, B. Logan, and K. Takadama (Eds.), *AAMAS 2004*, Volume 3415 of Lecture Notes in Computer Science, pp. 25–36. Berlin: Springer.

Low, M. Y. H., K. W. Lye, P. Lendermann, S. J. Turner, R. T. W. Chim, and S. H. Leo (2005). An agent-based approach for managing symbiotic simulation of semiconductor assembly and test operation. In *AAMAS '05: Proceedings of the Fourth International Joint Conference on Autonomous Agents and Multiagent Systems*, New York, NY, pp. 85–92. ACM Press.

Luck, M., N. Griffiths, and M. d'Inverno (1997). From agent theory to agent construction: A case study. In J. Mueller, M. Wooldridge, and N. Jennings (Eds.), *Third International Workshop on Agent Theories, Architectures, and Languages*, Volume 1193 of Lecture Notes in Artificial Intelligence, London, pp. 49–63. Berlin: Springer.

Melo, F., R. Choren, R. Cerqueira, C. Lucena, and M. Blois (2004). Deploying agents with the CORBA component model. In W. Emmerich and A. Wolf (Eds.), *CD 2004*, Volume 3083 of Lecture Notes in Computer Science, pp. 234–247. Berlin: Springer.

Minar, N., K. H. Kramer, and P. Maes (1999). Cooperating mobile agents for dynamic network routing. In A. L. G. Hayzelden and J. Bigham (Eds.), *Software Agents for Future Communication Systems*, pp. 287–304. Heidelberg, Germany: Springer.

Moro, G., S. Bergamaschi, and K. Aberer (Eds.) (2005). *Agents and Peer-to-Peer Computing*, Volume 3601 of Lecture Notes in Computer Science. Berlin: Springer.

Negnevitsky, M. (2002). *Artificial Intelligence: A Guide to Intelligent Systems*. Harlow, England: Addison-Wesley.

Odell, J. J. (2002). Objects and agents compared. *Journal of Object Technology 1*(1), 41–53.

O'Hare, G. M. P. and N. R. Jennings (Eds.) (1996). *Foundations of Distributed Artificial Intelligence*. New York, NY: Wiley.

Parunak, H. V. D. (1997). "Go to the ant": Engineering principles from natural agent systems. *Annals of Operation Research 75*, 69–101.

Parunak, H. V. D. (1999). From chaos to commerce: Practical issues and research opportunities in the nonlinear dynamics of decentralized manufacturing systems. In *Proceedings of Second International Workshop on Intelligent Manufacturing Systems*, pp. k15–k25.

Pew, R. W. and A. S. Mavor (Eds.) (1997). *Representing Human Behavior in Military Simulations: Interim Report*. Washington, DC: National Academy Press.

Pew, R. W. and A. S. Mavor (Eds.) (1998). *Modeling Human and Organizational Behavior: Application to Military Simulations*. Washington, DC: National Academy Press.

Philips, A. and J. Bresina (1991). NASA TILEWORLD manual. Technical Report TR_FIA_91-04, MASA Ames Research Center, Mountain View, CA.

Pollack, M. E., L. Brown, D. Colbry, C. E. McCarthy, C. Orosz, B. Peintner, S. Ramakrishnan, and I. Tsamardinos (2003). Autominder: An intelligent cognitive orthotic system for people with memory impairment. *Robotics and Autonomous Systems 44*, 273–282.

Pollack, M. E. and M. Ringuette (1990). Introducing the TILEWORLD: Experimentally evaluating agent architectures. In *AAAI-90*, Boston, MA, pp. 183–189.

Pretschner, A., O. Slotosch, E. Aiglstorfer, and S. Kriebel (2004). Model-based testing for real—The inhouse card case study. *International Journal on Software Tools for Technology Transfer (STTT) 5*(2–3), 140–157.

Rao, A. S. and M. P. Georgeff (1995). BDI agents: From theory to practice. Technical Report 56, Australian Artificial Intelligence Institute.

Rescue Technical Committee, T. R. (2000). RoboCup Rescue simulator manual. Ver. 0, Rev. 4.

Röhl, M. and A. M. Uhrmacher (2005). Controlled experimentation with agents—Models and implementations. In M.-P. Gleizes, A. Omicini, and F. Zambonelli (Eds.), *Post-Proceedings of the 5th Workshop on Engineering Societies in the Agents World*, Volume 3451 of Lecture Notes in Artificial Intelligence, pp. 292–304. Berlin: Springer.

Russell, S. J. and P. Norvig (2003). *Artificial Intelligence: A Modern Approach* (2nd ed.). New Jersey: Prentice-Hall.

Shoham, Y. (1993). Agent-oriented programming. *Artificial Intelligence 60*, 51–92.

Silverman, B. G. (2003). *Metrics and Methods in Human Performance Research toward Individual and Small Unit Simulation*, Chapter 9 Human Performance Simulation. Washington, DC, Human Systems Information Analysis Center.

Sim, S. E., S. Easterbrook, and R. C. Holt (2003). Using benchmarking to advance research: A challenge to software engineering. In *ICSE '03: Proceedings of the 25th International Conference on Software Engineering*, Washington, DC, USA, pp. 74–83. IEEE Computer Society.

Tambe, M., J. Adibi, Y. Alonaizon, A. Erdem, G. Kaminka, S. Marsella, and I. Muslea (1999). Building agent teams using an explicit teamwork model and learning. *Artificial Intelligence 110*(2), 215–239.

Tanenbaum, A. S. and M. van Steen (2002). *Distributed Systems: Principles and Paradigms*. New York: Prentice-Hall.

Tobias, R. and C. Hofmann (2004). Evaluation of free java-libraries for social-scientific agent based simulation. *Journal of Artificial Societies and Social Simulation 7*(1).

Uhrmacher, A., P. Fishwick, and B. Zeigler (Eds.) (2001). Special Issue: Agents in Modeling and Simulation; Exploiting the Metaphor, Volume 89 of *Proceedings of the IEEE*, pp. 127–213.

Uhrmacher, A. M. (1997). Concepts of object- and agent-oriented simulation. *Transactions on SCS 14*(2), 59–67.

Weiss, G. (Ed.) (2000). *Multiagent Systems: A Modern Approach to Distributed Artificial Intelligence*. Cambridge, MA: MIT Press.

Weyns, D., A. Helleboogh, and T. Holvoet (2005a). The packet-world: A test bed for investigating situated multi-agent systems. In R. Unland, M. Klusch, and M. Calisti (Eds.), *Software Agent-Based Applications, Platforms and Development Kits*, Whitestein Series in Software Agent Technology.

Weyns, D., K. Schelfthout, T. Holvoet, and T. Lefever (2005b). Decentralized control of E'GV transportation systems. In *Proceedings of AAMAS'05*, pp. 67–74. ACM Press.

Wise, B. P., M. McDonald, L. M. Reuss, and J. Aronson (2001). Task order (TO) 69: ATM human behavior modeling approach study. Technical report, National Air and Space Administration (NASA), Arlington, VA. Technical Advance in Air Transportation Concepts and Technologies (AATT).

Wooldridge, M. (2000). *Multiagent Systems: A Modern Approach to Distributed Artificial Intelligence*, Chapter Intelligent Agents. Cambridge, MA: MIT Press.

Wooldridge, M. (2002). *An Introduction to MultiAgent Systems*. Chichester, England: Wiley.

Wooldridge, M. and N. Jennings (1998). Pitfalls of agent-oriented development. In *Proceedings of the 2nd International Conference on Autonomous Agents (Agents-98)*, Minneapolis.

Wooldridge, M. and D. Vandekerckhove (1994). MyWorld: An agent-oriented testbed for distributed artificial intelligence. In S. Deen (Ed.), *CKBS-93 Proceedings of the 1993 Special Interest Group on Cooperating Knowledge-Based Systems*, Dake Centre, University of Keele, UK.

Yilmaz, L. and T. Ören (2005). Special issue on agent-directed simulation. *Simulation 81*(7), 463–464.

9
Distributed Modeling

Simon J. E. Taylor
Brunel University

9.1 Introduction 9-1
9.2 Modeling with COTS Simulation Packages 9-2
9.3 Distributed Simulation 9-3
9.4 CSP-Based Distributed Simulation 9-5
   The Problem of CSP-Based Distributed Simulation •
   CSP-Based Distributed Simulation: Current Progress
9.5 A Standards-Based Approach 9-7
   The Simulation Interoperability Standards Organization •
   Emerging Standards and the CSPI-PDG • The Type I
   Interoperability Reference Model • The Entity
   Transfer Specification
9.6 Case Study 9-13
   Illustrative Protocol
9.7 Conclusion 9-16

9.1 Introduction

Other chapters in this book discuss various aspects of dynamic systems modeling. The purpose of this chapter is to present an introduction to some of the contemporary innovations in the use of distributed computing techniques to support modeling. This is called *distributed simulation* and can be defined as the distribution of the execution of a single run of a simulation program across multiple processors (Fujimoto, 2000). The various motivations for this include: the reduction of the execution time of a single simulation run, the use of multiple computers to support the memory needs of the simulation when one computer cannot, and the linking of simulations sited in different locations (Fujimoto, 2003). A cursory examination of the many Winter Simulation Conferences (WSC) (www.informs-cs.org and www.acm.org/dl), the Principles of Advanced and Distributed Simulation (PADS) (www.acm.org/dl), and Simulation Interoperability Workshops (SIW) (www.sisostds.org) show the wide applications and issues of distributed simulation. By way of introduction, and to give focus to this work, we restrict our discussion to the use of distributed simulation in the field of modeling associated with the use of *Commercial-off-the-shelf (COTS) Simulation Packages* (CSPs). CSPs are widely used in industry. In this chapter, we term the combination of distributed simulation and CSPs, CSP-based distributed simulation.

In addition to the above general reasons to use distributed simulation, additional reasons to use CSP-based distributed simulation include the interoperation of discrete-event simulations across virtual organizations, extended enterprises, and supply chains; the reduction of the cost of model development by enabling the reuse of distributed model components; and the protection of intellectual property (information hiding in distributed models) (Gan et al., 2000; Mertins and Rabe, 2002; Paul and Taylor, 2002; Robinson et al., 2004). Additionally, variants of distributed simulation techniques can also reduce the time taken for simulation experimentation (distributed replication and experimentation) and reduce simulation project costs (remote model execution, group working) (Robinson, 2005; Taylor et al., 2005).

Although there are excellent examples of successful distributed simulations with CSPs (in particular, Boer et al. [2002a] and Mertins et al. [2000]), a general solution to this area is illusive. In general dynamic systems modeling this means that this potentially highly useful technology cannot be used without significant cost.

In this chapter we consider why this is the case and review some of the new standards-based approaches that are currently being developed. The chapter is structured as follows. First, the notion of the CSP is explored in more depth. Distributed simulation is then introduced. The problems of CSP-based distributed simulation and the current progress of research in this area is then considered. The chapter then introduces a standards-based approach to the solution of the problems faced by CSP-based distributed simulation. Within this the key roles of the Simulation Interoperability Standards Organization (SISO) and the COTS Simulation Package Interoperability Product Development Group (CSPI-PDG) in standardization are discussed. Detail is then given on one research "target" in this area, the Type I interoperability reference model (IRM). A case study outlining its use is then presented to show how progress in this area is being made.

9.2 Modeling with COTS Simulation Packages

Discrete-event simulation (DES) is a computer-based dynamic systems modeling technique typically used to model and investigate the behavior of complex, dynamic systems (Banks, 1998; Pidd, 1998; Robinson, 2004). *Discrete event* refers to the type of simulation that models a system in terms of state variables that change instantaneously at separated points in time (*events*) as opposed to continuous change (*continuous simulation*) (Law and Kelton, 2000). As with most modeling techniques, DES can be used to support system analysis, education and training, acquisition and system acceptance, research and planning, organizational change, and facilitation (Nance and Sargent, 2002; Robinson, 2002) in a range of diverse areas such as commerce (Bosilj-Vuksic et al., 2003), defense (Hofmann, 2004), health care (Eldabi et al., 2000), manufacturing (Bruzzone, 2003), supply chains (Goel et al., 2002), civil (Demirci, 2003), and maritime transportation (Lee et al., 2004). Visual interactive simulation has played an important role in DES for around 25 years (Bell, 1985; Bell and O'Keefe, 1987; Hurrion, 1998). We use the term *commercial-off-the-shelf discrete-event simulation packages* (CSPs) to describe commercially available software tools that have been developed from visual interactive simulation to facilitate the process of DES and to provide a distinction from other similar modeling approaches such as those based on Petri nets (Peterson, 1981) or systems dynamics (Lane, 1999). Examples of CSPs include ProModel (Harrell and Price, 2003), Arena (Bapat and Sturrock, 2003), AutoMod (Rohrer, 2003), Simul8 (www.simul8.com), Extend (Krahl, 2003), and Witness (www.lanner.com). Some of these packages support other modeling techniques; we restrict ourselves to DES in this discussion.

CSPs support environments use visual programming approaches that allow simulation modelers to build discrete-event models using drag and drop interfaces and provide a range of facilities for DES (e.g., 2/3D animation and visualization, replication control, experimentation and statistical analysis utilities, and optimization support) All support DES in that each CSP supports the building of models that change state at *events*. Generally, such DES models are typically composed of networks of alternating *queues* and *activities* that represent, for example, the series of buffers and operations composing a manufacturing system. *Entities*, consisting of sets of typed variables termed as *attributes*, represent the elements of the manufacturing system undergoing machining. Sometimes the term is used to refer to a class, i.e., an entity "Job" might refer to the class of jobs that require machining. Each individual entity, each individual "Job" can be distinguished by attributes. In this chapter, we use the term to refer to a collection of items (hence entities "Jobs" and entity "Job"). Entities are transformed as they pass through these networks and may enter and exit the model at specific points. Additionally, activities may compete for *resources* that represent, for example, the operators of the machines. Resources tend to be passive and elements of the model "compete" for them. For example, a resource might be used to represent a collection of operators—when a machine wishes to begin processing an entity it might request an operator from the operator resource. If there are any, an operator is assigned to the machine. If there are none, then the machine must wait

Distributed Modeling

for an operator to become available. To simulate a model a CSP will typically have a simulation executive, an event list, a clock, a simulation state, and a number of event routines. The simulation state and event routines are derived from the simulation model. The simulation executive is the main program that (generally) simulates the model by first advancing the simulation clock to the time of the next event and then performing all possible actions at that simulation time. For example, this may change the simulation state (e.g., ending a machining activity and placing an entity in a queue) or schedule new events (e.g., a new entity arriving in the simulation). This cycle carries on until some terminating condition is met.

Virtually every CSP has a different variant of the above. Each is based on a variant of a simulation *worldview*. A worldview, or conceptual framework, is "… a structure of concepts and views under which the simulationist [modeler] (developer) is guided for the development of a simulation model" (Balci, 1988). The most well known of these are event scheduling, activity scanning (Buxton and Laski, 1962), the three-phase approach (Tocher, 1963), and process interaction. In the 1960s, these gave rise to simulation programming languages such as GPSS, SIMAN, SIMSCRIPT, SIMULA, and SLAM. Many of these were the predecessors to the CSPs used today (Nance, 1996). In addition to these different worldviews, CSPs have widely differing terminology, representation, and behavior (Schriber and Brunner, 2003). For example, without reference to a specific CSP, in one CSP an entity as described above may be termed as an *item* and in another as *object*. In the first CSP, the datatypes might be limited to integer and string, while in the other the datatypes might be the same as those in any object-oriented programming language. The same observations are true for the other model elements of queue, activity, resource, and entry and exit point. Behavior is also important as the set of rules that govern the behavior of a network of queues and activities subtly differ between CSPs (e.g., the rules that govern behavior when an entity leaves a machine to go to a buffer). Indeed, even the representation of *time* can differ. This is also further complicated by variations in model elements over and above the "basic" set (e.g., transporters, conveyors, flexible manufacturing cells, and robots).

The consequence of this is a point that many researchers new to this area miss; it is entirely plausible to argue that there are now as many worldviews as there are CSPs. This presents a substantial challenge to the field of distributed simulation. Let us now consider general progress in distributed simulation and in particular CSP-based distributed simulation.

9.3 Distributed Simulation

The IEEE 1516 standard *The High Level Architecture* (HLA) (IEEE 1516, 2000) is a general standard that supports distributed simulation. This, and its predecessor the IEEE 1278 standard *Distributed Interactive Simulation* (DIS) (IEEE, 1995), came from the need of the US Department of Defense (DoD) to reduce the cost of training military personnel by reusing computer simulations linked via a network, i.e., through the creation of distributed simulations of real-time military applications.

The DIS standard described the format of data exchanged by simulators linked over a network for military applications. The limited domain of DIS (military, real-time applications) and technical problems, such as time management and limited bandwidth, led to the creation of the HLA. In the HLA, a distributed simulation is called a *federation*, and each individual simulator (in our case the combination of a CSP and its model) is referred to as a *federate*. A HLA *Runtime Infrastructure* (RTI) provides facilities to enable federates to interact with one another, as well as to control and manage the simulation. The HLA is composed of four parts: a set of rules (IEEE 1516.0, 2000), the Object Model Template (OMT) (IEEE 1516.1, 2000), the Federate Interface Specification (FIS) (IEEE 1516.2, 2000), and the federate development Process (FEDEP) (IEEE 1516.3, 2004). The rules are a set of 10 basic conventions, which define the responsibilities of federates and their relationship with the RTI. The FIS is an application interface standard for distributed simulation middleware, which defines how federates interact within the federation, and is implemented by an RTI. The OMT provides a common presentation format for HLA federates. Using the OMT, each federate defines, in its simulation object model (SOM), the data that it is willing to share (publish) with other federates and the data it requires from other federates (subscribe). The federation object model (FOM) combines the federate SOMs into a single object model for the federation and therefore defines the overall data to be exchanged (published and subscribed) between federates during a simulation execution. The FEDEP defines the recommended practice processes and procedures that should be followed by users of the HLA to develop and execute their federations.

Federates do not communicate with one another directly. Instead, they exchange information only using the services provided by the RTI. Each federate has an *RTIambassador* and a *FederateAmbassador*. A federate invokes an operation on the *RTIambassador* whenever it needs an RTI service (e.g., a request to advance simulation time). In the reverse direction, the RTI invokes an operation on the *FederateAmbassador* whenever it needs to pass data to the federate (e.g., to inform the federate that the request to advance simulation time has been granted). Thus, operations in the *FederateAmbassador* need to be implemented by the federate, as part of the federate code or as part of some interface service. As defined by the FIS, an RTI provides six classes of services:

- *Federation management*: These services allow federates to create and destroy federation executions, and join or resign from an existing federation.
- *Declaration management*: These services allow federates to publish federate data and to subscribe to updated data produced by other federates.
- *Object management*: These services allow federates to create and delete object instances, and produce and receive data.
- *Ownership management*: These services allow federates to transfer the ownership of object data during the federation execution.
- *Time management*: These services coordinate the advancement of simulation time of the federates.
- *Data distribution management* (DDM): These services can reduce unnecessary information transfer between federates by filtering out irrelevant data.

This overcame the shortcomings of DIS by being simulation-domain neutral (the OMT) and specifying functionality for time management and bandwidth control (in the FIS modules). In terms of heterogeneity, the HLA therefore provides facilities to describe any data exchange format as required. Specifically, the OMT provides neutral data representation types that are mapped to/from the RTI. These are the basic representation types of HLAinteger16/32/64BE/LE, HLAfloat32/64BE/LE, HLAoctetPairBE/LE, and HLAoctet (16/32/64 represents bit size, BE/LE represents big/little endian representation); the simple data representation types of HLAASCIIchar, HLAunicodeChar, and HLAbyte; user-defined enumerated types (including HLAboolean represented as a HLAinteger32BE with possible values of 0 and 1); the array datatype representation types of HLAASCIIstring, HLAunicodeString, and HLAopaqueData (uninterpreted); user-defined array types; user-defined fixed record datatypes; and user-defined variant record datatypes. Apart from datatypes, the OMT also provides 13 tables that are used to define various different aspects of the SOMs and FOM of a distributed simulation using the HLA. These are

- *Object model identification table*: This associates important identifying information with a HLA object model (SOM/FOM).
- *Object class structure table*: This records the namespace of all federate or federation object classes and describes their class–subclass relationships.
- *Interaction class structure table*: An interaction is a type of data exchange that models "(a)n explicit action taken by a federate that may have some effect or impact on another federate within a federation execution." This table records the namespaces of all federate or federation interaction classes and describes their class–subclass relationships.
- *Attribute table*: An attribute is a type of data exchange that models "(a) named characteristic of an object class or object instance" and is semantically different to attributes mentioned in our discussion of CSPs. This table specifies the object attributes in a federate or federation that can be exchanged.

- *Parameter table*: This specifies the parameters of interaction classes in a federate or federation.
- *Dimension table*: This specifies the dimensions used to filter instance attributes and interactions (used in DDM).
- *Time representation table*: This is used to specify the common representation of time values.
- *User-supplied tag table*: This specifies the representation of tags used in HLA services.
- *Synchronization table*: This specifies the representation and datatypes used in HLA synchronization services (typically used to synchronize the federation at the start and end of the simulation).
- *Transportation type table*: This table describes the transportation mechanisms used in the federation (essentially following UDP and TCP semantics).
- *Switches table*: This specifies the initial settings for parameters used by the RTI.
- *Datatype tables*: This specifies details of data representation in the object model (as described above).
- *Notes table*: This table expands explanations of any OMT table item as required.
- *FOM/SOM lexicon*: This defines all of the objects, attributes, interactions, and parameters used in the HLA object model.

What progress has been made in using this standard or other nonstandard approaches to support CSP-based distributed simulation? In other words, how do we use the complexity of the HLA?

9.4 CSP-Based Distributed Simulation

In this section we consider the specific problems that CSP-based simulation faces and the progress that has been made toward a general solution.

9.4.1 The Problem of CSP-Based Distributed Simulation

Consider the simple distributed simulation of Figure 9.1 In our discussion, a distributed simulation (federation) is composed of CSPs and their models (federates) that exchange data (interactions and attributes) via an RTI in a time synchronized manner. Two factories, F1 and F2, interact in various ways as denoted by the black double-headed arrow. Each model consists of an arrival source So_i, a queue Q_i, a workstation W_i, a resource R_i, and an exit sink Si_i (where i is the factory identifier). There are various types of model information that we might share. For example, entities might be passed between models (i.e., the two factories are linked together—entities leave F1 at Si1 and arrive in F2 at So2) and the resources R1 and R2 might be shared to reflect a shared set of operators that can operate workstations W1 and W2. If this was the case, factory F1 must publish and send information to the RTI in an agreed format and time synchronized manner and factory F2 must subscribe to and receive that information in the

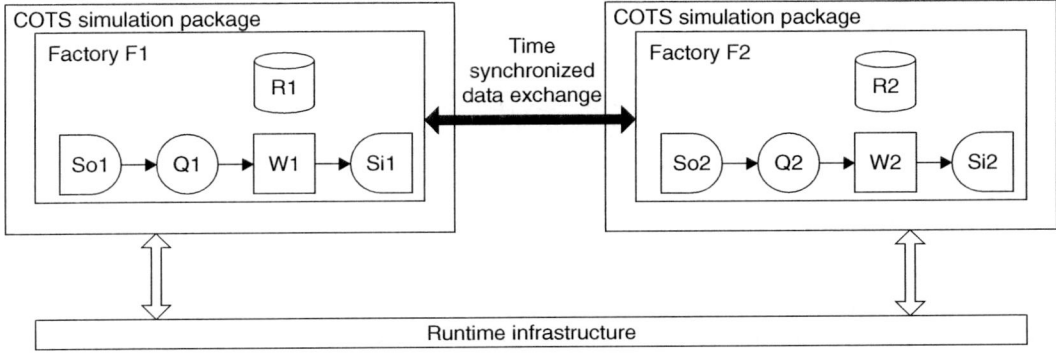

FIGURE 9.1 Simple distributed simulation.

same agreed certain format and time synchronized manner, i.e., both federates must agree on a common representation of data and both must use the RTI in a similar way. Further, the "passing" of entities and the sharing of resources require different distributed simulation protocols. In entity passing, the departure of an entity at a sink and the arrival of an entity at a source are effectively the same scheduled event in the two models—most distributed simulations represent this as a timestamped event message sent from one federate to another (with the timestamp typically equal to the time that the entity finished processing in the last workstation [W1 in our example]) (Boer et al., 2002b; Sudra et al., 2000). The sharing of resources cannot be handled in the same way. For example, when resource (R1) is released or an entity arrives in queue Q1, a CSP executing the simulation of F1 will determine if workstation W1 can start processing an entity. If resources are shared, then each time R1 or R2 changes state, a timestamped communication protocol is required to inform and update the changes of the shared resource state (Low et al., 2001).

Our heterogeneous distributed CSP integration problem therefore consists of several parts. These are

- what are the synchronization demands of data exchanged between federates?
- how should these be implemented through an RTI?
- what format should the data take and what relationship should this have to the CSPs and their models?

Let us now consider current progress to a solution to these questions.

9.4.2 CSP-Based Distributed Simulation: Current Progress

Although initial work on the use of the HLA to integrate heterogeneous distributed CSPs can be traced back to pioneering work done by Straßburger in the late 1990s (Straßburger, 2001), this area is still emerging (Taylor et al., 2003). Research has mainly focused on technological challenges using combinations of various CSPs and HLA-based and non-HLA-based approaches. Mertins et al. (2000), Rabe and Jäkel (2001, 2003), Hibino et al. (2002), McLean and Riddick (2000), and Linn et al. (2002) discuss the use of the HLA and the associated adapter technologies of the MISSION project to support the distributed simulation of manufacturing systems. Lenderman et al. (2001) and Straßburger et al. (2003) also discuss strategies for HLA use in the same domains. In terms of non-HLA approaches the following contributions have been made. Sudra et al. (2000) and Taylor et al. (2002) discuss the use of the *generic runtime infrastructure for distributed simulation* (GRIDS) to support the distributed simulation of supply chains and automotive engine production. Fuji et al. (2000) present an approach to the distributed simulation of virtual factories. Zülch et al. (2002) show how distributed simulations of manufacturing systems can be composed hierarchically. Gan et al. (2000) compare HLA against an MPI-based implementation extended from the protocol described in Gan and Turner (2000). Boer et al. (2002b) discuss the use of distributed simulation to link to real-time data sources using the FAMAS backbone. Finally, Boer et al. (2002b) also discuss the use of the same technology to support the distributed simulation of a port.

In terms of the questions posed above, what contributions have the above made? Let us summarize:

- what are the synchronization demands of data exchanged between federates?
  - Many!
- how could these be implemented through an RTI?
  - In many different, incompatible ways!
- what format could the data take and what relationship should this have to the CSPs and their models?
  - Many, different, incompatible forms!

In short, although much of the above work has led (to various degrees) to some successful solutions, the solutions themselves are incompatible. While each is excellent in its own context, the lack of a standardized approach means it is difficult for end users and CSP vendors to choose a solution. A standardized approach means that there is only (usually) one approach to select. Let us now consider how a standardized approach can be created in this area.

9.5 A Standards-Based Approach

In this section we first consider the organization responsible for the development of standards in the area of distributed simulation. We then consider the emerging standards being developed in this area before specifically considering two.

9.5.1 The Simulation Interoperability Standards Organization

It is the role of SISO to oversee the process of standards making in HLA-based distributed simulation. The HLA was initially developed in conjunction with the IEEE via the Simulation Interoperability Standards Committee (SISC) with US DoD support. In 2003, SISC was disbanded and replaced with (SISO's) Standards Activity Committee (SAC). SISO is led by its Executive Committee (EXCOM) that oversees work performed by SAC. The SAC in turn oversees the various product development groups (PDGs). A PDG is a working party responsible for developing a specific SISO product (a SISO term for *standard*). Products are developed according to a *balloted product development process* (BPDP) that describes a six-stage process: activity approval, product development, ballot product, product approval, distribution and configuration management, and periodic review. A prestage involves the development and submission of a document called a *product nomination*, a proposal for a PDG and the products intended for development. During the life of a PDG, members vote at various stages of the BPDP to accept, modify, or reject products under development. The process so seeks to reflect consensus agreements to create stable, well-understood, technically competent standards that have significant public support. An example of a SISO product is the RPR FOM, a HLA equivalent representation of DIS. Others are under development for military applications such as Link-16 and C4ISR by PDGs such as PDG-LINK16 and PDG-C4ISRTRM.

Essentially the main work of SISO's PDGs is to develop standards to support different distributed simulation domains. We now discuss emerging standards to support CSP-based distributed simulation.

9.5.2 Emerging Standards and the CSPI-PDG

In August 2002, the High Level Architecture COTS Simulation Package Interoperability Forum (HLA-CSPIF) was created in an attempt to produce a generalizable solution to the problem of distributed heterogeneous CSP integration. Over 2 years, discussions led by the Forum resulted in the splitting up of the integration problem into different requirements. The rationale is this. If we consider all possible distributed simulation requirements in this area three important observations can be made:

- not all distributed simulations need all integration requirements;
- some integration solutions are relatively straightforward and some are extremely complex; and
- not all integration requirements are known.

In the simple example of Figure 9.1, some distributed simulations only require entities to be passed between them. The problem of entity passing is somewhat simpler than the problem of synchronous shared state in the case of resource sharing. The issue of not being able to know all integration requirements has been demonstrated by the experiences of the Forum. Entity passing and resource sharing were the first requirements that were identified. However, the requirements to integrate models with shared (global) events, various data structures, and conveyors were later identified by members. It is expected that as simulation modelers use distributed simulation, more requirements will emerge.

The above requirements have been encapsulated into (currently) six IRMs (Taylor, 2003). These are

- Type I: Asynchronous entity passing
- Type II: Synchronous entity passing (bounded buffer)
- Type III: Shared resources
- Type IV: Shared events
- Type V: Shared data structures
- Type VI: Shared conveyor.

Briefly, the Type I IRM *asynchronous entity passing* deals with the common requirement of transferring entities between simulation models. The Type II IRM *synchronous entity passing* deals with the case where a receiving queue is *bounded*, i.e., in the above example queue Q2 has limited capacity. In this case, the requirement means that the federate containing the sending workstation W1 must, when the processing of an entity is complete, check to determine that there is space in Q2. If there is space available then the entity may be transferred. If there is none the federate must ensure that W1 is *blocked* until space becomes available. The Type III IRM *shared resources* deals with the sharing of resources across simulation models. For example, a resource R might be common between two models and represents a pool of workers. In this scenario, when a machine in a model attempts to process an entity waiting in its queue it must also have a worker. If a worker is available in R then processing can take place. If not then work must be suspended until one is available. The Type IV IRM *shared events* deals with the sharing of events across simulation models. For example, when a variable within a model reaches a given threshold value (a quantity of production, an average machine utilization, etc.) it should be able to signal this fact to all models that have an interest in this fact (to throttle down throughput, route materials via a different path, etc.). The Type V IRM *shared data structures* deals with the sharing of variables and data structures across simulation models that are semantically different to resources (e.g., a bill of materials or a shared inventory). Finally, the Type VI IRM *shared conveyors* deals with the problem of sharing transportation systems such as conveyor or barges across simulation models (as distinct to the representation of these in Type I IRMs). Note that not all IRMs will be applicable to all CSPs. For example, the Type I and II IRMs would only be applicable to CSPs that are capable of representing entities (such as those mentioned in Section 9.1).

The creation of the IRMs has proved to be a powerful tool in the development of standards in this area as it is now possible to create solutions for *specific* integration problems (rather than the general notion of integration as is currently the case). These have formed the basis for the creation of a new SISO-based standards group that arose from the HLA-CSPIF. Led by Taylor, this group is called the COTS Simulation Package Interoperability Product Development Group (CSPI-PDG) (www.cspi-pdg.com). They propose a suite of CSPI standards consisting of IRMs that outline different integration needs of CSPI, *Interoperability Frameworks* (IFs) that define the HLA-based solution to each IRM, appropriate data exchange representations to specify the data exchanged in an IF, and benchmarks termed *CSP Emulators* (CSPE) (see the Web site for recent versions of these). It has been noted that the creation of an efficient link between CSPs and an RTI is problematic as it requires investment by the vendor of the CSP (Taylor et al., 2003). While there are several good examples of this cited elsewhere in this chapter, it is difficult to judge the performance of the distributed simulation approach as the latency between a CSP and RTI is hidden. The use of a CSPE is intended to form a common platform to compare different proposed approaches to each IF. It is anticipated that there will be several data exchange formats to cover the possible needs of the IRMs. However, our concern in this chapter is a data exchange format specification that can deal with the passing of entities between federates and is relevant to Type I and II IRMs and their HLA-based IFs (as each of these IRM specifically deal with entities). We term this the *Entity Transfer Specification* (ETS). To discuss our ETS, we first present the Type I IRM in detail.

9.5.3 The Type I Interoperability Reference Model

Figure 9.2 shows the Type I IRM (asynchronous entity passing). This IRM represents models that interact on the basis of entities; models are linked together so that one model may "pass" an entity to another at a given timestamp. The reason why this is termed "asynchronous" is that there is no *immediate* or *direct feedback* when an entity is passed (this does not mean to say that no feedback can exist, just that it must happen at a different time to when an entity is passed—our case study shows an example of this). The model elements that have been placed in each model are there to indicate in a simple manner the relationships between models, i.e., the internal structure of a model can be far more complex—it is the relationship between the last workstation (W1), sink (S1), source (S2), and queue (Q2) that is important. Also, it is possible that models could have more that one set of links and that there could be more than two

Distributed Modeling 9-9

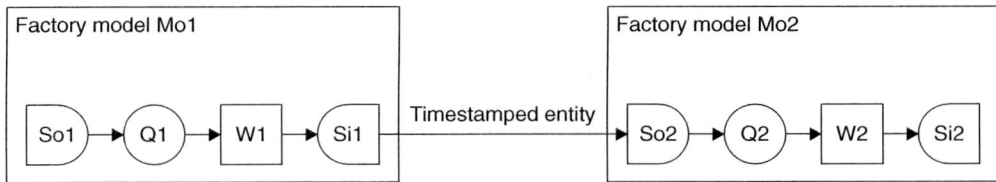

FIGURE 9.2 Type I interoperability reference model.

models connected in arbitrary topologies. Again, this IRM is intended to show the simplest relationship between models, one that can be extrapolated to many different scenarios.

In terms of minimum technological support of the logical link between the two models, all that is required is the transmission of timestamped entity information between model Mo1 and model Mo2 in such a way that model Mo2 receives the timestamped entity information in a correct order with its own events. The reason why this IRM has been termed "asynchronous" is that there is no synchronous message exchange needed to transfer the entity information between the two models (as is required in the Type II IRM). An IF solution to this Type I IRM must therefore be able to

- transfer timestamped entity information from one model to another via a timestamped message or such,
- allow a model to correctly receive timestamped entity messages from one or more models, and
- correctly coordinate this information with the receiving model events being processed by the COTS simulation package (Taylor and Mustafee, 2003).

We now discuss the latest contribution to this, the representation of entity information.

9.5.4 The Entity Transfer Specification

The ETS deals with the representation of entities in Type I and II IRMs. The reason for this is that both IRMs deal with the transfer of entities between CSPs. The difference between the IRMs is that Type II requires additional synchronization to deal with the bounded buffer problem (this is further discussed in Taylor [2003]). The following discussion is based on the current version of this emerging standard Version 1.1.1. Consider Figure 9.3. This shows the relationship between a CSP, interfacing software called the *CSP Handler* (CH) and an RTI (a candidate architecture for the CSPI IFs). We shall define a *source* model as being the model from which a timestamped entity leaves and a *destination* model as the model at which the timestamped entity arrives. These are necessary as there may be different possible routings between models (as defined by the model, not the RTI) and there must be enough information for this to be conveyed between a CH and RTI and vice versa for this model routing to be accomplished. Models may also have multiple entry points (i.e., the point at which an entity "arrives" at the model) and it is therefore important that there must be some way of indicating at which entry point an entity enters a model. In this version of the ETS, we assume there is only one receiving point in the destination model for a specific entity type from a specific source model, i.e., for different entity types there are different single receiving points. We shall define *time* as being the time when an entity leaves a source model and instantaneously arrives at the destination model (i.e., an event has occurred at *time* marking the departure of an entity from one subsystem to instantly arrive at another).

In terms of *entity representation*, as we are concerned with the transfer of a timestamped entity from a model in one federate to a model in another, our focus is a common data exchange format of the entity that has been prepared for transfer. We will assume that there is some translation mechanism between the heterogeneous CSP and CH to convert to and from our ETS representation. We shall also assume that *time* has been converted into the same units and resolution in both models. As with most distributed systems, the representation of an item must be marshaled (flat) so that it can be sent as a stream of bytes. We shall

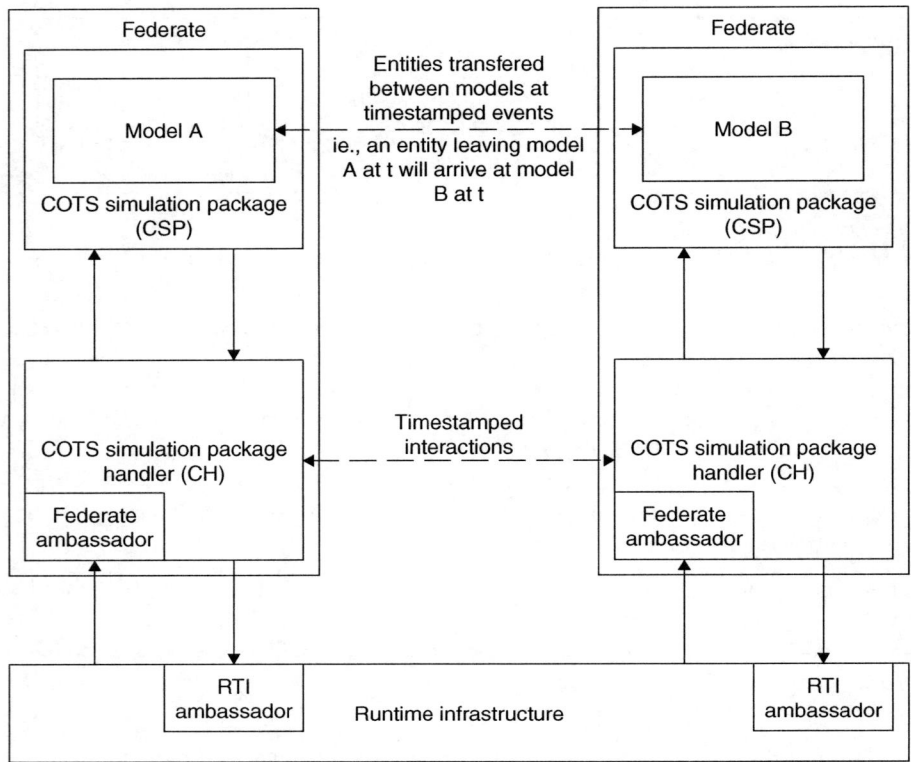

FIGURE 9.3 Entity transfer specification architecture.

therefore represent a mapped entity as a *name* and zero or more *attributes*. The form and type of the attributes are the result of the entity–entity mapping between the heterogeneous CSPs and their models.

An entity is therefore defined as follows

$$entity = \{entityName, attributes^*\}$$

for example,

$$widgetEntity = \{widgetEntity, 24, \text{``Acme''}\}$$

which represents a widget entity with attributes of (integer) 24 and (string) Acme.

When a CSP determines that an entity has left its model, the CSP must be able to deliver the following information to the CH:

$$output(entity, time, source, destination)$$

Similarly, when the CH is ready to pass an entity to the CSP, indicating that an entity has arrived, the CH must be able to deliver the following information:

$$input(entity, time, source)$$

where *entity* is the name of the entity *entityName* and zero or many attributes, *time* is the time at which the entity left the model, *source* is the name of the sending model, and *destination* is the name of the destination model.

On output, the *source* and *destination* are used by the CH to select the appropriate transfer mechanism. On input, the CSP uses *source* to determine the appropriate entry point in a model (i.e., where the entity has been transferred from). *time* is used to perform CSP time synchronization.

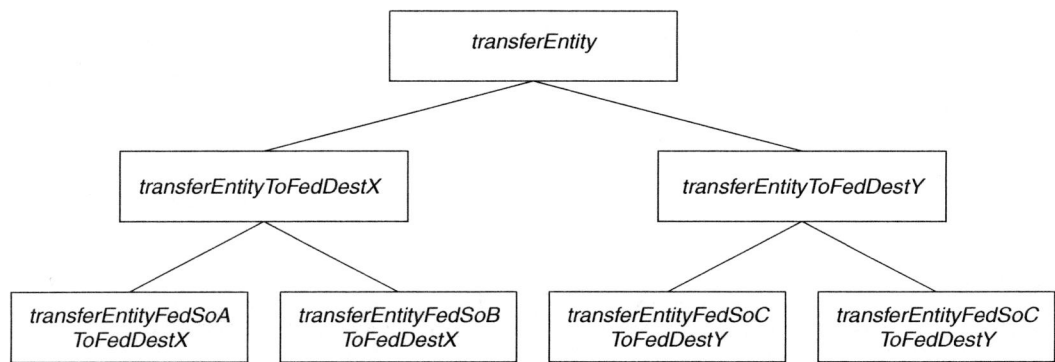

FIGURE 9.4 Interaction hierarchy.

In this specification, HLA *interactions* are used to represent the passing of an entity from one model to another at the RTI level. Figure 9.4 shows the ETS interaction class hierarchy. Features of this are

- *transferEntity*: the superclass. This allows a federate to conveniently subscribe to all instances of entity transfer (for purposes of monitoring, visualization, etc.).
- *transferEntityToFedDest*: a single subclass per receiving federate where *FedDest* is the name or abbreviation of the receiving federate's model. It exists for the convenience of the *FedDest* federate to subscribe to all instances of *transferEntity* bound to the destination federate without explicit naming.
- *transferEntityFedSoToFedDest*: subclasses for each entity transfer relation where *FedSo* is the name or abbreviation of the sending federate's model. It allows the source federate to send a timestamped interaction that represents the transfer of an entity from source to destination at a given time.

Note that in the above for an actual implementation *FedSo* and *FedDest* are replaced by the source and destination federate names and *Entity* is replaced by the name of the entity as appropriate. As we will see in our case study a wheel entity is transferred from the federate BAL to the federate WPL by the interaction *transferWheelEntityBALtoWPL*.

In a federate's SOM or the federation FOM the following tables are used in our exchange format.

- Interaction class table: This contains the interaction classes used to transfer the entities. These will be the interaction superclass *transferEntity*, its interaction subclasses *transferEntityToFedDest*, and their interactions subclasses *transferEntityFedSoToFedDest*.
- Parameter table: Each *transferEntityFedSoToFedDest* interaction will have a named parameter *Entity* with a named datatype *EntityType*. In the table, unless otherwise stated, *Available Dimensions* shall be NA (as data distributed management is not used), *Transportation* shall be HLAreliable (TCP semantics), and Order shall be timestamp, i.e., messages must arrive in timestamp order.
- Datatype table: A *fixed record datatype* table shall exist to represent the named *EntityType* and will consist of *entityName, source, destination,* and *attributes*. The datatypes of *entityName, source,* and *destination* will be of type *HLAASCIIstring*. The type of the attributes will be defined using the HLA datatype types as appropriate to best represent the type of the attribute.

These tables are shown in Figure 9.5. We assume that in any object model, these will be in addition to all other required tables (as described in the previous section). Additionally, as required by the OMT, the valid publish/subscribe options will be

- For an SOM:
  - *P (Publish)*: The federate is capable of publishing the interaction class.
  - *S (Subscribe)*: The federate is capable of subscribing to the interaction class.
  - *PS (PublishSubscribe)*: The federate is capable of publishing and subscribing to the interaction class.

Interaction class table

HLAinteraction Root(N)	TransferEntity(N/S)	TransferEntityToFedDest(N/S)	TransferEntityFedSoAToFedDestB(P/S)
			TransferEntityFedSoAToFedDestC(P/S)
			...

Parameter table

Interaction	Parameter	Datatype	Available dimensions	Transportation	Order
TransferEntityFedSoAToFedDestB	Entity	EntityType	NA	HLAreliable	TimeStamp

Fixed record datatype table

Record name	Field			Encoding	Semantics
	Name	Type	Semantics		
EntityType	EntityName	HLAASCIIString	Name of the entity	HLAfixedRecord	An entity
	Source	HLAASCIIString	FedSo		
	Destination	HLAASCIIString	FedDest		
	Attributes	Appropriate datatypes (4,12)	The various attributes of the entity		
	Attributes	Appropriate datatypes (4,12)	The various attributes of the entity		
	...	...	...		

FIGURE 9.5 OMT tables used for entity transfer specification.

- *N (Neither)*: The federate is incapable of either publishing or subscribing to the interaction class. HLAInteractionRoot is always this.
- In an SOM, *transferEntity* will be
  - *S* if a federate wishes to get all entity transfer interactions.
  - *N* if the federate is not interested in receiving this global information.
- In an SOM, *transferEntityToFedDest* will be
  - *S* if the federate's model is *FedDest*.
  - *N* if the federate's model is not *FedDest* (i.e., it is required to support the interaction class hierarchy for a publish-only *transferEntityFedSoToFedDest*).
- In an SOM, *transferEntityFedSoToFedDest* will be
  - *P* if the federate is *FedSo*, i.e., its model sends entities to *FedDest*'s model.
  - *S* if the federate is *FedDest*, i.e., its model receives entities from *FedSo*.
- For an FOM, these interactions will be
  - *transferEntity* will be *N* if there is no "monitor" federate or *S*.
  - *transferEntityToFedDest* will be *S*.
  - *transferEntityFedSoToFedDest* will always be *PS*.
- Classes designated as *Subscribe* or *Neither* are never sent, but they can have subclasses that are sent.
- It will be assumed that when an FOM is composed from SOMs there will be some kind of entity name resolution.
- HLAinteractionRoot is a superclass of all other interaction classes in an FOM or SOM and is mandatory.

The above is enough to define the representation of an entity transferred from one model to another via the RTI. The translation of the datatype of this representation and the internal type representation of the CSP must be performed by the CH according to the requirements of the CSP.

The interaction classes are meant to be used in the following way in a Type I IF. During initialization, a federate will

- indicate that it is capable of sending entities to various destination federates by publishing all *transferEntityFedSoToFedDest* interactions, and

Distributed Modeling 9-13

- indicate that it is capable of receiving entities from any other federate by subscribing to all *transferEntityToFedDest* interactions.

During runtime, when the CSP sends the message equivalent to

output(entity, time, source, destination)

the CH will use *destination* to select the appropriate interaction class to use. It will then parameterize an interaction instance with the details supplied in the output message. When the RTI passes an interaction instance to the CH, the CH will use the instance's details to pass the entity to the CSP in some input message with *source* to indicate which model the entity has arrived from.

We now present an illustrative case study showing the use of the ETS to support the integration of heterogeneous CSPs according to the Type I IRMs within the Type I IF.

9.6 Case Study

Consider the illustrative Type I IRM-based distributed simulation in Figure 9.6. A company manufactures bicycles. Three models exist in three possibly heterogeneous CSPs that represent the three main parts of the manufacturing system: a wheel production line (WPL), a frame production line (FPL), and a bicycle assembly line (BAL) that assembles two wheels to one frame to produce a bicycle. The BAL checks wheels for faults and can return them to the WPL for re-machining (an example of valid feedback for Type I IRMs). Frames have no such problems. To describe part of the simulation, raw materials for the WPL arrive every 20 min at entry point En1a and wait for processing in Q1a. When workstation W1a becomes free, raw materials are taken from queue Q1a, processed into wheels and released. This activity takes a fixed time of 20 min. The newly created wheels then take 100 min travel time to be transferred to the BAL's entry point En3a. We assume that the entry point, the queue, and the workstation are adjacent. Frame entities have the attributes "frame_size" which is of type integer and "frame_color" which is of type string. Wheels have a single attribute "wheel_size" of type integer. The rest of the distributed simulation can be described in a similar manner with the various times to perform actions shown on the models. Note that in our example all distributions are fixed. In a real simulation it is likely that these will be probabilistic distributions and that there will be a greater number of model elements. However, our simple model with its fixed distributions is appropriate for purposes of illustration as we are concerned with distributed simulation issues. Each simulation model in our example runs in a CSP or in different CSPs, with each CSP/model combination a federation in our approach (the models WPL, FPL, and BAL and their CSPs becoming federates Fd1, Fd2, and Fd3).

Figures 9.7(a)–9.7(c) show the SOMs for Fd1, Fd2, and Fd3. Figure 9.7(d) shows the composite FOM for the federation as a whole. As can be seen, these tables provide a neutral representation of data that the various heterogeneous CSPs are required to translate to and from as they send and receive entities. This illustrates our contribution to emerging standards in this area in support of the Type I and II IRMs and their IFs.

9.6.1 Illustrative Protocol

As part of the IF, the CH provides an interface consisting of a set of functions to be invoked by the CSP when needed. Through the interface, the CH invokes necessary calls to the RTI ambassador on behalf of the CSP and transfers the information received from the federate ambassador to the CSP. Figure 9.8 shows the basic communication protocol between the CSP, CH, and RTI and its relationship with the ETS *output* and *input*.

There are various different approaches to time management using a HLA RTI to support distributed discrete-event simulation (Fujimoto, 1998). The approach described here is based around *NextEventRequest* (others are currently under investigation as part of the work developing the Type I IF).

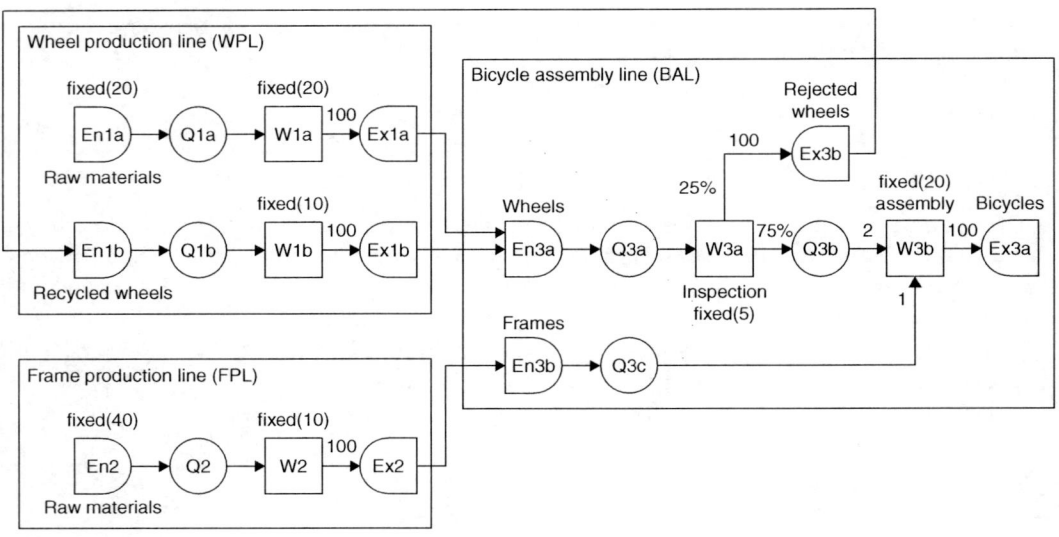

FIGURE 9.6 The bicycle factory model.

Interaction class table (4.3)

HLAinteractionRoot(N)	TransferWheelEntity(N)	TransferWheelEntityToBAL(N)	TransferWheelEntityWPLToBAL(P)
		TransferWheelEntityToWPL(S)	TransferWheelEntityBALToWPL(S)

Parameter table (4.5)

Interaction	Parameter	Datatype	Available dimensions	Transportation	Order
TransferWheelEntityBALToWPL	WheelEntity	WheelType	NA	HLAreliable	TimeStamp
TransferWheelEntityWPLToBAL	WheelEntity	WheelType	NA	HLAreliable	TimeStamp

Fixed record datatype table (4.12)

Record name	Field			Encoding	Semantics
	Name	Type	Semantics		
WheelType	Wheel	HLAASCIIString	Name of the entity	HLAfixedRecord	An Entity
	Source	HLAASCIIString	WPL or BAL		
	Destination	HLAASCIIString	WPL or BAL		
	Wheel_Size	HLAinteger16BE	Size of the wheel		

(a)

Interaction class table

HLAinteractionRoot(N)	TransferFrameEntity(N)	TransferFrameEntityToBAL(N)	TransferFrameEntityFPLToBAL(P)

Parameter Table

Interaction	Parameter	Datatype	Available dimensions	Transportation	Order
TransferFrameEntityFPLToBAL	FrameEntity	FrameType	NA	HLAreliable	TimeStamp

Fixed record datatype table

Record name	Field			Encoding	Semantics
	Name	Type	Semantics		
FrameType	Frame	HLAASCIIString	Name of the entity	HLAfixedRecord	An Entity
	Source	HLAASCIIString	FPL		
	Destination	HLAASCIIString	BAL		
	Frame_Size	HLAinteger16BE	Size of the frame		
	Frame_Color	HLAASCIIString	Color of the frame		

(b)

FIGURE 9.7 Entity transfer specifications. (a) Wheel production line SOM; (b) frame production line SOM.

Interaction class table

HLAinteractionRoot(N)	TransferWheelEntity(N)	TransferWheelEntityToBAL(S)	TransferWheelEntityWPLToBAL(S)
		TransferWheelEntityToWPL(N)	TransferWheelEntityBALToWPL(P)
	TransferFrameEntity(N)	TransferFrameEntityToBAL(S)	TransferWheelEntityFPLToBAL(S)

Parameter table

Interaction	Parameter	Datatype	Available dimensions	Transportation	Order
TransferEntityBALToWPL	WheelEntity	WheelType	NA	HLAreliable	TimeStamp
TransferEntityWPLToBAL	WheelEntity	WheelType	NA	HLAreliable	TimeStamp
TransferEntityFPLToBAL	FrameEntity	FrameType	NA	HLAreliable	TimeStamp

Fixed record datatype table

Record name	Field			Encoding	Semantics
	Name	Type	Semantics		
FrameType	Frame	HLAASCIIString	Name of the entity	HLAfixedRecord	An entity
	Source	HLAASCIIString	FPL		
	Destination	HLAASCIIString	BAL		
	Frame_Size	HLAinteger16BE	Size of the frame		
	Frame_Color	HLAASCIIString	Color of the frame		
WheelType	Wheel	HLAASCIIString	Name of the entity	HLAfixedRecord	An entity
	Source	HLAASCIIString	WPL or BAL		
	Destination	HLAASCIIString	WPL or BAL		
	Size	HLAinteger16BE	Size of the wheel		

(c)

Interaction class table

HLAinteractionRoot(N)	TransferWheelEntity(N)	TransferWheelEntityToBAL(S)	TransferWheelEntityWPLToBAL(PS)
		TransferWheelEntityToWPL(S)	TransferWheelEntityBALToWPL(PS)
	TransferFrameEntity(N)	TransferFrameEntityToBAL(S)	TransferFrameEntityFPLToBAL(PS)

Parameter table

Interaction	Parameter	Datatype	Available dimensions	Transportation	Order
TransferWheelEntityBALToWPL	WheelEntity	WheelType	NA	HLAreliable	TimeStamp
TransferWheelEntityWPLToBAL	WheelEntity	WheelType	NA	HLAreliable	TimeStamp
TransferFrameEntityFPLToBAL	FrameEntity	FrameType	NA	HLAreliable	TimeStamp

Fixed record datatype table

Record name	Field			Encoding	Semantics
	Name	Type	Semantics		
FrameType	Frame	HLAASCIIString	Name of the entity	HLAfixedRecord	An entity
	Source	HLAASCIIString	FPL		
	Destination	HLAASCIIString	BAL		
	Frame_Size	HLAinteger16BE	Size of the frame		
	Frame_Color	HLAASCIIString	Color of the frame		
WheelType	Wheel	HLAASCIIString	Name of the entity	HLAfixedRecord	An entity
	Source	HLAASCIIString	WPL or BAL		
	Destination	HLAASCIIString	WPL or BAL		
	Size	HLAinteger16BE	Size of the wheel		

(d)

FIGURE 9.7 Entity transfer specifications. (c) Bicycle assembly line SOM; (d) Bicycle manufacturing, system FOM.

When the CSP wishes to advance to the time of its next event, it issues an *advanceTime* request to the CH. The CH invokes the corresponding RTI service *nextEventRequest*. The response from the RTI is zero or many ETS interactions received via *receiveInteraction* and a new simulation time granted via *timeAdvanceGrant*. The interactions represent the arrival of entities at the time granted by *timeAdvanceGrant* and may be less than the time initially requested by the CSP (i.e., entities arrive before the time of the original next event—the new time of next event is that of the arriving entities). If no interactions appear, the time granted is exactly the requested time. Either way, this grant time is returned to the CSP with the entities received (if any) via *input(entity, time, source)*, the CSP advances its local simulation time and continues execution. If, as a consequence of this, any entities leave the simulation model, the CSP will send to CH as

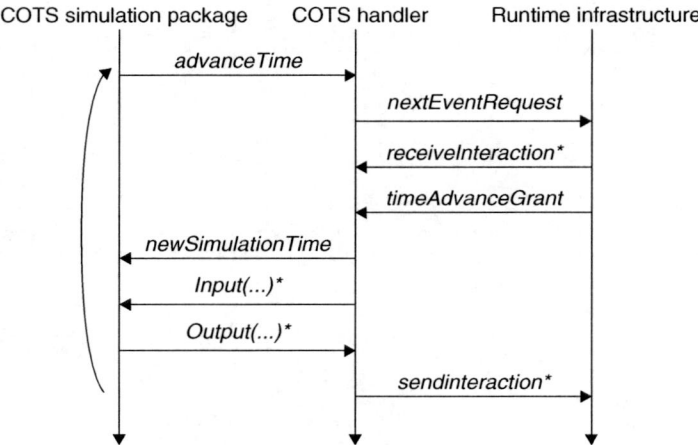

FIGURE 9.8 Illustrative protocol.

many *output(entity, time, source, destination)* as appropriate. CH will translate these into ETS interactions and then forward these to the RTI by invoking *sendInteraction*. This continues until some terminating condition is met.

9.7 Conclusion

The purpose of this chapter has been to introduce some of the contemporary innovations in the use of distributed computing techniques to support modeling. Our discussion has been centered around CSP-based distributed simulation and represents a rich area of research that continues to be very rewarding for those involved. As presented in the introduction, the development of a generalized solution to this area will open up many new opportunities for those involved in dynamic systems modeling. Indeed several successes have been achieved already. Brailsford et al. (to appear) describe work on the modeling of large systems in health care with a distributed version of the CSP Simul8. Lendermann et al. (to appear) describe progress in using a distributed version of the CSP AutoSched to model the rapidly changing demands of semiconductor manufacture.

Acknowledgments

Sections of this chapter are based on the paper: Taylor, S.J.E., Wang, X., Turner, S.J. and Low, M.Y.H. 2006. Integrating heterogeneous distributed COTS discrete-event simulation packages: An emerging standards-based approach. *IEEE Transactions on Systems, Man and Cybernetics: Part A,* 36, 1, 109–122. The author would also like to acknowledge the members of the CSPI-PDG for their hard work in developing the current range of standards offered by the group.

References

Balci, O., 1988. The implementation of four conceptual frameworks for simulation modeling in high-level languages. In *Proc. 1998 Winter Simulation Conference*, pp. 287–295.

Banks, J. (Ed.), 1998. *Handbook of Simulation: Principles, Methodology, Advances, Applications, and Practice,* New York, NY: Wiley.

Bapat, V. and D.T. Sturrock, 2003. The arena product family: Enterprise modeling solutions. In *Proc. 2003 Winter Simulation Conference*, pp. 210–217.

Bell, P.C., 1985. Visual interactive modeling in operational research: Successes and opportunities, *Journal of the Operational Research Society*, vol. 36, pp. 975–982.

Bell, P.C. and R.M. O'Keefe, 1987. Visual interactive simulation—History, recent developments, and major issues, *Simulation*, vol. 49, pp. 109–116.

Boer, C.A., A. Verbraeck, and H.P.M. Veeke, 2002a. Distributed simulation of complex systems: Application in container handling. In *Proc. 2002 EUROSIW*, 02E-SIW-03.

Boer, C.A., A. Verbraeck, and H.P.M. Veeke, 2002b. The possible role of a backbone architecture in real-time control and emulation. In *Proc. 2002 Winter Simulation Conference*, pp. 1675–1682.

Bosilj-Vuksic, V., M.I. Stemberger, J. Jarklic, and A. Kovacic, 2003. Assessment of E-business transformation using simulation modeling, *Simulation*, vol. 78, pp. 731–744.

Brailsford, S., K. Katsaliaki, N. Mustafee, and S.J.E. Taylor, Accepted. Modelling very large complex systems using distributed simulation: A pilot study in a healthcare setting. In *Proc. Simulation Workshop 2006*.

Bruzzone, A.G., 2003. Preface to modeling and simulation methodologies for logistics and manufacturing optimization, *Simulation*, vol. 80, pp. 119–120.

Buxton, J.N. and J.G. Laski, 1962. Control and simulation language, *The Computer Journal*, vol. 5, pp. 194–199.

Demirci, E., 2003. Simulation modelling and analysis of a port investment, *Simulation*, vol. 79, pp. 94–105.

Eldabi, T., R.J. Paul, and S.J.E. Taylor, 2000. Simulating economic factors in adjuvant breast cancer treatment, *Journal of the Operational Research Society*, vol. 51, no. 4, pp. 465–475.

Fujii, S., T. Kaihara, and H. Morita, 2000. A distributed virtual factory in agile manufacturing environment, *International Journal of Production Research*, vol. 38, no. 17, pp. 4113–4128.

Fujimoto, R.M., 1998. Time management in the high level architecture, *Simulation*, vol. 71, no. 6, pp. 388–400.

Fujimoto, R.M., 2000. *Parallel and Distributed Simulation Systems*, New York, NY: Wiley.

Fujimoto, R.M., 2003. Distributed simulation systems. In *Proc. 2003 Winter Simulation Conference*, pp. 124–134.

Gan, B.P., L. Li, S. Jain, S.J. Turner, C. Wentong, and J.H. Wen, 2000. Distributed supply chain simulation across enterprise boundaries. In *Proc. 2000 Winter Simulation Conference*, pp. 1245–1251.

Gan, B.P. and S.J. Turner, 2000. An asynchronous protocol for virtual factory simulation on shared memory multiprocessor systems, *Journal of Operational Research Society*, vol. 51, pp. 413–422.

Goel, S., D.R. Strong, N. Richards, and N.C. Goel, 2002. A simulation-based method for the process to allow continuous tracking of quality, cost, and time, *Simulation*, vol. 78, pp. 330–337.

Harrell, C.R. and R.N. Price, 2003. Simulation modeling using ProModel technology. In *Proc. 2003 Winter Simulation Conference*, pp. 175–181.

Hibino, H., Y. Fukuda, Y. Yura, K. Mitsuyuki, and K. Kaneda, 2002. Manufacturing adapter of distributed simulation systems using HLA. In *Proc. 2002 Winter Simulation Conference*, pp. 1099–1107.

Hofmann, M.A., 2004. Criteria for decomposing systems into components in modeling and simulation: Lessons learned with military simulations, *Simulation*, vol. 80, pp. 357–365.

Hurrion, R.D., 1998. Visual interactive meta-simulation using neural networks, *International Transactions in Operational Research*, vol. 5, pp. 261–270.

IEEE 1278, 1995. *Standard for Distributed Interactive Simulation (DIS)*, New York, NY: Institute of Electrical and Electronics Engineers.

IEEE 1516, 2000. *IEEE Standard for Modeling and Simulation (M&S) High Level Architecture (HLA)*, New York, NY: Institute of Electrical and Electronics Engineers.

IEEE 1516.0, 2000. *IEEE Standard for Modeling and Simulation (M&S) High Level Architecture (HLA)— Rules*, New York, NY: Institute of Electrical and Electronics Engineers.

IEEE 1516.1, 2000. *IEEE Standard for Modeling and Simulation (M&S) High Level Architecture (HLA)— Federate Interface Specification*, New York, NY: Institute of Electrical and Electronics Engineers.

IEEE 1516.2, 2000. *IEEE Standard for Modeling and Simulation (M&S) High Level Architecture (HLA)—Object Model Template (OMT) Specification*, New York, NY: Institute of Electrical and Electronics Engineers.

IEEE 1516.3, 2004. *IEEE Standard for Modeling and Simulation (M&S) High Level Architecture (HLA)—Federate Development Process (FEDEP)*, New York, NY: Institute of Electrical and Electronics Engineers.

Krahl, D., 2003. Extend: An interactive simulation tool. In *Proc. 2003 Winter Simulation Conference*, pp. 188–196.

Lane, D.C., 1999. Social theory and system dynamics practice, *European Journal of Operational Research*, vol. 113, pp. 501–527.

Law, A.M. and W.D. Kelton, 2000. *Simulation Modeling and Analysis* (3rd ed.), New York, NY: McGraw-Hill.

Lee, J.K., Y.H. Lim, and S.D. Chi, 2004. Hierarchical modeling and simulation environment for intelligent transportation systems, *Simulation*, vol. 80, pp. 61–76.

Lendermann, P., B.P. Gan, and L.F. McGinnis, 2001. Distributed simulation with incorporated APS procedures for high-fidelity supply chain optimization. In *Proc. 2001 Winter Simulation Conference*, pp. 1138–1145.

Linn, R.J., C.S. Chen, and J.A. Lozan, 2002. Development of distributed simulation model for the transporter entity in a supply chain Process. In *Proc. 2002 Winter Simulation Conference*, pp. 1319–1326.

Low, Y.H., B.P. Gan, J.J. Wei, X. Wang, S.J. Turner, and W. Cai, 2001. Implementation issues for shared state in HLA-based distributed simulation. In *Proc. 2001 European Simulation Symposium*, pp. 5–13.

Mertins, K. and M. Rabe, 2002. Inter-enterprise planning of manufacturing systems applying simulation with IPR protection. In *Proc. 5th International Conference on Design of Information Systems for Manufacturing (DIISM)*, pp. 149–156.

Mertins, K., M. Rabe, and F.W. Jäkel, 2000. Neutral template libraries for efficient distributed simulation within a manufacturing system engineering platform. In *Proc. 2000 Winter Simulation Conference*, pp. 1549–1557.

McLean, C. and F. Riddick, 2000. The IMS MISSION architecture for distributed manufacturing simulation. In *Proc. 2000 Winter Simulation Conference*, pp. 1539–1548.

Nance, R.E., 1996. A history of discrete event simulation programming languages, in *History of Programming Languages—II*, New York, NY: Association for Computing Machinery Press, pp. 369–427.

Nance, R.E. and R. Sargent, 2002. Perspectives on the evolution of simulation, *Operations Research*, vol. 50, no. 1, pp. 161–172.

Paul, R.J. and S.J.E. Taylor, 2002. What use is model reuse: Is there a crook at the end of the rainbow? In *Proc. 2002 Winter Simulation Conference*, pp. 648–652.

Peterson, J.L., 1981. *Petri Net Theory and the Modeling of Systems*, New Jersey,: Prentice-Hall.

Pidd, M., 1998. *Computer Simulation in Management Science* (4th ed.), Chichester, UK: Wiley.

Rabe, M. and F.W. Jäkel, 2001. Non military use of HLA within distributed manufacturing scenarios. In *Proc. Simulation und Visualisierung*, pp. 141–150.

Rabe, M. and F.W. Jäkel, 2003. On standardization requirements for distributed simulation, in *Production and Logistics. Building the Knowledge Economy*, Twente, The Netherlands: IOS Press, pp. 399–406.

Robinson, S., 2002. Modes of simulation practice: Approaches to business and military simulation, *Simulation Practice and Theory*, vol. 10, pp. 513–523.

Robinson, S., 2004. *Simulation: The Practice of Model Development and Use*, Chichester, UK: Wiley.

Robinson, S. 2005. Distributed simulation and simulation practice, *Simulation*, vol. 81, no. 1, pp. 5–13.

Robinson, S., R.E. Nance, R.J. Paul, M. Pidd, and S.J.E. Taylor, 2004. Simulation model reuse: Definitions, benefits and obstacles, *Simulation Modeling Practice and Theory*, vol. 12, pp. 479–494.

Rohrer, M.W., 2003. Maximizing simulation ROI with AutoMod. In *Proc. 2003 Winter Simulation Conference*, pp. 201–209.

Schriber, T.J. and D.T. Brunner, 2003. Inside discrete-event simulation software: How it works and why it matters. In *Proc. 2003 Winter Simulation Conference*, pp. 113–123.

Straßburger, S., 2001. *Distributed Simulation Based on the High Level Architecture in Civilian Application Domains*, Ghent, Belgium: Society for Computer Simulation International.

Straßburger, S., G. Schmidgall, and S. Haasis, 2003. Distributed manufacturing simulation as an enabling technology for the digital factory, *Journal of Advanced Manufacturing Systems*, vol. 2, no. 1, pp. 111–126.

Sudra, R., S.J.E. Taylor, and T. Janahan, 2000. Distributed supply chain management in GRIDS. In *Proc. 2000 Winter Simulation Conference*, pp. 356–361.

Taylor, S.J.E., 2003. HLA-CSPIF: The high level architecture—COTS simulation package interoperation forum. In *Proc. 2003 Fall Simulation Interoperability Workshop*, 03F-SIW-126.

Taylor, S.J.E., B.P. Gan, S. Strassburger, and A. Verbraeck, 2003. HLA-CSPIF technical panel on distributed simulation. In *Proc. 2003 Winter Simulation Conference*, pp. 881–887.

Taylor, S.J.E. and N. Mustafee, 2003. An analysis of internal/external event ordering strategies for COTS distributed simulation. In *Proc. 2003 European Simulation Symposium*, pp. 193–198.

Taylor, S.J.E., S. Robinson, and J. Ladbrook, 2005. An investigation into the use of net-conferencing groupware in simulation modeling, *Journal of Computing and Information Technology*, vol. 13, no. 1, pp. 1–10.

Taylor, S.J.E., R. Sudra, T. Janahan, G. Tan, and J. Ladbrook, 2002. *GRIDS-SCS: An Infrastructure for Distributed Supply Chain Simulation*, Simulation, vol. 78, pp. 312–320.

Tocher, K.D., 1963. *The Art of Simulation*, London, UK: English Universities Press.

Zülch, G., U. Jonsson, and J. Fischer, 2002. Hierarchical simulation of complex production systems by coupling models, *International Journal of Production Economics*, vol. 77, pp. 39–51.

10
Model Execution*

Kalyan S. Perumalla
Oak Ridge National Laboratory

10.1 Introduction 10-1
    Systems and Models • Elements of Execution • Execution Platforms • Generating Executables from Models • Executable Timelines • Pacing the Execution

10.2 Time-Stepped Execution 10-5
    Example: Heat Equation • Parallelizing Time-Stepped Execution

10.3 Discrete-Event Execution 10-7
    Execution Method • Example: ATM Multiplexer • Parallelizing Discrete-Event Execution

10.4 Summary 10-13

10.1 Introduction

A computer software-based model is typically designed to produce a trace of system evolution over time. The actual process of computing the model state and producing the state values as the simulation time is advanced is called model execution. Models could be designed with a specific execution technique in mind, or could be generally amenable to multiple execution techniques. Two popular methods that are used to execute models are time-stepped method and discrete-event method. Each of these methods could in turn be executed either sequentially (on a single processor), or in parallel (using multiple processors concurrently). In this chapter, we describe the time-stepped and discrete-event execution methods and outline some of the common approaches to their sequential and parallel execution. Execution concepts common to the methods are described followed by implementation details of the methods.

10.1.1 Systems and Models

Figure 10.1 shows the general architecture in which model execution is used. A system of interest is modeled at appropriate resolution and the resulting models are configured for given scenarios of interest. The execution is based on the given configuration of interest and its execution generates the required output. This chapter deals with different methods by which simulation models can be mapped to executable units and how their execution is controlled to obtain meaningful and useful output from the simulation. Model execution is multifaceted, involving issues such as execution order and synchronization.

---

*This manuscript has been authored by UT-Battelle, LLC, under contract DE-AC05-00OR22725 with the U.S. Department of Energy. The United States Government retains and the publisher, by accepting the article for publication, acknowledges that the United States Government retains a nonexclusive, paid-up, irrevocable, worldwide license to publish or reproduce the published form of this manuscript, or allow others to do so, for United States Government purposes.

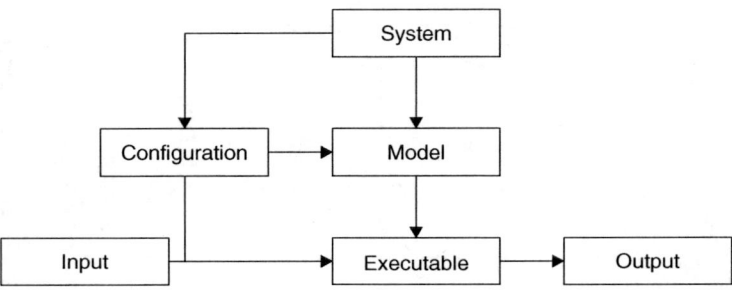

FIGURE 10.1 Relationships among modeling and simulation elements.

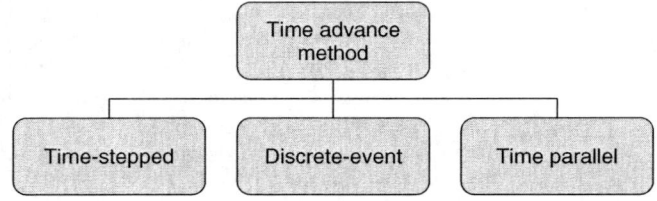

FIGURE 10.2 Time advance mechanisms.

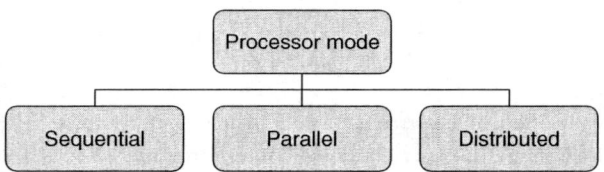

FIGURE 10.3 Simulation modes based on computation platforms.

10.1.2 Elements of Execution

Although simulation models are qualitatively different relative to each other, they share common underlying execution concepts (Tocher, 63). Essentially, the modeled entities need to be mapped on to suitable execution units, and the actions of the execution units have to be driven in simulation time order (Pidd, 92).

A key design element in model execution is the time advance mechanism (Nance, 81; Chandy, 89; Zeigler, 00), as illustrated in Figure 10.2. In the traditional *time-stepped* simulation method, time is advanced in fixed increments, and the system state is updated synchronously at each increment. *Discrete-event* simulation method, on the other hand, evolves different parts of system state at their own timescales, using the concept of events. Each event signals the specific instant in simulation time at which a particular part of the system is to be updated. Yet another method that is less commonly used is called the time parallel simulation, in which simulation time is partitioned into multiple segments, and each segment is executed independently from each other. Initial state of one segment is reconciled with the ending state of its previous segment and the process is iterated until convergence is reached. Time parallel simulation differs from time-stepped and discrete-event simulation methods in that time parallel simulation partitions the simulation across the simulation time dimension, whereas the latter two methods partition the simulation across the problem's spatial dimension.

The next important element is whether the execution is performed using one processor or using multiple processors, as shown in Figure 10.3.

Sequential simulation uses one processor for its execution. *Parallel simulation* can assume shared memory or high-speed interconnects among multiple processors. *Distributed simulation* is performed on loosely

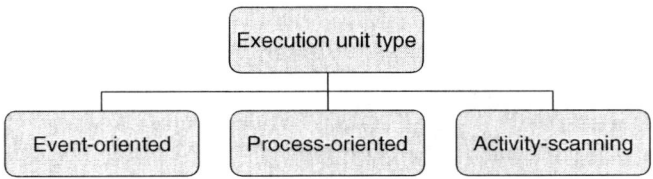

FIGURE 10.4 Classical simulation unit types, also called simulation worldviews.

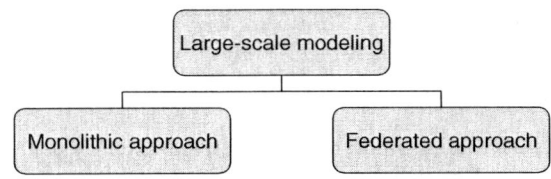

FIGURE 10.5 Approaches to simulating large-scale models/scenarios.

connected nodes, such as a cluster of workstations connected by a local network, or a geographically separated set of servers connected over a wide area network. Sequential execution is by far the most commonly used approach, primarily due to its simplicity of implementation and the availability of a large number of simulation tools that run on a single processor. Parallel execution using multiple processors is used when the execution speed needs to be increased and the model needs large amount of memory that cannot fit in a single processor's memory. Distributed execution employs multiple processors connected by a network and that do not share memory. The key distinction between parallel and distributed execution is that the former uses shared memory among processors while the latter executes on distributed memory platforms.

Another related element is the type of modeling framework (Balci, 88; Fishwick, 95; Banks, 96) used to map modeled entities to simulation units, as categorized in Figure 10.4. *Process-oriented* views are those that provide a full stack (thread) of context for each modeled entity, which typically results in more readable, maintainable, and shorter descriptions of the models (Banks, 96; Perumalla, 98). In contrast, the alternative *event-oriented* views are those that provide a bare-bones event-handler interface, for potentially better runtime performance than process-oriented views, but that are more complex to develop and maintain. Another simulation approach is called *activity scanning* in which execution of "a set of guarded actions" is enabled via continual scanning of predicated guards that prefix each action. Actions associated with the guards that evaluate to true are executed during each scan.

Finally, large-scale scenarios are developed and simulated using two distinct approaches, shown in Figure 10.5. In a *monolithic* approach, all models are developed and instantiated in a single simulator (sequential or parallel), and the entire scenario is configured and executed in the same simulator. In a *federated* approach, on the other hand, different models spanning multiple simulators are used to assemble a single scenario, and the simulators are integrated at runtime to exchange data across models (Nicol, 96) and synchronize their execution. While the monolithic approach is simpler and efficient to implement and use if all models are present in a single simulator, the federated approach is useful when no single simulator exists that contains all the desired functionality and, consequently, multiple simulators are needed to realize a large integrated scenario.

10.1.3 Execution Platforms

The most common environment for model execution is a desktop computer, which serves well for typical models that are small in size. A dual-processor or quad-processor workstation (called symmetric shared memory multiprocessor [SMP]) can be used to improve processing speed by employing parallel simulation techniques based on shared memory communication across processors. Cluster of workstations connected by local area network represents the next level of computation platform to leverage dozens of uniprocessor and SMP workstations. The highest levels of execution performance are afforded by supercomputing platforms containing hundreds to thousands of processors connected by fast processor interconnects.

In addition to these traditional simulation platforms, new nontraditional platforms are emerging that offer better performance to cost ratios. These include field programmable gate arrays (FPGA), application-specific integrated circuits (ASIC), and general purpose graphical processing units (GPGPU).

10.1.4 Generating Executables from Models

Models in general can be developed in various environments. Some of the most common forms are library-based, language-based, and integrated development environment (IDE)-based. In a library-based approach, runtime support is provided in the form of a library module that is linked into simulation model. Models are written using the interface provided by the library (e.g., process and event class hierarchies). The model, for example, invokes execution primitives to help coordinate time advances and message exchanges. Examples of library-based approach include Georgia Tech Time Warp (GTW) (Das, 94) and A Discrete Event Simulation (ADEVS) (Nutaro, 93) packages. In a language-based approach, the model is written in a high-level modeling language, and a compiler automatically transforms models into executable entities. The compiler also generates the runtime loop to invoke the generated executable entity code in appropriate order. Examples of language-based approach include Modelica and PARSEC (general purpose) (Bagrodia, 98), and telecommunications description language (TeD) (Perumalla, 98), VHDL, and Verilog (domain-specific). In an IDE-based approach, an IDE presents a graphical interface to compose models from model repositories, and transparently performs all the required translations and assemblage to execute the model. Examples of IDE-based systems include Simulink, Cadence, and OPNET. Another category might include single-use efforts commonly used to build simulators. Even spreadsheets such as Microsoft Excel can be used to quickly code certain simulations. These are not discussed here.

The generation of an executable model in a typical modeling system based on a hypothetical modeling language L is shown in Figure 10.6. A language-based approach uses such a translation system to generate the model executable. An IDE-based approach hides much of the internal operation from the user, but the internal operation roughly follows the language-based approach. A library-based approach can be viewed as the bottom half of the language-based process. In Figure 10.6, for example, the C++ code is generated by the user instead of by a separate translator.

A simulation language shields the user from details of simulation units (Nance, 93; Schriber, 74), and provides domain-specific constructs. For example, a modeling language for telecommunication networks provides network-specific constructs such as network packets and protocols, and shields the modeler from

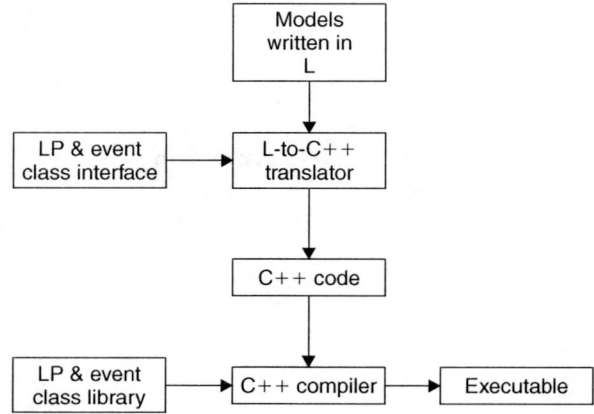

FIGURE 10.6 Illustration of how a hypothetical simulation language L is converted to an executable based on an intermediate conversion to C++.

simulation concepts such as logical processes and timestamped events. The modeler simply describes the model in the language constructs and then uses the compiler to generate equivalent simulation executable code. A system based on an IDE goes a bit further and hides much more of the model execution details from the modeler by performing compilation, linkage, and execution under the covers of the IDE.

Regardless of the approach, the fundamental units of simulation execution remain the same. These are the time advance method, the types of execution units, and whether the execution is sequential or parallel. Once a model is defined, it needs to be executed to compute the state of the system over time.

10.1.5 Executable Timelines

In model execution, there are generally three distinct time axes. The first corresponds to the *physical time*, which is the time in the physical system that is being modeled (e.g., 10–11 pm on January 1990). The second corresponds to the *simulation time*, which is a representation of the physical time for the purposes of simulation (e.g., number of seconds since 10 pm of January 1990, represented in floating point values in the range [0 ... 3600] corresponding to the simulated time period of the physical time). Finally, the last axis corresponds to *wallclock time,* which is the elapsed real time during execution of the simulation, as measured by a hardware clock (e.g., number of milliseconds of computer time during execution). For each, the notions of time axis and time instant can be defined such that the *time axis* is the totally ordered set of *time instant*s along the corresponding timeline.

10.1.6 Pacing the Execution

Almost always, there is a one-to-one mapping from physical time to simulation time. In contrast, there may or may not exist a specific relationship between simulation time and wallclock time. The mode of simulation execution determines this particular relationship. In an *as-fast-as-possible* execution, the simulation time is advanced as fast as computing speed can allow, unrelated to wallclock time. In *real-time* execution, on the other hand, advances in simulation time are performed in lockstep with wallclock time, such that one unit of simulation time is advanced exactly in one same unit of wallclock time. A variation of real-time execution is *scaled real-time* execution, in which simulation time period is some constant factor times an equivalent wallclock time period.

10.2 Time-Stepped Execution

By far the most commonly used mode of execution in scientific simulations is the time-stepped method. Time-stepped simulation is schematically illustrated in Figure 10.7. The horizontal bars represent timelines of each modeled entity, while the solid vertical lines represent points in simulation time at which the entities are updated.

The time-step value (simulation time period between successive updates to the state) is determined by model-specific means to ensure stability of numerical computation along with sufficient accuracy of results. The pseudocode for a generic sequential time-stepped algorithm is shown in Figure 10.8. It consists

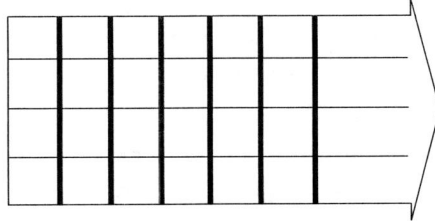

FIGURE 10.7 Schematic of a time-stepped simulation.

of a simple simulation loop which typically terminates when a certain simulation time is reached. Within the loop, each iteration consists of advancing the state of the entire set of modeled entities by a time step. The variable **t**$_{now}$ maintains the current simulation time, which is updated by time step in step 1.1. All elements are advanced using their **advance()** method to the new simulation time. From the loop, it can be seen that time-stepped execution is one of the easiest time advance methods to implement. Although this simple algorithm uses a constant time step during the entire execution, variants to this algorithm exist that vary the time step dynamically as permitted by the model.

10.2.1 Example: Heat Equation

We will illustrate the time-stepped algorithm using an example application. The application is a simulation of the diffusion process, such as heat transfer, which is a well-studied problem and has many applications. We will use the following one-dimensional version of the diffusion equation:

$$\frac{\partial Q}{\partial t} = \alpha_x \frac{\partial^2 Q}{\partial x^2} + \beta \tag{10.1}$$

For discretization of this continuous function, the *spatial* dimension is discretized by partitioning the space as a grid in the x dimension. Time is discretized into a grid with equidistant points, with the spacing fixed for all grid elements. Within each time step, the processing per grid element i in the 1-D grid can be performed by one of several known methods. In the following simple explicit method, q_i^{n+1} is the computed value of q_i at time step $n+1$.

$$q_i^{n+1} = q_i^n + \alpha_x \frac{q_{i-1}^n - 2q_i^n + q_{i+1}^n}{\Delta x^2} + \beta \tag{10.2}$$

This model can be executed to answer any of the several questions, such as (1) what is the value of a given q_i at some time t_n, and (2) what is the distribution of all values of q at a given time t_n. In the time-stepped algorithm, the **advance()** method updates q_i^n to q_i^{n+1} using the preceding equation. Of course, the new values are stored in temporary variables while all the values are updated, to correctly access the previous values for neighbors.

10.2.2 Parallelizing Time-Stepped Execution

We now turn to the question of how the time-stepped execution could be parallelized. This turns out to be quite easy. The sequential simulation loop of Figure 10.8 is modified to that in Figure 10.9. Grid elements are mapped to different processors, and the simulation loop is executed on every processor. Data exchange among neighboring grid elements is achieved by sending out a copy of the updated state to neighbors via interprocessor messages. Synchronization across processors is obtained by invoking a barrier call at the end of each time step, as shown in step 1.1 of the parallel algorithm. The barrier is a "flush barrier" which ensures that all messages destined to a processor are delivered before the barrier call returns. The use of the flush barrier ensures that all updates to neighboring grid elements from a previous iteration are incorporated before moving on to the next iteration. Note that the update messages are sent to other processors for only those grid elements that have neighbors outside their own processor.

```
1.    While not end of simulation
      /*Advance current simulation time*/
1.1   t_now += timestep
      /*Advance all entities to next timestep*/
1.2   For all (i): advance(i, t_now)
```

FIGURE 10.8 Time-stepped algorithm for sequential execution.

> 1. **While not end of simulation**
>   /*Synchronize with rest of processors to start the current timestep*/
> 1.1 **flush-barrier()**
>   /*Advance current simulation time*/
> 1.2 t_{now} += **timestep**
>   /*Advance all entities' state to next timestep*/
> 1.3 **For all (i on this processor): advance(i, t_{now})**
>   /*Send copy of new state to all neighboring entities*/
> 1.4 **For all (i on this processor having off-processor neighbors): send(state(i), neighbor-processors(i))**

FIGURE 10.9 Time-stepped algorithm for parallel execution.

10.3 Discrete-Event Execution

10.3.1 Execution Method

In discrete-event simulation, the system behavior is modeled in terms of actions at discrete points along the time axis. The evolution of the state of the system is fully expressed in terms of these actions. Each action, typically, is expressed as an instantaneous change in the system state, with causal dependencies among the actions. Thus, after effecting an instantaneous action, the model may dynamically determine a new set of future actions which are affected by this action. Each action is called an *event*. Since these events occur at discrete points in simulated time, this type of a simulation is called a *discrete-event simulation*.

Figure 10.10 shows a schematic of time advances occurring in discrete-event simulation. It illustrates the staggered time instants for updates to entities. The horizontal bands represent the timelines of entities in the model, and each vertical bar represents an event processed at a particular simulation time on an entity. Owing to the fact that update times are staggered, and also owing to the fact that future updates are scheduled while processing current updates, the simulation loop becomes slightly more complex than time-stepped simulation loop. The discrete-event simulation loop is shown in Figure 10.11. The simulation starts at time zero in step 1, and the entities are initialized in step 2, as part of which the entities schedule their initial set of events into the future. The main simulation loop starts in step 3. A priority queue data structure is used to store all scheduled events, such that the minimum timestamped event is always readily available to be dequeued. In each iteration of the loop, the minimum timestamped event is deleted from the event list. The event data structure contains the timestamp of the event along with the identity of the entity for whom the event is scheduled. The event is presented to the corresponding entity to process. As part of processing, the entity updates its state to the time of the event, and schedules new events, if any, to other entities (or to itself). The entity provides an event handler that processes events presented to the entity. Typically, there can be multiple event handlers per entity to handle different event types coming to the entity. In the algorithm in Figure 10.11, a generic handler called **process**() is assumed for the entity. Multiple event handlers can be added to the entity that are invoked within the **process**() method of that entity based on the type of event received.

10.3.2 Example: ATM Multiplexer

To illustrate the discrete-event execution method, we will use an example from the domain of telecommunications networks using asynchronous transfer mode (ATM) protocol. Consider a simple model shown in Figure 10.12 of a nonpreemptive ATM multiplexer (Fujimoto, 95) containing a buffer of size B (i.e., can hold at most B ATM cells). Suppose we are interested in measuring the cell loss probability (i.e., what fraction of incoming cells are to be discarded due to lack of space in buffer) and delay distributions on the

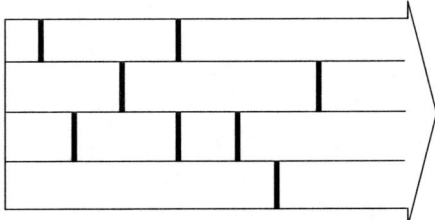

FIGURE 10.10 Schematic of updates in discrete-event simulation.

```
1.    t_now = 0
2.    For all entities (i) EventList.Insert(initial-event(i,dt_i))
3.    While EventList is not empty and t_now < end time
         /*Find the event with the earliest timestamp*/
3.1      E(i_min, t_min) = EventList.DeleteMin()
         /*Move global simulation time to the time of this event*/
3.2      t_now = t_min
         /*Let the entity process the event*/
3.3      Entity(i_min).process(E) /*This can schedule more events into EventList*/
```

FIGURE 10.11 Sequential discrete-event simulation algorithm.

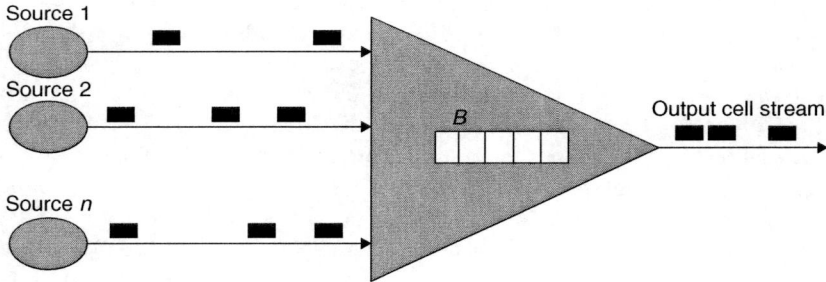

FIGURE 10.12 Schematic of a model of an ATM multiplexer.

queue (i.e., the amount of delay that each incoming cell experiences before being sent on output channel). This system can be modeled using multiple entities. The multiplexer itself is modeled as an entity, containing a model of the buffer and the output channel. One entity is used to model each ATM cell source. Cell source entities send events to the multiplexer entity to signal arrival of ATM cells from the source to the multiplexer on the corresponding input channel.

The state of the multiplexer entity is simplified in terms of four variables, as shown in Figure 10.13(a). All variables are initialized to zeros. The `qlen` variable is used to keep track of the current buffer occupancy; `sent` and `lost` are variables to accumulate statistics of the total number of cells transferred to the output link and the total number of cells dropped because of a full buffer respectively. The array `delays` measures the number of cells experiencing a given amount of delay, which in combination with the `sent` counter gives the cell delay distribution. Two event handlers are used to model actions of the multiplexer upon cell arrivals and departures. A cell arrival event occurs at the multiplexer entity to mark the arrival

```
Integer qlen;              if qlen < B                if qlen > 0
Integer sent;                     delays[qlen]++;           qlen--;
Integer lost;                     qlen++;                   sent++;
Integer delays [B];        else                             send(cell);
                                  lost++;
```

    (a)   (b)   (c)

FIGURE 10.13 Simple model of an ATM multiplexer. (a) State. (b) Cell arrival event handler. (c) Cell departure event handler.

of a cell from a source to the multiplexer. A cell departure event is scheduled by the multiplexer entity to signal the transfer of a cell from the buffer to the output channel.

 The code associated with the event handler in the multiplexer is shown in Figure 10.13(b) for processing a cell arrival event. If there is room in the buffer to add the cell, the `qlen` variable is incremented to mark that one more cell is added to the queue. The cell delay distribution is updated by noting that one more cell experienced a delay of `qlen` units of time. If there is no room to hold the new cell (i.e., the buffer is full), the cell is recorded as dropped by incrementing the `lost` counter. A cell departure event is scheduled by the multiplexer to periodically emit a cell, if there are any cells present in the buffer. The cell emission is marked by decrementing the `qlen` variable and incrementing the `sent` counter. The actual cell emission is performed in Figure 10.13(c) by sending a cell event on its output channel, which typically gets forwarded to the entity, if any, that is mapped to the output channel.

10.3.3 Parallelizing Discrete-Event Execution

How can discrete-event execution be parallelized? In other words, how can multiple processors be used to execute the same simulation such that it produces the same set of results as a sequential execution, albeit faster. A rich body of literature exists to address this question, and multiple approaches have been proposed (Fujimoto, 90b). In parallel simulation, model entities are mapped to different processors, and events among entities are exchanged via interprocessor communication (e.g., using shared memory or local area network). The crux of parallel discrete-event simulation (PDES) is the need to perform efficient synchronization such that the results from the parallel execution are the same as those from an equivalent sequential execution. This means that all events have to be processed in such a way that global timestamp order of processing is maintained for all events. This is achieved by processing all events local to each processor in strict order of nondecreasing timestamps and ensuring that no incoming events from other processors arrive in simulation past. Broadly, there are two main categories of PDES. A classification is shown in Figure 10.14. The main methods are (1) conservative parallel simulation and (2) optimistic parallel simulation.

 In conservative parallel simulation (Chandy, 89), at every processor, processing of the minimum timestamped local event is blocked until a guarantee is obtained that no event with a smaller timestamp will later arrive from other processors. This blocking can introduce idle time at the processor but will ensure that the event processing always strictly follows the timestamp order. Conservative parallel simulation, however, is constrained by an application-defined limitation called lookahead (explained later) which is necessary to permit concurrency.

 Optimistic parallel simulation (Fujimoto, 90a) is an approach by which the same timestamp ordered processing is ensured across all processors, but it is achieved as an asymptotic guarantee. In other words, the system might violate the timestamp ordered processing guarantee at certain times during execution, but uses corrective measures to undo the guarantee violations. Processors optimistically execute ahead and process locally minimum timestamped events without having to wait for absolute guarantees from other processors. When the other processors indeed end up generating lower timestamped events, the violation of the order is detected and the processor undoes the incorrect part of the optimistic computation, and resumes from the correct state.

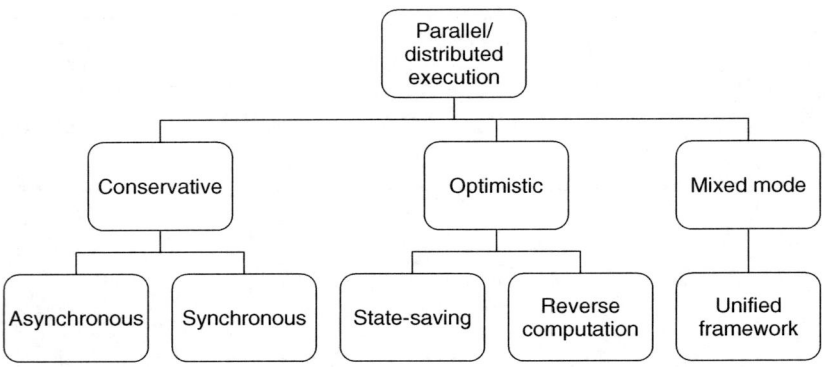

FIGURE 10.14 Classification of common parallel/distributed execution techniques.

10.3.3.1 Conservative Parallel Execution

A fundamental problem with conservative parallel simulation is concerned with the concept of *lookahead*. In the absence of the concept of lookahead, suppose any simulator that is processing an event with timestamp T can generate another event, whose timestamp is also equal to T, to another simulator. Moreover, this new event could be destined to any or all simulators. In such a scenario, to ensure timestamp-ordered processing, it is clear that there is little concurrency among federates. Only the event with the globally minimum timestamp in the entire system can be processed at its simulator, while all the rest of the simulators necessarily have to stay idle. Essentially, this degenerates to sequential execution, albeit with multiple simulators. Clearly, this is undesirable in interest of runtime performance. It becomes desirable to uncover concurrency among simulators to avoid such serialization. The concept of lookahead is defined to resolve this problem (Deelman, 01).

Lookahead is defined as the minimum increment in simulation time between an event and any new events generated during processing of that event. When this lookahead is greater than zero at all simulators, the parallel execution can experience concurrency. If the lookahead is zero for any federate (i.e., a simulator can generate events with zero delay), then the entire simulation suffers from serial execution (discounting unrelated events with equal timestamps at different simulators).

In simulation models, it is possible to extract lookahead by examining the minimum time for interactions to occur among entities. For example, signal transmission delays could be used to compute minimum propagation delays across the ATM source and multiplexer entities. In other models, it might be difficult to extract nonzero lookahead. Lookahead extraction is a topic of much research, and unfortunately remains a challenge in its generality.

A typical conservative parallel simulation algorithm is shown in Figure 10.15. A quantity called `lbts`, short for lower bound on incoming timestamps, is used to keep track of the smallest timestamp on events that can potentially arrive from other processors. If `lbts` equals infinity, it is clear that this loop simply degenerates to the sequential simulation loop in Figure 10.15. The complexity of the conservative algorithm is in the computation of `lbts`, in step 4.3.1. Assuming there are no events in transit in the interprocessor messaging network, it is easy to compute `lbts`, which is simply the minimum among all values passed to the **compute-lbts**() function by all processors. Once this value is computed, it should be corrected to take into account the events in transit, if any, among processors. This final value can now be used as the lower bound guarantee on incoming event timestamps, and the rest of the loop is a simple variant of the sequential loop.

An example scenario of conservative execution is shown in Figure 10.16. Let us assume each entity is mapped to a different processor, and consider the operation of the processor simulating the multiplexer entity. Suppose the multiplexer has just processed its events until time 8 (i.e., $t_{now}=8$). It now finds two events D@9 and A@10 in its event list. To determine if D@9 can be processed, it needs to compute the value of `lbts` and wait until `lbts` is at least 9. Assume that the modeler has specified a transmission

Model Execution

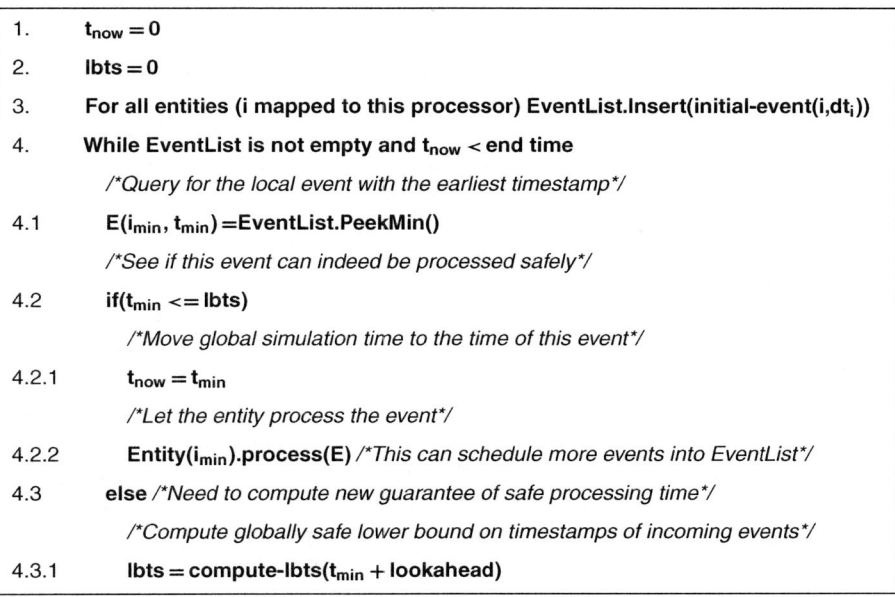

FIGURE 10.15 Conservative parallel discrete-event simulation algorithm.

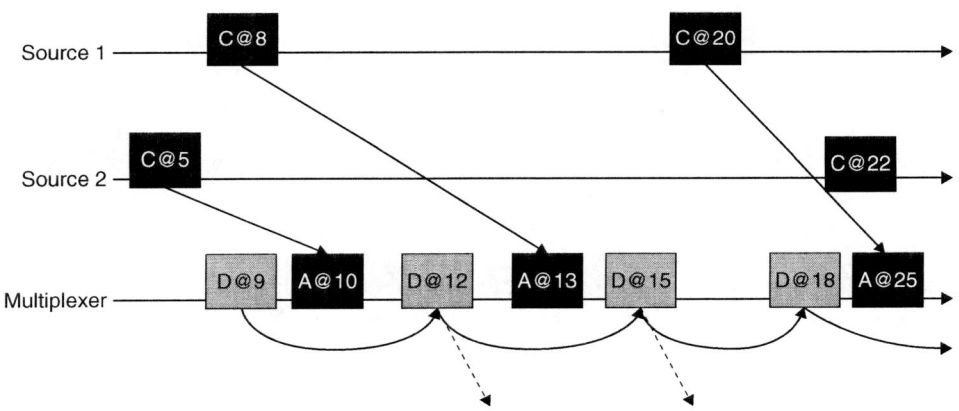

FIGURE 10.16 Example scenario of a two-source multiplexer model. Sources send cell arrival events A@t to the multiplexer, with a constant delay (lookahead) of 5 time units. Multiplexer schedules departure events D@t to itself every 3 time units.

delay (lookahead) of five time units from the sources to the multiplexer (the minimum time it takes for a cell to move from a source to the multiplexer). It is clear that the minimum timestamp on any cell arrival from source 1 or source 2 would be at least 13 (8+5). Thus, lbts at the multiplexer would be computed as 13. This enables the multiplexer to process all local events upto time 13, which allows it to safely process its events A@10 and D@12. Once the time reaches 13, this process repeats, by which a new lbts computation is initiated to determine a new (updated) guarantee on incoming timestamps.

10.3.3.2 Optimistic Parallel Execution

In optimistic parallel execution, local events are processed without having to wait for lbts guarantees, unlike conservative execution. Clearly, as a result, the system can sometimes find itself performing event processing in incorrect timestamp order. For example, in the scenario given in Figure 10.16, suppose the `lbts` value (13) is being computed. Conservative simulation simply waits until the `lbts` value is known. Optimistic simulation, on the other hand, goes ahead with processing the events A@10 and D@12 even while `lbts` is being computed. Suppose the `lbts` value is still being computed even after A@10 and D@12 are processed. The multiplexer continues its optimistic processing and processes D@15 as well. Later, the event A@13 from source 1 arrives and `lbts` is advanced to 13. Clearly, D@15 has been incorrectly processed ahead of A@13, and hence it needs to be corrected. The optimistic simulation then initiates a corrective action called rollback, which restores the state of the system back to the point of fault, namely to time 13. Once the rollback is completed, the incorrect effects of D@15 are fully undone, and hence the system is effectively in the same position as though it only processed until D@12 and never processed D@15. Event processing now resumes at A@13 correctly. The net gain is that the three events D@9, A@10, and D@12 have all been already processed (correctly) without having to wait for the actual `lbts` value. This overlap of computation with communication results in better concurrency in execution, thus increasing the overall speed of simulation.

The key element enabling this type of optimistic operation is the capability to rollback incorrect computation. For example, in our scenario, we assumed that the system state can be restored to its correct value before D@15 even though we have already processed D@15 and overwritten the state with newly computed values with D@15. How is it now possible to undo the effects of D@15? Also, D@15 scheduled a new event which should also be retracted. In general, optimistically processed events overwrite the entity state with newly updated values and also send out new events to other processors; all of these effects need to be carefully undone to restore the state to correct values for complete rollback. Let us consider the two distinct aspects of rollback: (1) state restoration and (2) event retraction.

For state restoration, several variants have been proposed, of which there are mainly two categories: (1) state saving and (2) reverse computation. In the state saving method, a copy of the system state is saved before it is modified in the optimistic event processing. In other words, a copy of the values of all system variables is saved before being overwritten by the event handler code. In the multiplexer example, a copy of all variables in Figure 10.17(a) is saved in a buffer before the departure event handler code of Figure 10.17(c) is invoked for the event D@15. A pointer to this buffer is stored in the event data structure of D@15 itself. When it is necessary to rollback D@15, the buffer is retrieved using the pointer, and the old values saved in the buffer are used to overwrite the multiplexer state variables, thus effectively restoring the state to correct values before D@15 was incorrectly processed.

An alternative method for state restoration is called reverse computation (Carothers, 99). In this approach, instead of using the copy-restore method, inverse event handlers are used to undo the effects of incorrectly executed event handlers, on a statement-by-statement basis of the handler code, in reverse order of invocation. For example, the inverse event handlers shown in Figure 10.17 are defined and used to undo incorrect invocations to event handlers.

A bit variable is added to the state to keep track of whether the "if" statement body has been evaluated in the forward (incorrect) execution. This single bit variable, stored in the event data structure, is sufficient to later restore the rest of the variables to their correct previous values. This is achieved by defining the

```
Integer qlen;
Integer sent;
Integer lost;
Integer delays[B];
bit b;
```

(a)

```
b = 0
if qlen > 0
    b = 1
    qlen--;
    sent++;
    send(cell);
```

(b)

```
if b == 1
    retract(cell);
    --sent;
    ++qlen;
```

(c)

FIGURE 10.17 Modified version of departure event handler of ATM multiplexer for reverse computation-based rollback. (a) Modified state (one bit variable added). (b) Departure event handler—Forward. (c) Departure event handler—Reverse.

reverse event handler shown in Figure 10.17(c). This is obtained by inverting the order of the statements in the forward event handler in Figure 10.17(b) and inverting the operations within each statement.

Event retraction within a processor is relatively easily achieved, simply removing the event from local priority queue data structure of event list. Retracting events that have been forwarded to other processors is slightly more involved, which requires forwarding a retraction request to the destination processor. Two cases arise: the retraction request arrives the destination processor after the original event already has arrived at that processor; the retraction request arrives at the destination processor even before the original event arrived at that processor. The first case is easy to handle, since the original event is already identifiable—the event is removed from the event list if it has not been processed, or it is first rolled back and then removed if it has been already processed. If the retraction arrives before the event arrives, the retraction is simply buffered and until the actual event is received and then both the retraction and the event are nullified.

10.3.3.3 Mixed-Mode Parallel Execution

In mixed-mode parallel execution, the system supports a mixture of optimistic and conservative execution *within* each processor. Entities can specify whether their events are to be processed using optimistic (rollback-based) synchronization or conservative (blocking-based) synchronization. As a result, the simulation algorithm becomes considerably more complex and hence not discussed here.

10.4 Summary

Model execution involves computing the evolution of state values for a set of interacting entities in a model over time. Computing the state evolution can be performed using different time advancing techniques, and executed sequentially or in parallel. Popular techniques include time-stepped execution and discrete-event execution. While sequential execution of time-stepped techniques can be parallelized in a relatively straightforward manner, parallelization of discrete-event execution involves resolution of important synchronization issues. Parallel discrete-event synchronization approaches broadly fall under two categories: conservative and optimistic methods. This chapter described the time-stepped and discrete-event methods with examples and presented algorithms for their implementation for sequential and parallel execution.

References

Bagrodia, R., R. Meyer, M. Takai, Y. Chen, X. Zeng, J. Martin, B. Park, and H. Song. "Parsec: A Parallel Simulation Environment for Complex Systems." *IEEE Computer*, 31(10) (1998): 77–85.

Balci, O. "The Implementation of Four Conceptual Frameworks for Simulation Modeling in High-Level Languages." Paper presented at the Winter Simulation Conference, 1988.

Banks, J., J. S. Carson II, and B. L. Nelson. *Discrete-Event System Simulation*. Upper Saddle River, NJ: Prentice-Hall, 1996.

Carothers, C., K. S. Perumalla, and R. M. Fujimoto. "Efficient Optimistic Parallel Simulations Using Reverse Computation." *ACM Transactions on Modeling and Computer Simulation*, 9(3) (1999): 224–253.

Chandy, K. M. and R. Sherman. "Space, Time, and Simulation." In *Proceedings of the SCS Multiconference on Distributed Simulation*, pp. 53–57, SCS Simulation Series, 1989.

Das, S., R. M. Fujimoto, K. Panesar, D. Allison, and M. Hybinette. "GTW: A Time Warp System for Shared Memory Multiprocessors." In *Proceedings of the 1994 Winter Simulation Conference*, pp. 1332–1339, 1994.

Deelman, E., R. Bagrodia, R. Sakellariou, and V. Adve. "Improving Lookahead in Parallel Discrete Event Simulations of Large-Scale Applications Using Compiler Analysis." In *Proceedings of the 15th Workshop on Parallel and Distributed Simulation*, pp. 5–13, 2001.

Fishwick, P. A. *Simulation Model Design and Execution: Building Digital Worlds*, 1st ed., Englewood Cliffs, NJ: Prentice-Hall, 1995.

Fujimoto, R. M. "Optimistic Approaches to Parallel Discrete Event Simulation." *Transactions of the Society for Computer Simulation*, 7(2) (1990a): 153–191.

Fujimoto, R. M. "Parallel Discrete Event Simulation." *Communications of the ACM*, 33(10) (1990b): 30–53.

Fujimoto, R. M., I. Nikolaidis, and A. C. Cooper. "Parallel Simulation of Statistical Multiplexers." *Journal of Discrete Event Dynamic Systems*, 5 (1995): 115–140.

Nance, R. E. "The Time and State Relationships in Simulation Modeling." *Communications of the ACM*, 24(4) (1981): 173–179.

Nance, R. E. "A History of Discrete Event Simulation Programming Languages." *ACM SIGPLAN Notices*, 28(3) (1993): 149–75.

Nicol, D. and P. Heidelberger. "Parallel Execution for Serial Simulators." *ACM Transactions on Modeling and Computer Simulation*, 6(3) (1996): 210–242.

Nutaro, J. 2003. Adevs: A Discrete Event System Simulator. In http://www.ece.arizona.edu/~nutaro (accessed 2006/06/01).

Perumalla, K. S. and R. M. Fujimoto. "Efficient Large-Scale Process-Oriented Parallel Simulations." In *Proceedings of the Winter Simulation Conference*, 459–466, 1998.

Pidd, M. "Object Orientation and Three Phase Simulation." Paper presented at the Winter Simulation Conference, Arlington, VA, 1992.

Schriber, T. J. *Simulation Using GPSS*. New York: Wiley, 1974.

Tocher, K. D. *The Art of Simulation*. London: English Universities Press, 1963.

Zeigler, B. P., H. Praehofer, and T. G. Kim. *Theory of Modeling and Simulation*, 2nd ed., New York: Academic Press, 2000.

11
Discrete-Event Simulation of Continuous Systems*

James Nutaro
Oak Ridge National Laboratory

11.1 Introduction 11-1
11.2 Simulating a Single Ordinary Differential
    Equation 11-2
11.3 Simulating Coupled Ordinary Differential
    Equations 11-6
11.4 DEVS Representation of Discrete-Event
    Integrators 11-8
11.5 The Heat Equation 11-13
11.6 Conservation Laws 11-16
11.7 Two-Point Integration Schemes 11-19
11.8 Conclusions 11-21

11.1 Introduction

Computer simulation of a system described by differential equations requires that some element of the system be approximated by discrete quantities. There are two system aspects that can be made discrete: time and state. When time is discrete, the differential equation is approximated by a difference equation (i.e., a discrete-time system), and the solution is calculated at fixed points in time. When the state is discrete, the differential equation is approximated by a discrete-event system. Events correspond to jumps through the discrete state space of the approximation.

The essential feature of a discrete time approximation is that the resulting difference equations map a discrete time set to a continuous state set. The time discretization need not be regular. It may even be revised in the course of a calculation. Nonetheless, the elementary features of a discrete time base and continuous state space remain.

The basic feature of a discrete-event approximation is opposite that of a discrete time approximation. The approximating discrete-event system is a function from a continuous time set to a discrete state set. The state discretization need not be uniform, and it may even be revised as the computation progresses.

These two different types of discretizations can be visualized by considering how the function $x(t)$, shown in Figure 11.1(a), might be reduced to discrete points. In a discrete time approximation, the value of the function is observed at regular intervals of time. This kind of discretization is shown in

---

*This chapter has been authored by UT-Battelle, LLC, under contract De-AC05-00OR22725 with the U.S. Department of Energy. The United States Government retains and the publisher, by accepting the article for publication, acknowledges that the Government retains a nonexclusive, paid-up, irrevocable, worldwide license to publish or reproduce the published form of this chapter, or allow others to do so, for the Government's purposes.

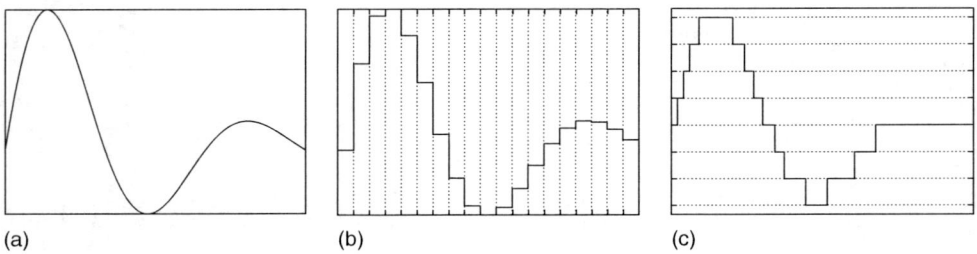

FIGURE 11.1 Time and state discretizations of a system. (a) Continuous, (b) Discrete time, and (c) Discrete state.

Figure 11.1(b). In a discrete-event approximation, the function is sampled when it takes on regularly spaced values. This type of discretization is shown in Figure 11.1(c).

From an algorithmic point of view, these two types of discretizations are widely divergent. The first approach emphasizes the simulation of coupled difference equations. Some distinguishing features of a difference equation simulator are nested "for" loops (used to compute function values at each time step), single instruction multiple data (SIMD) type parallel computing (using, e.g., vector processors or automated "for" loop parallelization, see Hwang, 1993), and good locality of reference.

The second approach emphasizes the simulation of discrete-event systems. The main computational features of a discrete-event simulation are very different from a discrete time simulation. Foremost among them are event scheduling, poor locality of reference, and multiple instruction multiple data (MIMD) type asynchronous parallel algorithms (see, e.g., Hwang, 1993). The essential data structures are different, too. Where difference equation solvers exploit a matrix representation of the system coupling, discrete-event simulations often require different, but structurally equivalent, data structures (e.g., influence graphs).

Mathematically, however, they share several features. The approximation of functions via interpolation and extrapolation are central to both. Careful study of error bounds, stability regimes, conservation properties, and other elements of the approximating machinery is essential. It is not surprising that theoretical aspects of differential operators, and their discrete approximations, have a prominent place in the study of both discrete time and discrete-event numerical methods.

This confluence of applied mathematics, mathematical systems theory, and computer science makes the study of discrete-event numerical methods particularly challenging. This chapter presents some basic results, while avoiding more advanced topics. The goal is to present essential concepts clearly, and so portions of this material will, no doubt, seem underdeveloped to a specialist. Pointers into the appropriate literature are provided for readers who want a more in-depth treatment.

The remainder of this chapter is organized as follows. In Section 11.2, discrete-event simulation of a single ordinary differential equation is introduced. This technique is expanded to systems of ordinary differential equations in Section 11.3. A general scheme for representing discrete-event integration schemes using the discrete-event system specification (DEVS) is described in 11.4, and the first-order method introduced in Sections 11.2 and 11.3 are recast in terms of DEVS. In Sections 11.5 and 11.6, the first-order DEVS integrator is applied to simulating partial differential equations using finite differences in space and discrete events in time. Second-order discrete-event integration schemes are discussed in Section 11.7. In Section 11.8, the chapter concludes with a brief, but broad, overview of related research.

11.2 Simulating a Single Ordinary Differential Equation

Consider an ordinary differential equation that can be written in the form

$$\dot{x}(t) = f(x(t)) \qquad (11.1)$$

Algorithm 11.1 Simulating a single ordinary differential equation.

$t \leftarrow 0$
$x \leftarrow x(0)$
while terminating condition not met **do**
    print t, x
    if $f(x) = 0$ **then**
        $h \leftarrow \infty$
    else
        $h \leftarrow \frac{D}{|f(x)|}$
    end if
    if $h = \infty$ **then**
        stop simulation
    else
        $t \leftarrow t + h$
        $x \leftarrow x + D\, sgn(f(x))$
    end if
end while

A discrete-event approximation of this system can be obtained in, at least, two different ways. To begin, consider the Taylor series expansion

$$x(t+h) = x(t) + h\dot{x}(t) + \sum_{n=2}^{\infty} \frac{h^n}{n!} x^{(n)}(t) \qquad (11.2)$$

If we fix the quantity $D = |x(t+h) - x(t)|$, then the time required for a change of size D to occur in $x(t)$ is approximately

$$h = \begin{cases} \frac{D}{|\dot{x}(t)|} & \text{if } \dot{x}(t) \neq 0 \\ \infty & \text{otherwise} \end{cases} \qquad (11.3)$$

This approximation drops the summation term in Eq. (11.2) and rearranges what is left to obtain h. Algorithm 11.1 uses this approximation to simulate a system described by Eq. (11.1). The procedure computes successive approximations to $x(t)$ on a grid in the state space of the system. The resolution of the state space grid is D, and h approximates the time at which the solution jumps from one state space grid point to the next.

The *sgn* function, on line 14 in Algorithm 11.1, is defined to be

$$sgn(q) = \begin{cases} -1 & \text{if } q < 0 \\ 0 & \text{if } q = 0 \\ 1 & \text{if } q > 0 \end{cases}$$

The expression $D\, sgn(f(x))$ on line 14 could, in this instance, be replaced by $hf(x)$ because

$$hf(x) = \frac{D}{|f(x)|} f(x) = D\, sgn(f(x))$$

However, the expression $D\, sgn(f(x))$ highlights the fact that the state space, and not the time domain, is discrete. Note, in particular, that the computed values of x are restricted to $x(0) + kD$, where k is an integer and D is the state space grid resolution. In contrast, the computed values of t can take any value.

TABLE 11.1 Simulation of $\dot{x}(t) = -x(t)$, $x(0) = 1$, Using Algorithm 11.1 with $D = 0.15$

t	x	$f(x)$	h
0.0	1.0	−1.0	0.15
0.15	0.85	−0.85	0.1765
0.3265	0.7	−0.7	0.2143
0.5408	0.55	−0.55	0.2727
0.8135	0.4	−0.4	0.3750
1.189	0.25	−0.25	0.6
1.789	0.1	−0.1	1.5
3.289	−0.05	0.05	3.0
6.289	0.1	−0.1	1.5
7.789	−0.05	0.05	3.0

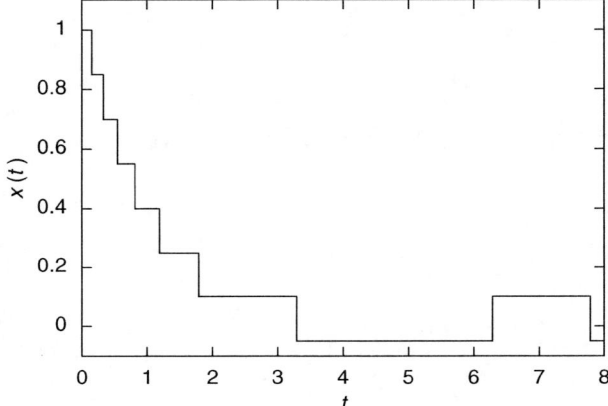

FIGURE 11.2 Computed solution of $\dot{x}(t) = -x(t)$, $x(0) = 1$ with $D = 0.15$.

The procedure can be demonstrated with a simulation of the system $\dot{x}(t) = -x(t)$, with $x(0) = 1$ and $D = 0.15$. Each step of the simulation is shown in Table 11.1. Figure 11.2 shows the computed $x(t)$ as a function of t.

The approximation given by Eq. (11.3) can be obtained in a second way. Consider the integral

$$\left| \int_{t_0}^{t_0+h} f(x(t)) \, dt \right| = D \qquad (11.4)$$

As before, D is the resolution of the state space grid and h is the time required to move from one point in the state space grid to the next. For time to move forward, it is required that $h > 0$. In the interval $(t_0, t_0 + h)$, the function $f(x(t))$ can be approximated by $f(x(t_0))$. Substituting this approximation into Eq. (11.4) and solving for h gives

$$h = \begin{cases} \frac{D}{|f(x(t_0))|} & \text{if } f(x(t_0)) \neq 0 \\ \infty & \text{otherwise} \end{cases}$$

This approach to obtaining h gives the same result as before.

There are two important questions that need answering before this can be considered a viable simulation procedure. First, can the discretization parameter D be used to bound the error in the simulation? Second, under what conditions is the simulation procedure stable? That is, under what circumstances can the error at the end of an arbitrarily long simulation run be bounded? Several authors (see, e.g., Zeigler et al., 2000; Kofman, 2004; Nutaro, 2003) have addressed these questions in a rigorous way. Happily, the answer to the first question is a "yes!" The second question, while answered satisfactorily for linear systems, remains (not surprisingly) largely unresolved for nonlinear systems.

The first question can be answered as follows: If $\dot{x}(t) = f(x(t))$ describes a stable and time-invariant system (see Szidarovszky and Bahill [1998], or any other introductory systems textbook), then the error at any point in a simulation run is proportional to D. The constant of proportionality is determined by the system under consideration. The time-invariant caveat is needed to avoid a situation in which the first derivative can change independently of $x(t)$ (i.e., the derivative is described by a function $f(x(t), t)$, rather than $f(x(t))$). In practice, this problem can often be overcome by treating the time-varying element of $f(x(t), t)$ as a quantized input to the integrator (see, e.g., Muzy et al., 2005).

The linear dependence of the simulation error on D is demonstrated for two different systems in Figure 11.3 and Figure 11.4. In these examples, $x(t)$ is computed until the time of next event exceeds a preset threshold. The error is determined at the last event time by taking the difference of the computed and known solutions. This linear dependency is strongly related to the fact that the scheme is exact when $x(t)$ is a line, or, equivalently, when the system is described by $\dot{x}(t) = k$, where k is a constant.

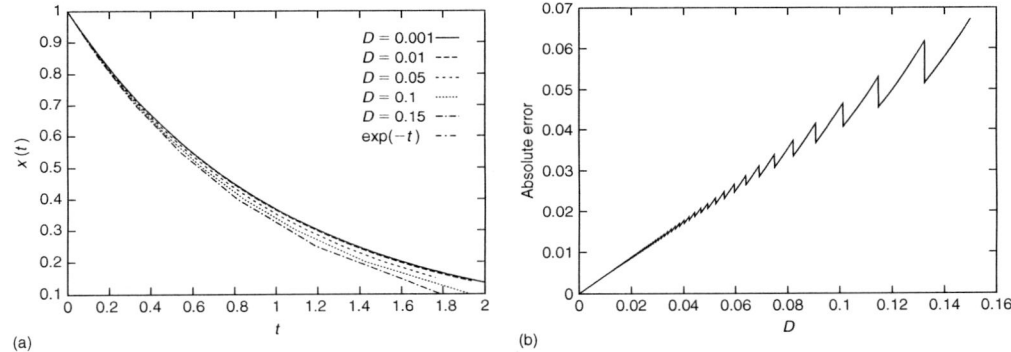

FIGURE 11.3 Error in the computed solution of $\dot{x}(t) = -x(t)$, $x(0) = 1$. (a) Comparison of computed and exact solutions. (b) Absolute error as a function of D.

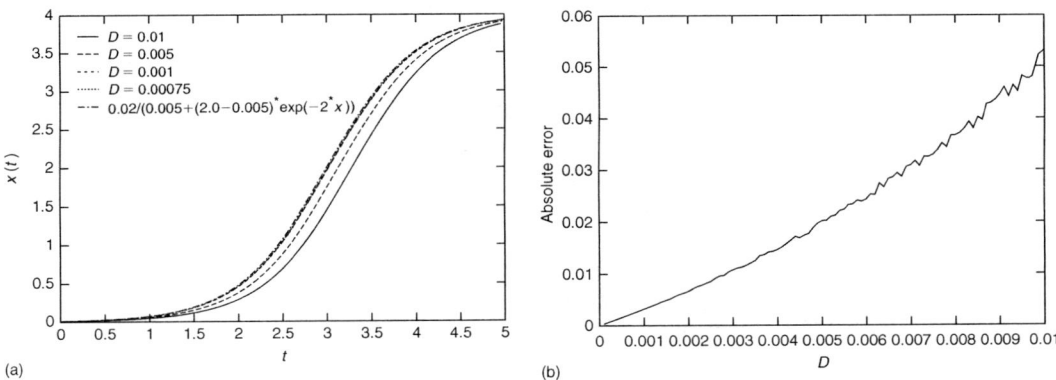

FIGURE 11.4 Error in the computed solution of $\dot{x}(t) = (2 - 0.5x(t))x(t)$, $x(0) = 0.01$. (a) Comparison of computed and exact solutions. (b) Absolute error as a function of D.

11.3 Simulating Coupled Ordinary Differential Equations

Algorithm 11.1 can be readily extended to sets of coupled ordinary differential equations. Consider a system described in the following form:

$$\dot{\bar{x}}(t) = \bar{f}(\bar{x}(t)) \tag{11.5}$$

where $\bar{x}(t)$ is the vector

$$[x_1(t), x_2(t), \ldots, x_m(t)]$$

and $\bar{f}(\bar{x}(t))$ is a function vector

$$[f_1(\bar{x}(t)), f_2(\bar{x}(t)), \ldots, f_m(\bar{x}(t))].$$

As before, we construct a grid in the m-dimensional state space. The grid points are regularly spaced by a distance D along the state space axes. To simulate this system, four variables are needed for each x_i, and so $4m$ variables in total. These variables are:

- x_i, the position of state variable i on its state space axis;
- tN_i, the time until x_i reaches its next discrete point on the ith state space axis;
- y_i, the last grid point occupied by the variable x_i; and
- tL_i, the last time at which the variable x_i was modified.

The x_i and y_i are necessary because the function $f_i(\cdot)$ is computed only at grid points in the discrete state space. Because of this, the motion of the variable x_i along its state space axis is described by a piecewise constant velocity. This velocity is computed using the differential function $f_i(\cdot)$ and the vector $\bar{y} = [y_1, \ldots, y_m]$. The value of y_i is updated when x_i reaches a state space grid point. The time required for the variable x_i to reach its next grid point is computed as

$$h = \begin{cases} \frac{D - |y_i - x_i|}{|f_i(\bar{y})|} & \text{if } f_i(\bar{y}) \neq 0 \\ \infty & \text{otherwise} \end{cases} \tag{11.6}$$

The quantity D is the distance separating grid points along the axis of motion, $|x_i - y_i|$ the distance already traveled along the axis, and $f_i(\bar{y})$ the velocity on the ith state space axis.

With Eq. (11.6), and an extra variable t to keep track of the simulation time, the behavior of a system described by Eq. (11.5) can be computed with Algorithm 11.2.

To illustrate the algorithm, consider the coupled linear system

$$\dot{x}_1(t) = -x_1(t) + 0.5 x_2(t) \tag{11.7}$$

$$\dot{x}_2(t) = -0.1 x_2(t)$$

with $x_1(0) = x_2(0) = 1$ and $D = 0.1$. Table 11.2 gives a step-by-step account of the first five iterations of Algorithm 11.2 applied to this system. The output values computed by the procedure are plotted in Figure 11.5 (note that the figure shows the results beyond the iterations listed in the table). Each row in the table shows the computed values at the end of an iteration (i.e., just prior to repeating the while loop). Blank entries indicate that the variable value did not change in that iteration. The blank entries, and the irregular time intervals that separate iterations, highlight the fact that this is a discrete-event simulation. An event is the arrival of a state variable at its next grid point in the discrete state space. Clearly, not every variable arrives at its next state space point at the same time, and so event scheduling provides a natural way to think about the evolution of the system.

Stability and error properties in the case of coupled equations are more difficult to reason about, but they generally reflect the one-dimensional case. In particular, the simulation procedure is stable, in the sense that the error can be bounded at the end of an arbitrarily long run, when it is applied to a stable

Discrete-Event Simulation of Continuous Systems

TABLE 11.2 Simulation of Two Coupled Ordinary Differential Equations on a Discrete State Space Grid

t	x_1	$\dot{x}_1$	y_1	tL_1	h_1	x_2	$\dot{x}_2$	y_2	tL_2	h_2
0	1	−0.5	1	0	0.2	1	−0.1	1	0	1
0.2	0.9	−0.4	0.9	0.2	0.25					
0.45	0.8	−0.3	0.8	0.45	0.3333					
0.7833	0.7	−0.2	0.7	0.7833	0.5					
1	0.6567	−0.25		1.0	0.2267	0.9	−0.09	0.9	1	1.111

Algorithm 11.2 Simulating a system of coupled ordinary differential equations.

$t \leftarrow 0$
for all $i \in [0, m]$ **do**
    $tL_i \leftarrow 0$
    $y_i \leftarrow x_i(0)$
    $x_i \leftarrow x_i(0)$
end for
while terminating condition not met **do**
    print $t, y_1, \ldots, y_m$
    for all $i \in [0, m]$ **do**
        $tN_i \leftarrow tL_i + h_i$, where h_i is given by Eq. (11.6)
    end for
    $t \leftarrow min\{tN_1, tN_2, \ldots, tN_m\}$
    Copy $\bar{y}$ to a temporary vector $\bar{y}_{tmp}$
    for all $i \in [0, m]$ such that $tN_i = t$ **do**
        $y_i \leftarrow x_i + h_i f_i(\bar{y}_{tmp})$
        $x_i \leftarrow y_i$
        $tL_i \leftarrow t$
    end for
    for all $j \in [0, m]$ such that a changed y_i alters the value of $f_j(\bar{y})$ and $tN_j \neq t$ **do**
        $x_j \leftarrow x_j + (t - tL_j) f_j(\bar{y}_{tmp})$
        $tL_j \leftarrow t$
    end for
end while

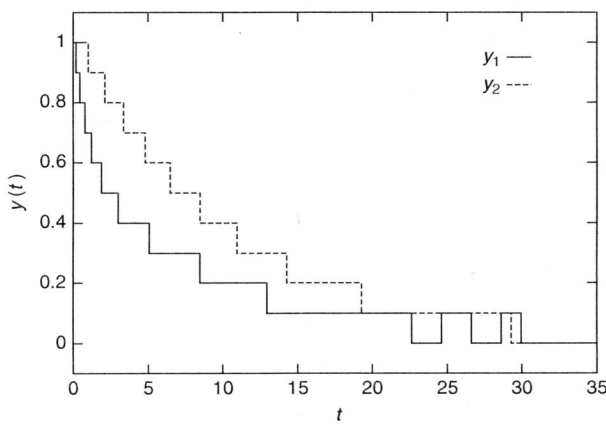

FIGURE 11.5 Plot of $y(t)$ for the calculation shown in Table 11.2.

and time invariant linear system (see Kofman, 2004; Zeigler et al., 2000). The final error resulting from the procedure is proportional to the state space grid resolution D (see Kofman, 2004; Zeigler et al., 2000).

11.4 DEVS Representation of Discrete-Event Integrators

It is useful to have a compact representation of the integration scheme that is readily implemented on a computer, can be extended to produce new schemes, and provides an immediate support for parallel computing. The DEVS satisfies this need. A detailed treatment of DEVS can be found in Zeigler et al. (2000). Several simulation environments for DEVS are available online, e.g., PowerDEVS (Kofman et al., 2003), adevs (Muzy and Nutaro, 2005), DEVSJAVA (Zeigler and Sarjoughian, 2005), CD++ (Wainer, 2002), and JDEVS (Filippi and Bisgambiglia, 2004) to name just a few.

DEVS uses two types of structures to describe a discrete-event system. Atomic models describe the behavior of elementary components. Here, an atomic model will be used to represent individual integrators and differential functions. Coupled models describe collections of interacting components, where components can be atomic and coupled models. In this application, a coupled model describes a system of equations as interacting integrators and function blocks.

An atomic model is described by a set of inputs, set of outputs, and set of states, a state transition function decomposed into three parts, an output function, and a time advance function. Formally, the structure is

$$M = <X, Y, S, \delta_{int}, \delta_{ext}, \delta_{con}, \lambda, ta>$$

where

X is a set of inputs,
Y is a set of outputs,
S is a set of states,
$\delta_{int}: S \to S$ is the internal state transition function,
$\delta_{ext}: Q \times X^b \to S$ is the external state transition function with $Q = \{(s,e) | s \in S \ \& \ 0 \leq e \leq ta(s)\}$ and X^b is a bag of values appearing in X,
$\delta_{con}: S \times X^b \to S$ is the confluent state transition function,
$\lambda: S \to Y$ is the output function, and
$ta: S \to \mathbb{R}$ is the time advance function.

The external transition function describes how the system changes state in response to input. When input is applied to the system, it is said that an external event has occurred. The internal transition function describes the autonomous behavior of the system. When the system changes state autonomously, an internal event is said to have occurred. The confluent transition function determines the next state of the system when an internal event and external event coincide. The output function generates output values at times that coincide with internal events. The output values are determined by the state of the system just prior to the internal event. The time advance function determines the amount of time that must elapse before the next internal event will occur, assuming that no input arrives in the interim.

Coupled models are described by a set of components and a set of component output to input mappings. For our purpose, we can restrict the coupled model description to a flat structure (i.e., a structure composed entirely of atomic models) without external input or output coupling (i.e., the component models cannot be affected by elements outside of the network). With these restrictions, a coupled model is described by the structure

$$N = <\{M_k\}, \{z_{ij}\}>$$

where

$\{M_k\}$ is a set of atomic models, and
$\{z_{ij}\}$ is a set of output to input maps $z_{ij}: Y_i \to X_j \cup \{\Phi\}$

where the i and j indices correspond to M_i and M_j in $\{M_k\}$ and Φ is the nonevent.

The output-to-input maps describe how atomic models affect one another. The output-to-input map is, in this application, somewhat overgeneralized and could be replaced with more conventional descriptions of computational stencils and block diagrams. The nonevent is used, in this instance, to represent components that are not connected. That is, if component i does not influence component j, then $z_{ij}(y_i) = \Phi$, where $y_i \in Y_i$.

These structures describe what a model can do. A canonical simulation algorithm is used to generate dynamic behavior from the description. In fact, Algorithms 11.1 and 11.2 are special cases of the DEVS simulation procedure. The generalized procedure is given as Algorithm 11.3. Its description uses the same variables as Algorithm 11.2 wherever this is possible. Algorithm 11.3 assumes a coupled model N, with a component set $\{M_1, M_2, \ldots, M_n\}$, and a suitable set of output-to-input maps. For every component model M_i, there is a time of last event and time of next event variable, tL_i and tN_i, respectively. There are also state, input, and output variables s_i, x_i, and y_i, in addition to the basic structural elements (i.e., state transition functions, output function, and time advance function). The variables x_i and y_i are bags, with elements taken from the input and output sets X_i and Y_i, respectively. The simulation time is kept in variable t.

Algorithm 11.3 DEVS simulation algorithm.

$t \leftarrow 0$
for all $i \in [1, n]$ **do**
  $tL_i \leftarrow 0$
  set s_i to the initial state of M_i
end for
while terminating condition not met **do**
  for all $i \in [1, n]$ **do**
    $tN_i \leftarrow tL_i + ta_i(s_i)$
    Empty the bags x_i and y_i
  end for
  $t \leftarrow min\{tN_i\}$
  for all $i \in [1, n]$ **do**
    if $tN_i = t$ **then**
      $y_i \leftarrow \lambda_i(s_i)$
      for all $j \in [1, n]$ & $j \neq i$ & $z_{ij}(y_i) \neq \Phi$ **do**
        Add $z_{ij}(y_i)$ to the bag x_j
      end for
    end if
  end for
  for all $i \in [1, n]$ **do**
    if $tN_i = t$ & x_i is empty **then**
      $s_i \leftarrow \delta_{int,i}(s_i)$
      $tL_i \leftarrow t$
    else if $tN_i = t$ & x_i is not empty **then**
      $s_i \leftarrow \delta_{con,i}(s_i, x_i)$
      $tL_i \leftarrow t$
    else if $tN_i \neq t$ & x_i is not empty **then**
      $s_i \leftarrow \delta_{ext,i}(s_i, t - tL_i, x_i)$
      $tL_i \leftarrow t$
    end if
  end for
end while

To map Algorithm 11.2 into a DEVS model, each of the x variables is associated with an atomic model called an integrator. The input to the integrator is the value of the differential function, and the output of the integrator is the appropriate y variable. The integrator has four state variables

- q_l, the last output value of the integrator;
- q, the current value of the integral;
- $\dot{q}$, the last known value of the derivative; and
- σ, the time until the next output event.

The integrator input and output events are real numbers. The value of an input event is the derivative at the time of the event. An output event gives the value of the integral at the time of the output.

The integrator generates an output event when the integral of the input changes by D. More generally, if Δq is the desired change, $[t_0, T]$ the interval over which the change occurs, and $f(x(t))$ the first derivative of the system, then

$$\int_0^T f(x(t_0 + t))dt = F(T) = \Delta q \qquad (11.8)$$

The function $F(T)$ gives the change in $x(t)$ over the interval $[t_0, T]$. Eq. (11.8) is used in two ways. If $F(T)$ and Δq are known, then the time advance of the discrete-event integrator is found by solving for T. If $F(T)$ and T are known, then the next state of the integrator is given by $q + F(T)$, where T is equal to the elapsed time (for an external event) or time advance (for an internal event).

The integration scheme used by Algorithms 11.1 and 11.2 approximates $f(x(t))$ with a piecewise constant function. At any particular time, the value of the approximation is given by the state variable $\dot{q}$. Using $\dot{q}$ in place of $f(x(t_0 + T))$ in Eq. (11.8) gives

$$\int_0^T \dot{q}\, dt = \dot{q}T$$

When $\dot{q}$ and T are known, then the function

$$\hat{F}(T, \dot{q}) = \dot{q}T \qquad (11.9)$$

approximates $F(T)$. Because T must be positive (i.e., we are simulating forward in time), the inverse of Eq. (11.9) cannot be used to compute the time advance. However, the absolute value of the inverse

$$\hat{F}^{-1}(\Delta q, \dot{q}) = \begin{cases} \frac{\Delta q}{|\dot{q}|} & \text{if } \dot{q} \neq 0 \\ \infty & \text{otherwise} \end{cases} \qquad (11.10)$$

is suitable.

The state transition, output, and time advance functions of the integrator can be defined in terms of Eq. (11.9) and (11.10). This gives

$$\delta_{int}((q_l, q, \dot{q}, \sigma)) = (q + \hat{F}(\sigma, \dot{q}), q + \hat{F}(\sigma, \dot{q}), \dot{q}, \hat{F}^{-1}(D, \dot{q}))$$
$$\delta_{ext}((q_l, q, \dot{q}, \sigma), e, x) = (q_l, q + \hat{F}(e, \dot{q}), x, \hat{F}^{-1}(D - |q + \hat{F}(e, \dot{q}) - q_l|, x))$$
$$\delta_{con}((q_l, q, \dot{q}, \sigma), x) = (q + \hat{F}(\sigma, \dot{q}), q + \hat{F}(\sigma, \dot{q}), x, \hat{F}^{-1}(D, x))$$
$$\lambda((q_l, q, \dot{q}, \sigma)) = q + \hat{F}(\sigma, \dot{q})$$
$$ta((q_l, q, \dot{q}, \sigma)) = \sigma$$

In this definition, $\hat{F}$ computes the next value of the integral using the previous value, the approximation of $f(x(t))$ (i.e., $\dot{q}$), and the time elapsed since the last state transition. The time that will be needed for the integral to change by an amount D is computing using $\hat{F}^{-1}$. The arguments to $\hat{F}^{-1}$ are the distance remaining (i.e., D minus the distance already traveled) and the speed with which the distance is being covered (i.e., the approximation of $f(x(t))$).

An implementation of this definition is shown in Figure 11.6. This implementation is for the adevs simulation library. The implementation is simplified by taking advantage of two facts. First, the output

```cpp
class Integrator: public atomic {
    public:
        /* Arguments are the initial variable value, variable index,
        integration quantum, and an array for storing output values. */
        Integrator(double q0, int index, double D, double* x):
        atomic(),index(index),q(q0),D(D),x(x) { x[index] = q; }
        /* Initialize the state prior to start of the simulation. */
        void init() {
            dq = f(index,x); compute_sigma();
        }
        /* DEVS state transition functions. */
        void delta_int() {
            q = x[index]; dq = f(index,x); compute_sigma();
        }
        void delta_ext(double e, const adevs_bag<PortValue>& xb) {
            q += e*dq; dq = f(index,x); compute_sigma();
        }
        void delta_conf(const adevs_bag<PortValue>& xb) {
            q = x[index]; dq = f(index,x); compute_sigma();
        }
        /* DEVS output function. */
        void output_func(adevs_bag<PortValue>& yb) {
            x[index] = q+ta()*dq;
            output(cell_interface::out,NULL,yb); // Notify influences of change.
        }
        /* Event garbage collection function. */
        void gc_output(adevs_bag<PortValue>& g){}
        /* Virtual derivative function. */
        virtual double f(int index, const double* x) = 0;
    private:
        /* Index of the variable associated with this integrator. */
        int index;
        /* Value of the variable, its derivative, and the integration quantum. */
        double q, dq, D;
        /* Shared output variable vector. */
        double* x;
        /* Sign function. */
        static double sgn(double z) {
            if (z > 0.0) return 1.0; if (z < 0.0) return -1.0; return 0.0;
        }
        /* Set the value of the time advance function. */
        void compute_sigma() {
            if (fabs(dq) < ADEVS_EPSILON) hold(ADEVS_INFINITY);
            else hold(fabs((D-fabs((q-x[index])))/dq));
        }
};
```

FIGURE 11.6 Code listing for the integrator class.

```cpp
/* Integrator for the two variable system. */
class TwoVarInteg: public Integrator {
    public:
        TwoVarInteg(double q0, int index, double D, double* x):
        Integrator(q0,index,D,x){}
        /* Derivative function. */
        double f(int index, const double* x) {
            if (index == 0) return -x[0]+0.5*x[1];
            else return -0.1*x[1];
        }
};
int main() {
    double x[2];
    TwoVarInteg* intg[2];
    // Integrator for variable x1
    intg[0] = new TwoVarInteg(1.0,0,0.1,x);
    // Integrator for variable x2
    intg[1] = new TwoVarInteg(1.0,1,0.1,x);
    // Connect the output of x2 to the input of x1
    staticDigraph g;
    g.couple(intg[1],1,intg[0],0);
    // Run the simulation for 3.361 units of time
    devssim sim(&g);
    while (sim.timeNext() <= 3.4) {
        cout << "t = " << sim.timeLast() << endl;
        for (int i = 0; i < 2; i++) {
            intg[i]->printState();
        }
        sim.execNextEvent();
    }
    // Done
    return 0;
}
```

FIGURE 11.7 Main simulation code for the two equation simulators.

values can be stored in a shared array that is accessed directly, rather than via messages. Second, the derivative value, represented as an input in the formal expression, can be calculated directly from the shared array of output values whenever a transition function is executed.

The integrator class is derived from the atomic model class, which is part of the adevs simulation library. The atomic model class has virtual methods corresponding to the output and state transition functions of the DEVS atomic structure. The time advance function of an adevs model is defined as $ta(s) = \sigma$, where s is a state variable of the atomic model, and its value is set with the $hold(\cdot)$ method. The integrator class adds a new virtual method, $f(\cdot)$, that is specialized to compute the derivative function using the output value vector $\bar{y}$.

A DEVS simulation of a system of ordinary differential equations, using Algorithm 11.3, gives the same result as Algorithm 11.2. This is demonstrated by a simulation of Eq. (11.7). The code used to execute the simulation is shown in Figure 11.7. The state transitions and output values computed in the course of the simulation are shown in Table 11.3. A comparison of this table with Table 11.2 confirms that they are identical.

TABLE 11.3 DEVS Simulation of Two Coupled Ordinary Differential Equations

t	q_1	$\dot{q}_1$	y_1	ta	Event Type	q_2	$\dot{q}_2$	y_2	ta_2
0	1	−0.5	1	0.2	Init	1	−0.1	1	1
0.2	0.9	−0.4	0.9	0.25	Internal				
0.45	0.8	−0.3	0.8	0.3333	Internal				
0.7833	0.7	−0.2	0.7	0.5	Internal				
1	0.6567	−0.25		0.2267	External	0.9	−0.09	0.9	1.111

11.5 The Heat Equation

In many instances, discrete approximations of partial differential equations can be obtained with a two-step process. In the first step, a discrete approximation of the spatial derivatives is constructed. This creates a set of coupled ordinary differential equations. The second step approximates the time derivatives. This step can be accomplished with the discrete-event integration scheme.

To illustrate this process, consider the heat (or diffusion) equation

$$\frac{\partial u(t,x)}{\partial t} = -\frac{\partial^2 u(t,x)}{\partial x^2} \tag{11.11}$$

The function $u(t,x)$ represents the quantity that becomes diffuse (temperature, if this is the heat equation). The spatial derivative can be approximated with a center difference, this giving

$$\frac{\partial^2 u(t, k\Delta x)}{\partial x^2} \approx \frac{u(t, (k+1)\Delta x) - 2u(t, k\Delta x) + u(t, (k-1)\Delta x)}{\Delta x^2} \tag{11.12}$$

where Δx is the resolution of the spatial approximation, and k are indices on the discrete spatial grid. Substituting Eq. (11.12) into Eq.(11.11) gives a set of coupled ordinary differential equations

$$\frac{du(t, k\Delta x)}{dt} = -\frac{u(t, (k+1)\Delta x) - 2u(t, k\Delta x) + u(t, (k-1)\Delta x)}{\Delta x^2} \tag{11.13}$$

that can be simulated using the DEVS integration scheme. This difference equation describes a grid of N integrators, and each integrator is connected to its two neighbors. The integrators at the end can be given fixed left and right values (i.e., fixing $u(t, -\Delta x)$ and $u(t, (N+1)\Delta x)$) equal to a constant, or some other suitable boundary condition can be used. For the sake of illustration, let $u(t, -\Delta x) = u(t, (N+1)\Delta x) = 0$. With these boundary conditions, two equivalent views of the system can be constructed. The first view, shown in Eq. (11.14), uses a matrix to describe the coupling of the differential equations in Eq.(11.13).

$$\frac{d}{dt}\begin{bmatrix} u(t,0) \\ u(t,\Delta x) \\ u(t,2\Delta x) \\ \ldots \\ u(t,(N-1)\Delta x) \\ u(t,N\Delta x) \end{bmatrix} = \frac{1}{\Delta x}\begin{bmatrix} -2 & 1 & 0 & 0 & 0 \\ 1 & -2 & 1 & 0 & 0 \\ 0 & 1 & -2 & 1 & 0 \\ 0 & \ldots & \ldots & \ldots & 0 \\ 0 & 0 & 1 & -2 & 1 \\ 0 & 0 & 0 & 1 & -2 \end{bmatrix}\begin{bmatrix} u(t,0) \\ u(t,\Delta x) \\ u(t,2\Delta x) \\ \ldots \\ u(t,(N-1)\Delta x) \\ u(t,N\Delta x) \end{bmatrix} \tag{11.14}$$

Because the kth equation is directly influenced only by the $(k+1)$st and $(k-1)$st equations, it is also possible to represent Eq. (11.13) as a cellspace in which each cell is influenced by its left and right neighbors. The discrete-event model favors this representation. The discrete-event cellspace, which is illustrated in Figure 11.8, has an integrator at each cell, and the integrator receives input from its left and right neighbors.

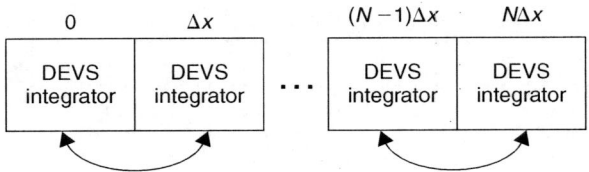

FIGURE 11.8 A cellspace view of the system described by Eq. (11.13).

Figure 11.9 shows the adevs simulation code for Eq. (11.13). The cellspace view of the equation coupling is implemented by the adevs Cellspace class.

The discrete-event approximation to Eq. (11.13) has two potential advantages over a similar discrete time approximation. The discrete time approximation is obtained from the same approximation to the spatial derivatives, but using the explicit Euler integration scheme to approximate the time derivatives (see, e.g., Strang, 1986). Doing this gives a set of coupled difference equations

$$u(t + \Delta t, k\Delta x) = u(t, k\Delta x) + \Delta t \left(\frac{u(t, (k+1)\Delta x) - 2u(t, k\Delta x) + u(t, (k-1)\Delta x)}{\Delta x^2} \right)$$

This discrete time integration scheme has an error term that is proportional to the time step Δt. In this respect, it is similar to the discrete-event scheme whose error is proportional to the quantum size D. However, there is an extra constraint in the discrete time formulation that is not present in the discrete-event approximation. This extra constraint is a stability condition on the set of difference equations (not the differential equations, which are inherently stable). For a stable simulation (i.e., for the state variables to decay rather than explode), it is necessary that

$$\Delta t \leq \frac{(\Delta x)^2}{2}$$

Freedom from the stability constraint is a significant advantage that the discrete-event scheme has over the discrete time scheme. For discrete time systems, this stability constraint can only be removed by employing implicit approximations to the time derivative. Unfortunately, this introduces a significant computational overhead because a system of equations, in the form $Ax = b$, must be solved at each integration step (see, e.g., Strang, 1986).

The unconditional stability of the discrete-event scheme can be demonstrated with a calculation. Consider a heat conducting bar with length 80. The ends are fixed at a temperature of 0. The initial temperature of the bar is given by $u(0, x) = 100 \sin(\pi x/80)$. Figure 11.10(a) and Figure 11.10(b) show the computed solution at $t = 300$ using $\Delta x = 0.1$ and different values of D. Even with large values of D, it can be seen that the computed solution remains bounded. Figure 11.10(c) shows the error in the computed solution for the more reasonable choices of D. From the figure, the correspondence between a reduction in D and a reduction in the computational error is readily apparent.

In many instances, the discrete-event approximation enjoys a computational advantage as well. An in-depth study of the relative advantage of a DEVS approximation to the heat equation over a discrete time approximation is described in Jammalamadaka (2003) and Zeigler (2004). This advantage is realized in a fire simulation described by Muzy et al. (2005), where a diffusive process is the spatially explicit piece of the fire model. In that report, the DEVS approximation is roughly four times faster than an explicit discrete time simulation giving the similar errors with respect to experimental data.

The reason for the performance advantage can be understood intuitively in two related ways. The first is to observe that the time advance function determines the frequency with which state updates are calculated at a cell. The time advance at each cell is inversely proportional to the magnitude of the derivative, and so cells that are changing slowly will have large time advances relative to cells that are changing quickly.

```cpp
class DiffInteg: public Integrator, public cell_interface {
    public:
        DiffInteg(double q0, int index, double D, double* x, double dx):
        Integrator(q0,index,D,x),cell_interface(){ dx2=dx*dx; }
        double f(int index, const double* x) {
            return (x[index−1]−2.0*x[index]+x[index+1])/dx2;
        }
    private:
        static double dx2;
};
double DiffInteg::dx2 = 0.0;

void print(const double* x, double dx, int dim, double t) {
    for (int i = 0; i < dim; i++) {
        double soln = 100.0*sin(M_PI*i*dx/80.0)*exp(−t*M_PI*M_PI/6400.0);
        cout << i*dx << " " << x[i] << " " << fabs(x[i]-soln) << endl;
    }
}

int main() {
    // Build the solution array and assign boundary and initial values
    double len = 80.0;
    double dx = 0.1;
    int dim = len/dx;
    double* x = new double[dim+2];
    // Half sine intial conditions with zero at boundaries
    for (int i = 0; i <= dim; i++) {
        x[i] = 100.0*sin(M_PI*i*dx/80.0);
    }
    x[0] = x[dim+1] = 0.0;
    // Create the DEVS model
    double D = 10.0;
    cellSpace cs(cellSpace::SIX_POINT,dim);
    for (int i = 1; i <= dim; i++) {
        cs.add(new DiffInteg(x[i],i,D,x,dx),i−1);
    }
    // Run the model
    devssim sim(&cs);
    sim.run(300.0);
    print(x,dx,dim+2,sim.timeLast());
    // Done
    delete [ ] x;
    return 0;
}
```

FIGURE 11.9 Code listing for the diffusion simulation.

This causes the simulation algorithm to focus effort on the rapidly changing portions of the solution, with significantly less work being devoted to portions that are changing slowly.

This is demonstrated in Figure 11.11. The state transition frequency at a point is given by the inverse of the time advance function following an internal event (i.e., $\dot{u}(i\Delta x,t)/D$, where i is the grid point

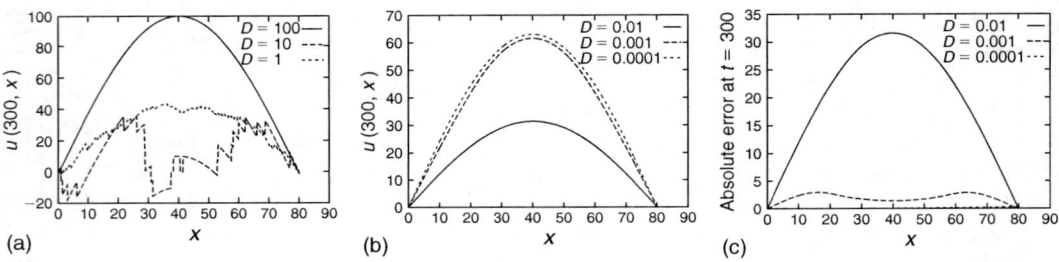

FIGURE 11.10 DEVS simulation of the heat equation with various quantum sizes. (a) Simulation with large D. (b) Simulation with small D. (c) Absolute errors.

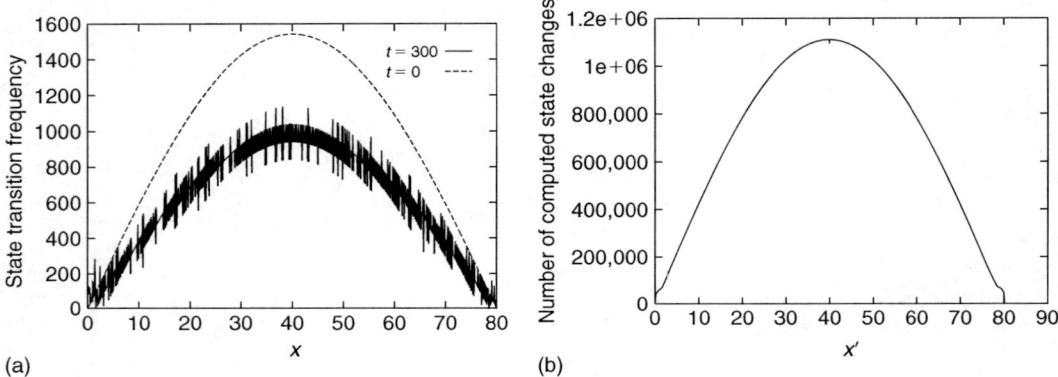

FIGURE 11.11 Activity tracking in the DEVS diffusion simulation using $D = 0.0001$. (a) State update frequency. (b) State transition count.

index). Figure 11.11(a) shows the state transition frequency at the beginning and end of the simulation. Figure 11.11(b) shows the total number of state changes that are computed at each grid point in the course of the calculation. It can be seen that the computational effort is focused on the center of the bar, where the state transition functions are evaluated most frequently.

A second explanation can be had by observing that the number of quantum crossings required for the solution at a grid point to move from its initial to final state is, approximately, equal to the distance between those two states divided by the quantum size. This gives a lower bound on the number of state transitions that are required to move from one state to another. It can be shown that, in many instances, the number of state transitions required by the DEVS model closely approximates this ideal number (see Jammalamadaka, 2003).

11.6 Conservation Laws

Conservation laws are an important application area where DEVS approximations of the time derivatives can be usefully applied. A DEVS simulation of Euler's fluid equations is presented in Nutaro et al. (2003). In that report, a significant performance advantage was obtained, relative to a similar time-stepping method, via the activity-tracking property described above. In this section, the application of DEVS to conservation laws is demonstrated for a simpler problem, where it is easier to focus on the derivation of the discrete-event model.

A conservation law in one special dimension is described by a partial differential equation

$$\frac{\partial u(t,x)}{\partial t} + \frac{\partial F(u(t,x))}{\partial x} = 0$$

The flux function $F(u(t,x))$ describes the rate of change in the amount of u (whatever u might represent) at each point x (see, e.g., Strang, 1986). To be concrete, consider the conservation law

$$\frac{\partial u(t,x)}{\partial t} + u(t,x)\frac{\partial u(t,x)}{\partial x} = 0 \tag{11.15}$$

Eq. (11.15) describes a material with quantity $u(t,x)$ that moves with velocity $u(t,x)$. In this equation, the flux function is $u(t,x)^2/2$. Eq. 11.15 is obtained by taking the partial derivative of the flux function with respect to x.

As before, the first step is to construct a set of coupled ordinary differential equations that approximates the partial differential equation. There are numerous schemes for approximating the derivative of the flux function with respect to x (see, e.g., Kroner, 1997). One of the simplest is an upwinding scheme on a spatial grid with resolution Δx. Applying an upwinding scheme to Eq. 11.15 gives

$$u(t, k\Delta x)\frac{\partial u(t, k\Delta x)}{\partial x} \approx -\frac{1}{2\Delta x}(u(t, (k-1)\Delta x)^2 - u(t, k\Delta x)^2) \tag{11.16}$$

Substituting Eq. (11.16) into Eq. (11.15) gives the set of coupled ordinary differential equations

$$\frac{du(t, k\Delta x)}{dt} = \frac{1}{2\Delta x}(u(t, (k-1)\Delta x)^2 - u(t, k\Delta x)^2) \tag{11.17}$$

It is common to approximate the time derivatives in Eq. (11.17) with the explicit Euler integration scheme using a time step Δt. This gives the set of difference equations

$$u(t + \Delta t, k\Delta x) = u(t, k\Delta x) + \frac{\Delta t}{2\Delta x}(u(t, (k-1)\Delta x)^2 - u(t, k\Delta x)^2)$$

that approximate the set of differential equations. The difference equations are stable provided that the condition

$$\frac{\Delta t}{\Delta x} max|u(i\Delta t, j\Delta x)| \leq 1$$

is satisfied at every time point i and every spatial point j.

Because Eq. (11.17) is nonlinear, it is not necessarily true that a discrete-event approximation will be stable regardless of the size of the integration quantum. However, it is possible to find a sufficiently small quantum for which the scheme works (see Nutaro, 2003). This remains an open area of research, but we will move recklessly ahead and try generating solutions with several different quantum sizes and observe the effect on the solution.

For this example, a space of 10 units in length is assigned the initial conditions

$$u(0,x) = \begin{cases} \sin(\pi x/4) & \text{if } 0 \leq x \leq 4 \\ 0 & \text{otherwise} \end{cases}$$

and the boundary conditions $u(t,0) = u(t,10) = 0$. The integrator implementation for this model is shown in Figure 11.12. The simulation main routine is identical to the one for the heat equation (except where DiffInteg is replaced by ClawInteg; see Figure 11.9). Figure 11.13 shows snapshots of the solution computed with $\Delta x = 0.1$ and three different quantum sizes: 0.1, 0.01, and 0.001. The computed solutions maintain important features of the real solution, including the shock formation and shock velocity (see Strang, 1986).

While the advantage of the discrete-event scheme with respect to stability remains unresolved (but looks promising!), a potential computational advantage can be seen. From the figure, it is apparent that the

```
class ClawInteg: public Integrator, public cell_interface {
    public:
        ClawInteg(double q0, int index, double D, double* x, double dx):
        Integrator(q0,index,D,x),cell_interface(){ ClawInteg::dx=dx; }
        double f(int index, const double* x) {
            return 0.5*(x[index−1]*x[index−1]−x[index]*x[index])/dx;
        }
    private:
        static double dx;
};
double ClawInteg::dx = 0.0;
```

FIGURE 11.12 Integrator for the conservation law simulation.

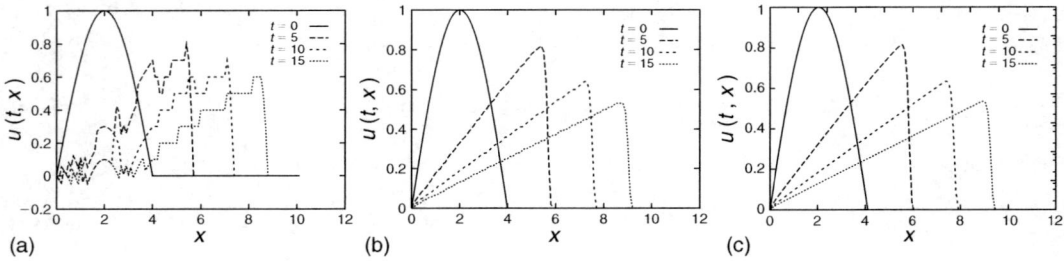

FIGURE 11.13 Simulation of Eq. (11.17) with various quantum sizes. (a) $D = 0.1$, (b) $D = 0.01$, and (c) $D = 0.001$.

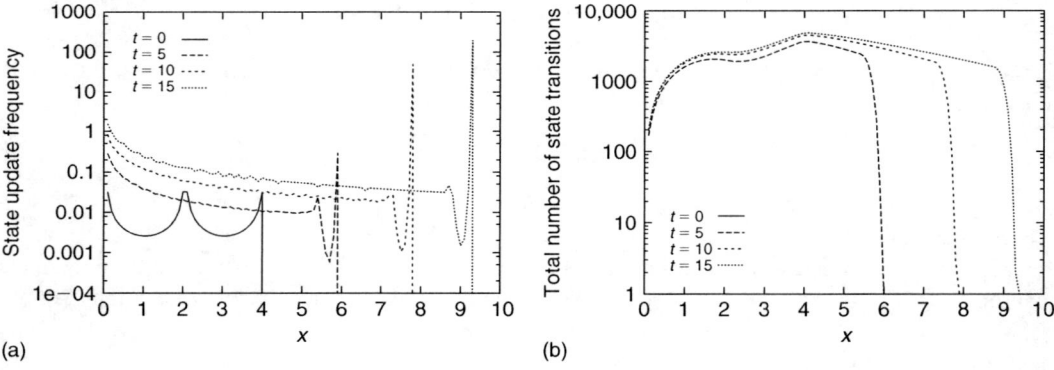

FIGURE 11.14 Front tracking in the DEVS simulation of Eq. (11.17) with $D = 0.001$. (a) Update frequency, and (b) State transition count.

larger derivatives follow the shock, with the area in front of the shock having zero derivatives and the area behind the shock having diminished derivatives. The DEVS simulation apportions computational effort appropriately. This is shown in Figure 11.14 for a simulation with $D = 0.001$. Figure 11.14(a) shows several snapshots of the cell update frequency (i.e., $\dot{u}(i\Delta x, t)/D$ following an internal event, where i is the grid point index) at times corresponding to the solution snapshots shown in Figure 11.13(c). Figure 11.14(b) shows the total number of state transitions computed at each cell at those times. The effect of this front-tracking behavior on the simulation running time can be significant. In Nutaro et al. (2003), it is responsible for a speedup of 35 relative to a discrete time solution for Euler's equations in one spatial dimension.

11.7 Two-Point Integration Schemes

The integration scheme discussed to this point is a single point scheme. It relies on a single past value of the function, and it is exact for the linear equation $\dot{x}(t) = k$, where k is a constant. Recall that the single point scheme for simulating a system $\dot{x}(t) = f(x(t))$ can be derived from the expression

$$\left| \int_{t_0}^{t_0+h} f(x(t))dt \right| = D \tag{11.18}$$

by approximating $f(x(t))$ with the value $f(x(t_0))$.

If the function $f(x(t))$ in Eq. (11.18) is approximated using the previous two values of the derivative, then the resulting method is called a two-point scheme. A DEVS model of a two-point scheme requires the state variables

q, the current approximation to $x(t)$;
q_l, the last grid point occupied by q;
σ, the time required to move from q to the next grid point;
$\dot{q}_1$ and $\dot{q}_0$, the last two computed values of the derivative; and, possibly,
h, the time interval between $\dot{q}_1$ and $\dot{q}_0$.

At least two different two-point methods have been described (see Kofman, 2004; Nutaro, 2005). The first method approximates $f(x(t))$ in Eq. (11.18) with the line connecting points $\dot{q}_1$ and $\dot{q}_0$. The distance moved by $x(t)$ in the interval $[h, h+T]$ can be approximated by

$$\int_{h}^{h+T} \frac{\dot{q}_1 - \dot{q}_0}{h} + \dot{q}_0 \, dt = \frac{\dot{q}_1 - \dot{q}_0}{2h} T^2 + \dot{q}_1 T = \Delta q$$

The functions

$$\hat{F}_1(T, \dot{q}_1, \dot{q}_0, h) = \frac{\dot{q}_1 - \dot{q}_0}{2h} T^2 + \dot{q}_1 T \tag{11.19}$$

and

$$\hat{F}_1^{-1}(\Delta q, \dot{q}_1, \dot{q}_0, h) = \Delta T \tag{11.20}$$

where ΔT is the smallest positive root of

$$\left| \frac{\dot{q}_1 - \dot{q}_0}{2h} T^2 + \dot{q}_1 T \right| = \Delta q$$

or ∞ if such a root does not exist, can be used to define the state transition, output, and time advance functions (which will be done in a moment). Eq. (11.19) and Eq. (11.20) are exact when $x(t)$ is quadratic.

The second method approximates $f(x(t))$ with the piecewise constant function

$$a\dot{q}_1 + b\dot{q}_0, \quad a+b = 1 \tag{11.21}$$

If $x(t)$ is the line $mt + b$, then $f(x(t)) = m$, $(am + bm) = (a+b)m = m$, and so this approximation is exact. Integrating Eq. (11.21) over the interval $[0, T]$ gives the approximating functions

$$\hat{F}_2(T, \dot{q}_1, \dot{q}_0) = (a\dot{q}_1 + b\dot{q}_0)T \tag{11.22}$$

$$\hat{F}_2^{-1}(\Delta q, \dot{q}_1, \dot{q}_0) = \frac{\Delta q}{|a\dot{q}_1 + b\dot{q}_0|} \tag{11.23}$$

This approximation does not require the state variable h.

For brevity, let $\bar{q}$ denote the state of the integrator, and $\bar{d}q$ denote the variables $\dot{q}_1, \dot{q}_0$ or $\dot{q}_1, \dot{q}_0, h$ as needed. Which is intended will be clear from the context in which it is used. The time advance function for a two-point scheme is given by

$$ta(\bar{q}) = \sigma$$

and the output function is defined by

$$\lambda(\bar{q}) = \hat{F}(\sigma, \bar{d}q)$$

If Eq. (11.19) and Eq. (11.20) are used to define the integration scheme, then the resulting state transition functions are

$$\delta_{int}(\bar{q}) = (q + \hat{F}_1(\sigma, \bar{d}q), q + \hat{F}_1(\sigma, \bar{d}q), q_1, q_1, \sigma, \hat{F}_1^{-1}(D, \dot{q}_1, \dot{q}_1, \sigma))$$

$$\delta_{ext}(\bar{q}, e, x) = (q_l, q + \hat{F}_1(e, \bar{d}q), x, q_1, e, \hat{F}_1^{-1}(D - |q + \hat{F}_1(e, \bar{d}q) - q_l|, x, \dot{q}_1, e))$$

$$\delta_{con}(\bar{q}, x) = (q + \hat{F}_1(\sigma, \bar{d}q), q + \hat{F}_1(\sigma, \bar{d}q), x, q_1, \sigma, \hat{F}_1^{-1}(D, x, \dot{q}_1, \sigma))$$

When Eq. (11.22) and Eq. (11.23) are used to define the integrator, then the state transition functions are

$$\delta_{int}(\bar{q}) = (q + \hat{F}_2(\sigma, \bar{d}q), q + \hat{F}_2(\sigma, \bar{d}q), q_1, q_1, \hat{F}_2^{-1}(D, \dot{q}_1, \dot{q}_1))$$

$$\delta_{ext}(\bar{q}, e, x) = (q_l, q + \hat{F}_2(e, \bar{d}q), x, q_1, \hat{F}_2^{-1}(D - |q + \hat{F}_2(e, \bar{d}q) - q_l|, x, \dot{q}_1))$$

$$\delta_{con}(\bar{q}, x) = (q + \hat{F}_2(\sigma, \bar{d}q), q + \hat{F}_2(\sigma, \bar{d}q), x, q_1, \hat{F}_2^{-1}(D, x, \dot{q}_1))$$

The scheme that is constructed using Eq. (11.19) and Eq. (11.20) is similar to the QSS2 method in Kofman (2004), except that the input and output trajectories used here are piecewise constant rather than piecewise linear.

The scheme constructed from Eq. (11.22) and Eq.(11.23) is nearly second-order accurate when a and b are chosen correctly. If $a = \frac{3}{2}$ and $b = -\frac{1}{2}$, then the error, with respect to Eq. (11.18), in the integral of Eq. (11.21) is

$$E = \left(f(x_1) - \frac{3f(x_1)}{2} + \frac{f(x_0)}{2}\right)T + \frac{1}{2}T^2\frac{d}{dt}f(x_1) + \sum_{n=3}^{\infty}\frac{1}{n!}\frac{d^{(n)}}{dt}f(x_1)T^n \quad (11.24)$$

For this scheme to be nearly second-order accurate, the terms that depend on T and T^2 need to be as small as possible. Let h be the time separating x_1 and x_0 (i.e., $x_1 = x(t_1)$, $x_0 = x(t_0)$, and $h = t_1 - t_0$), and let $\alpha = \frac{T}{h}$, the ratio of the current time advance to the previous time advance. It follows that $T = \alpha h$. The function $\frac{d}{dt}f(x_1)$ can be approximated by

$$\frac{d}{dt}f(x_1) \approx \frac{f(x_1) - f(x_0)}{h} \quad (11.25)$$

Substituting Eq. (11.25) into Eq.(11.24) and dropping the high-order error terms gives

$$E \approx \alpha h\left(\frac{f(x_1) - f(x_0)}{2} + \alpha\frac{f(x_0) - f(x_1)}{2}\right) \quad (11.26)$$

Eq. (11.26) approaches 0 as α approaches 1. It seems reasonable to assume T and h become increasingly similar as D is made smaller. From this assumption, it follows that the low-order error terms in Eq. (11.24) vanish as D shrinks.

Figure 11.15(a) and Figure 11.15(b) show the absolute error in the computed solution of $\dot{x}(t) = -x(t)$, $x(0) = 1$, as a function of D for these two integration schemes. The simulation is ended at $t = 1.0$, and

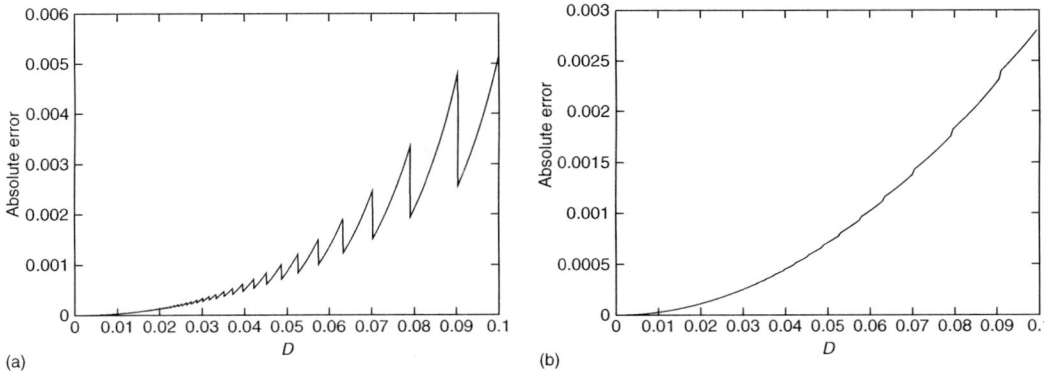

FIGURE 11.15 Simulation error as a function of D for the system $\dot{x}(t) = -x(t)$ with $x(0) = 1$. (a) Simulation error using Eq. (11.19) and Eq. (11.20). (b) Simulation error using Eq. (11.22) and Eq. (11.23).

α and the absolute error are recorded at that time. In both cases, it can be observed that the absolute error is proportional to D^2.

These two schemes use additional information to reduce the approximation error with respect to the single point scheme. Fortunately, these two schemes share the unconditional linear stability of the single point scheme (see Kofman, 2004; Nutaro, 2003), and so they represent a tradeoff between storage, execution time, and accuracy. When dealing with very large systems, the single point scheme has the advantage of needing less computer memory because it has fewer state variables per integrator. However, it will, in general, be less accurate than a two-point scheme for a given quantum size. If the quantum size is selected to obtain a given error, then the two-point scheme will generally use a larger quantum than the one-point scheme, and so the simulation will finish more quickly using the two-point scheme.

11.8 Conclusions

This chapter introduced some essential techniques for constructing discrete-event approximations to continuous systems. Discrete-event simulation of continuous systems is an active area of research, and the breadth of the field cannot be adequately covered in this short space. In the conclusion, some related research is summarized, and references are given, for the interested reader.

A side-by-side introduction of discrete time and discrete-event methods for continuous system simulation can be found in Cellier and Kofman (2006). Comparisons of nonadaptive discrete time and discrete-event methods can be found in Muzy et al. (2005), Nutaro et al. (2003), Zeigler (2004), and Kofman (2004). In-depth comparison of discrete-event schemes with adaptive time stepping, asynchronous, and other types of advanced numerical integration methods remains a topic for future research.

The construction of high-order discrete-event integration schemes is discussed in Nutaro (2005). A third-order scheme, and the only third-order method developed to date, is described in Kofman (2005a, 2005b).

In Bolduc and Vangheluwe (2003), an adaptive quantum scheme is introduced. This scheme allows the integration quantum to be varied during the course of the calculation to maintain an upper bound on the global error. An application of adaptive quantization to a fire-spreading model is discussed in Muzy et al. (2002).

A methodology for approximating general time functions as DEVS models is discussed in Giambiasi et al. (2000). The approximations introduced in that paper associate events with changes in the coefficients of an interpolating polynomial. An application of this methodology to partial differential equations is shown in Wainer and Giambiasi (2005).

Applying DEVS models to finite element methods for equilibrium problems is discussed in D'Abreu and Wainer (2003) and Saadawi and Wainer (2004). A steady state heat transfer problem is used to demonstrate the method.

Simulation of partial differential equations leads naturally to parallel computing. Parallel discrete-event simulation for the numerical methods presented in this chapter are discussed in Nutaro (2003) and Nutaro and Sarjoughian (2004). Specific issues that emerge when simulating DEVS models using logical process-based algorithms are described in Nutaro and Sarjoughian (2004). Parallel discrete-event simulation applied to particles in cell methods is discussed in Tang et al. (2005) and Karimabadi et al. (2005).

References

Bolduc, J.-S. and H. Vangheluwe (2003). Mapping ODEs to DEVS: Adaptive quantization. In *Proceedings of the 2003 Summer Simulation MultiConference (SCSC'03)*, Montreal, Canada, pp. 401–407.

Cellier, F. E. and E. Kofman (2006). *Continuous System Simulation*. New York: Springer.

D'Abreu, M. and G. Wainer (2003). Improving finite elements method models using cell-DEVS. In *Proceedings of the 2003 Summer Computer Simulation Conference*, Montreal, QC, Canada.

Filippi, J.-B. and P. Bisgambiglia (2004). JDEVS: An implementation of a DEVS based formal framework for environmental modeling. *Environmental Modelling & Software* 19(3), 261–274.

Giambiasi, N., B. Escude, and S. Ghosh (2000). GDEVS: A generalized discrete event specification for accurate modeling of dynamic systems. *Transactions of the Society Computer Simulation International* 17(3), 120–134.

Hwang, K. (1993). *Advanced Computer Architecture: Parallelism, Scalability, Programmability*. New York: McGraw-Hill.

Jammalamadaka, R. (2003). Activity characterization of spatial models: Application to the discrete event solution of partial differential equations. Master's thesis, University of Arizona, Tucson, Arizona, USA.

Karimabadi, H., J. Driscoll, Y. Omelchenko, and N. Omidi (2005). A new asynchronous methodology for modeling of physical systems: Breaking the curse of the Courant condition. *Journal of Computational Physics* 205(2), 755–775.

Kofman, E. (2004). Discrete event simulation of hybrid systems. *SIAM Journal on Scientific Computing* 25(5), 1771–1797.

Kofman, E. (2005a). A third order discrete event method for continuous system simulation. Part I: Theory. In *Proceedings of RPIC'05*.

Kofman, E. (2005b). A third order discrete event method for continuous system simulation. Part II: Applications. In *Proceedings of RPIC'05*.

Kofman, E., M. Lapadula, and E. Pagliero (2003). PowerDEVS: A DEVS-based environment for hybrid system modeling and simulation. Technical Report LSD0306, School of Electronic Engineering, Universidad Nacional de Rosario, Rosario, Argentina.

Kroner, D. (1997). *Numerical Schemes for Conservation Laws*. Chichester, NY: Wiley.

Muzy, A., E. Innocenti, A. Aiello, J.-F. Santucci, and G. Wainer (2002). Cell-DEVS quantization techniques in a fire spreading application. In *Proceedings of the 2002 Winter Simulation Conference*, Sandiego, CA.

Muzy, A. and J. Nutaro (2005). Algorithms for efficient implementations of the DEVS & DSDEVS abstract simulators. In *1st Open International Conference on Modeling & Simulation*, ISIMA/Blaise Pascal University, France, pp. 401–407.

Muzy, A., P.-A. Santoni, B. P. Zeigler, J. J. Nutaro, and R. Jammalamadaka (2005). Discrete event simulation of large-scale spatial continuous systems. In *IEEE International Conference on Systems, Man, and Cybernetics*, Vol. 4, Waikoloa, Hawaii, pp. 2991–2998.

Nutaro, J. (2003). Parallel discrete event simulation with application to continuous systems. Ph.D. thesis, University of Arizona, Tuscon, Arizona.

Nutaro, J. (2005). Constructing multi-point discrete event integration schemes. In *Proceedings of the 2005 Winter Simulation Conference*, Orlando, FL.

Nutaro, J. and H. Sarjoughian (2004). Design of distributed simulation environments: A unified system-theoretic and logical processes approach. *Simulation 80*(11), 577–589.

Nutaro, J. J., B. P. Zeigler, R. Jammalamadaka, and S. R. Akerkar (2003). Discrete event solution of gas dynamics within the DEVS framework. In P. M. A. Sloot, D. Abramson, A. V. Bogdanov, J. Dongarra, A. Y. Zomaya, and Y. E. Gorbachev (Eds.), *International Conference on Computational Science*, Volume 2660 of Lecture Notes in Computer Science, pp. 319–328. Berlin: Springer.

Saadawi, H. and G. Wainer (2004). Modeling complex physical systems using 2D finite elemental cell-DEVS. In *Proceedings of MGA, Advanced Simulation Technologies Conference 2004 (ASTC'04)*, Arlington, VA, USA.

Strang, G. (1986). *Introduction to Applied Mathematics*. Wellesley, MA: Wellesley-Cambridge Press.

Szidarovszky, F. and A. T. Bahill (1998). *Linear Systems Theory*, 2nd edition. Boca Raton, FL: CRC Press LLC.

Tang, Y., K. Perumalla, R. Fujimoto, H. Karimabadi, J. Driscoll, and Y. Omelchenko (2005). Parallel discrete event simulations of physical systems using reverse computation. In *ACM/IEEE/SCS Workshop on Principles of Advanced and Distributed Simulation (PADS)*, Monterey, CA.

Wainer, G. (2002). CD++: A toolkit to develop DEVS models. *Software: Practice and Experience 32*(13), 1261–1306.

Wainer, G. A. and N. Giambiasi (2005). Cell-DEVS/GDEVS for complex continuous systems. *Simulation 81*(2), 137–151.

Zeigler, B. P. (2004). Continuity and change (activity) are fundamentally related in DEVS simulation of continuous systems. In *Keynote Talk at AI, Simulation, and Planning 2004 (AIS'04)*, Jeju, Korea.

Zeigler, B. P., H. Praehofer, and T. G. Kim (2000). *Theory of Modeling and Simulation*, 2nd edition. New York: Academic Press.

Zeigler, B. P., H. Sarjoughian, and H. Praehofer (2000). Theory of quantized systems: DEVS simulation of perceiving agents. *Cybernetics and Systems 31*(6), 611–647.

Zeigler, B. P. and H. S. Sarjoughian (2005). Introduction to DEVS modeling and simulation with JAVA: Developing component-based simulation models. Unpublished manuscript.

III

Multiobject and System

12
Toward a Multimodel Hierarchy to Support Multiscale Simulation

Mark S. Shephard
Scientific Computation Research Center

E. Seegyoung Seol
Scientific Computation Research Center

Benjamin FrantzDale
Scientific Computation Research Center

12.1 Introduction 12-1
12.2 Functional and Information Hierarchies in Multiscale Simulation 12-4
  Mathematical Physics Description Transformations and Interactions • Domain Definitions, Transformations, and Interactions • Physical Parameter Definitions, Transformations, and Interactions
12.3 Constructing a Multimodel: Design of Functional Components to Support Multiscale Simulations 12-10
  Problem Definition • Equation Parameters • Geometric Domain • Discretized Geometric Domains • Tensor Fields • Scale-Linking Operators
12.4 Example Multimodel Simulation Procedures 12-14
  Automated Adaptive Mesh-Based Simulation • Adaptive Atomistic/Continuum Adaptive Multiscale Simulation
12.5 Closing Remarks 12-15

12.1 Introduction

There is a long history of developing mathematical representations capable of providing behavioral predictions of physical parameters on the atomic, molecular, microscopic, and macroscopic scales. Over the past half century, simulation programs have been developed to support the computerized solution of these mathematical representations which, in some cases, are discretized with billions of degrees of freedom (dofs) and solved on massively parallel computers with thousands of processors. Historically, scientists and engineers have applied these models (simulation programs) to solve problems on a single physical scale. However, in recent years it has become clear that to continue to make advances in the areas of nanotechnology and biotechnology, and to develop new products and treatments based on those advances, scientists and engineers must be able to solve sets of coupled models active over multiple interacting scales. For example, the development of new materials will require the design of structure and function across a hierarchy of scales, starting at the molecular scale to define nanoscale building blocks that will be used to construct mesoscale features that may be combined into micron-scale weaves that could be used in the manufacturing of complete parts (Figure 12.1). Such capabilities are clearly central to the development of nanoelectronics devices and future biomedical device design as well as many of other future products.

As an example of the potential impact of multiscale simulation on biomedical device design, consider a drug-eluting stent (Figure 12.2). Drug-eluting stents are hybrid device–drug medical products that release therapeutic drugs and provide a scaffold to maintain arterial lumen size after angioplasty. The design

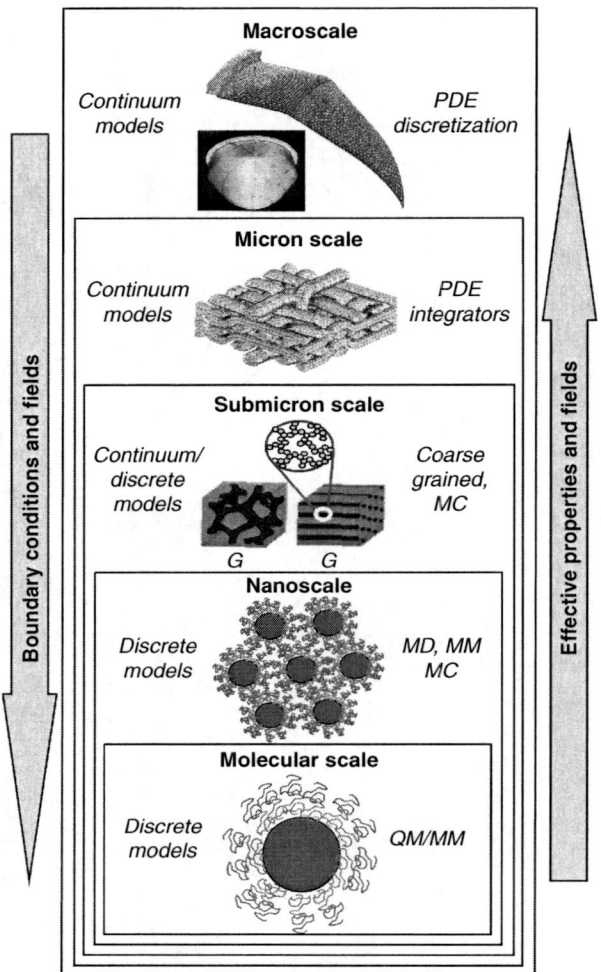

FIGURE 12.1 Multimodel hierarchy used in the design of a composite material system.

of these devices requires consideration of the mechanical function of holding open a diseased artery, and of the pharmacological function of delivering the appropriate drug in the appropriate concentration for the requisite length of time to prevent in-stent restenosis. The mechanical simulations involved with device deployment includes continuum-scale models of the stent and blood vessel, which employ complex material models. The complex nature of the blood vessel requires the application of multiscale methods to determine those material models. Consideration of drug delivery from the stent coating requires a model that includes continuum-level modeling of blood flow coupled with molecular-level diffusion and transport of drug molecules, which needs to be coupled to cellular- and molecular-level models of drug diffusion into blood vessels through cell membranes. Similar models and methods are central to many other applications. For example, similar models and model coupling are needed in the consideration of new automotive skins made of nanoreinforced materials in which the material interfaces are strong at strain rates consistent with normal use, while at high strain rates demonstrating substantial local damage, leading to high stiffness so that little dents are avoided, but providing high-energy absorption under impact loading to keep passengers safe.

There are many available models to solve various single-scale simulation problems, but while the development of multiscale methods is an active research area, there has been limited attention paid

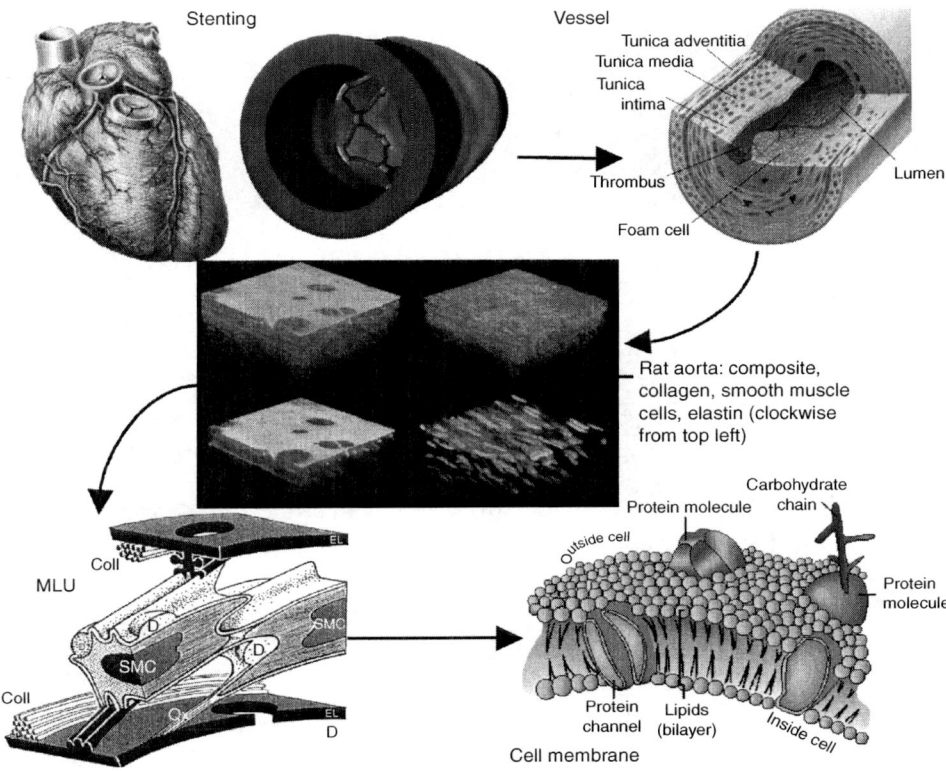

FIGURE 12.2 Multimodel hierarchy needed for a drug delivery system.

to the development of general multiscale modeling techniques. One procedure that does address the complex issue of bridging from atomistic to continuum physics, including the ability to adaptively control the model selection over the domain, is the quasicontinuum method (Knap and Ortiz, 2001; Miller and Tadmor, 2002). Since this procedure is based on a single method to define and link the physical scales, its development has not needed to address the inclusion of flexible methods or the insertion of alternative models and scale-linking methods. Other efforts have limited the range of models included to similar models such as the OCTA software, which includes four discrete mesoscale models (OCTA, 2006). Considering the thousands of person years of effort that has gone into the development of the existing single-scale models that operate at each of the physical scales needed, the effective development of multiscale simulations will be greatly facilitated by the development of methods that can easily integrate and use existing and developing single-scale models. This chapter outlines the overall structure of a component-based multimodel approach in which each model uses clearly defined interfaces and functionality for sharing information. Since these methods are early in the development phase, this chapter provides one view of how this complex problem could be addressed with the goal of opening a constructive dialog between the software engineers developing multimodel methods and the computational engineers developing multiscale methods.

The next section considers the key information and function hierarchies of a multiscale simulation. Section 12.3 discusses the overall design of a set of functional components defined to support the full set of interactions and transformations needed by multiscale simulation. It is the combination of these functional components with existing single-scale models that will provide an operational multimodel multiscale simulation system. Section 12.4 presents two example simulations combining existing single-scale models with prototypes of the interfaces described here.

12.2 Functional and Information Hierarchies in Multiscale Simulation

In abstracting multiscale simulation processes, one must consider the hierarchy of transformations required to go from a description of physical behavior to a set of mathematical and computational models capable of simulating the desired behaviors. The highest level in the hierarchy is an overall problem definition related to mathematical descriptions used to describe the behavior, typically coupled, at various scales, including equations relating parameters between the scales. The other two levels in the hierarchy are the discretizations of the mathematical models and the numerical algorithms used to solve the discretized mathematical models. A key to abstracting this process is to qualify the information needed to support the models and the transformations required as information is shared by models. The information used in the processes can be placed in the following three groups:

Mathematical models: The description of the mathematical equations used as a description of the physical behavior on the various scales, and mathematical equations that relate behaviors between scales.

Domains: The description of the domains over which the various mathematical equations apply. In the case of multiscale analysis, this includes appropriate definitions at each relevant scale of space and time, and the spatial and temporal interactions between them.

Physical parameters: The description of the various physical parameters used in the mathematical equations, defined over the appropriate domains as required to qualify the current instances of the governing equations to be solved.

The ability to properly support component-based multiscale simulation requires the specification of mathematical descriptions with associated domain and physical-parameter definitions at the highest possible level meaningful to the execution of the process so that the full range of solution methods interaction modes can be supported.

12.2.1 Mathematical Physics Description Transformations and Interactions

A mathematical physics description is a set of governing equations that are assumed to govern the behavior at a particular scale over a particular domain. The equations are written in terms of a set of dependent variables and given parameters, and are a function of the coordinates of the domain of the problem. To make this more concrete, consider the two most common forms of equations encountered in multiscale analysis: partial differential equations (PDEs), which are defined at various continuum scales, and molecular dynamics (MD), which is based on interatomic potentials that define the interactions between molecules and atoms at the small scales for which continuum equations over the domain are not applicable.

12.2.1.1 Partial Differential Equations

PDEs may be written in terms of multiple sets of dependent variables where each set can contain tensors of various orders that vary over the space–time domain. For the purposes of this discussion, consider the PDE

$$\mathcal{D}^m(u,\sigma) - f = 0 \quad \text{on } \Omega \tag{12.1}$$

subject to boundary conditions

$$\mathcal{D}^i(u,\sigma) - g_i = 0 \quad \text{on } \Gamma_i \quad \text{for } i = 0, 1, 2, \ldots, m-1 \tag{12.2}$$

where

$\mathcal{D}^m$ represents the appropriate mth-order differential operator;

$u(x, t)$ represents one or more dependent vector variables which are functions of the independent variables of space, x, and time, t;

σ represents one or more dependent scalar variables which are functions of the independent variables of space, x, and time, t;

f represents the forcing functions;

Ω represents the domain over which the equation is defined;

$\mathcal{D}^i$ are the appropriate ith-order differential operators;

g_i are the given boundary conditions;

Γ_i are the portions of the boundary over which the associated boundary conditions act.

Computerized models of the PDEs typically use mesh-based methods in which a two-part discretization process is used to transform the mathematical model into numerical systems which are solved. The first part of the discretization is the decomposition of the space–time domain into a set of mesh entities with simple shapes in space and time. The second part of the discretization is to discretize the shapes of the functions. A set of basis functions related to a "weak form" of the governing equation, and to difference relations for the differential operators, is used to discretize the dependent variables over the individual mesh entities in terms of a set of to-be-determined parameters, called dofs. The dofs can always be associated with a single mesh entity whereas the distribution functions (basis functions or difference relations) are associated with one or more mesh entities. In the case that the distribution is associated with multiple mesh entities, that set is defined by rules associated with the discretization operator and can be supported by using mesh adjacency information. Three common cases that employ different combinations of interactions among the mesh entities, the dofs, and the distributions are

Finite difference based on a vertex stencil, in which the dofs are typically values of the dependent variables at vertices of a mesh and the distribution functions are difference stencils defined in terms of vertex values for mesh vertices for the appropriate set of topologically adjacent mesh entities.

Finite volume methods are constructed in terms of distribution function written over individual mesh entities. In most cases the field being defined is discontinuous between elements (i.e., C^{-1}). Therefore, dofs are not shared between neighboring mesh entities. The coupling of the dofs from different mesh entities is through operators acting over boundary entities between neighboring mesh entities.

Finite element distribution functions are written over individual mesh entities called elements. In cases where C^m, $m \geq 0$ continuity is required, the distribution functions associated with neighboring elements are made C^m, $m \geq 0$ continuous by having common dofs associated with the bounding mesh entities common to the neighboring elements.

The application of the discretization operations over the mesh entities produces a local contributor which can be stated symbolically as

$$k^c d^c = f^c \tag{12.3}$$

where k^c is the discretized matrix for contributor c that multiplies the vector of dofs associated with the contributor, d^c.

These individual contributions are then assembled into a global algebraic system, $Kd = F$, based on an assembly operator defined by the relationships of the contributor-level dofs, d^c, with the assembled set of global dofs, d.

12.2.1.2 Molecular Dynamics

In MD, the mathematical model is a potential function describing the forces among interacting atoms and which depends on the relative position of the atoms (Frenkel and Smit, 2002). A common potential function is the Lennard–Jones potential which approximates the force between two atoms as

$$V_{\text{LJ}} = 4\epsilon \left[\left(\frac{\sigma}{r}\right)^{12} - \left(\frac{\sigma}{r}\right)^6 \right] \tag{12.4}$$

where σ and ϵ are Lennard–Jones parameters for a given material and r is the interatomic distance. The parameters in potential equations may be developed empirically or based on simulations performed on a finer ab initio scale using an electron model. Owing to the large number of atoms required to fill domains, MD simulation is typically performed over small subdomains where boundary conditions must be applied to the atoms on or near the boundary. Typical boundary conditions are free-surface, periodic, or fixed (Dirichlet). The direct outputs of an MD simulation, atom trajectories and forces on the atoms, are typically not of specific interest, but rather are needed to determine the meaningful higher-scale parameters of interest. The extraction of those higher-scale parameters is often done by taking statistical ensembles.

12.2.1.3 Interactions Between PDEs and MD

It is common for a simulation to require the solution of a set of coupled mathematical models where the coupling is defined by parameters assumed to be given in one model but which are actually the results of another model. In some cases, the coupling simply requires solving the models in a given order so that parameters are available when required. In other cases, parameters are shared in both directions, necessitating the application of an appropriate coupling method.

Coupling on a single scale occurs when multiple models are used to solve for different sets of the physical parameters of interest. A common example is fluid–structure interactions, in which the flow field is influenced by the geometry of the structure over which it flows and the geometry of the structure is a function of the forces on it caused by the flow field. The issues associated with the transfer of parameters between models depend on the portions of the domain over which the interactions occur and on how those portions have been discretized, both in terms of its geometry (mesh) and the distributions and dof used.

The interactions of parameters between models solved on multiple scales must account for differences of the domain representation at the different scales, for the models used to couple information between the scales, and for the relationships between the parameters passed between the models on the different scales. Two broad classes of scale-linking methods are "information-passing" and "concurrent-bridging" (Fish, 2006). With information-passing methods, fine scales are modeled and their gross response is infused into the coarse scale; the influences of coarse-scale fields on the fine scales are taken into account as boundary conditions and forcing functions on the fine scale. With concurrent bridging, the fine and coarse scales are simultaneously resolved. For nonlinear problems, the models at different scales are coupled in both directions and information continuously flows between the scales.

In many information-passing techniques, the fine-scale model is a representative unit cell subject to appropriate boundary conditions, and information passed to the larger scale is considered to be at a point on the larger scale. In concurrent techniques, the fine-scale model acts over some small finite portion of the coarse-scale domain and the parameters are passed through the common boundary between the domains, or through some overlap portion of the domains.

In multiscale methods, where entirely different models are used at each scale, the relationships of parameters between scales is usually not direct and care must be taken to define the appropriate operations to relate them. In some cases, these operations act as filters to remove information (e.g., the removal of high-frequency modes when up-scaling). In others, they must account for relating discrete and continuum models (e.g., relating atomic-level deformations defined by atomic positions to a continuum displacement field). In some cases, operations are needed to relate quantities with different forms of definition (e.g., atomic-scale forces to continuum stresses) or to define terms not defined at a given scale (e.g., defining continuum-level temperature in terms of atomic scale motions).

The complication of properly relating information between scales has led to the active development of methods for scale linking and to computer implementation of these methods. Representative information-passing methods include multiple-scale asymptotic techniques (Fish et al., 2002), variational multiscale methods (Hughes et al., 2000), heterogeneous multiscale methods (E and Enquist, 2002), multiscale enrichment schemes based on partition of unity (Fish and Yuan, 2005), discontinuous Galerkin discretizations (Hou and Wu, 1997), and equation-free methods (Kevrekidis et al., 2003). Spatially concurrent schemes are based on either multilevel (Fish and Belsky, 1995) or domain-bridging methods (Belytschko and Xiao, 2003; Broughton et al., 1999), while concurrent schemes in the time domain are typically based on multistep methods (Gravouil and Combescure, 2001).

12.2.2 Domain Definitions, Transformations, and Interactions

The domains considered here are space–time domains. Time is a linear progression that runs from an initial time to a final time and is typically discretized using a well-accepted set of methods based on time increments. In contrast, there are a number of general forms commonly used to provide a high-level representation of spatial domains. To meet the needs of multiscale simulation,

- The domain representation must support the transformation of an original domain definition into representations that can support a discretization of the governing equations over the domain. For example, the original definition of a domain may be a feature-based model of a domain over which a mesh-based simulation is to be performed. The process of creating the mesh in this case requires the transformation of the feature model into a nonmanifold geometric model upon which an automatic mesh generation procedure can be applied to generate the desired mesh (Shephard et al., 2004). In addition, the transformations needed to construct the required domain representations; it is necessary to maintain the relationship between the entities in each of the representations.
- The domain representation must support the definition of the physical parameters (attributes) associated with the equations to be solved and the proper transformation of that information into any derived representation to be used by the models. For example, the ability to map the components of a tensor with a given distribution onto a model entity, such as a surface of the geometric domain.
- The domain representation must support the ability to address any domain interrogation required during the execution of models involved with the simulation. Most of these interrogations can be limited to pointwise evaluations (e.g., determine the normal vector at a given point on a surface).
- The domain representation must support geometric interactions between related domains used in a multiscale simulation. For example, in a concurrent multiscale model, to determine the mesh entities in the continuum domain which overlap with the atomic region.

The definition of the domain is a function of the type of mathematical description used. For example, continuum domain definitions are needed in the case of PDEs while a discrete set of atomic positions is needed in MD.

12.2.2.1 Continuum Domains

There are multiple sources for domain definitions, the most common being CAD models, mesh models, and image data. CAD systems and mesh models employ a boundary representation. Image data are generally defined in terms of voxels. Except in cases of directly using the image data as the model, it is generally accepted that boundary representation is well suited for defining continuum domains. Common to all boundary representations is the use of topological entities and their adjacencies to represent the entities of various dimensions. Information defining the actual shape of topological entities can be thought of as information associated with each entity. The ability to interact with a domain definition in terms of the topological entities provides an effective means to develop abstract interfaces to a domain definition, allowing easy integration of multiple domain-definition sources.

An important consideration in selecting a boundary representation is its ability to represent the classes of domain needed. In the most general case, domains can be general combinations of 0-, 1-, 2-, and 3-D entities where lower-order entities are not required to bound higher-order entities. Figure 12.3 shows a typical analysis domain of this type, which would be appropriate for structural analysis. The boundary representations that can fully and properly represent such geometric domains are referred to as a nonmanifold boundary representations (Weiler, 1998).

In addition to the topological entities and associated shape information, geometric-modeling systems maintain numerical tolerance information on how well the entities fit together. The algorithms and methods within a geometric, modeling system are able to use such tolerance information to effectively

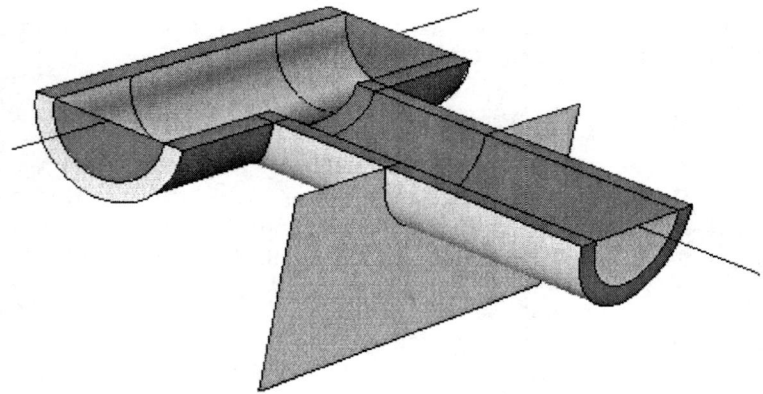

FIGURE 12.3 Example of a nonmanifold model.

define and maintain a consistent representation of the geometric domain. (The vast majority of what various geometry-based applications have referred to as "dirty geometry" is caused by a lack of knowledge of, or improper use of, the tolerance information (Beall et al., 2004).)

Abstracting topology is an effective way to allow the development of functional interfaces to boundary-based modelers that are independent of specific shape information. The developers of CAD systems have recognized the possibility of supporting geometry-based applications through general application program interfaces (APIs), where functions that provide entity adjacencies, calculate geometric information such as surface normals, etc. are keyed to topological entities. This has led to the development of geometric-modeling kernels such as ACIS (Spatial Inc.) and Parasolid (Parasolid, Inc.) which have been successfully used to develop automated finite element modeling processes (Shephard et al., 2005; Wan et al., 2005) and automatic mesh generators (Beall et al., 2004).

In the application of generalized numerical analysis processes, a meshed approximation must be created from a geometric domain. To support the full set of operations needed for reliable multiscale analysis, a mesh must maintain an association with its continuum-domain representation and with the distribution functions and number of dofs used in discretizing the PDEs (see Section 12.2.1.1). From the perspective of maintaining its relationship to the geometric domain, the use of an appropriate set of topological entities and their adjacency is ideal (Beall and Shephard, 1997).

A key component supporting mesh-based simulation is the association of a mesh to its geometric model (Beall and Shephard, 1997; Shephard and Georges, 1992), which indicates the mesh entities that represent particular model entities. This association is used for operations such as ensuring the mesh entities on the boundary of a model are properly curved when needed, associating boundary conditions defined at the model entity level with the appropriate mesh entities, etc. This association can be defined as follows:

Classification: The unique association of the ith mesh topological entity (with dimension d_i), $M_i^{d_i}$, to a topological entity of the geometric model of dimension d_j, $G_j^{d_j}$, on which it lies, where $d_i \leq d_j$. This is denoted $M_i^{d_i} \sqsubset G_j^{d_j}$ where the classification symbol, $\sqsubset$, indicates that the left-hand entity, or set, is classified on the right-hand entity.

Reverse Classification: For each model entity, G_j^d, the set of equal-order mesh entities classified on that model entity defines the reverse classification information for that model entity. Reverse classification is denoted as

$$\mathrm{RC}(G_j^d) = \{M_i^d \mid M_i^d \sqsubset G_j^d\} \tag{12.5}$$

Shape information can be effectively associated with the topological entities defining the mesh. In many cases this is limited to the coordinates of the mesh vertices and, if they exist, higher-order nodes associated with mesh edges, faces, or regions. In addition, it is possible to associate other forms of geometric information with the mesh entities. For example, the association of Bézier curves and surface definitions with mesh edges and faces for use in high-order curved finite elements (Luo et al., 2002). The mesh classification can be used to obtain other needed geometric information such as the coordinates of a new mesh vertex formed by splitting a mesh edge classified on a model face.

12.2.2.2 Discrete Domains

The domain definition for the discrete domains are the positions of the entities for which the potentials are written to relate. For example, in the case of MD this is the position of atoms. In many cases it is possible to define the full set of discrete entity positions from a higher-level construct with appropriate transformations. In this case the highest-level domain definition consists of the geometry of the domain to be included, parameters defining the distribution of the discrete positions, and the functions required to define those positions. The overall domain is often a representative volume that has portions of its boundary interior to a higher-level domain and may include knowledge of free surfaces.

The parameters and transformations used to define the atomic positions are a function of the type of material being defined. In the case of perfect crystals, the position of atoms within each crystal is defined by a set of lattice vectors, which provide information defining the positions of atoms. The definition of the geometric configuration of the crystal is a nontrivial process that can start with a statistical method to define an initial set of seed locations for crystals whose initial shape can then be defined as the Voronoi diagram of those points. To define more-realistic configurations, various grain-growth procedures can be applied which account for knowledge of the material system. There can be defects in the crystal systems (Hull and Bacon, 1965). With additional information about these defects and the total number of atoms, coordinates, and velocities of the atoms can have an initial adjustment applied to them. In the case of polyms, an atom's position must be defined by its position along its molecular chain, where there are strong bounds between neighboring units in the chain. Statistically based geometric constructs can be used to define these material-dependent chains in the simulation box. Methods like those just outlined, which can take a compact definition of discrete domains and produce a proper set of atomistic positions, are required. In some cases these methods will be purely geometric while in others will require the execution of a full atomistic relaxation model.

One approach to bridging scales is to interpolate the behavior of a large set of atoms in terms of a small subset of them. One such approach, well-suited to lattice structures, is the quasicontinuum method in which the position of atoms over simple shapes such as triangles and tetrahedra is described to vary linearly between known atom positions (Knap and Ortiz, 2001; Miller and Tadmor, 2002). In the case of polymer chains, atoms along a chain can be represented by a small number of "beads" placed along the chain (Mavrantzas et al., 1999; Padding and Briels, 2002).

12.2.2.3 Interactions of Domains

There are three general forms of domain interactions used in multiscale simulations. They are

Disjoint domains, which share information across a common boundary.
Overlapping domains, which have portions of the overall domain represented at more than one scale and the information is shared through the overlapped region.
Telescoping domains, which represent microstructure by many small-scale domains, which have essentially zero size with respect to the higher-scale domain. Thus, each small-scale domain passes information to a point in the higher-scale domain.

In each case, the operations used to transfer parameters between the scales must be consistent with the form of domain interaction.

12.2.3 Physical Parameter Definitions, Transformations, and Interactions

The physical parameters used in the mathematical equations are tensor quantities (Beju et al., 1983) defined over various portions of the domain that can be general functions of the independent variables of space and time as well as other dependent variables. Knowledge of the order of a tensor and the dimension of the spatial domain it is defined over defines the number of components needed to uniquely define the tensor. The symmetries, for tensors of order two or greater, define those components that are identical to, or negative of (antisymmetric), other components. The components of the tensor are, in general, functions of the domain parameters as well as other problem parameters. The ability to understand and use a tensor at any particular instant requires knowledge of the coordinate system in which the components are written. Tensors can be represented in other coordinate systems of equal or lower order through appropriate coordinate transformations.

To support the full range of simulation needs, the tensors used to define the equation's parameters must be related to the highest level of the geometric representation possible. For example, in the case of solving a PDE over continuum domains, the distribution of the given input tensors needs to be related to the entities in the geometric model. The model topological entities of regions, faces, edges, and vertices are ideally suited for supporting that specification in a general way.

The tensors associated with the dependent parameters are determined as part of the solution process. Therefore, these tensors, referred to as fields, are understood with respect to the spatial and equation discretizations used in the simulation process. Since the spatial discretizations are required to maintain the relationship to the original domain definition (see Section 12.2.2), the fields can also be related to the highest-level domain definitions.

In multiscale simulation, a single tensor field can be used by a number of different analysis routines that interact and the field may be associated with multiple spatial discretizations (e.g., meshes) having alternative relationships between them. In addition, different distributions can be used by a field to discretize its associated tensor. The ability to have a given tensor defined over multiple meshes or discretized in terms of multiple distributions can be handled by supporting multiple field instances.

12.3 Constructing a Multimodel: Design of Functional Components to Support Multiscale Simulations

In the design of a multimodel system to support multiscale simulations, it is important to determine the information required by the models and the transformations to be applied to provide the information in the needed form. Within the multimodel multiscale simulation environment, functional APIs are defined to support the various classes of information transformations needed. Employing the APIs provided by components makes it straightforward to combine various single-scale models to construct multimodels for multiscale simulations. Each of the various models interact with other models only through the component's API. For example, in a concurrent model, part of the scale-linking component would be a function linking atomistic to continuum using statistical averaging of atomic displacements on the boundary of the atomistic region thereby providing boundary deformations to the continuum model.

A key goal of this design is to build multiscale simulation procedures by using adaptive solution strategies to control existing time-tested single-scale models thereby ensuring the reliability of simulation in terms of providing predictions of the desired parameters to the required degree of accuracy. The only way to provide this reliability is to explicitly consider the approximation errors that can arise within each step executed by a model or transformation performed by a component. Since many of these errors cannot be controlled through a priori means, it is necessary to support adaptive feedback processes that use a posteriori information to control the execution of each model step and transformation.

A number of the models needed to perform specific simulation steps are well established and should be used. Two examples are generalized fixed-mesh continuum PDE solvers (finite element, finite volume, and finite difference) and discrete-level models for solving discrete-potential systems (ab initio, molecular statics, MD).

The majority of the mature and widely used software operates only through input and output files. In that case, the components will be a facade, crating the input files and interacting with the output files. Although this case does not take full advantage of the components, advantages gained are the easy substitution of other models, including ones that can more directly interact with the components.

Some programs support the addition of user-defined functionality. For example, ABAQUS (ABAQUS Inc.) supports user-defined material models and user-defined finite elements. Although limited, these two features facilitate the majority of the functionality needed for ABAQUS to be an effective model in a multiscale simulation environment.

Another area in which mature models exist to support multiscale modeling is the definition of geometric domains of 3-D parts using boundary representations. Most existing systems provide a functional API (Parasolid, Inc. [2006]; Spatial Inc. [2006]), which is ideal for creating a component for a multiscale simulation. These geometric-modeling APIs provide the capabilities needed to represent continuum-level domains in multiscale simulations. There are also many existing programs that can generate mesh-level discretizations of geometric domains, the interfaces of which range from file- to API-based (Beall et al., 2004). API-based interfaces have been used in the development of adaptive mesh modification procedures (Li et al., 2002) and complete adaptive PDE multimodel simulations (Shephard et al., 2005; Wan et al., 2005), and are well suited for the needs of multiscale simulation.

In designing a multimodel system to support multiscale simulation, we must identify appropriate levels of abstraction to support the flow of information between models such that information can be provided to procedures that execute any required transformations. The components defined to support multimodel multiscale simulations are

1. problem definition,
2. equation parameters,
3. geometric domains,
4. discretized geometric domains,
5. tensor fields, and
6. scale-linking operations.

A subset of similar functional components being defined to support the interoperability of simulation models is a topic of current development for mesh-based continuum-simulation methods both in terms of open-source code (Chand et al., 2007; TSTT Software, 2006) and commercial products (Simmetrix Inc., 2006).

Figure 12.4 illustrates the structure of a multiscale multimodel in which five functional components are used by the single-scale models in the multimodel of a given multiscale simulation. Additional scales can be added by adding another scale-linking component and model to either side of the diagram. Each instance of a component utilizes other components in the same scale to do its job. Information flow is indicated by the arrows. Furthermore, components will only share information through scale linking with

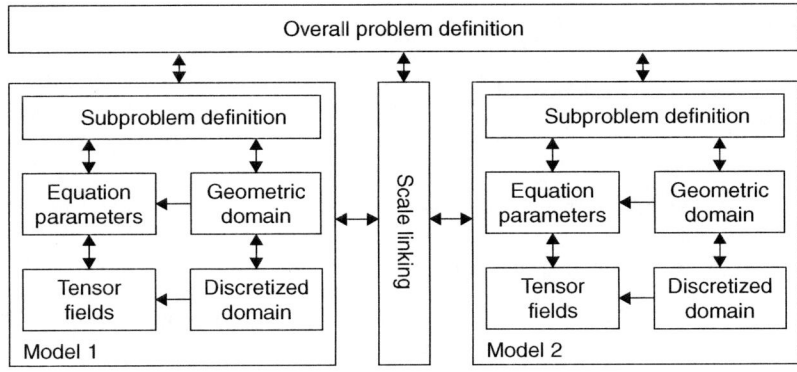

FIGURE 12.4 Interactions between components.

the component of the same type in the other model. Based on this, the functions within a component need to consider the following three modes of interaction:

1. Providing and modifying component-internal data (procedures which do not involve interactions with other components),
2. Providing information between components within a model, and
3. Providing information between like components for different models.

12.3.1 Problem Definition

To consider multiscale simulations, we need an overall problem definition. A problem definition must include all of the relevant physical parts, must define the relationships between the parts, and must facilitate the creation of alternate representations of the parts. In addition, a problem definition must provide any other information needed to construct and execute the desired solution process. A problem definition must also allow for the creation of viewpoint-specific interfaces as needed by the solution strategy and simulation models to be used (e.g., considering an atomic region as though it is a continuum).

Representations, including simulation viewpoints, are under development (Shephard et al., 2004). These representations can be defined in terms of graph structures similar to those used to define assembly and feature models in CAD systems (Bidarra and Bronsvoort, 2000; Hoffman and Joan-Arinyo, 1998), with the extensions to support hierarchal decompositions and multiple viewpoints (Bronsvoort and Jansen, 1993; Hoffman and Joan-Arinyo, 1998, 2000; Noort et al., 2002). In the single-scale case, a problem-definition component would then support the following interaction modes:

Component-internal: part definitions, relationships of parts, and mathematical equations governing part behavior.
Intercomponent: relationships of parts to domains, relationships of parts to parameters, relationships of parts to model functions, and model-level simulation strategy information.
Interacting with like components of other models: viewpoint construction rules, relation of mathematical equations on related parts in the different models, and multiscale simulation strategy information.

In the multiscale case, the overall problem definition is inherently multiscale. As such, it will depend on a scale-linking component, as well as single-scale models which compose it, complete with their own subproblem definitions.

12.3.2 Equation Parameters

The parameters in the mathematical equations representing physical quantities are, in general, tensors. The types of physical parameters these tensors define are material properties, loading functions, boundary conditions, and initial conditions. Although the execution of any model requires a specific set of these tensors associated with the mathematical equations of that model, the parameters can be stated abstractly in terms of the highest-level problem-definition entities. For example, a boundary heat flux of $1\,kW/m^2$ defined on surface 1 of mesh element 2 might be abstractly stated as the same heat flux applied to the bottom surface of a heat sink. This boundary conditions on the heat sink would be mapped onto a corresponding concrete solid model, and in turn onto a mesh. By defining conditions on the highest-possible level, it becomes easy to change the mesh, or even the solid model itself, without having to start from scratch. Generalized methods to define and manipulate such a structure of models have been defined (O'Bara et al., 2002, Shephard, 1985).

The equation parameters component must support

Component-internal: parameter information queries, parameter instance information queries, parameter coordinate transformations, and parameter reduction and modification functions.
Intercomponent: relation to problem parts and geometric domains, relation to model solution process, and relation to fields.

Interacting with like components of other models: dependencies between parameters for different models.

12.3.3 Geometric Domain

A geometric-domain component is a functional unit to describe a multiscale simulation domain at a particular scale (e.g., continuum domain or atomistic domain). Within a multiscale model, a geometric domain supports geometric interactions between other geometric domains.

Consider, first, a continuum geometric domain defined in a CAD modeler in terms of a boundary representation. This component must support

Component-internal: topological entity information, shape information, geometric-model tolerance information, and geometric-model modification.
Intercomponent: association with parts in the problem definition, association of equation parameters with the geometric domain (and, through that, association with domain discretizations—meshes), and association with scale linking.
Interacting with like components of other models: geometric interactions relating domains on different scales through boundaries or overlaps.

Atomic-scale models must support

Component-internal: the definition of atom layouts, and the geometric relationships among atoms.
Intercomponent: obtain potentials and provide forces.
Interacting with like components of other models: placement of the domain with respect to larger-scale domains, and geometric interactions relating domains of different scales through boundaries or overlaps.

12.3.4 Discretized Geometric Domains

A discretized geometric-domain component is a piecewise-geometric approximation of a corresponding geometric-domain component in terms of a mesh or idealized atomistic layout. This component must support

Component-internal: topological entity queries for meshes, geometric shape of mesh entities, and geometric queries such as position of atoms and distance between atoms or mesh entities.
Intercomponent: mesh Jacobian information, association with the geometric domain, and association with fields.
Interacting with like components of other models: mesh-to-mesh interaction, mesh-to-atomistic interactions, and discrete-to-atomistic interactions.

12.3.5 Tensor Fields

A tensor-field component is a discretization of a tensor field over a discretized domain. To support a tensor field defined over multiple discretized domains, the tensor-field component must support a collection of field instances for a single field where one field instance is defined over each of the discretized domains. The tensor-field component must support

Component-internal: field information queries, field coordinate transformation, and field reduction and modification.
Intercomponents: association of field with the discretized geometric-domain entities, association with quantities determined by model solution processes, and relation to parameters.
Interacting with like components of other models: field transfer between field instances, and field transformation or modification to meet the model needs.

12.3.6 Scale-Linking Operators

Scale-linking operators exist to transform parameters among different single-scale models. A scale-linking operator is defined in terms of

- the domains at each scale and the form of the domain interactions,
- the domain discretization used for the interacting fields,
- the distribution functions and number of dofs used to represent the interacting fields over the discretized domains, and
- the functional operations associated with transforming the field information on one scale to the other.

The methods used should allow scale-linking operations to be defined at the highest level of problem definition, with additional qualification as needed to account for specific forms of domain discretizations and field distributions used. As such it must support

Component-internal: the definition of the linking operations.
Intercomponent: the relationship to parts and domains, and to other fields and parameters.
Interacting with models: using single-scale components' interfaces to transfer information between scales.

By combining components as described, and as shown in Figure 12.4, each component will have minimal dependence on the rest of the system. This reduces software complexity and will allow the components to be easily interchanged.

12.4 Example Multimodel Simulation Procedures

The examples of automated adaptive simulation procedures presented in this section employ prototype implementations of the functional components outlined in Section 12.3. The first example is an automated adaptive single-scale procedure that is currently used in industry. The second is an adaptive atomistic–continuum multiscale procedure under development.

12.4.1 Automated Adaptive Mesh-Based Simulation

Many programs are used for the solution of PDEs on a given fixed mesh. Although they are capable of providing results to required levels of accuracy, the vast majority lack the ability to automatically control mesh discretization errors through adaptive methods. Using the interoperable components discussed in Section 12.3 in conjunction with existing fixed-mesh finite element models and a mesh-modification component (Li et al., 2002), multiple adaptive analysis procedures have been built.

One such example was created for 3-D forming simulations in which the deformable parts undergo large plastic deformations that result in major changes in the analysis domain geometry. The meshes of the deforming parts need to be frequently modified to continue the analysis owing to significant element distortions, mesh discretization errors, and geometric approximation errors. In these cases, it is necessary to replace the deformed mesh with an improved mesh that is consistent with the current configuration. Procedures using the two domain and field components are employed to determine a new mesh size field, which is provided to a local mesh modification procedure (Wan et al., 2005) which creates an adapted mesh. A tensor-field component is also used to transfer history-dependent field variables as each mesh modification is performed (Wan et al., 2005) so that the full set of information needed for the next set of analysis steps can be provided to the analysis model, the commercial finite element program DEFORM-3D (Fluhrer, 2004).

Figure 12.5 shows the setup, initial mesh, and final adapted meshes for a steering-link manufacturing problem solved using this multimodel capability. This simulation shows a total stroke of 41.7 mm. The initial mesh of the workpiece consists of 28,885 elements. The simulation was completed with 20 mesh-modification steps producing a final mesh with 102,249 elements.

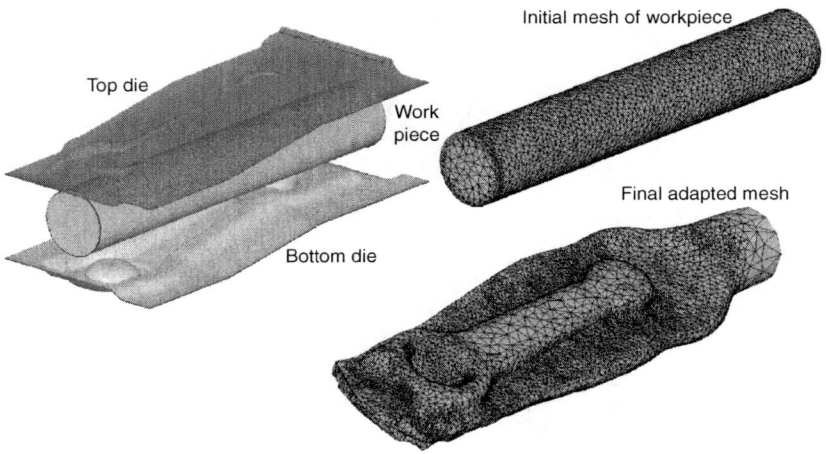

FIGURE 12.5 Adaptive forming simulation example where the left image shows the problem setup with geometry of the two dies and initial workpiece and the right two images show the initial (top) and final (bottom) meshes.

12.4.2 Adaptive Atomistic/Continuum Adaptive Multiscale Simulation

Concurrent adaptive multiscale simulation capabilities are being developed for modeling fracture in metallic structures (Datta et al., 2004). The key analysis engines for this multimodel application are nonlinear finite element models for the continuum level and molecular-statics models to address the atomistic aspects of dislocation formation and growth. Part of the simulation viewpoint in this case is the indication of the set of behaviors that can be associated with the parts which indicate that both linear- and nonlinear continuum behavior can be considered, and that atomistic regions can be superimposed at locations of dislocation formation such as crack tips. The equation parameters include the continuum material properties, loads and boundary conditions, and the atomistic potentials. The geometric domain includes the full part geometry and atomistic overlays, including defect locations for the locations that are adaptively determined to require an atomistic overlay. The computational representation of these two regions includes a finite element mesh and atomistic positions, taking account of the defects. The tensor fields include overall and local deformations as well as stresses at the continuum level, and atom positions and forces on the atomistic level. Since the atomistic and continuum levels overlap, the options for a scale-linking operator include relating local deformations and forces either through a common boundary or through an overlap region. In both cases, the atomistic deformations must be smoothed before being transferred to the continuum level and the discrete interatom forces must be transformed into stress-like quantities so they can be related to continuum-level stresses.

Figure 12.6 shows an example of an adaptive atomic continuum simulation for the definition and growth of dislocations at a crack tip. In this case, the cracked macrodomain was defined in a solid modeler and the finite elements were generated automatically. Atomic overlay regions are defined in the critical areas based on an error indicator; as defects form, the atomic domains automatically adjust.

12.5 Closing Remarks

The focus of this chapter has been an examination of the process of performing adaptive multiscale simulation with the goal of defining an appropriate set of high-level components to support the construction of multimodel simulations taking advantage of established models that can effectively address specific aspects of these simulations. Six functional components have been defined which support the transformation and transfer of information to the various models used by these multimodel simulations. The application

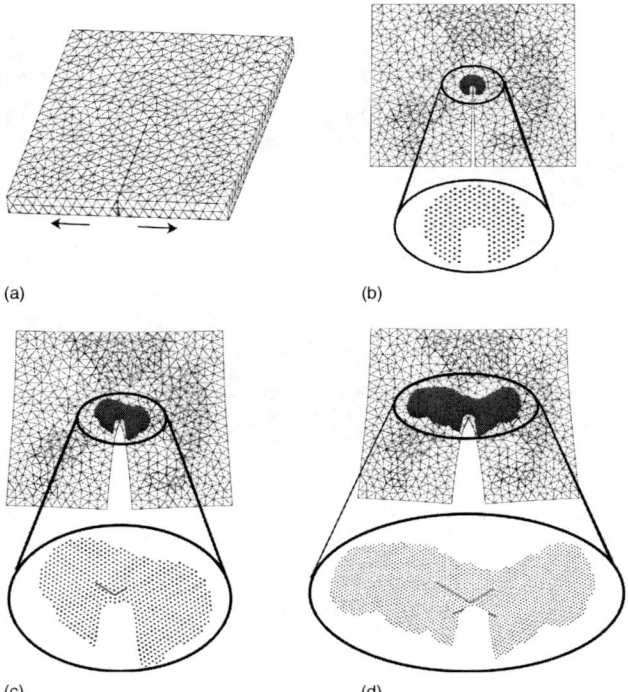

FIGURE 12.6 Adaptive molecular/continuum multiscale simulation. (a) 3-D domain with a crack and macroscale mesh; (b) adaptively superimposed atomic region (shaded); (c) adapted atomic region accounting for dislocation formation and growth; (d) continued adaptation of the atomic region with dislocation growth.

of these components has been demonstrated through two multimodel automated adaptive simulation examples building on an initial prototype of six functional components.

References

ABAQUS, Inc. (2006). http://www.abaqus.com.
Beall, M. and M. Shephard (1997). A general topology-based mesh data structure. *International Journal for Numerical Methods in Engineering* 40(9), 1573–1596.
Beall, M., J. Walsh, and M. Shephard (2004). Accessing cad geometry for mesh generation. *Engineering with Computers* 20(3), 210–221.
Beju, I., E. Soos, and Teodorescu (1983). *Euclidean Tensor Calculus with Applications*, Taylor & Francis, London, England.
Belytschko, T. and S. Xiao (2003). Coupling methods for continuum model with molecular model. *International Journal for Numerical Methods in Engineering* 1, 115–126.
Bidarra, R. and W. Bronsvoort (2000). Semantic feature modeling. *Computer-Aided Design* 32, 201–225.
Bronsvoort, W. and F. Jansen (1993). Feature modeling and conversion—key concepts to concurrent engineering. *Computers in Industry* 21(1), 61–86.
Broughton, J., F. Abraham, N. Berstein, and E. Kaxiras (1999). Concurrent coupling of length scales: Methodology and application. *Physical Review B* 60, 2391–2403.
Chand, K., L. Diachin, X. Li, C. Ollivier-Gooch, E. Seol, M. Shephard, T. Tautges, and H. Trease (2007). Toward interoperable mesh, geometry and field components for PDE simulation development. *Engineering With Computers*, to appear.

Datta, D., R. Picu, and M. Shephard (2004). Composite grid atomistic continuum method: An adaptive approach to bridge continuum with atomistic analysis. *Journal of Multiscale Computational Engineering* 2(3), 401–420.

E, W. and B. Enquist (2002). The heterogeneous multi-scale method. *Communications in Mathematical Sciences* 1, 87–132.

Fish, J. (2006). Discrete to continuum multiscale bridging. *Multiscaling in Molecular and Continuum Mechanics*, Springer, New York.

Fish, J. and V. Belsky (1995). Multigrid method for a periodic heterogeneous medium. *Computer Methods in Applied Mechanics and Engineering* 126, 1–38.

Fish, J., W. Chen, and G. Nagai (2002). Nonlocal dispersive model for wave propagation in heterogeneous media: One-dimensional case and multi-dimensional case. *International Journal of Numerical Methods in Engineering* 54(3), 331–363.

Fish, J. and Z. Yuan (2005). Multiscale enrichment based on the partition of unity. *International Journal of Numerical Methods in Engineering* 62(10), 1341–1359.

Fluhrer, J. (2004). *DEFORM-3D Version 5.0 User's Manual*. Scientific Forming Technologies Corporation.

Frenkel, D. and B. Smit (2002). *Understanding Molecular Simulations: From Algorithms to Applications* (Second ed.). Academic Press, San Diego, CA.

Gravouil, A. and A. Combescure (2001). Multi-time-step explicit method for nonlinear structural dynamics. *International Journal of Numerical Methods in Engineering* 50, 199–225.

Hoffmann, C. and R. Joan-Arinyo (1998). Cad and the product master model. *Computer-Aided Design* 30(11), 905–918.

Hoffmann, C. and R. Joan-Arinyo (2000). Distributed maintenance of multiple project views. *Computer-Aided Design* 32, 421–431.

Hou, T. and X. Wu (1997). A multiscale finite element method for elliptic problems in composite materials and porous media. *Journal of Computational Physics* 134, 169–189.

Hughes, T., L. Mazzei, and K. Jansen (2000). Large-eddy simulation and the variational multiscale method. *Computing and Visualization in Science* 3, 47–59.

Hull, D. and D. Bacon (1975). *Introduction to Dislocations*. Butterworth-Heinemann, Oxford, UK.

Kevrekidis, I. G., C. W. Gear, J. M. Hyman, P. G. Kevrekidis, O. Runborg, and C. Theodoropoulos (2003). Equation-free coarse-grained multiscale computation: Enabling microscopic simulators to perform system-level tasks. *Communications in Mathematical Sciences* 1(4), 715–762.

Knap, J. and M. Ortiz (2001). An analysis of the quasicontinuum method. *Journal of Mechanics and Physics of Solids* 49, 1899–1923.

Li, X., M. Shephard, and M. Beall (2002). Accounting for curved domains in mesh adaptation. *International Journal of Numerical Methods in Engineering* 58, 246–276.

Luo, X., M. Shephard, J.-F. Remacle, R. O'Bara, M. Beall, B. Szab, and R. Actis (2002). p-Version mesh generation issues. In *11th International Meshing Roundtable*, pp. 343–354.

Mavrantzas, V., T. Boone, E. Zervopoulou, and D. Theodorou (1999). End-bridging monte carlo: An ultra-fast algorithm for atomistic simulation of condensed phases of long polymer chains. *Macromolecules* 32, 5072–5096.

Miller, R. and E. Tadmor (2002). The quasicontinuum method: Overview, applications and current directions. *Journal of Computer-Aided Materials Design* 9, 203–209.

Noort, A., G. Hoek, and W. Bronsvoort (2002). Integrated part and assembly modeling. *Computer-Aided Design* 34, 899–912.

O'Bara, R., M. Beall, and M. Shephard (2002). Attribute management system for engineering analysis. *Engineering with Computers* 18(4), 339–351.

OCTA (2006). Meso-scale simulation programs. http://octa.jp/.

Padding, J. and W. Briels (2002). Time and length scales of polymer melts studied by coarse-grained molecular dynamics simulations. *Journal of Chemical Physics* 117(2), 925–943.

Parasolid, Inc. (2006). http://www.ugs.com/products/open/parasolid.

Shephard, M. (1985). Finite element modeling within an integrated geometric modeling environment: Part ii—Attribute specification, domain differences and indirect element types. *Engineering with Computers* 1, 72–85.

Shephard, M., M. Beall, R. O'Bara, and B. Webster (2004). Toward simulation-based design. *Finite Elements in Analysis and Design* 40, 1575–1598.

Shephard, M., J. Flaherty, K. Jansen, X. Li, X.-J. Luo, N. Chevaugeon, J.-F. Remacle, M. Beall, and R. O'Bara (2005). Adaptive mesh generation for curved domains. *Journal for Applied Numerical Mathematics* 50(2–3), 251–271.

Shephard, M. and M. Georges (1992). Reliability of automatic 3-D mesh generation. *Computational Methods in Applied Mechanics and Engineering* 101, 443–462.

Simmetrix Inc. (2006). http://www.simmetrix.com/.

Spatial Corp. (2006). 3D ACIS modeler. http://www.spatial.com/components/acis.

TSTT Software (2006). http://tstt-scidac.org/software/software.html.

Wan, J., S. Kocak, and M. Shephard (2005). Automated adaptive 3-D forming simulation process. *Engineering with Computers* 21(1), 47–75.

Weiler, K. (1998). The radial-edge structure: A topological representation for non-manifold geometric boundary representations. *Geometric Modeling for CAD Applications*, Elsevier Science Publisher, Amsterdam, The Netherlands, pp. 3–36.

13
Finite Elements

Marc Hoit
University of Florida

Gary Consolazio
University of Florida

13.1 Finite Element Theory 13-1
   Simple FEM Theory • Shape Functions •
   Tapered Extensional Example • Shape Function
   Accuracy • Numerical Integration • Mapping
   Errors • Available Elements
13.2 Membrane Elements 13-9
   Membrane Theory • 2-D Shape Functions •
   Shear Locking • Mesh Correctness and
   Convergence • Stress Difference to Indicate Mesh
   Accuracy • Element Meshing
13.3 Flat Plate and Shell Elements 13-12
   Plate Theory • Generalized Stress
13.4 Solid Elements 13-15
   Solid Element Behavior and DOF
13.5 Dynamics 13-16
   Single Degree of Freedom (SDOF) Dynamic Analysis •
   Multiple Degree of Freedom (MDOF) Dynamic Analysis
13.6 Summary 13-21

13.1 Finite Element Theory

The finite element method (FEM) allows complex, continuum problems to be modeled. The FEM was originally developed by structural engineers as an extension of matrix structural analysis. It has since been used in just about every field where differential equations are used to define problem behavior. Here we will use the physical behavior of stress flow to discuss how finite elements can be derived and how the derivation assumptions affect the results. This process can be abstracted to apply to other domains such as fluid flow, temperature flow, population changes, stellar physics, and electronic circuits. In these cases, the quantities developed here (stiffness, stress, displacement) need to be mapped into the problem domain described by the differential equation being used. The process of the FEM is to create a stiffness matrix (coefficients) and a set of loads (right-hand side). After that, the solution process is identical to that covered in any stiffness-based structural analysis textbook. Many excellent books covering the FEM exist; this section is intended only as a basic introduction (Bathe, 1995; McGuire et al., 2002; Zienkiewicz et al., 2000).

The basic physical idea of the FEM is to break up a continuum into a discrete number of smaller "elements." These elements are connected at discrete points called nodes. It is at these nodes where the solution is found. These elements can be modeled mathematically by a stiffness matrix and are connected by nodes that have degrees of freedom (DOFs) (allowed movements). This is identical to what is done for beam and truss elements. However, beams and trusses have natural locations at which to define nodes. In addition, the derivation of their stiffness matrices can be done on a physical basis.

13.1.1 Simple FEM Theory

More general finite elements require more complicated procedures than used for beams and trusses to derive the stiffness matrix. The basic procedure is to assume a shape function that describes how the nodal displacements (and sometimes stresses) are distributed throughout the element. From the differential equation, we form an operator matrix that will convert the displacements within the element into strains (derivatives of displacement). Next, the internal and external virtual work (an energy method) can be formed and equated to develop the stiffness matrix. The last step involves solving for the unknown displacements which is identical to that used for truss and bending elements.

As an example, we will develop the stiffness matrix for a truss element, an axial member. The truss element has two nodal displacements, v_1 and v_2, one at each end. For any given set of displacements at the ends, a function is required to convert these into displacements along the length of the element. The obvious selection for the functions is the linear set given below in Figure 13.1.

Note how the given functions distribute the end displacements throughout the element. These distribution functions are called the shape functions. The shape functions can be put into the matrix form along with the end displacements to form an equation that describes the displacement variation within the element. The displacement anywhere within the element is described by the following matrix equation:

$$u(x) = \left\langle 1 - \frac{x}{L} \quad \frac{x}{L} \right\rangle \left\{ \begin{array}{c} v_1 \\ v_2 \end{array} \right\} \quad (13.1)$$

Note that the displacement is a function of x, the position within the element. Also note that the displacement anywhere in the element, $u(x)$, is the sum of the displacements caused by both end displacements distributed throughout the element. Eq. (13.1) can be rewritten as

$$u(x) = \mathbf{H}(\mathbf{x}) * \mathbf{v} \quad (13.2)$$

The matrix $\mathbf{H}(\mathbf{x})$ is called the shape function matrix and $\mathbf{v}$ the vector of element nodal displacements. We now need a differential equation that converts the displacement into strain. For our one-dimensional (1-D) example the equation is

$$\varepsilon_x = \frac{\delta u(x)}{\delta x} \quad (13.3)$$

It is useful to rewrite Eq. (13.3) into the form of two matrices. This alternate form puts the differential operator into the form of an operator matrix, $\mathbf{D}$. Eq. (13.3) is rewritten as

$$\varepsilon_x = \mathbf{D} * u(x) \quad (13.4)$$

where the operator matrix has the form

$$\mathbf{D} = \left\langle \frac{\delta}{\delta x} \right\rangle \quad (13.5)$$

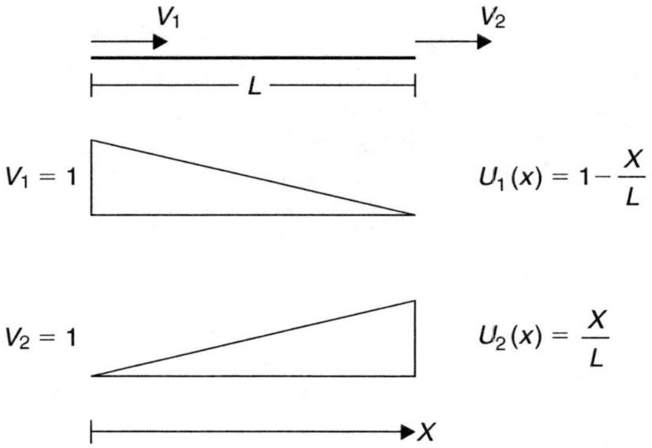

FIGURE 13.1 Unit displacement functions for axial member.

Finite Elements

If we apply this operator to the displacement, $u(x)$, given in Eq. (13.2), we can find the strain as a function of the element displacements, **v**, and the shape function matrix, **H(x)**. This gives the form for the strain in terms of the shape function matrix **H(x)**.

$$\varepsilon_x = \mathbf{D} * \mathbf{H(x)} * \mathbf{v} \tag{13.6}$$

Note that the nodal displacements, **v**, are constants (individual numbers for a given problem and actually the problem unknowns) with respect to x and need not be operated on, differentiated (as a result of the chain rule). Therefore, only the derivatives of the shape functions need to be taken. For our case of the axial element, the strain can then be written by substituting the shape functions, **H(x)**, into Eq. (13.6) and applying the **D** operator giving

$$\varepsilon(x) = \left\langle -\frac{1}{L} \quad \frac{1}{L} \right\rangle \begin{Bmatrix} v_1 \\ v_2 \end{Bmatrix} \tag{13.7}$$

Typically, the differential operator times the shape function matrix is called **B**, the strain–displacement matrix. The strain is then commonly written in the shorter form:

$$\varepsilon_x = \mathbf{B} * \mathbf{v} \tag{13.8}$$

where $\mathbf{B} = \mathbf{D} * \mathbf{H(x)}$. We also need the relationship of Hooke's law that converts strain into stress, given as

$$\sigma = \mathbf{E} * \varepsilon \tag{13.9}$$

Therefore, if we calculate the internal strain energy, a form of work energy, to develop a stiffness matrix, virtual strain times stress, with substitutions we have

$$\delta W_i = \int_{\text{volume}} \bar{\varepsilon}^T \sigma = \int_{\text{volume}} (\bar{v}^T * \mathbf{B}^T * \mathbf{E} * \mathbf{B} * v) \delta V \tag{13.10}$$

Equating internal to external virtual work (conservation of energy) and removing the arbitrary virtual displacements, $\bar{v}^T$, from both sides, we get

$$S = \int_{\text{volume}} (\mathbf{B}^T * \mathbf{E} * \mathbf{B}) \delta V * v \tag{13.11}$$

Looking at Eq. (13.11), we see that this is the familiar stiffness form of the element relationship between forces and displacements. As a result, we can see that the integral is just the element stiffness. Taking that portion out of the equation we have

$$K_e = \int_{\text{volume}} (\mathbf{B}^T * \mathbf{E} * \mathbf{B}) \delta V \tag{13.12}$$

This is the classic form of the finite element stiffness formulation. For our axial element example, we substitute for **B** from Eq. (13.7) and **E** is just the familiar Young's modulus. In the general case, **E** is the constitutive matrix. For the linear case, **E** is just the three-dimensional (3-D) representation of Hooke's Law. Integrating over the Y and Z coordinates for part of the volume integral we get the area of cross section. Multiplying the matrices after the partial integration for the area and removing the constants from the integral we get

$$K_e = AE * \int_{\text{length}} \begin{bmatrix} \frac{1}{L^2} & -\frac{1}{L^2} \\ -\frac{1}{L^2} & \frac{1}{L^2} \end{bmatrix} dx \tag{13.13}$$

Integrating the matrix term by term over the length, we get the familiar form for a truss stiffness as

$$K_e = \frac{AE}{L} * \begin{bmatrix} 1 & -1 \\ -1 & 1 \end{bmatrix} \tag{13.14}$$

This is the final result we are looking for. Of course, this is identical to the standard truss stiffness matrix developed by traditional stiffness methods. However, the described shape function process can be extended to other types of elements where the traditional stiffness-by-definition method is not possible.

13.1.2 Shape Functions

As seen in the previous section, one of the major variables in deriving a finite element is the choice of the shape function **H(x)**. Remember that it is the shape function that describes how the nodal displacements are distributed throughout the element. In the truss member, it was assumed that the end displacements are distributed linearly. This is an exact assumption for a uniform cross-section member.

The choice of the shape function is directly related to the number of unknowns in the element. The number of unknowns defines the order of approximation that the shape function can have. Again for the truss member, there were two unknowns (one at each end). This gave linear shape functions since two points give a straight line.

As a result, to increase the accuracy of an analysis you need to increase the number of unknowns. This can be handled by two methods: (1) you can increase the number of unknowns and use piecewise continuous linear shape functions or (2) you can increase the number of unknowns and increase the order of the shape function within an element. The next section will show the effects of these assumptions on a simple example.

13.1.3 Tapered Extensional Example

For a constant cross-section truss member, there is no need to increase the accuracy above linear shape functions. Instead, let us look at another simple problem. We want to see the effects of shape function selection on accuracy when trying to solve a tapered member subjected to axial load for different modeling conditions (see Figure 13.2).

The problem is simple enough that it can be solved exactly and the approximate finite element results compared with this exact result. The problem has a unit thickness and is subjected to an axial load of 20. Note that no units are given. Any consistent units are acceptable. The exact solution for the displacement is

$$u(x) = -0.0074074 * \ln(10 - 0.09 * x) + 0.0170561$$

The solution was achieved by integrating the strain, $\varepsilon = \frac{\sigma}{E}$. The exact solution for the stress, force over area, in the section is

$$\sigma(x) = \frac{20}{10 - 0.09 * x} \tag{13.15}$$

We will use the FEM and solve the problem using three different assumed shape functions. First, we will start by choosing a single unknown at the tip of the member. Second, we use two unknown displacements,

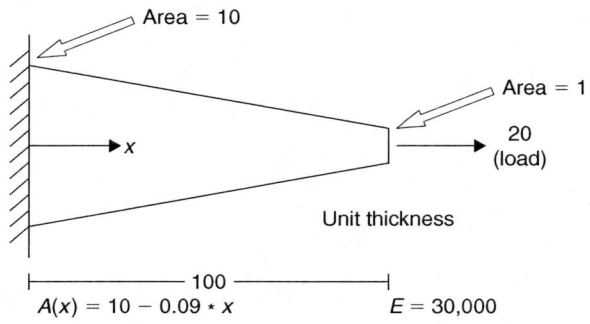

FIGURE 13.2 Axial loaded tapered problem.

Finite Elements **13**-5

one at the tip and the other at the midpoint. Using the two unknowns we have a choice, we can use a quadratic shape function or we can use two linear functions. We will perform the analysis using both. Figure 13.3 shows the finite element models and assumed shape functions.

Using the appropriate shape functions, forming the stiffness matrices using Eq. (13.12), and then solving for the displacements and stresses for each problem, we obtain the displacement and stress results shown in Figure 13.4 and Figure 13.5.

There are some important things to notice about the results. First let us look at the displacement plots (see Figure 13.4). The exact displacement is the natural log function given in Eq. (13.15). The assumed displacement functions try to approximate this function by a linear, bilinear, and a quadratic function, respectively. The FEM states that you can use the linear function, *but in the limit as the number of linear segments goes to infinity, the answer will be exact*. You can see that the bilinear is a better approximation than the linear. A trilinear would be even better, and so on. Also, the quadratic is even better than the other two at approximating the natural log function.

Next we will look at the stresses (see Figure 13.5); here we see an even more dramatic result. The stress in each linear element is *constant*. This is obvious since the stress is proportional to the strain and the strain is the *derivative* of the displacement. Another fact about finite elements is that it is a displacement-based method. This *assumes* that the displacements are *continuous*. However, it generally says nothing about the

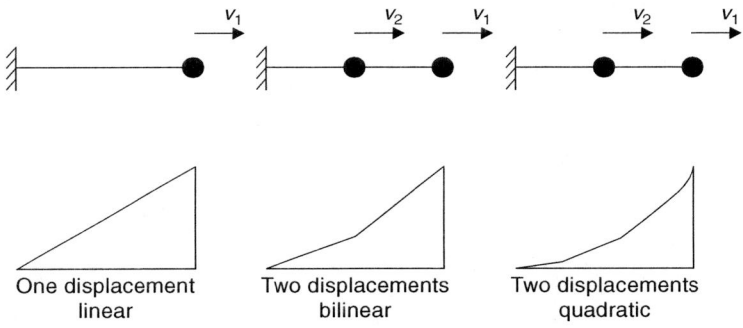

FIGURE 13.3 Three shape function models for tapered section.

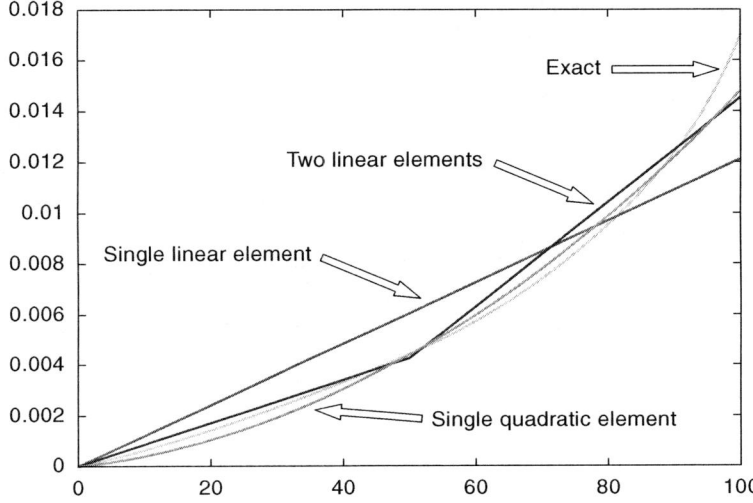

FIGURE 13.4 Axial displacement result for tapered element models.

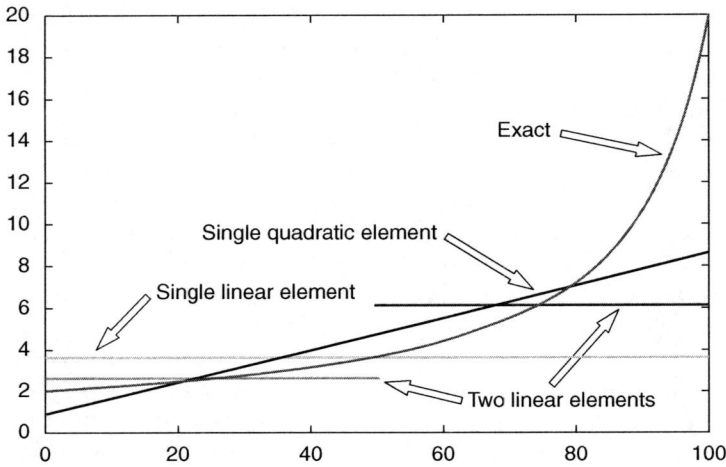

FIGURE 13.5 Axial stress for different tapered element models.

strain or stress. The displacement is *continuous* but the *stresses are not continuous*. This can be seen from the bilinear result. The stresses are constant but *different* between the two halves of the tapered element sections.

Finally, note that the stress from the quadratic element is linear. Again, this is due to the fact that the stress is proportional to the derivative of the displacement. Note how in all the cases, the finite element stress underestimated the true stress. Elements can be derived that also have stress continuity, but these are more complex.

13.1.4 Shape Function Accuracy

In summary, the accuracy of a finite element is dependent on the choice of the shape function for that element and the number of elements used in the model. *In the limit, as the number of elements approach infinity, the model will give exact results.* The displacements are assumed to be continuous between elements. However, *stresses are not continuous between elements*.

The best accuracy is achieved with many higher order elements. Equal accuracy can be achieved by a larger number of lower order elements. Clearly, the required number of elements depends on the displacement field you are trying to match. A linear displacement field only needs linear elements.

13.1.5 Numerical Integration

In the previous section, we saw how the choice of the number of unknowns dictated the order of the finite element shape function. The more nodes and unknowns used in the element, the higher the order. Once the shape function is chosen, the element can be formed by performing an integral over the element volume. In simple elements, this integration is easy to perform exactly. However, when we use irregular shaped elements in two and three dimensions, it is very difficult to perform the integration exactly.

As a result, most finite element programs rely on numerical integration techniques. Most engineers are familiar with Trapezoidal and Simpson's rule for numerical integration. While these methods are easy to understand, they are computationally inefficient. Instead, in finite element codes, the Gauss–Legendre quadrature method is often used.

Gauss–Legendre quadrature is *exact* for simple polynomials and requires very little computation. The basic formula for the method is

$$\int_{-1}^{1} F(x)\,dx = \sum_{i=1}^{\text{number of points}} A_i * F(\mu_i) \qquad (13.16)$$

This formula says that to get the integral for a function $F(x)$, you just need to *sum* the function at some given evaluation points, μ_i, times some weighting values, A_i, for the given number of evaluation points. *Both the evaluation points and the weights are known numbers!* It turns out that the weights (A_i) and the evaluation points (μ_i) never change. As an example, the two point Gauss–Legendre quadrature points and weights are

$$\mu_1 = \frac{1}{\sqrt{3}}$$
$$\mu_2 = -\frac{1}{\sqrt{3}} \quad (13.17)$$
$$A_1 = A_2 = 1.0$$

These formulas say that if the function to be integrated is evaluated at the two Gauss–Legendre points and the results are summed you get the exact integral (from -1 to 1).

The Gauss–Legendre quadrature procedure is *EXACT* depending on the number of points used for the sum. In the case above, for two-point integration, the method is exact for *up to* a cubic equation. The formula for exactness is $P = 2N - 1$, where P is the order of the polynomial that can be integrated exactly using N Gauss–Legendre sampling points.

Therefore, three Gauss–Legendre points ($N = 3$) is exact for up to a fifth-order polynomial. The Gauss–Legendre points do not have to be rederived once they are found. As a matter of fact, they can be looked up in many sources for up to 20 points.

It is clear that this method requires a very small number of function evaluations to get exact integration results. Also note that the method is defined as integrating a function from (-1 to 1). This restriction can be lifted by using the mapping formula:

$$\int_A^B F(x) = \frac{(B-A)}{2} \sum_{i=1}^{\text{number of points}} F\left(\frac{B-A}{2}x + \frac{B+A}{2}\right) \quad (13.18)$$

This formula takes the points defined on (-1 to 1) and shifts and scales (or maps) them into the new limits, A to B. Using the mapping method also has some additional benefits. Most finite element programs define their element on this -1 to 1 coordinate system. Then, whatever the actual shape of the element, it is *mapped* to the -1 to 1 system. This allows us to use *nonrectangular* shaped elements. The mapping of a nonrectangular element into a -1 to 1 coordinate system develops the *isoparametric* finite element.

This same mapping and integration procedures are used in two and three dimensions. When going to more dimensions, we need to extend the Gauss–Legendre quadrature scheme. The usual method is to use the same points as in the 1-D case but in all coordinate directions.

13.1.6 Mapping Errors

There can also be problems when using the mapping to shift to a -1 to 1 coordinate system. In isoparametric elements, the mapping procedure uses the displacement shape functions as mapping functions. Therefore, any real location X in the element can be found from its -1 to 1 coordinate by the formula:

$$X = \sum_{i=1}^{N} H_i(\mu) X_i \quad (13.19)$$

where $H(\mu)_i$ is the shape function for nodes i and X_i is the coordinate for node i. This mapping will work for (1-D, 2-D, 3-D) elements. If we look at the 1-D axial problem again, we can develop the coordinate mapping for the quadratic (two-node) element. If we allow the midpoint node to move between 1/4 from the left past the midpoint and 1/4 of the distance to the right end, we get the mapping as shown in Figure 13.6.

Note in Figure 13.6 that if the midpoint node is either larger than 3/4 or smaller than 1/4, the mapping goes outside the actual length of the element. This has the effect of saying that more than one location on the element maps to the *same* point in the real X coordinate.

Clearly, this is not a valid element configuration. It also means the element mapping is not valid because it is not invertible. The mapping from one coordinate to the other is handled by a transformation matrix called the *Jacobian*. The Jacobian matrix needs to be inverted to switch from one coordinate system to the other. The validity of a mapping can be determined by the invertibility of this mapping matrix. Therefore, if the *Jacobian* matrix is singular or noninvertible, then the mapping is not valid. This usually means that the nodes are not in the correct locations, outside the 1/4 points, or there are bad nodal coordinates.

13.1.7 Available Elements

Energy derivations (e.g., virtual work) are commonly used to form the stiffness for a variety of element types. The most common stiffness elements are the membrane (planar), plate, shell, and solid elements. Each of these elements has a given set of nodes and displacements associated with those nodes. The common forms of these elements are given in Figure 13.7.

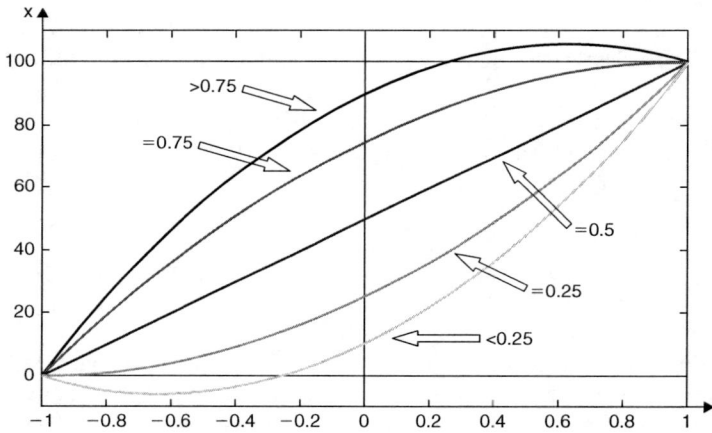

FIGURE 13.6 Mapping with midpoint node not at the center.

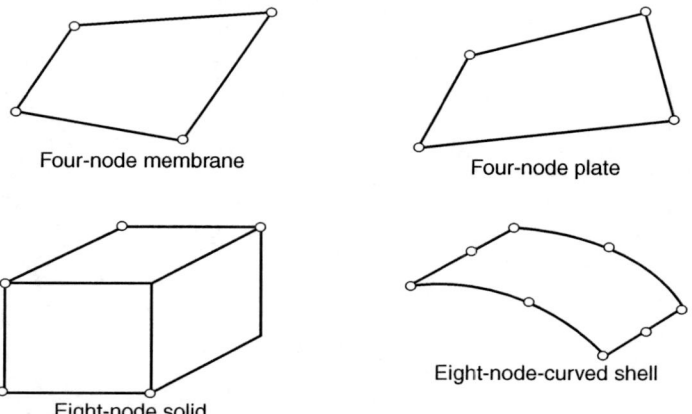

FIGURE 13.7 Common finite elements.

These elements have additional restrictions on their behavior that depend on their derivation. However, the result is always a stiffness matrix that can then be treated like any other stiffness matrix and may be rotated and transformed as desired. When combining these elements, the same concerns about boundary conditions and matching DOF at the nodes must be accounted for. Additional concerns are also generated since the shape function assumption can affect the accuracy of the results.

13.2 Membrane Elements

Membrane elements can be used to describe any continuum problem that is 2-D in nature. Either the plane stress or plane strain condition can be modeled using these elements. The *plane strain* condition says the out-of-plane strain is *zero*, and the *plane stress* condition says the out-of-plane stress is *zero*.

As an example, the following problems can be analyzed in 2-D using membrane elements (Figure 13.8).

13.2.1 Membrane Theory

The membrane element is a flat element. It is generally assumed to have constant thickness. It can be triangular, rectangular, four-sided polygonal, or have curved sides. The element is generally found in configurations of three, four, six, eight, nine, and variable 3–9 nodes. Whatever the shape or number of nodes, the element has *two* translational DOFs per node. These DOFs *must* lie in the plane of the element. The results from the element consist of two normal stresses and a shear stress in the plane of the element (see Figure 13.9). The stress results are generally given at each node in the element.

The difference in element behavior is dictated by the choice of the number of nodes and hence the number of DOFs for the element. The three-node triangle has *linear* shape functions and hence *constant* strain and stress. This element is referred to as the constant strain triangle. The four-node element has a slightly better response than the three-node element. The six-node triangle has quadratic shape functions and *linear* stress and strain. The eight- and nine-node element has better response than the six-node element.

13.2.2 2-D Shape Functions

Remember, the shape function describes how the nodal displacements are distributed throughout the element. In the membrane element, the X and Y displacements are assumed to be independent. As a result, the same shape functions are reused for the X and Y displacements. The three-node elements have linear shape functions for displacement. The four-node element adds an XY term to the shape function.

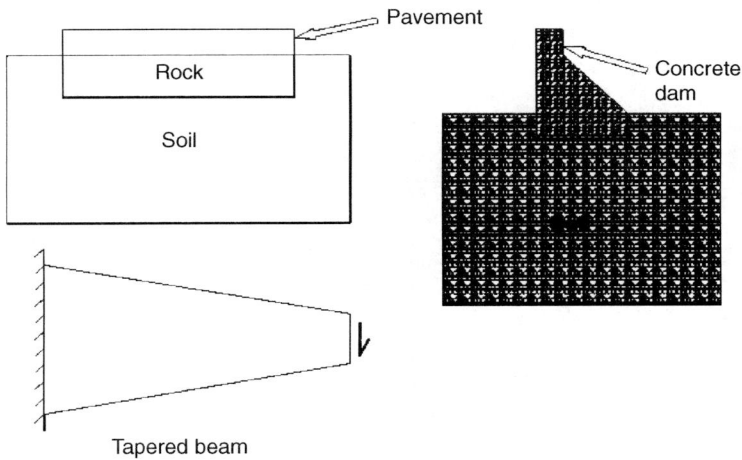

FIGURE 13.8   2-D problems using membrane.

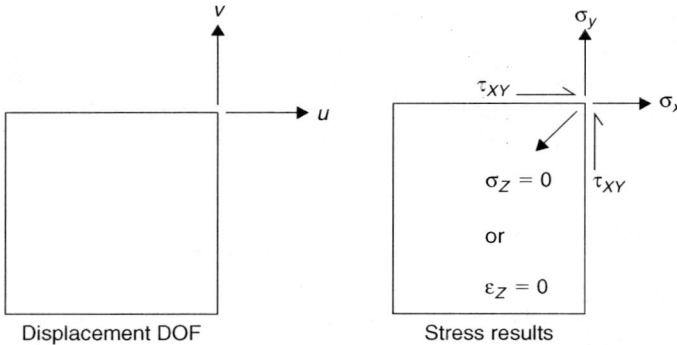

FIGURE 13.9 Membrane DOF and stress results.

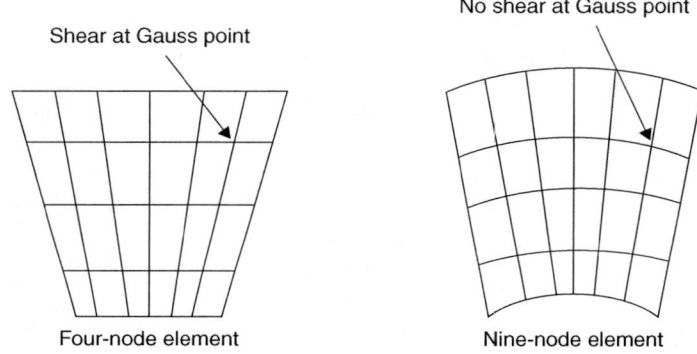

FIGURE 13.10 Shear locking of membrane.

The nine-node element has a quadratic variation of displacement. Remember the stress is the derivative of the shape function and hence it will be linear with some quadratic parts.

The order of the shape functions are given to reinforce the capabilities of the particular elements to model displacements. The more nonlinear the displacement field in a problem, the more elements required (like the tapered axial problem). In addition, higher order elements can approximate a more complex displacement field with fewer elements.

13.2.3 Shear Locking

When using or developing the FEM, the goal is to reduce the number of unknowns (nodes) and get the highest order displacement behavior possible. Often, modeling complex behavior using lower order elements can cause unexpected behavior. The problem with the four-node element (and all lower order elements) lies in how the element models a higher order displacement field like the bending (flexural) effect. Figure 13.10 gives the displaced shapes that each element uses to approximate beam bending.

The four-node element can only model bending (cubic displaced shape) by linear displacements. The nine-node element has quadratic capabilities. The linear displaced shape causes *shear* to occur in the element while in true bending *none exists*. The nine-node element does *not* generate this fictitious shear.

This phenomenon is called *shear locking*. Four-node elements exist that *do not* exhibit this problem. By changing the formulation, you can create an element that gives better behavior. The two most common methods are by creating a nonconforming element or by using reduced integration. A nonconforming element adds additional shape functions that contain the higher order (bending) shape (see Figure 13.11).

Finite Elements 13-11

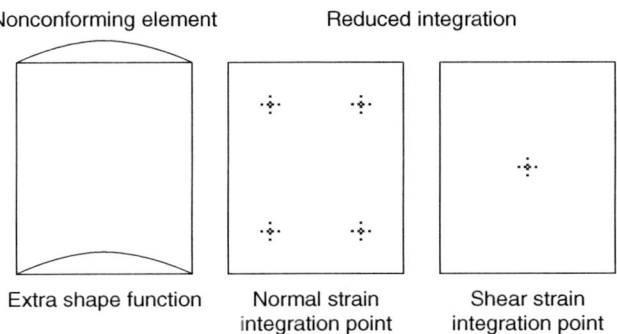

FIGURE 13.11 Methods for removing shear locking.

However, this additional shape function is *NOT* continuous across element boundaries and hence displacements are not continuous. Therefore, the effect is to allow gaps to open between elements. Reduced integration uses a different number of Gauss–Legendre points during the integration, a number less than that required for exact integration. The sampling points are chosen so that this shear is not included. For the current example, a single point in the center of the element is used for integrating the shear terms. As can be seen, this will neglect the shear developed in the element as a result of the linear displaced shape. There are other more complex constraint processes that remove the shear locking problem.

13.2.4 Mesh Correctness and Convergence

As discussed in Section 13.1, the accuracy of the solution depends on the number of elements and the order of the shape functions. As the number of elements increases, the piecewise displacement approximation approaches the true displacement field. Recall that two linear elements provided a better response than a single linear element. Also, a single quadratic element performed even better.

13.2.5 Stress Difference to Indicate Mesh Accuracy

The stress results also follow the same pattern. More elements provide better stress results. However, since we only guarantee the continuity of the displacements, the stresses are discontinuous. This means that at a node where two elements meet, the stresses do not match. However, as the number of elements increases, the difference in stresses between elements gets smaller. As an example, Figure 13.12 is a plot of the stress along the top of the cantilever beam. The results are plotted for 4-4, 2-9, and 40-4 node membranes.

Note that for the 4-4-node elements, the difference between the elements is 28%. This large percentage error indicates a poor mesh (or not enough elements). Looking at the two nine-node model we see a closer difference. Here the error is 14%. This indicates that the mesh is marginal but probably sufficient. Finally, we look at the 40-element model. Here the error is much better and only 3%. The 40-element model is very good. Note that many solution techniques perform another process of stress averaging to improve the final presented result.

The difference in element stresses at a node is an important measure of model correctness. In general, we do not have the exact displacements to compare and check our model. Hence, stress checks are necessary to verify convergence of our model. *If the difference in stresses between elements is small, the finite element mesh is good.*

13.2.6 Element Meshing

Defining a mesh is critical to finding a correct solution with a minimum number of elements. More elements are needed where the displacement field is highly nonlinear (or in a high stress *gradient* area).

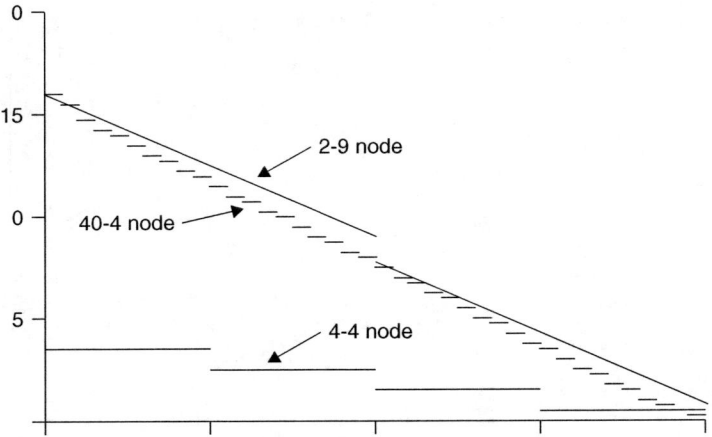

FIGURE 13.12 Stress error for different mesh and shape function.

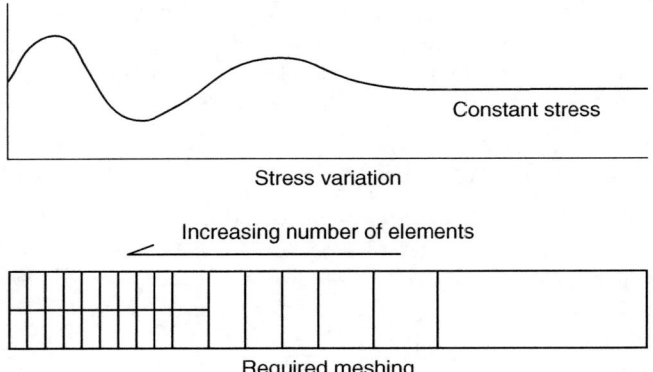

FIGURE 13.13 Mesh variation with stress gradient.

Fewer elements are needed as the response becomes linear and only a single element is required in a constant stress field. As an example, the following stress function could be modeled by the mesh given in Figure 13.13.

This change in the number of elements is handled by mesh changes from a single to multiple or higher order elements. This is where variable node elements are useful.

Another important principle is that stress concentrations are localized phenomena and do not affect the solution at a reasonable distance from the concentration. In other words, bad stress differences at one portion of a model do not necessarily affect the results in a well-modeled portion.

In summary, *a small stress difference between elements means a good mesh. Large stress differences in localized areas will not* necessarily *affect the result a reasonable distance away.*

13.3 Flat Plate and Shell Elements

The next finite element we examine is the flat plate bending element. This element can be thought of as a 2-D extension of a beam element. Beam elements provide both shear and bending resistance. Plate elements provide this same resistance but in two directions.

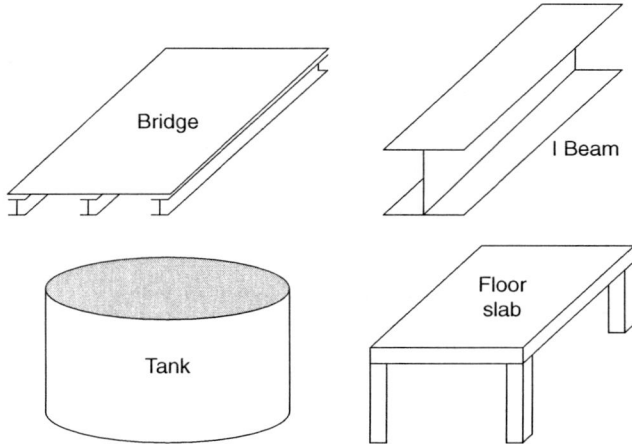

FIGURE 13.14 Structures where flat places can be used.

Plate bending elements can be used to model common structural elements such as floor slabs, floor diaphragms, bridge decks, and even I-beams. Whenever out-of-plane bending effects need to be considered, plate elements can be used. They are also useful for thin-walled structures like pipes and tanks. Most modeling situations require that both the out-of-plane plate bending and in-plane membrane effects be modeled. Some example structures are shown in Figure 13.14 that use plate and membrane elements.

True plate elements *do not include in-plane effects*. In-plane effects are handled by membrane elements. Similarly in a beam element the bending and axial effects are uncoupled. This is the same in two dimensions. These two elements are commonly merged to get a complete in-plane and out-of-plane element referred to as a *flat shell element*. A curved shell element would be needed to include the coupling of axial and bending effects. We will discuss a true plate element before discussing flat shell elements.

13.3.1 Plate Theory

There are two common versions of plate theory used in finite elements: Kirchoff and Mindlin. Kirchoff plate bending theory is derived in a similar fashion to beam bending but includes bending in both directions. The derivation assumes that the normal displacement, w, controls. In Kirchoff theory the rotation, Θ, in the plate is the derivative of w, the vertical displacement. This is the same as Euler–Bernoulli beam theory. In Mindlin theory, shear deformation is included and the rotation is the sum of the derivative of w and the shear deformation angle.

13.3.1.1 Kirchoff Theory

In Kirchoff theory, the *normal to the surface remains normal*. Hence, this theory *ignores* shear deformations (just like Euler–Bernoulli beam theory). To derive this type of finite element, a shape function that describes the distribution of the normal displacement $w(x, y)$ throughout the element is needed. This shape function has the property that its derivative is equal to the slope of the surface. An important implication of this is that the slope *is continuous* across elements! This is called a C^1 element, meaning that it has continuous first derivatives between elements. Figure 13.15 shows the relationship between w and Θ for Kirchoff theory.

13.3.1.2 Mindlin Theory

The second theory, Mindlin, *includes shear deformations*. As a result, a vector initially normal to the surface *does not* remain normal during deformation. The derivative of the shape function for the normal displacement $w(x, y)$ is *not* equal to the rotation. In Mindlin theory, the rotation of the surface is the sum

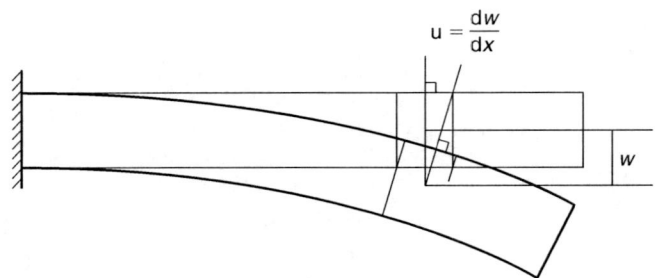

FIGURE 13.15 Kirchoff plate theory.

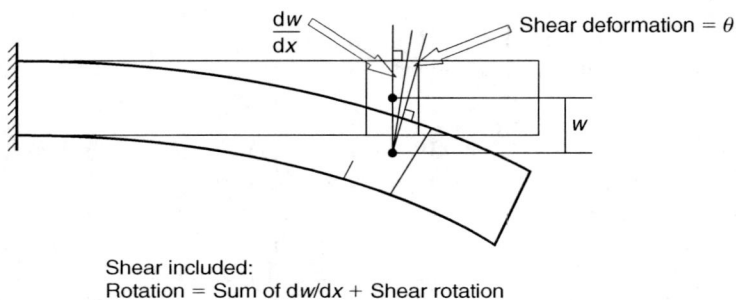

FIGURE 13.16 Mindlin plate theory.

of the derivative of $w(x, y)$ and the shear deformation angle change. Figure 13.16 shows the relationship between the displacement $w(x, y)$, shear angle γ, and the derivative of the displacement.

This sum of angles to get the total rotation implies that *independent shape functions* can be used for the displacement w and the rotations (Θ_x, Θ_y). Mindlin theory is the most common formulation found in flat plate and shell elements used in current computer programs. This means there will *not* be rotational continuity across elements boundaries (since shear exists). There is of course still rotational continuity at the nodes. Hence the elements are considered to be C^0 elements.

In both the Kirchoff and Mindlin formulations, the pure plate bending element has three DOFs per node: the normal displacement w and the out-of-plane rotations (Θ_x, Θ_y). These are shown in Figure 13.17.

13.3.2 Generalized Stress

In plate theory, most derivations refer to the equations for *generalized stress and strain*. This is because the equations for plate behavior can be converted to the form:

$$M(x, y) = \mathbf{E}^* * \Psi(\text{curvature}) \tag{13.20}$$

where $\mathbf{E}^*$ is a modified constitutive matrix. Note that this is just like the equation for stress and strain except we have moments replacing stresses and curvature replacing strain. In plates, the displacement unknowns are the normal displacement and the two rotations. Following the analogy of generalized stress, *moments* are equivalent to stress and *curvature* is equivalent to strain. This means that when using these elements in modeling, we treat the moment gradient like we would stress to determine the level of shape function and number of elements required for an accurate analysis. In addition, the difference in moment at a common node between two elements indicates the adequacy of the mesh.

Finite Elements

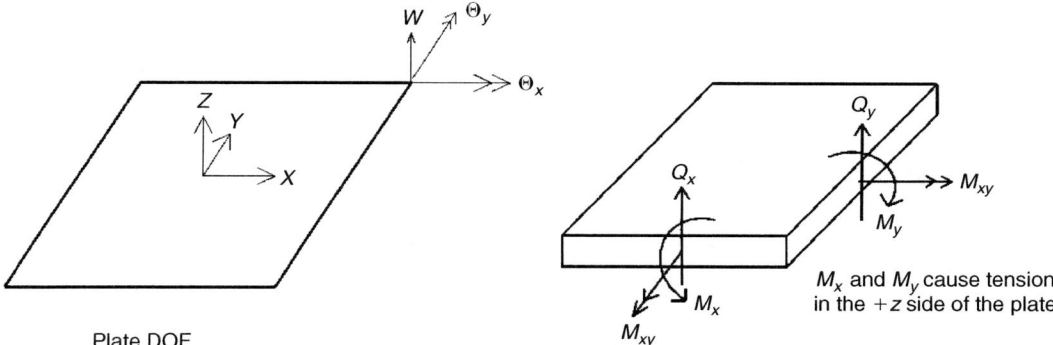

FIGURE 13.17 Plate DOF and stress results.

The results from all plate elements consist of moments and the transverse shear, Q. It is important to note that the moments and shear results are per unit length of plate.

Flat plate elements can be found in three- to nine-node versions, just like membrane elements. The same concepts of shape function order are true for plates as they were for the membranes. Three-node triangular plates model *constant moments* exactly. Nine-node elements model linear moments with some second-order effects. It is important to remember that in plates, *moments are equivalent to stress and curvature is equivalent to strain*, in terms of modeling. In other words, we need more elements in a high moment gradient area for plates.

Flat shell elements combine the effects of plate bending with in-plane (membrane) effects. There exist formulations for both flat and curved shell elements. The curved element formulation is a more complicated derivation. The flat shell however can be considered to be merely the addition of the membrane and flat plate elements. This is the most common form of shell element found.

13.4 Solid Elements

The final element we will look at is the 3-D solid element. This element is the most general of the finite elements. It fully represents a 3-D stress and strain state. It is a fundamental building block that can be used to create any shape structure. Clearly, as a structure becomes more complex, the effort required defining the geometry and mesh becomes time consuming. As a result of the required sophistication of many analyses, many people are moving to computer-based solid geometry modeling of structures. This especially includes mechanical components. Solid elements are a natural choice for this type of modeling since any solid object can be meshed by solid elements.

13.4.1 Solid Element Behavior and DOF

Solid elements are found in varied configurations analogous to membrane elements. They can be found in four-node tetrahedrons, eight-node bricks, 20-node bricks, and 27-node bricks. Of course, variable node versions exist that allow from 8 to 27 nodes.

The solid element is a 3-D analog of the membrane element. It has three DOFs per node. It does not have any rotational DOF or rotational stiffness at the nodes. The solid element is capable of representing a fully 3-D stress state. The available DOF and stress results for a typical eight node of a solid element are shown in Figure 13.18.

The properties required for the solid element consist of Young's modulus, E, and Poisson's ratio, v. Note no thickness is required as was the case for membrane element. Solid elements model the full 3-D effects of Hooke's Law.

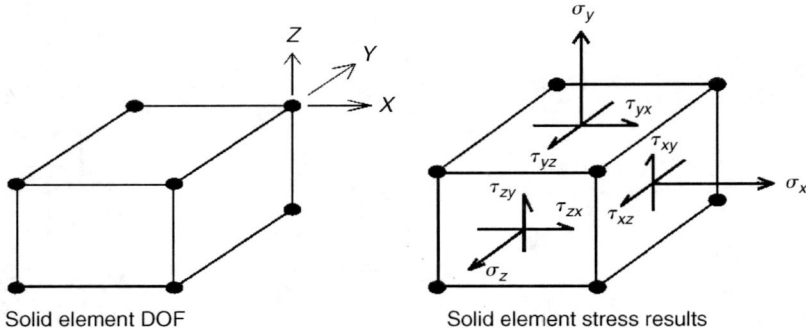

FIGURE 13.18 3-D solid DOF and stress results.

13.5 Dynamics

In the displacement-based linear static finite element analysis method described above, the governing equation to be solved is the static equilibrium equation:

$$Kq = F \qquad (13.21)$$

where K is the global stiffness matrix of the system, q a vector of global displacement DOF, and F a vector of global nodal forces (loads). Solving Eq. (13.21) yields a vector of spatially varying, but time-invariant, displacements q that may be used to recover (back-calculate) element quantities such as strains, stresses, and internal forces. Dynamic finite element analysis, in contrast, involves solution of a governing *equation of motion* that is both spatially varying and time-varying:

$$M\ddot{q} + C\dot{q} + Kq = F(t) \qquad (13.22)$$

In this equation, M is the global mass matrix of the finite element model under consideration, C the global damping matrix, and K the global stiffness matrix. Quantities $\ddot{q}$, $\dot{q}$, and q are vectors of time-varying nodal accelerations, velocities, and displacements, respectively, and $F(t)$ is a vector of time-varying forces acting at the system DOF. The equation of motion can be interpreted as a force balance between inertial force $M\ddot{q}$, damping force $C\dot{q}$, internal structural force Kq, and external load $F(t)$. Dynamic finite element analysis requires that the force balance described by Eq. (13.22) be satisfied at each point in time.

Formation of the global stiffness matrix K is carried out by assembling element stiffness matrices K_e, computed as indicated in Eq. (13.12), for all elements constituting the model. The global mass matrix M is similarly constructed as an assembly of element mass matrices. A commonly used method of forming element mass matrices M_e is referred to as the *consistent-mass* formulation and is given by

$$M_e = \int_{\text{volume}} \rho(H^T * H)\delta V \qquad (13.23)$$

where ρ is the mass density of the material and H a shape function matrix. Alternatively, in many applications is it sufficiently accurate to simply lump mass at the nodes of the model in accordance with tributary areas (or volumes) for each node. Such an approach produces a diagonal mass matrix and is referred to as a *lumped-mass* formulation.

Methods of formulating the global damping matrix C are more varied than those used to form K and M. Moreover, virtually all formulations of damping are simply mathematically convenient constructions that only approximately represent actual damping phenomena that occur in solid materials. Dynamic damping models are generally intended to model energy dissipation in structural materials. In Eq. (13.22),

Finite Elements

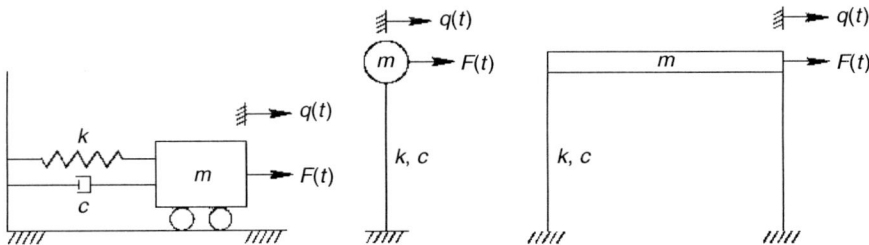

FIGURE 13.19 Schematic diagrams of single degree of freedom (SDOF) systems.

a *viscous damping* model is employed in which the energy dissipating damping forces $C\dot{q}$ are assumed to be proportional to the nodal (particle) velocities $\dot{q}$. A common and mathematically convenient means of formulating viscous damping matrices is referred to as Rayleigh damping (also called proportional damping), and is given by

$$C = \alpha M + \beta K \quad (13.24)$$

where α and β are scalar constants. If the Rayleigh damping model is employed at the element level, rather than at the global level indicated in Eq. (13.24), then element damping matrices are formed as $C_e = \alpha M_e + \beta K_e$ and are then assembled into a global damping matrix C in the same manner that element mass and stiffness matrices M_e and K_e are assembled.

13.5.1 Single Degree of Freedom (SDOF) Dynamic Analysis

SDOF dynamic systems are structures in which inertial, damping, and stiffness (internal structural) forces may be adequately described by a single displacement DOF q, and its time derivatives $\dot{q}$ and $\ddot{q}$. Schematic diagrams of example SDOF systems are shown in Figure 13.19.

For an SDOF system, the equation of motion, previously described in terms of matrices and vectors in Eq. (13.22), simplifies to

$$m\ddot{q} + c\dot{q} + kq = f(t) \quad (13.25)$$

where m, c, and k are scalar constants, and $\ddot{q}, \dot{q}, q$, and $f(t)$ are time-varying scalars. If no forcing (loading) function $f(t)$ is present, i.e., $f(t) = 0$, then *free vibration* of the system results. Oscillation of a system in free vibration may be initiated by nonzero initial conditions such as initial displacement and initial velocity. In Figure 13.20, free vibration displacement responses $q(t)$ for an undamped ($c = 0$) system are shown for cases of nonzero initial displacement $q_0 \neq 0$, nonzero initial velocity $\dot{q}_0 \neq 0$, and nonzero initial displacement and velocity $q_0 \neq 0$ and $\dot{q} \neq 0$.

In each case shown in Figure 13.20, the SDOF system oscillates at the same characteristic frequency, denoted the **circular natural frequency** of the system, and determined as

$$\omega = \sqrt{k/m} \text{ (rad/s)} \quad (13.26)$$

Parameters related to ω are the natural frequency of the system $f = \omega/2\pi$ (Hz), and the natural period of the system $T = 1/f$ (s). Because damping has been ignored in Figure 13.20, no dissipation of energy occurs, and oscillation continues without decay. For damped cases in which $c \neq 0$, energy dissipation during each cycle of oscillation causes the amplitude of free vibration to decay. The rate at which this decay occurs is a function of the level of damping c present in the structure. The level of damping may also be expressed as a fraction of a special level of damping called *critical damping*. The critical viscous damping coefficient for an SDOF structure is given by

$$c_{cr} = 2m\omega = 2\sqrt{mk} \quad (13.27)$$

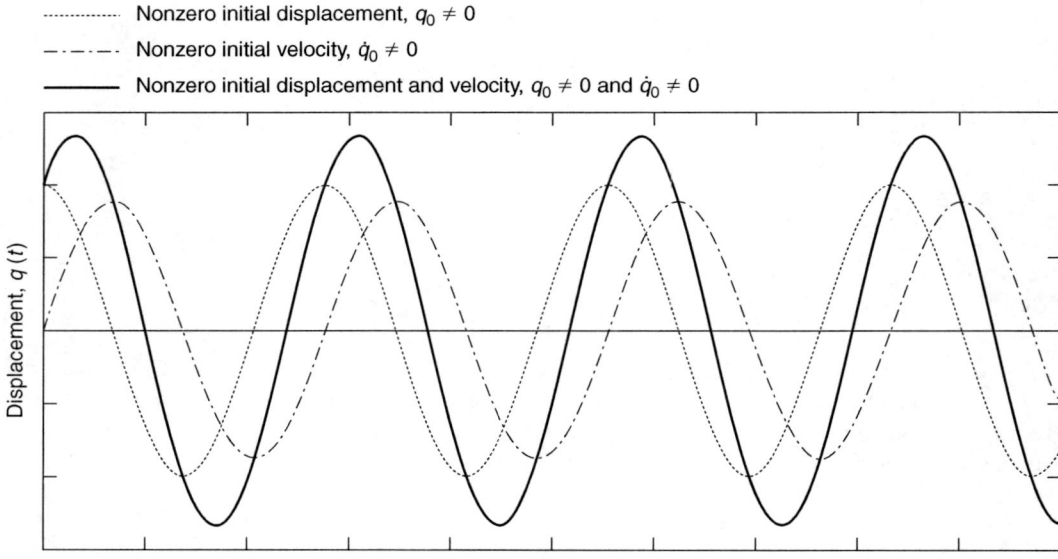

FIGURE 13.20 Free vibration of an undamped SDOF system under nonzero initial conditions.

Critical damping c_{cr} is the minimum level of damping that will prevent oscillation of an SDOF dynamic system in free vibration. For systems in which $c < c_{cr}$ (underdamped), free vibration oscillations will occur in response to nonzero initial conditions. However, for systems in which $c > c_{cr}$ (overdamped), free vibration oscillations will be completely suppressed. Damping levels in structures are therefore very often described in relation to c_{cr} by using the parameter:

$$\xi = c/c_{cr} \tag{13.28}$$

where ξ is called the *ratio of critical damping* or *fraction of critical damping*. Damping also generally affects the frequency and period of oscillation. Analogous to the undamped natural circular frequency ω, the *damped circular frequency* is defined as

$$\omega_d = \omega\sqrt{1-\xi^2} \text{ (rad/s)} \tag{13.29}$$

Following from this definition, the damped frequency is $f_d = \omega_d/2\pi$ (Hz) and the damped period is $T_d = 1/f_d$ (s). For most structural materials, natural damping levels fall within the range $\xi < 10\%$. For such conditions, $\omega_d \cong \omega$, $f_d \cong f$, and $T_d \cong T$. In Figure 13.21, displacement responses $q(t)$ for undamped and damped SDOF systems subjected to nonzero initial conditions are compared for various levels of damping.

When a nonzero forcing function $f(t)$ acts on a structure, the resulting response is referred to as *forced vibration*. Depending on the complexity of $f(t)$, one of a variety of different techniques may be employed to quantify the displacement time-history of the SDOF system. For relatively simple mathematical forms of $f(t)$, Duhamel's integral

$$q(t) = e^{-\xi\omega t}\left(q_0 \cos(\omega_d t) + \frac{\dot{q}_0 + \xi\omega q_0}{\omega_d}\sin(\omega_d t)\right)$$
$$+ \frac{1}{m\omega_d}\int_0^t f(\tau)e^{-\xi\omega(t-\tau)}\sin(\omega_d(t-\tau))d\tau \tag{13.30}$$

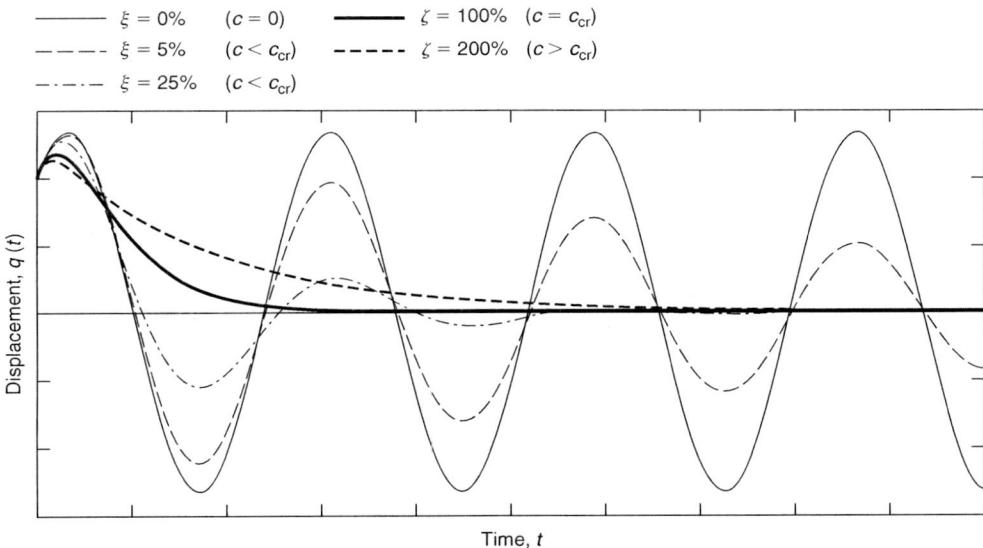

FIGURE 13.21 Free vibration of a damped SDOF system under non-zero initial displacement and velocity conditions.

may be used to compute the displacement response $q(t)$ of an SDOF system subjected to nonzero initial conditions q_0 and $\dot{q}_0$ and forcing function $f(t)$. For more complex forms of $f(t)$—as would arise, for example, in seismic (earthquake) analysis—numerical methods must instead be employed. Generally, *direct integration* techniques are used to time-integrate the equation of motion (Eq. [13.25]). All such methods fall under the general classification of *time-domain* analysis, because the equation of motion is solved in the time-domain, advancing one time-step at a time through the entire range of times of interest. Commonly employed direct integration methods include the average acceleration method, linear acceleration method, and central difference method. For additional details regarding these numerical procedures, the reader may consult Chopra (2000), Clough and Penzien (1993), Tedesco et al., (1999), and Weaver and Johnston (1987).

13.5.2 Multiple Degree of Freedom (MDOF) Dynamic Analysis

In MDOF time-domain analysis, direct time-integration procedures may be applied either to the coupled matrix equation of motion (Eq. [13.22]), or to a transformed form of that equation. For an MDOF finite element model having n global DOFs, the matrices M, C, and K, in Eq. (13.22) will be $n \times n$ in size and nondiagonal (generally) in form. The presence of off-diagonal terms in these matrices couples the various DOFs in the system together requiring that all n DOFs be time-integrated simultaneously. A variety of numerical methods are available for direct time integration of MDOF-coupled equations of motion. Examples include the average acceleration method, linear acceleration method, Wilson-θ method, Newmark-β method, and the central difference method.

An alternative time-domain analysis approach for MDOF systems is to transform the n-dimensional coupled equation of motion (Eq. [13.22]), into a set of n separate—and uncoupled—SDOF equations of motion that can be solved individually using SDOF analysis methods. MDOF methods based on this type of transformation are called *modal methods* because they generally employ normal (or natural) modes of vibration in the transformation process. Normal modes of vibration, and the corresponding frequencies at which these modes oscillate, are obtained by solving the eigen problem

$$K\phi = \lambda M\phi \qquad (13.31)$$

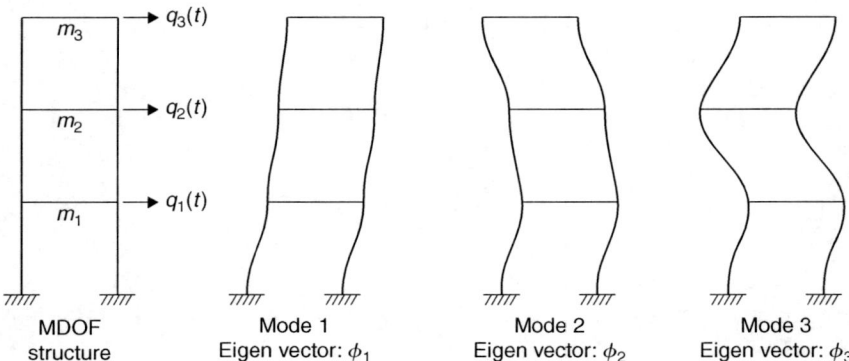

FIGURE 13.22   Normal mode shapes (eigen vectors) for an MDOF system.

for the n eigen vectors ϕ and eigen values λ of the system. Normal mode shapes of the structure (Figure 13.22) are described by the vectors ϕ whereas the corresponding frequencies are computed as $\omega = \sqrt{\lambda}$.

Normal modes of vibration exhibit a property called *orthogonality*, which is critically important in transforming (uncoupling) the coupled equations of motion. For two normal modes r and s of a structure, orthogonality with respect to the mass and stiffness matrices states that

$$\phi_r^T M \phi_s = 0 \quad \text{for } r \neq s \tag{13.32}$$

and

$$\phi_r^T K \phi_s = 0 \quad \text{for } r \neq s \tag{13.33}$$

If all n normal modes (eigen vectors) of a finite element model are collected into an eigen vector matrix $\Phi = [\phi_1 \phi_2 \ldots \phi_n]$, the coupled matrix equation of motion Eq. (13.22) can then be transformed into

$$\overline{M}\ddot{\overline{q}} + \overline{C}\dot{\overline{q}} + \overline{K}\overline{q} = \overline{F}(t) \tag{13.34}$$

where $\overline{M} = \Phi^T M \Phi$, $\overline{C} = \Phi^T C \Phi$, $\overline{K} = \Phi^T K \Phi$, and $\overline{F}(t) = \Phi^T F(t)$. In Eq. (13.34), $\overline{M}$, $\overline{C}$, and $\overline{K}$ are the modal mass, modal damping, and modal stiffness matrices, respectively; and $\ddot{\overline{q}}$, $\dot{\overline{q}}$, and $\overline{q}$ are the modal accelerations, modal velocities, and modal displacements, respectively. $\overline{F}(t)$ is the modal load vector. Each of the modal matrices $\overline{M}$, $\overline{C}$, and $\overline{K}$ is diagonal due to the orthogonality of the eigen vectors with respect to mass and stiffness (C must be a Rayleigh damping matrix for $\overline{C}$ to be diagonal). Because $\overline{M}$, $\overline{C}$, and $\overline{K}$ are diagonal, Eq. (13.34) then represents a set of n-separate, uncoupled SDOF modal equations of motion. For a given mode r, the corresponding SDOF modal equation of motion is then

$$\overline{m}_r \ddot{\overline{q}}_r + \overline{c}_r \dot{\overline{q}}_r + \overline{k}_r \overline{q}_r = \overline{f}_r(t) \tag{13.35}$$

where $\overline{m}_r = \phi_r^T M \phi_r$, $\overline{c}_r = \phi_r^T C \phi_r$, $\overline{k}_r = \phi_r^T K \phi_r$, and $\overline{f}_r(t) = \phi_r^T F(t)$. Each modal SDOF equation of motion given by Eq. (13.35) may be solved using SDOF time-integration methods. Once all of the SDOF modal equations of motion have been time-integrated to obtain time-histories of modal displacement $\overline{q}_r(t)$ for each mode r, time-histories of structural displacements may be recovered by superposition

$$q(t) = \sum_r \phi_r \overline{q}_r(t) \tag{13.36}$$

In many practical applications of MDOF dynamic analysis, only a relatively small percentage of the n modes of a system may participate substantially in dynamic response to loading. In such cases, the processes described above only need to be carried out for a truncated subset of p modes rather than making use of all n modes. Under such conditions, modal analysis techniques may be more computationally efficient than methods that time-integrate the n-dimensional coupled equation of motion directly.

13.6 Summary

The basics of the finite element process have been described using the simple virtual work formulation and physical elements as examples. There are numerous books and publications on finite elements that give more detailed descriptions, alternate formulations, and specific formulations for different domains. Whatever the domain or derivation method, the basics given in this chapter still hold and give the foundation for understanding more complex modeling.

References

Bathe, K. J., 1995, *Finite Element Procedures*, New Jersey, Prentice-Hall.
Chopra, A., 2000, *Dynamics of Structures: Theory and Applications to Earthquake Engineering* (2nd Edition), New Jersey, Prentice-Hall.
Clough, R. W. and Penzien, J., 1993, *Dynamics of Structures*, New York, McGraw-Hill.
McGuire, W., Gallagher, R., and Ziemian, R., 2002, *Matrix Structural Analysis, MATSTAN 2 V2.0* (2nd Edition), New Jersey, Wiley.
Tedesco, J. W., McDougal, W. G., and Ross, C. A., 1999, *Structural Dynamics: Theory and Applications*, Reading, MA, Addison-Wesley.
Weaver, W. and Johnston, P. R., 1987, *Structural Dynamics by Finite Elements*, Prentice-Hall.
Zienkiewicz, O. C. and Taylor R. L., 2000, *Finite Element Method: Volume 1, The Basis*, Oxford, Butterworth-Heinemann.
Zienkiewicz, O. C., 2000, *Finite Element Method: Volume 2, Solid Mechanics*, Oxford, Butterworth-Heinemann.
Zienkiewicz, O. C. and Taylor R. L., 2000, *Finite Element Method: Volume 3, Fluid Dynamics*, Oxford, Butterworth-Heinemann.

14
Multimodeling

Minho Park
Stephen F. Austin State University

Paul A. Fishwick
University of Florida

Jinho Lee
Samsung Electronics

14.1 Introduction 14-1
   Integrative Multimodeling • General Multimodeling • RUBE Framework
14.2 Scene Construction 14-5
   Ontology • Interaction Model Creation • Blender Interface
14.3 Multimodeling Exchange Language (MXL) 14-11
   Concepts of MXL • Multimodeling in MXL
14.4 Dynamic Exchange Language (DXL) 14-13
   DXL Concepts • Syntax of DXL • Semantics of DXL • Multimodeling
14.5 A Boiling Water Example 14-20
   2D Representation • Model Creation • Code Generation
14.6 Conclusion 14-27

14.1 Introduction

Multimodels are models that are composed of other models either through homogeneous or heterogeneous coupling. Multimodeling (Fishwick, 1995) is the process of engineering a model by combining different model types to form an abstraction network or hierarchy. Multimodeling, in a nutshell, endows the simulation modeler with the capability to blend different model types together to form hybrid models.

When we begin to understand a physical system by creating a model, we often find that the model is too limited: the model will answer only a very limited set of questions about system behavior. It is necessary, then, to create many models and link them, thereby maintaining a multilevel view of a system while permitting an analyst to observe system output at several abstraction levels. For example, consider this scenario: a region with several key military vehicles and targets, planes (both fighter as well as command and control center), surface-to-air missile (SAM) sites, and drones. A variety of models define the geometry and dynamics of these objects. Ideally, we can explore and execute these models within a 2D or 3D visualization environment by formalizing domain knowledge and providing a well-defined modeling methodology.

We will introduce two multimodeling approaches, integrative multimodeling (Fishwick, 2004; Park and Fishwick, 2004a, 2004b; Park, 2005) and general multimodeling (Fishwick, 1995; Lee, 2005). The purpose of integrative multimodeling is to provide a human–computer interaction environment that allows components of different model types to be linked to one another—specifically dynamic models used in simulation to geometry models for the phenomena being modeled. General multimodeling, however, describes a number of abstraction perspectives for a complex real-world system using simulation model types such as finite state model (FSM) and functional block model (FBM) (Fishwick, 1995).

To support the above multimodeling environments, we developed an XML-based modeling and simulation framework called RUBE (Kim et al., 2002; Kim and Fishwick, 2002a, 2002b; Fishwick, et al., 2003; Fishwick, 2002), which encompasses the modeling process of a real-world system as well as the simulation process of the model. The XML-based RUBE framework defines a formal approach to capture physical knowledge as well as the semantic information of the model and represent the information as separate XML documents.

We developed two XML-based languages: the multimodeling exchange language (MXL) (Kim et al., 2002; Kim and Fishwick, 2002a, 2002b; Fishwick et al., 2003; Fishwick, 2002; Damkjer, 2003) and the dynamic exchange language (DXL) (Lee, 2005; Lee and Fishwick, 2002), for the RUBE framework. MXL is an XML-based modeling language to support traditional heterogeneous model types such as FSM, FBM, and Petri net (Fishwick, 1995). DXL is an XML-based functional block language to support low-level simulation execution within the RUBE framework. We first discuss two multimodeling concepts, integrative multimodeling and general multimodeling as well as the overall structure of the XML-based RUBE framework. In Section 14.2, the process for constructing a multimodel is presented. The concepts and descriptions of MXL and DXL are explained in Sections 14.3 and 14.4. In Section 14.5, we demonstrate how the methodology is applied to a real-world application using an example.

14.1.1 Integrative Multimodeling

A real-world system can be embodied as a certain model type within a 2D or 3D visualization environment. It can be described by different perspectives depending on the modelers' viewpoint since the real-world system has the geometry or dynamics. Therefore, through the modeling process, the real world could be expressed in diverse model types, such as a geometry model or a dynamic model. Ideally, we can explore and execute these models within a unified 3D scene that integrates such models.

We present a novel method (i.e., integrative multimodeling) of visually merging two types of models with the intention of allowing the user to more easily, and contextually, associate dynamic model and scene model components. We need to define a formalized scene domain in which multiple model representations can exist together and a certain model type can be transformed into other model types via user interactions, by conceptualizing all objects, that the scene domain contains, and specifying properties (i.e., geometry and dynamics) of objects and relationships between objects. We employ the concept of ontology to formalize a certain scene domain. The purpose of integrative multimodeling is to provide a human–computer interaction environment that allows users to change model types within the same environment. Therefore, user interactions should be logically formalized and implemented to support the integrative multimodeling environment. We formalize the user interaction as an interaction model and derive the interaction model (Park and Fishwick, 2004a, 2004b; Park, 2005) based on first-order logic (FOL) rules. The concepts of ontology and the FOL rules along with the interaction model will be further discussed in Section 14.2.

14.1.2 General Multimodeling

In this section, we discuss a multimodel concept and two types of multimodels according to their model types, such as homogeneous and heterogeneous multimodels.

14.1.2.1 Intralevel and Interlevel Couplings

General multimodels are simulation models that are composed of heterogeneous simulation models. Since most simulation models like FSM and FBM represent only a part of the overall system behavior, they make only a subset of a solution for analyzing the prediction and diagnosis of the real world. General multimodels could, however, have a number of abstraction perspectives for a complex real-world system. Therefore, they can more correctly represent and analyze a complex real-world system. Different component models in a multimodel can define the activity at different stages of the simulation.

Figure 14.1 shows a general multimodel example having an abstraction hierarchy of heterogeneous simulation models. Double dotted lines in this figure mean the relation of its components to compose an upper-level model. This relation should ensure intralevel coupling. Intralevel coupling defines model components coupled to one another in the same model (Cubert and Fishwick, 1998). In Figure 14.1, there are two intralevel couplings. M_2 is composed of m_1, m_2, and m_3. M_3 is composed of s_1, s_2, and s_3. The intralevel coupling in M_2 defines how the submodels of M_2 are formed to represent a model M_2 and the intralevel coupling in M_3 defines how the submodels of M_3 are formed to represent a model M_3.

Multimodeling

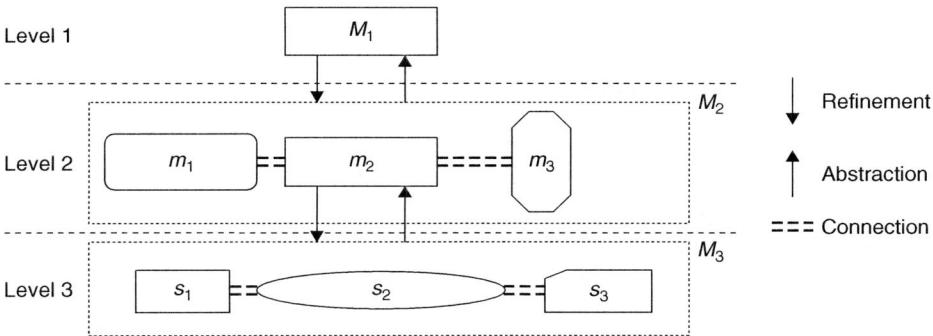

FIGURE 14.1 General multimodel structure.

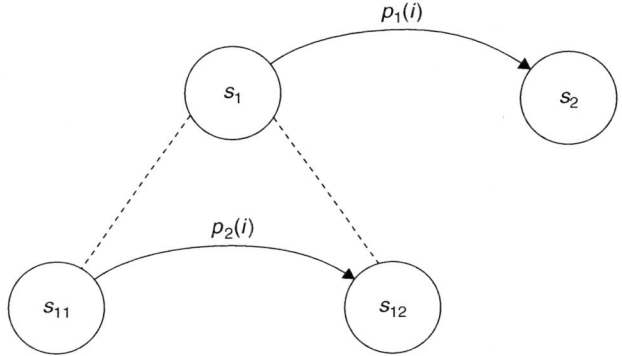

FIGURE 14.2 Homogeneous refinement: declarative → declarative.

In Figure 14.1, down-arrows mean refining a model and up-arrows mean the abstraction of a model. These two relations should ensure interlevel coupling. Interlevel coupling defines rules as to how model components from one model can be refined into models of different types (Cubert and Fishwick, 1998). In Figure 14.1, there are two interlevel couplings. M_2 refines a model M_1 and M_3 refines m_2. A model m_2 abstracts M_3, and M_1 abstracts M_2. To refine m_2 into M_3, we should ensure the intralevel coupling in M_2 when we change the submodel m_2 and M_3.

14.1.2.2 Homogeneous and Heterogeneous Multimodels

In Figure 14.2, the FSM shows top-down homogeneous decompositions (i.e., intralevel coupling), since the element of the upper model is defined by the same type of models. State s_1 is decomposed into the lower level of FSM. The predicates $p_1(i)$ and $p_2(i)$ involves external input variable i.

Figure 14.3 shows a two-level functional hierarchy, where function f is defined in terms of a composition of three other functions f_1, f_2, and f_3.

Heterogeneous decomposition of models in intralevel coupling describes the semantics of a model using different model's semantics. Figure 14.4 shows FSM that contains the internal state transitions associated with function f. The predicate $p(i)$ tests input variable i. Any FSM should be defined inside a functional block (i.e., function f in Figure 14.4) to represent explicit external input and output semantics.

Figure 14.5 shows a state-to-state space mapping, which is not immediately apparent from the figure. Specifically, most functional block models that represent some aspects of physical reality involve state transitions, which means f_1 and f_2 contain internal state transitions. There is a transition with two possible types of semantics coming out of state s in Figure 14.5. An external transition would be of the form $p_2(i)$ and an internal transition would be based on a variable o that is a component of internal state space of f_1 and f_2.

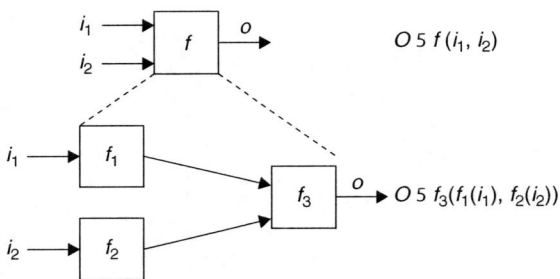

FIGURE 14.3 Homogeneous refinement: functional → functional.

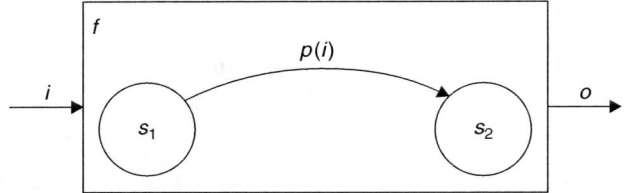

FIGURE 14.4 Heterogeneous refinement: functional → declarative.

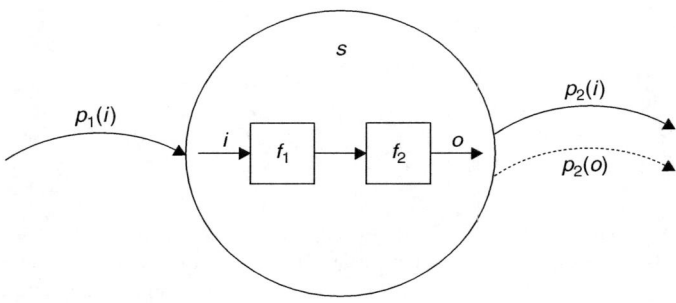

FIGURE 14.5 Heterogeneous refinement: declarative → functional.

14.1.3 RUBE Framework

The RUBE framework is an XML-based dynamic modeling and simulation framework permitting the users to specify and execute a dynamic model, with an ability to customize a model presentation using 2D or 3D visualization (Park and Fishwick, 2004a, 2004b; Park, 2005; Fishwick et al., 2003; Kim and Fishwick, 2002b). The purpose of RUBE is to facilitate a dynamic multimodel construction based on XML, and visualize and execute the model within a 2D environment (Fishwick et al., 2003) or a 3D immersive environment (Park and Fishwick, 2004a, 2004b; Park, 2005; Fishwick et al., 2003; Carey and Bell, 1997). The overall process of the XML-based RUBE framework is shown in Figure 14.6.

We will proceed using the architecture depicted in Figure 14.6, referencing each section by number as it appears in the figure. For each section, we will cover a description of that section.

A scene file contains 2D or 3D geometry objects that represent geometry and dynamic models. The RUBE framework has 2D- (i.e., Sodipodi, 2006) and 3D-based (i.e., Blender, 2006; Roosendaal and Selleri, 2004; Roosendaal and Wartmann, 2002) interfaces for representing 2D and 3D scenes, respectively. The scene files represent the appearance of geometric objects in the model and do not have any information about model behavior or dynamics. The scene could be represented either in standard 2D/3D XML documents

Multimodeling 14-5

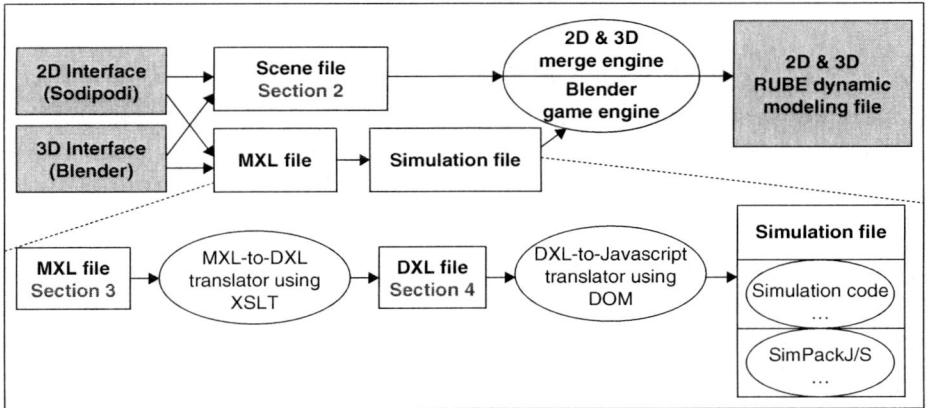

FIGURE 14.6 The RUBE framework.

(i.e., scalar vector graphics (SVG) (Eisenberg, 2002) or extensible 3D (X3D, 2006) or in Blender. Therefore, any 2D or 3D tools, which can generate SVG or X3D files, might be applied as a part of the RUBE framework.

For model creation there are two stages: model translation and model simulation. For model translation, a dynamic model is an actual model file that is represented in MXL. The MXL file describes the behavior of the model and represents the model file that includes the specification of a heterogeneous multimodel in an abstract level such as FBM and FSM. The MXL-to-DXL translator in extensible stylesheet language (XSL) (Kay, 2000) translates a model file written in MXL into a low-level functional specification language called DXL, which can be described with a homogeneous block diagrammatic presentation. For model simulation, the DXL is translated into an executable programming code for the model simulation using the DXL-to-simulation translator. The programming code either in JavaScript or in Python can be executed based on SimpackJ/S or Simpack Python (Fishwick, 1992; Park and Fishwick, 2002), which provides the underlying code foundation for libraries, classes, and objects for simulation.

For model merging we have two approaches: (1) using a 2D or 3D merge engine in XSL, we can merge a scene file with an actual model execution file in JavaScript or (2) within the Blender environment, we could naturally combine the 3D scene in Blender with simulation code in Python using Blender game engine. Then, finally we could visualize and execute the geometry and dynamic models within a 2D or 3D environment.

14.2 Scene Construction

14.2.1 Ontology

We need to formalize physical systems based on particular domain knowledge, since such formalism should help to build multiple, cooperative simulation models of certain physical systems. We can describe a certain target system using the concept of ontology. An ontology represents a formal conceptualization of a domain by clearly specifying meaning of terms and their interrelationships among concepts used in a particular domain (McGuinness, 2002; McGuinness and Harmelen, 2003; Berners-Lee et al., 2001). Ontologies consist of three general elements: classes, properties, and the relationships between classes. Any concepts in a target domain are represented as classes. Classes could be further generalized into specific categories using "is-a-kind-of" relationship (i.e., generalization in unified modeling language (UML). Also, we can express a structural relationship (i.e., association in UML) that specifies mappings between objects (i.e., instances of classes). Using "has-a"/"whole-part" relationship (i.e., aggregation in UML), structural dependency could be described. In addition, certain constraints, such as multiplicity and multiple associations, are specified along with the relationships between objects. We explain the conceptualization process by creating an ontology for a sample domain: a boiling water domain. Figure 14.7 shows an example scene ontology for boiling water (some concepts and relationships are not shown to avoid ontology complexity).

14.2.1.1 Classes and Relationships

Consider a pot of boiling water on a stovetop electric heating element. This domain contains a pot and an electric stove with a temperature knob. Initially, the pot is filled to some predetermined level with water. And this system has one input or control—the temperature knob. The knob is considered to be in one of the two states: on or off. On the basis of the given domain knowledge, we could create four basic classes, *Pot*, *Electric Stove*, *Knob*, and *Water*, as well as one abstract class called *Scene*. Because the overall scene (i.e., boiling water scene) "is-a-kind-of" scene, we could add one more abstract class called *Boiling Water Scene* as a subclass of the *Scene* class. And the electric stove "has a" knob. Therefore, aggregation relationship could be created between the *Electric Stove* and *Knob* classes. In addition, we could make aggregation relationships between the *Boiling Water Scene* class and *Pot*, *Electric Stove*, and *Water* classes, since the scene contains the three classes. If we consider system dynamics approach (Fishwick, 1995), which is a methodology for engineering simulation models, to examine the scene domain, we could derive the following simple causal diagrams:

- Knob_On → Water_Not_Cold
- Knob_Off → Water_Cold.

The graphs show that "turning on/off the stove causes a change of water phase from *cold/not cold* to *not cold/cold*." (We assume that initially the water is in the *cold* phase.) Also, we could further generalize the second nodes in the graphs. Hence, we could drive the following generalized causal graphs:

- Knob_On → Water_Heating (=Water_Getting_Hotter) → Water_Boiling
- Knob_Off → Water_Cooling (=Water_Getting_Colder) → Water_Cold.

By analyzing the causal graphs above, we could create a dependency relationship between *Knob* and *Water* classes. Also, the *Water* class could be further generalized into "cold" water and "not cold" water based on current water phase. Therefore, corresponding classes, such as *Cold* and *Not cold*, could be created as subclasses of the *Water* class. Besides, more generalized water phases (i.e., heating, cooling, and boiling phases) could be adopted as subclasses of the *Not cold* class. However, the following issue naturally arises: How do we express the sudden change in temperature within the heating and cooling phases? In other words, how do we naturally describe the continuous behaviors inside the phases? Therefore, we need to define reasonable functions that could describe more specific dynamic behaviors inside the phases. We could employ differential equations, since differential equations are the natural method of defining physical phenomena. Consider, for example, the heating phase. We could derive the following differential equation using Newton's law and the capacitance law:

$$T' = k(100 - T)$$

where T is the temperature ($\alpha \leq T \leq 100$, α is the ambient temperature) and k the thermal conductivity of water (i.e., rate constant).

In this case, one integrator, one multiplication, one subtraction, and one constant-number generation functions are needed. Therefore, corresponding functions are inserted into the *Heating* class as methods. Likewise, proper functions for cooling phase could be inserted into the *Cooling* class.

We define *Dynamic Model* and *Geometry Model* classes, since we want a formalized scene domain in which multiple model representations can exist together (i.e., integrative multimodeling) and a certain model type can be transformed into another model type via user interactions (i.e., geometry model to dynamic model and vice versa). In addition, *Boiling Water Dynamic Model* class is defined as a subclass of the *Dynamic Model* class. Similarly, *Boiling Water Geometry Model* class is defined as a subclass of the *Geometry Model* class. The *Boiling Water Dynamic Model* class could be composed of at least one dynamic model, such as FSM or FBM. Therefore, we define *FSM* and *FBM* classes as subclasses of the *Simulation*

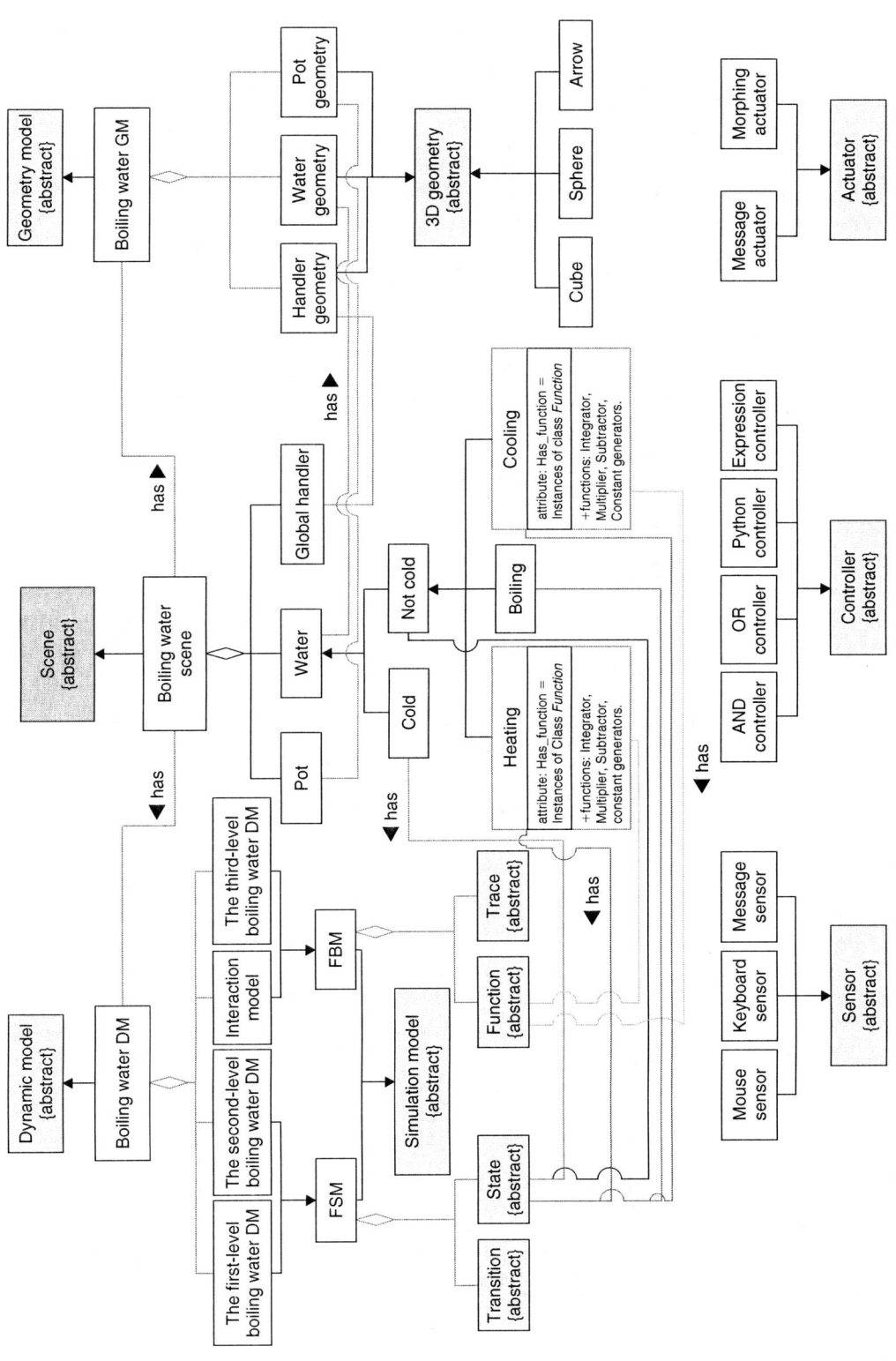

FIGURE 14.7 An Example scene ontology for the boiling water.

Model class. Because each simulation model has its model elements (i.e., transitions and states for FSM; traces and functions for FBM), we also define four classes such as *Transition*, *State*, *Trace*, and *Function*. Likewise, *Geometry* class is created, since 2D/3D objects are needed for explicitly representing geometric representations for a geometry model as well as a dynamic model in 2D or 3D space. Specific geometric types, such as primitive objects (i.e., cube, sphere, and arrow) and nonprimitive objects (i.e., stove, water, and pot), could be defined as subclasses of the *Geometry* class. By conceptualizing the simulation model type and 2D/3D geometry domains, we would provide syntactic and semantic mappings between objects in the boiling water domain and objects in the simulation model type domain, and objects in the boiling water domain and objects in the 2D/3D geometry domain.

To support the integrative multimodeling environment, we need to define additional classes, such as *Interaction Model*, *Scene Handler* (i.e., 3D icons for supporting integrative multimodeling), *Sensor*, *Controller*, and *Actuator*. We will cover these concepts in the following subsections.

14.2.1.2 Properties and Relationships

We assign two named properties (i.e., attributes), *Geometry Model* and *Dynamic Model*, to the *Boiling Water Scene* class. In addition, we generate two association relationships, named "has a geometry model," and "has a dynamic model," which connect to the *Boiling Water Geometry Model* and *Boiling Water Dynamic Model* classes. Therefore, an object (i.e., an instance) of class *Boiling Water Scene* could have two attributes whose values are objects of *Boiling Water Geometry Model* and *Boiling Water Dynamic Model* classes. Similarly, two attributes, *Geometry Object* and *Dynamic Model Element*, are declared for the five fundamental classes of the scene (i.e., *Pot*, *Electric Stove*, *Knob*, *Water*, and *Scene Handler* classes), since we assume that every object in the scene has a corresponding geometric object for presentation as well as a dynamic model element (i.e., state or function) for dynamics. Because *Cold*, *Not Cold*, *Boiling*, *Cooling*, and *Heating* classes are subclasses of the *Water* class, these generalized classes can have their own attribute values that are generated from *Geometry*, *State*, and *Function* classes. Therefore, we create association relationships, named "has a geometry object," and "has a dynamic model element," between the water domain and the geometry domain, and the water domain and the function/state domain. Because all classes in the water domain are generated based on the water phases, state-based dynamic model type (i.e., FSM) could be the reasonable choice to represent dynamics. Therefore, association relationships between the water domain and the state domain are generated.

We create another association relationship named "uses" between class *State/Function* and class *Geometry*, since 2D/3D objects are needed for explicitly representing geometric representations for a certain dynamic model in 2D/3D space. An attribute, *Interaction Model*, is defined inside *Geometry* class, since we attempt to induce an overall scene interaction model by taking an individual interaction model for each geometric object that the boiling water scene contains. In addition, we define three attributes, *Sensor*, *Controller*, and *Actuator*, in class *Interaction Model* so that an interaction model for a certain object could be inferred from its attribute values (i.e., sensor, controller, and actuator, which are components of an interaction model) through first-order logics. Likewise, the overall interaction model for the boiling water scene could be generated from each object's interaction model through first-order logics. We will explain the process for creating interaction models for the boiling water scene in next section. To represent relationships above, we create four association relationships, "has a sensor," "has a controller," "has an actuator," and "has an interaction model," between *Interaction Model* and *Sensor* classes, *Interaction Model* and *Controller* classes, *Interaction Model* and *Actuator* classes, and *Geometry* and *Interaction Model* classes, respectively.

14.2.2 Interaction Model Creation

The purpose of integrative multimodeling is to provide a human–computer interaction environment that allows users to change model types within the same environment. Therefore, user interactions should be logically formalized and implemented to support the integrative multimodeling environment. We formalize the user interaction as an interaction model and create the interaction model(s) through first-order logic rules. The interaction model consists of a sensor, a controller, and an actuator as model components as well as links between components, since we will utilize Blender game logic bricks (Roosendaal and

Wartmann, 2002), which have three logic components, sensors, controllers, and actuators, to execute the interaction model.

First, we induce all interaction models for the main model components through logic rules. Then an overall interaction model for a certain scene domain is generated from individual interaction models along with logic rules. The following are general first-order logic rules for creating interaction models used for an individual object and an overall scene domain.

- For an individual object
    1. There exists an object
       $\exists\, x\, Object(x)$
    2. There exists a sensor
       $\exists\, x\, Sensor(x)$
    3. There exists a controller
       $\exists\, x\, Controller(x)$
    4. There exists an actuator
       $\exists\, x\, Actuator(x)$
    5. Every Object has an interaction model
       $\forall\, x\, (Object(x) \rightarrow \exists\, y\, (InteractionModel(y) \wedge hasInteractionModel(x,y)))$
    6. Every Interaction Model has a sensor
       $\forall\, x\, (InteractionModel(x) \rightarrow \exists\, y\, (Sensor(y) \wedge hasSensor(x,y)))$
    7. Every Interaction Model has a controller
       $\forall\, x\, (InteractionModel(x) \rightarrow \exists\, y\, (Controller(y) \wedge hasController(x,y)))$
    8. Every Interaction Model has an actuator
       $\forall\, x\, (InteractionModel(x) \rightarrow \exists\, y\, (Actuator(y) \wedge hasActuator(x,y)))$
    9. If Interaction Model has a sensor and a controller, it has a link that connects the sensor with the controller
       $\exists\, x,\, y1,\, y2\, (InteractionModel(x) \wedge Sensor(y1) \wedge Controller(y2) \wedge hasSensor(x,y1) \wedge hasController(x, y2) \rightarrow Link(y1, y2) \wedge hasLink(x, Link(y1,y2)))$
    10. If Interaction Model has a controller and an actuator, it has a link that connects the controller with the actuator
        $\exists\, x,\, y1,\, y2\, (InteractionModel(x) \wedge Controller(y1) \wedge Actuator(y2) \wedge hasController(x, y1) \wedge hasActuator(x,y2) \rightarrow Link(y1, y2) \wedge hasLink(x, Link(y1,y2)))$
- For an overall scene domain
    11. There exists a scene
        $\exists\, x\, Scene(x)$
    12. The scene has an Interaction Model
        $\exists\, x, y\, ((Scene(x) \rightarrow (InteractionModel(y) \wedge hasInteractionModel(x,y))))$
    13. The Interaction Model includes all individual interaction models
        $\exists\, x\, (InteractionModel(x) \rightarrow \forall\, y\, (InteractionModel(y) \wedge includes(x, y)))$
    14. If there exists a "hasHandler" relationship between two objects and the objects have their interaction models, then the interaction model has a link that connects the actuator of the parent object with the sensor of the child object
        $\exists\, w, x1, x2, y1, y2, z1, z2\, (Object(x1) \wedge Object(x2) \wedge hasParent(x2, x1) \wedge hasInteractionModel(x1, y1) \wedge hasInteractionModel(x2, y2) \wedge hasActuator(y1, z1) \wedge hasSensor(y2, z2) \rightarrow Link(z1, z2) \wedge hasLink(InteractionModel(w), Link(z1,z2)))$
    15. If two objects are conceptually mapped (i.e., geometry and dynamic model components for a certain object), then the interaction model has a link that connects the *actuator* of the geometry object with *sensor* of the dynamic object
        $\exists\, v, w, x1, x2, x3, y1, y2, z1, z2\, (Object(w) \wedge hasDynamic(w, x1) \wedge hasGeometry(w, x2) \wedge uses(x1,x3) \wedge hasInteractionModel(x3, y1) \wedge hasInteractionModel(x2,y2) \wedge hasSensor(y1,z1) \wedge hasActuator(y2, z2) \rightarrow Link(z2,z1) \wedge hasLink(InteractionModel(v), Link(z2,z1)))$

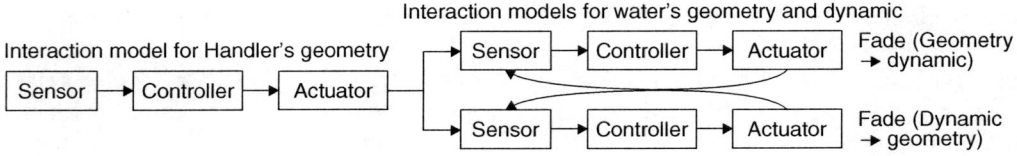

FIGURE 14.8 The overall interaction model for the boiling water scene.

16. If two objects are conceptually mapped (i.e., geometry and dynamic model components for a certain object), then the interaction model has a link that connects the *sensor* of the geometry object with *actuator* of the dynamic object
$\exists\ v,\ w,\ x1,\ x2,\ x3,\ y1,\ y2,\ z1,\ z2\ (Object(w) \wedge hasDynamic(w,\ x1) \wedge hasGeometry(w,\ x2) \wedge uses(x1,x3) \wedge hasInteractionModel(x3,\ y1) \wedge hasInteractionModel(x2,y2) \wedge hasSensor(y2,z2) \wedge hasActuator(y1,\ z1) \rightarrow Link(z2,z1) \wedge hasLink(InteractionModel(v),\ Link(z2,z1)))$.

Based on these rules, we could induce an interaction model for the boiling water scene. First, we need to define a user interaction scenario: (1) if an individual model object is touched, only the object is transformed into other model type's object (i.e., a geometry model object to a dynamic model object and vice versa) and (2) if a handler (i.e., 3D icon) is touched, all model objects are transformed into other model type's objects (i.e., all geometry model objects to all dynamic model objects and vice versa). According to the scenario, total three interaction models are needed: two for *Water* (i.e., geometry and dynamic objects) and one for *Handler* (i.e., geometry object). Additional association relationships, such as "has a link" and "has a hander," are inserted in the scene domain so that we could represent a topological connectivity between interaction model components and interaction models. Figure 14.8 shows the overall interaction mode for the boiling water scene.

The Protégé Ontology Editor (2006) is employed to create an actual scene ontology as well as first-order logic rules. Web ontology language (OWL) (Lacy, 2005) and semantic Web rule language (SWRL) (Horrocks et al., 2004) are utilized to construct an ontology and logic rules. To verify an ontology for a certain scene domain, we use RACER (2006) reasoner along with Protégé. Based on the ontological structures for the boiling water scene, we define all the classes and subclasses for the scene domain and modeling knowledge along with all necessary relationships based on Figure 14.7. An interaction model for the boiling water scene could be induced from the scene ontology using first-order logic rules. Therefore, using an inference engine such as Jess (2006), we could inference certain knowledge from domain knowledge in OWL through inference processes. In Protégé, we use SWRL to describe first-order logic rules, since SWRL tab is provided to allow users to create user-defined logic rules. There are a couple of approaches for inference in Protégé:

1. Create first-order logic in SWRL and then perform inference using a Jess rule engine.
2. Write and store logical constraints using Protégé Axiom Language (PAL) and then perform inference using PAL query statements.

We use the first approach since SWRL includes a high-level abstract syntax for Horn-like rules. However, the current version of Protégé does not provide inference capability for SWRL. Therefore, we manually generate an interaction model for the boiling water scene based on logic rules at this time.

14.2.3 Blender Interface

We developed a Python-based interface, *Blender Interface* (Park and Fishwick, 2004a, 2004b; Park, 2005), which can build 3D simulation models based on an ontology for a certain target domain. *Blender Interface* consists of two components, *Model Explorer* and *Simulation*. Figure 14.9 depicts the overall environment in the Blender software. The environment consists of Blender 3D Window (Scene Editor), *Blender*

Multimodeling

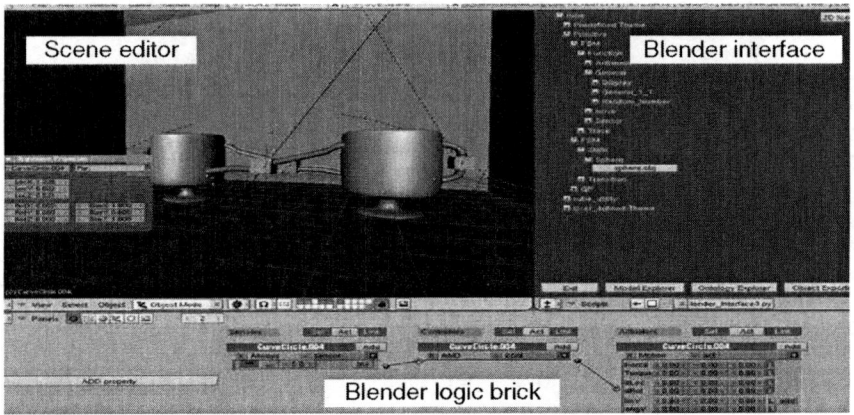

FIGURE 14.9 Blender environment containing the game engine.

Interface, and Blender Logic Brick. Geometry and dynamic models for a certain system are composed in the Scene Editor, while specifying the dynamic model types and styles for the system and generating Python simulation code for the dynamic model in the *Blender Interface*. The Python code is inserted into Blender Logic Brick to simulate the dynamic model. In addition, the interaction model, which could be induced through inference processes, is implemented and executed in the Blender Logic Brick for providing a human–computer interaction environment.

To use *Blender Interface*, RUBE must be installed. The RUBE has four folders: *primitive*, *predefined theme*, *user-defined theme*, and *rube_utility* folders. The *primitive* and *predefined theme* folders are given for users to provide libraries containing dynamic model objects and the corresponding MXL (we will explain it in detail in the next section) files and functions in Python. As the names imply, the *primitive folder* has primitive blender objects, such as cube and sphere, with the corresponding MXL file and function for each model type, such as FSM or FBM. The *predefined theme* folder contains prefabricated, customized and personalized blender objects, as well as MXL files and functions. If modelers want their own model representations, they can create an object and store it into a proper model-type folder under the *user-defined theme*.

Using *Model Explorer*, modelers are able to search model objects, which they want to import, within the RUBE folders. From the given dynamic model components in Blender 3D Editor using *Simulation* (we will explain it in detail in Section 14.5), we can generate an MXL file for the target system automatically, since we provide library systems that contain Blender objects as well as the corresponding MXLs and functions in a set of pairs.

14.3 Multimodeling Exchange Language (MXL)

14.3.1 Concepts of MXL

MXL is an XML-based dynamic modeling language used to represent traditional simulation model types within the RUBE framework. For MXL, the functional elements for each model type are clearly identified, and entry points of the functional elements are defined as ports.

The functional elements are model components that behave as functions that naturally fit as part of the model description. Figure 14.10 shows the multimodel structure of MXL. Multimodeling is permitted wherever the appropriate model component could be "extended" or "expanded" to another model whose outermost definition is a function. These couplings achieve multimodeling by inserting one function inside of another (interlevel coupling) or by connecting one function to another (intralevel coupling).

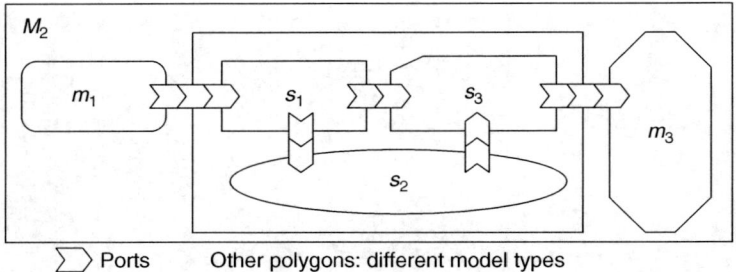

FIGURE 14.10 MXL multimodel structure.

```
<?xml version="1.0" encoding="utf-8"?>
<MXL>
    <fbm id="MXL">
        <block id="F1">
            <output id ="F1_outports_integer1"datatype="Integer" index="0"/>
            <script lang="Python" src="input.py" func="gen"/>
        </block>
        <block id="F2">
            <input id="F2_inports_integer1" datatype="Integer" index="0"/>
            <output id="F2_outports_string1" datatype="String" index="0"/>
            <fsm id="FSM_F2" src="FSM_F2.xml"/>
        </block>
        <trace from="F1_outports_integer1" to="F2_inports_integer1"/>
    </fbm>

    <simulation start_time="0" end_time="10" delta_time="0.1" cycle_time="0.1"/>
</MXL>
```

FIGURE 14.11 Functional block model in MXL.

14.3.2 Multimodeling in MXL

Suppose a system could be represented as multimodeling concepts. The system has an FBM with two blocks (i.e., functions) and one trace; the second block contains an FSM with two states and two transitions (i.e., interlevel coupling). Figure 14.11 and Figure 14.12 show the MXL representation for the system.

In Figure 14.11, the MXL contains *fbm* and *simulation* as subelements. *fbm* represents model type (i.e., FBM) and has two elements, *block* and *trace*. Each block element has an attribute, *id*, to specify the names of blocks. In addition, the element has port information, *input* or *output* as well as its functionality, *script*. Each *input* and *output* element contains *id*, *datatype*, and *index*, which define the name of *input/output*, data type of *input/output*, and the number of *input/output*, respectively. The script element has *lang*, *src*, and *func* to specify program language, the name of the script file, and function to be executed, respectively. If a block contains another model type (i.e., multimodel), the model type and its MXL can be specified in the block (i.e., Block F2 in Figure 14.11.) *trace* has from and to attributes denoting connectivity between blocks. *Simulation* has specific information for executing a certain dynamic model such as *start time*, *end time*, and *delta time*. Figure 14.12 depicts an internal dynamic model type which Block F2 contains. In this case, the MXL represents the FSM. Therefore, MXL has *state* and *transition* as subelements. Likewise, each *state* and *transition* has its *id*, *script* information, and topological connectivity.

```xml
<?xml version="1.0" encoding="utf-8"?>
<MXL>
  <fsm id="FSM_F2">
        <input id="F2_inports_integer1" datatype="Integer" index="0"/>
        <output id="F2_outports_string1" datatype="String" index="0"/>

        <state id="S1" start="true">
            <script lang="Python" src="input.py" func="off"/>
        </state>
        <state id="S2">
            <Script lang="Python"src="input.py" func="on"/>
        </state>
        <transition from="S1" to="S2">
            <script lang="Python" src="input.py" func="off2on"/>
        </transition>
        </transition from="S2" to="S1">
            <script lang="Python" src="input.py" func="on2off"/>
  </fsm>
</MXL>
```

FIGURE 14.12 Finite state model in MXL.

Using an MXL-to-DXL translator, the MXL file can be translated into DXL, which is a homogeneous assembly level block diagram modeling language consisting of *Connectors*, *Blocks*, and *Ports*. The concepts of DXL will be explained semantically and syntactically in the next section.

14.4 Dynamic Exchange Language (DXL)

14.4.1 DXL Concepts

DXL is a unified, low-level functional specification language, and associated diagrammatic presentation for simulation on the RUBE framework. In simple DXL models, models are defined by connection of input and output ports of primitive models such as multipliers or adders.

In DXL multimodels, the models may be abstracted upper-layer block models in the form of sublayer models. DXL combines the lower-level block models of subsystems to describe a complex system hierarchically (i.e., homogeneous multimodeling). Because each DXL model has its outermost block as a wrapper, it supports the important characteristics for multimodeling, composability, and reusability.

When each block plays a role as a leaf node in the structure of a multimodel, it can be encoded in one of the programming languages, such as JavaScript or Python. Multiple programming languages become the target codes for DXL-to-simulation translators, and an actual data flow for simulation is achieved by using XML data and schema.

In this section, a diagrammatic presentation and formalism for DXL, which operates like a circuit, is created. Ports constitute the interface that defines the boundary of components or subsystems in a system configuration. MXL and DXL ports are used for port coupling where all ports match in number and data type. MXL ports are used for conceptual multimodeling in the RUBE framework. DXL ports, however, are utilized to support the multimodel execution as well as to maintain the conceptual multimodeling. A DXL or MXL model supports not only streaming of simple data types but also XML "information streaming" since DXL ports are capable of encoding documents (i.e., XML documents) rather than data under the XML-based environment. XML schemata that define the typing structure accompany the "types."

14.4.2 Syntax of DXL

The right side of Figure 14.13 shows syntax elements of DXL. DXL elements are composed of blocks, input and output ports, and connects. Generally, a DXL model is composed of "block" and "connect" elements. A DXL block element is composed of "port" elements and "its actual computation codes." DXL block elements are connected by each connect element through input and output port elements, which can be included in a block element. Each port element is connected to the corresponding output and input port elements of other block elements for its data flow, which each connect element controls.

The left side of Figure 14.13 shows a simple DXL model consisting of two blocks and one connect. This model shows that the output becomes the input of "Block 2" after the computation of "Block 1" is finished. Therefore, "Block 2" executes its own computation using the output of "Block 1" as its input.

Each block becomes an object in object-oriented programming unlike a function in procedure programming. Input and output ports become the variables that are defined in the block objects according to their data types such as primitive and object data types. From the viewpoint of object-oriented programming, therefore, the DXL model of Figure 14.13 is explained in Figure 14.14.

The DXL block element might include another DXL model when it represents a homogeneous multi-model. In Figure 14.15, "Outer Block B" does not have "its actual computation codes," but it does contain another DXL model. Therefore, when "Outer Block B" starts, the input of "Outer Block B" becomes the input of "Inner Block a." In contrast, when "Outer Block B" finishes, the output of "Inner Block b" becomes the output of "Outer Block B." According to the flow of "connect" elements, after the inner blocks of "Outer Block B" are executed, the output of "Outer Block B" becomes the input of "Block C" in Figure 14.15.

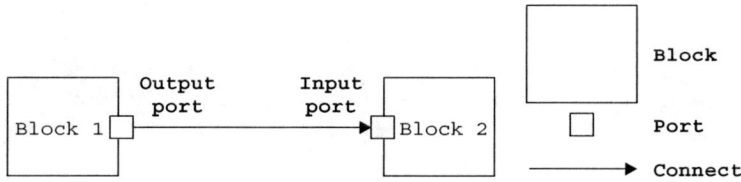

FIGURE 14.13 DXL syntax.

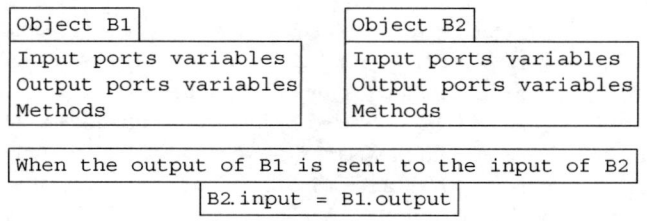

FIGURE 14.14 DXL programming semantics.

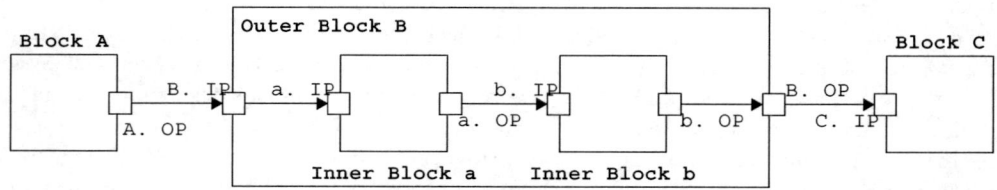

FIGURE 14.15 DXL multimodel syntax.

14.4.3 Semantics of DXL

14.4.3.1 Notation

Figure 14.16 defines a notation to describe the functionalities of DXL blocks. We describe algorithmic semantics of blocks using a set theory and some pseudocodes. IP and OP mean input port and output port of a block. For notation of DXL ports, the orders of the ports are expressed as an argument of IP or OP, and the identifiers of their parent blocks are subscripted. Therefore, the first and second input ports of block B1 become $IP_{B1}(1)$ and $IP_{B1}(2)$. The first and second output ports of block B1 become $OP_{B1}(1)$ and $OP_{B1}(2)$. OB means output blocks of current block. The input ports of these output blocks receive the output after the current block finishes its computation. For notation of DXL output blocks, the orders of output ports of the current block are presented as arguments of OB, and the identifier of the current block is subscripted. Therefore, the output blocks of the first and second output ports of the current block B1 become $OB_{B1}(1)$ and $OB_{B1}(2)$.

14.4.3.2 Information Stream Mechanism

Generally, a DXL model is composed of "block" and "connect" elements. A DXL block element is composed of "port" and "definition" elements. It might include another DXL model when it represents a homogeneous multimodel. The DXL block elements are connected by each connect element through the input and output port elements, which are included in a block element. The data flow between block elements is notated by the pseudocode SEND, which means different information streaming according to the various environment modes. For example, the pseudocode SEND means "generate next event" based on discrete-event scheduling methods (Fishwick, 1995) in sequential simulation, and also "send message" based on message-passing protocols in distributed simulation. This information streaming abstraction approach, which is used to generate different target codes on heterogeneous environments, makes it easy to create a DXL-to-simulation translator and integrate model components.

In Figure 14.17, the meaning of the SEND pseudocode is "transfer DATA in an output port OP of a block B1 to an input port IP of a block B2 and generate next event B2, after time t passes," based on discrete-event scheduling methods. If this model is based on message-passing protocols in distributed simulation, this

```
Notation
IPᵢ: a set of input ports for a block i
IPᵢ(n): the nth input port of a block i
OPᵢ: a set of output ports for a block i
OPᵢ(n): the nth output port of a block i
OBᵢ(j): a set of output blocks for the jth port of a block i
SEND(t, DATA(i), IPₖ(j)): Current block sends data of ith output
port to jth input port of block k after time t
```

FIGURE 14.16 DXL block notation.

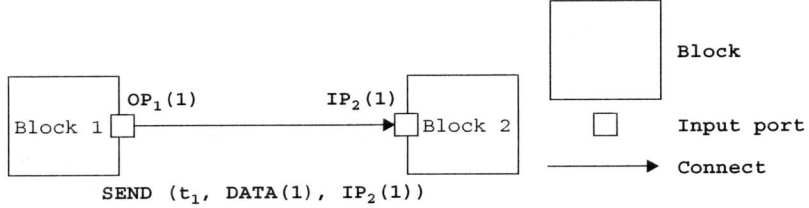

FIGURE 14.17 Augmented DXL syntax.

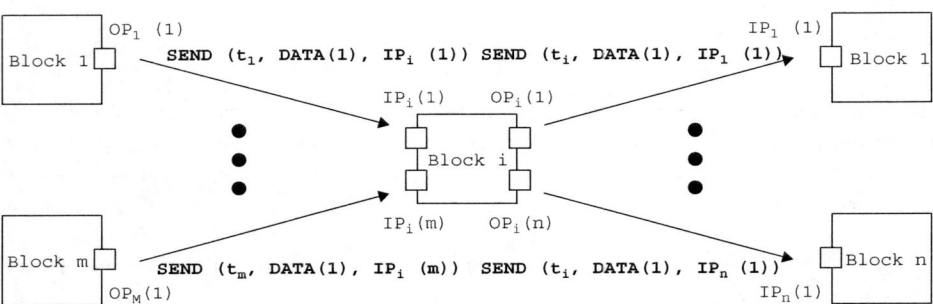

FIGURE 14.18 Examples of **SEND** pseudocode.

SEND pseudocode means "create DATA in an output port OP of a block B1 as a message and send the message to an input port IP of a block B2, after time t passes."

In Figure 14.18, the meanings based on discrete-event scheduling methods of SEND pseudocodes are as follows:

- **SEND (t_1, DATA(1), IP_i(1))**: transfer **DATA** in an output port **OP** of a block B_1 to an input port **IP1** of a block B_i and **generate next event B_i**, after time t_1 passes.
- **SEND (t_m, DATA(1), IP_i(m))**: transfer **DATA** in an output port **OP** of a block B_M to an input port **IPm** of a block B_i and **generate next event B_i**, after time t_m passes.
- **SEND (t_i, DATA(1), IP_1(1))**: transfer **DATA** in an output port **OP1** of a block B_i to an input port **IP** of a block B_1 and **generate next event B_1**, after time t_i passes.
- **SEND (t_i, DATA(1), IP_n(1))**: transfer **DATA** in an output port **OPn** of a block B_i to an input port **IP** of a block B_N and **generate next event B_n**, after time t_i passes.

In Figure 14.18, the meanings based on message-passing protocols in distributed simulation of SEND pseudocodes are as follows:

- **SEND (t_1, DATA(1), IP_i(1))**: create **DATA** in an output port OP_1 of a block B_1 as a message and **send the message to** an input port IP_1 of a block B_i, after time t_1 passes.
- **SEND (t_m, DATA(1), IP_i(m))**: create **DATA** in an output port OP_1 of a block B_m as a message and **send the message to** an input port IP_m of a block B_i, after time t_m passes.
- **SEND (t_i, DATA(1), IP_1(1))**: create **DATA** in an output port OP_1 of a block B_i as a message and **send the message to** an input port IP_1 of a block B_1, after time t_i passes.
- **SEND (t_i, DATA(1), IP_n(1))**: create **DATA** in an output port OP_n of a block B_i as a message and **send the message to** an input port IP_1 of a block B_n, after time t_i passes.

14.4.3.3 Synchronous Input Property

Block B3 in Figure 14.19 divides the output of block B1 with the output of block B2 and then sends its quotient to block B4 and remainder to block B5. To make the division, block B3 should have two inputs at the same time (i.e., a synchronous block in DXL). Note that the block has two output ports to support a different output at the same time unlike the functions of traditional programming languages, which have only one output. The right side of Figure 14.19 describes the processing algorithm of block B3.

14.4.3.4 Asynchronous Input Property

In Figure 14.20, block B3 plays a role in switching input data. It is not necessary to receive both inputs to relay its data. In other words, the block just relays its input to its next block whenever it receives the input. The block should therefore have an asynchronous input property, which is depicted with a gray color. The switch block algorithm is described on the right side of Figure 14.20. In DXL, the above combinations of simple blocks and their properties create an actual model for a system.

```
Algorithm for a block having synchronous inputs
if(for ∀x∈ IP_i, x has input data) then
    The computation of the function in a DXL block
    For ∀y∈ P_i, if(y has output data) then for z∈ OB_i(y), SEND(t,
DATA(y), z) elseif
```

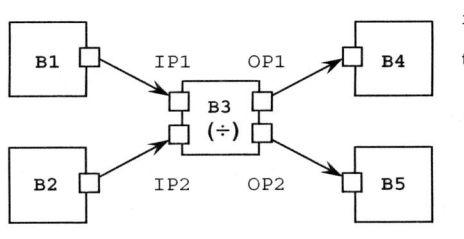

```
Algorithm for block B3
if (for ∀x∈ IP_B3, x has input data)
then
    OP_B3(1) ← IP_B3(1)/IP_B3(2)
    OP_B3(2) ← IP_B3(1)%IP_B3(2)
    SEND(t_B4, DATA(OP_B3(1)), B4)
    SEND(t_B5, DATA(OP_B3(2)), B5)
elseif
```

FIGURE 14.19 Example of synchronous inputs.

```
Algorithm for a block having asynchronous inputs
if(for ∃x∈ IP_i, x has input data) then
    The computation of the function in a DXL block
    For ∃y∈ OP_i, if(y has output data) then for z∈ OB_i(y),SEND(t,
DATA(y), z) elseif
```

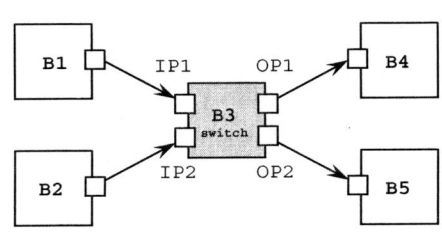

```
Algorithm for block B3
if (IP_B3(1) has input data) then
    OP_B3(2) ← IP_B3(1)
    SEND(t_B5, DATA(OP_B3(2)), B5)
elseif
if (IP_B3(2) has input data) then
    OP_B3(1) ← IP_B3(2)
    SEND(t_B4, DATA(OP_B3(1)), B4)
elseif
```

FIGURE 14.20 Example of asynchronous inputs.

14.4.4 Multimodeling

For DXL, multimodeling is defined by specifying a block circuit inside of a block, and continuing recursively as needed. Ports are coupled by ensuring matching data types for each connecting port on each connector. The general DXL methodology in transforming heterogeneous models (MXL models) into homogeneous models (DXL models) is as follows:

- Transform submodels in an MXL multimodel into DXL models and
- Incorporate each of the transformed models according to port coupling.

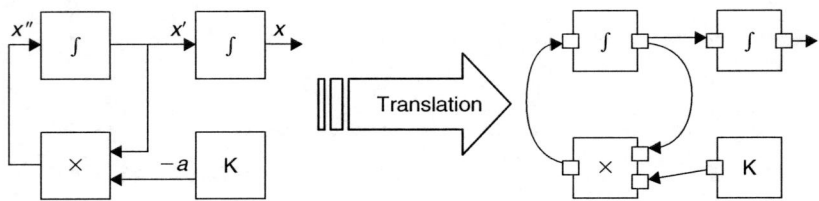

FIGURE 14.21 Translation of an FBM into a DXL model.

14.4.4.1 FBM-to-DXL Translation

The FBM is composed of functional elements where functions, along with inputs and outputs, are often depicted in a "block" form. An arbitrary number of blocks can be coupled to form an FBM. This FBM is similar to DXL except for synchronous and asynchronous inputs of DXL. Therefore, functions of FBM are translated into DXL blocks, and FBM's inputs and outputs are translated into DXL input and output ports. Because each block of traditional FBM has a pure function, all inputs of the function should be valid before the block is executed, meaning all FBM blocks have synchronous property for inputs.

An FBM for the differential equation $x'' = -ax'$ is described on the left side of Figure 14.21. This FBM is composed of four blocks: two integrators, one multiplier, and one constant. Because each block of a traditional FBM has a synchronous input property, these blocks are translated into blocks having synchronous inputs in DXL. Its arrows are translated into connectors of DXL. This FBM also has a continuous simulation property. To support the property, a start block of the DXL model is generated every simulation unit time.

14.4.4.2 FSM-to-DXL Translation

The FSM has states and transitions. A state represents the current condition or "snapshot" of a system for some length of time. Transitions enable the system to move from one state to another during the simulation while under the control of the system input. The basic rule of translating FSM into DXL is that all functional elements are translated into DXL blocks. The transitions of FSM are predicates under the system input and can, therefore, be translated into DXL blocks. These DXL blocks have the same type of input ports as the system input and a Boolean type of output ports that decide whether the predicates are true or false. Since states have the functional properties to access the system input, all states are translated into DXL blocks. In addition, we need the special block to control the system input.

The left side of Figure 14.22 shows an FSM modeling of a four-stroke gasoline engine with four phases: compression, ignition, expansion, and exhaustion. The key point that makes FSM different from FBM is a state-based model. To translate a state-based model into a function-based model, all states and transitions are translated into DXL blocks. Then DXL connectors and block properties control the semantics of the state-based model. The right side of Figure 14.22 shows a translated DXL model for the four-stroke gasoline engine. If we made a DXL code manually, we could create a simpler DXL code than the right side of Figure 14.22. But to make automatic MXL-to-DXL translation easy, we used the DXL model to include INPUT and OUTPUT blocks, as in Figure 14.22. Our translator generated the INPUT and OUTPUT blocks to control the FSM semantics. The right blocks next to INPUT blocks are translated transition blocks and the right blocks next to OUTPUT blocks are translated state blocks.

14.4.4.3 Multimodeling between Homogeneous Models

In Figure 14.23, the upper model is a two-level functional hierarchy, where function f is defined in terms of a composition of three other functions f_1, f_2, and f_3. The lower model shows multimodeling in DXL for the upper model. Because DXL is specifically based on a functional block diagram, FBM models in Figure 14.23 are refined only if the number and types of ports are matched.

Multimodeling 14-19

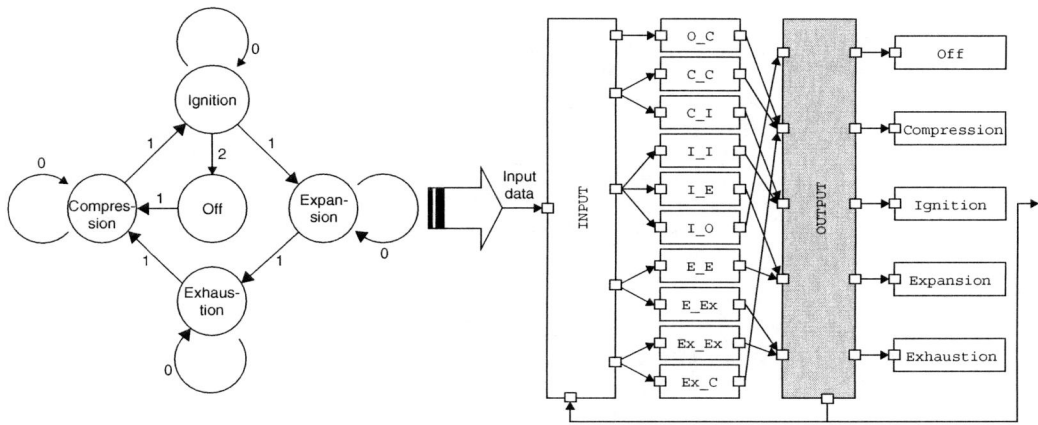

FIGURE 14.22 Translation of an FSM into a DXL model.

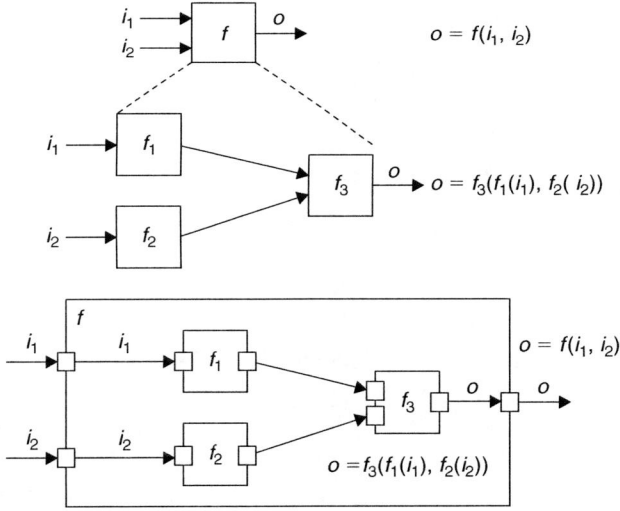

FIGURE 14.23 Multimodeling between functional block models in DXL.

In Figure 14.24, the upper model is a two-level finite state hierarchy, where state s_1 is defined in terms of another FSM. The lower model shows the transformed DXL model for the two-level finite state hierarchy. All FSMs should have this kind of functional box around them to represent external input and output semantics even though the functional box is not expressed explicitly in the FSM graphical representation.

Because FSMs represent behavior or dynamics of another component in multimodeling, an FSM is inserted in it to represent a state more specifically. In that case, multimodeling between FSMs should be supported, but there is a difficulty in integrating the two FSMs because of the implicit expression of input and output in FSMs.

However, in DXL, states and transitions in FSMs are transformed into the same blocks. MXL-to-DXL translator generates special blocks, which control state and transition blocks (i.e., INPUT and OUTPUT blocks). In Figure 14.24, after a submodel (i.e., the lower FSM model) is transformed into a DXL model, it is inserted in its upper state block and then connected with external input and output ports using connect elements in DXL.

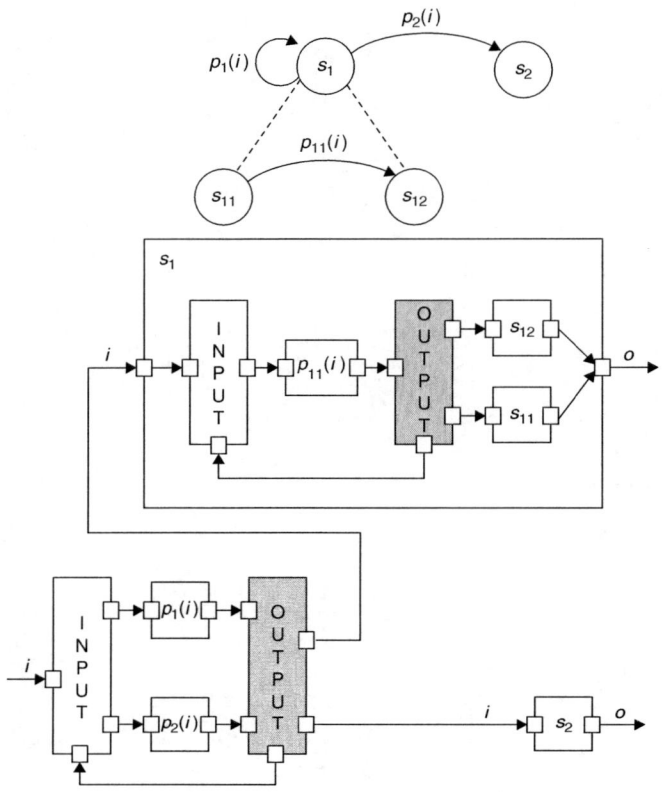

FIGURE 14.24 Multimodeling between FSM models in DXL.

14.4.4.4 Multimodeling between Heterogeneous Models in DXL

In Figure 14.25, the upper model is the same model as shown in Figure 14.4. The submodel in this figure is an FSM that defines the internal state transitions associated with function f. The generated DXL model for an FSM is inserted in function f. Because the number and types of input and output of the function f are the same as the input and output ports of the DXL model, the integration is accomplished through port matching.

In Figure 14.26, a state of an FSM decomposes into an FBM to represent internal functionality. The state in an FSM is different from an FBM from the viewpoint of a system. In an FSM, while input and output are not specified explicitly, control flow is unclear in an FBM. In a DXL model, however, the input and output of transformed FSM states are specified explicitly, which makes it easier for the integration of the components of an FSM and the components of an FBM. The lower model in Figure 14.26 indicates a transformed DXL multimodel. A sublayer of the model, a transformed DXL model for an FBM, is connected to the input and output ports of the outer state block s_1.

14.5 A Boiling Water Example

In this section, we explain how multimodeling environments (i.e., integrative and general multimodeling environments) can be implemented in the Blender 3D environment through our methodology, as proposed in the previous sections, using the example of the boiling water scene. The pot is filled to a predetermined level with water. A small amount of detergent is added to simulate the foaming activity that occurs

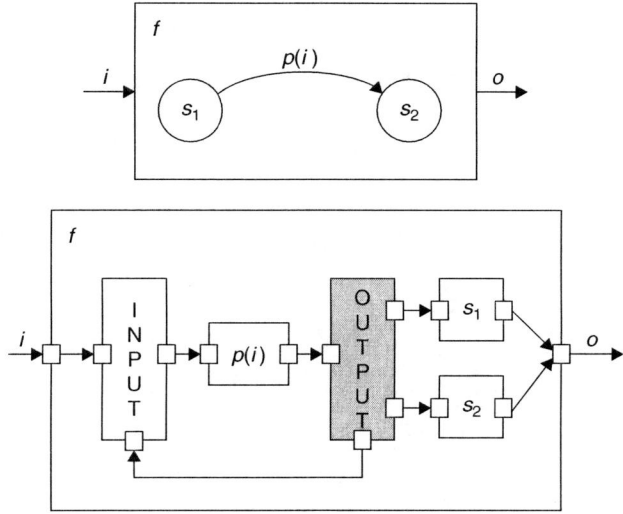

FIGURE 14.25 Multimodeling for an FBM including an FSM in DXL.

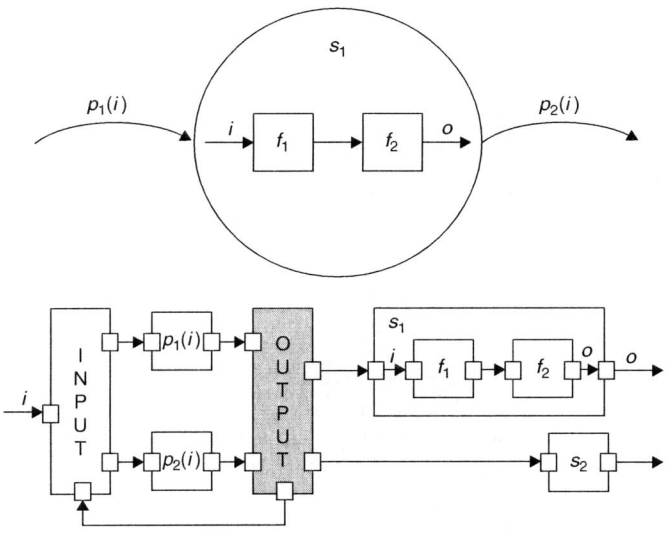

FIGURE 14.26 Multimodeling for an FSM including an FBM in DXL.

naturally when boiling certain foods. This system has a temperature knob. The knob is considered to be in one of two states: on or off. We first define the following conditions and assumptions in connection with this scene:

- External event: $I = \{\text{ON, OFF}\}$
- Internal event: $T = \{T = \alpha, \alpha < T < 100, T = 100\}$
- Input space: $\{(I = \text{ON}, T = \alpha), (I = \text{ON}, \alpha < T < 100), (I = \text{ON}, T = 100), (I = \text{OFF}, T = \alpha), (I = \text{OFF}, \alpha < T < 100), (I = \text{OFF}, T = 100)\}$
- Boiling state: $T = 100$
- Heating state: $T' = k_1(100 - T)$
- Cooling state: $T' = k_2(\alpha - T)$.

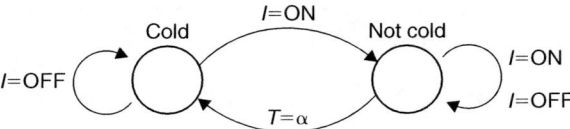

FIGURE 14.27   First-level FSM for the boiling water example.

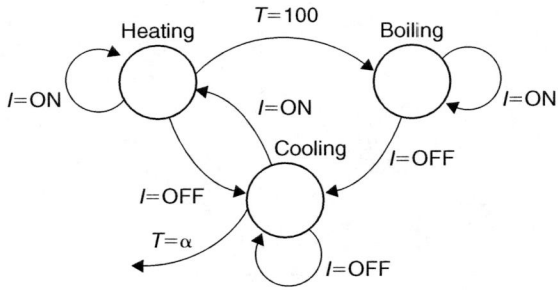

FIGURE 14.28   Second-level FSM for the boiling water example.

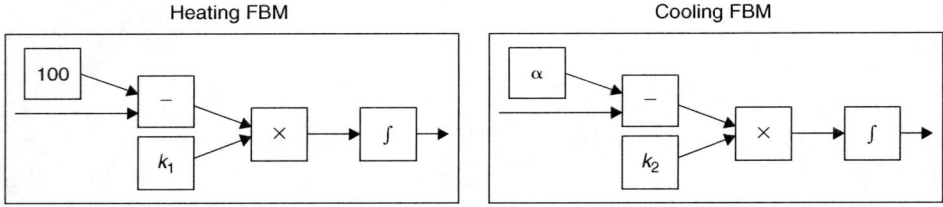

FIGURE 14.29   Third-level FBM for the boiling water example.

14.5.1   2D Representation

We begin the process by creating 2D representations of dynamic models for the boiling water scene based on the scene ontology, which is specified in the previous section, as well as conditions and assumptions above. Figure 14.27 shows the 2D representation for the highest level of FSM of the boiling water example. There are two states, "Cold" and "Not Cold." The input is either ON or OFF. Input can occur at any time and will facilitate a change in state. Because the "Not Cold" state could be further generalized, we could insert another FSM inside, as shown in Figure 14.28. To describe the continuous dynamic behaviors inside "Heating" and "Cooling" states, we need FBMs. Therefore, we could insert two FBMs, as shown in Figure 14.29, inside "Heating" and "Cooling" states. Figure 14.30 shows the overall 2D dynamic model representation for the boiling water example.

14.5.2   Model Creation
14.5.2.1   Geometry and Dynamic Models

We create the corresponding geometry and dynamic models for the scene. For geometry model, a teapot and a stove could be created within Blender Scene Editor (i.e., Blender 3D Window). And then using *Model Explorer*, which is provided by *Blender Interface*, we create dynamic models for the scene based on the 2D model representations. We employ a chemistry metaphor to describe the dynamic models instead of simple primitive objects (i.e., sphere and arrows), since the RUBE library system provides predefined chemistry objects in the *predefined theme* folder as well as corresponding MXLs and functions. To represent the "Cold"

Multimodeling

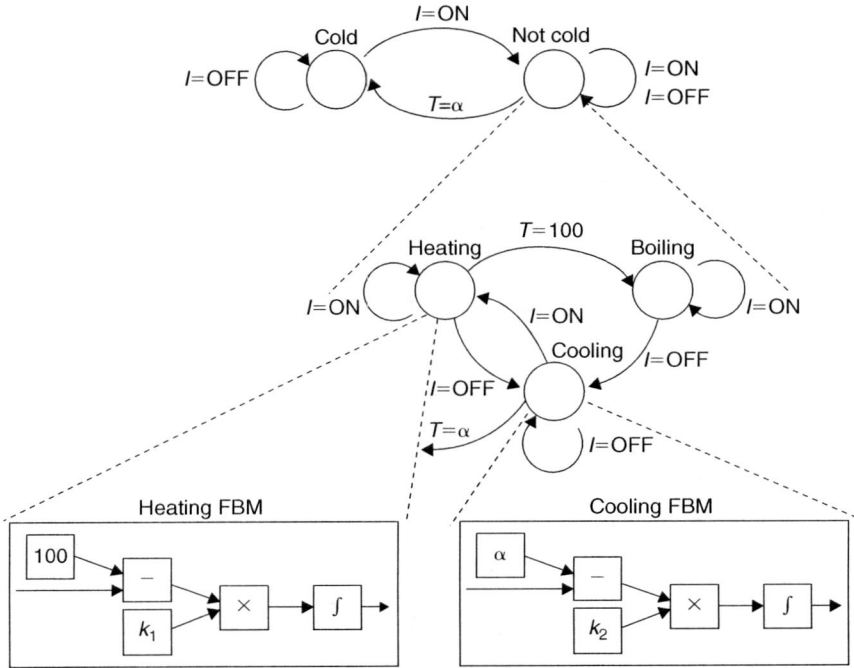

FIGURE 14.30 Multimodel for a boiling water example.

and "Not Cold" states as well as transitions, we use chemistry holders and chemistry pipes, respectively. And to describe the three states inside the "Not Cold" state and transitions, chemistry glasses and pipes are employed. In the case of two functional block models, which could be inserted into "Heating" and "Cooling" states, we use simple primitive objects (i.e., cubes and arrows).

14.5.2.2 Interaction Model

On the basis of the induced interaction model, which is discussed in Section 14.2, for the scene, we embody the interaction model within the Blender Game Logic. In the Blender Game Logic, there are sensors, controllers, and actuators. These are used to create "Logic Brick" graphs that drive the interaction. The Game Engine can cover most interaction behaviors since it has well-defined built-in sensors and actuators. If a user, however, wants more complicated interactions, the user can handle the interactions by Python scripting language. After creating a Python file, the user puts it in one of the controllers and connects the controller with any proper actuator.

14.5.3 Code Generation

From given dynamic model components in Blender 3D Editor, using the *Simulation* component, which is also provided by *Blender Interface*, we can generate a simulation code for a target system through the RUBE framework. The simulation code is Python scripting language. The process for generating a simulation code includes three steps:

1. Blender-to-MXL: First, an MXL file for a given scene dynamic model is created by gathering all segmented MXL files from the libraries, since the library system contains Blender objects as well as the corresponding MXLs and functions in a set of pairs.
2. MXL-to-DXL: Using XSLT, the MXL file is converted to a low-level XML language called DXL.
3. DXL-to-Python: The Python code for simulation is generated from DXL using DOM.

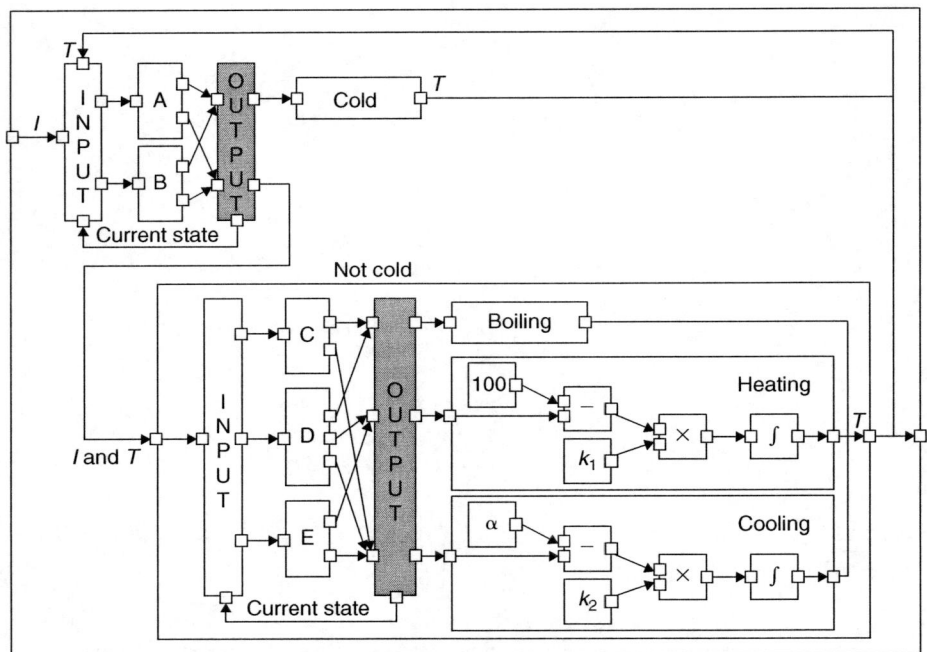

FIGURE 14.31 DXL Multimodel for the boiling water example.

The "Heating FBM" of Figure 14.29 describes the differential equation $T' = k_1(100 - T)$. This FBM is composed of five blocks: one integrator, one multiplier, one subtracter, and two constants. Because each block of this FBM has a synchronous input property, these blocks are translated into blocks having synchronous inputs in the "Heating DXL" of Figure 14.31. Its arrows are translated into connectors of the DXL. The "Cooling FBM" of Figure 14.29, which describes the differential equation $T' = k_2(\alpha - T)$, is translated into the "Cooling DXL" model according to the same procedure as the "Heating FBM."

The two FSMs of Figure 14.27 and Figure 14.28 are translated into Figure 14.31 except for the DXL models of "Heating" and "Cooling" blocks. Our translator generates the INPUT and OUTPUT blocks to control the FSM semantics. The right blocks next to INPUT blocks are translated transition blocks, and the right blocks next to OUTPUT blocks are translated state blocks, such as "Cold," "Not Cold," "Boling," "Heating," and "Cooling." Figure 14.31 shows the translated DXL multimodel for the boiling water example according to the MXL-to-DXL translation approach. In Figure 14.31, the output of state DXL blocks is T (the temperature of water). And the output of transition DXL blocks is I and T. We simplify blocks A–E because of the complexities in Figure 14.31. Figure 14.32 shows the specific descriptions of blocks A–E which can be found in Figure 14.31. Because DXL blocks are simple, the total DXL diagram could be complex. The simplification from Figure 14.31 to Figure 14.32 could be used for the optimization of DXL diagrams in our future research.

Figure 14.33 shows input and temperature trajectories. The input trajectory is displayed in a red line; turning the knob on and off over some time period. This was chosen at random to display phase switching. The temperature trajectory, which shows the execution results of simulation code in Python, is displayed in a blue line. The temperature rises to $T = 100$ in response to the step input and then it falls before reaching the ambient temperature of $T = 20$. The temperature levels off at $T = 100$.

Figure 14.34 and Figure 14.35 represent the 3D boiling-water scene that contains geometry and dynamic models. Through user interactions, we can convert a geometry model, which is shown in Figure 14.34, to a dynamic model, which is shown in Figure 14.35 (chemistry glasses and pipes used to describe states and transitions), and vice versa (i.e., integrative multimodeling).

Multimodeling

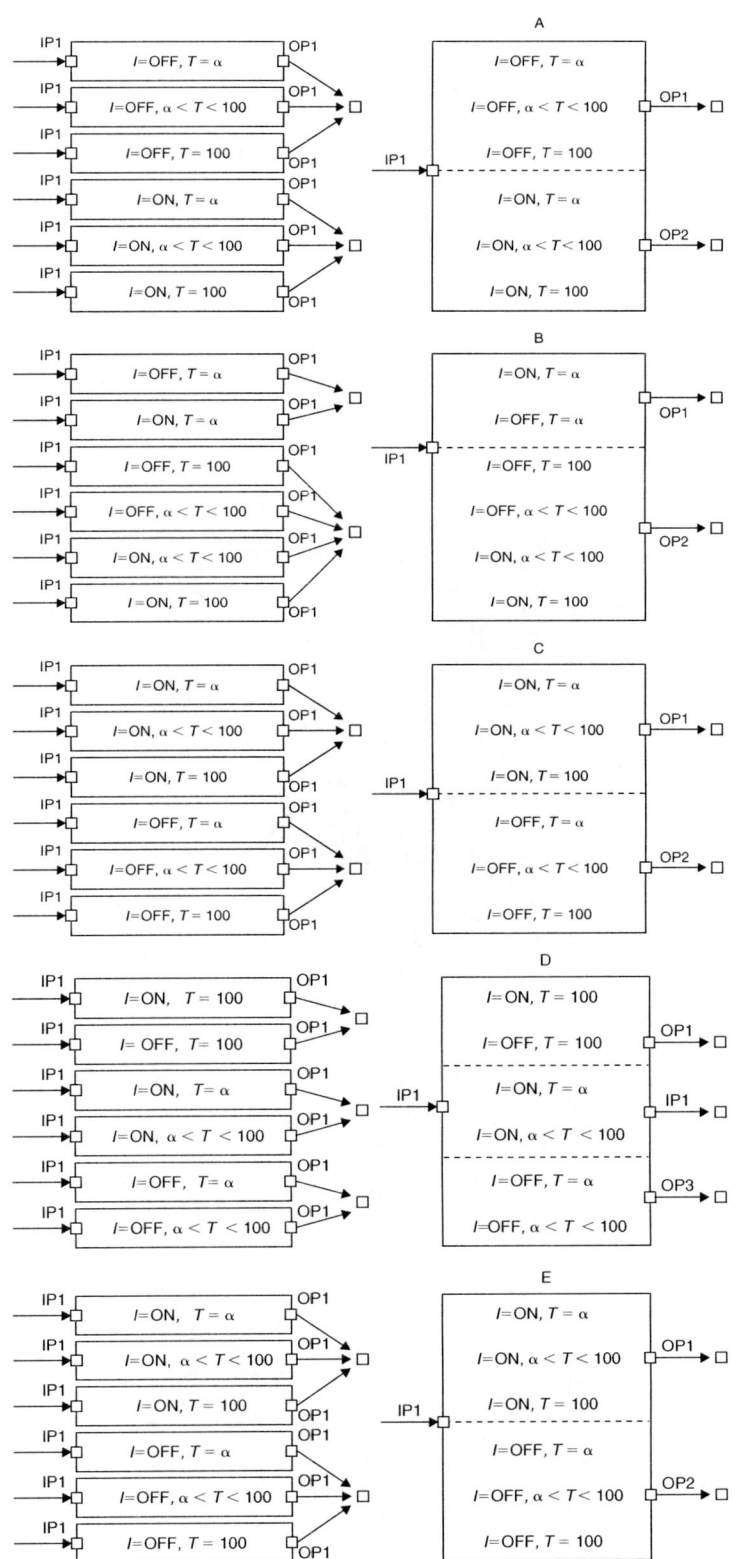

FIGURE 14.32 Descriptions of blocks A–E.

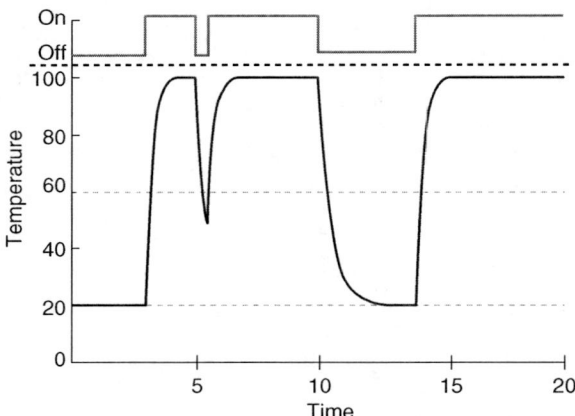

FIGURE 14.33 Simulation output.

FIGURE 14.34 Initial scene (geometry model).

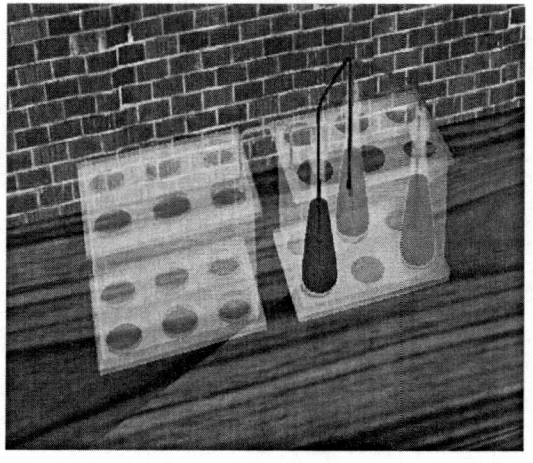

FIGURE 14.35 Dynamic model.

14.6 Conclusion

We have presented the approach and methodology for multimodeling. Also, we have explained and discussed the multimodeling concepts and environments. Using the example (i.e., boiling water), we have demonstrated how the multimodeling concepts and semantic Web technology are applied to a real-world system. The characteristics of this approach are summarized as follows:

- Web-based modeling and simulation through the use of XML to represent simulation models
- Multimodeling through ontology, MXL, DXL, and an MXL–DXL bridge
- Flexible diagrammatic presentation in terms of the functionality for each DXL "block"
- Port-based modeling and simulation capable of encoding documents rather than data
- Explicit support for discrete event and continuous simulation.

Several issues remain to be researched, such as model type extension, effective means of composability, ensuring complete mapping among models and ports, and expanding to remote distributed modeling and simulation.

References

Berners-Lee, T., Hendler, J., and Lassila, O. 2001. The semantic web. *Scientific American* 284(5): 34–43.

Blender, 2006. http://www.blender3d.org/.

Carey, R. and Bell, G. 1997. *The Annotated VRML 2.0 Reference Manual*. Reading, MA: Addison-Wesley.

Cubert, R. M. and Fishwick, P. A. 1998. MOOSE: An object-oriented multimodeling and simulation application framework. *Simulation* 70(6): 379–395.

Damkjer, K. L. 2003. Architecting RUBE worlds: A methodology for creating virtual analog devices as metaphorical representations of formal systems. M.S. Thesis, University of Florida.

Eisenberg, D. 2002. *SVG Essentials*. Sebastopol, CA: O'Reilly.

Fishwick, P. A. 1992. SimPack: Getting started with simulation programming in C and C++. In *Proceedings of the 1992 Winter Simulation Conference*: 154–162.

Fishwick, P. A. 1995. *Simulation Model Design and Execution: Building Digital Worlds*. Englewood Cliffs, NJ: Prentice-Hall.

Fishwick, P. A. 2002. RUBE: An XML-based architecture for 3D process model fusion. In *Proceedings of Enabling Technology for Simulation Science within SPIE '02 AeroSense Conference*: 330–335.

Fishwick, P. A. 2002. Using XML for simulation modeling. In *Proceedings of the 2003 Winter Simulation Conference*: 616–622.

Fishwick, P. A. 2004. Toward an integrative multimodeling interface: A human–computer interface approach to interrelating model structures. *Simulation* 80(9): 421–432.

Fishwick, P. A., Lee, J., Park, M., and Shim, H. 2003. RUBE: A customized 2D and 3D modeling framework for simulation. In *Proceedings of the 2003 Winter Simulation Conference*: 755–762.

Horrocks, I., Patel-Schneider, P. F., Boley, H., Tabet, S., Grosof, B., and Dean, M. 2004. SWRL: A semantic web rule language combining OWL and ruleML. http://www.daml.org/2004/04/swrl/.

Jess, 2006. http://herzberg.ca.sandia.gov/jess/.

Kay, M. 2000. *XSLT Programmer's Reference*. Birmingham, UK: Wrox Press.

Kim, T. 2002. A 3D XML-based modeling and simulation framework for dynamic models. Ph.D. Dissertation, University of Florida.

Kim, T. and Fishwick, P. A. 2002a. A 3D XML-based customized framework for dynamic models. In *Proceedings of the Seventh International Conference on 3D Web Technology*: 103–109.

Kim, T. and Fishwick, P. A. 2002b. An XML-based visualization and simulation framework for dynamic models. In *Proceedings of Enabling Technology for Simulation Science within SPIE'02 AeroSense Conference*: 336–347.

Kim, T., Lee, J., and Fishwick, P. A. 2002. A two-stage modeling and simulation process for web-based modeling and simulation. *ACM Transactions on Modeling and Computer Simulation*, 12(3): 230–248.

Lacy, L. 2005. *OWL: Representing Information Using the Web Ontology Language.* New Bern, NC: Trafford Publishing.

Lee, J. 2005. Architecture for a low-level functional specification language supporting multimodeling and simulation. Ph.D. Dissertation, University of Florida.

Lee, J. and Fishwick, P. A. 2002. A dynamics exchange language layer for *rube*. In *Proceedings of Enabling Technologies for Simulation Science within SPIE'02 AeroSense Conference*: 359–366.

McGuinness, D. L. 2002. Ontologies come of age. In D. Fensel, J. Hendler, H. Lieberman, and W. Wahlster, Eds, *Spinning the Semantic Web: Bringing the World Wide Web to Its Full Potential,* 171–194. Cambridge, MA: MIT Press.

McGuinness, D. L. and Harmelen, F. 2003. OWL web ontology language overview. http://www.w3.org/TR/owl-features.

Park, M. 2005. Ontology-based customizable 3D modeling for simulation. Ph.D. Dissertation, University of Florida.

Park, M. and Fishwick, P. A. 2002. SimPackJ/S: A web-oriented toolkit for discrete event simulation. In *Proceedings of Enabling Technology for Simulation Science within SPIE '02 AeroSense Conference*: 348–358.

Park, M. and Fishwick, P. A. 2004a. *An Integrated Environment Blending Dynamic and Geometry Models.* Lecture Notes in Computer Science Series, Volume 3397: 574–584. Berlin: Springer.

Park, M. and Fishwick, P. A. 2004b. A methodology for integrative multimodeling: Connecting dynamic and geometry models. In *Proceedings of Enabling Technology for Simulation Science within SPIE'04 AeroSense Conference*: 9–17.

The Protégé Ontology Editor and Knowledge Acquisition System, 2006. http://protege.stanford.edu/.

RACER, 2006. http://www.sts.tu-harburg.de/~r.f.moeller/racer/.

Roosendaal, T. and Selleri, S. 2004. *Blender 2.3 Guide.* Amsterdam, The Netherlands: Blender Foundation.

Roosendaal, T. and Wartmann, C. 2002. *The Official Blender Gamekit: Interactive 3D for Artists.* Amsterdam, The Netherlands: Blender Foundation.

Sodipodi, 2006. http://sodipodi.sourceforge.net/.

X3D, 2006. http://www.web3d.org/x3d/.

15
Hybrid Dynamic Systems: Modeling and Execution

Pieter J. Mosterman
The MathWorks, Inc.

15.1 Introduction 15-1
15.2 Hybrid Dynamic Systems 15-3
   Why Hybrid Dynamic Systems? • What Is a Hybrid Dynamic System • How Are Hybrid Dynamic System Models Designed?
15.3 Hybrid Dynamic System Behaviors 15-9
   An Operational Structure • Continuous-Time Behavior • Handling Mode Transitions • Reinitialization of State Variables
15.4 An Implementation 15-12
   Classes of Events • Classes of Temporal Behavior • Time-Driven Execution • Event-Driven Execution • Combining the Execution Types
15.5 Advanced Topics in Hybrid Dynamic System Simulation 15-17
   Zero-Crossing Detection • Mode Changes
15.6 Pathological Behavior Classes 15-22
15.7 Conclusions 15-23

15.1 Introduction

To remain competitive, organizations throughout industry are increasingly adopting Model-Based Design. Computational models take a central position in Model-Based Design as illustrated in Figure 15.1. The use of models throughout the design of engineered systems has a number of advantages. In Figure 15.1, four of these are shown:

- Early evaluation ensures the appropriateness, consistency, rigor, and unambiguity of requirements.
- Simulation allows quick design iterations while the use of models makes it possible to search a very large design space efficiently.
- Automatic code generation eliminates the expensive manual coding, which is labor-intensive and error-prone.
- Test vector sets can be composed in the early design stages, based on a system model, thus eliminating much of the lead time needed in traditional testing, which requires the real system to be available. Verification of completeness and parsimony of the design can be done before implementation.

Overall, from a business perspective, Model-Based Design allows a shorter design cycle so engineers can develop better products at a lower cost.

The models that are used throughout the product design have to address many different aspects. For example, structuring requirements may be best done by modeling scenarios and in an axiomatic manner,

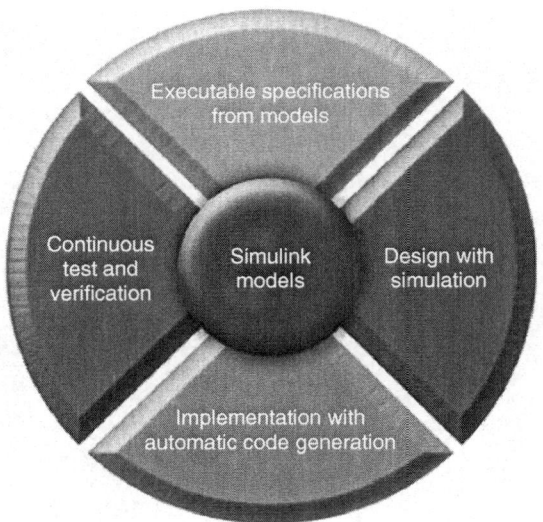

FIGURE 15.1 Model-based design.

FIGURE 15.2 A power window.

while the numerical models with an intensive data processing aspect are often best represented in a declarative manner. Meanwhile, execution models typically are best captured in an imperative manner.

Similarly, certain modeling tasks are best addressed by continuous-time models, while other tasks may be better performed with discrete-time or untimed models. For example, consider the power window in Figure 15.2 (Mosterman et al., 2004). Modeling the behavior of the window sliding up and down is often easiest using continuous-time differential equations, employing laws of physics such as Newton's Second Law of Motion that the acceleration, a, of a mass, m, is determined by the force, F, applied to it, $F = m \cdot a$, where the acceleration is the time-derivative of velocity, v, $a = \frac{dv}{dt}$.

In contrast, the desired behavior to start moving the window when the user presses a button is often best viewed as discrete in nature. An event *windowUp* then causes a state change from *neutral* to *moveUp*. At certain stages during the design, such an untimed model may be sufficient and even desired, as it leaves any implementation choices on how to achieve such control still open.

Hybrid Dynamic Systems: Modeling and Execution 15-3

To implement the control of the power window, an electronic control unit that operates at a certain sample rate can be selected. In this implementation, the untimed events can be modeled to occur at points in time. During the fixed intervals of time in between, no changes of the variable values in the model occur. This leads to a discrete time but periodic model.

Yet another task may consist of modeling the network that transmits the user command to move the window up to the controller that effects this command. In automobiles, these commands may be transmitted over a controller area network (CAN) (CAN specification, 1991), that is also used to transmit other commands such as moving the headlights up and down, adjusting the outside mirrors, and opening the sunroof. At a certain level, the network traffic can be thought of as discrete time but nonperiodic, since commands may be initiated with very different intervals of time in between.

When the integration of different parts of a system is studied, it becomes necessary to employ the different models in concert (Mosterman et al., 2005). This can be done in a number of ways. For example, the different types of models can be combined, integrated, and translated. Whichever approach is chosen, the result is likely to include both continuous-time and discrete-event behavior, and the overall system is often referred to as a *hybrid dynamic system*, or *hybrid system* for short.[1] It is important to stress, at this point, that an engineered system is not inherently a "hybrid system." A hybrid system is a term for a mathematical representation not a physical system. Whether an engineered system is modeled as a hybrid system depends on the problem that needs to be solved, the level of abstraction chosen, the phenomena the system embodies, and the background of the model designer (Mosterman and Vangheluwe, 2004).

This chapter intends to provide a bird's eye view of the modeling and execution of hybrid systems. Some of the uses of hybrid systems will be presented, and the different ways of simulating them and their idiosyncrasies will be investigated. In Section 15.2, hybrid dynamic systems are presented: what they are, how they come about, and what different modeling perspectives exist. In Section 15.3, the behavioral perspective is discussed and the different behavioral aspects are introduced. In Section 15.4, the two possible implementations of an execution engine, time- and event-driven implementation, and combinations are introduced. Section 15.5 then presents advanced topics with regards to numerical simulation. In Section 15.6 a number of pathological behaviors that may arise in simulation are discussed. Section 15.7 presents the conclusions of this work.

15.2 Hybrid Dynamic Systems

This section illustrates where the need for hybrid dynamic systems emerges from in embedded control systems design, what the mathematical notions are that hybrid dynamic systems involve, and the modeling concepts that require these notions.

15.2.1 Why Hybrid Dynamic Systems?

A power window will serve to illustrate the application of hybrid dynamic systems in the design of engineered systems. This example has been studied in more detail in other work (Mosterman et al., 2004; Mosterman and Vangheluwe, 2004) and is part of the Simulink® (Simulink, 2004) automotive demos. The elements of interest in this context are illustrated on the left-hand side in Figure 15.3, along with exemplary data as used in the design and analysis on the right-hand side. In the center near the top is the window lift mechanism. A dc motor translates the electrical signals that reflect the controller commands into movement in the mechanical domain. A current sensor is used to measure the current drawn by the dc motor, which is used by the controller to determine whether the top or bottom of the window frame is reached. Behavior of these components is often modeled as *continuous* in time.

The controller is often modeled as operating at a *periodic* sample rate and it may receive the user input over a CAN. This network provides communication between the electronic control units (ECUs) in an

---

[1] Note that omitting the adjective "dynamic" may lead to confusion with hybrid systems such as mixed neural network/fuzzy logic systems and combined electrical/mechanical drivetrains.

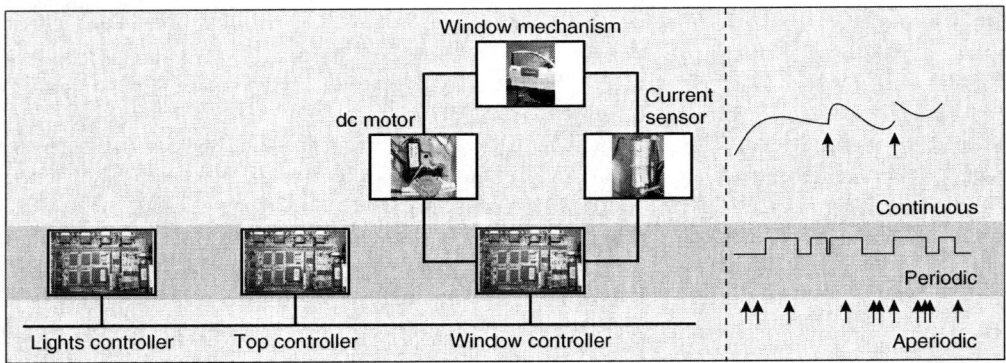

FIGURE 15.3 A hybrid dynamic system with different behavior classes.

automobile and connects the ECU that interfaces with the window up and down button to the ECU that interfaces with the window lift mechanism. Network traffic is often modeled as *aperiodic* events.

The design of such a system typically proceeds by determining a set of requirements that it has to satisfy. Among these requirements it may state that (Friedman and Ghidella, 2006)

1. The window should be closed within 5 s.
2. The passenger command should always be overruled by the driver command.
3. The driver and passenger commands should be acted upon within 200 ms.

To validate Requirement 1, a continuous-time model of the lift mechanism is best used. This could be a set of differential and algebraic equations (DAEs) that model how fast the window with mass, m_{window}, moves, v_{window}, given a voltage on the dc motor, u_{motor}, and the corresponding current that the motor draws, i_{motor}. The equations

$$
\begin{aligned}
F_{motor} &= r u_{motor} \\
F_{window} &= m_{window} \dot{v}_{window} \\
F_{lift} &= R_{lift} v_{window} \\
i_{motor} &= r v_{window} \\
F_{window} + F_{lift} &= F_{motor} \\
\dot{x}_{window} &= v_{window}
\end{aligned}
\tag{15.1}
$$

model the conversion of the dc motor voltage by a parameter, r, and include linear friction, R_{lift}, that causes a friction force component, F_{lift}, for the lift mechanism. Solving this system of equations over time shows whether the window can rise quickly enough for certain input voltages by evaluating the window position, x_{window}.

In contrast, the control structure that Requirement 2 addresses is best modeled by a state machine that has hierarchical structure of its states. This allows the passenger commands to execute when the driver commands are neutral. When the driver issues a command, the passenger control structure is departed and overruled by the driver command. This is illustrated by the Stateflow® (Stateflow, 2004) chart in Figure 15.4.

In this Stateflow chart, there are three states for the driver control, *driverNeutral*, *driverDown*, and *driverUp*. When *driverDown* is the active state, the window is moved down, and when the *driverUp* state is active, the window is moved up. The *driverNeutral* state is of a hierarchical nature, and contains the control structure for the passenger, analogous to the control structure for the driver. In this manner, the driver control takes precedence over the passenger control, and the passenger may command the window only if the driver does not issue a command (i.e., a neutral command). The initial states are indicated by

Hybrid Dynamic Systems: Modeling and Execution 15-5

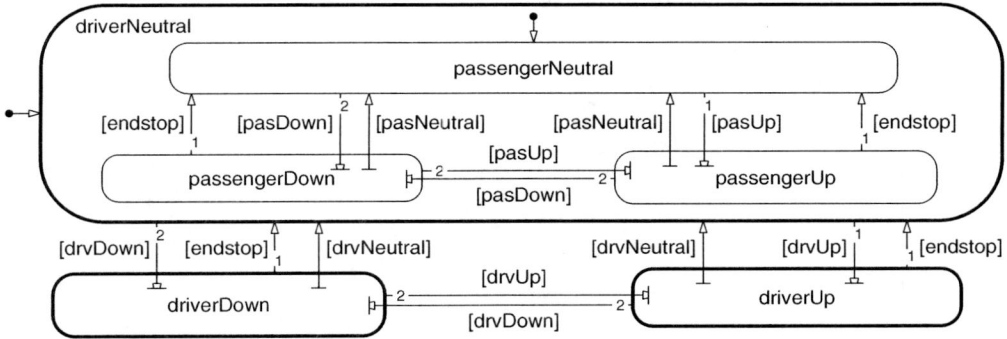

FIGURE 15.4 A finite state machine in the form of a Stateflow chart.

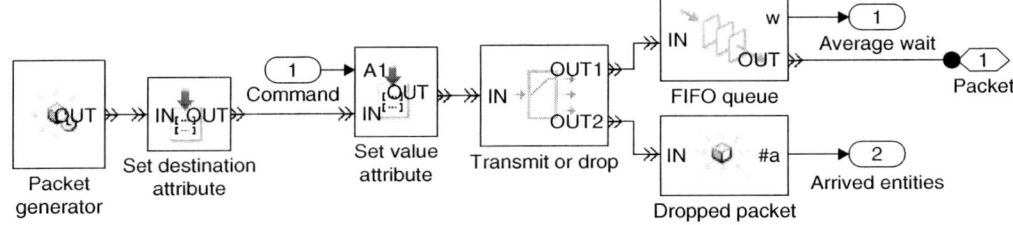

FIGURE 15.5 An entity network model of a transmitter.

a transition that is depicted by an arrow with a solid circle on one end and arrow head on the other end. Note that the ordering of transitions that could be activated simultaneously is indicated by the numbers associated with the respective transitions.

The events issued by the driver, *drvUp* and *drvDown*, command the window to go up or down. When any of the commands is released, the *drvNeutral* command is issued. The behavior is similar with the passenger-related commands *pasUp*, *pasDown*, and *pasNeutral*. The *endstop* command is issued when the window has reached the top or bottom of the door frame.

To validate the response time as put forward in Requirement 3, the communication network needs to be modeled. In this case, a server/queue type model is best used where the network is modeled as a server that aperiodically frees up and passes a packet from one ECU to another. A SimEvents™ (SimEvents, 2005) model of the packet transmitter is shown in Figure 15.5. Here, the command as issued by the driver is assigned as the attribute of packets that are generated at a fixed rate. The packets themselves have an attribute that identifies their destination. The system then attempts to put the packets into a first-in-first-out (FIFO) queue and if this fails, they are lost as dropped packets. The FIFO queue is connected to another subsystem, which models the CAN bus behavior.

Stochastic server times may be included to model the variable arrival times of network requests from control systems (such as the headlights control system) that communicate using the same CAN. This is a type of system integration test that is often performed in hardware because the different systems are provided by different suppliers, and the models used for the design of each system are not shared among these suppliers.

Combining all these different types of models that are needed to verify the requirements necessitates handling both continuous-time differential equation models as well as discrete-event state-transition models. This can be achieved, for example, by transforming the continuous-time models into discrete-event representations or the other way around. Alternatively, a computational model can be designed that can handle both the continuous-time and discrete-event behavior. This latter approach leads to the hybrid dynamic execution structure.

15.2.2 What Is a Hybrid Dynamic System?

A hybrid dynamic system is defined in behavioral terms, because it is model structure agnostic, and, therefore, does not impose unnecessary and undesired modeling assumptions.

In a geometric sense, a hybrid dynamic system evolves continuously in time in a mode, α_i, according to a field, f_{α_i} (Guckenheimer and Johnson, 1995; Mosterman and Biswas, 1997). This field defines a relation $f_{\alpha_i}(\dot{x}, x, u, t) = 0$ between the state, x, its time derivative, $\dot{x}$, the input u, and the time, t. This is illustrated in Figure 15.6, where the continuous-time behavior is shown as a solid directed line in the plane α_1.

The continuous-time behavior is often captured by an explicit representation as a set of differential equations

$$\dot{x} = f_{\alpha_i}(x, u, t) \tag{15.2}$$

where α_i is the mode of the model, x the continuous-time state vector, u the exogeneous input, and t the time. An alternative formulation is the implicit form

$$f_{\alpha_i}(\dot{x}, x, u, t) = 0 \tag{15.3}$$

often used in plant modeling, and the semi-explicit form that has an explicit representation of the time derivatives, $f_{\alpha_i}^d$, combined with a set of implicitly formulated algebraic constraints (van Dijk, 1994), $f_{\alpha_i}^a$,

$$\dot{x} = f_{\alpha_i}^d(x, u, t), \quad 0 = f_{\alpha_i}^a(x, u, t) \tag{15.4}$$

In Figure 15.6, the change to mode α_2 occurs when the state in mode α_1, x_{α_1}, reaches a threshold value. In general, a mode transition relation $\gamma_{\alpha_i}^{\alpha_{i+1}}(x, u, t) \geq 0$ defines the change from mode α_i to α_{i+1} when true.

The state space in a mode α_i consists of two parts: (i) the domain where f_{α_i} is properly defined and (ii) a *patch*, where $\gamma_{\alpha_i}^{\alpha_{i+1}}$ does not invoke a mode change. In Figure 15.6, the patches are shown as white areas in the state space. When the boundary of the patch in α_1 is reached, a mode transition as defined by $\gamma_{\alpha_1}^{\alpha_2}$ is invoked. In the new mode, a patch defined by $\gamma_{\alpha_2}^{\alpha_3}$ is entered in which the state can continue to evolve, now governed by the field f_{α_2}.

When a mode transition from α_i to α_{i+1} takes place, in general, the state x_{α_i} may change its value to $x_{\alpha_{i+1}}$. Without loss of generality it is first assumed that the explicitly defined state transition function, $x_{\alpha_{i+1}} = g_{\alpha_i}^{\alpha_{i+1}}(x_{\alpha_i}, u_{\alpha_i}, t)$, is the identity function, i.e., $x_{\alpha_{i+1}} = x_{\alpha_i}$, as shown in Figure 15.6 by the perpendicular dashed arrow.

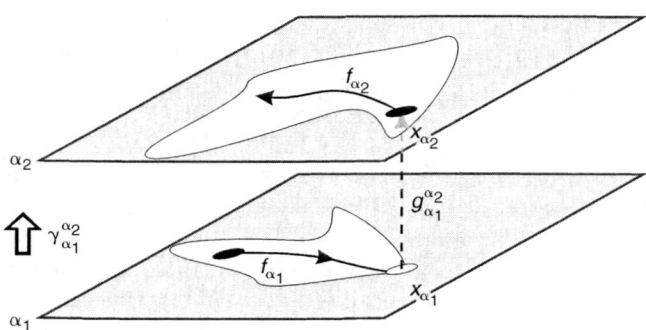

FIGURE 15.6 Geometric representation of a hybrid dynamic system. (From Mosterman, P. J. Mode transition behavior in hybrid dynamic systems. In *Proceedings of the 2003 Winter Simulation Conference*, pp. 623–631, New Orleans, LA, December 2003 (invited paper).)

15.2.3 How Are Hybrid Dynamic System Models Designed?

The behavioral definition of hybrid dynamic systems left open the specific underlying model structure that is employed. There are two basic approaches to the design of hybrid dynamic system models: (i) an implicit approach and (ii) an explicit approach.

15.2.3.1 Implicit Models

In modeling for analysis, often an implicit approach is favored. This is best illustrated by considering the modeling of a physical device under control, the *plant*, such as the power window mechanism of the system in Figure 15.3. The continuous-time behavior that models moving the window up is given in Eq. (15.1) in an implicit form. The equations are not converted in an explicit form that allows their solution yet. For example, the velocity of the window, v_{window}, is a state variable and so its time derivative needs to be computed by the system of equations in Eq. (15.1). However, applying Newton's Second Law of Motion leads to a formulation with the force, F_{window}, on the left-hand side. To arrive at an explicit computation of $\dot{v}_{\text{window}}$, the equation has to be reformulated into

$$\dot{v}_{\text{window}} = \frac{F_{\text{window}}}{m_{\text{window}}} \tag{15.5}$$

The reason for the implicit formulation is modeling convenience. It is often easier to model physical constraints in an implicit, constraint-based form such as Newton's Second Law of Motion. Another such constraint is, for example, conservation of momentum,

$$\sum m_i v_i^+ = \sum m_i v_i^- \tag{15.6}$$

which states that the some of the momentum, $m_i v_i^-$, of each body, m_i, with velocity, v_i^-, before a collision and after a collision, $m_i v_i^+$, should be the same. Which velocities are input to this constraint, and which ones are computed from the constraint, is left implicit.

In the same vein, mode changes in models of physical systems are often best represented in an implicit manner. For example, when the window reaches the top of its frame, further upward movement is restricted by a rapid increase of resistance to motion. This can be modeled in three different manners, illustrated in Figure 15.7.

- By a *nonlinear* resistance that is distance-dependent and rapidly increases when the window reaches the top of the frame position (Figure 15.7[a]).
- By a *piecewise linear* resistance that abruptly changes the resistance value when the window reaches the top of the frame position (Figure 15.7[b]).
- By an *ideal switch* of the resistance that enforces a hard stop when the window reaches the top of the frame position (Figure 15.7[c]).

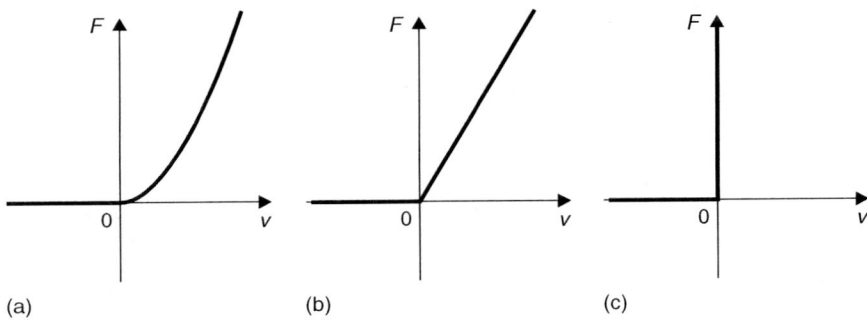

FIGURE 15.7 Abstraction classes for endstop models. (a) Nonlinear, (b) C^0 hybrid, and (c) ideal switch.

These three classes of modeling approaches are typical in the modeling of physical systems with perceived discontinuities. The latter two lead to hybrid dynamic systems.

To illustrate how the window reaching the top of the frame could be formulated as a mathematical relation that represents an ideal switch, the equation $F_{window} + F_{lift} = F_{motor}$ in Eq. (15.1) is replaced by

$$0 = L(F_{window} + F_{lift} - F_{motor}) + (L-1)v_{window} \tag{15.7}$$

where L is a logical variable that switches based on the window position, x_{window}, where $L = x_{window} \leq x_{top}$. Such "local" and "implicit" switching of equations is intuitive and convenient in the domain of plant modeling.

15.2.3.2 Explicit Models

Controller modeling typically takes a different perspective because the modeled behavior is the explicit objective of the design. As such, implicit modeling constructs such as algebraic loops that have a cyclic dependency (e.g., $x = -2x + 3$) are not desired or even illegal (such as for real-time code generation).

The corresponding continuous-time formalism that is often used in control system design is explicit ordinary differential equations (ODEs). Whereas the DAEs used for implicit modeling are of the form $0 = f(\dot{x}, x, u, t)$, the explicit ODE form is $\dot{x} = f(x, u, t)$. This form is especially popular because it is amenable to control law synthesis, in particular when it is linear. For example, when the behavior of the lift mechanism of the power window in Figure 15.3 is available in an explicit linear ODE form, root-locus methods can be applied to design a controller that can achieve the desired behavior, such as the position overshoot in response to a step input (Control System Toolbox, 2004).

The discrete-event behavior is often modeled by finite state machines. For example, *state transition diagrams* are graphical models that represent finite state machines. The different modes of behavior then correspond to the states of the transition diagram, and, consequently, are explicitly available. In addition, the imperative nature of state transition diagrams makes the state transition behavior become operational. The modeler has rendered the manner in which the transitions between states are made explicit. In the implicit formulation, constraints are provided on the modes, and a transformation to an operational form has to still be derived.

The combination of state transition diagrams and ODEs has been a popular modeling approach for control design as its explicit nature makes it amenable to reachability analyses (e.g., Lynch and Krogh, 2000). Such *hybrid automata* (Alur et al., 1993) consist of discrete states with associated ODEs. When in a state, the associated ODE governs continuous-time behavior. The transitions between states are enabled by *guards* and when enabled, the transition from one state to another *may* be taken. Each state has an *invariant* associated with it as well that cannot be violated while in that state. When the continuous behavior in a state reaches a point where it would violate the corresponding invariant, an enabled transition *must* be available and taken.

To illustrate, Figure 15.8(a) shows a hybrid automaton for the power window behavior that is modeled in an implicit form in the previous section. The state (also called *mode* as it captures the mode of operation of the entire system) on the left-hand side models the window moving up by the two equations for v_{window} and $\dot{x}_{window}$. The invariant of the state is $x_{window} < x_{top}$, which requires that the window can move up freely; that is, it is not pushing against the top of the frame.

When the top of the frame is reached, $x_{window} \geq x_{top}$ becomes true and the transition into the right-hand state is enabled. Because the invariant $x_{window} < x_{top}$ becomes false at the same time, the transition is forced to be taken, and the system moves into a state where the window does not move anymore ($v_{window} = 0$).

In case some bounce-back effect would be modeled because of the window colliding with the top of the frame, the continuous-time state v_{window} could be reinitialized with a value $-\eta v_{window}$, where η is a coefficient of restitution. This reinitialization can be included on the enabled transition as an action $v_{window} = -\eta v_{window}^{-}$, as shown in Figure 15.8(b). Note that the superscript is to indicate the left-hand limit value of the continuous-time behavior.

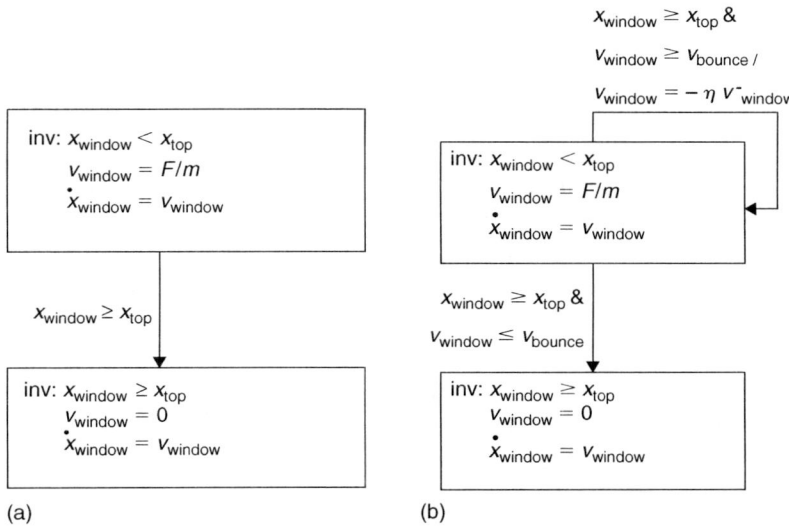

FIGURE 15.8 Hybrid automaton of power window endstop behavior. (a) without bounce back and (b) with bounce back.

15.3 Hybrid Dynamic System Behaviors

The different aspects of generating behaviors for hybrid systems are discussed and efficient methods to do so are presented.

15.3.1 An Operational Structure

To execute a hybrid dynamic system, three different types of behavior need to be handled (see Figure 15.9) (Mosterman and Biswas, 2002):

- Continuous-time behavior as specified by f needs to be generated.
- The crossing of thresholds of continuous-time variables needs to be detected and transitions between modes of behavior inferred, as specified by γ.
- The continuous-state variables need to be initialized and possibly reinitialized (think of the window that bounces back, reversing its velocity), as specified by g.

Each of these parts are reviewed here.

15.3.2 Continuous-Time Behavior

Continuous-time behavior is typically modeled by differential equations, either as ODEs or DAEs. To generate behaviors, a *numerical solver* such as differential/algebraic system solver (DASSL) (Petzold, 1982) can be used to integrate the time-derivative behavior.

Typically, the derivative with respect to time is computed at one or multiple points, after which a weighted mean is taken to infer the direction of behavior. In its most straightforward form, called a *forward Euler* integration scheme, the time derivative, $\dot{x}(t_k)$, at a time, t_k, is multiplied by the step in time, $h = t_{k+1} - t_k$, and the resultant is added to the current state, $x(t_k)$, to produce the state at t_{k+1}

$$x(t_{k+1}) = h\dot{x}(t_k) + x(t_k) \tag{15.8}$$

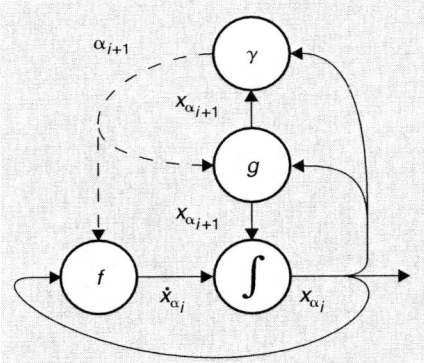

FIGURE 15.9 Functional model of hybrid dynamic systems.

More sophisticated solvers rely on continuity constraints on the continuous-time behavior. For example, a second-order Runge–Kutta algorithm is implemented by

$$m_1 = hf(x(t_k), t_k)$$
$$m_2 = hf(x(t_k) + 0.5m_1, t_k + 0.5h) \quad (15.9)$$
$$x(t_{k+1}) = x(t_k) + m_2$$

This scheme is based on a Taylor expansion

$$x(t_k + h) = x(t_k) + hf(t_k, x(t_k)) + \frac{h^2}{2}\left(\frac{\partial f(t_k, x(t_k))}{\partial t} + \frac{\partial f(t_k, x(t_k))}{\partial x} f(t_k, x(t_k))\right) + O(h^3) \quad (15.10)$$

and so it requires the integrated behavior to be continuous up to its third order for the error estimate to be valid. If this continuity constraint is not satisfied, the integration may still succeed, but without guaranteed error convergence behavior.

15.3.3 Handling Mode Transitions

Transitioning between two modes of continuous behavior requires: (i) detecting and locating the event that causes the mode change and (ii) inferring the new mode.

15.3.3.1 Event Detection and Location

The continuous-time behavior of a model typically affects the discrete-event part by events that are generated by continuous-time variables crossing threshold values. An example of this is when the power window in Figure 15.2 reaches the top of the door frame.

The exceeding of a threshold can be formulated as an inequality on 0 that either includes the boundary, $\gamma_{\alpha_i}^{\alpha_{i+1}}(x, u, t) > 0$, or not, $\gamma_{\alpha_i}^{\alpha_{i+1}}(x, u, t) \geq 0$. In general, multiple mode transition relations such as both a relation $\gamma_{\alpha_i}^{\alpha_{i+1}}(x, u, t) > 0$ as well as $\gamma_{\alpha_i}^{\alpha_{i+2}}(x, u, t) > 0$ may apply in any given mode. However, in case of deterministic models, only one of these should be active. Note that the behavior of the continuous-time states is best "left-closed". In other words, each of the abutting intervals of continuous behavior includes its starting point to satisfy causality requirements (Mosterman, 1999).

An implementation of the event generation requires two parts:

- First, it has to be *detected* whether the threshold has been exceeded when an integration step ΔT is about to be taken.

- Second, the values of the continuous states, input, and time for which the threshold was first exceeded have to be determined. In other words, the point in time at which the zero-crossing event occurs has to be *located*.

A robust implementation of the event location can be done by means of a bisectional search to find when the first event (in general there are multiple events) in the ΔT interval occurs. In this case, if an event is detected, the step size is reduced from ΔT to δt_m, where δt_m is computed based on whether an event occurs, $\sigma = 1$, or not, $\sigma = 0$, in the interval δt_i as follows

$$\delta t_{i+1} = \delta t_i + \Delta t_i (1 - \sigma)$$
$$\Delta t_{i+1} = \tfrac{1}{2} \Delta t_i \qquad (15.11)$$

The initial values for this iteration are $\delta t_0 = 0$ and $\Delta t_0 = \Delta T$, and the iteration terminates after a fixed number of a priori prescribed steps, m.

Other approaches to finding the point in time where the threshold is first exceeded such as *regular-falsi* and the *illinois* algorithm exist (Moler, 1997) and in practical simulation engines a combination of the different approaches tends to be employed.

Note that this requires the model not to change its mode until the zero crossing is located. For example, an absolute value computation may not be effected while the zero crossing is located. This implies that a negative value may indeed be computed to ensure that behavior is continuous and, in this case, smooth. In some cases, such a scheme may not work. For example, in case of a square root computation, a negative value of the argument is outside of the domain of the square root function (unless complex values are allowed).

15.3.3.2 Mode Transition Inferencing

The discrete events generated by the zero-crossing function may cause a mode change in the model. The relation $\gamma_{\alpha_i}^{\alpha_{i+1}}(x, u, t)$ transitions the model from a mode α_i to a mode α_{i+1}. It takes as arguments the continuous-time state vector, x, the exogenous input, u, and time, t. The mode change is typically implemented as an instantaneous transition, which means there is no passage of time. As such, it is best modeled by an untimed formalism.

The discrete-state transition behavior can be represented in two basic forms: (i) by *combinational* logic and (ii) by *sequential* logic. This is an important distinction as it makes quite a difference in the complexity of analyses of the hybrid dynamic systems.

Combinational Logic

An important approach that applies combinational logic is complementarity modeling, which has proven very successful in the domain of collision modeling (Lötstedt, 1981; Pfeiffer and Glocker, 1996) as well as, for example, power electronics (Kassakian et al., 1991). Complementarity formulations are further employed in the work on mixed logical dynamical (MLD) systems (Torrisi and Bemporad, 2004). A linear complementarity model is of the form

$$\begin{aligned} y &= Bz + b \\ yz &= 0 \\ y &\geq 0, \quad z \geq 0 \end{aligned} \qquad (15.12)$$

Only one of the variables y and z can be positive while the other has to be 0. This is an intuitive representation for points of contact in mechanical systems, where there is either some distance larger than 0 between two bodies and no force acting, or there is no distance and a force larger than 0 acting.

Sequential Logic

In case the logic contains memory, a combinational representation does not apply and sequential logic is needed. This holds true, for example, for certain models of sequences of collisions such as those found in Newton's Cradle (Mosterman, 2007).

Sequential logic can be represented by a state machine, ϕ, often with a finite number of states (Kohavi, 1978). Many graphical formalisms exist that are of a discrete state, sequential logic, nature. For example, there are state transition diagrams (Kohavi, 1978), statecharts (an example of these are given in Figure 15.4) (Harel, 1987), and Petri nets (David and Alla, 1992; Murata, 1989). Computationally, a finite state machine can be represented by a five tuple

$$\phi = \langle \alpha, \alpha_0, \sigma, \delta, \nu \rangle \qquad (15.13)$$

in which the state transition function, δ, changes the active state, α, in response to events, σ, while actions, ν, are generated. The initial state is given by α_0.

15.3.4 Reinitialization of State Variables

In response to a mode transition inferred by $\gamma_{\alpha_i}^{\alpha_{i+1}}$, the continuous-time state variables may be reinitialized, as governed by $g_{\alpha_i}^{\alpha_{i+1}}$. For example, in case of the power window bounce back, the window reverts its velocity upon impact with the frame and the corresponding state variable, ν_{window}, needs to be reinitialized from a positive to a negative value. An important implication of this is that the numerical solver may have to be reset. Sophisticated numerical solvers build up a history of time points and based on that history attempt to take larger steps in time to compute the next integration point. If an integrator state is reset, even if no mode transition occurs, this history becomes invalid and needs to be cleared. In this case, the integrator starts off with a minimal step size once continuous-time behavior resumes.

Note that to specify the reinitialization, semantics need to be defined for the two values around a discontinuity, the *a priori* and *a posteriori* values. In this work, if necessary, the *a priori* values are indicated by a "−" superscript, and the *a posteriori* values by a "+" superscript (see Eq. [15.6], for an example).

Finally, the number of continuous-time state variables may change between mode transitions. For example, while modeling a highway, vehicles may enter and leave, and, therefore, continuous-time states would be included or discarded. This, again, will require a reset of the numerical solver, depending on the integration algorithm that is being used.

15.4 An Implementation

To generate behaviors for a hybrid dynamic system, two basic approaches exist: (i) time-driven execution and (ii) event-driven execution. The former has the execution driven by moving time forward, often by means of a numerical solver. The latter jumps in time in response to discrete events.

15.4.1 Classes of Events

For purposes of discussion, it is convenient to first identify two classes of events (Cellier, 1979): (i) *time events* and (ii) *state events*. A time event is an event that occurs at a given point in time, independent of the continuous-time state of the model, x, and the forcing function, u. Therefore, a time event is predictable. A state event, however, is generated based on the values of the continuous-time state and the forcing function.

15.4.2 Classes of Temporal Behavior

Depending on the particular type of behavior that is generated, one or the other may be more efficient. For the purposes of execution analysis, three categories of behavior over time can be distinguished, as illustrated in Figure 15.10. These behaviors correspond to those shown in the power window example in Figure 15.3.

In Figure 15.10(a), a behavior is shown that evolves continuously in time. This evolution is typically modeled by differential equations and the traces are generated using numerical solvers. In Figure 15.10(a), there is a time event that occurs at the point in time that is marked by the dashed vertical. A state event is generated at the point in time when the continuous-time behavior, the solid line, exceeds the dashed horizontal. When the threshold is exceeded, it is backtracked to the earliest point in time where this occurred, indicated by the dashed arrow. The events are indicated by vertical solid arrows.

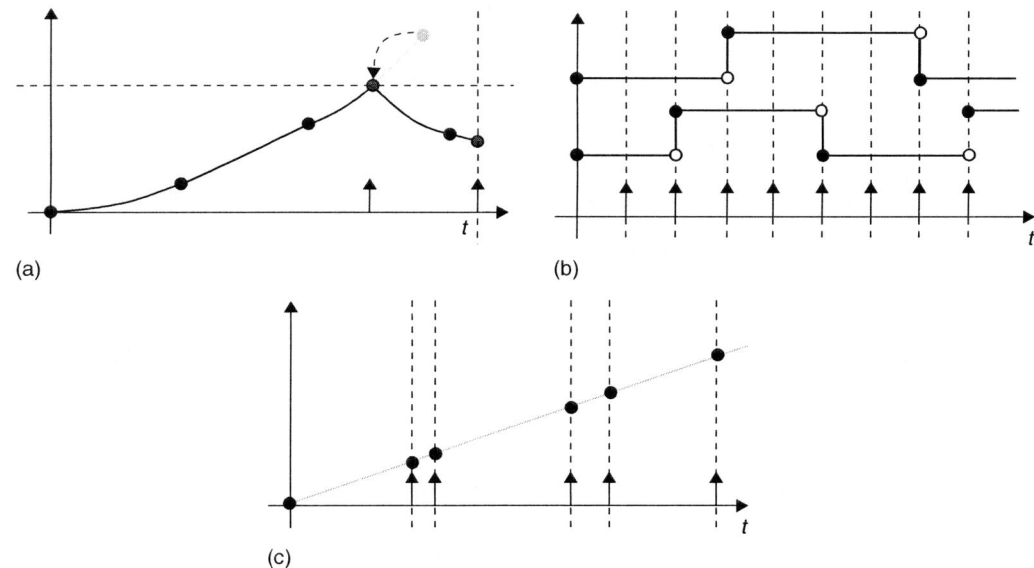

FIGURE 15.10 Different types of execution. (a) Continuous time (b) discrete time and (c) discrete event.

In Figure 15.10(b), two signals are shown: (i) the top signal has a period of 4 (ms) with an offset of 3 (ms) and (ii) the bottom signal has a period of 3 (ms) and an offset of 2 (ms). Such behaviors only contain time events that are of a periodic nature. This is typical in the design of an embedded controller, where the system operates at a fixed sample time, the *base rate*, and the different aspects of the control may execute at the fast base rate or have a sample time that is an integer multiple of the base rate. For the signals in Figure 15.10(b), the base rate is 1 (ms). Synchronous languages such as Lustre (Halbwachs et al., 1991) have been specifically designed to handle this class of systems.

In Figure 15.10(c), a behavior with only time events is shown, but the events are not periodic. Such behaviors are typical in network modeling, where the time in which a packet arrives is variable as is the time it takes to move the packet through a network. Another application is the modeling of the task scheduling on a microprocessor network, where events occur when a task is completed. This time is typically variable and the execution may even be preempted by higher priority tasks, resulting in further variability of the execution time.

15.4.3 Time-Driven Execution

In a time-driven execution, a numerical solver is applied that governs the advance of time. The numerical solver is typically provided with a set of differential equations, a start time, a set of initial states, a set of input values, and a stop time. It then attempts to solve the differential equations to generate a trace of numerical values for the states from the start time to the stop time. This trace may consist of any number of steps and corresponding integration points.

The trace is generated by the numerical solver computing the size of a step in time to take, based on the time gradients of the differential equations, as, for example, in Eq. (15.9). If there is a steep gradient because of the fast changes in the state values, small steps are taken, whereas when the behavior changes relatively slowly, larger steps are taken.

The stop time is either taken to be the end time of the requested simulation, or, in case there are time events, the first of such an event is provided as the stop time. If only time events are present, and they are statically known, then a schedule for execution can be precompiled. This is the case if all of the events are periodic in nature. The greatest common denominator of all the different sample times is determined and each of the events are executed at an integer multiple of this base rate.

TABLE 15.1 Event Calendar

Time (ms)	Event
20	*open_tonneau*
2020	*move_top_up_cmd*
2150	*move_down_window*
2250	*stop_moving_window*

For aperiodic events, this approach may not be applicable, as the accuracy with which the time at which events can be effected degenerates to the period of the base rate, which is typically too coarse.

In case state events are present, a numerical solver with zero-crossing detection built-in is desirable. Such a numerical solver returns before the final time is reached and reports the time at which a zero crossing was found.

15.4.4 Event-Driven Execution

Another approach to generating behavior takes an event-driven perspective. This is particularly efficient for time events that are aperiodic and often have a stochastic component to their time of occurrence.

Rather than having a numerical solver move time forward until each of the events occurs, time immediately jumps from one event time to another. To efficiently implement this, typically an event calendar is used to keep track of all events that are scheduled to occur. An example of an event calendar is presented in Table 15.1. It shows how the first upcoming event to occur is at 20 (ms), followed by another at 2020 (ms).

An event-driven approach is much more efficient in handling aperiodic time events as it does not require a time-driven solver to move time forward by means of integration over time. Typically, hundreds of thousands of events can be conveniently simulated in a matter of seconds.

To achieve such efficiency, it is critical that the event calendar be implemented in an efficient manner. In particular, efficient search of the event calendar needs to be facilitated, because new events must be inserted in the correct place, and events that were scheduled at one time may have to be located and retracted at a later time.

One implementation, called *calendar queue*, mimics a traditional calendar [3], which uses an array to index into a number of bins while the content of each bin is implemented as a doubly linked list. In this list, each item has a link to the next item as well as the previous item, which allows each item in the list to be accessed in a forward as well as backward manner. Deleting any item can then be done in constant time. The array that implements the number of bins has constant access time to each of the bins.

This is illustrated in Figure 15.11 where four bins are represented by the array at the top. The bin with index 0 contains all elements with a time stamp in the range from 0 to 1, with 1 not included; the bin with index 1 contains all elements with a time stamp in the range from 1 to 2, with 2 not included, etc. Events with a time stamp of 4 or more are distributed based on the result of the time stamp modulo 4.

The calendar queue combines the benefit of quick insertion of events in the doubly linked list where the array is more efficient for indexing a large event calendar. Further permutations of different data structures such as arrays, lists, and heaps can be devised depending on the characteristics of the event distributions particular to a problem under study. For example, if the number of scheduled events varies greatly during simulation, dynamically reconfiguring the data structure may become important.

Note that an event-driven implementation can be exploited to generate behaviors for continuous-time models as well. In this approach, the numerical solver is modeled as a discrete-event component and the computations at each time step are aperiodic events to be handled (e.g., D'Abren and Wainer, 2003).

15.4.5 Combining the Execution Types

In many applications of modeling and simulation, in particular for hybrid dynamic systems, it is common to have an extensive model component that is of an aperiodic discrete-event nature as well as a component that is of a continuous-time nature. For example, the power window system in Figure 15.3 may contain

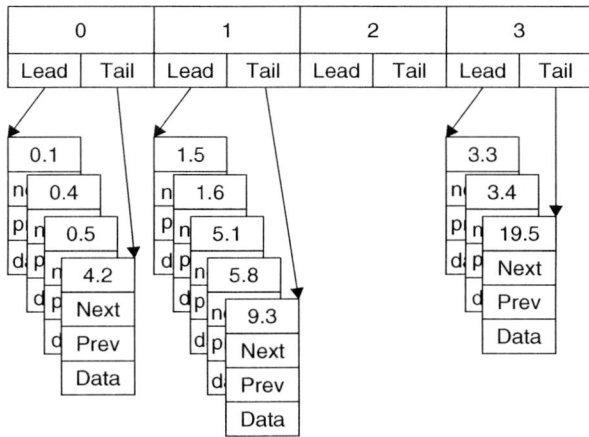

FIGURE 15.11 Calendar queue event list.

a detailed continuous-time model of the door with window, dc motor actuator, signal conditioning hardware, and the like, all of which may contain some discontinuities modeled by local finite state machines. Similarly, the controller may operate in a discrete-time manner because it executes with a given sample rate. The controller may implement an extensive signal processing component, data analysis computations, supervisory control, and further elements. The CAN behavior may be modeled in detail by a discrete-event model that performs behaviors such as capturing the generation of packets, routing them through the network, and possibly resending dropped packets.

To efficiently execute models that comprise all of these behaviors, it is desirable to support: (i) a numerical solver for continuous-time behavior, (ii) a scheduler for the discrete-time behavior, and (iii) an event calendar handler for the discrete-event behavior.

Because of the predictability of the discrete-time events and the fixed base rate, discrete-time behavior as handled by a scheduler can often be conveniently integrated into the numerical solver to comprise a time-driven execution engine. A *getNextEventTime* call is typically exploited to tell the solver the time until which it should solve for the continuous-time behavior.

The discrete-event part is a different matter, though. Because there is no fixed base rate, events may occur arbitrarily close in time. Halting the numerical solver at each of these points in time quickly becomes very inefficient. So an event-driven execution engine is preferred.

Rather than *integrating* the time-driven execution engine that uses a numerical solver and event schedule with the event-driven execution engine that relies on an event calendar, it is more efficient to *combine* the two different execution engines. When combining the two, it is important to first observe that the event-driven engine may process many events that have no bearing on the time-driven behavior. The time- and event-driven execution engine can then compute behavior independently until an event is generated that requires interaction between the two. This requires coordination between the time- and event-driven execution to ensure the two are synchronized upon communication (Nicolescu et al., 2006).

Two possibilities to implement this are illustrated in Figure 15.12. In Figure 15.12(a), the event-driven behavior shown at the bottom leads the time-driven behavior shown at the top. Behavior generation starts with processing the events (indicated as vertical arrows) that are registered on the event calendar. When the first event that has a bearing on the time-driven behavior is processed, a *time event* is set and the numerical solver starts integrating up to that point in time, shown at the top. A complication arises when the continuous-time behavior exceeds a threshold that causes a *state event* to occur that has a bearing on the event-driven behavior located in time before the time event that was set. This state event then occurs at a point in time that the event-driven execution has passed already. So, events that were already processed have to be retracted and the state of the event-driven behavior has to roll back.

The other possibility is illustrated in Figure 15.12(b), where the time-driven behavior leads the event-driven behavior. In this case, the state event is generated and registered on the event calendar before the event-driven behavior reaches this point in time and execution control is handed over to the event-driven behavior to catch up. Now, it may happen that the event-driven behavior sets a time event in the time-driven execution at a point in time before the state event occurred. This event may cause a change in continuous-time behavior, and the threshold that caused the state event may never be reached. So, the time-driven execution engine needs to roll back.

Efficient implementations can be developed based on, for example, employing an interpolation polynomial in the time-driven part to quickly obtain the previous continuous-time state values, so the numerical integration does not have to be restarted anew, which is an expensive proposition.

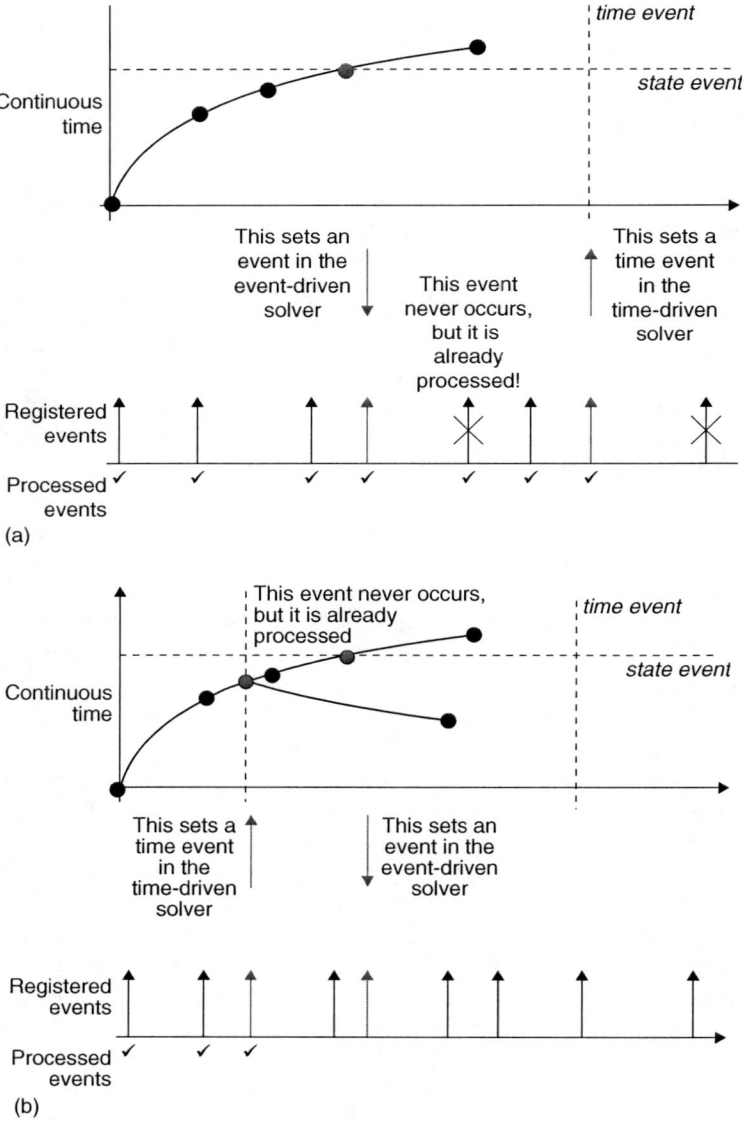

FIGURE 15.12 Combining event-driven and time-driven execution. (a) Event-driven leads and (b) time-driven leads.

Hybrid Dynamic Systems: Modeling and Execution

15.5 Advanced Topics in Hybrid Dynamic System Simulation

To build on the understanding of the basic elements of a hybrid system, some of the complications that arise when implementing and combining these elements are explored.

15.5.1 Zero-Crossing Detection

The typical approach to zero-crossing detection compares the sign of a function result and if it changes, it has crossed zero. This approach may fail if the zero-crossing function has an even number of zeros in the interval ΔT between the two evaluated points as determined by the numerical integration algorithm. This is illustrated in Figure 15.13(a). In general, the zero-crossing function, z, is a function of the model state, but it does not contribute to its continuous dynamics, f. Therefore, numerical integration can proceed without taking the dynamics of z into account, and when these are faster than the dynamics of f, the situation with even zeros may arise.

One solution to the even zeros problem is to include the dynamics of z in the model dynamics so the numerical solver adjusts its step size when too large an error in the zero-crossing function dynamics is found (Park and Barton, 1996). This does not, however, address the problem of moving outside of the domain of the zero-crossing function, such as outlined for the square root computation in Section 15.3.3.

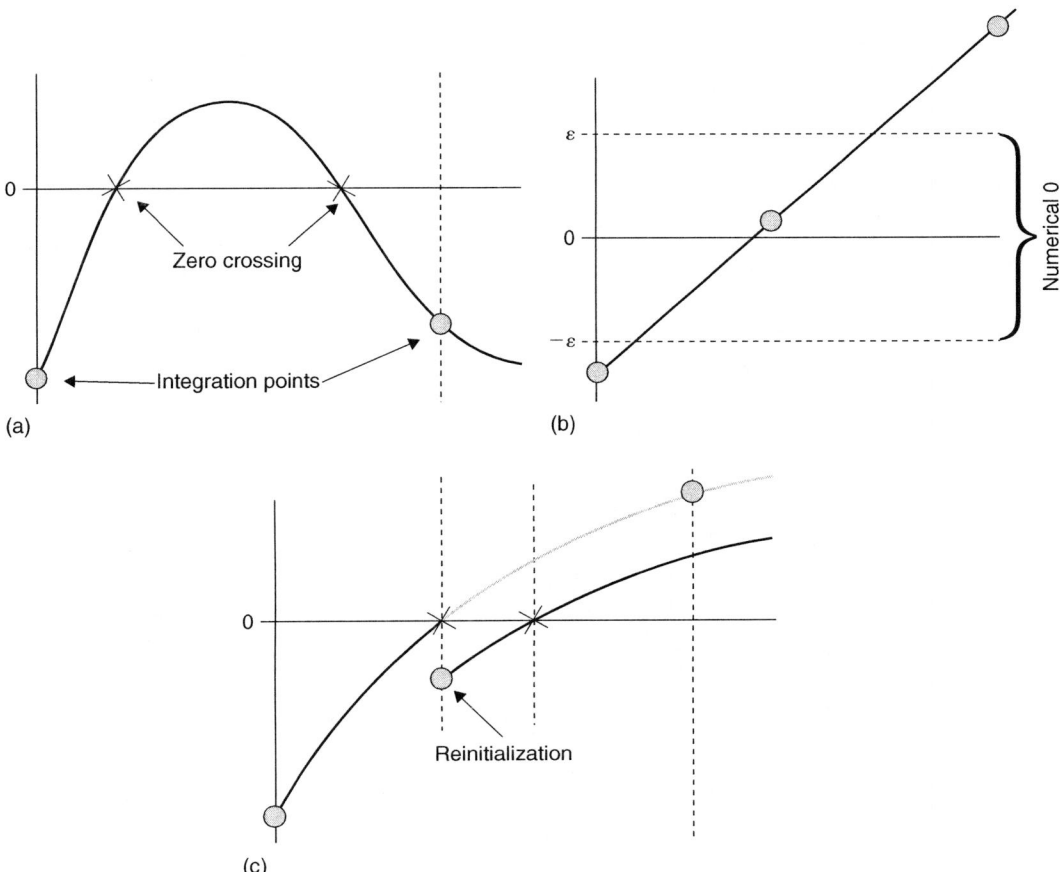

FIGURE 15.13 Difficulty in zero-crossing detection and location. (a) Even roots, (b) comparing a float, and (c) discontinuity sticking.

To avoid moving outside of a function domain, an alternate approach to event detection relies on a feedback-control concept. If k is the index of the points in time at which the numerical solver computes a value, t_k, then, if k is considered dense rather than discrete, the step size between consecutive points, h, becomes $h(k) = \frac{dt}{dk}$. This step size can be chosen as

$$h(k) = -\eta \frac{z(x)}{\frac{\partial z}{\partial x} f} \qquad (15.14)$$

to control the step size selection (Esposito et al., 2001). Given the sensitivity of the zero-crossing function, z, with respect to the step size

$$\frac{dz}{dk} = \left(\frac{\partial z}{\partial x} \frac{dx}{dt}\right) \frac{dt}{dk} = \left(\frac{\partial z}{\partial x} f\right) h(k) \qquad (15.15)$$

this results in z behaving as

$$\frac{dz}{dk} = -\eta z \qquad (15.16)$$

which gradually converges to 0 without actually crossing it. Instead, the solution to Eq. (15.16) is $z(k) = z(0)e^{-\eta k}$, and $z(k)$ approaches 0 exponentially as k tends to infinity. So, the function domain is never departed. However, this requires a number of additional computations during numerical integration such as the sensitivity, and, therefore, is computationally less attractive.

Other than exceeding the threshold value, another issue with zero-crossing detection arises when the zero-crossing function z returns 0 exactly. This does not constitute a *crossing*, and, therefore, is not detected as such. Root-finding facilities of numerical solvers may require the function z to actually change sign between two integration steps to report that a crossing has occurred. For example, when z starts off at 0 and then moves away from it, no zero-crossing event would be generated and consequently no mode change could occur.

One solution is the use of a zero-crossing function that is not at 0 exactly. Rather, it is chosen to be $-\epsilon$ when z is negative and ϵ when z is positive, with ϵ very small. When $-\epsilon < z < \epsilon$, both zero-crossing functions, $z - \epsilon$ and $z + \epsilon$, are used.

One effect of this implementation is the need for a "*numerical 0*"; a $\pm\epsilon$ band around 0 in which the zero-crossing function is considered to be the 0 value of the sign function.

$$\text{sign}(x) = \begin{cases} - & \forall_x (x < -\epsilon) \\ 0 & \forall_x (-\epsilon \leq x \leq \epsilon) \\ + & \forall_x (x > \epsilon) \end{cases} \qquad (15.17)$$

As long as the zero-crossing function evaluates to a value within the $\pm\epsilon$ band, its sign is considered to be 0.

This approach has an important implication for the analytical correctness of comparing for equality. If a simulation contains the comparison with zero, $x = 0$, when x is a continuous signal, a zero-crossing function $z(x, u, t)$ is used to find the value of x that satisfies this equality within a certain tolerance. For this value, x', the zero-crossing function has sign 0, whereas the strict equality $x = 0$ may not be satisfied. Figure 15.13(b) shows how the signal x may therefore cross 0 without having $x = 0$ evaluate to true because the analytical solution that satisfies this comparison is never evaluated by the numerical integration.

Another issue that is less critical but still causes inefficiencies to occur because of recomputation of the model variables after the zero crossing has been located. Because of numerical inaccuracies, even if no changes to the states are made that are used to compute a zero-crossing variable, the return value of the function may still differ. This is illustrated in Figure 15.13(c), which shows that restarting the simulation results in a function return value that is slightly different from the computed value immediately before the zero crossing. Continuing simulation leads to the same zero crossing being detected and located again. This phenomenon has been referred to as *discontinuity sticking* (Park and Barton, 1996).

15.5.2 Mode Changes

When a change between modes occurs, the continuous-time state variables may have to be reinitialized and an immediate consecutive mode change may occur as a result.

15.5.2.1 Reinitialization

An important phenomenon of the general DAE form $0 = f_\alpha(\dot{x}, x, u, t)$ is that the system may only be allowed to move in part of the *generalized* state space (Verghese et al., 1981). This is the case, for example, when the power window in Figure 15.2 collides with an object, m_{object}, in a perfectly nonelastic manner. After the collision, the window and object proceed to move with equal velocity, and this leads to the equations

$$m_{object}\dot{v}_{object} + m_{window}\dot{v}_{window} + R_{lift}v_{window} = ru_{motor}$$
$$v_{object} = v_{window}$$
$$\dot{x}_{window} = v_{window} \quad (15.18)$$
$$\dot{x}_{object} = v_{object}$$

to describe the dynamic behavior. In the form of a *matrix pencil* (Demmel and Kägström, 1986) with a forcing function, $E\dot{x} + Ax + Bu = 0$, this becomes

$$\begin{bmatrix} m_{window} & m_{object} & 0 & 0 \\ 0 & 0 & 0 & 0 \\ 0 & 0 & 1 & 0 \\ 0 & 0 & 0 & 1 \end{bmatrix} \begin{bmatrix} \dot{v}_{window} \\ \dot{v}_{object} \\ \dot{x}_{window} \\ \dot{x}_{object} \end{bmatrix} + \begin{bmatrix} R_{lift} & 0 & 0 & 0 \\ -1 & 1 & 0 & 0 \\ -1 & 0 & 0 & 0 \\ 0 & -1 & 0 & 0 \end{bmatrix} \begin{bmatrix} v_{window} \\ v_{object} \\ x_{window} \\ x_{object} \end{bmatrix} + \begin{bmatrix} -r \\ 0 \\ 0 \\ 0 \end{bmatrix} [u_{motor}] = 0 \quad (15.19)$$

Here, because of the constraint that $v_{window} = v_{object}$, only the part of the state space for which this constraint holds can be accessed. This implies that there is also a limitation on the reachability in the x_{window}, x_{object} space, but this is a nonholonomic constraint rather than it being disallowed.

In a general sense, the reduced state space can be represented in geometric terms as shown in Figure 15.14. Here, the system evolves in a mode α_1 until the boundary of a patch is reached. At this point in time, the system transitions into another mode, but the continuous-time state is not in the allowed space, which is marked by the thick solid line. To arrive at a consistent situation, the continuous-time state has to be in the allowed space, and the exact value is computed based on the *jump space*, which is the space in which the required instantaneous changes are allowed.

In the linear case, this computation in the jump space corresponds to a projection. This projection can be computed in several ways (Gantmacher, 1965; Griepentrog and März, 1986; Lewis, 1992; Mosterman, 2000, 2002; van der Schaft and Schumacher, 1996; Verghese et al., 1981). To illustrate the method in Mosterman (2001), the Weierstrass normal form is derived for the velocity part of the equations in Eq. (15.19). The

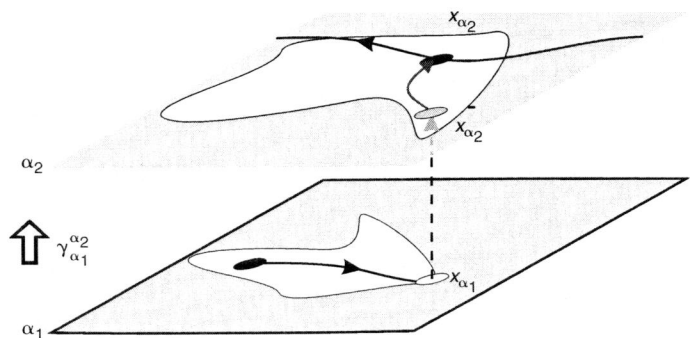

FIGURE 15.14 A projection. (From Mosterman, P. J. Mode transition behavior in hybrid dynamic systems. In *Proceedings of the 2003 Winter Simulation Conference*, pp. 623–631, New Orleans, LA, December 2003 (invited paper).)

positions are geometric states that do not change discontinuously, and, therefore, are irrelevant for the example. This leads to the system of equations

$$\begin{bmatrix} m_{\text{window}} & m_{\text{object}} \\ 0 & 0 \end{bmatrix} \begin{bmatrix} \dot{v}_{\text{window}} \\ \dot{v}_{\text{object}} \end{bmatrix} + \begin{bmatrix} R_{\text{lift}} & 0 \\ -1 & 1 \end{bmatrix} \begin{bmatrix} v_{\text{window}} \\ v_{\text{object}} \end{bmatrix} + \begin{bmatrix} -r \\ 0 \end{bmatrix} [u_{\text{motor}}] = 0 \quad (15.20)$$

After applying the following change of basis:

$$\begin{bmatrix} v_{\text{window}} \\ v_{\text{object}} \end{bmatrix} = \begin{bmatrix} 1 & -\frac{m_{\text{object}}}{m_{\text{window}}} \\ 0 & 1 \end{bmatrix} \begin{bmatrix} 1 & 0 \\ -\frac{m_{\text{window}}}{m_{\text{window}} + m_{\text{object}}} & 1 \end{bmatrix} \begin{bmatrix} \bar{v}_{\text{window}} \\ \bar{v}_{\text{object}} \end{bmatrix} \quad (15.21)$$

the following matrix pencil is arrived at:

$$\begin{bmatrix} m_{\text{window}} & 0 \\ 0 & 0 \end{bmatrix} \begin{bmatrix} \dot{\bar{v}}_{\text{window}} \\ \dot{\bar{v}}_{\text{object}} \end{bmatrix} + \begin{bmatrix} 0 & \frac{m_{\text{window}}}{m_{\text{window}} + m_{\text{object}}} R_{\text{lift}} \\ 0 & \frac{m_{\text{window}} + m_{\text{object}}}{m_{\text{window}}} \end{bmatrix} \begin{bmatrix} \bar{v}_{\text{window}} \\ \bar{v}_{\text{object}} \end{bmatrix} + \begin{bmatrix} -r \\ 0 \end{bmatrix} [u_{\text{motor}}] = 0 \quad (15.22)$$

The matrix $\begin{bmatrix} m_{\text{window}} & 0 \\ 0 & 0 \end{bmatrix}$ contains a finite space (top-left entry) $[m_{\text{window}}]$ and infinite space (bottom-right entry) $[0]$. Note that this is an *index 1* system of equations, as the infinite space is the null matrix. Therefore, its nilpotency is 1, and this is also referred to as the *index* of the system of equations (van Dijk, 1994). The nilpotency is an indicator of how many stages of substitution are required to compute all infinite variables. Eq. (15.22) is considered to be of index 1 because the infinite variable, $\bar{v}_{\text{object}}$, can be computed in one stage, i.e., $\bar{v}_{\text{object}} = 0$.

The finite part in Eq. (15.22) consists of $\bar{v}_{\text{window}}$ and this can be converted into a regular one-dimensional ODE by inverting the matrix $[m_{\text{window}}]$ and left-multiplying. Because it is a regular ODE, there is no discontinuous change in the variable $\bar{v}_{\text{window}}$, or

$$\bar{v}_{\text{window}} = \bar{v}_{\text{window}}^{-} \quad (15.23)$$

where the "−" superscript indicates the final value of $\bar{v}_{\text{window}}$ before the mode change (in case of the initialization before simulation starts, it is the user-supplied initial value).

Now, from v_{window}^{-} and v_{object}^{-}, the value $\bar{v}_{\text{window}}^{-}$ can be computed using the inverse change of basis. Straightforward computations yield

$$\bar{v}_{\text{window}}^{-} = v_{\text{window}}^{-} + \frac{m_{\text{object}}}{m_{\text{window}}} v_{\text{object}}^{-} \quad (15.24)$$

which equals $\bar{v}_{\text{window}}$. With $\bar{v}_{\text{object}} = 0$ the change of basis can now be applied to yield

$$v_{\text{window}} = \frac{1}{m_{\text{window}} + m_{\text{object}}} (m_{\text{window}} v_{\text{window}}^{-} + m_{\text{object}} v_{\text{object}}^{-}) \quad (15.25)$$

and from $v_{\text{window}} = v_{\text{object}}$, v_{object} can be computed. Details on the derivation are available in previous work (Mosterman, 2000).

15.5.2.2 Sequences of Mode Changes

After one mode change, $\gamma_{\alpha_i}^{\alpha_{i+1}}$, and computing the initial values of the continuous-state variables in the new mode, a new transition, $\gamma_{\alpha_{i+1}}^{\alpha_{i+2}}$, may follow immediately as shown in Figure 15.9 (Mosterman and Biswas, 1996). The mode change causes a new mode to be arrived at, after which reinitialization is performed again. This process repeats until no further mode changes occur and the system proceeds to evolve continuously again.

The values of the continuous-time state in modes where there is no continuous-time behavior can be of two types (Mosterman and Biswas, 1996):

- When the value does not change from the previous mode, and the mode is a so-called *mythical mode* (Figure 15.15(a)).
- When the reinitialization causes a change in value from the previous mode, which results in an isolated point, a so-called *pinnacle*, with no continuous behavior in that mode (Figure 15.15(b)).

In case of sequences of mode transitions, the left-closedness mentioned in Section 15.3.3 may (have to) be relaxed, though, in particular for sequences of pinnacles. For example, a pressure relief valve may move through a sequence of opening and closing cycles before the pressure has subsided to below the threshold for opening the relief valve. In a sufficiently detailed model, this sequence occurs over time, but when small

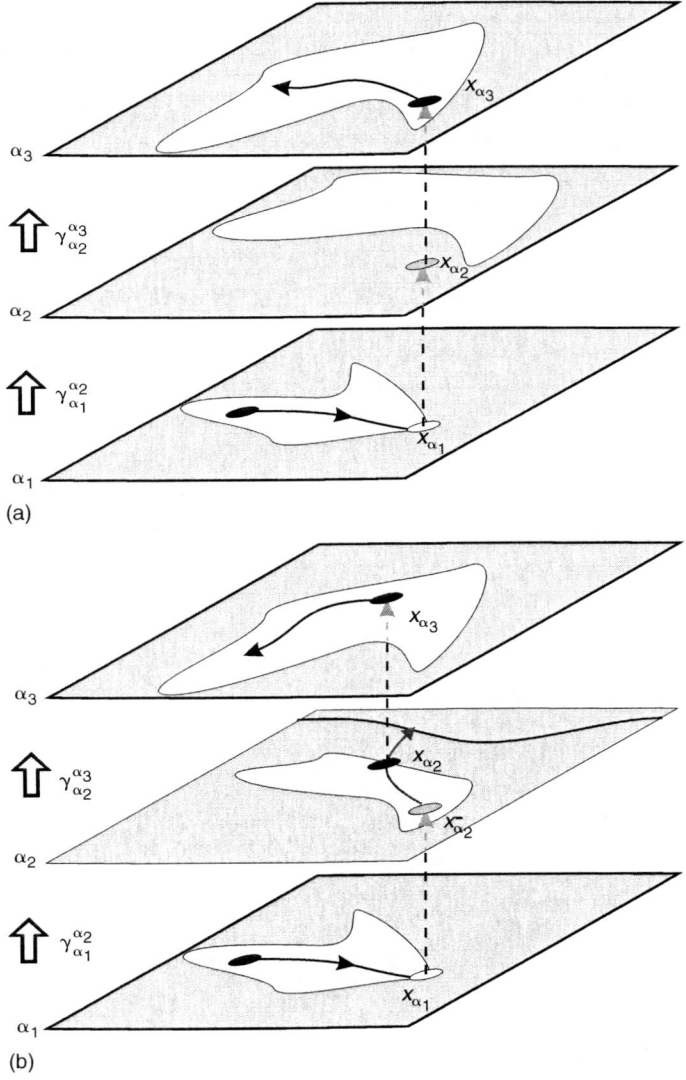

FIGURE 15.15 Sequences of mode transitions. (a) State invariance and (b) state reinitialization. (From Mosterman, P. J. Mode transition behavior in hybrid dynamic systems. In *Proceedings of the 2003 Winter Simulation Conference*, pp. 623–631, New Orleans, LA, December 2003 (invited paper).)

physical phenomena are not included in the model, for example, to simulate it in real time, this sequence may occur at one point in time (Mosterman, 2002). Note that this requires the "state" of the system to include an additional dimension to allow multiple values at one point in time. This can be done by using pairs that consist of the continuous-time state and an index (Guckenheimer and Johnson, 1995).

An important behavior that is not properly dealt with in simulation tools at present is the crossing of the patch boundary in an intermediate mode such as in mode α_2 in Figure 15.15(b). The proper value of the continuous state to be applied for initialization in mode α_3 appears to be the point at which the projection crosses the patch boundary. However, there may be physical phenomena that are best modeled with a different semantics. This is still a subject of research.

15.6 Pathological Behavior Classes

Once sequences of mode changes occur, models can be constructed that contain loops of mode changes, i.e., a previously visited mode is revisited, without continuous-time behavior evolving in between.

Two classes of behavior are illustrated in Figure 15.16. In Figure 15.16(a), the pathological case is shown that violates the *divergence of time* principle (Mosterman and Biswas, 1998). Here, the state is initialized inside of the patch in mode α_1. It evolves continuously until it reaches the patch boundary as defined by $\gamma_{\alpha_1}^{\alpha_2}$. When the state x_{α_1} is then transferred to mode α_2, it is outside of the patch as defined by $\gamma_{\alpha_2}^{\alpha_1}$ (note the exchange in subscripts of α). This causes the state to be transferred back to α_1 where it is outside of the patch as defined by $\gamma_{\alpha_1}^{\alpha_2}$. Thus, a loop of discrete changes between modes arises.[2] Because these are instantaneous, no time elapses, and, therefore, the model stops evolving in time. In other words, time does not diverge. Since this behavior is not observed in physical systems, such behavior is considered the result of anomalous models of physics.

Similar but different behavior is illustrated in Figure 15.16(b). Here, after reaching the patch boundary in α_1, the state transfers onto the patch boundary in α_2 as defined by $\gamma_{\alpha_2}^{\alpha_1}$ (note again the exchange in subscripts of α). Because it is the patch boundary, the state transfers back to α_1 after an infinitesimal step in time. This step results in a value x_{α_1} that may be immediately inside the patch in α_1 as defined by $\gamma_{\alpha_1}^{\alpha_2}$ and so another infinitesimal step will transfer the state back to α_2.[3]

Far-fetched and pathological as it may seem, this behavior, referred to as *chattering* or *sliding mode* behavior, is actually aimed for by robust control design methodologies (Utkin, 1992) (e.g., it is used in antilock braking systems), as it is relatively insensitive to plant model parameter variations. Unlike the behavior in Figure 15.16(a), here the state does continue to evolve in time and the divergence of time principle is satisfied. To efficiently derive the actual behavior along the *switching surface* as defined by the patches in mode α_1 and α_2, two methods can be applied: (i) equivalence of control (Utkin, 1992) and (ii) equivalence of dynamics (Filippov, 1960; Mosterman et al., 1999). Although there are classes of models for which these "regularizations" result in the same behavior, in general they may differ.

Finally, another class of pathological behaviors can be identified, namely Zeno behavior.[4] Behaviors that are Zeno do progress in time by a noninfinitesimal value each time a mode transition occurs. However, this time reduces upon each transition as a converging series. For example, in case the time is halved upon each transition, the transition series converges to a limit value in time

$$t_f = \Sigma_i \frac{1}{2^i} \qquad (15.26)$$

that is never exceeded. In case the bounce-back of the window in the hybrid automaton in Figure 15.8(b) does not include the threshold clause, the bounce transition would be taken indefinitely, with shorter

---

[2] Note that a loop may involve any finite number of modes.
[3] Note how left-closedness is violated in this particular instance of behavior. In general, an infinitesimal "hysteresis" effect may be present to guarantee left-closedness again.
[4] Named after the Greek philosopher Zeno who studied the relation between points and intervals, i.e., whether an interval is an infinite collection of points.

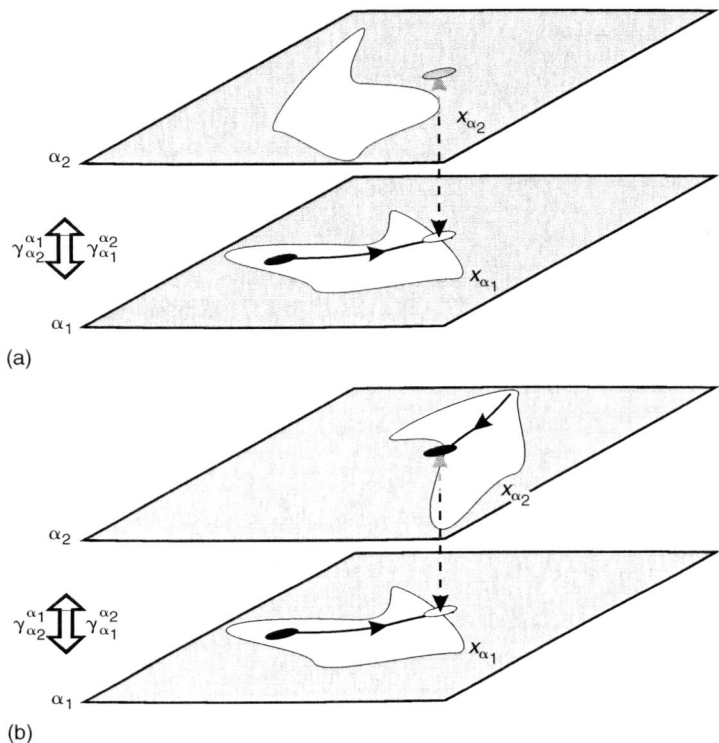

FIGURE 15.16 Recurring mode transitions. (a) Nondivergent and (b) sliding. (From Mosterman, P. J. Mode transition behavior in hybrid dynamic systems. In *Proceedings of the 2003 Winter Simulation Conference*, pp. 623–631, New Orleans, LA December 2003 (invited paper).)

and shorter intervals of time in between. Depending on the coefficient of restitution, η, the new interval between bounces would be a fraction of the present one. The increasingly smaller intervals would converge to a limit point in time beyond which time would not progress. Therefore, although time diverges locally, it does not do so globally.

It is now possible to compare the three mode revisiting behaviors.

- *Divergence of time*: infinitely many discrete steps in zero time. Time remains the same.
- *Chattering*: infinitely small time steps. Evolves past any value in time.
- *Zeno*: infinitely many time steps in a finite, nonzero, time interval. Does not evolve past a limit point in time.

Note that, although it is important to clearly distinguish between these three behaviors, often *Zeno* is used as an all-encompassing term that includes each of these (e.g., behavior that is locally not divergent in time is often called Zeno as well).

15.7 Conclusions

This chapter presented an overview of the field of numerical simulation for hybrid dynamic systems. It discussed the basic parts of continuous-time behavior generation, event detection and location, mode transitions, and reinitialization that are needed and how they combine together.

The implementation was shown to be potentially based on a numerical integrator scheme, use of a statically compiled schedule, and use of a dynamically kept event calendar. In some cases, it may be desirable to combine the different forms of implementation for efficiency and ways to achieve this were discussed.

The richness and complexity of mode transition behavior in hybrid dynamic systems was shown by presenting a number of mode transition behavior classes that require special attention. Pathological behaviors were classified.

Acknowledgments

The author wishes to acknowledge extensive discussions with Gautam Biswas and Feng Zhao on the topic.

MATLAB, Simulink, Stateflow, Handle Graphics, Real-Time Workshop, and xPC TargetBox are registered trademarks and SimBiology, SimEvents, and SimHydraulics are trademarks of The MathWorks Inc. Other product or brand names are trademarks or registered trademarks of their respective holders.

References

Alur, R., C. Courcoubetis, T. A. Henzinger, and P. Ho. Hybrid automata: An algorithmic approach to the specification and verification of hybrid systems. In R. L. Grossman, A. Nerode, A. P. Ravn, and H. Rischel, editors, *Lecture Notes in Computer Science*, volume 736, pp. 209–229. Springer, Berlin, 1993.

Brown, R. Calendar queues: A fast 0(1) priority queue implementation for the simulation event set problem. *Communications of the ACM*, 31(10): 1220–1227, 1988.

CAN specification. Technical Report, 1991. Robert Bosch GmbH.

Cellier, F. E. *Combined Continuous/Discrete System Simulation by Use of Digital Computers: Techniques and Tools*. PhD dissertation, ETH, Zurich, Switzerland, 1979.

Control System Toolbox. *Control System Toolbox User's Guide*. The MathWorks, Natick, MA, 2004.

D'Abreu, M. C. and G. A. Wainer. Defining hybrid system models using DEVS quantization techniques. In *Proceedings of the Winter Simulation Conference*, New Orleans, LA, December 2003.

David, R. and H. Alla. *Petri Nets & Grafcet*. Prentice-Hall, Englewood Cliffs, NJ, 1992.

Demmel, J. and B. Kågström. Stably computing the Kronecker structure and reducing subspaces of singular pencils $A - \lambda B$ for uncertain data. In J. Cullum and R. A. Willoughby, editors, *Large Scale Eigenvalue Problems*. Elsevier Science, 1986.

Esposito, J. M., V. Kumar, and G. J. Pappas. Accurate event detection for simulating hybrid systems. In M. D. Di Benedetto and A. Sangiovanni-Vincentelli, editors, *Hybrid Systems: Computation and Control*, Lecture Notes in Computer Science, pp. 204–217. Springer, Berlin, 2001.

Filippov, A. F. Differential equations with discontinuous right-hand sides. *Mathematicheskii Sbornik*, 51(1): 48–122, 1960.

Friedman, J. and J. Ghidella. Using model-based design for automotive systems engineering—Requirements analysis of the power window example. In *Proceedings of the SAE 2006 World Congress & Exhibition*, CD-ROM, Detroit, MI, April 2006.

Gantmacher, F. R. *Matrizenrechnung: Teil I Allgemeine Theorie*. Deutscher Verlag der Wissenschaften, Berlin, 1965.

Griepentrog, E. and R. März. *Differential-Algebraic Equations and Their Numerical Treatment*. BSB Teubner, Leipzig, 1986.

Guckenheimer, J. and S. Johnson. Planar hybrid systems. In P. Antsaklis, W. Kohn, A. Nerode, and S. Sastry, editors, *Hybrid Systems II*, Lecture Notes in Computer Science, volume 999, pp. 202–225. Springer, Berlin, 1995.

Halbwachs, N., P. Caspi, P. Raymond, and D. Pilaud. The synchronous dataflow programming language LUSTRE. *Proceedings of the IEEE*, 79(9): 1305–1320, 1991.

Harel, D. Statecharts: A visual formalism for complex systems. *Science of Computer Programming*, 8: 231–274, 1987.

Kassakian, J. G., M. F. Schlecht, and G. C. Verghese. *Principles of Power Electronics*. Addison-Wesley, Reading, MA, 1991.

Kohavi, Z. *Switching and Finite Automata Theory*. McGraw-Hill, New York, 1978.

Lewis, F. L. A tutorial on the geometric analysis of linear time-invariant implicit systems. *Automatica*, 28: 119–137, 1992.

Lötstedt, P. Coulomb friction in two-dimensional rigid body systems. *Zeitschrift fur, Angewandte Mathematik und Mechanik*, 61: 605–615, 1981.

Lynch, N. and B. Krogh, editors. *Hybrid Systems: Computation and Control*, Lecture Notes in Computer Science, volume 1790. Springer, Berlin, 2000.

Moler, C. Are we there yet? Zero crossing and event handling for differential equations. *EE Times*, pp. 16–17, 1997. Simulink 2 special edition.

Mosterman, P. J. An overview of hybrid simulation phenomena and their support by simulation packages. In F. W. Vaandrager and J. H. van Schuppen, editors, *Hybrid Systems: Computation and Control*, Lecture Notes in Computer Science, volume 1569, pp. 164–177. Springer, Berlin, March 1999.

Mosterman, P. J. Implicit modeling and simulation of discontinuities in physical system models. In S. Engell, S. Kowalewski, and J. Zaytoon, editors, *The 4th International Conference on Automation of Mixed Processes: Hybrid Dynamic Systems*, pp. 35–40. Dortmund, Germany, 2000 (invited paper).

Mosterman, P. J. HYBRSIM—A modeling and simulation environment for hybrid bond graphs. *Journal of Systems and Control Engineering*, 216: (special issue) 35–46, 2002.

Mosterman, P. J. Mode transition behavior in hybrid dynamic systems. In *Proceedings of the 2003 Winter Simulation Conference*, pp. 623–631, New Orleans, LA, December 2003 (invited paper).

Mosterman, P. J. On the normal component of centralized frictionless collision sequences. *ASME Journal of Applied Mechanics*, 2007, in press.

Mosterman, P. J. and G. Biswas. A formal hybrid modeling scheme for handling discontinuities in physical system models. In *AAAI-96*, pp. 985–990, Portland, OR, August 1996.

Mosterman, P. J. and G. Biswas. Principles for modeling, verification, and simulation of hybrid dynamic systems. In *Fifth International Conference on Hybrid Systems*, pp. 21–27, Notre Dame, IN, September 1997.

Mosterman, P. J. and G. Biswas. A theory of discontinuities in dynamic physical systems. *Journal of the Franklin Institute*, 335B(3): 401–439, 1998.

Mosterman, P. J. and G. Biswas. A hybrid modeling and simulation methodology for dynamic physical systems. *SIMULATION: Transactions of the Society for Modeling and Simulation International*, 178(1): 5–17, 2002.

Mosterman, P. J., J. Ghidella, and J. Friedman. Model-based design for system integration. In *Proceedings of the Second CDEN International Conference on Design Education, Innovation, and Practice*, CD-ROM, Kananaskis, Alberta, Canada, July 2005.

Mosterman, P. J., J. Sztipanovits, and S. Engell. Computer automated multi-paradigm modeling in control systems technology. *IEEE Transactions on Control System Technology*, 12(2): 223–234, 2004.

Mosterman, P. J. and H. Vangheluwe. Computer automated multi-paradigm modeling: An introduction. *SIMULATION: Transactions of the Society for Modeling and Simulation International*, 80(9): 433–450, 2004.

Mosterman, P. J., F. Zhao, and G. Biswas. Sliding mode model semantics and simulation for hybrid systems. In P. Antsaklis, W. Kohn, M. Lemmon, A. Nerode, and S. Sastry, editors, *Hybrid Systems V*, Lecture Notes in Computer Science, pp. 218–237. Springer, Berlin, 1999.

Murata, T. Petri nets: Properties, analysis and applications. *Proceedings of the IEEE*, 77(4): 541–580, 1989.

Nicolescu, G., F. Bouchhima, and L. Gheorghe. CODIS—A framework for continuous/discrete systems co-simulation. In *Proceedings of the 2nd IFAC Conference on Analysis and Design of Hybrid Systems*, pp. 274–275, Alghero, Italy, June 2006.

Park, T. and P. I. Barton. State event location in differential-algebraic models. *ACM Transactions on Modeling and Computer Simulation*, 6(2): 137–165, 1996.

Petzold, L. R. A description of DASSL: A differential/algebraic system solver. Technical Report SAND82-8637, Sandia National Laboratories, Livermore, CA, 1982.
Pfeiffer, F. and C. Glocker. *Multibody Dynamics with Unilateral Contacts*. Wiley, New York, 1996.
SimEvents. *SimEvents User's Guide*. The MathWorks, Natick, MA, 2005.
Simulink. *Using Simulink*. The MathWorks, Natick, MA, 2004.
Stateflow. *Stateflow User's Guide*. The MathWorks, Natick, MA, 2004.
Torrisi, F. D. and A. Bemporad. HYSDEL—A tool for generating computational hybrid models for analysis and synthesis problems. *IEEE Transactions on Control System Technology*, 12(2): 235–249, 2004.
Utkin, V. I. *Sliding Modes in Control and Optimization*. Springer, Berlin, 1992.
van der Schaft, A. J. and J. M. Schumacher. The complementary-slackness of hybrid systems. *Mathematics of Control Signals and Systems*, (9): 266–301, 1996.
van Dijk, J. On the role of bond graph causality in modelling mechatronic systems. PhD dissertation, University of Twente, The Netherlands, 1994.
Verghese, G. C., B. C. Lévy, and T. Kailath. A generalized state-space for singular systems. *IEEE Transactions on Automatic Control*, 26(4): 811–831, 1981.

16
Theory and Practice for Simulation Interconnection: Interoperability and Composability in Defense Simulation

Ernest H. Page
The MITRE Corporation

16.1 Introduction 16-1
16.2 The Practice of Simulation Interconnection—
 Simulation Interoperability 16-2
 Simulator Networking • Distributed Interactive Simulation • Aggregate Level Simulation Protocol • High Level Architecture • Summary Thoughts on the Practice of Simulation Interconnection
16.3 The Theory of Simulation Interconnection—
 Simulation Composability 16-6
 A Brief History of Composability • Composability and Complexity • Formalisms for Composability • Proposal to Restrict the Scope of Composability • Summary Thoughts on the Theory of Simulation Interconnection
16.4 Conclusions 16-9

If you nail two things together that have never been nailed together before, some schmuck will buy it from you.

George Carlin

16.1 Introduction

For the better part of the past quarter century, the defense simulation industry has invested significant resources in technologies and methods for making independently developed simulations work together at run-time. Many reasons for this activity exist. The initial impetus was the need for a common "synthetic environment" that could interconnect simulators in support of small-team training—such environments are now commonplace to today's online gamers. Subsequently, the notion of "moving electrons to the people" rather than "moving people to the electrons" led to a proliferation of geographically distributed simulation-based training environments. In addition, there was the belief that each Service (e.g., Army, Navy, Air Force, and Marines) could best model that Service's capabilities, and if you needed to access such capabilities, you should do so by interconnecting with that Service's "authoritative" simulations. And of course, the notion of cost savings through simulation *reusability* also drove the development of technologies for simulation interconnection. Today, simulation interconnection is pervasive in the defense simulation arena.

Issues in simulation interconnection are variously referred to as *integration, interoperation, composition, configuration*, and more. Today, the dominant notions are *interoperability* and *composability*. And while there is some ambiguity in their usage, for purposes of this chapter, interoperability is the realm of the practical aspects of simulation interconnection, and composability encompasses the theoretical work in simulation interconnection.

To a large extent, the simulation interconnection problem can be viewed as a computer science problem, tackled by the "modern miracles" of standards, middleware, distributed algorithms, data type coercion, and so forth. The practice of simulation interconnection (interoperability) is reviewed in these terms. But there is also the notion of what it *means* when you interconnect two simulations. And while this topic has received much less attention in the defense simulation arena than the more immediately tractable issues of aligning bits and bytes, some notable work has been done, and is covered under the theory of simulation interconnection (composability). A careful reader will no doubt observe a close connection between the issues confronted in the theory of simulation interconnection and the fundamental concerns of multiscale, multiresolution, and multiformalism modeling.

A tremendous volume of literature exists in this area. Simulation interconnection is, for practical purposes, an industry unto itself. There are several longstanding conferences and workshops devoted to the topic, most notably the Simulation Interoperability Workshops (see SISO, 2005). Entire texts are dedicated to aspects of the problem, e.g., the high-level architecture (HLA) (Kuhl et al., 1999), and simulation interconnection represents the primary business for many companies that support the defense simulation industry. The goal of this chapter is to provide a broad and gentle introduction to the topic, with lots of pointers to the more comprehensive, detailed sources.

And, just for the record: George Carlin was offering general advice to aspiring entrepreneurs. As far as this author is aware, Mr. Carlin had no connections with the defense simulation industry!

16.2 The Practice of Simulation Interconnection—Simulation Interoperability

The lineage of the practice of simulation interconnection is typically traced from the Defense Advanced Research Project Agency (DARPA) simulator networking (SIMNET) project, through the development of the distributed interactive simulation (DIS) protocols and aggregate level simulation protocol (ALSP), to the current approach defined by HLA for modeling and simulation. Each of these is briefly surveyed here. Interested readers should consult Voss (1993) and Miller and Thorpe (1995) for histories of SIM-NET and DIS, Miller and Zabek (1996) and Weatherly et al. (1996) for ALSP, and Kuhl et al. (1999) for HLA.

Interconnecting simulations over computer networks may be rightly viewed as an application of *distributed simulation*. Within the defense simulation arena, work in simulation interconnection has sometimes been referred to as advanced distributed simulation (ADS). Here, the distributed simulations are mostly (but not exclusively) used for training, and the purpose of distributed execution is enhanced functionality. A newcomer to this area should be aware that there is a separate body of work in distributed simulation whose objective is to reduce the execution time of a simulation by utilizing multiple processors. In his comprehensive text on parallel and distributed simulation, Fujimoto (2000) refers to distributed simulation in the defense community as distributed virtual environments (DVEs).

16.2.1 Simulator Networking

Prior to the 1980s, simulators were very expensive, special purpose devices used to train an individual in the essential skills needed to operate the real platform that the simulator represented. Although a few cases of one-to-one locally interconnected air combat simulators had been created, networking technology was inadequate to support the interconnection of large numbers of simulators or the interconnection of simulators at great distances. The evolution of the ARPAnet and its concomitant technologies (e.g., packet switching) changed things, however, and in 1983 DARPA launched the SIMNET project. The purpose of SIMNET was to investigate the capability of networked simulators to support group (or "collective") training in large scales and at great distances. The idea was that a large-scale, interactive, networked simulation created a "synthetic environment" that could be entered by any authorized combatant from anywhere on the network using his simulator as a porting device (Miller and Thorpe, 1995). The initial project scope was to develop a SIMNET testbed with at least four geographically distributed sites with 50–100 vehicle simulators each. Because it was believed that the networking technologies of the day would not support the demands (owing to speed and maneuverability) of aircraft simulators, the initial SIMNET testbed focused on slower moving ground-based platforms, e.g., tanks and armored personnel carriers.

SIMNET spurred technical advancements in both computer networking and image generation, and is the basis for a variety of successful Service programs, including the U.S. Army's Close Combat Tactical Trainer (CCTT). Most of the fundamental design principles underlying SIMNET have demonstrated lasting value for simulation interconnection (Miller and Thrope, 1995):

- *Selective fidelity*. To minimize simulator costs, a simulator should only contain high fidelity representations of those elements essential to the training task. All other elements should be represented at lower fidelities, or not at all.
- *Autonomous simulation nodes*. Each node is responsible for maintaining the state of at least one object in the synthetic environment, and for communicating to other nodes any events caused by its object(s). Each node receives event reports from other nodes and calculates the effects of those events on its objects. All events are broadcast on the simulation network, and are available to any node that is interested. There is no centralized controlling process. Nodes may join and leave the network without affecting other nodes. Each node advances simulation time according to a local clock (typically a hardware clock).
- *Transmission of "ground truth" data*. Each node transmits the absolute truth about the current state of the object(s) it represents. Alteration of data to suit simulation objectives is the responsibility of the receiving node. For example, the position of a vehicle is broadcast to the network with 100% accuracy. If an object in another simulator determines that it could perceive the vehicle through a particular sensor, but with an accuracy determined by the alignment of the sensor and current weather conditions, then the receiving simulator should degrade the reported position accordingly.
- *Transmission of state change information*. To minimize communications processing, nodes transmit state update information only. To accommodate late-joining nodes and networks with high packet loss, this rule is often relaxed. In these situations, nodes send periodic (but relatively infrequent) updates for each owned object regardless of whether or not their state changes. This update interval is known as the "heartbeat."
- *Dead reckoning*. Between state update messages, receiving nodes may extrapolate the last reported state of remote objects that are of interest. To keep the extrapolated values from becoming too far afield of the actual values, the sending node maintains the same approximation and transmits a state update whenever the true position (or orientation) of an object diverges from the calculated dead reckoned values by more than an agreed-upon threshold. Fujiomoto (2000, p. 206) discusses common dead reckoning algorithms.

Other mainstays of modern defense simulation introduced by SIMNET include semi-automated forces (SAF) and the "flying carpet" (or "stealth") display.

16.2.2 Distributed Interactive Simulation

Following the successful demonstrations of SIMNET in the late 1980s, the defense simulation community undertook an industry-wide effort to define a set of standard networking protocols for interconnecting simulations. This work was accomplished largely within a series of semiannual workshops, and the DIS protocols became an IEEE standard in the spring of 1993 (Pullen and Wood, 1995; Voss, 1993). The primary mission of DIS is (University of Central Florida, 1993, p. 3)

> ...to create synthetic, virtual representations of warfare environments by systematically connecting separate subcomponents of simulation which reside at distributed, multiple locations...The property of connecting separate subcomponents or elements affords the capability to configure a wide range of simulated warfare representations patterned after the task force organization of actual units...Equally important is the property of interoperability which allows different simulation environments to efficiently and consistently interchange data elements essential to representing warfighting outcomes.

The fundamental design principles for DIS follow directly from SIMNET. Most of the standardization effort focused on extending the basic SIMNET communication structure, the protocol data unit (PDU), a bit-encoded packet for communicating entity state, and other types of information identified as useful for distributed combat simulations, e.g., weapons fire and weapons detonation events.

Like SIMNET, DIS was primarily designed to support the interconnection of simulations that (1) run in real-time, and (2) have a significant visual component. A great deal of focus in the DIS arena dealt with minimizing network latencies for PDUs. The creation of DIS led to a burgeoning market in SAF. SAFs were used to populate synthetic environments with background objects that behaved in a "reasonable" way. They were "semiautomated" because human intervention was often required to make the modeled entities maintain their reasonable behavior. The power and utility of SAFs was recognized very quickly, and eventually DIS-supported simulation environments consisting entirely of SAFs emerged.

DIS has been used as the protocol underlying numerous warfighting experiments and advanced concepts technology demonstrations (ACTDs), most notably, the Synthetic Theater of War (STOW) family of experiments.

16.2.3 Aggregate Level Simulation Protocol

As noted by Page and Smith (1998), defense simulation has a vernacular that can be nonintuitive to simulationists from outside the defense arena. For example, a commonly applied taxonomy for defense simulation is the *virtual, live, constructive* taxonomy (U.S. Department of Defense, 1997):

- *Virtual simulation* refers to a simulation involving real people operating simulated systems. Virtual simulations inject human-in-the-loop in a central role by exercising motor control skills (e.g., flying an airplane), decision skills (e.g., committing fire control resources to action), or communication skills (e.g., as members of a C4I team).
- *Live simulation* refers to a simulation involving real people operating real systems.
- *Constructive simulation* refers to a simulation that involves simulated people operating in simulated systems. Real people stimulate (make inputs) to such simulations, but are not involved in determining the outcomes.

Essentially, virtual simulation refers to the use of simulators, live simulation to rehearsal, or practice with "go-to-war" systems, and constructive simulation refers to "classical" computerized simulation models. These classical simulation models are also categorized with respect to their inherent level of abstraction. If the simulation includes explicit representation of individual vehicles, it is referred to as an *entity-level simulation*. If, however, the basic unit of representation in the simulation corresponds to a military echelon, e.g., a platoon, company, brigade, or battalion, then the simulation is referred to as an *aggregate-level simulation*.

Like the DIS protocols, the ALSP is rooted in SIMNET, but ALSP was targeted toward support for the interoperation of aggregate-level simulations used within command post exercises (Page et al., 1997; Weatherly et al., 1993, 1996). In addition, ALSP supported the explicit representation and synchronization of simulation time using a variant of the Chandy–Misra–Bryant protocol (Chandy and Misra, 1979; Bryant, 1977).

Fielded in the spring of 1991, the ALSP Joint Training Confederation (JTC), currently known as the Joint Training Transformation Initiative Plus (JTTI+), has been successfully employed to support numerous major, large-scale, joint training exercises, including the annual Ulchi Focus Lens, Prairie Warrior, and Unified Endeavor exercises (Miller and Zabek, 1996; ALSP, 2005).

16.2.4 High Level Architecture

By 1995, SIMNET, DIS, and ALSP had each contributed to the demonstration that interconnecting simulations could be of practical value. SIMNET provided an efficient and effective mechanism for linking simulators. DIS extended SIMNET and provided scalability to many thousands of entities in SAF-based exercises. ALSP provided support for synchronization required to interconnect "logical time," e.g., discrete event, simulations. By 1995, many defense simulations had interconnection interfaces—some SIMNET, some DIS, some ALSP, some "home grown." To mitigate against the proliferation of homegrown interconnection standards, the U.S. Department of Defense (DoD) established a *unifying* standard for simulation interconnection known as the High Level Architecture (HLA). The HLA represents both a generalization and extension of SIMNET, DIS, and ALSP, and is defined by three components:

- *an object model template*—a common model definition and specification formalism,
- *an interface specification*—a collection of services describing the HLA run-time environment, and
- the *HLA rules*—governing compliance with the architecture.

The HLA is intended to have applicability across the full range of defense simulation applications, including those used to support training, analysis, mission rehearsal, and acquisition.

At the heart of the HLA is the notion of a *federation*. A federation is a collection of federates—simulations and other systems—that interoperate using the protocols described by the architecture. A federation object model (FOM) provides the model specification and establishes a contract between the federates regarding the nature of the activity taking place during federation run-time. Federation execution is accomplished through an HLA run-time infrastructure (RTI), which is an implementation of the infrastructure services defined by the architecture. In addition to defining services for the RTI, the HLA interface specification also defines services that must be implemented by federates.

In a typical federation execution, a federate joins the federation, indicates its operating parameters (e.g., information the federate will provide to the federation and information it will accept from the federation), and then participates in the evolution of federation state until the federate departs the federation or the simulation terminates. FOM data are provided to the RTI at run-time, enabling the infrastructure to provide a level of enforcement with respect to the "information contract" that the FOM represents.

16.2.5 Summary Thoughts on the Practice of Simulation Interconnection

It would be hard to argue that the pursuit of making independently developed systems work together at run-time has not been worthwhile for the defense simulation community. The technology has been used successfully too many times to condemn it. However, there is still the sense in the defense simulation community that simulations generally, and federations of simulations particularly, are expensive, difficult to use, and fragile. The community still has not achieved the level of interoperability that it would like. Although HLA was (is) a unifying standard, a great many DIS and a few ALSP applications still persist. The technology seems to be almost in a perpetual proof-of-concept mode. Organization after organization still spend nontrivial sums on the creation of federations with minimal utility. On the one hand, this seems like a bad sign. On the other, fundamental change within government organizations takes time, and 25 years is probably too soon to fully evaluate the impact of this technology.

16.3 The Theory of Simulation Interconnection—Simulation Composability

In recent years, the defense simulation community has begun to complement a robust practice of simulation interconnection with some regard to the development of supportive theories. Most of this work has arguably been accomplished under the rubric of simulation *composability*. Like many terms from the DoD lexicon, the notion of composability is vaguely and disparately applied, as evidenced by the history below, adopted from Page et al. (2004).

16.3.1 A Brief History of Composability

The earliest uses of the term composability within the defense simulation context date to the Composable Behavioral Technologies (CBT) project during the mid-1990s (see Courtemanche et al., 1997). The purpose of CBT was to give ModSAF users a convenient way to develop new entity behaviors without appealing to the underlying SAF source code. Shortly after the initiation of CBT, composability appeared as a system objective within the Joint Simulation System (JSIMS) Mission Needs Statement. A taxonomy and use case for JSIMS composability appears in (JSIMS Composability Task Force, 1997), and the impact of composability as a system objective on the JSIMS design is described in Pratt et al. (1999). Composability also appeared as a key system objective for OneSAF in 1999 (U.S. Army STRICOM, 2000).

In 1998, the DARPA Advanced Simulation Technology Thrust (ASTT)—which was chartered to develop technology in support of JSIMS—funded two separate studies on simulation composability: (1) the model-based simulation composition (MBSC) project, which developed a prototype composition environment for JSIMS (Aronson and Wade, 1998, 2000; Davis and Aronson, 1999; Wade and Aronson, 1999); and (2) a study by Page and Opper (1999) that investigated the composability problem from a computability and complexity theoretic perspective.

A focus-paper session at the 2000 Winter Simulation Conference addressed methodologies for composable simulation (Kasputis and Ng, 2000; Davis et al., 2000). Recently, the work of Petty, Weisel, and Mielke (Petty and Weisel, 2003a, 2003b; Petty et al., 2003, 2005; Weisel et al., 2003, 2005) provides a broad survey of the uses of the term composability, extends the work of Page and Opper, and examines the composite validation problem within the context of automata theory and computable functions.

The Defense Modeling and Simulation Office (DMSO) initiated a collection of studies as part of the Composable Mission Space Environments (CMSE) initiative in FY03 (DMSO, 2005). The comprehensive report by Davis and Anderson (2003) provides a broad survey of the topic of composability and suggests a wide-ranging investment strategy for the DoD in this area.

Related work outside the defense simulation community included the emergence of the topic of "Web-based simulation" in the mid-1990s, which included the concept of composing simulations via Web protocols (see Fishwick, 1996; Fishwick et al., 1998; Page et al., 2000). Most recently, the Extensible Modeling and Simulation Framework (XMSF) was initiated by the Naval Postgraduate School, George Mason University, SAIC, and Old Dominion University to develop a ubiquitous Web-based simulation environment (XMSF, 2005).

16.3.2 Composability and Complexity

Motivated by the high degrees of automated support for composability expressed within the JSIMS and OneSAF program requirements, Page and Opper (1999) consider composition from a computability and computational complexity theoretic perspective. The authors observe that prior work in analyzing simulation model specifications suggests that many of the problems attendant with simulation model development, verification, and validation are *fundamentally hard*, and that automation can only provide so much relief (Page and Opper, 1999, p. 554). For example, problems such as the following cannot be solved in the general case: (1) determining if a model is finite (i.e., will run to completion); (2) determining if a model specification is complete; and (3) determining if a model specification is minimal. Other analyses have shown that problems such as the following have no efficient solution: (1) determining whether any given state will occur during an execution of a model; (2) determining the existence of, or possibility for, simultaneous events; and (3) determining whether a model implementation satisfies a model specification.

Motivating the work of Page and Opper (1999) was the fairly simple observation that the class of *model specifications* must include the class of *model compositions*, and therefore all prior results in the computational complexity of model specifications also applied to model compositions. In support of this observation, the authors develop a formal analysis of a simple, generic methodology for composable simulation. They observe that building simulation models by composition implies not only identifying (via search) relevant candidates from (possibly massive) component repositories, but also answering the following: (1) does a combination of components exist that satisfies the modeling objectives, and (2) if so, can the "best" (or a "good enough") solution be identified in a reasonable time. If not, how closely can the objectives be met?

Determining whether a collection of components satisfies a modeling objective might be accomplished in any number of ways, including

- Determination made strictly on the basis of the descriptions of component capabilities (i.e., metadata).
- Determination made by modeling or approximating component interactions.
- Determination made by constructing the composed model and observing the result(s).

Page and Opper observe that a determination made on the basis of metadata is the least computationally intensive solution—assuming such a determination was possible. They suggest a simple formal model of composition based on set theory and describe a generic decision problem for composability as follows:

COMPOSABILITY
INSTANCE: A set O of objectives and a collection C of components.
QUESTION: Is there a valid composition that meets the objectives stated in O?

The authors conjecture that the decision problem COMPOSABILITY is NP-complete. In the development of a proof of this conjecture, the authors observe, however, that certain objectives may be undecidable on their face, e.g., the simulation terminates for a given set of circumstances. To accommodate this, Page and Opper suggest two variants of the COMPOSABILITY decision problem: (1) BOUNDED COMPOSABILITY—each objective in O is decidable; and (2) UNBOUNDED COMPOSABILITY—some objective in O is undecidable. Further, the authors observe that it may be possible for two components, A and B, to satisfy some objective O and that their ability to satisfy O could not be predicted based on any metadata for A and B. Page and Opper suggest that this characteristic is like the property of emergence in complex adaptive systems and suggests two more variants of the COMPOSABILITY decision problem: (1) EMERGENT COMPOSABILITY—composition cannot be evaluated based on metadata; and (2) NONEMERGENT COMPOSABILITY—the composition can be evaluated based on metadata.

The cross product of the variants yields four decision problems:

- UNBOUNDED EMERGENT COMPOSABILITY
- BOUNDED EMERGENT COMPOSABILITY
- UNBOUNDED NONEMERGENT COMPOSABILITY
- BOUNDED NONEMERGENT COMPOSABILITY (BNC).

From a complexity perspective, BNC is the simplest. Page and Opper provide a proof that BNC is NP-complete. This proof suggests that use cases for composability that imply automated support for determining valid combinations of components (e.g., Steps 2 and 5 from the JSIMS composability use case [JSIMS, 1997]) cannot have an efficient solution in the general case.

Petty et al. (2003) extend the work of Page and Opper and suggest another variant of the problem, ANTI-EMERGENT COMPOSABILITY (AC). In AC, two models A and B may each satisfy some objective O, but their combination does not. Petty, Weisel, and Mielke suggest a general form of the component selection problem that subsumes the variants and prove that it is NP-complete.

16.3.3 Formalisms for Composability

Petty and Wiesel (2003b) develop a formal model of simulation composition based on computable functions operating on vectors of integers. In this formalism, the composition of simulation models is isomorphic to the composition of computable functions. The authors suggest a mechanism to measure the validity of a composition by comparing its output vector to the output vector of a "perfect model," a model whose outputs perfectly correspond to the outputs of the system being modeled. Weisel et al. (2005) evaluate their formalism with respect to the DEVS formalism, noting their analytic equivalence despite the fact that DEVS is not explicitly restricted to computable functions.

16.3.4 Proposal to Restrict the Scope of Composability

Page et al. (2004) observe that despite the decades of successful practice in simulation interconnection and the solid theoretical work ongoing by Petty and others in the context of simulation composability, a refinement of the terminology in use by the defense simulation community could serve to better focus the community's research and development efforts.

As illustrated by the history of composability above, the term composability is used within the military simulation domain to imply a variety of notions, ranging from interoperability, to end-user tailorability, to any act of creation. Page et al. (2004) suggest that for the term to be most useful, it should be unambiguously differentiated from these other concepts.

The definition suggested by Petty and Weisel (2003a) loosely differentiates the notions of composability and interoperability as follows:

> Essentially, interoperability is the ability to exchange data or services at run-time, whereas composability is the ability to assemble components prior to run-time... It can be seen that interoperability is necessary but not sufficient to provide composability. Composability (engineering and modeling) does require interoperability (technical and substantive). Federates that are not interoperable can not be composed, so interoperability is necessary for composability. However, interoperability is not sufficient to provide composability, i.e. federates may be interoperable but not composable. Recall that an essential aspect of composability is the ability not just to combine federates but to combine and recombine federates into different simulation systems. Federates that are interoperable in one specific configuration or with one specific object model, and cannot be combined and recombined in other ways, are not composable... The matter of substantial effort is crucial to the distinction between interoperability and composability.

These distinctions seem somewhat problematic, because it is not immediately clear how a run-time characteristic (interoperability) could be a necessary condition to enable a prerun-time characteristic (composability). Further, the term "substantial effort" requires quantification.

Petty and Weisel (2003a) distinguish composability and integratability as follows:

> Integration is the process of configuring and modifying a set of components to make them interoperable and possibly composable. Essentially any federate can be integrated into any federation with enough effort, but composability implies that the changes can be made with little effort.

This distinction also seems somewhat problematic since the level of "effort" is also used to distinguish between composability and interoperability.

As an alternative, Page et al. (2004) suggest that composability be viewed as a property of a set of models. Specifically, it should be expressed as some function on the congruity of the objectives and assumptions underlying each model in the set. In this sense, they agree with the definition of Petty and Weisel. That is, composability is a property that may be assessed prior to run-time. However, in their view *composability is independent from interoperability*. Interoperability is a property of the software implementation of a set of models (or other systems). The objectives and assumptions underlying two models, A and B, may be wholly congruent and thus the models are composable, but their software implementations may utilize different programming languages, data types, marshaling protocols, and so forth such that the two implementations are not interoperable.

Informally, two models are composable if they share compatible objectives and assumptions. Quantifying and reasoning about the "compatibility" of objectives and assumptions should be the domain of research in algebras and calculi for composition. Thus, Page et al. (2004) suggest the following framework:

- *Composability*—realm of the model (e.g., two models are composable if their objectives and assumptions are properly aligned).
- *Interoperability*—realm of the software implementation of the model (e.g., are the data types consistent, have the little endian/big endian issues been addressed, etc.)
- *Integratability*—realm of the site the simulation is running at (e.g., have the host tables been set up, are the NIC cards working properly).

16.3.5 Summary Thoughts on the Theory of Simulation Interconnection

The defense simulation lexicon is generally underwhelming—although this may simply reflect the author's academic bias—and the community treatment of the terms surrounding the simulation interconnection problem is no exception. While the definitions promulgated by Petty et al., and currently embraced by the defense simulation community, are not as "crisp" as they could perhaps be, their theoretical work in the composition of models is a great service to the community. We do a reasonably good job, as a community, in getting the bits to flow between simulations. We do not have a good handle on how to reason about the semantics of the interoperating simulations once the bits start flowing. Algebras and calculi of composition are much in need here.

16.4 Conclusions

For the better part of the past quarter century, the defense simulation industry has been heavily invested in technologies and methods to make independently developed simulations work together. This pursuit has met with significant technological successes, notably: SIMNET, DIS, ALSP, and HLA. The pressures that drove the development of this technology were primarily fiscal. Defense simulations represent a significant investment, and mechanisms for their reuse must be defined to preserve the defense simulation community's capabilities in an era of ever-shrinking budgets. Like many areas where the realities of the day demand a solution, the practice of simulation interconnection has led the theory. But important theoretical work is beginning to take shape to quantify and reason about the semantics of interconnected simulations.

References

Aggregate Level Simulation Protocol (ALSP). 2005. Homepage. http://ms.ie.org/alsp.

Aronson, J. and Wade, D. 1998. Model based simulation composition. In *Proceedings of the Fall 1998 Simulation Interoperability Workshop*, 98F-SIW-055.

Aronson, J. and Wade, D.M. 2000. Benefits and pitfalls of composable simulation. In *Proceedings of the Spring 2000 Simulation Interoperability Workshop*, 00S-SIW-155.

Bryant, R.E. 1977. Simulation of packet communication architecture computer systems. Technical Report MIT-LCS-TR-188, Massachusetts Institute of Technology.

Chandy, K.M. and Misra, J. 1979. Distributed simulations: case study in design and verification of distributed programs, *IEEE Transactions on Software Engineering*, **SE-5**(5): 440–452.

Courtemanche, A.J., von der Lippe, S.R., and McCormack, J. 1997. Developing user-composable behaviors. In *Proceedings of the Fall 1997 Simulation Interoperability Workshop*, 97F-SIW-068.

Davis, D. and Aronson, J. 1999. Component selection techniques to support composable simulation. In *Proceedings of the Spring 1999 Simulation Interoperability Workshop*, 99S-SIW-043.

Davis, P.C., Fishwick, P.A., Overstreet, C.M., and Pegden, C.D. 2000. Model composability as a research investment: Responses to the featured paper. In *Proceedings of the 2000 Winter Simulation Conference*, pp. 1585–1591.

Davis, P.K. and Anderson, R.H. 2003. *Improving the Composability of DoD Models and Simulations*, RAND, National Defense Research Institute, Santa Monica, CA.

Defense Modeling and Simulation Office (DMSO). 2005. Composable mission space environments. Homepage. https://www.dmso.mil/public/warfighter/cmse/

Extensible Modeling and Simulation Framework (XMSF). 2005. Homepage. http://www.movesinstitute.org/xmsf/xmsf.html

Fishwick, P., Hill, D., and Smith, R. (Eds.). 1998. *Proceedings of the 1998 SCS International Conference on Web-Based Modeling and Simulation*, 11–14 January, San Diego, CA.

Fishwick, P.A. 1996. Web-based simulation: Some personal observations. In *Proceedings of the 1996 Winter Simulation Conference*, pp. 772–779.

Fujimoto, R.M. 2000. *Parallel and Distributed Simulation Systems*. New York: Wiley.

JSIMS Composability Task Force. 1997. Final Report.

Kasputis, S. and Ng, H.C. 2000. Composable simulations. In: *Proceedings of the 2000 Winter Simulation Conference*, pp. 1577–1584.

Kuhl, F., Weatherly, R., and Dahmann, J. 1999. *Creating Computer Simulation Systems—An Introduction to the High Level Architecture*. New Jersey: Prentice-Hall.

Miller, D.M. and Thorpe, J.A. 1995. SIMNET: The advent of simulator networking. *Proceedings of the IEEE*, **83**(8): 1114–1123.

Miller, G. and Zabek, A.A. 1996. The joint training confederation and the aggregate level simulation protocol. *MORS Phalanx*, **29**: 24–27.

Page, E.H., Briggs, R., and Tufarolo, J.A. 2004. Toward a family of maturity models for the simulation interconnection problem. In *Proceedings of the Spring 2004 Simulation Interoperability Workshop*, 04S-SIW-145.

Page, E.H., Buss, A., Fishwick, P.A., Healy, K.J., Nance, R.E., and Paul, R.J. 2000. Web-based simulation: Revolution or evolution? *ACM Transactions on Modeling and Computer Simulation*, **10**(1): 3–17.

Page, E.H., Canova, B.S., and Tufarolo, J.A. 1997. A case study of verification, validation and accreditation for advanced distributed simulation. *ACM Transactions on Modeling and Computer Simulation*, **7**(3): 393–424.

Page, E.H. and Opper, J.M. 1999. Observations on the complexity of composable simulation. In *Proceedings of the 1999 Winter Simulation Conference*, pp. 553–560.

Page, E.H. and Smith, R. 1998. Introduction to military training simulation: A guide for discrete event simulationists. In *Proceedings of the 1998 Winter Simulation Conference*, pp. 53–60.

Petty, M.D. and Weisel, E.W. 2003a. A composability lexicon. In *Proceedings of the Spring 2003 Simulation Interoperability Workshop*, 03S-SIW-023.

Petty, M.D. and Weisel, E.W. 2003b. A formal basis for a theory of semantic composability. In *Proceedings of the Spring 2003 Simulation Interoperability Workshop*, 03S-SIW-054.

Petty, M.D., Weisel, E.W., and Mielke, R.R. 2003. Computational complexity of selecting components for composition. In *Proceedings of the Fall 2003 Simulation Interoperability Workshop*, 03F-SIW-072.

Petty, M.D., Weisel, E.W., and Mielke, R.R. 2005. Composability theory overview and update. In *Proceedings of the Spring 2005 Simulation Interoperability Workshop*, 05S-SIW-063.

Pratt, D., Ragusa, C., and von der Lippe, S. 1999. Composability as an architecture driver. In *Proceedings of the 1999 I/ITSEC Conference*.

Pullen, J.M. and Wood, D.C. 1995. Networking technology and DIS. In *Proceedings of the IEEE*, **83**(8): 1157–1167.

Simulation Interoperability Standards Organization (SISO). 2005. Homepage. http://www.sisostds.org.

U.S. Army Simulation, Training and Instrumentation Command (STRICOM). 2000. OneSAF Operational Requirements Document, Version 1.0.

U.S. Department of Defense. 1997. DoD modeling and simulation glossary. DoD 5000-59-M, December.

University of Central Florida. 1993. Distributed interactive simulation standards development: Operational concept 2.3. TRI ST-TR-93-25, Institute for Simulation and Training, Orlando, FL.

Voss, L.D. 1993. *A Revolution in Simulation: Distributed Interaction in the '90s and Beyond*. Virginia: Pasha Publications.

Wade, D.M. and Aronson, J. 1999. Model based simulation composition: A unified scenario model to support composable simulation. In *Proceedings of the Spring 1999 Simulation Interoperability Workshop*, 99S-SIW-044.

Weatherly, R.M., Wilson, A.L., Canova, B.S., Page, E.H., Zabek, A.A., and Fischer, M.C. 1996. Advanced distributed simulation through the aggregate level simulation protocol. In *Proceedings of the 29th Hawaii International Conference on Systems Sciences*, Vol. 1, pp. 407–415.

Weatherly, R.M., Wilson, A.L., and Griffin, S.P. 1993. ALSP—Theory, experience, and future directions. In *Proceedings of the 1993 Winter Simulation Conference*, pp. 1068–1072.

Weisel, E.W., Petty, M.D., and Mielke, R.R. 2003. Validity of models and classes of models in semantic composability. In *Proceedings of the Fall 2003 Simulation Interoperability Workshop*, 03F-SIW-073.

Weisel, E.W., Petty, M.D., and Mielke, R.R. 2005. A comparison of DEVS and semantic composability theory. In *Proceedings of the Spring 2005 Simulation Interoperability Workshop*, 05S-SIW-153.

IV

Model Types

17
Ordinary Differential Equations

Francisco Esquembre
Universidad de Murcia

Wolfgang Christian
Davidson College

17.1 Introduction 17-1
  The Simple Pendulum • Dynamics of Chemical Reactions • Predator–Prey Population Dynamics • Planetary Motion
17.2 Numerical Solution 17-3
  Implementation Techniques
17.3 Taylor Methods 17-5
17.4 Runge–Kutta Methods 17-7
17.5 Implementation 17-8
17.6 Adaptive Step 17-11
17.7 Implementation of Adaptive Step 17-12
17.8 Performance and Other Methods 17-15
17.9 State Events 17-19
17.10 The OSP Library 17-20

17.1 Introduction

Right from the invention of calculus, ordinary differential equations (ODEs) have been used to model continuous systems in all scientific and engineering disciplines. The idea is simple in principle and applies to physical situations in which information can be obtained about the continuous rate of change of the state variables of the system. The hope is that we will be able to ascertain the evolution of the system using this information and the knowledge of its initial state.

We first consider some examples from different disciplines.

17.1.1 The Simple Pendulum

A simple pendulum is a physical abstraction in which a point mass m oscillates in a vertical plane at the end of a rod of length L with negligible mass. The motion of a simple pendulum can be modeled by the ODE:

$$mL^2\ddot{\theta}(t) + b\dot{\theta}(t) + mgL \sin \theta(t) = \tau_e(t) \tag{17.1}$$

where $\theta(t)$ is the angular position of the rod of the pendulum with respect to the vertical, b the friction coefficient, g the acceleration due to gravity, and $\tau_e(t)$ a time-dependent external torque which drives the motion. Typically, the pendulum starts at time $t = 0$ from a given angle θ_0 with zero initial angular velocity.

17.1.2 Dynamics of Chemical Reactions

The Brusselator was proposed in 1968 by R. Lefever and the Nobel Prize winner I. Prigogine, as a model for an autocatalytic, oscillating chemical reaction. The mechanism for the reaction is given by

$$A \xrightarrow{k_1} X \tag{17.2a}$$

$$B + X \xrightarrow{k_2} Y + C \tag{17.2b}$$

$$2X + Y \xrightarrow{k_3} 3X \tag{17.2c}$$

$$X \xrightarrow{k_4} D \tag{17.2d}$$

We are interested in the evolution of the intermediate products X and Y. If we assume that the concentration of the reactants A and B is kept constant, $a = [A]$ and $b = [B]$, and denote the concentrations of the products as $x = [X]$ and $y = [Y]$, the system is described by the differential equations:

$$\dot{x}(t) = k_1 a - k_2 b x(t) + k_3 x^2(t) y(t) - k_4 x(t) \tag{17.3a}$$

$$\dot{y}(t) = k_2 b x(t) - k_3 x^2(t) y(t) \tag{17.3b}$$

The initial values of x and y are zero. For appropriate values of the parameters, this system can exhibit oscillatory behavior.

17.1.3 Predator–Prey Population Dynamics

ODEs are frequently used in biology to model population dynamics. The famous Lotka–Volterra model describes the evolution of two species, one of which preys (feeds) on the other, using the ODEs:

$$\dot{x}_1(t) = a x_1(t) - b x_1(t) x_2(t) \tag{17.4a}$$

$$\dot{x}_2(t) = c x_1(t) x_2(t) - d x_2(t) \tag{17.4b}$$

where x_1 represents the number of prey and x_2 is the number of predators (in appropriate units so that they take continuous values in the interval $[0, 1]$), and a, b, c, and d are the parameters. The terms $a x_1$ and $-d x_2$ account for the reproduction rate of each species in the absence of interaction with the other and the nonlinear terms represent the effects of predation on the reduction of prey and the reproduction of the predators. This model was used in the mid-1920s to study how the intensity of fishing affected the different fish populations in the Mediterranean Sea.

17.1.4 Planetary Motion

Kepler discovered his three laws of planetary motion after a titanic analysis of years of astronomical observations by Tycho Brahe. Newton's inverse-square law of gravitation allows us to reformulate this motion in modern terms using the equations for a massless test particle about a particle of mass M located at the origin:

$$\ddot{x}(t) = -GM \frac{x(t)}{(x^2(t) + y^2(t))^{3/2}} \tag{17.5a}$$

$$\ddot{y}(t) = -GM \frac{y(t)}{(x^2(t) + y^2(t))^{3/2}} \tag{17.5b}$$

where M is the combined mass of the Sun and the planet, and G is the universal gravitational constant. The xy-coordinates for the massless test particle are the relative position of the planet with respect to the Sun in this two-body approximation of the solar system. The initial values for x and y and its derivatives are taken from direct astronomical observation.

Ordinary Differential Equations

The examples presented here are typical. Formulating the problem with the help of ODEs and stating the existence and uniqueness of a solution under reasonable regularity conditions are relatively easy tasks. But finding an analytic solution is often difficult and might be impossible. Although analytic solutions, such as in the case of linear ODEs, are very important theoretically, the vast majority of problems cannot be solved in this manner. In these cases, we must use numerical techniques to approximate the evolutionary behavior predicted by the model. This chapter introduces some of these techniques.

The numerical algorithms we will discuss can be implemented in many different ways, using almost any programming language. We have chosen Java to illustrate particular algorithmic implementations. Java is an object-oriented programming language that is designed to run on a virtual computer that can be implemented in any modern operating system. This promise of platform independence has become a reality, and we can now write programs that have attractive graphical user interfaces and support common tasks such as printing, disk access, and copy–paste data exchange with other applications.

Although these features can be implemented using the standard Java library, it usually requires much programming and a working knowledge of the Java application programmer interface (API). Learning the Java API is essential for software developers, but experienced programmers usually adopt or develop an add-on code library. To enable scientists and engineers to quickly begin writing their own programs, we have developed the Open Source Physics (OSP) library (Christian, 2007). This library enables users to quickly perform common visualization tasks such as creating a graph.

```
PlotFrame plot= new PlotFrame("f(t)","t","Sine Function");
for (int i = 0; i < 100; i++) {
   double t=i*2*Math.PI/100.0;
   plot.append(0,t,Math.sin(t));     // Add data to graph #0
}
```

You do not need to know much about Java to understand that these statements graph a sine function. The advantage of an object-oriented approach is that readers who are unfamiliar with or uninterested in Java may treat the code listings in this chapter as pseudocode while Java programmers can compile and run the examples. Source code for the pendulum, the chemical reaction, the predator–prey, and the planetary motion examples are available on the OSP website (http://www.opensourcephysics.org/CRC_examples). The website also has ready-to-run (executable) versions of these programs available for downloading.

An alternative approach to coding the examples in Java is to use a high-level modeling program. The *Easy Java Simulations* (Ejs) authoring tool uses the OSP library to provide a simple way to create your own simulations for continuous (and noncontinuous) systems. Ejs has a graphical drag-and-drop user interface that helps users build models and connect models to graphical elements to visualize the state of the system and its evolution. These simulations can then be used to explore the behavior of the system under different conditions. Examples of Ejs (including the examples in this chapter) can be found at http://fem.um.es/Ejs/CRC_examples.

17.2 Numerical Solution

The mathematical formulation most frequently used in modeling continuous systems is that of an *initial-value problem* for a first-order ordinary differential equation.[1] An initial-value problem is an expression of the form:

$$\dot{\mathbf{x}}(t) = \mathbf{f}(\mathbf{x}(t), t), \quad \mathbf{x}(a) = \mathbf{x}_0 \tag{17.6}$$

---

[1] ODEs can also appear in the form of *boundary value problems*, where the information about the state of the system is provided partly in one initial instant of time, and partly in one final instant of time (Press et al., 1992; Keller, 1992).

where t (which often represents the time) is called the independent variable and can take values in a given interval $[a, b]$ of the real line, $\mathbf{x}(t)$ represents a k-dimensional vector of real numbers that describes the state of the system at a given instant, and the vector function $\mathbf{f}$ conveys all the information that relates the rate of change of the state vector with the value of the independent variable and the state itself. The information provided at the initial instant of time, $\mathbf{x}(a) = \mathbf{x}_0$, is called the initial condition, and is necessary to completely determine a solution out of the many possible solutions.

ODEs in which the right-hand side has no explicit dependence on time, $\dot{\mathbf{x}}(t) = \mathbf{f}(\mathbf{x}(t))$, are called *autonomous*. Autonomous equations are of interest because they describe *dynamical systems* whose evolution depends only on their internal state independent of time. Nonautonomous equations contain an explicit time dependence, $\dot{\mathbf{x}}(t) = \mathbf{f}(\mathbf{x}(t), t)$. The simple pendulum in Section 17.1 is an example of a nonautonomous equation, while all the others describe dynamical systems.

Every nonautonomous ODE can be reformulated as a dynamical system by adding the time to the system of differential equations. That is, we add the independent variable t to create a new state vector $\hat{\mathbf{x}} = (\mathbf{x}(t), t)$ and we add the trivial time rate $dt/dt = 1$ to create a new rate $\hat{\mathbf{f}} = (\mathbf{f}(\mathbf{x}, t), 1)$. Although adding another variable makes the geometrical interpretation more difficult, it is inconsequential (and sometimes more convenient) from a computational point of view. We will use this approach in our implementation code.

We have chosen to represent the dynamical system state $(x_1, x_2, x_3, \ldots, x_k, t)$ as a vector because most interesting phenomena involve multiple state variables. Even in cases where the state can be described by a single number, derivatives of order higher than one usually appear. For instance, a typical physics model, such as the pendulum, obeys Newton's second law $\mathbf{F} = m\mathbf{a}$, which usually turns into a second-order differential equation. These cases are contained in our definition above, because any higher-order differential equation can be rewritten as a new first-order ODE. As an example, we can construct a system of first-order differential equations from Eq. (17.1) by introducing the angular velocity ω as the rate of change of the angle θ. The driven pendulum can then be written as the following coupled system of first-order differential equations for the variables (θ, ω, t):

$$\dot{\theta} = \omega \tag{17.7a}$$

$$\dot{\omega} = -\frac{g}{L}\sin(\theta) - \frac{b}{mL^2}\omega + \frac{1}{mL^2}\tau_e(t) \tag{17.7b}$$

$$\dot{t} = 1 \tag{17.7c}$$

The numerical solution of ordinary differential equations is a well-studied problem in numerical analysis. Most numerical techniques are based on *difference methods*. In these methods, we attempt to obtain approximate values of the solution at a sequence of mesh points $a = t_0 < t_1 < \cdots < t_N = b$. Although obtaining a solution at a finite number of mesh points appears restrictive, the idea is actually very useful. In many cases, scientists and engineers use models to study how systems evolve in time by simulating them with different parameters and initial conditions, plotting or displaying the state of the system at regular time steps. This is precisely what difference methods do.

Solving differential equations numerically is both a science and an art. There exist many authoritative books on the subject, both because the discipline is very mathematically advanced, and because different numerical techniques are developed for particular types of systems. A universal solving method for ordinary differential equations does not exist and practitioners should select a method based on requirements such as speed, accuracy, and the conservation of important physical properties such as energy.

The numerical methods most frequently used fall into the following categories: Taylor methods, Runge–Kutta methods, multistep methods, and extrapolation methods. We concentrate on Taylor and Runge–Kutta methods first because they work well in many situations and are easy to implement (particularly Runge–Kutta methods), and because they will help us illustrate the main concepts and details for the computer implementation of difference methods. We will discuss other methods in Section 17.8.

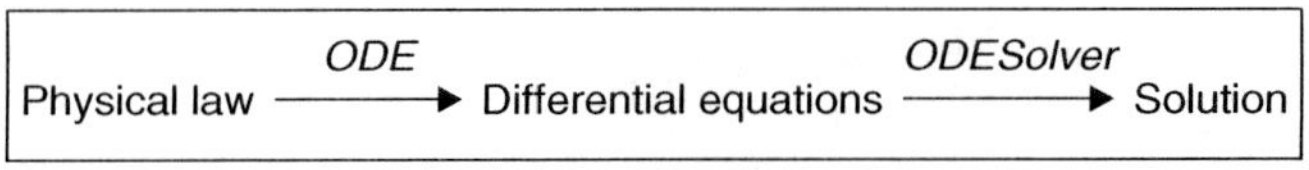

Ordinary Differential Equations

FIGURE 17.1 Differential equations can be solved using the `ODE` interface to define the equations and the `ODESolver` interface to define the numerical algorithm.

17.2.1 Implementation Techniques

The numerical solution of differential equations can be made simpler using object-oriented programming techniques that separate the differential equations from the numerical algorithm. The `ODE` and `ODESolver` interfaces shown in Figure 17.1 are code files that will be described in Section 17.5.

A Java interface is a list of methods that an object can perform. Note that Java allows us to define variables using an interface as a variable type. The `ODE` interface in the OSP library defines methods (similar to subroutines or functions) that enable us to encapsulate an initial value problem such as Eq. (17.6) in a Java class. The `ODESolver` interface defines methods that implement numerical algorithms for solving differential equations. The most important method in `ODESolver` is the `step` method that advances the state of the differential equation. This allows us to write a code such as

```
ODE ode = new Pendulum();           // creates ODE
ODESolver solver = new RK4(ode);    // creates numerical method
solver.initialize(0.01);            // each step advances by 0.01
for(int i=0; i<100; i++) {          // advances by 100 steps
  solver.step();
}
```

In the spirit of object-oriented programming, the details of how the differential equations are defined and solved are hidden in the `Pendulum` and `RK4` objects. This hiding is known as *encapsulation* and is a hallmark of good object-oriented design. A user can solve the pendulum problem using a different numerical algorithm by creating (instantiating) a different `ODESolver`. The OSP library contains many differential equation algorithms. The user assumes that these algorithms have been properly programmed and the library assumes that the user has obeyed the OSP API.

17.3 Taylor Methods

The first difference method we will study is based on the Taylor series expansion of the solution $\mathbf{x}(t)$. Namely

$$\mathbf{x}(t+h) = \mathbf{x}(t) + h\dot{\mathbf{x}}(t) + \frac{h^2}{2!}\ddot{\mathbf{x}}(t) + \frac{h^3}{3!}\mathbf{x}^{(3)}(t) + \cdots \quad (17.8)$$

If we know the value of $\mathbf{x}(t)$, we can compute $\dot{\mathbf{x}}(t)$ from the differential equation. The second and subsequent derivatives of $\mathbf{x}$ at t can be obtained by repeatedly differentiating Eq. (17.6). Thus,

$$\ddot{\mathbf{x}}(t) = \mathbf{f}_t(\mathbf{x}(t), t) + \mathbf{f}_x(\mathbf{x}(t), t)\mathbf{f}(\mathbf{x}(t), t) \quad (17.9)$$

and similarly for higher-order derivatives. Note, however, that the expression becomes more complicated as we compute higher-order derivatives.

Because we actually know from Eq. (17.6) the value of $\mathbf{x}$ at t_0, we can take $h_0 = t_1 - t_0$, and use a Taylor expansion to obtain $\mathbf{u}_1$, an approximation to $\mathbf{x}(t_1)$. We can now use $\mathbf{u}_1$ to repeat the process with $h_1 = t_2 - t_1$ and obtain an approximation of $\mathbf{x}(t_2)$, $\mathbf{u}_2$, and so on until we reach t_N.

Consider again the example of the simple pendulum (with no external torque, for simplicity). By repeatedly differentiating Eq. (17.7) and applying the described numerical scheme, we obtain the coupled recurrent sequences:

$$\theta_{n+1} = \theta_n + h_n \omega_n + \frac{h_n^2}{2!}\left[-\frac{g}{L}\sin(\theta_n) - \frac{b}{mL^2}\omega_n\right] + \cdots \tag{17.10a}$$

$$\omega_{n+1} = \omega_n + h_n\left[-\frac{g}{L}\sin(\theta_n) - \frac{b}{mL^2}\omega_n\right]$$
$$+ \frac{h_n^2}{2!}\left[-\frac{g}{L}\cos(\theta_n)\omega_n + \frac{b}{mL^2}\frac{g}{L}\sin(\theta_n) + \left(\frac{b}{mL^2}\right)^2\omega_n\right] + \cdots \tag{17.10b}$$

$$t_{n+1} = t_n + h_n \tag{17.10c}$$

for $n = 0, 1, \ldots, N-1$, where $t_0 = 0$, $\theta_0 = \theta(0)$, and $\omega_0 = 0$.

Implementing such an iterative procedure on a computer is straightforward. We only need to choose the number of terms in the Taylor expansion that we will use, and the appropriate values for the mesh points t_n. The typical choice is to first choose N (the number of steps), and then take all t_n equally spaced in the interval $[a, b]$. This choice leads to $h_n = h = (b-a)/N$, for all n.

We have implemented this algorithm and plotted the approximate solution of the pendulum for the algorithm including first-, second-, and third-order derivative terms of the Taylor expansion using $h = 0.1$. The result, shown in Figure 17.2, displays the typical behavior of Taylor methods. The first-order method (the largest plot in the figure), also known as the Euler method, provides a very poor approximation to the solution. A second-order expansion does a better job, but still not perfect. The solution that includes third-order terms (the darker one in the figure) is almost indistinguishable from the true solution.

The reason for this behavior can be explained using the error term for the Taylor expansion. If we include up to p order terms in this expansion, the remainder (the *local error*) has the form $O(h^{p+1})$. Thus, for the same value of h, we gain an order of magnitude in the approximation with each term. A method with $O(h^{p+1})$ local error is said to be of order p. Although the error in one step of the method is of the order of h^{p+1}, we need to take $(b-a)/h$ steps to go from a to b. Thus, the global error can be shown to be of the form $O(h^p)$.

The approximation described above can also be improved for any order by reducing the value of h. However, reducing h increases both the computational effort and the round-off error. Thus, we will need to balance the order of the method with the right value of h for our problem.

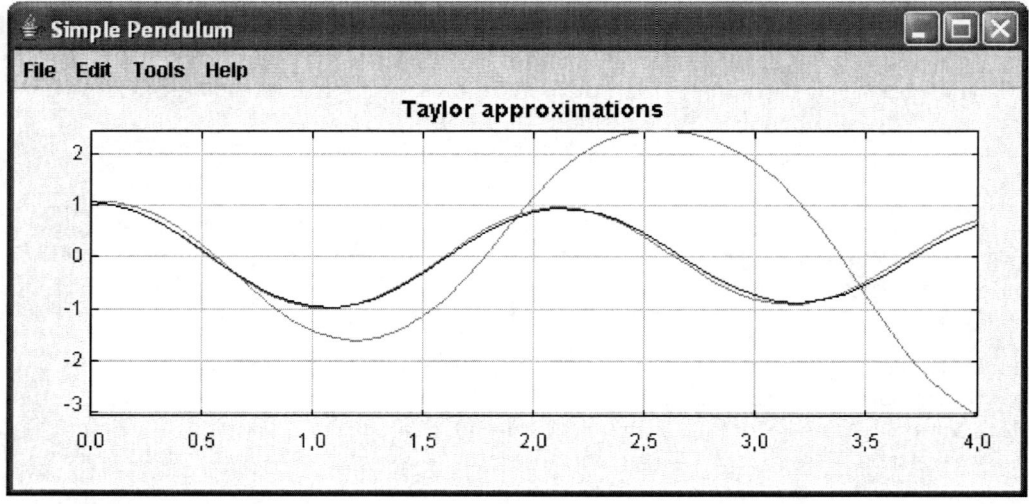

FIGURE 17.2 Taylor methods applied to the pendulum example.

17.4 Runge–Kutta Methods

Although methods based on a Taylor series expansion can be made very accurate by taking sufficiently many terms, computing higher-order derivatives becomes increasingly complicated and the resulting code cannot be reused for a different problem. For this reason, numerical analysts have developed methods with similar accuracy that are easier to implement and reuse. In particular, these methods only require evaluations of the function $\mathbf{f}$ that defines the differential equation. The Runge–Kutta methods described here are among the most popular.

Whereas Taylor methods advance the solution by evaluating $\mathbf{f}$ and its derivatives at a single point, Runge–Kutta methods advance the solution $\mathbf{u}_n$ by evaluating $\mathbf{f}$ at several intermediate points in the interval $[t_n, t_{n+1}]$. These intermediate results are combined in such a way as to match the Taylor expansion of the solution up to a given order. The precise formulation of an s-stage explicit Runge–Kutta method is the following:

$$\mathbf{k}_1 = \mathbf{f}(\mathbf{u}_n, t_n)$$
$$\mathbf{k}_2 = \mathbf{f}(\mathbf{u}_n + h_n a_{21} \mathbf{k}_1, t_n + c_2 h_n)$$
$$\cdots \qquad (17.11)$$
$$\mathbf{k}_s = \mathbf{f}(\mathbf{u}_n + h_n(a_{s1}\mathbf{k}_1 + \cdots + a_{s,s-1}\mathbf{k}_{s-1}), t_n + c_s h_n)$$
$$\mathbf{u}_{n+1} = \mathbf{u}_n + h_n(b_1 \mathbf{k}_1 + b_2 \mathbf{k}_2 + \cdots + b_s \mathbf{k}_s)$$

Note that the method describes a simple recurrent algorithm, which is easy to implement on a computer and requires only the evaluation of the function $\mathbf{f}$. The method is determined by the parameters, which are listed traditionally in the form of a table (Table 17.1). Usually, the c_i satisfies the condition: $c_i = \sum_{j=1}^{i-1} a_{ij}$.

For a given s, the algorithm consists of choosing appropriate values of the a, b, and c parameters that provide a good approximation of the solution. The case $s = 2$ illustrates the procedure. Consider the second-order Taylor approximation of the solution at t_n given by

$$\mathbf{x}(t_n + h_n) = \mathbf{x}(t_n) + h_n \mathbf{f} + \frac{h_n^2}{2!}(\mathbf{f}_t + \mathbf{f}_x \mathbf{f}) + O(h_n^3) \qquad (17.12)$$

where $\mathbf{f}$ and its derivatives are evaluated at $(\mathbf{x}(t_n), t_n)$. If we expand the value of the approximation produced by the method in Eq. (17.11) for $s = 2$ and use the Taylor expansion of the function $\mathbf{f}$, we find that

$$\mathbf{u}_{n+1} = \mathbf{u}_n + h_n[b_1 \mathbf{f}(\mathbf{u}_n, t_n) + b_2 \mathbf{f}(\mathbf{u}_n + h_n a_{21} \mathbf{f}(\mathbf{u}_n, t_n), t_n + c_2 h_n)]$$
$$= \mathbf{u}_n + h_n b_1 \mathbf{f} + h_n b_2(\mathbf{f} + h_n c_2 \mathbf{f}_t + h_n a_{21} \mathbf{f}_x \mathbf{f}) + O(h_n^3) \qquad (17.13)$$
$$= \mathbf{u}_n + h_n(b_1 + b_2)\mathbf{f} + h_n^2(b_2 c_2 \mathbf{f}_t + b_2 a_{21} \mathbf{f}_x \mathbf{f}) + O(h_n^3)$$

TABLE 17.1 Generic Runge–Kutta Table of Coefficients

0					
c_2	a_{21}				
c_3	a_{31}	a_{32}			
$\vdots$	$\vdots$	$\vdots$	$\ddots$		
c_s	a_{s1}	a_{s2}	$\cdots$	$a_{s,s-1}$	
	b_1	b_2	$\cdots$	b_{s-1}	b_s

TABLE 17.2 Coefficients of the Classical Fourth-Order Runge–Kutta Method

0				
1/2	1/2			
1/2	0	1/2		
1	0	0	1	
	1/6	2/6	2/6	1/6

Except where explicitly indicated, $\mathbf{f}$ and its derivatives are evaluated at $(\mathbf{u}_n, t_n)$. Hence, if $\mathbf{u}_n$ is a good approximation of $\mathbf{x}(t_n)$, the method will approximate the solution at $t_n + h_n$ with a third-order local error (which leads to a second-order method) if $b_1 + b_2 = 1$, $b_2 c_2 = 1/2$, and $b_2 a_{21} = 1/2$. Because we have four unknowns and only three equations, this expression gives a one-parameter family of methods for Runge–Kutta algorithms of order 2. The most popular are the following:

(Heun's method)

$$\mathbf{u}_{n+1} = \mathbf{u}_n + \frac{h_n}{2}\left[\mathbf{f}(\mathbf{u}_n, t_n) + \mathbf{f}(\mathbf{u}_n + h_n \mathbf{f}(\mathbf{u}_n, t_n), t_n + h_n)\right] \qquad (17.14)$$

(Midpoint method)

$$\mathbf{u}_{n+1} = \mathbf{u}_n + h_n \mathbf{f}\left(\mathbf{u}_n + \frac{h_n}{2}\mathbf{f}(\mathbf{u}_n, t_n), t_n + \frac{h_n}{2}\right) \qquad (17.15)$$

(Ralston's method)

$$\mathbf{u}_{n+1} = \mathbf{u}_n + \frac{h_n}{4}\left[\mathbf{f}(\mathbf{u}_n, t_n) + 3\mathbf{f}\left(\mathbf{u}_n + \frac{2}{3}h_n \mathbf{f}(\mathbf{u}_n, t_n), t_n + \frac{2}{3}h_n\right)\right] \qquad (17.16)$$

which correspond to $b_1 = 1/2$, $b_1 = 0$, and $b_1 = 1/4$, respectively. Among all second-order Runge–Kutta methods, Ralston's method[2] provides a minimum bound of a specified type for the local error (Ralston, 1962).

The same approach can be used to obtain families of higher-order Runge–Kutta methods, although the algebra becomes much more complicated and more sophisticated techniques must be used (Hairer et al., 2000; Butcher, 1987).

The best seller of all Runge–Kutta methods is the fourth-order *classical* method given in Table 17.2, which requires four rate evaluations per step. But this method is certainly not the end of the story, and we will give higher-order methods in Section 17.6. However, for orders 5 and above, all Runge–Kutta methods require a number of stages strictly greater than the order. This limitation is one of the *Butcher's barriers*.

17.5 Implementation

Runge–Kutta methods provide powerful, high-precision algorithms for solving initial value problems. In this section we show how these methods can be used to create computer algorithms that are effective, easy to program, and easy to reuse for different ODE problems. In particular, we will show how the OSP library implements some of these routines.

Experienced programmers divide a complex problem into several smaller parts that they later combine in a modular way. This division helps implement, test, and, if required, later replace or reuse code. The

---

[2]We name Eq. (17.16) as Ralston's method following Ralston (1962). Other authors define Ralston's second-order method using different coefficients.

Ordinary Differential Equations

object-oriented features of Java are very appropriate for such an approach. To show how it works in practice, OSP first defines a Java interface that encapsulates the mathematical definition of an ODE:

```
public interface ODE {
  public double[] getState();
  public void getRate(double[] state, double[] rate);
}
```

Encapsulation is an object-oriented concept that combines the data and behavior of an object, hiding the implementation from the user of that object. Encapsulation helps abstract in computer terms the characteristic features of an object. In this case, encapsulation means that to define a particular ODE a user needs to create a class that implements the two methods defined in the interface.[3]

The getState method returns the $(x_0, x_1, \ldots, x_k)$ array of state variables, while the getRate method evaluates the derivatives using the given state array and stores the result in the given rate array $(\dot{x}_0, \dot{x}_1, \ldots, \dot{x}_k)$. Since, as we have seen, Runge–Kutta methods evaluate the rate multiple times as they advance the system by h_n, the state given to getRate is usually not the current state of the system.

The OSP library uses the dynamical system approach and considers all ODEs as autonomous. Hence, the independent variable t is always included in the system of differential equations as a final trivial equation $\dot{t}(t) = 1$ (consequently, x_k represents the time). Also, in cases where second-order equations are coded, OSP uses the convention that the velocity coordinate follows the corresponding position coordinate when constructing the state and rate arrays. This ordering makes it possible to efficiently code certain numerical algorithms (such as the Verlet method in Section 17.8).

The second step is to abstract the general behavior of Runge–Kutta solvers. Independent of the number of stages it uses and the values of the parameters, a Runge–Kutta method (and many other methods as well) is encapsulated in the OSP library by the ODESolver interface:

```
public interface ODESolver {
  public void initialize(double stepSize);
  public double step();
  public void setStepSize(double stepSize);
  public double getStepSize();
}
```

Any implementing class also needs to provide a public constructor (a method that instantiates an object of the class), which takes an ODE class as a parameter. With this reference to the ODE class, the solver's initialize method sets the initial step size and allocates arrays to store temporary values. Typically, the initialize method needs to be called only once at the beginning of the program execution.

The step method is the heart of the algorithm. The ODESolver obtains a reference (pointer) to the state array using the ODE object's getState method and advances this state by applying the numeric method defined within the solver. Because adaptive algorithms (see Section 17.6) are free to change the step size, the method returns the value of the step size that was used. The interface also provides the setStepSize and getStepSize methods to modify and read the step size parameter.

Before showing how a particular Runge–Kutta solver class looks, let us see how the combination of these two interfaces works in practice to create clean, well-structured, and easy to reuse code. We first create a Java class that implements the ODE interface for the (undriven) pendulum example in Section 17.1:

```
public class Pendulum implements ODE {
  double[] state = new double[] {Math.PI/2.0, 0.0, 0.0};
  double m=1.0, L=1.0, g=9.8, b=0.1; // parameters
```

---

[3]*Method* is the term Java uses for both functions and routines. This usage should not be confused with its use in an expression that refers to an algorithm such as "Runge–Kutta method."

```
  public double[] getState() { return state; }
  public void getRate(double[] state, double[] rate){
    rate[0] = state[1];
    rate[1] = - g/L*Math.sin(state[0]) - b/(m*L*L)*state[1];
    rate[2] = 1; // time derivative
  }
}
```

If we assume that we have created an implementation of the ODESolver interface for the fourth-order Runge–Kutta method called RK4, a simple driver program would instantiate both the ODE and ODESolver classes and repeatedly invoke the solver's step method to solve the initial value problem:

```
public class PendulumMain {
  public static void main(String[] args){
      double max = 4; // solution range
      double h = 0.1;   // ode step size
      ODE ode = new Pendulum();
      ODESolver ode_solver = new RK4(ode);
      ode_solver.initialize(h);
      while(max>0) {
        double[] state = ode.getState();
        String xStr = "angular position = "+state[0];
        String vStr = ", angular velocity = "+state[1];
        String tStr = ", time = "+state[2];
        System.out.print(xStr+vStr+tStr+"\n");
        max -= ode_solver.step();
      }
  }
}
```

The important point is that the end user only needs to program the class that implements the ODE interface and thereby defines the initial value problem. Numerical algorithms, such as the RK4 class, are implemented in the OSP library. These algorithms can be written (and tested) once, perhaps by a professional numerical analyst, in an independent manner. Switching to a different numerical algorithm only requires instantiating a different ODESolver class (see Table 17.6 for a listing of solvers in the OSP library).

We now list, for completeness, a possible implementation of the step method for the RK4 method:

```
public double step() {
  double state[] = ode.getState();
  ode.getRate(state, k1);
  for(int i=0;i<numEqn;i++) {
    temp_state[i] = state[i]+stepSize*k1[i]/2;
  }
  ode.getRate(temp_state, k2);
  for(int i=0;i<numEqn;i++) {
    temp_state[i] = state[i]+stepSize*k2[i]/2;
  }
  ode.getRate(temp_state, k3);
  for(int i=0;i<numEqn;i++) {
    temp_state[i] = state[i]+stepSize*k3[i];
  }
```

```
    ode.getRate(temp_state, k4);
    for(int i=0;i<numEqn;i++) {
      state[i] = state[i]+stepSize*(k1[i]+2*k2[i]+2*k3[i]+k4[i])/6.0;
    }
    return stepSize;
}
```

17.6 Adaptive Step

An important aspect of solving ODEs numerically is that of choosing the best possible set of mesh points $t_0 < \cdots < t_N$, mentioned in Section 17.2. The *fixed step size* approach that we have used so far, that is, $h_n = h = (b-a)/N$, is not always appropriate. The solution of the ODE may vary rapidly in some parts of the $[a, b]$ interval (which requires a small step size), while it may be very smooth in other parts of it (which allows a larger step size). Taking the same step size for the entire interval might result in either loss of precision or in an unnecessary waste of computer resources, or both.

Making an appropriate choice of mesh points in advance can be difficult. A detailed analytical study of the local error at different points within the interval is required. Alternatively, there exist numerical techniques that allow the computer to estimate this error and automatically compute an appropriate value of h_n for each integration step. These techniques are, for Runge–Kutta methods, based on two approaches: interval-halving and embedded formulas.

Interval-Halving

The first approach consists in using, from each point $(\mathbf{u}_n, t_n)$, the same Runge–Kutta method in two parallel computations. The first one applies the method once with a given step size h, while the second uses the method twice, each with half this step size, i.e., $h/2$. We thus obtain two different approximations of the solution at $t_n + h$. Both values are then compared to estimate the error obtained with this step size, to accept the solution or not, and to compute a more appropriate step size (Schilling and Harris, 2000).

Although this extrapolation process can be easily programmed, the resulting code is inefficient. For the case of the fourth-order Runge–Kutta method, we would need 11 different evaluations of the rate function for each individual step and would achieve only a fifth-order approximation.

Embedded Runge–Kutta Formulas

A more efficient scheme was first discovered by Fehlberg, who found a pair of Runge–Kutta formulas that used the same set of coefficients to provide two approximations of different orders. The difference between both approximations can then be used to estimate the error of the lower-order formula, while taking the higher-order approximation as the final output of the method (this is called *local extrapolation*).

Since the original Fehlberg scheme, many other embedded formulas have been found. The most used ones are the 6-stage formulas of Cash and Karp (Table 17.3), and the 7-stage formulas of Dormand and Prince (table not provided), which attempt to minimize the error of the local extrapolation approximation. Table 17.3 contains a pair of Runge–Kutta–Fehlberg formulas of orders 5 and 4. The first row of b-coefficients in Table 17.3 corresponds to the higher-order approximation, while the second row gives the lower-order one. Both Cash–Karp and Dormand–Prince schemes are implemented in the OSP library.

Adapting the Step

Once the estimated error is found, the technique adapts the step size so that this error keeps within reasonable limits, decreasing the step size if the error is too large, and increasing it if the error is too small. A popular procedure to compute the new step size, $\tilde{h}$, consists in using the following formula (Press et al., 1992):

$$\tilde{h} = Sh \left| \frac{\Delta_0}{\Delta_1} \right|^\alpha \qquad (17.17)$$

TABLE 17.3 Coefficients for the Embedded Runge–Kutta Formulas Computed by Cash and Karp

0						
1/5	1/5					
3/10	3/40	9/40				
3/5	3/10	−9/10	6/5			
1	−11/54	5/2	−70/27	35/27		
7/8	1631/55296	175/512	575/13824	44275/110592	253/4096	
	37/378	0	250/621	125/594	0	512/1771
	2825/27648	0	18575/48384	13525/55296	277/14336	1/4

where S is a safety constant (such as 0.9) that prevents $\tilde{h}$ from actually reaching the limit for an acceptable step size, h the current step size, Δ_0 denotes the desired accuracy, and Δ_1 measures the current accuracy. Finally, α denotes a constant that (for a p-order method) is $1/(p+1)$ when the step size is increased, and $1/p$ otherwise.

There are different ways to interpret Δ_0 and Δ_1 in Eq. (17.17). Usually, Δ_1 is taken as the maximum of the absolute values of the components of the estimated error (for embedded formulas, of the difference of both approximations). For Δ_0, the simplest option is to let the user specify a suitable small tolerance, ϵ. For other possibilities such as including a scaling vector, see Press et al. (1992), Hairer et al. (2000), and Enright et al. (1995).

17.7 Implementation of Adaptive Step

Implementing an adaptive algorithm based on a pair of Runge–Kutta embedded formulas is straightforward. The following code shows how the `step` method can be implemented for the Cash–Karp formulas. Note that the code includes additional checks to avoid abrupt changes to the step size. In addition, the algorithm will return if the required precision cannot be attained.

```
public double step() {
  int iterations = 10;
  double currentStep = stepSize, error=0;
  double state[] = ode.getState();
  ode.getRate(state, k[0]); // gets the initial rate
  do {
    iterations--;
    currentStep = stepSize;
    // Compute the k's
    for (int s=1; s<numStages; s++) {
      for (int i=0; i<numEqn; i++) {
        temp_state[i] = state[i];
        for (int j=0; j<s; j++) {
          temp_state[i] = temp_state[i]+stepSize*a[s-1][j]*k[j][i];
        }
      }
      ode.getRate(temp_state, k[s]);
    }
    // Computes the error
    error = 0;
```

```java
    for (int i=0; i<numEqn; i++) {
      truncErr = 0;
      for (int s=0; s<numStages; s++) {
        truncErr = truncErr+stepSize*er[s]*k[s][i];
      }
      error = Math.max(error,Math.abs(truncErr));
    }
    if (error<=Float.MIN_VALUE) { // error too small
      error = tol/1.0e5; // actually increase stepSize x10
    }
    // find h step for the next try.
    if(error>tol) { // shrink, but by no more than x10
      double fac = 0.9*Math.pow(error/tol,-0.25);
      stepSize = stepSize*Math.max(fac,0.1);
    }
    else if(error<tol/10.0) { // grow, but no more than x10
      double fac = 0.9*Math.pow(error/tol, -0.2);
      if(fac>1) { // sometimes fac<1 if error/tol close to 1
        stepSize = stepSize*Math.min(fac, 10);
      }
    }
  } while(error>tol && iterations>0);
  // advance the state
  for (int i=0; i<numEqn; i++){
    for (int s=0; s<numStages; s++){
      state[i] = state[i]+currentStep*b5[s]*k[s][i];
    }
  }
  return currentStep; // step actually taken.
}
```

Independent of how the algorithm is implemented internally, Java's object-oriented features again help us construct easy-to-use and well-structured code for solving ODEs using adaptive methods. The OSP library defines the ODEAdaptiveSolver interface that encapsulates the methods for an adaptive solver:

```java
public interface ODEAdaptiveSolver extends ODESolver {
  public void setTolerance(double tol);
  public double getTolerance();
  public int getErrorCode();
}
```

Because the interface *extends* the previously defined ODESolver interface, it inherits all the methods defined in this interface. The only new methods are those that allow the user to specify or read the desired tolerance (the ϵ mentioned above) and to check for a possible error in the application of the algorithm. This interface allows the user to create a code such as the following:

```java
ODEAdaptiveSolver ode_solver = new CashKarp45(ode);
ode_solver.initialize(h);
ode_solver.setTolerance(1.0e-3);
```

Note that the only difference with the previous code is in the instantiation of the ODEAdaptiveSolver object and the addition of a line that sets the desired tolerance. All other parts of the driver program and the definition of the ODE remain unchanged.

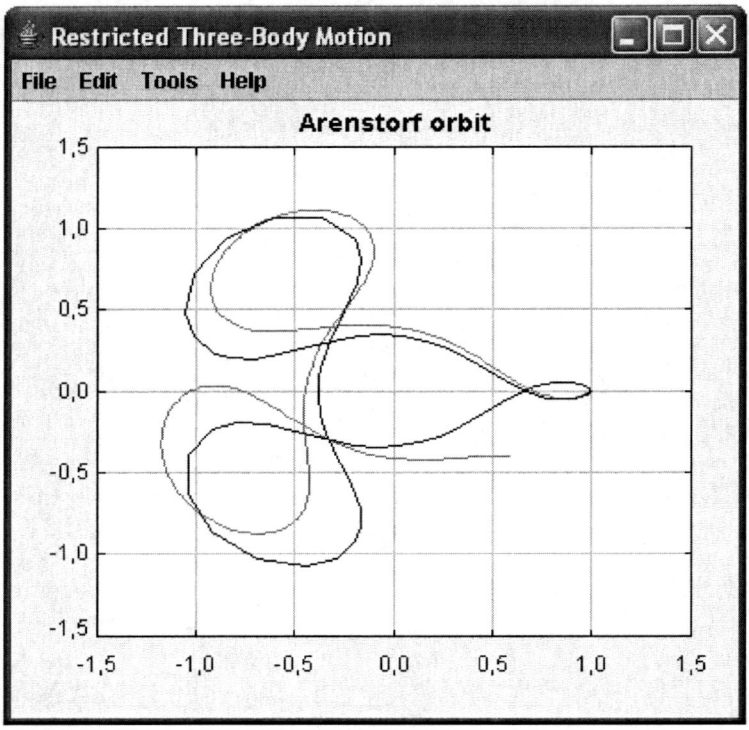

FIGURE 17.3 Arenstorf orbit computed with fixed step (gray) and adaptive (black) Runge–Kutta methods.

Using an adaptive algorithm can dramatically increase the performance of our programs in situations where the solution has regions of different behavior. A typical example is that of the computation of the *Arenstorf orbits*, closed trajectories of the restricted three-body problem (two bodies of masses μ and $1 - \mu$ moving in a circular orbit, and a third body of negligible mass moving in the same plane), such as a satellite–earth–moon system. The equations for the motion of the third body are given by Hairer et al. (2000):

$$\ddot{x}_1(t) = x_1(t) + 2\dot{x}_2(t) - (1-\mu)\frac{x_1(t)+\mu}{D_1} - \mu\frac{x_1(t)-(1-\mu)}{D_2} \quad (17.18a)$$

$$\ddot{x}_2(t) = x_2(t) - 2\dot{x}_1(t) - (1-\mu)\frac{x_2(t)}{D_1} - \mu\frac{x_2(t)}{D_2} \quad (17.18b)$$

where $D_1 = ((x_1(t)+\mu)^2 + x_2^2(t))^{3/2}$ and $D_2 = ((x_1(t) - (1-\mu))^2 + x_2^2(t))^{3/2}$.

The Arenstorf orbit for the initial values $x_1(0) = 0.994$, $\dot{x}_1(0) = 0$, $x_2(0) = 0$, $\dot{x}_2(0) = -2.03173262955733683573 02057924$, and $\mu = 0.012277471$ is displayed in Figure 17.3.[4] The gray trajectory has been computed using 3000 fixed step iterations of the fourth-order Runge–Kutta algorithm (the trajectory is not even closed), while the black, closed trajectory has been computed by the adaptive Cash–Karp algorithm with a tolerance of 10^{-5} in 123 steps with 98 steps accepted and 25 steps rejected. The fact that the second algorithm is of order 5 is secondary. Comparing the total number of rate evaluations (12,000 versus 713) shows that the adaptive method is far superior.

---

[4]This high accuracy is required because Arenstorf orbits are very sensitive to small changes in the initial conditions.

Ordinary Differential Equations

Multistepping

Given the relatively small additional effort required to compute two Runge–Kutta solutions and thereby to control the error, is there any reason to use a nonadaptive algorithm? Sometimes. Adaptive algorithms do not work if the rate contains discontinuous functions. A fixed step size is also convenient if the output is to have evenly spaced values.

However, a fixed step size can also be achieved by taking multiple steps. Because the step size is chosen so as to obtain the desired accuracy, the last step will almost always overshoot the desired endpoint. This overshoot can be eliminated by reducing the last step. The `ODEMultistepSolver` class in the OSP library combines multiple steps of any adaptive algorithm in this manner.[5]

Interpolation

Using an adaptive solver to compute the solution at fixed intervals has an obvious inconvenience, it prevents the adaptive algorithm from taking large steps in areas where the estimated error is small. A different technique, interpolation, can help improve performance.

The idea consists in letting the solver work internally at its own pace, taking as large a step as it sees fit and asking it to provide approximated values of the solution at equally spaced points. The algorithm is then assumed to interpolate the internally computed values to produce an approximation at the given points. The interpolation is expected to have a similar order of precision as the internal approximations of the solution.

Interpolation is a bit more complex to implement, since it actually involves the concept of *dense output* (Hairer et al., 2000). This means that, instead of providing an approximate solution to the ODE at a finite set of points of the interval $[a, b]$, the method provides a function that can approximate the solution of the ODE at *any* point of it. The user could then in principle use this function to produce solution points everywhere in the trajectory (a dense output). OSP has implemented this technique using the `ODEInterpolationSolver` interface for two of the most powerful, higher-order methods of the library: the Dopri5 and Dopri853 classes. The first one implements a pair of embedded optimized formulas of orders 5 and 4, found by Dormand and Prince, together with dense output interpolation of order 4. The second is based on a pair of embedded formulas of orders 8 and 5, with dense output interpolation of order 7.

These two ODE solvers are available in the OSP library and can be created using factory methods as the code below shows.

```
ODEAdaptiveSolver dopri5=ODEInterpolationSolver.Dopri5(ode);
ODEAdaptiveSolver dopri8=ODEInterpolationSolver.Dopri853(ode);
```

17.8 Performance and Other Methods

Up to this point, we have only used explicit Runge–Kutta methods for solving ordinary differential equations. Explicit Runge–Kutta methods work correctly in almost every situation. But they might not be the most efficient or convenient ones:

- When the ODE to solve is a *stiff* equation for which implicit algorithms have better stability properties and are therefore more efficient.
- When the solution of the ODE is very smooth or the rate function is very expensive to evaluate, a multistep algorithm can be preferred.
- When high accuracy is required (of the order of 10^{-12} or higher), and evaluating the rate function is not cheap in terms of CPU time, extrapolation techniques can be more efficient in this situation.

---

[5]Multistepping and the `ODEMultistepSolver` class should not be confused with the multistep methods described in Section 17.8.

- When the long-term behavior or the preservation of certain geometrical properties of the computed solutions of a Hamiltonian system are of interest, a *symplectic* algorithm is preferred.

We now briefly cover these four cases.

Implicit Algorithms and Stiff Equations
An ODE is said to be *stiff* when the solution comprises two or more terms that change at speeds which differ in several orders of magnitude. Consider the simple one-dimensional initial value problem (Chapra and Canale, 2006):

$$\dot{x}(t) = -1000x(t) + 3000 - 2000e^{-t}, \quad x(0) = 0 \qquad (17.19)$$

which has the exact solution $x(t) = 3 - \frac{997}{999}e^{-1000t} - \frac{2000}{999}e^{-t}$. The solution contains a slow-changing exponential together with a fast-changing one. Although the fast exponential quickly contributes only very small values, its presence forces a typical explicit algorithm to keep a small step size, even after the transient part of the solution becomes very small.

The stability of an algorithm refers to its capability of not exponentially propagating the (unavoidable) small errors that take place in the solving process, even when taking moderately large step sizes. Implicit algorithms typically show better stability properties and consequently perform better on stiff problems.

A more generic formulation of an s-stage Runge–Kutta algorithm,

$$\mathbf{k}_i = \mathbf{f}(\mathbf{u}_n + h_n \sum_{j=1}^{s} a_{ij}\mathbf{k}_j, t_n + c_i h_n), \quad i = 1, \ldots, s$$

$$\mathbf{u}_{n+1} = \mathbf{u}_n + h_n \sum_{i=1}^{s} b_i \mathbf{k}_i \qquad (17.20)$$

allows, when any of the $a_{ij}, j \geq i$, is not 0, for implicit algorithms, in which the values of the $\mathbf{k}_i$ may appear at both sides of the evaluation of some intermediate rate. If the rate function is nontrivial, finding the solution of the (possibly nonlinear) equations imposes an additional computation burden.

Implicit Runge–Kutta methods often possess better stability properties than their explicit counterparts given reasonably good behavior of the rate function $\mathbf{f}$. Their implementation is, however, more sophisticated, because it must include techniques to solve the nonlinear system of equations.

The OSP library implements an implicit Runge–Kutta method in the `Radau5` class, following the implementation proposed in Hairer and Wanner (2002). This class includes all the necessary routines to implement the Radau IIA fifth-order method given by Table 17.4, with step size control and dense output (interpolation).

Although implementing such an advanced algorithm is reserved to specialists, using it in the OSP framework is very easy. The `Radau5` class is created using the static (factory) method:

`ODEAdaptiveSolver radau5=ODEInterpolationSolver.Radau5(ode);`

and can be treated as any other `ODEAdaptiveSolver` object.

To show how the method compares to explicit Runge–Kutta methods, the classical fourth-order Runge–Kutta method requires 400 rate function evaluations (for a step size of 0.001) to achieve comparable results

TABLE 17.4 Coefficients for Radau IIA Fifth-Order Method

$(4-\sqrt{6})/10$	$(88-7\sqrt{6})/360$	$(296-169\sqrt{6})/1800$	$(-2+3\sqrt{6})/255$
$(4+\sqrt{6})/10$	$(296+169\sqrt{6})/1800$	$(88+7\sqrt{6})/360$	$(-2-3\sqrt{6})/255$
1	$(16-\sqrt{6})/36$	$(16+\sqrt{6})/36$	$1/9$
	$(16-\sqrt{6})/36$	$(16+\sqrt{6})/36$	$1/9$

Ordinary Differential Equations

in the small interval $[0, 0.1]$ when solving Chapra and Canale's initial value problem [Eq. (17.19)]. The adaptive Cash–Karp formulas required 217 evaluations for the same test (the number of steps growing with the length of the interval, not being restricted to the initial transient part of the solution). Finally, `Radau5` solves the problem with only 60 evaluations, showing no deviation from the analytical solution.

For nonstiff ODEs, explicit adaptive methods usually perform better than implicit ones. For instance, to obtain similar results for the Arenstorf orbit computation of Section 17.7, `Radau5` required 2.5 times the number of rate function evaluations of `CashKarp45`.

Multistep Methods

These methods take advantage of the information contained in previous steps of the solution. The general algorithm for a $(p+1)$-step method is

$$\mathbf{u}_{n+1} = \sum_{j=0}^{p} a_j \mathbf{u}_{n-j} + h \sum_{j=-1}^{p} b_j \mathbf{f}_{n-j}, \quad (n \geq p) \tag{17.21}$$

where $\mathbf{f}_j$ represents $\mathbf{f}(\mathbf{u}_j, t_j)$. Note that, if $b_{-1} \neq 0$, the method is implicit and must be solved by iteration. Simple multistep methods for initial value problems belong to two main families: Adams' methods and backwards differentiation formulas.

Adams' methods derive from numerical methods for the equivalent integral equation:

$$\mathbf{x}(t_{n+1}) = \mathbf{x}(t_n) + \int_{t_n}^{t_{n+1}} \mathbf{f}(\mathbf{x}(t), t) dt \tag{17.22}$$

Typical examples of Adams' method are the following fourth-order explicit and implicit methods (Atkinson, 1989):

$$\mathbf{u}_{n+1} = \mathbf{u}_n + \frac{h}{24}(55\mathbf{f}_n - 59\mathbf{f}_{n-1} + 37\mathbf{f}_{n-2} - 9\mathbf{f}_{n-3}) \tag{17.23a}$$

$$\mathbf{u}_{n+1} = \mathbf{u}_n + \frac{h}{24}(9\mathbf{f}_{n+1} + 19\mathbf{f}_n - 5\mathbf{f}_{n-1} + \mathbf{f}_{n-2}) \tag{17.23b}$$

Sometimes, explicit and implicit Adams' formulas are applied in pairs, in *predictor–corrector* formulas. The latter is a mixed method that tries to obtain some of the stability properties of implicit formulas with the simplicity of implementation of explicit ones. The scheme first uses an explicit method to *predict* the solution of the ODE at t_{n+1}. It then *corrects* the approximation by applying the implicit formula with the value of $\mathbf{f}_{n+1}$ computed from the prediction.

Backwards differentiation formulas are obtained by approximating the solution by a polynomial through a series of past $\mathbf{u}_{n-j}$ points and then taking $\mathbf{u}_{n+1}$ such that the polynomial satisfies the ODE at t_{n+1}. A typical result of this process is (Hairer et al., 2000)

$$\frac{25}{12}\mathbf{u}_{n+1} - 4\mathbf{u}_n + 3\mathbf{u}_{n-1} - \frac{4}{3}\mathbf{u}_{n-2} + \frac{1}{4}\mathbf{u}_{n-3} = h\mathbf{f}_{n+1} \tag{17.24}$$

Both Adams' methods and backwards differentiation formulas need the help of an auxiliary method to start the process (that is, to compute the first $p+1$ points of the solution) and require that the steps be equally spaced. This requirement makes it complicated (though not impossible) to implement adaptive step versions of the algorithms.

Extrapolation Methods

These algorithms effectively accelerate the convergence by using a power series to extrapolate to zero step size applying successively the algorithm with different step sizes. The Bulirsch–Stoer method chooses a sequence of increasing integer numbers N, and then uses, for each of them, a fixed step H, and $h = H/N$, the formula:

$$\mathbf{v}_1 = \mathbf{u}_n + h\mathbf{f}(\mathbf{u}_n, t_n) \tag{17.25a}$$

$$\mathbf{v}_{i+1} = \mathbf{v}_{i-1} + 2h\mathbf{f}(\mathbf{v}_i, t_n + ih), \quad \text{for } i = 1, 2, \ldots, N-1 \tag{17.25b}$$

$$\mathbf{u}_{n+1} = \frac{1}{2}[\mathbf{v}_N + \mathbf{v}_{N-1} + h\mathbf{f}(\mathbf{v}_N, t_n + H)] \tag{17.25c}$$

to advance from $(\mathbf{u}_n, t_n)$ to $(\mathbf{u}_{n+1}, t_n + H)$. The error term for this formula is an expansion with only even powers of h. Standard extrapolation techniques can then be used to obtain an arbitrarily high-order method that accurately approximates the solution in the given interval (Press et al., 1992).

Symplectic Integration Methods
These methods can be the preferred choice when studying Hamiltonian systems:

$$\dot{p}_i = -\frac{\delta H}{\delta q_i}(\mathbf{p}, \mathbf{q}), \qquad \dot{q}_i = \frac{\delta H}{\delta p_i}(\mathbf{p}, \mathbf{q}), \quad i = 1, \ldots, k \tag{17.26}$$

which define a $2k$-dimensional ODE with variables q_i and p_i representing the generalized coordinates of the system and the generalized momenta, respectively. Such problems frequently arise when modeling mechanical systems, where the Hamiltonian H is the energy function of the system.

The *symplectic* property of these problems states that the flow of the system preserves the differential *2-form*:

$$\omega^2 = \sum_{i=1}^{k} dp_i \wedge dq_i \tag{17.27}$$

The symbol $\wedge$ represents the exterior or *wedge* product between the two differential 1-forms (do Carmo, 1994). If $k = 1$ the 2-form above can be interpreted as an area. In this case, the symplectic property has the geometrical interpretation that the area (in phase space) is preserved by the flow. Numerical methods that also preserve this geometric property are called symplectic and are of interest when studying this type of problems.

Although no explicit Runge–Kutta method is symplectic, it is possible to find high-order so-called *partitioned* Runge–Kutta methods that are symplectic. These consist of a pair of Runge–Kutta methods with different sets of coefficients applied "separately" to the variables p_i and q_i. For separable systems, in which the Hamiltonian is of the form $H = T(\mathbf{p}) + V(\mathbf{q})$, such methods do exist (Hairer et al., 2000, 2002; Enright et al., 1995).

A simple lower-order symplectic method that can be applied to solve the equations of motion with acceleration $a(t) = f(x(t), v(t), t)$ for a system of particles is the Verlet algorithm. This is an easy-to-program multistep method that produces stable long-term trajectories. The method has the additional advantage that, although it does not preserve the energy of the system over short times, it does produce accurate averages over long times because the energy oscillates about the mean. It therefore reduces the computational effort for systems where statistical averages are more important than the accuracy of particular trajectories.

The partitioned Runge–Kutta coefficients for the Verlet method are given for (x, v) in Table 17.5. Because these coefficients define a simple implicit algorithm, it is convenient to solve for the values at t_{n+1}. This reformulation, which is included in the OSP library, averages the acceleration at the old and new positions to update the velocity. It is known as the velocity form of the Verlet method and is given by

$$x_{n+1} = x_n + v_n h + \frac{1}{2} a_n h^2 \tag{17.28a}$$

$$v_{n+1} = v_n + \frac{1}{2}(a_{n+1} + a_n)h \tag{17.28b}$$

Ordinary Differential Equations

TABLE 17.5 Partitioned Runge–Kutta Coefficients for the Verlet Method. The Left-Hand Table is Used to Advance the Position and the Right-Hand Table is Used to Advance the Velocity

0	0	0	1/2	1/2	0
1	1/2	1/2	1/2	1/2	0
	1/2	1/2		1/2	1/2

Important areas in which symplectic methods are preferred to other, perhaps more accurate, methods are molecular dynamics and astrophysics, where large-scale simulations involving thousands of particles are studied over very long time scales (Gray et al., 1994).

17.9 State Events

Sometimes the model of a continuous system needs to include discontinuities to reflect special situations. The typical case is that of a collision between two objects. Although the collision can be modeled at a microscopic scale, it is frequently much more convenient to stick to the macroscopic model and instruct the computer to detect that the collision has taken place, find the precise moment when it happened, compute explicitly the state of the system after the collision, and finally restart the continuous model from it. This is a particular case of what is known as a state event (Cellier and Kofman, 2006).

Because our numerical method solves the ODE at given step sizes (either fixed or adaptive), a particular event will almost always take place in between two successive computed states. A simple way to implement event detection is to provide a real valued function of the state, $h(\mathbf{x})$, which changes sign whenever an event takes place. The algorithm can then keep track of possible changes in the sign of h in each solution step and then apply a standard root-finding algorithm to find the precise instant of time when the event takes place. The user must provide a method (function) that changes the state of the system at that particular time. For example, a collision event would use conservation of energy and momentum to change particle velocities.

The OSP library implements a simple state event mechanism in the StateEvent interface. An event takes place whenever the event function h changes from positive to strictly negative (zero does not trigger an event). A sample implementation of this interface for a one-dimensional collision between two planar disks is

```
class OneDimensionalCollisionEvent implements StateEvent {

  public double getTolerance() { return 1.0e-4; } // precision

  public double evaluate(double[] state) { // the event function h
    return Math.abs(state[0]-state[2])-(radius1+radius2);
  }

  public boolean action() { // What to do on collision
    double v_temp = state[1];
    state[1] =
      (2*mass2*state[3]+(mass1-mass2)*state[1])/(mass1+mass2);
    state[3] =
      (2*mass1*v_temp+(mass2-mass1)*state[3])/(mass1+mass2);
    return true; // return state at the collision
  }
}
```

TABLE 17.6 List of ODE Solvers Implemented in the OSP Library (ERK, Explicit Runge–Kutta; IRK, Implicit Runge–Kutta)

Type of Solver	Solver Class Name
ERK fixed step	Euler, Butcher5, Felhberg8, Heun3, Ralston2, RK4
ERK adaptive	CashKarp45, DormandPrince45
ERK adaptive + multistepping	ODEMultistepSolver
ERK adaptive + interpolation	Dopri5, Dopri853
IRK adaptive + interpolation	Radau5
Predictor–corrector fixed step	Adams4, Adams5, Adams6
Symplectic	Verlet
Event-enabled fixed step	ODEBisectionSolver

Because the model is one-dimensional, the disks have the same y coordinate and the `state` array is $(x_1, vx_1, x_2, vx_2, t)$.

The `ODEBisectionEventSolver` class in the OSP library can use any `ODESolver` to advance the system. It checks for possible events and applies the bisection algorithm to find the instant of the event (up to the prescribed tolerance) if an event has occurred. Although more sophisticated techniques can be used, the bisection method is both simple and effective, reduces the number of the computations involved, and provides support for handling situations in which multiple events take place, perhaps simultaneously. Examples of systems displaying quite sophisticated event-handling are available on the Web links for this chapter.

17.10   The OSP Library

Table 17.6 summarizes the numerical algorithms implemented in the OSP library. The `Dopri5`, `Dopri853`, and `Radau5` methods were contributed to the library by Andrew Gusev and Yuri B. Senichenkov of Saint Petersburg Polytechnic University, Russia. These high-order differential equation solvers are being distributed in an optional `osp_ode.zip` archive. All other interfaces and algorithms are implemented in the numerics package (`org.opensourcephsics.numerics`) in the core OSP library and are being distributed in the `osp_core.zip` archive. Both code archives are available on the OSP website (`http://www.opensourcephysics.org`).

The OSP project is a synergy of curriculum development, computational physics, and physics education research. One goal of the project is to make a Java library and a large number of Java simulations available for education using the GNU Open Source model. You can redistribute it and/or modify it under the terms of the GNU General Public License (GPL),[6] as published by the Free Software Foundation either version 2 of the License or (at your option) any later version. Code that uses any portion of the code in the org.opensourcephysics package or any subpackage (subdirectory) of this package must also be released under the GNU GPL.

Programmers wishing to adopt OSP code for their projects are encouraged to do so, provided that they release their source code under the GNU Open Source GPL. Curricular material and developer resources are being distributed from the OSP server hosted at Davidson College and from other servers.

---

[6]The Free Software Foundation website http://www.gnu.org promotes the creation and distribution of open source software. The most common software license is the GNU General Public License (GNU GPL), which makes the software free to all users.

References

Atkinson, K. E. (1989). *An Introduction to Numerical Analysis, 2nd Ed.* Wiley, New York.

Butcher, J. C. (1987). *The Numerical Analysis of Ordinary Differential Equations. Runge–Kutta and General Linear Methods.* Wiley, Chichester.

Cellier, F. E. and E. Kofman (2006). *Continuous System Simulation.* Springer, New York.

Chapra, S. C. and R. P. Canale (2006). *Numerical Methods for Engineers.* McGraw-Hill, New York.

Christian, W. (2007). *Open Source Physics: A Users Guide with Examples.* Addison-Wesley, San Francisco.

do Carmo, M. P. (1994). *Differential Forms and Applications.* Springer, New York.

Enright, W. H., D. J. Higham, B. Owren, and P. Sharp (1995). A survey of the explicit Runge–Kutta method. Technical Report 94-291. University of Toronto.

Gray, S. K., D. W. Noid, and B. G. Sumpter (1994). Symplectic integrators for large scale molecular dynamics simulations: A comparison of several explicit methods. *Journal of Chemical Physics* 101(5) 4062–4072.

Hairer, E., C. Lubich, and G. Wanner (2002). *Geometric Numerical Integration, Structure-Preserving Algorithms for Ordinary Differential Equations.* Springer, Berlin.

Hairer, E., S. P. Nørsett, and G. Wanner (2000). *Solving Ordinary Differential Equations I (Nonstiff Problems),* 2nd Ed. Springer, Berlin.

Hairer, E. and G. Wanner (2002). *Solving Ordinary Differential Equations II (Stiff and Differential-Algenbaic Problems),* 2nd Ed. Springer, New York.

Keller, H. B. (1992). *Numerical Methods for Two-Point Boundary-Value Problems.* Dover, New York.

Press, W. H., S. A. Teukolsky, W. T. Vetterling, and B. P. Flannery (1992). *Numerical Recipes in C: The Art of Scientific Computing,* 2nd Ed. Cambridge University Press, New York.

Ralston, A. (1962). Runge–Kutta methods with minimum error bounds. *Mathematics of Computations* 16(80) 431–437.

Schilling, R. J. and S. L. Harris (2000). *Applied Numerical Methods for Engineers.* Brooks/Cole, Pacific Grove, CA.

18
Difference Equations as Discrete Dynamical Systems

Hassan Sedaghat
Virginia Commonwealth University

18.1 Introduction..18-1
18.2 Basic Concepts..18-2
18.3 First-Order Difference Equations..................................18-4
   Asymptotic Stability: Necessary and Sufficient
   Conditions • Cycles and Limit Cycles • Chaos • Notes
18.4 Higher Order Difference Equations..............................18-8
   Asymptotic Stability: Weak Contractions •
   Asymptotic Stability: Coordinate-Wise Monotonicity •
   Persistent Oscillations and Chaos • Semiconjugacy:
   First-Order Equations Revisited • Notes

18.1 Introduction

Interest in difference equations goes a long way back to the times before the discovery of differential and integral calculus. For instance, the famous sequence 0,1,1,2,3,5,8, ... that appeared in the work of Fibonacci (ca. 1202) is the solution of a difference equation, namely $x_n = x_{n-1} + x_{n-2}$ with given initial values $x_{-1} = 0, x_0 = 1$. This difference equation also generates the well-known Lucas numbers if $x_{-1} = 1$, $x_0 = 3$. Even after the invention of the concept of derivative until around the mid-twentieth century, difference equations found numerous applications in numerical analysis where they were used in the solution of algebraic and differential equations. Indeed, the celebrated Newton's method for finding roots of scalar equations is an example of a difference equation, as is the equally famous Euler's method for estimating solutions of differential equations through estimation of the derivative by a finite difference (Burden and Fairs, 1997). These are just two among many other and more refined difference methods for dealing with complex problems in calculus and differential equations. By the mid-twentieth century, the theory of linear difference equations had been developed in sufficient detail to rival, indeed parallel, its differential analog. This theory had already been put to use in the 1930s and 1940s by economists (Hicks, 1965; Samuelson, 1939) in their analyses of discrete-time models of the business cycle.

Interest of a different sort began to emerge in the 1960s and 1970s with important discoveries, such as the Mandelbrot and Julia sets, the Sharkovsky ordering of cycles, and the Li-Yorke "chaos theorem." A substantial amount of work by numerous researchers since then led to the creation of a qualitative theory of difference equations that no longer paralleled similar discoveries in differential equations. Although many analogs can be found between the two disciplines, there are also significant differences; for example, the Poincare–Bendixon theorem (Hirsch et al., 2004) establishes dimension 3 as the minimum needed for the occurrence of deterministic chaos in differential equations, whereas such behavior can appear even in simple, one-dimensional difference equations. The work in the latter part of the twentieth century inspired a further development of the qualitative theory of difference equations, which included the study of conditions for asymptotic stability of equilibria and cycles, and other significant aspects of nonlinear difference equations.

Solutions of difference equations are sequences and their existence is often not a significant problem, in contrast to differential equations. Furthermore, it is unnecessary to estimate solutions of difference equations. Because of their recursive nature, it is easy to generate actual solutions on a digital computer starting from given initial values. Therefore, modelers are quickly rewarded with insights about both the transient and the asymptotic behaviors of their equation of interest. A deeper understanding can then be had from the qualitative theory of nonlinear difference equations, which has now been developed sufficiently to make it applicable to a wide variety of modeling problems in the biological and social sciences. In studying nonlinear difference equations, qualitative methods are not simply things to use in the absence of quantitative exactitude. In the relatively rare cases, where the general solution to a nonlinear difference equation can be found analytically, it is often the case that such a solution has a complicated form that is more difficult to use and analyze than the comparatively simple equation that gave rise to it (see, e.g., Example 9). Thus, even with an exact solution at hand, it may not be easy to answer basic questions such as whether an equilibrium exists, or if it is stable, or if there are periodic or nonperiodic solutions.

In this chapter, we present some of the fundamental aspects of the modern qualitative theory of nonlinear difference equations of order one or greater. The primary purpose is to acquaint the reader with the outlines of the standard theory. This includes some of the most important results in the field as well as a few of the latest findings so as to impart a sense that a coherent area of mathematics exists in the discrete settings that is independent of the continuous theory. Indeed, there are no continuous analogs for many of the results that we discuss below. As it is not possible to cover so broad an area in a limited number of pages, we leave out all proofs. The committed reader may pursue the matters further through the extensive list of references provided. Entire topics, such as bifurcation theory, fractals and complex dynamics, and measure theoretic or stochastic dynamics had to be left out; indeed, each of these topics is quite extensive and it would be impossible to meaningfully include more than one of these within the confines of a single chapter.

18.2 Basic Concepts

A *discrete dynamical system* (autonomous, finite dimensional) basically consists of a mapping $F : D \to D$ on a nonempty set $D \subseteq \mathbb{R}^m$. We usually assume that F is continuous on D. We abbreviate the composition $F \circ F$ by F^2, and refer to the latter as an *iterate* of F. The meaning of F^n for $n = 3, 4, \ldots$ is inductively clear; for convenience, we also define F^0 to be the identity mapping. For each $\mathbf{x}_0 \in D$, the sequence $\{F^n(\mathbf{x}_0)\}$ of iterates of F is called a *trajectory* or *orbit* of F through $\mathbf{x}_0$ (more specifically, a *forward orbit* through $\mathbf{x}_0$). Sometimes, $\mathbf{x}_0$ is called the *initial point* of the trajectory. In analogy with differential equations, we sometimes refer to the system domain as the *phase space*, and call the plot of a trajectory in D a *phase plot*. Also, the plot of a scalar component of $F^n(\mathbf{x}_0)$ versus n is often called a *time series*.

Associated with the mapping F is the recursion

$$\mathbf{x}_n = F(\mathbf{x}_{n-1}) \tag{18.1}$$

which is an example of a *first-order, autonomous vector difference equation*. The vector equation (18.1) is equivalent to a *system* of scalar difference equations, analogously to systems of differential equations which are composed of a finite number of ordinary differential equations. If the map $F(\mathbf{x}) = A\mathbf{x}$ is linear, where $\mathbf{x} \in \mathbb{R}^m$ and A is an $m \times m$ matrix of real numbers, then Eq. (18.1) is called a *linear difference equation*. Otherwise, Eq. (18.1) is *nonlinear* (usually this excludes cases like the linear-affine map $F(\mathbf{x}) = A\mathbf{x} + B$, where B is an $m \times m$ matrix, since such cases are easy to convert to linear ones by a translation). Each trajectory $\{F^n(\mathbf{x}_0)\}$ is a *solution* of Eq. (18.1) with initial point $\mathbf{x}_0$ and may be abbreviated $\{\mathbf{x}_n\}$. Unlike a first-order nonlinear differential equation, it is clear that Eq. (18.1) always has solutions as long as F is defined on D, and that each solution of Eq. (18.1) may be recursively generated from some point of D by iterating F. A (*scalar, autonomous*) *difference equation of order m* is defined as

$$x_n = f(x_{n-1}, x_{n-2}, \ldots, x_{n-m}) \quad (18.2)$$

where the scalar map $f : \mathbb{R}^m \to \mathbb{R}$ is continuous on some domain $D \subset \mathbb{R}^m$. Given any set of m initial values $x_0, x_{-1}, \ldots, x_{-m+1} \in \mathbb{R}$, Eq. (18.2) recursively generates a solution $\{x_n\}$, $n \geq 1$. Eq. (18.2) may be expressed in terms of vector equations as defined previously. Associated with f or with Eq. (18.2) is a mapping

$$V_f(u_1, \ldots, u_m) \doteq [f(u_1, \ldots, u_m), u_1, u_2, \ldots, u_{m-1}]$$

of $\mathbb{R}^m$, which we call the *standard vectorization* or "unfolding" of f. Note that if we define

$$\mathbf{x}_n \doteq [x_n, \ldots, x_{n-m+1}], \quad n \geq 0$$

then

$$\begin{aligned} V_f(\mathbf{x}_{n-1}) &= [f(x_{n-1}, \ldots, x_{n-m}), x_n, \ldots, x_{n-m+1}] \\ &= [x_n, x_{n-1}, \ldots, x_{n-m+1}] \\ &= \mathbf{x}_n \end{aligned}$$

Hence, the solutions of Eq. (18.2) are known if and only if the solutions of the vector equation $\mathbf{x}_n = V_f(\mathbf{x}_{n-1})$ are known. The latter equation is the standard vectorization of Eq. (18.2) and a common way of expressing Eq. (18.2) as a system of difference equations.

A *fixed point* of F is a point $\bar{\mathbf{x}}$ such that $F(\bar{\mathbf{x}}) = \bar{\mathbf{x}}$. Clearly, iterations of F do not affect $\bar{\mathbf{x}}$, so $\bar{\mathbf{x}}$ is a *stationary point* or *equilibrium* of Eq. (18.2). For Eq. (18.2), $\bar{x}$ is a fixed point if and only if $\bar{\mathbf{x}} = (\bar{x}, \ldots, \bar{x})$ is a fixed point of V_f. A fixed point of F^k for some fixed integer $k \geq 1$ is called a *k-periodic point* of F. The orbit of a k-periodic point $\mathbf{p}$ of F is a finite set $\{\mathbf{p}, F(\mathbf{p}), \ldots, F^{k-1}(\mathbf{p})\}$ which is called a *cycle of F of length k*, or a *k-cycle* of F. A point $\mathbf{q}$ is *eventually periodic* if there is $l \geq 1$ such that $\mathbf{p} = F^l(\mathbf{q})$ is periodic. These definitions imply that a k-cycle of Eq. (18.2) is a finite set $\{p_1, \ldots, p_k\}$ of real numbers such that $p_{kn+i} = p_i$ for $i = 1, \ldots, k$ and $n \geq 1$. Evidently, the point $\mathbf{p} = (p_1, \ldots, p_k)$ is a period-k point of V_f.

A fixed point $\bar{\mathbf{x}}$ of F is said to be *stable* if for each $\varepsilon > 0$, there is $\delta > 0$ such that $x_0 \in B_\delta(\bar{\mathbf{x}})$ implies that $F^n(x_0) \in B_\varepsilon(\bar{\mathbf{x}})$ for all $n \geq 1$. Here, $B_r(\bar{\mathbf{x}})$ is the open ball of radius $r > 0$ with center $\bar{\mathbf{x}}$, which consists of all points in $\mathbb{R}^m$ that are within a distance r from $\bar{\mathbf{x}}$. Thus, $\bar{\mathbf{x}}$ is stable if trajectories starting near $\bar{\mathbf{x}}$ stay close to $\bar{\mathbf{x}}$. If a fixed point is not stable, then it is called *unstable*. If there is $r > 0$ such that for all $x_0 \in B_r(\bar{\mathbf{x}})$ the trajectory $\{F^n(x_0)\}$ converges to $\bar{\mathbf{x}}$, then $\bar{\mathbf{x}}$ is *attracting*. If $\bar{\mathbf{x}}$ is both stable and attracting, then $\bar{\mathbf{x}}$ is said to be *asymptotically stable*.

In the case of the scalar difference Eq. (18.2), linearization is more easily done than for Eq. (18.1) because the characteristic polynomial of the derivative is easily determined. The Jacobian of the vectorization V_f is given by the $m \times m$ matrix

$$\begin{bmatrix} \partial f/\partial x_1 & \partial f/\partial x_2 & \cdots & \partial f/\partial x_{m-1} & \partial f/\partial x_m \\ 1 & 0 & \cdots & 0 & 0 \\ \vdots & \vdots & \cdots & \vdots & \vdots \\ 0 & 0 & \cdots & 1 & 0 \end{bmatrix} \quad (18.3)$$

where the partial derivatives are evaluated at an equilibrium point $(\bar{x},\ldots,\bar{x})$. The characteristic polynomial of this matrix (and hence, also of the linearization of the scalar difference equation at the equilibrium) is computed easily as

$$P(\lambda) = \lambda^m - \sum_{i=1}^{m} \frac{\partial f}{\partial r_i}(\bar{x},\ldots,\bar{x})\lambda^{m-i} \quad (18.4)$$

The roots of this polynomial give the eigenvalues of the linearization of Eq. (18.2) at the fixed point $\bar{x}$. If all roots of $P(\lambda)$ have modulus less than unity (i.e., if all roots lie within the interior of the unit disk in the complex plane), then the fixed point $\bar{x}$ is *locally* asymptotically stable, i.e., it is stable and attracts trajectories starting in some usually small neighborhood of $\bar{x}$. If any root lies in the exterior of the unit circle, then $\bar{x}$ is unstable as some trajectories are repelled away from $\bar{x}$ because of the eigenvalues outside the unit disk.

If the mappings F or f above depend on the index n, then we obtain the more general *nonautonomous* versions of Eq. (18.1) and Eq. (18.2) as

$$\mathbf{x}_n = F(n, \mathbf{x}_{n-1}) \quad (18.5)$$

and

$$x_n = f(n, x_{n-1}, x_{n-2}, \ldots, x_{n-m}) \quad (18.6)$$

respectively. Eq. (18.5) and Eq. (18.6) are useful in modeling such things as periodic changes in the environment in the case of modeling discrete-time population growth or to account for nonhomogeneous media in the case of discrete spaces as in infarcted cardiac tissue.

For additional reading on fundamentals of difference equations, see Agarwal (2000), Alligood et al. (1996), Arrowsmith and Place (1990), Davies (1999), Demazure (2000), Devaney (1989, 1992), Drazin (1992), Elaydi (1999), Holmgren (1996), Jordan (1965), Kelley and Peterson (2001), Kocic and Ladas (1993), Lakshmikantham and Trigiante (1988), Sandefur (1993), Sedaghat (2003a), and Sharkovski et al. (1993).

18.3 First-Order Difference Equations

Eq. (18.1) and Eq. (18.2) are the same when $m=1$, i.e., the dimension of the phase space is 1, the same as the order of the equation. In this section, we present conditions for the asymptotic stability or instability of equilibria and cycles, the ordering of cycles for continuous mappings, and conditions for the occurrence of chaotic behavior for first-order equations. Eq. (18.5) and Eq. (18.6) are also the same when $m=1$, but they are not first-order equations, since the variable n in F or f adds an additional dimension (see, e.g., Drazin, 1992).

18.3.1 Asymptotic Stability: Necessary and Sufficient Conditions

Let $f: I \to I$ be a mapping of an interval I of real numbers. Here the *invariant interval* I may be any interval, bounded or unbounded as long as it is not empty or a singleton. Let $\bar{x}$ be a fixed point of f in I that is *isolated*, meaning that there is an open interval J such that $x \in J \subset I$ and J contains no other fixed points of f. The one-dimensional version of Eq. (18.3) is

$$|f'(\bar{x})| < 1 \quad (18.7)$$

In Eq. (18.7), it is assumed that f has a continuous derivative at the fixed point $\bar{x}$. If Eq. (18.7) holds, then $\bar{x}$ is stable and the orbit of each point x_0 will converge to $\bar{x}$ provided that x_0 is sufficiently near $\bar{x}$. Furthermore, the rate of convergence is *exponential*, i.e., $|f^n(x_0) - \bar{x}|$ is proportional to the quantity e^{-an} where $a > 0$ for all $n \geq 1$. However, if $|f'(\bar{x})| > 1$ then nearby points x_0 will be repelled by the unstable fixed point $\bar{x}$, also at an exponential rate.

Example 1

The equation

$$x_n = ax_{n-1}(1 - x_{n-1}) \qquad (18.8)$$

is called the one-dimensional *logistic equation* (with the underlying "logistic map" $f(x) = ax(1-x)$). For $a > 1$, this equation has a unique positive fixed point $\bar{x} = (a-1)/a$. Further, if $a \leq 4$ then f has an invariant interval $[0, 1]$. Since

$$f'(\bar{x}) = a - 2a\bar{x} = 2 - a$$

it follows that $\bar{x}$ is (1) asymptotically stable if $1 < a < 3$ or (2) repelling if $a > 3$. The origin 0 is the only other fixed point of the logistic equation and it is unstable because $|f'(0)| = a > 1$.

Despite its ease of use, Condition (18.7) that is based only on a linear approximation of f has some limitations: (1) it does not specify *how near* $\bar{x}$ the initial point x_0 needs to be (it may have to be quite near in some cases); (2) it requires that f be smooth; and (3) it conveys no information when $|f'(\bar{x})| = 1$; and (4) it is *not a necessary condition*.

We may remove one or more of these deficiencies by using more detailed information about the function f than its tangent line approximation can provide. In particular, *necessary and sufficient* conditions for asymptotic stability are given next.

Theorem 1. (Asymptotic Stability) *Let $f : I \to I$ be a continuous mapping and let $\bar{x}$ be an isolated fixed point of f in I. The following statements are equivalent*:

(a) $\bar{x}$ *is asymptotically stable.*
(b) *There is a neighborhood U of $\bar{x}$ such that for all $x \in U \subset I$ the following inequalities hold*:

$$f^2(x) > x \quad \text{if } x < \bar{x}, \qquad f^2(x) < x \quad \text{if } x > \bar{x} \qquad (18.9)$$

(c) *There is a neighborhood U of $\bar{x}$ such that for all $x \in U \subset I$ the following inequalities hold*:

$$f(x) > x \quad \text{if } x < \bar{x}, \qquad f(x) < x \quad \text{if } x > \bar{x} \qquad (18.10)$$

and the graph of f_r^{-1} (the inverse image of the part of f to the right of $\bar{x}$) lies above the graph of f.

Reversing the inequalities in Theorem 1 gives conditions that are necessary and sufficient for $\bar{x}$ to be repelling. This follows from the next result that characterizes all possible types of behavior at an isolated equilibrium for a mapping of the real number line.

Theorem 2. *Let $f : I \to I$ be a continuous mapping and assume that f has an isolated fixed point $\bar{x} \in I$. Then precisely one of the following is true*:

(i) $\bar{x}$ *is asymptotically stable*;
(ii) $\bar{x}$ *is unstable and repelling*;
(iii) $\bar{x}$ *is semistable (attracting from one side of $\bar{x}$ and repelling from the other side); or*
(iv) *there is a sequence of period-2 points of f converging to $\bar{x}$.*

Example 2

Let us examine the logistic Eq. (18.8) in cases $a = 1, 3$. If $a = 1$, then it is easy to verify that

$$f(x) = x(1-x) < x \quad \text{if } x \neq 0$$

It follows that 0 is a semistable fixed point. Now consider $a = 3$. This value of the parameter a represents the boundary between stability and instability. Linearization fails in this case so we apply Theorem 1. Note that $\bar{x} = 2/3$ when $a = 3$ and

$$f^2(x) = 3[3x(1-x)][1 - 3x(1-x)] = 9x(1-x)(1 - 3x + 3x^2)$$

The function $9(1-x)(1-3x+3x^2)$ is decreasing on $(-\infty, \infty)$ and has a value 1 at $x=2/3=\bar{x}$. Thus Eq. (18.9) is satisfied and the fixed point is asymptotically stable.

The situation in Theorem 2(iv) does not occur for the logistic map, but it does occur for maps of the line; a trivial example is $f(x)=-x$, which has a unique fixed point at the origin (see Sedaghat [2003a], p. 32] for a more interesting example).

18.3.2 Cycles and Limit Cycles

One of the most striking features of continuous mappings of the line is the way in which their periodic points can coexist. The following result establishes the peculiar manner and ordering in which the cycles of a continuous map of an interval appear.

Theorem 3. (Coexisting cycles) *Suppose that a continuous map $f : I \to I$ of the interval I has a cycle of length m, and consider the following total ordering relation $\triangleright$ of the positive integers:*

$$3 \triangleright 5 \triangleright 7 \triangleright 9 \triangleright 11 \triangleright \cdots$$

$$2 \times 3 \triangleright 2 \times 5 \triangleright 2 \times 7 \triangleright 2 \times 9 \triangleright \cdots$$

$$\vdots$$

$$2^i \times 3 \triangleright 2^i \times 5 \triangleright 2^i \times 7 \triangleright 2^i \times 9 \triangleright \cdots$$

$$\vdots$$

$$\cdots \triangleright 2^n \triangleright \cdots \triangleright 2^3 \triangleright 2^2 \triangleright 2 \triangleright 1$$

Then for every positive integer k such that $m \triangleright k$, there is a cycle of length k for f. In particular, if f has a cycle of length 3, then it has cycles of all possible lengths.

It may be shown that a continuous mapping f has a 3-cycle in the interval I if and only if there is $\alpha \in I$ such that

$$f^3(\alpha) \leq \alpha < f(\alpha) < f^2(\alpha) \quad \text{or} \quad f^3(\alpha) \geq \alpha > f(\alpha) > f^2(\alpha) \qquad (18.11)$$

These conditions have come to be known as the *Li–Yorke conditions* (Li and Yorke, 1975). The logistic map of Example 1 has a 3-cycle when $a \approx 3.38$.

It must be emphasized that even if a 3-cycle exists for a mapping f, most or even all of the cycles may be unstable. Therefore, in numerical simulations one will not see all of the above cycles. The question as to which of the cycles, if any, can be stable is the subject of the next theorem.

Theorem 4. (Limit cycles) *Let $f \in C^3(I)$ where $I = [a,b]$ is any closed and bounded interval, with Schwarzian*

$$Sf = \frac{f'''}{f'} - \frac{3}{2}\left(\frac{f''}{f'}\right)^2 < 0$$

on I. Here $Sf = -\infty$ is a permitted negative value.

(a) *If p is an asymptotically stable periodic point of f, then either a critical point of f or an end point of I converges to the orbit of p.*
(b) *If f has $N \geq 0$ critical points in I, then f has at most $N + 2$ limit cycles (including any asymptotically stable fixed points).*

Example 3

Consider the logistic map $f(x) = ax(1-x)$, $x \in [0,1]$, $1 < a \leq 4$. We know that $Sf < 0$ on $[0,1]$ since f is a quadratic polynomial and $f'''(x) = 0$ for all x. Since $f'(0) = a > 1$, it follows that 0 is unstable so it cannot attract the orbit of the unique critical point $c = 1/2$, unless $c \in f^{-k}(0)$ for some integer $k \geq 1$. If

$a < 4$, then the maximum value of f on $[0, 1]$ is $f(c) = a/4 < 1$, so that $f^{-k}(0) = \{0, 1\}$. Thus when $a < 4$, Theorem 4 implies that there can be at most one limit cycle, namely the one whose orbit attracts $1/2$. In fact, because Theorem 4 holds for nonhyperbolic periodic points also (Sedaghat, 2003a, p. 47), it follows that for the logistic map all cycles, except possibly one, must be repelling! If a limit cycle exists for some value of a, then certainly, we may compute $f^n(1/2)$ for sufficiently large n to estimate that limit cycle. If $a = 4$, then $f^2(1/2) = 0$, the unstable fixed point. Hence, by Theorem 4 there are no limit cycles in this case, even though a 3-cycle exists for $a = 4$ and therefore, by Theorem 3 there are (unstable) cycles of all lengths.

18.3.3 Chaos

Recall from Section 18.1 that by the Poincare–Bendixon theorem chaotic behavior does not occur in *differential* equations in dimensions 1 and 2 (i.e., first- and second-order differential equations), so dimension 3 is the minimum needed for the occurrence of deterministic chaos in the continuous case. However, even a first-order *difference* equation can exhibit complex, aperiodic solutions that are stable, in the sense that "most" solutions display such unpredictable behavior. The essential characteristic of these solutions is that they exhibit *sensitive dependence on the initial value* x_0; i.e., a trajectory starting from a point arbitrarily close to x_0 rapidly (at an exponential rate) diverges from the trajectory that starts from x_0. This unpredictability exists even though the underlying map f is completely known; therefore, this phenomenon has been termed "deterministic chaos." Since sensitive dependence on initial values exists near any repelling fixed point, usually deterministic chaos involves certain other features also, such as complicated orbits that may be dense in some subspace if not the whole space. The following famous result shows that "period 3 implies chaos."

Theorem 5. (Chaos: Period 3) *Let I be a bounded, closed interval and let f be a continuous function on I satisfying one of the inequalities in Eq. (18.11). Then the following are true:*

(a) *For each positive integer k, f has a k-cycle.*
(b) *There is an uncountable set $S \subset I$ such that S contains no periodic points of f and satisfies the following conditions:*
   (b1) *For every $p, q \in S$ with $p \neq q$,*

$$\limsup_{n \to \infty} |f^n(p) - f^n(q)| > 0 \tag{18.12}$$

$$\liminf_{n \to \infty} |f^n(p) - f^n(q)| = 0 \tag{18.13}$$

   (b2) *For every $p \in S$ and periodic $q \in I$,*

$$\limsup_{n \to \infty} |f^n(p) - f^n(q)| > 0 \tag{18.14}$$

The set S is called the *scrambled set* of f. Its existence characterizes chaotic behavior on the line in the sense of Li and Yorke. By Eq. (18.12), the orbits of two arbitrarily close points in S will be pulled a finite distance away under iterations of f so there is sensitivity to initial conditions. By Eq. (18.13), the orbits of any two points in S can get arbitrarily close to each other, and by Eq. (18.14) no orbit starting from within S can converge to a periodic solution. The logistic map with $3.83 < a \leq 4$ has a 3-cycle in $I = [0, 1]$ and thus satisfies the conditions of Theorem 5. If a is close to 3.83, then the 3-cycle is stable and it follows that S is a proper subset of I. In many cases, S is a "large" subset of I (e.g., when a is large enough then the 3-cycle becomes unstable) and for $a = 4$, in fact S is dense in I. However, S could be a fractal set of Lebesgue measure zero in some cases. Also see Lasota and Yorke (1973).

The logistic map displays erratic behavior for smaller values of a also and this suggests that the existence of period 3 may not be necessary for the occurrence of chaotic behavior. Indeed, for difference equations in dimension 2 or greater this is often the case. In particular, we can define a mapping of the unit disk that exhibits sensitive dependence on initial values and its trajectories are dense in the disk, but which has no k-cycles for $k > 1$ (Sedaghat, 2003a, p. 127). There is also a result that applies to higher dimensional maps and establishes a Li–Yorke type chaos in which Condition (a) in Theorem 5 is relaxed; under conditions of Theorem 12, chaotic behavior occurs in the logistic map for $a \geq 3.7$.

Chaotic behavior may occur even on unbounded intervals. The following example illustrates this feature.

Example 4

Consider the continuous, piecewise smooth mapping

$$\rho(r) = \left|1 - \frac{1}{r}\right|, \quad r > 0$$

This mapping does not leave the interval $(0, \infty)$ invariant since it maps 1 to 0, but it does have a scrambled set in $(0, \infty)$. We show this through an indirect application of Theorem 5 because ρ itself does not have a period-3 point. Remarkably, ρ does have period-p points for all positive integers $p \neq 3$ and in Sedaghat (2004b) these points were explicitly determined using the Fibonacci numbers. We can complete the list of periodic solutions for ρ by adding a "3-cycle that passes through ∞" as follows:

Let $[0, \infty]$ be the one-point compactification of $[0, \infty)$ and define ρ^* on $[0, \infty]$ as

$$\rho^*(0) = \infty, \quad \rho^*(\infty) = 1, \quad \rho^*(r) = \rho(r), \ 0 < r < \infty.$$

Note that ρ^* extends ρ continuously to $[0, \infty]$ and furthermore, ρ^* has a 3-cycle $\{1, 0, \infty\}$. The interval $[0, \infty]$ is homeomorphic to $[0, 1]$, so ρ^* is chaotic on $[0, \infty]$ in the sense of Theorem 5. To show that ρ is chaotic on $(0, \infty)$, we show that it has a scrambled set. Let S^* be a scrambled set for ρ^* and take out ∞ and the set of all backward iterates of 0, namely, $\cup_{n=0}^{\infty} \rho^{-n}(0)$ from S^*. The subset S that remains is contained in $(0, \infty)$. Further, S is uncountable because S^* is uncountable and each inverse image $\rho^{-n}(0)$ is countable for all $n = 0, 1, 2, \ldots$ (in fact $\cup_{n=0}^{\infty} \rho^{-n}(0)$ is the set of all nonnegative rational numbers; see (Sedaghat [2004b]). Further, since $\rho(S) \subset S$ and $\rho|_S = \rho^*|_S$ it follows that S is a scrambled set for ρ that has the properties stated in Theorem 5.

18.3.4 Notes

A proof of the equivalence of (a) and (b) in Theorem 1 first appeared in Sharkovski (1960) (also see Sharkovski et al. [1993]). A different proof of this fact, which also established the equivalence of Part (c), was first given in Sedaghat (1999). Theorem 1 is true only for maps on an invariant subset of the real number line. It is not true for invariant subsets of the Euclidean plane (see Sedaghat [1998b]). For complete proofs of Theorems 1 and 2, including other equivalent conditions not stated here and some further comments, see Sedaghat (2003a).

Theorem 3 was first proved in Sharkovski (1995) (also see Collet and Eckmann [1980], Devaney [1989], and Sedaghat [2003a]). Like Theorem 1 and 2, this result is peculiar to the maps of the interval and is not shared by the continuous mappings of a closed, one-dimensional manifold like the circle, or by continuous mappings of higher dimensional Euclidean spaces. See (Sedaghat (2004b)) for an example of a second-order difference equation all of whose solutions are either periodic of period 3 or else they converge to zero.

Theorem 4 was proved almost simultaneously in Allwright (1978) and Singer (1978) (also see Collet and Eckmann [1980], Devaney [1989], and Sedaghat [2003a]. Theorem 5 was first proved in Li and Yorke (1975) (also see Sedaghat [2003a]). Example 4 is extracted from Sedaghat (2004a).

18.4 Higher Order Difference Equations

Equations of type (18.2) or (18.6) with $m \geq 2$ are higher order difference equations. As noted above, these can be converted to first-order vector equations. Unfortunately, the theory of the preceding section largely fails for the higher order equations or for vector equations. Higher order equations provide a considerably greater amount of flexibility for modeling applications, but results that are comparable to those of the

Difference Equations as Discrete Dynamical Systems   18-9

previous section in power and generality are lacking. Nevertheless, there do exist results that apply to general classes of higher order difference equations and we shall be concerned with those results in this section. Because of space limitations we will not discuss results that are applicable to narrowly defined types of difference equations. Although many interesting results of profound depth have been discovered for such classes of difference equations, these types of equations do not offer the flexibility needed for scientific modeling.

18.4.1 Asymptotic Stability: Weak Contractions

A general sufficient condition for the asymptotic stability (nonlocal) of a fixed point is given next.

Theorem 6. (Asymptotic Stability) *Let f be continuous with an isolated fixed point $\bar{x}$, and let M be an invariant closed set containing $\bar{\mathbf{x}} = (\bar{x}, \ldots, \bar{x})$. If A is the containing $\bar{\mathbf{x}}$ and all $\mathbf{u} = (u_1, \ldots, u_m) \in M$ such that*

$$|f(\mathbf{u}) - \bar{x}| < \max\{|u_1 - \bar{x}|, \ldots, |u_m - \bar{x}|\}$$

then $\bar{x}$ is asymptotically stable relative to each invariant subset S of A that is closed in M; in particular, $\bar{x}$ attracts every trajectory with a vector of initial values $(x_{1-m}, \ldots, x_0) \in S$.

Further, if A is open and contains $\bar{x}$ in its interior, then $\bar{x}$ is asymptotically stable relative to $(\bar{x} - r, \bar{x} + r)$, where $r > 0$ is the largest real number such that $B_r(\bar{\mathbf{x}}) \subset A$. In particular, if $A = \mathbb{R}^m$, then $\bar{x}$ is globally asymptotically stable.

When f satisfies the inequality in Theorem 6 we may say that f is a *weak contraction* at $\bar{x}$ ("weak" because the vectorization V_f is not strictly a contraction in the usual sense).

Example 5

Consider the third-order equation

$$x_n = ax_{n-1} + bx_{n-3}\exp(-cx_{n-1} - dx_{n-3}), \quad a,b,c,d \geq 0, \ c+d > 0 \quad (18.15)$$

This equation was derived from a model for the study of observed variations in the flour beetle population (see Kuang and Cushing [1996]). For an entertaining account of the flour beetle experiments see Cipra (1997). We show that the origin is asymptotically stable (so that the beetles go extinct) if

$$a + b \leq 1, \quad b > 0 \quad (18.16)$$

We note that the linearization of Eq. (18.15) at the origin, i.e., the characteristic polynomial Eq. (18.4), has a unit eigenvalue $\lambda = 1$ when $a + b = 1$. Therefore, linear stability analysis is not applicable in this particular case.

Now, observe that if Eq. (18.16) holds, then for every $(x, y, z) \in [0, \infty)^3$,

$$ax + bz\exp(-cx - dz) \leq [a + b\exp(-cx - dz)]\max\{x, z\}$$
$$< (a+b)\max\{x, y, z\}$$
$$\leq \max\{x, y, z\}$$

so by Theorem 6 the origin is stable and attracts all nonnegative solutions of Eq. (18.15).

A generalization of Example 5 is the first part of the next corollary of Theorem 6, which in particular provides a simple tool for establishing the global stability of the zero equilibrium. The simple proof (showing that the map is a weak contraction) is omitted.

Corollary 1. (a) *Let $f_i \in C([0, \infty)^m, [0, 1))$ for $i = 1, \ldots, k$ and $k \geq 2$. If $\sum_{i=1}^k f_i(u_1, \ldots, u_m) < 1$ for all $(u_1, \ldots, u_m) \in [0, \infty)^m$, then the origin is the unique, globally asymptotically stable fixed point of the following equation*

$$x_n = \sum_{i=1}^{k} f_i(x_{n-1}, \ldots, x_{n-m})x_{n-i}$$

(b) *The origin is a globally asymptotically stable fixed point of the equation*

$$x_n = x_{n-k} g(x_{n-1}, \ldots, x_{n-m}), \quad 1 \leq k \leq m$$

where $g \in C(\mathbb{R}^m, \mathbb{R})$, *if* $|g(\mathbf{x})| < 1$ *for all* $\mathbf{x} \neq (0, \ldots, 0)$.

The following result gives an interesting and useful version of Theorem 6 for the nonautonomous equation (18.6).

Theorem 7. *Let f be the function in Eq. (18.6) and assume that there is a sequence $a_n \geq 0$ such that for all $u \in \mathbb{R}^m$ and all $n \geq 1$,*

$$|f(n, \mathbf{u})| \leq a_n \max\{|u_1|, \ldots, |u_m|\}$$

(a) *If* $\limsup_{n \to \infty} a_n = a < 1$ *then the origin is the globally exponentially stable fixed point of Eq. (18.6).*
(b) *If* $b_k = \max\{a_{mk}, a_{mk+1}, \ldots, a_{mk+k}\}$ *and* $\prod_{k=0}^{\infty} b_k = 0$, *then the origin is the globally asymptotically stable fixed point of Eq. (18.6).*

18.4.2 Asymptotic Stability: Coordinate-Wise Monotonicity

In case the function f in Eq. (18.2) or Eq. (18.6) is either nondecreasing or nonincreasing in all of its arguments, it is possible to obtain general conditions for asymptotic stability of a fixed point. There are several results of this type and we discuss some of them in this section along with their applications. The first result is the most general of its kind on attractivity of a fixed point within a given interval.

Theorem 8. *Assume that $f : [a, b]^m \to [a, b]$ in Eq. (18.2) is continuous and satisfies the following conditions:*

(i) *For each $i \in \{1, \ldots, m\}$ the function $f(u_1, \ldots, u_m)$ is monotone in the coordinate u_i (with all other coordinates fixed).*
(ii) *If (μ, ν) is a solution of the system*

$$f(\mu_1, \mu_2, \ldots, \mu_m) = \mu$$
$$f(\nu_1, \nu_2, \ldots, \nu_m) = \nu$$

then $\mu = \nu$, where for $i \in \{1., \ldots, m\}$ we define

$$\mu_i = \begin{cases} \mu & \text{if } f \text{ is nondecreasing in } u_i \\ \nu & \text{if } f \text{ is nonincreasing in } u_i \end{cases}$$

and

$$\nu_i = \begin{cases} \nu & \text{if } f \text{ is nondecreasing in } u_i \\ \mu & \text{if } f \text{ is nonincreasing in } u_i \end{cases}.$$

Then there is a unique fixed point $\bar{x} \in [a, b]$ for Eq. (18.2) that attracts every solution of Eq. (18.2) with initial values in $[a, b]$.

The following variant from Chan et al. (2006) has less flexibility in the manner in which f depends on variations in coordinates, but it does not involve a bounded interval and adds stability to the properties of $\bar{x}$.

Theorem 9. *Let r_0, s_0 be extended real numbers where $-\infty \leq r_0 < s_0 \leq \infty$ and consider the following hypotheses:*

(H1) $f(u_1, \ldots, u_m)$ *is nonincreasing in each* $u_1, \ldots, u_m \in I_0$ *where* $I_0 = (r_0, s_0]$ *if* $s_0 < \infty$ *and* $I_0 = (r_0, \infty)$ *otherwise.*
(H2) $g(u) = f(u, \ldots, u)$ *is continuous and decreasing for* $u \in I_0$.
(H3) *There is* $r \in [r_0, s_0)$ *such that* $r < g(r) \leq s_0$. *If* $r_0 = -\infty$ *or* $\lim_{t \to r_0^+} g(t) = \infty$, *then we assume that* $r \in (r_0, s_0)$.
(H4) *There is* $s \in [r, x^*)$ *such that* $g^2(s) \geq s$, *where* $g^2(s) = g(g(s))$.
(H5) *There is* $s \in [r, x^*)$ *such that* $g^2(u) > u$ *for all* $u \in (s, x^*)$.

Difference Equations as Discrete Dynamical Systems

Then the following is true:

(a) *If (H2) and (H3) hold, then Eq. (18.2) has a unique fixed point x^* in the open interval $(r, g(r))$.*
(b) *Let $I = [s, g(s)]$. If (H1)–(H4) hold, then I is an invariant interval for Eq. (18.2) and $x^* \in I$.*
(c) *If (H1)–(H3) and (H5) hold, then x^* is stable and attracts all solutions of Eq. (18.2) with initial values in $(s, g(s))$.*
(d) *If (H1)–(H3) hold, then x^* is an asymptotically stable fixed point of Eq. (18.2) and is an asymptotically stable fixed point of the mapping g; e.g., if g is continuously differentiable with $g'(x^*) > -1$.*

It may be emphasized that the conditions of Theorems 8 and 9 imply asymptotic stability over an interval and as such, they impart considerably greater information than linear stability results about the ranges on which convergence occurs. They also have the added advantage that if the extent of the interval I is not an issue, then we may reduce the amount of calculations considerably by using Theorem 9(d) instead of examining the roots of the characteristic polynomial (18.4).

Example 6

Consider the difference equation

$$x_n = \frac{\alpha - \sum_{i=1}^{m} a_i x_{n-i}}{\beta + \sum_{i=1}^{m} b_i x_{n-i}}, \quad \alpha, \beta > 0, \tag{18.17}$$

$$a_i, b_i \geq 0, \quad 0 < a = \sum_{i=1}^{m} a_i < \beta, \quad b = \sum_{i=1}^{m} b_i > 0$$

The function f for this equation is

$$f(u_1, \ldots, u_m) = \frac{\alpha - \sum_{i=1}^{m} a_i u_i}{\beta + \sum_{i=1}^{m} b_i u_i}$$

which is nonincreasing in each of its m coordinates if $u_i < \alpha/a$ for all $i = 1, \ldots, m$. Eq. (18.17) has one positive fixed point $\bar{x}$ which is a solution of the equation $g(t) = t$ where

$$g(t) = \frac{\alpha - at}{\beta + bt}$$

Eq. (18.17) satisfies (H1)–(H3) in Theorem 11 with $s = g^{-1}(\alpha/a)$ and

$$g'(\bar{x}) = \frac{-a\beta - \alpha b}{(\beta + b\bar{x})^2}$$

It is easy to verify that $\bar{x} \in [s, \alpha/a]$ with $|g'(\bar{x})| < 1$ so $\bar{x}$ is asymptotically stable according to Theorem 9(d).

We finally mention the following result for the nonautonomous equation:

$$x_n = f\left(\sum_{i=1}^{m} [a_{n-i} g(x_{n-i}) + g_i(x_{n-i})]\right), \quad n = 1, 2, 3, \ldots \tag{18.18}$$

which is of type (18.6) whose *autonomous* version (the coefficients a_{n-i} all have the same value) is a special case of Theorem 9. In Eq. (18.18), we assume that

$$a_{mn+i} = a_i \geq 0, \quad i = 1, \ldots, m, \quad n = 1, 2, 3, \ldots, \quad a = \sum_{i=1}^{m} a_i \tag{18.19}$$

The functions f, g, g_i are all continuous on some interval (t_0, ∞) of real numbers $\mathbb{R}$ and monotonic (nonincreasing or nondecreasing). It is assumed for nontriviality that all a_i and all g_i are not simultaneously zero. Define the function h as

$$h(t) = f\left(\sum_{i=0}^{m} g_i(t)\right), \quad g_0(t) = ag(t)$$

and assume that h satisfies the condition

$$h(t) \text{ is decreasing for } t > t_0 \geq -\infty \qquad (18.20)$$

Theorem 10. *Suppose that Eq. (18.19) and Eq. (18.20) hold and consider the following assumptions:*
(A1) *for some* $r > t_0 \geq -\infty$, $h(r) > r$;
(A2) *for some* $s \in (r, \bar{x})$, $h^2(s) \geq s$;
(A3) *for some* $s \in (r, \bar{x})$, $h^2(t) > t$ *for all* $t \in (s, \bar{x})$.
Then

(a) *If (A1) holds, then Eq. (18.18) has a unique fixed point* $\bar{x} > r$.
(b) *If (A1) and (A2) hold, then the interval $(s, h(s))$ is invariant for Eq. (18.18) and $\bar{x} \in (s, h(s))$.*
(c) *If (A1) and (A3) hold, then the fixed point $\bar{x}$ of Eq. (18.18) is stable and attracts every point in the interval $(s, h(s))$.*

Example 7

The propagation of an action potential pulse in a ring of excitable media (e.g., cardiac tissue) can be modeled by the equation

$$x_n = \sum_{i=1}^{m} a_{n-i} C(x_{n-i}) - A(x_{n-m}) \qquad (18.21)$$

if certain threshold and memory effects are ignored and if all the cells in the ring have the conduction properties. Models of this type can aid in gaining a better understanding of causes of cardiac arrhythmia. The ring is composed of m units (e.g., cardiac cell aggregates) and the functions A and C in Eq. (18.21) represent the restitutions of action potential duration (APD) and the conduction time, respectively, as functions of the diastolic interval x_n. The numbers a_n represent lengths of the ring's excitable units that are not generally constant, but the sequence $\{a_n\}$ is m-periodic since cell aggregate $m+1$ is the same as cell aggregate 1 and another cycle through the same cell aggregates in the ring starts (the reentry process).
The following conditions are generally assumed:

(C1) There is $r_A \geq 0$ such that the APD restitution function A is continuous and increasing on the interval $[r_A, \infty)$ with $A(r_A) \geq 0$.
(C2) There is $r_C \geq 0$ such that the conduction time or CT restitution function C is continuous and nonincreasing on the interval $[r_C, \infty)$ with $\inf_{x \geq r_C} C(x) \geq 0$.
(C3) There is $r \geq \max\{r_A, r_C\}$ such that $mC(r) > A(r) + r$.

Define the function $F = mC - A$ and note that by (C1)–(C3), F is continuous and decreasing on the interval $[r, \infty)$ and satisfies

$$F(r) > r \qquad (18.22)$$

Thus, (A1) holds and there is a unique fixed point $\bar{x}$ for Eq. (18.21). Now assume that the following condition is also true:

(C4) There is $s \in [r, \bar{x})$ such that $F^2(x) > x$ for all $x \in (s, \bar{x})$.

Then, (A1) and (A3) in Theorem 10 are satisfied with $h = F$, $f(t) = t$, $g = C$, and $g_i = -A/m$ for all i and the existence of an asymptotically stable fixed point for Eq. (18.21) is established. In the context of cardiac arrhythmia, this means that there is an equilibrium heart beat period of $A(\bar{x}) + \bar{x}$ that is usually shorter than the normal beat period by a factor of 2 or 3.

18.4.3 Persistent Oscillations and Chaos

It is possible to establish the existence of oscillatory solutions for Eq. (18.2) if the linearization of this equation exhibits such behavior; i.e., if some of the roots of the characteristic polynomial Eq. (18.4) are either complex or negative. Such linear oscillations occur both in stable cases where all roots of Eq. (18.4) have modulus less than 1 and in unstable cases where some roots have modulus greater than 1. Further, linear oscillations take place about the equilibrium (i.e., going past the fixed point repeatedly, infinitely often) and by their very nature; if linear oscillations are bounded and nondecaying, then they are not structurally stable or robust.

Our focus here is on a different and less familiar type of oscillation. This type of nonlinear oscillation is bounded and persistent (nondecaying), and unlike linear oscillations, it is structurally stable. Further, it need not take place about the equilibrium, although it is caused by the instability of the equilibrium.

We define a *persistently oscillating* solution of Eq. (18.2) simply as one that is bounded and has two or more (finite) limit points.

Theorem 11. (Persistent Oscillations) *Assume that f in Eq. (18.2) has an isolated fixed point $\bar{x}$ and satisfies the following conditions*:

(a) *For $i = 1, \ldots, m$, the partial derivatives $\partial f / \partial x_i$ exist continuously at $\bar{\mathbf{x}} = (\bar{x}, \ldots, \bar{x})$, and every root of the characteristic polynomial (18.4) has modulus greater than 1.*
(b) $f(\bar{x}, \ldots, \bar{x}, x) \neq \bar{x}$ *if* $x \neq \bar{x}$.

Then all bounded solutions of Eq. (18.2) except the trivial solution $\bar{x}$ oscillate persistently. If only (a) *and* (b) *hold, then all bounded solutions that do not converge to some $\bar{x}$ in a finite number of steps oscillate persistently.*

Example 8
The second-order difference equation

$$x_{n+1} = cx_n + g(x_n - x_{n-1}) \qquad (18.23)$$

has been used in the classical theories of the business cycle, where g is often assumed to be nondecreasing (also see Goodwin [1951], Hicks [1965], Puu [1993], Samuelson [1939], Sedaghat [1997, 2003a, 2003b], and Sedaghat and Wang [2000]). If $0 \leq c < 0$ and g is a bounded function, then it is easy to see that all solutions of Eq. (18.23) are bounded and confined to a closed interval I. If additionally g is continuously differentiable at the origin with $g'(0) > 1$, then by Theorem 11 for all initial values x_0, x_{-1} that are not both equal to the fixed point $\bar{x} = g(0)/(1-c)$, the corresponding solution of Eq. (18.23) oscillates persistently, eventually in the absorbing interval I.

We also mention that if $tg(t) \geq 0$ for all t in the interval I that contains the fixed point 0 of Eq. (18.23), then every eventually nonnegative and every eventually nonpositive solution of Eq. (18.23) is eventually monotonic (Sedaghat, 2003b, 2004c). This is true, in particular, if $g(t)$ is linear or an odd function. But in general it is possible for Eq. (18.23) to have oscillatory solutions that are eventually nonnegative (or nonpositive). For example, if

$$g(t) = \min\{1, |t|\}, \quad c = 0$$

then Eq. (18.23) has a period-3 solution $\{0, 1, 1\}$, which is clearly nonnegative and oscillatory. This type of behavior tends to occur when g has a global minimum (not necessarily unique) at the origin, including even functions (see Sedaghat [2003b] for more details).

With regard to chaotic behavior, it may be noted that persistently oscillating solutions need not be erratic. Indeed, they could be periodic as in the preceding example. The conditions stated in the next theorem are more restrictive than those in Theorem 11, enough to ensure that erratic behavior does occur. The essential concept for this result is defined next.

Let $F : D \to D$ be continuously differentiable where $D \subset \mathbb{R}^m$, and let the closed ball $\bar{B}_r(\bar{\mathbf{x}}) \subset D$ where $\bar{\mathbf{x}} \in D$ is a fixed point of F and $r > 0$. If for every $\mathbf{x} \in \bar{B}_r(\bar{\mathbf{x}})$, all the eigenvalues of the Jacobian $DF(\mathbf{x})$ have magnitudes greater than 1, then $\bar{\mathbf{x}}$ is an *expanding fixed point*. If, in addition, there is $\mathbf{x}_0 \in \bar{B}_r(\bar{\mathbf{x}})$ such that (a) $\mathbf{x}_0 \neq \bar{\mathbf{x}}$; (b) there is a positive integer k_0 such that $F^{k_0}(\mathbf{x}_0) = \bar{\mathbf{x}}$; and (c) $\det[DF^{k_0}(\mathbf{x}_0)] \neq 0$, then $\bar{\mathbf{x}}$ is a *snap-back repeller*. If $F = V_f$ is a vectorization so that $\bar{\mathbf{x}} = (\bar{x}, \ldots, \bar{x})$, then we refer to the fixed point $\bar{x}$ as a snap-back repeller for Eq. (18.2). Note that because of its expanding nature, a snap-back repeller satisfies the main Condition (a) of Theorem 11.

Theorem 12. (Chaos: Snap-back repellers) *Let $D \subset \mathbb{R}^m$ and assume that a continuously differentiable mapping F has a snap-back repeller. Then the following are true:*

(a) *There is a positive integer N such that F has a point of period n for every integer $n \geq N$.*
(b) *There is an uncountable set S satisfying the following properties:*
   (i) *$F(S) \subset S$ and there are no periodic points of F in S;*
   (ii) *For every $x, y \in S$ with $x \neq y$,*

$$\limsup_{n \to \infty} \|F^n(x) - F^n(y)\| > 0;$$

$$\limsup_{n \to \infty} \|F^n(x) - F^n(y)\| > 0$$

(c) *There is an uncountable subset S_0 of S such that for every $x, y \in S_0$*

$$\liminf_{n \to \infty} \|F^n(x) - F^n(y)\| = 0$$

The norm $\|\cdot\|$ in Parts (b) and (c) above may be assumed to be the Euclidean norm in $\mathbb{R}^m$. The set S in (b) is analogous to the similar set in Theorem 5, so we may also call it a "scrambled set". Unlike Theorem 5, Theorem 12 can be applied to models in any dimension (see Dohtani [1992] and Sedaghat [2003a] for applications to social science models).

18.4.4 Semiconjugacy: First-Order Equations Revisited

Given that there is a comparatively better-established theory for the first-order difference equations than for equations of order 2 or greater, it is of interest that certain classes of higher order difference equations can be related to suitable first-order ones. For these classes of equations, it is possible to discuss stability, convergence, periodicity, bifurcations, and chaos using results from the first-order theory. Owing to limitations of space, in this section we do not present the basic theory, but use specific examples to indicate how the main concepts can be used to study higher order difference equations. For the basic theory and additional examples, see Sedaghat (2002, 2003a, 2004b).

We say that the higher order difference equation (18.2) is *semiconjugate* to a first-order equation if there are functions h and φ such that

$$h(V_f(u_1, \ldots, u_m)) = \varphi(h(u_1, \ldots, u_m))$$

The function h is called a *link* and the function φ is called a *factor* for the semiconjugacy. The first-order equation

$$t_n = \varphi(t_{n-1}), \qquad t_0 = h(x_0, \ldots, x_{-m+1}) \tag{18.24}$$

is the equation to which Eq. (18.2) is semiconjugate. A relatively straightforward theory exists that relates the dynamics of Eq. (18.2) to that of Eq. (18.24) and bears remarkable similarity to the theory of Liapunov functions (LaSalle, 1976, 1986; Sedaghat, 2003a) and invariants (Ladas, 1995; Kulenovic, 2000). The link

Difference Equations as Discrete Dynamical Systems

function h also plays a crucial role and in some cases, two or more different factor and link maps can be found that shed light on different aspects of the higher order difference equation.

Example 9

Consider the second-order scalar difference equation

$$x_n = \frac{a}{x_{n-1}} + bx_{n-2} \qquad a, b, x_0, x_{-1} > 0 \tag{18.25}$$

It can be readily shown that this equation (see Magnucka-Blandzi and Popenda [1999]) is semiconjugate to the first-order equation

$$t_n = a + bt_{n-1} \tag{18.26}$$

with the link $H(x, y) = xy$ over the positive quadrant $D = (0, \infty)^2$. Indeed, multiplying both sides of Eq. (18.25) by x_{n-1} gives

$$x_n x_{n-1} = a + bx_{n-1} x_{n-2}$$

which has the same form as the first-order equation if $t_n = x_n x_{n-1}$. This semiconjugacy can be used to establish unboundedness of solutions for Eq. (18.25). Note that if $b \geq 1$ then all solutions of the first-order equation (18.26) diverge to infinity, so that the product

$$x_n x_{n-1} = t_n \tag{18.27}$$

is unbounded. It follows from this observation that every positive solution of Eq. (18.25) is unbounded if $b \geq 1$.

If $b < 1$, then every solution of the first-order equation converges to the positive fixed point $\bar{t} = a/(1 - b)$. It is evident that the curve $xy = \bar{t}$ is an invariant set for the solutions of Eq. (18.25) in the sense that if $x_0 x_{-1} = \bar{t}$ then $x_n x_{n-1} = \bar{t}$ for all $n \geq 1$. Hence, every solution of Eq. (18.25) converges to this invariant curve. Now since all solutions of the first-order equation $x_n x_{n-1} = \bar{t}$ are periodic with period 2, we conclude that if $b < 1$ then all nonconstant, positive solutions of Eq. (18.25) converge to period-2 solutions.

In this example, we saw how semiconjugacy allows a factorization of the second-order equation into two simple first-order ones. Subsequently, we can actually solve Eq. (18.25) exactly by first solving Eq. (18.26) to get

$$t_n = \alpha + \beta b^n, \quad \alpha = \frac{a}{1-b}, \quad \beta = t_0 - \alpha, \quad t_0 = x_0 x_{-1}$$

(for $b \neq 1$) and then using this solution in Eq. (18.27) to find an explicit solution for Eq. (18.25) as

$$x_n = x_0 \delta_n \prod_{k=1}^{n/2} \frac{1 + cb^{n-2k+2}}{1 + cb^{n-2k+1}} + x_{-1}(1 - \delta_n) \prod_{k=1}^{(n+1)/2} \frac{1 + cb^{n-2k+2}}{1 + cb^{n-2k+1}}$$

where $c = \beta/\alpha$ and $\delta_n = [1 + (-1)^n]/2$.

Example 10

Consider again the second-order difference equation from Example 8

$$x_n = cx_{n-1} + g(x_{n-1} - x_{n-2}), \qquad x_0, x_{-1} \in \mathbb{R}. \tag{18.28}$$

where g is continuous on $\mathbb{R}$ and $0 \leq c \leq 1$. In particular, in Puu (1993) the equation

$$y_n = (1-s)y_{n-1} + sy_{n-2} + Q(y_{n-1} - y_{n-2})$$

is considered where $0 \leq s \leq 1$ and Q is the "investment function" (taken as a cubic polynomial). We may rewrite this equation as

$$y_n = y_{n-1} - s(y_{n-1} - y_{n-2}) + Q(y_{n-1} - y_{n-2})$$

which has the form Eq. (18.28) with $c=1$ and $g(t) = Q(t) - st$. In the case $c=1$, Eq. (18.28) is semiconjugate with link function $h(x,y) = x - y$ and factor $g(t)$. In this case, since $x_n - x_{n-1} = t_n$, the solutions of Eq. (18.28) are just sums of the solutions of the first-order equation

$$t_n = g(t_{n-1}), \quad t_0 = x_0 - x_{-1}$$

i.e., $x_n = \sum_{k=1}^{n} t_k$. If g is a chaotic map (e.g., if it has a snap-back repeller or a 3-cycle), then the solutions of Eq. (18.28) exhibit highly complex and unpredictable behavior (see Puu [1993] and Sedaghat [2003a] for further details).

Example 11

Let a_n, b_n, d_n be given sequences of real numbers with $a_n \geq 0$ and $b_{n+1} + d_n \geq 0$ for all $n \geq 0$. The difference equation

$$x_{n+1} = a_n |x_n - cx_{n-1} + b_n| + cx_n + d_n, \quad c \neq 0 \tag{18.29}$$

is nonautonomous of type (18.6). Using a semiconjugate factorization, we show that its general solution is given by

$$x_n = x_0 c^n + \sum_{k=1}^{n} c^{n-k} \left(d_{k-1} + |t_0| \prod_{j=0}^{k-1} a_j + \sum_{i=1}^{k-1} (b_i + d_{i-1}) \prod_{j=i}^{k-1} a_j \right) \tag{18.30}$$

where $t_0 = x_0 - cx_{-1} + b_0$. To see this, note that Eq. (18.29) has a semiconjugate factorization as

$$x_n - cx_{n-1} + b_n = t_n, \quad t_n = a_{n-1}|t_{n-1}| + b_n + d_{n-1} \tag{18.31}$$

The second equation in Eq. (18.31) may be solved recursively to get

$$t_n = |t_0| \prod_{j=0}^{n-1} a_j + b_n + d_{n-1} + \sum_{i=1}^{n-1} (b_i + d_{i-1}) \prod_{j=i}^{n-1} a_j \tag{18.32}$$

Substituting Eq. (18.32) in the first equation in Eq. (18.31) and using another recursive argument yields Eq. (18.30). For additional details on the various interesting features of this and related equations, see Kent and Sedaghat (2005).

Example 12
As our final example we consider the difference equation

$$x_{n+1} = |ax_n - bx_{n-1}|, \quad a, b \geq 0, \quad n = 0, 1, 2, \ldots \tag{18.33}$$

An equation such as this may appear implicitly in smooth difference equations (or difference relations) that are in the form of, e.g., quadratic polynomials. We may assume that the initial values x_{-1}, x_0 in Eq. (18.33) are nonnegative and for nontriviality, at least one is positive. Dividing both sides of Eq. (18.33) by x_n we obtain a ratios equation

$$\frac{x_{n+1}}{x_n} = \left| a - \frac{bx_{n-1}}{x_n} \right|$$

which can be written as

$$r_{n+1} = \left|a - \frac{b}{r_n}\right|, \quad n = 0, 1, 2, \ldots \tag{18.34}$$

where we define $r_n = x_n/x_{n-1}$ for every $n \geq 0$. Thus, we have a semiconjugacy in which Eq. (18.34) is the factor equation with mapping

$$\phi(r) = \left|a - \frac{b}{r}\right|, \quad r > 0$$

and the link function is $h(x, y) = x/y$. Note that a general solution of Eq. (18.33) is obtained by computing the solution r_n of Eq. (18.34) and then using these values of r_n in the nonautonomous linear equation $x_n = r_n x_{n-1}$ to obtain the solution $x_n = x_0 \prod_{k=1}^{n} r_k$. However, closed forms are not known for the nontrivial solutions of Eq. (18.34). In fact, when $a = b = 1$ we saw in Example 4 that Eq. (18.34) has chaotic solutions. For a detailed analysis of the solutions of Eq. (18.33) and Eq. (18.34) see Kent and Sedaghat (2004) and Sedaghat (2004b).

18.4.5 Notes

Theorem 6 and related results were established in Sedaghat (1998a) (see also Sedaghat [2003a]). For some results concerning the rates of convergence of solutions in Theorem 6, see Stevic (2003). Theorem 6 has also been shown to hold in any complete metric space not just $\mathbb{R}$ (Xiao and Yang, 2003). For a proof of Theorem 7 and related results, see Berezansky et al. (2005).

Theorem 8 appeared without a proof in Grove and Ladas (2005); it is based on similar results for second-order equations that are proved in Kulenovic and Ladas (2001). This theorem in particular generalizes the main result in Hautus and Bolis (1979) where f is assumed to be nondecreasing in every coordinate. The requirement that the invariant interval $[a, b]$ be bounded is essential for the validity of Theorem 8, though it may be restrictive in some applications. Theorem 9 was motivated by the study of Eq. (18.21) in Example 7 and was established in different cases in Sedaghat et al. (2005) and Sedaghat (2005). The more complete version on which Theorem 9 is based is proved in Chan et al. (2006). Example 6 is based on results from Dehghan et al. (2005). Theorem 10 is proved in Sedaghat (2005) and Example 7 is based on results from Sedaghat (1998b). For additional background material behind the model in Example 7 also see Courtemanche and Vinet (2003), Ito and Glass (1992).

Theorem 11 is based on results in Sedaghat (1998b) where some classical economic models are also studied. Theorem 12 was first proved in Marotto (1978) (see Sedaghat [2003a] for various applications of this result to social science models).

References

Agarwal, R.P. (2000) *Difference Equations and Inequalities*, 2nd ed., Dekker, New York.
Alligood, K., T. Sauer, and J.A. Yorke (1996) *Chaos: An Introduction to Dynamic Systems*, Springer, New York.
Allwright, D.J. (1978) "Hypergraphic functions and bifurcations in recurrence relations," *SIAM J. Appl. Math.* **34**, 687–691.
Arrowsmith, D.K. and C.M. Place (1990) *An Introduction to Dynamical Systems*, Cambridge University Press, Cambridge.
Berezansky, L., E. Braverman, and E. Liz (2005) "Sufficient conditions for the global stability of nonautonomous higher order difference equations," *J. Difference Eq. Appl.* **11**, 785–798.
Burden, R.L. and J.D. Fairs (1997) *Numerical Analysis*, 6th ed., Brooks-Cole, Pacific Grove.
Chan, D., E.R. Chang, M. Dehghan, C.M. Kent, R. Mazrooei-Sebdani, and H. Sedaghat (2006) "Asymptotic stability for difference equations with decreasing arguments," *J. Difference Eq. Appl.* **12**, 109–123.
Cipra, B.A. (1997) "Chaotic bugs make the leap from theory to experiment," *SIAM News* **30**, July/August.
Collet, P. and J-P. Eckmann (1980) *Iterated Maps on the Interval as Dynamical Systems*, Birkhäuser, Boston.

Courtemanche, M. and A. Vinet (2003) "Reentry in excitable media," In: A. Beuter, L. Glass, M.C. Mackey, and M.S. Titcombe (Eds), *Nonlinear Dynamics in Physiology and Medicine*, Chapter 7, Springer, New York.

Davies, B. (1999) *Exploring Chaos: Theory and Experiment*, Perseus, New York.

Dehghan, M., C.M. Kent, and H. Sedaghat (2006) "Asymptotic stability for a higher order rational difference equation," *Proceedings of the Conference on Differential and Difference Equations and Applications*, pp. 335–339, Hindawi, New York.

Demazure, M. (2000) *Bifurcations and Catastrophes*, Springer, New York.

Devaney, R.L. (1989) *A Introduction to Chaotic Dynamical Systems*, 2nd ed., Addison-Wesley, Redwood City, CA.

Devaney, R.L. (1992) *A First Course in Chaotic Dynamical Systems*, Perseus, New York.

Dohtani, A. (1992) "Occurrence of chaos in higher dimensional discrete time systems," *SIAM J. Appl. Math.* **52**, 1707–1721.

Drazin, P.G. (1992) *Nonlinear Systems*, Cambridge University Press, Cambridge.

Elaydi, S.N. (1999) *An Introduction to Difference Equations*, 2nd ed., Springer, New York.

Goodwin, R.M. (1951) "The nonlinear accelerator and the persistence of business cycles," *Econometrica* **19**, 1–17.

Grove, E.A. and G. Ladas (2005) *Periodicities in Nonlinear Difference Equations*, Chapman & Hall/CRC, Boca Raton, FL.

Hautus, M.L.J. and T.S. Bolis (1979) "Solution to problem E2721," *Am. Math. Mon.* **86**, 865–866.

Hicks, J.R. (1965) *A Contribution to the Theory of the Trade Cycle*, 2nd ed., Clarendon Press, Oxford. (First ed. Oxford University Press, Oxford, 1950.)

Hirsch, M.W., S. Smale, and R.L. Devaney (2004) *Differential Equations, Dynamical Systems and An Introduction to Chaos*, Elsevier-Academic Press, San Diego.

Holmgren, R. (1996) *A First Course in Discrete Dynamical Systems*, Springer, New York.

Ito, I. and L. Glass (1992) "Theory of reentrant excitation in a ring of cardiac tissue," *Physica D* **56**, 84–106.

Jordan, C. (1965) *Calculus of Finite Differences*, Chelsea, New York.

Kelley, W. and A.C. Peterson (2001) *Difference Equations: An Introduction with Applications*, 2nd ed., Academic Press, San Diego.

Kent, C.M. and H. Sedaghat (2004) "Convergence, periodicity and bifurcations for the 2-parameter absolute-difference equation," *J. Difference Eq. Appl.* **10**, 817–841.

Kent, C.M. and H. Sedaghat (2005) "Difference equations with absolute values," *J. Difference Eq. Appl.* **11**, 677–685.

Kocic, V.L. and G. Ladas (1993) *Global Behavior of Nonlinear Difference Equations of Higher Order with Applications*, Kluwer, Dordrecht.

Kuang, Y. and J.M. Cushing (1996) "Global stability in a nonlinear difference delay equation model of flour beetle population growth," *J. Difference Eq. Appl.* **2**, 31–37.

Kulenovic, M.R.S. (2000) "Invariants and related Liapunov functions for difference equations," *Appl. Math. Lett.* **13**, 1–8.

Kulenovic, M.R.S. and G. Ladas (2001) *Dynamics of Second Order Rational Difference equations with Open Problems and Conjectures*, CRC, Boca Raton, FL.

Ladas, G. (1995) "Invariants for generalized Lyness Equations," *J. Difference Eq. Appl.* **1**, 209–214.

Lakshmikantham, V. and D. Trigiante (1988) *Theory of Difference Equations: Numerical Methods and Applications*, Academic Press, New York.

LaSalle, J.P. (1976) *The Stability of Dynamical Systems*, SIAM, Philadelphia.

LaSalle, J.P. (1986) *Stability and Control of Discrete Processes*, Springer, New York.

Lasota, A. and J.A. Yorke (1973) "On the existence of invariant measures for piecewise monotonic transformations," *Trans. Am. Math. Soc.* **183**, 481–485.

Li, T-Y. and J.A. Yorke (1975) "Period three implies chaos," *Am. Math. Monthly* **82**, 985–992.

Magnucka-Blandzi, E. and J. Popenda (1999) "On the asymptotic behavior of a rational system of difference equations," *J. Difference Eq. Appl.* **5**, 271–286.

Marotto, F.R. (1978) "Snap-back repellers imply chaos in $\mathbb{R}^n$," *J. Math. Anal. Appl.* **63**, 199–223.
Puu, T. (1993) *Nonlinear Economic Dynamics*, 3rd ed., Springer, New York.
Samuelson, P.A. (1939) "Interaction between the multiplier analysis and the principle of acceleration," *Rev. Econ. Stat.* **21**, 75–78.
Sandefur, J.T. (1993) *Discrete Dynamical Modeling*, Oxford University Press, Oxford.
Sedaghat, H. (1997) "A class of nonlinear second order difference equations from macroeconomics," *Nonlinear Anal. TMA* **29**, 593–603.
Sedaghat, H. (1998a) "Geometric stability conditions for higher order difference equations," *J. Math. Anal. Appl.* **224**, 255–272.
Sedaghat, H. (1998b) "Bounded oscillations in the Hicks business cycle model and other delay equations," *J. Difference Eq. Appl.* **4**, 325–341.
Sedaghat, H. (1999) "An Inverse map characterization of asymptotic stability on the line," *Rocky Mount. J. Math.* **29**, 1505–1519.
Sedaghat, H. (2002) "Semiconjugates of one-dimensional maps," *J. Difference Eq. Appl.* **8**, 649–666.
Sedaghat, H. (2003a) *Nonlinear Difference Equations: Theory with Applications to Social Science Models*, Kluwer, Dordrecht.
Sedaghat, H. (2003b) "The global stability of equilibrium in a nonlinear second-order difference equation," *Int. J. Pure Appl. Math.* **8**, 209–223.
Sedaghat, H. (2004a) "The Li-Yorke Theorem and infinite discontinuities," *J. Math. Anal. Appl.* **296**, 538–540.
Sedaghat, H. (2004b) "Periodicity and convergence for $x_{n+1} = |x_n - x_{n-1}|$," *J. Math. Anal. Appl.* **291**, 31–39.
Sedaghat, H. (2004c) "On the equation $x_{n+1} = cx_n + f(x_n - x_{n-1})$," *Fields Inst. Commun.* **42**, 323–326.
Sedaghat, H. (2005) "Asymptotic stability in a class of nonlinear, monotone difference equations," *Int. J. Pure Appl. Math.* **21**, 167–174.
Sedaghat, H., C.M. Kent, and M.A. Wood (2005) "Criteria for the convergence, oscillations and bistability of pulse circulation in ring of excitable media, *SIAM J. Appl. Math.* **66**, 573–590.
Sedaghat, H. and W. Wang (2000) "The asymptotic behavior of a class of nonlinear delay difference equations," *Proc. Am. Math. Soc.* **129**, 1775–1783.
Sharkovski, A.N. (1960) "Necessary and sufficient condition for convergence of one-dimensional iteration processes," *Ukrain. Math. Zh.* **12**, 484–489 (Russian).
Sharkovski, A.N. (1995) "Co-existence of cycles of a continuous mapping of the line into itself," *Int. J. Bifur. Chaos Appl. Sci. Eng.* **5**, 1263–1273. Original Russian edition in *Ukrain. Math. Zh.* **16**(1), 61–71, 1964.
Sharkovski, A.N., Yu.L. Maistrenko, and E.Yu. Romanenko (1993) *Difference Equations and their Applications*, Kluwer, Dordrecht.
Singer, D. (1978) "Stable orbits and bifurcations of maps of the interval," *SIAM J. Appl. Math.* **35**, 260–267.
Stevic, S. (2003) "Asymptotic behavior of a nonlinear difference equation," *Indian J. Pure Appl. Math.* **34**, 1681–1687.
Xiao, H. and X-S. Yang (2003) "Existence and stability of equilibrium points in higher order discrete time systems," *Far East J. Dyn. Syst.* **5**, 141–147.

19
Process Algebra

J.C.M. Baeten
Eindhoven University of Technology

D.A. van Beek
Eindhoven University of Technology

J.E. Rooda
Eindhoven University of Technology

19.1 Introduction 19-1
   Definition • Calculation • History • Hybrid Process Algebra
19.2 Syntax and Informal Semantics of the χ Process Algebra 19-4
   Controlled Tank • Assembly Line Example • Statement Syntax • Semantic Framework • Semantics of Atomic Statements • Semantics of Compound Statements
19.3 Algebraic Reasoning and Verification 19-11
   Introduction • Bottle Filling Line Example • Syntax and Semantics of the Recursion Scope Operator • Elimination of Process Instantiation • Syntax of the Normal Form • Elimination of Parallel Composition • Substitution of Constants and Additional Elimination • Tool-Based Verification
19.4 Conclusions 19-19

19.1 Introduction

19.1.1 Definition

Process algebra is the study of distributed or parallel systems by algebraic means. The word "process" here refers to the *behavior* of a *system*. A system is anything showing behavior, such as the execution of a software system, the actions of a machine, or even the actions of a human being. Behavior is the total of events, actions, or evolutions that a system can perform, the order in which these can be executed and maybe other aspects of this execution such as timing, probabilities, or continuous aspects. Always, the focus is on certain aspects of behavior, disregarding other aspects, so an abstraction or idealization of the "real" behavior is considered. Instead of considering behavior, we may consider an *observation* of behavior, where an action is the chosen unit of observation. As the origin of process algebra is in computer science, the actions are usually thought to be discrete: occurrence is at some moment in time, and different actions are separated in time. This is why a process is sometimes also called a *discrete-event system*. Please note that this is a less restrictive definition than, e.g., Cassandras and Lafortune (1999).

The word "algebra" denotes that the approach in dealing with behavior is algebraic and axiomatic. That is, methods and techniques of universal algebra are used. A process algebra can be defined as any mathematical structure satisfying the axioms given for the basic operators. A process is an element of a process algebra. By using the axioms, we can perform *calculations* with processes. Often, though, process algebra goes beyond the strict bounds of universal algebra: sometimes multiple sorts and/or binding of variables are used.

The simplest model of behavior is to see behavior as an input/output function. A value or input is given at the beginning of the process, and at some moment there is a value as outcome or output. This model was used to advantage as the simplest model of the behavior of a computer program in computer science, from the start of the subject in the middle of the twentieth century. It was instrumental in the development of (finite state) *automata theory*. In automata theory, a process is modeled as an automaton. An automaton has a number of *states* and a number of *transitions* going from a state to a state. A transition denotes the execution of an (elementary) action, the basic unit of behavior. Also, there is an initial state (sometimes, more than one) and a number of final states. A behavior is a run, i.e., a path from initial state to final state. An important aspect is when to consider two automata to be equal, expressed by a notion of equivalence. On automata, the basic notion of equivalence is "language equivalence," which considers equivalence in terms of behavior, where a behavior is characterized by the set of executions from an initial state to a final state. An algebra that allows equational reasoning about automata is the algebra of regular expressions (see, e.g., Linz, 2001).

Later on, this model was found to be lacking in several situations. Basically, what is missing is the notion of *interaction*: during the execution from initial state to final state, a system may interact with another system. This is needed in order to describe parallel or distributed systems, or the so-called *reactive* systems. When dealing with interacting systems, the phrase *concurrency theory* is used. Thus, concurrency theory is the theory of interacting, parallel, or distributed systems. When referring to process algebra, we usually consider it as an approach to concurrency theory, so that a process algebra usually (but not necessarily) has an operator (function symbol) to put things in parallel called parallel composition.

Thus, a usable definition is that process algebra is the study of the behavior of parallel or distributed systems by algebraic means. It offers means to describe or *specify* such systems, and thus it has means to specify parallel composition. Besides this, it can usually also specify alternative composition (put things in a choice) and sequential composition (sequencing, put things one after the other). Moreover, it is possible to reason about such systems using algebra, i.e., equational reasoning. By means of this equational reasoning, *verification* becomes possible, i.e., it can be established that a system satisfies a certain property.

What are these basic laws of process algebra? In this chapter, we do not present collections of such laws explicitly. Rather, it is shown how calculations can proceed. To repeat, it can be said that any mathematical structure with operators of the right number of arguments satisfying the given basic laws is a process algebra. Often, these structures are formulated in terms of *transition systems*, where a transition system has a number of states (including an initial state and a number of final states) and transitions between them. The notion of equivalence studied is usually not language equivalence. Prominent among the equivalences studied is the notion of *bisimulation*. Often, the study of transition systems, ways to define them, and equivalences on them are also considered as a part of process algebra, even in the case no equational theory is present.

19.1.2 Calculation

One form of calculation is verification by means of automated methods (called *model checking*, see e.g., Clarke et al., 2000) that traverse all states of a transition system and check that a certain property is true in each state. The drawback is that transition systems grow at a rate exponential in the number of components (in fact, due to the presence of parameters, often they become infinite). For instance, a system having 10 interacting components, each of which has 10 states, has a total number of 10,000,000,000 states. It is said that model checking techniques suffer from the *state explosion* problem.

In contrast, reasoning can take place in logic using a form of deduction. Also here, progress is made, and many *theorem proving* tools exist (Bundy, 1999). The drawback here is that finding a proof needs user assistance (as the general problem is undecidable), which requires a lot of knowledge about the system.

On the basis of an algebraic theory equational reasoning takes the middle ground. On the one hand, the next step in the procedure is usually clear, since it is more rewriting than equational reasoning. Therefore, automation can be done in a straightforward way. On the other, representations are compact and allow the presence of parameters, so that an infinite set of instances can be verified at the same time.

19.1.3 History

Process algebra started in the late seventies of the twentieth century. At that point, the only part of concurrency theory that existed was the theory of Petri nets, as discussed in Chapter 24.

The question was raised how to give semantics to programs containing a parallel composition operator. It was found that this was difficult using the semantic methods used at that time. The idea of a behavior as an input/output function needed to be abandoned. A program could still be modeled as an automaton, but the notion of language equivalence was no longer appropriate. This is because the interaction a process has between input and output influences the outcome, disrupting functional behavior. Secondly, the notion of *global* variables needed to be overcome. Using global variables, a state of an automaton used as a model was given as a valuation of the program variables, that is, a state was determined by the values of the variables. The independent execution of parallel processes makes it difficult or impossible to determine the values of global variables at a given moment. It turned out to be simpler to let each process have its own local variables and to denote exchange of information explicitly.

After some preliminary work by others, three main process algebra theories were developed. These are Calculus of Communicating Systems) (CCS) by Robin Milner (Milner, 1980, 1989), Communicating Sequential Processes (CSP) by Tony Hoare (Hoare, 1985), and Algebra of Communicating Processes (ACP) by Jan Bergstra and Jan Willem Klop (see Bergstra and Klop, 1984; Baeten and Weijland, 1990).

Comparing these best-known process algebras CCS, CSP, and ACP, we can say that there is a considerable amount of work and applications realized in all three of them. In that sense, there seem to be no fundamental differences between the theories with respect to the range of applications. Historically, CCS was the first with a complete theory. Different from the other two, CSP makes fewer distinctions between processes. More than the other two, ACP emphasizes the algebraic aspect: there is an equational theory with a range of semantic models. Also, ACP has a more general communication scheme; in CCS, communication is combined with abstraction, in CSP, there is also a restricted communication scheme.

Over the years, other process algebras were developed, and many extensions were realized. Most interesting for this book is the extension to hybrid systems. The language we consider in this chapter is most closely related to the ACP approach, as in this approach, there is most work and experience on hybrid extensions. For a taste of another approach, see He (1994).

19.1.4 Hybrid Process Algebra

Process algebra started out in computer science, and is especially geared to describing discrete-event systems such as computer programs and software systems. With the growing importance of embedded systems, which are software systems that are integrated in the machine or device that they control, it was considered to use process algebra also to model and reason about the controlled physical environment of the software. However, specifications of physical systems not only require discrete-event models (such as timed or untimed transition systems), but also continuous-time models (such as differential algebraic equations (Kunkel and Mehrmann, 2006)), leading to hybrid models.

In recent years, several attempts were made to incorporate such aspects into process algebra. In this chapter, we report on one of these based on the χ language. Other hybrid process algebras are HyPA (Cuijpers and Reniers, 2005), process algebra for hybrid systems ACP_{hs}^{srt} (Bergstra and Middelburg, 2005), and the ϕ-calculus (Rounds and Song, 2003). The history of the χ formalism dates back to quite some time. It was originally mainly used as a modeling and simulation language for discrete-event systems. The first simulator (Naumoski and Alberts, 1998) was successfully applied to a large number of industrial cases, such as integrated circuit manufacturing plants, breweries, and process industry plants (Beek et al., 2002). Later, the hybrid language and simulator were developed (Fábián, 1999; Beek and Rooda, 2000). Recently, the χ language has been completely redesigned. The result is a hybrid process algebra with a formal semantics as defined in Beek et al. (2006). This chapter informally defines the most important elements of the syntax and semantics of the χ process algebra. It also extends the formal definitions of Beek (2006) with a more user-friendly syntax including the specification of data types.

19.2 Syntax and Informal Semantics of the χ Process Algebra

In this section, the syntax and informal semantics of the χ process algebra is first illustrated by means of two examples: a controlled tank and an assembly line example. After this intuitive explanation, the syntax and semantics are more precisely defined.

19.2.1 Controlled Tank

Figure 19.1 shows a liquid storage tank with a volume controller VC. The incoming flow Q_i is controlled by the means of a valve n. The outgoing flow is given by the equation $Q_o = \sqrt{V}$. The volume controller maintains the volume V of the liquid in the tank between 2 and 10. The χ model of the controlled tank is as follows:

```
model Tank() =
|[ var n : nat = 0, cont V : real = 10, alg Qᵢ, Qₒ : real
:: V̇ = Qᵢ − Qₒ
 ‖ Qᵢ = n · 5
 ‖ Qₒ = √V
 ‖ *(V ≤ 2 → n := 1; V ≥ 10 → n := 0)
]|
```

Figure 19.2 shows the result of a simulation of the model for 7 time units. Initially, the volume in the tank equals 10, and the valve is closed ($n = 0$). The derivative of the volume equals the difference between the incoming and outgoing flows ($\dot{V} = Q_i - Q_o$). The specification of the controller is given

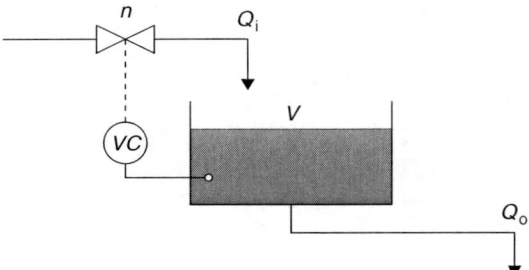

FIGURE 19.1 Controlled tank.

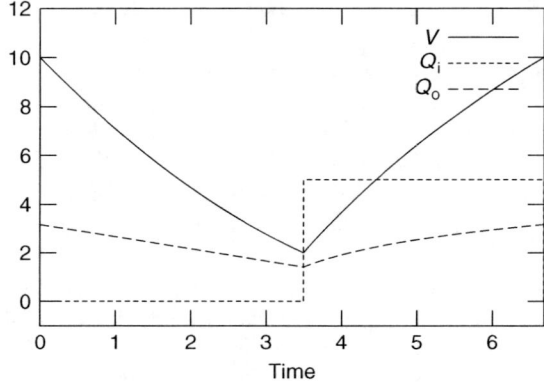

FIGURE 19.2 Simulation of the controlled tank.

by $*(V \leq 2 \rightarrow n := 1; V \geq 10 \rightarrow n := 0)$, where the loop statement $*(p)$ denotes the infinite repetition of statement p. The guard operator "$\rightarrow$" is used to specify conditional execution of a statement, by prefixing a condition (referred to as a guard) b to a statement p, which is written as $b \rightarrow p$. The sequential composition operator ";" is used to specify sequential execution of components, and the parallel composition operator "$\|$" is used to specify the parallel execution of components. In the example, the equations and the controller are all executed in parallel.

Initially, the three equations are enabled and the guard $V \leq 2$ is also enabled. Since the value of the guard is false initially ($V = 10$), the assignment $n := 1$ is disabled. The model executes by doing a sequence of delays, which involve passing of time, and actions, which are executed instantaneously, without passing of time. The model can do a delay of t time units when all enabled statements can simultaneously do a delay of t. A guard that is false, allows arbitrary delays until it becomes true (see Section 19.2.6.2), and equations allow a delay of t, when a solution of the equations exists that defines the values of the variables as a function of time (on domain $[0, t]$). At the end point of the delay(s), $V = 2$, and the guard becomes true. The assignment $n := 1$ is now enabled. The model can now no longer delay, since assignments cannot delay; an assignment is a so-called nondelayable statement (see Section 19.2.3). The model can do an action when any of the enabled statements can do an action. Assignments can do an action by executing the assignment. Therefore, the model executes the assignment $n := 1$, which models opening of the valve. The assignment causes the value of variable Q_i to immediately become 5, to satisfy the equation $Q_i = n \cdot 5$. This is referred to as the "consistent equation semantics": equations must be satisfied at all times. The value of the continuous variable V, however, is unchanged; only algebraic variables are allowed to change, to satisfy equations, when other variables are assigned. Execution of the assignment $n := 1$ causes the assignment to be disabled and the next statement ($V \geq 10 \rightarrow n := 0$) to be enabled. The guard $V \geq 10$ is false. Therefore, the model delays, while solving the equations, until the guard becomes true (volume in the tank equals 10). Now the assignment $n := 0$ is executed, modeling closing of the valve. As a result, the assignment is disabled and the first statement ($V \leq 2 \rightarrow n := 1$) of the repetition is reenabled.

The general form of a χ model is

$$\texttt{model}\ id(D_m) = \|[D :: p]\|$$

where id is an identifier that represents the name of the model, D_m denotes the model parameters that are not present in the example, and D denotes the declaration of variables or channels of the model. Channels are introduced in the assembly line example of Section 19.2.2. Finally, p denotes a statement also known as a process term. Notation $\|[D :: p]\|$ is in fact a scope operator, which is defined in Section 19.2.3, together with statement p. The following kinds of variable can be declared in D:

- "Discrete" variables, such as in $\texttt{var}\ n : \texttt{nat} = 0$. This declares a variable n with initial value 0. The name "discrete" is common in hybrid systems terminology, and refers to the fact that the variable takes only a limited number of values when the model is executed (in this case only 0 and 1). The value of a discrete variable remains constant when model time progresses. The value, in principle, changes only by means of assignments (e.g., $n := 1$). Discrete variables can be of type real, however.
- "Continuous" variables, such as in $\texttt{cont}\ V : \texttt{real} = 10$. Continuous variables are the only variables for which dotted variables (derivatives) can be used in models. Therefore, the declaration $\texttt{cont}\ V : \texttt{real}$ implies that V and its dotted version $\dot{V}$ can both be used in the model. The values of continuous variables may change according to a continuous function of time when model time progresses. The values of continuous variables are further restricted by equations (or in more general terms: delay predicates, defined in Section 19.2.5.2). The value of a continuous variable can also be changed by means of an assignment.
- "Algebraic" variables, such as in $\texttt{alg}\ Q_i, Q_o : \texttt{real}$. These variables behave in a similar way as continuous variables. The differences are that algebraic variables may change according to a discontinuous function of time, algebraic variables are not allowed to occur as dotted variables, and algebraic variables do not have a memory: the value of an algebraic variable is in principle determined by the enabled equations and not by assignments (e.g., $Q_o = \sqrt{V}$).

Finally, a predefined reserved global variable time, which denotes the model time, exists. Initially, the value of this variable is zero and it is incremented by t whenever the model does a delay of t.

19.2.2 Assembly Line Example

An assembly process A assembles three different parts that are supplied by three suppliers G. The order in which the parts are supplied is unknown, but each part should be received by the assembly process as soon as possible. When all the three parts have been received, assembly may start. Assembly takes t_A units of time. When the products have been assembled, they are sent to an exit process E. Figure 19.3 shows the iconic model of the assembly line, which is modeled as a discrete-event system. For the χ model of the assembly line, the first two types are declared. The type "part," representing a part as a natural number, and the type "assy," representing an assembled unit as a 3-tuple of parts:

```
type part = nat
   , assy = (part, part, part)
```

The χ model consists of parallel instantiations of the three generator processes G, the assembly process A, and the exit process E:

```
model AssemblyLine ( val t₀, t₁, t₂, tA : real) =
|[ chan a, b, c : part, d: assy
:: G(a, 0, t₀) || G(b, 1, t₁) || G(c, 2, t₂) || A(a, b, c, d, tA) || E(d)
]|
```

The channels $a, b, c,$ and d are used for communication and synchronization between the parallel processes. Each generator G sends a part n in every t time units:

```
proc G (chan a!: part val n: nat t: real) = |[ * (Δt; a!n)]|
```

The assembly process receives the parts by means of the parallel composition $(a?x \| b?y \| c?z)$. This ensures that each part is received as soon as possible. The parallel composition terminates when all parts have been received.

```
proc A (chan a?, b?, c? : part, d!: assy, val t: real) =
|[ var x, y, z : part
:: *((a?x || b?y | c?z); Δt; d!(x, y, z))
]|
```

The exit process is simply

```
proc E (chan a?: assy) = |[var x: assy :: *(a? x)]|
```

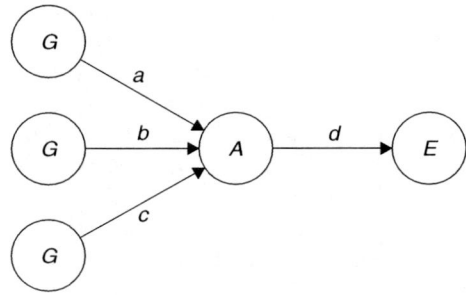

FIGURE 19.3 Iconic model of an assembly line.

To understand the meaning of the model, the process instantiations can be replaced by their definitions, as defined in (Beek et al. (2006), and the model parameters can be replaced by their values. Thus, the model instantiation *AssemblyLine*(5, 6, 7, 2) can be rewritten into the following equivalent form:

```
model AssemblyLine() =
|[ chan a, b, c : part d : assy
:: |[ var n : nat = 0, t : real = 5 :: *(Δt; a!n) ]|
 || |[ var n : nat = 1, t : real = 6 :: *(Δt; b!n) ]|
 || |[ var n : nat = 2, t : real = 7 :: *(Δt; c!n) ]|
 || |[ var x, y, z : part, t : real = 2 :: *((a?x ||| b?y ||| c?z); Δt; d!(x, y, z)) ]|
 || |[ var x : assy :: *(d?x) ]|
]|
```

Initially, the first statements of the repetitions are enabled. The first statement of the repetition of the assembly process is a parallel composition of three receive statements ($a?x \parallel b?y \parallel c?z$). Enabling a parallel composition enables its components. Therefore, initially, the statements $\Delta t, \Delta t, \Delta t, a?x, b?y, c?z$, and $d?x$ are enabled. Each of these statements can delay. A delay statement Δt behaves as a timer that can delay for at most t time units. After this, the timer is expired and can terminate by means of an action. The values of the three local variables of t are 5, 6, and 7, respectively. Therefore, initially, a (maximum) delay of 5 time units is possible. After this, the first timer terminates by means of an action, and the send statement $a!n$ is enabled. The enabled statements are now $a!n, \Delta t, \Delta t, a?x, b?y, c?z, d?x$, where the two timers modeled by Δt and Δt can delay for 1 and 2 remaining time units, respectively, before expiring. We now have an enabled pair of a send and a receive statement on the same channel that are placed in parallel: $a!n$ and $a?x$. This pair can simultaneously do a send and a receive action followed by joint termination. The result is comparable to the (distributed) execution of the assignment $x := n$, or $x := 0$, since the value of the first variable n is 0. After this, the send and receive statements are disabled. Disabling of $a!n$ enables the delay statement Δt again. The enabled statements are now: $\Delta t, \Delta t, \Delta t, b?y, c?z, d?x$, where the three timers need to delay for another 5, 1, and 2 time units, respectively, before expiring. After expiration of the second and third timer, communication of 1 via channel b and 2 via channel c takes place, respectively. Then, the parallel composition terminates, enabling the delay statement Δt of the assembly process. After this intuitive explanation of the χ language by means of examples, the next sections more precisely define the syntax and semantics.

19.2.3 Statement Syntax

This section defines the syntax of a considerable and representative subset of χ models using a Backus-Naur (BNF) like notation. The symbol | defines choice, and notation $\{Z\}^*$ denotes a sequence of zero or more Zs. Statements can be divided in two classes: the atomic statements that represent the smallest statement units and the compound statements that are constructed from one or more (atomic) statements by means of operators. The syntax of the atomic χ statements is as follows:

p_{atom} ::=	skip		nondelayable action
	x := **e**		nondelayable (multi)assignment
	[skip]		delayable action
	[**x** := **e**]		delayable (multi)assignment
	h ! e	h!	delayable send
	h ? x	h?	delayable receive
	Δd		delay
	u		delay predicate

where **x** and **e** denote comma separated variables $x_1, \ldots, x_n$ and expressions $e_1, \ldots, e_n$, respectively, for $n \geq 1$, h denotes a channel, and d denotes an expression of type real. Delay predicate u denotes a predicate over variables (including the variable `time`) and dotted continuous variables (derivatives). Delay predicates may occur in the form of differential algebraic equations, such as $\dot{x} = y, y = n$, or in the form of a constraint or invariant, such as $x \geq 1$.

The syntax of the compound χ statements is as follows:

$$
\begin{array}{rll}
p ::= & p_{\text{atom}} & \text{atomic} \\
 | & p; p & \text{sequential composition} \\
 | & b \to p & \text{guard operator} \\
 | & p \,[]\, p & \text{alternative composition} \\
 | & p \parallel p & \text{parallel composition} \\
 | & *p & \text{loop statement} \\
 | & b \stackrel{*}{\to} p & \text{while statement} \\
 | & |[D :: p]| & \text{variable and channel scope operator} \\
 | & id(\mathbf{e}) & \text{process instantiation} \\
 | & p_R & \text{recursion scope operator (see Sections 19.3.2 and 19.3.3),}
\end{array}
$$

where guard b denotes a predicate over variables. The operators are listed in descending order of their binding strength as follows: $\{*, \stackrel{*}{\to}, \to\}, ;, \{\parallel, []\}$. The operators inside the braces have equal binding strength. Parentheses may be used to group statements. For example, $x := 1; y := x \,[]\, x := 2; y := 2x$ means $(x := 1; y := x) \,[]\, (x := 2; y := 2x)$. To avoid confusion, parentheses are obligatory when alternative composition and parallel composition are used together. For example, $p \,[]\, q \parallel r$ is not allowed and should either be written as $(p \,[]\, q) \parallel r$ or as $p \,[]\, (q \parallel r)$.

19.2.4 Semantic Framework

In this chapter, the meaning (semantics) of a χ model is informally defined in terms of delay behavior and action behavior, based on the formal semantics as presented in Beek et al.(2006). Delay behavior involves passing of time, where the semantics defines for each variable how its value changes as a function of time. Action behavior is instantaneous: time does not progress, and the semantics defines for each variable the relation between its value before and after the action.

Atomic statements can be disabled or enabled. Actions and delays are done by *enabled atomic* statements, with one exception only: an enabled guarded statement $b \to p$, with a guard that is false can do any delay. Atomic statements terminate by doing an action. They never terminate by doing a delay. A statement that terminates becomes disabled by doing so.

Compound statements combine (sub)statements by means of operators. The operator defines the relation between enabling, disabling, and termination of the compound statement and its substatements. Enabling or disabling a compound statement is defined in terms of enabling or disabling its substatements. Enabling a compound statement implies enabling one or more of its substatements. For example, enabling a sequential composition $p_1; \ldots; p_n$ implies enabling the first statement p_1, whereas enabling a parallel composition $p_1 \parallel \ldots \parallel p_n$ implies enabling all statements $p_1 \ldots p_n$.

Execution of a χ model M, defined as $\mathtt{model}\ M(D_0) = |[D_1 :: p_0]|$, takes place by executing a sequence of delays and actions in the following way:

- At the start, statement p_0 is enabled.
- Any enabled skip statement or assignment statement (delayable or nondelayable) can do an action.
- An enabled pair of a send and a receive statement on the same channel that are placed in parallel can simultaneously do a send and a receive action followed by joint termination. The result, in terms of values of variables, is comparable to the (distributed) execution of a (multi)assignment. For example, execution of the communication action in $h\,!\,1 \parallel h\,?\,x$ is comparable to execution of the assignment $x := 1$.
- The model can do delays only when and for as long as:
  — All enabled statements can delay. The delayable versions of the skip statement, assignment, and send and receive statements can always delay (the nondelayable versions can never delay).

A delay statement Δd can delay for as long as its internal timer is not expired (see Section 19.2.5.3), and the set of all enabled delay predicates can delay for as long as they have a solution. Such a solution defines the values of the variables as a function of time for the period of the delay.

Note that the set of enabled statements may change while delaying. The reason for this is the guarded statement $b \rightarrow p$, because the value of the guard can change while delaying, owing to changes in the values of continuous or algebraic variables used in b.

— No parallel pair of a send and a receive statement on the same channel is enabled or becomes enabled. This is because, by default, channels in χ are urgent: communication or synchronization cannot be postponed by delaying.

- When different actions and/or delays are possible, any of these can be chosen. This is referred to as nondeterministic choice. Note that delays may always be shorter than the maximum possible length.

The values of the discrete and continuous variables are stored in memory. The values of the algebraic variables are not stored. This means that the starting point of the trajectory of a discrete or continuous variable equals its last value stored in memory. The starting point of the trajectory of an algebraic variables can be any value that is allowed by the enabled equations.

In models of physical systems, the delay behavior of the continuous and algebraic variables is usually uniquely determined: there is usually only one solution of the set of enabled differential algebraic equations. Multiple delays/solutions can be caused by underspecified systems of equations, where there are less equations than variables, or by delay predicates that allow multiple solution such as "true" or $\dot{x} \in [0, 1]$.

The action behavior of the discrete, continuous, and algebraic variables is as follows:

- The discrete and continuous variables do not change as a result of actions unless the change is explicitly specified, for example, by means of an assignment, or by receiving a value via a channel.
- The algebraic variables can, in principle, change arbitrarily in actions. In most models, their values are defined by equations.

19.2.5 Semantics of Atomic Statements

19.2.5.1 Skip and Multiassignment

An enabled skip statement can do an action, and then terminates. It corresponds to an assignment $x := x$, because the values of continuous and discrete variables are left unchanged. The skip statement can be used to make a choice in an alternative composition statement, because it executes an action (see process *Tank* in Section 19.3.2).

An enabled *multiassignment* statement $\mathbf{x}_n := \mathbf{e}_n$ for $n \geq 1$ can do an action that changes the values of the variables $x_1, \ldots, x_n$ in one step to the values of expressions $e_1, \ldots, e_n$, respectively, and then terminates. For $n = 1$, this gives a normal assignment $x := e$.

19.2.5.2 Delay Predicate

An enabled *delay predicate* u can perform delays but no actions. Delay predicates restrict the allowed trajectories of the variables while delaying in such a way that at each time point during the delay the delay predicate holds (its value must be true), when all variables and dotted variables in the predicate are replaced by their current value.

Delay predicates also restrict the action behavior of χ models, because the enabled delay predicates must also hold before and after each action. In fact, the enabled delay predicates of a χ model must hold at all times. This is referred to as the "consistent equation semantics."

The relation between the trajectory of a continuous variable x and the trajectory of its "derivative" $\dot{x}$ is given by the Caratheodory solution concept: $x(t) = x(0) + \int_0^t \dot{x}(s) ds$. This allows a nonsmooth (but continuous) trajectory for a differential variable x in the case that the trajectory of its "derivative" $\dot{x}$ is nonsmooth or even discontinuous, as in, for example, model $M() = \|[\text{cont } x : \text{real} = 0 :: \dot{x} = \text{step}(\text{time} - 1)]\|$, where step$(y)$ equals 0 for $y \leq 0$ and 1 for $y > 0$.

19.2.5.3 Delay Statement

A delay statement Δd behaves as a timer that can be in three modes: reset, running, or expired. A timer that is in mode running keeps track of the remaining time $t_{\exp}$ before expiring. Initially, timers are in mode reset. In modes reset and running, a timer can delay; in mode expired, it can terminate by means of an action. If the timer is enabled, its behavior is as follows:

- In mode reset, when the value c of expression d is bigger than zero, the timer can do a delay t for $t \leq c$. If $t < c$, the new mode after the delay is running with $t_{\exp} = c - t$. If $t = c$, the new mode is expired.
- In mode running, the timer can do a delay $t \leq t_{\exp}$ to mode running ($t < t_{\exp}$) or expired ($t = t_{\exp}$). It switches to mode reset when it is disabled as a result of a choice being made in an alternative composition (see Section 19.2.6.3).

  For example, in $x := 0; *(\Delta 3 \, [] \, \dot{x} = 1 \, [] \, x \geq 1 \rightarrow x := 0)$, when the delay statement/timer $\Delta 3$ becomes running, it switches to mode reset after 1 time unit, because of the execution of the (second, guarded) assignment $x := 0$, which enforces a choice in the alternative composition and disables the timer.
- In mode expired, or in mode reset when the value c of expression d equals zero, the timer can do an action, accompanied by termination to mode reset. It also switches to mode reset when it is disabled as a result of a choice being made in an alternative composition.

The mode of a timer remains unchanged when it is disabled as a result of the value of a guard becoming false. For example, in $\sin(2\pi \text{time}) \geq 0 \rightarrow \Delta 1$, the timer expires after two time units, that is after two periods of the sine function, because the timer only delays when the sine function is positive. As a final example, consider $*(h?d; \Delta d) \parallel *(h!1; h!2)$. The first delay of the timer is 1, the second delay is 2, and then the cycle is repeated.

19.2.6 Semantics of Compound Statements

19.2.6.1 Sequential Composition

In a *sequential composition* $p_1; \ldots; p_n$ ($n \geq 1$), only one statement p_i, $1 \leq i \leq n$, can be enabled at the same time. Enabling a sequential composition $p_1; \ldots; p_n$ implies enabling its first statement p_1. When statement p_i ($1 \leq i \leq n-1$) terminates (and is therefore also disabled), the next statement p_{i+1} becomes enabled. The sequential composition terminates upon termination of its last statement p_n.

19.2.6.2 Guard Operator

Enabling of a guarded statement enables its guard b. Behavior of a guarded statement $b \rightarrow p$ depends on the value of the guard b:

- Statement p is enabled while the guard is enabled and the value of the guard is true. Execution of the first action by p disables the guard. Thus, after this first action, the value of the guard becomes irrelevant.
- Statement p is disabled while the value of the guard is false. The guarded statement $b \rightarrow p$ can, in principle, do any delay while the guard is enabled and its value is false; only at the start point and end point of such a delay, the value of the guard may be true.

When a guarded statement occurs in parallel with another statement, as in $q \parallel b \rightarrow p$, the value of the guard can change owing to the actions of statement q, which may cause statement p to change from being disabled to enabled or vice versa. For example, $b := \text{false}; (\Delta 1; b := \text{true} \parallel b \rightarrow \text{skip})$.

When in $q \parallel b \rightarrow p$, the guard b contains continuous or algebraic variables, and q contains one or more enabled delay predicates, the value of the guard may change during a delay, causing statement p to change from being disabled to enabled or vice versa. For example, $\dot{x} = 1 \parallel x \geq 1 \rightarrow x := 0$.

19.2.6.3 Alternative Composition

Enabling $p_1 [] \ldots [] p_n$ enables the statements $p_1, \ldots, p_n$. Execution of an action by any one of the statements $p_1 \ldots p_n$ disables the other statements. In this way, execution of the first action makes a choice. When one of the statements $p_1, \ldots, p_n$ terminates, the alternative composition $p_1 [] \ldots [] p_n$ also terminates.

19.2.6.4 Parallelism

Enabling $p_1 \parallel \ldots \parallel p_n$ enables the statements $p_1, \ldots, p_n$. When a statement p_i, $1 \leq i \leq n$, executes an action, the other statements remain enabled. The parallel composition $p_1 \parallel \ldots \parallel p_n$ terminates when the statements $p_1, \ldots, p_n$ have all terminated.

Informally, we often refer to the statements $p_1, \ldots, p_n$ occurring in $p_1 \parallel \ldots \parallel p_n$ as parallel processes. Parallel processes interact by means of shared variables or by means of synchronous point-to-point communication or synchronization via a channel. Communication in χ is the sending of values of one or more expressions by one parallel process via a channel to another parallel process, where the received values are stored in variables. In case no values are sent and received, we refer to synchronization instead of communication.

19.2.6.5 Loop and While Statement

Loop statement $*p$ represents the infinite repetition of statement p. When $*p$ is enabled, p is enabled. Termination of p results in reenabling of p.

The while statement $b \xrightarrow{*} p$ can be interpreted as "while b do p." Enabling of $b \xrightarrow{*} p$ when b is true enables p (by means of an action), and enabling of $b \xrightarrow{*} p$ when b is false, leads to termination of the while statement (by means of an action).

19.2.6.6 Variable and Channel Scope Operator

A variable and channel scope operator may introduce new variables and new channels. Enabling of a variable and channel scope statement $|[D :: p]|$, where the local declaration part D introduces new variables and/or channels (see Sections 19.2.1 and 19.2.2), performs the variable initializations specified in D and enables statement p. Termination of p terminates the scope statement $|[D :: p]|$. Any occurrence of a variable or channel in p that is declared in D refers to that local variable or channel and not to any more global declaration of the variable or channel with the same name, if such a more global declaration should exist.

19.3 Algebraic Reasoning and Verification

19.3.1 Introduction

The χ process algebra has strong support for modular composition by allowing unrestricted combination of operators such as sequential and parallel composition, by providing statements for scoping, by providing process definition and instantiation, and by providing different interaction mechanisms, namely synchronous communication and shared variables.

The fact that the χ language is a process algebra with a wide range of statements potentially complicates the development of tools for χ, since the implementations have to deal with all possible combinations of the χ atomic statements and the operators that are defined on them. This is where the process algebraic approach of equational reasoning, that allows rewriting models to a simpler form, is essential.

To illustrate the required implementation efforts, consider the following implementations that are developed: a Python implementation for rapid prototyping; a C implementation for the fast model execution; and an implementation based on the MATLAB Simulink S-functions (The MathWorks, 2005), where a χ model is translated to an S-function block. Furthermore, there is an implementation for real-time control (Hofkamp, 2001). In Bortnik, et al. (2005) it has been shown that different model checkers each have their own strengths and weaknesses. Therefore, for verification, translations to several tools are defined. In particular, for hybrid models a translation to the hybrid I/O automaton-based PHAver (Frehse,

2005) model checker is defined. For timed models the following translations are defined: (1) a translation to the action-based process algebra μCRL (Groote, 1997), used as input language for the verification tool CADP (Fernandez et al., 1996); (2) a translation to PROMELA, a state-based, imperative language, used as input language for the verification tool SPIN (Holzmann, 2003); and (3) a translation to the timed automaton-based input language of the UPPAAL (Larsen et al., 1997) verification tool. In future, for verification of hybrid models, additional translations may be considered to tools such as HyTech (Alur et al., 1996), or one of the many other hybrid model checkers.

Instead of defining the implementations mentioned above on the full χ language as defined in Section 19.2.3, the process algebraic approach of equational reasoning makes it possible to transform χ models in a series of steps to a (much simpler) normal form, and to define the implementations on the normal form. The original χ model and its normal form are bisimilar, which ensures that relevant model properties are preserved. The normal form has strong syntactical restrictions, no parallel composition operator, and is quite similar to a hybrid automaton. Currently, correctness proofs are developed, and in the near future, implementations will be redesigned based on the normal form.

The steps to the normal form are as follows. First, the process instantiations are eliminated, by replacing them by their defining bodies, and replacing the formal parameters by actual arguments. Second, parallel composition is eliminated by using laws of process algebra, in particular a so-called *expansion law* (not given here). An example of a process algebra law in χ specifying that the guard distributes over alternative composition is $b \to (p \,[]\, q) = b \to p \,[]\, b \to q$. Finally, the normal form may be simplified further, taking advantage of the fact that it no longer contains parallel composition. Note that it is possible to construct models for which the normal form cannot be (easily) generated. These exceptions are not discussed in this chapter, since they do not restrict translation to the normal form for practical purposes. For a definition of the normal form see Section 19.3.5.

19.3.2 Bottle Filling Line Example

Figure 19.4 shows a bottle filling line, based on Baeten and Middelburg (2002), consisting of a storage tank that is continuously filled with a flow Q_{in}, a conveyor belt that supplies empty bottles, and a valve that is opened when an empty bottle is below the filling nozzle, and is closed when the bottle is full. When a bottle has been filled, the conveyor starts moving to put the next bottle under the filling nozzle, which takes one unit of time. When the storage tank is not empty, the bottle filling flow Q equals Q_{set}. When the storage tank is empty, the bottle filling flow equals the flow Q_{in}. The system should operate in such a way that overflow of the tank does not occur. We assume $Q_{in} < Q_{set}$.

Figure 19.5 shows an iconic representation of the model of the filling line. It consists of the processes *Tank* and *Conveyor* that interact by means of the channels *open* and *close*, and shared variable Q. The

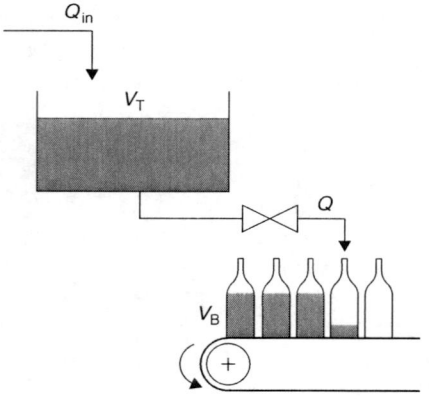
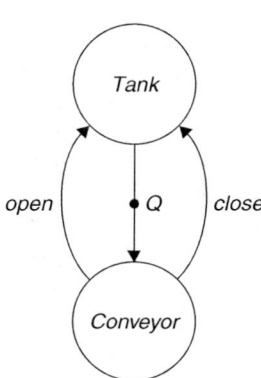

FIGURE 19.4 Filling line.   **FIGURE 19.5** Iconicmodel of the filling line.

model is defined below. It has two parameters: the initial volume V_{T0} of the storage tank, and the value Q_{in} of the flow that is used to fill the storage tank. The constants Q_{set}, V_{Tmax}, and V_{Bmax} define the maximum value of the bottle filling flow Q, the maximum volume of the storage tank, and the filling volume of the bottles, respectively. The model *FillingLine* consists of the algebraic variable Q, the channels *open* and *close*, and the parallel composition of the process instantiations for the tank and the conveyor.

```
const Q_set : real = 3
    , V_Tmax : real = 20
    , V_Bmax : real = 10

model FillingLine (val V_T0, Q_in : real) =
|[ alg Q : real, chan open, close : void
:: Tank(Q, open, close, V_T0, Q_in) || Conveyor(Q, open, close)
]|
```

The tank process has a local continuous variable V_T that is initialized to V_{T0}. Its process body is a recursion scope consisting of three modes: closed, opened, and openedempty that correspond to the valve being closed, the valve being open, and the valve being open while the storage tank is empty. The syntax and semantics of recursion scopes is defined in Section 19.3.3. In the mode opened, the storage tank is usually not empty. When the storage tank is empty in mode opened, the delayable skip statement [skip] may be executed causing the next mode to be openedempty. Owing to the consistent equation semantics, the skip statement can be executed only if the delay predicate in the next mode openedempty holds. This means, among others, that $V_T = 0$ must hold. Therefore, the transition to mode openedempty can be taken only when the storage tank is empty. Note that the comma in delay predicates denotes conjunction. For example, $\dot{V}_T = Q_{in}, Q = 0$ means $\dot{V}_T = Q_{in} \wedge Q = 0$.

```
proc Tank(alg Q : real, chan open?, close? : void, val V_T0, Q_in : real) =
|[ cont V_T : real = V_T0
:: |[ mode closed =
        ( V̇_T = Q_in, Q = 0, V_T ≤ V_Tmax [] open?; opened )
    , mode opened =
        ( V̇_T = Q_in − Q, Q = Q_set, 0 ≤ V_T ≤ V_Tmax
          [] [skip]; openedempty
          [] close?; closed
        )
    , mode openedempty =
        ( V_T = 0, Q = Q_in [] close?; closed )
   :: closed
   ]|
]|
```

Process *Conveyor* supplies an empty bottle in 1 unit of time ($V_B := 0; \Delta 1$). Then it synchronizes with the storage tank process by means of the send statement *open!*, and it proceeds in mode filling. When the bottle is filled in mode filling ($V_B \geq V_{Bmax}$), the process synchronizes with the storage tank to close the valve and returns to mode moving. The initial mode is moving.

```
proc Conveyor(alg Q : real, chan open!, close! : void) =
|[ cont V_B : real = 0
:: |[ mode moving = ( V_B := 0; Δ1; open!; filling )
    , mode filling = ( V_B ≥ V_Bmax → close!; moving )
   :: moving
   ]|
|| V̇_B = Q
]|
```

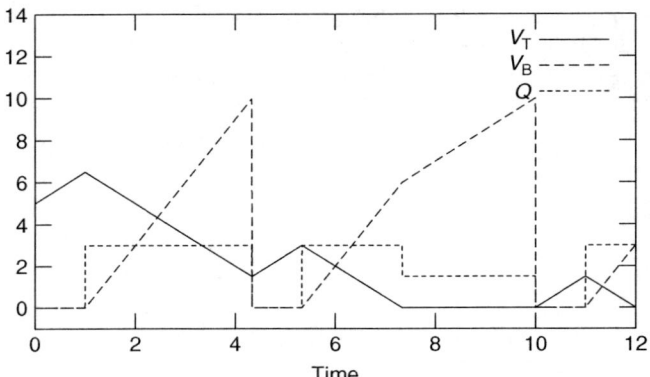

FIGURE 19.6 Simulation results of model *FillingLine*.

Figure 19.6 shows the results of the first 12 time units of a simulation run of the model *FillingLine*(5, 1.5), that is with model parameters $V_{T0} = 5$ and $Q_{in} = 1.5$. The graph shows that the first bottle is filled from time point 1 until time point $1 + 10/3 \approx 4.33$. Filling of the second bottle starts 1 time unit later, and somewhat after 7 time units, the storage tank becomes empty, so that filling continues at the reduced flow rate.

19.3.3 Syntax and Semantics of the Recursion Scope Operator

The syntax of the recursion scope operator statement p_R, that was introduced in Section 19.2.3, and first used in Section 19.3.2 is defined as

$$p_R ::= |[\text{mode } X = p^+ \{, \text{mode } X = p^+\}^* :: X]|$$

where X denotes a recursion variable, and statements p^+ consist of statements p (see Section 19.2.3) to which recursion variables X are added

$$p^+ ::= p \mid p; X \mid p^+ \,[]\, p^+ \mid b \rightarrow p^+ \mid p; p^+$$

The syntax enforces any recursion variable X to occur only at the end of a sequential composition. An additional restriction is that each recursion scope operator must be "complete." This means that in

$$|[\text{mode } X_1 = p_1^+, \ldots, \text{mode } X_n = p_n^+ :: X_k]|$$

all occurrences of free recursion variables in p_i^+ ($1 \leq i \leq n$) must be defined in the recursion scope operator itself. These restrictions enforce structured use of recursion: only one recursion variable X_i with corresponding statement p_i^+ can be executed at the same time, and termination of any of the statements p_i terminates the scope operator itself. This structured use of recursion simplifies analysis of χ models, it simplifies the translation to the normal form as discussed in Section 19.3, and it simplifies tool support for χ.

The meaning of recursion scope operators is as follows. Enabling the recursion scope operator $|[X_1 = p_1^+, \ldots, X_n = p_n^+ :: X_i]|$, enables the statement X_i ($1 \leq i \leq n$). When a recursion variable X_j ($1 \leq j \leq n$) is enabled (or disabled), its defining statement p_j is enabled (or disabled) instead. When a defining statement p_j terminates, the recursion scope operator terminates.

Process Algebra   19-15

19.3.4 Elimination of Process Instantiation

Elimination of the process instantiations for the *Tank* and *Conveyor* processes by replacing the process instantiations by their definitions, as defined in Beek et al. (2006), leads to the following model:

```
model FillingLine(val V_T0, Q_in : real) =
|[ alg Q : real, chan open, close : void
:: |[ cont V_T : real = V_T0^L
   , var V_T0^L : real = V_T0, Q_in^L : real = Q_in
   :: |[ mode closed =
           ( V̇_T = Q_in^L, Q = 0, V_T ≤ V_Tmax [] open?; opened )
        , mode opened =
           ( V̇_T = Q_in^L − Q, Q = Q_set, 0 ≤ V_T ≤ V_Tmax
             [] [skip]; openedempty
             [] close?; closed
           )
        , mode openedempty =
           ( V_T = 0, Q = Q_in^L [] close?; closed )
      :: closed
      ]|
   ]|
|| |[ cont V_B : real = 0
   :: |[ mode moving = ( V_B := 0; Δ 1; open!; filling )
        , mode filling = ( V_B ≥ V_Bmax → close!; moving )
      :: moving
      ]|
   || V̇_B = Q
   ]|
]|
```

To avoid naming conflicts between the formal parameters V_{T0} and Q_{in} declared in the process definition for process *Tank*, and the actual arguments V_{T0} and Q_{in} in the process instantiation *Tank* (Q, *open*, *close*, V_{T0}, Q_{in}), the newly defined local discrete variables that are used to hold the values of the last two parameters of the process instantiation, are renamed to V_{T0}^L and Q_{in}^L.

19.3.5 Syntax of the Normal Form

A slightly simplified syntax for the normal form in χ is given by a model with on the outer level a global variable and channel declaration D (see Sections 19.2.1 and 19.2.2), on the inner level a local variable and channel declaration D, and one recursion scope operator statement:

$$\chi_{norm} ::= \text{model } id(D_m) = |[\, D :: |[\, D :: |[\, X = p_{norm} \{, X = p_{norm}\}^* :: X \,]|\,]|\,]|$$

The normalized statements p_{norm}, used to define the recursion variables X, may consist of undelayable normalized atomic statements p_{na} (defined below). Such a normalized atomic statement may be prefixed by a guard b, and/or it may be made delayable (e.g., $b \to p_{na}$ and $[p_{na}]$). Sequential composition is allowed only in the form of such (guarded or delayable) atomic statements followed by a recursion variable. Finally, all of these statements may be part of alternative composition:

$$
\begin{array}{lll}
p_{norm} ::= & p_{nga} & \text{(guarded) atomic action} \\
| & u & \text{delay predicate} \\
| & p_{nga}; X & \text{atomic action followed by recursion variable} \\
| & p_{norm} \,[]\, p_{norm} & \text{alternative composition}
\end{array}
$$

where the normalized guarded atomic action statements p_{nga} are defined by

$$p_{nga} ::= p_{na} \qquad \text{nondelayable atomic action statement}$$
$$| \; b \to p_{na} \qquad \text{guarded nondelayable atomic action statement}$$
$$| \; [p_{na}] \qquad \text{delayable atomic action statement}$$
$$| \; b \to [p_{na}] \qquad \text{guarded delayable atomic action statement}$$

and the normalized atomic action statements p_{na}, that are all nondelayable, are defined by

$$p_{na} ::= \text{skip} \qquad \text{skip statement}$$
$$| \; \mathbf{x} := \mathbf{e} \qquad \text{multiassignment}$$
$$| \; h!? \qquad \text{synchronization via channel } h$$
$$| \; h!? \, \mathbf{x} := \mathbf{e} \qquad \text{communication via channel } h$$

The synchronization statement $h!?$ and communication statement $h!? \; \mathbf{x} := \mathbf{e}$ are required because of the fact that there is no parallel composition in the normalized form. The parallel composition $h! \parallel h?$ is normalized to $h!?$, and $h!\mathbf{e} \parallel h?\mathbf{x}$ is normalized to $h!?\mathbf{x} := \mathbf{e}$. The statement $h!?$ is comparable to the skip statement, and the statement $h!?\mathbf{x} := \mathbf{e}$ is comparable to the multiassignment statement $\mathbf{x} := \mathbf{e}$. The effect on the values of the variables is the same. There is only a small difference with respect to the occurrence of channel h, possibly accompanied by the value of $\mathbf{e}$, in the transition system.

As an example, that clarifies how the delay statement is eliminated in the translation to the normal form, consider the statement $x := 2; \Delta 1$ is first rewritten as $x := 2; |[\, \text{cont } t : \text{real} = 1 :: \dot{t} = -1 \,[]\, t \leq 0 \to \text{skip}\,]|$ and then normalized to

|[cont t: real
:: |[$X_0 = (\, x, t := 2, 1; X_1 \,)$
  , $X_1 = (\, \dot{t} = -1 \,[]\, t \leq 0 \to \text{skip} \,)$
  :: X_0
]|
]|

The normal form makes it easy to analyze system behavior and it simplifies tool implementations in the following way. When a model is defined as

```
model M(val x:t) =
|[ D_0
:: |[ D_1   :: |[ X_1 = p_{norm_1}, ..., X_n = p_{norm_n} :: X_i ]| ]|
]|
```

$M(\mathbf{c})$ defines a particular model instantiation. At each point of execution of this model instantiation, exactly one recursion variable X_i is enabled, so that the set of all possible next steps is determined by the term p_{norm_i} only. In addition, the term p_{norm_i} defines for each action the recursion variable (if any) that is enabled after execution of the action. Process definition, process instantiation, parallel composition, send and receive statements, the loop statement, while do statement, and delay statement are no longer present. Also scoping has been eliminated, apart from one top-level variable and channel scope operator, and one top-level recursion scope operator.

Process Algebra

19.3.6 Elimination of Parallel Composition

Elimination of parallel composition and translation to the normal form, as discussed in Section 19.3.5, leads to the model

```
model FillingLine (val V_T0, Q_in : real) =
|[ alg Q: real, chan open, close : void
:: |[ cont V_T : real = V_T0^L, V_B : real = 0
   , cont t : real, var V_T0^L : real = V_T0, Q_in^L : real = Q_in
   :: |[ moving_closed =
```
$$(\dot{V}_T = Q_{in}^L, Q = 0, V_T \leq V_{Tmax}, \dot{V}_B = Q$$
$$[] \; V_B, t := 0, 1; moving_0\_closed$$
$$)$$
```
      , moving_0_closed =
```
$$(\dot{V}_T = Q_{in}^L, Q = 0, V_T \leq V_{Tmax}, \dot{V}_B = Q, \dot{t} = -1$$
$$[] \; t \leq 0 \rightarrow skip; moving_1\_closed$$
$$)$$
```
      , moving_1_closed =
```
$$(\dot{V}_T = Q_{in}^L, Q = 0, V_T \leq V_{Tmax}, \dot{V}_B = Q$$
$$[] \; open!?; filling\_opened$$
$$)$$
```
      , filling_opened =
```
$$(\dot{V}_T = Q_{in}^L - Q, Q = Q_{set}, 0 \leq V_T \leq V_{Tmax}, \dot{V}_B = Q$$
$$[] \; [skip]; filling\_openedempty$$
$$[] \; V_B \geq V_{Bmax} \rightarrow close!?; moving\_closed$$
$$)$$
```
      , filling_openedempty =
```
$$(V_T = 0, Q = Q_{in}^L, \dot{V}_B = Q$$
$$[] \; V_B \geq V_{Bmax} \rightarrow close!?; moving\_closed$$
$$)$$
```
   :: moving_closed
   ]|
  ]|
]|
```

19.3.7 Substitution of Constants and Additional Elimination

The model below is the result of substitution of the globally defined constants by their values. Furthermore, the discrete variables Q_{in}^L and V_{T0}^L, that were introduced by elimination of the process instantiations, are eliminated. Also, the presence of the undelayable statements $V_B, t := 0, 1$ and $open!?$ in modes moving_closed and moving$_1$_closed, respectively, allows elimination of the differential equations in these modes.

Most hybrid automaton-based model checkers, such as PHAver (Frehse, 2005) and HyTech (Henzinger et al., 1995), do not (yet) have urgent transitions that can be combined with guards. Therefore, the urgency in the guarded statements is removed by making the statements that are guarded delayable, and adding the closed negation of the guard as an additional delay predicate (invariant). For example, $t \leq 0 \rightarrow skip$ is rewritten as $t \geq 0 \; [] \; t \leq 0 \rightarrow [skip]$.

```
model FillingLine (val V_T0, Q_in : real) =
|[ alg Q: real, chan open, close : void
:: |[ cont V_T : real = V_T0, V_B : real = 0, t: real
:: |[ moving_closed =
```

$$
\begin{aligned}
&\quad (\ V_T \leq 20, Q = 0 \\
&\quad [] \ V_B, t := 0,1; \ moving_0\_closed \\
&\quad) \\
&, moving_0\_closed = \\
&\quad (\ \dot{V}_T = Q_{in}, Q = 0, V_T \leq 20, \dot{V}_B = 0, \dot{t} = -1, t \geq 0 \\
&\quad [] \ t \leq 0 \rightarrow [skip]; \ moving_1\_closed \\
&\quad) \\
&, moving_1\_closed = \\
&\quad (\ V_T \leq 20, Q = 0 \\
&\quad [] \ open!?; \ filling\_opened \\
&\quad) \\
&, filling\_opened = \\
&\quad (\ \dot{V}_T = Q_{in} - 3, Q = 3, 0 \leq V_T \leq 20, \dot{V}_B = 3, V_B \leq 10 \\
&\quad [] \ [skip]; \ filling\_openedempty \\
&\quad [] \ V_B \geq 10 \rightarrow [close!?]; \ moving\_closed \\
&\quad) \\
&, filling\_openedempty = \\
&\quad (\ V_T = 0, Q = Q_{in}, \dot{V}_B = Q, V_B \leq 10 \\
&\quad [] \ V_B \geq 10 \rightarrow [close!?]; \ moving\_closed \\
&\quad) \\
&:: moving\_closed \\
&] | \\
&] | \\
&] |
\end{aligned}
$$

Figure 19.7 shows a graphical representation of the model. By means of straightforward mathematical analysis of the model, it can be shown that overflow never occurs if $Q_{in} \leq 30/13$.

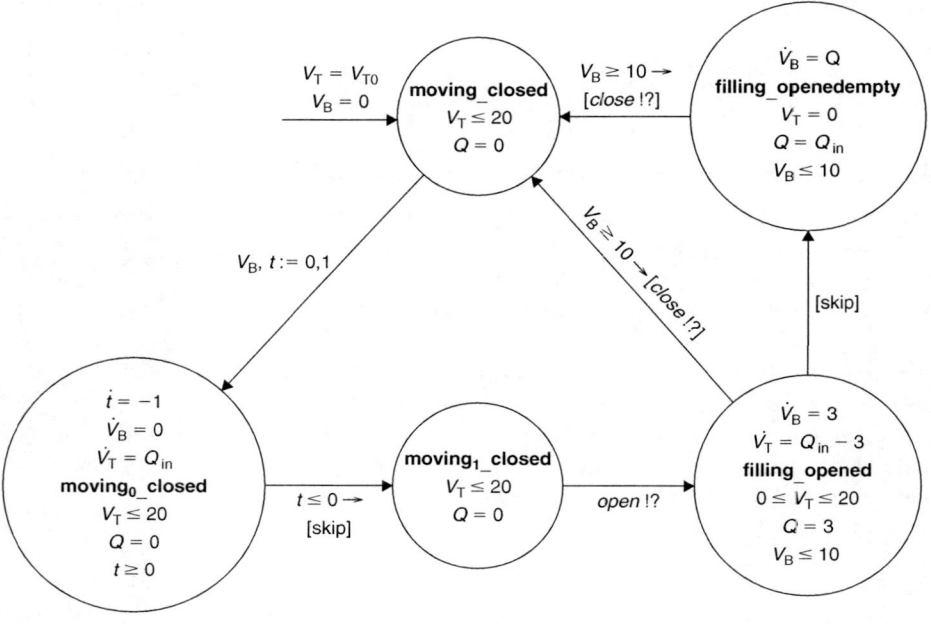

FIGURE 19.7 Graphical representation of the normalized χ model.

19.3.8 Tool-Based Verification

As a final step, for the purpose of tool-based verification, the model is translated to the input language of the hybrid IO automaton-based tool PHAVer (Frehse, 2005). Since most hybrid automata, including PHAVer, do not know the concept of an algebraic variable, first the algebraic variables are eliminated from the χ model. Because of the consistent equation semantics of χ, each occurrence of an algebraic variable in the model can simply be replaced by the right-hand side of its defining equation. The urgency owing to unguarded undelayable statements is in principle translated by defining the corresponding flow clause as false. The resulting PHAVer model follows below. Note that an additional variable x is introduced and the derivatives of Vb and Vt need to be defined in all locations, because of the current inability of PHAVer to define false as flow clause.

```
automaton filling_line
  state_var: Vt,Vb,t,x;
  parameter: Vt0,Qin;
  synclabs : open,close,tau;
  loc moving_closed:
    while Vt <= 20 & x==0 wait {x'==1 & Vb'==0 & Vt'==0};
    when true sync tau do {Vt'==Vt & Vb'==0 & t'==1 & x'==0}
      goto moving0_closed;
  loc moving0_closed:
    while Vt <= 20 & t >= 0 wait {Vb'==0 & t'==-1 & Vt==30/13};
    when t <= 0 sync tau do {Vt'==Vt & Vb'==Vb & t'==t & x'==0}
      goto moving1_closed;
  loc moving1_closed:
    while Vt <= 20 & x==0 wait {x'==1 & Vb'==0 & Vt'==0};
    when true sync open do {Vt'==Vt & Vb'==Vb & t'==t}
      goto filling_opened;
  loc filling_opened:
    while Vt >= 0 & Vt <= 20 & Vb <= 10 wait {Vb'==3 & Vt'==30/13-3};
    when Vt==0 sync tau do {Vt'==Vt & Vb'==Vb & t'==t}
      goto filling_openedempty;
    when Vb >= 10 sync close do {Vt'==Vt & Vb'==Vb & t'==t & x'==0}
      goto moving_closed;
  loc filling_openedempty:
    while Vt == 0 & Vb <= 10 wait {Vb'==30/13};
    when Vb >= 10 sync close do {Vt'==Vt & Vb'==Vb & t'==t & x'==0}
      goto moving_closed;
  initially moving_closed & Vt == Vt0 & Vb==0 & x==0;
end
```

The following properties were derived: if $Q_{in} = 30/13$ and $0 \leq V_{T0} \leq V_{Tmax} - 30/13$, overflow does not occur, and the storage tank does not become empty when filling a bottle. The volume of the storage tank then remains in the region $V_{T0} \leq V_T \leq V_{T0} + 30/13$. If $Q_{in} > 30/13$, eventually overflow occurs. If $Q_{in} < 30/13$, eventually the container becomes empty every time a bottle is filled. In this small example, these properties can also be derived by means of straightforward mathematical analysis of the χ models of Section 19.3.6 or 19.3.7.

19.4 Conclusions

Process algebra originated in the domain of theoretical computer science, where it was designed for the purpose of reasoning about the behavior of concurrent discrete-event systems. Recently, process algebra theory has been extended to also include continuous-time systems, and combined discrete-event/

continuous-time, or hybrid systems. The χ process algebra, that has been used as an example in this chapter, illustrates that process algebra is not only suited to verification, but also very well suited to high-level modeling and simulation of complex dynamical systems. The compositional semantics of a process algebra facilitates modular composition of processes and statements not only using parallel composition but also sequential composition, and in fact any kind of combination of statements by means of the process algebra operators. The equational reasoning, that is characteristic of process algebra, allows rewriting of complex specifications to a straightforward normal form, where parallel composition and many other operators and statements have been eliminated. For the χ process algebra, the normal form is very similar to a hybrid automaton, and thus simplifies the use and development of tools for simulation and verification.

Acknowledgments

The authors thank Albert Hofkamp for providing the main functionality of the χ toolset, and for many helpful comments on drafts of this text. They thank Rolf Theunissen for his preparative work on the bottle filling example, and for analysis of the properties of the resulting hybrid automaton using PHAVer. Finally, they thank Ramon Schiffelers for enabling hybrid simulation of χ models.

References

Alur, R., T. A. Henzinger, and P. H. Ho (1996). Automatic symbolic verification of embedded systems. *IEEE Transactions on Software Engineering 22*(3), 181–201.

Baeten, J. C. M. and C. A. Middelburg (2002). *Process Algebra with Timing*. EACTS Monographs in Theoretical Computer Science. New York: Springer.

Baeten, J. C. M. and W. P. Weijland (1990). *Process Algebra*, Volume 18 of Cambridge Tracts in Theoretical Computer Science. Cambridge, UK: Cambridge University Press.

Beek, D. A. v., K. L. Man, M. A. Reniers, J. E. Rooda, and R. R. H. Schiffelers (2006). Syntax and consistent equation semantics of hybrid Chi. *Journal of Logic and Algebraic Programming 68*(1/2), 129–210.

Beek, D. A. v. and J. E. Rooda (2000). Languages and applications in hybrid modelling and simulation: Positioning of Chi. *Control Engineering Practice 8*(1), 81–91.

Beek, D. A. v., A. van den Ham, and J. E. Rooda (2002). Modelling and control of process industry batch production systems. In *15th Triennial World Congress of the International Federation of Automatic Control*, Barcelona.

Bergstra, J. A. and J. W. Klop (1984). Process algebra for synchronous communication. *Information and Control 60*(1/3), 109–137.

Bergstra, J. A. and C. A. Middelburg (2005). Process algebra for hybrid systems. *Theoretical Computer Science 335*(2/3), 215–280.

Bortnik, E. M., N. Trčka, A. J. Wijs, B. Luttik, J. M. van de Mortel-Fronczak, J. C. M. Baeten, W. J. Fokkink, and J. E. Rooda (2005). Analyzing a Chi model of a turntable system using Spin, CADP and UPPAAL. *Journal of Logic and Algebraic Programming 65*(2), 51–104.

Bundy, A. (1999). A survey of automated deduction. In M. Wooldridge and M. Veloso (Eds.), *Artificial Intelligence Today: Recent Trends and Developments*, Volume 1600 of Lecture Notes in Computer Science, pp. 153–174. Berlin: Springer.

Cassandras, C. G. and S. Lafortune (1999). *Introduction to Discrete Event Systems*. Springer International Series on Discrete Event Dynamic Systems. Berlin: Springer.

Clarke, E. M., O. Grumberg, and D. A. Peled (2000). *Model Checking*. Cambridge, MA: MIT Press.

Cuijpers, P. J. L. and M. A. Reniers (2005). Hybrid process algebra. *Journal of Logic and Algebraic Programming 62*(2), 191–245.

Fábián, G. (1999). *A Language and Simulator for Hybrid Systems*. Ph.D. thesis, Eindhoven University of Technology.

Fernandez, J. C., H. Garavel, A. Kerbrat, L. Mounier, R. Mateescu, and M. Sighireanu (1996). CADP—A Protocol validation and verification toolbox. In *Proceedings 8th Conference on Computer Aided*

Verification (CAV'96), Volume 1102 of Lecture Notes in Computer Science, pp. 437–440. Berlin: Springer.
Frehse, G. (2005). PHAVer: Algorithmic verification of hybrid systems past HyTech. In M. Morari and L. Thiele (Eds.), *Hybrid Systems: Computation and Control, 8th International Workshop*, Volume 3414 of Lecture Notes in Computer Science, pp. 258–273. Berlin: Springer.
Groote, J. F. (1997). The syntax and semantics of timed μCRL. Technical Report SEN-R9709, CWI, The Netherlands.
He, J. (1994). From CSP to hybrid systems. In A. W. Roscoe (Ed.), *A Classical Mind: Essays in Honour of C.A.R. Hoare*, pp. 171–189. Hertfordshire, UK: Prentice-Hall.
Henzinger, T. A., P.-H. Ho, and H. Wong-Toi (1995). A user guide to HyTech. In *First International Conference on Tools and Algorithms for the Construction and Analysis of Systems TACAS*, Volume 1019 of Lecture Notes in Computer Science, pp. 41–71. Berlin: Springer.
Hoare, C. A. R. (1985). *Communicating Sequential Processes*. Englewood-Cliffs, NJ: Prentice-Hall.
Hofkamp, A. T. (2001). *Reactive Machine Control: A Simulation Approach Using χ*. Ph.D. thesis, Eindhoven University of Technology.
Holzmann, G. J. (2003). *The SPIN Model Checker: Primer and Reference Manual*. Boston, MA: Addison-Wesley.
Kunkel, P. and V. Mehrmann (2006). *Differential-Algebraic Equations: Analysis and Numerical Solution*. Zürich: EMS Publishing House.
Larsen, K. G., P. Pettersson, and W. Yi (1997). UPPAAL in a Nutshell. *International Journal on Software Tools for Technology Transfer* 1(1/2), 134–152.
Linz, P. (2001). *An Introduction to Formal Languages and Automata*. Sudbury, MA: Jones and Bartlett.
Milner, R. (1980). *A Calculus of Communicating Systems*, Volume 92 of Lecture Notes in Computer Science. Berlin: Springer.
Milner, R. (1989). *Communication and Concurrency*. Englewood-Cliffs, NJ: Prentice-Hall.
Naumoski, G. and W. Alberts (1998). *A Discrete-Event Simulator for Systems Engineering*. Ph.D. thesis, Eindhoven University of Technology.
Rounds, W. C. and H. Song (2003). The ϕ-calculus: A language for distributed control of reconfigurable embedded systems. In O. Maler and A. Pnueli (Eds.), *Hybrid Systems: Computation and Control, 6th International Workshop*, Volume 2623 of Lecture Notes in Computer Science, pp. 435–449. Berlin: Springer.
The MathWorks Inc. (2005). *Writing S-functions*, version 6. (http://www.mathworks.com).

20
Temporal Logic

Antony Galton
University of Exeter

20.1 Propositional Logic 20-1
20.2 Introducing Temporal Logic 20-3
20.3 Syntax and Semantics 20-5
20.4 Models of Time 20-6
  Transitivity • Reflexivity • Irreflexivity • Linearity • Boundedness • Density and Discreteness • Combinations of Properties
20.5 Further Extensions to the Formal Language 20-11
20.6 Illustrative Examples 20-11
  A System with an Attractor • Application to Reasoning about Programs • Other Applications
20.7 Conclusion 20-14
20.8 Further Reading 20-14

In this chapter, we introduce temporal logic (TL), a logical formalism for reasoning about domains in which the truth or falsity of statements can vary over time. These statements could include formulae representing states of a dynamical system, which brings TL clearly within the scope of this handbook. Since TL is an extension of propositional logic, we believe that the reader will best be in a position to appreciate the former if we first provide a brief introduction to the latter; readers already familiar with propositional logic can skip directly to Section 20.2.

20.1 Propositional Logic

Propositional logic is a basic system of formal logic that underlies most logical systems. Its purpose is to enable the formalization of certain types of reasoning, such as

> If the power supply is on, and the light-switch is down, and the bulb is not blown, then the light is on.
> The light-switch is down and the light is not on.
> Therefore, either the bulb is blown or the power supply is not on.

This piece of reasoning takes the form of an inference or argument; it consists of a set of premises (the first two statements) followed by a conclusion (flagged by "therefore"), the idea being that the conclusion is supposed to follow logically from the premises. An inference is said to be valid if the conclusion does indeed logically follow from the premises; this means that *the truth of the premises is sufficient to guarantee the truth of the conclusion*, i.e., in any situation in which the premises are all true, the conclusion must be true as well. Propositional calculus (PC) is a formal system by which the process of determining whether an inference is valid or not may be effectively mechanized.

PowerOn	SwitchDown	BulbBlown	LightOn	Premise 1	Premise 2	Conclusion
T	T	T	T	T	F	T
T	T	T	F	T	T	T
T	T	F	T	T	F	F
T	T	F	F	F	T	F
T	F	T	T	T	F	T
T	F	T	F	T	F	T
T	F	F	T	T	F	F
T	F	F	F	T	F	F
F	T	T	T	T	F	T
F	T	T	F	T	T	T
F	T	F	T	T	F	T
F	T	F	F	T	T	T
F	F	T	T	T	F	T
F	F	T	F	T	F	T
F	F	F	T	T	F	T
F	F	F	F	T	F	T

FIGURE 20.1 Truth table used for determining validity of an inference.

Writing *PowerOn*, *SwitchDown*, *BulbBlown*, and *LightOn* to stand for "The power supply is on," "The light-switch is down," "The bulb is blown," and "The light is on" respectively, the above inference can be formalised as

$$\frac{\begin{array}{c} PowerOn \wedge SwitchDown \wedge \neg BulbBlown \rightarrow LightOn \\ SwitchDown \wedge \neg LightOn \end{array}}{BulbBlown \vee \neg PowerOn}$$

Here, the symbol "$\wedge$," known as conjunction, is read "and"; "$\vee$" (disjunction) as "or," "$\rightarrow$" (material implication) as "if …, then …," and "$\neg$" (negation) as "not" (or "it is not the case that"). An additional symbol commonly used is "$\leftrightarrow$" (material equivalence), read "if and only if."[1] The specific interpretations of these symbols are provided by the following semantic rules, in which α and β stand for arbitrary propositions:

- $\alpha \wedge \beta$ is true if and only if both α and β are true.
- $\alpha \vee \beta$ is true if and only if at least one of α and β is true.
- $\alpha \rightarrow \beta$ is true if and only if it is not the case that α is true and β is false.
- $\alpha \leftrightarrow \beta$ is true if and only if α and β are either both true or both false.
- $\neg \alpha$ is true if and only if α is false.

Because the meanings of these symbols are defined purely in terms of truth values, they are known as truth-functional connectives; they are also known as Boolean connectives, in honor of George Boole who first introduced a system recognizable as the PC in his *Laws of Thought* (1854).

Using these rules, we can show that the inference above is valid. A mechanical, if sometimes unnecessarily complicated, way of doing this is through the construction of a truth table, as shown in Figure 20.1. Here, every possible combination of truth values for the four atomic propositions occurring in the inference is listed; for each combination, we use the semantic rules for the connectives to compute the truth values of

---

[1] Commonly used alternative versions of some of these symbols include $\supset$ for material implication, $\equiv$ for material equivalence, and $\sim$ for negation. In some contexts, these symbols are used in addition to the ones given in the main text, with different meanings. Note also the bracket-dropping convention: $\alpha \wedge \beta \rightarrow \gamma$ is understood to be implicitly grouped as $(\alpha \wedge \beta) \rightarrow \gamma$, and in general both $\wedge$ and $\vee$ are regarded as binding their arguments more closely than $\rightarrow$ and $\leftrightarrow$, unless this is explicitly overridden by brackets, as in $\alpha \wedge (\beta \rightarrow \gamma)$.

the premises and conclusion of the inference. That the inference is valid is shown by the fact that every combination of truth values which makes both the premises true also makes the conclusion true (lines 2, 10, and 12 of the table).

Within PC it is possible to establish the validity or invalidity of any pattern of inference that can be expressed using the truth-functional connectives. Many types of reasoning, however, cannot be expressed in this way, and for this reason the PC has been extended to produce various more expressive formalisms. The most well known of these is the first-order predicate calculus, or first-order logic (FOL).[2] However, the TL described in this chapter represents an alternative extension of PC.

20.2 Introducing Temporal Logic

In the PC inference discussed above, no reference is made to *time*, and this is true of the classical forms of logic generally. The truth or falsity of each of the propositions is implicitly referred to a particular, unstated time, so it is not possible to use PC to reason about propositions which refer to different times. TL was introduced to remedy this deficiency.

An inference requiring some form of TL for its formulation is

> The light is only ever on if the battery is charged.
> Whenever the battery is charged, at a later time it will not be charged.
> Therefore, the light will not always be on.

A TL formulation of this inference is as follows:

$$\frac{\begin{array}{c}\Box(LightOn \to BatteryCharged)\\ \Box(BatteryCharged \to \Diamond \neg BatteryCharged)\end{array}}{\neg \Box LightOn} \quad (20.1)$$

Here, the symbols "$\Box$" and "$\Diamond$" may be read "always" and "eventually," respectively. More exactly, their meanings are expressed by the following semantic rules:

- $\Box \alpha$ is true at time t_0 if and only if α is true at every time $t \geq t_0$.
- $\Diamond \alpha$ is true at time t_0 if and only if α is true at at least one time $t \geq t_0$.

Note that if $\Box \neg \alpha$ is true at t_0, then $\neg \alpha$ must be true at every time $t \geq t_0$, which means that there cannot be any time $t \geq t_0$ at which α is true; hence $\Diamond \alpha$ must be false at t_0. Conversely, if $\Box \neg \alpha$ is false then $\Diamond \alpha$ must be true. For this reason, we can affirm that the formulae $\neg \Box \neg \alpha$ and $\Diamond \alpha$ are always equivalent, and this is usually captured by *defining* $\Diamond$ in terms of $\Box$ as follows:

$$\Diamond \alpha =_{\text{def}} \neg \Box \neg \alpha \quad (\text{Def}\Diamond)$$

The validity of inference (Eq. [20.1]) may be demonstrated informally by a *reductio ad absurdum* argument. Suppose that, at t_0, the premises are both true but the conclusion is false. At t_0, therefore, we have $\Box(LightOn \to BatteryCharged)$, $\Box(BatteryCharged \to \Diamond \neg BatteryCharged)$, and $\Box LightOn$. By the semantic rule for $\Box$, this means that at t_0 (and indeed at every time $t \geq t_0$), we have

$$LightOn \to BatteryCharged \quad (20.2)$$

$$BatteryCharged \to \Diamond \neg BatteryCharged \quad (20.3)$$

$$LightOn \quad (20.4)$$

---

[2]An account of FOL can be found in almost any textbook on formal logic, for example, Hodges (2001) and Jeffrey (2006).

By PC reasoning, from Eq. (20.4) and Eq. (20.2) we also have

$$BatteryCharged \qquad (20.5)$$

at t_0 and hence also from Eq. (20.5) and Eq. (20.3)

$$\Diamond \neg BatteryCharged \qquad (20.6)$$

By the semantic rule for $\Diamond$, this means that there is a time $t \geq t_0$ at which $\neg BatteryCharged$ is true (note that since $BatteryCharged$ is true at t_0, we must in fact have $t > t_0$). Moreover, since $t \geq t_0$, we know from our earlier deductions that both Eq. (20.2) and Eq. (20.4) are true at t; and these imply that $BatteryCharged$ is true at t as well. This contradicts our earlier deduction that $\neg BatteryCharged$ is true then. Hence, the supposition that the premises could be true and the conclusion false is absurd, and we conclude that if the premises are true, the conclusion must be true too—i.e., the inference is valid.

Although we referred to individual times (t_0, t) in this reasoning, TL allows us to work purely at the level of the $\Box$, $\Diamond$ language used in the formulae, with no explicit reference to times. To illustrate this, we shall prove the above inference in a TL in which the following formulae are posited as axioms:[3]

$$\Box(\alpha \rightarrow \beta) \rightarrow (\Box\alpha \rightarrow \Box\beta) \qquad (\text{AxK})$$

$$\Box\alpha \rightarrow \alpha \qquad (\text{AxT})$$

and the following rules of inference are introduced:[4]

$$\text{From } \vdash \alpha \text{ and } \vdash \alpha \rightarrow \beta \text{ infer } \vdash \beta \qquad (\text{MP})$$

$$\text{From } \vdash \alpha \text{ infer } \vdash \Box\alpha \qquad (\text{R}\Box)$$

$$\text{If } \alpha \text{ is a theorem of PC, infer } \vdash \alpha \qquad (\text{PCT})$$

Here "$\vdash \alpha$" means that α is a theorem, i.e., considered proved. All the axioms count as theorems, and anything that can be inferred from the axioms using the rules of inference is also a theorem.

To establish the validity of the inference above, we make use of three PC theorems, not proved here:

$$(\alpha \rightarrow \beta) \wedge (\beta \rightarrow \gamma) \rightarrow (\alpha \rightarrow \gamma) \qquad (20.7)$$

$$(\alpha \rightarrow \beta) \wedge (\neg\alpha \rightarrow \beta) \rightarrow \beta \qquad (20.8)$$

$$(\alpha \rightarrow \beta) \leftrightarrow (\neg\beta \rightarrow \neg\alpha) \qquad (20.9)$$

Starting with the two premises

$$\Box(LightOn \rightarrow BatteryCharged)$$

$$\Box(BatteryCharged \rightarrow \Diamond \neg BatteryCharged)$$

we can use (PCT), Eq. (20.7), and (AxK) to give $\Box(LightOn \rightarrow \Diamond \neg BatteryCharged)$, and hence by (AxT)

$$LightOn \rightarrow \Diamond \neg BatteryCharged \qquad (20.10)$$

Premise 1 also gives, by (AxK), $\Box LightOn \rightarrow \Box BatteryCharged$. Using Eq. (20.9), this can be rewritten as $\neg \Box BatteryCharged \rightarrow \neg \Box LightOn$, and hence, by (Def$\Diamond$), as

$$\Diamond \neg BatteryCharged \rightarrow \neg \Box LightOn \qquad (20.11)$$

---

[3] K and T are the labels conventionally given to these axioms; as far as I am aware, these letters do not "stand for" anything.

[4] See Section 20.3 for a general explanation of axioms and rules of inference. "MP" here stands for *Modus Ponens*, the Latin name given to this pattern of inference in the logical tradition.

From Eq. (20.10) and Eq. (20.11), using Eq. (20.7) and (AxK), we now have

$$LightOn \to \neg \Box LightOn \qquad (20.12)$$

By (AxT), we also have $\Box LightOn \to LightOn$, which using Eq. (20.9) gives

$$\neg LightOn \to \neg \Box LightOn \qquad (20.13)$$

Finally, from Eq. (20.12) and Eq. (20.13), using Eq. (20.8) we infer the desired conclusion $\neg \Box LightOn$.

20.3 Syntax and Semantics

The preceding section introduced a particular, rather simple, variety of TL, and in the next section we will examine some other varieties. To prepare the way for this, it is necessary to say a few words about what a logical system consists of and how it is structured.

At the heart of the system is a formal language. This is defined as a set of well-formed formulas (wffs), each of which is a legitimate concatenation of symbols taken from some predefined vocabulary. Which concatenations are legitimate is determined by the rules of formation of the language, which define its syntax (or formal grammar).

For the TL in Section 20.2, the vocabulary consists of the logical constants "$\neg$," "$\wedge$," "$\vee$," "$\to$," "$\leftrightarrow$," "$\Box$," and "$\diamond$" together with a set of literals to represent elementary propositions; these may be tailored to the application context by taking a form suggestive of their intended meanings (e.g., *LightOn*, *BatteryCharged*), but in more general application-independent treatments it is convenient to use simple letters (e.g., A, B, C, ...) instead. Finally, parentheses "(" and ")" are used to help articulate the structure of formulae.

The wffs of this system are defined by the following rules of formation:

1. Any literal is a wff.
2. If α and β are wffs then so are $\neg \alpha$, $(\alpha \wedge \beta)$, $(\alpha \vee \beta)$, $(\alpha \to \beta)$, $(\alpha \leftrightarrow \beta)$, $\Box \alpha$, and $\diamond \alpha$.

This is a recursive definition; by repeated applications of these rules we can build up complex wffs such as $\Box(\Box A \to B) \vee \Box(\Box B \to A)$.[5]

Thus far, all we have defined is a set of symbol-concatenations which we have dignified with the name of wff; these wffs do not, on their own, possess any meaning. Wffs can be made meaningful in two ways:

1. Explicitly, by specifying how they are to be interpreted.
2. Implicitly, by specifying their logical relationships to other wffs.

The former method gives rise to what is known as model theory, and the latter to proof theory.

Consider the symbol "$\wedge$." Its semantic rule, given above, states that a wff $\alpha \wedge \beta$ is true if and only if both α and β are true. But a formula such as α can be given many different interpretations, in some of which it stands for a true statement, in others for a false one; an interpretation in which α stands for a true statement is said to satisfy α. The meaning of "$\wedge$" is specified by stipulating that an interpretation satisfies $\alpha \wedge \beta$ if and only if it satisfies both α and β. The other semantic rules in Section 20.1 are to be understood analogously.

Given a collection of wffs, we may or may not be able to find an interpretation, consistent with the semantic rules, in which all the wffs are true; any such interpretation is called a model for that set of wffs. Model theory is thus the way of specifying the meanings of formulae by laying down what counts as an interpretation of the formal language, and what it is for an interpretation to satisfy a formula. For our TL, this was what is provided, in part, by the semantic rules given above.

---
[5]Strictly speaking this should be $(\Box(\Box A \to B) \vee \Box(\Box B \to A))$, but it is conventional to omit the outer pair of brackets in cases such as this (see also footnote 1).

Recall the semantic rule for "$\Box$": $\Box\alpha$ is true at time t_0 if and only if α is true at every time $t \geq t_0$. This is not telling us about the truth of $\Box\alpha$ absolutely, but its truth *at a particular time*; moreover, the rule specifies the truth of $\Box\alpha$ at one time in terms of the truth of α at other times—thus t_0 is just one of a set of times, elements of which can be compared using the relation $\geq$ (where $t_1 \geq t_0$ means that t_1 is later than or equal to t_0). From these observations we see that an interpretation of a TL formula must be considered as relative to a time taken from some ordered system of times.

Let T be the set of times, and let $\prec$ denote the ordering relation on T (this may be read as "precedes," with the caveat that in some models, a time may precede itself). Together, these constitute a *temporal frame* $(T, \prec)$. We write

$$(T, \prec, t) \models \alpha$$

to mean that the wff α is true at time t in the temporal frame $(T, \prec)$. With this notation, the semantic rules for $\Box$ and $\Diamond$ can be reformulated as[6]

- $(T, \prec, t) \models \Box\alpha$ if and only if, for every $t' \in T$, if $t \prec t'$ then $(T, \prec, t') \models \alpha$.
- $(T, \prec, t) \models \Diamond\alpha$ if and only if, for at least one $t' \in T$, $t \prec t'$ and $(T, \prec, t') \models \alpha$.

Proof theory offers a completely different approach to specifying meanings. We characterize the meaning of a logical symbol by stating logical properties of formulae containing that symbol. One way of doing this is in terms of *inference rules*. For conjunction, "$\wedge$," the inference rules most commonly given are

($\wedge$-**introduction**) From $\vdash \alpha$ and $\vdash \beta$, derive $\vdash \alpha \wedge \beta$.
($\wedge$-**elimination**) From $\vdash \alpha \wedge \beta$, derive both $\vdash \alpha$ and $\vdash \beta$.

Introduction and elimination rules, some more complex than these, can be given for the other connectives. We have already met some inference rules for TL (MP, R$\Box$, and PCT). A "natural deduction" system provides a proof theory consisting entirely of inference rules; an axiom system includes in addition a set of formulae, called axioms, which are stipulated to be true (e.g., AxK and AxT).

Logic is concerned with making valid inferences. The *meaning* of validity is generally defined in terms of a model theory: an inference is valid if and only if every model for the premises satisfies the conclusion. This corresponds to the intuitive idea that in a valid inference, if the premises are true then the conclusion cannot fail to be true as well. The model-theoretic definition sets the standard which the proof theory, as a practical set of inferential procedures, should live up to. A proof theory is *sound* with respect to a given model theory if every inference validated by the former is indeed valid according to the latter; and it is *complete* with respect to the model theory if every inference valid according to the latter is validated by the former. The ideal is a proof theory that is both sound and complete with respect to the model theory corresponding to the intended application of the logic. In some cases this can be achieved, but in others it cannot.

20.4 Models of Time

An interpretation of TL is defined by specifying a temporal frame $(T, \prec)$, a reference time t_0 ("now"), and, for each literal L in the language, its truth value at each time $t \in T$. The semantic rules for the logic then determine the truth value of each formula at each time, and in particular at the reference time t_0.

The temporal frame obviously plays an essential role in all this, and the truth values of formulae in the logic can depend critically on the properties of the frame. The frame in effect encapsulates a *model* of time. In this section, we look at some of the key frame properties that are used in different applications, and the impact they can have on the truth values of different formulae. As a side effect of this survey, we shall also be led to indicate extensions to the logical vocabulary which have been introduced to enhance the expressive power of the logic with respect to different models of time.

---

[6]The semantic rules for the Boolean connectives, when considered as forming part of TL, can likewise be written in this style, e.g., for conjunction: $(T, \prec, t) \models \alpha \wedge \beta$ if and only if $(T, \prec, t) \models \alpha$ and $(T, \prec, t) \models \beta$.

Temporal Logic

Of the two components required to specify a frame, i.e., the set T and the relation $\prec$, the latter will play a major role in what follows. As far as the former is concerned, nothing depends on what the elements of T actually are: essentially they can be regarded as abstract elements that are taken to represent "times," but there is little need here for a philosophical discussion on what we actually mean by "a time." The only exception to this concerns whether we regard times as durationless instants or extended intervals. Most developments of TL opt for the former understanding, but the latter also has had its advocates, especially when TL is used to specify the semantics of temporal expressions in natural language.[7] Here we shall assume that times are effectively point-like, so that no change can take place *within* a time; thus each formula has a single truth value at each time. That apart, the only feature of T that is relevant for TL is its cardinality, e.g., whether it is finite, countably infinite, or uncountable.

We now embark on the survey of frame properties. These will be expressed in both English and FOL. Readers unfamiliar with the latter should be able to understand the properties from the English description alone.

20.4.1 Transitivity

It is natural to regard the ordering of times as transitive, as implied by the use of the word "ordering." It means that if one time precedes a second, which in turn precedes a third, then the first time must also precede the third; in symbols

$$\forall t \forall t' \forall t''(t \prec t' \wedge t' \prec t'' \to t \prec t'') \tag{Trans}$$

Transitive temporal frames can be characterized by the fact that in any interpretation over such a frame, for any TL formula α, the formula

$$\Box \alpha \to \Box \Box \alpha \tag{Ax4}$$

is true at every $t \in T$. This is often postulated as an axiom, thereby restricting the allowed interpretations to transitive frames.

One way of describing transitivity is that the future of the future already counts as future now—this corresponds to the formula $\Diamond \Diamond \alpha \to \Diamond \alpha$ which can be proved to be equivalent to (Ax4). A nontransitive frame thus requires the notion of "future" to be qualified in some way which leads to a breakdown in transitivity, e.g., the "near" future ("within the next 10 years," say).

20.4.2 Reflexivity

A temporal order is reflexive if every time is regarded as "preceding" itself, i.e.,

$$\forall t(t \prec t) \tag{Refl}$$

In effect, this means that the precedence relation $\prec$ is interpreted not as "precedes" but as "precedes or equals" (i.e., "not later than" rather than "earlier than"). This was, in fact, the interpretation used in our example in Section 20.2, where the semantic rules used "$t \geq t_0$" for "$t_0 \prec t$."

Reflexive temporal frames are characterized by the schema (AxT) which we discussed in Section 20.2. An equivalent formulation is $\alpha \to \Diamond \alpha$.

20.4.3 Irreflexivity

A nonreflexive frame satisfies $\neg \forall t(t \prec t)$ ("not every time precedes itself"), but for a frame to count as irreflexive it must satisfy the stronger condition

$$\forall t \neg (t \prec t) \tag{Irref}$$

---

[7]Examples are, in the context of reasoning about programs (Moszkowski, 1986) and in the context of analyzing natural language (Dowty, 1979).

i.e., no time precedes itself. This is satisfied by our normal understanding of "precedes" as "earlier than." However, there is no TL formula that exactly characterizes irreflexive frames in the way (AxT) characterizes reflexive ones.

20.4.4 Linearity

The reader has probably assumed in the foregoing that the temporal structure is linear, i.e., that the times are as it were strung out along a single line, in keeping with our ordinary notion of a "time line." However, such linearity is by no means entailed by any of the properties mentioned above, all of which are compatible with a wide range of branching or network-like structures. If we wish our time line to be linear, then this has to be built in explicitly.

The property of linearity is expressed by the formula

$$\forall t \forall t' (t \prec t' \vee t = t' \vee t' \prec t) \tag{Lin}$$

which says that either the times t and t' are the same time ($t = t'$) or one precedes the other ($t \prec t' \vee t' \prec t$). The TL we have used so far cannot express this notion of linearity. The best we can do is

$$\Diamond \alpha \wedge \Diamond \beta \rightarrow \Diamond(\alpha \wedge \beta) \vee (\alpha \wedge \Diamond \beta) \vee (\Diamond \alpha \wedge \beta)$$

which captures the idea that of any two *future* times, either they coincide or one precedes the other. This characterizes time as "linear in the future,"[8] but does not rule out "alternative pasts" whose times do not stand in any direct temporal relation to each other. Future-linearity is expressed by the formula

$$\forall t \forall t' \forall t'' (t'' \prec t \wedge t'' \prec t' \rightarrow t \prec t' \vee t = t' \vee t' \prec t) \tag{FLin}$$

For linear time we also require past-linearity (or "left-linearity"), i.e.,

$$\forall t \forall t' \forall t'' (t \prec t'' \wedge t' \prec t'' \rightarrow t \prec t' \vee t = t' \vee t' \prec t) \tag{PLin}$$

and this cannot be expressed using the operators $\square$ and $\Diamond$, since these operators are future-directed, describing the present in terms of the future. To express past-linearity, and hence full linearity, we need past-directed operators as well.

The usual notation, when both sets of operators are used, is to write **F** and **G** for the future-directed operators (i.e., our earlier $\Diamond$ and $\square$), and **P** and **H** for the past-directed ones, as captured by the following mnemonics:

    FA A will be true sometime in the **F**uture
    GA A is always **G**oing to be true
    PA A was true sometime in the **P**ast
    HA A **H**as always been true

The operators **F**, **G**, **P**, and **H** are called tense operators, and a TL that uses both sets of operators is called a tense logic. Historically, tense logics were the first kind of TL to be developed, arising from the work of the philosopher A. N. Prior in the 1950s and 1960s (Prior, 1967, 1968).

Backward and forward linearity are expressed by the tense logic formulae

$$\mathbf{P}\alpha \wedge \mathbf{P}\beta \rightarrow \mathbf{P}(\alpha \wedge \beta) \vee (\alpha \wedge \mathbf{P}\beta) \vee (\mathbf{P}\alpha \wedge \beta)$$
$$\mathbf{F}\alpha \wedge \mathbf{F}\beta \rightarrow \mathbf{F}(\alpha \wedge \beta) \vee (\alpha \wedge \mathbf{F}\beta) \vee (\mathbf{F}\alpha \wedge \beta)$$

but with both past- and future-tense operators available the same properties can be more simply expressed by

$$\mathbf{FP}\alpha \rightarrow \mathbf{P}\alpha \vee \alpha \vee \mathbf{F}\alpha$$
$$\mathbf{PF}\alpha \rightarrow \mathbf{P}\alpha \vee \alpha \vee \mathbf{F}\alpha$$

---

[8]Often called "right-linear," presupposing a picture in which time flows from left to right!

The first formula says that anything that will be past is either already past, present, or still future. This rules out the possibility of separate past time lines converging on a common time. Similarly the second formula likewise rules out separate future time lines diverging from a common time.

One might expect that (PLin) and (FLin) together would suffice for full linearity (Lin), but this is not so, since they allow two or more "parallel" time lines between which there are no temporal relations. Indeed, it is not possible to rule out this latter kind of model by means of any tense logic formula.

While for many purposes it is natural to demand that the model of time is linear, this is not always the case. If TL is used to model the evolution of a nondeterministic dynamical system, one way of incorporating the indeterminacy is to model the full set of possible histories by means of a temporal frame which branches into the future—and hence lacks future-linearity. It is usual even in this case to retain past-linearity, so as to rule out convergent time lines. Then each point in time has a unique past but may have more than one future. This kind of structure was introduced in the context of reasoning about nondeterministic computations by Lamport (1980).

For reasoning about such future-branching models, an extension to TL called computation tree logic (CTL) has been introduced. In addition to temporal operators such as $\Box$ and $\Diamond$, CTL uses the "path operators" $\mathbf{A}$ and $\mathbf{E}$ to describe properties as true for respectively all or some of the futures diverging from the time at which a formula is evaluated. For example, the CTL axiom $\mathbf{A}\Box\alpha \to \neg\mathbf{E}\Diamond\neg\alpha$ says that if α is true throughout all possible futures then there is no possible future in which α is at any time false. CTL was introduced (Emerson and Clarke, 1982) as a method for deriving the "synchronization skeleton" of a concurrent computer program from a high-level specification. The specification, in CTL, expresses the temporal constraints which must be satisfied by any execution of the program, and the synchronization skeleton that is derived from it is "an abstraction of the actual program where detail irrelevant to synchronization is suppressed." Subsequently, more expressive extensions of CTL such as CTL*, ECTL, and ECTL$^+$ were introduced (see (Emerson, 1990), for references).

20.4.5 Boundedness

A model of time is bounded in the past if it has a start time, that is, a time which is not preceded by any other time; and it is bounded in the future if it has an end time which is not followed by any other time. So long as time is linear, there can be at most one of each, but in a nonlinear temporal frame there can be many possibilities, for example, some branches might have an end time but others not. For reasoning about a dynamic system that is set up at a particular time and then allowed to evolve, it is natural to include a start time, but not an end time, in the temporal model. For reasoning about natural systems which may be regarded as having histories extending indefinitely far into the past as well as into the future, it may be better to drop the start time as well. In that case, we have an unbounded temporal frame. The first-order formulae

$$\forall t \exists t'(t' \prec t) \tag{PUnb}$$

$$\forall t \exists t'(t \prec t') \tag{FUnb}$$

express unboundedness in the past and future, respectively. Corresponding tense logic axioms are $\mathbf{H}\alpha \to \mathbf{P}\alpha$ and $\mathbf{G}\alpha \to \mathbf{F}\alpha$. The first says that a formula true at *every* past time is true at *some* past time, and hence there must *be* at least one past time; for this to hold universally, it must be that for every time there is an earlier time, as expressed by (PUnb). The second axiom may be explained similarly. The conjunction of (PUnb) and (FUnb) may be designated (Unb).

20.4.6 Density and Discreteness

The transitivity axiom, discussed in Section 20.4.1, says that whenever t precedes t' and t' precedes t'', then t precedes t''. An important question for a model of time is whether the converse holds, i.e., if t precedes t'' must there be a time t' such that t precedes t' and t' precedes t''. A temporal frame that satisfies this condition is called dense. The first-order formulation of density is

$$\forall t \forall t'(t \prec t' \to \exists t''(t \prec t'' \wedge t'' \prec t')) \tag{Dens}$$

and the corresponding TL formulation is $\Box\Box\alpha \to \Box\alpha$. Note the relationship between this and (Ax4), reflecting the status of transitivity and density as mutually converse. An equivalent formulation of density is $\Diamond\alpha \to \Diamond\Diamond\alpha$.

A dense model of time corresponds well to our normal practice of identifying positions in the time series by means of numbers. Given times t_1 and t_2 identified as, say, exactly 10 and 11 s, respectively after some agreed "zero" time t_0, then if we assume the existence of a time t corresponding to 10.5 seconds after t_0 we are in effect making use of the density property. Further applications of density then allow us to bring in times at 10.25 and 10.75 s, then 10.125, 10.375, 10.625, 10.875 s, and so on, there being no limit, on this picture, to how finely the time line can be subdivided.

For some purposes, however, it is more convenient to regard time as proceeding in discrete steps rather than a continuous flow. An example is when modeling the execution of a computer program, where nothing significant, from the point of view of the program, occurs between successive steps of the execution. Another example would be a game such as chess, where what matters is the configuration of the board after each successive move, forming a discrete series of times that are relevant. Such discrete time series can still be marked off using numbers, only now the appropriate number system to use is the integers ($\mathbb{Z}, <$) rather than the real numbers ($\mathbb{R}, <$).

A discrete-temporal frame has the two properties

$$\forall t \forall t'(t' \prec t \to \exists t''(t'' \prec t \wedge \neg \exists u(t'' \prec u \wedge u \prec t))) \quad \text{(PDisc)}$$

$$\forall t \forall t'(t \prec t' \to \exists t''(t \prec t'' \wedge \neg \exists u(t \prec u \wedge u \prec t''))) \quad \text{(FDisc)}$$

Thus (FDisc) says that if t is followed by any time at all, then it has an immediate successor t'' characterized by the fact that there is no time between t and t''. Similarly, (PDisc) says that any point with a predecessor has an immediate predecessor. We use (Disc) to denote the conjunction of these two formulae.

It can be shown that there is no tense logic formula that exactly characterizes discreteness (see Van Benthem, 1983).[9] However, the discreteness of the temporal frame can be implicitly incorporated into a TL by extending the language. A new temporal operator $\bigcirc$, called the *Next-time operator*, is introduced, interpreted so that $\bigcirc\alpha$ is true at time t just if α is true at the immediate successor time, which we may denote $t + 1$ on the assumption that integers are used to label the times in the model.

20.4.7 Combinations of Properties

Certain combinations of the above-mentioned properties are noteworthy. The combination (Trans, Irref, Lin, Unb) corresponds to the intuition of time as a line extending indefinitely far back into the past ("no beginning") and indefinitely far into the future ("no end"). We may add to this combination either (Dens) or (Disc), the resulting combinations being labeled DE and DI, respectively (Van Benthem, 1983). In neither case do we end up specifying the flow of time completely. DE, for example, holds for both a real ($\mathbb{R}, <$) and a rational ($\mathbb{Q}, <$) time line; to identify ($\mathbb{R}, <$) uniquely we need a *second*-order formula (expressing Dedekind continuity). Gabbay et al. (1994) give the following pair of TL formulae to express this:

$$\mathbf{FG}\alpha \wedge \mathbf{F}\neg\alpha \wedge \mathbf{G}(\neg\alpha \to \mathbf{F}\neg\alpha) \to \mathbf{F}(\mathbf{G}\alpha \wedge \neg\mathbf{PG}\alpha)$$
$$\mathbf{PH}\alpha \wedge \mathbf{P}\neg\alpha \wedge \mathbf{H}(\neg\alpha \to \mathbf{P}\neg\alpha) \to \mathbf{P}(\mathbf{H}\alpha \wedge \neg\mathbf{FH}\alpha)$$

Likewise, while the most natural model for DI is ($\mathbb{Z}, <$), others are possible, e.g., two copies of $\mathbb{Z}$, one after the other. To identify ($\mathbb{Z}, <$) uniquely, we need a second-order condition (given (Disc), Dedekind continuity suffices again).

---

[9]Note, however, that if the temporal frame is assumed to be linear, then the formulae $\mathbf{GP}\alpha \to \alpha \vee \mathbf{P}\alpha$ and $\mathbf{HF}\alpha \to \alpha \vee \mathbf{F}\alpha$ together secure discreteness as well. For, to take the second formula, if t does not have an immediate predecessor, then letting α be true at all predecessors of t and at no other times, we see that $\mathbf{HF}\alpha$ is true at t (since every point in the past of t has a point in its future which is still in the past of t), but $\alpha \vee \mathbf{F}\alpha$ is false at t.

20.5 Further Extensions to the Formal Language

An important addition to the expressive power of TL was made by Kamp (1968) who introduced new temporal operators **S** and **U**, read "since" and "until." Syntactically, these are binary operators, acting on a pair of propositions rather than a single one as in the case of **P** and **F**. As binary operators, they may be treated either as prefixes (e.g., "**S**pq") or as infixes (e.g., "p**S**q")—both conventions have been used; here we use infix notation to emphasize the reading of "p**S**q" as "p since q" (and likewise with **U**).
The semantic rules for the new operators are

- $(T, \prec, t) \models \alpha \mathbf{S} \beta$ if and only if, for some $t' \in T$, $t' \prec t$, $(T, \prec, t') \models \beta$ and, for every $t'' \in T$, if $t' \prec t'' \preceq t$ then $(T, \prec, t'') \models \alpha$
- $(T, \prec, t) \models \alpha \mathbf{U} \beta$ if and only if, for some $t' \in T$, $t \prec t'$, $(T, \prec, t') \models \beta$ and, for every $t'' \in T$, if $t \preceq t'' \prec t'$ then $(T, \prec, t'') \models \alpha$.

Thus, $\alpha \mathbf{S} \beta$ says that β was true at some past time, and α has been true ever since then, up to and including the present. Similarly, $\alpha \mathbf{U} \beta$ says that β will be true at some future time, but until then, α will always be true.

A weak variant of **U** that is sometimes used is the operator $\mathbf{U}_W$, perhaps misleadingly read "unless," defined so that $\alpha \mathbf{U}_W \beta$ is equivalent to $\alpha \mathbf{U} \beta \vee \mathbf{G} \alpha$, meaning that α will be true until such time as β becomes true—there being no requirement for this ever to happen. In the case that β never becomes true, α must remain true forever, as indicated by the second disjunct.

The significance of these new operators is that, subject to certain assumptions about the model of time, they bring the expressive power of the logic up to that of the FOL of time. For example, over linear discrete time, or linear continuous time, for any first-order temporal formula $\phi(t)$, there is an S, U-TL formula α which holds at all and only those times t for which $\phi(t)$. That we need **S** and **U** follows from the fact that these operators cannot be defined in terms of **P**, **F**, **G**, and **H**—that is, no formula constructed using these four operators is equivalent to either $p\mathbf{S}q$ or $p\mathbf{U}q$.

In contrast the other operators *can* be defined in terms of **S** and **U**. To do this, it is convenient to introduce formulae $\bot$ and $\top$ that are defined to be always true (i.e., equivalent to $\alpha \vee \neg \alpha$) and always false, respectively (equivalent to $\alpha \wedge \neg \alpha$). Then we can define **P**α as $\top \mathbf{S} \alpha$, and **F**α as $\top \mathbf{U} \alpha$, with **H** and **G** then defined as $\neg \mathbf{P} \neg$ and $\neg \mathbf{F} \neg$, respectively. The formula $\top \mathbf{S} \alpha$ says that α was true at some past time, since when $\top$ has always been true. Since the latter clause must hold in any case, the statement reduces to an assertion that α was true at some past time—i.e., **P**α; and likewise with **U** and **F**.

20.6 Illustrative Examples

In this section, we provide some brief illustrations of a range of possible applications of TL. The area in which it has been most widely used, and which has provided the main stimulus to its development, is in the specification and verification of concurrent computer programs, as illustrated in Section 20.6.2. However, similar considerations, concerning sequencing, scheduling, and temporal constraint satisfaction, can arise in the specification of any system which evolves through time, and in Section 20.6.3 we mention a couple of examples to illustrate this.

20.6.1 A System with an Attractor

Suppose we have a system with just three states; these can be represented by three mutually exclusive formulae P, Q, and R, with the understanding that each of the formulae is true just when the state it represents holds. The dynamics of the system are such that whenever it is in state P, it will eventually enter state Q, and whenever it is in state Q it will eventually enter state R. Moreover, state R is an attractor state, i.e., once the system has entered it, it can never leave it. With a little reflection it seems obvious that that the system will eventually enter state R and stay there. One might, naively, illustrate this with a state sequence such as

$$PPPPPPPQQQQQQQQQQQQQRRRRRRRRRRRRRRR \ldots$$

but this is only illustrative since in reality there is an infinite number of possible state sequences for this system, another one being, for example

$$PPPPQPPQPPQQQQPQPPPQQRRRRRRRRRRRRRRR \ldots$$

The state sequences illustrated here implicitly presuppose a discrete linear time, but is either of these properties necessary for the desired result to follow? TL can help us to determine exactly what underlying assumptions are required for the reasoning to go through correctly.

To establish the desired conclusion, we can formulate the problem in TL as follows. Our description of the system dynamics corresponds to the following four formulae (we use $\Box$, $\Diamond$ notation since we are reasoning forward in time):

$$\Box(P \vee Q \vee R) \tag{20.14}$$

$$\Box(P \rightarrow \Diamond Q) \tag{20.15}$$

$$\Box(Q \rightarrow \Diamond R) \tag{20.16}$$

$$\Box(R \rightarrow \Box R) \tag{20.17}$$

From these we have to derive the formula $\Diamond \Box R$. Without writing out the full formal derivation, we can indicate the lines along which it proceeds as follows.

We know from Eq. (20.14) that at any time in the system's evolution we have $P \vee Q \vee R$. Several applications of Eq. (20.15), Eq. (20.16), and Eq. (20.17) enable us to derive from this the formula $\Diamond \Diamond \Box R \vee \Diamond \Box R \vee \Box R$. To get from here to $\Diamond \Box R$ we need the following assumptions:

$$\Diamond \Diamond \alpha \rightarrow \Diamond \alpha \tag{20.18}$$

$$\Box \alpha \rightarrow \Diamond \Box \alpha \tag{20.19}$$

Assumption Eq. (20.18) is equivalent to (Ax4), the transitivity axiom. Given transitivity, Eq. (20.19) may be derived from $\Box \Box \alpha \rightarrow \Diamond \Box \alpha$, which in turn is an instance of (FUnb). We have thus laid bare the assumptions underlying our intuitive reasoning: the flow of time is transitive and unbounded in the future. There is *no* need to assume linearity, density, discreteness, or any of the other conditions we might suppose to be required (and which may be present in the mental models we construct in the course of our intuitive reasoning).

20.6.2 Application to Reasoning about Programs

TL has been considered the most appropriate logic for reasoning about the behavior of reactive programs— that is, programs which maintain an ongoing interaction with the environment rather than delivering a final output and then halting. Commonly cited examples of reactive programs include computer operating systems, airline reservation systems, and many e-business systems. One area in which it is important to be able to reason correctly about the behavior of such systems is correctness: how can we be sure that the program as implemented actually behaves in the manner required by its specification? Among the correctness properties of programs we may distinguish *safety* properties, informally characterized as ensuring that "nothing bad happens," and *liveness* properties, ensuring that "something good will happen."[10] In TL terms, safety properties are typically expressed using formulae of the form $\Box \alpha$, while liveness properties are expressed using $\Diamond \alpha$.

---

[10]These terms, and the associated slogans, are due to Lamport (1977).

Temporal Logic

To illustrate these ideas with a simple example, suppose that we have a number of concurrent processes which have to apply to some resource allocator for the use of various resources.[11] We introduce the following primitive propositions to describe this situation:

> $Available(i)$ says that resource i is unallocated.
> $Granted(i, j)$ says that process i is granted use of resource j.
> $Requesting(i, j)$ says that process i is requesting use of resource j.
> $Using(i, j)$ says that process i is using resource j.

The following are all safety properties:

- A process cannot request a resource which it is already using
  $\Box(Using(i, j) \rightarrow \neg Requesting(i, j))$
- A process cannot use a resource which it has not been granted
  $\Box(Using(i, j) \rightarrow Granted(i, j))$
- Two processes cannot simultaneously use the same resource
  $\Box \neg(Using(i, j) \land Using(i', j) \land i \neq i')$

The following are liveness properties:

- If a process requests use of a resource, it is eventually granted it
  $\Box(Requesting(i, j) \rightarrow \Diamond Granted(i, j))$
- If a process is granted a resource, it will eventually use it
  $\Box(Granted(i, j) \rightarrow \Diamond Using(i, j))$
- If a process is using a resource, it will eventually release it
  $\Box(Using(i, j) \rightarrow \Diamond \neg Using(i, j))$

These are just a sample of the system requirements that can be expressed as TL formulae. A complete set of such formulae may constitute a formal specification for a system to exhibit the desired behavior. This can be used in various ways: it provides a standard to which any implementation of the program must adhere, and temporal reasoning may be used to verify that a given implementation does indeed meet the specification. To do this, it is necessary to show that the formulae in the specification hold for every possible execution sequence of the program (see Manna and Pnueli, 1981). A related goal is to use the specification as a starting point from which to synthesize the synchronization skeleton of a program, as indicated in (Section 20.4.4).

20.6.3 Other Applications

The extremely general nature of TL formalism means that in principle it should be applicable to a wide range of situations in which it is necessary to specify or control the behavior of systems evolving through time. Here we mention just two examples.

In artificial intelligence, TL has been used by Bacchus and Kabanza (1996) as a means for controlling search in a forward-chaining planner, thereby at least mitigating the combinatorial explosion to which such planners are inevitably prone. Their approach is implemented in a system called TLPlan, which uses a first-order linear temporal logic, LTL for expressing strategies tailored to the chosen planning domain—an example being the formula

$$\forall [x : object(x)] \Box (polish(x) \land \bigcirc \neg polish(x) \rightarrow \bigcirc \Box \neg polish(x))$$

which "prohibits action sequences where an object is polished twice."

In a very different setting, Wood (1989) used TL to specify the operation of a bank of identical elevators servicing a number of floors in a building. This was approached in the spirit of a case study to investigate the suitability of TL for this kind of work; the author's conclusion is largely positive, including the

---

[11] This example is adapted from Pnueli (1985).

statement that TL specification is "a splendid vehicle for communicating precisely the behavior of the elevator system." Examples of temporal formulae used are

$$\forall i, m \Box (\neg \mathit{flDoorsClosed}(i,m) \leftrightarrow \neg \mathit{elDoorsClosed}(i) \land \mathit{elPosition}(i,m))$$

which states that at all times, the door for accessing elevator i on the mth floor will be open if and only if elevator i is at the mth floor and its doors are open; and

$$\forall i, m (\mathit{dormant}(i,m) \land \mathit{elDestLit}(i,m) \rightarrow \bigcirc \neg \mathit{elDoorsClosed}(i))$$

which says that if elevator i is in a dormant condition at the mth floor and the user pushes the destination button for the mth floor in that elevator, then the elevator doors will open immediately afterwards (note the use of the "next time" operator $\bigcirc$ to express this, implying a discrete model of time).

20.7 Conclusion

We have introduced TL, a system for reasoning about the dynamical properties of the world by manipulating formulae expressed in a formal language specifically designed to allow different properties to hold at different times. We described the formalism from both a syntactic and a semantic point of view, and introduced a range of different varieties of TL, adapted to working with a range of different models of time. Some applications were briefly described. Owing to the introductory nature of the chapter, much of importance has inevitably been left out, and the reader interested in pursuing the topic is encouraged to consult the references recommended in the next section.

20.8 Further Reading

There is a voluminous literature on TL, emanating from several distinct (though overlapping) research communities, notably computer science, mathematics, logic, and philosophy, with a correspondingly wide range of focus. Much of this literature appears in the form of journal articles or conference proceedings, but over the years a number of books have been produced on the subject, some of which have been very influential.

In the philosophical tradition, the seminal work (Prior, 1967) still retains much of interest to the modern reader, although its readability is somewhat compromised by the use of "Polish" prefix notation for the logical operators. This notation is largely unfamiliar to present-day readers, who may have the feeling of having to decipher the formulae rather than read them; but this has been remedied in the later work (Prior, 1968), which has been recently reissued with all the formulae rewritten in the more familiar infix notation (Prior, 2003). An important work exploring the logical and mathematical properties of models of time and their logics, with many interesting philosophical sidelights, is Van Benthem (1983).

For the computer science perspective, some useful works to consult are Galton (1987), Goldblatt (1987), and Bolc and Szałas (1995). A detailed survey, with many references, is provided by Emerson (1990). More technical, but containing a wealth of useful material and many stimulating observations, is Gabbay et al. (1994). A sample of recent research, covering many aspects of TL, is Barringer et al. (2000). A still more recent compilation with an emphasis on applications to artificial intelligence—but with much of interest in a wider setting—is Fisher et al. (2005).

For the reader interested in the history of TL, Øhrstrøm and Hasle (1995) is recommended, although it should be noted that the applications in computer science are largely neglected.

An important forum for the dissemination of research in TL is provided by a series of international symposia on Temporal Representation and Reasoning entitled simply TIME, first held in Florida in 1994 and still going strong at the time of writing.

References

Bacchus, F. and F. Kabanza (1996). Using temporal logic to control search in a forward-chaining planner. In M. Ghallab and A. Milani (Eds.), *New Directions in AI Planning*, pp. 141–153. Amsterdam: IOS Press.

Barringer, H., M. Fisher, D. Gabbay, and G. Gough (Eds.) (2000). *Advances in Temporal Logic*, Volume 16 of Applied Logic Series. Dordrecht: Kluwer.

Bolc, L. and A. Szałas (Eds.) (1995). *Time and Logic: A Computational Approach*. London: UCS Press.

Dowty, D. (1979). *Word Meaning and Montague Grammar*. Dordrecht: D. Reidel.

Emerson, E. A. (1990). Temporal and modal logic. In J. van Leeuwen (Ed.), *Handbook of Theoretical Computer Science (Volume B: Formal Models and Semantics)*, pp. 995–1072. Amsterdam: Elsevier.

Emerson, E. A. and E. M. Clarke (1982). Using branching time temporal logic to synthesize synchronisation skeletons. *Science of Computer Programming 2*, 241–266.

Fisher, M., D. Gabbay, and L. Vila (Eds.) (2005). *Handbook of Temporal Reasoning in Artificial Intelligence*. New York: Elsevier.

Gabbay, D. M., I. Hodkinson, and M. Reynolds (Eds.) (1994). *Temporal Logic: Mathematical Foundations and Computational Aspects*. Oxford: Oxford University Press.

Galton, A. (Ed.) (1987). *Temporal Logics and Their Applications*. London: Academic Press.

Goldblatt, R. (1987). *Logics of Time and Computation*. Stanford, CA: Center for the Study of Language and Information.

Hodges, W. (2001). *Logic*. London: Penguin Books.

Jeffrey, R. C. (2006). *Formal Logic: Its Scope and Limits* (Fourth edition). Indianapolis: Hackett Publishing.

Kamp, J. A. W. (1968). *Tense Logic and the Theory of Linear Order*. Ph.D. thesis, University of California, Los Angeles.

Lamport, L. (1977). Proving the correctness of multiprocess programs. *IEEE Transactions on Software Engineering SE-3*, 125–143.

Lamport, L. (1980). "Sometime" is sometimes "not never": On the temporal logic of programs. In *Proceedings of the 7th ACM Symposium on Principles of Programming Languages*, pp. 174–85.

Manna, Z. and A. Pnueli (1981). Verification of concurrent programs: The temporal framework. In R. S. Boyer and J. S. Moore (Eds.), *The Correctness Problem in Computer Science*, pp. 215–273. London: Academic Press.

Moszkowski, B. (1986). *Executing Temporal Logic Programs*. Cambridge: Cambridge University Press.

Øhrstrøm, P. and P. F. V. Hasle (1995). *Temporal Logic: From Ancient Ideas to Artificial Intelligence*. Amsterdam: Kluwer.

Pnueli, A. (1985). Applications of temporal logic to the specification and verification of reactive systems: A survey of current trends. In J. W. de Bakker, W.-P. de Roever, and G. Rozenberg (Eds.), *Current Trends in Concurrency*, Volume 224 of Lecture Notes in Computer Science, pp. 510–584. Berlin: Springer.

Prior, A. N. (1967). *Past, Present, and Future*. Oxford: Clarendon Press.

Prior, A. N. (1968). *Papers on Time and Tense*. Oxford: Clarendon Press.

Prior, A. N. (2003). *Papers on Time and Tense*. Oxford: Oxford University Press. (New edition, edited by Per Hasle, Peter Øhrstrøm, Torben Bräuner, and Jack Copeland.)

Van Benthem, J. F. A. K. (1983). *The Logic of Time* Second edition, 1991. Dordrecht: Kluwer.

Wood, W. G. (1989). Temporal logic case study. Technical Report CMU/SEI-89-TR-024, Carnegie Mellon Software Engineering Institute.

21

Modeling Dynamic Systems with Cellular Automata

Peter M.A. Sloot
University of Amsterdam

Alfons G. Hoekstra
University of Amsterdam

21.1 Introduction...21-1
21.2 A Bit of History..21-2
21.3 Cellular Automata to Model Dynamical Systems..........21-3
21.4 One-Dimensional CAs..21-3
21.5 Lattice Gas Cellular Automata Models of
     Fluid Dynamics..21-5
     The Road to Lattice Gas Cellular Automata • LGCA and
     Fluid Dynamics • Simulating an LGCA • Some
     Applications • Lattice Boltzmann Method

21.1 Introduction

In modeling dynamic systems, one of the first questions to be answered is whether the involved processes can be viewed to be discrete in state, time, space, or continuous. The model choice should be robust with respect to the chosen space–time–state framework. Table 21.1 gives a selective overview of the various modeling approaches.

In this chapter, we focus on complete discrete model systems: cellular automata (CAs). CAs are decentralized spatially extended systems consisting of large numbers of simple and identical components with local connectivity. Such systems have the potential to perform complex computations with a high degree of efficiency and robustness as well as to model the behavior of complex systems from nature. CAs have been studied extensively in the natural sciences, mathematics, and computer science. They have been considered as mathematical objects about which formal properties can be proved and have been used as parallel computing devices, both for high-speed simulation of scientific models and for computational

TABLE 21.1 Mathematical and numerical modeling approaches to spatio-temporal processes.

Model/Variable	State	Space	Time
PDEs	C	C	C
Integro-difference equations	C	C	D
Coupled ODEs	C	D	C
Interacting particle systems	D	D	C
Coupled map lattices, systems of difference equations, LBE models	C	D	D
Cellular automata and lattice gas automata	D	D	D

PDE, partial differential equation; ODE, ordinary differential equation; LBE, lattice Boltzmann equation. For more details see Berec (2002) and Deutsch and Dormann (2004).

tasks such as image processing. CAs have also been used as abstract models for studying "emergent" cooperative or collective behavior in complex systems (e.g., Sloot, 2001b). In addition, CAs have been successfully applied to the simulation of a large variety of dynamical systems such as biological processes including pattern formation, earthquakes, urban growth, galaxy formation, and most notably in studying fluid dynamics. Their implicit spatial locality allows for very efficient high-performance implementations and incorporation into advanced programming environments. For a selection of the numerous papers in all of these areas, see, e.g., Bandini (2002), Burks (1970), Deutsch and Dormann (2004), Farmer et al. (1984), Forrest (1990), Frisch et al. (1986), Ganguly et al. (2003), Gutowitz (1990), Jesshope et al. (1994), Kaandorp et al. (1996), Mitchell (1998), Naumov (2004), Sloot (1999), Sloot and Hoekstra (2001), Sloot et al. (1997, 2001c, 2002, 2004), and Wolfram (1986a, 1986b, 2002).

In this chapter, we will give some background on CA modeling and simulation of dynamical systems with an emphasis on simulating fluid dynamics.

21.2 A Bit of History

In 1948, on the occasion of the Hixon Symposium at Caltech, John von Neumann gave a lecture entitled "The General and Logical Theory of Automata" (von Neumann, 1951, 1966), where he introduced his thoughts on universal, self-reproducing machines, trying to develop an abstract model of self-reproduction in biology, a topic that had emerged from investigations in cybernetics (Wolfram, 2002, 876 ff). von Neumann himself said to have been inspired by Stanislaw Ulam (1952, 1962) and Turing's theory of universal automata, which dates back another 10 years (Turing, 1936). Some scientists regard the paper by Wiener and Rosenblueth (1946) as the start of the field (Wolf-Gladrow, 2000), or mathematical work that was done in the early 1930s in Russia.

So we see that the roots of CA may be traced back to biological modeling, computer science, and (numerical) mathematics. From the early days of von Neumann and Ulam up to the recent book of Wolfram, CAs have attracted researchers from a wide variety of disciplines. It has been subjected to rigorous mathematical and physical analysis for the last 50 years, and its application has been proposed and explored in almost all branches of science. A large number of research papers are published every year. Specialized conferences, such as Sloot et al. (2004), Automata (2005), and NKS (2005), and special issues of various journals on CA have been initiated in the last decades. Several universities started offering courses on CA. The reason behind the popularity of CA can be traced to their simplicity and to the enormous potential they hold in modeling complex systems, in spite of their simplicity. Or in the words of R. May: "We would all be better off if more people realized that simple dynamical systems do not necessarily lead to simple dynamical behavior" (May, 1976). This has led to some very remarkable claims and predictions by renowned researchers about the potential of CAs. In this respect, we came across the following statements that are worth mentioning:

> *The entire universe is being computed on a computer, possibly a cellular automaton.*
> Konrad Zuse, as he referred to this as "Rechnender Raum" (Zuse, 1967, 1982)

> *I am convinced that CA, in one form or another will eventually be found lurking at the very heart of how the universe really works*
> Andrew Ilachinski (2001)

> *The view of the Universe as a cellular automaton provides the (same) perspective, (i.e.,) that reality ultimately is a pattern of information.*
> Ray Kurzweil (2002)

> *I have come to view [my discovery] as one of the more important single discoveries in the whole history of theoretical science*
> Stephan Wolfram: Talking about his CA work in his NKS book (Wolfram, 2002)

The remainder of this chapter is organized as follows, we start with a formal description of CA and some of their spatio-temporal properties and we will briefly discuss their capacity to model dynamical systems. Next we will focus on the use of CAs to model fluid flow, starting from Lattice Gas CAs up to recent developments in the related Lattice Boltzmann Method for fluid flow.

21.3 Cellular Automata to Model Dynamical Systems

In general, a CA is specified by the following four characteristics:

- A *discrete lattice* $\mathcal{L}$: This is the discrete lattice of cells (nodes and sites) upon which the CA dynamics unfolds. $\mathcal{L} \subset \mathcal{R}^{\mathcal{D}}_\bullet$ consists of a set of cells that homogeneously cover a $\mathcal{D}$-dimensional Euclidian space. $\mathcal{L}$ can have any dimension "$\mathcal{D}$" (normally 1, 2, or 3), with well-defined boundary conditions.
- A *finite state space*: Each cell can assume only one of a finite number of different values: $\sigma_{i \in \mathcal{L}}(t) \in \Sigma \equiv \{0, 1, 2, \cdots, k-1\}$, where σ_i is the value of the ith cell at time t, and Σ is usually taken to be the set of integers modulo k, $\mathcal{Z}_k$ (formally any finite commutative ring will do). For a finite lattice of $\mathcal{N}$ cells, the total number of global states is also finite and given by $k^{\mathcal{N}}$.
- *Boundary conditions*: Boundary conditions play an important role in CA dynamics. Although CA are defined on infinite large lattices, computer simulations impose finite sets. Common boundary conditions are *periodic* (i.e., the lattice is repeated periodically in each direction, in effect wrapping the boundaries onto each other in each direction), *reflecting* (i.e., boundary values are reflected back into the lattice), and *fixed* (i.e., the boundary values have a prescribed fixed value).
- *Dynamical update and transition Rule* ϕ:
  ϕ: $\underbrace{\Sigma \times \Sigma \times \cdots \times \Sigma}_{n} \to \Sigma$, where n is the number of cells that defines the "neighborhood" of a given cell i. With $\mathcal{S}_i$ to be the sublattice neighborhood about cell i, the transition rule is given by $\sigma_i(t+1) = \phi(\sigma_j(t) \in \mathcal{S}_i)$.

The spatial arrangement of the cells is specified by the nearest neighbor connection links, obtained by joining pairs of cells. State transitions are local in both space and time. Individual cells evolve iteratively according to a fixed (often deterministic) function of the current state of that cell and its neighboring cells. One iteration step of the dynamical evolution is achieved after synchronous (i.e., simultaneous in time) application of the rule ϕ to each cell in the lattice $\mathcal{L}$.

21.4 One-Dimensional CAs

The general form of a one-dimensional (1D) CA rule ϕ with an arbitrary range 'r' is given by $\sigma_i(t+1) = \phi(\sigma_{i-r}(t), \ldots, \sigma_i(t), \ldots, \sigma_{i+r}(t))$; with ϕ: $\Sigma^{2r+1} \to \Sigma$, where $\sigma_j \in \{0, 1, \ldots, k-1\}$, and ϕ is explicitly defined by assigning values to each of the k^{2r+1} possible $(2r+1)$-tuples of possible configurations for a given sublattice neighborhood $\mathcal{S}_i$. From this we see that we have a total of $k^{k^{2r+1}}$ possible rules in a 1D CA. So for a binary state 1D CA, with nearest neighbor interaction, there are 256 (2^{2^3}) possible rules.

The boundary conditions imposed in a 1D CA can be

- *Periodic*: When the left boundary cell is kept identical to the most right (normal) cell, and the right boundary cell is kept identical to the most left (normal) cell.
- *Reflecting*: When the left boundary cell is kept identical to the most left (normal) cell, and the right boundary cell is kept identical to the most right (normal) cell.
- *Fixed*: When the boundary cells are set to a fixed value.

The system dynamics of the CA is determined by the local transition rule ϕ, which can be *spatial homogeneous* (independent of cell position), or *inhomogeneous* (for instance in the case of fixed boundary conditions shown in Figure (21.1). Furthermore, the update can be *time dependent* or *time independent*,

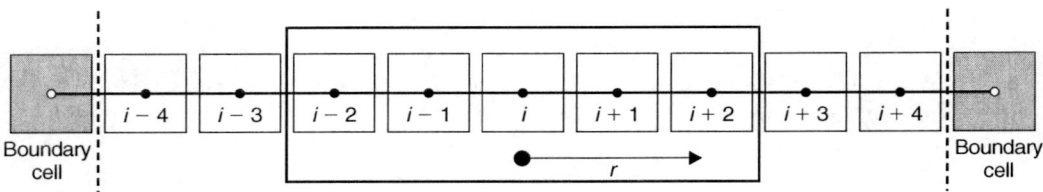

FIGURE 21.1 A 1D CA with range $r = 2$.

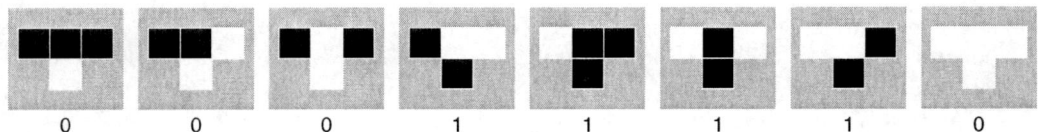

FIGURE 21.2 Updates for rule 30. The upper three blocks denote the cell that must be update, together with its left and right neighbors, and the lower block shows the outcome of applying rule 30. Black denotes for a state 1 and white for a state 0.

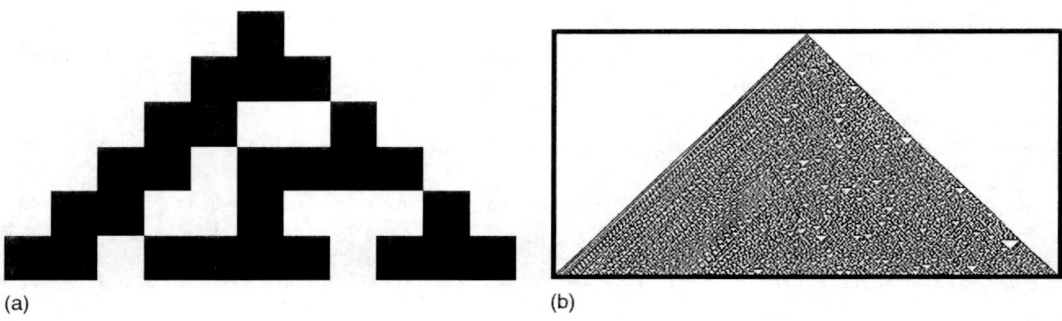

FIGURE 21.3 (a) Space–time diagram of Wolfram's elementary ($k = 2, r = 1$) CA rule 30 starting from a single seed. The sequence of updates "time" runs from top to bottom. (b) same rule for 300 updates on the 1D CA.

synchronous or *asynchronous*, *deterministic* or *stochastic*. This gives the modeler a plethora of possibilities to capture the dynamics of the system to be modeled. In the paragraph on lattice gas CAs, we will see more examples of how to design and apply such update rules. For the binary state 1D CA with nearest neighbor interaction, Wolfram introduced a convenient rule-code $\mathcal{R}$ for these important CAs, which uniquely identifies the update mechanism:

$$\mathcal{R}[\phi] = 2^7 \cdot \phi_{1,1,1} + 2^6 \cdot \phi_{1,1,0} + 2^5 \cdot \phi_{1,0,1} + 2^4 \cdot \phi_{1,0,0} + 2^3 \cdot \phi_{0,1,1} + 2^2 \cdot \phi_{0,1,0} + 2^1 \cdot \phi_{0,0,1} + 2^0 \cdot \phi_{0,0,0}$$

The number ϕ is an eight-bits number that encodes all possible 256 rules for this 1D CA and $\phi_{i,j,k}$ are the bits of the binary notation of ϕ. The rule should be read as follows. The outcome of the rule is determined by the current state of a cell (0 or 1) and the current state of the left and right neighbor of the cell. The binary number that is formed by concatenating the state of the left neighbor, the cell itself, and the right neighbor is a number between 0 and 7. The outcome of rule $\mathcal{R}[\phi]$ is then the bit of the binary representation of ϕ at the position of the number encoded by the input states. So, for example, $\mathcal{R}[30] = 00011110$, since the decimal value of $00011110 = 30$. The update rules in terms of black ($=1$) and white ($=0$) blocks are graphically depicted in Figure 21.2.

The first five updates (*spatial homogeneous, time independent, synchronous, and deterministic*) of this $\mathcal{R}[30]$, starting from a single seed is shown in Figure 21.3.

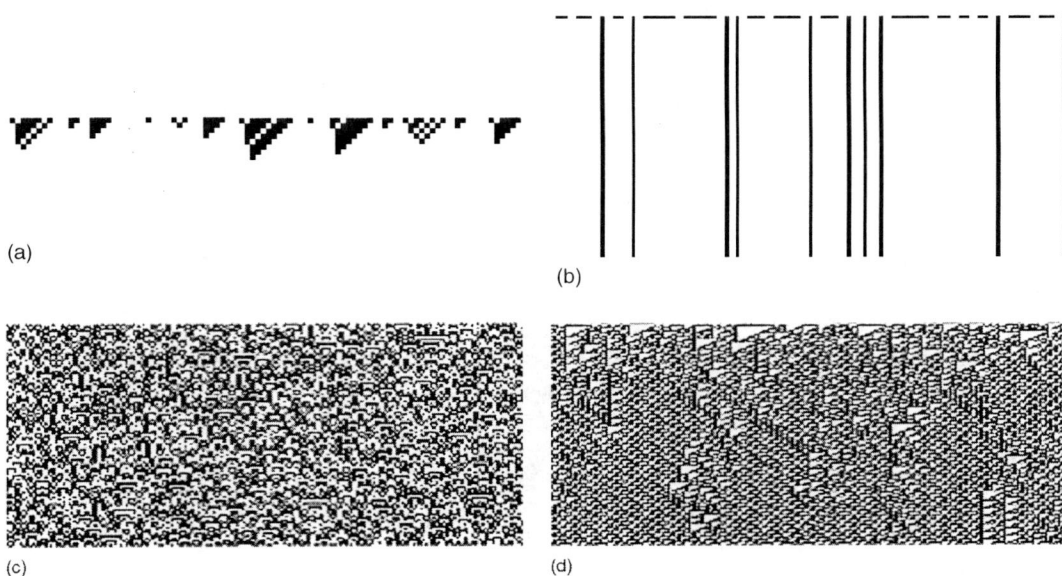

FIGURE 21.4 (a) Class 1 (e.g., Rules 0, 8, 128, 136, 160, 168): evolution leads to a homogeneous state, in which all cells eventually attain the same value (continues analog: attractive fixed limit point), shown is rule 168. (b) Class 2 (e.g., Rules 4, 37, 56, 73): evolution leads to inhomogeneous state; either simple stable states or periodic and separated structures (continues analog: limit cycle), shown is rule 4. (c) Class 3 (e.g., Rules 18, 45, 105, 126): evolution leads to chaotic nonperiodic patterns (continues analog: strange attractor), shown is rule 105. (d) Class 4 (e.g., rules 30, 110): evolution leads to complex, localized propagating structures (no continuous analog), shown is rule 110.

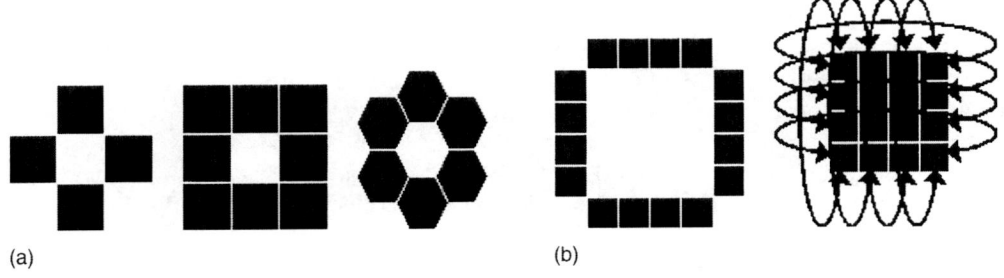

FIGURE 21.5 (a) von Neumann, Moore, and Hexagonal neighborhoods. (b) Fixed and periodic boundary conditions.

In his seminal paper on CA classification, Wolfram (1986a) identified four classes of CAs, linking them to analogs in continuous system dynamics (see Figure 21.4).

The behavior in two-dimensional (2D) CAs is much more complex and less well understood. In this case, we need again to define the update neighborhood and the boundary conditions; this is shown in Figure 21.5.

21.5 Lattice Gas Cellular Automata Models of Fluid Dynamics

Perhaps the most successful practical application of CAs as computing devices has been in the field of fluid dynamics. Coined lattice gas cellular automata (LGCA), this class of CA mimics a fully discretized fictitious fluid. Both the positions and velocities of the fluid's "molecules" are discrete, and tightly coupled to the

LGCA's discrete lattice $\mathcal{L}$. Moreover, the dynamics of the molecules is highly simplified and completely synchronous. All molecules perform free streaming from one lattice node to a neighboring one in a time period δt. Next, particles arriving at a node collide with each other, thus exchanging momentum in some deterministic or stochastic way. The collisions on the nodes all happen at the same time and the duration of a collision is assumed to take zero time. By enforcing conservation of mass, momentum, and energy in a collision, we have created a model gas with a fully discrete and simplified, yet physically correct micro dynamics. With this LGCA dynamics, we may then investigate macroscopic variables, i.e., averaged quantities such as fluid density or momentum, which vary over time and length scales much larger than those of the micro dynamics, and hope that they behave as a real fluid. In fact, we know that if Λ is sufficient isotropic (to be defined later) an LGCA, when operated in the right limits, reproduces the incompressible Navier–Stokes equations and therefore is a model for fluid dynamics.

The most complete account of LGCA (including a highly useful "guide to further reading") is the book by Rivet and Boon (2001). Other influential monographs on LGCA are Rothman and Zaleski (1997), Wolf-Gladrow (2000), and Chopard and Droz (1998). Finally, Boghosian (1999) provides a nice overview of lattice gases and cellular automata.

21.5.1 The Road to Lattice Gas Cellular Automata

As suggested by Rivet and Boon (2001), LGCA can be traced back to discrete kinetic models, in which a gas is modeled as a collection of particles with continuous position and time variables, but with a (small) discrete set of velocities. Such discrete kinetic models were studied intensively starting in the sixties of the previous century. LGCA would then be one step further down the road to minimalist models, in which also space and time are discrete. Indeed, in 1972 the point of departure for Hardy and Pomeau, who 1 year later introduced the first real LGCA, was the discrete velocity Maxwell model (Hardy and Pomeau, 1972). In contrast, Boghosian (1999) suggests a connection between early minimalist discrete models in statistical physics (such as the Ising spin, Creutz, and Kawasaki models) in the sense that LGCAs are comparable minimalist models, but on top of that they are also truly dynamic and have conserved quantities (mass, momentum, and energy) whose dynamics (approach to equilibrium) are of interest. Boghosian also points out that a first step toward LGCA probably was the Kadanoff–Swift model (Kadanoff and Swift, 1968). Strictly speaking this was not a CA, but it had a number of ingredients that are close to LGCA, such as fictitious particles living on a 2D Cartesian lattice with discrete velocities oriented along the diagonals of the lattice. The dynamics would then be performed sequentially on randomly selected particles. From a statistical physics point of view, this model already had many features of real fluids that made it quite interesting to study. Interestingly, the basic papers on LGCA have never referenced the Kadanoff–Swift paper.

The first real LGCA was introduced by Hardy, Pomeau and de Pazzis in 1973 (Hardy et al., 1973), and its hydrodynamics were studied in detail in Hardy et al. (1976). The HPP model, as we now call it, is a real CA. It has an underlying 2D Cartesian lattice. On each node, particles with unit mass are defined that can have one one of the four discrete velocities c_i, $i \in \{1, 2, 3, 4\}$ (see also Figure 21.6(a)):

$$c_1 = \begin{pmatrix} 1 \\ 0 \end{pmatrix}; \quad c_2 = \begin{pmatrix} 0 \\ 1 \end{pmatrix}; \quad c_3 = \begin{pmatrix} -1 \\ 0 \end{pmatrix}; \quad c_4 = \begin{pmatrix} 0 \\ -1 \end{pmatrix} \quad (21.1)$$

At each node only one particle can have a velocity c_i. This exclusion principle allows us to denote the state of a node as a binary 4-vector $\mathbf{n}$. A value "TRUE" (or 1) of element i of the state vector (denoted as n_i) encodes for the presence of a particle with velocity c_i, and a value "FALSE" (or 0) denotes the absence of a particle with velocity c_i. So, the vector (1,0,0,1) would represent a state as shown in Figure 21.6(b). The number of different states per node in HPP is 16. Owing to the exclusion principle, the maximum number of particles on a node is 4. The dynamics is very straightforward. First, incoming particles at a node collide, and next particles perform a free streaming, in which they move from their node to the neighboring node in the direction of their velocity. We assume that the time for this streaming $\delta t = 1$, so

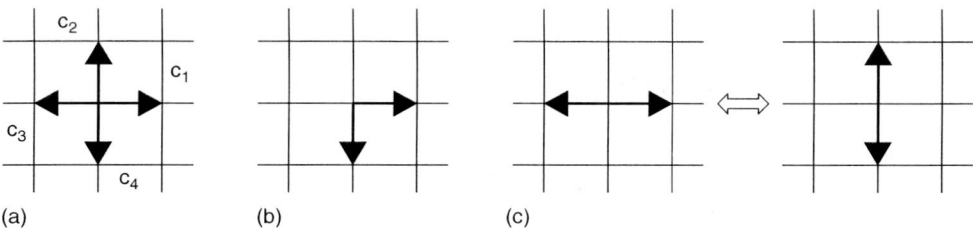

FIGURE 21.6 The HPP model: (a) shows the four velocities of the model, (b) an example of a node with state (1,0,0,1), and (c) the only possible collisions in HPP, (1,0,1,0) ⇔ (0,1,0,1).

that with the velocities as defined in Eq. (21.1), particles move exactly from one node to a neighboring node during the streaming phase. Assigning a position vector $\mathbf{r}$ to the state vector of the node at $\mathbf{r}$, and a time stamp t, the streaming step can be mathematically expressed as

$$n_i(\mathbf{r} + \mathbf{c}_i, t + 1) = n_i(\mathbf{r}, t) \qquad (21.2)$$

In words, the content of the element n_i of the state vector at position $\mathbf{r}$ at time t is copied in the next time step to the neighbor at position $\mathbf{r} + \mathbf{c}_i$. If $n_i(\mathbf{r}, t) = 1$, this means that a particle streams from $\mathbf{r}$ to $\mathbf{r} + \mathbf{c}_i$. If $n_i(\mathbf{r}, t) = 0$, it just means that at $t + 1$ $n_i(\mathbf{r} + \mathbf{c}_i, t + 1)$ is correctly set to 0.[1]

The collision is also very straightforward. In a collision, the velocities of the particles are redistributed. During a collision we must conserve mass, momentum, and energy. Since for all particles the magnitude of velocity and their mass is 1, their kinetic energy is always 1/2. Therefore, mass conservation immediately implies energy conservation.[2] Mass conservation means preservation of the number of particles in the collision. The momentum of a particle is just $\mathbf{c}_i$. So, during a collision we must preserve the total momentum on a node, i.e., $n_1 \mathbf{c}_1 + n_2 \mathbf{c}_2 + n_3 \mathbf{c}_3 + n_4 \mathbf{c}_4$. By systematically going through all possible 16 states, it turns out that only *one* type of collision is possible (see Figure 21.6[c]). These are the "head-on" collisions, where two particles arriving from north-south (or east-west) are scattered over 90° and after the collision propagate to east-west (or north-south), or in terms of the state vector, (1,0,1,0) ⇔ (0,1,0,1).

The collision can also be described more formally. We define a collision operator $\Delta_i(\mathbf{n})$, which can take the values $\{-1, 0, 1\}$. If before the collision a particle with $\mathbf{c}_i$ is present, and after the collision this particle is scattered into another direction, we must have $\Delta_i(\mathbf{n}) = -1$ (i.e., the particle is removed from velocity channel i). In the reverse case, when as the result of a collision, an empty channel i is filled with a particle, we have $\Delta_i(\mathbf{n}) = 1$. Finally, if the collision does nothing to channel i, $\Delta_i(\mathbf{n}) = 0$. To find an expression for $\Delta_i(\mathbf{n})$, we must therefore have a trick to select certain states that undergo a collision (the two head-on cases) and then assign the correct value to $\Delta_i(\mathbf{n})$. Here we take advantage of the binary notation that we introduced using the symbols "0" and "1." In the collision operator, we assume these symbols are in fact the integer numbers 0 or 1 and we compute with them. So, to select say the state (1,0,1,0) we could formulate a logical expression as n_1 AND (NOT n_2) AND n_3 AND (NOT n_4), which returns TRUE if the state (1,0,1,0) is present at a node and FALSE in all other cases. We could also write $n_1(1 - n_2)n_3(1 - n_4)$ and fill in the number 1 and 0 depending on the state. This expression will evaluate to 1 for the state (1,0,1,0) and to 0 in all other cases. In the case of a precollision state (1,0,1,0), we know that the result of the collision must be (0,1,0,1), so for, e.g., channel 1 we must have $\Delta_1(\mathbf{n}) = -1$. This we can achieve by writing $\Delta_1(\mathbf{n}) = -n_1(1 - n_2)n_3(1 - n_4)$. However, this is not the complete expression as we must also accommodate the reverse situation, i.e., that the precollision state is (0,1,0,1), in which case a particle will appear in channel 1, and $\Delta_1(\mathbf{n}) = 1$. We achieve this by adding another term, i.e., $(1 - n_1)n_2(1 - n_3)n_4$ resulting in the full expression $\Delta_1(\mathbf{n}) = -n_1(1 - n_2)n_3(1 - n_4) + (1 - n_1)n_2(1 - n_3)n_4$, and likewise for

---

[1] Physicists would say that in this case a "hole" is streaming from $\mathbf{r}$ to $\mathbf{r} + \mathbf{c}_i$.
[2] For this reason, the HPP model (and other "homokinetic" models) have no thermal effects.

the other channels 2–4. By introducing the shorthand notation $\bar{n}_i = 1 - n_i$, we can finally write the HPP collision operator in the following compact form:

$$\Delta_i(\mathbf{n}) = -n_i \bar{n}_{i+1} n_{i+2} \bar{n}_{i+3} + \bar{n}_i n_{i+1} \bar{n}_{i+2} n_{i+3} \tag{21.3}$$

where the index i is taken modulo 4 (so if $i = 2, i + 3 = 1$). With an expression for the collision operator, we can now write the full equation for the dynamics of the HPP model:

$$n_i(\mathbf{r} + \mathbf{c}_i, t + 1) = n_i(\mathbf{r}, t) + \Delta_i(\mathbf{n}(\mathbf{r}, t)) \tag{21.4}$$

In fact, Eq. (21.4) expresses the micro dynamics for all LGCAs. However, for each specific model, the total number and definitions of the $\mathbf{c}_i$ may be different, and the details of the collision operator are different.

We can now formulate the following CA rule for LGCA.

```
for each node in the Lattice do
    1. Perform a collision step, i.e., redistribute the state
       vector n such that n_i := n_i + Δ_i (n)
    2. Perform a streaming step, i.e.,
       for all i
           copy n_i to n_i at position = my_position + c_i.
       update the time t := t + 1
```

Having defined the structure, collision operator, and the dynamics as expressed in the CA rule, we are now in the position to execute the HPP LGCA. Next, we must define observables. The total number of particles and total momentum on a node are obtained by summing n_i and $n_i \mathbf{c}_i$ over all i, respectively. However, these instantaneous observables are very noisy, they fluctuate strongly as a function of time and position. Although these fluctuations contain a wealth of interesting physical information (Rivet and Boon, 2001), we want to observe smooth hydrodynamic fields such as the density or momentum of the fluid. To achieve this we must first take ensemble averages of n_i, yielding $f_i = <n_i>$. The f_i values are real numbers between 0 and 1 and should be interpreted as the probability to find a particle with velocity $\mathbf{c}_i$. In an LGCA simulation, we can compute the ensemble average by, e.g., taking spatial or temporal averages of the $n_i(\mathbf{r},t)$. We can now define the fluid density ρ and fluid velocity $\mathbf{u}$ as follows:

$$\begin{aligned} \rho(\mathbf{r}, t) &= \sum_{i=1}^{b} f_i(\mathbf{r}, t) \\ \rho(\mathbf{r}, t) \mathbf{u}(\mathbf{r}, t) &= \sum_{i=1}^{b} \mathbf{c}_i f_i(\mathbf{r}, t) \end{aligned} \tag{21.5}$$

where b is the total number of velocity vectors (for HPP $b = 4$). With these definitions, and the full machinery of statistical mechanics and kinetic theory, one can work out the equations that govern ρ and $\mathbf{u}$. Although the resulting equations for HPP have a strong resemblance to the Navier–Stokes equations that govern macroscopic fluid flow, there is a major flaw. It turns out that the resulting macroscopic equations are not isotropic, meaning that the flow properties depend on the orientation with respect to the underlying lattice. This problem can be traced back to isotropy properties of the underlying HPP lattice and its four discrete velocities.[3] This anisotropy is of course unacceptable for a model of fluid dynamics, and therefore HPP did not catch the attention of people interested in doing fluid dynamics.

In 1986 the big breakthrough came for LGCA. Frish et al. (1986) introduced the first LGCA that produces isotropic macroscopic equations for the density and velocity, reproducing the Navier–Stokes equations for an incompressible fluid. The main innovation in this FHP model was to change the underlying lattice

---

[3]Technically, the HPP model has a crystallographic isotropy of order 3, which is too low to obtain isotropic macroscopic equations (for details on this issue, see Rivet and Boon [2001]).

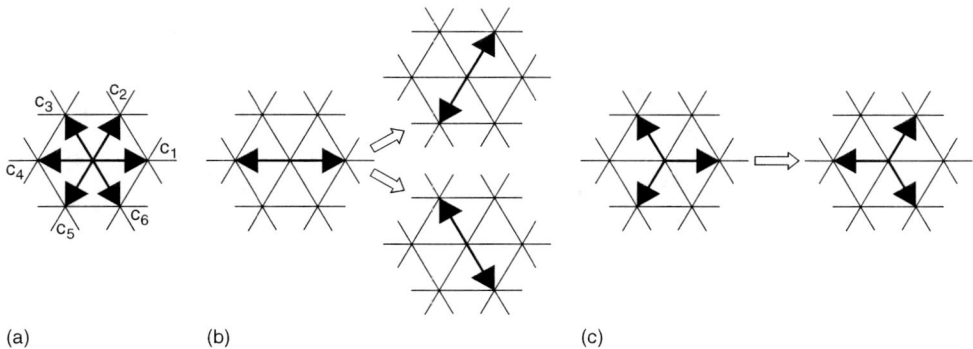

FIGURE 21.7 The FHP model: (a) shows the six velocities of the model, (b) the head on collision, and (c) the triplet collision.

to a 2D triangular lattice, and define six discrete velocities along the six directions of the lattice (see Figure (21.7[a]). This lattice has sufficient isotropy[4] to reproduce isotropic Navier–Stokes equations for an incompressible fluid. In this version of the FHP model (coined FHP-I),[5] the magnitude of the six discrete velocities are the same and equal to 1. Therefore, mass conservation and energy conservation are equivalent and FPH-I is also an a-thermal model. The micro dynamics are governed by Eq. (21.4), but we need to adapt the collision operator. In FHP-I, we only consider two types of collisions (see Figure 21.7[b] and Figure 21.7[c]). The first type is the head-on collision (as in HPP). However, in this case two possible postcollision states are possible (rotated $+60°$ or $-60°$ with respect to the incoming direction). FHP-I randomly selects one of those two postcollision states with a probability of 0.5. Because of this, FHP-I is no longer a deterministic CA, but has become probabilistic. The second type of collision is the triplet state, where three particles arrive with mutual angles of $120°$. The postcollision state is the same triplet rotated over $60°$. Using the same procedure as for HPP, we can express the collision operator for FHP-I as follows

$$\begin{aligned}\Delta_i^{FHP-I}(\mathbf{n}) = &- n_i \bar{n}_{i+1} \bar{n}_{i+2} n_{i+3} \bar{n}_{i+4} \bar{n}_{i+5} \\ &+ \xi \bar{n}_i n_{i+1} \bar{n}_{i+2} n_{i+3} \bar{n}_{i+4} \bar{n}_{i+5} \\ &+ (1-\xi) \bar{n}_i \bar{n}_{i+1} n_{i+2} \bar{n}_{i+3} \bar{n}_{i+4} n_{i+5} \\ &- n_i \bar{n}_{i+1} n_{i+2} \bar{n}_{i+3} n_{i+4} \bar{n}_{i+5} \\ &+ \bar{n}_i n_{i+1} \bar{n}_{i+2} n_{i+3} \bar{n}_{i+4} n_{i+5}\end{aligned} \quad (21.6)$$

where ξ is a Bernouilli random variable (i.e., it randomly takes the values 0 or 1) with mean 0.5. On each time step and at each lattice node ξ is evaluated. On the right-hand side, the first three terms represent the head-on collisions and the last two terms the triplet collision. The variable i is now taken modulo 6.

We can now proceed to execute the FPH-I LGCA and compute observables using Eq. (21.5), where $b=6$. In Figure 21.8, an example is shown, demonstrating the need to perform ensemble averaging before computing the observables. In Figure 21.8(a) we show the results of a single iteration of FHP-I, in fact we have assumed that $f_i = n_i$ (i.e., no ensemble averaging is performed). Clearly, the resulting flow field is very noisy. To observe smooth flow lines one should really compute $f_i = <n_i>$. Because the flow is static, we compute the ensemble average f_i by averaging the Boolean variables n_i over a large number of FHP iterations. The resulting flow velocities are shown in Figure 21.8(b).

---

[4]It has crystallographic isotropy of order 5, so sufficient for the required fourth-order isotropy needed for isotropic large-scale dynamics.

[5]A number of extensions to FHP exist, including the so-called rest particles (i.e., particles on a node with zero velocity, $c_0 = \{0,0\}$), and with extended collision operators, taking into account all possible collisions (including the rest particle). These extensions will not be further treated here, but see Rivet and Boon (2001).

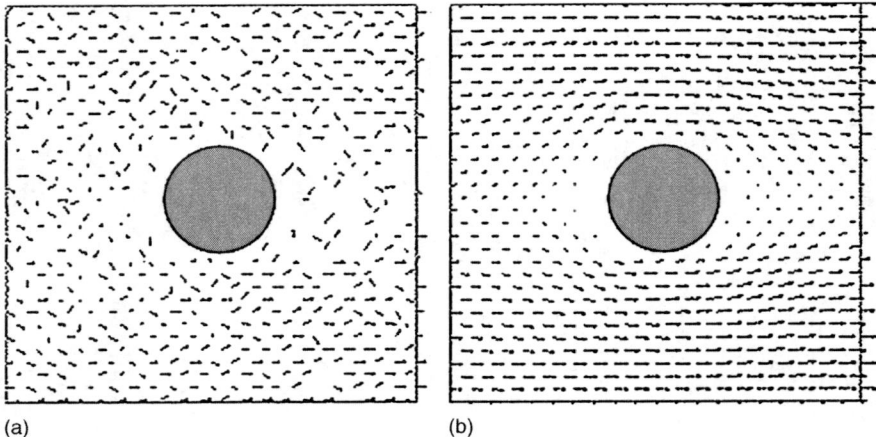

FIGURE 21.8 (a) FHP simulation of flow around a cylinder, the result of a single iteration of the LGCA is shown (i.e., no averaging). The arrows are the flow velocities, the length is proportional to the absolute velocity. The simulations were done on a 32 × 64 lattice, the cylinder has a diameter of 8 lattice spacing, only a 32 × 32 portion of the lattice is shown; periodic boundary conditions in all directions apply. (b) As in (a), the velocities are shown after averaging over 1000 LGCA iterations.

The year 1986 marked the beginning of an enormous research activity on LGCAs. Chapter 11 (guide for further reading) from Rivet and Boon (2001) provides numerous references to this literature. Here we will only touch upon a few interesting developments of practical interest. First, we mention the extension to three-dimensional (3D) models, clearly a necessity for a serious model of hydrodynamics. It turns out that a 3D LGCA is far from trivial. In fact, the LGCA community rapidly realized that all 3D regular (Bravais) lattices do not have sufficient isotropy ("the magic didn't work," as Rivet and Boon put it). However, by taking a detour into four-dimensional (4D) space and projecting back to three dimensions, d'Humières, Lallemand, and Frisch were able to build a 3D LGCA with all required isotropy properties (d'Humières et al., 1986). This model is based on a 4D face-centered hyper cube (FCHC), and has as many as 24 velocity channels per node. The sheer amount of possible states in this model ($2^{24} = 16,777,261$) and the number of possible collisions (18,736 or 10,805, depending on assumption put on the model) make it completely impossible to write down an explicit equation for the collision operator, and one must resort to a more general approach. Moreover, an efficient implementation of such a complicated collision operator requires new algorithmic strategies (see, e.g., Hénon, 1987), clever lookup table strategies (see, e.g., Rivet and Boon, 2001), or a combination of both.

An example of a thermal LGCA (i.e., a model where energy conservation is nontrivial) is the 2D model proposed by (Grosfils et al. 1992). This GBL model, like FHP, is defined on the 2D triangular lattice, but now has 19 velocities. It has one rest particle, six particles connected to nearest neighbors ($\|c_i\| = 1$), six connected to next-nearest neighbors ($\|c_i\| = \sqrt{3}$), and six connected to next-next-nearest neighbors ($\|c_i\| = 2$). The GBL model has $2^{19} = 524,288$ possible states, of which 517,750 can undergo collisions that change the state while preserving mass, momentum, and energy. Again, the complexity of the collision operator requires efficient implementations (see, e.g., Dubbeldam et al., 1999). GBL is a true thermo-hydrodynamic model for 2D fluid dynamics.

LGCAs are easily extended to multiple species models by coloring the particles. Besides the normal collision rules, one also demands a color conservation. After collisions, colors are then randomly redistributed, while preserving total color. In this way nonreacting mixed fluids can be modeled. Moreover, by adding reactions one can create reaction–diffusion LGCAs (see, e.g., Boon et al., 1996) or by adding an interaction term between differently colored particles, one can model multiphase immiscible fluids (see, e.g., Rothman and Zaleski, 1997).

21.5.2 LGCA and Fluid Dynamics

The Navier–Stokes equations for an incompressible fluid read

$$\nabla \cdot \mathbf{u} = 0 \tag{21.7}$$

$$\frac{\partial \mathbf{u}}{\partial t} + \mathbf{u} \cdot \nabla \mathbf{u} = -\nabla P + \nu \nabla^2 \mathbf{u} \tag{21.8}$$

where P is the pressure and ν the viscosity of the fluid. Eq. (21.7) expresses mass conservation, and Eq. (21.8) momentum conservation. LGCAs with sufficient isotropy (e.g., FHP, GBL, and FCHC) can reproduce these Navier–Stokes equations under the assumption that the velocities $\mathbf{u}$ are small. This can be demonstrated by explicit simulations and by theory. In this section, we just outline such theory, for all details we refer to Frisch et al. (1987), Rivet and Boon (2001), Rothman and Zaleski (1997), Chopard and Droz (1998), and Wolf-Gladrow (2000).

The starting point is the LGCA micro dynamics (see Eq. [21.4]). The mass and momentum conservation of the collision operator can be expressed as

$$\sum_{i=1}^{b} \Delta_i(\mathbf{n}(\mathbf{r},t)) = 0 \tag{21.9}$$

$$\sum_{i=1}^{b} \mathbf{c}_i \Delta_i(\mathbf{n}(\mathbf{r},t)) = 0 \tag{21.10}$$

One can ask if the evolution equation (21.4) is also valid for the averaged particle densities f_i. It turns out that this is true, but only under the Boltzmann molecular chaos assumption, which states that particles that collide are not correlated before and after collisions, or, that for any number of particles k, $\langle n_1 n_2 \ldots n_k \rangle = \langle n_1 \rangle \langle n_2 \rangle \cdots \langle n_k \rangle$. In this case, one can show that $\langle \Delta_i(\mathbf{n}) \rangle = \Delta_i(\mathbf{f})$, where $\mathbf{f}$ is the vector containing all f_i. By averaging Eq. (21.4) and applying the molecular chaos assumption we find

$$f_i(\mathbf{x} + \mathbf{c}_i, t+1) - f_i(\mathbf{x},t) = \Delta_i(\mathbf{f}(\mathbf{x},t)) \tag{21.11}$$

A first-order Taylor expansion of $f_i(\mathbf{x}+\mathbf{c}_i, t+1)$, substituted into Eq. (21.11), results in

$$\partial_t f_i(\mathbf{x},t) + \partial_\alpha c_{i\alpha} f_i(\mathbf{x},t) = \Delta_i(\mathbf{N}(\mathbf{x},t)) \tag{21.12}$$

Note that the shorthand ∂_t means $\partial/\partial t$; the subscript α denotes the α-component of a D-dimensional vector, where D is the dimension of the LGCA lattice; and we assume the Einstein summation convention over repeated Greek indices (e.g., in two dimensions $\partial_\alpha c_{i\alpha} f_i(\mathbf{x},t) = \partial_x c_{ix} f_i(\mathbf{x},t) + \partial_y c_{iy} f_i(\mathbf{x},t)$). Next, we sum Eq. (21.12) over the index i and apply Eq. (21.5), Eq. (21.9), and Eq. (21.10), thus arriving at $\partial_t \rho + \partial_\alpha \rho \mathbf{u} = 0$, or

$$\frac{\partial \rho}{\partial t} + \nabla \cdot \rho \mathbf{u} = 0 \tag{21.13}$$

which is just the equation of continuity that expresses conservation of mass in a compressible fluid. One can also first multiply Eq. (21.12) with $\mathbf{c}_i$ and then summate over the index i. In this case, we arrive at

$$\partial_t \rho \mathbf{u} + \partial_\beta \Pi_{\alpha\beta} = 0 \tag{21.14}$$

with

$$\Pi_{\alpha\beta} = \sum c_{i\alpha} c_{i\beta} f_i \tag{21.15}$$

The quantity $\Pi_{\alpha\beta}$ must be interpreted as the flow of the α-component of the momentum into the β-direction; $\Pi_{\alpha\beta}$ is the momentum density flux tensor. To proceed, one must be able to find expressions for the particle densities f_i. This is a highly technical matter that is described in detail in, e.g., Frisch et al. (1987) or Rivet and Boon (2001). The bottom line is that one first calculates the particle densities for an LGCA in equilibrium, f_i^0, and then substitute them into Eq. (21.15). This results in an equation that is almost similar to the Euler equation, i.e., the expression of conservation of momentum for an inviscid fluid. Next, one proceeds by taking into account small deviations from equilibrium, resulting in viscous effects. Then, after a lengthy derivation one is able to derive the particle densities, substitute everything into Eq. (21.15) and derive the full expression for the momentum conservation of the LGCA, which very closely resembles the Navier–Stokes equations for an incompressible fluid. The viscosity and sound speed of the LGCA are determined by its exact nature (i.e., the lattice, the discrete velocities, and the exact definition of the collision operator).

We must stress that the derivations in this section are very loose, in the sense that we ignored many important details. For instance, the Taylor expansion, which resulted in Eq. (21.12), was only accurate up to first order. In fact one can show that an LGCA obeys the Navier–Stokes equations up to second order. Also, we introduced very loosely the concept of equilibrium distributions, and small deviations from equilibrium that give rise to viscous effects. By using a very powerful technique, known as the Chapmann–Enskog expansion, one is able to solve Eq. (21.11) and derive expressions for mass and momentum conservation of an LGCA, which turn out to be almost equal to the equations for a real, incompressible fluid.

To be complete, we note that the derivation of the Navier–Stokes equations for the LGCA is correct in the limit of small Mach and small Knudsen numbers. The first restriction means that the flow velocities must be much smaller than the sound speed of the LGCA, and the second limit demands that the mean free path of the particles must be much smaller than some macroscopic dimensions of the LGCA, i.e., the particle density cannot be too small.

Finally, we must mention one last technical detail. As stated above, the momentum conservation equations of the LGCA are almost equal to the Navier–Stokes equations of a real fluid. The difference lies in a factor $g(\rho)$ in the advection term (the $\mathbf{u} \cdot \nabla \mathbf{u}$ term in Eq. [21.8]), which leads to the breakdown of Galilean invariance. In the low velocity limit, however, this is not a real problem, because the fluid becomes incompressible and $g(\rho)$ a constant. A rescaling of the velocity and time with $g(\rho)$ allows to fully recover the exact Navier–Stokes equations for an incompressible fluid.

21.5.3 Simulating an LGCA

Although LGCA simulations can benefit from generic CA environments, there are a few typical aspects of LGCA that will be discussed here. First, most 2D LGCAs are defined on triangular lattices. Such lattice should be represented by some 2D array. To denote each node in the triangular lattice with coordinates (x,y) by integer numbers, we multiply the coordinates by a factor $(2, 2/\sqrt{3})$. To avoid awkward diamond-shaped grids representation, the streaming step is different for even and odd parity of the lattice (see Figure 21.9). This same mapping can be used for the 19 velocity GBL model.

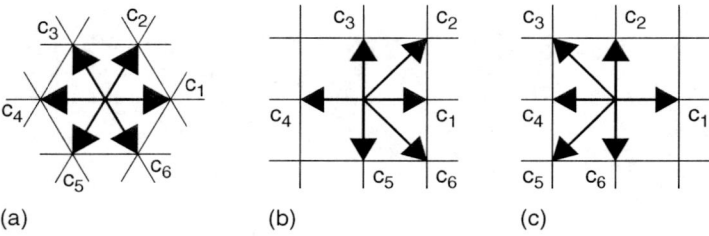

FIGURE 21.9 The mapping of the original triangular lattice (a) to a square domain. The mapping and streaming is based on the parity of the lattice. On even lines we propagate using (b) and, for odd lines we propagate using (c).

The state vector per node of an LGCA is a b-vector containing bits. So, the total LGCA lattice can be stored with $N \times b$ bits with N being the total number of nodes in the lattice. The streaming step can be accomplished by using two different lattice grids. The second lattice is used to calculate the new state at the next time step from the first lattice. After updating the lattice, we swap the pointers to the lattices. If memory size is a problem (e.g., large 3D problems), then it would be possible to do the streaming step in-place, at the cost of accessing each lattice point several times instead of just once.

The collision can be handled in two ways. If the collision is not very complicated (as with HPP or FHP-I), then it can be implemented as logical operations on the state bits. For instance, to implement the HPP collision operator (Eq. [21.3]), one could first define a selection operator S_i that returns TRUE if a head-on collision state is present. For instance, for channel 1 the selection operator S_1 would be

$$S_1 = (n_1 \text{ AND } (\text{NOT } n_2) \text{ AND } n_3 \text{ AND } (\text{NOT } n_4)) \text{ OR } ((\text{NOT } n_1) \text{ AND } n_2 \text{ AND } (\text{NOT } n_3) \text{ AND } n_4).$$

If S_1 returns 1, a collision occurs, and the state n_1 must flip (from 1 to 0 or vice versa). If S_1 returns 0, the state n_1 must stay the same. The postcollision state $(n_1 + \Delta_1(\mathbf{n}))$ can be obtained by applying the exclusive OR operation on n_1 with S_1, i.e., $n_1 + \Delta_1(\mathbf{n}) = n_1 \text{ XOR } S_1$.

This postcollision value is then streamed using the procedures as sketched above.

Another approach is to use lookup tables. If the collision becomes too complex to explicitly write down the Boolean expression (as with FCHC or GBL) this is the only possibility. However, one may also want to resort to lookup tables for HPP or FHP as this may be faster. We give an example of the use of lookup tables in the implementation of the GBL model (for details see Dubbeldam et al., 1999). More discussions on the lookup tables can be found in Rivet and Boon (2001). The first step is to group all 2^{19} states of GBL in equivalence classes with the same total mass, momentum, and energy. The collision then amounts to randomly selecting a state from the equivalence class to which the input state belongs. It is clear that the input state is also among them (meaning no collision when selected as output state), but since most equivalence classes are quite large this has little influence. For the 19-bits GBL model we create a collision table of 2^{19} indexes, followed by the equivalence classes (see Figure 21.10). Every index of an element in a class points to the start of the class (for instance, in Figure 21.5, 138, 273, and 41,024 all point to $2^{19} + X$). The left 12 bits are used to indicate the number of collision outcomes. If the number is zero, then the outcome is equal to the input state. Otherwise, the value of the right 20 bits is an index pointing to the start point of an array of possible postcollision states, of which we choose one at random (using the information on the size of the equivalence class).

To have solid boundaries or solid objects in the flow, one must be able to set boundary conditions. In LGCA this is almost trivial using the bounce-back rule, where particles that hit the boundary are reversed and sent back into the direction they came from. To start an LGCA simulation the Boolean field must be initialized. Usually one knows some initial values of the macroscopic fields (the density and the fluid velocity). Based on these values, the initial Boolean field must be computed. This is done by using the equilibrium distribution f_i^0 that is explicitly known as a function of ρ and $\mathbf{u}$ in the limit of small $\mathbf{u}$ (see Rivet and Boon, 2001, Chapter 4). This equilibrium distribution gives the average of the Boolean field in equilibrium (i.e., the probability that a particle is present in channel i), so, the actual field is computed from the equilibrium distribution using a random number generator that delivers random numbers between 0.0 and 1.0. If the random number is smaller than f_i^0, then n_i is initialized to 1, otherwise it is set to 0.

With the initialization and the boundary conditions in place, the LGCA simulation can be started. If the flow is driven by some pressure gradient, one can apply a body force that, after each collision, effectively adds some momentum to the nodes. This was done in the simulations presented in Figure 21.8. Finally, to extract the wanted macroscopic fields, an ensemble average must be computed. This is typically done using time- or space-averaging, or a combination of both. In Figure 21.8(b) time averaging was applied.

LGCA simulations have been executed on every type of computers, from desktop PCs to massively parallel supercomputers, and even dedicated CA and LGCA machines (including programmable FPGA hardware). On current state-of-the-art computers, one can easily simulate quite large 2D and 3D LGCAs (see Dubbeldam et al., 1999) for an example of running the GBL model on a parallel computer).

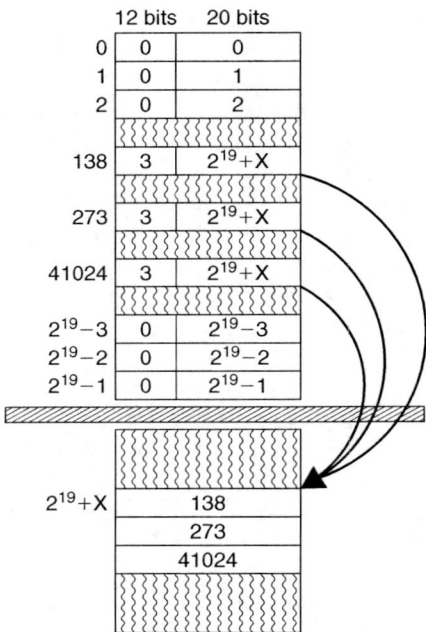

FIGURE 21.10 Collision table. The first 2^{19} indices are divided into 12 and 20 bits. The left 12 bits denote the number of collision outcomes, the rightmost 20 bits denote an index number from where the collision outcomes are stored in the table. These collision outcomes start from index 2^{19}. The figure shows an example for an equivalence class of size 3.

21.5.4 Some Applications

Lattice gases have been and are still being applied[6] to challenging problems of multiphase fluids in complex (e.g., porous) flow domains. These typically include fluid mixtures and colloids, reaction diffusions systems, immiscible fluids and phase separation, multiphase flows, fluids with free interfaces, magneto hydrodynamics, nonideal fluids, etc. Because of the ease to define boundary conditions using the bounce back rule, all such complex fluids have been studied in porous media. For details we refer to the LGCA books and references therein (Rivet and Boon, 2001; Rothman and Zaleski, 1997; Chopard and Droz, 1998).

21.5.5 Lattice Boltzmann Method

Immediately after the discovery of LGCA as a model for hydrodynamics in 1986, it was criticized on three points. LGCA have noisy dynamics, lack Galilean invariance, and have an exponential complexity of the collision operator. The noisy dynamics is clearly illustrated in Figure 21.8 and the lack of Galilean invariance was discussed above. Adding more velocities in an LGCA leads to increasingly more complex collision operators, exponentially in the number of particles (remember the numbers for GBL and FCHC). Therefore, another model, the lattice Boltzmann method (LBM), was introduced. This method is reviewed in detail in Succi (2001).

The basic idea is that one should not model the individual particles n_i, but immediately the particle densities f_i, i.e., one iterates the lattice Boltzmann equation (Eq. [21.11]). This means that particle densities are streamed from cell to cell, and particle densities collide, immediately solving the problem of noisy

---

[6]Although it must be said that the Lattice Boltzmann method, which will be mentioned in the next section, currently is the preferred method over LGCAs, although in special situations LGCAs are still to be preferred.

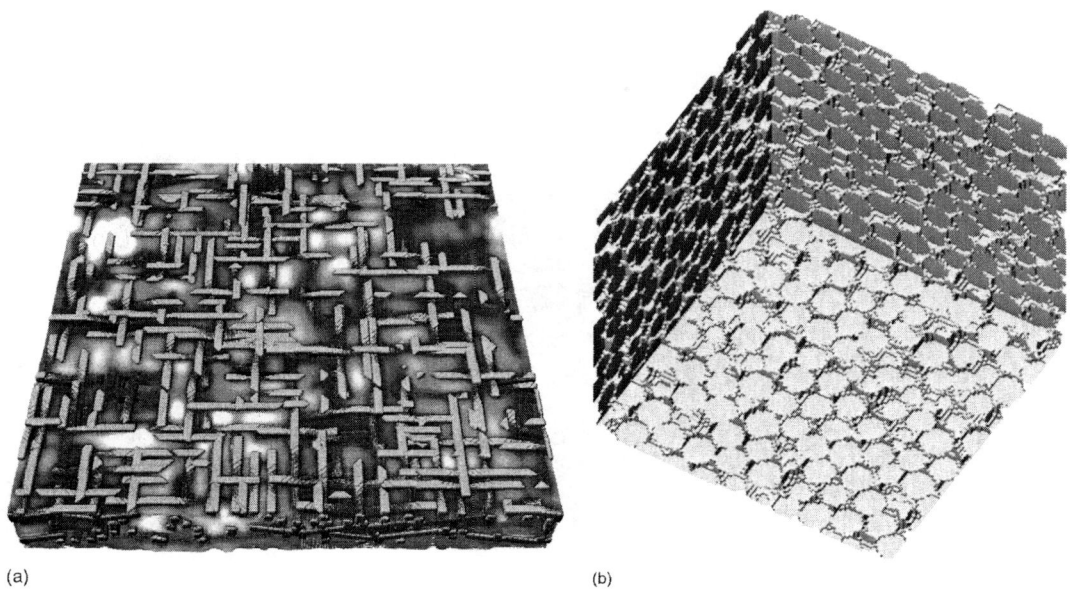

FIGURE 21.11 (a) Flow in a porous fiber mat; and (b) a porous medium composed of a dense random packing of spheres.

dynamics. However, in a strict sense, we no longer have a CA with a Boolean state vector (in fact, the state is now a vector of real numbers, so the state space per node is infinite). However, we can view LBM as a generalized CA. By a clever choice of the equilibrium distributions f_i^0, the model becomes isotropic and Galilean invariant, thus solving the second problem of LGCA. Finally, a very simple collision operator can be introduced. This so-called BGK collision operator models the collisions as a single-time relaxation τ, toward equilibrium, i.e.,

$$\Delta_i^{BGK}(\mathbf{N}) = -\frac{1}{\tau}\left(f_i - f_i^0\right) \tag{21.16}$$

Eq (21.11) and Eq. (21.16) together with a definition of the equilibrium distributions result in the Lattice-BGK (L-BGK) model. The L-BGK model leads to correct hydrodynamic behavior in two and three dimensions. The L-BGK not only applies to the triangular lattice, but also correctly works for other lattices, e.g., 2- or 3D cubical lattices with nearest and next-nearest neighbor interactions. The LBM and especially the L-BGK have found widespread use in simulations of highly complex fluid dynamical problems, including turbulence, multiphase flows, spinal decomposition, etc. To get a flavor we refer to Succi (2001). Below we present a few examples of successful L-BGK applications. All these L-BGK simulations were routinely executed on parallel computers, using a highly efficient parallel implementation of the L-BGK method (Kandhai et al., 1998).

The L-BGK method has played a significant role in studying flow in porous media. In Figure 21.11(a), we show an example of flow in a porous medium that consists of randomly placed fibers. This fiber mat is a model system for paper. Through L-BGK simulations, one can compute the permeability of such fiber mats, as a function of the porosity (Kopenen, et al., 1998). The results showed a remarkable agreement with experimental results. Since the permeability could be studied over a very large range of volume fractions, much larger than accessible through experiments, the authors were able to check the quality of theoretical approximations for this particular problem.

Another example of successful L-BGK simulations of flow in porous media was the transient dispersion in homogeneous porous media. Figure 21.11(b) shows an example of a medium of a random distribution

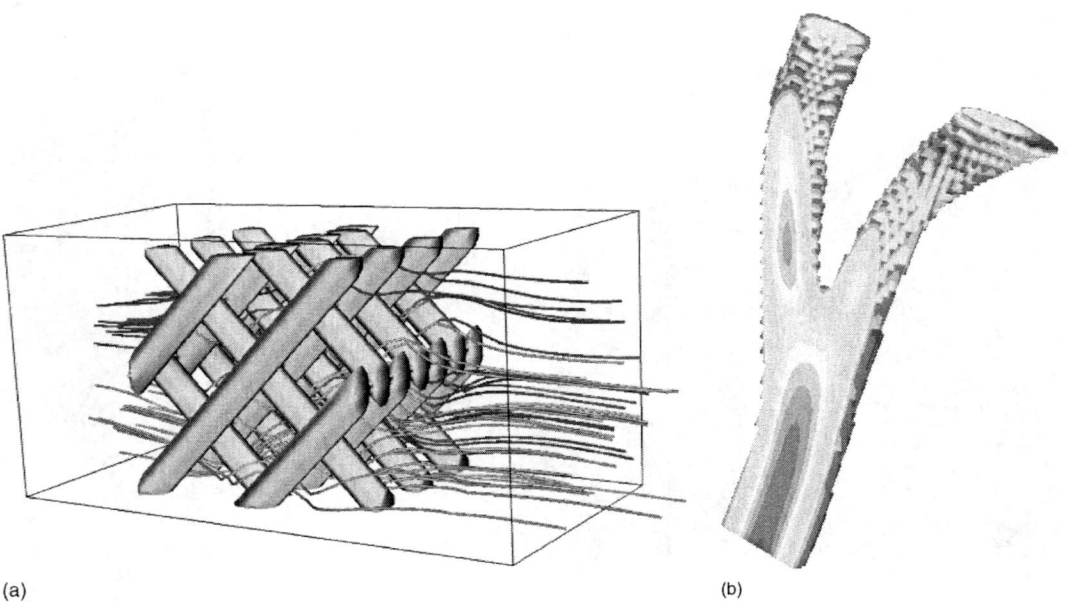

FIGURE 21.12 (a) Flow in a static mixer reactor; and (b) flow in the lower abdominal aortic bifurcation at peak systole.

of densely packed spheres. Here the flow around the spheres is computed, as well as advection-diffusion of tracer particles in the flow, and diffusion of the tracer particles through a microporous structure of the spheres. From these simulations, the authors could derive transient tracer dispersion curves, which showed a very good agreement with propagators that were measured using nuclear magnetic resonance (Kandhai et al., 2002a, 2002b).

L-BGK simulations have been compared extensively with traditional computational fluid dynamics (CFD) codes. As an example, we mention flow through a static mixer reactor (see Figure 21.12[a]). The flow fields as computed with L-BGK were in very good agreement with results stemming from a commercial CFD package (FLUENT). The computed fluid flux through the system, as a function of the applied pressure gradient, was in very good agreement with experimental data (Kandhai et al., 1999). It turned out that preparing the computational grid for L-BGK is trivial (just a Cartesian grid), whereas creating a body-fitted finite element grid for this geometry is very tedious.

As a final example we mention the application of L-BGK for simulation of blood flow. In Figure 21.12(b), we show the results of L-BGK simulations of time periodic systolic flow through the bifurcation of the lower abdominal aorta. The geometry is taken from MRI images of real patients. The results are in good agreement with previously published studies of systolic flow in region of the arterial tree (Artoli et al., 2005).

References

Artoli, A.M.M., A.G. Hoekstra, and P.M.A. Sloot. 2005. Mesoscopic simulations of systolic flow in the human abdominal aorta. *Journal of Biomechanics* 39: 873–884.
Automata. 2005. *11th Workshop on Cellular Automata*, Gdansk, 3–5 September 2005 http://iftia9.univ.gda.pl/~CA2005/
Bandini, S. 2002. Special issue on cellular automata. *Future Generation Computing Systems* 18: 871–1004.
Berec, L. 2002. Techniques of spatially explicit individual-based models: Construction, simulation and mean-field analysis. *Ecological Modelling* 150: 55–81.

Boghosian, B.M. 1999. Lattice gas and cellular automata. *Future Generation Computing Systems* 16: 171–185.
Boon, J.P., D. Dab, R. Kapral, and A. Lawniczak. 1996. Lattice gas automata for reactive systems. *Physics Reports* 273: 556–647.
Burks, A.W. 1970. *Essays on Cellular Automata*. Urbana, IL: University of Illinois Press.
Chopard, B. and M. Droz. 1998. *Cellular Automata Modeling of Physical Systems*. Cambridge: Cambridge University Press.
Deutsch, A. and S. Dormann. 2004. *Cellular Automaton Modeling of Biological Pattern Formation*. Basel: Birkhauser.
d'Humières, D., Lallemand, P., and Frisch, U. 1986. Lattice gas models for 3-D hydrodynamics. *Europhysics Letters* 2: 291–297.
Dubbeldam, D., A.G. Hoekstra, and P.M.A. Sloot. 1999. Computational aspects of multi-species lattice-gas automata. In P.M.A. Sloot, M.T. Bubak, A.G. Hoekstra, and L.O. Hertzberger (Eds.), *High-Performance Computing and Networking*, Amsterdam, The Netherlands, Lecture Notes in Computer Science, 1593: 339–349. Berlin: Springer-Verlag.
Farmer, D., T. Toffoli, and S. Wolfram. 1984. *Cellular Automata: Proceedings of an Interdisciplinary Workshop*. Amsterdam: North-Holland.
Forrest, S. 1990. Emergent computation: Self-organizing, collective, and cooperative phenomena in natural and artificial computing networks. *Physica D* 42: 1–11.
Frisch, U., D. d'Humières, B. Hasslacher, P. Lallemand, P. Pomeau, and J.-P. Rivet. 1987. Lattice gas hydrodynamics in two and three dimensions. *Complex Systems* 1: 649–707.
Frisch, U., B. Hasslacher, and Y. Pomeau. 1986. Lattice-gas automata for the Navier–Stokes Equation. *Physical Review Letters* 56: 1505–1508.
Ganguly, N., B.K. Sikdar, A. Deutsch, G. Canright, and P.P. Chaudhuri. 2003. A survey on cellular automata. Technical Report, Centre for High Performance Computing, Dresden University of Technology. http://www.cs.unibo.it/bison/publications/CAsurvey.pdf
Grosfils, P., J.P. Boon, and P. Lallemand. 1992. Spontaneous fluctuation correlations in thermal lattice gas automata. *Physical Review Letters* 68: 1077–1080.
Gutowitz, H.A. 1990. *Cellular Automata*. Cambridge, MA: MIT Press.
Hardy, P., O. de Pazzis, and Y. Pomeau. 1976. Molecular dynamics of a classical lattice gas: Transport properties and time correlation functions. *Physical Review A* 13: 1949–1961.
Hardy, P. and Y. Pomeau. 1972. Thermodynamics and hydrodynamics for a modeled fluid. *Journal of Mathematical Physics* 13: 1042–1051.
Hardy, P., Y. Pomeau, and O. de Pazzis. 1973. Time evolution of a two-dimensional model system. I. Invariant states and time correlation functions. *Journal of Mathematical Physics* 14: 1746–1759.
Hénon, M. 1987. Isometric collision rules for the 4-D FCHC lattice gas. *Complex Systems* 1: 475–494.
Ilachinski, A. 2001. *Cellular Automata, a Discrete Universe*. Singapore: World Scientific.
Jesshope, C., V. Jossifov, and W. Wilhelmi. 1994. *International Workshop on Parallel Processing by Cellular Automata and Arrays* (Parcella'94). Berlin: Akademie Verlag.
Kaandorp, J.A., C.P. Lowe, D. Frenkel, and P.M.A. Sloot. 1996. The effect of nutrient diffusion and flow on coral morphology. *Physical Review Letters* 77: 2328–2331.
Kadanoff, L.P. and J. Swift. 1968. Transport coefficients near the critical point: A master-equation approach. *Physical Review* 165: 310–322.
Kandhai, B.D., D. Hlushkou, A.G. Hoekstra, P.M.A. Sloot, H. van As, and U. Tallarek. 2002a. Influence of stagnant zones on transient and asymptotic dispersion in macroscopically homogeneous porous media. *Physical Review Letters* 88: 234501.
Kandhai, B.D., A. Koponen, A.G. Hoekstra, M. Kataja, J. Timonen, and P.M.A. Sloot. 1998. Lattice Boltzmann hydrodynamics on parallel systems. Computer Physics Communications 111: 14–26.
Kandhai, B.D., U. Tallarek, D. Hlushkou, A.G. Hoekstra, P.M.A. Sloot, and H. van As. 2002b. Numerical simulation and measurement of liquid hold-up in biporous media containing discrete stagnant zones. *Philosophical Transactions: Mathematical, Physical & Engineering Sciences* 360: 521–534.

Kandhai, B.D., D.J.-E. Vidal, A.G. Hoekstra, H.C.J. Hoefsloot, P. Iedema, and P.M.A. Sloot. 1999. Lattice-Boltzmann and finite element simulations of fluid flow in a SMRX mixer. *International Journal for Numerical Methods Fluids* 31: 1019–1033.

Koponen, A., D.B. Kandhai, E. Héllen, M. Alava, A.G. Hoekstra, M. Kataja, K. Niskanen, P.M.A. Sloot, and J. Timonen. 1998. Permeability of three-dimensional random fibre webs. *Physical Review Letters* 80: 716–719.

Kurzweil, R. 2002. Reflections on Stephen Wolfram's 'A New Kind of Science'. http://www.kurzweilai.net/meme/frame.html?main=/articles/art0464.html

May, R.M. 1976. Simple mathematical models with very complicated dynamics. *Nature* 261: 459–467.

Mitchell, M. 1998. Computation in cellular automata: A selected review. In T. Gramss, S. Bornholdt, M. Gross, M. Mitchell, and T. Pellizzari (Eds.), *Nonstandard Computation*, pp. 95–140. Weinheim: VCH Verlagsgesellschaft.

Naumov, L. 2004. CAMEL—Cellular Automata Modeling Environment & Library. Cellular Automata. 6th International Conference on Cellular Automata for Research and Industry. *Lecture Notes in Computer Science*, 3305: 735–744. Heidelberg: Springer.

NKS Conference. 2005. http://www.cs.indiana.edu/~dgerman/2005midwestNKSconference/index.html

Rivet, J.-P. and J.P. Boon. 2001. *Lattice Gas Hydrodynamics*. Cambridge: Cambridge University Press.

Rothman, D.H. and S. Zaleski. 1997. *Lattice-Gas Cellular Automata, Simple Models of Complex Hydrodynamics*. Cambridge: Cambridge University Press.

Sloot, P.M.A. 1999. High performance simulation with cellular automata. In N. Piskunov (Ed.), Proceedings of HiPer'99, pp. 169–207. Computer Centre University of Tromso, Tromso, Sweden.

Sloot, P.M.A., F. Chen, and C.A. Boucher. 2002. Cellular automata model of drug therapy for HIV infection. *Lecture Notes in Computer Science* 2493: 282–293.

Sloot, P.M.A., B. Chopard, and A.G. Hoekstra. 2004. Cellular Automata: 6th International Conference on Cellular Automata for Research and Industry, ACRI 2004, Amsterdam, The Netherlands, October 2004. *Lecture Notes in Computer Science*, 3305. Heidelberg: Springer.

Sloot, P.M.A. and A.G. Hoekstra. 2001. Cellular automata as a mesoscopic approach to model and simulate complex systems. *Lecture Notes in Computer Science*, 2074 (II): 518–527. Heidelberg: Springer.

Sloot, P.M.A., J.A. Kaandorp, A.G. Hoekstra, and B.J. Overeinder. 2001a. Distributed cellular automata: Large scale simulation of natural Phenomena. In A.Y. Zomaya, F. Ercal, and S. Olariu (Eds.), *Solutions to Parallel and Distributed Computing Problems: Lessons from Biological Sciences*, pp. 1–46. Wiley: New York.

Sloot, P.M.A., B.J. Overeinder, and A. Schoneveld. 2001b. Self-organized criticality in simulated correlated systems. *Computer Physics Communications* 142: 76–81.

Sloot, P.M.A., A. Schoneveld, J.F. de Ronde, and J.A. Kaandorp. 1997. Large scale simulations of complex systems Part I: Conceptual framework. SFI Working Paper: 97-07-070, *Santa Fe Institute for Complex Studies*.

Succi, S. 2001. *The Lattice Boltzmann Equation, for Fluid Dynamics and Beyond*. Oxford: Clarendon Press.

Turing, A.M. 1936. On computable numbers, with an application to the Entscheidungsproblem. *Proceedings of the London Mathematical Society*, Ser. 2, Vol. 42; and extract On computable numbers, with an application to the entscheidungsproblem. A correction. *Proceedings of the London Mathematical Society*, Ser. 2, Vol. 43, 1937.

Ulam, S. 1952. Random processes and transformations. *Proceedings of the International Congress on Mathematics* 2: 264–275.

Ulam, S. 1962. On some mathematical problems connected with patterns of growth of figures. *Proceedings of Symposia in Applied Mathematics*, Vol. 14, pp. 215–224. Providence: American Mathematical Society.

von Neumann, J. 1951. The general and logical theory of automata. In L.A. Jeffress (Ed.), *Cerebral Mechanisms and Behavior: The Hixon Symposium*, pp. 1–41. Wiley: New York.

von Neumann, J. 1966. *Theory of Self-Reproducing Automata*. Urbana, IL: University of Illinois Press.

Wiener, N. and Rosenbleuth, A. 1946. The mathematical formulation of the problem of conduction of impulses in a network of connected excitable elements, specifically in cardiac muscle. *Archivos del Instituto de Cardiologica de Mexico* 16: 202–265.

Wolf-Gladrow, D.A. 2000. *Lattice-Gas Cellular Automata and Lattice Boltzmann Models.* Heidelberg: Springer

Wolfram, S. 1986a. *Theory and Applications of Cellular Automata.* Singapore: World Scientific.

Wolfram, S. 1986b. *Journal of Statistical Physics* 45: 471–526.

Wolfram, S. 2002. *A New Kind of Science.* Champaign, IL: Wolfram Media.

Zuse, K. 1982. The computing universe. *International Journal of Theoretical Physics* 21: 589–600.

Zuse, K. 1967. Rechnender Raum. *Elektronische Datenverarbeitung* 8: 336–344. See also: http://www.idsia.ch/~juergen/digitalphysics.html

22
Spatio-Temporal Connectionist Networks

Stefan C. Kremer
University of Guelph

- 22.1 Introduction 22-1
- 22.2 Connectionist Networks (CNs) 22-2
  Basic Approach • Function Approximation • Learning
- 22.3 Spatio-Temporal Connectionist Networks 22-4
  Basic Approach • Specific Architectures
- 22.4 Representational Power 22-6
- 22.5 Learning 22-6
- 22.6 Applications 22-8
- 22.7 Conclusion 22-9

22.1 Introduction

In this chapter, we look at a popular class of dynamical systems that are based on simplified models of how the brain processes information. These systems fall under the heading of artificial neural networks called spatio-temporal connectionist networks (SCNs). SCNs are particularly interesting because they are capable of representing all dynamical systems and all models of computation. This means that they are as powerful (in terms of what they can do) as all other dynamical and computational systems. In addition, these systems also typically embody adaptive mechanisms. That means they are capable of changing over time to suit a particular task. In particular, the systems incorporate algorithms that allow them to be *trained* to match a dataset of example behavior. Furthermore, they are able to both interpolate and extrapolate. This means that they can perform well, not only on the data that was used to train the system, but also on other data that has never been seen. In short, regularization can serve to address data points beyond those available during the adaptation process and SCNs are thus able to generalize to novel situations.

The chapter is organized as follows. In Section 22.2, we begin with a brief exposition of connectionist (also called *neural*) networks, the basic model underlying all SCNs. We also examine the role of SCNs within this paradigm. We follow this, in Section 22.3, with a generalized and comprehensive SCN model to formalize the definition of SCNs and also present some specific examples popularized in the literature. Then, in Section 22.4, we examine the representational power of this class of dynamical systems. Next, in Section 22.5, we address the issue of adapting SCNs including a discussion of the challenges involved in this type of inference. In Section 22.6, we present a brief survey of some applications to which SCNs have been applied. Finally, in Section 22.7 we give some concluding remarks and avenues for future research. Throughout the chapter, we present references to the source literature so that the reader can follow up with a more extensive examination of the materials.

22.2 Connectionist Networks (CNs)

Computer models of how neurons in the brain process information date back to the 1940s (McCulloch and Pitts, 1943). More recently these models have been used for two different (though not necessarily exclusive) purposes: either (1) to provide a model of brain function designed to correlate with actual brain behavior (in other words, to model real brains) or (2) to provide a computational architecture for systems designed to mimic *intelligent* behavior (or, to build "artificial intelligence"). While biologists and neuroscientists typically focus on the former goal, cognitive scientists, computer scientists, and engineers usually focus on the latter.

In an attempt to build a computational system designed to exhibit intelligent behavior, a number of models have been developed. These models focus on analog computation, parallel implementation and distributed information representation and processing. This means that they process real-valued (continuous) numbers as opposed to the binary representations common in most computer systems. They also process multiple streams of information in parallel and store pieces of information, not just in one location, but spread across many venues. We shall refer to these models as CNs as this reflects their essential character though the terms *artificial neural networks*, *PDP systems*, and others are also in popular use.

22.2.1 Basic Approach

The key to CNs as a computational paradigm is their difference from the typical von Neumann architecture. Specifically, CNs consist of many small and simple processing elements (PEs) that have an internal state and communicate with each other through connections to achieve complex tasks. The basic operation of a PE can be summarized by the following equation:

$$a_i = f\left(\sum_j w_{ij} \cdot a_j\right) \quad (22.1)$$

where, a_i represents the state or *activation* of PE i, $f()$ is a transfer function, j an index over all PEs connected to PE i, w_{ij} the weight of the connection from j to i, and a_j the activation value of PE j. The transfer function is typically a sigmoid function such as tanh or the logistic: $1/(1+e^{-x})$. Basically, we compute a weighted average of the activations of units a_j and then feed the average to an s-shaped transfer function, which limits the output of each process into the range from zero to one.

To prevent circular definitions of the activation (where an activation directly or indirectly depends on itself), we enforce a specific constraint. This constraint says that the nodes in the network must be ordered, and that no higher-ordered node can connect to any lower-ordered node, i.e., we requiring that $w_{ji} = 0$ $\forall j \leq i$. In mathematical terms, this ensures that the connectivity is limited to a directed acyclic graph (i.e., connections have a direction and there are no loops).

In practice, connection schemes more restrictive than directed acyclic graphs are implemented in the form of layered architectures. That is, the PEs of the CN are organized into layers with connections only existing between successive layers. Input information to the CN is encoded in a special layer (usually called the *input layer*) that has its activation values explicitly set rather than obeying Eq. 22.1 like layers of PEs. Also one special element in the network, called a *bias*, has its activation permanently set to a value of 1. This provides a zero-offset to the summation in Eq. 22.1. Without this offset every hidden unit would produce a value of 0.5 whenever the input values are all zeros. The inclusion of the bias with a trainable weight makes it possible to generate any activation in the hidden units when the inputs are all zero.

By far the most common connectionist network consists of an input layer of elements that encode the input information, connected to a layer of hidden (or internal) elements, connected to a layer of output elements which provide the resulting data. In such networks, each input unit is connected to each hidden unit and each hidden unit is connected to each output unit (see Figure 22.1).

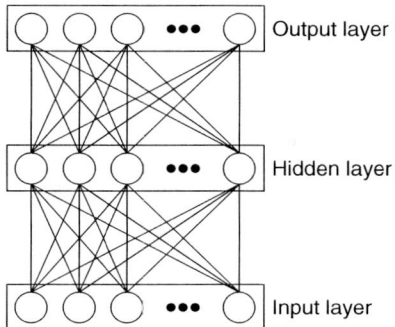

FIGURE 22.1 A typical connectionist network (CN).

There are a few variations on the basic PE presented here, one of the more notable of which is the use of second-order connections replace the linear summation in Eq. 22.1 by a quadratic (Giles and Maxwell, 1987):

$$a_i = f\left(\sum_{j,k} w_{kji} \cdot a_k \cdot a_j\right) \qquad (22.2)$$

Because this equation multiplies two activations together, it can be viewed as a generalization of the AND function in the sense that input activations of 0 and 1 elicit responses of 0 and 1 that are consistent with the Boolean AND function.

22.2.2 Function Approximation

Another way to view a CN is as a continuous function, mapping from input data to output data. Since both input and output data are expressed as real-valued activation values of PEs, the operation can be expressed as a function mapping a vector in the n-dimensional input space to a vector in the m-dimensional output space: $\Re^n \to \Re^m$. Under this formulation, it is natural to ask, "for a given network architecture (defined by connectivity scheme), what is the class of vector functions that can be implemented"?

It turns out that simple networks of PEs are surprisingly powerful. For example, simple CNs with a single layer of PEs between an input and an output layer are universal function approximators (Hornik et al., 1989). This speaks to the computational versatility of this approach and suggests that more elaborate connection schemes, PEs, are not necessary.

22.2.3 Learning

One of the most exciting features of CNs is that many define an adaptation mechanism in addition to their computational specification. These are typically referred to as *learning algorithms*. Such algorithms can be broadly classified into two categories: (1) unsupervised and (2) supervised. Adaptation in unsupervised learning algorithms is based solely on their input data and typically performs tasks such as clustering, nearest neighbor identification, and data completion. In contrast, supervised learning algorithms rely on an external mechanism to provide additional information about the desired output values. Often the exact output vector for each input vector is provided. (One other alternative, that of semisupervised training methods, has recently received some attention in the CN paradigm, but has thus far not had much influence in the spatio-temporal versions we will discuss below.)

By far the most common learning algorithms for CNs are based on approaches that compute an error gradient in weight space. By defining the error as the Euclidean difference between target, $\vec{a}^*$, and actual, $\vec{a}$, output vectors,

$$E = |\vec{a}^* - \vec{a}| \qquad (22.3)$$

and defining the error gradient based on that,

$$\nabla_{ij}E = \frac{\partial E}{\partial w_{ij}} \tag{22.4}$$

It is possible to use a simple steepest descent algorithm to find a local minimum:

$$\Delta w_{ij} = -\eta \nabla_{ij}E \tag{22.5}$$

where η is a scaling constant. The gradient in this equation can be computed by a simple equation

$$\Delta w_{ij} = -\eta \frac{\partial E}{\partial net_i} a_j \tag{22.6}$$

where

$$\frac{\partial E}{\partial net_i} = \begin{cases} a_i(1-a_i)(a_i^* - a_i) & \text{if } i \text{ is an output element} \\ a_i(1-a_i) \sum_k w_{ki} \frac{\partial E}{\partial net_k} & \text{otherwise} \end{cases} \tag{22.7}$$

This technique (proposed by Werbos [1994] and popularized by Rumelhart et al. [1986]) will lead the CN to adopt a weight configuration that represents a local minimum in the error space, E. While many theoretical approaches to finding global optima instead of local ones have been proposed, in practice, most applications rely on restarting the system with a different set of random initial weights (i.e., a different starting point in the weight space) as a simple method to avoid local minima.

22.3 Spatio-Temporal Connectionist Networks

SCNs are CNs with a temporal delay associated with their connections and therefore no need of restrictions on cyclical connectivity. By using the activation values of PEs at one time step to compute the activation values of PEs at another step, a form of feedback is created that makes SCNs dynamical systems. With proper connectivity, the feedback can be used to serve as an internal memory of old states and inputs, to induce oscillatory activation patterns, and even implement chaotic dynamics. In fact, by the same approach that allows CNs to serve as universal function approximators, SCNs can be universal dynamical systems.

22.3.1 Basic Approach

The operation of an individual PE in an SCN can be summarized by the following general equation:

$$a_i(t) = f\left(\sum_j \int_{t'=0}^{t} w_{ij}(t') \cdot a_j(t-t') dt' \right) \tag{22.8}$$

where $a_i(t)$ represents the activation of PE i at time t, $f()$ is a transfer function, j an index over the PEs connecting to i, t' a variable integrated over time from $t' = 0$ to $t' = t$, $w_{ij}(t')$ a function giving the weight of influence of the activation of element j as a function of time, and $a_j(t-t')$ the activation of element j at time $t-t'$. Note that this equation replaces the simple weight in Eq. 22.1 with a temporal weight kernel.

In most applications, time is assumed to be sampled at discrete unit time intervals so the integration in Eq. 22.8 can be replaced by a summation. In other words, we implement a discrete time simulation of a continuous time model with synchronous updates. Further, while there have been a number of studies of varying weight kernels (see, e.g., Mozer, 1994), most SCNs rely on the simplest of kernels: an impulse of delay 1. In this popular, degenerate case

$$a_i(t) = f\left(\sum_j w_{ij} \cdot a_j(t-1) \right) \tag{22.9}$$

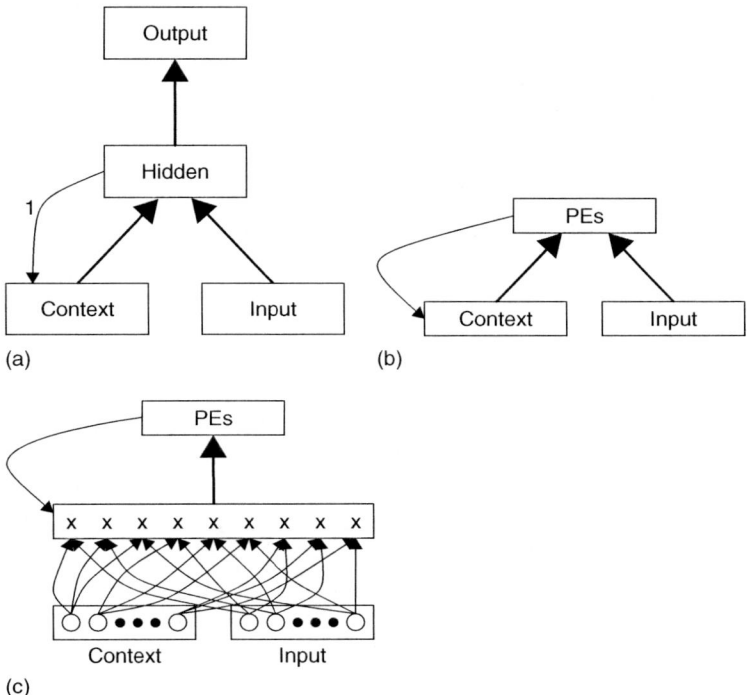

FIGURE 22.2 The most popular SCNs. (a) Elman network (From Elman, J. L. (1990). *Cognitive Science 14*, 179–211; Elman, J. L. (1991). *Machine Learning 7*(2/3), 195–226.) (b) Williams and Zipser network (From Williams, R. J. and D. Zipser (1989). *Neural Computation 1*(2), 270–280.) (c) Giles et al. network.

22.3.2 Specific Architectures

A number of specific architectures have been proposed and, as in the acyclic CN, they are organized into layers. Owing to the cyclical nature of the connections in SCNs, it is useful to illustrate them by showing the same layer at a given time step and also at a previous time step. By convention, the layer's previous time step is referred to as a *context layer*. The context layer is a virtual layer that does not involve any computation and merely serves to better illustrate the operation of the network. Some of the most popular SCN architectures are: the Elman network, the Williams and Zipser network, and the Giles et al. network.

22.3.2.1 Elman

In the Elman network (Elman, 1990, 1991), Figure 22.2(a), an input layer is connected to a hidden layer which in turn is connected to an output layer. A virtual context layer is really the hidden layer at the previous time step. The context layer also connects to the hidden layer, providing a simple feedback loop from the activation vector of the hidden layer at the previous time step to the current activation vector of the hidden layer.

22.3.2.2 Williams and Zipser

In the Williams and Zipser network (Williams and Zipser, 1989), Figure 22.2(b), the input layer is directly connected to all activation computing PEs, which in turn have time-delayed connections back to themselves. A subset of the PEs are used to represent the output of the network.

22.3.2.3 Giles et al.

In the Giles et al. network (Giles et al., 1990), second-order multiplicative elements are used. Thus, the activation of each PE is computed based on a sum of products. Each product is computed by multiplying the activation value of a PE with the activation value of an input element. Thus, the governing equation is

$$a_i(t) = f\left(\sum_{j,k} w_{ijk} \cdot a_j(t-1) \cdot a_k\right) \quad (22.10)$$

where j represents an input element and k a regular PE.

22.3.2.4 Other Architectures

A plethora of other architectures and variants on these three approaches have been developed, but a comprehensive survey is beyond the scope of this chapter. Interested readers are referred to Kremer (2001) and Barreto et al. (2003).

22.4 Representational Power

One issue of particular interest to researchers working with SCNs has been an examination of their computational power and their representation ability relative to other systems previously studied. The question asked is: as a class of systems, what can SCNs do? It is easiest to formulate an answer to this question in the context of other systems and two spring immediately to mind: (1) formal models of computation and (2) dynamical systems. The former are used to model digital computers and include finite-state automata, pushdown automata, Turing machines, and similar mathematical models of computers and their programs. The latter are generalized mathematical models involving time.

The computational power of SCNs has been extensively studied. The paradigm is inherently capable of universal dynamical system approximation (by a trivial corollary of the universal function approximation capability of CNs). This means that since CNs can implement any mathematical function, SCNs can also implement any dynamical system that can be specified mathematically to an arbitrarily close approximation. Beyond this, much focus has been placed on comparisons of specific SCN models with formal models of computation. These formal models rely on discrete input symbols. By encoding these inputs to vectors, SCNs can be applied to the same classes of models as finite-state automata, pushdown automata, and Turing machines.

Not surprisingly (given the universal dynamical system approximation), it can be shown that SCNs can exhibit the same or superior computational powers to their conventional counterparts. For example, Kremer (1995) shows that Elman networks are capable of implementing arbitrary finite-state automata, Wiles et al. (2001) prove that SCNs are capable of implementing pushdown automata models in their internal dynamics, and Siegelmann (2001) showed that not only can Turing machines be represented but they also may even be able to compute more efficiently when implemented in these continuous dynamical systems thus exhibiting super-Turing capabilities (though the existence of hypercomputation in itself is debatable).

A particularly interesting approach to relating SCN power to the conventional spectrum of computational formalisms is that of Giles et al. (1990). The Giles network is designed in such a way that the weights between the products of input/context elements and subsequent PEs correspond exactly to transitions in a traditional finite-state automaton. This correspondence allows the authors to naturally encode finite-state automata into these networks (Giles and Omlin, 1992) an even extract automata from trained SCNs (Giles et al., 1992). The extraction of automata from SCNs is somewhat controversial, however, and some have argued that the effort is futile (Kolen, 1994).

22.5 Learning

The impressive representational powers of SCNs gives confidence in their use on a broad range of applications. This leads one naturally to consider the matter of learning. For this, we are typically interested in mapping, not one vector (like in CNs), but rather a sequence of input vectors presented at each time step into not one vector but rather a sequence of output vectors. Thus, the system is trained to act as a vector-sequence to

Spatio-Temporal Connectionist Networks

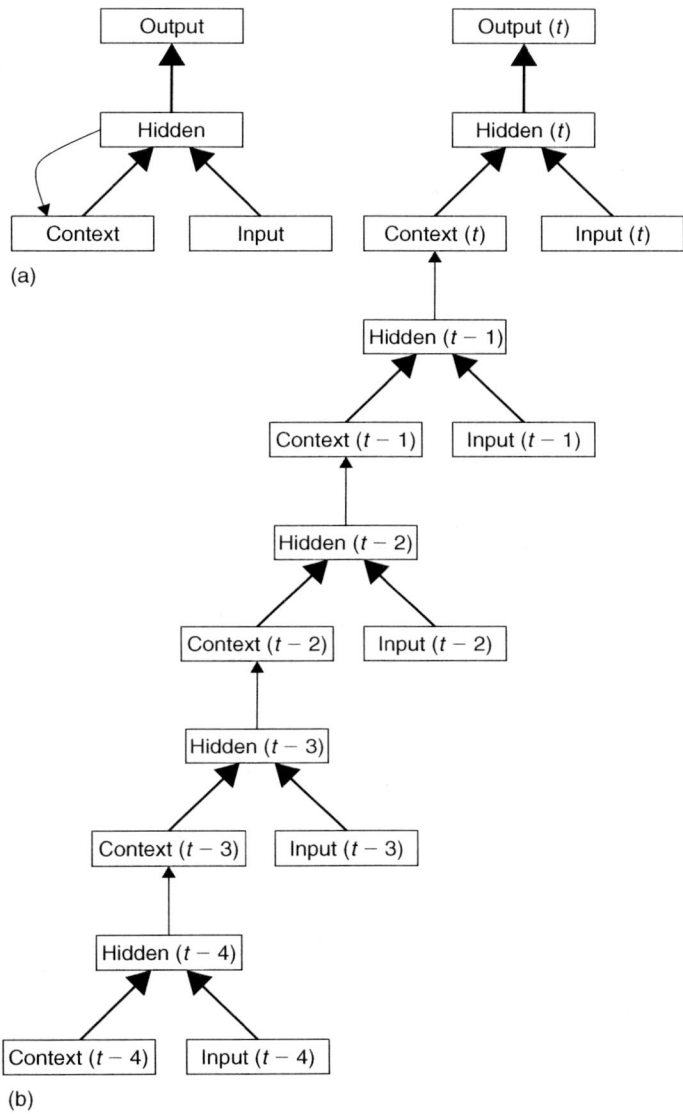

FIGURE 22.3 Unfolding an SCN.

vector-sequence mapping system. In this case, we define the output error as an error over the output vectors over time (usually summing the error values of the output over all the time steps). It turns out that computing error derivatives in the weight space of an SCN is a simple extension of the CN training algorithm.

One of the simplest methods for computing such derivatives involves unrolling a network in time using the context element trick over and over again to trace activation dynamics back to the very first time step. For example, consider the Elman network in Figure 22.3(a). We can unfold this network for four time steps to produce the unfolded network shown in Figure 22.3(b). In the resulting network, the connections from hidden to context and input to hidden at all levels of unfolding are identical. Thus, the chain rule for derivatives allows the weight changes to each virtual set of connections to be summed to compute the gradient.

$$\Delta w_{ij} = -\eta \sum_{t} \frac{\partial E}{\partial net_i(t+1)} a_j(t) \qquad (22.11)$$

where

$$\frac{\partial E}{\partial net_i(t+1)} = \begin{cases} a_i(t+1)(1-a_i(t+1))(a_i^*(t+1)-a_i(t+1)) & \text{if } i \text{ is an output element} \\ a_i(t+1)(1-a_i(t+1))\sum_k w_{ki}\frac{\partial E}{\partial net_k(t+2)} & \text{otherwise} \end{cases} \quad (22.12)$$

This method is a quick way of computing the error gradient in weight space, but requires memory to store all the successive activation values and partial derivatives in the unfolded network. This implies a memory requirement in the order of the number of time steps.

Interestingly, Elman himself does not unroll the network in the way described. Instead, he uses a truncated version of gradient descent that ends the gradient computation at the virtual context layer. This effectively results in an approximation to the gradient, which ignores distal effects in favor of more proximal ones. This algorithm uses a fixed amount of memory and greatly reduces the computational time since only one step of backpropagation needs to be performed.

Williams and Zipser (1989) propose an alternative approach to gradient computation where memory usage is constant, but the computational time increases by a factor proportional to the number of PEs. Another algorithm (Schmidhuber, 1992) combines the unrolling approach with that of Williams and Zipser to get the best of both worlds. Some interesting dynamical systems can be induced using these methods, for an extensive set of examples, the reader is referred to Pearlmutter (2001).

Unfortunately, however, the story does not end there. It turns out that computing the gradient is not an effective method for training SCNs. Specifically, for problems where information must be retained for long time periods, gradient approaches break down. Consider a problem where a network must remember the first bit in a long binary sequence. To store this bit, the information must be effectively latched within the PEs of the network. This requires the system to adopt a stable dynamic. Such a dynamic, however, requires the PEs to operate in the flat outer regions of the transfer function $f()$. But, in these flat outer regions the error gradient approaches zero. As a consequence, when an SCN faces a problem requiring the stable storage of information over long-time periods, it typically is unable to learn such behavior because the closer the PEs come to stable operation, the slower the learning gets and shorter-term effects dominate the training process. This phenomenon is known as the *shrinking gradients* problem (Hochreiter et al., 2001).

Interestingly, this limitation was not discovered for some time and these networks were applied to a number of problems without this limitation ever becoming apparent. The reason for this is that the limitation described here, while mathematically significant, does not occur in many problems to which these networks are typically applied. In fact, many of the models of human language actually benefit from such a limitation as it can help to describe the limitations of human performance in comparison to formal grammar models.

Nonetheless, a number of remedies have been suggested to the shrinking gradients problem. These include the long short-term memory network (Hochreiter and Schmidhuber, 1997) and other training algorithms (Palmer and Kremer, 2005). The long short-term memory approach relies on the use of linear memory elements fed by gating elements to ensure that gradients do not deteriorate to zero. The linear activation function guarantees that gradients do not deteriorate, while the gating elements ensure a nonlinear dynamics which would limit the operation of the system to linearly separable problems only.

22.6 Applications

A number of applications of SCNs have been proposed and developed. For example, in Prokhorov et al. (2001), an SCN is presented as a general purpose controller. This controller is applied to a plant that is a third-order system with two inputs and two outputs, and to the problem of financial portfolio optimization.

Tabor (2001) has shown that the domain of linguistics can be tackled by an adaptive SCN. Specifically, he uses dynamical systems theory embodied in the SCN to provide effective formal models of structures in natural languages. His work provides empirically reproducible predictions of human language performance.

Bakker and Schmidhuber (2004) used SCNs to implement learning and subgoal discovery in robots. This work applies to problems such as robot soccer and other tasks, where machines must dynamically adapt strategies to a changing environment.

Eck and Schmidhuber (2002) applied SCNs to learning long-term structure in music. This can be applied to both analysis and composition.

22.7 Conclusion

In this chapter, we have seen how spatio-temporal connectionist networks are a type of dynamical system. These models are loosely based on abstractions of neuronal processing and typically incorporate a learning mechanism. It is easy to extol the inherent computational capabilities of these systems, as they can be proven to be just as powerful as the best digital computers, capable of computing anything that is computable by a Turing machine (the formal definition of computable) and capable or representing any dynamical system to an arbitrary degree of precision. There are a number of learning algorithms proposed for these systems. The simplest of these suffer from an interesting limitation called the shrinking gradients problem. This problem identifies a mathematical limitation to what can be learned, but does not always apply to practical problems. A few solutions to the shrinking gradients have been recently proposed. SCNs have been successfully applied to a number of interesting real-world problems.

References

Bakker, B. and J. Schmidhuber (2004). Hierarchical reinforcement learning based on subgoal discovery and subpolicy specialization. In F. Groen, N. Amato, A. Bonarini, E. Yoshida, and B. Krse (Eds.), *Proceedings of the 8-th Conference on Intelligent Autonomous Systems, IAS-8*, Amsterdam, The Netherlands, pp. 438–445.

Barreto, G. A., A. F. R. Araújo, and S. C. Kremer (2003). A taxonomy for spatio-temporal connectionist networks revisited: The unsupervised case. *Neural Computation* 15(6), 1255–1320.

Eck, D. and J. Schmidhuber (2002). Learning the long-term structure of the blues. In J. Dorronsoro (Ed.), *Artificial Neural Networks — ICANN 2002 (Proceedings)*, pp. 284–289. Berlin: Springer.

Elman, J. (1990). Finding structure in time. *Cognitive Science* 14, 179–211.

Elman, J. L. (1991). Distributed representations, simple recurrent networks and grammatical structure. *Machine Learning* 7(2/3), 195–226.

Giles, C., G. Sun, H. Chen, Y. Lee, and D. Chen (1990). Higher order recurrent networks & grammatical inference. In D. S. Touretzky (Ed.), *Advances in Neural Information Processing Systems 2*, San Mateo, CA, pp. 380–387. Morgan Kaufmann.

Giles, C. L. and T. Maxwell (1987). Learning, invariance, and generalization in high-order neural networks. *Applied Optics* 26(23), 4972–4978.

Giles, C. L., C. B. Miller, D. Chen, G. Z. Sun, H. H. Chen, and Y. C. Lee (1992). Extracting and learning an unknown grammar with recurrent neural networks. In J. E. Moody, S. J. Hanson, and R. P. Lippmann (Eds.), *Advances in Neural Information Processing Systems 4*, San Mateo, CA, pp. 317–324. Morgan Kaufmann.

Giles, C. L. and C. Omlin (1992). Inserting rules into recurrent neural networks. In S. Kung, F. Fallside, J. A. Sorenson, and C. Kamm (Eds.), *Neural Networks for Signal Processing II, Proceedings of the 1992 IEEE Workshop*, Piscataway, NJ, pp. 13–22. IEEE Press.

Hochreiter, S., Y. Bengio, P. Frasconi, and J. Schmidhuber (2001). Gradient flow in recurrent nets: The difficulty of learning long-term dependencies. In J. F. Kolen and S. C. Kremer (Eds.), *A Field Guide to Dynamical Recurrent Networks*, pp. 237–244. Piscataway, NJ: IEEE Press.

Hochreiter, S. and J. Schmidhuber (1997). Long short-term memory. *Neural Computation* 9(8), 1735–1780.

Hornik, K., M. Stinchcombe, and H. White (1989). Multilayer feedforward networks are universal approximators. *Neural Networks* 2(5), 359–366.

Kolen, J. F. (1994). Fool's gold: Extracting finite state machines from recurrent network dynamics. In J. D. Cowan, G. Tesauro, and J. Alspector (Eds.), *Advances in Neural Information Processing Systems 6*, Volume 6, pp. 501–508. San Mateo, CA: Morgan Kaufmann.

Kremer, S. C. (1995). On the computational power of elman-style recurrent networks. *IEEE Transactions on Neural Networks* 6(4), 1000–1004.

Kremer, S. C. (2001). Spatio-temporal connectionist networks: A taxonomy and review. *Neural Computation* 13(2), 249–306.

McCulloch, W. and W. Pitts (1943). A logical calculus of ideas immanent in nervous activity. *Bulletin of Mathematical Biophysics* 5, 115–133.

Mozer, M. C. (1994). Neural net architectures for temporal sequence processing. In A. Weigend and N. Gershenfeld (Eds.), *Time Series Prediction*, pp. 243–264. Reading, MA: Addison-Wesley.

Palmer, J. and S. Kremer (2005). Learning long-term dependencies with reusable state modules and stochastic correlation. In *Intelligent Engineering Systems Through Artificial Neural Networks*, Volume 15, pp. 103–109. ASME.

Pearlmutter, B. (2001). Gradient calculations for dynamic recurrent networks. In J. F. Kolen and S. C. Kremer (Eds.), *A Field Guide to Dynamical Recurrent Networks*. pp. 179–206. Piscataway, NJ: IEEE Press.

Prokhorov, D., G. Puskorius, and L. Feldkamp (2001). Dynamical recurrent networks in control. In J. F. Kolen and S. C. Kremer (Eds.), *A Field Guide to Dynamical Recurrent Networks*. pp. 143–152. Piscataway, NJ: IEEE Press.

Rumelhart, D., G. Hinton, and R. Williams (1986). Learning internal representation by error propagation. In J. L. McClelland, D. Rumelhart, and the P. D. P. Group (Eds.), *Parallel Distributed Processing: Explorations in the Microstructure of Cognition, Volume 1: Foundations*. Cambridge, MA: MIT Press.

Schmidhuber, J. H. (1992). A fixed size storage $o(n^3)$ time complexity learning algorithm for fully recurrent continually running networks. *Neural Computation* 4(2), 243–248.

Siegelmann, H. (2001). Universal computation and super-turing capabilities. In J. F. Kolen and S. C. Kremer (Eds.), *A Field Guide to Dynamical Recurrent Networks*, pp. 143–152. Piscataway, NJ: IEEE Press.

Tabor, W. (2001). Sequence processing and linguistic structure. In J. F. Kolen and S. C. Kremer (Eds.), *A Field Guide to Dynamical Recurrent Networks*, pp. 291–310. Piscataway, NJ: IEEE Press.

Werbos, P. J. (1994). *The Roots of Backpropagation: From Ordered Derivatives to Neural Networks and Political Forecasting*. New York: Wiley.

Wiles, J., A. Blair, and M. Bodén (2001). Representation beyond finite states: Alternatives of push-down automata. In J. F. Kolen and S. C. Kremer (Eds.), *A Field Guide to Dynamical Recurrent Networks*, pp. 129–142. Piscataway, NJ: IEEE Press.

Williams, R. J. and D. Zipser (1989). A learning algorithm for continually running fully recurrent neural networks. *Neural Computation* 1(2), 270–280.

23
Modeling Causality with Event Relationship Graphs

Lee Schruben
University of California

23.1 Introduction 23-1
23.2 Background and Definitions 23-2
 Discrete-Event Systems and Models • Discrete-Event System Simulations • The Basic Event Relationship Graph Modeling Element • Verbal Event Graphs • Reading Event Relationship Graphs
23.3 Enrichments to Event Relations Graphs 23-7
 Parametric Event Relationship Graphs • Building Large and Complex Models • Variations of Event Relationship Graphs
23.4 Relationships to Other Discrete-Event System Modeling Methods 23-10
 Stochastic Timed Petri Nets • Mapping Petri Nets into Event Relationship Graphs • Process Interaction Flows • Generalized Semi-Markov Processes • Mathematical Optimization Programs
23.5 Simulation of Event Relationship Graphs 23-16
23.6 Event Relationship Graph Analysis 23-16
23.7 Experimenting with ERGs 23-17

23.1 Introduction

Events are any potential change in the state of a dynamic system. Event relationship graphs (ERGs) explicitly model the ways in which one system event may cause another system event to occur. The cause and effect relationships between events modeled in an ERG, along with simple rules for execution and initial conditions, completely specify all possible sample paths (state trajectories) of a dynamic system model. Continuous dynamics systems have been modeled as ERGs, but they are most commonly used to model discrete-event system dynamics. The ERG for a queueing system is typically a system of simple difference equations analogous to a system of differential equations used in modeling continuous time system dynamics. ERGs are completely general in that any dynamic system can be modeled as an ERG (Savage et al, 2005). They are easy to develop and understand and facilitate the design of efficient simulation models. ERGs also have analytical representations that aid in systems analysis, specifically when the potential system trajectories for ERG model are represented as the solutions to mathematical optimization problems.

23.2 Background and Definitions[1]

23.2.1 Discrete-Event Systems and Models

It will be sufficient for our purposes to define a "system" as *a collection of entities that interact with a common purpose according to sets of laws and policies.* A system may already exist, or it may be hypothetical or proposed. Here, we intentionally do not define a system by the specific elements in it or its boundaries. Rather, we define a system by its purpose. Thus, we speak of a communications system, a health care system, and a production system. Using a functional definition of a system helps avoid thinking of a system as having a preconceived structure. Consequently, a system is viewed in terms of how it ought to function rather than how it currently functions. To design better systems it is important to think beyond the status quo.

The "entities" making up the system may be either physical or mathematical. A physical entity might be a patient in a hospital or a part in a factory; a mathematical entity might be a variable in an equation. When developing models of queueing systems, it is often useful to classify entities as being either *resident* entities or *transient* entities. Resident entities remain part of the system for long intervals of time, whereas transient entities enter into and depart from the system with relative frequency. In a factory, a resident entity might be a machine; a transient entity might be a job or a part. Depending on the level of detail desired, a factory worker might be regarded as a transient entity in one model and a resident entity in another. The states of resident entities can often be modeled sufficiently on a computer using simple fixed-dimension integer arrays, while transient entities often require creating and maintaining dynamic records or objects. Entities are described by their characteristics (referred to here as *attributes*). Attributes can be quantitative (represented in a computer by numeric codes) or qualitative. Moreover, they can be static and never change (the speed of a machine), or they can be dynamic and change over time (the length of a waiting line). Dynamic attributes can further be classified as deterministic or stochastic depending on whether the changes in their values can be predicted with certainty or not.

The rules that govern the interaction of entities in a system that are not under our control are called "laws." Similar laws are grouped in families, members of which are distinguished by *parameters*. Rules that are under our control are called "policies"; a family of similar policies may be distinguished by the values of their *factors*.

We will define a *model* simply as *a system used as a surrogate for studying another system.* In this chapter, when we use the word *system*, without qualification, we are referring to a real or hypothetical system that is the subject of a modeling analysis or simulation study. In typical computer simulation models, systems of mathematical equations and computational objects are used as a surrogate for a real or proposed system of physical entities.

The *state* of a system is a complete description of the system and includes values of all attributes of entities, parameters of laws, factors for its policies, time, and what might be known about the future. The *state space* is the set of all possible system states. A *process* is an indexed sequence of system states; typically, the index is time, but it might be the index of jobs in a queueing system or some other system characteristic.

A discrete-event model of a dynamic system is one where the state of the system changes at particular instants of time. Examples include queueing system models where a job arriving or leaving the system is a discrete change in the system state. Changes in the system state for a discrete-event system are called *events*. In a production system, for example, events might include the following:

1. Completion of a machining operation; such an event might be called "finish" and the state of the machine involved would change from "busy" to "idle."
2. Failure of a machine; a "failure" event would change the machine state to "broken."
3. Arrival of a repair crew; an event called, say, "start_repair" where the machine state would change to "under_repair."

---

[1] *Source*: Figures, examples, and model descriptions in this chapter are adapted from Schruben, D. L. and L. W. Schruben, *Event Relationship Graph Modeling Using SIGMA*. ©Custom Simulations, used with permission.

4. Arrival of a part at a machining center; this "arrival" event would, if an appropriate machine were idle and in good working condition, immediately cause another event, a "start_processing" event where the machine will again become "busy."

The ability to identify and abstract system events in a discrete-event model are important skills that take practice to acquire. The following simple steps identify system events: first, identify all the *dynamic* attributes of the system entities; then, identify the circumstances (state conditions and/or time delays) that may cause the values of these attributes to change. These circumstances are the system events and they will in turn schedule or cancel other system events.

23.2.2 Discrete-Event System Simulations

We *model* (verb) the dynamic behavior of a discrete-event system by describing the events and the relationships between these events. When we use the word *model* (noun), without qualification, we will be referring to a graphical description of a system called an ERG. *Simulations* will refer to computer programs developed from ERGs. Simulations will be our methodology for studying the model.

The building blocks of a discrete-event simulation program are *event procedures*. Each event procedure makes appropriate changes in the state of the system and, perhaps, may schedule a sequence of further events to occur. Event procedures might also cancel previously scheduled events. An example of event canceling might occur when a busy computer breaks down. End-of-job events that might have been scheduled to occur in the future must now be canceled (these jobs will not end in the normal manner as originally expected).

The event procedures describing the state changes in a discrete-event system simulation are executed by a main control program that operates on a master appointment list of scheduled events. This list is called the *pending events list* and contains all of the events that are scheduled to occur in the future. The main control program will advance the simulated time to the time for the next scheduled event. The corresponding event procedure is executed, typically changing the system state and perhaps scheduling or canceling further events. Once this event procedure has finished executing, the event is removed from the future events list. Then the control program will again advance time to the next scheduled event and execute the corresponding event procedure. The simulation operates in this way, successively calling and executing the next scheduled event procedure until some condition for stopping the simulation run is met. The operation of the main simulation event scheduling and execution loop is shown in Figure 23.1.

23.2.3 The Basic Event Relationship Graph Modeling Element

ERGs are a way of explicitly expressing all the relationships between events in a discrete-event dynamic system model. Some early references include Schruben (1983), Sargent (1988), Som and Sargent (1990), and Wu and Chung (1991). ERG models of discrete-event system dynamics have been presented in many textbooks on simulation, stochastic processes, and manufacturing systems engineering (Pegden, 1986; Hoover and Perry, 1990; Law and Kelton, 2000; Askin and Standridge, 1993; Nelson, 1995; Seila et al., 2003).

The three elements of a discrete-event system model are the state variables, the events that change the values of these state variables, and the relationships between the events (one event causing another to occur, or preventing it from occurring). An ERG organizes sets of these three objects into a simulation model. In the graph, events are represented as nodes (circles) and the relationships between events are represented as arcs (arrows) connecting pairs of event nodes. The basic unit of an ERG is an arc connecting two nodes. Suppose the arc represented in Figure 23.2 is part of an ERG.

We interpret the arc between A and B as follows: *whenever event A occurs, it might cause event B to occur.* Arcs between event nodes are labeled with the conditions under which one event will cause another event to occur, perhaps after a time delay. The state changes associated with each event are in braces next to the event node.

ERGs may look similar to flow graphs, but they are very different. While one might sometimes think of these graphs as modeling the flow of, say jobs, through a queueing system network, these graphs are actually

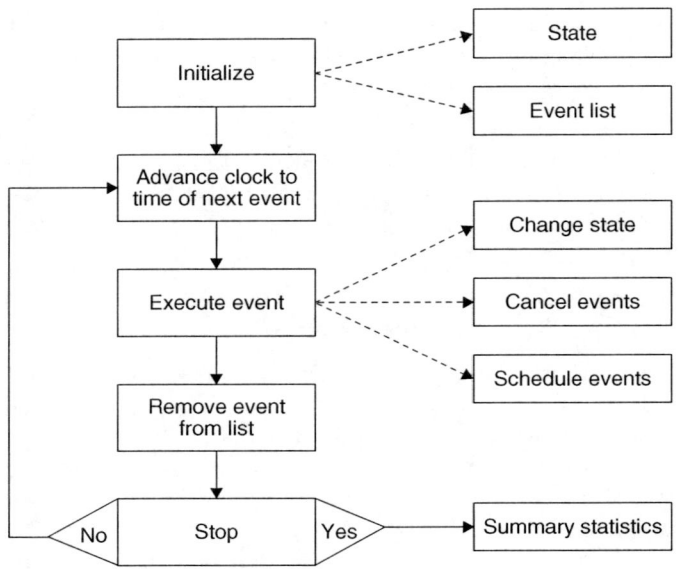

FIGURE 23.1 Main discrete-event execution algorithm.

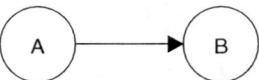

FIGURE 23.2 Event relationship: Event A might cause event B to occur, perhaps in the future.

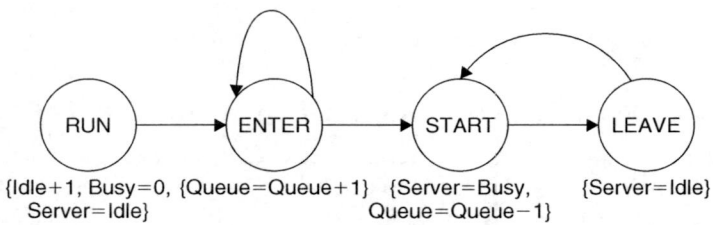

FIGURE 23.3 Event relationship graph for a single server queue.

modeling the scheduling and canceling causal *relationships* between pairs of events that may change the state of a system entity. Limiting the interpretation of ERGs to flow processes can lead to errors or model inefficiencies. ERGs are related to state machines or state transition diagrams much like a derivative is related to a function. In a state machine or transition diagram, each node represents the *value* of a state; in an ERG, each node represents a *change* in state.

Using this notation, we can build a model that simulates a simple waiting line with one server (e.g., a ticket booth at a theater and the drive-in window at a fast-food restaurant). An ERG for this system is represented in Figure 23.3.

We will begin our examination of this graph by discussing each node. The RUN node models the initialization of the simulation, the ENTER node models the event when a new job enters the system, the

Modeling Causality with Event Relationship Graphs

START node models the start of service event, and the LEAVE node models the end of service when a job will leave the system. The state variables chosen to describe this system are

SERVER = the status of the server (busy, idle), initially set idle

QUEUE = the number of jobs waiting in line, initially set equal to zero

To make our model more readable, we also define the constants IDLE=1 and BUSY=0.

Next, we will focus on the changes in the state variables shown in braces. The simulation RUN is started by making the server available for use {IDLE=1, BUSY=0, SERVER=IDLE}. (Instead of using an initial event, the initially scheduled events are indicated by a broken arrows in the ERGs in Law and Kelton [2000].) Each time a job ENTERs the line, the length of the waiting line is incremented {QUEUE=QUEUE+1}. When service STARTs, the server is made busy {SERVER=BUSY} and the length of the line is decremented {QUEUE=QUEUE−1}. Whenever a job has been finished and LEAVEs the system, and the server is again made available {SERVER=IDLE} to serve other jobs.

The dynamics of an ERG model are captured in the arcs of the graph. We read an ERG simply by describing the arcs *exiting* each node (out-arcs). In-arcs take care of themselves. Continuing with our example, we look at each arc in Figure 23.3 (event names are in italics).

The simulation *run* is started by having the first job *enter* the system (arc from *run* to *enter*). If the *enter*ing job finds the server idle, service will *start* immediately (arc from *enter* to *start*). Each time a job *enters* the system, the next job will be scheduled to *enter* sometime in the future (arc from *enter* to *enter*). The *start* service event will always schedule a job to *LEAVE* after that job has been served (arc from *start* to *leave*). Finally, if there are jobs waiting in line when a job *leaves*, the server will *start* servicing the next job right away (arc from *leave* to *start*).

The self-scheduling arc (the loop) on the ENTER event is the conventional way of perpetuating successive customer arrivals to the system. There will typically be some random time delay between customer arrivals.

The state changes for an event node for a queueing system are typically very simple. Most of the action occurs on the arcs of the graph. The conditions and delays associated with the arcs of the ERG are very important; it is on the graph arcs that the logical flow and dynamic behavior of the model are defined. For each arc in the graph we will need to define under what conditions and after how long one event might schedule another event to occur.

To make the event relationships explicit, we label each arc with the conditions that must be true for an event to be scheduled. Also associated with each arc will be a label that is the delay time equal to the interval until the scheduled event occurs. Time will be measured in minutes for our examples. In Figure 23.4, the basic ERG is enriched to include arc conditions and arc delay times using the notation of Askin and Standridge (1993).

This arc is interpreted as follows: *if and only if condition (i) is true right after event A occurs, then event B will be scheduled to occur t minutes later*. If the condition is not true, nothing will happen, and the arc can be ignored until the next time event A occurs. Arcs are stochastic: an arc does not exist unless its condition is true. If the condition for an arc is always true, the condition label is left off the graph. We will call arcs with conditions that are always true *unconditional* arcs. Zero time delays for arcs are also not labeled on the graph.

Our queueing system model with arc conditions and delay times is shown in Figure 23.5.

The state variables SERVER and QUEUE are now denoted by R and Q, respectively, and the status of S is indicated by 1 or 0 (1 if the server is idle and 0 if busy). In addition, the time between successive job arrivals (often random) is denoted by t_a and the service time required to process a job is denoted by t_s.

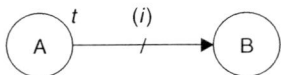

FIGURE 23.4 Event relationship arc with condition (*i*) and delay *t*.

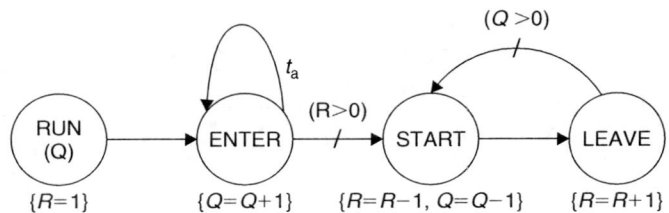

FIGURE 23.5 Event relationship graph for a single server queue.

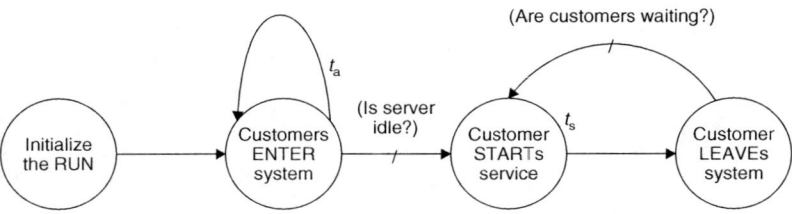

FIGURE 23.6 Verbal event relationship graph for a single server queue.

When values of t_a are actually needed, they might be obtained from a data file or generated by random variable generation algorithms. The graph in Figure 23.5 represents a well-defined dynamic model. To simulate this model, only the starting and ending boundary conditions for the run need to be specified.

23.2.4 Verbal Event Graphs

Before designing a simulation model, it is important to first develop a verbal description of the system dynamics. This can be done using a verbal ERG. This description would include state changes associated with each node along with a verbal description of each arc condition and delay time on the graph. A verbal ERG for a generic single server queueing system is shown in Figure 23.6.

Developing a verbal description of a system is a reasonable first step toward building a realistic and accurate simulation model. It will help to conceptualize the major components in the system, determine the key events and their interrelationships, and identify the state variables, arc conditions, and time delays necessary for the model. It is easy to determine if a variable is required to define the state of a system: the state variables needed for an ERG are those that permit testing all arc conditions in a verbal ERG. Once a detailed verbal description has been developed, an ERG model is easier to build.

The visual modeling power of ERGs is most appreciated after one recognizes the complicated details involved in a discrete-event simulation. The fundamental concept in ERG modeling is to use a directed graph as a picture of the relationships among the *elements* in sets of expressions characterizing the dynamics of the system. Each node of the graph is identified with a set of expressions for the state changes that result when the corresponding event occurs. Each arc in the graph identifies sets of logical and temporal relationships between a pair of events.

23.2.5 Reading Event Relationship Graphs

A concise description of the dynamics of an ERG model can be obtained by paraphrasing the arc definitions given for Figure 23.4. The description will have roughly one sentence per arc. As you read the following description, identify a single arc in Figure 23.5 with each sentence, again event names are in italics.

At the start of the simulation *run*, the first job will *enter* the system. Successive jobs *enter* the system every t_a minutes. If *enter*ing jobs find that the server is available ($S>0$), they can *start* service. Once jobs start service, t_s minutes later they can *leave*. Whenever a job *leaves*, if the queue is not empty ($Q>0$), the server will *start* with the next job.

Now reread the above paragraph without looking at Figure 23.5. You will see that it is a concise description of the dynamic behavior of this queueing system. A verbal system description can be read easily from the arcs of very complicated ERGs. Simply define each arc in the graph in a sentence—the graph itself will connect these sentences into a complete description of the system's dynamics. This is an excellent way to communicate the essential features of a simulation model and a good step in model validation. With experience in reading ERGs, it becomes easier to detect modeling errors.

23.3 Enrichments to Event Relations Graphs

23.3.1 Parametric Event Relationship Graphs

One of the most powerful enrichments of ERGs is parameters for the event nodes. Using event parameters, a basic ERG can become an element in a multiple-dimensional array of ERGs without increasing the size or complexity of the graph. For modeling large-scale systems, such as huge networks of queues, the ERG represents a generic element in an array of ERGs for each component of the system. The events in the ERG are parameterized (subscripted) with the particular station in the queueing network to which it applies by using event parameters. For this, the basic definition of an ERG is expanded to allow the values of a string of expressions computed after an event is executed to be passed to a string of state variables when an event is scheduled. The values of expressions are included in boxes on the arcs.

To be explicit: define a string of state variables for each event, these are actually arguments for the state change method for an event object. The values of these parameters are computed by a string of arc expressions called arc attributes. The definition of a parametric arc in a simulation ERG is in Figure 23.7.

This arc is read as follows: *Immediately after executing the state change for Event A, the elements of an array of expressions, k, are evaluated. Before B is executed, the array of state variables, j, are assigned the values previously computed for k.* The arc condition and delay time, if any, are defined as before. This uses a "pass-by-value" argument passing convention like in the C and Java programming languages.

Treating event parameters as subscripts, the graph becomes an element in an array of ERGs that model a network composed of a large number of similar systems (Chan and Schruben, 2005). For example, Figure 23.8 is an ERG for m parallel server queues in tandem. (The state changes are omitted in Figure 23.8 since these merely increment and decrement state variables $Q(k)$ and $R(k)$ for station k in the obvious fashions.) The only change to the graph structure to obtain Figure 23.8 from Figure 23.5 is an additional arc from the Finish(k) event for station k to the Arrival(k) event for the next station, $k+1$ (and the arc attributes).

The ERG in Figure 23.8 can be easily extended to model more general queueing network systems. For example, a generic queueing network can be modeled by adding a parameter, j, to the objects in the graph that indicates job type. The attribute, $k+1$, on arc from Finish(k, j) to Arrival(k, j) in the resulting ERG can be replaced by a general routing function. Transportation resources and move times could be added using conditions and delays on this arc. The ERG then becomes an element of a two-dimensional array (kth station by jth job type) of ERGs. See Schruben and Schruben (2005) where a general queueing network of arbitrary size is modeled with an ERG having only two nodes.

We conclude this section with an example of an ERG for a more complex system. This is a model of a tool commonly found in semiconductor manufacturing called a cluster tool (a configuration of isolated processing chambers served by one or more dedicated robots). In recent years, the use of cluster tools in semiconductor manufacturing has increased rapidly, causing the performance of cluster tools to become more and more important (see Perkinson et al., 1994; Chan and Schruben, 2004; Ding and Yi, 2004). The ERG for this system is given by Figure 23.9, which is an element in an array of ERGs each of which defines the relationships among events that simulate wafer j as it is processed or moved between chambers m and n by a robot.

Chamber 0 is a load lock for loading or removing wafers from the cluster tool. In Figure 23.9, f is the next wafer on the robot's move schedule; $x(j)$ the current location of wafer j, $y(j)$ its next destination, and z the current location of the robot. The time for the robot to move from chamber n to chamber m is denoted as $t_{n,m}$ and the time for chamber n to process wafer j is $p_{j,n}$.

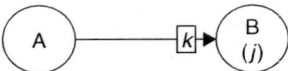

FIGURE 23.7 Parametric event relationship graph arc.

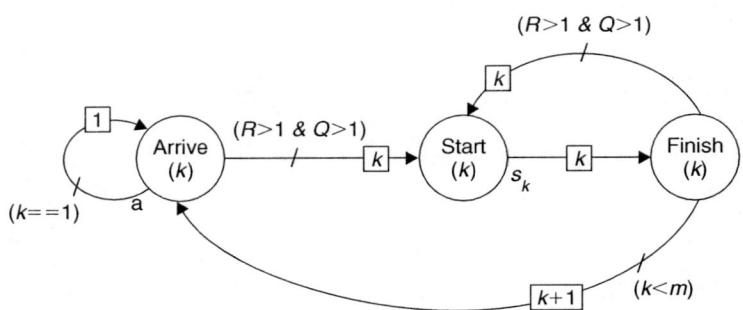

FIGURE 23.8 A system of m multiple-server tandem queues.

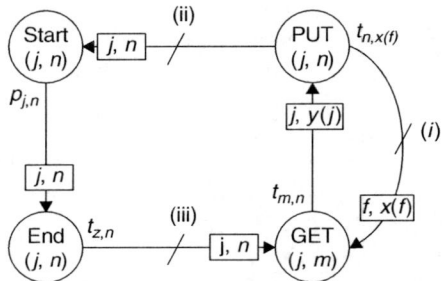

FIGURE 23.9 An ERG model for a semiconductor cluster tool.

In the ERG of Figure 23.9, wafers move from chamber to chamber around the center rectangle in the graph and the robot cycles between the GET and the PUT events. The events are defined as follows: Start (j, n) = Start processing wafer j in module n; End(j, n) = End processing wafer j in module n; Get (j, n) = Robot gets wafer j from module n and Put (j, n) = Robot loads wafer j into module n. The relationships between these events are defined by their delay times and their arc conditions. Condition (i) is true if there is a wafer (f) waiting to be picked up at chamber $x(f)$ and the destination chamber, $y(f)$, has been reserved. Condition (ii) is true when the requirements to start processing job j in chamber n are satisfied (these might include conditions such as having a full batch for a batch process). Condition (iii) is true if the robot is idle and has no backlog of wafers waiting to be moved. Many ERG simulations of cluster tools have been developed (Pederson and Trout, 2002). Two of the more elegant ERGs for a generic cluster tool, both modeled with only three events, are in Nehme and Pierce (1994) and Ding and Yi (2004).

23.3.2 Building Large and Complex Models

There is an important distinction between complicated ERG models and ERG models that are simply large. Very large systems can merely be a large number of simple components. The ERG for such a large model of similar components is the same as that for one of its components, but with parameter values indicating to which components a particular event applies. Complex models, however, have different types of ERG components, each of which might be quite complicated.

In this section, a moderately complex component is used to illustrate how ERGs can be developed and enriched into a large system of such components. We start off using the graph in Figure 23.5 to model a

Modeling Causality with Event Relationship Graphs

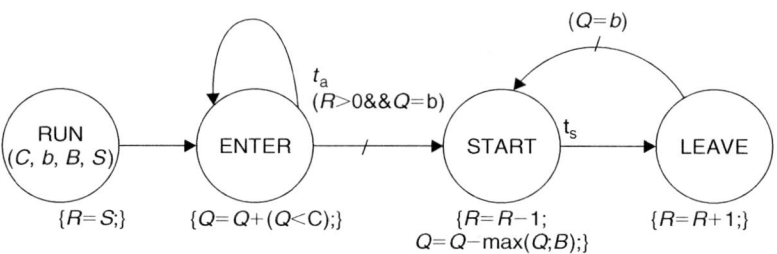

FIGURE 23.10 ERG for queue with s servers, capacity c, and batch size between b and b.

system with S identical servers, a queue capacity of C, and a flexible batch size between b and B. A batch of at least b jobs is required and at most B jobs can be processed together in a batch. The ERG for this system is given in Figure 23.10. Boolean variables (conditional expressions that equal 1 if true and 0 if false) are used. For example, the state change $Q=Q+(Q<C)$ will increment Q by 1 if and only if the expression $(Q<C)$ is true, modeling a queue with capacity C.

Reading the arcs of this graph as before (the event node names are in italics).

A simulation *run* starts by setting the number of idle servers equal to S and the first job *enters* the system. Jobs *enter* at the system every t_a time units, but join the queue only if the Q size is less than the capacity C. When new jobs *enter*, if they find an idle server and the minimum batch size of jobs is waiting, then a batch of at most B jobs will *start* service. After a service time of t_b, the batch of jobs will *leave* the system. When a batch of jobs *leaves*, if a minimum batch of b jobs is now waiting, the server will *start* on the next batch of jobs.

The arc from ENTER to START is tested every time a new job enters the system regardless of whether there is room in the queue for it to join or not. This does not cause any logical problems since the conditions for the START event must still be satisfied.

The ERG model in Figure 23.10 can be used as the fundamental element in an array of ERGs to model an arbitrarily large network of queues with different job types, each with a different routing among many different server types. This is done by attaching four parameters to each event and arc in Figure 23.10 that indicate the job type involved in an event, the step in that job's route, the server type, and the batch size being processed. The only structural change in the graph is a single additional arc from LEAVE to ENTER so batches of jobs can go from one server type on their routes to the next. This arc will cause a LEAVE (job, step, server, batch) event to schedule an ENTER (job, step, server, batch) node with attribute values given by the expressions: job, route (job, step+1), step+1, batch. Here route(job, step) is a function that specifies the server type for the next step on the route of a job. Complicated batching rules (jobs processed only in batches with certain other types of jobs) can be modeled with the same graph structure; the state changes and arc conditions can become as complex as necessary. Modeling resource failures or assembly operations also does not require the graph structure to change (see Schruben and Schruben, 2005 for examples).

23.3.3 Variations of Event Relationship Graphs

Specializations of ERGs include resource cycle graphs (RCG) for simulating queueing networks (Hyden et al., 2001). In an RCG, the state variables are all integer arrays and the resource state changes associated with every event are expressed as one or more integer difference equations. Simulating these models involves increasing and decreasing the values of elements in the state arrays when specific events occur (e.g., the number of idle servers of a particular class would decrease whenever a start_service event occurs for this class). ERG modeling of such systems has certain advantages in terms of simplicity and efficiency in simulation (Schruben and Schruben, 2005). For example, a simulated RCG for an actually semiconductor factory ran orders of magnitude faster than the most popular commercial simulator (Schruben and Roeder, 2003).

Object-oriented ERGs (called listening event graph objects, LEGOs) have been developed by Buss and Sanchez (2002). LEGO is a public domain collection of powerful Java applets for modeling general system objects.

Qualitative ERGs have been developed for the analysis of qualitative decisions (Ingalls, et al., 2000). These ERGs capture the relationships between various qualitative variables, modeling uncertainty using ranges rather than distributions and get sensitivities to various aspects of a problem. Qualitative ERGs have been applied to project management simulations where there are many unknowns that cannot easily be quantified.

23.4 Relationships to Other Discrete-Event System Modeling Methods

23.4.1 Stochastic Timed Petri Nets

In Schruben and Yucesan (1994), a method for representing a general stochastic timed Petri net (STPN) model as an ERG is presented. Basically, the nodes in a Petri net become arcs in an ERG. It has been shown that ERGs have equivalent representations using only zero-delay conditional arcs and nonzero delay unconditional arcs (Yucesan and Schruben, 1992; Schruben and Yucesan, 1994). These two classes of arcs are referred to here simply as conditional and timed arcs. Timed stochastic Petri nets can be mapped directly into ERGs by representing the transition nodes in a Petri net as timed arcs in an ERG and the place nodes in a Petri net as conditional arcs in an ERG.

Using the method for eliminating redundant events in Som and Sargent (1990), the resulting ERG can determine the smallest number of essential events that captures the Petri net dynamics for efficient simulation and analysis. ERGs are more general than Petri nets, so the reverse mapping of ERGs into STPNs is not possible without additional restrictions.

A Petri net is a directed graph with two classes of vertices, places (balls), and transitions (bars). The graph is bipartite, meaning that no two vertices in the same class are adjacent, places are connected only to transitions and transitions are connected only to places. The state of the system is represented by a marking of tokens in each place. The number of tokens at a place typically represents the value of some associated state variable.

The dynamics of the STPN are modeled by a transition firing rule. For convenience in coding a simulation, we separate state transitions into two sequential operations. A transition first becomes *enabled* when all of its input places are marked with at least a single token. Enabling a transition involves removing a single token from each of its input places. Next, after a delay associated with the transition, the transition *fires*, depositing a token in each of its output places. There is no restriction on the conservation of the total number of tokens. Note that we remove tokens from input places when a transition is enabled and then we deposit tokens when the transition fires as done in Fishwick (1995) rather than removing tokens when a transition is fired as is more common. Removing tokens when transitions are enabled often results in a simpler dynamic model. One can think of a token as residing in a transition while it is active. If the STPN is simulated, the number of tokens in a transition is the number of firing "events" for this transition that are currently scheduled to occur on the pending events list.

An STPN model for the multiple server queue is given in Figure 23.11.

In Figure 23.11, the number of idle servers is the number of tokens in place, R, and the number of jobs in queue is the marking of the place, Q. The marking shown is the initial marking for a system with three identical servers. The time between job arrivals is t_a and the time for service of a job is t_s.

Next we consider a situation where jobs are processed in a batch of size, B; this model is in Figure 23.12. Numerous enrichments of Petri net have been suggested for modeling dynamic systems. One we will use in Figure 23.12 is a token count which is a number placed on an arc with the number of tokens that are to be removed on enabling a transition or deposited on firing (if different from one).

A slight modification of the system in Figure 23.12, where there is a minimum batch size, b, and maximum batch size, B, probably cannot be represented as a general STPN where the STPN remains the

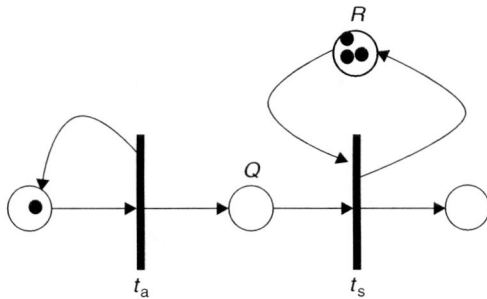

FIGURE 23.11 A multiple server queue with stochastic service times, t_s, and job arrival times, t_a.

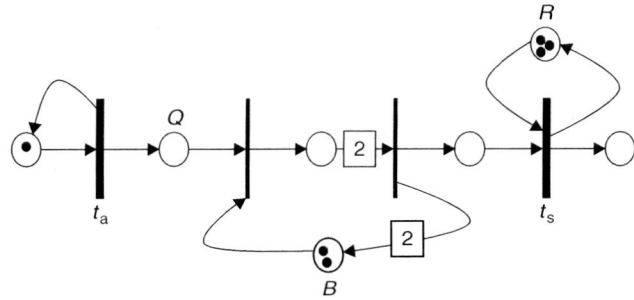

FIGURE 23.12 Initial marking for an STPN for a 3-server queue with batch size of 2.

same for all ranges of batch sizes with only the initial markings to differentiate the different batch size limits. A flexible batch size such as this for arbitrary minimum and maximum batch sizes, b and B, is modeled easily as an ERG in Figure 23.10.

STPNs provide an intuitive methodology for visualizing the dynamics of some discrete-event systems. Using simple tokens, STPNs graphically represent the dynamic relationships between system entities, modeling concurrency and contention for system resources. However, STPNs can easily become unwieldy when modeling large or complex systems. Direct simulations of such dynamic models result in codes that are often very slow to execute. Also, a simple method for parametric Petri nets is needed before highly redundant, large-scale systems can be modeled easily. Event graphs however can model very large and complex systems easily.

23.4.2 Mapping Petri Nets into Event Relationship Graphs

Let P denote the set of places and T the set of transitions in the STPN. We also define

$d(t)=$ Delay time (possibly random) for each transition $t \in T$
$\{I_p(t), O_p(t)\}=$ Set of input and output places for $t \in T$
$\{I_t(p), O_t(p)\}=$ Set of all input and output transitions for $p \in P$

The algorithm for translating an STPN into an ERG is as follows:

Step 0. $\forall\, p \in P$: define an integer state variable, $X(p)$.
Step 1. $\forall\, t \in T$: create two event vertices, $O(t)$ and $D(t)$ (denoting the origin and destination of a transition) and an arc $(O(t), D(t))$ with delay $d(t)$.
Step 2. $\forall\, p \in P$ with *unique* $(I_t(p), O_t(p))$ pair: create the arc $(D(I_t(p)), O(O_t(p)))$ with the condition that all $p \in I_p(O_t(p))$ are marked. (For inhibitor arcs, the arc condition is that the input places must *not* be marked.)

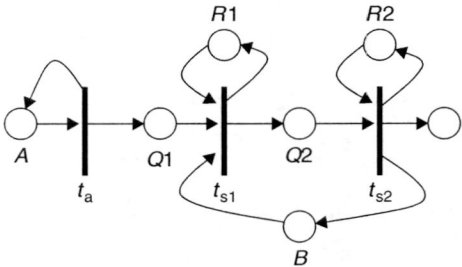

FIGURE 23.13 An STPN for a communications channel with limited packet capacity.

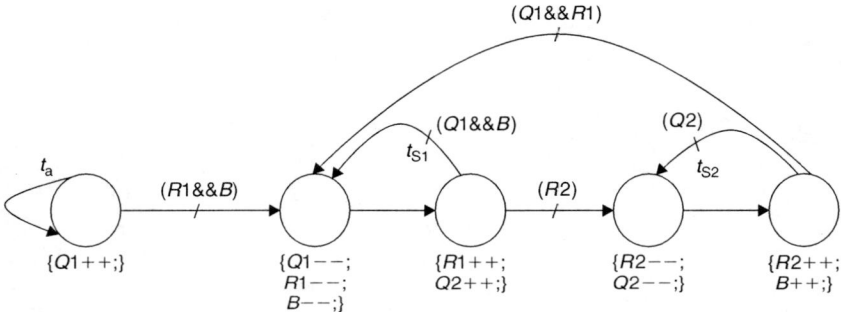

FIGURE 23.14 An ERG for a tandem communications channel with limited message packet capacity.

Step 3. Add state changes:
for $O(t)$, decrement $X(p) \; \forall \, p \in I_p(t)$;
for $D(t)$, increment $X(p) \; \forall \, p \in O_p(t)$.
Step 4. Initialize the values of state variables in the ERG to the initial markings of the STPN.

This algorithm generalizes some of the more intuitive ideas in Schruben and Yucesan (1994). The algorithm applied to the STPN in Figure 23.11 results in the ERG in Figure 23.5 after elimination of some redundant events needed by the Petri net (Yucesan and Schruben, 1992).

Our second example is a buffered tandem queue with limited buffer size (channel bandwidth) between the two resources—a sending resource $R1$ and a receiving resource R2. Here, we model communications blocking with limited concurrent message packet capacity B between the two resources. The STPN model for this system is given in Figure 23.13.

Applying the translation algorithm given earlier to this model results in the ERG given in Figure 23.14. The arc conditions in Figure 23.14 are not shown; these conditions simply require that all state variables decremented in the arc's destination vertex are positive. In Figure 23.14, the "++" notation from C is used: X++ means X is incremented by 1 ($X=X+1$) and X−− means X is decremented by 1 ($X=X-1$).

ERGs are more general and typically more parsimonious than basic STPNs for modeling discrete-event dynamic systems. Petri nets have an intuitive graphical representation of resource contention. The two modeling paradigms are most effective when used together. Basic STPNs can be used to develop an intuitive high-level model of resource contention for a single component. These STPNs should then be translated to ERGs for efficient simulation. In cases where STPNs are impossibly cumbersome, parametric ERGs can be very compact and effective in capturing the system dynamics.

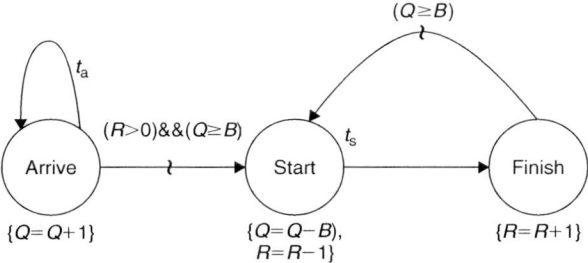

FIGURE 23.15 Batch processing with parallel resources (b, batch size; r, idle resources; t_s service times; t_a, interarrival times).

23.4.3 Process Interaction Flows

It is easy to model process flows by considering the subgraph of an ERG that models the action of a resource (e.g., the START and LEAVE nodes in Figure 23.5) as a single "process" node. When these nodes are combined into a single node then the ERG becomes a flow graph for modeling transient entity flows.

23.4.4 Generalized Semi-Markov Processes

A Generalized semi-Markov Processes (GSMP) is a mathematical construction, consisting of sets of events, event clocks, and states for modeling discrete-event dynamic systems. GSMPs have been used extensively to prove various mathematical properties of models and analysis methodologies. GSMPs are not general, but the class of systems they model is large and includes many interesting systems. However, GSMPs are abstract and provide very little insight into the systems being modeled. Like Petri nets, GSMPs can be mapped into ERGs using the algorithm in Schruben and Savage (1996). Representing GSMPs as ERGs allows visualization of a system that can be used to give intuitive interpretations to some of the mathematical properties for particular GSMPs. An example is the conditions on a GSMP for asymptotic consistency of infinitesimal perturbation analysis gradient estimators for simulation response optimization. In Freimer and Schruben (2001) these properties are expressed as intuitive structural properties of their ERG representations.

23.4.5 Mathematical Optimization Programs[2]

The dynamics of continuous systems are often modeled by a system of differential equations that express the relationships between changes in the values of system state variables. Given an initial state (boundary condition) these equations completely specify a model of the system's dynamic behavior. When this system of differential equations is particularly simple or has some special properties, it can be solved analytically to find the system's path of motion (trajectory). However, many interesting models are too complex and must be simulated by numerically solving (integrating) the set of differential equations. If the system is modeled using random processes, then the simulations can be used to generate sample paths for statistical analysis.

In an analogous manner, the relationships between changes in the values of state variables in a discrete-event system (events) can be modeled with an ERG. The vertices of the graph represent changes in state and the arcs the dynamic and logical relationships between these changes. An ERG, along with initial conditions, completely specifies the discrete-event system dynamics. Also, like for continuous systems, the dynamics of most discrete-event systems are complex and must be numerically simulated. In this section, we look at some properties of discrete-event models that allow us to solve, analytically, a system of difference equations represented in their ERGs.

A simple ERG models will be used as examples. Figure 23.15 is a simple batch processing system with parallel resources.

---

[2] *Source*: This section is largely adapted from Schruben 2000, used with permission.

The initial value of Q is assumed to be zero and the initial value of R is the number of parallel servers. Of course, there are several equivalent ERGs for this system (in fact, special cases of this system can be simulated with only one "Finish" event vertex).

We will first "read" the ERG in Figure 23.15. As discussed earlier, ERGs are read by substituting for A, B, (i), and t and paraphrasing the arc definition in Figure 23.4. An accurate and concise description of the system dynamics can typically be given with only one sentence for each arc in the graph. The four arcs of the ERG in Figure 23.15 give a four sentence full description of the system dynamics (event names are in italics).

Jobs *arrive* every t_a time units. Jobs that *arrive* to find an idle resource and complete a full batch will immediately *start* service. A resource will *finish* processing a batch of jobs t_s time units after it *starts* processing. When a resource *finishes* a batch, if a full batch of jobs is waiting, the resource can immediately *start* on the next batch.

Again, for networks of queues, the ERG represents a generic element of an array of ERGs by using event parameters k and j in the basic arc definition.

We will start by specifying the trajectory (say, waiting times or queue sizes) of n jobs processed in a G/G/1 FIFO queue as the solution to a linear program. The G/G/1 queue has general random time between job arrivals, general service times, and a single server. The ERG for a G/G/1 queue is Figure 23.15 with a batch size of $B=1$ and the number of idle resources R initially equal to 1. For this system, the linear program that specifies the dynamic system trajectory is almost obvious. The nonnegative decision variables in our linear program will be the event times. Here

A_i = the time of the ith Arrival event

S_i = the time of the ith Start event

F_i = the time of the ith Finish event.

The objective or the optimization will be to execute each event as soon as possible subject to the constraints imposed by the ERG,

$$\text{Min } Z = \Sigma(A_i + S_i + F_i)$$

Other objective functions will work for this model. In fact, we do not need to include the job arrival times in our objective function since they are not scheduled by conditional arcs. Also, if we knew that the N jobs occurred in the same busy period, then the simple objective of minimizing the length of the busy period, F_N, would suffice.

Each of the arcs of the ERG imposes a constraint on how events can be executed. For a given input process of arrival times and service times $(t_A(i), t_S(i): i = 1, 2, \ldots n)$. Simple linear constraints are enforced by each arc in the ERG.

$$A_{i+1} - A_i = t_A(i) \quad \text{(Arrival-Arrival arc)}$$
$$F_i - S_i = t_S(i) \quad \text{(Start-Finish arc)}$$
$$A_i \leq S_i \quad \text{(Arrival-Start arc)}$$
$$F_i \leq S_{i+1} \quad \text{(Finish-Start arc)}$$

$$A_i, S_i, \text{ and } F_i \geq 0$$

The analytic solution to this model is the dynamic system trajectory. If, for example, we can compute the sequence of customer waiting times as, $W_i = S_i - A_i$ either after the model is solved or by adding it as a constraint. The queue length process $Q(t) = C_A(t) - C_S(t)$ can also easily be computed from the trajectory.

We now illustrate the details with the simulation of multiple parallel resources and batch processing in Figure 23.15. A general algorithm for expressing ERGs as linear programs is given in Chan and

Modeling Causality with Event Relationship Graphs

Schruben (2005). We will use the well-known relationship between an event and its counting point process (Nelson, 1995),

$$E_i \leq t \Leftrightarrow C_E(t) \geq i$$

This relationship simply enforces the forward progression of time: if E occurs i times at or before time t then it will have occurred at least that many times by time t.

The number of jobs in line at time t, $Q(t)$, is equal to the number of Arrival events (incrementing Q by 1) that have occurred minus B times the number of Start events (each decrementing Q by B) that have occurred, or,

$$Q(t) = C_A(t) - B * C_S(t)$$

At any time in the simulation $Q(t)$ must be greater than zero. Consider the instant, $t = S_i$, at which the ith batch service starts, then $C_S(S_i) = i$.

Therefore,

$$0 \leq Q(S_i) = C_A(S_i) - B * C_S(S_i)$$
$$\Rightarrow C_A(S_i) \geq B * C_S(S_i)$$
$$\Rightarrow C_A(S_i) \geq B * i$$
$$\Rightarrow A_{B*i} \leq S_i$$

This constraint, $A_{B*i} \leq S_i$ simply says that, since jobs are processed B at a time, then B times as many jobs must arrive as have started service.

Consider again any instant, t, at which the ith service can start. For service to start, there must be at least one idle resource at time t. The number of idle resources at any time is equal to a count of the initial number of idle resources, R, less than the count of the number of Start events (decrementing resources) plus the count of the number of Finish events (incrementing resources). Therefore, at the time of the ith Start event, S_I, there must be a positive number of idle resources,

$$R + C_F(S_i) - C_S(S_i) \geq 1$$
$$\Rightarrow C_F(S_i) \geq C_S(S_i) + 1 - R$$
$$\Rightarrow C_F(S_i) \geq i + 1 - R$$
$$\Rightarrow F_{i+1-R} \leq S_i$$

This constraint, $F_{i+1-R} \leq S_i$ simply says that, since there are only R resources, the number of Start events cannot exceed the number of Finish events by more than $R+1$. This constraint is enforced in the simulation by the arc from the Finish event to the Start event in the ERG of Figure 23.15.

In general, the number of occurrences of events that decrement the availability of a limited resource can never exceed the number of events that increment that resource by the more than the number of such resources. Sets of vertices in the resource-driven simulation that relate to limited resources will have such constraints even if they do not share an arc in the ERG.

To summarize the ERG in Figure 23.15 translates into the following linear program:

(Events occur as soon as feasible)

$$\text{Min } \Sigma(A_i + S_i + F_i)$$

The two unconditional timed arcs provide the constraints

$$A_{i+1} = A_i + t_A(i) \text{ and}$$
$$F_k = S_i + t_S(i)$$

While the two conditional zero-delay arcs provide the constraints

$$S_i \geq A_{B*i} \quad \text{and} \quad S_i \geq F_{i+1-R}$$

The subscripts on the last constraints reflect the bounds on the number of resources, R, in our system and the Batch size.

Applications of modeling ERG trajectories as the solutions to linear programs include developing optimal resource scheduling models, using the duals of the linear programs for sensitivity analysis, using properties of the linear programs to prove system properties such as tandem queue reversibility, and determining stochastic orderings of systems (Chan, 2005).

23.5 Simulation of Event Relationship Graphs

ERG models can be developed graphically and simulated easily with the software package SIGMA (Schruben and Schruben, 2005), This software provides a simple but powerful and practical method for ERG simulation modeling. SIGMA, the Simulation Graphical Modeling and Analysis system, is an integrated, interactive approach to building, testing, animating, and experimenting with discrete-event simulations. SIGMA supports the full simulation ERG model life cycle from model building and testing to output analysis, animation, documentation, and report writing.

For speed and portability, SIGMA can automatically translate a model (with a mouse click) into a fast C code. Not only does this code allow models to run thousands of times faster, but also the compiled models can be run from a spreadsheet interface using preprogrammed Excel Visual Basic Templates (Schruben and Schruben, 2005) with multiple experimental runs batched together. SIGMA also can write a verbal description of a model in English to aid in debugging and model verification.

SIGMA graphically models the events taking place within a system and the cause and effect relationships among these events. One of SIGMA's more useful features is that simulation models can be created, enriched, and edited *while they are running*. Events can be added, altered, executed, canceled, or deleted during a simulation run. Logic can be changed and errors corrected without stopping a run to change code and recompile, so the modeler can interactively "replay" interesting events.

Animation support is fundamentally different in SIGMA from most other simulation modeling environments. Animations are not created from simulation models using separate add-on software. In SIGMA, the *animation and the simulation code is identical*. In addition to graphical modeling, analysis, and animation, SIGMA also includes graphical data tracking tools and allows pictures, graphs, plots, and data to be pasted into spreadsheets and word processors.

Multiple SIGMA sessions can be run concurrently. Objects can be copied and pasted from one modeling session to another. In fact, models can be developed in one SIGMA session and then graphically integrated into another simulation model while that model is executing.

23.6 Event Relationship Graph Analysis

There is a rich literature on the analysis of ERGs. This includes mappings of GSMPs and Petri nets into ERGs as mentioned earlier allowing the rich analytical methodologies of Petri nets and GSMPs to be applied to ERGs. There are also ERG-specific analytical tools such as in Som and Sargent (1990), where rules for elimination and consolidation of events for more efficient simulation are given. They also identify events that "interact" and might cause logical errors when they occur at the same time and are executed in different orders.

Earlier it was shown that how the set of all possible sample paths from an ERG as the solutions to optimization programs (Schruben, 2000; Chan and Schruben, 2005). This is done by algorithmic mapping of the ERG models into linear and mixed integer programs. These programs are often simple linear programs which have duals that can be used for sensitivity analysis. For example, the dual of a G/G/s queueing system is what is known in operations research as the production lot sizing problem. These optimization model representations for ERGs also provide optimal resource scheduling models and alternative simple proofs of complicated system properties, such as the reversibility of tandem queues (Chan and Schruben, 2003a, 2003b; Chan, 2005). This mapping allows the rich algorithms and theory of optimization to be applied to ERGs.

23.7 Experimenting with ERGs

Parametric ERGs can be used for running experiments where simultaneous replications of the same or competing systems are done in a single run. An event parameter is used to designate to which system each event execution belongs.

It is also possible to use different timescales for the systems corresponding to different design points. In this manner, the run can focus on factor settings or systems that are likely to be optimal and feasible rather than spend time simulating systems that are not contenders. This idea can be generalized as illustrated in Hyden (2000) where a large-scale experiment to optimize a production system was solved orders of magnitude faster than the commercial simulation optimization software available at that time.

In Schruben (1997) an example is presented using a penalty function to dilate event times to find the cycle-time constrained capacity of a queue. The cycle-time constrained capacity of a queueing system is defined as the maximum job arrival rate to the system so that the average job delay time is below some target.

Determining the cycle-time constrained capacity of a queueing system is straightforward when the cycle-time function is known from queueing theory. However, small errors in approximating this function can cause large errors in capacity estimation. This is particularly true for the short cycle-times found in highly competitive industries; the trade-off curve in this region is flat. Standard simulation experiments are almost useless in estimating the asymptotic upper bound on queueing system capacity. At high arrival rates, observed cycle-times are highly correlated and run initialization bias is a serious concern. These factors combine to give simulation estimators of heavy-traffic cycle-times both high bias and high variance. For example, simulation of millions of wafer flows are needed to estimate semiconductor factory capacities.

An experimental strategy for simulation optimization is to assign parameters to the "arrival" from a grid of interesting arrival rates. A range of arrival rates can then be simultaneously simulated during a single run. This grid can be refined during the run if the initial grid is found to be too coarse.

As the simulation progresses, we want to spend more and more time running events that are near the solution. To do this we will penalize rates that are not performing well or appear to be infeasible. The "penalty function" takes the form of time dilation for events associated with arrival rates that are unlikely to be near the capacity. When their relative timescales are increased, events will naturally tend to be scheduled near the end of the pending events list. If the events list is very large, these penalized events become essentially irrelevant and have no detrimental effect on execution speed. The number of event executions devoted to a particular design point reflects the likelihood that the design point is optimal. Hence, the simulation run is concentrated on those experimental points where success is most likely. This has the positive effect of minimizing the estimator variance at exactly the right place. Events corresponding to uninteresting parameter values will occur occasionally as in simulated annealing. This is necessary if there is to be any theoretical hope of global optimization.

For illustration, this experimental technique was tested with a simulated single queue like that in Figure 23.5. The simultaneous replication strategy and time dilation techniques in Schruben (1992) were used. For this test system, exponential interarrival times and service times were used so the true cycle-time constrained capacity rate is known. To put this problem in perspective, conventional replication methods would require hundreds of thousands of simulated jobs to get a reasonably good estimate of the cycle-time constrained capacity of this simple system.

Without using any of the information about the system (or even the fact that cycle-times increase with arrival rate), we chose a grid of 40 arrival rates from 0.03125 to 1.25 is run. Therefore, there were

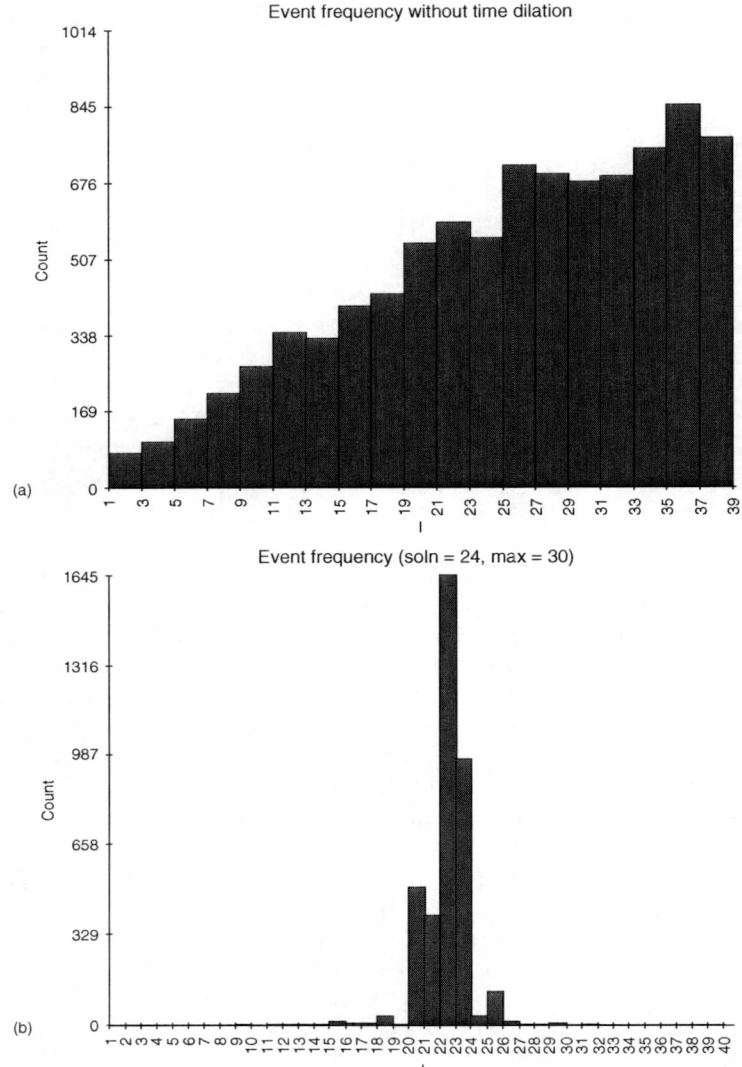

FIGURE 23.16 Event execution frequency. (a) Without time dilation and (b) with dilation.

40 different possible parameter values for each executed event, depending on which arrival rate spawned the event. We will set the average processing time at 1 so systems with arrival rates over 1 are unstable. We did not use this information so some of the systems included in our run are unstable.

Let W_i denote the current average job delay corresponding to the ith arrival rate being used. We will estimate the $W_i < 4$ capacity of this system with one simulation run. For time dilation, the timescale for the arrival event at rate i was multiplied by $(W_i - 4)^2$. The $W_i < 4$ capacity corresponds to event parameter $i = 24$. All event indices over 30 correspond to unstable arrival rates.

After an initialization period of 5,000 jobs, the relative frequency that events corresponding to the different arrival rates appear in Figure 23.16(a).

For the next 5,000 jobs, a quadratic time dilation penalty was then invoked (Figure 23.16[b]). After a total of 10,000 simulated jobs shared across all 40 systems, the event index at the correct solution of

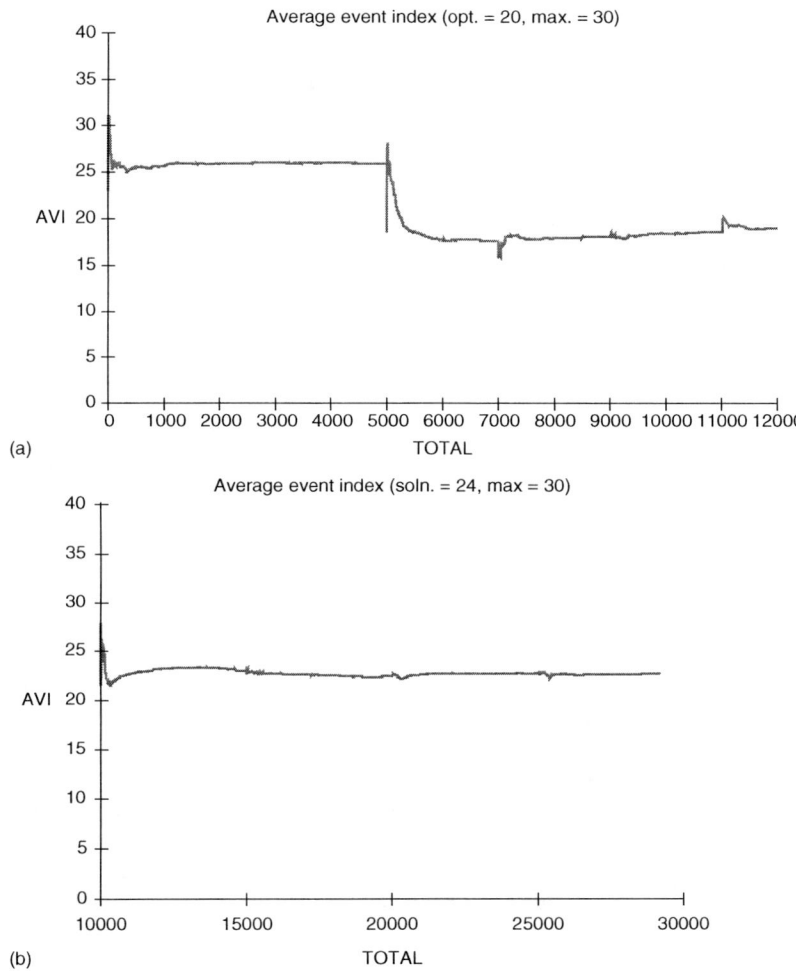

FIGURE 23.17 Average event indices for (a) $W_i < 2$ and (b) $W_i < 4$ capacity experiments.

24 is clearly indicated. Events with unstable arrival rates received early penalties, effectively excluding them from the experiment.

In Figure 23.17(a) and Figure 23.17(b), average event indices are plotted for both the $W_i < 2$ and $W_i < 4$ capacities. Periodically, these averages are reset causing a jump in the figures. These averages quickly converged to the correct indices for ERGs having the optimal arrival rates as their event parameters.

Summary

This chapter introduces ERGs and illustrates some of their applications and properties. ERGs are a minimalist approach to modeling discrete-event system dynamics that are completely general (having the full power of Turing machines). ERGs express the logical and temporal relationships between changes in system state, analogous to systems of differential equations for modeling continuous state system dynamics. Parametric ERGs can model huge and complex systems using very small finite graphs. Specification of the initial and terminating states for an ERG completely and unambiguously (perhaps using arc priorities) determine the state trajectory for a discrete-event model. Such models can be easily and efficiently simulated using software such as SIGMA.

Some analytical properties of ERGs are also presented, most notably their representation as linear programming or mixed integer optimization models. These linear program (LP) representations allow the rich set of analytical and algorithmic methodologies from optimization to be applied to the study of discrete-event dynamic systems as well as define explicit dual for discrete-event systems.

Acknowledgments

The author appreciates the support of the National Science Foundation through grant DMI0323765 to the University of California, Berkeley. He is also grateful for the contributions of his students and colleagues, in particular, those of former students W. K. Chan, P. Hyden, M. Oman, D. Pederson, V. Peterson, and T. Roeder.

References

Askin, R. G. and C. R. Standridge (1993). *Modeling and Analysis of Manufacturing Systems*, New york: Wiley.

Buss, A. H. and P. J. Sanchez (2002). Modeling very large scale systems: Building complex models with LEGOs (listener event graph objects). *Proceedings of the 2002 Winter Simulation Conference*, pp. 732–737.

Chan, W. K. (2005). Mathematical programming models for discrete event systems dynamics PhD thesis, Department of Industrial Engineering and Operations Research, University of California, Berkeley.

Chan, W. K. and L. W. Schruben (2003a). Properties of discrete event systems from their mathematical programming representations. *Proceedings of the 2003 Winter Simulation Conference*, pp. 496–502.

Chan, W. K. and L. W. Schruben (2003b). Reversibility of tandem queueing networks from their mathematical programming representations. Technical Report, University of California-Berkeley, Berkeley, CA.

Chan, W. K. and L. W. Schruben (2004). Generating scheduling constraints for discrete event dynamic systems. *Proceedings of the 2004 Winter Simulation Conference Proceedings*, pp. 568–574.

Chan, W. K. and L. W. Schruben (2005). Optimization representations of discrete-event system dynamics, unpublished manuscript (submitted for publication).

Ding, S. and J. Yi (2004). An event relationship graph based simulation and analysis of multi-cluster tools. *Proceedings of the 2004 Winter Simulation Conference Proceedings*, pp. 1915–1920.

Fishwick, P. (1995) *Simulation Model Design and Execution: Building Digital Worlds*, Englewood Cliffs, NJ: Prentice-Hall.

Freimer, M. and L. W. Schruben (2001). Graphical representation of IPA estimation. *Proceedings of the 2001 Winter Simulation Conference*, pp. 422–427.

Hoover, S. and R. Perry (1990). *Simulation: A Problem Solving Approach*, Reading, MA: Addison-Wesley.

Hyden, P. (2000). Designing simultaneous simulation experiments, PhD thesis, Department of Operations Research and Industrial Engineering, Cornell University.

Hyden, P., L. W. Schruben and T. K. Roeder (2001). Resource graphs for modeling large-scale, highly congested systems. *Proceeding of the 2001 Winter Simulation Conference*, pp. 523–529.

Ingalls, R. G., D. J. Morrice, and A. B. Whinston (2000). The implementation of temporal intervals in qualitative simulation graphs. *ACM Transactions on Modeling and Computer Simulation (TOMACS)* 10(3), pp. 215–240.

Law, A. and W. D. Kelton (2000). *Simulation Modeling and Analysis (3rd ed.)*, New York: McGraw-Hill.

Nehme, D. A. and N. G. Pierce (1994). Evaluating the throughput of cluster tools using event-graph simulations. *IEEE/SEMI 1994 Advanced Semiconductor Manufacturing Conference and Workshop. Theme—Manufacturing Excellence: A Global Challenge. ASMC'94 Proceedings*, pp.189–192

Nelson, B. (1995). *Simulation and Stochastic Processes*, New York: Wiley.
Pederson, D. and C. Trout (2002) Demonstrated benefits of cluster tool simulation. *Proceedings of the 2002 International Conference on Modeling and Analysis of Semiconductor Manufacturing*, pp. 84–89.
Pegden, C. D. (1986). *Introduction to SIMAN* (2nd ed.), State College, PA: Systems Modeling Corp.
Perkinson, T. L., P. K. McLarty, R. S. Gyurcsik, and R. K. Cavin III (1994). Single-wafer cluster tool performance: An analysis of throughput. *IEEE Transactions on Semiconductor Manufacturing* 7(3), pp. 369–373.
Sargent, R. G. (1988). Event graph modeling for simulation with an application to flexible manufacturing systems. *Management Science* 24(10), pp. 1231–1351.
Savage, E. L., L. W. Schruben and E. Yucesan (2005). On the generality of event relationship graph models. *INFORMS Journal on Computing* 17(1), pp. 3–9.
Schruben, L. (1983). Simulation modeling with event graphs. *Communications of the Association of Computing Machinery* 26(11), pp. 957–963.
Schruben, L. and E. Savage (1996). Visualizing generalized semi-Markov processes. *Proceedings of the 1996 Winter Simulation Conference*, Orlando, FL, December 11–14, 1994, pp. 560–565.
Schruben, L. W. (1997). Simulation optimization using simultaneous replications and event time dilation. *Proceedings of the 1997 Winter Simulation Conference*, pp. 177–180.
Schruben, L. W. (2000). Mathematical programming models of discrete event system dynamics. *2000 Winter Simulation Conference Proceedings*, vol. 1, pp. 381–385.
Schruben, L. W. and T. M. Roeder (2003). Fast simulations of large-scale highly-congested systems. *Simulation: Transactions of the Society for Modeling and Simulation International* 79(3), pp. 1–11.
Schruben, D. and L. W. Schruben (2005). *Graphical Simulation Modeling Using SIGMA*, Custom Simulations, www.customsimulations.com.
Schruben, L. W. and E. Yucesan (1994). Transforming Petri nets into event graph models. *Proceedings of the 1994 Winter Simulation Conference*, Orlando, FL, December 11–14, 1994, pp. 560–565.
Seila, A. F., V. Ceric, and P. Tadikamalla (2003). *Applied Simulation Modeling*, Belmont, CA: Thomson.
Som, T. K. and R. G. Sargent (1990). A formal development of event relationship graphs as an aid to structured and efficient simulation programs. *ORSA Journal on Computing* 1(2), pp. 107–125.
Wu, J.-H. and C.-N. Chung (1991). Timed finite automata as the theoretical foundation for simulation modeling with event relationship graphs. Technical Report, Department of Dec. Sci. and Inf. Sys., University of Kentucky, Lexington, KY.
Yucesan, E. and L. W. Schruben (1992). Structural and behavioral equivalence of simulation models. *ACM Transactions on Modeling and Computer Simulation* 2(1), pp. 82–103.

24
Petri Nets for Dynamic Event-Driven System Modeling

Jiacun Wang
Monmouth University

24.1 Introduction 24-1
24.2 Petri Net Definition 24-1
24.3 Transition Firing 24-3
24.4 Modeling Power 24-4
24.5 Petri Net Properties 24-5
   Reachability • Safeness • Liveness
24.6 Analysis of Petri Nets 24-7
   Reachability Analysis • Incidence Matrix and State Equation • Invariant Analysis • Simulation
24.7 Colored Petri Nets 24-10
24.8 Timed Petri Nets 24-12
   Deterministic Timed Petri Nets • Stochastic Timed Petri Nets
24.9 Concluding Remark 24-16

24.1 Introduction

Petri nets were introduced in 1962 by Dr. Carl Adam Petri (Petri, 1962). Petri nets are a powerful modeling formalism in computer science, system engineering, and many other disciplines. Petri nets combine a well-defined mathematical theory with a graphical representation of the dynamic behavior of systems. The theoretical aspect of Petri nets allows precise modeling and analysis of system behavior, while the graphical representation of Petri nets enables visualization of the modeled system state changes. This combination is the main reason for the great success of Petri nets. Consequently, Petri nets have been used to model various kinds of dynamic event-driven systems such as computer networks (Ajmone Marsan et al., 1986), communication systems (Merlin and Farber, 1976), manufacturing plants (Venkatesh et al., 1994; Zhou and DiCesare, 1989; Desrochers and Ai-Jaar, 1995), command and control systems (Andreadakis and Levis, 1988; Wang et al., 2000), real-time computing systems (Mandrioli et al., 1996; Tsai et al., 1995), logistic networks (van Landeghem and Bobeanu, 2002), and workflows (van der Aalst and van Hee, 2000; Lin et al., 2002) to mention only a few important examples. This wide spectrum of applications is accompanied by wide spectrum different aspects, which have been considered in the research on Petri nets.

24.2 Petri Net Definition

A Petri net is a particular kind of bipartite directed graphs populated by four types of objects. These objects are *places*, *transitions*, *directed arcs*, and *tokens*. Directed arcs connect places to transitions or transitions to places. In its simplest form, a Petri net can be represented by a transition together with an input place and an output place. This elementary net may be used to represent various aspects of the modeled systems. For example, a transition and its input place and output place can be used to represent a data processing event, its input data and output data, respectively, in a data processing system. To study the dynamic behavior of a Petri net modeled system in terms of its states and state changes, each place may contain zero or a positive number of tokens. Tokens are a primitive concept for Petri nets in addition to places and transitions. The presence or absence of a token in a place can indicate whether a condition associated with this place is true or false, for instance.

Denote by N the set of nonnegative integers. A Petri net is formally defined as a five-tuple $N = (P, T, I, O, M_0)$, where

(1) $P = \{p_1, p_2, \ldots, p_m\}$ is a finite set of places.
(2) $T = \{t_1, t_2, \ldots, t_n\}$ is a finite set of transitions. $P \cup T \neq \emptyset$ and $P \cap T = \emptyset$.
(3) $I: T \times P \rightarrow N$ is an *input matrix* that specifies directed arcs from places to transitions; its entry $I(t_i, p_j)$ represents the number of arcs connecting place p_j to transition t_i.
(4) $O: T \times P \rightarrow N$ is an *output matrix* that specifies directed arcs from transitions to places; its entry $O(t_i, p_j)$ represents the number of arcs connecting transition t_i to place p_j.
(5) $M_0: P \rightarrow N$ is the *initial marking*.

A *marking* in a Petri net is an assignment of tokens to the places of a Petri net. Tokens reside in the places of a Petri net. The number and position of tokens may change during the execution of a Petri net. The tokens are used to define the execution of a Petri net.

Most theoretical work on Petri nets is based on the formal definition of Petri nets. However, a graphical representation of a Petri net is much more useful for illustrating the concepts of Petri net theory. A Petri net graph is a Petri net depicted as a bipartite directed multigraph. Corresponding to the definition of Petri nets, a Petri net graph has two types of nodes: a *circle* that represents a place, and a *bar* or *box* that represents a transition. Directed arcs (arrows) connect places and transitions, with some arcs directed from places to transitions and other arcs directed from transitions to places. An arc directed from a place p_j to a transition t_i defines p_j to be an *input place* of t_i, denoted by $I(t_i, p_j) = 1$. An arc directed from a transition t_i to a place p_j defines p_j to be an *output place* of t_i, denoted by $O(t_i, p_j) = 1$. If $I(t_i, p_j) = k$ (or $O(t_i, p_j) = k$), then there exist k directed (parallel) arcs connecting place p_j to transition t_i (or connecting transition t_i to place p_j). Usually, in the graphical representation, parallel arcs connecting a place (transition) to a transition (place) are represented by a single directed arc labeled with its multiplicity, or weight k. A circle containing a *dot* represents a place contains a token (Peterson, 1981).

Example 1
A simple Petri net.

Figure 24.1 shows a simple Petri net. In this Petri net, we have

$P = \{p_1, p_2, p_3, p_4\}$;
$T = \{t_1, t_2, t_3\}$;
$I(t_1, p_1) = 2, I(t_1, p_i) = 0$ for $i = 2, 3, 4$;

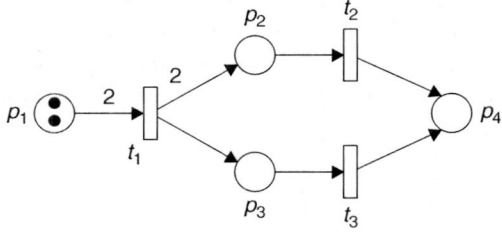

FIGURE 24.1 A simple Petri net.

$I(t_2, p_2) = 1, I(t_2, p_i) = 0$ for $i = 1, 3, 4$;
$I(t_3, p_3) = 1, I(t_3, p_i) = 0$ for $i = 1, 2, 4$;
$O(t_1, p_2) = 2, O(t_1, p_3) = 1, O(t_1, p_i) = 0$ for $i = 1, 4$;
$O(t_2, p_4) = 1, O(t_2, p_i) = 0$ for $i = 1, 2, 3$;
$O(t_3, p_4) = 1, O(t_3, p_i) = 0$ for $i = 1, 2, 3$;
$M_0 = (2\ 0\ 0\ 0)$.

24.3 Transition Firing

The execution of a Petri net is controlled by the number and distribution of tokens in the Petri net. By changing distribution of tokens in places, which may reflect the occurrence of events or execution of operations, for instance, one can study the dynamic behavior of the modeled system. A Petri net is executed by *firing* transitions. We now introduce the enabling rule and firing rule of a transition, which govern the flows of tokens:

(1) *Enabling Rule*: A transition t is said to be *enabled* if each input place p of t contains at least the number of tokens equal to the weight of the directed arc connecting p to t, i.e., $M(p) \geq I(t, p)$ for all p in P. If $I(t, p) = 0$, then t and p are not connected, so we do not care about the marking of p when considering the firing of t.

(2) *Firing Rule*: Only enabled transitions can fire. The firing of an enabled transition t removes from each input place p the number of tokens equal to $I(t, p)$, and deposits in each output place p the number of tokens equal to $O(t, p)$.

Mathematically, firing t at M yields a new marking

$$M'(p) = M(p) - I(t, p) + O(t, p) \quad \text{for all } p \text{ in } P$$

Note that since only enabled transitions can fire, the number of tokens in each place always remains nonnegative when a transition is fired. Firing a transition can never try to remove a token that is not there.

A transition without any input place is called a *source transition*, and one without any output place is called a *sink transition*. Note that a source transition is unconditionally enabled, and that the firing of a sink transition consumes tokens, but does not produce tokens.

A pair of a place p and a transition t is called a *self-loop*, if p is both an input place and an output place of t. A Petri net is said to be *pure* if it has no self-loops.

Example 2
Transition firing.

Consider the simple Petri net shown in Figure 24.1. Under the initial marking, $M_0 = (2\ 0\ 0\ 0)$, only t_1 is enabled. Firing of t_1 results in a new marking, say M_1. It follows from the firing rule that

$$M_1 = (0\ 2\ 1\ 0)$$

The new token distribution of this Petri net is shown in Figure 24.2. Again, in marking M_1, both transitions of t_2 and t_3 are enabled. If t_2 fires, the new marking, say M_2, is

$$M_2 = (0\ 1\ 1\ 1)$$

If t_3 fires, the new marking, say M_3, is

$$M_3 = (0\ 2\ 0\ 1)$$

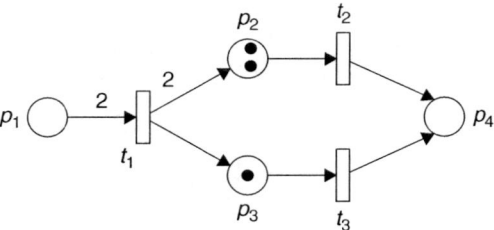

FIGURE 24.2 Firing of transition t_1.

24.4 Modeling Power

The typical characteristics exhibited by the activities in a dynamic event-driven system, such as concurrency, decision making, synchronization, and priorities, can be modeled effectively by Petri nets.

1. *Sequential Execution.* In Figure 24.3(a), transition t_2 can fire only after the firing of t_1. This imposes the precedence constraint "t_2 after t_1." Such precedence constraints are typical of the execution of the parts in a dynamic system. Also, this Petri net construct models the causal relationship among activities.
2. *Conflict.* Transitions t_1 and t_2 are in conflict in Figure 24.3(b). Both are enabled but the firing of any transition leads to the disabling of the other transition. Such a situation will arise, for example, when a machine has to choose among part types or a part has to choose among several machines. The resulting conflict may be resolved in a purely nondeterministic way or in a probabilistic way, by assigning appropriate probabilities to the conflicting transitions.
3. *Concurrency.* In Figure 24.3(c), the transitions t_1 and t_2 are concurrent. Concurrency is an important attribute of system interactions.
4. *Synchronization.* It is quite normal in a dynamic system that an event requires multiple resources. The resulting synchronization of resources can be captured by transitions of the type shown in Figure 24.3(d). Here, t_1 is enabled only when each of p_1 and p_2 receives a token. The arrival of a token into each of the two places could be the result of a possibly complex sequence of operations elsewhere in the rest of the Petri net model. Essentially, transition t_1 models the joining operation.
5. *Mutually exclusive.* Two processes are mutually exclusive if they cannot be performed at the same time due to constraints on the usage of shared resources. Figure 24.3(e) shows this structure. For example, a robot may be shared by two machines for loading and unloading. Two such structures are parallel mutual exclusion and sequential mutual exclusion.
6. *Priorities.* The classical Petri nets discussed so far have no mechanism to represent priorities. Such a modeling power can be achieved by introducing an *inhibitor arc*. The inhibitor arc connects an input place to a transition, and is pictorially represented by an arc terminated with a small circle. The presence of an inhibitor arc connecting an input place to a transition changes the transition-enabling conditions. In the presence of the inhibitor arc, a transition is regarded as enabled if each input place, connected to the transition by a normal arc (an arc terminated with an arrow), contains at least the number of tokens equal to the weight of the arc, and no tokens are present on each input place connected to the transition by the inhibitor arc. The transition firing rule is the same for normally connected places. The firing, however, does not change the marking in the inhibitor arc connected places. A Petri net with an inhibitor arc is shown in Figure 24.3(f). t_1 is enabled if p_1 contains a token, while t_2 is enabled if p_2 contains a token and p_1 has no token. This gives priority to t_1 over t_2: in a marking in which both p_1 and p_2 have a token, t_2 would not be able to fire until t_1 is fired.
7. *Resource constraint.* Petri nets are well suited to model and analyze systems that are constrained by resources. For instance, Figure 24.4 depicts the Petri net model of a queue with two servers. The transition *a* models the arrival of clients, *b* and *c* indicate the start and end of the service, whereas *d* models the departure. The place *p* indicates clients that are waiting to be served, *q* models clients

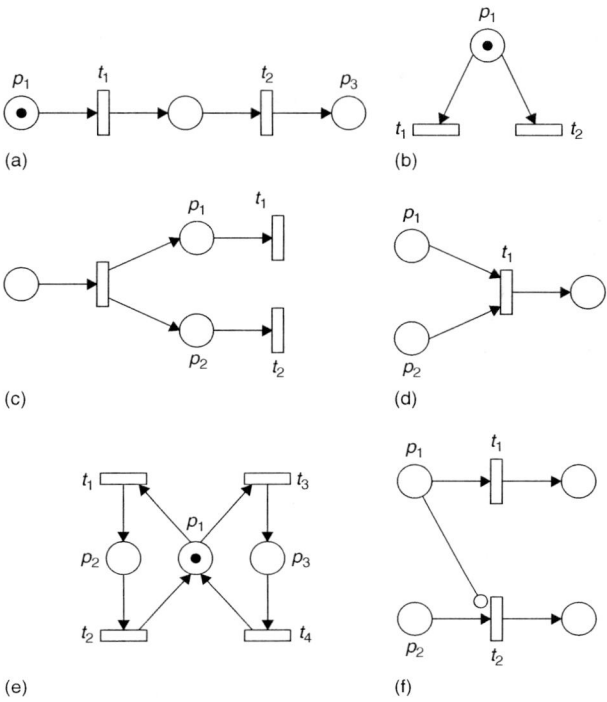

FIGURE 24.3 Petri net primitives to represent system features. (a) Sequential, (b) conflict, (c) concurrent, (d) synchronization, (e) mutual exclusive, and (f) priority.

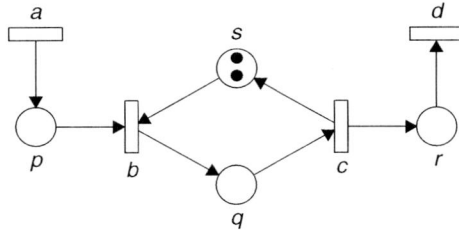

FIGURE 24.4 A queue with two servers.

that are being served. A client being served occupies a resource, the *server*. Place s indicates free resources; initially there are two resources (servers) available. To fire b, a waiting client in place p and a free server in place s have to be available. When the first client arrives, a server becomes busy. When the second client arrives before the first one has finished, the second server becomes busy as well. When the third arrives before the other two have finished, he has to wait until one of the two servers becomes available. The modeling of situations like the one sketched, where behaviors are concurrent up to a certain degree, can be done very naturally by means of Petri nets.

24.5 Petri Net Properties

As a mathematical tool, Petri nets possess a number of properties. These properties, when interpreted in the context of the modeled system, allow system designer to identify the presence or absence of the application domain-specific functional properties of the system under design. Two types of properties can be distinguished, behavioral and structural. The behavioral properties are those which depend on the initial state or marking of a Petri net. In contrast, the structural properties do not depend on the initial marking of a Petri net. They depend on the topology, or net structure, of a Petri net (Murata, 1989). Here we provide an overview of some of the most important, from the practical point of view, behavioral properties: reachability, safeness, and liveness.

24.5.1 Reachability

An important issue in designing event-driven systems is whether a system can reach a specific state, or exhibit a particular functional behavior. In general, the question is whether the system modeled with a Petri net exhibits all desirable properties as specified in the requirement specification, and no undesirable ones.

To find out whether the modeled system can reach a specific state as a result of a required functional behavior, it is necessary to find such a transition firing sequence that would transform its Petri net model from the initial marking M_0 to the desired marking M_j, where M_j represents the specific state, and the firing sequence represents the required functional behavior. In general, a marking M_j is said to be *reachable* from a marking M_i if there exists a sequence of transition firings that transforms M_i to M_j. A marking M_j is said to be *immediately reachable* from M_i if firing an enabled transition in M_i results in M_j. The set of all markings reachable from marking M is denoted by $R(M)$. We will explain how to get $R(M)$ later.

24.5.2 Safeness

In a Petri net, places are often used to represent information storage areas in communication and computer systems, product and tool storage areas in manufacturing systems, etc. It is important to be able to determine whether proposed control strategies prevent from the overflows of these storage areas. The Petri net property, which helps to identify the existence of overflows in the modeled system, is the concept of *boundedness*.

A place p is said to be *k-bounded* if the number of tokens in p is always less than or equal to k (k is a nonnegative integer number) for every marking M reachable from the initial marking M_0, i.e., $M \in R(M_0)$. It is *safe* if it is 1-bounded.

A Petri net $N = (P, T, I, O, M_0)$ is k-bounded (safe) if each place in P is k-bounded (safe). It is *unbounded* if k is infinitely large. For example, the Petri net of Figure 24.1 is 2-bounded, but the net of Figure 24.4 is unbounded.

24.5.3 Liveness

The concept of liveness is closely related to the *deadlock* situation, which has been situated extensively in the context of computer operating systems.

A Petri net modeling a deadlock-free system must be *live*. This implies that for any reachable marking M, any transition in the net can eventually be fired by progressing through some firing sequence. This requirement, however, might be too strict to represent some real systems or scenarios that exhibit deadlock-free behavior. For instance, the initialization of a system can be modeled by a transition (or a set of transitions) that fires a finite number of times. After initialization, the system may exhibit a deadlock-free behavior, although the Petri net representing this system is no longer live as specified above. For this reason, different levels of liveness are defined. Denote by $L(M_0)$ the set of all possible firing sequences starting from M_0. A transition t in a Petri net is said to be

(1) L0-live (or dead) if there is no firing sequence in $L(M_0)$ in which t can fire.
(2) L1-live (potentially firable) if t can be fired at least once in some firing sequence in $L(M_0)$.
(3) L2-live if t can be fired at least k times in some firing sequence in $L(M_0)$ given any positive integer k.
(4) L3-live if t can be fired infinitely often in some firing sequence in $L(M_0)$.
(5) L4-live (or live) if t is L1-live (potentially firable) in every marking in $R(M_0)$.

For example, all the three transitions in the net of Figure 24.1 are $L1$-live because t_1 and t_3 can only fire once each while transition t_2 can fire twice. However, all transitions in the net of Figure 24.4 are $L4$-live, because they are all $L1$-live in every reachable marking.

24.6 Analysis of Petri Nets

We have introduced the modeling power of Petri nets in the previous sections. However, modeling by itself is of little use. It is necessary to *analyze* the modeled system. This analysis will hopefully lead to important insights into the behavior of the modeled system.

There are four common approaches to Petri net analysis: (1) reachability analysis, (2) the matrix-equation approach, (3) invariant analysis, and (4) simulation. The first approach involves the enumeration of all reachable markings, but it suffers from the state-space explosion issue. The matrix-equations technique is powerful but in many cases it is applicable only to special subclasses of Petri nets or special situations. The invariant analysis determines sets of places or transitions with special features, as token conservation or cyclical behavior. For complex Petri net models, discrete-event simulation is an option to check the system properties.

24.6.1 Reachability Analysis

Reachability analysis is conducted through the construction of reachability tree if the net is bounded. Given a Petri net N, from its initial marking M_0, we can obtain as many "new" markings as the number of the enabled transitions. From each new marking, we can again reach more markings. Repeating the procedure over and over results in a tree representation of the markings. Nodes represent markings generated from M_0 and its successors, and each arc represents a transition firing, which transforms one marking to another.

The above tree representation, however, will grow infinitely large if the net is unbounded. To keep the tree finite, we introduce a special symbol ω, which can be thought of as "infinity." It has the properties that for each integer n, $\omega > n$, $\omega + n = \omega$, and $\omega \geq \omega$. Generally, we do not know if a Petri net is bounded or not before we perform the reachability analysis. However, we can construct a *coverability tree* if the net is unbounded or a *reachability tree* if the net is bounded according to the following general algorithm:

1. Label the initial marking M_0 as the root and tag it "new."
2. For every new marking M:
   2.1. If M is identical to a marking already appeared in the tree, then tag M "old" and go to another new marking.
   2.2. If no transitions are enabled at M, tag M "dead-end" and go to another new marking.
   2.3. While there exist enabled transitions at M, do the following for each enabled transition t at M:
      2.3.1. Obtain the marking M' that results from firing t at M.
      2.3.2. On the path from the root to M if there exists a marking M'' such that $M'(p) \geq M''(p)$ for each place p and $M' \neq M''$, i.e., M'' is coverable, then replace $M'(p)$ by ω for each p such that $M'(p) > M''(p)$.
      2.3.3. Introduce M' as a node, draw an arc with label t from M to M', and tag M' "new."

If ω appears in a marking, then the net is unbounded and the tree is a coverability tree; otherwise, the net is bounded and the tree is a reachability tree. Merging the same nodes in a coverability tree (reachability tree) results in a coverability graph (*reachability graph*).

Example 3

Reachability analysis.

Consider the Petri net shown in Figure 24.1. All reachable markings are $M_0 = (2, 0, 0, 0)$, $M_1 = (0, 2, 1, 0)$, $M_2 = (0, 1, 1, 1)$, $M_3 = (0, 2, 0, 1)$, $M_4 = (0, 0, 1, 2)$, $M_5 = (0, 1, 0, 2)$, and $M_6 = (0, 0, 0, 3)$. The reachability tree of this Petri net is shown in Figure 24.5(a), and the reachability graph is shown in Figure 24.5(b).

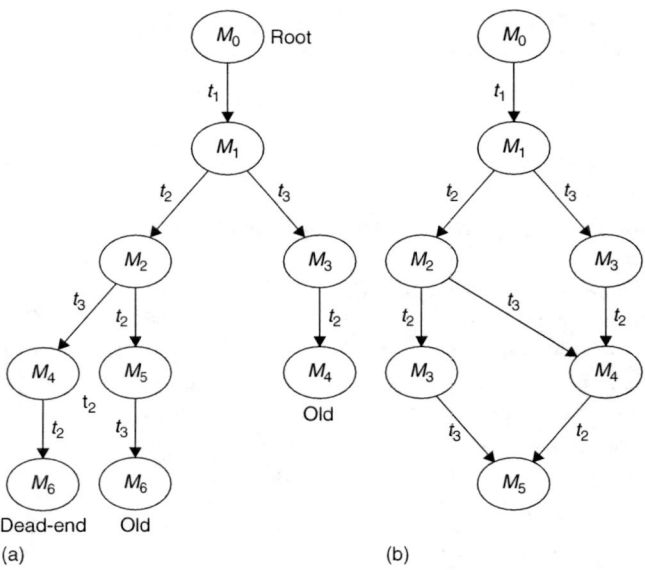

FIGURE 24.5 (a) Reachability tree. (b) Reachability graph.

24.6.2 Incidence Matrix and State Equation

For a Petri net with n transitions and m places, the incidence matrix $A = [a_{ij}]$ is an $n \times m$ matrix of integers and its entry is given by

$$a_{ij} = O(t_i, p_j) - I(t_i, p_j)$$

In writing matrix equations, we write a marking M_k as an $m \times 1$ column vector. The jth entry of M_k denotes the number of tokens in place j immediately after the kth firing in some firing sequence. The kth firing or *control vector* u_k is an $n \times 1$ column vector of $n-1$ zeroes and one nonzero entry, a 1 in the ith position indicating that transition i fires at the kth firing. Since the ith row of the incidence matrix A denotes the change of the marking as the result of firing transition i, we can write the following state equation for a Petri net:

$$M_k = M_{k-1} + A^T u_k, \quad k = 1, 2, \ldots$$

Suppose that a destination marking M_d is reachable from M_0 through a firing sequence $\{u_1, u_2, \ldots, u_d\}$. Writing the state equation for $k = 1, 2, \ldots, d$ and summing them, we obtain

$$M_d = M_0 + A^T \sum_{k=1}^{d} u_k$$

This state equation specifies a necessary condition for marking M_d being reachable from M_0, which is that there is a nonnegative solution of the firing vector $\sum_{k=1}^{d} u_k$.

24.6.3 Invariant Analysis

In a Petri net, arcs describe the relationships among places and transitions. They are represented by two matrices I and O. By examining the linear equations based on the execution rule and the matrices, one can find subsets of places over which the sum of the tokens remains unchanged. One may also find that a transition firing sequence brings the marking back to the same one. The concepts of *S-invariant* and *T-invariant* are introduced to reflect these properties.

Mathematically, an S-invariant is an integer solution y of the homogeneous equation

$$Ay = 0$$

and a T-invariant is an integer solution x of the homogeneous equation

$$A^T x = 0$$

The nonzero entries in an S-invariant represent weights associated with the corresponding places so that the weighted sum of tokens on these places is constant for all markings reachable from an initial marking. These places are said to be covered by an S-invariant. The nonzero entries in a T-invariant represent the firing counts of the corresponding transitions, which belongs to a firing sequence transforming a marking M_0 back to M_0. Although a T-invariant states the transitions comprising the firing sequence transforming a marking M_0 back to M_0, and the number of times these transitions appear in this sequence, it does not specify the order of the transition firings.

Invariant findings may help in the analysis of some Petri net properties. For example, if each place in a net is covered by an S-invariant, then it is bounded. However, this approach is of limited use since invariant analysis does not include all the information of a general Petri net.

The set of places (transitions) corresponding to nonzero entries in an S-invariant $y \geq 0$ (T-invariant $x \geq 0$) is called the *support of an invariant* and is denoted by $\|x\|$ ($\|y\|$). A support is said to be *minimal* if there is no other invariant y_1 such that $y_1(p) \leq y(p)$ for all p. Given a minimal support of an invariant, there is a unique minimal invariant corresponding to the minimal support. We call such an invariant a *minimal-support invariant*. The set of all possible minimal-support invariants can serve as a generator of invariants. That is, any invariant can be written as a linear combination of minimal-support invariants (Memmi and Roucairol, 1980).

Example 4

Figure 24.6 shows a simple manufacturing system with a single machine and a buffer. The capacity of the buffer is 1. A raw part can enter the buffer only when it is empty, otherwise it is rejected. As soon as the part residing in the buffer gets processed, the buffer is released and can accept another coming part. Fault may occur in the machine when it is processing a part. After being repaired, the machine continues to process the uncompleted part. The places and transitions in this Petri net are as follows:

- p_1: The buffer available.
- p_2: A part in the buffer.
- p_3: The machine available.
- p_4: The machine processing a part.
- p_5: The machine failed.
- t_1: A part arrives.
- t_2: The machine starts processing a part.
- t_3: The machine ends processing a part.

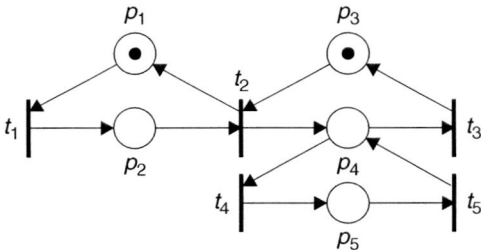

FIGURE 24.6 A simple manufacturing system with a single machine and a buffer.

t_4: The machine fails.
t_5: Repair the machine.

We are going to find the S-invariants of the net and use them to see how this simple manufacturing system works. First, we obtain the incidence matrix directly from the model:

$$A = \begin{bmatrix} -1 & 1 & 0 & 0 & 0 \\ 1 & -1 & -1 & 1 & 0 \\ 0 & 0 & 1 & -1 & 0 \\ 0 & 0 & 0 & -1 & 1 \\ 0 & 0 & 0 & 1 & -1 \end{bmatrix}$$

Then, by solving $Ay = 0$ we get two minimal-support S-invariants, $y_1 = (1\ 1\ 0\ 0\ 0)^T$ and $y_2 = (0\ 0\ 1\ 1\ 1)^T$, where $\|y_1\| = \{p_1, p_2\}$ and $\|y_2\| = \{p_3, p_4, p_5\}$ are corresponding minimal supports. Since the initial marking is $M_0 = (1\ 0\ 1\ 0\ 0)^T$, then $M_0^T y_1 = 1$ and this leads to the fact that

$$M(p_1) + M(p_2) = 1$$

It shows that the buffer is either free or busy. Similarly, it results from $M_0^T y_2 = 1$ that

$$M(p_3) + M(p_4) + M(p_5) = 1$$

It shows how the machine spends its time. It is either up and waiting, or up and working, or down.

24.6.4 Simulation

For complex Petri net models, simulation is another way to check the system properties. The idea is simple, that is, using the execution algorithm to run the net. Simulation is an expensive and time-consuming technique. It can show the presence of undesirable properties but cannot prove the correctness of the model in general case. Despite this, Petri net simulation is indeed a convenient and straightforward yet effective approach for engineers to validate the desired properties of a discrete-event system. The algorithm is given as follows:

(1) Initialization: decide the initial marking and the set of all enabled transitions in the marking.
(2) If the number of preset simulation steps or certain stopping criteria is met, stop. Otherwise, if there is no transition enabled, report a deadlock marking and either stop or go to Step 1.
(3) Randomly pick a transition to fire. Remove the same number of tokens from each of its input places as the number of arcs from that place to the transition and deposit the same number of tokens to each of its output places as the number of arcs from the transition to that place.
(4) Remove all disabled transitions from the enabled transition set, and add all newly enabled ones to the enabled transition set. Go to Step 2.

The above algorithm can be modified to simulate extended Petri nets such as timed ones. The advantage of the simulation methods is to allow one to derive the temporal performance for a system under very realistic assumptions. A list of Petri net simulation tools along with feature descriptions can be found in the following *Petri Nets World* website: http://www.informatik.uni-hamburg.de/TGI/PetriNets/.

24.7 Colored Petri Nets

In a standard Petri net, tokens are indistinguishable. Because of this, Petri nets have the distinct disadvantage of producing very large and unstructured specifications for the systems being modeled. To tackle this issue, high-level Petri nets were developed to allow compact system representation. Colored

Petri nets (CPNs) (Jensen, 1981) and Predicate/Transition (Pr/T) nets (Genrich and Lautenbach, 1981) are among the most popular high-level Petri nets. We will introduce colored Petri nets in this section.

Introduced by Kurt Jensen in 1981, a CPN has its each token attached with a color, indicating the identity of the token. Moreover, each place and each transition has a set of colors attached. A transition can fire with respect to each of its colors. By firing a transition, tokens are removed from the input places and added to the output places in the same way as that in original Petri nets, except that a functional dependency is specified between the color of the transition firing and the colors of the involved tokens. The color attached to a token may be changed by a transition firing and it often represents a complex data-value. CPNs lead to compact net models by using of the concept of colors. This is illustrated by Example 5.

Example 5

A manufacturing system.

Consider a simple manufacturing system comprising two machines M1 and M2, which process three different types of raw parts. Each type of parts goes through one stage of operation, which can be performed on either M1 or M2. After the completion of processing of a part, the part is unloaded from the system and a fresh part of the same type is loaded into the system. Figure 24.7 shows the (uncolored) Petri net model of the system. The places and transitions in the model are as follows:

$p_1(p_2)$: Machine M1 (M2) available.
$p_3(p_4, p_5)$: A raw part of type 1 (type 2, type 3) available.
$p_6(p_7, p_8)$: M1 processing a raw part of type 1 (type 2, type 3).
$p_9(p_{10}, p_{11})$: M2 processing a raw part of type 1 (type 2, type 3).
$t_1(t_2, t_3)$: M1 begins processing a raw part of type 1 (type 2, type 3).
$t_4(t_5, t_6)$: M2 begins processing a raw part of type 1 (type 2, type 3).
$t_7(t_8, t_9)$: M1 ends processing a raw part of type 1 (type 2, type 3).
$t_{10}(t_{11}, t_{12})$: M2 ends processing a raw part of type 1 (type 2, type 3).

Now let us take a look at the CPN model of this manufacturing system, which is shown in Figure 24.8. As we can see, there are only 3 places and 2 transitions in the CPN model, compared at 11 places and 12 transitions in Figure 24.7. In this CPN model, p_1 means machines are available (corresponding to places p_1 and p_2 in Figure 24.7), p_2 means parts available (corresponding to places p_3–p_5 in Figure 24.7), p_3 means processing in progress (corresponding to places p_6–p_{11} in Figure 24.7), t_1 means processing starts (corresponding to transitions t_1–t_6 in Figure 24.7), and t_2 means processing ends

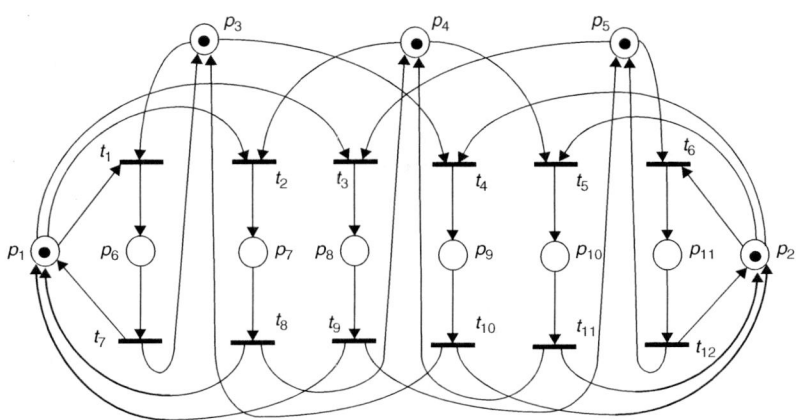

FIGURE 24.7 Petri net model of a simple manufacturing system.

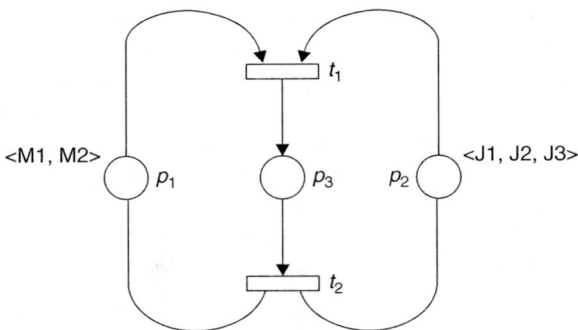

FIGURE 24.8 Colored Petri net model of the manufacturing system.

(corresponding to transitions t_7–t_{12} in Figure 24.7). There are three color sets: SM, SP, and $SM \times SP$, where $SM = \{M1, M2\}$, $SP = \{J1, J2, J3\}$. The color of each node is as follows:

$C(p_1) = \{M1, M2\}$
$C(p_2) = \{J1, J2, J3\}$
$C(p_3) = SM \times SP$
$C(t_1) = C(t_2) = SM \times SP$

CPN models can be analyzed through reachability analysis. As for ordinary Petri nets, the basic idea behind reachability analysis is to construct a reachability graph. Obviously, such a graph may become very large, even for small CPNs. However, it can be constructed and analyzed totally automatically, and there exist techniques that make it possible to work with condensed occurrence graphs without losing analytic power. These techniques build upon equivalence classes. Another option to the CPN model analysis is simulation. Readers are referred to Jensen (1997) for a detailed description of the concepts, analysis methods, and practical use of colored Petri nets.

24.8 Timed Petri Nets

The need for including timing variables in the models of various types of dynamic systems is apparent since these systems are real time in nature. In the real world, almost every event is time related. When a Petri net contains a time variable, it becomes a *timed Petri net* (Wang, 1998). The definition of a timed Petri net consists of three specifications:

- the topological structure,
- the labeling of the structure, and
- firing rules.

The topological structure of a timed Petri net generally takes the form that is used in a conventional Petri net. The labeling of a timed Petri net consists of assigning numerical values to one or more of the following things:

- transitions,
- places, and
- arcs connecting the places and transitions.

The firing rules are defined differently depending on the way the Petri net is labeled with time variables. The firing rules defined for a timed Petri net control the process of moving the tokens around.

The above variations lead to several different types of timed Petri nets. Among them, deterministic timed Petri nets (DTPNs) (Ramchandani, 1974) and stochastic timed Petri nets (STPNs) (Molloy, 1982;

Bause and Kritzinger, 2002), in which time variables are associated with transitions, are the two most widely used extended Petri nets.

24.8.1 Deterministic Timed Petri Nets

The introduction of deterministic time labels into Petri nets was first attempted by Ramchandani (1974). In his approach, the time labels were placed at each transition, denoting the fact that transitions are often used to represent actions, and actions take time to complete. The obtained extended Petri nets are called *deterministic timed Petri nets* (DTPNs). (Ramamoorthy and Ho, 1980) used such an extended model to analyze system performance. The method is applicable to a restricted class of systems called *decision-free nets*. This class of nets involves neither decisions nor nondeterminism. In structural terms, each place is connected to the input of no more than one transition, and to the output of no more than one transition.

A DTPN is a six-tuple (P, T, I, O, M_0, τ), where (P, T, I, O, M_0) is a Petri net and $\tau: T \to R^+$ a function that associates transitions with deterministic time delays.

A transition t_i in a DTPN can fire at time τ if and only if

(1) for any input place p of this transition, there have been the number of tokens equal to the weight of the directed arc connecting p to t_i in the input place continuously for the time interval $[\tau - \tau_i, \tau]$, where τ_i is the associated firing time of transition t_i;
(2) after the transition fires, each of its output places, p, will receive the number of tokens equal to the weight of the directed arc connecting t_i to p at time τ.

An important application of DTPN is to calculate the cycle time of a Petri net model. For a decision-free Petri net where every place has exactly one input arc and one output arc, the minimum cycle time (maximum performance) C is given by

$$C = \max \left\{ \frac{T_k}{N_k} : k = 1, 2, \Lambda, q \right\}$$

where $T_k = \sum_{t_i \in L_k} \tau_i$ is the sum of the execution times of the transitions in circuit k; $N_k = \sum_{p_i \in L_k} M(p_i)$ the total number of tokens in the places in circuit k; and q the number of circuits in the net.

Example 6

A communication protocol.

Consider the communication protocol between two processes, one indicated as the sender and the other as the receiver. The sender sends messages to a buffer, while the receiver picks up messages from the buffer. When it gets a message, the receiver sends an acknowledgment (ACK) back to the sender. After receiving the ACK from the receiver, the sender begins processing and sending a new message. Suppose that the sender takes 1 time unit to send a message to the buffer, 1 time unit to receive the ACK, and 3 time units to process a new message. Then, the receiver takes 1 time unit to get the messages from the buffer, 1 time unit to send back an ACK to the buffer, and 4 time units to process a received message. The DTPN model of this protocol is shown in Figure 24.9. The legends of places and transitions and timing properties are as follows:

p_1: The sender ready.
p_2: Message in the buffer.
p_3: The sender waiting for ACK.
p_4: Message received.
p_5: The receiver ready.
p_6: ACK sent.
p_7: ACK in the buffer.
p_8: ACK received.

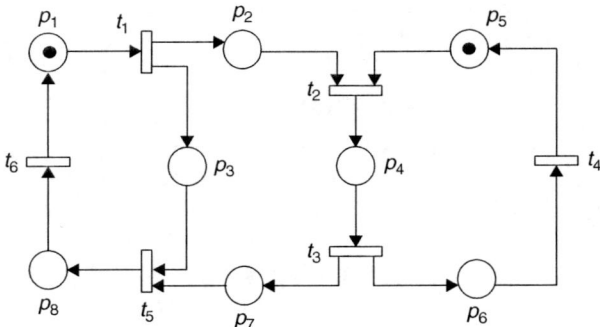

FIGURE 24.9 Petri net model of a simple communication protocol.

t_1: The sender sends a message to the buffer. Time delay: 1 time unit.
t_2: The receiver gets the messages from the buffer. Time delay: 1 time unit.
t_3: The receiver sends back an ACK to the buffer. Time delay: 1 time unit.
t_4: The receiver processes the message. Time delay: 4 time units.
t_5: The sender receives the ACK. Time delay: 1 time unit.
t_6: The sender processes a new message. Time delay: 3 time units.

There are three circuits in the model. The cycle time of each circuit is calculated as follows:

$$\text{circuit } p_1 t_1 p_3 t_5 p_8 t_6 p_1 : C_1 = \frac{T_1}{N_1} = \frac{1+1+3}{1} = 5$$

$$\text{circuit } p_1 t_1 p_2 t_2 p_4 t_3 p_7 t_5 p_8 t_6 p_1 : C_2 = \frac{T_2}{N_2} = \frac{1+1+1+1+3}{1} = 7$$

$$\text{circuit } p_5 t_2 p_4 t_3 p_6 t_4 p_5 : C_3 = \frac{T_3}{N_3} = \frac{1+1+4}{1} = 6$$

After enumerating all circuits in the net, we know the minimum cycle time of the protocol between the two processes is 7 time units.

24.8.2 Stochastic Timed Petri Nets

STPNs are Petri nets in which stochastic firing times are associated with transitions. An STPN is essentially a high-level model that generates a stochastic process. STPN-based performance evaluation basically comprises modeling the given system by an STPN and automatically generating the stochastic process that governs the system behavior. This stochastic process is then analyzed using known techniques (Haas, 2002). STPNs are a graphical model and offer great convenience to a modeler in arriving at a credible, high-level model of a system.

The simplest choice for the individual distributions of transition firing times is negative exponential distribution. Because of the memoryless property of this distribution, the stochastic process associated with the STPN is a continuous-time homogeneous Markov chain (Ethier, 2005) with state space in one-to-one correspondences with marking in $R(M_0)$, the set of all reachable markings. The transition rate matrix of the Markov chain can be easily constructed from the reachability graph given the firing rates of the transitions of the STPN. Exponential timed stochastic Petri nets, often called *stochastic Petri nets* (SPNs), were independently proposed by Natkin (1980) and Molloy (1981), and their capabilities in the performance analysis of real systems have been investigated by many authors.

An SPN is a six-tuple $(P, T, I, O, M_0, \Lambda)$, where (P, T, I, O, M_0) is a Petri net and $\Lambda : T \to R$ a set of firing rates whose entry λ_k is the rate of the exponential individual firing time distribution associated with transition t_k. Natkin and Molloy have shown that the marking process of an SPN is a continuous-time Markov chain. The state space of the Markov chain is the reachable set $R(M_0)$. Suppose there are s markings in $R(M_0)$, and the underlying Markov chain is ergodic, then the steady-state probability distribution $\Pi = (\pi_0, \pi_1, \ldots, \pi_s)$ can be obtained by resolving the following linear system:

$$\Pi Q = 0 \qquad \sum_j \pi_j = 1, \quad \pi_j \geq 0, \quad j = 0, 1, 2, \ldots$$

where Q is a transition rate matrix whose elements outside the main diagonal are the rates of the exponential distributions associated with the transitions from state, while the elements on the main diagonal make the sum of the elements of each row equal to zero. Denote by $E(M_i)$ the set of all enabled transition at marking M_i, and T_{ij} the set of enabled transitions at marking M_i whose firings lead the SPN to another marking M_j. Then Q is determined as follows:

$$q_{ij} = \sum_{t_k \in T_{ij}} \lambda_k \qquad q_i = q_{ii} = - \sum_{t_k \in E(M_i)} \lambda_k$$

The probability of marking M_i changing to M_j is the same as the probability that one of the transitions in the set T_{ij} fires before any of the transitions in the set $T \setminus T_{ij}$. Since the firing times in an SPN are mutually independent exponential random variables, it follows that the required probability has the specific value given by

$$\alpha_{ij} = q_{ij}/q_i$$

In the expression for α_{ij} deduced above, note that the numerator is the sum of the rates of those enabled transitions in M_i, the firing of any of which changes the marking from M_i to M_j; whereas the denominator is the sum of the rates of all the enabled transitions in M_i. Also note that $\alpha_{ij} = 1$ if and only if $T_{ij} = E(M_i)$.

Example 7

A stochastic Petri net.

Figure 24.10 shows a simple SPN model with its reachable markings and its reachable graph. The linear system of steady-state probabilities is

$$\pi_0 + \pi_1 + \pi_2 + \pi_3 + \pi_4 = 1$$

Let $\Lambda = (1\ 1\ 1\ 1)$, then solution to this system is

$$\pi_0 = \pi_4 = 2/7, \qquad \pi_1 = \pi_2 = \pi_3 = 1/7$$

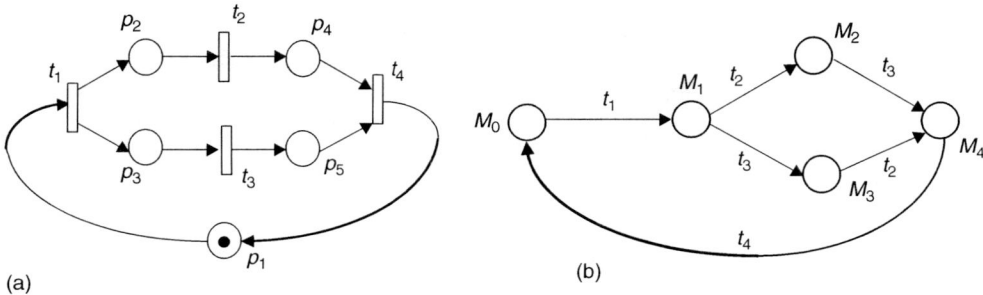

FIGURE 24.10 (a) SPN model. (b) Reachability graph.

The analysis of an SPN model is usually aimed at the computation of more aggregate performance indices than the probabilities of individual markings. Several kinds of aggregate results are easily obtained from the steady-state distribution over reachable markings. In this section, we quote some of the most commonly and easily computed aggregate steady-state performance parameters (Ajmone Marsan, 1990).

- The probability of an event defined through place markings (e.g., no token in a subset of places, or at least one token in a place while another one is empty), can be computed by adding the probabilities of all markings in which the condition corresponding to the event definition holds. Thus, for example, the steady-state probability of the event A defined through a condition that holds for the markings $M_i \in H$ is obtained as

$$P\{A\} = \sum_{M_i \in H} \pi_i$$

- The average number of tokens in a place can be obtained by computing the individual probabilities as those of the event "place p_i contains k tokens."
- The frequency of firing a transition, i.e., the average number of times the transition fires in unit time, can be computed as the weighted sum of the transition firing rate:

$$f_i = \sum_{i: t_j \in E(M_i)} \lambda_j(M_i) \pi_i$$

where f_j is the frequency of firing t_j and $\lambda_j(M_i)$ the firing rate of t_j at M_i.
- The average delay of a token in traversing a subnet in steady-state conditions can be computed using Little's formula

$$E(T) = \frac{E(N)}{E(\gamma)}$$

where $E(T)$ is the average delay, $E(N)$ the average number of tokens in the process of traversing the subnet, and $E(\gamma)$ the average input (or output) rate of tokens into (or out of) the subnet. This procedure can be applied whenever the interesting tokens can be identified inside the subnet (which can also comprise other tokens defining its internal condition, but these must be distinguishable from those whose delay is studied) so that their average number can be computed, and a relation can be established between input and output tokens (e.g., one output token for each input token).

24.9 Concluding Remark

Petri nets have been proven to be a powerful modeling tool for various types of dynamic event-driven systems. Since Petri nets were introduced in 1962, numerous research papers have been published. The research has been conducted in many branches, with each branch exploring a promising application aspect of this formalism. Given the rich research results from the Petri net society, it is hard to cover all of them in such a short book chapter. Therefore, this chapter only aims at briefly introducing the most basic concepts, theory and applications of ordinary Petri nets, and a few of the most popular extended Petri nets.

References

Ajmone Marsan, M. 1990. Stochastic Petri nets: An elementary introduction. *Advances in Petri Nets*, Lecture Notes in Computer Science 424.

Ajmone Marsan, M., M.G. Balbo, and G. Conte. 1986. *Performance Models of Multiprocessor Systems*. Cambridge, MA: MIT Press.

Andreadakis, S.K. and A.H. Levis. 1988. Synthesis of distributed command and control for the outer air battle. *Proceedings of the 1988 Symposium on C^2 Research*, SAIC, McLean, VA.

Bause, F. and P. Kritzinger. 2002. *Stochastic Petri Nets—An Introduction to the Theory*. Germany: Vieweg Verlag.

Desrochers, A. and R. Ai-Jaar. 1995. *Applications of Petri Nets in Manufacturing Systems: Modeling, Control and Performance Analysis*. New York: IEEE Press.

Ethier, S. N. 2005. *Markov Processes*. New York: Wiley.

Genrich, J.H. and K. Lautenbach. 1981. System modeling with high-level Petri nets. *Theoretical Computer Science* 13: 109–136.

Haas, P. 2002. *Stochastic Petri Nets: Modelling, Stability, Simulation*. New York: Springer.

Jensen, K. 1981. Colored Petri nets and the invariant-method. *Theoretical Computer Science* 14: 317–336.

Jensen, K. 1997. *Coloured Petri Nets. Basic Concepts, Analysis Methods and Practical Use* (3 volumes). London: Springer.

Lin, C., L. Tian, and Y. Wei. 2002. Performance equivalent analysis of workflow systems. *Journal of Software* 13(8): 1472–1480.

Mandrioli, D., A. Morzenti, M. Pezze, P. San Pietro, and S. Silva. 1996. A Petri net and logic approach to the specification and verification of real time systems. In: *Formal Methods for Real Time Computing* (C. Heitmeyer and D. Mandrioli, eds.). New York: Wiley.

Memmi, G. and G. Roucairol. 1980. Linear algebra in net theory. *Lecture Notes in Computer Science*, 84: 213–223.

Merlin, P. and D. Farber. 1976. Recoverability of communication protocols—Implication of a theoretical study. *IEEE Transactions on Communications* 1036–1043.

Molloy, M. 1981. *On the Integration of Delay and Throughput Measures in Distributed Processing Models*. Ph.D. Thesis, UCLA, Los Angeles, CA.

Molloy, M. 1982. Performance analysis using stochastic Petri nets. *IEEE Transactions on Computers* 31(9): 913–917.

Murata, T. 1989. Petri nets: Properties, analysis and applications. *Proceedings of the IEEE* 77(4): 541–580.

Natkin, S. 1980. *Les Reseaux de Petri Stochastiques et Leur Application a l'evaluation des Systems Informatiques*. These de Docteur Ingegneur, Cnam, Paris, France.

Peterson, J.L. 1981. *Petri Net Theory and the Modeling of Systems*. New Jersey: Prentice-Hall.

Petri, C.A. 1962. *Kommunikation mit Automaten*. English Translation, 1966: *Communication with Automata*, Technical Report RADC-TR-65-377, Rome Air Dev. Center, New York.

Ramchandani, C. 1974. *Analysis of Asynchronous Concurrent Systems by Petri Nets*. Project MAC, TR-120, MIT, Cambridge, MA.

Ramamoorthy, C. and G. Ho. 1980. Performance evaluation of asynchronous concurrent systems using Petri nets. *IEEE Transaction on Software Engineering* 6(5): 440–449.

Tsai, J., S. Yang, and Y. Chang. 1995. Timing constraint Petri nets and their application to schedulability analysis of real-time system specifications. *IEEE Transactions on Software Engineering* 21(1): 32–49.

van der Aalst, W. and K. van Hee. 2000. *Workflow Management: Models, Methods, and Systems*. Cambridge, MA: MIT Press.

van Landeghem, R. and C.-V. Bobeanu. 2002. Formal modeling of supply chain: An incremental approach using Petri nets. *14th European Simulations Symposium and Exhibition*, Dresden, Germany.

Wang, J. 1998. *Timed Petri Nets, Theory and Application*. Boston, MA: Kluwer.

Wang, J., Y. Deng and M. Zhou, 2000. Compositional time Petrinets and reduction rules. *IEEE Transactions on Systems, Man and Cybernetics*, Part B, 30(4): 562–572.

Venkatesh, K., M.C. Zhou, and R. Caudill. 1994. Comparing ladder logic diagrams and Petri nets for sequence controller design through a discrete manufacturing system. *IEEE Transaction on Industrial Electronics* 41(6): 611–619.

Zhou, M.C. and F. DiCesare. 1989. Adaptive design of Petri net controllers for error recovery in automated manufacturing systems. *IEEE Transactions on Systems, Man, and Cybernetics* 19(5): 963–973.

25
Queueing System Models

Christos G. Cassandras
Boston University

25.1 Introduction 25-1
25.2 Specification of Queueing System Models 25-2
25.3 Performance of a Queueing System 25-4
25.4 Queueing System Dynamics 25-6
25.5 Little's Law 25-7
25.6 Simple Markovian Queueing Models 25-8
   The $M/M/1$ Queueing System
25.7 Markovian Queueing Networks 25-11
   The Departure Process of the $M/M/1$ Queueing System • Open Queueing Networks • Closed Queueing Networks • Product Form Networks
25.8 Non-Markovian Queueing Systems 25-17

25.1 Introduction

In a separate chapter, models for discrete-event systems (DES) were introduced and discussed. To briefly recap, the behavior of such systems is governed by *discrete events* occurring asynchronously over time and solely responsible for generating state transitions. In between event occurrences, the state of such systems is unaffected. Examples include computer and communication networks, automated manufacturing systems, air traffic control systems, command-control systems, advanced monitoring and control systems in automobiles or large buildings, intelligent transportation systems, distributed software systems, and so forth. DES are also referred to as *event-driven* systems to distinguish them from *time-driven* systems. In the latter, the state of the system generally changes as time changes: with every "tick" of an underlying clock the state is expected to change and differential (or difference) equations are the standard modeling framework one uses in such cases. In contrast, in an event-driven system state transitions are the result of combining asynchronous concurrent event processes. Modeling frameworks for DES include state automata and Petri nets, discussed elsewhere in the book.

An important class of DES is that of *queueing systems*. The term "queueing" is associated with the fact that the resources we need to use in most systems we design and build (as well as in our daily life) are not always accessible: to use them, we have to wait. For example, to use the resource "bank teller" in a bank, people form a line and wait for their turn. Sometimes, the waiting is not done by people, but by discrete objects or more abstract "entities." For example, to use the resource "CPU" in a computer, various "tasks" wait somewhere until they are given access to it through potentially complex mechanisms. There are three basic elements that comprise a queueing system: (i) The *entities* that do the waiting in their quest for resources; these are traditionally referred to as *customers*. (ii) The *resources* for which the waiting is done; since resources typically provide some form of service to the customers, we shall generically call them *servers*; and (iii) The space where the waiting is done, which we shall call a *queue*.

The study of queueing systems is motivated by the simple fact that resources are not unlimited; if they were, no waiting would ever occur. This fact gives rise to obvious problems of resource allocation and related tradeoffs so that (i) customer needs are adequately satisfied, (ii) resource access is provided in fair

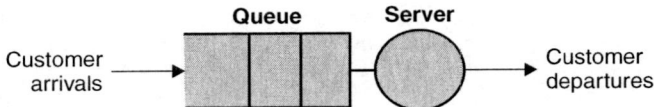

FIGURE 25.1 A simple queueing system.

and efficient ways among different customers, and (iii) the cost of designing and operating the system is maintained at acceptable levels.

Graphically, we will represent a simple queueing system as shown in Figure 25.1. A circle represents a server, and an open box represents a queue preceding this server. The slots in the queue are meant to indicate waiting customers. Customers are thought of as *arriving* at the queue, and *departing* from the server. The process of serving customers takes a positive amount of time (otherwise there would be no waiting). Thus, a server may be thought of as a "delay block," which holds a customer for some amount of service time.

Queueing theory is a subject to which many books have been devoted. It ranges from the study of simple single-server systems modeled as birth–death Markov chains to the analysis of arbitrarily complex networks of queues. In this chapter, we limit ourselves to the essential ideas and techniques that are used to analyze simple queueing systems. Queueing theory has as its main goal the determination of a system's performance under certain operating conditions, rather than the determination of the operating policies to be used to achieve the best possible performance. Thus, its mission has been largely to develop "descriptive" tools for studying queueing systems, rather than "prescriptive" tools for controlling their behavior in an ever-changing dynamic and uncertain environment. Still, we must stress that queueing theory has made some of the most important contributions to the analysis of stochastic DES where resource contention issues are predominant.

25.2 Specification of Queueing System Models

There are three basic aspects of a queueing model that require specification: (i) *stochastic models* for the arrival and service processes; (ii) *structural parameters* of the system, e.g., the storage capacity of a queue and the number of servers; and (iii) *Operating policies*, e.g., conditions under which arriving customers are accepted and preferential treatment (prioritization) of some types of customers by the server.

Stochastic Models for Arrival and Service Processes. Viewed as a DES, the simple model of Figure 25.1 has an event set $E = \{a, d\}$, where a denotes a customer arrival and d a departure following a service completion. We associate with arrival events a a stochastic sequence $\{Y_1, Y_2, \ldots\}$, where Y_k is the kth *interarrival time*, i.e., the time elapsed between the $(k-1)$th and kth arrival, $k = 1, 2, \ldots$. For simplicity, we always set $Y_0 = 0$, so that Y_1 is the random variable describing the time of the first arrival. In most simple queueing models, it is assumed that the stochastic sequence $\{Y_k\}$ is iid (i.e., $Y_1, Y_2, \ldots$ are independent and identically distributed). Therefore, a single probability distribution

$$A(t) = P[Y \leq t] \tag{25.1}$$

completely describes the interarrival time sequence. In Eq. (25.1), the random variable Y is often thought of as a "generic" interarrival time that does not need to be indexed by k. The mean of the distribution function $A(t)$, $E[Y]$, is particularly important and it is customary to use the notation

$$E[Y] \equiv \frac{1}{\lambda} \tag{25.2}$$

to represent it. Thus, λ is the *arrival rate* of customers.

Similarly, we associate to the event d a stochastic sequence $\{Z_1, Z_2, \ldots\}$, where Z_k is the kth service time, i.e., the time required for the kth customer to be served, $k = 1, 2, \ldots$. If we assume that the stochastic sequence $\{Z_k\}$ is also iid, then we define

$$B(t) = P[Z \leq t] \qquad (25.3)$$

where Z is a generic service time. Similar to Eq. (25.2), we use the following notation for the mean of $B(t)$:

$$E[Z] \equiv \frac{1}{\mu} \qquad (25.4)$$

so that μ is the *service rate* of the server in our model.

Structural Parameters. The most common structural parameters of interest in a queueing system are (i) The *storage capacity* of the queue, usually denoted by $K = 1, 2, \ldots$. By convention, we normally agree to include in this storage capacity the space provided for customers in service. (ii) The *number of servers*, usually denoted by $m = 1, 2, \ldots$. In the simple system of Figure 25.1, we have $K = \infty$ and $m = 1$.

Operating Policies. Even for the simple system of Figure 25.1, there are various schemes one can adopt in handling the queueing process. Here are just some of the natural questions that arise about the operation of the system: (i) Is the service time distribution the same for all arriving customers? (ii) Are customers differentiated on the basis of belonging to different "classes," some of which may have a higher priority than others in requesting service? (iii) Are all customers admitted to the system? (iv) Are customers allowed to leave the queue before they get served, and, if so, under what conditions? (v) How does the server decide which customer to serve next (if there are more than one in the queue); for example, the first one in queue, anyone at random, and so forth? (vi) Is the server allowed to preempt a customer in service to serve a higher-priority customer that just arrived?

Clearly, operating policies can cover a wide range of possibilities. In categorizing them, the most common issues we consider are the following: (i) *Number of customer classes*. In the case of a *single-class* system, all customers have the same service requirements and the server treats them all equally. This means that the service time distribution is the same for all customers. In the case of a multiple-class system, customers are distinguished according to their service requirements and/or the way in which the server treats them. (ii) *Scheduling policies*. In a multiple-class system, the server must decide upon a service completion which class to process next. For example, the server may always give priority to a particular class, or it may preempt a customer in process because a higher priority customer just arrived. (iii) *Queueing disciplines*. A queueing discipline describes the order in which the server selects customers to be processed, even if there is only a single class. For example, first-come-first-served (FCFS), last-come-last-served (LCFS), and random order. (iv) *Admission policies*. Even if a queue has infinite storage capacity, it may be desirable to deny admission to some arriving customers. In the case of two arriving classes, for instance, higher priority customers may always be admitted, but lower priority customers may only be admitted if the queue is empty or if some amount of time has elapsed since such a customer was admitted. (v) *Batching*. It is possible that arrivals occur in "batches," i.e., customers may arrive one at a time or in groups of size $n > 1$. Similarly, it is possible that service occurs in batches of size $n > 1$ customers processed at a time.

In the simple case of Figure 25.1, we assume a single-class system with all arriving customers admitted and served one at a time and the queueing discipline is FCFS.

Notation. It is customary in queueing theory to employ a particular type of notation to succinctly describe a system. This notation (attributed to Kendall) is $A/B/m/K$, where A is the interarrival time distribution, B the service time distribution, m the number of servers present, $m = 1, 2, \ldots$, and K the storage capacity of the queue, $K = 1, 2, \ldots$. Thus, the infinite queueing capacity single-server system of Figure 25.1 is described by $A/B/1/\infty$. To simplify the notation, if the K position is omitted it is understood that $K = \infty$. Therefore, in our case, we have $A/B/1$. Furthermore, there is some common notation used to represent the distributions A and B. In particular, G stands for a *General* distribution when nothing else is known about the arrival/service process, GI for a *General* distribution in a *renewal* arrival/service process (i.e., all interarrival/service times in that process are iid), D for the *Deterministic* case, (i.e., the interarrival/service times are fixed), and M for the *Markovian* case, i.e., the interarrival/service times are exponentially distributed; using our notation in Eqs. (25.1)–(25.4), this means that $A(t)=1-e^{-\lambda t}$ and $B(t)=1-e^{-\mu t}$.

These are the most often encountered cases. Here are some examples to illustrate the $A/B/m/K$ notation: the $M/M/1$ queueing system has a single server and infinite storage capacity with both interarrival and service times exponentially distributed; the $M/M/1/K$ system is the same but with storage capacity given by some $K<\infty$ (including space at the server); the $M/G/2$ system has two servers and infinite storage capacity, with exponentially distributed interarrival times, while the service times have an arbitrary (general) distribution.

Note that the $A/B/m/K$ notation does not specify the operating policies to be used. It also does not describe the case where two or more customer classes are present, each with different interarrival and service time distributions. Finally, note that when $A(t)=1-e^{-\lambda t}$ the underlying arrival process is Poisson. Thus, the statement "the arrival process is Poisson" is the same as "interarrival times are independent and exponentially distributed." In the case of departure events generated by a Poisson process, we normally use the statement "service times are exponentially distributed," because these events are only feasible when the server is busy.

The $A/B/m/K$ notation can be extended to include one more characteristic of a queueing system: whether it is open to an infinite population of customers who can request service at any time or whether the system is limited to a finite population of customers, usually denoted by N. In the latter case, a customer completing service at some server is always routed to another queue and never leaves the system. We refer to the former as an *open* queueing system and the latter as a *closed* one. In such cases, the notation $A/B/m/K/N$ is used, where N represents the customer population residing in the system. As in the case of infinite storage, omitting N implies an open system (i.e., $N=\infty$). Note that if $K=\infty$, but $N<\infty$, we normally write $A/B/m//N$. Closed queueing systems should not be thought of as strictly consisting of a particular set of fixed customers. Instead, the population N may indicate a number of resources limiting access to more customers. A typical example arises in modeling a computer system with N access points. Here, the total number of users is limited to N, but users certainly come and go replacing each other at various access points. Similarly, in a manufacturing system production parts are often carried in pallets whose number is limited to N. When a finished part leaves the system, it relinquishes its pallet to a new part, so that the effective number of customers in the system is limited to N.

25.3 Performance of a Queueing System

In addition to the random variables already introduced, i.e., the interarrival time Y_k and the service time Z_k, let us define A_k to be the *arrival time* of the kth customer, D_k its *departure time*, W_k its *waiting time*, and S_k its *system time* (from arrival instant until departure), also referred to as *response time*, *sojourn time*, or *delay*. Note that

$$S_k = D_k - A_k = W_k + Z_k \tag{25.5}$$

and

$$D_k = A_k + W_k + Z_k \tag{25.6}$$

In addition, we define the random variables $X(t)$ to denote the *queue length* at time t, $X(t)\in\{0,1,2,\ldots\}$ and $U(t)$ to denote the *workload* (or *unfinished work*) at time t, i.e., the amount of time required to empty the system at t.

The stochastic behavior of the waiting time sequence $\{W_k\}$ provides important information regarding the system's performance. The probability distribution function of $\{W_k\}$, $P[W_k\leq t]$, generally depends on k. We often find, however, that as $k\to\infty$ there exists a stationary distribution, $P[W\leq t]$, independent of k, such that

$$\lim_{k\to\infty} P[W_k \leq t] = P[W \leq t] \tag{25.7}$$

If this limit indeed exists, the random variable W describes the waiting time of a typical customer at steady state. Intuitively, when the system runs for a sufficiently long period of time (equivalently, the system has processed a sufficiently large number of customers), every new customer experiences a "stochastically identical" waiting process described by $P[W \leq t]$. The mean of this distribution, $E[W]$, represents the *average waiting time* at steady state. Similarly, if a stationary distribution exists for the system time sequence $\{S_k\}$, then its mean, $E[S]$, is the *average system time* at steady state.

The same idea applies to the stochastic processes $\{X(t)\}$ and $\{U(t)\}$. If stationary distributions exist for these processes as $t \to \infty$, then the random variables X and U are used to describe the queue length and workload of the system at steady state. We will use the notation π_n, $n = 0, 1, \ldots$, to denote the stationary queue length probability, that is,

$$\pi_n = P[X = n], \quad n = 0, 1, \ldots \tag{25.8}$$

Accordingly, $E[X]$ is the *average queue length* at steady state, and $E[U]$ the *average workload* at steady state.

In general, we want to design a queueing system so that a typical customer at steady state waits as little as possible (ideally, zero). In contrast, we also wish to serve as many customers as possible by keeping the server as busy as possible, that is, we try to maximize the server's utilization. To do so, we must constantly keep the queue nonempty; in fact, we should make sure there are always a few customers to serve in case several service times in a row turn out to be short. However, this is directly contrary to our objective of achieving zero waiting time for customers. This informal argument serves to illustrate a fundamental performance tradeoff in all queueing systems. To keep a server highly utilized we must be prepared to tolerate long waiting times; conversely, to maintain low waiting times we have to tolerate some server idling. With this observation in mind, the main measures of performance (at steady state) that we are interested in are (i) the *average waiting time* of customers, $E[W]$, (ii) the *average queue length*, $E[X]$, (iii) the *utilization* of the system, i.e., the fraction of time that the server is busy, and (iv) the *throughput* of the system, i.e., the rate at which customers leave after service. Our objective is to keep the first two as small as possible, while keeping the last two as large as possible.

To gain some more insight on the utilization of a queueing system, we define the *traffic intensity* ρ as

$$\rho \equiv \frac{[\text{arrival rate}]}{[\text{service rate}]}$$

Then, by the definitions of λ and μ in Eq. (25.2) and Eq. (25.4), we have

$$\rho = \frac{\lambda}{\mu} \tag{25.9}$$

In the case of m servers, the average service rate becomes $m\mu$, and, therefore,

$$\rho = \frac{\lambda}{m\mu} \tag{25.10}$$

In a single-server system at steady state, the probability π_0, defined in Eq. (25.8), represents the fraction of time the system is empty, and hence the server is idle. It follows that for a server at steady state:

$$[\text{utilization}] \equiv [\text{fraction of time server is busy}] = 1 - \pi_0$$

Since a server operates at rate μ and the fraction of time that it is actually in operation is $(1 - \pi_0)$, the throughput of a single-server system at steady state is

$$[\text{throughput}] \equiv [\text{departure rate of customers after service}] = \mu(1 - \pi_0)$$

At steady state, the customer flows into and out of the system must be balanced, that is,

$$\lambda = \mu(1 - \pi_0)$$

It then follows from Eq. (25.9) that

$$\rho = 1 - \pi_0 \tag{25.11}$$

Thus, the traffic intensity, which is defined by the parameters of the service and interarrival time distributions, also represents the utilization of the system. This relationship holds for any single-server system with infinite storage capacity. Note that if $\pi_0 = 0$, the system is permanently busy, which generally leads to an instability in the sense that the queue length grows to infinity. Thus, the values of ρ must be such that $0 \leq \rho < 1$.

A typical design problem for queueing systems is the selection of parameters such as the service rate and number of servers to achieve some desirable performance in terms of the measures above. In controlling queueing systems, our task is to select operating policies that help us achieve such performance.

25.4 Queueing System Dynamics

Consider once again the queueing system of Figure 25.1, operating on an FCFS basis. Using the notation we have established, a typical sample path of this system is shown in Figure 25.2. In this example, the first arriving customer at time A_1 finds an empty queue. During the interval starting at A_1 and ending with D_3 the server remains busy. Such an interval is termed a *busy period* of the queueing system. During the interval starting with D_3 and ending with the next arrival at A_4 the server remains idle. We term this an *idle period* of the system. We can see that one way to view this system is as a sequence of alternating cycles each consisting of a busy period followed by an idle period.

Taking a closer look at Figure 25.2 helps us identify the basic dynamic mechanism of this queueing system. When the kth customer arrives, two cases are possible. In the first case, the system is empty, therefore $W_k = 0$. The system can only be empty when $D_{k-1} \leq A_k$, i.e., the previous customer departed before the current customer arrived. Thus

$$D_{k-1} - A_k \leq 0 \Leftrightarrow W_k = 0 \tag{25.12}$$

This is clearly seen in Figure 25.2 with the case $W_4 = 0$, which is a result of $D_3 < A_4$.

In the second case, the system is not empty, therefore, $W_k > 0$ and the kth customer is forced to wait until the previous, i.e., $(k-1)$th, customer departs. Thus,

$$D_{k-1} - A_k > 0 \Leftrightarrow W_k = D_{k-1} - A_k \tag{25.13}$$

This situation arises with $W_2 = D_1 - A_2 > 0$ in Figure 25.2, as well as with W_3 and W_5.

Combining Eq. (25.12) and Eq. (25.13), we obtain

$$W_k = \max\{0, D_{k-1} - A_k\} \tag{25.14}$$

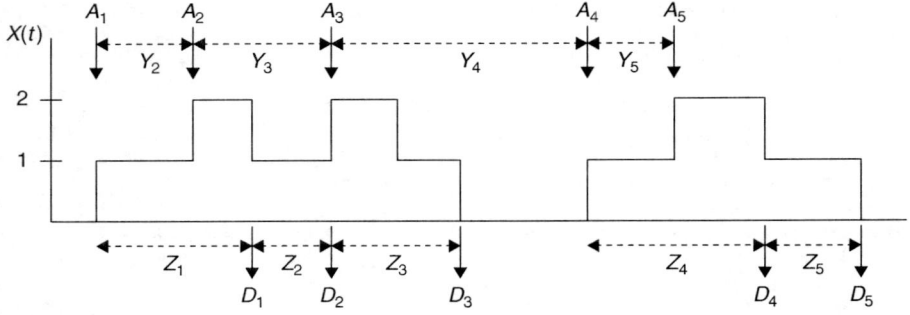

FIGURE 25.2 A typical queueing system sample path.

Now using Eq. (25.6) for the $(k-1)$th customer on a sample path, and recalling that $A_k - A_{k-1} = Y_k$, we get the following recursive expression for waiting times:

$$W_k = \max\{0, W_{k-1} + Z_{k-1} - Y_k\} \tag{25.15}$$

Similarly, we can rewrite this relationship for system times:

$$S_k = \max\{0, S_{k-1} - Y_k\} + Z_k \tag{25.16}$$

Finally, we can obtain a similar recursive expression for departure times using the fact that $W_k = D_k - A_k - Z_k$ and we get

$$D_k = \max\{A_k, D_{k-1}\} + Z_k \tag{25.17}$$

These relationships capture the essential dynamic characteristics of a queueing system. Eq. (25.15) is referred to as *Lindley's equation*, after D. V. Lindley who studied queueing dynamics in the 1950s (Lindley, 1952). Also note that these relationships are very general, in the sense that they apply on any sample path of the system regardless of the distributions characterizing the various stochastic processes involved. There are few results we can obtain for queueing systems, which are general enough to hold regardless of the nature of these distributions. Aside from the relationships derived above, there is one more general result which we discuss in the next section.

25.5 Little's Law

Consider once again the queueing system of Figure 25.1 and define $N_a(t)$ and $N_d(t)$ to count the number of arrivals and departures, respectively, in the interval $(0, t]$. Assuming that the system is initially empty, it follows that the queue length $X(t)$ is given by

$$X(t) = N_a(t) - N_d(t) \tag{25.18}$$

Let $U(t)$ be the total amount of time all customers have spent in the system by time t. Thus, the average system time per customer by time t, denoted by $\bar{S}(t)$, is

$$\bar{S}(t) = \frac{U(t)}{N_a(t)} \tag{25.19}$$

Similarly, dividing $U(t)$ by t, we obtain the average number of customers present in the system over the interval $(0, t]$, i.e., the average queue length along this sample path, $\bar{X}(t)$:

$$\bar{X}(t) = \frac{U(t)}{t} \tag{25.20}$$

Finally, dividing the total number of customers who have arrived in $(0, t]$, $N_a(t)$, by t, we obtain the arrival rate $\lambda(t)$:

$$\lambda(t) = \frac{N_a(t)}{t} \tag{25.21}$$

Combining Eq. (25.19)–(25.21) gives

$$\bar{X}(t) = \lambda(t)\bar{S}(t) \tag{25.22}$$

We now make two assumptions. We assume that as $t \to \infty$, $\lambda(t)$ and $\bar{S}(t)$ both converge to fixed values λ and $\bar{S}$, respectively, i.e., the following limits exist: $\lim_{t \to \infty} \lambda(t) = \lambda$, $\lim_{t \to \infty} \bar{S}(t) = \bar{S}$. These values represent the steady-state arrival rate and system time, respectively, for a given sample path. If these limits exist, then, by Eq. (25.22), $\bar{X}(t)$ must also converge to a fixed value $\bar{X}$. Therefore,

$$\bar{X} = \lambda \bar{S} \qquad (25.23)$$

This relationship applies to a particular sample path we selected. Suppose, however, that the limits we have assumed exist for all possible sample paths and for the same fixed values of λ and $\bar{S}$, and hence $\bar{X}$. In other words, we are assuming that the arrival, system time, and queue length processes are all ergodic. In this case, $\bar{X}$ is actually the mean queue length $E[X]$ at steady state, and $\bar{S}$ the mean system time $E[S]$ at steady state. We may then rewrite Eq. (25.23) as

$$E[X] = \lambda E[S] \qquad (25.24)$$

This is known as *Little's Law*. It is a powerful result in that it is independent of the stochastic features of the system, that is, the probability distributions associated with the arrival and departure events. This derivation is not a proof of the fundamental relationship [Eq. (25.24)], but it does capture its essence. In fact, this relationship was taken for granted for many years even though it was never formally proved. Formal proofs finally started appearing in the literature in the 1960s (e.g., Little, 1961; Stidham, 1974).

It is important to observe that Eq. (25.24) is independent of the operating policies employed in the queueing system under consideration. Moreover, it holds for an arbitrary configuration of interconnected queues and servers. This implies that Little's Law holds for a single queue (server not included) as follows:

$$E[X_Q] = \lambda E[W] \qquad (25.25)$$

where $E[X_Q]$ is the mean queue content (without the server) and $E[W]$ the mean waiting time. Similarly, if our system is defined by a boundary around a single server, we have

$$E[X_S] = \lambda E[Z] \qquad (25.26)$$

where $E[X_S]$ is the mean server content (between 0 and 1 for a single server) and $E[Z]$ the mean service time.

25.6 Simple Markovian Queueing Models

Most interesting performance measures can be evaluated from the stationary queue length probability distribution $\pi_n = P[X = n]$, $n = 0, 1, \ldots$, of the system. Obtaining this distribution (if it exists) is therefore a major objective of queueing theory. This is generally an extremely hard problem, even when the interarrival and service time distributions are relatively simple. The Markovian case, where they are both exponential, is of particular interest because it captures many practical situations and it is analytically tractable under certain assumptions regarding the structure of the queueing system.

The key observation is that the state transition diagram of a simple queueing model, such as the one in Figure 25.1, is that of a birth–death Markov chain as seen in Figure 25.3: in such a DES, there are two events, a "birth" and a "death" with a state $X(t)$ and an underlying state space $\{0, 1, \ldots\}$. The time between births is exponentially distributed and its (generally state-dependent) *birth rate* parameter is λ_i, $i = 0, 1, \ldots$. Similarly, the time between deaths (defined only when $X(t) > 0$) is exponentially distributed and its *death rate* parameter is μ_i, $i = 1, 2, \ldots$. It is easy to see that this model also represents a queueing process where births correspond to customer arrivals and deaths correspond to customer departures. Birth–death Markov chains have been extensively studied in the stochastic process literature. Of particular interest for our purposes is the fact that the steady-state probabilities π_n, $n = 0, 1, \ldots$ of such a chain are given by (see, e.g., Cassandras and Lafortune, 1999; Kleinrock, 1975).

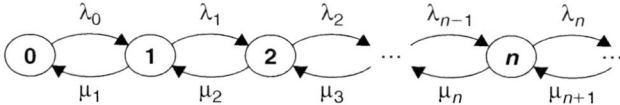

FIGURE 25.3 State transition rate diagram of a birth–death Markov chain.

$$\pi_n = \left(\frac{\lambda_0 \cdots \lambda_{n-1}}{\mu_1 \cdots \mu_n} \right) \pi_0, \quad n = 1, 2, \ldots \quad (25.27)$$

$$\pi_0 = \frac{1}{1 + \sum_{n=1}^{\infty} \left(\frac{\lambda_0 \cdots \lambda_{n-1}}{\mu_1 \cdots \mu_n} \right)} \quad (25.28)$$

where λ_n and μ_n are the birth and death rates, respectively, when the state is n.

Before proceeding with applications of this general result to Markovian queueing models, let us discuss an important property of the Poisson process, which has significant implications to our analysis. Looking once again at the single-server system of Figure 25.1, consider the event [arriving customer at time t finds $X(t) = n$] that is to be compared to the event [system state at some time t is $X(t) = n$]. The distinction may be subtle, but it is critical. In the first case, the observation of the system state (queue length) occurs at specific time instants (arrivals), which depend on the nature of the arrival process. In the second case, the observation of the system state occurs at random time instants. In general,

$$P[\text{arriving customer at time } t \text{ finds } X(t) = n] \neq$$
$$P[\text{system state at some time } t \text{ is } X(t) = n]$$

However, equality does hold for a Poisson arrival process (Kleinrock, 1975), regardless of the service time distribution, as long as the arrival and service processes are independent. This is also known as the Poisson Arrivals See Time Averages (PASTA) property. Using the notation $\pi_n(t) = P[\text{system state at some time } t \text{ is } X(t) = n]$ and $\alpha_n(t) \equiv P[\text{arriving customer at time } t \text{ finds } X(t) = n]$, the PASTA property asserts that $\pi_n(t) = \alpha_n(t)$.

Theorem 1. *For a queueing system with a Poisson arrival process independent of the service process, the probability that an arriving customer finds n customers in the system is the same as the probability that the system state is n, i.e., $\pi_n(t) = \alpha_n(t)$.*

Using Eq. (25.27)–(25.28) above, as well as Theorem 1, it is possible to analyze a number of Markovian queueing systems of practical interest. We will limit ourselves here to the $M/M/1$ queueing system and refer the reader to the queueing theoretic literature where more such systems are analyzed extensively (e.g., Asmussen, 2003; Kleinrock, 1975; Trivedi, 1982).

25.6.1 The $M/M/1$ Queueing System

Using the notation we have established, this is a single-server system with infinite storage capacity and exponentially distributed interarrival and service times. It can therefore be modeled as a birth–death chain with a state transition rate diagram as shown in Figure 25.3 with birth and death parameters $\lambda_n = \lambda$ for all $n = 0, 1, \ldots$ and $\mu_n = \mu$ for all $n = 1, 2, \ldots$. It follows from Eq. (25.28) that

$$\pi_0 = \frac{1}{1 + \sum_{n=1}^{\infty} (\lambda/\mu)^n}$$

The sum in the denominator is a simple geometric series that converges as long as $\lambda/\mu < 1$. Under this assumption, we get

$$\sum_{n=1}^{\infty} \left(\frac{\lambda}{\mu}\right)^n = \frac{\lambda/\mu}{1 - \lambda/\mu}$$

and, therefore,

$$\pi_0 = 1 - \frac{\lambda}{\mu} \tag{25.29}$$

Let us set $\rho = \lambda/\mu$, which is the traffic intensity defined in Eq. (25.9). Thus, we see that

$$\pi_0 = 1 - \rho \tag{25.30}$$

which is in agreement with Eq. (25.11). Note also that $0 \leq \rho < 1$, since Eq. (25.29) was derived under the assumption $\lambda/\mu < 1$.

Next, using Eq. (25.30) in Eq. (25.27), we obtain $\pi_n = \left(\frac{\lambda}{\mu}\right)^n (1 - \rho)$ or

$$\pi_n = (1 - \rho)\rho^n, \quad n = 0, 1, \ldots \tag{25.31}$$

Eq. (25.31) gives the stationary probability distribution of the queue length of the $M/M/1$ system. The condition $\rho = \lambda/\mu < 1$ is the *stability condition* for the $M/M/1$ system. We are now in a position to obtain explicit expressions for various performance measures of this system.

Utilization and Throughput. The utilization is immediately given by Eq. (25.30), since $1 - \pi_0 = \rho$. The throughput is the departure rate of the server, which is $\mu(1 - \pi_0) = \lambda$. This is to be expected since at steady state the arrival and departure rates are balanced. Thus, for a stable $M/M/1$ system, the throughput is simply the arrival rate λ. In contrast, if we allow $\lambda > \mu$, then the throughput is simply μ, since the server is constantly operating at rate μ.

Average Queue Length. This is the expectation of the random variable X whose distribution is given by Eq. (25.31). Thus,

$$E[X] = \sum_{n=0}^{\infty} n\pi_n = (1 - \rho) \sum_{n=0}^{\infty} n\rho^n \tag{25.32}$$

We can evaluate the preceding sum by observing that

$$\frac{d}{d\rho} \left(\sum_{n=0}^{\infty} \rho^n \right) = \sum_{n=0}^{\infty} n\rho^{n-1} = \frac{1}{\rho} \sum_{n=0}^{\infty} n\rho^n$$

Since $\sum_{n=0}^{\infty} \rho^n = \frac{1}{1-\rho}$ and $\frac{d}{d\rho}\left(\frac{1}{1-\rho}\right) = \frac{1}{(1-\rho)^2}$, we get

$$\sum_{n=0}^{\infty} n\rho^n = \frac{\rho}{(1-\rho)^2}$$

Then, Eq. (25.32) gives

$$E[X] = \frac{\rho}{1 - \rho} \tag{25.33}$$

Note that as $\rho \to 1$, $E[X] \to \infty$, that is, the expected queue length grows to ∞. This clearly reveals the tradeoff we already identified earlier: As we attempt to keep the server as busy as possible by increasing the utilization ρ, the quality of service provided to a typical customer declines, since, on the average, such a customer sees an increasingly longer queue length ahead of him.

Average System Time. Using Eq. (25.33) and Little's Law in Eq. (25.24), we get

$$\frac{\rho}{1 - \rho} = \lambda E[S]$$

or, since $\lambda = \rho\mu$,

$$E[S] = \frac{1/\mu}{1-\rho} \qquad (25.34)$$

As $\rho \to 0$, we see that $E[S] \to 1/\mu$, which is the average service time. This is to be expected, since at very low utilizations the only delay experienced by a typical customer is a service time. We also see that as $\rho \to 1, E[S] \to \infty$. Once again, this is a manifestation of the tradeoff between utilization and system time: the higher the utilization (good for the server), the higher the average system time (bad for the customers). Owing to the nonlinear nature of this relationship, the issue of sensitivity is crucial in queueing systems. Our day-to-day life experience (traffic jams, long ticket lines, etc.) suggests that when a queue length starts building up, it tends to build up very fast. This is a result of operating in the range of ρ values where the additional increase in the arrival rate causes drastic increases in $E[S]$.

Average Waiting Time. It follows from Eqs. (25.5)–(25.6) at steady state that $E[S] = E[W] + E[Z] = E[W] + 1/\mu$. Then, from Eq. (25.34) we get

$$E[W] = \frac{1/\mu}{1-\rho} - \frac{1}{\mu}$$

or

$$E[W] = \frac{\rho}{\mu(1-\rho)} \qquad (25.35)$$

As expected, we see once again that as $\rho \to 1, E[W] \to \infty$, that is, increasing the system utilization toward its maximum value leads to extremely long average waiting times for customers.

Before leaving the $M/M/1$ system, we briefly discuss the issue of determining the transient solution for the queue length probabilities $\pi_n(t) = P[X(t) = n], n = 0, 1, \ldots$. This requires solving a set of flow balance equations obtained from Figure 25.3 with $\lambda_n = \lambda$ and $\mu_n = \mu$:

$$\frac{d\pi_n(t)}{dt} = -(\lambda + \mu)\pi_n(t) + \lambda\pi_{n-1}(t) + \mu\pi_{n+1}(t), \quad n = 1, 2, \ldots \qquad (25.36)$$

$$\frac{d\pi_0(t)}{dt} = -\lambda\pi_0(t) + \mu\pi_1(t) \qquad (25.37)$$

Obtaining the solution $\pi_n(t), n = 0, 1, \ldots$, of these equations is a tedious task. We provide the final result below to give the reader an idea of the complexity involved even for the simplest of all interesting queueing systems we can consider (see also Asmussen, 2003; Kleinrock, 1975):

$$\pi_n(t) = e^{-(\lambda+\mu)t}\left[\rho^{(n-i)/2}J_{n-i}(at) + \rho^{(n-i-1)/2}J_{n+i+1}(at) + (1-\rho)\rho^n \sum_{j=n+i-2}^{\infty} \rho^{-j/2}J_j(at)\right]$$

where the initial condition is $\pi_i(0) = P[X(0) = i] = 1$ for some given $i = 0, 1, \ldots$, and $a = 2\mu\rho^{1/2}$,

$$J_n(x) = \sum_{k=0}^{\infty} \frac{(x/2)^{n+2k}}{(n+k)!k!}, \quad n = -1, 0, 1, \ldots$$

Here, $J_n(x)$ is a modified Bessel function, which makes the evaluation of $\pi_n(t)$ particularly complicated.

25.7 Markovian Queueing Networks

The queueing systems we have considered thus far involve customers requesting service from a single service-providing facility (with one or more servers). In practice, however, it is common for two or more servers to be connected so that a customer proceeds from one server to the next in some fashion.

In communication networks, for instance, messages often go through several switching nodes followed by transmission links before arriving at their destination. In manufacturing, a part must usually proceed through several operations in series before it becomes a finished product. This leads to models referred to as *queueing networks*, where multiple servers and queues are interconnected. In such systems, a customer enters at some point and requests service at some server. Upon completion, the customer generally moves to another queue or server for additional service. In the class of *open networks*, arriving customers from the outside world eventually leave the system. In the class of *closed networks*, the number of customers remains fixed.

In the simple systems considered thus far, our objective was to obtain the stationary probability distribution of the state X, where X is the queue length. In networks, we have a system consisting of M interconnected nodes, where the term "node" is used to describe a set of identical parallel servers along with the queueing space that precedes it. In a network environment, we shall refer to X_i as the queue length at the ith node in the system, $i = 1, \ldots, M$. It follows that the state of a Markovian queueing network is a vector of random variables

$$\mathbf{X} = [X_1, X_2, \ldots, X_M] \tag{25.38}$$

where X_i takes on values $n_i = 0, 1, \ldots$ just like a simple single-class stand-alone queueing system. The major objective of queueing network analysis is to obtain the stationary probability distribution of X (if it exists), i.e., the probabilities

$$\pi(n_1, \ldots, n_M) = P[X_1 = n_1, \ldots, X_M = n_M] \tag{25.39}$$

for all possible values of $n_1, \ldots, n_M, n_i = 0, 1, \ldots$.

In the next few sections, we present the main results pertinent to the analysis of *Markovian* queueing networks. This means that all external arrival events and all departure events at the servers in the system are generated by processes satisfying the Markovian (memoryless) property, and are therefore characterized by exponential distributions. A natural question that arises is: "what about internal arrival processes?" In other words, the arrival process at some queue in the network is usually composed of one or more departure processes from adjacent servers; what are the stochastic characteristics of such processes? The answer to this question is important for much of the classical analysis of queueing networks and is presented in the next section.

25.7.1 The Departure Process of the *M/M/*1 Queueing System

Let us consider an $M/M/1$ queueing system. Recall that Y_k and Z_k denote the interarrival and service time, respectively, of the kth customer, and that the arrival and service processes are assumed to be independent. Now let us concentrate on the departure times $D_k, k = 1, 2, \ldots$, of customers, and define Ψ_k to be the kth *interdeparture time*, that is, a random variable such that $\Psi_k = D_k - D_{k-1}$ is the time elapsed between the $(k - 1)$th and the kth departure, $k = 1, 2, \ldots$, where, for simplicity, we set $\Psi_0 = 0$, so that Ψ_1 is the random variable describing the time of the first departure. As $k \to \infty$, we will assume that there exists a stationary probability distribution function such that

$$\lim_{k \to \infty} P[\Psi_k \leq t] = P[\Psi \leq t]$$

where Ψ describes an interdeparture time at steady state. We will now evaluate the distribution $P[\Psi \leq t]$. The result, stated below as a theorem without proof (see also Buzen, 1973) is quite surprising:

Theorem 2. *The departure process of a stable stationary $M/M/1$ queueing system with arrival rate λ is a Poisson process with rate λ, i.e., $P[\Psi \leq t] = 1 - e^{-\lambda t}$.*

This fundamental property of the $M/M/1$ queueing system is also known as *Burke's theorem* (Burke, 1956): a Poisson process supplying arrivals to a server with exponentially distributed service times results in a Poisson departure process with the exact same rate. This fact also holds for the departure process of an $M/M/m$ system. Burke's theorem has some critical ramifications when dealing with networks of Markovian queueing systems, because it allows us to treat each component node independently, as long as there are no customer feedback paths. When a node is analyzed independently, the only information required is the number of servers at that node, their service rate, and the arrival rate of customers (from other nodes as well as the outside world).

25.7.2 Open Queueing Networks

We will consider a general open network model consisting of M nodes, each with infinite storage capacity. We will assume that customers form a single class, and that all nodes operate according to an FCFS queueing discipline. Node i, $i = 1, \ldots, M$, consists of m_i servers each with exponentially distributed service times with parameter μ_i. External customers may arrive at node i from the outside world according to a Poisson process with rate r_i. In addition, internal customers arrive from other servers in the network. Upon completing service at node i, a customer is routed to node j with probability p_{ij}; this is referred to as the *routing probability* from i to j. The outside world is usually indexed by 0, so that the fraction of customers leaving the network after service at node i is denoted by p_{i0}. Note that $p_{i0} = 1 - \sum_{j=1}^{M} p_{ij}$.

In this modeling framework, let λ_i denote the total arrival rate at node i. Thus, using the notation above, we have

$$\lambda_i = r_i + \sum_{j=1}^{M} \lambda_j p_{ji}, \quad i = 1, \ldots, M \tag{25.40}$$

where the first term represents the external customer flow and the second term represents the aggregate internal customer flow from all other nodes.

Before discussing the general model, let us first consider the simplest possible case, consisting of two single-server nodes in tandem, as shown in Figure 25.4. In this case, the state of the system is the two-dimensional vector $\mathbf{X} = [X_1, X_2]$, where X_i, $i = 1, 2$, is the queue length of the ith node. Since all events are generated by Poisson processes, we can model the system as a Markov chain whose state transition rate diagram is shown in Figure 25.5.

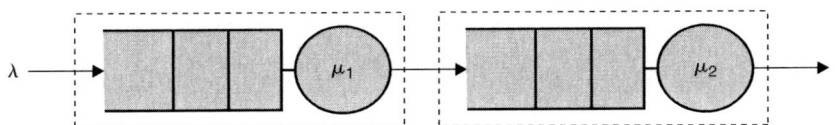

FIGURE 25.4 A two-node open queueing network. (From Cassandras, C.G., and Lafortune, S., *Introduction to Discrete Event Systems*, Springer, Berlin, 1999, pp. 485–486.)

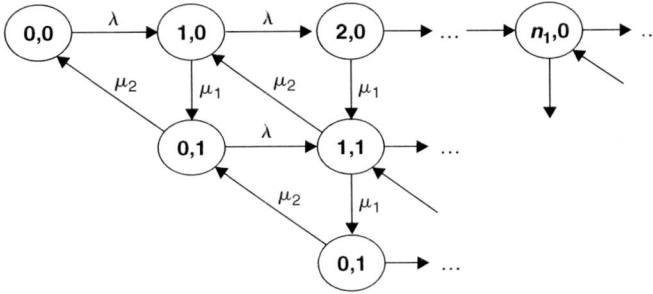

FIGURE 25.5 State transition rate diagram for a two-node open queueing network. (From Cassandras, C.G., and Lafortune, S., *Introduction to Discrete Event Systems*, Springer, Berlin, 1999, pp. 485–486.)

We can now write down flow balance equations by inspection of the state transition rate diagram. First, for any state (n_1, n_2) with $n_1 > 0$ and $n_2 > 0$ we have

$$\lambda \pi(n_1 - 1, n_2) + \mu_1 \pi(n_1 + 1, n_2 - 1) + \mu_2 \pi(n_1, n_2 + 1) - (\lambda + \mu_1 + \mu_2) \pi(n_1, n_2) = 0 \quad (25.41)$$

Similarly, for all states $(n_1, 0)$ with $n_1 > 0$ we have

$$\lambda \pi(n_1 - 1, 0) + \mu_2 \pi(n_1, 1) - (\lambda + \mu_1) \pi(n_1, 0) = 0 \quad (25.42)$$

and for states $(0, n_2)$ with $n_2 > 0$:

$$\mu_1 \pi(1, n_2 - 1) + \mu_2 \pi(0, n_2 + 1) - (\lambda + \mu_2) \pi(0, n_2) = 0 \quad (25.43)$$

and finally for state $(0, 0)$:

$$\mu_2 \pi(0, 1) - \lambda \pi(0, 0) = 0 \quad (25.44)$$

In addition, the state probabilities must satisfy the normalization condition:

$$\sum_{i=0}^{\infty} \sum_{j=0}^{\infty} \pi(i, j) = 1 \quad (25.45)$$

The set of Eqs. (25.41)–(25.45) can readily be solved to give

$$\pi(n_1, n_2) = (1 - \rho_1) \rho_1^{n_1} \cdot (1 - \rho_2) \rho_2^{n_2} \quad (25.46)$$

where

$$\rho_1 = \frac{\lambda}{\mu_1}, \quad \rho_2 = \frac{\lambda}{\mu_2}$$

with the usual stability conditions $0 \leq \rho_1 < 1$ and $0 \leq \rho_2 < 1$. Clearly, ρ_1 is the traffic intensity at the first node. Moreover, ρ_2 is the traffic intensity of node 2, since the throughput (departure rate) of node 1, which is λ, is also the arrival rate at node 2. Observe that if we view each of the two nodes as separate $M/M/1$ systems with stationary state probabilities $\pi_1(n_1)$ and $\pi_2(n_2)$, respectively, we get

$$\pi_1(n_1) = (1 - \rho_1) \rho_1^{n_1}, \quad \pi_2(n_2) = (1 - \rho_2) \rho_2^{n_2}$$

We see, therefore, that we have a simple *product form solution* for this two-node network:

$$\pi(n_1, n_2) = \pi_1(n_1) \cdot \pi_2(n_2) \quad (25.47)$$

In fact, the product form in Eq. (25.47) is a consequence of Burke's theorem, which also asserts that the queue length at time t in an $M/M/1$ system is independent of its departure process prior to t. It allows us to decouple the two nodes, analyze them separately as individual $M/M/1$ systems, and then combine the results as in Eq. (25.47). It is straightforward to extend this solution to any open Markovian network with no customer feedback paths. Let us briefly indicate why customer feedback may create a problem. Suppose that in a simple $M/M/1$ system customers completing service are returned to the queue with probability p, or they depart with probability $(1-p)$. The difficulty here is that the process formed by customers entering the queue, consisting of the superposition of the external Poisson process and the feedback process, is *not* Poisson (in fact, it can be shown to be characterized by a hyperexponential distribution). Remarkably, the departure process of this system is still Poisson (see also Walrand, 1988).

Returning to the general open network model described above, it was established by Jackson (1963) that a product form solution still exists even if customer feedback is allowed. This type of model is also referred to as a *Jackson network*. What is interesting in this model is that an individual node need not have

Poisson arrivals (due to the feedback effect), yet it behaves as if it had Poisson arrivals and can therefore still be treated as an $M/M/1$ system. Thus, we have the general product form solution:

$$\pi(n_1, n_2, \ldots, n_N) = \pi_1(n_1) \cdot \pi_2(n_2) \cdots \pi_M(n_M) \tag{25.48}$$

where $\pi_i(n_i)$ is the solution of an $M/M/1$ queueing system with service rate μ_i and arrival rate given by λ_i in the solution of Eq. (25.40). It is worth mentioning that the same result also holds for $M/M/m_i$ queueing systems, i.e., systems with $m_i \geq 1$ servers at node i. To guarantee the existence of the stationary probability distribution in Eq. (25.48), we impose the usual stability condition for each node:

$$\lambda_i = r_i + \sum_{j=1}^{M} \lambda_j p_{ji} < m_i \mu_i \tag{25.49}$$

The open network model we have considered can be extended to include state-dependent arrival and service processes. In particular, Jackson considered the case where an external arrival process depends on the total number of customers $(n_1 + \cdots + n_N)$, and the ith service rate depends on the queue length at node i (Jackson, 1963). Although the analysis becomes considerably more cumbersome, the stationary state probability can still be shown to be of the product form variety.

25.7.3 Closed Queueing Networks

A closed queueing network is one with a finite population of N customers. From a modeling standpoint, a closed network may be obtained from the open network model of the previous section by setting

$$r_i = 0 \quad \text{and} \quad \sum_{j=1}^{M} p_{ij} = 1 \quad \text{for all } i = 1, \ldots, M$$

In this case, no external arrivals occur and no customers can leave the system. Under these conditions, the state variables of the system, $X_1, \ldots, X_M$, must always satisfy $\sum_{i=1}^{M} X_i = N$. Thus, the state space is finite and corresponds to the number of placements of N customers among M nodes, given by the binomial coefficient

$$\binom{M+N-1}{M-1} = \frac{(M+N-1)!}{(M-1)!N!}$$

In addition, if we balance customer flows as in Eq. (25.40) we get

$$\lambda_i = \sum_{j=1}^{M} \lambda_j p_{ji}, \quad i = 1, \ldots, M \tag{25.50}$$

There is an important difference between the set of equations (25.40) and that in Eq. (25.50). In Eq. (25.40), we have M linearly independent equations, from which, in general, a unique solution may be obtained. In contrast, the absence of external arrival rate terms r_i in Eq. (25.50) results in $(M-1)$ linearly independent equations only. Thus, the solution of Eq. (25.50) for $\lambda_1, \ldots, \lambda_M$ involves a free constant. For instance, suppose we choose λ_1 to be this constant. We may then interpret $\lambda_i, i \neq 1$, as the relative throughput of node i with respect to the throughput of node 1.

It turns out that this class of networks also has a product form solution for its stationary state probabilities $\pi(n_1, \ldots, n_M)$, with the values of $n_1, \ldots, n_M$ constrained to satisfy $\sum_{i=1}^{M} n_i = N$ (Gordon and Newell, 1967). For simplicity, we limit our discussion here to single-server nodes, although the result also applies to the more general case where node i consists of m_i servers. Solving the flow balance equations in this case gives:

$$\pi(n_1, \ldots, n_M) = \frac{1}{C(N)} \rho_1^{n_1} \cdots \rho_M^{n_M} \tag{25.51}$$

where $\rho_i = \lambda_i/\mu_i$ with λ_i obtained from the solution of Eq. (25.50) with a free constant (arbitrarily chosen) and $C(N)$ is a constant dependent on the population size N, which is obtained from the normalization condition

$$\frac{1}{C(N)} \sum_{n_1,\ldots,n_M} \rho_1^{n_1} \cdots \rho_M^{n_M} = 1$$

Thus, to obtain the stationary state probability distribution in Eq. (25.51) we need to go through several steps of some computational complexity. First, the solution of the linear equations Eq. (25.50) to determine the parameters $\lambda_1, \ldots, \lambda_M$ must be obtained. This solution includes a free constant, which must next be selected arbitrarily. This selection will not affect the probabilities in Eq. (25.51), but only the values of the parameters ρ_i in the product form. Lastly, we must compute the constant $C(N)$, which is usually not a trivial computational task. There exist several computationally efficient algorithms to carry out this computation, e.g., see Buzen (1973) where the following recursive relationship is exploited:

$$C_i(k) = C_{i-1}(k) + \rho_i C_i(k-1), \quad i = 2, \ldots, M, \quad k = 2, \ldots, N \qquad (25.52)$$

with initial conditions $C_1(k) = \rho_1^k$, $k = 1, \ldots, N$ and $C_i(1) = 1$, $i = 1, \ldots, M$, from which $C(N)$ is obtained as $C(N) = C_M(N)$.

It can also be shown that the utilization of node i when the population size is N is given by

$$\mu_i[1 - \pi_i(0)] = \rho_i \frac{C(N-1)}{C(N)}$$

Expressions for other performance measures can similarly be derived in terms of the parameters ρ_i and $C(k), k = 1, \ldots, N$.

25.7.3.1 Mean Value Analysis

If we are only interested in obtaining performance measures such as the network throughput and the mean values of the queue length and system time distributions at nodes, then *Mean Value Analysis* (MVA), developed in Reiser and Lavenberg (1980), bypasses the need for computing the normalization constant $C(N)$. MVA exploits a simple relationship between the average customer system time at a network node and the average queue length at that node. Specifically, consider a customer arriving at node i, and let $\bar{S}_i$ be the average system time the customer experiences at i. Moreover, let $\bar{X}_i$ be the average queue length seen by that arrival. Observe that

$$\bar{S}_i = \frac{1}{\mu_i} + \bar{X}_i \frac{1}{\mu_i}$$

where $1/\mu_i$ is the mean service time at node i. In other words, the customer's system time consists of two parts: his own service time, and the total time required to serve all customers ahead of him. It can be shown that in a closed queueing network with N customers, $\bar{X}_i$ is the same as the average queue length at i in a network with $(N-1)$ customers. Intuitively, what a typical customer sees is the network without that customer. Therefore, if we denote by $\bar{X}_i(N)$ and $\bar{S}_i(N)$ the average queue length and average system time at node i, respectively, when there are N customers in the network, we obtain the following recursive equation:

$$\bar{S}_i(N) = \frac{1}{\mu_i}[1 + \bar{X}_i(N-1)], \quad i = 1, \ldots, M \qquad (25.53)$$

with initial condition $\bar{X}_i(0) = 0, i = 1, \ldots, M$. In addition, we can use Little's Law (25.24) twice. First, for the whole network, we have

$$N = \sum_{i=1}^{M} \Lambda_i(N) \bar{S}_i(N) \qquad (25.54)$$

where $\Lambda_i(N)$ is the node i throughput when there are N customers in the network and N the fixed number of customers in the network. Note that $\Lambda_i(N)/\Lambda_j(N) = \lambda_i/\lambda_j$ for all i, j where λ_i, λ_j are obtained from Eq. (25.50), so that $\Lambda_j(N)$ is obtained from Eq. (25.54) as

$$\Lambda_j(N) = N \left[\sum_{i=1}^{M} \left(\frac{\lambda_i}{\lambda_j}\right) \bar{S}_i(N) \right]^{-1} \tag{25.55}$$

The second use of Little's Law is made for each node i and we get

$$\bar{X}_i(N) = \Lambda_i(N) \bar{S}_i(N), \quad i = 1, \ldots, M \tag{25.56}$$

The four equations (25.53)–(25.56) define an algorithm through which $\bar{X}_i(N)$, $\bar{S}_i(N)$, and $\Lambda_i(N)$ can be evaluated for various values of $N = 1, 2, \ldots$.

25.7.4 Product Form Networks

The open and closed queueing network models considered thus far are referred to as *product form* networks. Obviously, this is because the stationary state probabilities $\pi(n_1, \ldots, n_M)$ can be expressed as products of terms as in Eq. (25.48) or Eq. (25.51). The ith such term is determined by parameters of the ith node only, which makes a decomposition of the network easy and efficient. Even though we have limited ourselves to Markovian networks, it turns out that there exist significantly more complicated types of networks, which are also of the product form variety. The most notable type of product form network is one often referred to as a BCMP network after the initials of the researchers Baskett, Chandy, Muntz, and Palacios who studied it (Baskett et al., 1975). This is a closed network with K different customer classes. Class $k, k = 1, \ldots, K$, is characterized by its own routing probabilities and service rates and four different types of nodes are allowed: (i) A single-server node with exponentially distributed service times and $\mu_i^k = \mu_i$ for all classes. In addition, an FCFS queueing discipline is used. This is the simplest node type which we have used in our previous analysis as well. (ii) A single-server node and any service time distribution, possibly different for each customer class, as long as each such distribution is differentiable. The queueing discipline here must be of the *processor-sharing* (PS) type (i.e., each customer in queue receives a fixed "time slice" of service in round-robin fashion and then returns to the queue to wait for more service if necessary; the PS discipline is obtained when this time slice is allowed to become vanishingly small). (iii) The same as before, except that the queueing discipline is of the Last-Come-First-Served (LCFS) type with a preemptive resume (PR) capability. This means that a new customer can preempt (i.e., interrupt) the one in service, with the preempted customer resuming service at a later time. (iv) A node with an infinite number of servers and any service time distribution, possibly different for each customer class, as long as each such distribution is differentiable.

In this type of network, the state at each node is a vector of the form $\mathbf{X}_i = [X_{i1}, X_{i2}, \ldots, X_{iK}]$, where X_{ik} is the number of class k customers at node i. The actual system state is the vector $\mathbf{X} = [\mathbf{X}_1, \mathbf{X}_2, \ldots, \mathbf{X}_K]$ and, assuming the population size of class k is N_k, we must always satisfy the condition $\sum_{j=1}^{M} X_{ik} = N_k$ for all k. The actual product form solution can be found in Baskett et al. (1975) or several queueing theory textbooks such as Trivedi (1982).

25.8 Non-Markovian Queueing Systems

Beyond queueing systems whose arrival and service processes are modeled through exponential distributions, analytical results are very limited. There are several sophisticated techniques for approximating the distributions of these processes using combinations of exponential ones or by approximating interesting performance measures by means of the first two moments only. These and related techniques for analyzing complex queueing models are beyond the scope of this chapter and the reader is referred to specialized books, including Asmussen (2003), Bremaud (1981), Chen and Yao (2000), Kelly (1979), Kleinrock (1975),

Neuts (1981), Walrand (1988). We limit ourselves to one well-known and very useful analytical result that applies to the $M/G/1$ queueing system. In particular, it is possible to derive a simple expression for the average queue length $E[X]$ of such a system. This is known as the *Pollaczek–Khinchin formula* (or PK formula):

$$E[X] = \frac{\rho}{1-\rho} - \frac{\rho^2}{2(1-\rho)}(1 - \mu^2 \sigma^2) \tag{25.57}$$

where $1/\mu$ and σ^2 are the mean and variance, respectively, of the service time distribution and $\rho = \lambda/\mu$ is the traffic intensity as defined in Eq. (25.9), λ being the Poisson arrival rate. We can immediately see that for exponentially distributed service times, where $\sigma^2 = 1/\mu^2$, (Eq. 25.57) reduces to the average queue length of the $M/M/1$ system, $E[X] = \rho/(1-\rho)$, as in Eq. (25.33).

Finally, it is worth mentioning that many software tools have been developed over the years to facilitate the process of modeling queueing systems and of estimating performance measures such as throughput, mean queue lengths, and mean delays. Some are based on discrete event simulation (Extend and SimEvents are two of the most recent commercial products of this type), while others are based on approximation methods (RESQ and QNA are such examples). More specialized tools have also been developed for particular application areas such as communication networks (e.g., Opnet) and manufacturing systems (e.g., MPX).

References

Asmussen, S., *Applied Probability and Queues*, 2nd ed., Wiley, New York, 2003.
Baskett, F., Chandy, K. M., Muntz, R. R., and Palacios, R. R., Open, closed, and mixed networks of queues with different classes of customers, *J. ACM*, 22(2), 248–260, 1975.
Bremaud, P., *Point Processes and Queues*, Springer, New York, 1981.
Burke, P. J., The output of a queueing system, *Oper. Res.*, 4, 699–704, 1956.
Buzen, J. P., Computational algorithms for closed queueing networks with exponential servers, *Commun. of ACM*, 16(9), 527–531, 1973.
Cassandras, C.G., and Lafortune, S., *Introduction to Discrete Event Systems*, Kluwer, Norwell, MA, 1999.
Chen, H., and Yao, D.D., *Fundamentals of Queueing Networks*, Springer, New York, 2000.
Gordon, W. J., and G. F. Newell, Closed queueing systems with exponential servers, *Oper. Res.*, 15(2), 254–265, 1967.
Jackson, J. R., Jobshop-like queueing systems, *Manage. Sci.*, 10(1), 131–142, 1963.
Kelly, F. P., *Reversibility and Stochastic Networks*, Wiley, New York, 1979.
Kleinrock, L., *Queueing Systems, Volume I: Theory*, Wiley, New York, 1975.
Lindley, D. V., The theory of queues with a single server, *Proc. Cambridge Philos. Soc.*, 48, 277–289, 1952.
Little, J. D. C., A proof of $L = \lambda W$, *Oper. Res.*, 9(3), 383–387, 1961.
Neuts, M. F., *Matrix-Geometric Solutions in Stochastic Models: An Algorithmic Approach*, Dover, New York, 1981.
Reiser, M. and Lavenberg, S. S., Mean-value analysis of closed multichain queueing networks, *J. of ACM*, 27(2), 313–322, 1980.
Stidham, S., Jr., A last word on $L = \lambda W$, *Oper. Res.*, 22, 417–421, 1974.
Trivedi, K. S., *Probability and Statistics with Reliability, Queuing and Computer Science Applications*, Prentice-Hall, Englewood Cliffs, NJ, 1982.
Walrand, J., *An Introduction to Queueing Networks*, Prentice-Hall, Englewood Cliffs, NJ, 1988.

26
Port-Based Modeling of Engineering Systems in Terms of Bond Graphs

Peter Breedveld
University of Twente

26.1	Introduction	26-1
	Port-Based Modeling versus Traditional Modeling • Dynamic Models of Engineering Systems	
26.2	Structured Systems: Physical Components and Interaction	26-4
26.3	Bond Graphs	26-5
	Appearance • Elementary Behaviors • Causality	
26.4	Multiport Generalizations	26-21
	Multiport Storage Elements	
26.5	Conclusion	26-28

26.1 Introduction

26.1.1 Port-Based Modeling versus Traditional Modeling

To be concise, port-based modeling of dynamic behavior will be introduced from the point of view that the reader is well aware of traditional modeling techniques as well as the common pitfalls of modeling, viz.:

- confusion of physical components with ideal concepts (conceptual elements), and
- confusion of modeling with model manipulation, where modeling is the mere decision process of which aspects should be taken into account to arrive at a competent model for a specific problem context, while model manipulation is any kind of transformation of the representation of a model into a form that allows better insight in the particular aspect under study (none of these transformations should change the physics properties of the model).

The key idea in any modeling approach that starts from the a priori knowledge about the physics properties of the system to be modeled is that a conceptual separation is made between various fundamental behaviors that are (considered to be) relevant for a given problem context. Although this step is always present, it is often preceded by a step in which the system to be modeled is subdivided into subsystems on the basis of aspects of function or configuration. A pump (drive) system, for example, can be seen as an interconnection of an electric power source (an amplifier), an electric motor, a transmission, and a mechanism for displacement of a fluid (a gear box and a load). This is the first conceptual level of observable, functional components. The total behavior consists not only of the observable dynamic interaction between these subsystems, but also of the conceptual interaction between the fundamental behaviors that constitute the behavior of these subsystems. Already when modeling at this level, it is useful to consider both forms of interaction from the point of view of *bilateral* relations, instead of the unidirectional input–output relations that are often used, thus implicitly assuming no "back-effect" (Figure 26.1). This bilateral relation

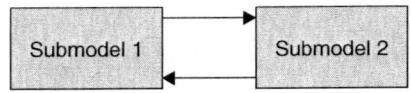

FIGURE 26.1 Bilateral relation.

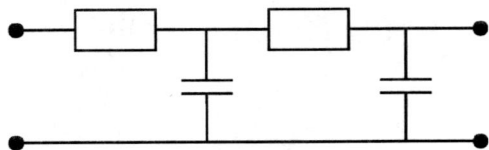

FIGURE 26.2 Double RC network.

and its relation to the concept of energy is the key idea behind a port-based modeling approach, as will be one of the main objectives of this chapter.

To give a rather simple example: when modeling the behavior of a double resistor-capacitor (RC) network (Figure 26.2) assuming connection to a fit source, e.g., a sine generator, the first step is to identify the components (two resistors and two capacitors). Unlike many other domains of physics, the fabrication of components applied in electrical circuits is given so much attention that, in a context of normal operation (i.e., neglecting radiation and high frequencies), the components map during modeling in a one-to-one way to the conceptual elements with the same name (resistor R and capacitor C) that represent two basic physical concepts, viz. Ohm's law for dissipation and electrical storage (of charge and electrostatic energy), respectively. It is discussed later that the use of this conceptual resistor implies an implicit assumption that the variations of the temperature are such that they may be neglected in the given context, such that "dissipatable" free energy can be used instead of globally conserved "true" energy. If one would identify each RC network as a subsystem and consider the transfer function between the input and the output voltages, the total transfer function is not equal to the product of these transfer functions, unless one includes a so-called separation amplifier. This is due to the bilateral relation between all these elements and these two subsystems in particular.

Note that the adjective "physical" (physical system, physical model, and physical modeling) is not used in the sense of "concrete" within the context of dynamic modeling, and thus this text, but in the sense of "obeying the laws of physics." As of now the terminology "physical properties" will be used to address what has been called "physics properties" earlier. Consequently, a physical model in this sense can be rather abstract.

In most classical modeling approaches, most of the relations are thought to be unilateral (input–output), in other words, the back-effect is implicitly considered negligible. In those cases where the back-effect cannot be neglected, it is thought of as a separate, unilateral relation, often addressed as the feedback.

Next all conceptual behaviors are commonly linked at the conceptual level into a set of differential and algebraic equations using common variables and balance equations. In principle, the algebraic equations can be used to eliminate a number of variables, such that a set of (partial) differential equations remains. In some simple cases (e.g., linear systems), such a set may have an analytical solution. However, in most cases such an analytical solution does not exist and numerical integration (simulation) is used to obtain an approximate solution. This is discussed extensively in Chapter 17.

However, if the model is written in terms of equations, an important change in the nature of the model representation is implicitly made: from a *simultaneous*, by definition graphical, representation of a system model in terms of interconnected subsystems and interconnected conceptual elements, the representation becomes *sequential*, in the sense that the equations can only be read one after the other. As the parts of the actual system that is being described are present simultaneously and take part in the behavior simultaneously, simultaneous representations generally provide more insight, in particular during the conceptual phase of the modeling process. Furthermore, when algebraic equations are eliminated, not only some physically relevant variables are lost but also the (conceptual) structure of the relations, such that local changes in the model structure are not possible anymore, even though such elimination may be very useful for analysis or computational purposes.

Several examples of domain-dependent, simultaneous model representations exist and are frequently used, in particular in engineering, although at the same time they represent the wish to take the structure apart for a local analysis per submodel or element. For example, a free-body diagram is used to compute the net contribution to the rate of change of momentum (dp/dt) on the basis of the forces acting on a body. For the analysis of electrical circuits similar techniques exist, like (modified) nodal analysis, mesh analysis, and incidence matrices, see any standard textbook on circuit analysis (e.g., Hayt et al., 2002).

The bilateral interaction between conceptual elements requires the concept of a port, introduced in the context of electrical networks by Harold Wheeler (Wheeler and Dettinger, 1949) and extended by Henry Paynter to all relevant physical domains (Paynter, 1961). Some authors have taken this concept even beyond physical systems as any relation can be considered bilateral. As a bilateral interaction can be seen as a combination of two opposite unilateral interactions or signals, *two* relevant quantities are required to characterize such a port. This aspect is discussed later in more detail.

Important at this point is that port-based modeling can be distinguished from classical modeling techniques not only by its use of bilateral relations a priori in a simultaneous representation (i.e., graphical), but also in particular by the fact that no decisions are made about the input–output nature of the constituting signals of this relation. The relations between the ports are said to be acausal, a terminology to be explained later and somewhat different from the commonly used meaning of causality and acausality related to the "arrow of time," even though it is related in case of the so-called storage ports (cf., the concept of preferred causality in Section 26.3.3.1). This property of a port not only enables easy reuse of model parts within other contexts, but also gives immediate feedback to the modeler about his modeling decisions as the so-called causal properties of ports to be connected not only determine the final computational structure, but may also indicate that certain physical phenomena are not taken into account. This feedback stimulates the modeler in taking a deliberate decision about what he should include in his model and what not.

26.1.2 Dynamic Models of Engineering Systems

To be precise, the aspects of reality to be modeled are "limited" herein to behavior in time, i.e., dynamics, of systems that obey the basic principles of classical physics, in the sense that quantum effects play no role. This means that the dynamic behavior is modeled of macroscopic phenomena that are all supposed to obey the basic principles of thermodynamics, viz. (global) conservation of energy and positive entropy production, independent of some specific domain. The adjective "macroscopic" refers to the *nature* of the scope, viz. a thermodynamic approach, rather than the *scale* of the phenomena: phenomena at the nanoscale may for instance still be modeled using a macroscopic viewpoint, which combines the kinetic energy of the random motions into thermal energy using entropy not only as an energy state but also as a property to express the increase of random motion in terms of the positive entropy production principle. In contrast, the microscopic view does not collect any motion into thermal energy and this kind of view may even be applied to objects at a relatively large scale like queueing behavior.

The particular problem context in which the model is made determines the criteria for the selection of basic phenomena that may determine the relevant behavior of each of the functional or tangible components. As discussed earlier, the structured set of physical *components* that constitutes a model, has to be further abstracted into a conceptually interconnected set of basic behaviors, like storage and transformation, which will be discussed in more detail later in the form of ideal *elements* (Section 26.3.2). It will turn out that this process may be supported by the use of multiport generalizations of the most basic conceptual elements, in particular, if the relevant phenomena take place in various physical domains. Such systems are called "multidomain," "multidisciplinary," "multiphysics," and "mechatronic."

A clear distinction is required between intended behavior and realized behavior, where intended behavior is analyzed by functional modeling during design, while during design evaluation as well as during troubleshooting the behavior that will be or is actually realized should be analyzed by physical modeling of all realized aspects, in other words, including the undesired behaviors from a functional point of view. It can be often observed during troubleshooting that the functional models that served their purpose well in the design phase, fail to provide the required insight for troubleshooting. This means that when a designer is involved in either concept validation or troubleshooting, he has to make the difficult step of reducing the priority of what he envisioned on a functional basis, while trying to capture all physical behaviors (to be) actually realized, even when there is no direct link with the intended function. Making a clear separation between functional modeling and physical modeling supports this step considerably.

26.2 Structured Systems: Physical Components and Interaction

In this section, a special focus of the port-based approach to modeling is put on structured systems, i.e., systems that can be seen as a set of physical or functional components in a structure. Key element of such a structure is that it represents the relations between the constituting parts or subsystems. As these relations are bilateral, the (virtual) "points" where these relations can be "attached" to a subsystem have a bilateral nature too: the *ports*. In other words, the relations interconnect subsystems into a system via the ports of the subsystems. The properties of ports related to behavior in terms of classical physics are discussed as well as the basic concepts that are required to describe the subsystem behavior: *state* and *change* (of state). In contrast with domain-specific simultaneous representations like free-body diagrams, the analysis of the model in the port-based approach is not based on tearing the structure apart and performing local analyses: the structural elements are considered subsystems (nodes of a graph) as well and called *junctions* (cf., Section 26.3.2.5).

Within the context of dynamic systems, where the core of the dynamics itself lies in the process of "storage," the two relevant quantities or variables in a bilateral relation discussed later in this section are strongly related to the concept of equilibrium, the balance between the two parts in which a (locally) conserved quantity or so-called extensive state can be stored: they can either *establish* the equilibrium by changing the stored quantity ("equilibrium establishing variable" or "rate of change of extensive state") or they *determine* the equilibrium by forming a balance independent of the extent of the communicating parts ("equilibrium determining variable" or "intensive state"). For example, in case of diffusion processes, the flows of matter of the participating species are the equilibrium-establishing variables (called generalized fluxes in nonequilibrium thermodynamics) that become zero when the equilibrium is reached. The concentrations of the participating species are the equilibrium-determining variables (called generalized forces in nonequilibrium thermodynamics) that become equal (zero difference or gradient) when equilibrium is reached.

The conjugation between these two types of variables that characterize a port is called *dynamic conjugation*. Diffusion phenomena are commonly studied at constant pressure and temperature, thus allowing energy exchange with the environment that is not considered relevant for the dynamics of the diffusion process. However, as soon as such assumptions cannot be made and more than one domain (i.e., type of conserved quantity) needs to be considered, the energy bookkeeping becomes an important means for consistent modeling, and the dynamic conjugation has to be further constrained to *power conjugation* in the sense that the relation between two ports that dynamically interact, describes the power, in other words, the energy exchange linked to the relation between two connected *power ports* (in the sequel "port" will refer to a "power port" unless otherwise indicated). In other words, the power of a port has to be a function of the two power conjugated variables. The shape of this function is not dictated by nature, but a common choice is that the *product* of the two power conjugate variables forms the power of the interconnection between two power ports called *bond*. This is a natural consequence of the mathematical property that a change of energy E, being a function of one or more stored quantities $E(q_1, \ldots, q_i, \ldots, q_n)$, can be written as the inner product of the partial derivatives with respect to the stored quantities and the changes of those stored quantities: $dE(q_1, \ldots, q_i, \ldots, q_n) = \sum_{i=1}^{n} \frac{\partial E}{\partial q_i} dq_i$. Domain-dependent examples of power conjugate variables are voltage and current, force and velocity, pressure and volume flow, and temperature and entropy flow. The domain-independent terminology for the equilibrium-determining variable is *effort*, and *flow* is used for the equilibrium-establishing variable. In case of common (power)

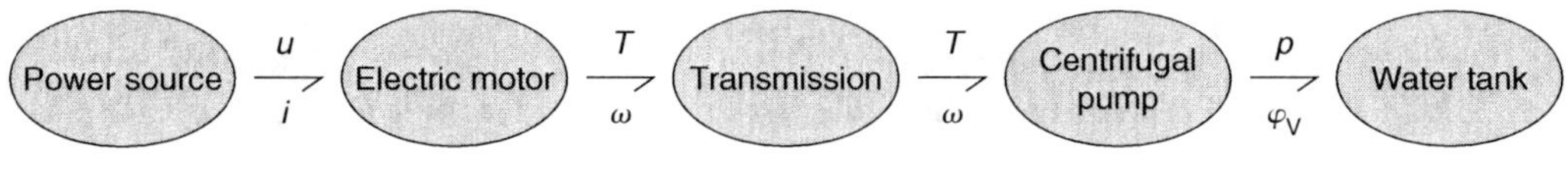

FIGURE 26.3 Word bond graph of a pump driven by a DC motor.

FIGURE 26.4 Causal stroke.

bonds, the product of effort and flow is assumed to represent the power, in case of pseudobonds effort and flow may be related to the power in another manner. For instance, in case of the common variables used to model thermal behavior, viz. temperature and heat flow, the heat flow itself is the (thermal) power already. However, it only represents a rate of change of the stored energy as long as the dynamics of other domains play no role; otherwise, the amount of heat cannot play the role of a proper state variable, as discussed at length in many introductions to thermodynamics (e.g., Callen, 1960).

Figure 26.3 demonstrates how specific functional or material parts of an actual system can be represented in the form of submodels by a short text enclosed by some closed line and identifying them in an insightful way (often an ellipse is used to distinguish them from the common "blocks" of a block diagram of which the interface consists of inputs and outputs and where the blocks themselves represent mathematical operations). These ellipses form the nodes of a graphical representation, a so-called *word bond graph*, of which the edges represent the so-called *bonds* between the ports of these subsystems (Paynter, 1961). The word bond graph can be seen as the highest level description of a dynamic model, close to the functional and structural level. The direction of the edges in the form of the *half-arrow* that is typical for a bond represents the positive orientation of both flow and power related to that bond.

Another aspect that can be added to a bond is its so-called causal stroke, a little stroke attached to one end of the bond and perpendicular to it. It represents the computational direction (when chosen) of the effort signal and consequently also that of the flow signal as the signals of the two conjugate variables are opposite by definition (Figure 26.4).

26.3 Bond Graphs

Readers who have been exposed previously to analog models and the bond graph notation in particular are warned that many introductions start from the misconception that analogies are merely based on similarity of the underlying differential equations. In this treatise this is considered as an exchange of cause and effect. The actual source of analogies that lead to the domain independence of both differential equations and bond graph notation is the mere fact that in human reasoning about dynamic behavior only a rather limited set of elementary concepts is exploited. This set of ideal concepts is chosen as the backbone of Section 26.3.2 and maps directly onto the bond graph symbolism. In Section 26.3.3, it is shown how the concept of computational causality that augments the representation of the physical structure by a bond graph with a representation of the computational structure, leads to the formulation of a computable mixed set of differential and algebraic equations.

In models that are solved analytically, it is commonly preferable to eliminate the algebraic equations as much as possible, as only the structure of the differential equations determines the nature of the described behavior. In mathematics, this is often taken one step further by scaling or normalizing the variables, which takes the model a step further away from its physical interpretation. In the days of analog computing, scaling and reduction to a minimal form was a necessity and in the early days of digital computing "superfluous" computations were to be avoided. However, the algebraic part describes the interconnection structure and maintaining this model information, particularly in a graphical representation, allows rapid model modification, quick interpretation of simulation results and thus enhances insight, in particular if the variables continue to represent common physical concepts. Owing to the increase of computational power and robustness of modern computing maintaining this information that is not essential for solving the dynamics is less costly, in particular due to automatic symbolic manipulation.

Starting from the word bond graph, this section explains the basic structure of a bond graph and the most important semantics and grammar of this graphical language.

26.3.1 Appearance

The mathematical structure of a word bond graph is that of a labeled, directed graph or digraph of which the directed edges are the bonds with their orientation (half-arrows) and of which the labeled nodes (ellipses with text) represent the subsystems characterized by some (functional) description (Figure 26.3). The transition to a regular bond graph is relatively simple: basic behaviors with respect to energy are represented by a mnemonic code (acronym) consisting of up to a few capital letters. In this chapter, these basic behaviors will be discussed at different levels, which lead to different categorizations. However, the key elementary behaviors of macroscopic physics, viz. *conservation* (in fact expressing the need for any useful model to satisfy *time translation symmetry*) and *positive entropy production* (expressing the need for most macroscopic models to satisfy *time reflection asymmetry*, in other words, the "arrow of time") will play a key role at all times.

26.3.2 Elementary Behaviors

26.3.2.1 Storage

We start with a bottom-up discussion that assigns all reversible storage to those nodes of the graph that are called storage elements. Note that the adjective "reversible" is in fact a tautology when identifying this ideal process and will be omitted in the sequel. In principle, these storage nodes may have an arbitrary number of ports and are called multiport storage elements in that generic case. For reasons of simplicity, the one-port version will be described first. The storage element or capacitive element stores a specific, (locally) conserved quantity that, by definition, is extensive in nature, i.e., it is proportional to the spatial or material extent of the object to be modeled. Examples are electric charge, amount of moles, entropy, momentum, and magnetic flux (linkage). Some of these quantities, like charge and momentum, are also globally conserved, but others, like entropy, elastic displacement, and flux linkage, are not. Such a locally conserved quantity q is by definition a state variable, in the sense that its cyclic integral is zero:

$$\oint dq = 0 \qquad (26.1)$$

which means that it describes the state of the system in a unique way, independent of the history of the system. One could also say that the history of the system is uniquely reflected in its current state, in other words, the states can be considered the "memory" of the system.

The locally conserved quantity also determines the nature of the so-called physical domain and will require some reference value. The process of storage of such a locally conserved quantity can be represented by the time integral of its rate of change, which is the already identified equilibrium establishing variable or flow variable f

$$\int_{q(t_0)}^{q(t)} dq(t) = \int_{t_0}^{t} \frac{dq(\tau)}{d\tau} d\tau = \int_{t_0}^{t} f(\tau) d\tau = q(t) - q(t_0) \qquad (26.2)$$

where $q(t) - q(0)$ is the increase of the stored quantity in a time interval $t - t_0$. Later it is shown that the powerful—in particular in the conceptual modeling phase—feature of the relation of the effort and flow variables with the concept of equilibrium is commonly lost due to a symmetrization of the role of the power conjugate variables. However, the concept of the symplectic gyrator to be discussed later, allows an alternative approach that maintains this asymmetry and can be converted into the traditional form.

Independent of the domain, the storage element is represented by the label **C** that stands for *capacitor*. The conjugate effort or equilibrium determining variable has to be an intensive state, i.e., a state that is independent of the extent of the described system, as equilibrium between two systems does not depend on their individual extents. Obviously, intensive states are related to the extensive states. In case of physical properties with a spatial orientation, like force and electrical current, it is clear that the concept of extent has to be related to the nature of the spatial dependence of the physical quantity, so the role of the intensity is played by the linear vector and the role of the extensity by the axial vector. Without going in to more detail, it is noted that this creates an additional constraint on how these vectorial extensities can be combined in case of equilibrium.

The relation between the intensive state and extensive state is described by the *constitutive relation* of the C-element, which depends both on constituting matter and geometry. In the common, linear, one-port case the constitutive relation reduces to a proportionality constant

$$e(t) = \frac{q(t)}{C} = \frac{1}{C}\int_{t_0}^{t} f(\tau)d\tau + q(t_0) \tag{26.3}$$

The inverse of this proportionality constant is called the *capacitance C*, which, as a natural consequence of the above discussion, should be extensive in nature. Examples are the capacitance of an electric capacitor and the compliance of a spring (inverse of a spring constant). Such constitutive parameters (also in the nonlinear case) are always dependent on material parameters that are not related to the extent, like mass density and geometric parameters like length, width, and height that are by definition related to the spatial extent. Sometimes they also contain global natural constants. Sometimes the spatial extent is combined with the mass density into a material extent expressed in amount of moles or in kilograms.

For example, an electric flat-plate capacitor of which fringing and losses can be neglected can be described by a linear constitutive relation:

$$u(t) = \frac{q(t)d}{A\varepsilon_0\varepsilon_r} = \frac{q(t)}{C} = \frac{1}{C}\left(\int_{t_0}^{t} i(\tau)d\tau + q(t_0)\right) = \frac{1}{C}\int_{t_0}^{t} i(\tau)d\tau + u(t_0) \tag{26.4}$$

where the capacitance is $C = \frac{A\varepsilon_0\varepsilon_r}{d}$, A (plate area), and d (distance between plates) are the geometrical parameters, ε_r (relative permittivity or dielectric constant) the material parameter, and ε_0 the global natural constant (permittivity).

Another example of a linear storage element is an ideal, nonrelativistic mass m (product of mass density ρ and volume V), storing momentum p:

$$v(t) = \frac{p(t)}{m} = \frac{p(t)}{\rho V} = \frac{1}{m}\int_{t_0}^{t} F(\tau)d\tau + v(t_0) \tag{26.5}$$

Although the association with a capacitance may seem somewhat awkward, it is emphasized that so far only one type of storage has been identified on the basis of the asymmetric role of effort and flow. The kinetic equilibrium determining variable (kinetic effort) is the velocity v, while the equilibrium establishing variable is the rate of change of momentum (kinetic flow) or net force F. The paradox with common modeling approaches that treat a force as an effort and a velocity as a flow will be resolved later.

26.3.2.2 Environment, Sources, Boundary Conditions, and Constraints

If the system to be modeled cannot be isolated in an energetic sense from the rest of the world, its so-called environment, the environment should "store" too as this allows exchange of the stored conserved quantity with this environment. However, the criterion for selecting the system boundary that separates the system from its environment is that the environment does not influence the dynamic characteristic of the system. In other words, the intensive state of the environment does not depend on its extensive state, such that the extent of the environment becomes irrelevant, which coincides with the concept of an environment. This means that the intensive state or effort is also independent of the rate of change of its extensive state, i.e., its conjugate flow. This kind of imposed intensity at the boundary is often called a *source*, a *constraint*, or a *boundary value*. In bond graph terminology, such an environmental influence becomes a source of effort with acronym **Se**. Not only from a mathematical point of view it can be considered a **C**-type element of which the capacitance has become infinite, such that it becomes independent of a change in extent and just depends on its initial value (or "initial condition"), but also from a modeling point of view, since storage in the environment can only be considered as such when the capacitances of all capacitors in the system itself are relatively small compared to the capacitance of the environment.

The energy of a system is the sum of all partial energies stored in the capacitors. In other words, the energy is a state that is a function of all stored quantities (independent extensive states) and as such an extensive state too. Since the environment is not related to the extent, the reference of each conserved state is also made at the intensive level and can be either absolute or relative in nature, for instance, temperature and pressure have absolute zero points, but voltage and velocity (the effort of the kinetic domain) have no absolute zero point.

In many cases these types of boundary conditions, viz. sources of effort, are combined with certain types of transducers that are considered ideal in such a way that the constant effort imposed to the transducer (ideal pump or motor) is translated into an imposed flow variable in another domain. Such combinations of effort sources with interdomain transducers are called flow sources with acronym **Sf**. Both types of sources can also be deliberately approximated by means of external energy supply and feedback control. In those cases, the conceptual connection to some "infinite" storage becomes much less elucidating.

In the degenerate case of zero-valued sources, the power is zero too (also called Dirichlet and von Neumann type of boundary conditions), which means that the modeled system is energetically isolated and its energy becomes an invariant. Many analyses of systems described by partial differential equations (such as finite-element analyses) are based on these kinds of boundary conditions. If such models are considered submodels to be combined with other submodels the description of this interaction is not straightforward (Ligterink et al., 2006).

26.3.2.3 Power Continuous Structure

Now that all storage in the system (**C**-type ports) and in the environment (**Se**- and **Sf**-type ports) has been described, the principle of energy conservation means that all other conceptual elements in the system need to be *power continuous* accordingly. In other words, they require an instantaneous power balance between all the ports

$$\sum_i \varepsilon_i P_i = \sum_i \varepsilon_i e_i f_i = 0 \qquad (26.6)$$

where $\varepsilon_i = \pm 1$ depending on the orientation of the half-arrow of the connecting bond.

This condition means that all other elements need to have at least two ports. This seems a quite unnatural conclusion, because the average reader will immediate think of the resistor as a basic one-port element. To resolve this paradox, the ideal resistor is treated first.

26.3.2.4 Resistor

Indeed, this basic element in its general form is a two-port, irreversible transducer, where one of the ports is by definition a thermal port with the flow of irreversibly produced entropy as conjugate flow variable of the thermal effort, the temperature. However, in most cases, it is *implicitly* assumed that the temperature variations can be neglected at the timescale of interest. This means that this thermal port can be considered to be connected to an effort source and that the conjugate flow is irrelevant for the dynamics of the system. In fact, this is also the case for the storage elements and, as will be explained later in more detail, this means that the stored energy is replaced by one of its Legendre transforms (cf., Section 26.4.1.1), the free energy, as a generating function of the constitutive relations. The free energy does not satisfy a conservation principle and can thus be *dissipated* by a one-port R-element.

Many constitutive relations of these resistive elements, like Ohm's law in the electric case, are considered linear algebraic relations between the two conjugate variables of the remaining port:

$$u - Ri = 0 \tag{26.7}$$

The power into the electric port is the product of effort and flow, voltage u and current i, and is considered dissipated when the (absolute) temperature T is considered constant:

$$P_{\text{diss}} = ui = Ri^2 = \frac{u^2}{R} = Tf_{S_{\text{irr}}} = P_{\text{thermal}} \tag{26.8}$$

In contrast, if the temperature cannot be considered constant, the thermal port needs to be explicitly modeled. Owing to the power continuity constraint the flow of irreversibly produced entropy $f_{S_{\text{irr}}}$ can always be computed, independent of the form of the constitutive relation of the resistive port, by dividing the input power that equals the thermal output power, i.e., the heat flow, by the absolute temperature:

$$f_{S_{\text{irr}}} = \frac{P_{\text{thermal}_{\text{out}}}}{T} = \frac{P_{\text{in}}}{T} \tag{26.9}$$

This relation cannot be inverted as the flow of produced entropy is zero in equilibrium.

Furthermore, it shows that the irreversible transducer that represents the positive entropy production in the system is by definition nonlinear, even if the constitutive relation of the nonthermal port is linear, like in the case of an ohmic resistor. From another perspective to be discussed later, it will become clear that all linear power continuous two-ports are *reversible* transducers, which confirms the conclusion that an *irreversible* transducer needs to have at least one *nonlinear* constitutive relation. To distinguish the two-port irreversible transducer from the one-port dissipator of free energy **R**, the symbol **RS** is used, emphasizing the explicit representation of the production (source) of entropy.

Any nonlinear constitutive relation of the nonentropy producing port of the **R(S)** should be a relation that lies in the first and third quadrant owing to the second law of thermodynamics. This means that all resistive constitutive relations cross the origin and that all linear resistors are positive. Only differential resistances in an operating point that is kept out of equilibrium by external energy input can be considered negative. Living systems are in such a state.

26.3.2.5 Generalized Junction Structure: Transduction and Interconnection

Now that all energy storage and irreversible production of entropy (dissipation of free energy) has been given a conceptual location in the **C** & **Se** (**Sf**) and **R(S)**, respectively, the rest of the model can be seen as a power continuous multiport that is called *generalized junction structure* (GJS). However, this GJS can be usefully substructured into elements with specific properties that can be globally split into *transducers* and *interconnections*.

The common choice for the relation between the two power conjugate variables and the power is a product operation, which can be observed from the common choices for effort and flow. However, as a simple linear transformation transforms effort- and flow-type variables into so-called *(wave) scattering variables* of which the relation with the power is a sum operation, this illustrates that this is just the modeler's choice (Paynter and Busch-Vishniac, 1988). The transformation to scattering variables provides quite some insight if the properties of power continuous n-ports ($n \geq 2$) are studied, thus providing a means to create a substructure in the GJS.

When the *interconnection* aspect of power continuous n-ports is studied, the key property is that one should be able to make *arbitrary connections*, in other words, all ports should behave in the same way and cannot be mutually distinguished. This means that the interconnection n-port is not only power continuous but also port-symmetric, which can be easily translated into the form of the scattering matrix that represents the constitutive relations in terms of scattering variables. By means of this scattering approach it is relatively straightforward to prove that there are only two possible solutions when these two constraints are applied (Hogan and Fasse, 1988). These two solutions are described by linear constitutive relations without any characteristic parameter and can be seen as a combination of a generalized form of a Kirchhoff's law for one of the conjugate variables combined with the identity of the other conjugate variable:

$$\sum_{i=1}^{n} \varepsilon_i e_i = 0 \quad \varepsilon_i = -1, +1 \text{ and } f_i = f_j \quad \forall i, j : i \neq j$$

or (26.10)

$$\sum_{i=1}^{n} \varepsilon_i f_i = 0 \quad \varepsilon_i = -1, +1 \text{ and } e_i = e_j \quad \forall i, j : i \neq j$$

In all the cases, these kind of analyses start from the assumption that the inbound orientation is positive in principle, which means that $\varepsilon_i = +1$ for inbound port orientations and $\varepsilon_i = -1$ for outbound port orientations. These interconnection multiports are called *junctions*, where the **0**-junction describes the instantaneous balance of the flow variables combined with the identity of the effort variable and where the dual case, i.e., interchanged role of efforts and flows, is called a **1**-junction. No other assumption about the form of the constitutive relations was made than power continuity and port symmetry, such that this linear, parameter-free result holds for any domain in which power and ports are meaningful concepts.

Applying scattering variables to power continuous two-ports without the port symmetry constraint to get more insight in the *transformation* aspect of the GJS, it turns out that these two-ports are characterized by *multiplicative* constitutive relations

$$e_1 = -\varepsilon n(.)e_2 \quad f_2 = \varepsilon n(.)f_1 \quad \varepsilon = -1, +1 \text{ and } e_1 = -\varepsilon r(.)f_2 \quad e_2 = \varepsilon r(.)f_1 \quad \varepsilon = -1, +1$$

that in the linear case reduce to one parameter:

$$e_1 = -\varepsilon n e_2 \quad f_2 = \varepsilon n f_1 \quad \varepsilon = -1, +1 \text{ and } e_1 = -\varepsilon r f_2 \quad e_2 = \varepsilon r f_1 \quad \varepsilon = -1, +1.$$

If the two-port orientations are chosen unequal (one inbound and one outbound), these relations reduce to the common relations for a transformer **TF** and a gyrator **GY**, respectively: $e_1 = n e_2$, $f_2 = n f_1$ and $e_1 = r f_2$, $e_2 = r f_1$. The nonlinear case (including the time-variant case) is constrained to *modulation* by the multiplier, which is expressed by adding the letter **M** to the acronyms **MTF**: $e_1 = n(.)e_2$, $f_2 = n(.)f_1$ and **MGY**: $e_1 = r(.)f_2$, $e_2 = r(.)f_1$, where (.) stands for any modulating signal. If these signals are external, this does not require additional attention. In contrast, internal signals mean that system variables or functions of system variables may become part of the constitutive relation, which may result in constitutive relations of other types. An example is the gravitational force on a pendulum mass (Figure 26.5[a]). In principle, this force is constant and independent of the velocity (**Se**-type source). However, the kinematic constraint has the nature of an **MTF** between the translational domain and the rotational domain with the angular position as a modulating signal (Figure 26.5[b]). Owing to this state modulation, the **Se**-port acts via the **MTF** as a **C**-type port on the rotational domain, for small deflections even as a linear **C** that results in an oscillator when combined with the kinetic storage in the mass (Figure 26.5[c] and Figure 26.5[d]).

Both the **MTF** and **MGY** can be generalized to modulated multiport versions, where the generic multiport **MTF** is an $m + n$-port, characterized by an $m \times n$-matrix **T**(.)

$$e_1 = \mathbf{T}^t(.)e_2 \quad f_2 = \mathbf{T}(.)f_1 \tag{26.11}$$

where e_1 and f_1 are the power conjugate variables of an m-dimensional inbound multiport, and e_2 and f_2 are the power conjugate variables of an n-dimensional outbound multiport. The transposition of the matrix follows from the power continuity constraint:

$$\left. \begin{array}{l} f_2 = \mathbf{T}(.)f_1 \\ P_1 = P_2 : e_1^t f_1 = e_2^t f_2 \end{array} \right\} \quad e_1^t f_1 = e_2^t \mathbf{T}(.)f_1 \quad e_1^t = e_2^t \mathbf{T}(.) \quad e_1 = \mathbf{T}^t(.)e_2 \tag{26.12}$$

The generic multiport **MGY** is an arbitrary n-port characterized by a $n \times n$-matrix **G**(.):

$$e = \mathbf{G}(.)f \tag{26.13}$$

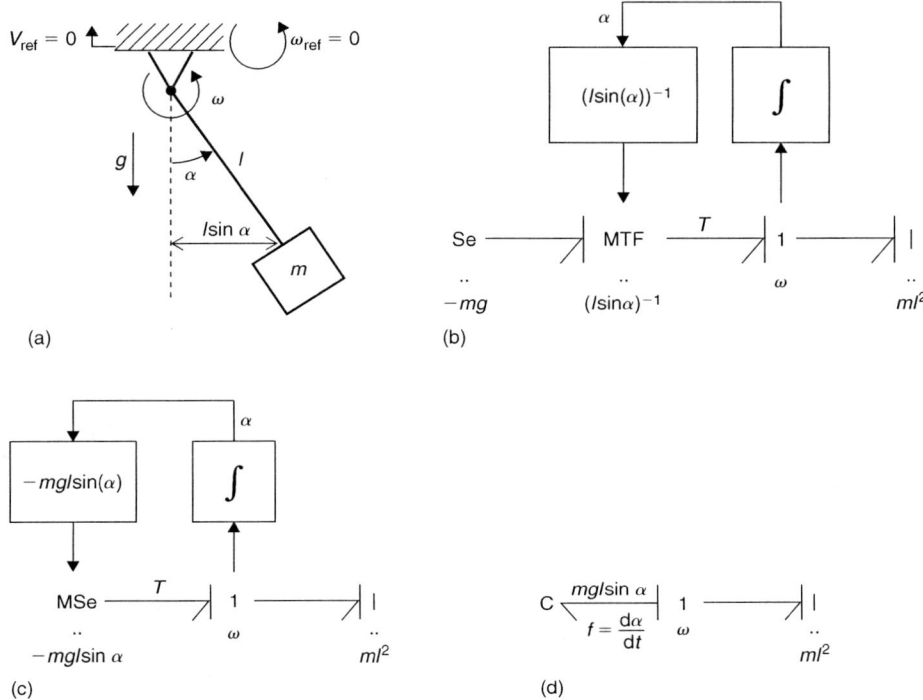

FIGURE 26.5 Pendulum: **Se** via position-modulated **MTF** results in C-type storage.

The power continuity constraint results in this case in the antisymmetry of the matrix **G**:

$$P = e^t f = f^t \mathbf{G}^t(.)f = 0 \quad \mathbf{G}(.) = -\mathbf{G}^t(.) \tag{26.14}$$

By mixing efforts and flows in the input and output vectors and by taking all ports inbound again $(-f_2)$, the multiport **MTF** can be written in a similar manner

$$P = e^t f = \begin{bmatrix} e_1^t & -f_2^t \end{bmatrix} \begin{bmatrix} f_1 \\ e_2 \end{bmatrix} = \begin{bmatrix} f_1^t & e_2^t \end{bmatrix} \begin{bmatrix} 0 & -\mathbf{T}^t(.) \\ \mathbf{T}(.) & 0 \end{bmatrix} \begin{bmatrix} f_1 \\ e_2 \end{bmatrix} = 0 \tag{26.15}$$

The GJS is a combination of all possible forms and thus leads to an antisymmetric matrix too. The quadratic form of an arbitrary matrix representing a multiplicative, algebraic relation between power conjugate efforts and flows of an arbitrary multiport represents the power generated or absorbed by such a multiport. The power continuity constraint thus requires this quadratic form to be zero. Indeed, this corresponds to the antisymmetry of the constitutive matrix:

$$P = e^t f = \begin{bmatrix} e_1^t & -f_2^t \end{bmatrix} \begin{bmatrix} f_1 \\ e_2 \end{bmatrix} = \begin{bmatrix} f_1^t & e_2^t \end{bmatrix} \begin{bmatrix} \mathbf{G}(.) & -\mathbf{T}^t(.) \\ \mathbf{T}(.) & \mathbf{H}(.) \end{bmatrix} \begin{bmatrix} f_1 \\ e_2 \end{bmatrix} = 0 \quad \mathbf{G} = -\mathbf{G}^t, \quad \mathbf{H} = -\mathbf{H}^t \tag{26.16}$$

where **H** is an "inverse" gyration matrix ($f_2' = \mathbf{H}(.)e_2$), not necessarily generated though by taking some inverse, in other words, **H** does not need to have an inverse. This results in the GJS decomposition in Figure 26.6, where the double-lined bonds represent multibonds; i.e., arrays of bonds, similar to the common notation for signal arrays in block diagrams.

The categorization of the junction structure elements used here initially is not based on the asymmetry between effort and flow pointed out earlier on a physical basis, as this type of categorization corresponds to the conventional viewpoint provided by the existing mathematical analyses for engineering systems like

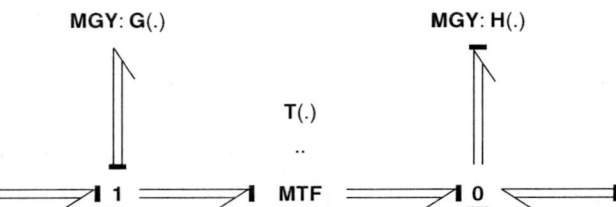

FIGURE 26.6 GJS decomposition.

electrical networks. Even though a voltage (potential difference) or a current (rate of change of charge/flow of moving electrons) are physically quite different, they are treated in a symmetric, dual way in Kirchhoff's equations.

In the sequel, we return to the physical differences, in which an effort can be considered a point-like variable (scalar potential, intensive state) when related to storage and a line vector (negative gradient) when related to a potential difference. In contrast, a flow can be seen as a rate of change of an extensive stored quantity (proportional to available volume or amount of matter) or one of the axial vectors that contribute to this rate of change in the balance equation of the stored quantity (divergence).

Given this distinction, it is straightforward to conclude that the discrete versions of the (negative) gradient and divergence operations are represented by the summing port relations of the **1**- and **0**-junctions, respectively, keeping in mind that the junctions themselves represent the respective conjugate flow and effort, which are common at all ports.

When elaborating this distinction based on operators in a continuum approach, it should be noted here already that a unit gyrator, called *symplectic gyrator* (**SGY**) for reasons to be discussed later, represents a spatially discretized version of two rotation operators between the efforts on the one hand and the two nonconjugate flows on the other. This **SGY** is indeed similar to the junctions in the sense that it is linear without any constitutive parameter. Later, this **SGY** will turn out to be useful for explicitly representing two interdomain couplings that are not self-evident, despite the fact that they are often considered that way and structurally eliminated by partial dualization accordingly. Dualization here means the inversion of the roles of effort and flow in the constitutive relations and "partial" means that not all ports are dualized. Allowing dualization immediately leads to a full symmetrization of the roles of effort and flow in the constitutive relations, thus not only going back to dual **0**- and **1**-junctions, but also giving every other type of port its dual, viz. **Se** and **Sf**, a storage element in which the effort is integrated into a conserved quantity (generalized momentum) called **I**-type port (generalized "inertia" or "inductance"). Inverting the roles of effort and flow in a constitutive relation of an **R**-type port (e.g., in Ohm's law) does not change its nature, although the relation is inverted and a resistance parameter becomes a conductance parameter. Finally, it should be noted that **TF** and **GY** are each others *partial* dual, i.e., they are obtained from each other by dualization of only one of the two ports. However, the **S**-type port of the **RS** cannot be dualized, as temperature and entropy production are by definition asymmetric. In the thermal domain, dual storage is not a viable concept as it would violate the second law of thermodynamics (Breedveld, 1982).

The full symmetrization of effort and flow thus leads to the common *nine* basic port types:

C & I, (M)Se & (M)Sf, (M)R(S), (M)TF & (M)GY, (X)0 & (X)1

where the letter **M** is added in case the constitutive relations can be modulated and the combination of the letter **X** with a junction means that the junction can be switched by a (Boolean) condition:

$$X1: \quad e_j = \text{if condition then} -\varepsilon_j \sum_{i \neq j}^{n} \varepsilon_i e_i \text{ else } 0 \text{ end}; \quad \varepsilon_i = -1, +1 \text{ and}$$

$$f_i = \text{if condition then } f_j \text{ else } 0 \text{ end}; \quad \forall i, j : i \neq j$$

or (26.17)

$$\mathbf{X0}: \quad f_j = \textit{if condition then} - \varepsilon_j \sum_{i \neq j}^{n} \varepsilon_i f_i \textit{ else } 0 \textit{ end;} \quad \varepsilon_i = -1, +1 \textit{ and}$$

$$e_i = \textit{if condition then } e_j \textit{ else } 0 \textit{ end;} \quad \forall i,j : i \neq j$$

Modulation of storage ports is omitted in principle as it violates energy conservation and contradicts the concept of storage (change of content without a rate of change). At a level where it can be accepted that the energy exchange via one of the ports of a multiport **C** can be neglected with respect to the other ports, one might choose to represent this kind of so-called bond activation by a modulated **C**. However, it should be clear that this condition exists and it should be checked that it is satisfied.

These are the nine basic node types in common bond graphs that are frequently used to describe electric circuits and simple mechanical systems. Trying to fit other domains in this approach often leads to complications although the hydraulic domain can be relatively easily incorporated as long as incompressibility is assumed, which allows description of a mass flow by a volumetric flow.

This is not a coincidence: in the descriptions of both electrical networks and mechanical systems, some implicit assumptions are made by default that lead to a simplification of a coupling that is an interdomain coupling in principle. In case of electrical circuits containing capacitors and coils, both the electrical domain (storage and conservation of charge, dielectric displacement) and the magnetic domain (storage and conservation of magnetic flux, more particular flux linkage of a coil) play a role. In its most general form this coupling is described by Maxwell's equations, in particular the third equation (Faraday's law of induction)

$$rotE = \nabla \times E = -\frac{\partial B}{\partial t} \tag{26.18}$$

and the fourth equation (Ampere's law)

$$rotH = \nabla \times H = \frac{\partial D}{\partial t} + J \tag{26.19}$$

where E is the electric field strength, D the dielectric displacement, B the magnetic induction, H the magnetic field strength, J the electric current density, and ρ the electric charge density.

However, if one assumes that electrical networks do not radiate, which is identical to assuming that changes in the EM fields take place quasistationary and if simple geometries are considered (e.g., an ideal coil with n windings), such that these relations can be written in terms of standard efforts and flows, this coupling (based on two rotation operators, cf., Eq. [26.18] and Eq. [26.19]) reduces to two simple identities:

$$E' \cdot l' = u = \frac{nd(B' \cdot A')}{dt} = \frac{nd\Phi}{dt} = \frac{d\lambda}{dt}, \text{ i.e., } u = e_{\text{elec}} = \frac{d\lambda}{dt} = f_{\text{mag}} \tag{26.20}$$

the identity between the electric effort (voltage u) and the magnetic flow (rate of change of flux, for one current loop) or flux linkage λ (for a coil with n windings; change of sign due to choice of orientation) and

$$H' \cdot l' = MMF = n(J' \cdot A') = ni = ne_{\text{mag}}, \text{ i.e., } e_{\text{mag}} = i = \frac{dq}{dt} = f_{\text{elec}} \tag{26.21}$$

the identity between the magnetic effort and the electrical flow (electric current). The quotes refer to the fact that the configuration is chosen such that the required spatial integrations are simplified. These two identities are exactly represented by a unit gyrator called **SGY** between the two domains, which can be eliminated next by dualization of the magnetic storage into the common **I**-type storage element. Magnetic dissipation is not commonly assumed relevant for electric circuits and neither are the magnetic sources. The only element that is commonly considered relevant, the magnetic storage in a coil or solenoid, is thus

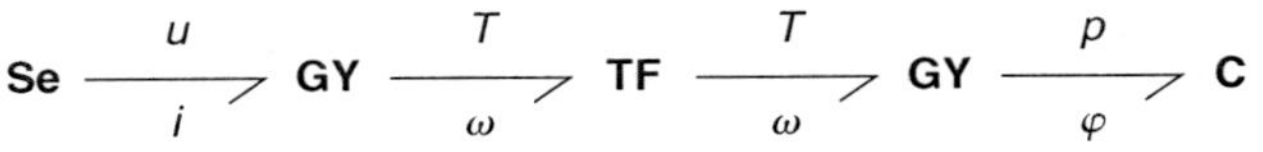

$$\textbf{Se} \xrightarrow[i]{u} \textbf{GY} \xrightarrow[\omega]{T} \textbf{TF} \xrightarrow[\omega]{T} \textbf{GY} \xrightarrow[\varphi]{p} \textbf{C}$$

FIGURE 26.7 Initial bond graph of the pump system.

considered part of the electric domain as a *dual* type of storage. This explains the common modeling difficulties that occur if permanent magnets play a role.

Although this may be harder to accept, a similar situation exists between the potential (or elastic) domain and the kinetic domain that are commonly considered one domain, viz. the mechanical domain. In the mechanical domain, the implicit **SGY** coupling between the potential domain (storage of displacement) and the kinetic domain (storage of momentum) is commonly considered unconditionally present. However, the implicit assumptions that are made by default here are that motions are described with respect to an *inertial* reference frame and in *inertial* coordinates.

Only in that case Newton's second law states that the rate of change of momentum (kinetic flow) is an effort of the potential domain (force). Obviously, in such a situation the velocity (kinetic effort) serves as a rate of change of elastic or gravitational displacement (potential flow).

In recent work (Golo et al., 2000), the link between the spatially discrete, quasistationary network approach, and the continuum approach viz. 1-junction related to gradient, 0-junction related to divergence, and SGY related to two rotations (Breedveld, 1984a), has been generalized into the so-called Dirac TransFormer or DTF, which can be seen as a condensed notation for the fact that for numerical solution (simulation) of the underlying partial differential equations a nontrivial spatial discretization has to be performed that commonly cannot be seen independent from the required time discretization. The DTF represents the so-called Dirac-structure (Maschke et al., 1995). An extension that includes the rotation operators unconditionally is called the Stokes–Dirac structure, which is represented by the acronym stokes-dirac transformer (SDTF) or stokes-dirac structure (SDS) (Maschke and van der Schaft, 2001).

Before making the relation between the physical structure of the bond graph and the computational structure that can be added via causality assignment, the example of the pump system in Figure 26.3 will be converted into an initial bond graph by translating the components into their most dominant elementary behavior. From this perspective the electrical power supply, e.g., the power grid voltage, can be considered a voltage source (**Se**). The dominant behavior of the electric motor, i.e., the Lorentz force that relates the torque with the same ratio to the motor current as the voltage (rate of change of flux linkage) to the angular velocity, can be considered a gyrator (**GY**). The transmission basically relates two angular velocities with the same ratio as the two torques (**TF**). The momentum balance in a centrifugal pump relates the pressure difference to the angular velocity and the volume flow to the torque. Although these relations will generally be nonlinear, this dominant behavior can be captured by a gyrator (**GY**). Finally, the water tank primarily stores water, where the pressure at the bottom of the tank (assuming that the inlet is there too), depends on the stored volume, i.e., the integral of the net volume flow. This means that Figure 26.3 can be converted into the initial bond graph in Figure 26.7, assuming that the environmental pressure can be taken as the zero reference pressure.

26.3.3 Causality

26.3.3.1 Causal Port Properties

Each of the nine basic elements (**C, I, R(S), TF, GY, Se, Sf, 0, 1**) introduced above has its own causal port properties, which can be categorized as follows: fixed causality of the first kind, fixed causality of the second kind, preferred causality, arbitrary causality, and causal constraints between ports. The representation by means of the causal stroke has been introduced already (cf., Figure 26.4).

Fixed Causality of the First Kind

It needs no explanation that a source of effort always has an effort as output signal, in other words, the causal stroke is attached to the end of the bond that is connected to the rest of the system (Figure 26.8 and

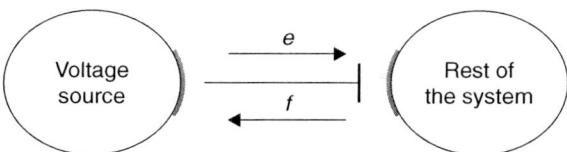

FIGURE 26.8 Fixed effort-out causality of an effort (voltage) source.

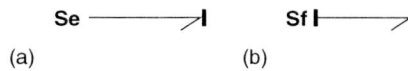

FIGURE 26.9 Fixed causality of sources.

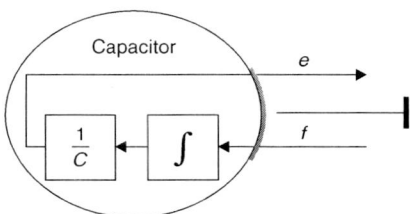

FIGURE 26.10 Preferred integral causality of a capacitor.

Figure 26.9[a]). Mutatis mutandis the causal stroke of a flow source is connected at the end of the bond connected to the source (Figure 26.9[b]). These causalities are called "fixed causalities" accordingly.

Fixed Causality of the Second Kind

Apart from the fundamentally fixed causalities of the first kind, all ports of elements that may become nonlinear and noninvertible, i.e., all but the regular (i.e., nonswitched) junctions, may become computationally fixed due to the fact that the constitutive relation may only take one form that cannot be inverted.

Preferred Causality

A less strict causal port property is that one of the two possibilities is, for some reason, preferred over the other. Commonly, this kind of property is assigned to storage ports, as the two forms of the constitutive relation of a storage port require either differentiation with respect to time or integration with respect to time (Figure 26.10). Owing to the amplification of numerical noise by numerical differentiation the integral form is preferred from a computational perspective, but there are more fundamental arguments as well. A first indication is that the integral form requires an initial condition, while the differential form does not. Obviously, an initial state or content of some storage element is a physically relevant property that supports the statement that "integration exists in nature, whereas differentiation does not." Although one should be careful with the concept of "existence" when discussing modeling, this statement seeks to emphasize that differentiation with respect to time requires information about future states in principle, in contrast with integration with respect to time. The discussion of causal analysis will make clear that violation of a *preferred integral causality* gives important feedback to the modeler about his modeling decisions. Some forms of analysis, e.g., finding the rank of the system matrix, may require that the differential causal form is preferred at some point in the analysis too, but this requirement is never used as a preparation of the equations for numerical simulation.

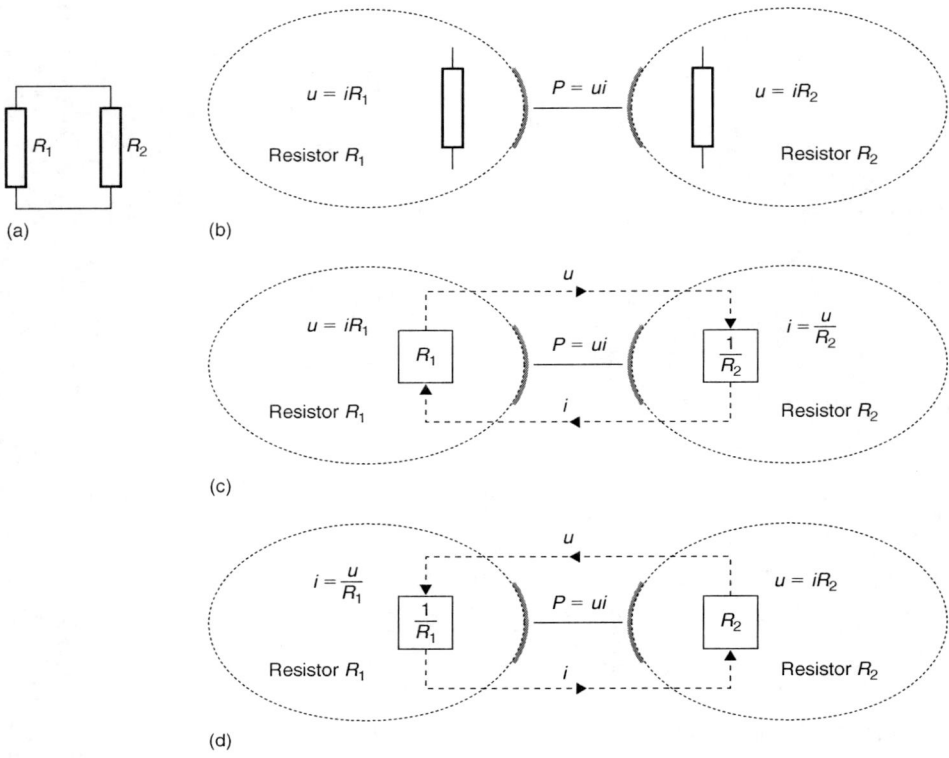

FIGURE 26.11 Arbitrary causality of two resistors causing an algebraic loop.

Arbitrary Causality

The expected next possibility in the sequence is that the causality of a port is neither fixed nor preferred, thus arbitrary. Examples of arbitrary port causality are linear, resistive ports that are consequently invertible and algebraic, respectively. The acausal form of the constitutive relation of an ohmic resistor is $u - Ri = 0$, the effort-out causal form is $u = Ri$, while the flow-out causal form is $i = u/R$ (cf., Figure 26.11).

Causal Constraints

Causal constraints only exist for basic multiports, i.e., elements with two or more ports like the transducers (**TF**, **GY**) and the junctions (**0**, **1**). For instance, if the constitutive relation of the two-port transducers is linear (the junctions are intrinsically linear), the first port to which causality is assigned is arbitrary, but the causality of the second port is immediately fixed. For instance, the two-port transformer always has one port with effort-out causality and one with flow-out causality. In contrast, the causalities of the ports of a two-port gyrator always have the same type of causality. In graphical terms: a **TF** has only one causal stroke directed to it, while a **GY** has either both causal strokes directed to it or none.

The fundamental feature of the junctions that either all efforts are common (**0**) or all flows are common (**1**) shows that only one port of a **0**-junction can have "effort-in causality," i.e., flow-out causality, viz. the result of its flow-balance. In contrast, only one port of a **1**-junction can have "flow-in causality," i.e., effort-out causality, viz. the result of its effort-balance. In graphical terms, only one causal stroke can be directed toward a **0**-junction, while only one open end can be directed toward a **1**-junction.

26.3.3.2 Causal Analysis: Feedback on Modeling Decisions

Causal analysis, also called *causality assignment* or *causal augmentation*, is the algorithmic process of putting the causal strokes at the bonds on the basis of the causal port properties induced by the nature

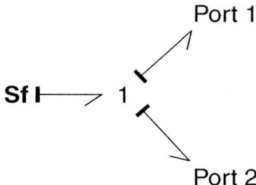

FIGURE 26.12 Propagation of a fixed causality via a 1-junction.

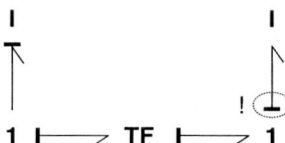

FIGURE 26.13 Dependent inertias via de causal constraints of 1-junctions and transformer.

of the constitutive relations. Not only the final result, but also the assignment process itself provides immediate feedback on modeling decisions.

Fixed Causality of the First Kind

The first step is to assign fixed causalities of the first kind and immediately propagate them via the causal constraints, as the latter cannot be violated. For instance, if a flow source is connected to a 1-junction, the source-port flow-out causality by definition, which in turn means that the corresponding port at the 1-junction gets flow-in causality, which means that all other ports of the 1-junction get flow-out causality (Figure 26.12), unless one can apply numerical iteration to impose the same flow at the junction from more than one of its ports, but this is not anticipated in regular causal analysis that is focused on feedback nor is it possible if only sources are involved. Conflicts between fixed causalities of the first kind via the constraints indicate that the problem is ill-posed, e.g., two voltage sources in parallel or two force sources trying to impose a different value to the same force. The causality propagation may lead to violation of preferred causalities, e.g., a voltage source in parallel to a capacitor or a velocity source on a mass. This violation feeds back to the modeler that no independent state is related to the storage element as its content is imposed by a source, which also means that it is dynamically inactive as its rate of change will only influence the source, which is insensitive to it by definition. Not only storage ports, but also resistive ports that get their causality imposed by a source are dynamically inactive, as they cannot form signal loops (so-called causal paths) via other ports. A causal path can be found to be following either a causal stroke or an open end from one **R**-, **C**-, or **I**-type port to another. At a **GY** the causality reverses, so if one follows the stroke one should continue with the open end or vice versa.

The adjective "dynamically inactive" does not mean that the port-variables cannot change, but that the port does not contribute to the dynamic characteristics (like time constants and eigen frequencies) of the system, which will become more clear after the discussion of preferred causality below. Conflicts with fixed causalities of the second kind inform the modeler that the model has to be changed in such a way that either the constitutive relation is changed such that it can be inverted or the structure is changed such that the conflict disappears. If this is impossible, numerical iteration has to be applied to solve the noninvertible relation.

Preferred Causality

After the assignment of all fixed causalities of the first kind, the preferred causalities are similarly assigned and immediately propagated via the causal constraints. Conflicts at this stage indicate that a port may get differential causality as a result of another port getting preferred integral causality. Figure 26.13 shows the bond graph of two rigidly linked inertias, e.g., the motor inertia and the load inertia in a servo system

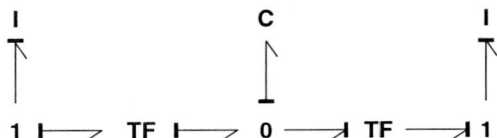

FIGURE 26.14 Independent inertias by adding the elasticity of the transmission (e.g., belt-drive).

model, including a transmission (**TF**), but without any compliance. The causality of this bond graph shows the modeler that he has chosen a model in which two storage ports depend on each other and form a signal loop (causal path) with an integration that is compensated by a differentiation, i.e., a net algebraic loop. The computational problem may be solved by

- the application of implicit numerical integration,
- changing the model (the sequence of putting the causal strokes hints the modeler where a model change should be made, e.g., adding the compliance of the transmission between the two rigid bodies (see Figure [26.14]), or
- symbolic manipulation (either manually or automatically) of the model.

Preferred integral causalities that impose other preferred integral causalities give immediate feedback on the existence of *second-order* signal loops by identifying the *causal paths* in the bond graph, i.e., loops containing two integrations that lead to behavior as described by second-order differential equations, viz. potentially oscillatory behavior. A similar kind of feedback on the dynamic properties of the model is obtained by a port with arbitrary causality that is assigned its causality via propagation of an integral causality. The resulting *first-order causal path* informs the modeler about relaxation type of behavior. Finally, if a preferred causality creates a conflict with a fixed causality of the second kind, then the source of the fixation has to be reconsidered and the problem solved by either changing the constitutive relation or the model structure or by implementing numerical iteration.

Fixed Causality of the Second Kind

As discussed above, fixed causalities of the first kind and preferred causalities are given a higher priority than fixed causalities of the second kind, unless a physical meaning can be assigned to the noninvertibility, in which case the model needs reconsideration in the sense that the model becomes ill-posed or that the number of free-to-choose initial conditions may be reduced. Accordingly, fixed causalities of the second kind can only propagate to ports with arbitrary causality. In that case, mostly an algebraic loop will occur and, if not solved symbolically beforehand, this requires numerical iteration during simulation too. This shows that an explicit ODE model can only be obtained if fixed causalities of the second kind obtain their proper causalities via propagation of fixed causalities of the first kind and preferred causalities. The modeler should consider adapting the model if this is not the case.

Arbitrary Causality

Commonly all ports in a bond graph are causal after assigning and propagating fixed and preferred causalities, but if this is not the case, it means that at least two ports with arbitrary causality are present. If an arbitrary choice is made for one of these ports, this means that at least one other port will obtain its causality as a result of propagation via the causal constraints (cf., Figure 26.11). The dual choice would have the same effect. This shows the modeler that this situation always results in an algebraic loop (or its reverse form corresponding to the dual choice of causality) that requires numerical iteration during simulation. Similar to other causal conflicts, e.g., generated by differential causality, the assignment procedure itself hints the modeler how to change the model to prevent the algebraic loop. The causality assignment process is completely algorithmic and more advanced variations on this algorithm exist and are implemented that can handle all possible situations in an automated way (van Dijk and Breedveld, 1991; Golo, 2002). As a result, it can be used without using the notation itself, e.g., by replacing the bond graph with the more common iconic diagram representation or the linear graph notation. However, this largely reduces the amount of feedback that can be given to the modeler about his modeling decisions and the effect of model modifications becomes less obvious. Nevertheless, if one is merely interested in converting simple iconic diagrams into code ready for simulation, this is a powerful option.

The support offered by causal analysis in the common trade-off between conceptual and computational complexity of a model is illustrated by the simple example of a rigid constraint between two rigid bodies. Conceptual simplicity leads to a causal problem (a so-called dependent inertia with differential causality) and consequently to numerical complexity. The example in Figure 26.13 already showed that a loop emerges containing an integration and a differentiation that cancel each other, i.e., a "net" algebraic loop, similar to the situation where two resistors form such an algebraic loop (Figure 26.11). Direct equation generation does not generate a set of ordinary differential equations (ODE), but a mixed set of differential and algebraic equations (DAE). A set of DAE cannot be solved straightforwardly by means of explicit numerical integration (e.g., with the common Runge–Kutta fourth-order method). However, the way in which the causal problem emerges in the model during causal analysis of the bond graph clearly suggests how the model can be modified to prevent the causal problem. In this example, the rigid constraint can be replaced by an elastic element, i.e., a finite rigidity. Although this gives the model some more conceptual complexity, the numerical (structural) complexity is reduced due to the fact that the resulting equations are a set of ODE that can be solved by explicit numerical integration schemes (Ascher and Petzold, 1998).

The resulting model needs a rather stiff constraint and thus introduces dynamics at a timescale that is not of interest. This not only means that both options to formulate the model can be a solution depending on the problem context, the available tools, etc., but also that a third solution can be obtained, viz. a symbolic transformation of the model as to eliminate the dependent inertia. In other words, two rigidly connected rigid bodies may be considered as one rigid body. This possibility is directly induced by the causal analysis of the bond graph model.

Example of Causal Analysis

If causality analysis is applied to the bond graph in Figure 26.7, the fixed causality of the **Se** propagates all the way through the graph via the causal constraints of the two-ports and imposes differential causality on the storage element (Figure 26.15).

It shows that the state of the system is imposed by the source, which means that the model cannot be used to capture any physical behavior. At best, it may serve to analyze a stationary situation, although it is not likely in a practical situation that none of the components has any losses. This immediate feedback on modeling decisions is obtained without writing any equation. If dynamic behavior is to be captured, one needs to modify the graph in such a way that it contains at least one independent state, e.g., the content of the water tank. To that end it will need to have integral causality. For instance, if one would decide that the friction in the bearings in the mechanical part should be modeled, the causal graph immediately shows that this does not change the causality of the storage element (Figure 26.16).

FIGURE 26.15 Causally augmented initial bond graph of the pump system: no dynamics.

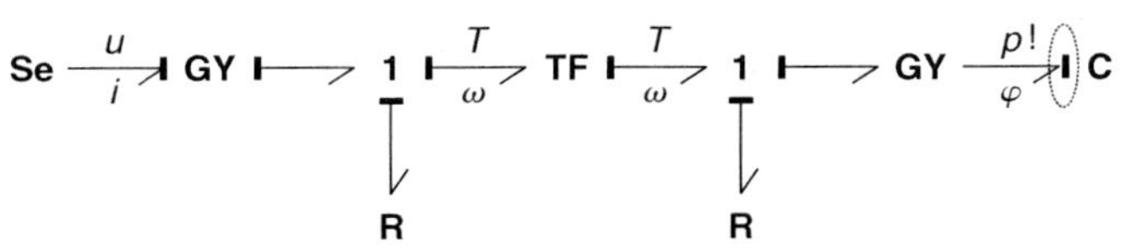

FIGURE 26.16 Addition of friction to the pump model: still no dynamics.

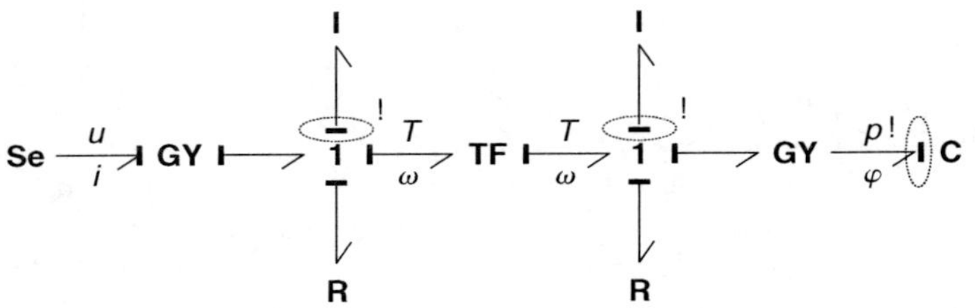

FIGURE 26.17 Addition of inertias to the pump model: still no dynamics.

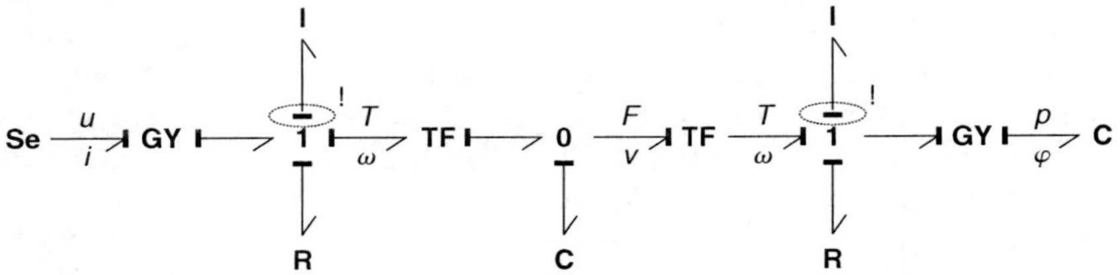

FIGURE 26.18 Addition of inertias to the pump model: some dynamics, but not the expected dominant behavior.

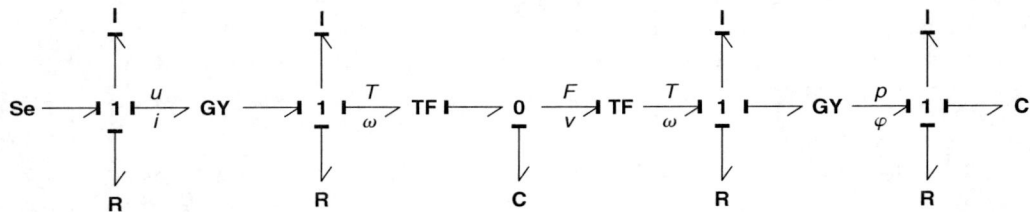

FIGURE 26.19 Addition of motor inductance and fluid inertia solves the causal conflicts: ample dynamics.

The same holds for adding the inertias of the rotating parts (Figure 26.17). The differential causality of the **I**-ports demonstrates that the imposed voltage not only imposes the hydraulic pressure, but also all velocities.

The bond graphs shows that the only way to "break" the causal propagation between voltage source, velocities, and pressure is to add a junction that breaks the causal path. This can be either a **0**-junction in the mechanical connections, i.e., allowing some relative speed or a nonstiff connection, or a **1**-junction in the electrical or hydraulic domain, i.e., allowing a voltage difference between source and motor or a pressure difference between pump and tank. For instance, if the transmission is a belt drive, the elasticity of the belt can be modeled (Figure 26.18).

However, this only gives the tank and the belt independent states: the two Cs form a second-order loop via the **GY**-action of the pump. Furthermore, it forms a first-order loop (causal path via this **GY** with the mechanical R at the right-hand side. The first inertia still gets its velocity imposed by the source, while the state of the second depends on the state of the tank. The only way the resolve the first dependency is to add an electric **1**-junction connected to the resistance or the inductance of the current loop. Similarly, the second dependency can be resolved by adding a hydraulic **1**-junction connected to the resistance or the inertia of the hydraulic line (Figure 26.19).

The process of eliminating all dependencies resulted in a relatively high-order system, viz. a sixth-order system ($4 \times \mathbf{I}, 2 \times \mathbf{C}$). If we assume for instance at this point that the elasticity of the belt can be neglected after all, the two mechanical Is become dependent again, but these can be symbolically combined into one. If the hydraulic lines are relatively short, the hydraulic I can be omitted too (Figure 26.20), such that a

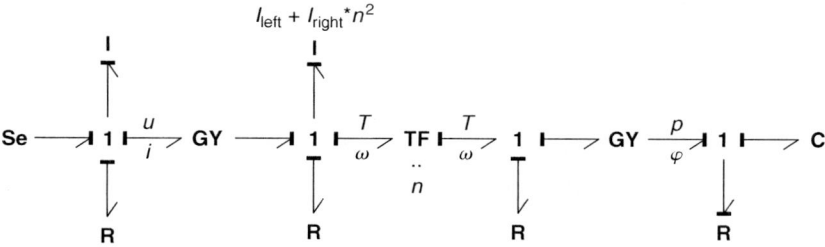

FIGURE 26.20 Simplification of the model by symbolic manipulation and maintaining key behavior.

third-order model (2 × I, 1 × C) remains to capture the observed dynamics. Further modeling iterations require a more detailed study of problem context and measurement data.

The above example demonstrates that it is possible to make far-reaching modeling decisions quickly, because no detailed equations have to be written yet. Causality assignment thus proves to be not only a means to create a model in a form ready for numerical simulation, but also provides direct feedback on modeling decisions. The nine basic elements can be used to model many different dynamic problems. However, in the next section it will be shown how the generalization of the basic elements can lead to even more powerful modeling concepts.

26.4 Multiport Generalizations

All basic elements can be generalized to a multiport form and put into a multibond notation of which the most important aspects are quite intuitive, cf., Figure 26.6 (Breedveld, 1985). Sources just form arrays without any coupling owing to their nature, i.e., no relations between their conjugate variables. Junctions already are arbitrary multiports. The multiport forms of the (modulated) **TF** and **GY** were already introduced in Section 26.3.2.5. Herein, it is merely mentioned that position modulated multiport transformers (**MTF**) are a powerful tool in modeling planar and spatial mechanisms (Tiernego and van Dixhoorn, 1979; Bos and Tiernego, 1985; Bos, 1986; Stramigioli, 1998).

The generalization of a resistor, or rather the resistive port of an irreversible transducer, is also straightforward as it concerns an algebraic relation between the port variables. Assuming that the generic algebraic form $\Phi(\mathbf{e}, \mathbf{f}) = 0$, where $\mathbf{e}$ and $\mathbf{f}$ are the conjugate effort and flow vector of the resistive multiport can be written $\mathbf{e} = \mathbf{e}(\mathbf{f})$ (or $\mathbf{f} = \mathbf{f}(\mathbf{e})$) with variations $de_i = \sum_j \frac{\partial e_i(\mathbf{f})}{\partial f_j} df_j$, where $\frac{\partial e_i}{\partial f_i}$ is the "self resistance" of the ith port (diagonal element of the Jacobian) and $\frac{\partial e_i}{\partial f_j} \forall j \neq i$ are the nondiagonal coupling terms. If the Jacobian is constant, the relations are linear $e_i = \sum_j \frac{\partial e_i}{\partial f_j} f_j = \sum_j B_{ij} f_j$ or $\mathbf{e} = \mathbf{B}\mathbf{f}$. In that case, the irreversibly produced entropy flow $f_{S_{irr}}$ conjugate to temperature T can be written $f_{S_{irr}} = T^{-1} \mathbf{e}^t \mathbf{f} = T^{-1} \mathbf{f}^t \mathbf{B}^t \mathbf{f}$, i.e., a quadratic form in the flows. As only the symmetric part $\mathbf{R} = (\mathbf{B} + \mathbf{B}^t)/2$ of the constitutive matrix contributes to the entropy production, only this part $\mathbf{R}$ is commonly considered to describe the resistive multiport. Its symmetry corresponds to what is called Onsager symmetry or reciprocity in irreversible thermodynamics (Callen, 1960), as the multiport $\mathbf{R}$ can be considered to represent a domain-independent form of description common in irreversible thermodynamics of the relations between generalized forces and fluxes. The antisymmetric part $\mathbf{G} = (\mathbf{B} - \mathbf{B}^t)/2$ is readily identified as a power continuous multiport gyrator as introduced already from another perspective in Section 26.3.2.5. It corresponds to Casimir's extension of the Onsager relations (Breedveld, 1982). Although it is quite interesting to see that this extension only occurs between ports of domains that were identified earlier as having a symplectic interdomain coupling on different grounds, further discussion goes beyond the scope of this treatise. Finally, it is noted that the multiport resistor, just like the elementary one-port can be modulated without changing its fundamental nature and that the symmetric matrix $\mathbf{R}$ has to be positive-definite where all diagonal elements have to be positive as well as the determinants of all subblocks to satisfy the positive entropy production principle at all times. Finally, only the multiport generalization of a storage element adds a new behavior, viz. transduction. For this reason, multiport storage will be analyzed in more detail in the following section.

For all the multiports mentioned above (canonical), decompositions (Breedveld, 1984b) can be found that not only enhance insight into the potential dynamic behavior of a model, but also allow recognition of substructures as specific multiports (composition), of which an alternative decomposition may improve numerical solution properties, as is the case for the well-known Euler junction structure (Karnopp and Rosenberg, 1968; Breedveld, 1999).

26.4.1 Multiport Storage Elements

The generalization of a one-port storage element into an n-port storage element, where the ports may belong to different domains, requires an approach based on the commonly conserved property, the energy. It turns out that a new type of behavior will emerge from this extension, viz. transduction by a cycle process. In other words, power can be transformed from one port to another and thus from one domain into another by performing a cycle process with a multiport storage element with at least two ports. As much of the literature is rather inaccurate about the use of the concept of energy, the distinction between coenergies (often incorrectly considered energies when they have the same numerical value as the energy) and the globally conserved energy will be made clear first by a discussion of Legendre transformations of homogeneous functions.

26.4.1.1 Coenergy and Legendre Transforms

A concise overview of homogeneous functions and Legendre transforms is given and applied to energy functions, thus introducing the concept of coenergy. Various properties of energy and coenergy related to multiport storage are shortly discussed and related to a physical interpretation. Finally, some domain-specific forms of coenergy are discussed.

Math Background: Homogeneous Functions and Euler's Theorem
A function $F(\underline{x})$ with $\underline{x} = x_1, \ldots, x_k$ is *homogeneous* of order n if $F(\alpha \underline{x}) = \alpha^n F(\underline{x})$. If $y_i(\underline{x}) = \partial F/\partial x_i$, then $y_i(\alpha \underline{x}) = \partial F(\alpha \underline{x})/\partial \alpha x_i = \frac{\alpha^n}{\alpha} \cdot \partial F(\underline{x})/\partial x_i = \alpha^{n-1} y_i(\underline{x})$, in other words, $y_i(\underline{x})$ is homogeneous of order $(n-1)$. For such homogeneous functions *Euler's theorem* holds

$$\sum_{i=1}^{k} \frac{\partial F}{\partial x_i} \cdot x_i = n \cdot F(\underline{x}) \quad \text{or} \quad F(\underline{x}) = \frac{1}{n} \sum_{i=1}^{k} y_i \cdot x_i = \frac{1}{n} \cdot \underline{y}^T \cdot \underline{x} \quad (26.22)$$

By definition the variation of F can be written as

$$dF = \sum_{i=1}^{k} \frac{\partial F}{\partial x_i} \cdot dx_i = \sum_{i=1}^{k} y_i dx_i = \underline{y}^T \cdot d\underline{x} \quad (26.23)$$

but also using Eq. (26.22)

$$dF = d\left(\frac{1}{n} \underline{y}^T \cdot \underline{x}\right) = \frac{1}{n} \underline{y}^T \cdot d\underline{x} + \frac{1}{n} (d\underline{y})^T \cdot \underline{x} \quad (26.24)$$

Combination of Eq. (26.23) and Eq. (26.24) gives

$$(d\underline{y})^T \cdot \underline{x} = (n-1) \underline{y}^T \cdot d\underline{x}$$

and for $n=1$: $(d\underline{y})^T \cdot \underline{x} = 0$, for $n \neq 1$: $dF = \frac{1}{n-1}(d\underline{y})^T \cdot \underline{x}$.

Homogeneous Energy Functions
The *energy* of a system with k state variables $\underline{q}$ is $E(\underline{q}) = E(q_1, \ldots, q_k)$. If q_i is an "extensive" state variable, this means that $E(\alpha \underline{q}) = \alpha E(\underline{q}) = \alpha^1 E(\underline{q})$. Hence $E(\underline{q})$ is *first* order ($n=1$) homogeneous, so $e_i(\underline{q}) = \partial E/\partial q_i$ is *zeroth* order ($n-1=0$) homogeneous, which means that $e_i(\underline{q})$ is an "intensive" variable, i.e., $e_i(\alpha \underline{q}) = \alpha^0 e_i(\underline{q}) = e_i(\underline{q})$.

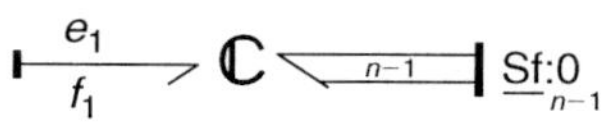

FIGURE 26.21 Bond graph of a "one-port storage element."

This means also that in case $n=1$ and $k=1$ $e(q)$ is constant, i.e., $\partial e/\partial q = de/dq = 0$, which changes the behavior of this element into that of a source. This means that storage elements should in principle be multiports ($k > 1$). A one-port storage element is in principle an $n-$ port storage element of which the flows of $n-1$ ports are kept zero. Accordingly, the corresponding $n-1$ states are constant and commonly not recognized as states. Such states are often considered parameters: if $E(q_1, q_2, \ldots, q_n)|_{dq_i=0, \forall i \neq 1} = E'(q_1)$, then $E'(q_1)$ not necessarily first-order homogeneous in q_1 (cf., Figure 26.21).

For $n=1$ and k independent extensities there are only $k-1$ independent intensities, because for $n=1$ we find *Gibbs' fundamental relation*: $E(q) = \underline{e}^T \cdot \underline{q}$. As by definition holds that $dE = \underline{e}^T \cdot d\underline{q}$, this results in the *Gibbs–Duhem relation*: $(d\underline{e})^T \cdot \underline{q} = 0$, which expresses that one of the intensities depends on the others.

Math Background: Legendre Transforms

A *Legendre transform* L of $F(\underline{x})$ with respect to x_i is by definition $L\{F(\underline{x})\}_{x_i} = L_{x_i} = F(\underline{x}) - y_i \cdot x_i$, where $y_i = \partial F/\partial x_i$ and the *total Legendre transform* of $F(\underline{x})$ is $L\{F(\underline{x})\} = L = F(\underline{x}) - \sum_{i=1}^{k} y_i x_i$. Eq. (26.22) shows that $L=0$ for $n=1$. Now $dL_{x_i} = dF - d(y_i x_i) = dF - y_i dx_i - x_i dy_i = \sum_{j \neq i} y_j dx_j - x_i dy_i$ or $L_{x_i} = L_{x_i}(x_1, \ldots, x_{i-1}, y_i, x_{i+1}, \ldots, x_k)$, which means that x_i is replaced by y_i as *independent* variable or "coordinate!" Hence $L = L(\underline{y})$; $dL = -\sum_{i=1}^{k} x_i dy_i = -(d\underline{y})^T \cdot \underline{x}$.

Coenergy Functions

The complimentary energy or *coenergy* $E^*_{q_i}$ of $E(\underline{q})$ of with respect to q_i is by definition: $E^*_{q_i} = -L_{q_i} = E^*_{q_i}(q_1, \ldots, q_{i-1}, e_i, q_{i+1}, \ldots, q_k)$. Hence $E(\underline{q}) + E^*_{q_i}(\ldots, e_i, \ldots) = e_i \cdot q_i$. For the *total coenergy* $E^*(\underline{e})$ of $E(\underline{q})$ holds that $E^* = -L$, hence $E(\underline{q}) + E^*(\underline{e}) = \underline{e}^T \cdot \underline{q}$, which explains the terminology. For $n=1$: $E^*(\underline{e}) = 0$, confirming the earlier conclusion that there are only $k-1$ independent intensities e_i. For $n=2$: $E(\underline{q}) = E^*(\underline{e}) = \frac{1}{2} \underline{e}^T \cdot \underline{q}$ (i.e., equal in value, but different in nature as is the case for linear constitutive relations that result in quadratic energy functions!) and for $n=3$: $E(\underline{q}) = \frac{1}{3} \underline{e}^T \cdot \underline{q}$ while $E^*(\underline{e}) = \frac{2}{3} \underline{e}^T \cdot \underline{q}$. The following relations for coenergy functions can be useful:

$$dE^*_{q_i} = de_i \cdot q_i - \sum_{j \neq 1} e_j \cdot dq_j \quad dE^* = \sum_{i=1}^{k} de_i \cdot q_i = (d\underline{e})^T \cdot \underline{q} = (n-1)\underline{e}^T \cdot d\underline{q} = (n-1)dE$$

$$E^* = (n-1)E = \frac{n-1}{n} \cdot \underline{e}^T \cdot \underline{q} = \left(1 - \frac{1}{n}\right) \underline{e}^T \cdot \underline{q}.$$

Next, several domain dependent manifestations of the Legendre transforms will be discussed.

Legendre Transforms in Thermodynamics

In thermodynamic systems of "simple" systems with internal energy U that is a function of the extensities entropy S, volume V, total mole number N, and mole number per species i N_i, the conjugate intensities are temperature $T = \partial U/\partial S$, pressure $p = \partial U/-\partial V = -\partial U/\partial V$ (energy increases in case of compression $-\partial V$), total material potential $\mu^{tot} = \partial U/\partial N$, and chemical potential $\mu_i = \partial U/\partial N_i$, respectively. The Legendre transforms are the *free energy F*, the *enthalpy H*, and the *Gibbs free enthalpy G*. The *free energy F* is a Legendre transform of the energy with respect to the entropy S: $L_S = F(T, V, \underline{N}, N) = U - TS =, -pV + \sum_{i=1}^{m-1} \mu_i N_i + \mu^{tot} \cdot N$ ($= N \cdot f(T, v, \underline{c})$) where $\underline{c}$, is the vector of molar fractions and with variation $dF = -SdT - pdV + \sum_{i=1}^{m-1} \mu_i dN_i + \mu^{tot} \cdot dN$, which means that the thermal domain does not influence the free energy at constant temperature. The *enthalpy H* is a Legendre transform of the energy with respect to the available volume V: $L_V = H(S, p, \underline{N}, N) = U - (-pV) = U + pV$ ($= N \cdot h(s, p, c)$) with variation $dH = TdS + Vdp + \sum_{i=1}^{m-1} \mu_i \cdot dN_i + \mu^{tot} \cdot dN$, which means that the mechanical domain does not influence the enthalpy at constant pressure. Finally, the *Gibbs free enthalpy G* is a Legendre transform of the energy with respect to both the entropy S and the available volume V: $L_{S,V} = G(T, p, \underline{N}, N) = U - TS - (-pV) = \mu^{\text{tot}} \cdot N + \sum_{i=1}^{m-1} \mu_i N_i \, (= N \cdot g(T, p, \underline{c}))$ with variation $\mathrm{d}G = -S\mathrm{d}T + V\mathrm{d}p + \sum_{i=1}^{m-1} \mu_i \mathrm{d}N_i + \mu^{\text{tot}} \cdot \mathrm{d}N$, which means that both the mechanical domain and the thermal domain do not influence the Gibbs free enthalpy at constant pressure and temperature. For one constituent ($m = 1$) it holds that $g = \mu^{\text{tot}}(T, p)$, which means that the equilibrium of the material is determined by pressure and temperature.

Legendre Transforms in Mechanics

In mechanical systems with kinetic energy T and potential energy V, i.e., of which the total energy is the *Hamiltonian* H that is a function of the extensities displacements vector $\underline{x}$ and the momenta vector $\underline{p}$: $E(q) = H(\underline{x}, \underline{p}) = T + V$, while the conjugate intensities are the velocities vector $\underline{v} = \partial H/\partial \underline{p}$ and the vector of forces $\underline{F} = \partial H/\partial \underline{x}$, respectively. The Legendre transforms of the *Hamiltonian* $\mathbf{H}$ are the *Lagrangian* $\mathbf{L}$, the *co-Hamiltonian* and the *co-Lagrangian* or *Hertzian*. The *Lagrangian* $\mathbf{L}$ is a *negative* Legendre transform (coenergy) of the Hamiltonian with respect to the momenta $\underline{p}$: $\mathbf{H}_{\underline{p}}^* = -\mathbf{L}_{\underline{p}} = \underline{v}^T \cdot \underline{p} - \mathbf{H} = (T + T^*) - (T + V) = T^* - V = \mathbf{L}(\underline{x}, \underline{v})$, the *co-Hamiltonian* is a *negative* Legendre transform (coenergy) of the Hamiltonian with respect to both the displacements $\underline{x}$ and the momenta $\underline{p}$: $\mathbf{H}_{\underline{x},\underline{p}}^* = \underline{v}^T \cdot \underline{p} + \underline{F}^T \cdot \underline{q} - \mathbf{H} = (T + T^*) + (V + V^*) - (T + V) = T^* + V^* = \mathbf{H}^*(\underline{F}, \underline{v})$ and the *co-Lagrangian* or *Hertzian* is a *negative* Legendre transform (coenergy) of the Hamiltonian with respect to the displacements $\underline{x}$: $\mathbf{H}_{\underline{x}}^* = \underline{F}^T \cdot \underline{q} - \mathbf{H} = (V + V^*) - (T + V) = V^* - T = -\mathbf{L}^*(\underline{F}, \underline{p})$.

Note that adding the symplectic coupling between the kinetic and potential domains leads to the common Hamiltonian formulation of the equations of motion, where $\mathbf{J}$ is the symplectic matrix:

$$\frac{\mathrm{d}}{\mathrm{d}t}\begin{bmatrix} q \\ p \end{bmatrix} = \begin{bmatrix} 0 & +1 \\ -1 & 0 \end{bmatrix} \begin{bmatrix} \dfrac{\partial \mathbf{H}}{\partial q} \\ \dfrac{\partial \mathbf{H}}{\partial p} \end{bmatrix} + \begin{bmatrix} 0 \\ F_{\text{external}} \end{bmatrix} = \mathbf{J} \begin{bmatrix} \dfrac{\partial \mathbf{H}}{\partial q} \\ \dfrac{\partial \mathbf{H}}{\partial p} \end{bmatrix} \tag{26.25}$$

while after the Legendre transform into the Lagrangian the equations of motion are written

$$\frac{\mathrm{d}}{\mathrm{d}t}\begin{bmatrix} q \\ \dfrac{\partial \mathbf{L}}{\partial v} \end{bmatrix} = \begin{bmatrix} 0 & +1 \\ -1 & 0 \end{bmatrix} \begin{bmatrix} \dfrac{\partial \mathbf{L}}{\partial q} \\ v \end{bmatrix} + \begin{bmatrix} 0 \\ F_{\text{external}} \end{bmatrix} \text{ or, by eliminating } \frac{\mathrm{d}q}{\mathrm{d}t} = v: \frac{\mathrm{d}}{\mathrm{d}t}\frac{\partial \mathbf{L}}{\partial v} + \frac{\partial \mathbf{L}}{\partial q} = F_{\text{ext}}$$

Note that in both common formulations velocity sources are excluded.

Legendre Transforms in Electrical Circuits

In electrical circuits with capacitor charges $\underline{q}$ and conjugate voltages $\underline{u}$ and coil flux linkages $\underline{\lambda}$ and conjugate currents $\underline{i}$, the total stored energy is $E(\underline{q}, \underline{\lambda}) = E_C(\underline{q}) + E_L(\underline{\lambda})$. The total coenergy is $E^*(\underline{u}, \underline{i}) = \underline{u}^T \cdot \underline{q} + \underline{i}^T \cdot \underline{\Phi} - E = E$. Note that the last equal sign only holds in case of circuit elements with linear constitutive relations. Nevertheless, this assumption is seldom made explicit when coenergy is used as the "energy" of an electric circuit.

Legendre Transforms and Causality

The above similarities between the Legendre transforms in these different domains are often not made explicit. For instance, some properties of and processes with multiports that are quite common in thermodynamics are not common in mechanics and electrical circuit theory, although they may be used to enhance insight in similar ways.

In the domain independent approach as denoted in bond graphs one can conclude that if an effort is "forced" on a port of a **C**-element ("derivative causality" or "flow-out causality"), this means that the roles of e and q are interchanged in the set of independent variables, which means that the energy has to be *Legendre transformed* to maintain a generating function for the constitutive relations. Such a transformation is particularly useful when the *effort* e is *constant*. For example, an electrical capacitor in an

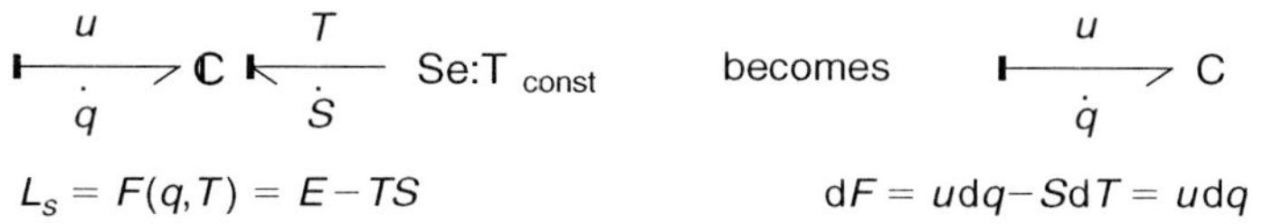

FIGURE 26.22 Use of free energy at constant temperature: no thermal port required.

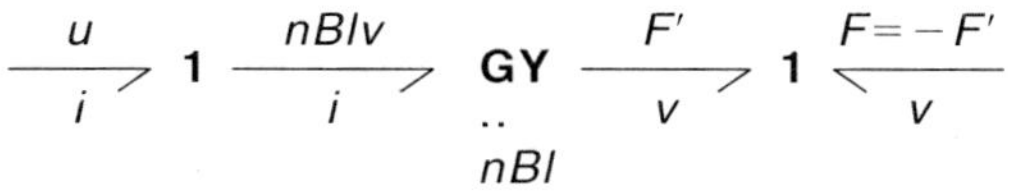

FIGURE 26.23 Basic transduction in a loudspeaker.

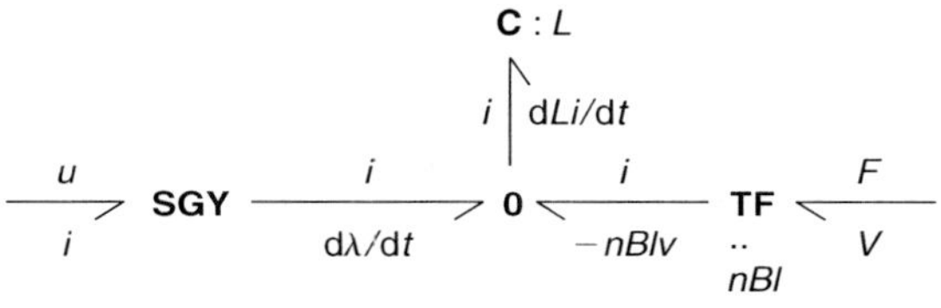

FIGURE 26.24 Addition of magnetic storage with an explicit magnetic domain.

isothermal environment with $T = T_{\text{const}}$ can be characterized by the free energy: $dF = udq - SdT = udq$ and the thermal port can be omitted as it does not result in changes of the free energy F (Figure 26.22):

$$P = u \cdot \dot{q} = \frac{dF}{dt} \qquad (26.26)$$

Independent of the domain the function $e_i(q)$ that characterizes a storage port is called *constitutive relation* or *constitutive equation, constitutive law, state equation*, and *characteristic equation*. If $e_i(q)$ is *linear*, i.e., first-order homogeneous, then $E(q)$ is second-order homogeneous, i.e., $E(q)$ is *quadratic*. In this case, and only in this case: $E(\alpha q) = \alpha^2 E(q)$ $E(q) = \frac{1}{2} \underline{e}^T \cdot \underline{q} = E^*(\underline{e})$ and $dE^* = (d\underline{e})^T \cdot \underline{q} = \underline{e}^T \cdot d\underline{q} = dE$, i.e., the value of energy and coenergy are equal.

Another property of constitutive relations of storage elements that is commonly used in thermodynamics is *Maxwell reciprocity* or *Maxwell symmetry*. From the principle of energy conservation can be derived that the mixed second derivatives of the energy should be equal $\partial^2 E / \partial q_i \partial q_j = \partial^2 E / \partial q_j \partial q_i$. This means that $\partial e_j / \partial q_i = \partial e_i / \partial q_j$, i.e., the *Jacobian* matrix of the constitutive relations is *symmetric*. Maxwell symmetry requires the "true energy" form (integral causality) as Legendre transforms generally destroy the symmetry of the Jacobian. Yet another property that is commonly used in thermodynamics (le Chatelier–Braun principle) to check the *intrinsic stability* of a system that can be considered a multiport storage element is that the *Jacobian* of and all its subblocks are also *positive-definite*: det $(\partial \underline{e} / \partial \underline{q}) > 0$ and that the *diagonal elements* of the Jacobian are positive: $\partial e_i / \partial q_i > 0\ \forall i$.

26.4.1.2 Loudspeaker Example

The dominant behavior of a loudspeaker, in particular the electromechanical transduction based on the Lorentz force, $F = n(i \times B)l$ and on $d\lambda/dt = u = n(v \times B)l$, can in the given configuration be approximated by a gyrator, $F = (nBl)i$ and $u = (nBl)v$ (Figure 26.23).

Making the relation between electric and magnetic domain explicit by means of an SGY shows that the storage of magnetic energy in the voice coil can be represented by a magnetic C (Figure 26.24) for

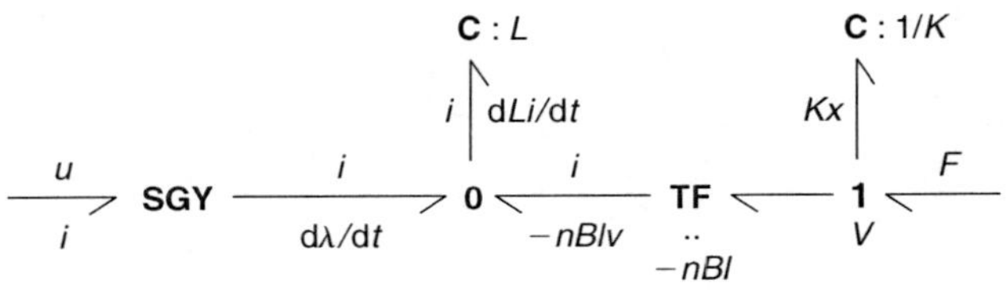

FIGURE 26.25 Addition of elastic cone suspension resulting in the decomposition of a two-port storage element.

which can be written $d\lambda/dt = dLi/dt + nBlv = dLi/dt + nBldx/dt$, after integration: $\lambda = Li + nBlx$. If this constitutive relation is put in preferred integral causality $i = 1/L\lambda - (nBl)/(L)x$ and combined with the relation of the mechanical port $F = -nBli = (-nBl/L)\lambda + ((nBl)^2/L)x$ into a relation of a two-port C in matrix form:

$$\begin{bmatrix} i \\ F \end{bmatrix} = \begin{bmatrix} \dfrac{1}{L} & -\dfrac{nBl}{L} \\ -\dfrac{nBl}{L} & \dfrac{(nBl)^2}{L} \end{bmatrix} \begin{bmatrix} \lambda \\ x \end{bmatrix} \qquad (26.27)$$

it can be concluded that this two-port satisfies Maxwell symmetry, but that it is singular $(1/L((nBl)^2/L) - (-nBl/L)^2 = 0)$, such that a positive mechanical spring constant K representing the connection to the frame of the moving voice coil, has to be added to make it intrinsically stable: $1/L(((nBl)^2/L) + K) - (-nBl/L)^2 > 0$ (Figure 26.25).

The junction structure with the transformer and the two Cs coincides with the so-called congruent canonical decomposition of a linear two-port C of which the energy can be used as a generating function of the constitutive relations (Breedveld, 1984b). When starting again without a mechanical spring, the energy of this two-port should be written in terms of λ and x:

$$dE(\lambda, x) = id\lambda + Fdx = \left(\dfrac{\lambda}{L} - \dfrac{(nBl)x}{L}\right) d\lambda + \left(-\dfrac{(nBl)\lambda}{L} + \dfrac{(nBl)^2 x}{L}\right) dx = d\dfrac{(\lambda - (nBl)x)^2}{2L}$$

such that $E(\lambda, x) = (\lambda - (nBl)x)^2/2L$, but is commonly mistaken by what is actually the coenergy in terms of i and x: $E^*(i, x) = Li^2/2 + nBlix$.

If this is the case and the force is incorrectly derived by taking the partial derivative of this coenergy as if it were an energy $F \stackrel{\text{incorrect!}}{=} \partial E^*(i,x)/\partial x = (i^2/2)dL/dx + nBli$, a sign error is obtained, as it should be

$$F = \dfrac{\partial E(\lambda, x)}{\partial x} = \dfrac{\partial}{\partial x}\dfrac{(\lambda - (nBl)x)^2}{2L} = -(nBl)\left(\dfrac{\lambda}{L} - \dfrac{(nBl)x}{L}\right) - \dfrac{1}{2}\left(\dfrac{\lambda}{L} - \dfrac{(nBl)x}{L}\right)^2 \dfrac{dL}{dx} = -(nBl)i - \dfrac{1}{2}i^2\dfrac{dL}{dx}$$

Note that the mere difference in the form of a minus sign between these results only occurs for the particular case in which the current i depends linearly on the flux linkage λ. In case of a nonlinear relation, the error is larger than "just" a sign error that is commonly not noticed as it is compensated by an implicit change in orientation: in contrast with the global convention, the mechanical port is then taken positively outbound.

If a spring is added to satisfy intrinsic stability the energy increases with the potential energy $E(x) = Kx^2/2$ and similar constitutive relations are found from the energy function:

$$\begin{bmatrix} i \\ F \end{bmatrix} = \begin{bmatrix} \dfrac{\partial i}{\partial \lambda} & \dfrac{\partial i}{\partial x} \\ \dfrac{\partial F}{\partial \lambda} & \dfrac{\partial F}{\partial x} \end{bmatrix} \begin{bmatrix} \lambda \\ x \end{bmatrix} = \begin{bmatrix} \dfrac{\partial^2 E}{\partial \lambda^2} & \dfrac{\partial^2 E}{\partial x \partial \lambda} \\ \dfrac{\partial^2 E}{\partial \lambda \partial x} & \dfrac{\partial^2 E}{\partial x^2} \end{bmatrix} \begin{bmatrix} \lambda \\ x \end{bmatrix} = \begin{bmatrix} \dfrac{1}{L} & -\dfrac{nBl}{L} \\ -\dfrac{nBl}{L} & \dfrac{(nBl)^2}{L} + K \end{bmatrix} \begin{bmatrix} \lambda \\ x \end{bmatrix} \qquad (26.28)$$

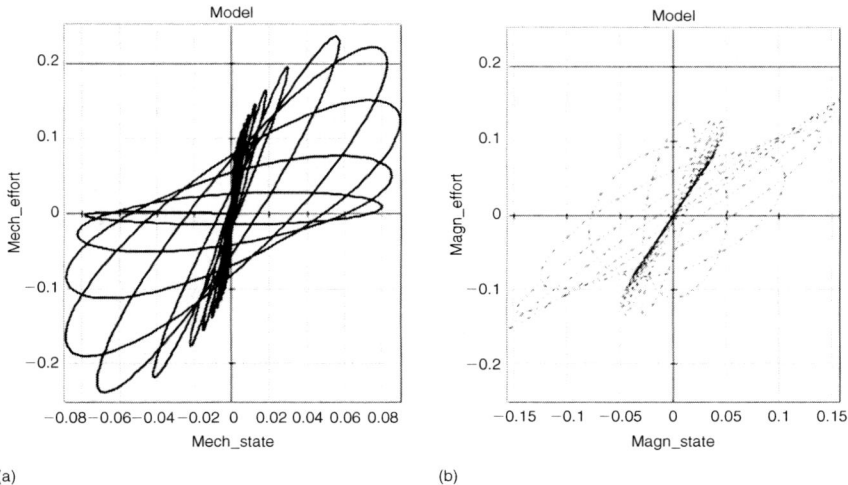

FIGURE 26.26 Simulation of the mechanical and magnetic cycle with frequency sweep input.

which is intrinsically stable as long as $(nBl)^2/L^2 + K/L - (nBl)^2/L^2 = K/L > 0$. Figure 26.26 shows the simulation results of a loudspeaker model containing this two-port, with a sinusoidal frequency sweep input at the electric port. Both ports indeed perform a cycle process in which the magnetic port cycles clockwise and the mechanical port counterclockwise (large loops, i.e., more energy transferred per cycle, represent a simulation at the resonance frequency of the speaker), which demonstrates that indeed per cycle a net amount of magnetic (electric) energy is transduced into mechanical energy. The areas of the loops are equal in the conservative case.

Obviously, if the coenergy is explicitly identified as being different from the energy, the derivations lead to the correct results too

$$E^*(i,x) = -(E(\lambda,x) - i\lambda) = i\lambda - E(\lambda,x) = \frac{Li^2}{2} + nBlix - \frac{Kx^2}{2} \qquad (26.29)$$

$$dE^*(i,x) = \lambda di - Fdx = Lidi + nBlidx - Kxdx \qquad (26.30)$$

Hence

$$E^*(i,x) = \frac{Li^2}{2} + nBlix - \frac{Kx^2}{2}\lambda = \frac{\partial E^*(i,x)}{\partial i} = Li + nBlx \qquad (26.31)$$

so $u = d\lambda/dt = d/dt(\partial E^*(i,x)/\partial i) = L(di/dt) + nBlv = u_{\text{self ind}} + u_{\text{Lorentz}}$ and $F = -\partial E^*(i,x)/\partial x = -nBli + Kx$, in matrix form:

$$\begin{bmatrix} \lambda \\ F \end{bmatrix} = \begin{bmatrix} \frac{\partial \lambda}{\partial i} & \frac{\partial \lambda}{\partial x} \\ \frac{\partial F}{\partial i} & \frac{\partial F}{\partial x} \end{bmatrix} = \begin{bmatrix} \frac{\partial^2 E^*}{\partial i^2} & \frac{\partial^2 E^*}{\partial i \partial x} \\ -\frac{\partial^2 E^*}{\partial x \partial i} & -\frac{\partial^2 E^*}{\partial x^2} \end{bmatrix} = \begin{bmatrix} L & nBl \\ -nBl & K \end{bmatrix} \begin{bmatrix} i \\ x \end{bmatrix} \qquad (26.32)$$

i.e., no symmetry of the Jacobian if energy and coenergy are confused.

This example demonstrates that the concept of multiport storage, in particular two-port storage, may not only lead to conceptual models of transduction phenomena, but also that reversible transduction requires a cycle process in principle. This means that a transducer cannot continuously transduce a DC input. By changing its configuration into that of an electric motor, the configuration takes care of the cycle of the conductors in the magnetic field.

Hence, pure "DC"-type transduction can only be achieved by some form of "carrier" that performs a cycle: gears carrying teeth (with elastic deformation!) during contact, a cooling fluid, and a rolling wheel that cycles its interaction point, which is a useful insight during conceptual design of transducers.

26.5 Conclusion

In this chapter, the basics of the port-based approach were introduced as well as their natural notation, viz. bond graphs. The main advantages of the use are

(1) the *domain-independence* of the elementary behaviors and their graphical notation that allow quick analysis of dynamic interaction across domain boundaries;
(2) that *not* all ports have an a-priori *fixed* causality, thus allowing flexible reuse of submodels;
(3) the *combination* of physical and computational structure *in one notation*, thus allowing

   a. *direct physical interpretation* of required changes in model structure, e.g., for controller design,
   b. *direct feedback* on modeling decisions,
   c. *direct graphical input* for simulation software.

All these features contribute to rapid insight and the ability of efficient iteration during the modeling process.

It was also demonstrated that various domains of physics that use some form of energy-based model formulation technique are all strongly related, even though terminology and a loss of conceptual distinction between conserved energy and its nonconserved Legendre transforms commonly obstructs this insight.

References

Ascher, U. M. and Petzold, L. R. 1998. *Computer Methods for Ordinary Differential Equations and Differential-Algebraic Equations*. Philadelphia: SIAM.

Bos, A. M. 1986. *Modelling Multibody Systems in Terms of Multibond Graphs*. Ph.D. Thesis, Electrical Engineering, University of Twente, Enschede, Netherlands.

Bos, A. M. and Tiernego, M. J. L. 1985. Formula manipulation in the multibond graph modelling and simulation of large mechanical systems. *J. Franklin Inst.*, Vol. 319, No. 1/2, pp. 51–65.

Breedveld, P. C. 1982. Thermodynamic bond graphs and the problem of thermal inertance. *J. Franklin Inst.*, Vol. 314, No. 1, pp. 15–40.

Breedveld, P. C. 1984a. *Physical Systems Theory in Terms of Bond Graphs*. ISBN 90-9000599-4. University of Twente, Enschede, The Netherlands. Distributed via the author.

Breedveld, P. C. 1984b. Decomposition of multiport elements in a revised multibond graph notation. *J. Franklin Inst.*, Vol. 318, No. 4, pp. 253–273.

Breedveld, P. C. 1985. Multibond graph elements in physical systems theory. *J. Franklin Inst.*, Vol. 319, No. 1/2, pp. 1–36.

Breedveld, P. C. 1999. Insight in rigid body motion stability via an alternative for the Eurlerian junction structure. *Proceedings 1999 International Conference on Bond Graph Modeling and Simulation* (ICBGM'99), San Francisco, USA, Western Multi Conference Simulation Series, Vol. 31, No. 1, pp. 269–274.

Callen, H. B. 1960. *Thermodynamics*. New York: Wiley.

Golo, G. 2002. *Interconnection Structures in Port-Based Modelling: Tools for Analysis and Simulation*. Ph.D. Thesis, University of Twente, Twente University Press.

Golo, G., Breedveld, P. C., Maschke, B. M. and van der Schaft, A. J. 2000. Input–output representations of Dirac structures and junction structures in bond graphs. *Proceedings 14th International Symposium on Mathematical Theory of Networks and Systems*, MTNS 2000, Perpignan, France, June 19–23.

Hayt, W. H., Jr., Kemmerly, J. E. and Durbin, S. M. 2002. *Engineering Circuit Analysis*, 6th edition. New York: McGraw-Hill.

Hogan, N. J. and Fasse, E. D. 1988. Conservation principles and bond graph junction structures. *Proceedings of the ASME 1988 WAM*, DSC-Vol. 8, pp. 9–14.

Karnopp, D. C. and Rosenberg, R. C. 1968. *Analysis and Simulation of Multiport Systems: The Bond Graph Approach to Physical System Dynamics*. Cambridge, MA: MIT Press.

Ligterink, N. E., Breedveld, P. C. and van der Schaft, A. J. 2006. Physical model reduction of interacting, continuous systems. *Proceedings 17th International Symposium on Mathematical Theory of Networks and Systems,* Kyoto, Japan, July 24–28.

Maschke, B. M. and van der Schaft, A. J. 2001. Canonical interdomain coupling in distributed parameter systems: An extension of the symplectic gyrator. *Proceedings of the International Mechanical Engineering Congress and Exposition*, New York, USA, November 11–16.

Maschke, B. M., van der Schaft, A. J. and Breedveld, P. C. 1995. An intrinsic Hamiltonian formulation of the dynamics of LC-circuits. *Trans. IEEE on Circuits and Systems, I: Fundamental Theory and Applications*, Vol. 42, no. 2, pp. 73–82.

Paynter, H. M. 1961. *Analysis and Design of Engineering Systems*. Cambridge, MA: MIT Press.

Paynter, H. M. and Busch-Vishniac, I. J. 1988. Wave-scattering approaches to conservation and causality. *J. Franklin Inst.*, Vol. 325, No. 3, pp. 295–313.

Stramigioli, S. 1998. *From Differentiable Manifolds to Interactive Robot Control*. Ph.D. Thesis, Delft University of Technology, Netherlands.

Tiernego, M. J. L. and van. Dixhoorn, J. J. 1979. Three-axis platform simulation: Bond graph and Lagrangian approach. *J. Franklin Inst.*, Vol. 308, No. 3, pp. 185–204.

Van Johannes, Dijk and Breedveld, P. C. 1991. Simulation of system models containing zero-order causal paths—part I: Classification of zero-order causal paths and part II: Numerical implications of class-1 zero-order causal paths. *J. Franklin Inst.*, Vol. 328, No. 5/6, pp. 959–979, 981–1004.

Wheeler, H. A. and Dettinger, D. 1949. Wheeler Monograph 9.

27
System Dynamics Modeling of Environmental Systems

Andrew Ford
Washington State University

27.1 Introductory Examples 27-1
27.2 Comparison of the Flowers and Sales Models 27-4
27.3 Background on Daisy World 27-6
27.4 The Daisy World Model 27-6
27.5 The Daisy World Management Flight Simulator 27-9

This chapter provides a short tutorial on the system dynamics approach to computer simulation modeling. The modeling usually begins when managers face a dynamic pattern that is causing a problem. The modeling is based on the premise that we can improve our understanding of the dynamic behavior by the construction and testing of models that focus on the information feedback. The approach was pioneered by Forrester (1961) and is explained in recent texts by Ford (1999) and Sterman (2000). The models are normally implemented with visual software such as

Stella (http://www.iseesystems.com),
Vensim (http://www.vensim.com/), and
Powersim (http://www.powersim.com/).

These programs use "stock and flow" icons to help us see the accumulation in the system. The programs also help one to see the information feedback in the simulated system. The programs use numerical methods to show the dynamic behavior of the simulated system.

This chapter begins with simple models to demonstrate that different systems can exhibit the identical pattern of growth over time. One model shows the growth in flowered area; the second shows the growth in a sales company. These systems show the same dynamic behavior because their growth is governed by the same feedback loop structure. The flower model is then extended to simulate the imaginary Daisy World created by Watson and Lovelock (1983). Daisy World provides an additional example of feedback loop structure. It also provides a convenient way to illustrate interactive "flight simulators," one of the several methods to promote communication and learning from system dynamics models. The chapter concludes with a list of readings for those who wish to learn more about system dynamics modeling of environmental systems.

27.1 Introductory Examples

Figure 27.1 shows a flow diagram of a model to simulate the growth in the area covered by flowers. (This diagram is in Stella.) In this example, we start with 10 acres of flowers located within a suitable area of 1000 acres. The area of flowers is called a stock variable; the growth and decay variables are called flows.

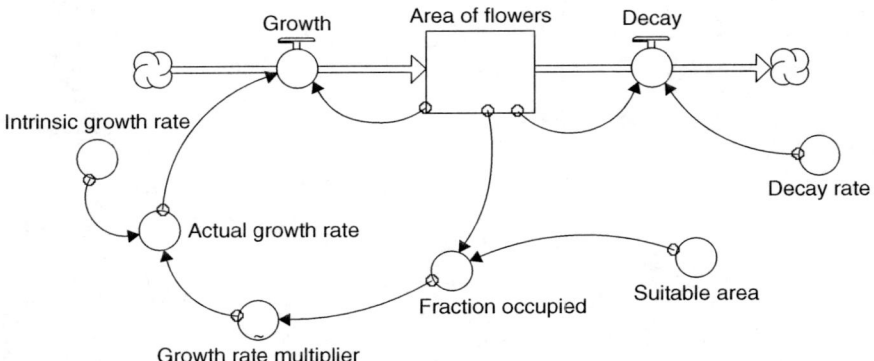

FIGURE 27.1 Stella diagram of a model to simulate growth in the area of flowers (from Island Press, 1999).

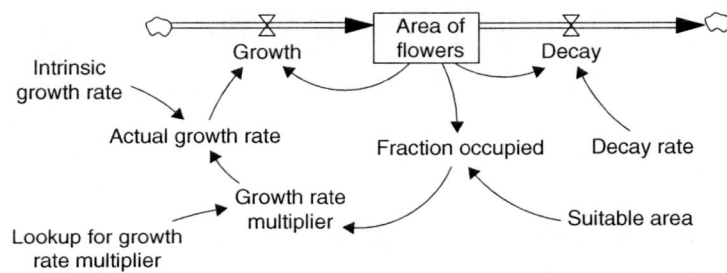

FIGURE 27.2 Vensim diagram of the flowers model.

The growth flow increases the area of flowers over time. The decay flow reduces the flowered area over time. The remaining variables in the Stella diagram are called converters. The converters are used to help explain the flows. Stella depicts the action of the flows by the double lines leading into or out of the stocks. Stella represents the information connections by single lines. For example, the single lines connecting to the decay flow indicate that the decay is influenced by the area of flowers and the decay rate. The variables in system dynamics models are normally assigned long names to make their meaning clear. When this is done, the reader can usually guess the equation for each variable. For example, you can probably guess that the equation for the decay is the product of the area of flowers and the decay rate.

Figure 27.2 shows the Vensim version of the flow diagram for the flowers model. The two programs use somewhat different icons, but the stock and flows are clear to see. Figure 27.2 shows that the area of flowers is a stock which is increased by growth and reduced by decay. Additional variables (such as the decay rate) appear in Figure 27.2 by their names. Vensim allows the user to assign a variety of symbols to the variables, but the convention is to avoid extra symbols to minimize clutter in the diagram.

System dynamics models may be viewed as a coupled set of differential equations. In the flower model, we have only one stock variable, so the model may be represented by a single differential equation, as shown in Table 27.1. This format will be familiar to many handbook readers who have studied calculus and differential equations. These readers will appreciate that an analytic solution to the differential equation is difficult because of the nonlinear relationship for the growth rate multiplier.

System dynamics models are filled with nonlinear relationships, so the software programs are designed to make it easy to find a numerical solution to the equations. Most simulations are performed with Euler's method. The model user is responsible for setting a sufficiently short step size to ensure accurate results. Accuracy is usually checked by simply repeating the simulation with the step size cut in half. If we see the same results, we know the results are numerically accurate.

TABLE 27.1 Flower Model in the Form of a Differential Equation

Let	
A = area of flowers	$dA(t)/dt = G(t) - D(t)$
G = growth	$A(0) = 10$
D = decay	
	$D(t) = A(t)^{*}d$
	$G(t) = A(t)^{*}g$
d = decay rate	$g(t) = ig^{*}gm(t)$
g = actual growth rate	$gm = f(FO)$
ig = intrinsic growth rate	$FO(t) = A(t)/SA$
SA = suitable area	$SA = 1000$
gm = growth rate multiplier	$ig = 1.0$
FO = fraction occupied	$d = 0.2$

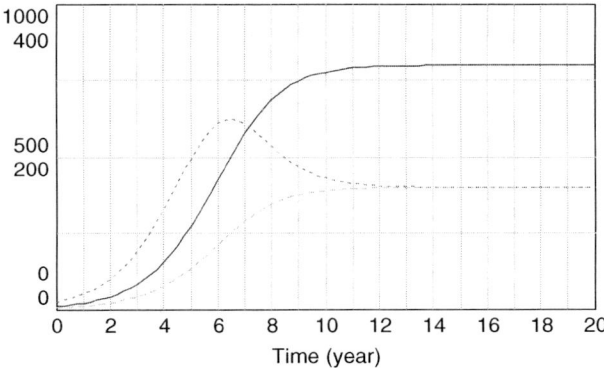

FIGURE 27.3 Vensim graph of flower simulation results. (The area is scaled to 1000 acres. The growth and decay are scaled to 400 acres/year.)

Figure 27.3 shows a Vensim time graph of a 20-year simulation. The graph shows rapid growth in the space occupied by flowers during the first 5 years as the growth far exceeds the decay. As the flowers fill up the suitable area, their growth rate falls below the intrinsic growth rate that applied at the start of the simulation. The growth gradually declines until it matches the decay. At this point, the system has reached a dynamic equilibrium with 800 acres of flowers. The flowered area is maintained by a growth and decay of 160 acres/year.

Figure 27.4 shows a somewhat more complicated model. This is a Vensim flow diagram of a model to simulate the growth in a sales company. A stock variable represents the size of the sales force. The stock is increased by the new hires, and it is reduced by the departures. Departures are controlled by the exit rate, which is set to 20% per year. The new hires are controlled by the budgeted size of the sales force. To illustrate, consider a numerical example with the sales force initially at 50 people. With a small sales force, we assume that each person can sell 2 widgets per day. If the widget price is $100, the 50 people could generate $3.65 million in annual revenues. If the company assigns half of the revenues to the sales department budget, the company could budget for 73 sales persons paid an annual salary of $25,000. In this situation, the new hires would be calculated to build the stock toward the goal of 73 persons. The pace at which the new hires builds the stock is controlled by the hiring fraction, which is measured as a fraction per year. When new hires exceed departures, the sales force will grow toward saturation. With so many sales persons working the market, their effectiveness will fall below the initial value of 2 widgets per day.

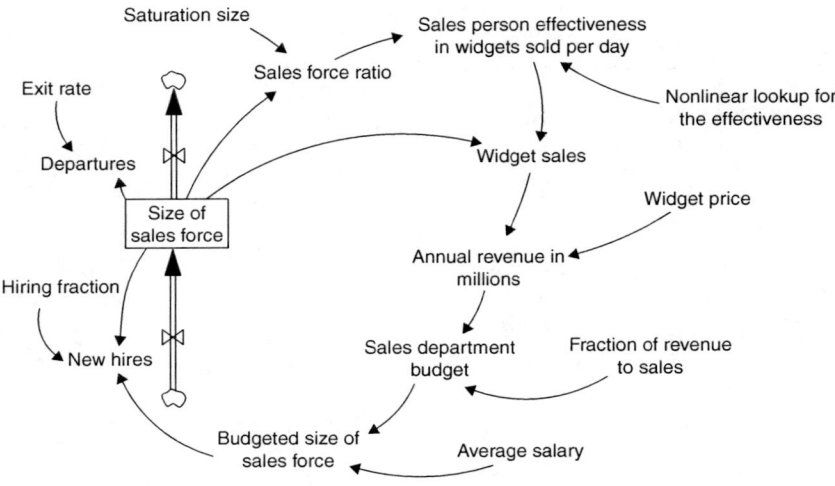

FIGURE 27.4 Vensim diagram of a model to simulate the growth of a sales company.

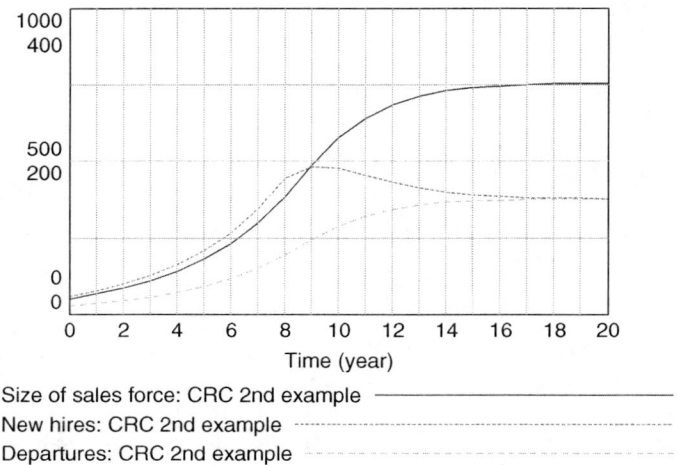

FIGURE 27.5 Vensim time graph for the sales model. (The sales force is scaled to 1000 persons. The new hires and departures are scaled to 400 persons per year.)

Figure 27.5 shows the simulated growth in the sales force when the model is simulated over a 20-year period. The graph shows rapid growth in the size of the sales force during the first 9 years as the new hires far exceed the departures. However, as the sales force climbs past 500 persons, the new hires begins to decline. New hires gradually decline until the company achieves dynamic equilibrium with around 150 new hires and departures per year. The equilibrium size of the company is around 750 persons.

27.2 Comparison of the Flowers and Sales Models

Figure 27.3 and Figure 27.5 show that the sales company exhibits essentially the same pattern of behavior as the flowers system. Both simulations begin with rapid growth. The growth is powered by a high growth rate of flowers and by the high effectiveness of the sales persons. But both systems face limits, so they cannot grow forever. As the systems encounter these limits, their growth declines in a gradual manner.

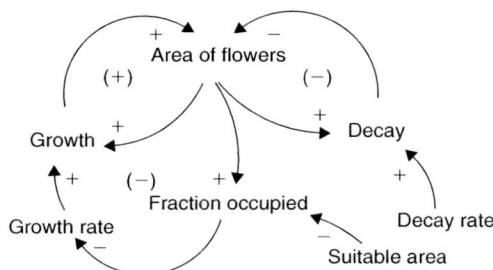

FIGURE 27.6 Feedback loop structure of the flowers model (from Island Press, 1999).

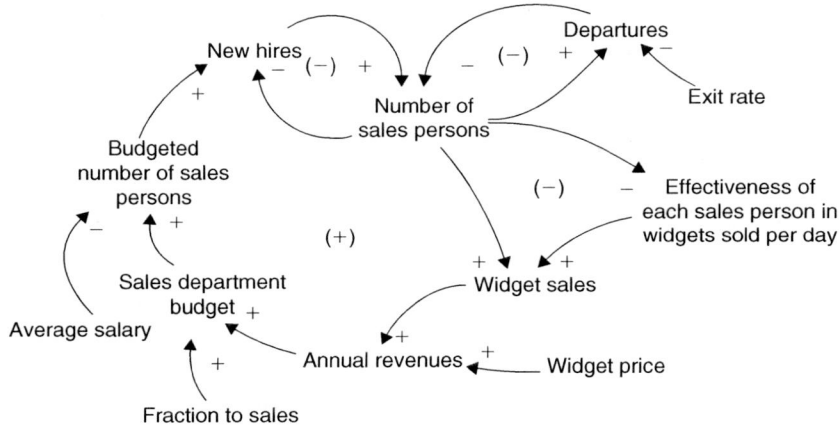

FIGURE 27.7 Feedback loop structure of the sales model (from Island Press, 1999).

Eventually, they reach a state of dynamic equilibrium, an equilibrium which could be maintained year after year. The similarity in these patterns is no coincidence. It arises from the similarity in their underlying structure. You can see some similarity by comparing the model diagrams. But the similarity will be even more apparent when we draw causal loop diagrams to focus our attention on the feedback in the systems. We expect to see positive feedback when a system is able to grow on its own; we expect to see negative loops that limit or shape the growth of the system. These loops come into clearer view when we show variables and their interconnections as in Figure 27.6 and Figure 27.7.

Figure 27.6 shows the three feedback loops in the flowers model. The rapid growth is powered by the positive feedback loop in the upper left corner of the diagram: more flowered area leads to more growth, which leads to more flowered area. The convention is to label the positive feedback loops with a (+). These loops will be found whenever a system has the power to exhibit exponential growth on its own. Figure 27.6 shows two negative loops, both of which are marked with a (−). One negative loop involves the decay of flowered area over time. The strength of this loop is controlled by the decay rate, which is assumed to remain constant in this example. The other negative loop involves the reduction in the growth rate as the flowers occupy a larger and larger fraction of the suitable area. This is a negative loop that becomes stronger and stronger over time. (Such feedback is sometimes called density-dependent feedback in environmental systems.) The negative feedback on flower growth eventually reaches the point where growth and decay are in dynamic equilibrium.

Figure 27.7 shows a somewhat similar set of loops in the sales model. Since the sales force grows rapidly on its own, we know that there should be positive feedback in the system. This loop involves the size of the sales force and the annual revenues: a larger sales force leads to greater sales, increased revenues, a larger budget for the sales department, a larger budgeted number of sales persons, more new hires, and a still larger sales force. This positive feedback loop serves the same general function as the positive loop in the flower model. If the positive feedback is sufficiently strong, the sales company will be able to grow. However, the positive feedback in the sales company is different in that the company has a clearly identified goal associated with the budgeting process. (Such a goal is not apparent in the flowers model.) Figure 27.7 shows a negative feedback loop associated with the company's attempt to bring the size of the sales force into balance with the budgeted goal. Such loops are to be expected whenever a system is designed with goal-seeking behavior. An additional negative loop is associated with the departures of sales persons. This loop is controlled by the exit rate and is similar in function to the decay loop in the previous model.

The fourth loop in Figure 27.7 provides the density-dependent feedback that we would expect to see if the sales company is to eventually come into dynamic equilibrium. To understand the impact of the loop, imagine there is a major increase in the size of the sales force. This would reduce the effectiveness of each sales person which tends to reduce the widget sales. Lower sales lead to lower revenues, a reduction in the budgeted size of the sales force, fewer new hires, and a smaller size of the sales force. By working through the loop in a step-by-step fashion, we see that the loop acts to negate the effect of the original change. This is a characteristic effect of negative feedback.

27.3 Background on Daisy World

Daisy World is an imaginary world invented by Watson and Lovelock (1983) to illustrate a system with close coupling between the biota and the global environment. This imaginary world is inhabited by white daisies and black daisies. The white daisies have a high albedo, which means they are highly reflective. When the planet surface is covered by a large area of white daisies, the planet will tend to reflect much of the incoming solar luminosity. The black daisies have a low albedo, so their surface tends to absorb much of the incoming luminosity. Consequently, the mix of daisies on the planet influences the absorbed luminosity and the planet's temperature. Now imagine that the planet's temperature influences the rate of growth of the daisies. If the temperature is close to the optimum value for flower growth, the flowered areas will spread across the planet. But if the temperature is too high or too low, the flowered areas will recede over time.

This imaginary world was created as a concrete example of system with homeostatic properties, like those described for the human body by Cannon (1932). According to Levine (1993, p. 89), Lovelock's conclusion on a living planet followed from the principle of physiology: "that the living body strives to maintain the constancy of its internal environment." Lovelock believed that "Earth behaves in the same way: its living and nonliving parts collaborate to hold temperature and other conditions at reasonably constant levels." The idea of a "living earth" needed a name, and Lovelock was reluctant to invent a "barbarous acronym such as Biocybernetic Universal Systems Tendency/Homeostasis." Instead, he chose the name "Gaia Hypothesis" with the hope that Gaia would remind us of The Greek Earth goddess, and "hypothesis" would encourage scientists to ask a different set of questions about the world.

The Gaia Hypothesis is a subject of much discussion (Capra, 1996; Dawkins, 1982; Joseph, 1990; Levine, 1993; Lovelock, 1988, 1991, 1995), as is the example of Daisy World (Capra, 1996; Joscourt, 1992; Kirchner, 1989; Lovelock, 1991). My purpose in this chapter is to use Daisy World as a third example of system dynamics modeling. This new model builds from the simple flower model shown previously so that the reader can appreciate how system dynamics models can be created by building on previous work. The Daisy World model will also illustrate how causal loop diagrams can help one understand and anticipate the dynamic behavior. Finally, the model is used to illustrate a "management flight simulator" a model designed to promote learning through interactive experimentation.

27.4 The Daisy World Model

The Daisy World model was created by adapting the original differential equations by Watson and Lovelock (1983) into a system dynamics model implemented in Stella (Ford, 1999, Chapter 21). Figure 27.8 shows part of the model in the form of an equilibrium diagram. (An equilibrium diagram is a stock-and-flow

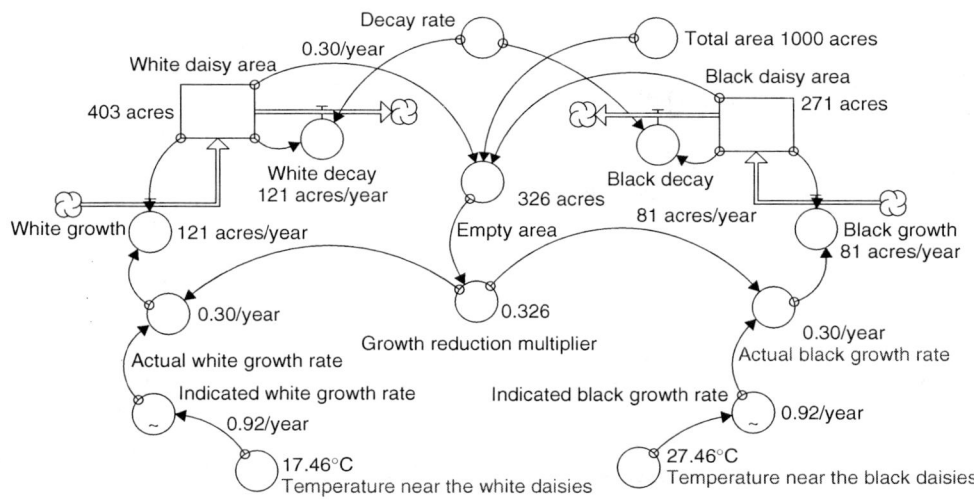

FIGURE 27.8 Equilibrium diagram for the daisy areas in the Daisy World model (from Island Press, 1999).

diagram with the equilibrium values for each variable on display.) The model represents a world with 1000 acres. When the system reaches equilibrium, the white daisies occupy 403 acres, and the black daisies occupy 271 acres. Both flowered areas are subject to a fixed decay rate of 30%/year, so the system finds its way to an equilibrium with both growth rates at 30%/year. With this mix of areas, the average albedo is 0.53 which means the plant would absorb 47% of the incoming solar luminosity. Lovelock and Watson set the incoming luminosity at 1.0, but their main interest was how the world would respond to changes in the solar luminosity. They calculated the planet's average temperature as a nonlinear function of the absorbed luminosity. In this case, the average temperature turns out to be 21.8°C. They then assumed that local effects would lead to somewhat higher temperatures in the area covered by black daisies and somewhat lower temperatures in the area covered by white daisies. Figure 27.8 shows the equilibrium temperatures at 17.46°C near the white daisies and 27.46°C near the black daisies. Both flowers grow best (at 100%/year) when their local temperature is 22.5°C. Figure 27.8 shows that the white flowers are experiencing temperatures cooler than the optimum, and their indicated rate of growth would be 92% if there were no limits based on the empty area. The black flowers are experiencing temperatures hotter than the optimum, and their indicated rate of growth is also 92%/year (if there were no limits based on the empty area). Both flowers experience a reduction in the actual growth rates from the space limitations. By the time all of these effects are sorted out, the two areas grow at the rate of 30%/year, just what is required to counter the decay rate of 30%/year. The preceding description skips over some of the equations, but the interested reader can learn the missing details from *Modeling the Environment* or by downloading the model (http://www.wsu.edu/~forda/GuidetoRest.html).

It is useful to now turn to the feedback loops that will help us anticipate how this world would respond to a change in solar luminosity. Figure 27.9 shows a collection of eight feedback loops. This is a complex diagram, so it is helpful to find a familiar place to start. Let us begin with the black area in the upper right corner of the diagram and look for the feedback loops that remind us of the three loops shown previously in Figure 27.6. This comparison shows that the spread of the black daisies is controlled by the same set of loops. There is a positive loop to power the growth in a flowered area; a negative loop based on a fixed rate of decay, and a negative loop whose strength changes with the amount of empty area. If we turn to the white area in the upper left part of the diagram, we see a similar set of three loops that control the spread of the white daisies.

The remaining two loops in Figure 27.9 involve the plant's absorbed luminosity and average temperature. Each of these loops provides negative feedback, and it is these feedback effects that give the world the homeostatic properties to illustrate the Gaia Hypothesis. To understand these loops, let us trace the world's reaction to a major change in solar luminosity. To begin, let us assume a large increase in incoming luminosity and trace the reaction of the left-side loop involving the growth of the white daisies. With an increase in solar luminosity, we would see an increase in the absorbed luminosity, an increase in the planet's average temperature, and an increase in the local temperature for the white daisies. The white flowers would then experience temperatures that are closer to their optimum growing conditions. This increases the white growth rate and the area covered by white flowers. With a greater area covered by white daisies, the planet has a higher albedo and a lower absorbed luminosity. By tracing the cause and effect reactions around the loop, we learn that the loop acts to negate the impact of the assumed increase in solar luminosity. We have learned that this loop acts to protect the world against the threat of increases in the solar luminosity.

The remaining loop in Figure 27.9 operates through the black daisy area and the average temperature. To understand the effect of this loop, let us imagine that the world experiences a major drop in the incoming solar luminosity. This would lower the absorbed luminosity and the planet's average temperature. The local temperature near the black daisies would be lowered, so the black flowers would then experience temperatures closer to their optimum growing conditions. This increases the black growth rate and the area covered by black flowers. With a greater black area, the planet has a lower albedo and a greater absorbed luminosity. The overall reaction of this loop is to negate the impact of the assumed drop in solar luminosity. We have learned that this loop acts to protect the world against the threat of reduced solar luminosity.

Some readers may look at the temperature-dependent loops in Figure 27.9 and think of the actions of a heating/cooling system that protects our home against changes in the external temperature. But this analogy was not what Watson and Lovelock had in mind for Daisy World. They intentionally avoided creating a world with anything resembling the target temperature on our home thermostat. Given their view of the natural world, a better analogy is the human body which has evolved over the centuries to achieve homeostatic regulation of body temperature. This is accomplished through involuntary physiological

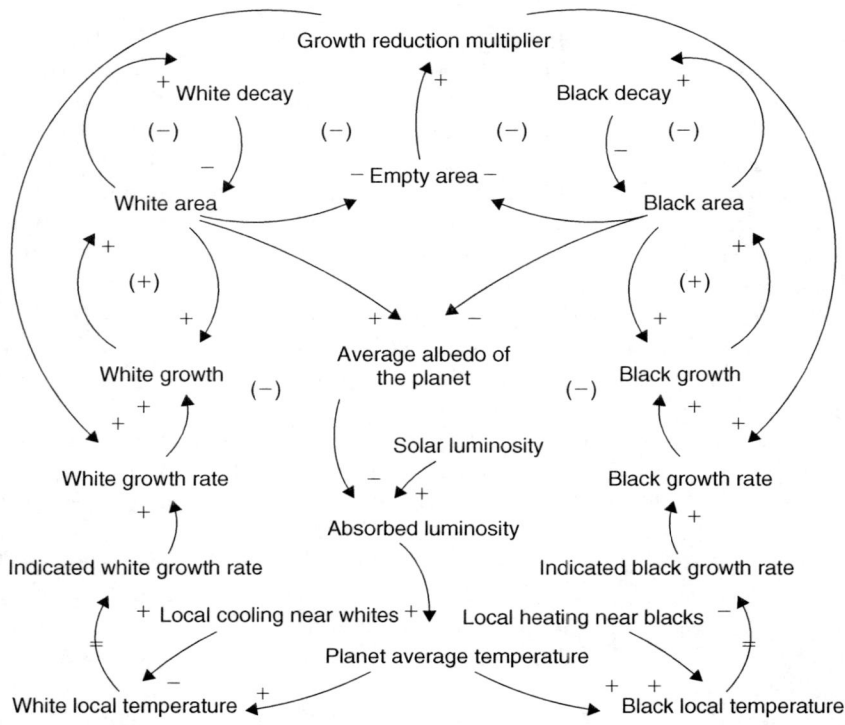

FIGURE 27.9 Feedback loop control of Daisy World temperatures (from Island Press, 1999).

reactions such as shivering and sweating. Shivering acts to protect the body's core temperature under cold conditions, so one might think of shivering as analogous to the spread of black daisies on Daisy World. Sweating acts to protect the body's core temperature under hot conditions, so one might think of sweating as analogous to the spread of white daisies on Daisy World.

27.5 The Daisy World Management Flight Simulator

Figure 27.10 concludes the Daisy World example with a view of an interface to make it easy to experiment with the model in interactive fashion. Such models are sometimes called "management flight simulators" because they remind us of a pilot flight simulator, an electromechanical model of an airplane. A pilot flight simulator is a simplified representation of an airplane equipped with a sophisticated interface resembling the cockpit. Pilots experiment with the simulated airplane to improve their flying instincts. A management flight simulator has a similar purpose. It is designed to help managers improve their business instincts. Such simulations have proven useful in several business systems including health care, airlines, insurance, real estate, and project management (Morecroft, 1988; Sterman, 1992; Senge, 1990). Environmental examples include simulators for management of fees and rebates (to promote sale of cleaner vehicles) and the management of a salmon fishery (Ford, 1999).

The simulator in Figure 27.10 provides information buttons to explain that the user is to experiment with strategies to help Daisy World support life in the face of massive uncertainties in the solar luminosity. In the example shown in Figure 27.10, the user has elected to test the simulated world's response to a heat shock scenario in which the solar luminosity jumps from 1.0 to 1.25 in the year 2010. The user may select from a variety of daisies with different albedos and optimum temperatures. In this example, the user has selected the 5th variety (which corresponds to the flowers described by Watson and Lovelock). The graph in Figure 27.10 shows the first 30 years of a 40-year simulation. (The simulation is advanced 2 years at a time when the user clicks on the "Run for 2 years" button.) The graph shows that the planet's average temperature has declined to around 22°C by the year 2010. This is the year of the sudden increase in solar

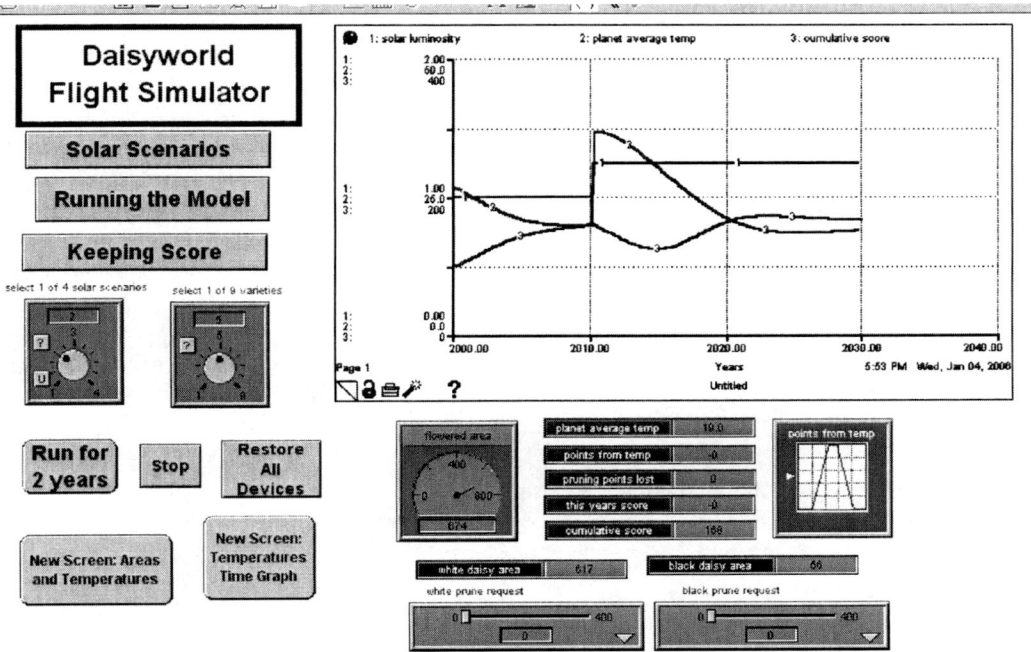

FIGURE 27.10 Part way through a simulation of a heat shock scenario for Daisy World.

luminosity, and the planet's average temperature jumps immediately to around 37°C. But the world is not fixed at 37°C even though the solar luminosity is fixed at 1.25. Figure 27.10 shows that the planet's average temperature declines to around 22°C within a decade after the onset of the heat shock conditions. This decline is made possible by the spread of white daisies and the retreat of the black daisies, both of which act to increase the planet's average albedo. By the 30th year of the simulation, the world appears to have found a new equilibrium condition with the average temperature close to the value well suited for the flowers. This is the response that Lovelock and Watson were looking for since it shows the biota reacting with the nonliving part of the system to maintain conditions suitable for life—the central feature of the Gaia Hypothesis.

Figure 27.10 shows two distinctive features that are crucial to the use of system dynamics models for interactive experimentation as management flight simulators. Since system dynamics focuses on dynamic behavior, the first and most important feature is the display of simulated behavior over time. In this case, the time graph shows the jump in temperature in 2010 followed by the decline in temperature as the planet's feedback mechanisms negate the effect of the heat shock scenario. The display of dynamic behavior over time is present in all system dynamics models and is often a key feature that distinguishes the approach from other modeling methods. (In the case of Watson and Lovelock's Daisy World, for example, the original set of differential equations was not simulated over time. Rather, the authors found the equilibrium properties of the world through analytical methods.) The second distinguishing feature is the use of models in a highly interactive fashion to promote learning. Interactive simulation is encouraged by a "score keeper" which gives a quantitative measure of the success in maintaining a world with homeostatic properties that are conducive to life. The score keeper is a convenient way to promote friendly competition and increased discussion within a group of students or managers.

Further Reading

The preceding examples provide the reader with simple, concrete examples of system dynamics modeling of environmental systems. To learn more, the reader may turn to a variety of texts and articles. Starting with texts, two of the early books are Forrester's (1961) *Industrial Dynamics* and Richardson and Pugh's (1981) *Introduction to System Dynamics Modeling with Dynamo*. In my view, the most comprehensive and useful textbook is *Business Dynamics* by Sterman (2000). Readers interested in environmental systems can learn from *Modeling the Environment* (Ford, 1999).

System dynamics has been applied to a wide variety of environmental and resources systems. The most widely known application is probably the collection of models associated with the study of population growth in a world with finite resources (Forrester, 1971; Meadows et al., 1972, 1973, 1974, 1992). Some of the recent applications are described in special issues of the *System Dynamics Review*. A special edition edited by Sterman (2002) presents historical reflections and recent applications, all organized to celebrate the life of Dana Meadows. A special issue edited by Ford and Cavana (2004) presents the articles summarized in Table 27.2.

Readers may learn of other applications from the citations listed in the bibliography of the System Dynamics Society (2003). Figure 27.11 reveals the distribution of citations with keywords "environmental" or "resource" over the time period from 1960 to 2002. There are 635 citations, approximately 10% of the total publications in the 2003 bibliography. The number of publications increased in the early 1970s, probably as a result of the tremendous interest in the *Limits to Growth* (Meadows et al., 1972). There were around 10 citations per year during the 1980s and early 1990s followed by a dramatic increase at the turn of the century. The appearance of 70–80 annual citations in the past few years is certainly a dramatic development that might be attributed to growth in the System Dynamics Society and a growing awareness of environmental problems.

Figure 27.12 provides further analysis of the number of environmental and resource publications during the past four decades. This chart shows that energy and resource applications dominate the frequency of citations. Resource applications cover a lot of ground, so it is reasonable for this keyword to extract a large

System Dynamics Modeling of Environmental Systems

TABLE 27.2 Summary and Comparison of Papers in the Special Issue of the *System Dynamics Review* (Ford and Cavana, 2004)

	Dudley	Martinez Fernandez and Esteve Selma	Moxnes	Faust, Jackson, Ford, Earnhardt and Thompson	Arquitt and Johnstone
Author(s)					
Model	Log export ban	Irrigated landscapes	Misperception of feedback	Wildlife population management (spectacled and grizzly bears)	Blue-green algae blooms
Geographic location	Forests in Indonesia	Irrigated lands in SE Spain	Reindeer rangelands of Norway	AZA zoos, and Greater Yellowstone Ecosystem, USA	Coastal waters of Queensland, Australia
Issue(s)	Substantial increases in illegal logging and deforestation	New irrigated lands leading to overexploitation of available water resources	Overuse of renewable resources owing to misperceptions of the dynamics	Improved management and conservation of wildlife populations in captivity and natural habitats	Algae blooms threatens water quality, coastal ecosystems, and harmful to humans
Model purpose	To understand the effects of a log export ban on the forestry sector	To analyze the key socioeconomic and environmental factors leading to overexploitation.	Simulators to test decision making of reindeer management with limiting resources	To assess the impact of management actions, help guide data collection and allocation of conservation resource	Scoping and consensus model to understand the dynamics of algae blooms and for developing research directions
Client/sponsor/ decision maker	Centre for International Forestry Research, Public and private forestry managers	Public policy makers (Spanish National Water Plan), local decision makers, general public, and technical people	Public and private managers of renewable resource systems	Population biologists, zoo managers, national park administrators and advisors	Committee est. by Environmental Protection Agency, Queensland from Government departments, academics and technical people, and community organizations

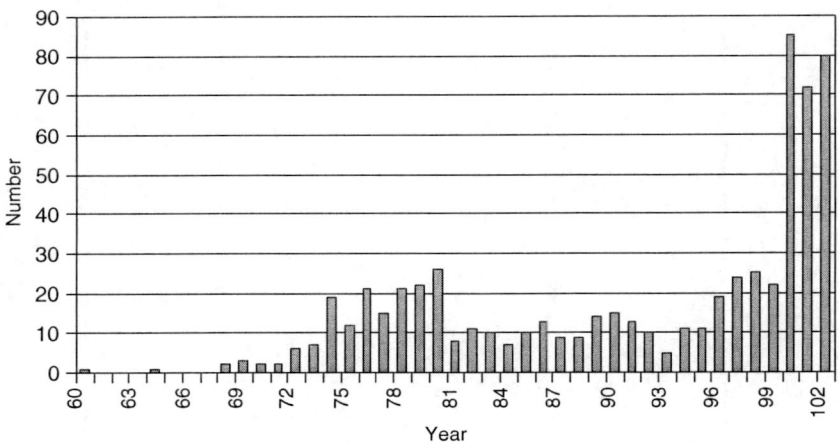

FIGURE 27.11 Number of publications listed in the System Dynamics Society bibliography extracted using the keywords "environmental" or "resource" (from Cavana and Ford, 2004).

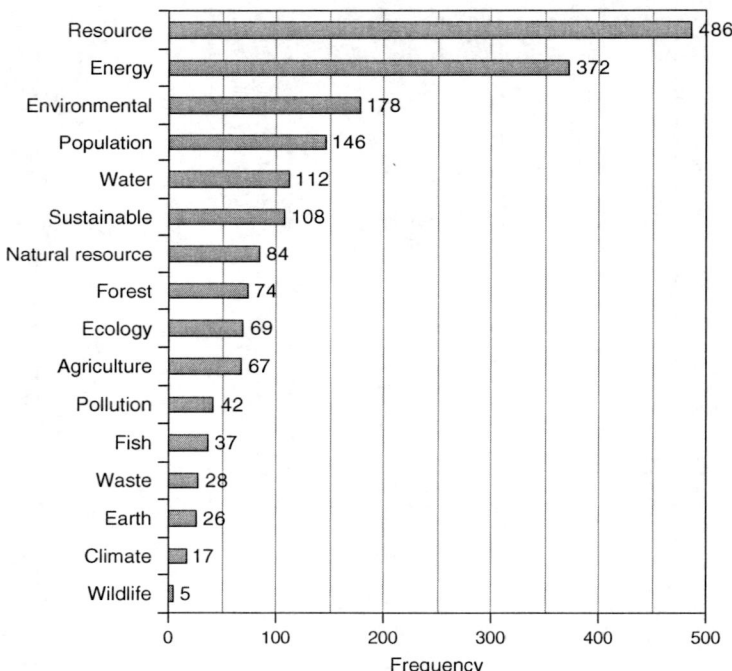

FIGURE 27.12 Analysis of System Dynamics Society 2003 bibliography by keyword frequency (from Cavana and Ford, 2004).

number of citations. Energy issues have received major attention of system dynamics practitioners, as explained by Ford (1999) and by Bunn and Larsen (1997). Figure 27.12 shows a significant number of citations for more narrowly defined keywords such as fisheries and earth systems.

The citation analysis shows a tremendous quantity of work, and these citations are mostly limited to those who elect to publish their models in the system dynamics literature. The number of citations would be even greater if we were to count the many modeling applications using system dynamics software such as Stella. These citations would include work by Costanza and Ruth (1997), Grant et al. (1997), Costanza (1998), Deaton and Winebrake (1999) and the recent book on *Mediated Modeling* by Van Den Belt (2004). These researchers do not place the same emphasis on feedback structure that is common in system dynamics studies advocated by Forrester (1961), Ford (1999), and Sterman (2000). Nevertheless, their use of computer simulation helps one understand the reasons for dynamic behavior. Counting such applications would reinforce the upward trends in the citation analysis. Clearly, there is a major surge in interest in the application of system dynamics methods to environmental systems.

References

Bunn, D.W. and Larsen, E.R., *Systems Modeling for Energy Policy*, Wiley, Chichester, 1997.
Cannon, W., *The Wisdom of the Body*, W.W. Norton, New York, 1932.
Capra, F., *The Web of Life*, Doubleday, New York, 1996.
Cavana, R. and Ford, A., Environmental and resource systems: editors' introduction, *Sys. Dyn. Rev.*, 20(2), 2004.
Costanza, R., Ed., *Ecological Modeling* (special issue on modeling in Stella), Vol. 112, Elsevier Science, 1998.
Costanza, R. and Ruth, M., Dynamic systems modeling for scoping and consensus building, in *Sustainability and Global Environmental Policy: New Perspectives*, Dragun, A.K. and Jakonson, K.M., Eds., Edward Elgan, Cheltenham, UK, 1997.
Dawkins, R., *The Extended Phenotype*. Oxford University Press, New York, 1982.
Deaton, M. and Winebreake, J., *Dynamic Modeling of Environmental Systems*, Springer, New York, 1999.
Ford, A., *Modeling the Environment: An Introduction to System Dynamics Modeling of Environmental Systems*, Island Press, Washington, DC, 1999.
Ford, A., and Cavana, R., Eds., Special issue on environmental and resource systems, *Syst. Dyn. Rev.*, 20(2), 2004.
Forrester, J., *Industrial Dynamics*, Pegasus Communications, Waltham, MA, 1961.
Forrester, J., *World Dynamics*, Pegasus Communications, Waltham, MA, 1971.
Grant, W., Pedersen, E., Marin, S., *Ecology & Natural Resource Management: Systems Analysis and Simulation*, Wiley, New York, 1997.
Joscourt, S., Comments on 'Chaos in Daisyworld', *Tellus*, 44B, 243–346, 1992.
Joseph, L., *Gaia: The Growth of an Idea*, St. Martin's Press, New York, 1990.
Kirchner, J., The Gaia hypothesis: can it be tested? *Rev. Geophys.*, 27(2), 223–235, 1989.
Levine, L., GAIA: goddess and idea, *BioSystems*, 31, 85–92, 1993.
Lovelock, J., *The Ages of Gaia: A Biography of Our Living Earth*, W.W. Norton, New York, 1988.
Lovelock, J., *Healing Gaia*, Harmony Books, New York, 1991.
Lovelock, J., *Gaia: A New Look at Life on Earth*, Oxford University Press, New York, 1995.
Meadows, D.H., Meadows, D.L., and Randers, J., *Beyond the Limits: Confronting Global Collapse, Envisioning a Sustainable Future*, Chelsea Green, Post Mills, VT, 1992.
Meadows, D.H., Meadows, D.L., Randers, J., and Behrens, W.W., III, *The Limits to Growth*, Universe Books, New York, 1972.
Meadows, D.L., Behrens, W.W., III, Meadows, D.H., Naill, R.F., Randers, J., and Zahn, E., *Dynamics of Growth in a Finite World*, Pegasus Communications, Waltham, MA, 1974.
Meadows, D.L. and Meadows, D.H., Eds., *Toward Global Equilibrium: Collected Papers*, Pegasus Communications, Waltham, MA, 1973.
Morecroft, J., System dynamics and microworlds for policy makers, *Eur. J. Oper. Res.*, 59, 9–27, 1988.
Richardson, G., and Pugh, A., *Introduction to System Dynamics Modeling with Dynamo*, Pegasus Communications, Waltham, MA, 1981.
Senge, P., *The Fifth Discipline*, Doubleday Currency, New York, 1990.
Sterman, J.D., Teaching takes off: flight simulators for management education, *OR/MS Today*, October, 1992.

Sterman, J.D., *Business Dynamics: Systems Thinking and Modeling for a Complex World*, Irwin McGraw-Hill, Boston, 2000.

Sterman, J.D., Ed., Special issue. The global citizen: celebrating the life of Dana Meadows. *Syst. Dyn. Rev.*, 18(2), 2002.

System Dynamics Society, *System Dynamics Bibliography* (User Notes Version 03B, July 2003), System Dynamics Society, Albany, USA (available from web site: http://www.albany.edu/cpr/sds/biblio/sdbib.html), 2003.

Van Den Belt, M., *A System Dynamics Approach to Environmental Consensus Building*, Island Press, Washington, DC, 2004.

Watson, A.J. and Lovelock, J.E., Biological homeostasis of the global environment: the parable of daisy world. *Tellus Series B*, 35, 284–289, 1983.

28
Dynamic Simulation with Energy Systems Language

Clay L. Montague
University of Florida

28.1 Introduction 28-1
28.2 Reading an Energy Systems Language Diagram 28-4
 Layout of a Diagram and its Connection to Energy Theory • Symbols within the System Boundary • Stylistic Diagramming Features not Directly Involved in Equation Formulation
28.3 Translating a Diagram to Dynamic Equations 28-8
 Equation Naming Convention Used in this Chapter • Basic Equation Forms Indicated in Energy Systems Language Diagrams
28.4 Calibration of Model Constants 28-21
28.5 Preparation for Simulation 28-22
 Simulation Software and Energy Systems Language • Timescales and Numerical Integration
28.6 Dynamic Output of the Marsh Sector Model.............28-27
 Model Output Analysis • Model Validation
28.7 A Brief Comparison with Forrester's Systems Dynamics Approach 28-29
28.8 Conclusions..28-31

28.1 Introduction

Energy Systems Language is a tool for identifying self-organized environmental systems and exploring their development and possible responses to environmental change. It is a diagrammatic language not a computer language. The meaning of the resulting diagrams is so specific, however, that a simulation model can be directly derived from a diagram. The process of reading a diagram and deriving the simulation model is the main focus of this chapter. In addition, some background history and theory of the method will be mentioned along with some general guidelines for system identification through diagramming. With these tools, the reader may begin to find appropriate uses of Energy Systems Language in their own work.

At its core, Energy Systems Language incorporates both an energy-based philosophy about natural organization of systems, and a methodology for developing and testing complex hypotheses about ecosystem organization and response to change. The philosophy and methodology were developed by H.T. Odum (1924–2002), one of the most influential intellectual leaders in the field of ecology and a primary developer of the discipline of systems ecology. The philosophy includes a broad theory of self-organization of natural systems such as ecosystems and other open energy systems of humanity and nature.

No plans can be consulted for the design of self-organized systems, however, basic laws of nature must be followed. Energy is the primary natural resource for self-organized systems. A constant supply of energy is essential both to maintain each living individual as well as to obtain all other resources needed for life.

A process of natural selection is at work that replaces components that become less competitive for the energy sources with more productive ones that accomplish a similar function. As long as a constant source of energy is available, and changes are not too great and sudden, the self-organization of a natural system, once begun, is self-perpetuating.

Odum's systems philosophy and methodology therefore focused on energy. The theoretical basis combines the principles of natural selection and the laws of thermodynamics to account for the development and structure of natural systems and so their response to environmental change. Out of the theory and some massive field studies of energy flow through entire ecosystems, came complex ideas that were difficult to communicate to other ecologists at a time when few incorporated energy concepts in their research (mid-1950s).

Energy Systems Language was then developed to address the need for precise communication of (a) the likely structure of naturally organized energy systems; and (b) the consequences of a given arrangement of energetic components and connections on the response of a natural system to environmental change. The language consists of symbols that can be arranged in a diagram for precise communication of the energy and resource network that defines a given model. The methodology is accomplished through a process of diagramming a network of flows and storages, followed by numerical simulations and analyses derived directly from the diagram, given limited empirical estimates with which to calibrate the model.

An example diagram, Figure 28.1, is used in this chapter to illustrate the process. The diagram was developed originally to help describe salient features of the ecological and economic system associated with marsh estuaries of the Gulf of Mexico coast (Montague and Odum, 1997). At first glance, the diagram may seem overwhelming and cluttered, but the rules for diagram layout are strict and the symbols have precise meaning. The diagram is not primarily a visual aid. Rather, it communicates specific ideas in a systematic way. Both the choice of symbols and their layout in the diagram have specific meanings. This is the reason for calling the method a language. The symbols are simply words and sentences. The sense of the document is contained in the arrangement of these on the page. Increasing familiarity with the rules and uses of Energy Systems Language will demystify such diagrams. The process can begin with this chapter.

As can be seen in Figure 28.1, an Energy Systems Language diagram connects energy resources, plants, animals, ecological processes, people, mineral resources, economic products, and socioeconomic processes in a complex network of flow and control based on careful thought about the nature of each individual flow, storage, and interaction in the set thought to define a given system. The diagram then is an expression of a complex hypothesis of system structure that will produce specific dynamic responses to change. Compared with a verbally stated or mental model, it is a very rigorous representation of the hypothesis. The computer is required because the brain cannot fathom the consequences of the many influences thought to be operating simultaneously in the real system. Like many scientific hypotheses, complex models of poorly known self-organized systems cannot be proven correct, but they are capable of disproof. Often the disproof comes with the output from the first simulation model!

An important aspect of Energy Systems Language is the fixed relationship between a sufficiently detailed diagram and a set of differential and algebraic equations that describe dynamic change in all storages and flows. The unique feature of the approach is the use of relatively few types of control and storage processes to represent a complex system by arranging them in a highly specific, information rich diagram. Accordingly, a few basic types of equation are sufficient to represent most of the individual flows, storages, and interactions in an Energy Systems Language diagram. The diagram consists of a complex arrangement of relatively simple symbols that corresponds to a complex system of relatively simple equations.

In this chapter, a portion of Figure 28.1 (the Marsh sector) will be translated into equations and simulated to illustrate this process and the highly dynamic output that can result even in a constant environment. *A caveat*: The experienced user of Energy Systems Language has intimate knowledge of the equations implied by the diagram, and chooses to draw subtle details in a certain way so that the proper process occurs. Because a sufficiently detailed diagram can be translated nearly by rote, beginners are often encouraged to draw a diagram and see what simulation model it produces. This speeds up the process of getting interesting model output to discuss; however, it postpones the need to recognize how to represent known processes adequately by combining model components appropriately. Experimentation

Dynamic Simulation with Energy Systems Language

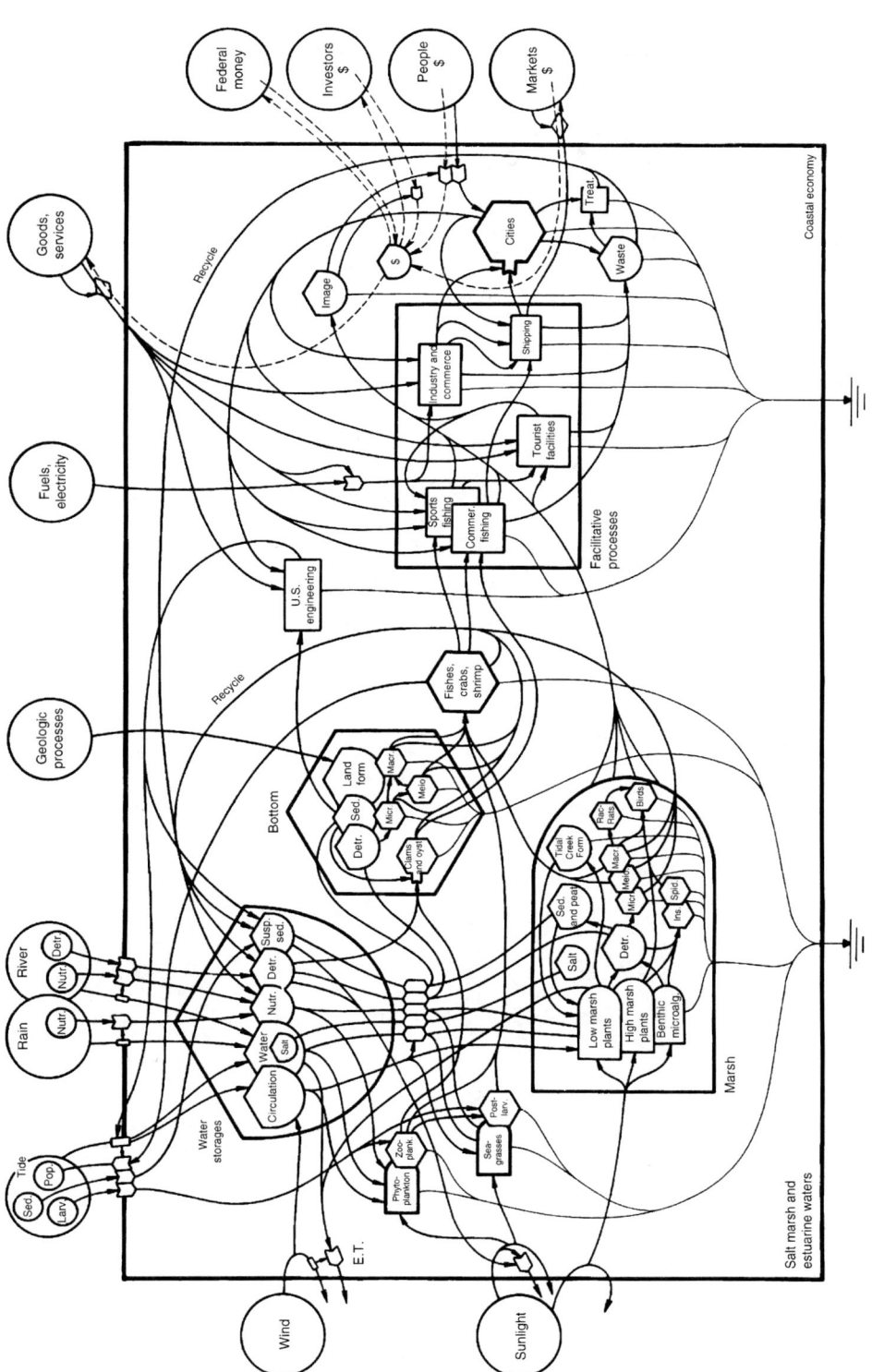

FIGURE 28.1 Energy Systems Language diagram of a marsh estuary system of the Gulf of Mexico coast of the United States (Reproduced with permission from Montague, C.L. and H.T. Odum, 1997. Introduction: The intertidal marshes of Florida's Gulf coast. In Coultas, C.L. and Y.-P. Hsieh (eds.), *Ecology and Management of Tidal Marshes: A Model from the Gulf of Mexico*. St Lucie Press, Delray Beach, Florida, pp. 1–7, 355pp.).

with various model structures is helpful, along with advanced reading on the subject. A good introduction to the approach along with many simple model examples can be found in Odum and Odum (2000). The most complete description of the behavior of combinations of components is in Odum (1983, revised and retitled as Odum, 1994).

28.2 Reading an Energy Systems Language Diagram

The diagram in Figure 28.1 includes the nine most commonly used symbols of Energy Systems Language. For reference, these symbols are listed in Figure 28.2. Lines throughout the diagram connect various symbols to one another. Flows of energy and materials are represented by solid lines. Dashed lines represent flows of money. Money always flows in the opposite direction of the products and sources it purchases.

Reading the diagram involves the symbols, their combination in connected pairs, and the overall layout scheme that together provide a wealth of information about the system under study. Complex Energy Systems Language diagrams, such as Figure 28.1, communicate several levels of detail about system organization. Translating the diagram involves the finest level of detail. At the highest level, a sense of system self-organization is imparted. Reading Energy Systems Language diagrams is easier once major features of diagram layout are recognized. With experience, the richness of the diagram becomes more apparent, while at the same time, the complexity of it seems less daunting.

28.2.1 Layout of a Diagram and its Connection to Energy Theory

The layout process imparts distinctive left–right, up–down, clockwise, and counterclockwise facets of meaning to the diagram. Lines of energy and material flow generally from left to right and from up to

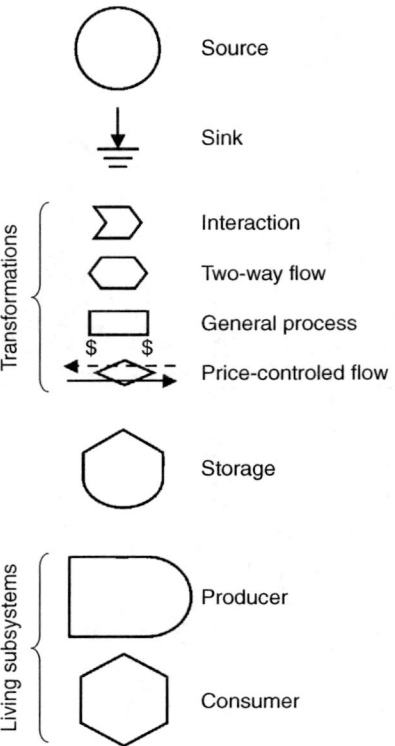

FIGURE 28.2 The nine most common symbols used in Energy Systems Language diagrams.

down. Toward the center and further right, however, some lines bend upward and back to the left creating a counterclockwise impression of flow. Conversely, dashed lines representing money flows enter the system generally from the right and bend upward and back to the right in a clockwise manner.

In the theory behind Energy Systems Language, the primary function of a persistent self-organizing system is to combine and upgrade energy and material sources into products that reproduce and maintain its lasting components. The products include the living components and concentrated stores of nonliving resources. The processes of upgrading and maintaining the system components use large amounts of energy. Most of the energy that enters the system is consumed and degraded into relatively dilute heat no longer able to do useful work. Relatively little of the energy is left behind in the form of upgraded products. Some of the upgraded products leave the system, either by leakage, migration, or by export in trade for high-quality inputs.

Diagram layout incorporates the distinction between energy sources, upgraded energy, and downgraded energy. The diagram is drawn so that sources enter each symbol from the left or top, and upgraded products exit to the right. Upgraded products also cross the right side of the system boundary. Lines representing heat losses, however, leave the bottom of each energy-using component and terminate in a heat sink symbol outside the lower boundary. In general, the lines in the diagram that curve continuously downward are heat losses. In Figure 28.1, these lines converge at one of two heat sinks at the bottom of the diagram.

Diagram layout itself is not critical to equation translation, but knowing the scheme helps to organize the process. Understanding the overall principle behind the layout allows the diagram to be read and studied at telescoping levels of detail. For those fluent in Energy Systems Language and indoctrinated into its theory, a diagram can begin to impart a sense of system self-organization and of the response to be expected with a given change.

28.2.1.1 Transformity Principle of Diagram Layout

The concept of *transformity*—central to Odum's energy systems theory of self-organization—provides the principle for layout of Energy Systems Language diagrams. Essentially, transformity is the amount of basic source energy (usually sunlight) used in the necessary network of flows and transformations to place a single unit of source energy into a given upgraded product. Each component of the network depicted in Figure 28.1, for example, has its own transformity, which can be calculated from amounts of the various sources that ultimately lead to it. To do this, each source must be weighted by its own transformity.

Each source represented by a circle is itself a product of energetic processes that occur in a network. This network must be identified to determine the transformity of the source. The progression then extends to the sources for the source until ultimately each can be traced back to primary sources for all processes on Earth (sunlight, the Earth's internal heat, and the planetary motions that create tides, seasons, etc.).

Transformity analysis is beyond the scope of this chapter, but it is a second major use of Energy Systems Language diagrams. The procedures are described in detail in Odum (1996).

28.2.1.2 Arrangement of Sources in Order of Transformity

In Energy Systems Language diagrams, circles are arranged outside the sides and top of the system boundary in order of the transformity of the indicated source. The arrangement is clockwise beginning with sunlight at the lower left, which, as the basis for the approach, has a transformity of 1. All other sources have higher transformities. In Figure 28.1, the next one is wind energy, a joule of which, according to transformity analysis, requires on the order of 1500 J of sunlight.

Third is tide, which is a primary energy source. Part of transformity analysis requires establishing a sunlight energy equivalence for the work of tide. Extrapolation from the solar tide would not yield the required equivalence to light energy; however, sunlight energy can lift water through evaporation, which is the usual basis for establishing a sunlight equivalence for tide.

In any case, the layout of sources continues clockwise. Those above the middle of Figure 28.1 change from natural energy sources of high transformity (geological processes that form land) to sources that have been transformed into commodities in part through human work. In Figure 28.1 these sources include electricity, fuels, goods and services, and others arranged around the diagram. In Figure 28.1, the last source is labeled markets. It has such a high transformity because markets are based largely on information. According to transformity analysis, useful information is the highest transformity product of society (Odum, 1996).

28.2.1.3 Overall Effect of Transformity on Diagram Organization

Inside the diagram, the various symbols are arranged according to their proximity to the resources needed for their production. Given the layout of sources and the rules of connecting flows to symbols, a progression occurs from left to right across the diagram that reflects a general increase in transformity of the various components as the number of preceding transformations increases, and higher transformity sources are used. The layout procedure results in nonhuman processes and the products of nature generally appearing to the left side of a diagram, while those involving greater amounts of human labor and education are further to the right.

High transformity components of systems are often involved with the production of lower transformity products, thereby forming positive feedback loops. Animals, for example, regenerate nutrients that influence rates of plant production, which ultimately provides more food for more animal production. Likewise, people farm foods and direct natural resources for their own use. Feedback from higher transformity components to lower ones is distinguished by the counterclockwise return flows that sweep upward and back to the left in the diagram.

28.2.2 Symbols within the System Boundary

Two categories of symbol are used in Energy Systems Language: basic symbols and composite symbols. Composite symbols represent commonly occurring sets of basic symbols. The two most common composite symbols are for living components: the bullet-shaped symbol for primary producers and the hexagon for consumers.

The basic symbols include the circles that denote sources and the heat sinks that are generally shown outside the system boundaries. Five other basic symbols occur within the system boundary and within various composite symbols. These include the storage symbol (reminiscent of a water storage tank), and the five transformation symbols listed in Figure 28.2.

In Figure 28.1 storages include water, salt, nutrients, detritus, circulation of water, land formations, coastal image, money, and waste. A storage is part of the two composite symbols as well. A total of 43 separate storages are shown in Figure 28.1. The amount of material or energy within each storage changes according to the net effect of all inflows and outflows connected to them.

A variety of symbols represent specific energy transformation processes that have standard mathematical representations. The most common one is the interaction symbol shown in Figure 28.2. Figure 28.1 also includes several two-way flows, a price-controlled interaction, and several rectangular general process symbols. The general process remains unspecified in the diagram. It may be a simple transformation, or a composite. In any case, the rectangle is the only transformation symbol that requires further specification than appears in the diagram before it can be used in an equation.

When the interaction symbol is used, the primary source of energy to be used in the transformation process enters from the left and the product exits from the right and usually enters a storage symbol. Smaller amounts of upgraded energy from elsewhere in the system enter the symbol from the top. These assist or subsidize the transformation. In this way, the main energy source appears to pass through the interaction, while the interacting subsidy appears to control or assist the rate of transformation.

By far the most frequently used function associated with the interaction symbol is multiplication. Multiplication is so common that an unspecified interaction is assumed to be multiplicative. Otherwise, a mathematical character, such as a division sign, will appear within the interaction symbol to indicate its function. The rationale for this frequent use stems from its similarity to the second-order reaction kinetics in physical chemistry. A rate of reaction of two chemicals is often adequately represented by the multiple of the concentrations of the two reactants times a rate constant. The rate declines as the reactants are converted to products. As in chemical kinetics, multiplication is thought to adequately represent many kinds of energy transformation that combine several necessary forms of energy and materials.

All of the flows entering and leaving the interaction symbol have the same equation form. This applies to the formation rate of the product, the loss of raw materials from the main source and the assisting variables, and heat loss associated with the transformation. The equation form consists of the indicated combination of source and assisting variable multiplied by a constant. Only the value of the constant will differ for the associated flows.

When two or more variables assist a transformation, adjacent interaction symbols will occur in series, and the terms of the equation are built accordingly. One multiplicative interaction gives a rate equation with a second-degree term. Likewise, a series of three multiplicative interactions will produce a fourth-degree term. Again all the reactant and product flows will have the same equation form.

A two-way interaction symbol denotes a flow of energy and materials back and forth created by an interacting source of energy. The tide is such an energy source in Figure 28.1. The tide moves water and the materials it contains between the marsh and the estuary. The rate of movement depends also on the relative amounts of material stored on each side of the interaction relative to an equilibrium ratio, and on various fixed conditions that along with the tide determine the rate at which equilibrium can be restored. The specific equation structure is given in the section on equation writing.

The price-controlled flow is a symbol that converts a flow of purchased products to a flow of money. The flow of product may be caused by a variety of events. The price controls the flow of money in return. The supply of money and products may also feed back and influence the price. When the price is so influenced, lines will enter the price-controlled flow from the top and the details of the equation must be specified. Often the effect on price is simply proportional to the supply of product or to the ratio of product and available money. Figure 28.1 shows a situation where the markets outside the system influence the price. If no explicit price controls are specified, a fixed price is implied that will be represented in the model by a constant.

The rectangle is reserved for unspecified processes. This symbol can represent a single transformation equation, or a combination of symbols. Rectangles are used to reduce clutter when the process involved is complex. In any case, the detail must be provided before a simulation model can be built. In Figure 28.1, the rectangle is used for many of the processes associated with the economic system on the right side of the diagram.

Rectangles and price controls are the only symbols for which no direct translation to equations is possible. Before simulation can be done, the process within each rectangle must be represented either in a special diagram of its own, or by a suitable input–output relationship already in equation form (a regression equation, for example).

A variety of other symbols were introduced by Odum over the years to represent specific recurring processes found in open energy systems; however, these are not as often used as those listed in Figure 28.2. Among others, they include ratio interactions, switches, backflow interactions, and saturation limits; all are described in Odum (1983, 1994).

28.2.3 Stylistic Diagramming Features not Directly Involved in Equation Formulation

Certain features of Energy Systems Language diagrams are primarily stylistic, but help organize like components, indicate relative importance, and reduce clutter. These include the use of different sizes for symbols and line thickness, grouping related symbols within sector boundaries, and stacking or clustering related symbols. Symbol size and line thickness indicate the relative sizes of various storages and flows of upgraded materials. Lines of degraded energy are often thinner than other lines even though most of the energy travels through them. This is to reduce clutter and provide an additional way to distinguish an upgrade from a downgrade in a transformation process.

28.2.3.1 Sector Boundaries

Sector boundaries are lines that encompass a group of related symbols within the system boundary. Sector boundaries can be rectangles or, when the overall effect of the sector on the larger system is similar to that of a given symbol, sector boundaries may take the shape of the relevant symbol.

Several sectors are identified by boundaries in Figure 28.1. On the right side of Figure 28.1, a grouping labeled "Facilitative processes" is outlined with a rectangle. The sector contains some of the infrastructure of the coastal economic system. Near the top of the left side of Figure 28.1 is a group of "Water storages" enclosed in a sector boundary resembling a storage tank. Below that is a sector called "Marsh." Because in general more organic matter is produced in coastal marshes than is consumed there, the sector boundary is bullet shaped like the primary producer symbol. Finally in Figure 28.1 is a hexagonal sector boundary labeled "Bottom" because the living organisms and biochemical reactions within the estuarine bottom consume more organic matter than they produce.

28.2.3.2 Nested Symbols

Sometimes basic symbols contain smaller and differently labeled symbols of the same type inside them. Nested within the tide source in Figure 28.1, for example, are source symbols for larvae, adult animal populations, and sediments that move into the coastal system from nearshore waters with the tide. The rain and river symbols contain source symbols for nutrients and detritus that are delivered in significant amounts along with the freshwater that enters the system.

In the water storage sector, the symbol for the water storage itself includes a storage of salt. The sector boundary already points out the collection of important constituents that it contains. Salt is treated differently to highlight an important difference in its effect from the other constituents. The saltiness of coastal water is represented by its salinity. Production in estuaries is often higher at lower salinity. The nesting of salt within the water represents salinity.

28.2.3.3 Vertical and Horizontal Stacks of Symbols

Throughout Figure 28.1 are symbols that touch or partially overlap one another. The proximity suggests tight coupling of the symbols. In the marsh sector, for example, is a vertical stack of three primary producer symbols. The stack shows that all the three are influenced by the set of five variables entering the top of the stack. The order in the stack is also important. Low and high refer to elevation relative to sea level. By being closer to the estuarine water, the low marsh plants receive the effects of the water constituents first and then the high marsh plants.

At the bottom of the stack, the smaller size of the symbol for benthic microalgae indicates their lower biomass and the inset of the symbol under the other two indicates their existence in the shade of the other plants. This latter issue does affect the primary production equation for benthic microalgae. The part of the diagram that dictates this for the equation is found on the line that represents the flow of sunlight energy to primary producers. The benthic microalgae receive light last. This will be clarified below when the marsh sector is redrawn for equation writing.

Also in the marsh sector is an illustration of a horizontal stacking of the community of detritivores. In this case, the stack means that all three consume detritus, meiofauna also consume microbes, and macrofauna, which are the last consumers in the stack, consume all three. To write equations from a diagram requires that such details be explicitly shown. The flow lines were left out to reduce clutter.

28.2.3.4 Converging and Diverging Flow Lines

Figure 28.1 includes a variety of convergent and divergent flow lines to reduce clutter. The line labeled recycle, for example, originates from the three components of the detritivore community in the marsh sector. Three separate lines converge into one at the marsh sector boundary. Divergent lines are also used in Figure 28.1. A single line from the source labeled "Goods, Services" splits into several as it crosses the right side of the upper system boundary.

28.3 Translating a Diagram to Dynamic Equations

From a sufficiently detailed diagram, a mathematical system of first-order, nonlinear differential equations can be derived. The set of differential equations represents the net rate of change over time for each material and energy storage in the model system. The rates are based on cause and effect influences from the storages themselves and various environmental inputs to the system. The equations are solved on a computer to reveal the dynamic patterns produced in each model variable.

The set of equations that can be developed from an Energy Systems Language diagram will at a minimum consist of one equation for each flow and one integrating equation for each storage that combines the flows into and out of it. In total, Figure 28.1 contains 43 storages and nearly 120 explicitly represented flows, plus a number of flows implied in the stacking of symbols. A complete translation of Figure 28.1 for simulation would have close to 200 equations for dynamic variables, and another 200 that specify inputs, constants, and initial conditions. Figure 28.1, however, is incompletely specified and the resulting model would provide an unwieldy demonstration at best! So to illustrate the process, a portion of the diagram (the marsh sector) is redrawn, the corresponding equations derived, and the resulting model simulated.

Equation writing is easier than it may appear from a complex diagram, especially one that includes all necessary details. However, only a few types of equation are actually needed. Like the repeated use of the same symbols in the diagram, the same equation form is repeated many times.

Equations are based upon symbol connections in Energy Systems Language diagrams. The individual symbols provide only a part of the equation. In general, a combination of two components connected by a line are necessary to determine the form of each equation used.

The necessary details for deriving model equations for the Marsh sector are given in the diagram shown in Figure 28.3. A close examination of the marsh sector in Figure 28.1 will reveal a few important differences. First, the effect of Tidal Creek Form is excluded because the intended effect is not clearly understood. Second, the line from high marsh plants to birds is misplaced in Figure 28.1. In the revised

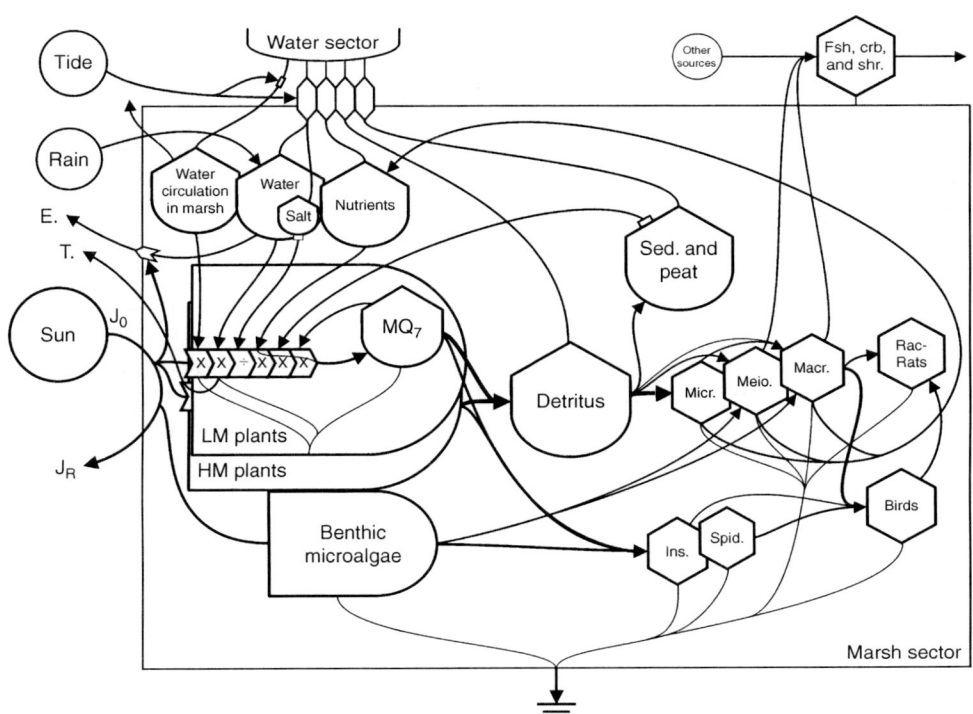

FIGURE 28.3 Detailed Energy Systems Language diagram of the marsh sector of the coastal system diagram. The diagram is sufficiently detailed for the derivation of model equations.

marsh sector diagram of Figure 28.3, a line from high marsh plants goes to detritus and another line from insects to birds is added. The line from rats and raccoons to birds is also reversed (rats and raccoons eat birds eggs but few rats and raccoons are eaten by birds in the marsh). Finally, a missing line is added from salt to the primary producers.

Only the low-elevation portion of the marsh sector has been translated into equations and simulated in this chapter. The equation list for this simulation is given in Table 28.1. Equations for the high marsh portion would differ little other than in the calibration of constants to reflect high marsh conditions. The correspondence between equations in Table 28.1 and the symbols on the diagram (Figure 28.3) will become apparent by cross-referencing the diagram labels with the column of Table 28.1 labeled "Meaning."

The 16 storages involved in the low marsh subsector are listed first in Table 28.1 (Eqs. [001]–[016]). These are followed by the rate equations, then model inputs, model constants, and the time parameters used in the simulation. A total of 141 entries are made in Table 28.1.

28.3.1 Equation Naming Convention Used in this Chapter

The equation list uses a naming convention for variables that helps to track the relationship among model variables. In general, storage i is indicated by Q_i and rates of flow from storage j to storage k by J_{jk}. Marsh variables are preceded by an M and estuarine water variables by a W. For example, MQ_7 represents the marsh plants (Eq. [007]), and WQ_4 the nutrients in estuarine water (Eq. [077]). MJ_{71} represents the flow from plants to detritus (MQ_1) in the marsh (Eq. [048]). J_{W4M4} represents the two-way interaction (upper left of the diagram in Figure 28.3) that defines exchange of nutrients between the water and the marsh. By listing W_4 as the first subscript, the positive direction is toward the marsh for this two-way nutrient exchange. Equally, a negative flow represents export of nutrients from the marsh.

Constants in rate equations are indicated by a K. An A denotes an auxiliary variable that must first be calculated to assign a value to a flow rate. Constants and auxiliaries are prefixed and subscripted like the rate to which they apply. MA_{20}, for example, denotes the light available for evaporation of water (Eq. [031]), which is used to compute evaporation (MJ_{20}, Eq. [040]). The latter equation also uses the proportionality constant denoted as MK_{20}.

A subscript of 0 indicates a location outside the system boundary. A flow from a source outside the boundary to a given storage i is labeled J_{0i}. Sunlight from outside the system is assimilated by benthic microalgae (MQ_9), for example, so their rate of gross primary production is labeled MJ_{09} (Eq. [034]).

Conversely, flows that leave the system from storage i are denoted by J_{i0}. The flow MJ_{a0}, for example, identifies the metabolic heat loss from microbes (MQ_a). The equation number for MJ_{a0} is 054 in Table 28.1. Note that the constant in the equation, MK_{a0}, has the same subscript. If more than one flow from a given source connects to the same sink, the flows are distinguished by a third character appended to the name. Evaporation and transpiration, which are two ways in which water leaves the marsh system boundary, are represented by MJ_{20} (Eq. [040]) and MJ_{207} (Eq. [041]), respectively. The added 7, in this case, indicates plants (MQ_7) that are involved in the rate. Additional characters are also used to distinguish the same flow converted to different units. Burial of detritus in sediment requires energy units when subtracted from the detritus storage as MJ_{15} (Eq. [035]), but requires mass units when added to the sediment as MJ_{15D} (Eq. [036]).

Likewise, if more than one auxiliary variable is calculated for the same flow rate, a third subscript is used. To calculate gross primary production of low marsh plants (MJ_{07}, Eq. [033]), for example, requires calculation of available light (MA_{07a}, Eq. [027]), and a multiplier based on other environmental conditions (MA_{07b}, Eq. [028]).

Model constants associated with a flow have the same subscript as the flow. For example, the equation for assimilation of meiofauna by macrofauna, MJ_{bc}, has a constant MK_{bc} (Eq. [060]). As before, where a given equation has more than one constant, a third character is appended to distinguish the two. The equation for the two-way exchange of sediment between estuary and marsh (J_{W6M5}, Eq. [025]) has a rate constant K_{W6M5} for restoring equilibrium, and an equilibrium constant K_{W6M5n}.

Dynamic Simulation with Energy Systems Language

TABLE 28.1 Equations Corresponding to the Energy Systems Language Diagram of the Marsh Sector (Low Marsh Portion). The Table is Organized into Five Sections: Storage Equations, Rates, Inputs, Model Constants, and Simulation Time Parameters

No.	Storage Equation[a]	Units	Meaning	Used in Eq. No.
(001)	MQ1 = INTEG(JW5M1+MJ71−MJ15−MJ1a−MJ1b−MJ1c, Initial value: 9.18000e+014)	J	Organic detritus in marsh	(024), (035), (036), (037), (038), and (039)
(002)	MQ2 = INTEG(JRain+JW2M2−MJ20−MJ207, Initial value: 5.43750e+006)	kg-H$_2$O	Water in marsh	(021), (027), (028), (030), (031), and (040)
(003)	MQ3 = INTEG(JW3M3, Initial value: 2.56217e+005)	kg-salt	Salt in marsh	(022), (028), and (030)
(004)	MQ4 = INTEG(JW4M4+MJa4+MJb4+MJc4−MJ47−MJ49, Initial value: 2.52000e+004)	kg-N	Inorganic nutrients in marsh	(023), (028), and (030)
(005)	MQ5 = INTEG(JW6M5+MJ15D, Initial value: 1.59030e+010)	kg-sediment	Sediment and peat in marsh	(025)
(006)	MQ6 = INTEG(MJ06−MJ60−MJ607, Initial value: 1.47000e+011)	J	Water circulation energy in marsh	(028), (030), (044), and (045)
(007)	MQ7 = INTEG(MJ07−MJ70−MJ706−MJ71−MJ7e, Initial value: 9.18000e+014)	J	Low marsh plants	(027), (031), (033), (045), (046), (048), and (049)
(008)	MQ9 = INTEG(MJ09−MJ90−MJ9b−MJ9c−MJ9e, Initial value: 2.52000e+012)	J	Benthic microalgae	(029), (034), (050), (051), (052), and (053)
(009)	MQa = INTEG(MJ1a−MJa0−MJab−MJac, Initial value: 2.52000e+011)	J	Microbes on detritus	(037), (054), (056), and (057)
(010)	MQb = INTEG(MJ1b+MJ9b+MJab−MJb0−MJbc−MJbFCS, Initial value: 2.52000e+012)	J	Meiofauna in marsh	(038), (051), (056), (058), (060), and (061)
(011)	MQc = INTEG(MJ1c+MJ9c+MJac+MJbc−MJc0−MJcd−MJcFCS−MJcg, Initial value: 2.52000e+012)	J	Macrofauna in marsh	(039), (052), (057), (060), (062), (064), (065), and (066)
(012)	MQd = INTEG(MJcd+MJgd−MJd0, Initial value: 1.26000e+012)	J	Raccoons and rats in marsh	(064), (067), and (074)
(013)	MQe = INTEG(MJ7e+MJ9e−MJe0−MJef−MJeg, Initial value: 2.52000e+012)	J	Insects in marsh	(049), (053), (068), (069), and (070)
(014)	MQf = INTEG(MJef−MJf0−MJfg, Initial value: 2.52000e+011)	J	Spiders in marsh	(069), (071), and (072)
(015)	MQg = INTEG(MJcg+MJeg+MJfg−MJg0−MJgd, Initial value: 2.52000e+012)	J	Birds in marsh	(066), (070), (072), (073), and (074)
(016)	QFCS = INTEG(J0FCS+MJbFCS+MJcFCS−JFCS0−JYield, Initial value: 4.83000e+013)	J	Fishes, crabs, and shrimps in marsh creeks and estuary	(019), (061), and (065)

(*Continued*)

TABLE 28.1 Continued

No.	Rate (or Auxiliary) Equation	Units	Meaning	Used in Eq. No.
(017)	J0FCS = K0FCS	J/y	Nonmarsh food assimilation by fishes, crabs, and shrimps	(016)
(018)	J0LM = 1.42009e+017	J/y	Light influx	(027) and (031)
(019)	JFCS0 = KFCS0 * QFCS	J/y	Metabolic heat losses by fishes, crabs, and shrimps	(016)
(020)	Jrain = Rain	kg-H$_2$O/y	Rain on marsh	(002)
(021)	JW2M2 = Tide * KW2M2 * (WQ2 − KW2M2n * MQ2)	kg-H$_2$O/y	Water exchange with estuary	(002)
(022)	JW3M3 = KW3M3 * (WQ3 − KW3M3n * MQ3)	kg-salt/y	Salt exchange with estuary	(003)
(023)	JW4M4 = KW4M4 * (WQ4 − KW4M4n * MQ4)	kg-N/y	Inorganic nutrient exchange with estuary	(004)
(024)	JW5M1 = KW5M1 * (WQ5 − KW5M1n * MQ1)	J/y	Organic detritus exchange with estuary	(001)
(025)	JW6M5 = KW6M5 * (WQ6 − KW6M5n * MQ5)	kg-sediment/y	Sediment exchange with estuary	(005)
(026)	Jyield = Yield	J/y	Catch of fishes, crabs, and shrimps in estuary and creeks	(016)
(027)	MA07a = J0LM / (1 + (MK07 * MA07b * MQ7) + (MK20E * MK20 * MQ2))	J/y	Light influx available to low marsh plants	(029) and (033)
(028)	MA07b = (MQ2 * MQ4 * MQ6)/MQ3	Complex	Environmental conditions for low marsh plant photosynthesis	(027), (031), and (033)
(029)	MA09a = MA07a / (1 + (MK09 * MA09b * MQ9))	J/y	Light influx available to benthic microalgae	(034)
(030)	MA09b = (MQ2 * MQ4 * MQ6)/MQ3	Complex	Environmental conditions for benthic microalgal photosynthesis	(034) and (029)
(031)	MA20 = J0LM / (1 + (MK07 * MA07b * MQ7) + (MK20E * MK20 * MQ2))	J/y	Light influx available to evaporate water from marsh	(040)
(032)	MJ06 = Tide	J/y	Influx of tidal energy to marsh	(006)
(033)	MJ07 = MK07 * MA07b * MQ7 * MA07a	J/y	Gross primary production by low marsh plants	(007), (041), (042), and (047)
(034)	MJ09 = MK09 * MA09b * MQ9 * MA09a	J/y	Gross primary production by benthic microalgae	(008) and (043)
(035)	MJ15 = MK15 * MQ1	J/y	Burial of detritus in sediment (in energy units)	(001)
(036)	MJ15D = MK15D * MQ1	kg-sediment/y	Burial of detritus in sediment (in sediment units)	(005)
(037)	MJ1a = MK1a * MQ1 * Mqa	J/y	Assimilation of detritus by microbes	(001) and (009)
(038)	MJ1b = MK1b * MQ1 * Mqb	J/y	Assimilation of detritus by meiofauna	(001) and (010)
(039)	MJ1c = MK1c * MQ1 * Mqc	J/y	Assimilation of detritus by macrofauna	(001) and (011)
(040)	MJ20 = MK20 * MQ2 * MA20	kg-H$_2$O/y	Evaporation of water in marsh	(002)
(041)	MJ207 = MK207 * MJ07	kg-H$_2$O/y	Transpiration of water by low marsh plants	(002)
(042)	MJ47 = MK47 * MJ07	kg-N/y	Uptake of inorganic nutrients by low marsh plants	(004)
(043)	MJ49 = MK49 * MJ09	kg-N/y	Uptake of inorganic nutrients by benthic microalgae	(004)
(044)	MJ60 = MK60 * MQ6	J/y	Baseline dissipation of water circulation energy in marsh	(006)
(045)	MJ607 = MK607 * MQ6 * MQ7	J/y	Dissipation of circulation energy by low marsh plants	(006)
(046)	MJ70 = MK70 * MQ7	J/y	Baseline metabolic heat losses by low marsh plants	(007)
(047)	MJ706 = MK706 * MJ07	J/y	Metabolic heat losses by low marsh plants due to production processes	(007)

No.	Inputs		Meaning	Used in Eq. No.
(048)	MJ71 = MK71 * MQ7	J/y	Death of low marsh plant tissue and flow to detritus	(001) and (007)
(049)	MJ7e = MK7e * MQ7 * MQe	J/y	Assimilation of low marsh plants by insects	(007) and (013)
(050)	MJ90 = MK90 * MQ9	J/y	Metabolic heat losses by benthic microalgae	(008)
(051)	MJ9b = MK9b * MQ9 * MQb	J/y	Assimilation of benthic microalgae by meiofauna	(008) and (010)
(052)	MJ9c = MK9c * MQ9 * MQc	J/y	Assimilation of benthic microalgae by macrofauna	(008) and (011)
(053)	MJ9e = MK9e * MQ9 * MQe	J/y	Assimilation of benthic microalgae by macrofauna	(008) and (013)
(054)	MJa0 = MKa0 * MQa	J/y	Metabolic heat losses by detritus microbes	(009) and (055)
(055)	MJa4 = MKa4 * MJa0	kg-N/y	Nutrient regeneration by detritus microbes	(004)
(056)	MJab = MKab * MQa * MQb	J/y	Assimilation of microbes by meiofauna	(009) and (010)
(057)	MJac = MKac * MQa * MQc	J/y	Assimilation of microbes by macrofauna	(009) and (011)
(058)	MJb0 = MKb0 * MQb	J/y	Metabolic heat losses by meiofauna	(010) and (059)
(059)	MJb4 = MKb4 * MJb0	kg-N/y	Nutrient regeneration by meiofauna	(004)
(060)	MJbc = MKbc * MQb * MQc	J/y	Assimilation of meiofauna by macrofauna	(010) and (011)
(061)	MJbFCS = MKbFCS * MQb * QFCS	J/y	Assimilation of meiofauna by fishes, crabs, and shrimps	(010) and (016)
(062)	MJc0 = MKc0 * MQc	J/y	Metabolic heat losses by macrofauna	(011) and (063)
(063)	MJc4 = MKc4 * MJc0	kg-N/y	Nutrient regeneration by macrofauna	(004)
(064)	MJcd = MKcd * MQc * MQd	J/y	Assimilation of macrofauna by raccoons and rats	(011) and (012)
(065)	MJcFCS = MKcFCS * MQc * QFCS	J/y	Assimilation of macrofauna by fishes, crabs, and shrimps	(011) and (016)
(066)	MJcg = MKcg * MQc * MQg	J/y	Assimilation of macrofauna by birds	(011) and (015)
(067)	MJd0 = MKd0 * MQd	J/y	Metabolic heat losses by raccoons and rats	(012)
(068)	MJe0 = MKe0 * MQe	J/y	Metabolic heat losses by insects	(013)
(069)	MJef = MKef * MQe * MQf	J/y	Assimilation of insects by spiders	(013) and (014)
(070)	MJeg = MKeg * MQe * MQg	J/y	Assimilation of insects by birds	(013) and (015)
(071)	MJf0 = MKf0 * MQf	J/y	Metabolic heat losses by spiders	(014)
(072)	MJfg = MKfg * MQf * MQg	J/y	Assimilation of spiders by birds	(014) and (015)
(073)	MJg0 = MKg0 * MQg	J/y	Metabolic heat losses by birds	(015)
(074)	MJgd = MKgd * MQd * MQg	J/y	Assimilation of birds eggs by raccoons and rats	(012) and (015)

No.	Inputs	Units	Meaning	Used in Eq. No.
(075)	Rain = 4.5e+010	kg-H$_2$O/y	Annual rainfall on marsh	(020)
(076)	Tide = 1.03752e+014	J/y	Tidal energy influx to marsh	(021) and (032)
(077)	WQ2 = 1e+011	kg-H$_2$O	Water in estuary and tidal creeks	(021)
(078)	WQ3 = 2.5641e+009	kg-salt	Salt in estuary and tidal creeks	(022)

(Continued)

TABLE 28.1 Continued

No.	Model constants	Units	Meaning	Used in Eq. No.
(079)	WQ4 = 5600	kg-N	Nutrients in estuary and tidal creeks	(023)
(080)	WQ5 = 5.1e+013	J	Organic detritus in estuary and tidal creeks	(024)
(081)	WQ6 = 7e+006	kg-sediment	Sediment in estuary and tidal creeks	(025)
(082)	K0FCS = 2.898e+012	J/y	Nonmarsh food assimilation by fishes, crabs, and shrimps	(017)
(083)	KFCS0 = 0.38375	1/y	Fraction of biomass energy used per year in metabolic processes by fishes, crabs, and shrimps	(019)
(084)	KW2M2 = 5.86735e−014	1/y	Rate constant for water exchange with estuary	(021)
(085)	KW2M2n = 18934.6	dmls	Equilibration constant for water exchange with estuary	(021)
(086)	KW3M3 = 3.04375	1/y	Rate constant for salt exchange with estuary	(022)
(087)	KW3M3n = 10007.5	dmls	Equilibration constant for salt exchange with estuary	(022)
(088)	KW4M4 = 8.11667	1/y	Rate constant for inorganic nutrient exchange with estuary	(023)
(089)	KW4M4n = −1.09205	dmls	Equilibration constant for inorganic nutrient exchange with estuary	(023)
(090)	KW5M1 = 1	1/y	Rate constant for organic detritus exchange with estuary	(024)
(091)	KW5M1n = 0.0555556	dmls	Equilibration constant for organic detritus exchange with estuary	(024)
(092)	KW6M5 = 1.33333	1/y	Rate constant for sediment exchange with estuary	(025)
(093)	KW6M5n = 0.000492821	dmls	Equilibration constant for sediment exchange with estuary	(025)
(094)	MK07 = 1.1316e−033	complex	Rate constant for primary production by low marsh plants	(027), (031), and (033)
(095)	MK09 = 1.03014e−031	complex	Rate constant for primary production by benthic algae	(029) and (034)
(096)	MK15 = 0.0206749	1/y	Rate constant for detritus burial in sediment (detritus outflow)	(035)
(097)	MK15D = 1.21617e−009	1/y	Rate constant for detritus burial in sediment (sediment inflow)	(036)
(098)	MK1a = 2.37925e−012	1/(J-y)	Rate constant for assimilation of detritus by microbes	(037)
(099)	MK1b = 2.18782e−014	1/(J-y)	Rate constant for assimilation of detritus by meiofauna	(038)
(100)	MK1c = 5.46954e−015	1/(J-y)	Rate constant for assimilation of detritus by macrofauna	(039)
(101)	MK20 = 9.51857e−014	1/J	Fraction of water evaporated per unit of available light energy	(027), (031), and (040)
(102)	MK207 = 1.26752e−005	kg-H$_2$O/J	Water transpired per unit of energy assimilated by low marsh plants	(041)
(103)	MK20E = 1.36889e+007	J/kg-H$_2$O	Energy required to evaporate water	(027) and (031)
(104)	MK47 = 2.25e−008	kg-N/J	Inorganic nutrients assimilated per unit energy assimilated by low marsh plants	(042)
(105)	MK49 = 2.5e−008	kg-N/J	Inorganic nutrients assimilated per unit energy assimilated by benthic microalgae	(043)
(106)	MK60 = 635.217	1/y	Fraction of water circulation energy dissipated per unit time	(044)
(107)	MK607 = 7.68842e−014	1/(J-y)	Fraction of circulation energy dissipated year per unit of low marsh plants	(045)
(108)	MK70 = 0.773471	1/y	Fraction of biomass energy used in metabolic processes by low marsh plants	(046)
(109)	MK706 = 0.005	dmls	Fraction of low marsh plant production lost as heat	(047)
(110)	MK71 = 0.689162	1/y	Fraction of low marsh plant biomass that dies per year	(048)
(111)	MK7e = 3.03863e−014	1/(J-y)	Fraction of low marsh plants consumed per year per unit insect	(049)

Dynamic Simulation with Energy Systems Language

No.		Units	Meaning	
(112)	MK90 = 69.0035	1/y	Fraction of biomass energy used per year in metabolic processes by benthic microalgae	(050)
(113)	MK9b = 1.36912e−011	1/(J-y)	Fraction of benthic microalgae consumed per year per unit meiofauna	(051)
(114)	MK9c = 1.09529e−011	1/(J-y)	Fraction of benthic microalgae consumed per year per unit microfauna	(052)
(115)	MK9e = 2.73823e−012	1/(J-y)	Fraction of benthic microalgae consumed per year per unit insect	(053)
(116)	MKa0 = 1747.32	1/y	Fraction of biomass energy used per year in metabolic processes by detritus microbes	(054)
(117)	MKa4 = 5e−008	kg-N/J	Regenerated nutrients per unit energy metabolized by detritus microbes	(055)
(118)	MKab = 1.04007e−010	1/(J-y)	Fraction of microbes consumed per year per unit meiofauna	(056)
(119)	MKac = 6.93382e−011	1/(J-y)	Fraction of detritus microbes consumed per year per unit macrofauna	(057)
(120)	MKb0 = 72.7162	1/y	Fraction of biomass energy used per year in metabolic processes by meiofauna	(058)
(121)	MKb4 = 6e−008	kg-N/J	Regenerated nutrients per unit energy metabolized by meiofauna	(059)
(122)	MKbc = 1.60309e−012	1/(J-y)	Fraction of meiofauna consumed per year per unit macrofauna	(060)
(123)	MKbFCS = 8.36395e−014	1/(J-y)	Fraction of meiofauna consumed per year per unit fishes, crabs, and shrimps	(061)
(124)	MKc0 = 48.7219	1/y	Fraction of biomass energy used per year in metabolic processes by macrofauna	(062)
(125)	MKc4 = 6e−008	kg-N/J	Regenerated nutrients per unit energy metabolized by macrofauna	(063)
(126)	MKcd = 1.71895e−012	1/(J-y)	Fraction of macrofauna consumed per year per unit raccoons and rats	(064)
(127)	MKcFCS = 4.48327e−014	1/(J-y)	Fraction of macrofauna consumed per year per unit fishes, crabs, and shrimps	(065)
(128)	MKcg = 4.29646e−013	1/(J-y)	Fraction of macrofauna consumed per year per unit birds	(066)
(129)	MKd0 = 4.93012	1/y	Fraction of biomass energy used per year in metabolic processes by raccoons and rats	(067)
(130)	MKe0 = 31.3155	1/y	Fraction of biomass energy used per year in metabolic processes by insects	(068)
(131)	MKef = 6.90377e−012	1/(J-y)	Fraction of insects consumed per year per unit spiders	(069)
(132)	MKeg = 6.90377e−013	1/(J-y)	Fraction of insects consumed per year per unit birds	(070)
(133)	MKf0 = 15.6578	1/y	Fraction of biomass energy used per year in metabolic processes by spiders	(071)
(134)	MKfg = 6.90377e−013	1/(J-y)	Fraction of spiders consumed per year per unit birds	(072)
(135)	MKg0 = 2.69679	1/y	Fraction of biomass energy used per year in metabolic processes by birds	(073)
(136)	MKgd = 2.37812e−013	1/(J-y)	Fraction of birds eggs consumed per year per unit raccoons and rats	(074)
(137)	Yield = 0	J/y	Capture of fishes, crabs, and shrimps in estuary and creeks	(026)

No.	Simulation time parameters[b]	Units	Meaning
(138)	FINAL TIME = 4	y	The final time for the simulation
(139)	INITIAL TIME = 0	y	The initial time for the simulation
(140)	TIME STEP = 2e−007	y	The time step for the simulation
(141)	SAVEPER = 0.001	y	The frequency with which output is stored

[a] INTEG(inflows − outflows) refers to the integration of the net flow rate to determine the storage value. The initial values given for each storage are the assumed steady-state values used for model calibration. These values yield a flat-line output. To obtain dynamic output, halve the initial value for MQd.

[b] Vensim simulation software was used with the fourth-order Runge–Kutta routine selected. The model was run with identical results both on a MacIntosh PowerBook G4 (1 Ghz PowerPC G4 (3.3) processor, OS: Apple Classic 9.2.2 under OS10.4.6) and on an Abit KT7-RAID PC motherboard with an AMD Duron 700 processor (OS: Windows 2000 Professional).

28.3.2 Basic Equation Forms Indicated in Energy Systems Language Diagrams

Each storage in an Energy Systems Language diagram has lines indicating one or more inflows and outflows. The equation for each storage is simply the integral of its net flow rate over time, that is to say, the integral of the sum of all inflows into it minus the sum of outflows from it from all time past to the present moment. Because the concept of "all time past" is indefinite, an initial starting time is specified (Time Zero), at which point each storage is represented by an initial value. Eq. (001)–(016) in Table 28.1 show the rates contained in an integration function and the initial value supplied for each of the model storages.

The form of each rate equation, however, is tied to specific details in the diagram. Each rate arises from a symbol combination. The translation rules can be illustrated best with simpler diagrams of processes that repeat many times in Figure 28.3. By applying a few simple rules, the entire equation list can be constructed.

The composite symbols for primary producers (bullet shapes) and consumers (hexagons) have the same standard symbol combination to represent a reproduction process called "autocatalytic growth." This standard structure is given in Figure 28.4. The only difference between primary producers and consumers is the type of energy assimilated: usually sunlight for primary producers and some form of organic matter for consumers.

In either case, the designated energy source R passes at rate J_1 through a multiplicative interaction to be assimilated and stored in biomass Q_1. As shown in Figure 28.4, energy from the recipient storage (Q_1) feeds back and intersects the top of the interaction symbol, and then dissipates as heat to the heat sink symbol.

The rate J_1 is a net assimilation rate. Assimilation measurements for most living organisms are usually of net assimilation. As the curly bracket in Figure 28.4 indicates the energy of Q_1 fed back and used specifically in the production process is rarely separated explicitly from the amount assimilated. In ecological literature, the rate J_1 is known simply as "assimilation" for consumers. For primary producers, it is the gross primary production. The net production rate is $J_1 - J_2$, which is net primary production in the case of plants and net secondary production for consumers. Old age death, or other passive loss of biomass is indicated by rate J_3.

The equations for the relevant energy flows and their combined effect on the storage are indicated in the top panel of Figure 28.4. Consider first the net assimilation rate J_1, the form of which arises from the multiplicative interaction symbol involved. According to the translation rules of Energy Systems Language, all flows (J_i) into and out of an interaction term have the same equation form. Equations differ only by the numerical value of an associated constant (k_i). In the case of multiplicative interaction, the resource (R) and the biomass (Q_1) multiply together to produce the indicated equation for net assimilation rate $J_1 = k_1 R Q_1$. In other words, the rate of production is proportional to the multiple of the amount of energy available for transformation in resource R and the amount already available in the autocatalytic biomass Q_1. The biomass acquires resources to form additional biomass. That is to say, in the presence of its resources, growth is autocatalytic. The rate constant k_1 is analogous to the second-order rate constant in chemical reaction kinetics. The dimensional units for k_1 are fractions of R assimilated per unit Q_1 per unit time.

The outflows J_2 and J_3 in Figure 28.4 do not pass through an interaction symbol as drawn. If they did, the form of the equation would follow interaction rules like those for J_1. When the rate of outflow is simply a constant proportion of the donor storage, the outflow line does not pass through an explicit interaction symbol downstream. Hence, the equations for J_2 and J_3 in Figure 28.4 are constant proportions of Q_1. The dimensional units of the rate constants involved are simply the fraction of Q_1 that is degraded per unit time for k_2, or in the case of k_3, the fraction of upgraded material (Q_1) that dies from old age, leaks out, or is otherwise passively lost (per unit time).

In the complete marsh sector diagram (Figure 28.3), all of the metabolic heat losses from living organisms are constant proportions of the energy in biomass. These flows are identified in the equation list (Table 28.1) as MJx0, where x is the subscript associated with the standing stock Q of a living component x. Donor proportional outflows are used elsewhere in Figure 28.3 as well. The flow from plants to detritus (MJ$_{71}$, Eq. [048]) is proportional to plant biomass (MQ$_7$) and the burial of detritus in sediments as peat (MJ$_{15}$, Eq. [035]) is proportional to the detritus accumulation (MQ$_1$).

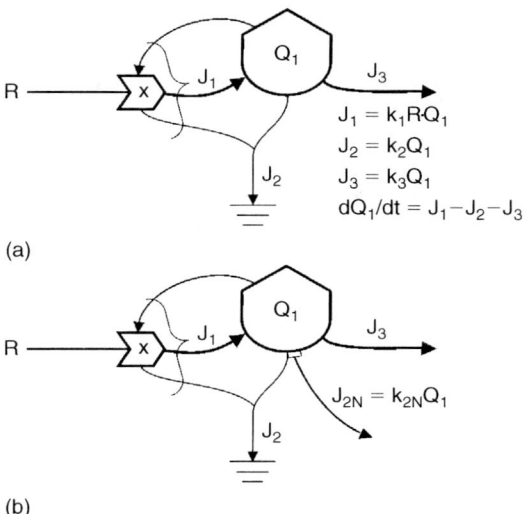

FIGURE 28.4 Basic symbolic representation and equations for the autocatalytic process usually used in the composite symbols for primary producers and consumers: (a) without nutrient regeneration indicated; (b) with nutrient regeneration indicated (J_{2N}). The curly bracket pointing to J_1 indicates combined flows (see text). R is a variable that represents the energy resource to be transformed, J's are for the various flows, k's are rate constants, Q_1 represents the storage of biomass, and dQ_1/dt is the rate of change of biomass with time.

Although not illustrated in Figure 28.4, some of the biomass of one living component may be captured and consumed by predators or other consumers. The amount is called the yield of biomass, and it becomes the resource for the autocatalytic growth of the recipient consumers. The equation for such an outflow from Q_1 to a downstream consumer would have the same multiplicative form as used for J_1. If, for example, Q_1 was the food of consumer Q_2, the flow between them—call it J_4—would be $J_4 = k_4 Q_1 Q_2$. The constant k_4 would represent the fraction of Q_1 assimilated per unit time by each unit of consumer Q_2.

A food Web can be represented by linking primary producer symbols to a network of consumer symbols. Consumers may represent trophic levels, feeding guilds, or other functionally similar combinations of species. Figure 28.3 includes a food Web based on marsh plant detritus, live marsh plants, and benthic microalgae that after passing through a number of intermediate consumers, culminates in the birds, raccoons, and rats of the salt marsh and the fishes, crabs, and shrimps of the tidal creeks and adjacent estuary (Q_{FCS}). Each equation for an energy transfer from resource to consumer uses the two-way multiplicative interaction described above.

The composite symbol for consumers sometimes has a flow of regenerated nutrients indicated in conjunction with the metabolic energy loss. One way to handle this is illustrated by the flow labeled J_{2N} in the bottom panel of Figure 28.4. The flow line emanates from a small rectangle attached close to the bottom of the storage tank. The small rectangle, called a "sensor," denotes that no loss of the storage of energy occurs, in this case because the flow is in units of something else besides the energy that is stored in biomass (nutrients). Nutrient regeneration is a product of metabolism. Nutrients accompany the energy losses denoted by the flow to the sink, but they must be tracked in a separate flow stream. Regenerated nutrients flow from the consumer to a nutrient storage elsewhere. The lower panel of Figure 28.4 shows the equation for nutrient regeneration (J_{2N}). It has the same form as the metabolic energy loss equation (J_2), but the dimensional units for the constant (k_{2N}) are nutrient units released per unit time per unit of energy stored, rather than fraction of energy per time.

The marsh sector diagram (Figure 28.3) shows nutrient regeneration by microbes (MQ_a), meiofauna (MQ_b), and macrofauna (MQ_c). The nutrient regeneration flow lines arise from the vicinity of the metabolic heat losses for these three consumers and cycle up and to the left terminating in the nutrients storage (MQ_4). The tiny sensor symbols are left out of the diagram in Figure 28.3, and in the equation list (Table 28.1) an alternate equation form is used for MJ_{a4}, MJ_{b4}, and MJ_{c4} (Eq. [055], Eq. [059], and Eq. [063], respectively). In this form, nutrient regeneration is a constant proportion of the respective metabolic heat losses rather than to the amount of energy stored in each consumer's biomass. The constants therefore define the amount of nutrients regenerated per unit of energy metabolized, which may be conceptually more familiar to some. Either way is consistent with the diagram.

In either case, the flow of regenerated nutrients is not subtracted from the accumulations of energy (MQ_a, MQ_b, and MQ_c) because the associated energy subtraction takes place with the respective flows to the heat sink (see Eqs. [009]–[011]). The nutrient flows are, however, added to the accumulation of nutrients (Eq. [004]).

28.3.2.1 The Flow-Limited Source

A curved line entering the system boundary from a source and then passing back out, while being tapped along the way inside the boundary, means that the equation for a "flow-limited source" is to be used. In the coastal system and marsh sector diagrams (Figure 28.1 and Figure 28.3), sunlight is tapped by evaporation of water and by the various primary producers in the system. Unused light is reflected by the system (the system's albedo), as represented by the line that curves back out of the system boundary.

Some resources, such as fossil fuels, are limited by the amount available in storage. For these resources, use rate exceeds the regeneration rate. Once the storage is depleted, further use is limited by the rate of regeneration. The resource then becomes flow-limited. Water flowing in a hillside stream is limited by the flow from upstream. From its first diversion for irrigation or drinking water, a stream is recognized as flow-limited. After diversion only the remaining flow is available for additional withdrawals. Likewise, the flow of sunlight ultimately limits photosynthesis. Sunlight used in evaporating water is not available for photosynthesis, and only the sunlight remaining in the shade and sunflecks beneath the tallest plants can be used by the shorter ones.

When storage is the resource transformed for the production of an autocatalytic component, the associated flow rate equations are based on the multiple of the donor and recipient storages. The production rates of all consumers of the food Web illustrated in the marsh sector diagram (Figure 28.3) are of this type. For primary producers, however, the main energy resource is flow-limited. Hence, the equation for gross primary production is based on the remaining unused flow of the resource, rather than a donor storage. The remaining flow is known generally as J_R. A simple diagram of this process is shown in Figure 28.5. In this case, the only tap is for gross primary production (J_1), so the remaining light (J_R) is simply the solar input (J_0) less that already incorporated into production. Substituting and rearranging this set of implicit equations yields the explicit relationship suitable for models. The equation for J_R is hyperbolic and declines to zero as the storage grows. The first equation for J_1 is simply the multiple of J_R and Q_1. The dimensional units of the constant k_1 are the inverse of the storage units. The constant represents the fraction of the remaining light assimilated by each unit of biomass.

The equation derived for J_1 from the definitions of J_1 and J_R (shown in Figure 28.5) is the Monod function (or the Michaelis–Menten equation), often used in ecological models to represent flow saturation at high levels of a resource and limitation at low levels. In the case illustrated, the maximum gross primary production rate is equal to J_0 and the level of biomass energy (Q_1) at which the rate is half the maximum is the inverse of k_1. The practical maximum is actually considerably less than J_0 for two reasons: (a) other factors also limit photosynthesis; and (b) other processes compete for sunlight, such as water evaporation, conversion to heat on dark surfaces, and reflection.

In the marsh sector diagram (Figure 28.3), the taps of sunlight for evaporation and for marsh plant production originate from the same point, and the tap for benthic microalgae occurs afterwards. Usually, tapping from the same point means the processes are all competing for the same remaining light, however, this is not the case when the point refers to different areas. The vertically overlapping primary producer symbols for low and high marsh plants indicate this distinction. The various grasses, rushes, and sedges that comprise each of the two elevation regions exist literally side by side in the marsh, so they receive the same

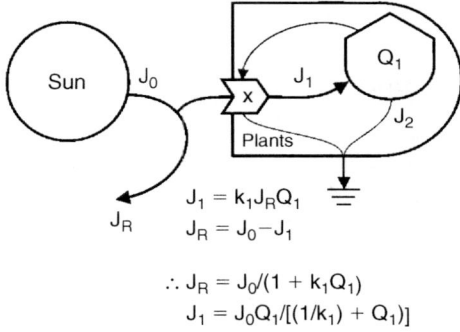

FIGURE 28.5 Flow-limited source coupled with an autocatalytic primary production process. Equations for J_1 and J_R are given two ways.

amount of light from overhead. A complete model of the marsh sector would have two subsectors: one for the low elevation marsh and the other for the high marsh. The equations of Table 28.1 show only the low marsh subsector. The high marsh would be represented by an identical set of equations calibrated for high marsh conditions. The total marsh sector would consist of the sum of each component in each subsector.

The independent process of water evaporation shown at the same tap point as plant production, however, is a competing process, separately competing in both the low marsh and the higher elevation marsh. Processes competing for the same flow-limited resource use the same calculation of remaining light for its functions. Water evaporates through an interaction with sunlight, as illustrated in the diagram. The equation for the evaporation rate (MJ_{20}, Eq. [040]) is based on the multiple of the storage of water (MQ_2), and an auxiliary that provides the remaining light available (MA_{20}, Eq. [031]). The right-hand side of the corresponding auxiliary equation for plants is identical (MA_{07a}, Eq. [027]). The form of each equation for light remaining at the combined tap is similar to that given for a single tap in the simple diagram of Figure 28.5, except that the denominator is based on the sum of the competing uses: evaporation of water and production by low marsh plants. Note also the conversion of evaporation from water to energy using the constant MK_{20E} (specified in Eq. [103]) for dimensional consistency.

Benthic microalgae exist in water-saturated soil often covered by water, and they grow in the shade of the emergent vascular plants of the marsh. Accordingly, the diagram shows the sunlight for benthic microalgae drawn downstream of the simultaneous tap for the marsh plant production and water evaporation. The source light available to benthic microalgae (analogous to J_0 in Figure 28.5) is the light remaining after evaporation of water and assimilation by the plants that shade them. This was already determined by MA_{07a} in Eq. (027) (or equivalently by MA_{20} in Eq. [031]). Otherwise the calculation of J_R for benthic microalgae (MA_{09a}) follows the form given in Figure 28.5, as shown in Eq. (029).

28.3.2.2 Multiple Simultaneous Interactions in Energy Systems Language

According to Figure 28.1 five other factors influence the transformation of sunlight by plants. These are illustrated more precisely in the marsh sector diagram (Figure 28.3). Water, nutrients, sediment, and circulation energy (caused by winds and tide) all positively influence gross primary production. In Energy Systems Language, these combine in a series of multiplicative interactions to determine the rate of gross primary production. Salt negatively influences gross production, however, so it is represented by a divisor. The resulting autocatalytic production equation for plants consists of a constant multiplied by the four positive influence variables, divided by salt, and all multiplied as before by the remaining light and the energy stored in plant biomass. A total of seven state variables are combined with one constant creating a seventh-degree equation for gross photosynthesis.

The multiplication of all these values to achieve a given rate of gross photosynthesis means that the single constant involved will have very complex dimensional units. Because production units and light flux units are both in energy units per time, a production function based only on the multiple of remaining light and biomass involves a constant with dimensions of inverse biomass units (as shown in Figure 28.5). If, however, two additional variables (say A and B) multiply and one divides (C), then for dimensional consistency, the implied units for the constant are C units per A unit per B unit per biomass unit. In the case given by the marsh sector diagram in Figure 28.3, the constant must cancel units from a total of six variables.

An additional practical detail concerning the constant is the small numerical value that must be assigned to it. The multiple of all the variables involved in a complex interaction produces a very large number that is proportional to the much smaller number that represents the resulting production rate. A very small constant does the conversion. In fact, it can be too small to represent on a computer!

The need for a tiny constant became a significant practical issue in the marsh sector simulation, so the multiplier auxiliaries for the gross photosynthesis equations, MA_{07b} and MA_{09b} (Eq. [028] and Eq. [030]), do not include the influence of sediment (MQ_5). To include the sediment influence, the constant needed was on the order of 10^{-44}. The number was too small to be represented precisely enough to run a required model check on the available computer. Sediment was potentially the least dynamic influence on gross production. The timescale of sediment dynamics in the model was on the order of 10,000 y, whereas the next longest time scale (that of salt) was around 3 y. Sediment was unlikely to influence model output spanning less than a few decades. By leaving out the sediment influence on production, the order of magnitude for the gross production constant was increased to 10^{-33}, large enough to allow the simulation check (the check involved reproducing the steady-state condition used to calibrate the model).

28.3.2.3 Feedback Effect of Production Processes on the Environmental Variables Involved

Some of the environmental influences on production are significantly consumed in the process. Nutrients, for example, are used at a rate proportional to gross production of low marsh plants (MJ_{47}, Eq. [042]) and benthic microalgae (MJ_{49}, Eq. [043]). Likewise, water is transpired by low marsh plants (MJ_{207}, Eq. [041]), and circulation energy is dissipated by marsh plant biomass (MJ_{607}, Eq. [045]). The accumulations of salt and sediment, however, are not changed directly by gross primary production. This lack of effect is denoted in Figure 28.3 by the use of the sensor symbol (small rectangle attached to the storage symbol where the line of influence originates).

Because gross production by benthic microalgae is not explicitly drawn on the marsh sector diagram, the influencing variables and any influences on them are unspecified. The same variables involved in low marsh plant production are involved with benthic microalgae. An examination of the equation list (Table 28.1) shows that with the exception of nutrient uptake (MJ_{49}, Eq. [043]), gross primary production by benthic microalgae does not consume any of the variables that influence it, at least for this model.

28.3.2.4 Two-way Interactions

In coastal systems, water circulating with the tides creates an exchange of materials between marshes and adjacent estuarine waters. Five exchanges are illustrated at the top left of the marsh sector diagram using two-way interaction symbols acted upon as a group by the tides. These are water, salt, nutrients, detritus, and sediment. All of these exchanges are driven by tidal energy. A two-way interaction indicates a balance of forces arising from storages of the same material on the two sides of the interaction. In the case of Figure 28.3, the storages are in the marsh and in the adjacent waters. A tendency toward dynamic equilibrium of materials is expected. The rate equation involves the difference between the two storages. Its general form is

$$J = k_r(Q_S - k_e Q_R)$$

where Q_S is the quantity stored in the designated source location, and Q_R is that stored in the receiving location. A positive difference in parentheses sends the flow toward the recipient, and a negative difference establishes flow in the opposite direction. For the marsh sector model, a positive flow represents an import into the marsh from the estuary. The constant k_e is the dimensionless equilibrium ratio of $Q_S:Q_R$ expected once a dynamic equilibrium is established. The other constant k_r is a rate constant: the fraction of the difference within the parentheses that can be closed per unit time. It is inversely proportional to the half-life of the difference (half-life $= \ln 2/k_r$).

The above rate equation was applied to the two-way flow for water, salt, inorganic nutrients, organic detritus, and sediment in Eqs. (021)–(025) in Table 28.1. Note that Eq. (021) for exchange of water (J_{W2M2}) includes the multiplication by tide. This is not the case for the other four variables. In those cases, the effect of tide is implicit in the selection of a value for the rate constant. As long as tidal influx is constant, this distinction is unimportant to the model output; however, before tidal energy influx can be tested as a variable, all the two-way exchanges must be converted to the form of Eq. (021). The conversion can be done simply by dividing each rate constant by the constant tidal influx used now (Eq. [086]). The values in question are given in Eq. (086), Eq. (088), Eq. (090), and Eq. (092), respectively.

28.3.2.5 Passive Inputs and Environmental Conditions

A basic input from the environment passively received by storage within the system is perhaps the simplest concept used to create a rate equation. Unlike with flow- and storage-limited sources from outside the system boundary, nothing within the system feeds back to limit the availability of such inputs. Examples in the marsh sector model are the input of rain to marsh water (Rain, Eq. [075]) and the input of tidal power to circulation (Tide, Eq. [076]).

When modeling a sector of a larger system, inputs from other sectors become basic environmental conditions similar to passive inputs. For example, inputs to the marsh sector include the storages of water, salt, nutrients, organic detritus, and sediment in the adjacent estuarine water (Eqs. [077]–[081]). These storages would become dynamic model variables of the water sector in a complete model of the coastal system diagrammed in Figure 28.1. Instead, they are set as passive inputs that do not change in response to the flows in and out of the marsh.

A passive input may be an average represented as a constant, a test pattern (such as a sine wave), a statistical distribution, or a complex time variable based on real data. In the early stages of model analysis, such variables are often set to average conditions represented by constants so as to not confound dynamics generated by the model's own feedback structure with those that are simply responding to variable driving forces. Holding the environmental inputs constant allows the internally created dynamics to be revealed first.

28.3.2.6 An Unspecified Process (Rectangle)

The interaction of tide with the exchange of water circulation is represented by a small rectangle, indicating a general process that must be made explicit in the documentation. In Figure 28.3, the rectangle merely signifies that water circulation energy is affected by the tide and by other forces provided through the water sector. Wind effects, for example, are shown in the coastal system diagram of Figure 28.1. The two kinds of forces are additive, but the marsh sector model only uses tidal power to drive circulation.

28.4 Calibration of Model Constants

All of the types of rate equation indicated or implied in the marsh sector diagram and given in the equation list have now been defined. The next step is to provide numerical values for all constants. In Energy Systems Language, this is nearly always accomplished by back calculation from quantitative estimates of all storages, rates, and system inputs. To do this, the rate equations are algebraically rearranged to solve for the constants in terms of the rates, storages, and inputs involved. The process of back-calculating constants in this manner is called model calibration.

To illustrate, consider the detailed diagram for a living composite symbol given in Figure 28.4(a). With estimates of the amount storage (Q_1), the amount of resources (R), and flow J_1, an estimate for the constant k_1 can be calculated. Likewise, with estimates of the storage Q_1 and each of the other flows (J_2, and J_3), the constants k_2 and k_3 can be made.

Obtaining a complete set of applicable estimates for model calibration can be a time-consuming task; however, it is easier than obtaining independent estimates for the constants themselves. Many of the constants, especially those with complex units, are not easily measured. By comparison, measurements of the needed rates of production, metabolism, resource use and regeneration, and measurements of the standing stock and energy content of biomass are commonly made in ecological studies. Even so, a sufficiently complete set of measurements about the same system is very rare. A broad range of experience and literature must be explored to come up with a set of preliminary numbers that can be used for model calibration.

Moreover, the estimates must apply to the same time period. To correctly compute a constant, the corresponding rates and storages must be in proper proportion. Average values for the rates and storages are usually used because they integrate long time periods, so are thought to be in correct proportions. A snapshot of synoptic measurements would provide a more precise set of estimates, but a specially coordinated field study would be required. Such a study would be very involved for a snapshot of the 80 or so variables involved in the marsh model given in Table 28.1.

Table 28.2 provides the list of nominal steady-state values for each storage and rate that were used in calibrating the constants for the model in Table 28.1. The nominal values were estimated from many kinds of information. The basis for each estimate is provided and any formula involved is also given in the Table 28.2. The nominal values so provided are for illustrative purposes only. They are plausible, based on general literature and experience of the author, but they may not apply adequately to a given marsh. Nevertheless, an analysis of model output sensitivity to changes in constants can be done to help identify the estimates that have the most significant influence on the dynamics produced by the model.

28.5 Preparation for Simulation

Once in equation form with all constants specified, the process of simulation is no different from that commonly used for nonlinear causal models in a wide variety of engineering and science fields. The same issues of initial value selection, choice of integration procedure, specification of simulation step size (time interval), and evaluation of roundoff error must be addressed. The same heuristic analytical procedures can be used, and the same limitations and pitfalls are present.

Because the resulting differential equations are seldom solvable analytically with calculus, Energy Systems Language models are represented and solved numerically on a computer by whatever means available. Historically, this effort began with analog computers, the use of which is now all but forgotten by Energy Systems Language practitioners. Today, Energy Systems Language models are often represented in programming languages for scientists and engineers such as FORTRAN, BASIC, PASCAL, and C.

28.5.1 Simulation Software and Energy Systems Language

Simulation software, such as Vensim®, STELLA®, and EXTEND® are valuable tools for conceptualizing the system, documenting flow equations, writing and integrating storage equations, and plotting output.[1] Both Vensim and STELLA were developed specifically for the systems dynamics approach begun by Jay Forrester (1961, 1968), but these programs work equally well for many other approaches to dynamic simulation modeling, including Energy Systems Language models. Vensim Personal Learning Edition was used for the model given in Table 28.1. To date, this version is free for educational use and comes with well-written and informative documentation.

When writing simulation models in a programming language such as BASIC, the order of equations is very important. Constants and initial values must be given before the iterative loop that represents the passage of time. Within the loop, flow rates must be calculated before time is advanced, and storages must be calculated immediately afterwards. Simulation languages free the modeler from the necessity to properly sequence model equations, to program an integration procedure, and to plot output data.

---

[1] Vensim is a registered trademark of Ventana Systems Inc., Harvard, MA; STELLA is a registered trademark of Isee Systems Inc., Lebanon, NH; and EXTEND is a registered trademark of Imagine That Inc., San Jose, CA.

Dynamic Simulation with Energy Systems Language

TABLE 28.2 Assumptions and Formulas Employed to Set the Nominal Steady-State Values of Storages, Flows, and Auxiliaries so that Model Constants could be Calibrated by Back-Calculation with the Model Simulation Equations. Variables in Table are given in Alphabetical Order. Refer to the List of Model Equations for the Meaning of each Variable

Variable and formula for nominal steady-state value	Nominal value	Units	Basis for value used
JOFCS = (0.6 * 46.0e+3) * 1.05e+8	2.89800e+12	J/y	0.6 g-C/m²/y production, in 10,000 ha estuary + 500 ha of tidal creeks, 46 kJ/g-C
JOLM = (150 * 3.1557e+7) * 3.0e+7	1.42009e+17	J/y	Sunlight (150 W/m²) × low marsh area (3000 ha)
JFC50 = (JOFCS+MJbFCS+MJcFCS−JYield)	1.85351e+13	J/y	All food assimilated by fishes, crabs, and shrimps (but not caught) is used in their own metabolism
JW2M2 = −0.4 * Rain	−1.80000e+10	kg-H$_2$O/y	Net export equal to 40% of rainfall
JW3M3 = 0	0.00000e+0	kg-salt/y	Salt is at equilibrium, so net exchange = 0
JW4M4 = (MJ47+MJ49) − (MJa4+MJb4+MJc4)	2.68822e+5	kg-N/y	Net nitrogen import from estuary equals uptake in marsh less regeneration in marsh
JW5M1 = MJ15+(MJ1a+MJ1b+MJ1c)−MJ71	0.00000e+0	J/y	Net detritus exchange with estuary is detritus buried plus that assimilated minus that produced in the marsh
JW6M5 = −MJ15D	−1.11644e+6	kg-sediment/y	Net sediment export equals the detrital burial rate in marsh
Jyield = 0	0.00000e+0	J/y	No fishes, crabs, or shrimps are caught in first model run
MA07a = J0LM * (0.12/(1−0.02)	1.73889e+16	J/y	Light input × portion (P) of light available for low marsh plants (albedo 12%) adjusted for fraction assimilated by benthic microalgae (2%).
MA07b = (MQ2 * MQ4 * MQ6/MQ3)	7.86156e+16	Complex	Multiple of all production effects indicated for low marsh plants except light and standing stock.
MA09a = 0.12 * J0LM	1.70411e+16	J/y	Light input × marsh total albedo (12%)
MA09b = (MQ2 * MQ4 * MQ6/MQ3)	7.86156e+16	Complex	Multiple of all production effects indicated for benthic microalgae except light and standing stock
MA20 = MA07a	1.73889e+16	J/y	Same as light available for low marsh plants (=MA07a)
MJ06 = Tide	1.03752e+14	J/y	Circulation energy equals the tidal energy influx
MJ07 = 0.01 * J0LM	1.42009e+15	J/y	Light input × fraction assimilated by plants (1%)
MJ09 = 0.02 * MA07a	3.47778e+14	J/y	2% of light assimilated by benthic microalgae after interception by water and plants
MJ15 = 0.03 * MJ71	1.89795e+13	J/y	3% of detritus input is buried in sediments
MJ15D = MJ15/17e+6	1.11644e+6	kg-sediment/y	Energy content of amount buried divided by 17 kJ/g
MJ1a = 0.87 * MJ71	5.50406e+14	J/y	87% of detritus input is consumed by microbes
MJ1b = 0.08 * MJ71	5.06121e+13	J/y	8% of detritus input is consumed by meiofauna
MJ1c = 0.02 * MJ71	1.26530e+13	J/y	2% of detritus input is consumed by macrofauna
MJ20 = 0.2 * Rain	9.00000e+9	kg-H$_2$O/y	Evaporation equivalent to 20% of rainfall (30 cm/y)
MJ207 = 0.4 * Rain	1.80000e+10	kg-H$_2$O/y	Transpiration equivalent to 40% of rainfall (60 cm/y)
MJ20E = J0LM * (1−0.01−((0.12/(1−0.02)))	1.23200e+17	J/y	Light for water evaporation is the portion neither assimilated by low marsh plants (1%), by benthic microalgae (2% after plant assimilation), nor reflected (12%)
MJ47 = (MJ07−MJ70) * 4.5e−08	3.19521e+7	kg-N/y	0.045 g-N is assimilated by low marsh plants for every kJ of gross production
MJ49 = (MJ09−MJ90) * 5.0e−8	8.69444e+6	kg-N/y	0.05 g-N is assimilated by benthic microalgae for every kJ of gross production

(Continued)

TABLE 28.2 Continued

Variable and formula for nominal steady-state value	Nominal value	Units	Basis for value used
MJ60 = 0.9 * MJ06	9.33770e+13	J/y	90% of circulation energy influx does not affect low marsh plants
MJ607 = 0.1 * MJ06	1.03752e+13	J/y	10% of circulation energy influx is intercepted by plants
MJ70 = 0.5 * MJ07	7.10046e+14	J/y	50% of low marsh plant gross production is used in the plant's own metabolism
MJ706 = 0.01 * (MJ07−MJ70)	7.10046e+12	J/y	1% of the net primary production of low marsh plants (before transpiration) is used in transpiration
MJ71 = 0.9 * (MJ07−MJ70−MJ706)	6.32651e+14	J/y	90% of the net primary production of low marsh plants becomes detritus
MJ7e = 0.1 * (MJ07−MJ70−MJ706)	7.02946e+13	J/y	10% of the net primary production of low marsh plants is assimilated by insects
MJ90 = 0.5 * MJ09	1.73889e+14	J/y	50% of benthic microalgal gross production is used in the plant's own metabolism
MJ9b = 0.5 * (MJ09−MJ90)	8.69444e+13	J/y	50% of benthic microalgal net production assimilated by meiofauna
MJ9c = 0.4 * (MJ09−MJ90)	6.95555e+13	J/y	40% of benthic microalgal net production assimilated by macrofauna
MJ9e = 0.1 * (MJ09−MJ90)	1.73889e+13	J/y	10% of benthic microalgal net production assimilated by insects
MJa0 = 0.8 * MJ1a	4.40325e+14	J/y	80% of the detritus assimilated by microbes is used in microbial metabolism
MJa4 = 5.0e−8 * MJa0	2.20163e+7	kg-N/y	0.05 g-N is regenerated for every kJ of energy metabolized by microbes
MJab = 0.6 * (MJ1a−MJa0)	6.60488e+13	J/y	60% of the net production of microbes is assimilated by meiofauna
MJac = 0.4 * (MJ1a−MJa0)	4.40325e+13	J/y	40% of the net production of microbes is assimilated by macrofauna
MJb0 = 0.9 * (MJ1b+MJ9b+MJab)	1.83245e+14	J/y	90% of the food assimilated by meiofauna is used in meiofaunal metabolism
MJb4 = 6.0e−8 * MJb0	1.09947e+7	kg-N/y	0.06 g-N is regenerated for every kJ of energy metabolized by meiofauna
MJbc = 0.5 * (MJ1b+MJ9b+MJab−MJb0)	1.01803e+13	J/y	50% of the net production of meiofauna is assimilated by macrofauna
MJbFCS = 0.5 * (MJ1b+MJ9b+MJab−MJb0)	1.01803e+13	J/y	50% of the net production of meiofauna is assimilated by fishes, crabs, and shrimps in the estuary and tidal creeks
MJc0 = 0.9 * (MJ1c+MJ9c+MJac+MJbc)	1.22779e+14	J/y	90% of the food assimilated by macrofauna is used in macrofaunal metabolism
MJc4 = 6.0e−8 * MJc0	7.36675e+6	kg-N/y	0.06 g-N is regenerated for every kJ of energy metabolized by macrofauna
MJcd = 0.4 * (MJ1c+MJ9c+MJac+MJbc−MJc0)	5.45685e+12	J/y	40% of the net production of macrofauna is assimilated by rats and raccoons
MJcFCS = 0.4 * (MJ1c+MJ9c+MJac+MJbc−MJc0)	5.45685e+12	J/y	40% of the net production of macrofauna is assimilated by fishes, crabs, and shrimps in the estuary and tidal creeks
MJcg = 0.2 * (MJ1c+MJ9c+MJac+MJbc−MJc0)	2.72843e+12	J/y	20% of the net production of macrofauna is assimilated by birds
MJd0 = MJcd+MJgd	6.21195e+12	J/y	All food assimilated by rats and raccoons is metabolized (they are top carnivores)
MJe0 = 0.9 * (MJ7e+MJ9e)	7.89151e+13	J/y	90% of the food assimilated by insects is used in insect metabolism
MJef = 0.5 * (MJ7e+MJ9e−MJe0)	4.38417e+12	J/y	50% of the net production of insects is assimilated by spiders
MJeg = 0.5 * (MJ7e+MJ9e−MJe0)	4.38417e+12	J/y	50% of the net production of insects is assimilated by birds
MJf0 = 0.9 * MJef	3.94575e+12	J/y	90% of the food assimilated by spiders is used in spider metabolism
MJfg = MJef−MJf0	4.38417e+11	J/y	All of the net production of spiders is assimilated by birds
MJg0 = 0.9 * (MJcg+MJeg+MJfg)	6.79591e+12	J/y	90% of the food assimilated by birds is used in bird metabolism
MJgd = MJcg+MJeg+MJfg−MJg0	7.55102e+11	J/y	All of the net production of birds is consumed (as eggs) by rats and raccoons

Variable	Value	Units	Description
MQ1 = MQ7	9.18000e+14	J	Organic detritus similar in quantity to live plant biomass
MQ2 = (0.25 * 0.125 + 0.30 * 0.50) * 3.0e+7	5.43750e+6	kg-H$_2$O	Average water over marsh (inundated 25% of the time, 12.5 cm deep) plus water in root zone (30 cm deep, 50% water content)
MQ3 = 45 * MQ2/(1000−45)	2.56217e+5	kg-salt	Average marsh salinity of 45 g/kg-seawater
MQ4 = (200.0e−6 14) * (0.3 * 3.0e+7)	2.52000e+4	kg-N	200 micromoles per liter of wet sediment in root zone
MQ5 = 1767 * (0.3 * 3.0e+7)	1.59030e+10	kg-sediment	Sediment bulk density of 1767 kg/m^3 in root zone
MQ6 = ((WQ2 * 9.8) * 0.5) * 0.3	1.47000e+11	J	Estuarine water force against gravity × tidal amplitude (0.5 m) prorated for low marsh (30% of total marsh area)
MQ7 = (1.8e+3 * 17.0e+3) * 3.0e+7	9.18000e+14	J	1.8 kg/m^2 at 17 kJ/g × marsh area
MQ9 = (4.0 * 21.0e+3) * 3.0e+7	2.52000e+12	J	4 g/m^2 at 21 kJ/g × marsh area
MQa = 0.10 * MQ9	2.52000e+11	J	Microbial standing stock ~10% that of benthic microalgae
MQb = MQ9	2.52000e+12	J	Meiofauna standing stock similar to that of benthic microalgae
MQc = MQb	2.52000e+12	J	Macrofauna standing stock similar to that of meiofauna
MQd = MQg/2	1.26000e+12	J	Raccoons and rats standing stock half that of birds
MQe = Mqc	2.52000e+12	J	Insect standing stock similar to that of macrofauna
MQf = 0.10 * MQe	2.52000e+11	J	Spider standing stock ~ 10% that of insects
MQg = Mqc	2.52000e+12	J	Bird standing stock similar to that of macrofauna
QFCS = (20 * 23.0e+3) * 1.05e+8	4.83000e+13	J	20 g/m^2 at 23 kJ/g in 10,000 ha of coastal water and 500 ha of tidal creeks
Rain = (1.50 * 3.0e+7) * 1000	4.50000e+10	kg-H$_2$O/y	Rainfall (150 cm/y) × water density (1000 kg/m^3) × marsh area (3.0e+7 m^2)
Tide = 0.5 * (1e+11 * 9.8) * 705.797	1.03752e+14	J/y	Tidal amplitude (0.5 m) × water force against gravity (9.8e+11 N) × tidal cycles/y (705.797)
WQ2 = (1.0 * 1e+8) * 1000	1.00000e+11	kg-H$_2$O	Water volume (depth of 1 m, area of 10,000 ha) × density
WQ3 = 25 * WQ2/(1000−25)	2.56410e+9	kg-salt	Average salinity of 25 g salt per kg seawater
WQ4 = (4e−6 * 14) * (1e+8)	5.60000e+3	kg-N	Inorganic nitrogen of 4 μm N × water volume (1e+8 m^3)
WQ5 = (30 * 17e+3) * 1.0e+8	5.10000e+13	J	30 g/m^3 suspended organic matter at 17 kJ/g × water volume
WQ6 = 70e−3 * 1.0e+8	7.00000e+6	kg-sediment	70 g/m^3 suspended inorganic matter × water volume

Some features offered in simulation software may not be appropriate for Energy Systems Language simulation. In particular, storages should always be free to exceed physically impossible limits of the real world should the model produce such behavior in a given simulation. Artificially preventing a storage from going negative or from creating overflow errors will mask serious errors in the *flow* equations, and it can mask the use of an incorrect integration procedure.

The storage equations themselves must not be manipulated. They are derived from a mathematical integration process. Manipulating the storage in ways other than through changes in flow rates alters the integration procedure, not the hypothesis. The greatest power of Energy Systems Language modeling is in its ability to reject an erroneous hypothesis. Preserving this power must be done if the output is to be used to evaluate the plausibility of the hypothesis.

The intensive focus on special diagramming symbols, layout, and equation-writing rules, and the necessity for numerical integration procedures, when combined with the idea that ecologists and environmental scientists should be the best ecological modelers, creates a compelling incentive for developing a simulation language specifically to represent HT Odum's Energy Systems Language. As noted by those who developed DYNAMO, STELLA, and Vensim for easing simulation with Forrester's Systems Dynamics approach, a special simulation language can eliminate programming effort and numerical integration mistakes so that those most knowledgeable about ecological systems need not also become experts in computer simulation technique to test complex ideas. To date, however, no special simulation language has been forthcoming for Odum's Energy Systems Language.

28.5.2 Timescales and Numerical Integration

Energy Systems Language diagramming of complex systems may produce models with components that operate over many timescales. This can lead to practical difficulties of numerical integration on a digital computer, such as a tradeoff between roundoff error and truncation error, and misleading parasitic solutions. Long central processing times are required for the computer, and output sampling is necessary if available computer storage media is insufficient to hold all the data generated by the model.

The timescale of a given storage component is represented by the order of magnitude of its turnover time, which is the amount stored divided by the sum of the inflows to it or the outflows from it. Turnover time can be determined for any instant or averaged over various time periods. The steady-state turnover times for the various components of the example model are given in Table 28.3. They span eight orders of magnitude.

TABLE 28.3 Turnover Times at Steady State (in Years and Days) for the 16 Storages Included in the Low Marsh Subsector Model. Turnover Times Span Eight Orders of Magnitude

Storage	Model Symbol	Steady-State Value	Units	Years	Days
Water in marsh	MQ2	5.44 E6	kg-H_2O	2.01E−4	0.074
Detrivorus microbes	MQa	2.52 E11	J	4.58E−4	0.17
Nutrients in marsh	MQ4	2.52 E4	kg-N	5.88E−4	0.22
Water circulation energy	MQ6	1.47 E11	J	1.42E−3	0.52
Benthic microalgae	MQ9	2.52 E12	J	5.79E−3	2.1
Marsh meiofauna	MQb	2.52 E12	J	1.12E−2	4.1
Marsh macrofauna	MQc	2.52 E12	J	1.63E−2	5.9
Marsh insects	MQe	2.52 E12	J	2.74E−2	10
Marsh spiders	MQf	2.52 E11	J	5.48E−2	20
Raccoons and rats	MQd	1.26 E12	J	1.80E−1	66
Marsh birds	MQg	2.52 E12	J	3.09E−1	110
Low marsh plants	MQ7	9.18 E14	J	6.46E−1	240
Detritus in marsh	MQ1	9.18 E14	J	1.45E0	530
Fishes, crabs, and shrimp	QFCS	4.83 E13	J	2.37E0	866
Salt in marsh	MQ3	2.56 E5	kg-salt	3.04E0	1100
Sediment and peat in marsh	MQ5	1.59 E10	kg-sediment	1.42E4	5,200,000

The model was successfully simulated using the Vensim Personal Learning Edition simulation language with the fourth-order Runge–Kutta integration procedure selected and a tiny time step of 2.0e−7 y. Owing to insufficient computer storage media, output from every model variable was sampled every 0.001 y and stored. Many trials of time steps were required before a satisfactory one was achieved. Solutions converged below 5e−7 y and above 5e−8 y. Presumably, divergence with shorter time steps was caused by roundoff error, and with larger time steps by truncation error.

At the chosen time step, a 12 y continuous run was not possible (the CPU kept running, but the data recording to screen and disk stopped and the Vensim program became unresponsive and had to be rebooted). To achieve a 12-y simulation, the model was run for 4 y and the ending values for storages were set as the initial values for another 4-y run. This was repeated a second time to achieve 12 y of data.

28.6 Dynamic Output of the Marsh Sector Model

When all constants and initial conditions are set to six significant digits, the model produces a flat line steady-state response in all model variables. This steady-state check helps to verify that the equations are operating as intended. However, if any one of the initial conditions is set slightly off of the steady state, oscillations begin within a simulated year or two. Larger changes in any initial condition cause the oscillations to begin sooner, but the patterns that eventually result are similar in frequency, amplitude, and irregular appearance. The oscillations occur because of the feedback structure of the model. All the environmental conditions are constant.

Figure 28.6 provides an example set of patterns for 14 of the 16 storages when the model is simulated for 12 y. These dynamics result when the initial value for the storage of raccoons and rats is halved from its steady-state value, while the other initial values remain calibrated for steady state. The dynamic patterns of the 16 storages differ widely. Salt and sediment, the two storages with the highest turnover times, did not change at all during the 12-y simulation, so they are not plotted in Figure 28.6. The others oscillate over a wide range of frequencies and amplitudes. Most of the patterns include some amplitude modulation. The storages with higher frequency oscillations show frequency modulation.

Many of the storages cycle several times per year; the larger animals of the marsh cycle more slowly. Rats and raccoons are at the top of the marsh food Web. Their quantities are generally low for 3 y and then spike every fourth year or so. Birds exhibit a complex set of cycles. A twice a year cycle seems superimposed on a 5-y cycle. The mean and amplitude of the higher frequency cycles build for about 3 y and then birds suddenly decline and remain low for a couple of years.

Between years 8 and 10, several interesting differences in the output occur: not only do high densities of raccoons, rats, and birds occur, but also the oscillations of the high-frequency variables cycle even faster, benthic microalgae all but disappear, the oscillating spider density seems to skip a peak and then produces the highest peak of the entire 12 y run, and finally, the highest peak level of insects occurs. A longer simulation is needed to determine whether such behavior recurs periodically.

The interesting mix of patterns is typical of complex models developed with Energy Systems Language. Such patterns could in fact exist in nature and go undetected. Ecological field measurements are notoriously variable, but are usually made too infrequently to capture the details of patterns such as those generated by the model. Instead, the variation among repeated field measurements is most often used to describe a statistical variance around a sample mean. Yet average values resulting from field data in this manner become the data to which the model is calibrated in the first place. The model output shows that significant nonrandom variation can result from deterministic processes like those represented in the marsh model.

The output from this preliminary model, however, is unlikely to be valid in any more than a general sense. Some of the output is highly suspicious. Marsh macrofauna, for example, precipitously decline to near zero after the first five simulated years and remain in unrealistically low quantities for the remainder of the 12-y simulation. Marsh macrofauna include fiddler crabs and marsh snails that are easily noticed by anyone visiting a salt marsh in the southeastern United States. If the macrofauna of tidal marshes

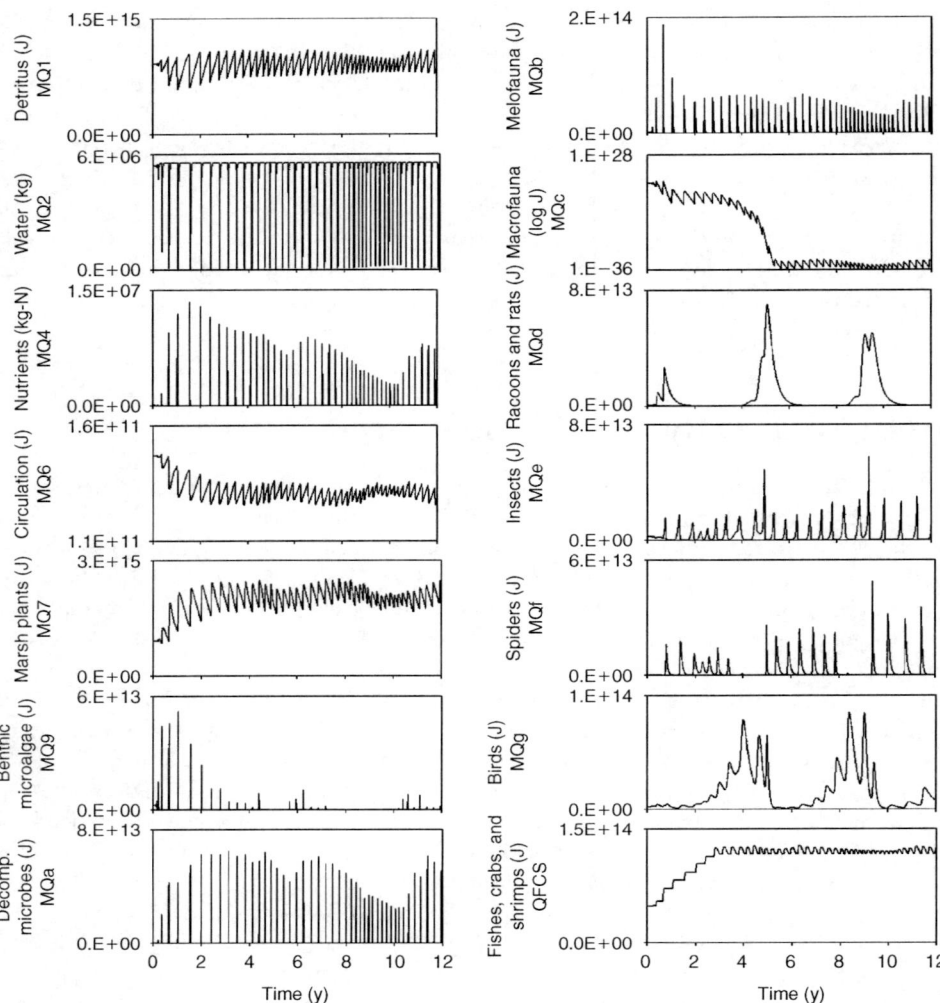

FIGURE 28.6 Dynamic output over 12 simulated years for 14 of the 16 storages represented in the marsh sector model. The run resulted from a single deviation from steady-state conditions caused by halving the initial number of raccoons and rats. The set of outputs includes a rich array of frequencies, amplitudes, and modulations.

periodically disappeared for years at a time, the loss would be noticed and widely reported, but no such reports are on record.

28.6.1 Model Output Analysis

Additional work with the model may be justified whether or not the output from the model is plausible. A thorough analysis will reveal the reasons for the model's behavior, whether it is plausible or not. Model output analysis involves systematic experimentation with altered initial conditions, test inputs, and model constants, and sometimes with alternate equations or even alternate system identifications.

A sensitivity analysis of a model consists of running the model repeatedly, each time with a standardized adjustment to a different model constant. Halving and doubling is a common adjustment. Normally, only one constant is changed on each such rerun. The constant is reset to its original nominal value before the next one is halved and doubled. In this way, a standardized test is made in which all changes are based on an arbitrary but identical type of change, and all tests are independent of one another. The concern is not the realism of the adjustment, just that it is standardized so that relative influence of each coefficient on specific features of the model output can be identified and ranked. Further measurement of the real system can be directed toward obtaining better understanding of the most influential parameters first.

The ranking of constants by their relative influence on model output can be usefully compared to a ranking based on the confidence the modeler has in the nominal value used for the constant. Better estimates would be sought first for those constants with the highest combination of effectiveness in creating change in the model output and uncertainty in the original estimate.

28.6.2 Model Validation

A model is an abstraction of the cause-and-effect mechanisms believed to operate in the system under study. A complex simulation model must be treated as a scientific hypothesis about cause and effect that is too complex to be evaluated wholly in the mind. Computer representation allows the consequences of an imagined system to be revealed and evaluated against what is known to be true about the system under study. The adequacy of the abstraction must be defended in all of its aspects, including its level of detail, the validity of each mathematical representation, and the interpretation of output.

Confidence is gained as the model passes repeated attempts to falsify it—a process called validation. If the model produces plausible output on the computer, further examination is warranted using an interplay between models, field observations, and experimentation with the real system. This process of validation will either cause the hypothesis to eventually be rejected, or will bolster confidence in it. Model results often lead to testable ideas that are explored experimentally or by careful observation of nature. Most published simulation models in ecology await this process.

The utility often derives from model failure in spite of state-of-the-art thinking represented in it. Hypothesis rejection is a powerful tool that defines the scientific method. Computer simulation of complex hypotheses is a natural extension of the scientific method in the age of computers. Taken in conjunction with observation and experimentation, computer simulation allows complex ideas to be tested.

With continued recycling between the marsh model, its output, and measurements from nature, the knowledge and predictability of the marsh ecosystem will improve. The intriguing output and the obvious failures stimulate thinking, identify key variables, and guide the planning of additional field measurements and experimentation. It is through the interaction of these facets that the understanding arises.

28.7 A Brief Comparison with Forrester's Systems Dynamics Approach

The many components and connections within Energy Systems Language models continually respond to one another in a complex system of feedback loops. The loops account for the dynamic behavior of a naturally self-organized system. Control and alteration of the self-organization process is often desired to direct more of the energy into the human sector, or sometimes to restore the direction of self-organization back to an earlier state, and to do so without unintended consequences.

The above description of the nature of the systems studied with Energy Systems Language would seem to provide an ideal setting for the Systems Dynamics approach of Jay Forrester (1961, 1968). In Forrester's approach, a specific cause of a dynamic pattern is sought. Odum's approach takes the opposite approach: Energy System Language is used to propose a system structure and discover the dynamics that would result if a system was organized as described. The two are reverse processes of one another.

The method of systems identification involved in Energy Systems Language, while it can lead to a simulation model filled with feedback processes, does not focus on feedback structure. In fact, the focus is somewhat reversed from Forrester's approach. Where the Systems Dynamics approach seeks loops to account for existing dynamic patterns, simulation with Energy Systems Language links all essential energy transformation and control processes thought to be present and identifies the resulting dynamics that could be produced.

The two methods can complement one another. They do not replace one another, nor are they at philosophical odds. The two are simply focused differently. The founding difference of Odum's approach is the primary focus on energy flow rather than on identifying feedback loops that account for observed dynamics. Energy tracing, rather than an analysis of dynamics, identifies the basic flow and control system. Instead of using feedback theory to provide the evidence for controls on flows, generalizations about the responses of organisms to the environment and to other organisms determine the rates of flows in ecosystem models.

When dealing with environmental systems, the two approaches can be used together to considerable advantage, but few analysts are adequately trained in the practical aspects of both methods. A reason for this separation may relate to the different contexts within which the methods developed—System Dynamics from the perspective of industrial management, and Energy Systems Language from the theory of ecosystem recognition and ecosystem development.

By examining industrial systems with wild inventory fluctuations that could not be controlled through traditional methods of industrial management, Forrester recognized that an unidentified system for operating the industrial plant had arisen on its own. It had self-organized, and no longer operated as designed. Identifying this system within the confines of the industry included looking at how orders were received, filled, and delivered, given the decisions made by various managers and other employees involved. Forrester's concept, originally called Industrial Dynamics, was to use features of the patterns of fluctuation, such as frequency, amplitude, and damping rates, as clues that would help uncover the self-organized feedback loops that were operating. Once this was done, the system could be modeled on a computer, and the likely causes of the problem demonstrated. Furthermore, the effects of different decisions could be explored in the model, discussed among the managers, and eventually tried in the real system.

To bring this approach into the hands of industrial managers, Forrester encouraged the development of special computer software that would be easier for the industrial manager to use than a higher level programming language such as FORTRAN. In response, Alexander Pugh created the simulation language DYNAMO. The award-winning success of this approach in industry caused it to be generalized to many other types of self-organized systems. Forrester then changed the name of this approach from Industrial Dynamics to Systems Dynamics and began a philosophy of general principles of system self-organization in society and nature.

At the same time and completely separately, H.T. Odum, an ecologist who had studied global strontium cycling, was writing about general principles of self-organization from an ecosystem perspective. His observations of ecosystems were not so much based on noticing curious dynamic patterns, but instead on the more basic work of identifying the details of a self-organized system by tracing energy flows from sunlight through plants, then animals, and ultimately the decomposers that consumed the remaining energy in dead matter and recycled its component elements.

Like Forrester's interest in easing the process of modeling for industrial managers, Odum wanted a better way for ecologists to relate to the mathematical relationships that could be developed to represent energy flows through ecosystems and how they might be controlled by amounts of various living and nonliving components, and the genetically programmed responses of organisms. Out of this effort came a diagrammatic language, originally called Energy Circuit Language, and now called Energy Systems Language by most of its main practitioners.

In its overall goal, Odum's approach is similar to Forrester's for its attempt to identify and formally represent the dynamic behavior of self-organized systems of humanity and nature so that the consequences of change can be better assessed. Odum began with an interest in ecosystems and, like Forrester, soon generalized his theory to all open energy systems, including the systems of humanity and nature that have led to many of the environmental concerns of today. Through Energy Systems Language as tool to express a theory and philosophy of natural system organization and behavior, he established considerable evidence for a strong relationship between the natural environment and the economy (Odum and Odum, 2001).

Aside from the wide difference in focus between their fields of study, an important difference between Odum's and Forrester's approaches can perhaps be traced to the nature of data available for observing industrial inventory problems versus that for ecosystems. First, owing to the difficulties of field measurement, ecosystem data are rarely obtained at a frequency that can distinguish temporal patterns caused by feedback processes from random variation. Second, even when sufficiently frequent data are collected, it is often impossible to separate the pattern generated by feedback processes within the ecosystem itself from patterns generated by responses to regular changes in environmental factors. In general, the most obvious dynamic patterns in ecological data are dominated by daily and seasonal changes in light, temperature, rainfall, and other exogenous variables (forcing or driving variables). Unlike an out-of-control industrial inventory, ecosystem data do not inspire particular curiosity about the feedback processes involved. Most of the obvious dynamics are caused by feedback within a larger astronomical system that is not the focus of ecological research.

Deviations from perfect tracking of exogenous variables might be found in the dynamics of key variables of a system that could provide clues to feedback structure. This possibility, however, seems less likely to occur to someone who examines the relatively uninspiring, exogenously driven, incompletely recorded temporal variation found in most ecosystem data records. Such a notion is rather more likely to occur to someone exposed both to systems ecology and to feedback analysis. With that combination comes the imperative to monitor ecosystems frequently enough to recognize interesting patterns like those suggested by the marsh model developed in this chapter.

28.8 Conclusions

The interesting dynamics produced by the marsh model could stimulate a closer look at the fluctuations of animal and plant biomass in field or experimental settings. Conversely, a close look at actual dynamics from the real system may force a completely new idea about system organization. In these alternate outcomes lies the true value of modeling in environmental management. Models do not replace the need for real system measurement and understanding. Instead they demand it and direct it.

Dynamic phenomena are the *patterns* of change we think happen through time. They are signals from self-organized systems that we can analyze and attempt to explain. In contrast, a simulation model of a self-organized system is a detailed hypothesis of how and why we think a given dynamic phenomenon happens, and why certain changes may occur in response to new conditions. A computer provides a mechanism for tracking complex networks of influences thought to be involved in causing change. The output dynamics allow visualization of the dynamic consequences of thinking that way.

Energy Systems Language guides the imagination toward plausible constructs that follow reasonable principles of self-organization. It also simplifies equation specification and provides a way to use scarce information to estimate rate constants. The Energy Systems Language procedure does not replace the judgment and real system understanding required to identify the system and evaluate the adequacy of a given model.

Increased familiarity with the diagramming rules eases the process of representing an environmental system with Energy Systems Language. Nevertheless, even among those most familiar with the theoretical basis and instructions, to settle issues of how to diagram a process often requires intense thought and discussion with others familiar with the process. Although equations can be developed from the diagram, the complex reasoning for drawing it in a particular way can rarely be appreciated by those who were not involved in its production. A lot of detail is represented in Energy Systems Language diagrams. The diagram must be accompanied by considerable explanation even when presented to the most accomplished users of this approach. Complete justification of the logic behind each line of Figure 28.1, however, is beyond the scope here. For greater insight into the derivation of the diagram in Figure 28.1, the interested reader may refer to the source publication and the other contributions to the book edited by Coultas and Hsieh (1997).

A simulation model of a self-organized system built with Energy Systems Language is a rigorous statement of a complex hypothesis composed of many relational and quantitative estimates. It is unlikely to be completely correct, but when used in conjunction with real system measurements, it forms a basis for tracking, testing, and improving understanding of the environmental system that the model represents. An Energy Systems Language model can become a record of the evolving state-of-the-art understanding about a given system that can transcend generations of environmental managers. Each generation can improve the model as the interplay between measured phenomena and model prediction proceeds.

Flaws in understanding and difficulty in detecting dynamic phenomena each interfere with successful environmental systems management. Environmental data in the absence of an explanation may add description, but does little for understanding. Simulation models, in the absence of dynamic phenomena from the real system, are untested hypotheses that produce a prediction of dynamics waiting to be tested. With continued cycling between model formulation and measurement of phenomena, the high level of understanding of self-organized environmental systems required to manage such systems will be forthcoming. It is the interplay between models and measurement of dynamics that can guide understanding toward management principles that work.

A system model can be built that shows impossible dynamics, sometimes intriguing ones, and perhaps less often accurate ones. Nevertheless, Energy Systems Language opens the door to a rigorous energy theory of nature that, perhaps unlike the models themselves, continues to have an enormous practical impact on the observation, analysis, and management of ecosystems today.

Acknowledgments

This chapter is dedicated to the memory of Professor H.T. Odum with whom the author was privileged to work and learn as a colleague and friend for over 20 years. This chapter reflects the author's understanding of Professor Odum's methods. Thanks to Dr. Mark T. Brown, also a longtime colleague of Dr. Odum, and to Elizabeth C. Odum, Dr. Odum's spouse and frequent coauthor, for reviewing the first draft manuscript of this chapter. A third in-house reviewer, Mr. Mike Lemmons, provided in-depth review from a novice's point of view. Considerable revision resulted from these reviews, the comments of three anonymous reviewers, and the editor, Paul Fishwick. The author owes a debt of gratitude to each of these reviewers for a greatly improved, clarified, and more substantive presentation of Energy Systems Language.

References

Coultas, C.L. and Y.-P. Hsieh (eds.), *Ecology and Management of Tidal Marshes: A Model from the Gulf of Mexico*. St Lucie Press, Delray Beach, FL, 355pp.

Forrester, J. 1961. *Industrial Dynamics*. MIT Press, Cambridge, MA.

Forrester, J. 1968. *Principles of Systems*. Wright-Allen Press, Cambridge, MA.

Montague, C.L. and H.T. Odum. 1997. Introduction: The intertidal marshes of Florida's Gulf coast. In Coultas, C.L. and Y.-P. Hsieh (eds.), *Ecology and Management of Tidal Marshes: A Model from the Gulf of Mexico*. St Lucie Press, Delray Beach, FL, pp. 1–7, 355pp.

Odum, H.T. 1983. *Systems Ecology: An Introduction*. Wiley, New York, 644pp.

Odum, H.T. 1994. *Ecological and General Systems: An Introduction to Systems Ecology*. University Press of Colorado, Boulder, CO, 644pp. (revised edition of Odum, 1983).

Odum, H.T. 1996. *Environmental Accounting: Energy and Environmental Decision Making*. Wiley, New York, 370pp.

Odum, H.T. and E.C. Odum. 2000. *Modeling for All Scales: An Introduction to System Simulation*. Academic Press, London, 458pp.

Odum, H.T. and E.C. Odum. 2001. *A Prosperous Way Down*. University Press of Colorado, Boulder, CO, 326pp.

29
Ecological Modeling and Simulation: From Historical Development to Individual-Based Modeling

David R.C. Hill
Blaise Pascal University

P. Coquillard
Université de Nice-Sophia Antipolis

29.1	Introduction	29-1
29.2	An Old Story?	29-2
29.3	Determinism or Probability?	29-5
29.4	Modeling Techniques	29-5
29.5	The Use of Models in Ecology	29-6
29.6	Models are Scientific Instruments	29-7
29.7	Levels of Organization and Methodological Choices	29-8
29.8	Individual-Based Models	29-9
29.9	Applications	29-12
29.10	Conclusion	29-15

29.1 Introduction

Ecology studies the relationships between living organisms (individuals)—vegetal or animal—and the environment in which they live. The biosphere in which we live is made up of the whole set of terrestrial, marine, and aerial ecosystems. A scientist who endeavors to explain the relationships and the interactions within an ecosystem must often be in contact with colleagues more specialized in other fields like chemistry, genetics, oceanography, geography, hydrology, ethology, and climatic sciences. Considering the complexity of this multidisciplinary study, modeling is a fundamental tool to understand ecosystems. A good modeling practice is to design a model with a precise goal in mind; a modeler expects that his model outputs (results) will help understanding the real system under study. The model final objective also helps selecting the model simplifying assumptions and the level of realism, taking into account the limits of our knowledge and the limits of our modeling techniques. Even if we had precise descriptions and observations, estimating future trends will always be considered very risky. Considering the lack of biological data, we see in Begon et al. (1990) that many ecologists focus on the following levels of organization: individuals, populations of individuals, and ecosystems.

In this chapter, we will first focus on the history of ecological modeling, starting in the twelfth century. A discussion will present the controversy opposing deterministic and probabilistic approaches in ecological modeling. We follow this discussion by an overview of the model types employed by ecological modelers. The main focus is then given to individual-based models (IBMs). They appeared in the seventies (Kaiser, 1976, 1979; Lomnicki, 1978) and have been more intensively used since the remarkable paper from Huston et al. (1988). Individual-based modeling is a bottom-up approach focusing on the individuals (i.e., the parts) of an ecological population (i.e., the system). Then, the modeler tries to understand how the properties of the population can emerge from the interaction among individuals. The strengths and weakness of IBMs will be presented. Despite their advantages, bottom-up approaches are not sufficient to build theories at the population level (Grimm, 1999). Indeed, more traditional mathematical models, with state variables, described "top-down" and used like black boxes, provide integrated views replying to relevant questions at the population level.

29.2 An Old Story?

The great adventure of modeling ecological processes began at the end of the twelfth century with regard to a story about rabbits, in Pisa, Italy. In this thriving commercial town, young Leonardo of Pisa (1175–1240) is very rapidly confronted with practical problems of arithmetic. Leornardo, son of Guilielmo and a member of the Bonacci family, is trying to solve what appears at first sight to be quite a simple problem: a couple of young rabbits will become adult and then will produce at each mating season, a new couple of young rabbits. After a new season, the new couple has grown up and is in turn able to produce another couple of rabbits. The question is how many rabbits will there be after "n" seasons?

In Table 29.1, it can be seen that the number of adult couples at season n is equal to the total number of couples at season $n-1$, and that the number of young couples at season n is equal to the total number of couples at season $n-2$. A little reflection convinces us very quickly that the total number of rabbit couples at season n can be computed with the following sequence:

$$T_n = T_{n-1} + T_{n-2} \tag{29.1}$$

This formula generates a series of numbers whose characteristic is that each number is equal to the sum of the two previous numbers. Generally, this sequence is known as Fibonacci numbers: 1, 1, 2, 3, 5, 8, 13, 21, 34, 55, 89, 144, 233, 377, 610... The Fibonacci name was given in the nineteenth century by Guillaume Libri, a mathematics historian (Leonardo of Pisa, was son of Bonacci: Filius Bonacci—which gave the Fibonacci nickname). Naturally, the Fibonacci sequence is not very realistic for this problem, since for instance it takes no account of mortality and running out of food resources. However, this sequence has numerous other interesting applications for basic ecological models. Indeed, Fibonacci numbers are interesting in understanding honeybees family trees, petals on flowers, seed heads, pinecones, leaf arrangements, and even shell spirals. All the previously cited natural elements use what mathematicians call the golden ratio. Curiously, we can perceive that the ratio T_n/T_{n-1} is converging to a particular value, $21/13 \simeq 1.615$; $34/21 \simeq 1.619$; $144/89 \simeq 1.617$; $610/377 \simeq 1.618$. This value is called the golden ratio or the

TABLE 29.1 Number of Couples at Season n

Season n	Total Number of Couples $(= T_n)$	Number of Adult Couples $(= T_{n-1})$	Number of Young Couples $(= T_{n-2})$
1	1	0	1
2	1	1	0
3	2	1	1
4	3	2	1
5	5	3	2
6	8	5	3
7	13	8	5
8	...	...	...

golden number (∼1.618034 and often represented by a Greek letter Phi [ϕ]) and it has also been used by humans in many artistic creations, including architecture.

Thereafter, several centuries go by before mathematicians take an interest in biological and ecological problems: Bernouilli in the eighteenth century proposes a mathematical theory of smallpox epidemic; then, in the nineteenth, Malthus and Quetelet take an interest in the dynamics of human populations. But it is Pierre François Verhulst (Verhulst, 1845), a Belgian mathematician, who invents in 1844 the famous S curve known as the logistic curve:

$$\frac{dy}{dt} = ry\left(\frac{K-y}{K}\right) \tag{29.2}$$

This is the foundation of mathematical modeling applied to biological sciences, strongly influencing the modeling of system dynamics throughout the twentieth century. This equation has been studied in depth, notably in its discrete form. It will also be integrated in the famous double equation system, attributed independently to Lokta (1925) and Volterra (1926), where they propose a mathematical formalism for the relationship between consumers and resources.

Finally, mathematical modeling of sciences, apart from the strict circle of Physics, does not take off until the twentieth century. In the nineteenth century, biological sciences were not particularly favorable to using mathematical formalism. According to Giorgio Israel this can be explained by "The old ambition, still alive and kicking, to achieve a unified mechanical and reductionistic description of the world, an ambition resisting all difficulties and all failures." Again, according to the same author: "The fundamental method of modeling in the 20th, is mathematical analogy (where the fragment of mathematics unifies all the phenomena it is supposed to represent), and no longer mechanical analogy, which has been for a very long time, the principal mathematical method."

With a mathematical analogy, Leslie (1945) presented to an adaptation of Markov Chains to the dynamics of populations. This technique is interesting for modeling the changes in age-structured populations (Figure 29.1).

In 1965, Lefkovitch introduced an alternative approach. A similar matrix model considers population growth in organisms grouped by stages instead of age (Lefkovitch, 1965) (stage-specific survival rates). Kent Holsinger gives the following reasons to explain why Lefkovitch models are often preferred to Leslie matrix:

- It's often difficult or impossible to age animals and plants accurately.
- In some organisms, especially perennial plants, survivorship and fecundity are more related to size (or some other variable by which a population might be stage-classified) than to age.
- In some organisms, especially herbaceous perennial plants, individuals may actually revisit stages they already left, e.g., they may get smaller from one season to the next.
- Focusing on life-cycle stages helps to focus attention on identifying the critical transitions that may provide opportunities for management. (Holsinger, 2005)

At the end of the eighties, Caswell (1989) also adapted Leslie's matrix to model the development of a species of a vegetal species (*Dipsacus sylvestris*) in six development stages (from seed to flowering plant). Finally,

$$P = \begin{bmatrix} F_1 & F_2 & \ldots & F_{n-1} & F_n \\ p_1 & 0 & \ldots & 0 & 0 \\ 0 & p_2 & \ldots & 0 & 0 \\ \vdots & \vdots & & \vdots & \vdots \\ 0 & 0 & \ldots & p_{n-1} & 0 \end{bmatrix}$$

FIGURE 29.1 A Leslie matrix. The first row contains the fertilities (F_i) of each the n class of age. p_i represents the transition probability (i.e., the survival probability) from one class of age to the next.

$$M_{AS}(p) = \begin{bmatrix} F & F & \ldots & F & F \\ T & 0 & \ldots & 0 & 0 \\ 0 & T & \ldots & 0 & 0 \\ \ldots & \ldots & \ldots & \ldots & \ldots \\ 0 & 0 & \ldots & T & T \end{bmatrix}$$

(a)

Where the submatrices F (fertilities) and T (transition) are for instance:

$$F = \begin{bmatrix} f_1 & f_2 & f_3 \\ 0 & 0 & 0 \\ 0 & 0 & 0 \end{bmatrix} \quad \text{and} \quad T = \begin{bmatrix} p_1 & 0 & 0 \\ q_1 & p_2 & 0 \\ 0 & q_2 & q_3 \end{bmatrix}$$

(b)

FIGURE 29.2 (a) A stage-by-age model or multistate model. (b) Fertilities and transition matrix.

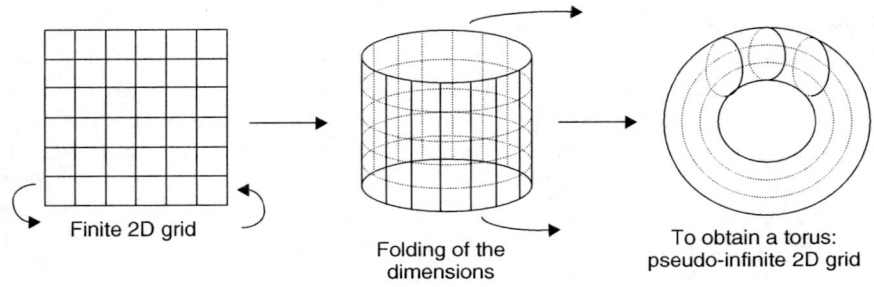

Finite 2D grid → Folding of the dimensions → To obtain a torus: pseudo-infinite 2D grid

FIGURE 29.3 Folding of a finite 2D grid to obtain a pseudo-infinite grid.

Lebreton (2005) proposed a stage-by-age models (or a multistate model) uniting the two characteristics state and time (Figure 29.2[a] and Figure 29.2[b]).

After the Second World War, the original thoughts on self-reproducing cellular automata were introduced. "This introduction can be traced back to 1948 when John von Neumann gave a talk entitled "The General and Logical Theory of Automata" (Von Neumann, 1951). More than a decade later, Von Neuman, Ulam, and John Conway introduced self-reproducing cellular automata (Von Neumann, 1966; Burks, 1970) and the Game of Life (Gardner, 1970). Von Neumann and Ulam were modeling a minimal biological self-reproduction and the Conway mathematical game was introduced to a wide public thanks to columns written by Martin Gardner in Scientific American. The result was a famous synchronous spatial model, centered on cells, and with a time-based evolution. The future of a cell depends on rules relying on the other cells in its neighborhood. Each cell on a pseudo-infinite 2D grid (obtained by folding, Figure 29.3) has two potential states: dead or alive. The generated cells are clones, exactly the same as the parent. From these basic elements, more complex models can take into account heredity, variability, fitness and other characteristics, which can be useful to model for instance the evolution by means of natural selection. Here are the basic transition rules applied at each time step in the first issue of the Game of Life:

- If two or three neighbors of a cell are alive and the cell itself is currently alive, its next state is alive.
- If three neighbors of a cell are alive and the cell itself is currently not alive, its next state is alive.
- Otherwise the next cell state is dead.

Those simple rules were sufficient to give rise to order from an apparent "chaos." Scientists rapidly identified patterns like blinkers, gliders, guns, which grow and evolve for very long periods. Some other patterns lead to dying or stagnating configurations. Since the theory of cellular automata was introduced, many enhancements were proposed and the theory was simplified in the 1980s by Langton and Byl (Wolfram,

1986). Cellular automata have also been recently embedded in the powerful Cell-DEVS environment, which has been successfully applied to different ecological models (Wainer and Giambiasi, 2001). We would not stay too long on cellular automata since they are detailed in a specific chapter of this book, proposed by Peter Sloot.

29.3 Determinism or Probability?

Evidently, from what has just been said, the mathematical approach, i.e., deterministic, represents the base on which modeling and simulation in environmental sciences has been developed. The introduction of randomness with Monte Carlo simulation and the question of using models with probabilistic components provoke fierce scientific controversy. For example, Papoulis explains: "The controversy of determinism and causality versus randomness and probability has been the topic of extensive discussions […] the phenomenon is thus inherently deterministic, and probabilistic considerations are necessary only because of our ignorance" (Papoulis, 1965). Up to the early seventies, the majority of models concerning ecology are deterministic. The position of Norman E. Kowal (1971, pp. 123–171) sums up the aversion of scientists at that time to use probabilistic processes. Concerning ecological processes, which present: "Changes in Space, or in Space and Time," he writes: "Since there is more than one independent variable (two or three for space, and, perhaps, one for time), the required mathematical theory becomes very complex and difficult to work analytically. The most useful mathematical structures to use as models of such systems are partial differential equations […] these models will probably always be solved by numerical approximation on digital computers." In the last sentence of his paragraph he also states that: "Probability density functions may also be used." In the Coda Volume I of System Analysis and Simulation in Ecology, Bernard C. Patten writes: "Simulation models do not have to reproduce dynamic behavior realistically to be useful […] the thought that goes into them may be their greatest value" (Patten, 1971). Patten preferred to reduce the realism of the model rather than include stochastic aspects. From our point of view, the final modeling goal has to be borne in mind when deciding whether or not a modeler can abandon the reproduction of realistic behavior.

Thanks to the steady increase in memory capacity and calculation speed, to the emergence of procedural and object-oriented programming languages, the limits of ecosystem modeling have rapidly extended. Dealing with spatial systems and processes dependant on time, scientists found themselves confronted with such a complexity that deterministic mathematics alone could not resolve (Jorgensen, 1994).

29.4 Modeling Techniques

Whether deterministic or not, there is a plethora of modeling techniques. Using the "Science direct" database from Elsevier, we have built a classification of the most employed techniques found in papers published in the Ecological Modeling journal since 1975 (Table 29.2). Most of them are not specific to ecological modeling and they are detailed in other chapters of this book.

Two expressions in Table 29.2 do not belong to usual simulation vocabulary: Individual Based Model and Gap Model. It is an example of vocabulary introduced by ecologists to qualify some kinds of simulation models. IBMs have already been introduced, so we will present the Gap models. They are dedicated to the simulation of vast forest spaces, discretized into small units (a few m^2) on which the number of trees of different categories and species, the transition probabilities and the reproduction success probabilities are known. The term Gap model comes from the fact that these models were developed originally to simulate the behavior of a forest area in which, over the course of time, a natural process of clearings healed up more or less rapidly depending on characteristics of their immediate environment.

If we analyze more deeply the published databases in ecological modeling, we can see that ecological specialists have now been using discrete simulation and IBMs more intensively particularly over the last two decades (Grimm et al., 1999). Multiagent models can be considered as a special case of IBMs where

TABLE 29.2 Results Extracted using the Science Direct Database on the Ecological Modeling Journal (between 1975 and 2005). Expressions were Researched in the Titles, Keywords, and Abstracts. In Brackets the Results Obtained by Extending the Search to Journals Filed in Environmental Sciences

Expression	Number of Articles
Model OR Modeling	3576 (36308)
Simulation	1361 (8146)
Mathematical model	298 (9396)
Individual-based model	106
Neural network	104 (1801)
Markov OR Leslie	91 (2273)
Cellular automata	46 (441)
Bayesian statistics	42 (120)
Mixed model	25 (561)
Gap model	26
Multiagent	6 (28)
Multimodel(ing)	4 (96)

individuals have shown a social behavior. This modeling technique is not specific to ecological modeling, it has been introduced in Artificial Intelligence and more precisely in Distributed Artificial Intelligence. Adelinde Uhrmacher is devoting a chapter of this book to agent-oriented modeling. In ecological modeling, this approach is particularly interesting for of ethological systems. We have for instance studied herbivorous animals, and also the memory of ewes at pasture using this approach (Dumont and Hill, 2001, 2004).

This development of individual-based modeling and its derivative can be explained first by the availability of powerful desktop computers, of user friendly software development environments, and of prototyping design methods. With such tools and techniques, ecological modelers can study the complexity of many systems they are interested in, in terms of number of animals, plants (or trees), strips of land, volumes of air, flow of energy, interacting in time and in three dimensions.

This quick review of techniques in ecological models shows that a long road is ahead to use multiagents, coupled models, mixed models and multimodels with different formalisms. The formalization of models, using DEVS (Zeigler et al., 2000) or other more formal specification techniques, is still extremely rare in ecological modeling. In our opinion, their use can be very promising, the introduction of Cell-DEVS mentioned above, and other recent applications demonstrate the interest of the simulation community in applying such advances to ecological and environmental models (Muzy et al., 2005).

29.5 The Use of Models in Ecology

In ecological modeling, the knowledge acquired these last decades experienced a spectacular growth correlatively with the mastering of new sampling techniques (satellite and space imagery; radiogoniometric follow-up of animals; automation of the physicochemical data acquisition of air and water), of numerical analysis techniques (statistical analysis of multidimensional data and analysis of time series), and of data-processing tool (hardware and software). At the same time, this amount of data made decision makers aware of a wiser management of human activities. Table 29.3 presents several ecological topics discussed in the Ecological Modeling Journal (still between 1975 and 2005).

In Table 29.3, we notice that the predominant subject of interest is incontestably the problem of biodiversity, followed, but at a considerable distance, by global change, forestry problems, and population dynamics. We therefore find here subjects covered by the media in the news, even though fundamental research subjects are not neglected, as is proved by the 334 articles on population dynamics or the prey–predator relationship. The greenhouse effect is almost not covered by the Ecological Modeling journal. This can also be noticed in the "Environmental Sciences" section where only 197 references are found on this subject, which, in reality, is dealt with by climatology specialists in collaboration with ecologists.

TABLE 29.3 Number of Articles found in the Ecological Modeling Review (1975–2005) Related to Several Ecological Topics

Expression	Number of Articles
Organic matter	100
Nitrogen OR nitrogen cycle	295
Carbon OR carbon cycle	290
Global change	344
Greenhouse effect	2
Fish model	122
Fishing model	83
Trees OR forest	636
Populations dynamics	344
Predator–Prey	97
Productivity	187
Behavioral model	192
Biodiversity	2947
Plant architecture	34

TABLE 29.4 Number of Articles found in the Ecological Modeling Review (1975–2005) Related to Ecosystems

Expression	Number of Articles
Taïga	44
Tundra	145
Temperate forest	130
Tropical forest	198
Meadows	206
Arid ecosystem	18
Lake	141
River	227
Desert	275
Marine ecosystem	187
Estuarine ecosystem	29
Coral reef	10
Alga(e)	39

From Table 29.4, we observe that the ecosystems mostly studied by ecologists are mainly terrestrial ecosystems, notably forest, tropical or not, where an attempt is made to predict their evolution in terms of production, population structure, and biodiversity for future decades with a supposed changing climate. Lakes and rivers are also the subject of numerous studies. One could advance the hypothesis that the management of fresh and drinkable water resources arises today as one of the major challenges over the next decades. Again from Table 29.4 we note that marine ecosystems are less studied, in spite of their extremely dominant situation on the globe, and their capital importance in regulating climate. We have developed models in oceanography and we could explain this small number of studies by the fact that oceanographic model parameters are difficult to measure, involving costly and occasionally dangerous operations (Hill et al, 1998; Coquillard et al., 2000).

29.6 Models are Scientific Instruments

From our simulation experience in various application fields, Ecological models basically do not differ—not even by their complexity—from models developed in other disciplines such as meteorology and nuclear physics, or from the models developed for manufacturing systems.

The incredible complexity of the operation of an ecosystem cannot be seized by the simple acquisition of the whole set of parameters that characterizes it, as this has been the case for a long time when ecology was confined within a descriptive approach. The latter was a necessary stage, but the multiple interactions and feedbacks within ecosystems reveal behaviors which one could not seize by the interpretation of data collected (were they exhaustive). Even if we do not know all the ecosystem parts, studies in complexity showed that the behavior of a complex system is not equivalent to the sum of the behaviors of the parts (this is known as the "system effect"). Only a software, modeling the interaction of the various parts of a system, can reveal the emerging behaviors. Even if models could also show impossible behaviors that only field experts will be able to detect, they are more and more used for a better comprehension of ecosystems. In his state-of-the-art textbook, Jorgensen (1994) summarizes in four points the advantages of modeling:

1. Models have their utility in the *monitoring of complex systems*.
2. Models can be used *to reveal the properties of ecological systems*.
3. Models can *show deficiencies in our knowledge* and can be used to define priorities in research.
4. Models are useful *to test scientific assumptions*, insofar as the model can simulate the reactions of the ecosystem, which can be compared with the observations.

29.7 Levels of Organization and Methodological Choices

In practice, the development of a model is based on the two following constraints:

1. *The objectives to be reached* (which types of results are we expecting?).
2. *The state of our knowledge concerning the studied system and the data at our disposal* (or at least what it is reasonable to hope to acquire in an assigned time).

These two constraints will define the *level of organization* (or the *scale of the study*). The choice of this scale is in direct relationship with the complexity of the model. Will we be interested in the individuals, in parts of individuals, in sets of individuals, in whole populations even in sets of populations? It is also necessary to have in mind that the level of organization will also influence the choice of the modeling technique that will be implemented. For instance, if we organize our model around the population level, an individual-based modeling is completely unnecessary. In addition, there must be a close consistency between the objectives (first point) and the available data (second point). If we only have data at the molecular level, will it be really consistent to expect results at the population level? The identification of the level of organization imposes a thorough examination of the data: the objectives being fixed and thus the working scale determined, it is still necessary to have available data relating to this level of organization. It is a frequent case that for a given level it is advisable to obtain data concerning a lower scale. Even if the level of a population is retained, it could be necessary to collect data relating to the individuals. Indeed, the behavior of the population results from the individual interactions. Biological individuals themselves are not free from influences coming from the higher levels of the organization hierarchy: other populations, environmental factors. How many factors do we have to take into account? Which one among them can be regarded as negligible? There is no absolute rule in this field; a wise approach would be to bear in mind the main goal of the modeling (i.e., the principal expected results). This will help to simplify the making of choices.

The level of organization being selected according to the objectives (Figure 29.4), it is appropriate to specify what is the level of detail. Thus, for a model of forest growth (with a level of organization corresponding to a population), will we have to take into account, or not, the following nonexhaustive set of parameters?

- Seasonal variation of the light intensity.
- CO_2 partial pressure of the atmosphere.
- Competition for water resources and ground nutrients.
- Competition for space.

Ecological Modeling and Simulation

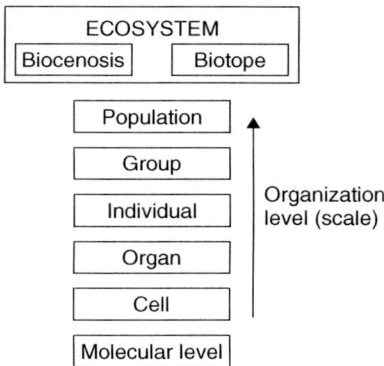

FIGURE 29.4 Organization levels.

- Density of the settlement.
- Competition for light.
- Competition with the shrubby and herbaceous species.

One quickly sees that objectives and data will interact; the choices will be in accordance with the experimental framework. Thus, if we want to model the growth of a local forest over the next 5 years, it is certainly not necessary to take into account the possible infinitesimal variations of the CO_2 on this interval of time. In the same way, the seasonal variation of the light intensity received by the settlement does not appear crucial in the modeling, the whole settlement being subjected to the same light conditions. Lastly, the competition for resources with the shrubby/herbaceous species will have to be neglected if we consider that trees escape from this competition. We can also suppose that the forest ecosystem is relatively homogeneous from this point of view.

A systematic increase in the model complexity by addition of variables will often not make a substantial improvement in terms of model validation. Beyond a certain point, the addition of new variables does nothing but increase the complexity of the model and accumulate uncertainties. It even occurs that these increases in uncertainty make the model radically diverge from the real system under study. It is often advisable to prefer holistic variables to several elementary variables whose uncertainties can only obliterate the quality of the model. It goes without saying that an increase in model complexity will add to the implementation difficulties, the ability to check the internal consistency of the software is reduced, and computing speed problems can also occur even with our fastest computers. Unless we have specific goals requiring the implementation of many details, simplification is a virtue in modeling.

Now that we have presented some methodological elements, we will give our main focus to IBMs. We have seen in the previous sections that many modeling techniques used in ecological modeling are common to other research fields, most of them are presented in this textbook and our preference goes to individual-based modeling. This choice is not a way to champion this technique among others and we will also discuss its drawbacks. However, this technique has met with an interesting development linked to the evolution of computer technologies. In the next sections we will not distinguish IBMs from individually oriented models (IOM), though the latter can be more powerful, see Fishwick et al. (1998).

29.8 Individual-Based Models

The IBM approach completes the set of mathematical methods that are still interesting for many applications (Grimm, 1994; Sultangazin, 2004). For instance, differential equations or partial differential equations are very efficient to give a rough estimation of the spatial evolution of large areas. However, if we have previously seen how the latter were limited in simulating biological processes (Grimm and Uchmanski, 1994) when we face questions requiring a high level of details, ecological modeling often has to take into account:

- the diversity of individuals,
- the spatial heterogeneity of the environment,
- the changing interaction network (and changes in biotic structures),
- the discrete and distant interactions,
- some random aspects and behavior (i.e., random spatial interactions).

At the end of the 1980s, biologists and computer scientists highlighted the convergence between biological concepts and object model concepts (Huston et al., 1988). Indeed, IBMs are often implemented by object-oriented models (Coquillard and Hill, 1997) or by multiagent models when there is a need to represent an autonomous social behavior of individuals involved in a common goal (Ferber, 1999). It is by the way remarkable that the Simula language (Dahl and Nygaard, 1966), which introduced the essence of object-orientation in computer science, is still providing one of the most convenient ways to implement IBMs without additional libraries. Even if the authors of this chapter have abandoned this language, they want to emphasize the work done by some of their pioneer colleagues in Germany (Kiel Ecology Center) who built various models dealing with a wide range of delicate ecological interactions in Simula (Breckling et al., 1998).

An object-oriented simulation will be based on individual organisms rather than on aggregated variables. Classes and their relationships can define the biological taxonomy retained for the model (Baveco, 1997). A domain analysis will produce an object model tightly associated with the biological model and helping the communication process between both scientific communities (Computer Science/Biology) (Hill, 1996). Coupling object-oriented concepts with individual-based modeling has the main following advantages:

- It avoids difficult or impossible mathematical modeling (i.e., a competition between three different species including spatial constraints). It is also possible to enrich a model using inheritance, which recovers all its original meaning. In addition, object classes can take into account a part of mathematical modeling to obtain combined simulations if needed (mixing the discrete and continuous approach as explained in the methodological chapters of this book).
- It avoids keeping scrupulous track of stands over longtime periods, which are necessary to feed Markovian analysis. However, a lot of fieldwork is necessary as well as a substantial knowledge of the species modeled by object classes. Class attributes can reflect the current expert knowledge; the detail level will depend on the expected model result.
- It takes into account the spatial aspects of ecosystems, which is hardly possible with partial differential equations (compartment models), or with classical Markovian analysis except with time-dependent matrix (nonhomogeneous chains) which sets (i) the problem of the investigation time cost and (ii) initialization difficulties in variable conditions (i.e., combinatorial exploration).
- It provides the possibility to manage, for each individual, the set of all the parameters which the biologist decides to integrate in his model. The management of individuals, and correlatively their physiological variations, enables the refinement of the model to close reality with the detailed level the user wants.

We think that the two last points are essential. Even if spatial diffusion processes can be modeled by partial differential equations, it is indeed impossible to take into account spatial constraints and distant stochastic interactions (propagation of cuttings, of seeds, of sparks hundreds of meters away and sometimes a few kilometers away). For instance, Figure 29.5(b) presents the spreading of *Caulerpa taxifolia* in the harbor of "Villefranche sur Mer," France, this model will be presented in details in the next section. *C. taxifolia* is a green alga of tropical origin introduced by mistake in the Mediterranean Sea (Meinesz and Hesse, 1991), it is currently colonizing the North Mediterranean (Hill, 1997) and we quote this application here since it will give a concrete example of the importance of spatial constraints in some ecological models. In this

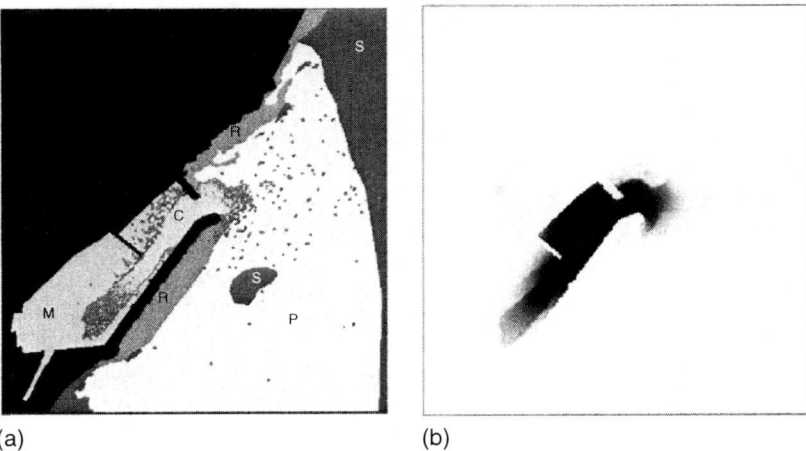

FIGURE 29.5 (a) Map provided by the simulation model coupled with a GIS; C = *Caulerpa taxifolia*; R = rocks; M = harbor mud; S = sediment; P = *Posidonia oceanica* beds. Inside the harbor: C = *C. taxifolia*, it appears in a two gray scale; light gray corresponds to densities higher than 80%. (b) Result of a spectral analysis with 255 replicates.

model, the individual behavior including the spreading of cuttings and the stolon growth are important colonization factors. In Figure 29.5, inside the harbor, denoted by a "C," *C. taxifolia* appears mainly in light gray, and spotting outside the harbor correspond to the spreading of *Caulerpa* cuttings. To obtain colonization maps, we have to use spatial constraints, including the shape of the harbor, the bathymetry, the position of undersea substrates, which are key elements. Such constraints impose the coupling of an individual-based discrete-event simulation with a Geographical Information System and also the design of discrete geometry algorithms (Hill et al., 1998).

As opposed to what can be modeled with IBMs, Huston, DeAngelis, and Post (Huston et al., 1988) note that when we use formal models we often combine many individual organisms to describe a single variable and each individual is assumed to have an equal effect on every other individual. This kind of modeling is often mandatory because of the mathematical tools used. The resulting problem is that it violates two main biological principles:

1. Each individual of a given population is different, with regard to behavior and physiology.
2. Interactions between individuals are inherently local. Spatial aspects are often key factors in biology.

IBMs do not require the same level of simplifying assumptions used for state-variable models. This fact mainly explains the rise of this modeling approach observed by Judson, Breckling, and Müller at the beginning of the 1990s (Breckling and Müller, 1994; Judson, 1994). The object-oriented approach enables us to describe individual organisms that change their inner states and their environment. In Figure 29.6 we see again how an IBM model can take into account spatial constraints by coupling the simulation model with a geographical information system (GIS). Figure 29.6 presents a modeling of herbivorous behavior in the French "Massif Central." GIS combined with GPS devices assigned to horses and cows were used to obtain precious field data (such as animal behavior, animal pasturing choices, and location).

In an IBM, individuals are able to react to their changing environmental state and to reproduce in their "virtual" environment. This could sound like a definition of an "artificial life," but the scientific community studying "artificial life" has the following objective: they want to give access to the domain of *life-as-it-could-be* by extending the limits of our current biology knowledge described as *life-as-we-know-it*. For more details, the interested reader should consult the official Journal of the International Society of Artificial Life edited by MIT Press. A common interest of both disciplines is the study of emergent properties. With an IBM it is possible to understand how local interactions between individuals contribute

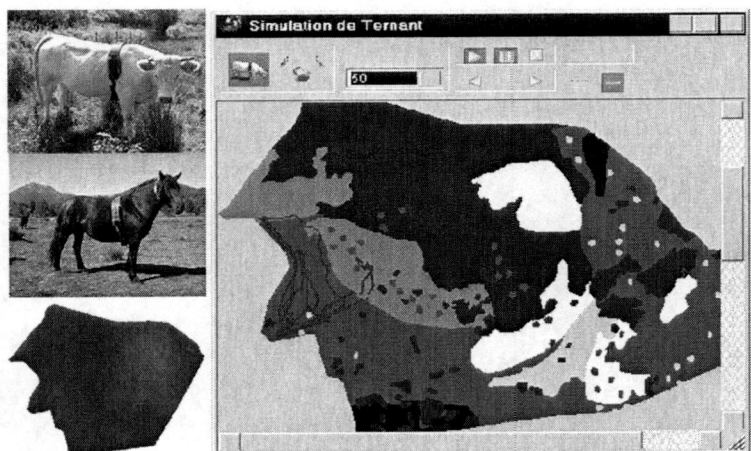

FIGURE 29.6 Modeling of herbivore followed by GPS devices (horses and cows).

to structural changes at higher levels (DeAngelis and Gross, 1992; Grimm and Railsback, 2005). In an IBM, we try to set up a model where local interactions work as far as possible as they do in nature (according to field observation). Local interactions enable the emergence of properties observed in an ecological context (Breckling et al., 2005). Scientists, working in Social Sciences, study the use of primitive rules assigned to individuals to observe a more global behavior; they often use multiagent systems. As we said, we can consider a multiagent system as an extension of an IBM in which social interactions have a significant role to explain the system behavior.

IBMs do also present many limits; most of them can be rapidly reached depending on our computing capacity. The main limit is that IBMs do not scale well in space, they are best suited to study ecosystems at a local or regional scale. In addition, the number of individuals and individual interactions is also limited (linked computing capacity). This is not crucial when we work on the scale of an ecosystem (millions of individuals can be simulated over a 100 years even on a desktop personal computer).

To work on different scales, we have to face the challenge of simulating scale transfers. The team of scientists running the Earth Simulator supercomputer in Japan introduced what they called Holistic Simulation to explore complex interdependencies between micro- and macroscale processes (Sato, 2003). On such a supercomputer, it is possible to simulate a reasonable number of different scales with the major interactions between the most significant processes. Many scientists are trying to develop simulations of the interactions between the main processes at different scales to help understanding the behavior of complex systems (not only ecosystems). However, only Japan possesses a supercomputer dedicated to this purpose, and it has been the World's fastest public computer for two successive years (2003 and 2004). Multiscale modeling methods are presented in this book by Mark Sheppard.

29.9 Applications

We will now give more details for a couple of linked applications where the development of an IBM was necessary. The first application was already briefly introduced, it deals with the spreading of *C. taxifolia* in the North Mediterranean. *Posidonia oceanica*, a protected Mediterranean species, and other more common species are endangered by this colonization. The second application deals with the simulation of *C. taxifolia* biocontrol using the *Elysia subornata* (a marine slug). This biocontrol simulation was achieved using a multimodel embedding various models of different formalisms. A chapter of this book is devoted to multimodeling (by Minho Park, Paul Fishwick, and Jinho Lee).

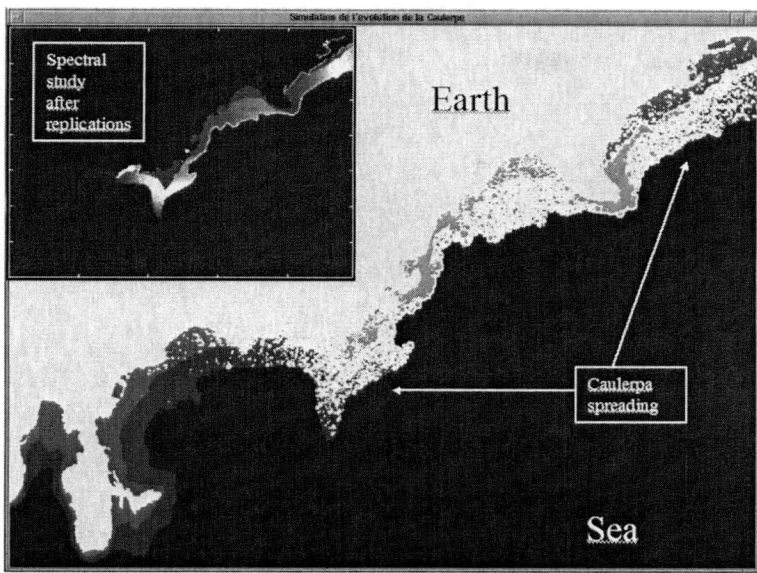

FIGURE 29.7 Simulation of *C. taxifolia* spreading around Monaco 12 years after its introduction. A spectral result obtained for the same geographical site is also presented on the top left of the figure.

In 1984, the French coast of the Mediterranean Sea near Monaco was the initial site of the development of *C. taxifolia*. Twelve years latter, this species had colonized several thousand hectares of the French and Italian coasts and was detected in numerous places of the northwestern Mediterranean coast, from Croatia (Adriatic sea) to the Balearic Islands (Spain). Figure 29.7 presents the simulated situation around Monaco 12 years after the alga introduction and a spectral result that will be discussed a few paragraphs later. This development has locally induced an intense alteration of the coastal ecosystems both on endogenous species distribution (alga, cnidaria, sponges, echinoderms, fishes, etc.) as well as on the ecosystem functioning. To predict the development of *C. taxifolia*, a simulation study was undertaken through an interdisciplinary joint venture between marine ecologists, biologists, and computer scientists. An IBM had been implemented to take into account spatial interactions and anthropic dispersion (dispersion by people) or activities such as eradication. The attentive reader will recall that this model was based on a coupling of a Geographical Information System with a stochastic discrete-event simulation, and was able to deal with distant stochastic interactions. Even though the model had to cope with incomplete data and sampling difficulties encountered in the hostile environment of the sea, interesting calibration and validation procedures specified by oceanographers have been successfully achieved. The calibration has been achieved on various sites and the model results matched a satisfying level of prediction at different spatial scales (from a few centimeters to a few kilometers). Our calibration dealt with the local and spatial patterns of expansion, the increase of *C. taxifolia* biomass, the increase in covered surfaces, and the invasive behavior toward existing communities. The fact that we knew the initial settlements and that we had been following the maps of *Caulerpa* evolution since 1988 was of great help in the calibration process. Indeed, concrete evolution maps have been successfully compared with simulated maps obtained by a spectral analysis of stochastic spatial results (Hill, 1997). Figure 29.5(b) presents on the right a spectral analysis of simulation results over 5 years, using 255 replicates on Villefranche-sur-Mer harbor (La Darse) site. This kind of result can be understood as a map of colonization probabilities; the darker the colors, the higher the frequency of settlement. Spatial aspects in the spreading of cuttings did take into account geographical data available such as harbor walls, bathymetry, and different substrates. More details on how the model determines whether or not a specific place was frequently or infrequently settled are given in Hill et al. (1998).

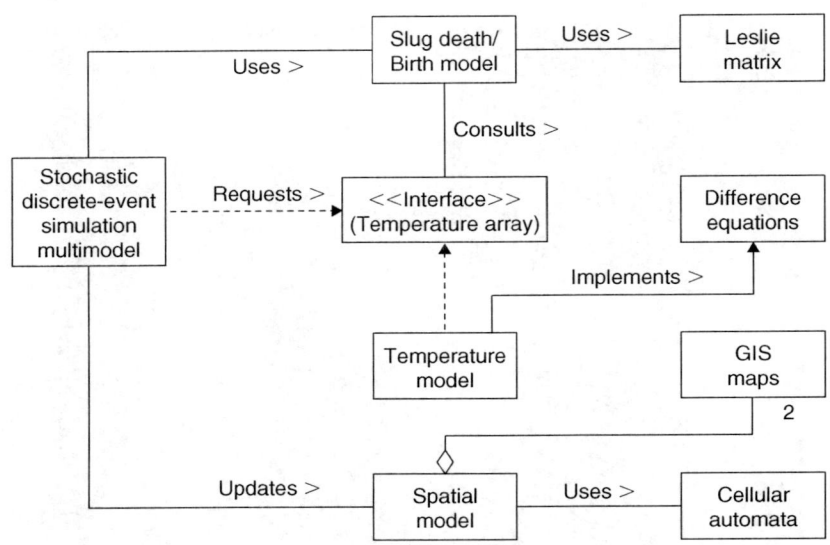

FIGURE 29.8 Excerpt of the UML class diagram for the biocontrol multimodel.

After an intensive use of this first model, we developed at the end of the 1990s a multimodeling simulation to assess the potential of a biocontrol of the alga *C. taxifolia* by means of a small marine slug of tropical origin: *Elysia subornata*. Biological and ecological parameters were considered as key factors related to the behavior of *E. subornata* toward *C. taxifolia* and toward the Mediterranean conditions. To this end, growth, survival, reproduction, feeding on *C. taxifolia*, and foraging of *E. subornata* were studied through laboratory experiments, details are given in Coquillard et al. (2000). Simulations, taking into account spatial effects, were in agreement with the laboratory experiments, the results demonstrated that the greatest impacts on *Caulerpa* were obtained using either just some adults or mixing adults and juvenile slugs. In any case, better results were obtained by a scattering of slugs on isolated spots rather than on clusters (with constant surfaces). Lastly, the choice of a suitable date for scattering increases the weak consumption of *Caulerpa* resulting from the scattering of juvenile slugs. To set up experiments in large mesocosm, an E.E.C. agreement is now necessary. Figure 29.8 presents a metamodel of this simulation application using the UML (unified modeling language). The metamodel describes the multimodel used for this biocontrol application; it integrates submodels of different formalisms, and can still be considered as an IBM, since at its main level we do consider individual differences and local interactions between individuals.

Many other applications can be found in which IBM or multiagent simulations are of greatest interest. For instance, the simulation of spatial interactions was essential in the modeling of the reproduction of the fine-scale horizontal heterogeneity of a pure grass sward with its dynamics (Lafarge et al., 2005). We also tried other formalisms like the metamodeling approach using artificial neural networks (ANN) trained by the simulation results (Aussem and Hill, 1999). After a long training, this approach was able to give a rough prediction of surface colonization but it was impossible to obtain colonization maps. More formal approaches have also been studied though they are still using discrete-event simulation. In a recent paper, we show how initial ideas and work proposed by Vasconcelos and Zeigler (1993) for fire spreading and simulation can benefit from the recent advances in the theory of modeling and simulation (Muzy et al., 2005). The last point we would like to highlight before ending this application section is the impact of visualization tools. In the domain of ecological modeling or environmental modeling these tools are often necessary for communication between specialists. Not only do they provide substantial gains in model validation, but as in many other domains, the involvement of decision makers will also depend on the kind

Ecological Modeling and Simulation

FIGURE 29.9 Snapshot of the postprocessed visualization tool designed for a coupling with an IBM model of forest fire simulations.

and variety of presented results. A snapshot of a visualization tool we developed for fire simulation is given in Figure 29.9. Decision makers should, however, be warned that visualization can lead to wrong decisions (either intentionally or unintentionally). In the case of stochastic simulations under spatial constraints, end users have to be aware that the animation of simulation results they are watching is only a very small sample of what could really occur. However, maps presenting spectral results do integrate a sufficient number of replications and are thus a better tool for decision-making.

29.10 Conclusion

In this chapter, we have exposed a short history of ecological modeling followed by a survey of current modeling practices using the best journal for publications in this research field. We have given some general indications concerning methodological choices including the importance of stochastic modeling. Design problems such as the selection of an organization level have also been discussed. We hope that the reader has seen that ecological modeling could benefit from many modeling formalisms, most of them are presented in different chapters of this book. Among the various formalisms we wanted to explain why more ecological modelers started to build IBMs. This choice was retained since this technique is not presented in other chapters of this book and because IBMs are of great help when we want to study the role of individual difference, space, and diversity. With such models, we use what computer scientist would call a bottom-up approach, since the emergence of ecological properties can be observed from the activity of individual behavior. Two concrete applications have been presented dealing with stochastic simulations under spatial constraints (*C. taxifolia* spreading and biocontrol in the North Mediterranean). However, we must also warn people that taking this approach, most IBMs rapidly lead to complex simulation programs (Lorek and Sonnenschein, 1999) and they are only interesting for limited spatial scales (local and regional). In addition, IBMs need a large amount of attention for the design, development, and debugging (model verification). Software programming and statistical analysis experience is required when dealing with stochastic simulation programs; interesting tools are presented in Kleijnen and Groenendaal (1992). Lorek and Sonnenschein present software frameworks designed to help the building of IBMs, most of the time they are built on top of general purpose programming languages, some rely on simulation languages. IBMs can be extended and linked with many other formalisms in a multimodel (Fishwick et al., 1998; Coquillard et al., 2000). Whatever the modeling technique, it is important to remember that serious calibration, as a first stage of model validation, can be achieved only if we have collected enough ecological information, often supposing an extensive fieldwork.

References

Aussem, A. and Hill, D.R.C. 1999. Wedding connectionist and algorithmic modelling towards forecasting *Caulerpa taxifolia* development in the north-western Mediterranean sea. *Ecological Modelling*, 120 (2–3): 225–236.

Baveco, H. 1997. Population dynamics in object-oriented and individual-based models. PhD Thesis, University of Amsterdam.

Begon, M., Harper, J.L., and Townsend, C.R. 1990. *Ecology—Individuals, Populations, Communities*. Birkhäuser, Basel.

Breckling, B. and Müller, F. 1994. Current trends in ecological modeling. *Ecological Modelling*, 75–76: 667–675.

Breckling, B., Reuter, H., Jopp, F., Hölker, F., Middelhoff, U., and Eschenbach, C. 1998. Simula meets ecology II: Object oriented model applications. *ASU News Letter*, 27 (2): 1–38.

Breckling, B., Müller, F., Reuter, H., Hölker, F., and Fränzle O. 2005. Emergent properties in individual-based ecological models—Introducing case studies in an ecosystem research context. *Ecological Modelling*, 186 (4): 376–388.

Burks, A.W. 1970. *Essays on Cellular Automata*. University of Illinois Press, Urbana, IL.

Caswell, H. 1989. *Matrix Population Models*. Sinauer Associates, Inc. Pub., Sunderland, MA.

Coquillard, P. and Hill, D.R.C. 1997. *Modélisation et simulation d'écosystèmes*. Masson Ed., Paris.

Coquillard, P., Thibaut, T., Hill, D.R.C., Gueugnot, J., Mazel, C., and Coquillard, Y. 2000. Simulation of the mollusc Ascoglossa Elysia subornata population dynamics: Application to the potential biocontrol of *Caulerpa taxifolia* growth in the Mediterranean Sea. *Ecological Modelling*, 135 (1): 1–16.

Dahl, O.J. and Nygaard, K. 1966. Simula—an ALGOL based simulation language. *Communication of the ACM*, 9 (9): 671–678.

DeAngelis, D.L. and Gross, L. 1992. *Individual-Based Models and Approaches in Ecology*. Chapman & Hall, New York.

Dumont, B. and Hill, D.R.C. 2001. Multi-agent simulation of group foraging in sheep: effects of spatial memory, conspecific attraction and plot size. *Ecological Modelling*, 141 (1–3): 201–215.

Dumont, B. and Hill, D. 2004. Spatially explicit models of group foraging by herbivores: What can Agent Based Models offer? *Animal research*, 53: 419–428.

Ferber, J. 1990. *Multi-Agent Systems: An Introduction to Distributed Artificial Intelligence*. Addison-Wesley, Reading, MA.

Fishwick, P.A., Sanderson, J.G., and Wolff, W.F. 1998. A Multimodeling basis for across-trophic-level ecosystem modeling: The Florida everglades example. *SCS Transactions on Simulation*, 15 (2): 76–89.

Gardner, M. 1970. The fantastic combinations of John Conway's New Solitaire Game of Life. *Scientific American*, 23 (4): 120–123.

Grimm V. and Railsback S.F. 2005. *Individual-Based Modeling and Ecology*, Princeton University Press. 428 p.

Grimm, V. 1994. Mathematical models and understanding in ecology. *Ecological Modelling*, 75–76: 641–651.

Grimm, V. 1999. Ten years of individual-based modelling in ecology: What have we learned and what could we learn in the future? *Ecological Modelling*, 115 (2–3): 129–148.

Grimm, V. and Uchmanski, J. 1994. Ecological systems are not dynamic systems: Some consequences of individual variability. *Predictability and Nonlinear Modelling in Natural Sciences and Economics*, pp. 248–259. Kluwer, Dordrecht.

Grimm, V., Wyszomirski, T., Aikman, D., and Uchmanski, J. 1999. Individual-based modelling and ecological theory: Synthesis of a workshop. *Ecological Modelling*, 115 (2–3): 275–282.

Hill, D.R.C. 1996. *Object-Oriented Analysis and Simulation*. Addison-Wesley, Wokingham.

Hill, D.R.C. 1997. Modélisation des processus d'expansion: application à Caulerpa taxifolia. *Proceedings of "Dynamics of invading Species"*, International Conference by invitation only, March 13–15, Paris, French Science Academy, Ed. Tec & Doc, pp. 219–230.

Hill, D.R.C., Coquillard, P., de Vaugelas, J., and Meinesz, A. 1998. An algorithmic model for invasive species: Application to Caulerpa taxifolia (Vahl) C. Agardh development in the North-Western Mediterranean Sea. *Ecological Modelling*, 109 (3): 251–266.

Holsinger, K. 2005. Population Viability analysis. (http://darwin.eeb.uconn.edu/eeb310/lecture-notes/pva.pdf) Creative Common Licence.

Huston, M., DeAngelis, D.L., and Post, W. 1988. New computer models unify ecological theory. *BioScience*, 38 (10): 682–691.

Jorgensen, S.E. 1994. *Fundamentals of Ecological Modeling* (2nd Edition). Elsevier, Amsterdam.

Judson, O. 1994. The rise of individual based models in ecology. *Trends in Ecology and Evolution*, 9: 9–14.

Kaiser, H. 1976. Quantitative description and simulation of stochastic behavior in dragonflies (*Aeschna cyanea*, Odonata). *Acta Biotheoretica*, 25: 163–210.

Kaiser, H. 1979. The dynamics of populations as result of the properties of individual animals. *Fortschritte der Zoologie*, 25: 109–136.

Kleijnen J. and Groenendaal W.V. 1992. *Simulation a Statistical Perspective*. Wiley, New York.

Kowal, N.E. 1971. A rationale for modeling dynamic ecological systems. *System Analysis and Simulation in Ecology*, pp. 123–171. Academic Press, New York and London.

Lafarge, M., Mazel, C., and Hill, D.R.C. 2005. A modelling of the tillering capable of reproducing the fine-scale horizontal heterogeneity of a pure grass sward and its dynamics. *Ecological Modelling*, 183 (1): 125–141.

Lebreton, J.-D. 2005. Dynamical and statistical models for exploited populations. *Australian and New Zealand Journal of Statistics*, 47 (1): 49–63.

Leslie P.H., 1945. On the use of matrices in certain population mathematics. *Biometrica*, 35, 183–212.

Lefkovitch, L.P. 1965. The study of population growth in organisms grouped by stages. *Biometrics*, 21, 1–18.

Lokta, A.J. 1925. *Elements of Physical Biology*. Williams & Wilkins, Baltimore.

Lomnicki, A. 1978. Individual differences between animals and the natural regulation of their numbers. *Journal of Animal Ecology*, 47: 461–475.

Lorek, H. and Sonnenschein, M. 1999. Modelling and simulation software to support individual-based ecological modeling. *Ecological Modelling*, 115: 199–216.

Meinesz, A. and Hesse, B. 1991. Introduction et Invasion de l'algue *Caulerpa taxifolia* en Méditerranée nord-occidentale, *Oceano, Acta*, 14 (4): 415–426.

Muzy, A., Innocenti, E., Aïello, A., Santucci, J.F., Santoni, P., and Hill, D.R.C. 2005. Modelling and simulation of ecological propagation processes: Application to fire spread. *Environmental Modelling & Software*, 20 (7): 827–842.

Papoulis, A. 1965. *Probability, Random Variables and Stochastic Processes*. McGraw-Hill, New York.

Patten, C. 1971. Coda Chapter of *System Analysis and Simulation in Ecology*. Academic Press, New York & London.

Sato, T. 2003. The earth simulator. *Holistic Simulation and Science Evolution*, 58 (2): 79–85.

Sultangazin, U., Ed. 2004. Mathematical modeling of ecological systems. *Mathematics and Computers in Simulation*, 67 (4–5): 275–500 (Special Issue).

Vasconcelos, M. and Zeigler, B.P. 1993. Discrete-event simulation of forest landscape response to fire disturbances. *Ecological Modelling*, 65 (3–4): 177–198.

Verhulst, P.F. 1845. Recherches mathématiques sur la loi d'accroissement de la population. *Mémoires de l'Académie Impériale et Royale des Sciences et Belles-Lettres de Bruxelles*, 18: 1–38.

Volterra, 1926. *Nature*, 118: 558–560.

Von Neumann, J. 1951. The general and logical theory of automata. In L.A. Jeffress, Ed., *Cerebral Mechanisms and Behavior: The Hixon Symposium*, pp. 1–32. Wiley, New York.

Von Neumann, J. 1966. *Theory of Self-Reproducing AutomataI*. University of Illinois Press, Urbana, IL.

Wainer, G. and Giambiasi, N. 2001. Application of the Cell-DEVS paradigm for cell spaces modeling and simulation. *Simulation*, 76 (1): 22–39.

Wolfram, S., Ed. 1986. *Theory and Application of Cellular Automata*. World Scientific, Singapore.

Zeigler, B.P., Praehofer, H., and Kim, T.G. 2000. *Theory of Modeling and Simulation: Integrating Discrete Event and Continuous Complex Dynamic Systems*. Academic Press, New York.

30
Ontology-Based Simulation in Agriculture and Natural Resources

Howard Beck
University of Florida

Rohit Badal
University of Florida

Yunchul Jung
University of Florida

30.1 Introduction 30-1
30.2 Ways in Which Ontologies can be Applied to Simulation 30-2
  Model Base • System Structure (Logical and Physical) • Representing Equations and Symbols • Connecting Models with Data Sources • Integrating Documentation and Training Resources • Reasoning
30.3 How to Build an Ontology-Based Simulation—Bioprocessing Example 30-6
  Collection of Relevant Documents • Define Model in Terms of Elements • Identifying Classes, Individuals, and Properties • Define Equations • Initial Values of State Variables, Constants, and Database Access • Generating Program Code for Implementing the Simulation • Simulation Execution
30.4 Tools for Ontology-Based Simulation 30-10
  Ontology Editor • EquationEditor • SimulationEditor
30.5 Conclusions 30-12

30.1 Introduction

An ontology is a formal representation of concepts and relationships among concepts within a particular domain. In this chapter, the domain is models of agricultural and natural resource processes. There are a vast number of concepts in this domain including abstract entities such as energy, mass, organism, and mathematical operators (addition and subtraction) and specific entities such as soil chemicals, plant components, or a particular plant disease. There are many relationships among these concepts (plant–water, pest–host). A formal representation is a data structure that describes a concept or relationship and which is based on a well-defined language. Many different languages have been developed for building ontologies including the Web Ontology Language known as OWL (OWL, 2005), a W3 standard.

Ontologies languages are object-oriented languages, but they are not programming languages. Ontologies include objects called individuals that represent particular things in the world (e.g., a particular soybean crop). Classes categorize similar individuals, and classes can be generalized to superclasses (i.e., soybean is a legume) creating a taxonomy of concepts. Individuals have properties that can be primitive values such as numbers (a soybean crop that has a size of 35 acres) or relationships to other individuals (the soybean crop is host to several pests).

Because of the formal nature of ontology languages such as OWL, reasoners can be built that perform inferences over the concepts in an ontology. Such inferences include automatically determining what class an individual belongs to, determining whether a certain class is a special case (subclass) of some other class, and determining how two concepts are similar or different. Reasoners can be used for many purposes, including checking for consistency and querying the ontology.

An ontology-based approach to simulation, in which a model is represented using ontology concepts, can help to address several problems with current methodology used to develop simulations within the domain of agriculture and natural resources. The general goal is to better communicate knowledge about models, model elements, and data sources among different modelers and between different computers. This is achieved through the ontology's ability to explicitly represent and thus define concepts used in models.

Various researchers create simulations within a particular domain to address a specific problem. There is an overlap of the concepts and interactions used in these simulations. Frequently, different modelers use different symbols for the same concept. The use of different programming languages makes communication even more difficult (Reitsma and Albrecht, 2005). Typically, a model is implemented in a particular programming language like FORTRAN, C++, or Java. However, the meaning of the model is lost when it is represented using program code (Furmento et al., 2001). Researchers must understand the programming language to understand the model. While such models are usually documented using papers and manuals, this documentation is physically separate from the model implementation itself. It is difficult to maintain both the model and the documentation, and often the documentation is not an accurate description of the model implementation. All the details of program code are difficult to describe in written documentation, so that ultimately it is necessary to read computer code to truly understand how the model works. These issues need to be addressed, so that the knowledge in a simulation can be made explicit (Lacy and Gerber, 2004; Cuske et al., 2005).

Typically, many different yet similar models are available for a particular domain. The challenge lies in knowing precisely how two models are similar or different and selecting the one most suitable for a particular task (Yang and Marquardt, 2003). When a particular model is encoded in a conventional programming language, it is very difficult to do comparisons between models, and impossible to conduct comparisons using automated techniques.

Most of the simulations in agriculture and natural resources use databases as a source of input data. A simulation requires input data in a particular format, which is defined inside the simulation and which is usually different from the format of the data stored in the database. The input data required for a simulation must be matched with a database, and to do this matching, knowledge of the internals of the simulation as well as the database is required. The matching is traditionally done manually, which is certainly tedious if not error prone. There is a need for a technology that can represent and interpret diverse data sources and support integration of these sources (Altman et al., 1999). The interoperability of data can be solved by information integration (Miled et al., 2002; Altman et al., 1999).

Utilizing ontologies for managing model and simulation knowledge facilitates representing this knowledge in an explicit manner. An ontology provides the model semantics that allows machines to interpret concepts in an automated manner (Lacy and Gerber, 2004). The construction of ontologies encourages the development of conceptually sound models, more effectively communicates these models, enhances interoperability between different models, and increases the reusability and sharing of model components (Reitsma and Albrecht, 2005). It also provides assistance in computation by structuring data (Altman et al., 1999).

In this chapter, we discuss several important issues and problems addressed by ontology-based simulation. We give an example of how to build a simulation using ontology techniques in a model of sequential batch anaerobic composting (SEBAC). Tools for building ontology-based simulation are presented based on a system we have developed that uses an ontology as a database management system.

30.2 Ways in Which Ontologies can be Applied to Simulation

The notion of combining ontologies with simulation has received much attention in recent years (Fishwick and Miller, 2004; Lacy and Gerber, 2004; Miller et al., 2004; Raubal and Kuhn, 2004). This chapter explores several different ways in which ontologies can be applied to simulation, and in particular how ontologies can solve some problems in current methods of building simulations for agriculture and natural resources.

30.2.1 Model Base

Many physical processes in agriculture and natural resources are fundamental and well studied. For example, basic crop physiology such as respiration and photosynthesis, soil/water dynamics, and crop–pest interactions have been modeled extensively (Lu et al., 2004). Many different yet similar models have been developed in each of these areas. One reason for the diversity of models is the diversity of environments in which models need to be applied. There are hundreds of crops grown commercially, and while as physical and biological systems they all share commonality, there is variability of climate and geography as well as individualized crop characteristics that lead to differences in models. Furthermore, each researcher describes a real-world problem based on *different* perspectives using his/her distinct modeling environment (Park and Fishwick, 2005).

There are many crop models but there is no comprehensive management system for managing all these models. Research is being done to develop a suite of crop models for a variety of crops and integrate these with models for weeds and insects (Agriculture Research Service, 2005). Many other crops can be modeled by assembling components from available models and changing parameters and rate equations. However, having so many different yet similar models causes problems in managing models and in sharing model components among developers. There is unnecessary redundancy resulting from poor communication among developers. For example, there may be as many as two dozen irrigation models that all basically operate on the principles of water balance. They may use similar ways of calculating processes such as evapotranspiration, or they may use different equations to achieve the same results. Unfortunately, the traditional methods for creating these models make it very difficult to compare the models to see how they are similar or different.

An ontology can be used to build a database of models, that is, a "model base" that can help to classify different but similar models and that can be searched to locate models and model components suitable for some application. Each specific model can be represented by an instance in the ontology, and abstract model structure and behaviors represented as classes. Similar models can be grouped together into a class, and related classes grouped together to form superclasses. At the top of the resulting taxonomy would be generic modeling approaches. If an ontology is also used to represent the internal structure of a model, then model internals can be compared in an automated fashion to determine which parts of the models are similar and which are different.

The vast collection of models and model components resulting from this analysis would create a large but organized taxonomy. This taxonomy could be searched using query processors based on ontology reasoners to locate models and model components of interest. It can also be used to compare and contrast two models and explicitly identify how they are different or similar.

30.2.2 System Structure (Logical and Physical)

System structure can take many forms including a geometric structure, a chemical structure, or a physiological structure. The use of object-oriented design for analysis of system structure is well known and one of the first applications of object-oriented programming dating back to the 1960s. The biological and physical systems in agriculture and natural resources are also analyzed in this fashion by decomposing a complex system into simpler interconnected parts and subparts, and modular, object-oriented designs are widely used (Beck et al., 2003; Kiker, 2001). Of course traditional object-oriented design uses programming languages such as Java or C++ as a representation language. Using an ontology is the next step in this approach (Fishwick and Miller, 2004). There are several advantages to elevating the objects comprising the system to the status of ontology objects. For one, the model description and behavior is forced to be done in an entirely declarative fashion. Ontologies do not utilize methods or program code to represent behavior. Instead the representation is based entirely on concepts and relationships. Also, by using ontology objects, model components can be classified and interrelated based on their meaning.

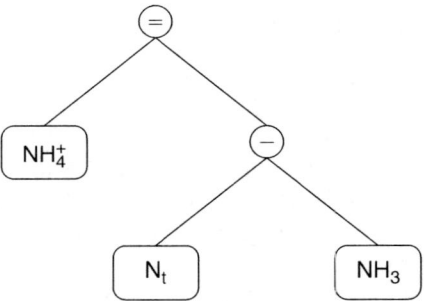

FIGURE 30.1 Representation of equation as a tree structure.

System structure is made explicit in a way that can be exploited by ontology reasoners to compare and contrast model structures.

30.2.3 Representing Equations and Symbols

Model behavior can be described entirely using mathematical equations (Cuske et al., 2005). Equations are composed of symbols, and each of these symbols can be represented as a concept in the ontology. This enables the symbol's meaning to be more exposed and accessible to analysis and manipulation. Whereas equations describe the quantitative behavior of variables, the variables are also symbols, and the things the symbols represent can be made explicit. Furthermore, the basic mathematical operators can also be treated as symbols and described in the same fashion.

Equations can be stored in the ontology by representing them as tree structures. For example, the formula

$$NH_4^+ = N_t - NH_3$$

can be expressed using the tree structure in Figure 30.1. The tree is rooted on the equal symbol, and equal has a left side and right side which are the first two branches in the tree. Operators, such as minus, are nodes in the tree with subtrees for each of the operator arguments. Each node in the tree, including operators and variables, become concepts in the ontology. Each concept includes associations to related concepts, for example, "minus" contains associations to the concepts being subtracted.

The advantage of better defining symbols appearing in equations is improved interoperability of concepts and associated symbols appearing in different models. In addition, with the inclusion of basic operators, the ontology can classify groups of equations and organize them taxonomically from generic forms to specific applications. This will lead to discovery of similarities in forms of equations used in different models, and will help to communicate among different modelers (Altman et al., 1999).

There is one obvious limitation of using ontologies to store and organize equations. While an ontology is a valuable tool for representing the meaning of the symbols, including operators, appearing in equations, it has no facilities for solving equations or even performing simple arithmetic operations needed to do simulation. Although, it is possible that an ontology could be extended to support analytical equation solving, this area has not been explored and goes beyond the scope of ontology reasoners. Instead, whereas the ontology acts as a library for organizing equations and their symbols, external facilities are needed to solve the equations. An external code generator can take equation structures that are stored in the ontology and automatically produces XML, or program code in C++ or Java (or other languages) that can implement the simulation.

30.2.4 Connecting Models with Data Sources

Most simulations in agriculture and natural resources require access to data. Basic physical data, such as weather data, soil parameters, and records of production operations including nutrient, pesticide, and irrigation applications, are all stored in databases (Beck et al., 2004). These databases can come in many forms and are geographically distributed. Unfortunately, there are many and different databases implemented on different database management systems.

There are several problems associated with attaching these data sources to the simulations that need the data to run. These include logical considerations, such as determining if the database has the necessary data in the correct form to provide the inputs required by the simulation. There is also a physical connection problem as the software used for the simulation must be capable of attaching to the database and querying it to extract the necessary information. There are also issues of resolution, in time and space, such that the data provided is on the same resolution or can be converted to the same resolution as that required by the model.

Establishing the logical and physical connection between a model and database can be a tedious process. To establish this connection, a developer must become very familiar with both the model and database. The input requirements for the model must be fully specified (this is typically done with text-based documents) and understood, and likewise the database schema and the meaning of all the attributes available in the database must be studied in detail. Then the person establishing the connection must match the model inputs and the database attributes. Seldom is this a perfect match, and often the model requires inputs that the database cannot provide. In this case, the database must be expanded to include the necessary inputs or it cannot be used to drive the simulation. Creating the physical connection is also tedious, as usually code in the form of SQL statements must be written to extract the information from the database needed by the simulation. Depending on how the database is published, additional work may be needed to connect the simulation software with the data source using technologies such as ODBC, RMI, or Web services (Mills-Tettey et al., 2002).

It would be possible for a simulation to automatically search for and attach to a data source. This would require having a sufficiently rich description, in the form of an ontology, of both the simulation input symbols and the entities and attributes in the database. A particular input symbol in the simulation could match automatically to a database attribute. This would be accomplished by matching the ontology descriptions of the model symbol and the database attribute to prove that they are compatible. Databases and models could be published on the Internet as a Web service (Knutson and Kreger, 2002). The Web service registry could be queried to determine the contents of databases contained in the service. The physical connection between the model and database would be automatic using XML as the method of exchanging data.

30.2.5 Integrating Documentation and Training Resources

If the ontology is part of a complete database management system, the ontology can store and organize any content, including multimedia content in the form of rich text, images, 2D/3D animations, and video. In the context of simulation, this creates a complete environment for all information associated with the simulation. In particular, all research materials (experimental procedures, raw data, statistical analysis, technical reports, and journal articles) and educational resources (training-based simulations, scenario training, and case studies) can be integrated.

30.2.6 Reasoning

The power of ontologies lies not only in their ability to provide declarative representations of concepts and their relationships, but also in the ability to automatically reason about those concepts. Basic reasoning facilities include ontology validation, automatically determining subsumption relationships (determining if class A is a subclass of class B), and classification (automatically determining the location of a new class within the class taxonomy). Extended facilities included automatic clustering (conceptual clustering) of concepts, and analogical reasoning or similarity-based queries and case-based reasoning. These facilities can be applied to simulation to automatically classify models, model components, and the equations and symbols used in the models. Query facilities based on reasoning can help to locate simulation elements within a large collection. Clustering techniques can compare the structure of two models and tell how they are similar or different.

For example, the knowledge in an ontology in the bioprocessing domain can be used for automatically generating equations based on physiochemical equilibrium laws. A particular law can be applied based on the specific property of an individual. In the SEBAC simulation (Section 30.3) fatty acids dissociate into fatty acid ions based on a physiochemical equilibrium law, and that law is represented by an equation stored in the ontology. The reasoner can automatically instantiate an equation corresponding to the law when it finds that an individual of the fatty acid class has a property called "in equilibrium with" and the range of the property is fatty acid ion. It would use the particular properties of the individuals involved to parameterize the equation.

30.3 How to Build an Ontology-Based Simulation— Bioprocessing Example

SEBAC is an anaerobic digestion process that decomposes organic matter into methane and carbon dioxide by a series of reactions in the presence of several microorganisms. It is used for treating the organic fraction of municipal solid waste. A mathematical model of SEBAC has been developed to understand the SEBAC system and to study the response of the system for various feed conditions (Annop et al., 2003). The model consists of a set of differential equations, which have been constructed based on mass balance and physiochemical equilibrium relationships. The steps in building an ontology-based SEBAC model are as follows.

30.3.1 Collection of Relevant Documents

The first step in building an ontology-based simulation is to collect all relevant documents such as technical papers of the system and any existing related models. In the case of SEBAC, an existing model had already been implemented using MATLAB (Lai, 2001). We were able to obtain a graduate thesis describing the variables and equations used in the model (Lai, 2001), a research publication describing the implementation of the mathematical model (Annop et al., 2003), and source code of the MATLAB implementation.

It would have been useful to have access to a conceptual model for understanding the conceptual schema of the system. A simple conceptual model of the process was sketched for understanding the SEBAC domain. Figure 30.2 shows a conceptual model with nine concepts (owl: Class) and three types of interactions (owl: ObjectProperty). These classes have individuals that can be mapped to the variables used in the simulation. There are six individuals of bacteria and six individuals of fatty acids in the SEBAC system that are mapped to the state variables of the model.

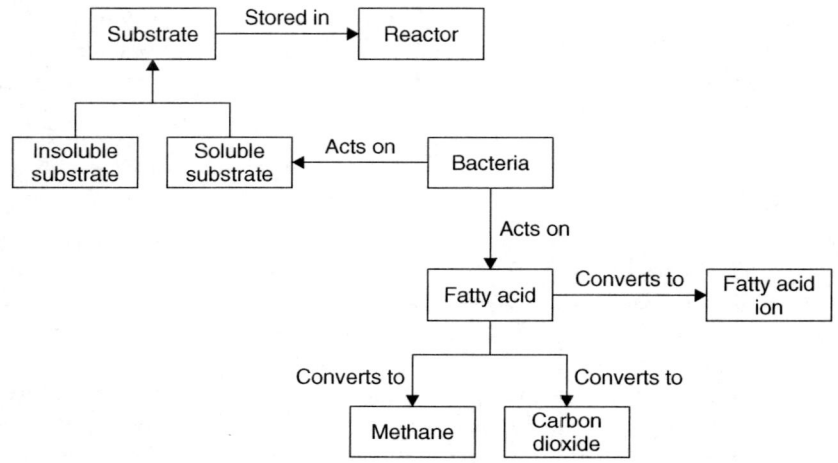

FIGURE 30.2 Conceptual model of SEBAC system.

30.3.2 Define Model in Terms of Elements

The next step is to define the model in term of elements. Elements are used to modularize the model into logical units. Related classes, individuals, properties, and equations are entered in a particular element. The description of the model in terms of elements is helpful in understanding the structure of the model. Typically, a modeler designs a particular model by creating a graph containing elements and links indicating the information flow between elements. Our tools allow modeler to use a SimulationEditor (Section 30.4) to build the module structure in the form of an element graph.

The SEBAC simulation involves biological process. The simulation is described in terms of elements that capture the important processes like bioconversion of fatty acids and substrate and dissociation of fatty acids. Figure 30.3 shows the elements of the SEBAC model and gives an overview of the SEBAC process including various transformations that occur during the process.

30.3.3 Identifying Classes, Individuals, and Properties

After defining the general elements of the model, specific concepts in the model are identified. For the SEBAC system, the concepts were identified from the list of variables used in the model (Lai, 2001). From these, we created the following classes with the corresponding properties:

- Reactor—liquid volume, gas head space, and reactor temperature
- Fatty acid ion—equilibrium constant for dissociation and conversion factor
- Fatty acid
- Bacteria—biomass death rate, half velocity constant and maximum growth constant
- Methane
- Carbon dioxide
- Soluble substrate and insoluble substrate

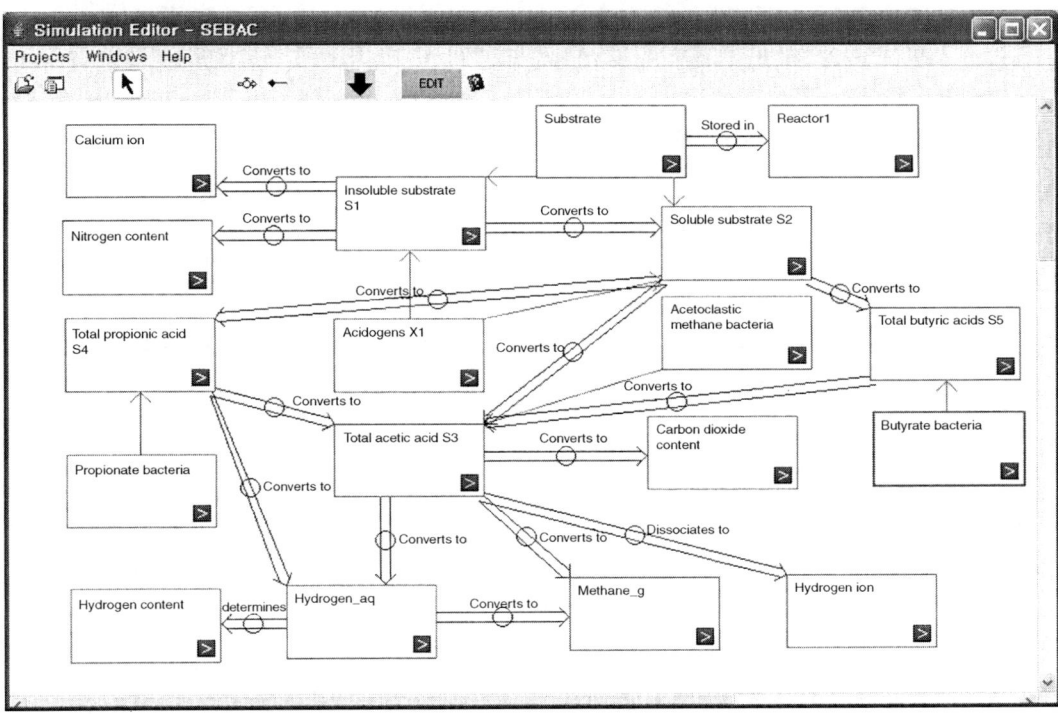

FIGURE 30.3 SimulationEditor diagram for SEBAC process showing elements of SEBAC simulation and showing various transformations that occur during the process.

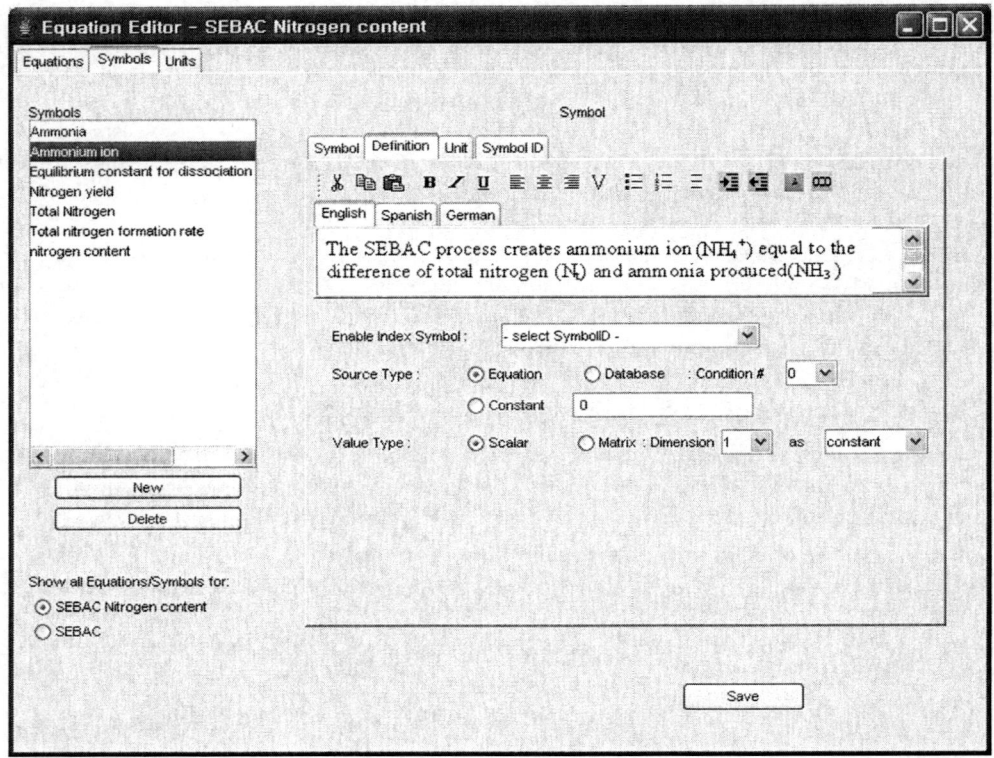

FIGURE 30.4 Interface of EquationEditor to input the concepts in a particular element of the simulation.

Some of these classes have several individuals. There are three individuals of fatty acid ion, and each fatty acid ion has a specific value of equilibrium constant for dissociation and conversion factor. Relevant classes, individuals, and properties are entered in a particular element. Figure 30.4 shows how an individual called "Ammonium ion" is entered into the ontology database. The other classes, individuals and properties are entered into the database in a similar fashion.

In conventional modeling languages, the meaning of the symbols and the relationships between the symbols are not defined explicitly. The SEBAC model has symbols for various forms of nitrogen such as ammonia, nitrate, and ammonium ion, but the simulation written in MATLAB does not explicitly specify relationship between these forms of nitrogen or the meaning of each form. The meaning of the symbols and relationships can be defined explicitly using an ontology. Figure 30.5 shows a portion of the ontology for different forms of nitrogen.

In the SEBAC model, total dissolved nitrogen is found in the form of ammonia, which in turn can be found in two forms: ammonium ion (NH_4^+) or dissolved ammonia gas (NH_3). In Figure 30.5, there is a relationship called "consists of" with a domain of total dissolved nitrogen and a range of forms of ammonia (NH_4^+, NH_3). Ammonium ion concentration is calculated by the difference of total dissolved nitrogen and ammonia. NH_4^+ and NH_3 are in equilibrium and their concentration is given by the equation:

$$NH_4^+ \rightleftharpoons NH_3 + H^+$$

Figure 30.5 contains a property called "in equilibrium with" having NH_4^+ as a domain and NH_3 as a range that models reversible conversion between these two forms of ammonia. Ammonium ion and hydrogen ion are specific kinds of ion, and thus are shown as subclasses of the class ion. Ammonia is defined as a specific kind of gas, so it is also subclass of the class gas. There is a relationship called "converts to" between NH_4^+ and NH_3.

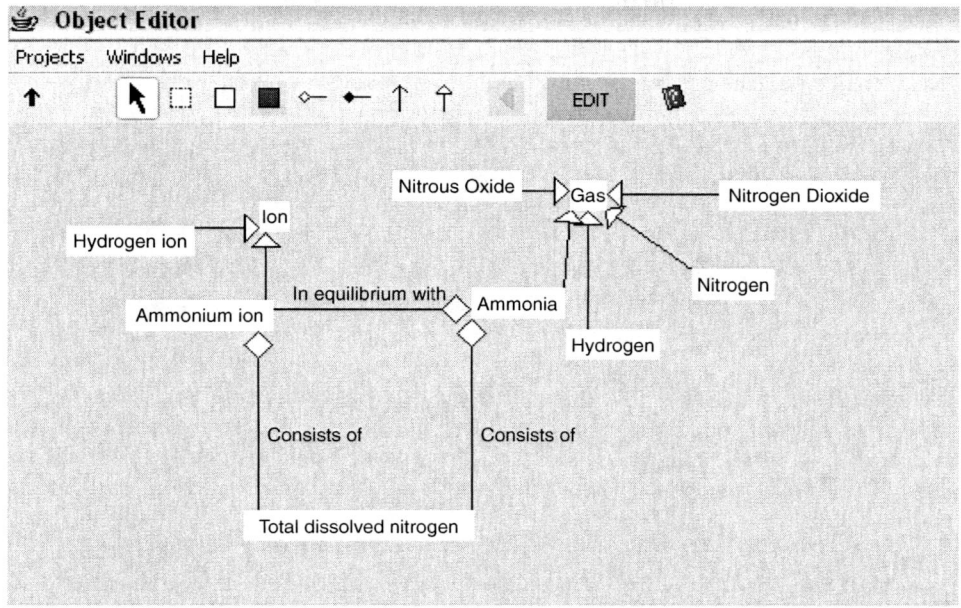

FIGURE 30.5 Ontology for different forms of nitrogen.

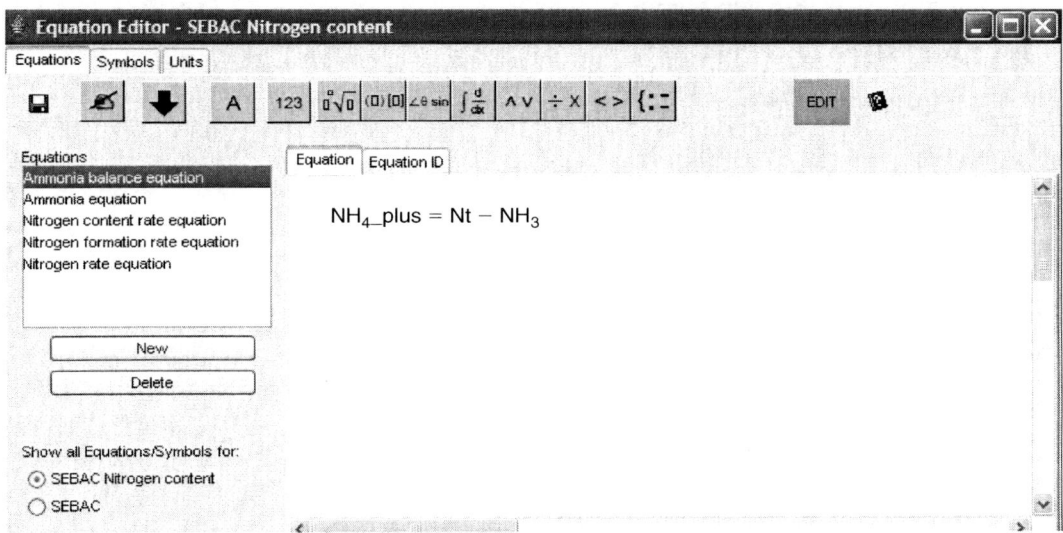

FIGURE 30.6 Interface of the EquationEditor for entering equation.

30.3.4 Define Equations

The equations describing dynamic behavior are entered in the system after entering the classes, individuals, and properties of symbols that are used in the equation. Figure 30.6 shows the EquationEditor for entering an equation that represents the relationship between total dissolved nitrogen, ammonia, and ammonium ion concentration.

An equation models the dynamic relationship between concepts (classes) and represents a statement of a specific law. The Michaelis–Menten equation (Heidel and Maloney, 2000) models a relationship between acid and bacteria. Acetic acid is an individual of the acid class and acetolistic methane bacteria is an individual of the bacteria class. The acetolistic methane bacteria acts on acetic acid, and this relationship can be modeled by Michaelis–Menten equation. These relationships can be explicitly shown in ontology as properties. It is also possible to store the specific laws in the ontology based on the relationship between classes so that the equations can be automatically generated based on the specific relationships between individuals by using an ontology reasoner as described in Section 30.2.6.

30.3.5 Initial Values of State Variables, Constants, and Database Access

Initial values of state variables are entered manually using an input form which is generated automatically based on the logic that each differential equation has a state variable and that an initial value is required for each state variable. The SEBAC simulation has 21 state variables so the input form has 21 text fields. The value of constants (like the universal gas constant) and other parameters used in the simulation are entered as properties in individuals representing these constants as described in Section 30.3.3.

Many simulations obtain inputs from databases, such as weather databases or database of farm production practices. The simulation can input values from the database by matching individuals representing the inputs with individuals representing the database attributes as discussed in Section 30.2.4.

30.3.6 Generating Program Code for Implementing the Simulation

Program code for running the simulation is automatically generated by processing the descriptions of model structure and behavior (equations) stored in the ontology. Currently, our system generates Java code, but other languages can be supported. Code generation involves retrieving equations and symbols belonging to each element in the ontology database and making a reference list of symbols having the hierarchical structure of operators in each equation. A Java class is generated for each element of the simulation (mainly to partition the code into logical modules). The symbols for variables belonging to an element are generated as member variables in the Java class while the equations are generated as Java methods. Each method returns a value for a particular variable based on an equation defined for that variable. For example, a Java method is generated corresponding to the ammonia balance equation shown in Figure 30.6 that returns the value of NH_4^+.

30.3.7 Simulation Execution

After generating the Java code, the code is compiled and the simulation is executed. The simulation results are presented in the form of charts and tables. To enhance interpretation, the results of the simulation can also be presented as an animation. The dynamics of the SEBAC simulation is shown in terms of reactors that change colors based on pH and other chemical properties of the system (Figure 30.7). The ontology facilitates creating these animated interfaces by storing graphic objects that can be used to render an animation along with the associated model concepts.

30.4 Tools for Ontology-Based Simulation

This section describes some new Web-based authoring tools that we have developed for facilitating construction of ontology-based simulations. These tools are Java applets that can be accessed online using standard Web browsers that have the Java plug-in installed. Wherever possible we designed these tools to be visually similar to traditional tools such as equation editors and editors for building system component diagrams. However, our tools utilize the ontology as the back-end for representing all concepts in the model.

30.4.1 Ontology Editor

There exists a variety of ontology editors such as KAON, which is a popular ontology editor extending RDFS (Volz et al., 2003), and (Knowledge Base Editor) (KBE), also known as Zeus and Protégé (Noy et al., 2001). These editors have a rich set of features for building ontologies and support exporting their ontology model to OWL. They include common functionalities and support several plug-in.

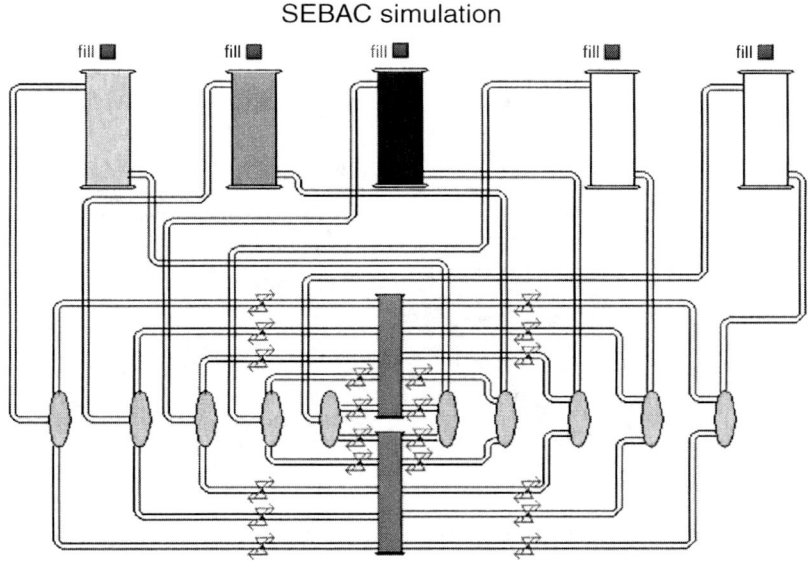

FIGURE 30.7 Interface for presenting results of SEBAC simulation using animation.

Although they have excellent features, these tools can be difficult to use because they require high level skills and understanding of ontologies and ontology modeling languages. These ontology editors are general purpose tools and are not designed specifically for developing simulation models. To facilitate developing ontologies for simulation, we created new editors which not only have more familiar graphic interfaces, but also are based on the features of these traditional ontology editors as a back-end. They have a dual-modality in which the modeler can either view the ontology-based simulation using traditional visual modeling tools (Figure 30.6) or view the ontology concepts in a graph-based ontology editor (Figure 30.8).

For the classic ontology view, our ontology editor adopts the dynamic graph layout from TouchGraph (TouchGraph Website, 2005) as illustrated in Figure 30.8. The editor includes tools for creating concepts and describing properties. Two types of concepts are displayed in this editor: concepts representing the symbols that appear in equations and concepts representing entire equations. The dynamic graph layout integrates these two as one graphic view such as shown in the figure. Each node represents a symbol, and the relation sign represents relationships between symbols in an equation.

30.4.2 EquationEditor

The process of building an equation starts with representing the equation as a mathematical expression of symbols, including numbers and operators. The equation is rendered visually using classic mathematical notation, but internally a hierarchical data structure (tree) is used for storing operators and symbols. Each node in the equation tree is also a node in the ontology. For a particular equation, the equal operator is the root node of the equation tree. Operators (like + and −) used in the equation become a node in the tree with child nodes being additional operators or symbols.

The EquationEditor (Figure 30.4) includes a symbol dictionary for entering symbols appearing in the equations along with their definitions and units. Each symbol is defined both using a gloss (this is a brief definition in English) and by relationships to other symbols as expressed in the ontology. Equations can automatically be converted to a sharable format, such as OpenMath or MathML, or to Java code using code generators.

It is very important to carefully track the units associated with symbols. Different models may use the same symbol but having different units. For example, mass per unit volume can be expressed with kg/m^3

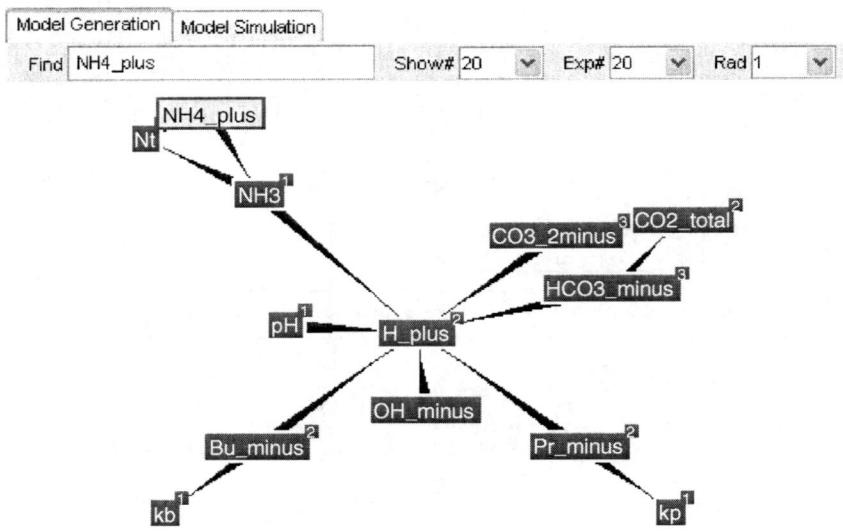

FIGURE 30.8 Classic graph-based ontology editor showing simulation symbols and their relationships.

or with g/cm^3. In the ontology, units are expressed not as simple strings but as a composition of other units and operators (kg, /, m, power operator, 3) each of which are also symbols in the ontology. Such composite units can be used to analyze the dimension of symbols appearing in equations. This facilitates automatic unit conversions and helps determine the correctness of the equation and the dimension and magnitude of the symbols.

30.4.3 SimulationEditor

SimulationEditor incorporates the EquationEditor described in the previous section and is used for specifying overall model structure in the form of elements. SimulationEditor also contains facilities for automatically generating and running simulations and generating reports. SimulationEditor is designed to represent models of dynamic systems using graphic elements such as source, sink, storage, and flow. We adopt this concept from the compartmental model (Peart and Curry, 1998) and Forrester notation (Forrester, 1971), which is widely used in agriculture and natural resource models.

Figure 30.9 shows a mathematical equation that has been converted to various different languages including MathML, OpenMath, and Java. For program code generation, each equation in the model is converted to a Java method returning the result of solving the equation. The code is compiled and executed after it is generated. Independent threads are used for generating and compiling code. The threads check and report the consistency of the equation and the correspondence of unit and value. Simulation results are shown as graphs and tables created with a report generator that is integrated into the SimulationEditor.

30.5 Conclusions

This chapter explored several ways in which ontologies can be applied to simulation in the agriculture and natural resource domains. As an example, an ontology-based simulation was developed in the bioprocessing domain. Development involved seven steps including collection of relevant documents, defining the model in terms of elements, identifying classes, individuals, and properties, encoding equations, entering initial values of state, constant, and other parameters, generating code, and executing the simulation.

We have developed an ontology database management system where the database management capabilities are built entirely around an ontology language rather than a traditional relational or object database

[Source Equation]
A = B + C

[generated documents]

```
<!-- MathML -->
<math display= 'block' xmlns = 'http://www.w3.org/1998/Math/MathML'>
<mrow>
   <mi> A </mi>
   <mo>=</mo>
   <mrow>                      <!-- OpenMath---->
    <mi> B </mi>               <OMOBJ xmlns = "http://www.openmath.org/OpenMath"
                                  version = "2.0" cbase = "http://www.openmath.org/cd">
    <mo> + </mO>               <OMA>
    <mi> C </mi>                 <OMS cd= "relation1 " name= "eq" />
   </mrow>                       <OMV name = "A" />
</mrow>                          <OMA>
</math>                            <OMS cd= "arith1" name= "plus" />      <!--- Java--->
                                   <OMV name= "B" />
                                   <OMV name= "B" />                     public double A()
                                  < /OMA>                                {
                                 < /OMA>                                 "return B() + c()" ;
                                < /OMBJ>                                 }
```

FIGURE 30.9 MathML, OpenMath, and Java code generated automatically from an equation.

model. Authoring tools for developing models are provide in this environment which have interfaces similar to traditional tools such as conventional equation editors. They utilize an ontology as the back-end for representing all concepts in the model.

The development of an ontology for simulation models explicitly exposes knowledge contained in models at a higher level. This knowledge can be further used for constructing conceptual models, simulations of similar systems, and educational and training materials. The construction of an ontology will allow better communication of knowledge about models, model elements, and data sources among different modelers, enhance interoperability between different models, and increase the reusability and sharing of model components.

References

Agriculture Research Service, The in-house research arm of United State Department of Agriculture, 2005, http://www.ars.usda.gov/research/projects/projects.htm?ACCN_NO=407227.

Altman, R. B., Xiaoqian, M.B., Chai, J., Carillo, M. W., Chen, R. O., and Abernethy, N. F., RiboWeb: An ontology-based system for collaborative molecular biology, *IEEE Intell. Syst.*, 14(5), 68–76, 1999.

Annop, N., Pullammanappallil, P. C., and Clarke, W. P., A dynamic mathematical model for sequential leach bed anaerobic digestion of organic fraction of municipal solid waste, *Biochem. Eng. J.*, 13, 21–33, 2003.

Beck, H.W., Albrigo, L. G., and Kim, S., DISC citrus planning and scheduling program. in *Proceedings of the Seventh International Symposium on Modelling in Fruit Research and Orchard Management*, P. Braun (Ed.,) pp. 25–32, 2004.

Beck, H.W., Papajorgji, P., Braga, R., Porter, C., and Jones, J.W., Object-oriented approach to crop modeling: concepts and issues. *Proc. of Congress on Computers in Agriculture and Natural Resource*, ASAE Publication Number 701P0301.

Cuske, C., Dickopp, T., and Seedorf, S., JONTORISK: an ontology-based platform for knowledge-based simulation modeling in financial risk management, *Proc. of 2005 ESM Conference*, 2005.

Fishwick, P.A., and Miller, J. A., Ontologies for modeling and simulation: issues and approaches. *Proc. of 2004 Winter Simulation Conference*, 2004.

Forrester, J.W., *Principles of Systems*, Wright Allen, Cambridge, MA, 1971.

Furmento, N., Mayer, A., McGough, S., Newhouse, S., Field, T., and Darlington, J., Optimization of component-based applications within a grid environment. *Proc. of the 2001 ACM/IEEE Conference*, 2001.

Heidel, J., and Maloney, J., An analysis of a fractal Michaelis–Menten curve, *J. Aust. Math. Soc. Ser. B*, 41, 410–422, 2000.

Kiker, G.A. Development of JAVA based, object oriented modeling system for simulation of South African hydrology, *Proc. of Am. Soc. of Agricultural Engineers*, ASAE Publication Number 01-2030, 2001.

Knutson, J., and Kreger, H., Web Services for J2EE, Version 1.0. Public Draft v0.3, IBM, 2002.

Lacy, L., and Gerber, W., Potential modeling and simulation applications of the web ontology language—OWL, *Proc. of 2004 Winter Simulation Conference*, 2004.

Lai, T.E. Rate Limiting Factors of the Anaerobic Digestion of Municipal Solid Waste in Bioreactor Landfills, PhD dissertation, Department of Chemical Engineering, The University of Queensland, Brisbane, 2001.

Lu, H.Y., Lu, C.T., Wei, M.L., and Chan, L.F., Comparison of different models for nondestructive leaf area estimation, *Taro. Agron. J.*, 96, 448–453, 2004.

Miled, Z. B., Webster, Y. W., Li, N., Bukhres, O., Nayar, A. K., Martin, J., and Oppelt, R., BAO, a biological and chemical ontology for information integration, *Online J. BioInformat.*, 1, 60–73, 2002.

Miller, J. A., Fishwick, P.A., Baramidze, G. T., and Sheth, A. P., Investigating ontologies for simulation modeling, *Proc. of 37th Annual Simulation Symposium (ANSS'04)*, 2004.

Mills-Tettey, G. A., Johnston, G., Wilson, L. F., Kimpel, J. M., and Xie, B., A security framework for the agent-based environment for linking simulations (ABELS), *Proceeding of the 2002 Winter Simulation Conference*, 2002.

Noy, N. F., Sintek, M., Decker, S., Crubezy, M., Fergerson, R. W., and Musen, M. A., Creating semantic web contents with protege-2000, *IEEE Intell. Syst.*, 16(2), 60–71, 2001.

OWL, *Web Ontology Language Guide*, http://www.w3.org/TR/owl-guide, 2005.

Park, M., and Fishwick, P. A., Ontology-Based Customizable 3D Modeling for Simulation, PhD dissertation, Department of Computer Science Engineering, The University of Florida, Gainesville, 2005.

Peart, R., and Curry, R. B., *Agricultural Systems Modeling and Simulation*, Marcel Dekker, New York, 1998.

Raubal, M., and Kuhn, W., Ontology-based task simulation, Spat. Cogn. Comput., 4(1), 15–37, 2004.

Reitsma, F., and Albrecht, J., Modeling with the semantic web in the geosciences, *IEEE Intell. Syst.*, 20(2), 86–88, 2005.

TouchGraph Website, http://www.touchgraph.com, 2005.

Volz, R., Oberle, D., Staab, S., and Motik, B., KAON SERVER—a semantic web management system in alternate track, *Proc. of the Twelfth International World Wide Web Conference*, 2003.

Yang, A., and Marquardt, W., COGents: applying software agent technology to support process modeling, Presentation LPT–pre–2003–06, 2003.

31
Modeling Human Interaction in Organizational Systems

Stewart Robinson
University of Warwick

31.1 Introduction 31-1
31.2 Systems and Human Interaction 31-2
31.3 Why Model Human Interaction? 31-3
31.4 Modeling Human Interaction: Research and Practice 31-4
   Modeling Human-to-System Interaction • Modeling Human-to-Human Interaction
31.5 The KBI Methodology 31-5
   Stage 1: Understanding the Decision Making Process • Stage 2: Data Collection through Simulation • Stage 3: Determining the Decision Makers' Decision Making Strategies • Stage 4: Determining the Consequences of the Decision Making Strategies • Stage 5: Seeking Improvements
31.6 A Case Study: Modeling Human Decision Making at Ford Motor Company 31-8
   KBI Stage 1 • KBI Stage 2 • KBI Stage 3 • KBI Stage 4 • KBI Stage 5 • Findings about KBI
31.7 Conclusion 31-12

31.1 Introduction

Most systems involve some level of human interaction. Therefore, if we are to model such systems it seems necessary to have at least some understanding of the nature of that interaction. It would also seem necessary to be able to model the interaction of humans to obtain a proper understanding of the system and its performance.

This chapter explores the idea of modeling the interaction of humans with organizational systems. First, the concept of system and the way in which humans interact with systems, and each other, are discussed. The motivation for modeling human interaction is then addressed, before reviewing some examples of modeling human-to-system and human-to-human interaction. A key focus of the chapter is to outline the knowledge-based improvement (KBI) methodology. This methodology focuses on modeling a specific form of human interaction, namely human decision making. An example of the methodology in practice at the Ford Motor Company is then described.

31.2 Systems and Human Interaction

Checkland (1981) identifies four main types of system. *Natural systems* have their inception with the origins of the universe and range from atomic systems, through living systems on earth, to galaxies. Although many of these systems appear to arise from natural processes, there is evidence that some are the result of (intelligent) design (Dembski, 1999). *Designed physical systems* are the result of conscious design and exist to serve some human purpose. Examples are a wrench, a manufacturing plant, and a car. Meanwhile, *designed abstract systems*, such as mathematics and literature, are similarly the product of conscious design, but are abstract in form. Finally, there are *human activity systems*. These less tangible, but observable, systems are more or less consciously ordered and exist for some purpose. Examples include a political system, a health system, a charitable organization, and a football team.

All of these systems can, and often do, involve some level of human interaction. This is most obviously the case for human activity systems, since they exist as systems of human interaction. Designed abstract systems are developed and exist through human consciousness and interaction between human actors. Designed physical systems are developed by humans and often involve human interaction in their use. A car, for instance, exists as a designed physical system, but only becomes useful when a human starts to interact with the system, that is, drives the car. Even natural systems involve some level of human interaction. The climate is affected by human activity through emissions of CO_2, and as I write, humans are crashing a spaceship into a comet having some (small) effect on our galaxy.

To one degree or another, the nature and degree of human activity determines the performance of these systems. Place a hesitant driver in a Formula 1 car and its performance is going to be very different from that expected when a professional is behind the wheel. The performance of a regional health system is hugely dependent on the human activity that defines that system. Indeed, as we move from natural systems through to human activity systems, human interaction has an increasing impact on the performance of the system.

In this chapter, the focus is on organizational systems primarily as they exist in business or public sector organizations. These can primarily be classified as designed physical systems or human activity systems. Indeed, rather than seeing these as bipolar extremes, there exists a continuum from systems that are largely designed physical systems through to systems that mainly exist as human activity. This is illustrated in Figure 31.1 with some examples of organizational systems. On the far left are designed physical systems that involve little or no human interaction in their daily operation. Although an automated warehouse requires human activity for its initial design and build, there is little or no human activity in its daily operation with the exception of occasional manual overrides and maintenance.

To the far right are systems that exist purely as human activity. In between these two extremes are a whole range of systems that are in part designed physical systems and in part human activity systems. A retail bank, for instance, is designed with a layout and facilities. It also acts as a human activity system with customers interacting with staff in some purposeful activity. The performance of the bank is in part determined by the design and in part by the human activity.

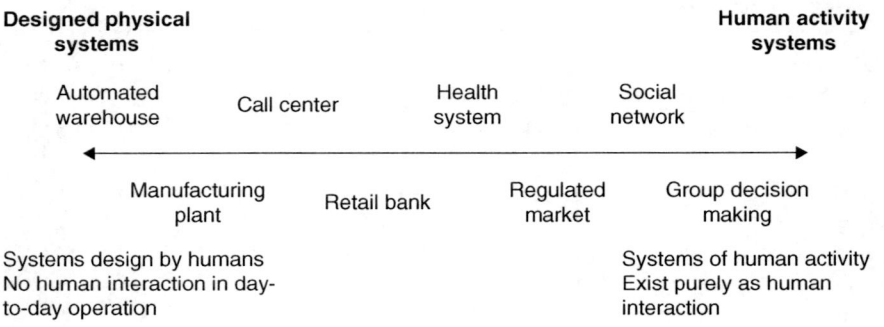

FIGURE 31.1   Systems and human interaction: a continuum.

Within these organizational systems there are two types of human interaction. The first involves an interaction between the human and the designed physical system (human-to-system interaction). Taking the example of the bank, a customer interacts with the layout of the bank in choosing a route to a service counter. The customer may also interact with an automated teller. A member of staff interacts with the equipment at the service counter.

Secondly, there is human-to-human interaction. In the bank customers interact with staff when receiving service. Customers also interact with one another (e.g., collision avoidance) as do staff. As we move further to the right in Figure 31.1, human-to-human interaction gradually predominates over human-to-system interaction.

31.3 Why Model Human Interaction?

Put simply, it is important to model human interaction because systems involve human activity and because that activity helps determine the performance of a system. Balanced against this is the difficulty of modeling human interaction, not least because humans are complex. A particular difficulty is being able to obtain data on how humans behave. Customer behavior in a bank, for instance, is complex. What determines their choice of service? How do customers choose which queue to join? When do customers balk, jockey, or leave a queue? How do customers decide how to move from location to location in the bank? What are the determinants of the time it takes to serve customers? Added to the difficulty of answering these questions is the recognition that every customer behaves differently.

While human behavior is complex and important in determining system performance, expert modelers advise that models should be kept as simple as possible (Robinson, 1994; Pidd, 1999). There are a number of advantages to simpler models (Innis and Rexstad, 1983; Ward, 1989; Salt, 1993; Chwif et al., 2000; Thomas and Charpentier, 2005):

- Faster development
- More flexible
- Less onerous data requirements
- Faster execution speed
- Easier interpretation of the results

These advantages are lost as the complexity of a model increases. This does not mean that complex models should not be developed. However, it does mean that the aim should be to develop the simplest model possible to achieve the objectives of a modeling study.

Robinson (2004) illustrates the need for simple models with the diagram in Figure 31.2. As complexity (the scope and level of detail of the model) increases, the accuracy of the model also increases, but with diminishing returns. Eventually a point is reached where the accuracy of the model may diminish with

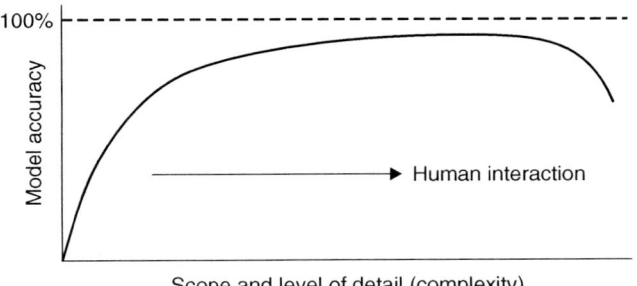

FIGURE 31.2 Model accuracy and increasing complexity. (Adapted with permission from Robinson, S., *Simulation: The Practice of Model Development and Use*, p. 68, 2004.)

increasing complexity. This is because the data required to support such models are not available. The modeling of human interaction in organizational systems certainly involves increasing the complexity of a model, and possibly to the point where the accuracy begins to diminish.

There is something of a paradox when it comes to modeling human interaction in organizational systems. At one level it is important to include human factors because they affect the performance of a system, but at another, the implied complexity suggests that human factors should be excluded from any model of a system. The paradox is resolved to an extent by noting that the requirement is to design the model with respect to the objectives of the modeling study (Robinson, 2004). The desire to model human interaction should, therefore, be driven by a specific requirement to study human interaction. Among the many reasons for this might be:

- Understanding the effects of human interaction on the performance of the system
- Improving the performance of human actors in the system
- Improving system performance

None of these necessarily require highly detailed or complex models. High model fidelity is not always required to make a model useful (Hodges, 1991; Robinson, 2001). Instead, it is argued here that the key aim is to provide a learning environment, in which the model is used to better understand the interaction between human actors and the system. Through this improved understanding, actions can be taken to improve the performance of the human actors and/or to improve the performance of the system.

31.4 Modeling Human Interaction: Research and Practice

Before describing a specific example of modeling human interaction, it is worth pausing to obtain an understanding of the range of research and practical work that is being carried out in this area. That said, it should be noted that the modeling of human interaction is a very underdeveloped field. Much of the work remains at the level of research on examples, with only some practical applications. It is also a diverse field with very little cohesion. As a result, the aim here is not to give a complete review of the field, but to provide some examples of the type of work that has been, and is being, performed. This work is described under two headings: modeling human-to-system interaction and modeling human-to-human interaction.

31.4.1 Modeling Human-to-System Interaction

Perhaps the first place to start in discussing the modeling of human-to-system interaction is to recognize that simulation games incorporate human interaction with a system. A business game, for instance, involves human actors entering policies and decisions on a periodic basis, which then determine the performance of the organization being modeled. Simulators also incorporate human interaction with a system. The best known example is probably a flight simulator where a human interacts with a simulation as a means of practicing flight procedures. Neither of these strictly model human interaction since they rely on interaction between real people and the game or simulator rather than on a model of the human actors. What they do provide is a means of assessing the impact of human actors on a system.

In terms of specifically modeling human interaction, one area with a reasonably long history of research has been the modeling of human decision making. Flitman and Hurrion (1987), Williams (1996), and Lyu and Gunasekaran (1997) all describe examples of linking simulation models with artificial intelligence (AI) representations of human decision makers. Flitman and Hurrion link a discrete-event simulation model of a coal yard with a rule-based expert system, enabling human actors to input control decisions. The expert system learns from the control actions taken. Williams uses a simulation to elicit knowledge from experts concerning the scheduling of replenishment at sea by the Royal Navy. The knowledge is used to train a rule-based expert system, which is then linked back to the simulation model to test the performance of the scheduling strategy. Lyu and Gunasekaran represent managers' decisions concerning unloading operations at a wharf using a rule-based expert system, and then test the performance of decision strategies using a simulation. Similar ideas are discussed by Robinson et al. (1998) and Standridge and Steward (2000). In all these cases, rule-based expert systems are used for representing the decision making strategies.

As an alternative to the use of rule-based expert systems for representing human decision making strategies, both Liang et al. (1992) and Curram (1997) use neural networks. Meanwhile, Moffat (2000) discusses the use of Bayesian inference for modeling command and control in simulation models of military combat.

Outside of modeling human decision making in systems, Harmon et al. (2006) provide a general discussion on human behavior representation in simulation with a focus on the validation of such representations. A specific example of modeling worker behavior in manufacturing simulation models is given by Baines and Kay (2002). They develop a pilot model, which includes the effects of worker stereotype (e.g., action oriented or thinking person) and the environment (e.g., noise level, temperature, and cleanliness) on worker performance. The model utilizes a neural network to evaluate the relationship between these factors and performance. In a similar vein, Brailsford and Schimdt (2003) model patients' attendance behavior for diabetic retinopathy. Using Schmidt's PECS framework (Schmidt, 2000), the model takes into account the patients' physical state, emotions, cognitions, and social status. Both of these projects incorporate ideas taken from cognitive psychology.

31.4.2 Modeling Human-to-Human Interaction

Agent-based modeling provides concepts and constructs for modeling human-to-human interaction. Unlike traditional simulation approaches which aim to model a system as a defined set of relations between system elements (top-down approach), agent-based models adopt a bottom-up approach. Agents with given behaviors interact with other agents, leading to a system with emergent properties. The prisoners' dilemma illustrates how complexity can arise from what appears to be a simple two-person human-to-human interaction (Axelrod, 1997). Silverman et al. (2006) discuss the state-of-the-art in human behavior modeling and identify the key challenges in improving models of socially intelligent agents.

Two specific examples of agent-based simulation in an organizational context are given by Baxter et al. (2003) and Schelhorn et al. (1999). In the former, an agent-based model is developed to understand how word of mouth influences the adoption of products and services by customers. Meanwhile, Schelhorn et al. simulate pedestrian movements in an urban area based on the influence of spatial configuration, predefined activity schedules and different combinations of land use.

Another branch of work in modeling human-to-human interaction is in evacuation simulations. These models not only involve interactions between evacuees, but also interactions between people and the system (e.g., a building). A number of examples of this type of simulation exist, for instance, Kim et al. (2004) describe progress toward a simulation for evacuation in marine accidents, while Learmount (2005) summarizes work on simulating evacuation from aircraft.

Social network analysis (SNA) provides a means for modeling and understanding human-to-human interaction in organizations. Cross and Parker (2004) discuss the use of SNA in business organizations, while Carley (2003) focuses on their use for intelligence work in identifying and understanding terrorist networks.

What is apparent in this work is that a key challenge in modeling both human-to-system and human-to-human interaction is developing valid models of human behavior. Further to this, it is difficult to obtain data to populate these models. What now follows is a description of a methodology that aims to address both these needs within the context of modeling human decision making. The methodology is illustrated by a case study example in a Ford manufacturing plant.

31.5 The KBI Methodology

KBI uses visual interactive simulation (VIS) (Pidd, 2004; Robinson, 2004) and AI to develop an understanding of human decision making within organizational systems, primarily manufacturing systems. VIS is used to elicit knowledge from decision makers and AI is used to learn and represent different decision

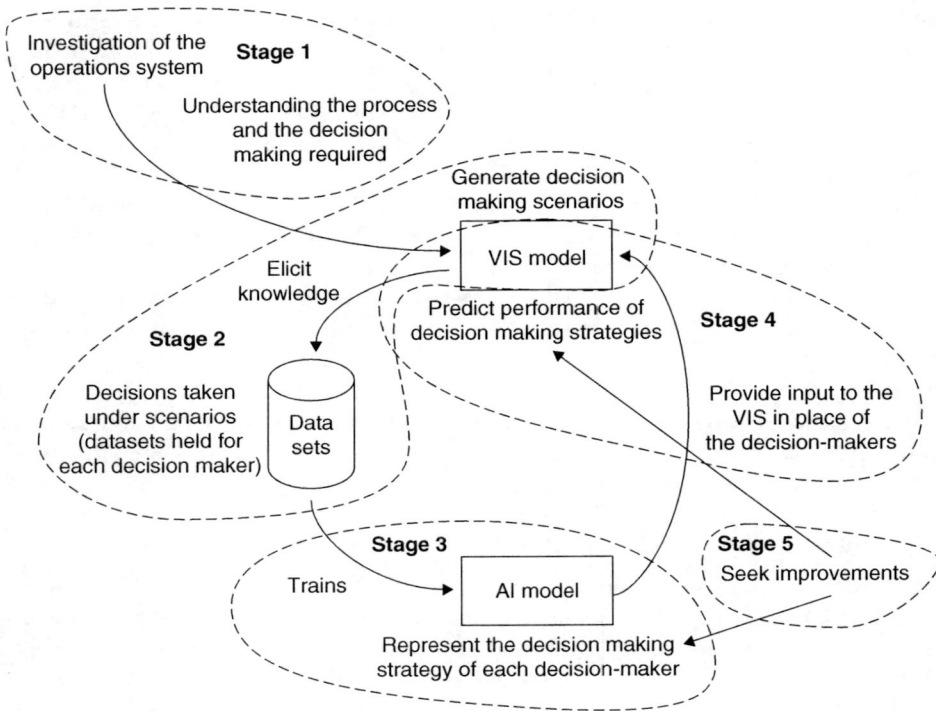

FIGURE 31.3 The Knowledge Based Improvement methodology. (From Robinson et al. 2005. *Journal of the Operational Research Society* 56: 913. With permission.)

making strategies. The VIS and AI are then linked to predict the performance of the system under different decision making strategies.

As discussed above, the prime motivation of KBI is not to develop a more accurate simulation, but to create a learning environment in which good practice can be identified and shared among decision makers. KBI consists of five stages (Figure 31.3):

- Understanding the decision making process
- Data collection through simulation
- Determining the decision makers' decision making strategies
- Determining the consequences of the decision making strategies
- Seeking improvements

Each stage is summarized below, with a more complete discussion to be found in Robinson et al. (2005).

In the KBI methodology the problem is conceived as follows. A decision making process consists of two vectors:

$$\mathbf{D}_{i,j} = [d_1, d_2, \ldots]$$
$$\mathbf{A}_i = [a_1, a_2, \ldots] \tag{31.1}$$

$\mathbf{D}_{i,j}$ is a set of decisions made by decision maker j at decision point i. Each decision made is represented by *decision variables* $(d_1, d_2, \ldots)$. The values that these decision variables take are known as the *decision options*. Vector $\mathbf{A}_i$ lists the *decision attributes* $(a_1, a_2, \ldots)$ at decision point i, described as the *decision scenario*. The decision attributes can take on various *attribute levels*.

The aim of the KBI methodology is to identify and improve the function f_j, which represents the decision making strategy of decision maker j:

$$\mathbf{D}_{i,j} = f_j(\mathbf{A}_i) \tag{31.2}$$

Function f_j is itself a model, in this case of a human decision making strategy. As such, the discussion above concerning models as simplifications applies. Indeed, KBI does not set out with the notion that a complete model of a decision making strategy can be obtained, but instead that only a partial explanation is possible. This is similar to Zeigler's (1976) view that the base model (a complete explanation of a system) is unknowable, and that instead a lumped model (a partial and appropriate explanation) should be utilized. The aim of KBI, therefore, is to identify a partial model of a decision making strategy based on a partial understanding of the decision variables and decision attributes. As a consequence, the decision making strategy as identified by KBI can be expressed as

$$\hat{\mathbf{D}}_{i,j} = \hat{f}_j(\hat{\mathbf{A}}_i) \tag{31.3}$$

denoting that the two vectors and the function $\hat{f}_j$ are all approximate.

This does leave the question of what is the appropriate level of approximation for the function $\hat{f}_j$. It is supposed that a relation similar to that set out in Figure 31.2 applies. Similarly, the determination of how far to move along the curve should be driven by the objectives of the study. If the aim is to devise an AI system to largely replace a human decision maker in a real system, then a much greater level of accuracy is required than if the aim is to gain a better understanding of the effects of human decision making on system performance. The selection of an appropriate simplification in modeling is itself a field for further research. For discussions on this topic see Zeigler (1976), Ward (1989), Nance (1994), Powell (1995), Brooks and Tobias (1996), Pidd (1999) and Pace (2000).

31.5.1 Stage 1: Understanding the Decision Making Process

In the first stage of KBI, the aim is to identify the decision variables, decision options, decision attributes, and attribute levels. Although interviews and discussion with the decision makers can reveal some information about the decision making process, usually a decision maker cannot explicitly identify and list all the decision making components. Various other approaches can be used to develop a better understanding of the decision variables and attributes:

- Observing the decision makers as they take decisions.
- Hypothesizing about likely decision variables and attributes.
- Interaction with a simulation model (stage 2 of the methodology).

31.5.2 Stage 2: Data Collection through Simulation

This step involves collecting example decisions from the decision makers to obtain the decision matrices:

$$\hat{\mathbf{D}}_j = \begin{bmatrix} d_{1,1} & d_{1,2} \\ \cdot & \cdot \\ d_{i,1} & d_{i,2} \\ \cdot & \cdot \\ d_{I,1} & d_{I,1} \end{bmatrix} \quad \hat{\mathbf{A}} = \begin{bmatrix} a_{1,1} & a_{1,2} \\ \cdot & \cdot \\ a_{i,1} & a_{i,2} \\ \cdot & \cdot \\ a_{I,1} & a_{I,1} \end{bmatrix} \tag{31.4}$$

where $\hat{\mathbf{D}}_j$ is the series of decisions taken by the decision maker j under the decision attributes represented by $\hat{\mathbf{A}}$. A total of I scenarios are presented to each decision maker. These data are collected through a VIS model. The decision maker interacts with a VIS of the system in question entering his/her decisions to the model. The model records the value of each decision variable and decision attribute to a data file. As a result, a set of values for the matrices $\hat{\mathbf{D}}_j$ and $\hat{\mathbf{A}}$ are collected. Separate $\hat{\mathbf{D}}_j$ matrices can be generated for each decision maker by presenting them with the same set of scenarios, $\hat{\mathbf{A}}$.

There are various advantages of using VIS over observing decision makers in the real system:

- The speed with which decision scenarios can be presented to the decision makers.
- The ability to control the nature of the decision scenarios.
- The ability to replicate the same decision scenarios with different decision makers.
- The ability to easily record every attribute of the simulated system at each decision point.

Of course, certain problems with the use of VIS also exist:

- The ability of a simulation to contain all the decision attributes of a decision scenario.
- The extent to which a decision maker makes realistic decisions in a simulated environment.
- The monotony of working in a simulated environment.

31.5.3 Stage 3: Determining the Decision Makers' Decision Making Strategies

AI methods are used to determine the decision making strategies ($\hat{f}_j$) of the different decision makers from whom data have been collected. Various AI approaches can be used, for instance, rule-based expert systems, case-based reasoning, and artificial neural networks, as well as statistical pattern recognition techniques (e.g., logistic regression). The data collected in stage 2 are used to train an AI representation using, for instance, the ID3 algorithm (Quinlan, 1986) for determining a decision tree in a rule-based system. Since rule-based systems provide a decision tree they are favored. This is because they not only model a decision making strategy, but they also have the greatest explanatory power. A separate AI representation (e.g., decision tree) should be created for each decision maker.

31.5.4 Stage 4: Determining the Consequences of the Decision Making Strategies

In this step, the performance of the decision making strategies ($\hat{f}_j$) are determined by linking the AI representation of each decision maker with the simulation model. Instead of requesting input from the decision maker, the simulation now requests input from the AI models. This enables long simulation runs to be performed so that the effect of different decision making strategies on the performance of the system can be determined.

31.5.5 Stage 5: Seeking Improvements

Improved decision making strategies can be sought by combining the best strategies or by taking a good strategy and discussing with the decision makers how improvements can be made. More formal heuristic search methods could also be employed to look for better strategies. In each case, the alternative strategies can be tested by running them with the VIS to determine their effectiveness.

31.6 A Case Study: Modeling Human Decision Making at Ford Motor Company

The Ford engine assembly plant at Bridgend (South Wales) is one of the main production plants for the Zetec petrol engine. The plant consists of three main assembly lines (Line A, Line B, and the Head Subassembly Line) followed by a Hot Test area and an After Test Dress line for finishing operations. The KBI methodology was applied to the problem of machine failures on the main assembly lines. When a machine fails, the machine is first inspected and then the maintenance supervisors determine what courses of action to take. The research project focused on the decision making strategies of the maintenance supervisors. To reduce the scope of the work, only the maintenance activities in the first of the four areas on the line were considered, representing about a quarter of the machines on the line.

31.6.1 KBI Stage 1

Initial investigations in stage 1 of the methodology were carried out by observing maintenance engineers at work and through discussion with the engineers and plant manager. These revealed that there were three decision variables surrounding the failure of a machine (decision options are shown in brackets):

- What action should be taken? (repair immediately [RI], "standby"—operate manually [SB], switch-off the machine [SO])
- Who should act? (group leaders [L1, L2], second skilled engineer [L3], semiskilled engineer [L4], unskilled engineer [L5], ask the production manager [APM])
- Should planned maintenance be performed on the machine? (perform planned repair [PR], schedule planned repair [SPR])

The decision vector in this context is therefore expressed as a set of binary variables:

$$\hat{D}_{i,j} = [\text{RI SB SO L1 L2 L3 L4 L5 APM PR SPR}] \quad (31.5)$$

Following these investigations (and some use of the VIS in stage 2 of the methodology), the decision attributes were found to be:

- Type of fault (TF)
- Estimated repair time (ERT)
- Machine number (MN)
- Time of day (TD)
- Number of engines produced so far this shift (NEP)
- Engines waiting on the conveyor before the machine (EWC)
- Number of heads in the buffer (NHB)
- Number of breakdowns on this machine today (NBMT)
- Number of breakdowns on this machine this month (NBMM)
- Number of breakdowns of this type on this machine today (NBTMT)
- Number of breakdowns of this type on this machine this month (NBTMM)

These attributes could each take on various levels. The vector of decision attributes is expressed as follows:

$$\hat{A}_i = [\text{TF ERT MN TD NEP EWC NHB NBMT NBMM NBTMT NBTMM}] \quad (31.6)$$

31.6.2 KBI Stage 2

A VIS model of the assembly line already existed, but this did not contain any complex decision making logic for maintenance decisions; it defaulted to a repair immediately decision. The model was adapted to enable decisions to be entered by an expert user as machines failed and an interface was devised for data collection (Figure 31.4 and Figure 31.5). The interface, written in Visual Basic®,[1] reported the status of the simulation (the decision attributes) and asked the user to enter a decision (the decision variables). Data collection sessions were run with three decision makers; the maintenance supervisor for each of three shifts on the first area of the line. Each was presented with the same set of 63 scenarios ($\hat{A}$) via the simulation. These scenarios were selected from a historic trace of data on machine failures obtained from the plant. The trace was adjusted, removing repetitive and simple cases, to ensure that there was a wide range of decision scenarios. Data collection sessions took around 1 h with each decision maker. This was found to be both as long as the maintenance supervisors could spare for this exercise and as long as their concentration when using the model could last.

---

[1] Registered trademark of Microsoft.

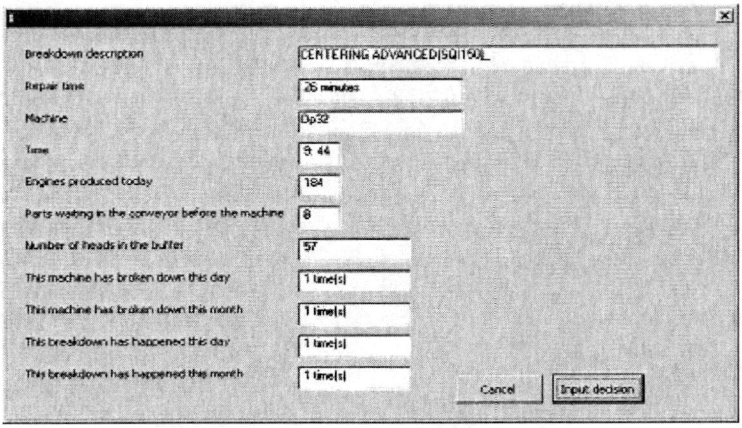

FIGURE 31.4 Visual Basic interface for VIS model: reporting decision attributes.

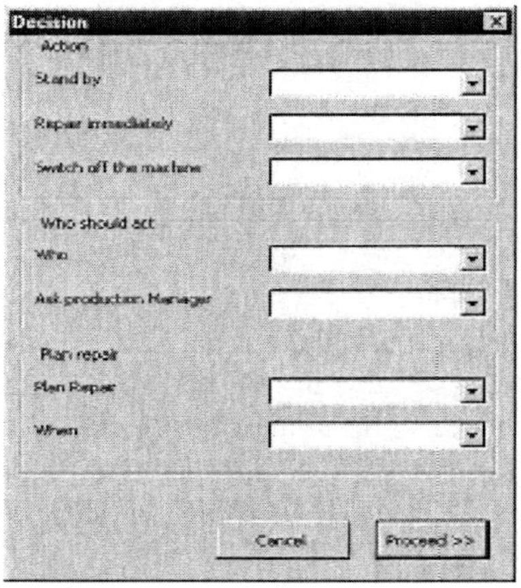

FIGURE 31.5 Visual Basic interface for VIS model: entering decision variables.

31.6.3 KBI Stage 3

Based on the data obtained ($\hat{D}_j$), three approaches were used to learn the decision making strategies ($\hat{f}_j$) of the three decision makers: artificial neural networks, logistic regression, and rule-based expert systems. MATLAB®,[2] SPSS®,[3] and the XpertRule®[4] software were used, respectively, for implementing these methods. All provided a reasonable fit to the original decisions. As stated previously, the rule-based expert system was preferred due to its explanatory power in that it provides a decision tree showing each decision making strategy.

An example extract of a decision tree generated using this approach is shown in Figure 31.6. If machine Op1025 breaks down, the decision maker will repair immediately if the estimated repair time is less than 20 min, otherwise standby operation will be used. For Op1060 a different repair time threshold applies.

---
[2] Registered trademark of The Math Works Inc., Natick, MA.
[3] Registered trademark of SPSS Inc., Chicago, IL.
[4] Registered trademark of XpertRule Software, Leigh, UK.

Modeling Human Interaction in Organizational Systems

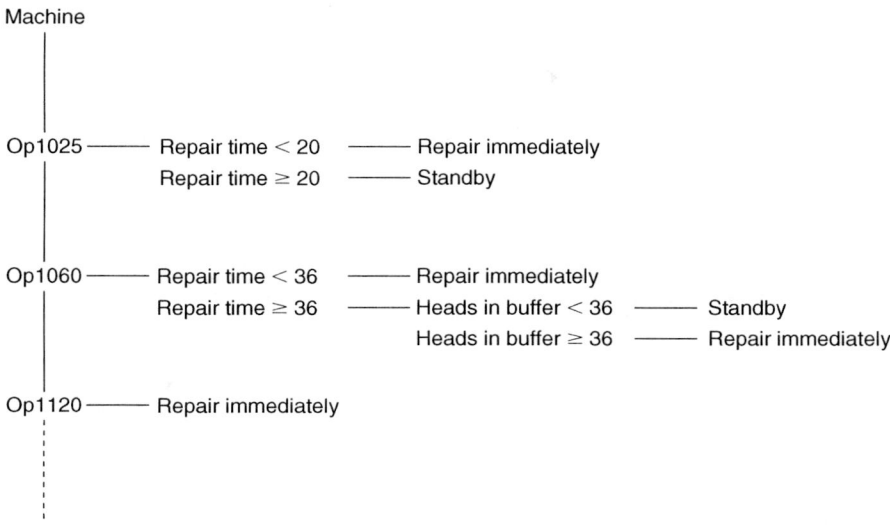

FIGURE 31.6 Extract from the decision tree derived for one of the decision makers. (From Robinson et al. 2005. *Journal of the Operational Research Society* 56: 918. With permission.)

TABLE 31.1 Performance of Decision Making Strategies in Terms of Assembly Line Daily Throughput

	DM1	DM2	DM3
Mean daily throughput	325.53	325.15	318.87
Standard deviation	36.28	28.54	34.34

Further to this, the decision maker takes account of the number of engine heads that are available in the buffer. For Op1120 the decision is always to repair immediately.

Notice that only a few of the decision attributes are required to determine what decisions would be taken. The same is true for all the decision makers, although some may take account of different attributes. There is something of a disconnect then between the attributes that a decision maker suggests he/she uses in forming decisions and those that are actually used. This helps confirm the view that simply asking decision makers what decisions they take and how is unlikely to yield accurate information. Very often experts simply do not know how they reach decisions or even what decisions they might take, since much knowledge is tacit.

31.6.4 KBI Stage 4

The rule-based expert system was then linked to the VIS to perform a long simulation run of over 100 days. As decision points were reached during the model run, the expert system provided decisions in place of the human decision maker. The model then continued its run simulating the consequences of each decision.

This simulation runs showed some variations in performance between the decision makers as measured by the throughput of the assembly line (Table 31.1). Note that a multiplier has been applied to the figures for reasons of confidentiality. Decision makers 1 and 2 achieve a similar level of throughput, but decision maker 2 achieves a lower variation on a day-to-day basis. Decision maker 3 achieves a mean throughput of between 6 and 7 less engines a day.

As previously stated, these decision makers only supervise maintenance activities on the first area of the assembly line. There are three further areas. In the simulations, it was assumed that for the rest of the facility the decision is always to repair immediately. As a result, the effect of the decision makers studied on overall assembly line performance is limited. To understand the full effect of maintenance decisions on assembly line performance, a further study of the other three line areas would have to be carried out.

31.6.5 KBI Stage 5

In the Ford project stage 5 was not performed explicitly. However, it is notable that at one point in the project a new supervisor used the model as a training device. When he was uncertain about which decision to take for a specific scenario, he sought advice from a colleague. In this way the methodology acted as a catalyst for improving the supervisor's decision making strategy.

31.6.6 Findings about KBI

This project demonstrated the validity of the KBI approach, showing that a VIS could be used to elicit knowledge from decision makers and an AI tool could then learn their (approximate) decision making strategies. Further to this, the VIS and AI models could be linked to determine the performance of the manufacturing system.

Various issues concerning the methodology also emerged:

- *Stage 1*: the use of formal problem structuring methods might aid this stage, e.g., cognitive mapping (Eden and Ackermann, 2001) or soft systems methodology (Checkland, 1981).
- *Stage 2*: the use of VIS needs further investigation, particularly the design of the human–computer interface, e.g., level of visual representation, reporting of decision attributes, and selection of decision scenarios to obtain sufficient valid data rapidly.
- *Stage 3*: further exploration of AI approaches is required to identify those with greatest learning and explanatory power.
- *Stage 4*: validation of the AI representations is problematic and requires methodological developments. How can we be sure that an AI representation of a decision making strategy is close to the actual strategy used in the real system?
- *Stage 5*: use of informal and formal methods for improving decision making strategies need to be explored.

A second project is focusing specifically on the issues surrounding stage 2 (Lee et al. 2005).

31.7 Conclusion

The KBI methodology provides an approach for obtaining data on, and representing, human decision making strategies. It appears to do this with some success, but also a number of limitations. This suggests the need for further development to the methodology.

Meanwhile, human decision making is just one form of human interaction with a system. There are many other forms of interaction between humans and systems (e.g., speed of work, requests for information, and environmental effects on humans) and between humans within systems (e.g., communication and physical interaction). Since the performance of many organizational systems is in part determined by the interaction of humans, it would be useful to better understand the nature of that interaction and its effects on system performance. The modeling of systems and human interaction provides a means for achieving this.

At present examples of modeling systems and human interaction are quite limited and the field requires further development with emphasis on two key areas. First, there is the need for modeling frameworks to represent human interaction. KBI provides one such framework. Other approaches might be found in fields such as AI, agent-based methods, game theory, evolutionary algorithms, and Bayesian belief networks. A second requirement is for data on human behavior and human interaction to be able to populate the modeling frameworks. This is a complex task. Further development of the field surely requires a multidisciplinary approach with input from modeling experts, computer scientists, and cognitive psychologists.

Acknowledgement

Sections of this chapter are based on the paper: Robinson, S., J.S. Edwards and E.P.K. Lee. 2005. Simulation based knowledge elicitation. In *International Conference on Human–Computer Interface*: *Advances for Simulation Modeling*, pp. 18–24. San Diego, CA: Society for Modeling and Computer Simulation.

References

Axelrod, R. 1997. *The Complexity of Cooperation: Agent-Based Models of Competition and Collaboration.* Princeton, NJ: Princeton University Press.

Baines, T.S. and J.M. Kay. 2002. Human performance modelling as an aid in the process of manufacturing system design: a pilot study. *International Journal of Production Research* 40(10): 2321–2334.

Baxter, N., D. Collings and I. Adjali. 2003. Agent-based modelling—intelligent customer relationship management. *BT Technology Journal* 21(2): 126–132.

Brailsford, S. and B. Schmidt. 2003. Towards incorporating human behaviour in models of health care systems: an approach using discrete event simulation. *European Journal of Operational Research* 150: 19–31.

Brooks, R.J. and A.M. Tobias. 1996. Choosing the best model: level of detail, complexity and model performance. *Mathematical and Computer Modelling* 24(4): 1–14.

Carley, K.M. 2003. Estimating vulnerabilities in large covert networks. In *Dynamic Social Network Modeling and Analysis: Workshop Summary and Papers*. Washington, DC: The National Academies Press.

Checkland, P.B. 1981. *Systems Thinking, Systems Practice.* Chichester, UK: Wiley.

Chwif, L., M.R.P. Barretto and R.J. Paul. 2000. On simulation model complexity. In *Proceedings of the 2000 Winter Simulation Conference*, pp. 449–455. Piscataway, NJ: IEEE.

Cross, R. and A. Parker. 2004. *The Hidden Power of Social Networks: Understanding How Work Really Gets Done in Organizations.* Boston, MA: Harvard University Press.

Curram, S. 1997. *Representing Intelligent Decision Making in Discrete Event Simulation: A Stochastic Neural Network Approach*. PhD Thesis, Coventry, UK: University of Warwick.

Dembski, W.A. 1999. *Intelligent Design: The Bridge between Science and Theology.* Downers Grove, IL: InterVarsity Press.

Eden, C. and F. Ackermann. 2001. SODA—The Principles. In *Rational Analysis for a Problematic World Revisited*, pp. 21–42. Chichester, UK: Wiley.

Flitman, A.M. and R.D. Hurrion. 1987. Linking discrete-event simulation models with expert systems. *Journal of the Operational Research Society* 38(8): 723–734.

Harmon, S.Y., C.W.D. Hoffman, A.J. Gonzalez, R. Knauf and V.B. Barr. 2006. Validation of human behavior representations. www.dmso.mil/public/library/projects/vva/found_02/sess_papers/b3_harmon.pdf (accessed June 2006).

Hodges, J.S. 1991. Six (or so) things you can do with a bad model. *Operations Research* 39: 355–365.

Innis, G. and E. Rexstad. 1983. Simulation model simplification techniques. *Simulation* 41(1): 7–15.

Kim, H., J.-H. Park, D. Lee and Y.-S. Yang. 2004. Establishing the methodologies for human evacuation simulation in marine accidents. *Computers and Industrial Engineering* 46(4): 725–740.

Learmount, D. 2005. Evacuation tool given green light. *Flight International* 167(4982): 12.

Lee, E.P.K., S. Robinson and J.S. Edwards. 2005. Improving the use of simulation-assisted AI in eliciting operations knowledge. In *Third European Conference on Intelligent Management Systems in Operations*, pp. 105–115. Birmingham, UK: The Operational Research Society.

Liang, T.-P., H. Moskowitz and Y. Yuehwern. 1992. Integrating neural networks and semi-Markov processes for automated knowledge acquisition: an application to real-time scheduling. *Decision Sciences* 23(6): 1297–1314.

Lyu, J. and A. Gunasekaran. 1997. An intelligent simulation model to evaluate scheduling strategies in a steel company. *International Journal of Systems Science* 28(6): 611–616.

Moffat, J. 2000. Representing the command and control process in simulation models of conflict. *Journal of the Operational Research Society* 51(4): 431–439.

Nance, R.E. 1994. The conical methodology and the evolution of simulation model development. *Annals of Operations Research* 53: 1–45.

Pace, D.K. 2000. Simulation conceptual model development. In *Proceedings of the 2000 Spring Simulation Interoperability Workshop*. www.sisostds.org. (accessed July 2005).

Pidd, M. 1999. Just modeling through: a rough guide to modeling. *Interfaces* 29(2): 118–132.

Pidd, M. 2004. *Computer Simulation in Management Science*, 5th ed. Chichester, UK: Wiley.

Powell, S.G. 1995. Six key modeling heuristics. *Interfaces* 25(4): 114–125.

Quinlan, J.R. 1986. Induction of decision trees. *Machine Learning* 1(1): 81–106.

Robinson, S. 1994. Simulation projects: building the right conceptual model. *Industrial Engineering* 26(9): 34–36.

Robinson, S. 2001. Soft with a hard centre: discrete-event simulation in facilitation. *Journal of the Operational Research Society* 52(8): 905–915.

Robinson, S. 2004. *Simulation: The Practice of Model Development and Use*. Chichester, UK: Wiley.

Robinson, S., T. Alifantis, J.S. Edwards, J. Ladbrook and T. Waller. 2005. Knowledge based improvement: simulation and artificial intelligence for identifying and improving human decision making in an operations system. *Journal of the Operational Research Society* 56(8): 912–921.

Robinson, S., J.S. Edwards and W. Yongfa. 1998. An expert systems approach to simulating the human decision maker. In *Proceedings of the Winter Simulation Conference 1998*, pp. 1541–1545. Piscataway, NJ: IEEE.

Salt, J. 1993. Simulation should be easy and fun. In *Proceedings of the 1993 Winter Simulation Conference*, pp. 1–5. Piscataway, NJ: IEEE.

Schelhorn, T., D. O'Sullivan, M. Hakley and M. Thurstain-Goodwin. 1999. Streets: an agent-based pedestrian model. Working Paper Series: Paper 9. University College London: Centre for Advanced Spatial Analysis.

Schmidt, B. 2000. *The Modelling of Human Behaviour*. Erlangen, Germany: SCS Publications.

Silverman, B.G., M. Johns, J. Cornwell, and K. O'Brien. 2006. Human behavior models for agents in simulators and games. Part 1: Enabling science with PMFserv. *Presence* 15(2): 139–162.

Standridge, C.R. and D. Steward. 2000. Using expert systems for simulation modeling of patient scheduling. *Simulation* 75(3): 148–156.

Thomas, A. and P. Charpentier. 2005. Reducing simulation models for scheduling manufacturing facilities. *European Journal of Operational Research* 161(1): 111–125.

Ward, S.C. 1989. Arguments for constructively simple models. *Journal of the Operational Research Society* 40(2): 141–153.

Williams, T. 1996. Simulating the man-in-the-loop. *OR Insight* 9(4): 17–21.

Zeigler, B.P. 1976. *Theory of Modelling and Simulation*. New York: Wiley.

32
Military Modeling

Roger Smith
Modelbenders LLC

32.1 Introduction 32-1
32.2 Applications 32-1
32.3 Representation 32-2
 Engineering • Virtual • Constructive • Live • Environment
32.4 Dynamics 32-4
 Movement • Perception • Exchange • Engagement • Reasoning • Dynamic Environment
32.5 Modeling Approach 32-8
 Physics • Mathematic • Stochastic • Logical Process • Artificial Intelligence
32.6 Military Simulation Systems 32-11
32.7 Conclusion 32-12

32.1 Introduction

The military has always been a very heavy user and innovative developer of modeling techniques and technologies. The nature of military missions requires that they rehearse missions to better understand their complex interactions and to estimate outcomes. This need has led them to apply modeling and simulation to a number of different activities over the last 300 years. In this chapter, we will explore the major applications of military modeling and will discuss the most common forms of dynamic modeling.

32.2 Applications

The U.S. military has made its own unique definitions of the terms "modeling" and "simulation." For their purposes, modeling is often defined as, "a descriptive, functional, or physical representation of a system" (National Simulation Center, 2000). These representations may take the form of a mathematic equation, a logical algorithm, a three-dimensional digital image, or a partial physical mock-up of the system. Models are applied so widely that the variety of systems of interest is almost without bounds. In these systems, military weapons systems are usually very prominently represented, to include land, air, and sea vehicles; communications and radar equipment; handheld weapons; and individual soldiers. But models also represent the decision-making process and automated information processing that occurs inside the human brain and within battlefield computers. They extend to representations of the environment that is made up of terrain, vegetation, cultural features, the atmosphere, ocean, and radio frequency (RF) environment. Different combinations of all of these are needed to accurately represent potential military situations.

One military definition of simulation is, "a system or model that represents activities and interactions over time. A simulation may be fully automated, or it may be interactive or interruptible" (NSC, 2000).

This definition attempts to encompass human-in-the-loop simulators for training as well as systems that serve as analytical tools for computing outcomes without the aid of a human participant.

The official categorization of the use of models and simulation within the military is to divide them into three large application groups.

The first is for use in "research, development, and acquisition." In these applications, models are used to provide insight into the cost and performance of military equipment, processes, or missions that are planned for the future. These use scientific inquiry to discover or revise facts and theories of phenomena, followed by transformation of these discoveries into physical representations.

The second category is in exploring "advanced concepts and requirements." These models present military systems and situations in a form that allows the military to conduct concept exploration and trade studies into alternatives. These trade studies often explore multiple variations on a new weapon or tactic and attempt to measure the effectiveness of each of them. The result is a general appreciation for the different options available and some rough measure for ranking them. The models may be used to understand physical weapons or equipment, but they may also explore different processes for organizing and executing a mission. These require an understanding of processes and the interactions that occur between the different steps in the processes. The models assist the military in creating a doctrine of operations, constructing an internal organization, and selecting materials for acquisition.

The third category is in "training, education, and military operations." Models that are embedded in a simulation system are used to stimulate individuals and groups of personnel with specific military scenarios. The goal is to determine the degree to which they have learned to execute the doctrines they have been taught. It also gives them the opportunity to experiment with new ideas and to determine how useful these might be in a real warfare situation. All of this can be done in a controlled environment that is free of the life-threatening situations that are part of real combat operations.

Finally, it should be noted that military modeling and simulation has always been the basis for a large segment of entertainment products. Many of the modeling concepts behind paper board-wargaming in the 1950s were developed simultaneously by the RAND Corporation for serious military training and by Charles Roberts at the Avalon Hill game company for popular entertainment (Perla, 1990). This trend has continued for over 50 years and can be seen today in comparing realistic three-dimensional military training systems and the product of the very popular computer gaming industry. Systems like America's Army provide an environment for experimentation and training in the military, a device to enhance Army recruitment and education about the military lifestyle, and a game for use by anyone looking for a little excitement in their free time (America's Army, 2006).

32.3 Representation

Models, by their very nature are an abstraction or simplification of the real world. Therefore, it is possible to create an almost infinite number of variations on the representation of objects, actions, and events in a simulation. Over the past several decades, a number of different types of models have been developed for representing a military system or mission. These have gradually converged into commonly recognized categories of representation. These categories have significantly improved the ability of military modelers to communicate with each other and to exchange models with a better understanding of the differences between the products being created.

32.3.1 Engineering

Engineering models focus on the details of what a system does. These capture the physical properties of materials, liquids, aerodynamics, servomechanisms, and computer control of specific systems. They also include interactions between two physical objects or between an object and its environment. An engineering model attempts to understand the physical capabilities of the system at a level that is accurate enough to be used to design the system. Historically, physical prototypes were used to conduct these experiments. However, advanced computer technologies and modeling techniques have allowed us to create digital models of systems that are nearly as predictive as are live physical tests. These models offer many advantages over their physical counterparts. They are almost infinitely malleable so that experiments can be conducted on many thousands of variations rather than just a few physical prototypes. They are nearly infinitely instrumentable. It is possible to collect data from all points in space and time around the event of interest. When using physical prototypes we are often limited by our ability to place sensor, communication, and recording equipment at the precise place and time of interest. Engineering models become a more prominent part of creating or studying military systems because of the accessibility of the required computers and more mature methods for representing real systems.

32.3.2 Virtual

A "virtual model" often refers to a three-dimensional representation of a system that is operating in a digital three-dimensional environment. The focus is usually on the visual appearance of the object and the environment, more than on the properties of physics that are the focus of engineering models. Because of its visual focus, the objects most often represented are military vehicles and humans that would appear on a battlefield. This category is closely aligned with the more popularly recognized term "virtual reality."

A virtual model and environment are usually constructed to simulate individual soldiers who are immersed in a system that generates visual, aural, and tactile stimuli. The goal is usually to train, test, or measure the ability of the human to respond in a desirable manner to the stimuli. Flight simulators are the most popularly recognized form of these models and systems.

32.3.3 Constructive

A "constructive model" represents objects that are separate from the human user or player, but which are under the control of this person. The user sends commands to these objects, but is not immersed in the middle of the battle as he or she would be in a virtual environment. Historically, constructive models have often been aggregated as well. Rather than representing individual vehicles or people, the model represents groups of these in an attempt to reduce the number of details about each and to make it possible for the computer and the human to control many more of them. More recently, constructive simulations that represent individual objects have become very popular and very powerful. These are often referred to as semiautomated forces (SAF) systems because of the way that control of the objects is shared between a human user of the system and intelligent models of the human behavior embedded in the software. A human may provide the overall mission and direction, but the SAF will supplement this with detailed control of activities like movement and engagement.

A constructive model may represent a flight of four aircraft as a single item in the simulation or it may represent each aircraft individually. What separates the constructive from the virtual is usually the method of human interaction, the lack of a three-dimensional representation of the object, and the number of objects that are controlled by a single user. A constructive may also group several hundred vehicles, humans, and equipment into a single object model. This model must then represent the aggregated behaviors of its many different constituent parts. There are a number of motivations for this type of modeling. First, it allows the simulation system developers to capture the operations of a much broader battlefield in a form that can be run on a reasonable computer suite. Second, in many cases the behavior of groups of objects are not understood at the engineering or virtual level, but can be represented as a higher-level aggregate. Third, this type of model mimics the organization, representation, and information that are used in the real military organizational hierarchy.

Very basic constructive models of military operations can be seen in many board and computer games, such as Chess, Stratego, and Risk. Constructive simulation systems differ from virtual systems in that the human operator or player is often positioned outside of the battle. Engagements are not usually targeted at the human player, so they are in a position to think more strategically about the situation and are not required to react to individual events that appear to threaten them personally, as would occur in a virtual system.

32.3.4 Live

Though a "live model" appears to be an inappropriate description, the term has been adopted to refer to activities in which live humans, vehicles, and equipment engage in mock combat. The combat events do not involve real munitions and attempt to avoid situations that could have lethal outcomes. Using computer, communication, navigation, and laser technologies, training areas have been constructed in which combatants can use their real weapons in a form that is as physically realistic as possible. Laser beams and radio often replace bullets, and radio messages indicate where bombs are dropped.

Live modeling allows humans to train in the real environment, to experience the physical hardships of traversing rough terrain, operating in the desert sun, and experiencing the effects of dirt and water on the equipment. The humans and vehicles become living models in a living simulation. In many cases, these live participants are also supplemented with virtual and constructive models to enrich the entire training experience. The largest, and in many ways, the definitive live exercise, was the Louisiana Maneuvers that were conducted in 1941 and used to prepare U.S. forces for entry into World War II. These maneuvers involved over half a million soldiers operating over an area of 3400 square miles of terrain in Louisiana (Sanson, 2006).

32.3.5 Environment

The model of the environment has typically been a static representation of terrain, vegetation, roads, rivers, wind, clouds, rain, ocean waves, salinity, ocean bottom, and any number of other features. This environment provides a medium within which the above models could operate. The environment impedes the movement of objects, obstructs sensor visibility, and changes the outcomes of all types of operations (Mamaghani, 1998). However, in the midst of all of this activity, the environment itself usually remained static and unchanged. A bomb dropped on a truck may destroy the truck, but makes no change to the underlying terrain or the surrounding vegetation.

Recently, this has been changing. Military simulation systems have included dynamic models of the interaction between military systems and environmental features. Simulated objects are able to knock down trees, crater roads, dig holes, build barriers, and destroy buildings. To support this, a new form of environmental model has evolved which understands the physical effects of vehicles and weapons on dirt, trees, and masonry block structures. Representing these changes has required better understanding of the physical properties of environmental features, especially as they relate to military operations. It is also driving advances in the representation of environmental features as both data structures and three-dimensional rendered scenes. As with all models, those of the environment contain almost an uncountable number of variations on how objects and events are represented. It is not possible to identify or enumerate all of these, but an interested reader is encouraged to explore this area more at the SEDRIS website given at the end of the chapter.

32.4 Dynamics

To this point, we have focused on defining and categorizing military modeling according to its application. Those categorizations were meant to illustrate the unique situations, problems, and interests of the developers and customers for military models and simulation systems. In this section, we will describe the most dominant forms of dynamic modeling that are used in the community. Because military systems and problems are so diverse and such a large investment has been made in exploring them, there are many more unique forms of dynamic modeling than can be captured in a single chapter or an entire book. However, the forms that are described here are some of the most dominant in military systems.

Dynamic modeling of military systems and missions often focuses on activities like

- movement,
- perception,

- exchange,
- engagement,
- reasoning, and
- dynamic environment.

In this section, we describe the dynamics that are included in each of these categories. This is followed by a section that explores multiple approaches to modeling these dynamics.

32.4.1 Movement

Dynamic representation of movement captures the change in an object's position over time. Models may represent position as a coordinate in two-space, three-space, or as a velocity vector. Two-space coordinates usually include a position in X and Y, such as latitude and longitude. For models that represent only ground-based vehicles like trucks, tanks, and foot soldiers, this can be sufficient. The object may have no variation in elevation, or the elevation may come from the underlying elevation of the terrain on which it sits. Position may also include orientation, which in two-space would be limited to a 360° angle around the vehicle. A common reference system for this angle is with the zero point being aligned with true north and proceeding clockwise with 90° being east, 180° being south, and 270° being west.

In three-space, the coordinate system includes a representation of elevation. This third dimension may be height above the local terrain, elevation above mean sea level, or distance from the center of a sphere that represents the Earth. The latter measurement evolved during the creation of distributed heterogeneous simulation systems. When networking multiple simulations, differences in the terrain representation within each system led to significant differences in vehicle position with respect to the terrain. Therefore, a nonterrain referenced coordinate system was needed to overcome these differences. When a three-space orientation is added to this model, it includes the pitch, roll, and yaw of the object, creating a six degree-of-freedom (6-DOF) model. When represented as a vector, this may also include the velocity of the vehicle along the axis of orientation.

In their basic form, movement models change the position and orientation coordinates according to a logical or physical representation of movement, as described in the next section. However, most implementations go further to include the effects that movement has on the object and the environment. The movement model may be linked to a model of the fuel consumption of the vehicle. This adds a limiting factor that can stop movement when the fuel is depleted. The inclusion of a fuel model leads to the need for the simulation system to represent a process for replenishing fuel as well. Otherwise, the objects in the simulation will eventually grind to a halt as fuel is depleted and there is no mechanism to refuel. In military modeling, the addition of each detail often leads to the need for more models to drive the additional variables that are added. Systems can grow far larger than can be developed, funded, or hosted on a computer if there is not a strict management of the details that are included in the models. Many authors have warned against this gradual creep in features that leads to the eventual failure of the system being developed (Law and Kelton, 1991). This type of growth is not limited to movement modeling, but can occur throughout the system if the designers do not control it.

A movement model may also calculate the number of hours of operation that the object has been used. This information is the root of most system failure and maintenance representations, and drives the mean time between failure (MTBF), repair (MTBR), or other similar models.

The interaction of object movement with the terrain can generate environmental changes that trigger yet another model, such as the generation of smoke or dust clouds in the wake of a vehicle. If these changes to the environment are represented, then they call for specific environmental models that can calculate the size and density of the cloud created as well as its drift and dispersion over time.

32.4.2 Perception

Military objects move about the environment to interact with other objects. One of the first steps in this interaction is to perceive or detect the existence, position, and identification of the other object. Sensor models capture the signatures of those objects, as when a visual sensor captures reflected light from an object to the sensor. In most cases, the sensor model does not actually represent the path of a light vector, but instead considers the range and orientation between the target object and the sensor, and calculates whether the target is potentially detectable based on the effective range and field-of-view of the sensor. A sensor model may also include information about the environment in which the detection is being attempted. For a visual sensor, atmospheric factors like the presence of smoke, dust, fog, and lighting may be used to diminish the possibility of detection. Also, environmental features like hills, trees, and buildings may be interposed between the target and the sensor and impact the detection of an object. The physical characteristics of the target may also be considered. Its size, in contrast with the background, rate of movement, and material composition, may significantly impact its detectability. Larger targets may be easier to see than smaller ones. Targets may have a higher or lower degree of camouflage, changing the ability of the sensor to separate them from the background image.

In military simulations, visual sensors are just one of a large variety that are available. Many systems include sensor models that collect signature information in the infrared spectrum, sound, emitted radio and radar signals, magnetic properties, and movement and vibrations. Models of each of these can be constructed at a number of different levels of detail, but each must determine whether to include the properties of the sensor, sensing platform, paired geometry, environment, target, and external interference. As illustrated earlier, as the sensor model becomes more complex, it drives the complexity of the entire system. Including all of the categories just listed would trigger the need for additional detail in the sensor model, but also the need for additional details in all target objects and the environment. Often the limitation in creating a high-fidelity sensor model is not driven by our understanding of the sensor, but, rather, by our ability to represent the characteristics of the target and environment that are needed to implement such a model. In a military simulation system, the detail included in a model may be limited both by the needs of the customer and by the desire to keep the entire system balanced, not allowing one model to drive others to a level of detail that is not necessary or affordable (Pritsker, 1990).

32.4.3 Exchange

After moving and detecting, models are needed to allow objects to exchange materials and information with each other. Battlefield operations often lead to the depletion of materials like fuel, ammunition, food, medical supplies, vehicles, and people. A logistics model may be used to represent the ability of the military to constantly deliver these materials to units and objects in operation. Such models are often based on an understanding of the rates of consumption, the predeployment of supplies to locations that are close to the operation, and the constant replenishment of supplies through a network of supply nodes. Replenishing supplies within an object on the battlefield is the culminating model of a much more complex representation of the logistics infrastructure that can stretch across an entire country or even around the world. The logistics model must also include mistakes and interference that cause it to breakdown and deprive the military objects of the supplies that keep them operating. A logistics model may be driven by textbook ratios of consumption or it may include specific messages from the military objects about the levels and rates of consumption. In the latter case, a communications model is needed to carry information about what materials are being consumed, by whom, and where they are located.

Communication is another model of exchange. The thing being exchanged is information rather than physical items. In the modern military, the amount of information that is carried around in a physical form, such as a book, letter, or paper map, is quite small compared to the amount that is transmitted in digital form. Therefore, modern models focus on communications in the form of digital computers and networks as well as analog radio networks. A model of radio communications, like that of a sensor, may include the characteristics of the transmitter, transmitting platform, environment, the receiver, the geometry between the sender and receiver, and interference by other objects. Details in the representation of the radio or the signal it generates call for corresponding details in the receivers, environment, and countermeasures.

Military models of digital computer communications are similar to the tools used to study Internet traffic. They can represent the senders, receivers, relay nodes, interference from competing traffic, multiple paths for the information to travel, and the loss of a message or the failure of a network. Modeling how people, objects, and units respond to the receipt of this information is included in the section on reasoning.

32.4.4 Engagement

Strictly speaking, engagement is another form of exchange. However, it is listed separately because it has been the central focus on military simulations for decades. The item being exchanged is a weapon and the effect is the degradation of the operational capabilities of the target. Most military simulations perform movement, perception, and exchange specifically so they can put themselves in a position to engage enemy targets. Engagement has historically been the pivotal centerpiece of a simulation system and one of the most important models in the system. Certainly, not all objects engage the enemy, but those that do not are often referred to as support elements whose mission is to make engagement possible for combat equipped units (Smith, 2000).

An engagement model typically includes the exchange of weapons or firepower from a shooter to a target. This exchange decrements the capability of the shooter by expending ammunition in one of its many forms (e.g., bullets, missiles, bombs, rockets, grenades, and artillery rounds). Just as in the perception and communication models described above, this exchange is usually impacted by the geometry between the shooter and the target. Environmental features like trees, terrain, water, and buildings may interfere with the optimal delivery of the weapon and reduce its impact on the target. The target may also contain defensive systems that counter the effects of the engagement. A defensive model may represent the effects of flares or chaff in deceiving and misleading a guided missile or the protective effects of armor to deflect the weapon.

If the weapon successfully impacts the target and is powerful enough to overcome any interference or defenses, then a level of attrition must be calculated for the target. Different approaches to modeling attrition are described in the next section. Attrition is usually directed at the model state variables that control its ability to perform its primary functions. These may include health or strength, fuel levels, communications capabilities, and mobility. Models may also make a binary decision about whether a vehicle, human, or unit is completely destroyed or not. Models of engagement have been of great interest to both the training and analysis communities for decades. A great deal in study and a number of publications dedicated to these exist. Interested readers may consult sources like Ball (1985), Epstein (1985), Parry (1995), and Shubik (1983). They should also visit the website of the Military Operations Research Society given later.

The attrition model may be linked to communications and medical models. Communications models propagate the outcome of an engagement so that units or operators are aware that an engagement has occurred. These communications may trigger a medical model that will attempt to conduct extraction and provide medical treatment to simulated humans that are wounded. It may also trigger the logistics model to extract and repair vehicles.

32.4.5 Reasoning

Within large military simulation systems, there are usually many models of human decision making and behaviors. These have become more prevalent as systems have grown in both the breadth of coverage and the depth of battlefield detail that are represented. Representing human thinking and even some computer reasoning are some of the most challenging parts of the current practice of military modeling. This type of information processing is largely not understood and requires general approximations and simplifications in models.

Reasoning models often rely on the techniques developed within the Artificial Intelligence field. Techniques like finite state machines (FSMs), expert systems, rule-based systems, case-based reasoning, neural networks, fuzzy logic, means-ends analysis, and others are used to organize information and create decisions that are similar to those of living humans. FSMs are currently the most widely used approach to modeling reasoning in both military models and commercial games. These reasoning models are challenged to perform a wide array of operations, to include commanding subordinate units, decomposing and acting on commands from higher level units, reacting to enemy attacks, selecting maneuver routes, identifying threats and opportunities for engagement, fusing sensor data, and extracting meaning from intelligence reports. Each of these functions can be extremely complicated and require significant computing resources to execute. Reasoning models must balance their level of realism between robotic reactions to stimuli and detailed consideration of the situation prior to selecting an action.

The variety of reasoning models that are required on a battlefield cannot be fit to a single modeling technique. In practice, multiple techniques are required, each applied to a reasoning problem for which it is best suited (Russell and Norvig, 1995).

32.4.6 Dynamic Environment

Earlier, we described the evolution of the simulated environment from static state structures to dynamic representations of features and their interactions with military objects. Military objects interact with the environment both through direct intention and through accidental collocation. An engineering unit may be tasked to destroy a bridge or a road. This is an operation in which the effects on the environment are the specific intent of the action. In another case, an aircraft may bomb a convoy of trucks moving on a road. In this case, the trucks are the primary targets, but the road may sustain damage because of its collocation with the trucks.

Until recently, military simulations seldom included impacts on the environment. However, with the current focus on precision operations, there is much more interest in destroying specific buildings, roads, bridges, communications equipment, and pieces of the social infrastructure. Since this data is usually found in the environmental database, models that accurately modify environmental information are needed.

For decades, military organizations have worked on models that accurately represent the engagements that take place between two tanks, soldiers, airplanes, or ships. It is becoming necessary for those models to also impact the trees, terrain, and roads in the vicinity of the targeted objects. This means that information on the effects of weapons on trees is necessary as well as their effects on buildings, roads, bridges, and a host of other types of surrounding terrain.

Though the type and level of damage done to a tree is seldom the focus of the experiment or exercise that is being conducted, similar damage to buildings, power grids, and command facilities may be the focus of an experiment. As a result, the military modeling and simulation community is pursuing new methods for accurately representing these types of engagements and doing so within the constraints of available computer systems.

32.5 Modeling Approach

In the previous section, we discussed many patterns of relationships that exist between multiple models and described in very general terms what would be represented in those models. However, we did not explore specific mathematic or logical algorithms that could be used in those models. In practice, the number of techniques, algorithms, and equations that are used in military models is close to uncountable. It is not possible to describe all of them or even those that might be considered "the best." So many different problems are studied with military models that there is no "best" approach that can applied universally when representing a specific vehicle, human, or unit. However, the techniques that are used do exhibit characteristics that allow us to talk about them in terms of general categories. However, it is not unusual for a model or a simulation system to combine techniques from any of these categories to create the effects that they need for training or experimentation. The categories aid us in explanation, but should not be considered a universal ontology of approaches. Figure 32.1 illustrates this with a missile that can be modeled using any one of the four modeling categories that will be described.

32.5.1 Physics

Physics-based models are most often found in engineering and virtual simulation systems. For example, a missile pursuing a target would be represented by the physics of motion, momentum, mass, and aerodynamics. Changes in the fin positions would drive aerodynamic equations and change the vector of

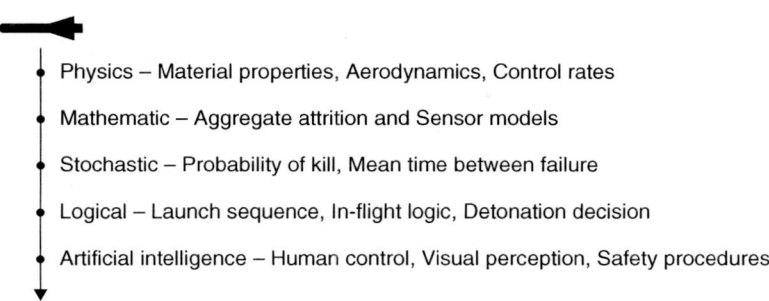

FIGURE 32.1 Five different approaches to modeling the behavior of a missile.

the missile based on the forces at work on the mass of the missile. Similarly, the seeker head in the missile would scan the environment electronically using the same pattern, revisit rates, and sampling rates of the real missile. This behavior would allow the simulated missile to collect data about a target in the same way that the real missile does.

Physics-based models are most often used to analyze the behavior of an existing weapon or to assist in the design of a new weapon. Understanding exactly how the pieces of the system will behave is an important part of exploring the design space to find optimum capabilities and combinations of capabilities that are optimum for the entire system.

Physics models require a great deal of data and mathematics. The data must be available for the system being modeled, the environment in which it is operating, and any other objects that it will interact with. Mathematic equations are required to represent a number of different behaviors of the system, interactions that occur within the system and interactions that occur with other objects. Given this need, it is not sufficient to collect data and build equations only for the missile that is to be studied. The model builders must do the same for the environment and for any objects that will interact with the missile.

Because of the volume of data, and the number and complexity of the equations that are required, physics models are necessarily reserved for smaller scenarios that involve only a few objects. Once constructed, the models can be computationally intensive, requiring high-powered computers or accepting extremely long simulation times. The budget of the project limits the former and the schedule limits the latter. Therefore, the models can literally be a compromise of what the project can afford in time, money, and skilled staff. These types of limitations are one of the primary causes of the diversity of military modeling solutions.

32.5.2 Mathematic

Though a physics model is certainly mathematic, there are a number of modeling techniques that are based in mathematics, but which neither represent the physics of the situation nor employ stochastic methods of representing aggregate behaviors. In military modeling, the classic example of this is the Lanchester equations. In his 1912 and 1916 publications, F.W. Lanchester attempted to represent the attrition experienced by large military forces in combat using differential equations. These assume that the combat power of each side can be represented accurately with fire power scores and that the weapons of each side can be brought to bear equally on all targets and under all conditions. This creates a model which will "grind down" both sides as they engage in combat over time. Each side loses capabilities at a rate proportional to the size of the enemy that is attacking it, and in some cases, also incorporates the size of the targeted unit. Lanchester equations have been used widely since their introduction and have only been displaced relatively recently as the military has sought to represent combat situations that are not symmetric between the attacker and the defender (Davis, 1995a and 1995b).

Lanchester's differential equations may be a useful way of representing a large barrage of missiles engaging targets. Instead of modeling the engagement and attrition of every individual missile, these equations would model the overall impact of a large number of such engagements and determine the attrition to the target as a result of all engagements. There have been many variations to and criticisms of Lanchester's equations. But they remain a foundational part of military simulation techniques.

32.5.3 Stochastic

Stochastic processes, probability and statistics, are most often found in virtual and constructive models. As simulation systems grow larger in their scope of representation, there is a need to capture many more activities and interactions in models. Lacking the detailed knowledge, breadth of expertise, access to data, time to build, and compute power to run a pure physics-based system, modelers have often resorted to a statistical representation of objects and interactions. In this case, the models capture the behavior of many iterations of an event and represent individual event results using a probability function and the results of a pseudo-random number generator. This type of modeling was introduced to the military modeling community by Stanislaw Ulam when he was working on the design of atomic weapons during World War II. Ulam encountered a number of problems for which the specific physical behaviors were not known, but where the pattern of outcomes had been measured. Therefore, he chose to use the statistical properties of the event and rely on multiple simulation runs to arrive at an accurate behavior for the entire system (Metropolis, 1987).

The previous missile example lends itself well to stochastic models. Instead of representing all of the minute physical interactions, a modeler could choose to represent the outcome of a missile engagement given a limited number of input variables governing each event and recourse to a probability distribution. The use of a pseudorandom number in decision making means that no one engagement contains all of the details of the event as in a purely physics-based model. However, if the model is run a number of times, the randomness of multiple replications will blend together and create an accumulated result that is representative of the system behavior that emerges from all of the interacting models.

Stochastic modeling has proven to be extremely useful because it allows modelers to study problems that were previously beyond our understanding of the physics of an event and perhaps beyond the computational capability of accessible computers. This has led to the creation of very large simulation systems capable of representing tens of thousands of events and objects on a battlefield. However, these models also require that their creators understand both the physical behavior of the system and the statistical aggregation of those behaviors to create accurate stochastic models.

32.5.4 Logical Process

A logical model of the missile's behavior may capture the sequential steps and the branching decisions that are used to control the flight of a missile. This model represents the programmed logic within the missile's computers, allowing scientists to explore all of the possible branches and to match logical decisions with the environmental stimuli that the missile will encounter.

When an object is controlled by a simulation system rather than a human operator, most of the time it is following a logical set of defined processes. These instructions tell it when to move, which direction to go, how fast to proceed, which objects to focus on, and which to ignore. These may be very complex processes, but they do not necessarily involve equations of physics or random decision points. In situations when an object should follow some form of "textbook" operation, logical models are an excellent method of encoding this.

FSMs are often used to assist in organizing very complex sets of behaviors. FSMs allow the modeler to capture hierarchical behaviors, set triggers for changing from one behavior to another, and encapsulate behaviors that can be reused in multiple FSMs. These structures are so useful that they often form the framework in which models of all types are organized. As mentioned earlier, SAF systems and computer games are dominated by FSMs for decision making.

32.5.5 Artificial Intelligence

Many military simulations require the representation of complex human decision making that goes beyond the capabilities of logical models. These attempt to model the behavior of individual soldiers, groups, and commanders. The community has turned to artificial intelligence (AI) as a source of unique and powerful methods for representing human behavior. Adopted techniques include FSMs, expert systems, case-based reasoning, neural networks, means-ends analysis, constraint satisfaction, learning systems, and any other technique that shows promise in accurately capturing the complex reasoning process of humans.

To illustrate this category, the missile guidance and navigation example that we have been using needs to be augmented with a simulated human-in-the-loop as it pursues a target. Though a missile model may use an FSM to represent its movement, it is not attempting to create a model of human intelligence; rather it represents a logical process that is followed robotically by the weapon. If the missile were being controlled remotely by a human who was viewing the target on a computer screen, then the behavior of the human might be represented using an AI technique. A neural network may represent the human's ability to discriminate a target in the scene and means-ends analysis may represent the human's decision process in selecting a target, leading its position, and switching from one target to another opportunistically.

AI techniques usually focus on processing information in a human-like manner. Using databases or rule sets, the algorithms attempt to make deductions that lead to behavior selection. These models may incorporate deterministic or stochastic methods in representing human behavior (Russell and Norvig, 1995). As we pointed out at the beginning of this section, these categories are not necessarily mutually exclusive; they are simply useful for explanation and understanding.

32.6 Military Simulation Systems

Modeling is one part of creating a military simulation system. Within any one of these systems there can be a large number of models. Using the major categories of models described above, Figure 32.2 illustrates the relationships that often exist between these models to create a working simulation system. This figure includes only the major categories. For a specific system, the number of models would be much larger and the relationships between them would be more complex. This figure illustrates many of the causal relationships that were described in the earlier sections.

The movement model is a good place to begin tracing the models in the figure. When this model calculates the new location, orientation, and velocity for an object, it may also trigger the dynamic environment model to represent the creation of smoke, dust clouds, or tracks in the sand. Once completed, the objects are in a position to execute perception models that can detect other objects from their newly achieved position. The perception model provides the necessary information to allow the objects to engage

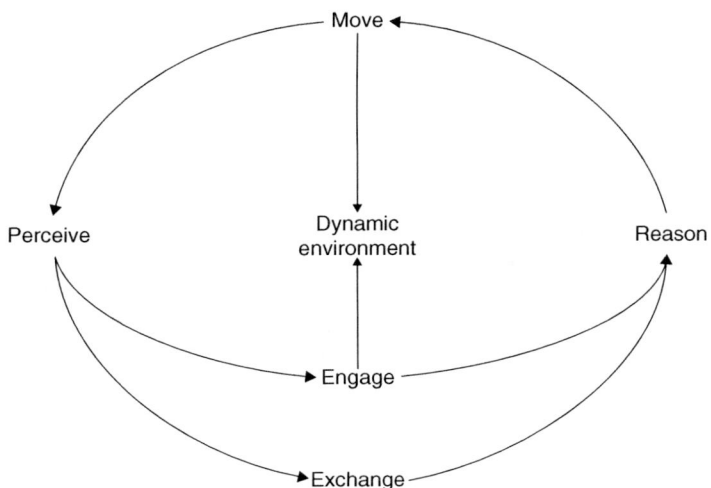

FIGURE 32.2 Model relationships for a simple battlefield simulation system.

each other or to exchange information or material with each other. Engagement also triggers the dynamic environment models to create effects like road craters and destroyed buildings in the environment. An engagement between two objects may create collateral damage to surrounding trees, buildings, and roads. In some cases, the engagement is actually targeted at an environmental feature like a road or a bridge. The exchange model calculates functions like refueling an aircraft or transmitting a message. Following the sequence of move–perceive–engage, the system may allow the objects to reason over what has just happened. This reasoning can take into account the results of each of the engagement, exchange, and perception, integrating them to enable the reasoning model to select the next action to be taken. Once completed, the cycle can begin again with a new objective that is received from a human user or from the reasoning models.

This cyclic diagram is a simplification of a real system. In actual implementation, the reasoning model may be activated at the completion of each of the other models, providing much finer control over the decision-making process. That is characteristic of virtual-level simulations in which the reasoning component is providing very detailed control of a computer-controlled entity. The reasoning model may also be triggered much less frequently than the other models. This occurs in constructive-level simulations where the reasoning is at a much higher level of command and decisions are made infrequently with respect to the rate of activities in the other models.

32.7 Conclusion

This chapter has provided a high-level overview of the dynamic modeling necessary to create military simulation systems. The very large number and variety of military systems that have been created, makes it impossible to describe the most common or "best" approach to modeling. Existing military systems focus heavily on movement, perception, and engagement. But, they may also include models of medical operations, communications, intelligence processing, military engineering, logistics networks, and command and control.

The mission of military organizations changes in response to the political situation in the world. The changes that have occurred in world politics are influencing the types of things that are being modeled in military systems. Newer simulation systems are focusing more on communications, social influence, police actions, one-on-one interactions with noncombatants, and urban environments. These call for scenarios that study smaller interactions between competing military forces or between the military and the civilian populace, rather than large theater-level models involving thousands of combatants on each side.

Models of the threat or opposing forces are also changing significantly. New models are being created that represent suicide bombers, improvised explosive devices, riots and protests, and active avoidance of direct engagement.

The future of military modeling will include increasing level of dynamics in the modeled world. Rather than focusing only on the combat-relevant activities of an object, we will be creating objects that have a much more extreme range of dynamic properties. These "extreme dynamic" models will create a more realistic world in which the human users and the automated objects will be able to interact with the virtual world in all of the ways that a real person would. This could include being able to assemble primitive objects into more complex ones, breaking objects into multiple pieces, tapping into the electrical systems of buildings, digging holes in the terrain, or interfering with the normal operations of a vehicle by flattening its tires or inserting rocks in its gun barrel. Such a dynamic representation of the world is far beyond our current capabilities due to limitations in both our modeling capabilities and the processing capacity of current computers. But, within a decade or two, military models will represent a world that is "McGuiver-ready." This means that the modeling is so rich that a user will be able to do almost anything he can imagine in the world and a model will be there ready to represent those actions realistically.

Computer games like The Sims illustrate some of the richness that we are looking toward in military simulation systems. These games often focus on mundane activities like creating a meal, painting a house, mowing the grass, and reading a book. Though these activities will probably never be the primary focus of military simulations, they can play an important part in creating a realistic world in which to rehearse or experiment with military actions. During the Cold War, the primary military problem changed very little and this had a direct impact on the evolution of military models and simulations. In today's more chaotic and ever changing environment, the military is being forced to look for ways to represent a much wider variety of objects and interactions. This will lead to significant changes in the dynamics that are modeled in future simulation systems.

References

America's Army. 2006. America's Army Web Site. Accessed May 28, 2006 at http://www.americasarmy.com/.
Ball, R. 1985. *The Fundamentals of Aircraft Combat Survivability Analysis and Design.* New York: AIAA Press.
Davis, P. 1995a. Aggregation, Disaggregation, and the 3:1 Rules in Ground Combat. RAND Monograph Report, MR-638. Available at http://www.rand.org/pubs/monograph_reports/MR638/.
Davis, P. 1995b. "Distributed interactive simulation in the evolution of DoD warfare modeling and simulation." *Proceedings of the IEEE* 83(8), 1138–1155.
Epstein, J. 1985. *The Calculus of Conventional War: Dynamic Analysis without Lanchester Theory.* Washington, DC: Brookings Institution.
Law, A. and Kelton, W.D. 1991. *Simulation Modeling & Analysis*, 2nd edition. New York: McGraw-Hill.
Mamaghani, F. 1998. "Creation and use of synthetic environments in realtime networked interactive simulation." *Digital Illusion: Entertaining the Future with High Technology.* New York: ACM Press.
Metropolis, N. 1987. "The beginning of the Monte Carlo Method." *Los Alamos Science.* Special Issue dedicated to Stanislaw Ulam, pp. 125–130. Accessed May 28, 2006 at http://www.fas.org/sgp/othergov/doe/lanl/pubs/00326866.pdf.
National Simulation Center. 2000. *Training with Simulations: A Handbook for Commanders and Trainers.* Fort Leavenworth, KS: National Simulation Center.
Parry, S., Ed. 1995. *Military OR Analyst's Handbook: Conventional Weapons Effects.* Alexandria, VA: Military Operations Research Society.
Perla, P. 1990. *The Art of Wargaming: A Guide for Professionals and Hobbyists.* Annapolis, MD: Naval Institute Press.
Pritsker, A. 1990. *Papers, Experiences, Perspectives.* West Lafayette, IN: Systems Publishing Corp.
Russell, S. and Norvig, P. 1995. *Artificial Intelligence: A Modern Approach.* Reading, MA: Prentice-Hall.
Sanson, J.P. 2006. Studying War: Central Louisiana and the Training of United States Troops 1939–1945. Accessed May 28, 2006 at http://www.crt.state.la.us/tourism/lawwii/maneuvers/Studying_War.htm
Shubik, M., Ed. 1983. *Mathematics of Conflict*, Editor. Amsterdam: Elsevier.
Smith, R. 2000. Simulation. In *Encyclopedia of Computer Science*, 4th edition, Ralston, A., Reilly, E.D. and Hemmendinger, D., Ed. New York: Grove's Dictionaries, pp. 1578–1587.

Further Information

Conferences

Interservice/Industry Training, Simulation, and Education Conference—http://www.iitsec.org/
Simulation Interoperability Workshop—http://www.sisostds.org/
Computer Generated Forces and Behavioral Representations Conference—http://www.sisostds.org/
Winter Simulation Conference—http://www.wintersim.org/
Military Operations Research Society Conferences—http://www.mors.org/
Society for Computer Simulation Conferences—http://www.scs.org/
Game Developers Conference—http://www.gdconf.com/
Parallel and Distributed Simulation Workshop—http://www.acm.org/sigsim/

Journals and Trade Magazines

ACM Transactions on Modeling and Computer Simulation (TOMACS)—http://www.acm.org/tomacs
NTSA Training Industry News—http://www.ntsa.org/ and http://www.ndia.org/
National Defense Magazine—http://www.ndia.org/
Military Training Technology Magazine—http://www.mt2-kmi.com/
Training and Simulation (Annual Issue). Published by the Armed Forces Journal International—http://www.afji.com/
Phalanx and Military Operations Research Journal—http://www.mors.org/
SCS Journal of Defense Modeling and Simulation—http://www.scs.org/pubs/jdms/jdms.html

Web Sites

ACM Special Interest Group on Simulation—http://www.acm.org/sigsim/
Army Modeling and Simulation Office—http://www.amso.army.mil/
Center for Army Lessons Learned—http://call.army.mil/call.html
Defense Modeling and Simulation Office—http://www.dmso.mil/
Institute for Simulation and Training—http://www.ist.ucf.edu/
MSRR—http://www.msrr.dmso.mil/
National Simulation Center—http://leav-www.army.mil/nsc/
National Training Center—http://www.irwin.army.mil/
Program Executive Office for Simulation, Training, and Instrumentation—http://www.peostri.army.mil/
SEDRIS—http://www.sedris.org/
Simulation Interoperability Standards Organization—http://www.sisostds.org/
Simulation Technology Magazine—http://www.simulationtechnology.org/

33
Dynamic Modeling in Management Science

Michael Pidd
Lancaster University Management School

33.1 Introduction 33-1
  Models to Support Decision Making • Dynamic Systems Modeling in Management Science
33.2 An Approach to Dynamic Systems Modeling in Management Science 33-4
  Project Management Skills • Conceptual Modeling Skills • Technical Skills
33.3 Discrete Event Simulation 33-7
  What Types of Problem are Well Suited to Discrete-Event Simulation? • Common Application Areas for Discrete-Event Simulation in Management Science • Discrete-Event Simulation Terminology • Activity Cycle Diagrams in Conceptual Modeling • Implementing a DES Model in Computer Software • Discrete Event Simulation using a VIMS
33.4 System Dynamics in Management Science 33-16
  System Structure and System Behavior • Qualitative System and Dynamics: Causal Loop Diagrams • Quantitative System Dynamics: Level-Rate Diagrams
33.5 Model Validation 33-21
33.6 Chapter Summary 33-22

33.1 Introduction

There was a time when even large businesses were staffed by many clerical workers whose main role was to process data and turn it into information. The data included sales invoices, payments, orders, production schedules, and the like. It had to be processed so that the managers of the organization, whether public or private, could make sound decisions. Decision making is a fundamental part of any manager's job; sometimes the decisions are trivial and have little impact, but others are so significant that they could open up new opportunities or ruin the business. Nowadays, the serried ranks of clerical workers have gone in advanced economies and information processing is automated in computer systems, some of which offer decision support. The importance of decision making and the use of information on which to base those decisions led to the development of management science: sometimes known as operations research in the USA or as operational research in the UK.

This chapter focuses on dynamic modeling to support decision making and planning as it has developed in management science. A glance at the catalog of any academic publisher will show that there are many, many texts that cover the techniques associated with management science. Examples include Hillier and Lieberman (2005), Daellenbach and McNickle (2004), and Taha (2003) and the list of such books is constantly changing. This chapter is not intended to provide a condensed version of these books, but instead discusses dynamic systems modeling in management science and provides practical tips on how to go about doing this. Clearly, not everything can be covered in the space available, and much more detail can be found in Pidd (2004a).

Models are used in decision making and planning to explore the consequences of decisions before they are made. The decisions may be one-off: for example, a business may wish to plan its distribution network for the next few years. Getting this right can save much money. A model that carefully explores the various options against likely demand and cost scenarios can be very useful in this regard. Other decisions are repetitive, for example airlines make routine use of dynamic pricing systems when selling tickets over the Internet. As most of us know, two passengers sitting in adjacent seats on the same flight may pay very different fares. Often these differences are due to the time at which the flight was booked, with early bookers paying the lowest prices. As the flight fills, so the ticket prices rise, though they may fall again if demand is lower than expected. This dynamic pricing, which is relatively simple for point-to-point airlines, is executed through a set of interlinked models that forecast likely demand through the booking period and adjust the prices so as to maximize revenue or profit from the flight. Similar systems are used for other "perishable" goods (ones with no value after a sell-by date) such as hotel rooms. An airline that operates many flights could not use dynamic pricing without automated decision models.

33.1.1 Models to Support Decision Making

The obvious first question to be faced is: Why use models to support good decision making and planning? Aren't some managers just very good, intuitive decision makers? Why go through all the rigmarole of modeling to support a decision? While it may well be true that some managers are good intuitive decision makers, a look at the longer term financial performance of almost any organization, whether public or private, makes it clear that these are rare beasts. The trick is not to make a single, good decision, but to go on making sensible choices time after time. For present purposes a model to support decision making will be defined as follows (taken from Pidd, 2003):

> an external and explicit representation of part of reality as seen by the people who wish to use that model to understand, to change, to manage and to control that part of reality.

This may seem a little vague, so it will now be unpacked in a little more detail.

As Sterman (2000) and others in the system dynamics community point out, we all use mental models to understand our experiences and to plan for the future. For example, I live on the west coast of Britain near the sea, where the weather is very changeable and can switch from sunshine to rain rather quickly. Over time, I have learned to "read" the sky and the wind to forecast the weather in the next few hours or so. Though this mental model can be improved by experience, it is really not much use beyond an immediate prediction and is unlikely to work for locations on the east or south coasts; let alone on another continent. Early weather forecasting was based on folklore, which embodied experience: "Red sky at night, shepherd's delight, red sky at dawning, fisherman's warning." However, things are very different now, and it is possible to get a 5-day weather forecast for almost anywhere in the world by logging on to one of the many Web sites that provide this service. How is this done? Data from terrestrial weather stations and from orbiting satellites is fed into data centers that use mathematical models of the weather systems to predict the weather. This is obviously easier in some locations (for example in a desert) than others, but the same generalized models are used. These were developed by meteorologists over many years and are external and explicit—that is, anyone who understands the science and mathematics can explore, criticize, and improve them. The models used in management science to support decision making and planning are likewise external and explicit and open to scrutiny.

The second part of the definition stresses that such models are representations of part of reality. This is important, because models are always simplifications. Any model that is as complicated and complex as the system it represents would be as expensive to build and as difficult or dangerous to use. Models are used in place of real experimentation, which was often known as "suck it and see" by old-style engineers—an approach that seems attractive, unless the system being studied contains the economic equivalent of strychnine. It is important to realize that the simplification needed to build a model is actually a good thing and should not be regarded as a limitation. It forces the modeler and decision makers to focus on the important aspects of the system being studied. Thus, models that forecast sales in a supermarket may be based on loyalty card data and it is usually enough to divide customers into categories for targeting purposes, rather that considering them individually. That is, a simplified model of purchases by different categories of customer may be enough.

The third part of the definition stresses that people are involved in the modeling and the use of the model and that they have purposes in mind. This provides guidance on the appropriate simplification, as discussed in the previous paragraph, and is also important when assessing whether a model is valid. In these terms, a valid model is one that is useful enough for the purpose in hand—even if it is not good enough for some other, possibly related, purpose. For example, a model to support the daily scheduling of jobs in a manufacturing plant will need to be very detailed and will probably incorporate the characteristics of jobs in the order book and of the available manufacturing resources. In contrast, a model to plan capacity over the next 3 years may need to focus on trends in orders and likely changes in technology but will not incorporate details of individual jobs. Thus the second, longer term, model may be very useful for its intended purpose but may be hopeless for daily job scheduling.

It is also important to realize that models are used for decision support and planning rather than to totally replace human decision making. Thus, an important maxim for modeling is: *model simple, think complicated*. The models are used to free humans for tasks that they are rather good at—making sense of unstructured situations—by performing the rapid calculations and inferences that most people do relatively poorly. If the model turns out to be too simple, then it can always be made more complicated. In one-off decision making, this essential link between the human decision maker and the model is obvious, but what of routine decision making such as dynamic pricing systems? Though these systems, if well-designed, are adaptive to changes in the environment such as increases in demand or costs, they have their limits. For example, intense competition on a route may lead an airline to introduce a wholly new fare structure to win customers. This will require human intervention and decisions; supported, of course, by appropriate decision models.

The other human component in model building and use, which is often sadly ignored, concerns the nature of "reality"—a term used in the definition. Sometimes, we are not modeling reality, but are investigating something that does not exist—the future—and a model that is valid for now may not be valid for then. Recognizing this is important. Also, we must ask what we mean by reality. This is an issue considered by philosophers for millennia and it has practical import in modeling. First, people differ in what they regard as "real." Sometimes this is because they are psychotic, but this is rarely the case. It is well known in the law courts that two people can provide quite different accounts of the same incidents and neither may be lying. They have seen the incident from different perspectives and may have different experiences with which they make sense of what they have seen. That is, each person's knowledge is partial and personal. Sharing and exposing that knowledge to criticism makes it more likely that the knowledge will be more complete. This is well illustrated in the following story that seems common in a number of cultures:

> Six blind men are led into a circus ring, not knowing where they are and are asked, using touch, to say where they are. The crowd watches with interest as an animal is brought slowly and quietly into the ring. The men touch it at different places. One cries out, 'I'm sure I've found a tree, it's rough like bark and I can put my arms around it'. Another tries the same and agrees; then another and a fourth. The fifth man reports that he's holding something long, narrow and sinuous—which must be a tree creeper. The sixth man makes the same claim. Together they confer and announce that they've been led into a tropical forest. Meanwhile the crowd is stifling its laughter, knowing that the blind men have been feeling their way around an elephant.

The point of the story is that no human is omniscient like the crowd but we all, like the blind men, have only partial knowledge. Sharing this incomplete knowledge makes it more likely that we will establish the truth, but there is no human guarantee. Reality and its modeling is an elusive concept, since modeling forms part of the investigation of that reality.

Thus we arrive at the final part of the definition, which again stresses the purposes for which the models are built and used. Though this section has focused on decision making, models are used in management science for other purposes, too. For example, most police forces operate control rooms and these act as contact points for members of the public who may phone in to report an incident or to ask for help. In the UK, each police force is required to meet national standards that specify, for example, the proportion of calls that are answered within 15 s. If a police force finds that it is not achieving these standards, it might choose to model the control room operations to investigate what is happening and why targets are not being achieved. That is, models are used to explore the "as is" situation as well as to conduct "what if" experiments.

33.1.2 Dynamic Systems Modeling in Management Science

There are two approaches to dynamic system modeling that find broad use in management science to support decision making and planning. Each of these is discussed in much more detail later in this chapter, but a quick overview is as follows:

- *Discrete-event simulation* (DES): in which a system to be modeled is treated as a set of discrete entities (people, machines, vehicles, and orders) that change state from time to time. The states are discrete: for example a machine may move between states such as busy, idle, under repair, and warming up. Thus, the states are modeled as discrete attributes of the system entities. The dynamic behavior of the system is thus the result of the interactions between entities as they change state. The simulation model consists of sets of rules that define how different entity classes change state, given particular conditions within the simulation. To build a DES, the modeler must decide which entities to include and must find some way to represent the logic of their state changes. Most organizational systems include variable behavior (e.g., the number of calls received per hour at a contact center) and DES models often use random sampling methods to cope with this variation. In DES models are usually developed using software known variously as simulators, or visual interactive modeling systems (VIMSs). Examples include Arena, ProModel, AutoMod, Witness, Simul8, and Micro Saint, descriptions of which can be found in the annual Proceedings of the Winter Simulation Conference.
- *System dynamics*: in which the focus is not on individual entities and their state changes, but rather on populations of objects and the rules that govern their behavior through time. It is normal to think in terms of flow rates (e.g., orders received per day) and accumulations (e.g., the order backlog), often abbreviated as rates and levels. These are used to model the structure of business processes so as to understand the effect that information feedback has on their behavior. Thus, process structure is seen as important in system dynamics modeling and it is assumed that behavior follows from this structure. Most system dynamics models are developed using dedicated VIMS and popular examples are Stella (also known as IThink), Vensim and Powersim.

33.2 An Approach to Dynamic Systems Modeling in Management Science

A number of texts, notably Robinson (2004) and Pidd (2004a), provide advice on how to participate in and manage a simulation modeling project so as to ensure that it adds value. It is important to realize that a successful simulation project requires more than a set of technical skills—though these are obviously essential. A successful study is usually based on three different sets of skills, which can all be developed with experience.

33.2.1 Project Management Skills

Within management science, most projects have a life cycle of their own, beginning with the need to understand what the client wants or needs and ending either with further work, or with the end of the current project. Figure 33.1 is a grossly simplified version of this cycle and shows 3 phases, each of which could, in turn, be regarded as a further cycle of activity.

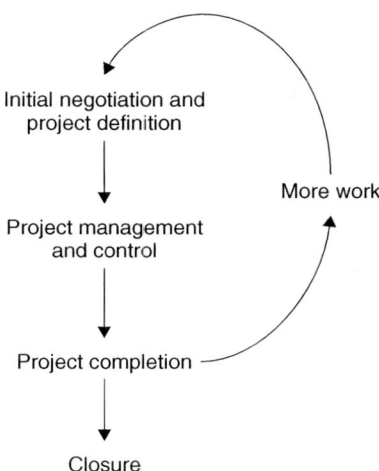

FIGURE 33.1 Management of a simulation project.

1. *Initial negotiation and project definition*: in which the terms of reference for the work are discussed and agreed, usually within some defined budget. It is often very difficult to do this for two reasons. First, though loath to admit it, most clients do not know precisely what they want from such a study and therefore cannot define what must be done. This is not because the clients are incompetent but because there are several ways in which models are used and the actual use may only emerge after some time. In addition, most clients are not modeling experts and may under-estimate or over-estimate the work that is needed. It is also true that the modeler is unlikely to be an expert in managing the system she is studying. Hence, the project definition is something that is negotiated between two parties with different interests and with partial knowledge.
2. *Project management and control*: projects rarely run smoothly and slippage against dates and budgets is highly likely unless the project is properly managed. This requires the project manager, who may be the analyst engaged in the technical work, to keep in close touch with the client. Disappearing off to another site for weeks on end is not an ingredient in a recipe for success. Instead, it is normal to agree a series of milestones for the work and to review progress against them. This may require considerable interpersonal skills as well as technical ones. Milestones may not be met for very good reasons, at least from the analyst's viewpoint. These include a changing view of what is required as both modeler and client start to realize what is needed. They also include shortages of data, which were assumed available when the project was defined. Sadly, technical incompetence by the analyst is another reason for slippage. If there are good reasons for the slippage, then client and project manager need to revise the milestones as appropriate.
3. *Project completion*: projects typically end in three ways. The first is that client and analyst agree that the work has been satisfactorily completed and that there is no need for further work, at least at this time. The second is that both client and analyst agree that further work is needed, possibly in refining the same model or in doing other related work. If this happens, it is wisest to manage the new work as a new project. Third, things may have gone so wrong that the project ends in failure and this sometimes happens despite great effort and good practice.

33.2.2 Conceptual Modeling Skills

As noted earlier, a management science model is a simplification and abstraction that includes factors believed to be relevant to some defined purpose. This raises two obvious questions: what should be included and how do we decide this? In answering these questions it is helpful to distinguish between a conceptual model and the implementation of that model. In the case of computer simulation, the conceptual model is implemented in computer software so that the model may mimic the dynamic behavior of the system being modeled. There is no widely accepted definition of the term "conceptual model," so for present purposes, it is defined as "an external statement of the factors and relationships that should be included in an implemented model." As earlier, the reason for stating that this is an external statement is to ensure that it is open to scrutiny and debate by other people. A purely mental and unexpressed conceptualization cannot be scrutinized or properly debated by other people.

Used in this sense, a conceptual model is a statement of the elements of a system that are thought important enough for inclusion in some implemented model. Needless to say, such a conceptual model will develop through time and is part of a learning cycle in which the modeler develops ideas about the elements of a system that are important enough to feature in the model. Developing a conceptual model is rarely straightforward and is an aspect of practice that distinguishes novices from experts. It requires a process of investigation in which the modeler tries to tease out what the client requires in the model while trying to understand the main features and behavior of the system to be modeled.

Consider, for example, the task of developing a simulation of a call center. It would be a strange simulation model of this that did not include some aspects of the behavior of the customers who phone the call center. However, which of these aspects should be included and in what detail? For example, are all callers to be treated as if from the same population, or should they be classified into subgroups? This will depend on the nature of those callers but also on the purpose of the simulation project. The overall population of callers may include some who need advice and others who wish to purchase a product. If the main aim is to establish an appropriate staffing level for current demand it may not be important to distinguish between them. However, if the aim is to consider what would happen if the business generated more new customers, then it is probably important to distinguish between them. Similarly, in what detail should the behavior of the call center operatives be modeled? Is it enough to develop a model based on probability distributions of call durations, or is it important to model worsening performance toward the end of a shift (if this occurs)?

Hence, a conceptual model must capture the important elements of the system to be modeled in a form that is independent of the computer software used for the model's final implementation. There are several approaches that can be used and one, activity cycle diagrams, is discussed later in the chapter.

33.2.3 Technical Skills

Developing and using dynamic models in management science requires the modeler to develop skills in two areas of work: computing and statistical methods. However, this does not mean that the modeler must be a professional computer scientist or statistician. It was once true that dynamic models could only be implemented on a computer if the modeler were a skilled programmer or had access to someone with those skills. Hence, early books on computer simulation provided specific advice about how particular programming languages could be used. Indeed, the very earliest such simulation models were implemented before languages such as FORTRAN (let alone Visual Basic or Java) were in widespread use. Hence, the early modelers had to learn how to program a computer in binary, machine code or assembler. Thankfully, those days have long gone.

Most simulation applications in management science are implemented using VIMSs in which the model is developed interactively by selecting objects from a palette and linking them on-screen. To do this requires the usual skills with a PC that children acquire at home or at school. Hence, the vendors of these systems are fond of implying that using their products requires no skill beyond this, but it is not true. As will be clear later, using a VIMS almost always requires the modeler to write some code, though not a full program. Some training and experience in computer programming is therefore desirable.

The dynamic systems modeled in management science are rarely deterministic: that is, they are not wholly predictable. A business that can be certain of the precise demand for its products, with customers that always do the same things, and whose production technologies are utterly reliable is very fortunate indeed. Most organizational systems exhibit considerable variation. Some of this can be understood by analyzing root causes but some can only be understood by expressing it in statistical terms. Hence, we might conclude that the number of calls per hour arriving at a call center can be modeled as a Poisson distribution. Therefore, though we can never say how many calls will arrive in the next 60 min, we can have some idea of the likely number if we apply correct statistical principles. Modelers must develop skills in data analysis (to fit distributions to observed data sets), statistical inference (to see if observed variation is significant), and in probability theory to underpin the other two areas.

33.3 Discrete Event Simulation

Within management science, two types of simulation modeling find widespread use: DES and system dynamics; the latter is considered later in the chapter. There are many excellent introductions to DES and the annual Winter Simulation Conference (www.wintersim.org) includes a range of tutorials, some aimed at beginners and others at people who wish to further develop their skills and knowledge. The approach taken here is based on the lengthier coverage in Pidd (2004a). Perhaps the most thorough coverage of DES is to be found in Law and Kelton (2000). Other chapters in this book covering DEVS, Event Graphs, and ARENA also provide valuable insights into the methods and practice of DES.

33.3.1 What Types of Problem are Well Suited to Discrete-Event Simulation?

Many applications of DES involve queuing systems of one kind or another. The queuing structure may be obvious, as in a queue of jobs waiting to be processed on a machine, or in a stack of aircraft waiting for landing space at an airport or of phone calls arriving at a call center. As time progresses, queues build up or are reduced as demand varies and as service patterns change. Mathematical queuing theory (see Hlynka, 2004, for a survey of available books) can be used to model some simpler queuing systems, but is ineffective when demand is nonstationary and when customers baulk (refuse to join a queue because it looks too long) or renege (enter a queue but leave it before being served)—which is common in most real applications. Sometimes, the queuing structure may be less obvious, as in the deployment of fire appliances in a large city. In this case, the customers are the fires needing attention and the servers are the fire-fighters together with their associated equipment.

DES is worth considering for systems with the following characteristics:

- They are *dynamic*: that is they display distinctive behavior which is known to vary through time. This variation might be due to factors which are not well understood and may therefore be amenable to statistical analysis—for example, the apparently random failures of equipment. Or they might be due to well-understood relationships which can be captured in equations—for example, the flight of a missile through a nonturbulent atmosphere.
- They are *interactive*: that is, the system is made up of a number of components that interact with one another and this interaction produces the distinctive behavior of the system. For example, the observed behavior of aircraft under air traffic control will be due to the performance characteristics of the individual aircraft, the intervention of the air traffic controllers, the weather, and any routing problems. This mix of factors will be varying all the time and their interaction will produce the observed behavior of the air traffic.
- They are *complicated*: that is, there are many objects in the system of interest, and their individual dynamics need careful consideration and analysis.

Despite the availability of excellent computer software for developing DES models, it is still true that building, testing, and using such a model is surprisingly time-consuming. Hence, it is sometimes better to try a very approximate mathematical model before resorting to simulation. If a discrete-event model is being considered, then a number of characteristics are likely to be present in the system to be modeled.

- *Individual entities*: the behavior of the system comes from the behavior of individual objects of interest, usually known as entities. The simulation program tracks the behavior of each of these entities through simulated time and will be minutely concerned with their individual logics. The entities could be truly individual objects (e.g., machines, people, and vehicles) or could be a group of such objects (e.g., a crowd, a machine shop, and a convoy of vehicles).
- *Discrete events*: each entity's behavior is modeled as a sequence of events, where an event is a point of time at which the entity changes state. For example, a customer in a shop may arrive (an event), may wait for a while, their service may begin (an event), their service will end (an event), and so on. The task of the modeler is thus to capture the distinctive logic of each of these events (e.g., what conditions must hold if a begin service event is to occur?). Though real-time flows smoothly, the flow of simulation time in a DES is not smooth, as it moves from event time to event time and these intervals may be irregular.
- *Stochastic behavior*: the interval between successive events is not always predictable, for example the time taken to serve a number of customers in a shop will vary. There may sometimes be obvious and entirely predictable reasons for this (the server may speed up as the queue of waiting customers increases) or there may be no obvious reason to explain things. In the latter case, the varying intervals between events could be represented in a probability distribution and modeled by taking samples from that distribution.

33.3.2 Common Application Areas for Discrete-Event Simulation in Management Science

There are many different application areas for DES and surveys of the use of MS/OR methods usually place simulation as one of the top three most frequently used approaches. The Winter Simulation Conference (ref) is a fertile hunting ground for those seeking case studies of successful applications.

The last 20 years have seen increased competition between manufacturing companies and this has led to a range of strategies including cost reduction and product differentiation. Both of these have relied on increased automation, which in turn depends on large-scale investments. It clearly makes sense to ensure that the investment is a wise one before making it and DES is often used for this purpose. The idea is to simulate the operation of a manufacturing system before installing it or to simulate an existing manufacturing system to seek performance improvements.

Many of the early simulation applications within MS/OR were in primary industries such as coal and steel. Since then, simulation has been used in many industries including food manufacture, microelectronics, aerospace, and automobile assembly. The latter is especially significant, since some of the commonly used VIMS (most notably Witness) were developed within the automotive industry, which is where they find most of their users. Hence, the terminology of many VIMS is geared toward that application area, using terms like work station, conveyor, and cycle time.

A second major application area for DES is health care. Worldwide expenditure on health care has been rising steeply in recent years. Current estimates suggest that health care expenditure in the USA amounts to about 17% of GDP, of which about 50% is funded through taxation. In the UK, where most health care is funded through taxation, it will soon account for about 10% of GDP. Why these large sums? First, because people's expectations are rising and, second, because of technological developments. Given the scale of this activity and the risk of adverse outcomes from medical interventions, it is clear that improvements in health care delivery are needed. Simulation methods have been successfully used in many areas of health care. Example include: models of disease transmission for AIDS and hepatitis; models of treatments and intervention for glaucoma and cardiac shunts; and models of hospital operations such as clinics and emergency departments. Despite this level of activity there are curiously few VIMS dedicated to health care modeling.

A third major area can be summarized under the heading of business process improvement; once known, rather unfortunately, as business process re-engineering (BPR). Many organizations have tried to streamline their business processes so as to make efficiency gains and, at the same time, to make

Dynamic Modeling in Management Science

their operations increasingly customer focused. For example, buying car insurance 20 years ago was a long, drawn-out process involving forms completed by hand, visits to agents, repeated correspondence, and laborious comparisons of quotations. Now we take it for granted that we can phone an insurance company and arrange the insurance in a few minutes in a single conversation. Alternatively, we can use price comparison Web sites and then enact the purchase using Web forms. A transaction that once took a couple of weeks, end-to-end, now takes a few minutes. A similar story applies to the purchase of airline tickets. These customer-focused business processes depend on IT systems that enable prices to be quickly generated, credit standing to be quickly checked, and transactions to be completed with the minimum of fuss. However, even the best IT in the world is a waste of time unless the processes are properly designed and managed. Hence, computer simulation models are commonly used in the design and improvement of call centers, workflow management systems, and document production systems.

A fourth major application area is transport and physical logistics—the movement of people and goods and the associated systems that support this. Airlines and airports have long been major users of simulation models to design and improve terminal buildings and baggage systems. Likewise, options for air traffic control are usually simulated before any attempt at implementation, it being much safer if a simulated aircraft runs out of fuel in a simulated flight. Physical distribution and supply chain management are two more major application areas for computer simulation in transport and physical logistics. Large retailers commonly own regional distribution centers to which suppliers deliver goods and from which sorted supplies are sent to retail outlets. These centers are large complex operations that must run in a highly synchronized manner if supplies are to reach the retail stores in time to meet customer demand. Simulation methods allow supply chain managers to see how the centers will operate and to experiment with policies that will enable proper synchronization.

Finally, and possibly the largest application area of all, is defence. War gaming and battle planning have long been a feature of military training and the increasing expense of high tech battle equipment makes simulation an essential part of most military operations. Applications are not restricted to gaming, however, and range from studies equipment, of logistics operations, of deployment options through to battle simulations. Not surprisingly, the majority of defence simulations are not reported in the open literature.

33.3.3 Discrete-Event Simulation Terminology

Some writers have tried to develop precise definitions of the terms used in DES modeling, notably Zeigler (1976) whose concept of DEVS is discussed in Chapter 6 of this book. However, the use of terms in the DES community is not always consistent. This section defines terms as they are normally used. The first set of terms refers to the objects within the system and the second set refers to their state changes and activity. When referring to the objects in a DES model it is normal to divide them into two main types.

1. *Simulation entities*: These are the individual elements of the system that are being simulated and whose behavior is being explicitly tracked. Examples might include machines in a factory, patients in a hospital, or aircraft at an airport. When the simulation is implemented in computer software, the computer program maintains information about each entity and therefore each one can be individually identified. As an entity changes state in the simulation, the computer program keeps track of these state changes. The overall system state is a result of the interaction of the simulation entities and resources (see below) and the number of entities in a model gives some idea of how fast it will run. Simulation entities are, conceptually at least, divided into *classes* which display similar behavior. Examples might include a class of aircraft, of customers of doctors, or whatever. Taken together, the number of entities and the number of classes partially determine the complexity of a DES model.
2. *Simulation resources*: These are also system elements but they are not modeled individually. Instead, they are treated as countable items whose individual behavior is not tracked in the computer program. Examples might be the number of passengers waiting at a bus stop or the number of boxes of a product available in a warehouse. Thus, a resource consists of identical items and the

program keeps a count of how many are available, but their individual states are not tracked. Resource contention occurs when the same resources are needed by different entities at the same simulation time. The number of resource types in a DES model is another clue to its complexity, but this is not the same as the number of resource items.

Whether a system element should be treated as an entity or as a resource is a decision for the modeler and it is normal for either representation to be possible for one of more classes of object in the system. In management science, DES models are built for particular purposes, typically to support decision making and planning, and this purpose will often be what determines the representation. Consider the simulation of a telephone call center in which callers dial in, wait for an operator, hold a conversation, and then end the call. If we need to know the performance of each operator within the call center, then we will represent these as entities whose precise behavior and quirks will be modeled. However, if we are only interested in the general performance of the population of operators, we may choose to represent them as a resource pool. In this second case, when a new call arrives, we check to see if the pool of available operators is greater than zero; if not, the call will not be taken until an operator becomes free. If the current pool size is greater than zero, it will be reduced by one (if a single operator is required) at the start of the conversation. At the end of the conversation, the operator is now free and the pool size will be increased by one to represent this.

As well as representing the entities and resources, we also need to capture how they interact and change state through time. This leads to the second set of terminology.

3. *Simulation events*: These are instants of time at which significant state changes occur in the system, such as when a customer arrives, a machine breaks down, or some operation begins. Note that the analyst must define whether an event is significant or not in the context of the objectives of the simulation. For example, when simulating an airport baggage system it may not matter when a rain shower begins and ends. However, when modeling the use of airspace around an airport, this might be an important consideration when trying to establish separation rules for landing aircraft.

4. *Simulation activities*: Entities move from state to state because of the operations in which they engage. The operations and procedures that are initiated at each event are known as activities and these transform the state of the entities. For example, a phone conversation in a call center requires a caller and an operator to co-operate for a period while the conversation proceeds. The start and end of the conversation may be modeled as simulation events, and the conversation itself, which takes time, is a simulation activity. Its start changes the state of caller and operator, and so does its end. It can only begin if the caller is waiting and the operator is currently idle. Hence, starting a simulation activity depends on the system state and also changes that system state. Ending a simulation activity also changes the system state.

5. *Simulation processes*: Sometimes it is useful to group together a sequence of events in the chronological order in which they will occur. Such a sequence is known as a process and is often used to represent all or part of the life of an entity class. For example, a phone call arrives at a call center, may join a queue, may renege and leave the queue, may be answered by an operator and will then leave the system. Hence simulation processes may include logical decision points.

6. *Simulation clock*: Since a DES model is concerned with the dynamic behavior of a system through time, this movement of time must itself be simulated. As mentioned earlier, time does not progress smoothly within a DES, but jumps from event to event in a chronological sequence. The simulation clock time is the point reached by current simulated time in a simulation. Hence in a simulation where the time unit is minutes, the test "is clock = 240?" might be used to test whether a lunch break is due. If so, appropriate activity could then be initiated in the simulation.

33.3.4 Activity Cycle Diagrams in Conceptual Modeling

There are many different approaches to conceptual modeling in DES, each offering some support to the modeler trying to represent the important elements of a system that will be simulated using a DES model. Examples include event graphs (Schruben, 1983; also in Chapter 23), control flow graphs (Cota and

Dynamic Modeling in Management Science

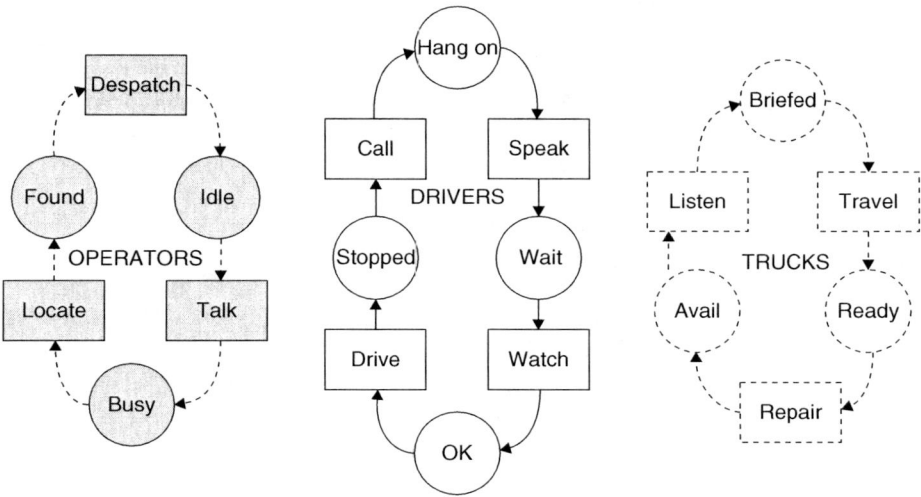

FIGURE 33.2 Separate activity cycles for the *CallOut* simulation.

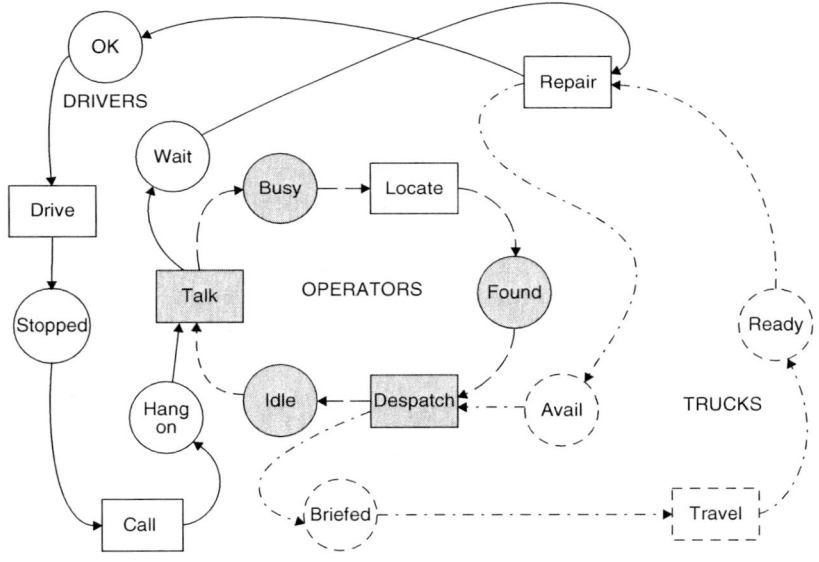

FIGURE 33.3 Combined activity cycle diagram for the *CallOut* simulation.

Sargent, 1990), Petri nets (Zimmerman, 2004; also in Chapter 24), and activity cycle diagrams, which are described here. Activity cycle diagrams are very simple and require the modeler to represent the interaction of entities and resources as a network that uses two symbols (See Figure 33.2 and Figure 33.3).

The rectangles represent active states, which may involve the co-operation of entities and resources. The states of those entities and the size of the resource pools may change at the start and end of the activity. Active states take simulation time (which may include zero time—for example, two customers may phone simultaneously) and we assume that this time can be computed when the activity begins. That is, when it begins, the start time of an active state is known and its end can be scheduled at some future simulation time. When simulating a call center, a conversation between an operator and caller is usually represented as an active state.

The circles are used to represent dead states, sometimes known as queues, and these have durations that cannot be known in advance. Though we know the simulation time at which an entity enters a dead state, such a state is defined as one in which the entity waits until some event provides a resource or co-operating entity that allows it to proceed from that dead state. Hence, we do not know the time at which it will end. A dead state is always preceded by an active state and is always followed by an active state. When simulating a call center, the time a caller spends waiting for an operator to answer is usually modeled as a dead state; as is the time that the operator is idle, waiting for another call.

To illustrate the use of activity cycle diagrams, consider the operation of the *CallOut* response center. *CallOut* provides a breakdown and recovery service to car drivers who pay an annual subscription. It operates several call centers, one of which supports motorists whose car has broken down while on the road. To use the service, motorists phone the call center—in most cases they do so from their cell phones using the number for on-road breakdown assistance. When they phone the call center, their call is taken by the first available operator whose main task is to check that the caller has an up-to-date subscription to *CallOut*, to note the precise location of the vehicle, its make and model details, and a summary of the problem. At the end of the call, the operator checks a real-time location system to find the nearest free breakdown truck (the trucks carry GPS locators), to whom she allocates the call. The truck driver then travels to the stricken vehicle, provides whatever aid is required, and, at the end of the job, reports back by phone that he is free for another job.

There are three important simulation object classes which, at this stage, we can think of as entity classes:

- *Drivers*: whose vehicles breakdown and who phone *CallOut* asking for assistance.
- *Operators*: who take the calls from drivers and despatch a breakdown truck to the location of the stricken car.
- *Trucks*: which are despatched by the operator to the scene and offer aid.

Figure 33.3 shows an activity cycle for each of these entity classes.

The operators are initially *Idle*, they then *Talk* to a caller, after which they *Locate* and *Despatch* appropriate resources, and return to an *Idle* state. Three of these are active states (*Talk*, *Locate*, and *Despatch*) and each is separated by a dead state from the next active state. One such dead state, *Idle*, is genuine, the others are there for convenience and to keep the rules of activity cycle diagrams.

The drivers are initially *OK*, and *Drive* a working vehicle until they break down (and enter a dead state, *Stopped*) and *Call* the call center and *Hang on* until they *Speak* to an operator. They then *Wait* until a breakdown truck arrives and then *Watch* while a repair is executed; after which they are *OK* again. Following similar logic to the activity cycle for the operators, *Wait* is clearly a dead state and *Phone* is an active state. Keeping to the alternative live and dead state rule means that *OK* is a dead state and *Watch* is an active state.

Finally, consider the activity cycle of the trucks, which are initially ready for work in an *Avail* state, from which they enter a *Listen* state in which they receive instructions from an operator, after which they are *Briefed*. They then *Travel* to the location of the stricken vehicle, where they execute a *Repair*, after which they are then once again in an *Avail* state. *Instruct*, *Travel*, and *Repair* are all active states and *Avail*, *Briefed*, and *Ready* are dead states. Applying the alternate active: dead state rule, leads to the third activity cycle in Figure 33.3.

As independent cycles, these diagrams are of little use. Their power comes when they are combined to show the interaction of the entity classes; for it is this interaction that causes the dynamic behavior of the system. An activity cycle diagram that combines the three cycles is shown in Figure 33.3, this is a little messy, but can be very useful in conceptualizing a possible simulation model. Some of the active states require the co-operation of two classes of entity: when an operator is in the active state *Talk*, this is the same state labeled as *Speak* in the drivers' activity cycle. This combined active state is labeled as *Talk* in Figure 33.4. The operators' active state *Despatch* requires the co-operation of a truck that is previously in the *Avail state*. On the truck cycle of Figure 33.3 this is labeled as *Instruct* and the operators' label *Despatch* is used for the combined active state. Finally, the active state *Repair* from the trucks' cycle is clearly the

Dynamic Modeling in Management Science

same as the active state *Watch* from the drivers' cycle. In combining the two, the trucks' state name of *Repair* is used.

Examining the combined activity cycles of Figure 33.3 provides some useful insights that will serve the basis of the implementation of a suitable DES model using computer software, independent of the type of software used. First, we can write down the conditions that govern the start of the active states that involve the co-operation of two or more entity classes. The *Talk* active state can only begin if two conditions hold:

- There must be a driver who is in the dead state *Hang on*.
- AND an operator in the dead state *Idle*.

Likewise, for the *Despatch* active state to begin, two conditions must both be satisfied:

- There must be a truck in the dead state *Avail*.
- AND an operator in the dead state *Found*.

The start of the *Repair* active state is also governed by two conditions:

- There must be a truck in the dead state *Ready*.
- AND a driver/car must be in the dead state *Wait*.

Notice that an unspoken approximation has crept into the model: we are not distinguishing between drivers and cars. This is reasonable, unless the purpose for which the model is being built requires us to do otherwise.

The other active states (*Drive*, *Locate*, and *Travel*) do not require the co-operation of two entity classes to begin. These are active states in which their relevant entities (drivers for *Drive*, operators for *Locate*, and trucks for *Travel*) are bound to occur once the preceding active is complete. For example, the truck will automatically enter the *Travel* active state once the *Despatch* active state is complete—which means that it will spend zero time in the intervening dead state, *Briefed*. The same applies to drivers in the *OK* dead state and to operators in the dead state, *Busy*. These dummy dead states are needed so as to allow the separation of the life cycles (or processes) of the co-operating entities at the end of the active state in which they have co-operated. It is tempting, but inadvisable, to ignore these dead states; and doing so may lead to problems later.

In this way, we can use devices such as activity cycle diagrams to represent the logical interactions between the entities and resources of the system. Rather more detail will be needed if the conceptual model is to be implemented in computer software, but modeling in too much detail at the start of the simulation project is almost always a mistake.

33.3.5 Implementing a DES Model in Computer Software

There are many different DES computer packages available on the market. The OR/MS Today magazine of INFORMS publishes occasional surveys of available software and the annual Winter Simulation Conference (www.wintersim.org) usually includes presentations from the major vendors. Hence there is plenty of choice and gaining information has been made much easier by the Internet. Broadly speaking, there are two ways in which a DES model may be implemented in computer software: write a program or use a VIMS.

Whatever the mode of implementation, it is important to distinguish between the model logic and the simulation executive (sometimes known as the simulation engine). The model logic is an expression of the rules that govern the system being simulated—its active and dead states and the rules that govern the state changes. Each different DES model will have its own application logic, as captured in the conceptual model. By contrast, the simulation executive is general purpose and can control any DES model whose logic is expressed in a suitable form. The executive exists to ensure that the entities and resources change state at the right simulation times and as required by the conditions in the simulation. In most circumstances, the modeler is concerned only to ensure that the model logic is correctly expressed; that is, the simulation executive is provided by the software vendor and is usually hidden from the modeler.

33.3.6 Discrete Event Simulation using a VIMS

As stated earlier, most simulation applications in management science are implemented using VIMS, which require little in the way of computer programming. These are shrink-wrapped software systems such as Arena, Witness, Simul8, MicroSaint, and Automod in which models are built by point and click using predefined objects for which the user must supply properties. Though there is no proper survey evidence to support this assertion, the proportion of discrete simulation applications that use a VIMS may be as high as 90%, with many of the remaining being large-scale models developed in the defence sector. The latter, by contrast, usually involve significant programming and may require explicit management of the events that comprise the dynamic behavior of the simulation. The pros and cons of the two approaches are discussed at length in Pidd (2004b).

Why are VIMS so popular? Because they offer the prospect of rapid application development by people who are not computer professionals. There is no doubt that a VIMS can be easy to learn, at least for straightforward applications, as most simulation instructors will testify. However, it is also true that a VIMS can run out of power when faced with large and complex applications. This, of course, may not matter for routine business applications in which the 80:20 rule may apply: that is, a simple model may be good enough. Few senior managers would invest very large sums in a one-off capital development based on a complicated stochastic analysis with wide and overlapping confidence intervals. All models are approximations; very approximate models are often good enough in management science, and VIMS are good enough for very approximate models.

Figure 33.4 shows the logical composition of a typical VIMS. It presents a user interface that exploits the API provided by the operating system (usually a version of Windows™). Working with the user interface, models may be developed, edited, and run and experiments may be conducted. The latter usually requires at least some statistical analysis, and VIMS usually allow the export of results files in some suitable format for a spreadsheet or statistical package. Models are constructed by selecting icons that represent features of the system being simulated and these are linked together on-screen, and parameterized using property sheets. This is fine if the objects provided are a good fit for the application. However, the default logic provided by the simulator may be inadequate to model the particular interactions of specific business processes. To allow customization, most VIMS provide a coding language in which interactions can be programmed. Some do this with a general-purpose programming language (e.g., Visual Basic or C#). Others incorporate simulation quasi-languages that permit little beyond the assignment of attributes, the definition of if statements, loops, and limited access to component properties. However, the modeler need not write a full program, merely code sections to represent specific logic.

Inside every VIMS is a generic simulation model that is not usually available to the user, who works with a network diagram that represents the activity and interactions of the model. Figure 33.5 shows such a network for an accident and emergency department of a hospital using Micro Saint Sharp (Micro Analysis and Design, 2005), a DES VIMS. In essence, the generic model takes the network diagram as data.

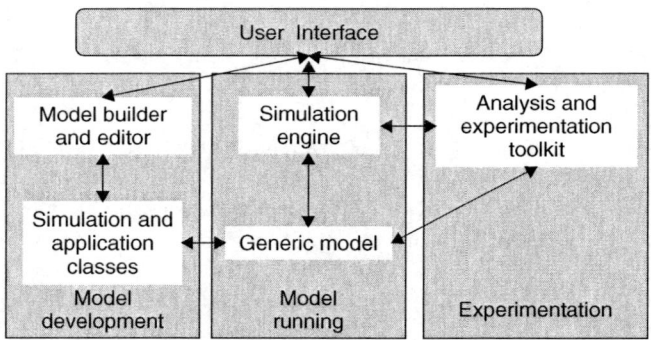

FIGURE 33.4 Internal organization of Visual Interactive Modeling Systems.

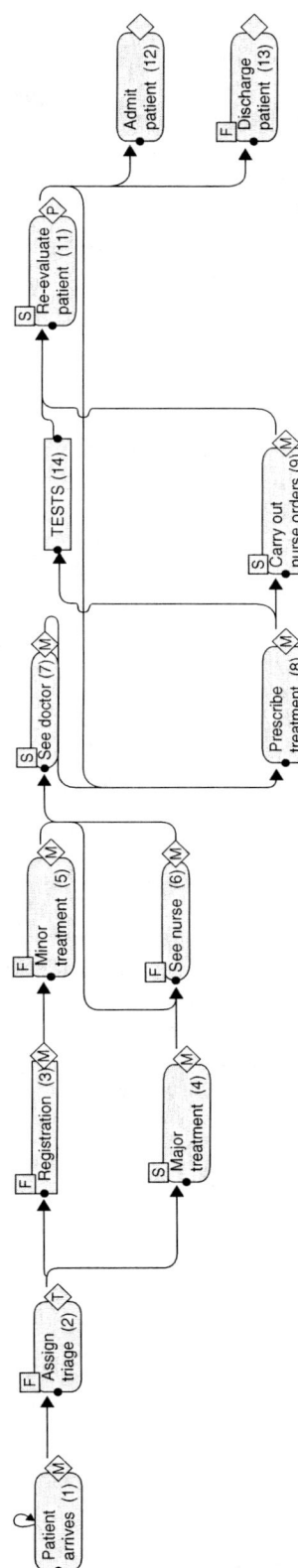

FIGURE 33.5 Micro Saint Sharp task diagram for an accident and emergency department.

Thus, if it were thought desirable, a VIMS on-screen network could be replaced by a series of verbal commands each of which carries attribute data to represent the property sheets. The generic model assumes that simulation entities change state and it reads the network description to define the sequence of those states and the conditions that govern them. The code fragments, in whatever simulation language, are then used to modify the standard generic model in some way or other. Once suitably parameterized, the generic model is run by a simulation executive, which sequences and schedules the tasks that define the application model. Like the generic model, the simulation executive is also hidden from the user. Hence, the form of the simulation executive is rarely revealed by the VIMS vendors, who are keen to present their software as easy to use and also powerful. How the simulation runs is deemed to be of little concern to the users.

33.4 System Dynamics in Management Science

DESs are based on the dynamic interactions of individual simulation entities, using rules to represent this logic. System dynamics, an approach introduced in Chapter 27, is the other dynamic modeling approach that is commonly used in management science. In system dynamics, the aim is to model system behavior at a more macroscopic level by considering how system structure may affect system behavior. Since a systems perspective underpins system dynamics, this has led some enthusiasts to claim the name "systems thinking" for their approach. This is a shame, since there are many different varieties of systems thinking, as discussed, for example in Jackson (2003). Further damage is done by overblown claims such as those of Senge (1999), whose popular book "The fifth discipline" shows how the ideas of system dynamics can be used to support organizational learning. This is fine, but the claim that this represents a whole new way of thinking that is revolutionary seems somewhat exaggerated. Instead, we should regard system dynamics as providing a way to support thinking about decisions and plans.

When first developed by Forrester and his colleagues in the early 1960s, system dynamics, then known as industrial dynamics (Forrester, 1961), had a strong engineering "feel" about it. Indeed, Forrester acknowledged that he had taken ideas used in control engineering in which feedback structures are modeled as differential equations (see Chapter 17 and Chapter 18). Forrester replaced the differential equations with first-order difference equations (see Chapter 27) and simple Euler–Cauchy integration. The result was an approach that could be used by people who were not trained engineers. The industrial dynamics approach stressed system simulation in which models consisting of linked sets of difference equations were run through time to provide a system simulation. This quantitative system dynamics was joined, some years later, by qualitative approaches that stress the use of system dynamics to develop understanding rather than as a simulation tool. Both approaches will be briefly covered in this chapter. For a thorough description of both approaches see Sterman (op cit), for a quantitative treatment see Coyle (1996) and see Wolstenholme (1990) for a more qualitative treatment.

33.4.1 System Structure and System Behavior

To illustrate the difference between system structure and system behavior, consider the following example. Suppose we notice that the sales of a product have increased and we start to think about why this has happened. That is, we try to develop a causal argument that links the observed increase in sales with some other influences and events. In effect, this is a set of informal hypotheses such as: sales have increased because the weather is good or sales have increased because we had the stocks in the shops when the customers wished to buy them. This is, of course, what economic modelers aim to do when trying to understand economic behavior—though in their case, the models are usually based on equations that rely on statistical methods for their parameters. In economic models and in system dynamics, the aim is to show how changes in one factor affects other variables. The intention is to help people to understand the effect of their actions and those of others.

The question is, what causes the observed behavior? A fundamental assumption in system dynamics is that behavior is a result of structures—both inside the system and in its environment. However, these are not structures as might often be understood—such as the number of staff employed or the layout of a factory. Instead, they are the underlying, general features that can be observed across many types of system. At their most basic, these are the feedback loops and delays that are present in most organizational systems. System dynamics provides a way to model those feedback structures and delays so as to understand how they affect the behavior of the system.

With this in mind, it is important to understand the distinction between content and structure. Structure defines how variables interact, and content expresses the meaning of those variable and interactions—rather like considering syntax and semantics in language. Two systems may have similar structures but quite different content. For example, a supermarket and a telephone call center can both be analyzed in terms of their queuing structure. Both systems have customers who are served, but the meaning (and importance) of the customers differ. In the call center, the customers are calls awaiting response, whereas in the supermarket the customers are the shoppers. Coyle (op cit) points out that the management scientist has to maintain two views of a system at the same time. To model it, the system must be seen in terms of its structure, but to consider making changes it is crucial to keep in mind the meaning of the variables. Only then will it be clear which changes are feasible.

33.4.2 Qualitative System and Dynamics: Causal Loop Diagrams

As an example, consider the parent company of the *CallOut* response center introduced earlier in this chapter. Suppose that the managers of this business are concerned about customer churn: that is, though they gained new customers they also lost customers during the year. Because of this concern they have called a meeting and as the discussion proceeds it becomes clear that most people think that the best way to reduce this problem is to increase their marketing effort. This thinking is captured in the diagram of Figure 33.6, which shows that they regard marketing effort as having two components: marketing expenditure and the effectiveness of their advertising. Figure 33.6 is an example of a causal loop diagram in which the arrows are intended to show causal links between two factors. Each arrow carries a sign: a positive sign indicates that an increase in the factor at the tail of the arrow will also lead to an increase in the factor at its head. A positive sign also means that any decrease in the factor at the tail of the arrow is expected to lead to a decrease in the factor at its head. Conversely, a negative sign indicates that an increase in the factor at the tail of the arrow is thought likely to lead to a decrease in the factor at the head of the arrow. Likewise, a negative sign could indicate that a decrease in the factor at the tail of the arrow is thought likely to lead to an increase in the factor of its head.

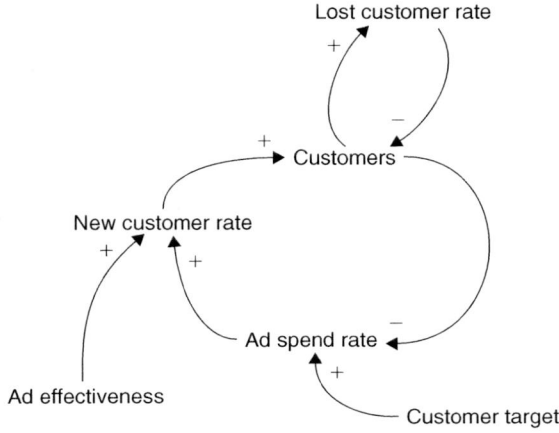

FIGURE 33.6 Causal loop diagram for advertising to gain new customers.

Thus the main loop of the diagram reflects straightforward beliefs and also what we might think of as the "physics" of the system. That is, an increase in the number of new customers will lead to an increase in the number of customers. This in turn will lead to a decrease in the expenditure on advertising, since as they edge closer to their target number of customers, they can reduce advertising. It is also believed, quite reasonably, that increasing advertising expenditure will lead to an increase in the number of new customers. This loop is known as a balancing loop in the terminology of system dynamics and it reflects negative feedback control. That is, the number of customers will not continually increase because, as the number of customers increase, advertising expenditure will decrease, which will decrease the number of new customers. This, of course, only works because an increase in the number of customers leads to an increase in the number of lost customers that will, in turn, lead to a decrease in the number of customers. Finally, since they believe that advertising effectiveness can vary, the diagram shows a positive link between advertising effectiveness and of the number of new customers recruited. There is, of course, a problem with the worldview shown in the causal loop diagram. It assumes that they can do nothing about the number of customers that are lost each period.

Once they realize that this is wrong, they can construct a rather better causal loop diagram as shown in Figure 33.7. This time two new factors have been included: marketing spend and expenditure on incentives to existing customers. The idea of this is to retain existing customers since it is usually the case that doing so is cheaper than spending money to gain new ones. There is a positive link shown between customers and total marketing expenditure, since the more customers they have the more they can afford to spend on marketing. In turn, marketing expenditure is split between incentives and advertising and it is clear that a major decision facing the company is how to divide this expenditure between the two types of marketing. There is a negative link from incentive expenditure to the number of lost customers, since it seems reasonable to believe that the more they spend on these incentives the less likely customers are to leave. This leads to a second balancing loop on the diagram which connects the customers, incentive expenditure and lost customers.

Causal loop diagrams of this type are used in two ways. First, they are at the core of what has come to be known as qualitative system dynamics. In this mode, the idea is to encourage individuals and, especially, groups of people to sketch out their thinking using causal loop diagrams. Even if the diagrams do not lead to a simulation of some kind they can be very useful devices for encouraging people to think about the factors to be considered when tackling a difficult issue. In addition, skilled users of system dynamics can recognize features of these diagrams that are based on structures found in many different types of system.

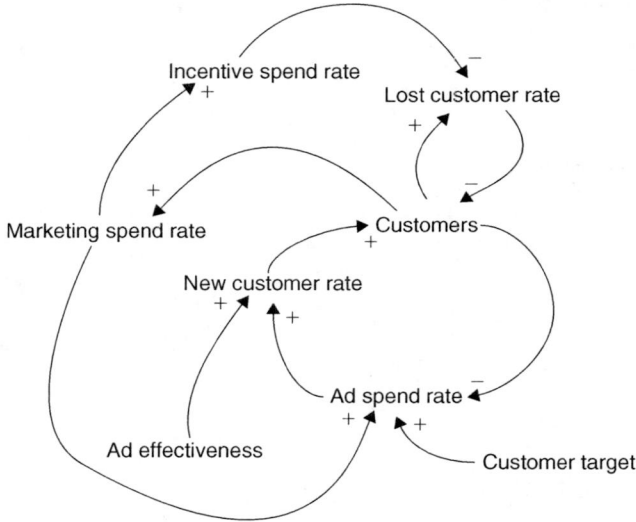

FIGURE 33.7 Causal loop diagram for customer retention expenditure.

Richmond and Petersen (1994) and Sterman (op cit) provide several examples of these system archetypes and discuss the effects that they are likely to have on system behavior.

33.4.3 Quantitative System Dynamics: Level-Rate Diagrams

System dynamic models are based on two fundamental concepts as discussed in Chapter 27: levels (sometimes known as stocks) and rates (sometimes known as flow rates). Levels represent accumulations within a system and will persist even if all activity within the system ceases for a while. For example, in Figure 33.6 and Figure 33.7, customers are clearly a level since, even if the company ceased to recruit any new customers, the customers that they do have will remain, at least for a while. Rates or flow rates represent activity within the system and these dropped to zero when activity ceases. It should be obvious that the number of new customers recruited in any period is a rate, as is the number of customers lost in a period. Advertising expenditure is also a rate and it will vary from period to period depending on the policy of the company.

Figure 33.8 shows a level-rate diagram based on the causal loop diagram in Figure 33.6 using the symbols provided by the Powersim software (www.powersim.com). Similar symbols are used in other system dynamics software: levels are shown as rectangles and rates are shown as pipes with valves on them. New customers are seen as arising from a cloud and this is to indicate that this source is not included in the model; that is, the cloud represents the system boundary. Likewise lost customers disappear into a cloud since we are not intending to model their destination. The main physical flow of the model represents customers and it should be clear that if the inflow of new customers exceeds the outflow of lost customers, the number of customers will increase. In a system dynamics model, a level will always have at least one outflow or at least one inflow. In our example the level has one inflow and one outflow. Advertising expenditure is also modeled as a flow rate and this is seen as an emerging from a cloud and disappearing to a cloud because the model is not concerned with where this money comes from, nor where it goes to. The rate of expenditure on advertising is a function of the actual number of customers and the target number of customers. In addition, like all businesses, there is a limit to their expenditure and so Figure 33.8 has an extra parameter—the maximum advertising spend each month. Finally, the diagram also includes two diamond shapes of which one represents the number of customers they would like to have (customer target) and the other represents the effectiveness of a particular advertising campaign. The diamonds indicate that these are parameters for the model.

As discussed in Chapter 27, a system dynamics model consists of two main types of equations: level equations and rate equations. When a modern system dynamics package such as Powersim, Stella/IThink

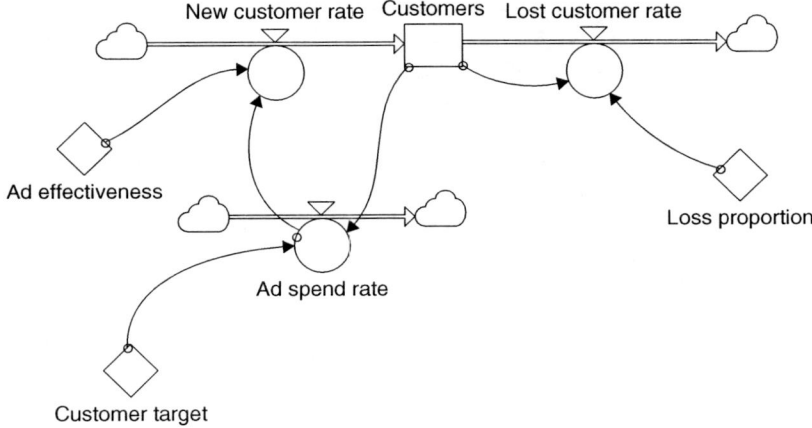

FIGURE 33.8 Powersim diagram for advertising to gain new customers.

or Vensim is used, the modeller draws the diagram on screen by selecting appropriate icons from a palette. Hence, the software knows which objects are levels which are rates and which are parameters. In a more complex model there might also be objects on the diagram to represent the delay is of different kinds. Since the software knows which objects are levels and which are rates it can determine the level equations. In our example there is only one level, customers, and the value that it takes at any time depends on its previous value and the difference between the incoming rate and the outgoing rate. Most system dynamics models are time slicing; that is, they move simulated time forwards in equal steps. This time increment is usually known as dt. Hence the software can determine that the level equation for customers will take the following form:

```
new customer level = old customer level + (new customer rate - lost
customer rate)*dt
```

where this assumes that the new customer rate and lost customer rate are both constant over the interval dt. In essence, the level equations are determined by the "physics" of the system, in that matter is neither created nor destroyed, but is conserved.

However, the software cannot determine the rate equations, for these depend on the policies chosen by the company and also by factors outside of their control. For example, they may have analyzed the customer losses over the last few years and discovered a regular pattern in which they lose a proportion of their customers each month. Similarly, they may represent the effectiveness of an advertising campaign by a constant which indicates the amount that should be spent to gain a new customer. Unless the model is provided with this information, the simulation software has no way of knowing what these values are, nor can it know what form the relationships take. Hence, in all system dynamics software, the user must define each rate equation and the links between the system parameters and the various equations in the model.

Given what is said above, a possible set of rate equations might be as follows:

```
AdSpendRate = MIN(IF(CustomerTarget-Customers > 0,
            50*(CustomerTarget-Customers), 0), MaxAdSpend)
LostCustomerRate = Customers*LossProportion
NewCustomerRate = AdSpendRate/AdEffectiveness
```

The first rate equation represents a policy of spending £50 per missing customer, where the number of missing customers is the difference between the customer target and the actual current number of customs. Since it is possible for the number of customers to be higher than the target, the calculation is wrapped in an if statement that checks whether or not the actual number of customers is less than the target. If the number of customers is greater than the target then it makes no sense at all to spend money on advertising and hence the equation returns the value of zero. The other two rate equations are self-explanatory.

We can now run a simulation using this model if we provide values for the parameters and an initial value for the number of customers. Suppose that the initial number of customers is 5000, the target number of customers is 6000, the proportion of customers lost each period is 0.05, and the advertising effectiveness takes the value of 45. Note that this value of 45 compares with a unit expenditure per missing customer 50, implying that advertising is not 100% effective. Running a simulation for 100 time units, where a time unit might be a week or a month, leads to the behavior shown in the graphs of Figure 33.9. The bottom graph shows the effect of fixing *MaxAdSpend* at 10,000—they continually lose customers. The top graph shows that raising this to 12,000 means that they will gradually approach their target. Hence, by varying the parameter values the managers can experiment with different policies to see what effect they may have. Similarly, they could take the causal loop diagram shown in Figure 33.7 and develop a set of system dynamics equations to represent it. If they wished they could then run further simulations to see what effect different options might have given the slightly more complicated representation of this second model.

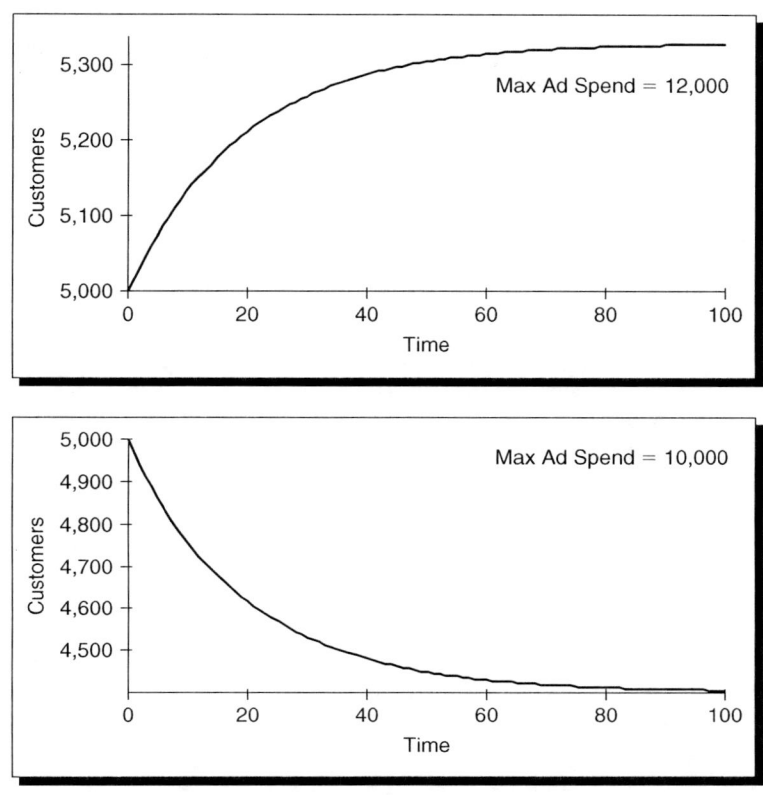

FIGURE 33.9 Output from the Powersim model of Figure 33.8.

33.5 Model Validation

In any modeling application it is important to ask whether the model is valid for its intended purpose. Sadly this seems to be an area that is often neglected by management scientists engaged in simulation modeling, whether using DES or system dynamics. It is rarely possible to fully validate any model, especially if the model is to be used to investigate new situations. This is because using a model to investigate new situations means that any data on which it was built may have to be stretched to represent the situation being investigated. Even if we are clear that the model is valid for the current situation it may not be valid for the new one being investigated. Nevertheless, some formal validation is always possible and the various approaches are discussed in detail in Chapter 12 of Pidd (2004b). Not attempting any model validation is risky because it may lead to wrong conclusions being drawn and wrong decisions being made.

The simplest form of model validation is often referred to as Black Box validation. In this we run the model under well-defined conditions and collect output. We select the conditions under which the model run is to be, as far as possible, identical to conditions that have occurred in the system being simulated. We then tune the simulation model, that is we modify its values, as appropriate, in an attempt to ensure that the model output is indistinguishable from the output of the real system. This is called Black Box validation because we are only concerned with the inputs and outputs of the model and of the system rather than the internal workings. We should, though, be aware that coincidences do happen and that they could be several reasons why the model and the real system appear to behave in the same way. We cannot be sure that this occurs because of the model adequately represents the operation of the real system. If we wish to assure ourselves on this count, then we must attempt what is often known as White Box validation.

A better name for White Box validation would be transparent box validation, since the idea is to look at the detailed internal workings of the model to see if they adequately represent the operation of the system being simulated. For example, in simulating a telephone call center we may wish to assure ourselves that the way that one operator hands over to another at the end of a shift is correctly represented in the model. This is not a question of looking at the model output but rather at the way the logic has been programmed into the model. We must, though, notice that even a White Box validation can never be complete because we may wish, as mentioned above, to simulate situations which as yet do not exist. In such cases we can only check that the logic of the simulation model is the logic that is expected to hold in the real system.

Sadly, though, the known failings of both Black Box and White Box validation are unlikely to be the reasons why many management science simulation models are not validated. In many cases they are not properly validated because the analyst or the client have not allowed enough time for this to be done. Shortage of time and other resources often leads to corners being cut, and proper validation is often the victim of this. Fortunately, this may not matter in at least some situations, for it is very unlikely that most organizations will make significant investments on the basis of small probabilities. That is, they are only likely to make investments if there is a really significant benefit to be gained. It is possible that even a partially valid model may be able to demonstrate this.

33.6 Chapter Summary

A core feature of management science, or operational research, is the use of models to aid in decision making. Models help decision makers and planners to think through the possible consequences of decisions before taking any action. There are many different modeling approaches used in management science and the most commonly used dynamic approaches are DES and system dynamics. A DES model captures the detailed logical interactions of the entities that compose the system of interest. In management science, DES models are usually built using VIMSs, and it is rare for such applications to be programmed from scratch. To build a system dynamics model we must take a rather different focus. Instead of modeling the detailed interaction of individual entities we are concerned with the changes in populations of entities. These are represented in first-order difference equations which may then be used, with a numerical integration method, to simulate the operation of the system. Contemporary system dynamics in management science may be qualitative or quantitative and in the former case, causal loop diagrams offer a useful way to understand how systems operate. Sad to say, though, model validation is clearly very important, it is probably true that this is not well executed in management science for either DES or system dynamics.

References

Cota, B.A. and Sargent, R.G. (1990) Control flow graphs: a method of model representation for parallel discrete event simulation. CASE Center Technical Report 9026, CASE Center, Syracuse University.
Coyle, R.G. (1996) System dynamics modelling: a practical approach. Chapman & Hall, London.
Daellenbach, H. and McNickle, D. (2004) Management science: decision-making through systems thinking. Palgrave-MacMillan, London.
Forrester, J.W. (1961) Industrial dynamics. MIT Press, Cambridge, MA.
Hillier, F.S. and Lieberman, G.J. (2005) Introduction to operations research (8th Ed). McGraw-Hill, New York.
Hlynka, M. (2004) Queueing theory books. http://www2.uwindsor.ca/~hlynka/qbook.html
Jackson, M.C. (2003) Systems thinking: creative holism for managers. Wiley, Chichester.
Law, A.M. and Kelton, W.D. (2000) Simulation modeling and analysis (3rd Ed). McGraw-Hill, New York.
Micro Analysis and Design (2005) Micro Saint Sharp simulation software. www.maad.com
Pidd, M. (2003) Tools for thinking: modelling in management science. Wiley, Chichester.
Pidd, M. (2004a) Computer simulation in management science. Wiley, Chichester.

Pidd, M. (2004b) Simulation worldviews—so what? In: Proceedings of the 2004 Winter Simulation Conference. R.G. Ingalls, M.D. Rossetti, J.S. Smith, and B.A. Peters, eds. 5–8 December 2004, Washington Hilton and Towers, Washington, DC.

Richmond, B. and Petersen, S. (1994) Stella II: an introduction to systems thinking. High Performance Systems Inc., Hanover, NH.

Robinson, S.L. (2004) Simulation: the practice of model development and use. Wiley, Chichester.

Schruben, L. (1983) Simulation modeling with event graphs, Communications of the ACM, 26(11), 957–963.

Senge, P. (1999) The fifth discipline: the art and practice of the learning organization. Random House, London.

Sterman, J.D. (2000) Business dynamics: systems thinking and modeling for a complex world. Irwin McGraw-Hill, New York.

Taha, H.A. (2003) Operations research: an introduction (7th Ed). Prentice-Hall, Englewood Cliffs, NJ.

Winter Simulation Conference web site: www.wintersim.org

Wolstenholme, E.F. (1990) System enquiry: a system dynamics approach. Wiley, Chichester.

Zeigler, B.P. (1976) Theory of modelling and simulation. Wiley, New York.

Zimmerman, A. (2004) Petri nets. http://pdv.cs.tu-berlin.de/~azi/petri.html

34
Modeling and Analysis of Manufacturing Systems

E. Lefeber
Eindhoven University of Technology

J.E. Rooda
Eindhoven University of Technology

34.1 Introduction 34-1
34.2 Preliminaries 34-2
34.3 Analytical Models for Steady-State Analysis 34-3
   Mass Conservation (Throughput) • Queueing Relations (Wip, Flow Time)
34.4 Discrete-Event Models 34-7
34.5 Effective Process Times 34-8
34.6 Control of Manufacturing Systems: A Framework ... 34-10
34.7 Standard Fluid Model and Extensions 34-12
   A Common Fluid Model • An Extended Fluid Model • An Approximation to the Extended Fluid Model • A Hybrid Model
34.8 Flow Models 34-16
   Introduction to Traffic Flow Theory: The LWR Model • A Traffic Flow Model for Manufacturing Flow
34.9 Conclusions 34-18

34.1 Introduction

The dynamics of manufacturing systems has been a subject of study for several decades (Forrester, 1961; Hopp and Spearman, 2000). Over the last years, manufacturing systems have become more and more complex and therefore a good understanding of their dynamics has become even more important.

The goal of this chapter is to introduce a large variety of models for manufacturing systems. By means of examples it is illustrated how certain modeling techniques can be used to derive models that can be used for analysis or control. In addition to references that can be used as a starting point for further inquiry, recent developments in the modeling, analysis, and control of manufacturing systems are presented.

Since no familiarity with manufacturing systems is assumed, in Section 34.2 some terminology and basic properties of manufacturing systems are introduced. Section 34.3 provides some analytical modeling techniques and methods for analyzing steady-state behavior. Section 34.4 is concerned with deriving discrete-event models, which yield a more detailed insight in the dynamics of a manufacturing system. To reduce the complexity of discrete-event models, the concept of effective process times (EPTs) is introduced in Section 34.5, which results in modeling a manufacturing system as a large queueing network. This way of modeling a manufacturing system is a first step in a larger control framework, which is introduced in Section 34.6. This control framework makes it possible to study problems of controlling the dynamics of manufacturing systems by means of the available inputs. An important role in this control framework is played by approximation models. The most commonly used approximation models are presented in Section 34.7. Recently, a new class of approximation models has been proposed, which is presented in Section 34.8. Section 34.9 concludes this chapter.

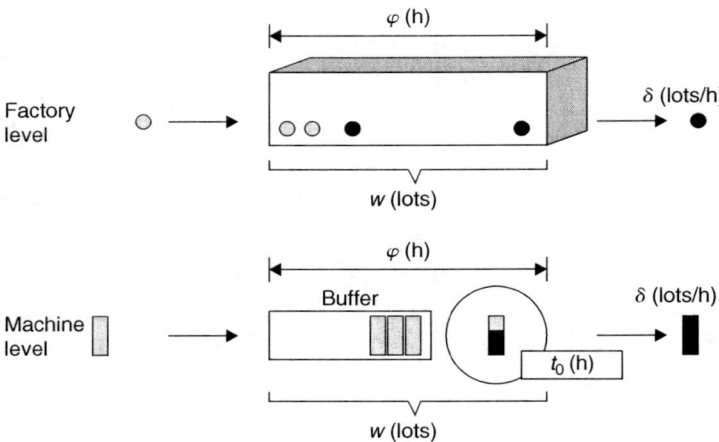

FIGURE 34.1 Basic quantities for manufacturing systems.

34.2 Preliminaries

First a few basic quantities are introduced as well as the main principles for manufacturing system analysis. The items produced by a manufacturing system are called *lots*. Also the words product and job are commonly used. Other important notions are throughput, flow time, wip, and utilization. These notions are illustrated in Figure 34.1 at factory and machine level.

Raw process time t_0 of a lot denotes the net time a machine needs to process the lot. This process time excludes additions such as setup time, breakdown, or other sources that may increase the time a lot spends in the machine. The raw process time is typically measured in hours or minutes.

Throughput δ denotes the number of lots per unit time that leaves the manufacturing system. At a machine level, this denotes the number of lots that leave a machine per unit time. At a factory level it denotes the number of lots that leave the factory per unit time. The unit of throughput is typically lots/hour.

Flow time φ denotes the time a lot is in the manufacturing system. At a factory level this is the time from the release of the lot into the factory until the finished lot leaves the factory. At a machine level this is the time from entering the machine (or the buffer in front of the machine) until leaving the machine. Flow time is typically measured in days, hours, or minutes. Instead of flow time the words cycle time and throughput time are also commonly used.

Work in process (wip) w denotes the total number of lots in the manufacturing system, i.e., in the factory or in the machine. Wip is measured in lots.

Utilization u denotes the fraction a machine is not idle. A machine is considered idle if it could start processing a new lot. Thus process time as well as downtime, setup time, and preventive maintenance time all contribute to the utilization. Utilization has no dimension and can never exceed 1.0.

Ideally, a manufacturing system should both have a high throughput and a low flow time or low wip. Unfortunately, these goals are conflicting (cf. Figure 34.2) and both cannot be met simultaneously. If a high throughput is required, machines should always be busy. As from time to time disturbances like machine failures happen, buffers between two consecutive machines are required to make sure that the second machine can still continue if the first machine fails (or vice versa). Therefore, for a high throughput many lots are needed in the manufacturing system, i.e., wip needs to be high. As a result, if a new lot starts in the system it has a large flow time, since all lots that are currently in the system need to be completed first.

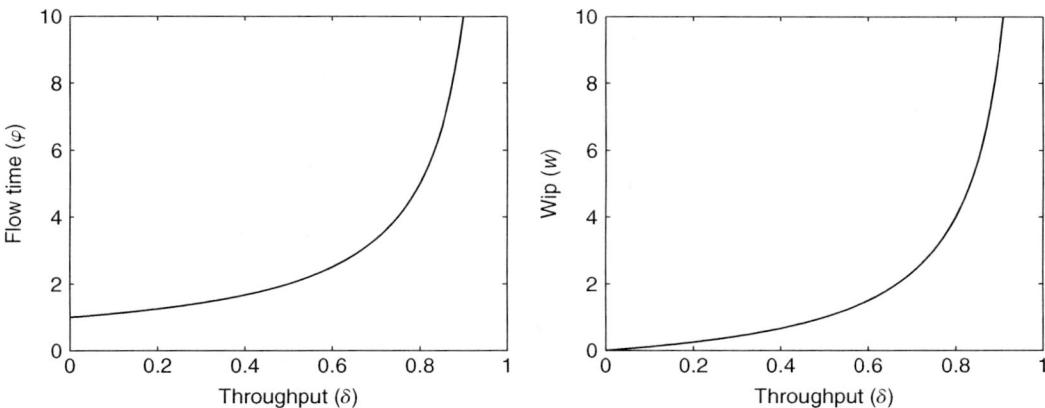

FIGURE 34.2 Basic relations between basic quantities for manufacturing systems.

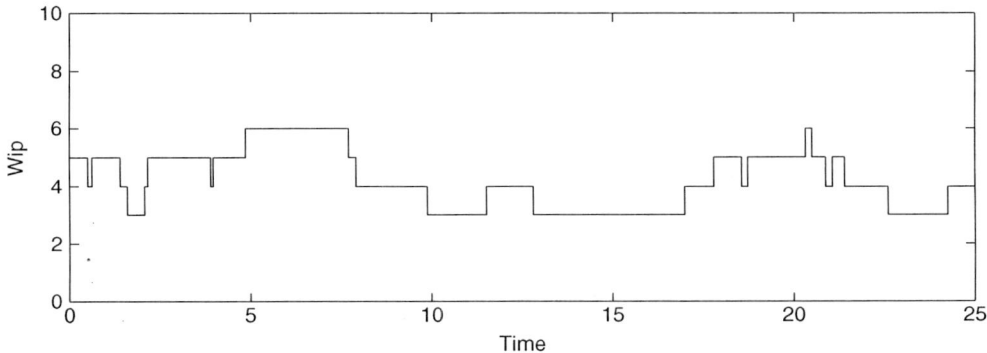

FIGURE 34.3 A characteristic time-behavior of wip at a workstation.

Conversely, the least possible flow time can be achieved if a lot arrives at a completely empty system and never has to wait before processing takes place. As a result, the wip level is small. However, for most of the time machines are not processing, yielding a small throughput.

When trying to control manufacturing systems, a trade-off needs to be made between throughput and flow time, so the nonlinear (steady-state) relations depicted in Figure 34.2 need to be incorporated in any reasonable model of manufacturing systems.

A final observation of relevance for modeling manufacturing systems is the nature of the system signals. In Figure 34.3 a characteristic graph of the wip at a workstation as a function of time is shown. Wip always takes integer values with arbitrary (nonnegative real) duration. One could consider a manufacturing system to be a system that takes values from a finite set of states and jumps from one state to the other as time evolves. This jump from one state to the other is called an *event*. As we have a countable (discrete) number of states, it is clear that discrete-event models are often used in modeling manufacturing systems. Discrete-event models for manufacturing systems are considered in Section 34.4. But first some analytical models for analyzing steady-state behavior of manufacturing systems are presented in the next section.

34.3 Analytical Models for Steady-State Analysis

To get some insights in the steady-state performance of a given manufacturing system simple relations can be used. In this section, we deal with mass conservation for determining the mean utilization of

34.3.1 Mass Conservation (Throughput)

Using mass conservation the mean utilization of workstations can easily be determined.

Example 1

Consider the manufacturing system with rework and bypassing in Figure 34.4. The manufacturing system consists of three buffers and four machines. Lots are released at a rate of λ lots/h. The numbers near the arrows indicate the fraction of the lots that follow that route. For instance, of the lots leaving buffer B_1 90% goes to machine M_1 and 10% goes to buffer B_3. The process time of each machine is listed in the table in Figure 34.4.

Let δ_{M_i} and δ_{B_i} denote the throughput of machine M_i $(i=1,2,3,4)$ and buffer B_i $(i=1,2,3)$, respectively. Using mass conservation we obtain

$$\delta_{M_1} = 0.9\delta_{B_1} \qquad \delta_{B_1} = \lambda$$
$$\delta_{M_2} = 0.2\delta_{B_2} \qquad \delta_{B_2} = \delta_{M_1} + \delta_{M_2}$$
$$\delta_{M_3} = 0.8\delta_{B_2} \qquad \delta_{B_3} = \delta_{M_3} + 0.1\delta_{B_1}$$
$$\delta_{M_4} = \delta_{B_3} \qquad \delta = \delta_{M_4}$$

Solving these linear relations results in:

$$\delta_{M_1} = 0.9\lambda \qquad \delta_{B_1} = \lambda$$
$$\delta_{M_2} = 0.225\lambda \qquad \delta_{B_2} = 1.125\lambda$$
$$\delta_{M_3} = 0.9\lambda \qquad \delta_{B_3} = \lambda$$
$$\delta_{M_4} = \lambda \qquad \delta = \lambda$$

Using the process times of the table in Figure 34.4, we obtain for the utilizations:

$$u_{M_1} = 0.9\lambda \cdot 2.0/1 = 1.8\lambda \qquad u_{M_3} = 0.9\lambda \cdot 1.8/1 = 1.62\lambda$$
$$u_{M_2} = 0.225\lambda \cdot 6.0/1 = 1.35\lambda \qquad u_{M_4} = \lambda \cdot 1.6/1 = 1.6\lambda$$

Clearly, machine M_1 is the bottleneck and the maximal throughput for this line is $\lambda = 1/1.8 = 0.56$ jobs/h.

Using mass conservation, utilizations of workstations can be determined straightforwardly. This also provides a way for determining the number of machines required for meeting a given throughput. By modifying the given percentages the effect of rework or a change in product mix can also be studied.

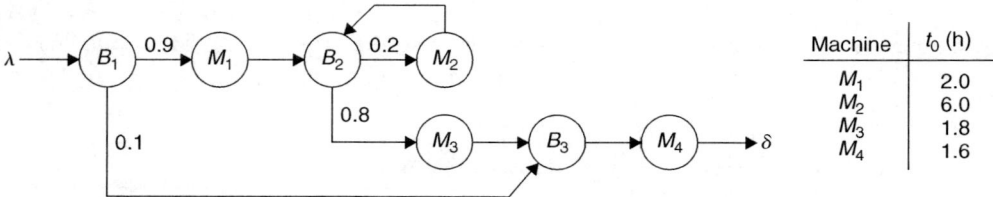

FIGURE 34.4 Manufacturing system with rework and bypassing.

34.3.2 Queueing Relations (Wip, Flow Time)

For determining a rough estimate of the corresponding mean flow time and mean wip, basic relations from queueing theory can be used.

Consider a single machine workstation that consists of infinite buffer B_∞ and machine M (see Figure 34.5). Lots arrive at the buffer with a stochastic interarrival time. The interarrival time distribution has mean t_a and a standard deviation σ_a, which we characterize by the coefficient of variation $c_a = \sigma_a/\mu_a$. The machine has stochastic process times, with mean process time t_0 and coefficient of variation c_0. Finished lots leave the machine with a stochastic interdeparture time, with mean t_d and coefficient of variation c_d. Assuming independent interarrival times and independent process times, the mean waiting time φ_B in buffer B can be approximated for a stable system by means of Kingman's equation (Kingman, 1961):

$$\varphi_B = \frac{c_a^2 + c_0^2}{2} \frac{u}{1-u} t_0 \qquad (34.1)$$

with the utilization u defined by: $u = t_0/t_a$. Eq. (34.1) is exact for an $M/G/1$ system, i.e., a single machine workstation with exponentially distributed interarrival times and any distribution for the process time. For other single machine workstations it is an approximation.

For a stable system, we have $t_d = t_a$. We can approximate the coefficient of variation c_d by Kuehn's linking equation (Kuehn, 1979):

$$c_d^2 = (1 - u^2)c_a^2 + u^2 c_0^2. \qquad (34.2)$$

This result is exact for an $M/M/1$ system. For other single machine workstations it is an approximation. Having characterized the departure process of a workstation, the arrival process at the next workstation has been characterized as well. As a result, a line of workstations can also be described.

Example 2 (Three workstations in line)

Consider the three workstation flow line in Figure 34.6. For the interarrival time at workstation 0 we have $t_a = 4.0$ h and $c_a^2 = 1$. The three workstations are identical with respect to the process times: $t_{0,i} = 3.0$ h for $i = 0, 1, 2$ and $c_{0,i}^2 = 0.5$ for $i = 0, 1, 2$. We are interested to determine the mean total flow time per lot.

Since $t_a > t_{0,i}$ for $i = 0, 1, 2$, we have a stable system and $t_{a,i} = t_{d,i} = 4.0$ h for $i = 0, 1, 2$. Subsequently, the utilization for each workstation is $u_i = 3.0/4.0 = 0.75$ for $i = 0, 1, 2$.

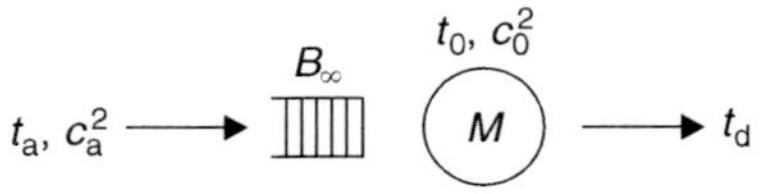

FIGURE 34.5 Single-machine workstation.

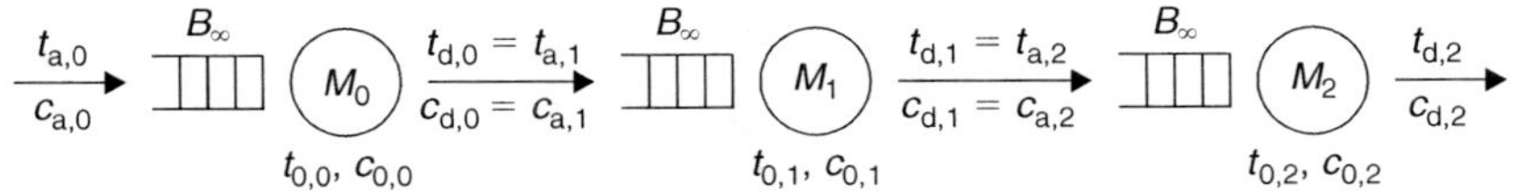

FIGURE 34.6 Three workstation flow line.

Using Eq. (34.1) we calculate the mean flow time for workstation 0

$$\varphi_0 = \varphi_B + t_0 = \frac{c_a^2 + c_0^2}{2} \frac{u}{1-u} t_0 + t_0 = \frac{1 + 0.5}{2} \frac{0.75}{1 - 0.75} 3.0 + 3.0 = 9.75\,\text{h}$$

Using Eq. (34.2), we determine the coefficient of variation on the interarrival time $c_{a,1}$ for workstation W_1

$$c_{a,1}^2 = c_{d,0}^2 = (1 - u^2)c_a^2 + u^2 c_0^2 = (1 - 0.75^2)1 + 0.75^2 \cdot 0.5 = 0.719$$

and the mean flow time for workstation 1

$$\varphi_1 = \frac{0.719 + 0.5}{2} \frac{0.75}{1 - 0.75} 3.0 + 3.0 = 8.49\,\text{h}$$

In a similar way, we determine that $c_{a,2}^2 = 0.596$, $\varphi_2 = 7.93\,\text{h}$. We then calculate the mean total flow time to be

$$\varphi_{\text{tot}} = \varphi_0 + \varphi_1 + \varphi_2 = 26.2\,\text{h}$$

Note that the minimal flow time without variability ($c_a^2 = c_{0,i}^2 = 0$) equals 9.0 h.

Eq. (34.1) and Eq. (34.2) are particular instances of a workstation consisting of a single machine. For workstations consisting of m identical machines, in parallel the following approximations can be used:

$$\varphi_B = \frac{c_a^2 + c_0^2}{2} \frac{u^{\sqrt{2(m+1)}-1}}{m(1-u)} \cdot t_0 \tag{34.3}$$

$$c_d^2 = (1 - u^2)c_a^2 + u^2 \frac{c_0^2 + \sqrt{m} - 1}{\sqrt{m}} \tag{34.4}$$

Note that in case $m = 1$ these equations reduce to Eq. (34.1) and Eq. (34.2).

Once the mean flow time has been determined, a third basic relation from queueing theory, Little's law (Little, 1961), can be used for determining the mean wip level. Little's law states that the mean wip level (number of lots in a manufacturing system) w is equal to the product of the mean throughput δ and the mean flow time φ, provided the system is in steady state

$$w = \delta \varphi \tag{34.5}$$

An example illustrates how Kingman's equation and Little's law can be used.

Example 3

Consider the system of Example 2 as depicted in Figure 34.6. From Example 34.3.2 we know that the flow times for the three workstations are, respectively,

$$\varphi_0 = 9.75\,\text{h}, \quad \varphi_1 = 8.49\,\text{h}, \quad \varphi_2 = 7.93\,\text{h}$$

Since the steady-state throughput was assumed to be $\delta = 1/t_a = 1/4.0 = 0.25$ lots/h, we obtain via Little's law

$$w_0 = 0.25 \cdot 9.75 = 2.44\,\text{lots}$$
$$w_1 = 0.25 \cdot 8.49 = 2.12\,\text{lots}$$
$$w_2 = 0.25 \cdot 7.93 = 1.98\,\text{lots}$$

The above-mentioned relations are simple approximations that can be used for getting a rough idea about the possible performance of a manufacturing system. These approximations are fairly accurate for high degrees of utilization but less accurate for lower degrees of utilization. A basic assumption when using these approximations is the independence of the interarrival times, which in general is not the case, e.g., for merging streams of jobs. Furthermore, using these equations only steady-state behavior can be analyzed. For studying things like ramp-up behavior or for incorporating more details like operator-behavior, more sophisticated models are needed, as described in the next section.

34.4 Discrete-Event Models

In the previous section simple methods have been introduced for analyzing steady-state behavior of manufacturing systems. For analyzing the dynamics of manufacturing systems, more sophisticated models are required. Using Figure 34.3 in Section 34.2 it was illustrated that typical models of manufacturing systems are the so-called discrete-event models. In this section, we present examples how to build a discrete-event model of a manufacturing system using the specification language χ, explained in more detail in Chapter 19 of this handbook.

The way to build a discrete-event model is to consider the manufacturing system as a network of concurrent processes through which jobs and other types of information flows. For example, a basic machine can be modeled as a process which repeatedly tries to receive a lot, waits for the period of time (the process time), and tries to send a lot. Using χ, we can write

$$\text{proc } M\,(\text{chan } a?b!: \text{lot}, \text{var } t_e: \text{real}) = \|[\text{var } x: \text{lot} :: (a?x;\ \Delta t_e;\ b!x)]\|$$

The machine is able to receive lots via external channel a, is able to send lots via external channel b, and the process time of the machine is given by parameter t_e. Repeatedly, the machine tries to receive a lot over external channel a and store this lot in discrete variable x. Next, the machine waits for t_e, after which the machine tries to send x via external channel b.

A buffer can be modeled as a process that may receive new lots if it is not full and may send lots if it is not empty. Using χ, a finite first in, first out (FIFO) buffer with a maximal buffer size n can be modeled as

```
proc B (chan a?b!: lot, var n: nat) =
|[ disc xs: [lot] = [], x: lot
::(len(xs) < n → a?x; xs := xs++[x]
[] len(xs) > 0 → b!hd(xs); xs : = tl(xs)
)
]|
```

This process can receive lots via external channel a, send lots via external channel b, and has its maximal buffer size n as a parameter. Repeatedly, two alternatives can be executed:

- Trying to receive a lot via channel a (only if the length of the list xs is less than n) into discrete variable x and consecutively adding it to list xs of lots (using a concatenation of lists).
- Trying to send the head of the list (its first element) via channel b (only if the list is not empty) and consecutively reducing list xs to its tail (everything but the first element).

These two processes can be used to model a workstation that consists of a 3-place buffer and a machine with process time t_e by simply executing the two previously specified processes in parallel:

$$\text{proc } W(\text{chan } a?b!: \text{lot}, \text{var } t_e: \text{real}) = \|[\text{chan } c: \text{lot} :: B(a,c,3) \| M(c,b,t_e)]\|$$

We assume that lots arrive at this workstation according to a Poisson arrival process with mean arrival rate of λ jobs per unit time. This can be modeled by means of the generator process

$$\text{proc } G\,(\text{chan } a!: \text{lot}, \text{var } \lambda: \text{real}) =$$
$$\|[\text{disc } u: \rightarrow \text{real} = \text{exponential}(1/\lambda) :: (a!\tau;\ \Delta\sigma u;\ b!x)]\|$$

Here the type lot is a real number which contains the time this lot entered the system. Generator G is able to send lots via external channel a and has a mean departure rate, which is given by parameter λ. The discrete variable u contains an exponential distribution with mean $1/\lambda$. Repeatedly, the generator tries to send a lot over external channel a, where at departure it gets assigned the current time τ. Next, the generator waits for a period which is given by a sample from the distribution u.

Once a lot has been served by workstation W it leaves to the exit process E:

proc E(chan a? : lot) = |[var x : lot :: (a?x)]|

This process repeatedly tries to receive a lot via external channel a.

For an arrival rate of $\lambda = 0.5$ and a process time of $t_e = 1.5$, the specification of the discrete-event model can be completed by

model $GWE()$ = |[chan a,b : lot :: $G(a,0.5)$ ‖ $W(a,b,1.5)$‖$E(b)$]|

In this way a manufacturing system can be modeled as a network of concurrent processes through which jobs and other types of information flows. The presented model is rather simple, but clearly many more ingredients can be added. For example, to include an operator for the processing of the machine we can modify the process M into

proc $\bar{M}$ (chan a?b! : lot, c?, d! : operator, var t_e : real) =
|[var x : lot, y : operator :: (c?y; a?x; Δt_e; b!x; d!y)]|

Highly detailed models of manufacturing systems can be made in this way, even before the system has been build. The influence of parameters can be analyzed by running several experiments with the discrete-event model using different parameter settings. This is common practice when designing a several billion wafer fab. However, since in practice manufacturing systems are changing continuously, it is very hard to keep these detailed discrete-event models up-to-date.

Fortunately, for a manufacturing system in operation it is possible to arrive at more simple/less detailed discrete-event models by using the concept of EPTs as introduced in the next section.

34.5 Effective Process Times

As mentioned in the previous section, for the processing of a lot at a machine, many steps may be required. It could be that an operator needs to get the lot from a storage device, set up a specific tool that is required for processing the lot, put the lot on an available machine, start a specific program for processing the lot, wait until this processing has finished (meanwhile doing something else), inspect the lot to determine if all went well, possibly perform some additional processing (e.g., rework), remove the lot from the machine and put it on another storage device, and transport it to the next machine. At all of these steps something might go wrong: the operator might not be available, after setting up the machine the operator finds out that the required recipe cannot be run on this machine, the machine might fail during processing, no storage device is available anymore so the machine cannot be unloaded and is blocked, etc.

It is impossible to measure all sources of variability that might occur in a manufacturing system. While some of the sources of variability could be incorporated into a discrete-event model (tool failures and repairs, maintenance schedules), not all sources of variability can be included. This is clearly illustrated in Figure 34.7, obtained from Jacobs et al. (2003).

The left graph contains actual realizations of flow times of lots leaving a real manufacturing system, whereas the right graph contains the results of a detailed deterministic simulation model and the graph in the middle contains the results of a similar model including stochasticity. It turns out that in reality flow times are much higher and much more irregular than simulation predicts. So, even if one tries hard to capture all variability present in a manufacturing system, still the outcome predicted by the model is far from reality.

The term EPT has been introduced by Hopp and Spearman (2000) as the time seen by lots from a logistical point of view. To determine the EPT they assume that the contribution of the individual sources of variability is known.

Modeling and Analysis of Manufacturing Systems 34-9

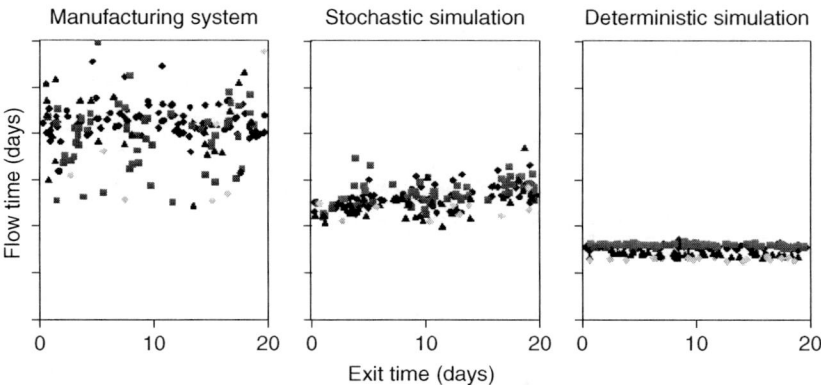

FIGURE 34.7 A comparison.

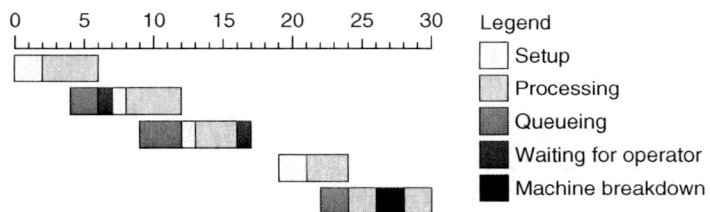

FIGURE 34.8 Gantt chart of five lots at a single machine workstation.

A similar description is given in Sattler (1996) where the effective process time has been defined as all flow time except waiting for another lot. It includes waiting owing to machine down time and operator availability and a variety of other activities. In Sattler (1996) it was also noticed that this definition of effective process time is difficult to measure.

Instead of taking the bottom-up view of Hopp and Spearman, a top-down approach can also be taken, as shown by Jacobs et al. (2003), where algorithms have been introduced that enable determination of effective process time realizations from a list of events.

We consider a single machine workstation and assume that the Gantt chart of Figure 34.8 describes a given time period.

- At $t = 0$ the first lot arrives at the workstation. After a setup, the processing of the lot starts at $t = 2$ and is completed at $t = 6$.
- At $t = 4$ the second lot arrives at the workstation. At $t = 6$ this lot could have been started, but apparently there was no operator available, so only at $t = 7$ the setup for this lot starts. Eventually, at $t = 8$ the processing of the lot starts and is completed at $t = 12$.
- The fifth lot arrives at the workstation at $t = 22$, processing starts at $t = 24$, but at $t = 26$ the machine breaks down. It takes until $t = 28$ before the machine has been repaired and the processing of the fifth lot continues. The processing of the fifth lot is completed at $t = 30$.

From a lot's point of view we observe:

- The first lot arrives at an empty system at $t = 0$ and departs from this system at $t = 6$. Its processing took 6 units of time.
- The second lot arrives at a nonempty system at $t = 4$ and needs to wait. At $t = 6$, the system becomes available again and hence from $t = 6$ on there is no need for the second lot to wait. At $t = 12$ the

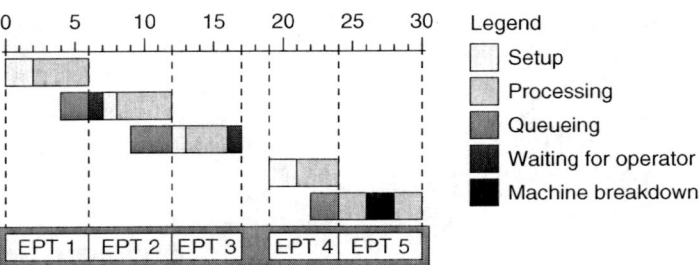

FIGURE 34.9 EPT realizations of five lots at a workstation.

second lot leaves the system, so from the point of view of this lot, its processing took from $t = 6$ till $t = 12$; the lot does not know whether waiting for an operator and a setup is part of its processing.
- The third lot sees no need for waiting after $t = 12$ and leaves the system at $t = 17$, so it assumes to have been processed from $t = 12$ till $t = 17$.

Following this reasoning, the resulting effective process times for lots are as depicted in Figure 34.9. Note that only arrival and departure events of lots to a workstation are needed for determining the effective process times. Furthermore, none of the contributing disturbances needs to be measured. In highly automated manufacturing systems, arrival and departure events of lots are being registered, so for these manufacturing systems, effective process time realizations can be determined rather easily. These EPT realizations can be used in a relatively simple discrete-event model of the manufacturing system. Such a discrete-event model only contains the architecture of the manufacturing system, buffers, and machines. The process times of these machines are samples from their EPT distribution as measured from real manufacturing data. There is no need for incorporating machine failures, operators, etc., as this is all included in the EPT-distributions. Furthermore, the EPTs are utilization independent. That is, EPTs determined collected at a certain throughput rate are also valid for different throughput rates. Also, machines with the same EPT-distribution can be added to a workstation. This makes it possible to study how the manufacturing system responds in case a new machine is added, or all kinds of other what-if-scenarios. Finally, since EPT-realizations characterize operational time variability, they can be used for performance measuring. For more on this issue, the interested reader is referred to Jacobs et al. (2003, 2006). What is most important is that EPTs can be determined from real manufacturing data and yield relatively simple discrete-event models of the manufacturing system under consideration. These relatively simple discrete-event models can serve as a starting point for controlling manufacturing systems dynamically.

34.6 Control of Manufacturing Systems: A Framework

In the previous section, the concept of effective process times has been introduced as a means to arrive at relatively simple discrete-event models for manufacturing systems, using measurements from the real manufacturing system under consideration. The resulting discrete-event models are large queueing networks which capture the dynamics reasonably well. These relatively simple discrete-event models are not only a starting point for analyzing the dynamics of a manufacturing system, but can also be used as a starting point for controller design. If one is able to control the dynamics of the discrete-event model of the manufacturing system, the resulting controller can also be used for controlling the real manufacturing system.

Even though control theory exists for controlling discrete-event systems, unfortunately none of it is appropriate for controlling discrete-event models of real-life manufacturing systems. This is mainly due to the large number of states of a manufacturing system. Therefore, a different approach is needed.

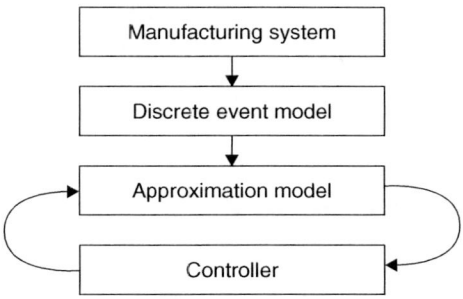

FIGURE 34.10 Control framework (I).

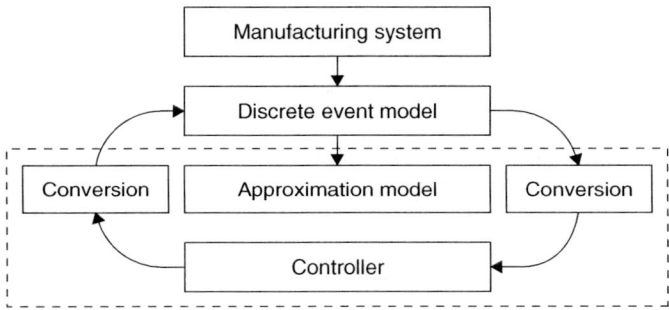

FIGURE 34.11 Control framework (II).

If we concentrate on mass production, the distinction between lots is not really necessary and lots can be viewed in a more continuous way. Therefore, instead of the discrete-event model we consider an approximation model.

Using the approximation model, we can use standard control theory to derive a controller for the approximation model (cf. Figure 34.10).

When the closed-loop system of the approximation model and the controller behaves as desired, the controller can be connected to the discrete-event model. However, since the derived controller is not a discrete-event controller its control actions still need to be transformed into events. It might very well be that the optimal control action would be to produce 2.75 lots during the next shift. One still needs to decide how many jobs to really start (2 or 3), and also when to start them. This is the left conversion block in Figure 34.11. Similarly, a conversion is needed from discrete-event model to controller: a simple conversion would be to sample the discrete-event model once every shift, but other sampling strategies might also be followed. For example, if at the beginning of a shift a machine breaks down it might not be such a good idea to wait until the end of the shift before setting new production targets.

Once the two conversion blocks have been properly designed a suitable discrete-event controller for the discrete-event model is obtained, as illustrated in Figure 34.11 (dashed).

Eventually, as a final step, the designed controller can be disconnected from the discrete-event model, and attached to the real manufacturing system.

In the presented control framework two crucial steps can be distinguished. First, the discrete-event model should be a good enough approximation of the real manufacturing system, i.e., the model needs to be validated and if found unsatisfactory it needs to be improved. Second, the approximation model should be a good enough approximation of the discrete-event model, or actually, of the discrete-event model and conversion block(s), since that is the system that needs to be controlled by the continuous controller. Depending on the variables of interest, a valid approximation model needs to be used. An overview of common used approximation models, assuming mass production, is provided in the next two sections. In Section 34.7 approximation models are presented that mainly focus on throughput. In Section 34.8 approximation models are presented that incorporate both throughput and flow time, taking into account the nonlinear relations as depicted in Figure 34.2.

34.7 Standard Fluid Model and Extensions

The analytical approximations models of Section 34.3 are only concerned with steady state, no dynamics is included. This disadvantage is overcome by discrete-event models as discussed in Section 34.4. However, since they model each job separately and stochastically long simulation times are required for obtaining satisfactory results. Using an approximation model where jobs are viewed in a continuous way we can overcome these long simulation times.

34.7.1 A Common Fluid Model

The current standard way of deriving fluid models is most easily explained by means of an example. Consider a simple manufacturing system consisting of two machines in series, as displayed in Figure 34.12. Let $u_0(t)$ denote the rate at which jobs arrive at the system at time t, $u_i(t)$ the rate at which machine M_i produces lots at time t, $y_i(t)$ the number of lots in buffer B_i at time t ($i \in \{1,2\}$), and $y_3(t)$ the number of lots produced by the manufacturing system at time t. Assume that machines M_1 and M_2 have a maximum capacity of μ_1 and μ_2 lots per time unit, respectively.

The rate of change of the buffer contents is given by the difference between the rates at which lots enter and leave the buffer. Under the assumption that the number of lots can be considered continuously, we get

$$\dot{y}_1(t) = u_0(t) - u_1(t),$$
$$\dot{y}_2(t) = u_1(t) - u_2(t), \quad (34.6)$$
$$\dot{y}_3(t) = u_2(t)$$

which can be rewritten as

$$\dot{x}(t) = \begin{bmatrix} 0 & 0 & 0 \\ 0 & 0 & 0 \\ 0 & 0 & 0 \end{bmatrix} x(t) + \begin{bmatrix} 1 & -1 & 0 \\ 0 & 1 & -1 \\ 0 & 0 & 1 \end{bmatrix} u(t) \quad (34.7a)$$

$$y(t) = \begin{bmatrix} 1 & 0 & 0 \\ 0 & 1 & 0 \\ 0 & 0 & 1 \end{bmatrix} x(t) + \begin{bmatrix} 0 & 0 & 0 \\ 0 & 0 & 0 \\ 0 & 0 & 0 \end{bmatrix} u(t) \quad (34.7b)$$

where $u = [u_0, u_1, u_2]^\top$ and $y = [y_1, y_2, y_3]^\top$. We also have capacity constraints on the input as well as the constraint that the buffer contents should remain positive, i.e.,

$$0 \leq u_1(t) \leq \mu_1, 0 \leq u_2(t) \leq \mu_2 \quad \text{and} \quad y_1(t) \geq 0, y_2(t) \geq 0, y_3(t) \geq 0 \quad (34.8)$$

This system is a controllable linear system of the form $\dot{x} = Ax + Bu, y = Cx + Du$, extensively studied in control theory.

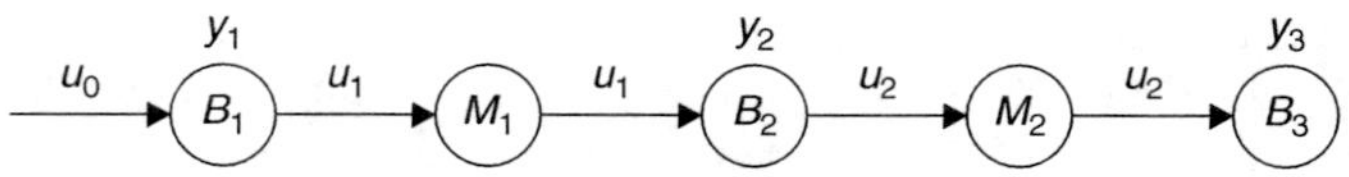

FIGURE 34.12 A simple manufacturing system.

Note that instead of a description in continuous time, a description in discrete time can also be used:

$$\dot{x}(k+1) = \begin{bmatrix} 1 & 0 & 0 \\ 0 & 1 & 0 \\ 0 & 0 & 1 \end{bmatrix} x(k) + \begin{bmatrix} 1 & -1 & 0 \\ 0 & 1 & -1 \\ 0 & 0 & 1 \end{bmatrix} u(k) \quad (34.9a)$$

$$y(k) = \begin{bmatrix} 1 & 0 & 0 \\ 0 & 1 & 0 \\ 0 & 0 & 1 \end{bmatrix} x(k) + \begin{bmatrix} 0 & 0 & 0 \\ 0 & 0 & 0 \\ 0 & 0 & 0 \end{bmatrix} u(k) \quad (34.9b)$$

Also, the description of Eq. (34.7) is not the only possible input/output/state model that yields the input/output behavior Eq. (34.6). To illustrate this, consider the change of coordinates

$$x(t) = \begin{bmatrix} 1 & -1 & 0 \\ 0 & 1 & -1 \\ 0 & 0 & 1 \end{bmatrix} \bar{x}(t) \quad (34.10)$$

which results in the following input/output/state model:

$$\dot{\bar{x}}(t) = \begin{bmatrix} 0 & 0 & 0 \\ 0 & 0 & 0 \\ 0 & 0 & 0 \end{bmatrix} \bar{x}(t) + \begin{bmatrix} 1 & 0 & 0 \\ 0 & 1 & 0 \\ 0 & 0 & 1 \end{bmatrix} u(t) \quad (34.11a)$$

$$y(t) = \begin{bmatrix} 1 & -1 & 0 \\ 0 & 1 & -1 \\ 0 & 0 & 1 \end{bmatrix} \bar{x}(t) + \begin{bmatrix} 0 & 0 & 0 \\ 0 & 0 & 0 \\ 0 & 0 & 0 \end{bmatrix} u(t) \quad (34.11b)$$

Note that in this description, the state $\bar{x}$ denotes the cumulative production at each workstation.

We would like to study the response of the output of the system Eq. (34.7), or equivalently Eq. (34.11). Assume we initially start with an empty production line (i.e., $x(0) = 0$), that both machines have a capacity of 1 lot per unit time (i.e., $\mu_1 = \mu_2 = 1$) and that we feed the line at a rate of 1 lot per time unit (i.e., $u_0 = 1$). Furthermore, assume that machines produce at full capacity, but only in case something is in the buffer in front of it, i.e.,

$$u_i(t) = \begin{cases} \mu_i & \text{if } y_i(t) > 0 \\ 0 & \text{otherwise} \end{cases} \quad i \in \{1, 2\} \quad (34.12)$$

Under these assumptions, the resulting contents of buffer B_3 are as displayed in Figure 34.13. Note that immediately lots start coming out of the system. Clearly, this is not what happens in practice. Since both machines M_1 and M_2 need to process the first lot, it should take the system at least $(1/\mu_1) + (1/\mu_2)$ time

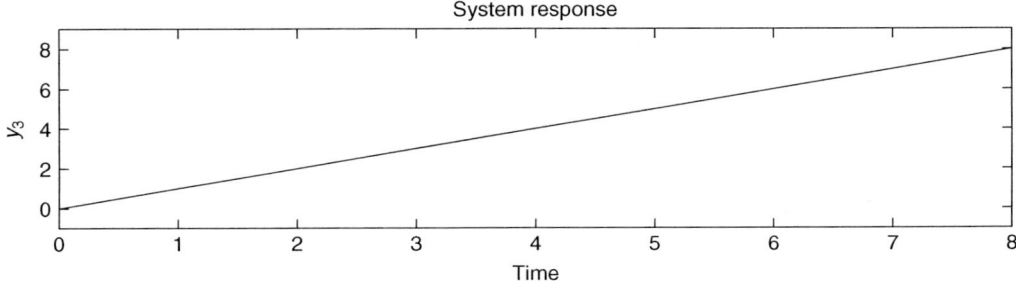

FIGURE 34.13 Output of the manufacturing system using model Eq. (34.6).

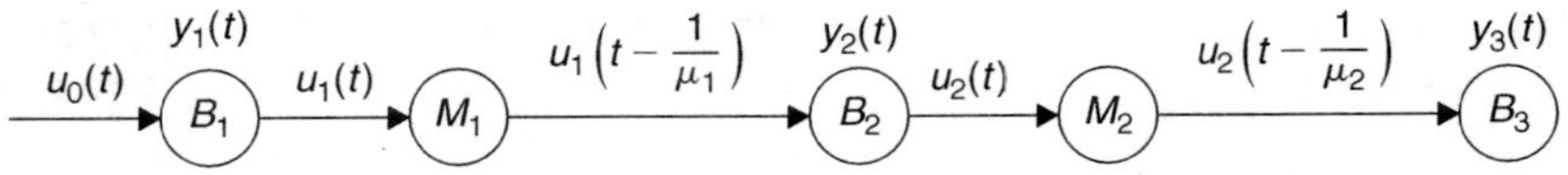

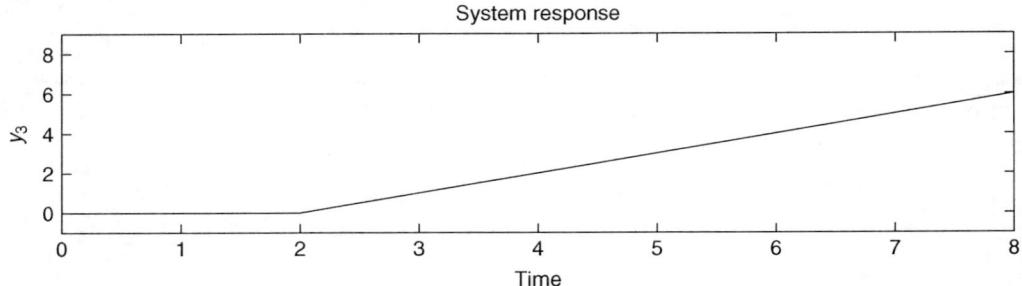

FIGURE 34.14 A simple manufacturing system revisited.

FIGURE 34.15 Output of the manufacturing system using model Eq. (34.13).

units before lots can come out. This illustrates fluid models as given by Eq. (34.7) or Eq. (34.11) do not incorporate flow times.

34.7.2 An Extended Fluid Model

In the previous subsection, we noticed that in the standard fluid model lots immediately come out of the system, once we start producing. A way to overcome this problem is to explicitly take into account the required delay. Whenever we decide to change the production rate of machine M_1, buffer B_2 notices this $1/\mu_1$ time units later. As a result, the rate at which lots arrive at buffer B_2 at time t is equal to the rate at which machine M_1 was processing at time $t - 1/\mu_1$. This observation results in the following model (see also Figure 34.14):

$$\dot{y}_1(t) = u_0(t) - u_1(t)$$
$$\dot{y}_2(t) = u_1\left(t - \frac{1}{\mu_1}\right) - u_2(t) \quad (34.13)$$
$$\dot{y}_3(t) = u_2\left(t - \frac{1}{\mu_2}\right)$$

Clearly, the constraints of Eq. (34.8) also apply to the model given by Eq. (34.13).

Figure 34.15 shows the response of the system given by Eq. (34.12) to the ramp-up experiment that lead to Figure 34.13. Comparing these two figures we see that no products enter buffer B_3 during the first 2.0 time units for the extended fluid model. Clearly, the extended fluid model produces more realistic results than the standard fluid model.

34.7.3 An Approximation to the Extended Fluid Model

In the previous subsection, we proposed an extended version of the standard fluid model. Although the model of Eq. (34.13) still is a linear model, standard linear control theory is not able to deal with this model, due to the time delay. Instead we have to rely on control theory of infinite-dimensional linear systems (see, e.g., Curtain and Zwart, 1995).

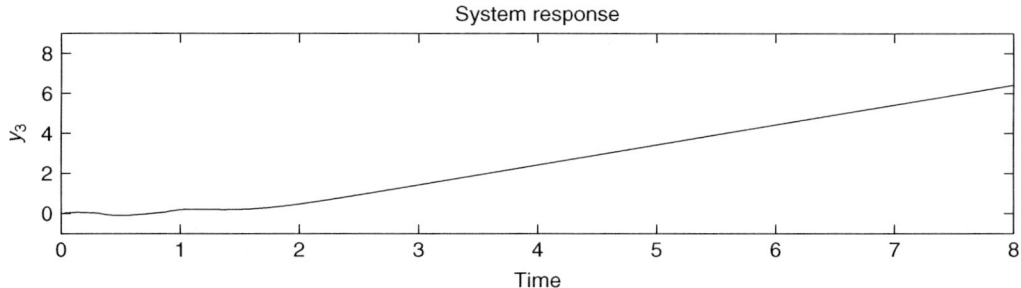

FIGURE 34.16 Output of the manufacturing system using model Eq. (34.14).

Another possibility would be to approximate the time delays by means of a Padé approximation (Baker, 1965). Using a second-order Padé approximation, the model of Eq. (34.13) can be approximated as:

$$\dot{x} = \begin{bmatrix} \mathbf{0} & \mathbf{0} & 0 & 0 & 0 & 0 & 0 \\ \mathbf{0} & \mathbf{0} & 0 & 0 & 0 & 0 & 0 \\ 0 & 0 & 4 & 6 & -3 & 0 & 0 \\ 0 & 0 & 0 & 4 & 0 & 0 & 0 \\ \mathbf{0} & \mathbf{0} & 0 & 0 & \mathbf{0} & 0 & 0 \\ 0 & 0 & 0 & 0 & 4 & 6 & -3 \\ 0 & 0 & 0 & 0 & 0 & 4 & 0 \end{bmatrix} x + \begin{bmatrix} 1 & 0 & 0 \\ 0 & 1 & 0 \\ 0 & 0 & 0 \\ 0 & 0 & 0 \\ 0 & 0 & 1 \\ 0 & 0 & 0 \\ 0 & 0 & 0 \end{bmatrix} u \qquad (34.14\text{a})$$

$$y = \begin{bmatrix} 1 & -1 & 0 & 0 & \mathbf{0} & 0 & 0 \\ 0 & 1 & -3 & 0 & -1 & 0 & 0 \\ 0 & 0 & 0 & 0 & 1 & -3 & 0 \end{bmatrix} x + \begin{bmatrix} 0 & 0 & 0 \\ 0 & 0 & 0 \\ 0 & 0 & 0 \end{bmatrix} u \qquad (34.14\text{b})$$

Note the structure in Eq. (34.14). In bold face we can easily recognize the dynamics of Eq. (34.11). The additional dynamics results from the Padé approximation.

If we initiate the system of Eq. (34.14) from $x(0) = 0$ and feed it at a rate $u_0 = 1$ while using Eq. (34.12), we obtain the system response as depicted in Figure 34.16. It is clear that we do not get the same response as in Figure 34.15, but the result is rather acceptable from a practical point of view. At least it is closer to reality than the response displayed in Figure 34.13.

34.7.4 A Hybrid Model

In the previous subsections, we provided extensions to the standard fluid model by taking into account the time delay lots encounter owing to the processing of machines. We also mentioned the constraints of Eq. (34.8) that have to be obeyed. These are constraints that we have to take into account when designing a controller for our manufacturing system. The way we dealt with these constraints in the previous subsections was by requiring the machines to produce only in case the buffer contents in front of that machine were positive (cf. Eq. 34.12).

A way to extend the standard fluid model Eq. (34.7) is to think of these constraints in a different way. As illustrated in Subsection 34.7.1, when we turn on both machines, immediately lots start coming out of the system. This is an undesirable feature that we would like to avoid. In practice, the second machine can only start producing when the first machine has finished a lot. Keeping this in mind, why do we allow machine M_2 to start producing as soon as the buffer contents of the buffer in front of it are positive? Actually, machine M_2 should only start producing as soon as a whole product has been finished by the machine M_1.

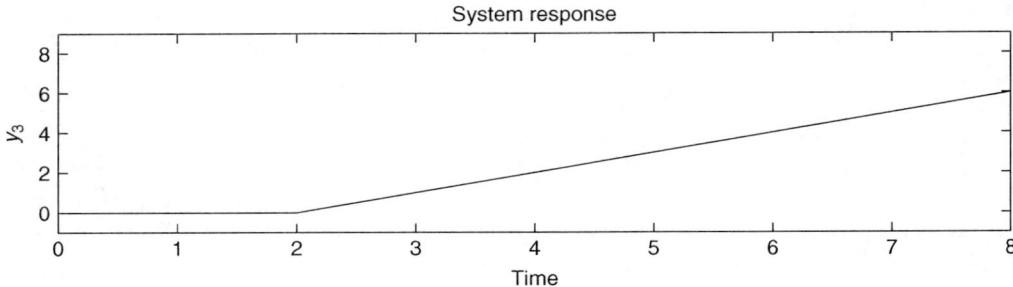

FIGURE 34.17 Output of the manufacturing system.

In words, machine M_2 should only start producing as soon as the buffer contents of the buffer in front of it becomes 1. Therefore, we should not allow for a positive u_2 as soon as $y_2 > 0$, but only in case $y_2 \geq 1$.

When we consider the initially empty system Eq. (34.7), i.e., $x(0) = 0$, and assume

$$u_i(t) = \begin{cases} \mu_i & \text{if } y_i(t) \geq 1 \\ 0 & \text{otherwise} \end{cases} \quad i \in \{1, 2\} \tag{34.15}$$

the resulting system response to an input of $u_0 = 1$ is shown in Figure 34.17. Note that we obtain exactly the same response as in Figure 34.15.

While this hybrid model has an acceptable behavior when we ramp up our manufacturing system, it stops producing at buffer level $y_i = 1$ when we ramp down. This is not what we would like to have. Therefore, in case $u_1 = 0$, machine M_1 should be allowed to produce until $y_1 = 0$.

Under these conditions, we could also think of our model operating in different modes. For the manufacturing system under consideration, we can distinguish the following modes:

mode 1: $0 \leq y_1 \leq 1$, $0 \leq y_2 \leq 1$, $u_0 = 0$, $u_1 \geq 0$, $u_2 = 0$.
mode 2: $0 \leq y_1 \leq 1$, $0 \leq y_2 \leq 1$, $u_0 \geq 0$, $u_1 = 0$, $u_2 \geq 0$.
mode 3: $1 \leq y_1$, $\quad\quad\quad\;\; 0 \leq y_2 \leq 1$, $\quad\quad\quad\;\; u_1 = 0$, $u_2 \geq 0$.
mode 4: $1 \leq y_1$, $\quad\quad\quad\;\; 0 \leq y_2 \leq 1$, $\quad\quad\quad\;\; u_1 \geq 0$, $u_2 = 0$.
mode 5: $0 \leq y_1 \leq 1$, $1 \leq y_2$, $u_0 = 0$, $u_1 \geq 0$.
mode 6: $0 \leq y_1 \leq 1$, $1 \leq y_2$, $u_0 \geq 0$, $u_1 = 0$.
mode 7: $1 \leq y_1$, $\quad\quad\quad\;\; 1 \leq y_2$.

In all of these modes, the system dynamics is described by Eq. (34.7).

The hybrid model just presented is also known as a piecewise affine (PWA) system (Sontag, 1981). Other well-known descriptions are linear complementarity (LC) systems (Heemels et al., 2000; Schaft and Schumacher, 1998) and mixed logical dynamical (MLD) systems (Bemporad and Morari, 1999). In Bemporad et al. (2000b) and Heemels et al. (2001) it was shown that (under certain assumptions like well-posedness) these three descriptions are equivalent. This knowledge is useful, as each modeling class has its own advantages. Stability criteria for PWA systems were proposed in DeCarlo et al. (2000) and Johansson and Rantzer (1998), and control and state-estimation techniques for MLD hybrid models have been presented in Bemporad et al. (2000a, 1999) and Bemporad and Morari (1999). These results can now be applied for controlling the hybrid systems model of our manufacturing system.

34.8 Flow Models

The fluid models presented in the previous section are (still) not satisfactory. While they do not suffer from the problem that lots come out of the system as soon as we start producing, flow times are not truly present in these models. It is not possible to determine the time it takes lots to leave once they have entered the system. Furthermore, according to these models any throughput can be achieved by means of zero inventory, whereas in Section 34.2 we already noticed that the nonlinear (steady-state) relations depicted in Figure 34.2 should be incorporated in any reasonable model of manufacturing systems.

In this section, we present approximation models that do incorporate both throughput and flow time. These dynamic models are inspired by the continuum theory of highway traffic. Therefore, before presenting this dynamic model we first present some results from traffic theory.

34.8.1 Introduction to Traffic Flow Theory: The LWR Model

In the mid-1950s Lighthill and Whitham (1955) and Richards (1956) proposed a first-order fluid approximation of traffic flow dynamics. This model nowadays is known in traffic flow theory as the LWR model.

Traffic behavior for a single one-way road can be described using three variables that vary in time t and space x: flow $u(x, t)$, density $\rho(x, t)$, and speed $v(x, t)$. Flow is the product of speed and density:

$$u(x, t) = \rho(x, t)v(x, t) \quad \forall x, t \tag{34.16}$$

For a highway without entrances or exits, the number of cars between any two locations x_1 and x_2 ($x_1 < x_2$) needs to be conserved at any time t, i.e., the change in the number of cars between x_1 and x_2 is equal to the flow entering via x_1 minus the flow leaving via x_2:

$$\frac{\partial}{\partial t} \int_{x_1}^{x_2} \rho(x, t) dx = u(x_1, t) - u(x_2, t) \tag{34.17a}$$

or in differential form:

$$\frac{\partial \rho}{\partial t}(x, t) + \frac{\partial u}{\partial x}(x, t) = 0 \tag{34.17b}$$

The two relations Eq. (34.16) and Eq. (34.17) are basic relations that any model must satisfy. As we have three variables of interest, a third relation is needed. For this third relation, several choices can be made. The LWR model assumes in addition to the relations Eq. (34.16) and Eq. (34.17) that the relation between flow and density observed under steady-state conditions also holds when flow and density vary with x and/or t; i.e., for a homogeneous highway:

$$u(x, t) = S(\rho(x, t)) \tag{34.18}$$

The model given by Eqs. (34.16)–(34.18) can predict some traffic phenomena rather well. To overcome some of the deficiencies of the LWR model, in the early 1970s higher order theories have been proposed where Eq. (34.18) has been replaced by another partial differential equation, containing diffusion or viscosity terms. Unfortunately, these extended models experience some undesirable properties, as made clear in (Daganzo, 1995). The most annoying of these properties is the fact that in these second-order models cars can travel backward. Second-order models that do not suffer from this deficiency have been presented in Jiang et al. (2002) and Zhang (2002).

34.8.2 A Traffic Flow Model for Manufacturing Flow

In the previous subsection, we introduced the LWR model from traffic flow theory. This model describes the dynamic behavior of cars along the highway at a macroscopic level and contains information both about the number of cars passing a certain point and about the time it takes cars to go from one point to the other. The LWR model can not only be used for describing the flow of cars along the highway, but also for describing the flow of products through a manufacturing line.

Consider, instead of a homogeneous highway, a homogeneous manufacturing line, i.e., a manufacturing line that consists of a lot of identical machines. Let t denote the time and x the position in the manufacturing line. The behavior of lots flowing through the manufacturing line can also be described by three variables that vary with time and position: flow $u(x,t)$ measured in unit lots per unit time, density $\rho(x,t)$ measured in unit lots per unit machine, and speed $v(x,t)$ measured in unit machines per unit time. Now we can relate these three variables by means of Eqs. (34.16)–(34.18), where in Eq. (34.18) the function S describes the relation between flow and density observed under steady-state conditions.

To make this last statement more explicit, consider a manufacturing line consisting of m machines with exponentially distributed process times and an average capacity of μ lots per unit time. Furthermore, consider a Poisson arrival process where lots arrive at the first machine with a rate of λ lots per unit time ($\lambda < \mu$), and assume that buffers have infinite capacity. Then we know from queueing theory (Kleinrock, 1975) that the average number of lots N in each workstation (consisting of a buffer and a machine) in steady-state is given by

$$N = \frac{\frac{\lambda}{\mu}}{1-\frac{\lambda}{\mu}} = \frac{\lambda}{\mu - \lambda} \tag{34.19}$$

In words, in steady-state we have $\rho(x,t)$ is constant and

$$\frac{1}{m}\rho(x,t) = \frac{u(x,t)}{\mu - u(x,t)} \tag{34.20}$$

from which we can conclude that in steady-state

$$u(x,t) = \frac{\mu \rho(x,t)}{m + \rho(x,t)} \tag{34.21}$$

For this example, this is the mentioned function $S(\rho)$.

With this information, we can conclude that the dynamics of this manufacturing line might be described by means of the partial differential equation

$$\frac{\partial \rho}{\partial t} + \mu \frac{\partial}{\partial x}\left(\frac{\rho}{m+\rho}\right) = 0 \tag{34.22a}$$

Together with the relations

$$u = \frac{\mu \rho}{m+\rho} \quad \text{and} \quad v = \frac{u}{\rho} \quad \text{or} \quad v = \frac{\mu}{m+\rho} \tag{34.22b}$$

this completes our model.

Note that contrary to the fluid models presented in the previous section, the dynamic model of Eq. (34.22) is able to incorporate the stochasticity as experienced in manufacturing lines. If the manufacturing line would be in steady-state, the throughput and flow time as predicted by the model of Eq. (34.22) is exactly the same as those predicted by queueing theory. However, contrary to queueing theory, the model of Eq. (34.22) is not a steady-state model, but also incorporates dynamics. Therefore, the model Eq. (34.22) is a dynamic model that incorporates both throughput and flow time. Furthermore, given the experience in the field of fluid dynamics, the model is computationally feasible as well. For more on these flow models, the interested reader is referred to Armbruster et al. (2004, 2005) and Armbruster and Ringhofer (2005).

34.9 Conclusions

In this chapter, we presented some of the models used in the modeling, analysis, and control of manufacturing systems. In Section 34.3 some analytical modeling techniques and methods for analyzing steady-state behavior of manufacturing systems have been introduced. To get a more detailed insight in the dynamics of a manufacturing system discrete-event models, as introduced in Section 34.4 can be used. A disadvantage of discrete-event models is their complexity. To reduce the complexity of discrete-event models, EPTs have been introduced in Section 34.5. This enables the modeling of a manufacturing system as a large queueing network.

Once the dynamics of manufacturing systems can be well described by a relatively simple discrete-event model, the problem of controlling the dynamics of manufacturing systems becomes of interest. In Section 34.6 a control framework has been presented. A crucial role in this framework is played by approximation models of manufacturing systems. In Section 34.7 the most common approximation models, fluid models, have been introduced, together with some extensions of these models. These fluid models mainly focus on throughput and do not contain information on flow times. Finally, in Section 34.8, flow models have been presented that do incorporate both throughput and flow time information.

References

Armbruster, D., P. Degond, and C. Ringhofer (2005). Continuum models for interacting machines. In D. Armbruster, K. Kaneko, and A. Mikhailov (Eds.), *Networks of Interacting Machines: Production Organization in Complex Industrial Systems and Biological Cells*. Singapore: World Scientific Publishing.

Armbruster, D., D. Marthaler, and C. Ringhofer (2004). Kinetic and fluid model hierarchies for supply chains. *SIAM Journal on Multiscale Modeling and Simulation* 2(1), 43–61.

Armbruster, D. and C. Ringhofer (2005). Thermalized kinetic and fluid models for re-entrant supply chains. *SIAM Journal on Multiscale Modeling and Simulation* 3(4), 782–800.

Baker Jr., G. A. (1965).The theory and application of the Pade approximant method. In K. A. Brueckner (Ed.), *Advances in Theoretical Physics*, Volume 1, pp. 1–58. New York: Academic Press.

Bemporad, A., F. Borrelli, and M. Morari (2000a, June). Piecewise linear optimal controllers for hybrid systems. In *Proceedings of the 2000 American Control Conference*, Chicago, IL, pp. 1190–1194.

Bemporad, A., G. Ferrari-Trecate, and M. Morari (2000b, October). Observability and controllability of piecewise affine and hybrid systems. *IEEE Transactions on Automatic Control* 45(10), 1864–1876.

Bemporad, A., D. Mignone, and M. Morari (1999, June). Moving horizon estimation for hybrid systems and fault detection. In *Proceedings of the 1999 American Control Conference*, San Diego, CA, pp. 2471–2475.

Bemporad, A. and M. Morari (1999). Control of systems integrating logic, dynamics, and constraints. *Automatica* 35, 407–427.

Curtain, R. F. and H. Zwart (1995). *An Introduction to Infinite-Dimensional Linear Systems Theory*. Berlin, Germany: Springer.

Daganzo, C. F. (1995). Requiem for second-order fluid approximations of traffic flow. *Transportation Research Part B* 29(4), 277–286.

DeCarlo, R. A., M. Branicky, S. Petterson, and B. Lennartson (2000). Perspectives and results on the stability and stabilizability of hybrid systems. *Proceedings of the IEEE* 88(7), 1069–1082.

Forrester, J. W. (1961). *Industrial Dynamics*. Cambridge, MA: MIT Press.

Heemels, W. P. M. H., J. M. Schumacher, and S. Weiland (2000). Linear complementarity systems. *SIAM Journal on Applied Mathematics* 60(4), 1234–1269.

Heemels, W. P. M. H., B. d. Schutter, and A. Bemporad (2001). Equivalence of hybrid dynamical models. *Automatica* 37(7), 1085–1091.

Hopp, W. J. and M. L. Spearman (2000). *Factory Physics*, second ed. New York: Irwin/McGraw-Hill.

Jacobs, J. H., P. P. v. Bakel, L. F. P. Etman, and J. E. Rooda (2006). Quantifying variability of batching equipment using effective process times. *IEEE Transactions on Semiconductor Manufacturing* 19(2), 269–275.

Jacobs, J. H., L. F. P. Etman, E. J. J. v. Campen, and J. E. Rooda (2003). Characterization of the operational time variability using effective processing times. *IEEE Transactions on Semiconductor Manufacturing* 16(3), 511–520.

Jiang, R., Q. S. Wu, and Z. J. Zhu (2002). A new continuum model for traffic flow and numerical tests. *Transportation Research. Part B, Methodological 36*, 405–419.

Johansson, M. and A. Rantzer (1998). Computation of piece-wise quadratic Lyapunov functions for hybrid systems. *IEEE Transactions on Automatic Control 43*(4), 555–559.

Kingman, J. F. C. (1961). The single server queue in heavy traffic. *Proceedings of the Cambridge Philosophical Society 57*, 902–904.

Kleinrock, L. (1975). *Queueing Systems, Volume I: Theory*. New York: Wiley.

Kuehn, P. J. (1979). Approximate analysis of general queueing networks by decomposition. *IEEE Transactions on Communication 27*, 113–126.

Lighthill, M. J. and J. B. Whitham (1955). On kinematic waves. I: Flow movement in long rivers. II: A theory of traffic flow on long crowded roads. *Proceedings of the Royal Society A 229*, 281–345.

Little, J. D. C. (1961). A proof of the queueing formula $l = \lambda w$. *Operations Research 9*, 383–387.

Richards, P. I. (1956). Shockwaves on the highway. *Operations Research 4*, 42–51.

Sattler, L. (1996, November). Using queueing curve approximations in a fab to determine productivity improvements. In *Proceedings of the 1996 IEEE/SEMI Advanced Semiconductor Manufacturing Conference and Workshop*. Cambridge, MA, pp. 140–145.

Schaft, A. J. v. d. and J. M. Schumacher (1998). Complementarity modelling of hybrid systems. *IEEE Transactions on Automatic Control 43*, 483–490.

Sontag, E. D. (1981). Nonlinear regulation: the piecewise linear approach. *IEEE Transactions on Automatic Control 26*(2), 346–358.

Zhang, H. M. (2002). A non-equilibrium traffic model devoid of gas-like behavior. *Transportation Research. Part B, Methodological 36*, 275–290.

35
Sensor Network Component-Based Simulator

Boleslaw K. Szymanski
Rensselaer Polytechnic Institute

Gilbert Gang Chen
Rensselaer Polytechnic Institute

35.1 The Need for a New Sensor Network Simulator 35-1
  Why a New Simulator • Features of SENSE • Currently Available Components and Simulation Engines
35.2 Component Simulation Toolkit 35-3
  Motivation: From Object to Component • Functor • Inport and Outport Class • Simulation Time and Port Index • Timer
35.3 Wireless Sensor Network Simulation 35-7
  Running the Simulation
35.4 Conclusions 35-15

35.1 The Need for a New Sensor Network Simulator

The emergence of wireless sensor networks brought many open issues to network designers. Traditionally, the three main techniques for analyzing the performance of wired and wireless networks are *analytical methods*, *computer simulation*, and *physical measurement*. However, because of many constraints imposed on sensor networks, such as energy limitation, decentralized collaboration, and fault tolerance, algorithms for sensor networks tend to be quite complex and usually defy analytical methods that have been proved to be fairly effective for traditional networks. Furthermore, few sensor networks have come into existence, for there are still many unsolved research problems, so measurement is virtually impossible. It appears that simulation is the only feasible approach to the quantitative analysis of sensor networks.

35.1.1 Why a New Simulator

A good simulator possesses two essential features. First, it must support reusable models. A model written for one simulation should be amenable to effortless embedding into other simulations that require the same kind of a model. Second, the model should be easy to be built from scratch. Interestingly, we observe that most existing simulators do not possess these two features simultaneously. Most commercial simulators provide a reusable model library, often coming with a friendly graphical user interface, but adding new models to the library is always a painful task. However, most freely available simulators follow a bottom-up approach; writing models from scratch is straightforward, but the reusability is severely limited.

ns2 (ns2, 1990), perhaps the most widely used network simulator, has been extended to include some basic facilities to simulate sensor networks. However, one of the problems of ns2 is its object-oriented design that introduces much unnecessary interdependency between modules. Such interdependency sometimes makes the addition of new protocol models extremely difficult, only mastered by those who have intimate familiarity with the simulator. Being difficult to extend is not a major problem for simulators targeted at traditional networks, for there, the set of popular protocols is relatively small. For example, Ethernet is widely used for wired LAN, IEEE 802.11 for wireless LAN, and TCP for reliable transmission over unreliable media. For sensor networks, however, the situation is quite different. There are no such dominant protocols or algorithms and there will unlikely be any, because a sensor network is often tailored for a particular application with specific features, and it is unlikely that a single algorithm can always be the optimal one under various circumstances.

Many other publicly available network simulators, such as JavaSim (Javasim, 2004), SSFNet (ssfnet, 2000), Glomosim (GloMoSim, 2004), and its descendant Qualnet (Qualnet, 2004), attempted to address problems that were left unsolved by ns2. Among them, JavaSim developers realized the drawback of object-oriented design and tried to attack this problem by building a component-oriented architecture. However, they chose Java as the simulation language, inevitably sacrificing the efficiency of the simulation. Moreover, C++ with Standard Template Library (Musser and Saini, 1996) can easily achieve high efficiency while maintaining a high level of code reuse, which matched our design goal better than Java. SSFNet and Glomosim designers were more concerned about parallel simulation, with the latter more focused on wireless networks. They are not superior to ns2 in terms of design and extensibility.

35.1.2 Features of SENSE

SENSE is designed to be an efficient and powerful sensor network simulator that is also easy to use. We identify three most critical factors as

- Extensibility: The enabling force behind the fully extensibility network simulation architecture is our progress on component-based simulation. We introduced a *component-port model* that frees simulation models from interdependency usually found in an object-oriented architecture, and then proposed a *simulation component classification* that naturally solves the problem of handling simulated time. The component-port model makes simulation models extensible: a new component can replace an old one if they have compatible interfaces, and inheritance is not required. The simulation component classification makes simulation engines extensible: advanced users have the freedom to develop new simulation engines that meet their needs.
- Reusability: The removal of interdependency between models also promotes reusability. A component developed for one simulation can be used in another if it satisfies the latter's requirements on the interface and semantics. There is another level of reusability made possible by the extensive use of C++ template: a component is usually declared as a template class so that it can handle different types of data.
- Scalability: Unlike many parallel network simulators, especially SSFNet and Glomosim, parallelization is provided as an option to the users of SENSE. This reflects our belief that completely automated parallelization of sequential discrete-event models, however tempting it may seem, is impossible, just as automated parallelization of sequential programs. Even if it is possible, it is doomed to be inefficient. Therefore, parallelizable models require larger effort than sequential models do, but a good portion of users are not interested in parallel simulation at all. In SENSE, a parallel simulation engine can only execute components of compatible components. If a user is content with the default sequential simulation engine, then every component in the model repository can be reused.

35.1.3 Currently Available Components and Simulation Engines

- Battery Model: Linear Battery, Discharge Rate Dependent and Relaxation Battery
- Application Layer : Random Neighbor; Constant Bit Rate
- Network Layer: Simple Flooding; a simplified version of ADOV without route repairing, a simplified version of DSR without route repairing
- MAC Layer: NullMAC; IEEE 802.11 with DCF

- Physical Layer: Duplex Transceiver; Wireless Channel
- Simulation Engine: CostSimEng (sequential)

35.2 Component Simulation Toolkit

Component-oriented simulation toolkit (COST) is a general-purpose discrete-event simulator (Chen and Szymanski, 2001). The main design purpose of COST is to maximize the reusability of simulation models without losing efficiency. To achieve this goal, COST adopts a component-based simulation worldview based on a component-port model. A simulation is built by configuring and connecting a number of components, either off the shelf or fully customized. Components interact with each other only via input and output ports, thus the development of a component becomes completely independent of others. The component-port model of COST makes it easy to construct simulation components from scratch. Implemented in C++, COST also features a wide use of templates to facilitate language-level reuse.

COST is a library of several classes that facilitates the development of discrete-event simulation using CompC++, a component-oriented extension to C++. It differs from many other tools in the simulation worldview it adopts. There are primarily two worldviews that are widely used in the discrete-event simulation community: event scheduling and process interaction. Both have their strengths and weaknesses. The event scheduling is much more efficient, but it is hard to program. Process interaction technique requires less programming effort. However, it is difficult to implement using imperative programming languages and many implementations based on special simulation languages are not efficient.

COST adopts a component-oriented worldview, which is a variation of the event scheduling worldview. Using this technique, a discrete-event simulation is viewed as a collection of *components* that interact with each other by exchanging messages through communication *ports*. Besides *components*, the simulation contains a *simulation engine* that is responsible for synchronizing *components*. An event-oriented view is adopted to model individual *components*, i.e., the *component* has one or more event handlers each of which performs corresponding actions upon the arrival of a certain type event. Events are divided into two categories. Synchronous events are the messages arriving at the input *ports*, which are sent by its neighboring *components*. Asynchronous events are associated with *timers*, a special kind of ports lying between the *components* and the *simulation engine*. *Components* receive and schedule asynchronous events through *timers*.

COST takes advantage of component-oriented features that are only available in CompC++.

35.2.1 Motivation: From Object to Component

Convenient and powerful as object-oriented programming is, it has its limits. One of these is that it often imposes unnecessary interobject dependence on the deployment of objects that prevents objects from being reusable. As a small example, in Figure 35.1, an object A calls a method g() of another object B.

Object A must keep a pointer (or a reference) to object B to make such a method call. Let us assume that these two objects have been set up correctly in one program. The difficulty arises when A is to be *reused*

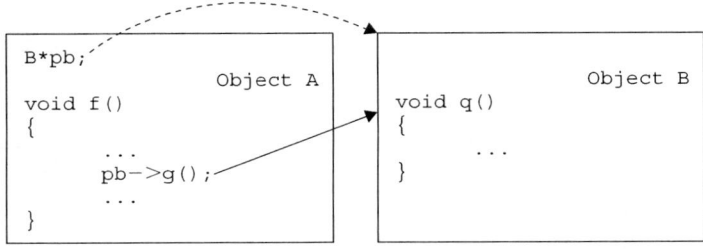

FIGURE 35.1 Object dependencies in object-oriented languages.

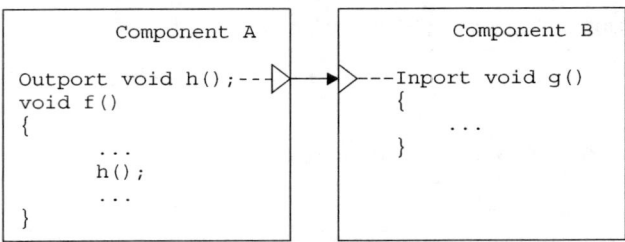

FIGURE 35.2 Component dependencies in component-oriented simulation.

in another program. Obviously, in such a case, either B must also be moved over to the new program, or there is another object that is derived from B available for A to access. If it is the former, things will become worse if some of B's methods are dependent on another object C. As a result, any object that A depends upon, either directly or indirectly, must be available in the new program.

Yet we only need, in the above example, a method that provides the same functionality as B.g() does. We should not be concerned with which object can provide such a functionality, whether it be B or a different, completely unrelated object D. In object-oriented programming, once you make an interobject method call, you not only introduce explicit interobject dependence manifested by the method call (represented by the solid arrow in the Figure 35.1), but also implicit dependence that is hard to trace and maintain (represented by the dashed arrow).

The solution is to introduce *inports* and *outports*. The use of inports and outports has been introduced in DEVS formalism (Zeigler et al., 1997; DEVS, 2004); however, in our approach their role have expanded to become, next to component themselves, an integral part of composing simulations. An inport defines what functionality an object (now it is becoming a component) provides, and an outport defines what functionality it needs. In Figure 35.2, it is apparent that the implicit dependence has been removed. Another benefit of having inports and outports is that any interaction a component may have with other components can be deduced only from its interface (which is composed of declarations of inports and outports). In contrast, for an object, the same information is clear only after scanning through the entire source code.

The central idea of CompC++ is to add inports and outports to objects to make them look and function like components. The extension to the standard C++ language is minimal: only four new keywords (*component, inport, outport,* and *connect*) and four new syntactic rules are needed. The addition of these, nevertheless, opens up a whole new programming paradigm, which is referred to as *component-oriented programming*.

35.2.1.1 Implementation of COST

The first issue of implementing the aforementioned simulation component model is the choice of the implementing language. Discrete-event simulators can be roughly divided into two groups: those based on a special simulation language, such as GPSS and SIMSCRIPT (Law and Kelton, 1982), and those based on a general programming language, such as SIMPACK (Fishwick, 1992) and SIMKIT (Gomes et al., 1995) Simulation languages contain abundant semantics designed for simulation, but requires a steep learning curve. General programming languages are more familiar to programmers, but lack the essential simulation constructs.

We chose C++ as the implementation language for two reasons. First, general programming languages always have good compiler support, and thus their execution speed is generally faster after optimization. Second, language-level reusability is a factor as important as component-level reusability, and C++ is one of the few languages that support code reuse well. With STL (Austern, 1999; Musser and Saini, 1996), C++ programs can easily achieve high efficiency while maintaining a high level of code reuse, which matches our design goal.

Sensor Network Component-Based Simulator

However, with C++ we ran into a problem. As mentioned in the last section, input ports are equivalent to functions, so it is natural to define them as member functions of the component. But how can we represent output ports? C++ language standard requires that the address of an object must be provided when the member function is being called. This conflicts with the requirement that component development should be completely independent. The classical solution for such a problem is a functor, which is the generalization of the function pointer.

35.2.2 Functor

A functor, or a function object, is an object "that can be called in the same way that a function is" (Austern, 1999; Musser and Saini, 1996). A functor class overloads the operator *()* so that it appears as a function pointer. For instance, the following is declaration of a functor class that takes one function argument.

```
template <class T> class functor {
public:
  typedef funct_t bool (*f)(T );
  functor (funct_t _f): f(_f) {}
  virtual bool operator (T t) { return f(t); }
private:
  funct_t f;
};
```

The class *functor* is a helper class that wraps a function pointer of type *funct_t*. Upon invocation, it calls the actual function pointer and returns the result. The syntax of using the functor is exactly the same as that of a function pointer.

The same idea can be applied to member functions as well. In C++, a member function of a class always takes an implicit parameter *this*, which is a pointer to the object upon which the member function will be invoked. As a result, two member functions that belong to different classes but take the same explicit parameters are treated as functions of different types. In the component level, however, they should be viewed as interchangeable. A *mem_functor* declared below can hide the class type as well as the implicit parameter *this*.

```
template <class C, class T>
class mem_functor : public functor {
public:
  typedef funct_t
  bool (C::*f)(T);
  mem_functor (C* _c, funct_t _f)
    : c(_c), f(_f) {}
  virtual bool operator(T t){return c->f(t);}
private:
  C* c;
  funct_t f;
};
```

With these two classes, *functor* and *mem_functor*, it is now straightforward to implement input and output ports. An input port could be simply an instantiation of the *mem_functor* class. Since an output port does not know the component(s) to which it will be connected, it could be represented as a pointer to a *functor*. When connecting an input port to an output port implemented in this way, the address of the *mem_functor* object corresponding to the input port is assigned to the functor pointer corresponding to the output port, because the class *mem_functor* is derived from the *functor* class. When the output port is invoked, the operator *()* of the *mem_functor* class is called, because it is declared as virtual.

35.2.3 Inport and Outport Class

The method of implementing input and output ports directly on top of two functor classes should work well, but there are some practical considerations. For instance, a port should have a name for the purpose of debugging and a port must be set up properly before it can be used to initialize its member variables. Moreover, one to multiple connections would make topology generation more convenient. It is easy to connect an input port to multiple output ports by passing its address to each of them, but when connecting an output port to multiple input ports, the output port must store the addresses of all connected input ports. These reasons are the main motivation for building the *inport* and *outport* classes on top of functor classes.

The *outport* class is declared to be a class with a template parameter that is the type of the events that can be handled by this output port. The function *Setup()* gives the port a string name. The function *Write()* is invoked by the component that outputs a message. *ConnectTo()* connects an input port to the output port.

```
template <class T>
class outport {
public:
 void Setup(typeii* c, const char* name);
 bool Write(T t);
 void ConnectTo(inport& port);
private:
 std::vector<functor<T>*> inports;
};
```

Similarly, the *inport* class takes one template parameter that is the type of the function argument. It must be bound to a member function of a component, therefore the type of the component is passed as the template parameter of the member template function *Setup()*, as shown below.

```
template <class T>
class inport {
public:
 template <class C>
 void Setup( typeii* c,
   mem_functor<C,T>::funct_t _f,
   const char* name);
 bool Write(T t) { return (*f)(t); };
private:
 functor<T>* f;
};
```

Since the type of the member function bound to the input port must be passed to the *Setup()* function, we need to find a way to construct this type from two template parameters, C and T. Fortunately, this type is declared publicly in the class *mem_functor<C,T>* as *funct_t*.

35.2.4 Simulation Time and Port Index

Until now, functors in COST take only one function argument, which is the message exchanged between components. However, two more arguments are necessary. First, all the components in COST are time-dependent components, so messages should be timestamped. Hence, an extra argument is needed to denote the simulation time at which the message is generated. Another extra argument is for arrays of input ports, which are convenient if a number of input ports are of the same type. All elements in an input port array share the member function bound to them. Therefore, it is necessary to have an extra argument to distinguish between them by their indices. The index of an input port that is an element of an array is always zero. The resulting *functor* class could be like (other classes must be modified accordingly):

```
template <class T> class functor {
public:
 typedef funct_t bool (*f)(T,double,int );
 functor (funct_t _f): f(_f) {}
 virtual bool operator (T t, double time) {
  return f(t,time,index); }
private:
 funct_t f;
 int index;
};
```

35.2.5 Timer

The timer class requires two different functor classes, *t_functor* and *mt_functor*, because a time event has empty content, so the binding function of a timer only takes the timestamp argument and the index argument. A *timer* object is actually an array of timers, each of which is identified with a unique integer number, as in the input port arrays. The timer class has two methods: *Set()* to schedule an event and *Cancel()* to cancel an event.

```
class timer {
public:
 void Setup( typeii*,
  mt_functor<C>::funct_t, const char* name);
 void Set(double time, int index=0);
 void Cancel(int index=0);
private:
 t_functor * f;
};
```

So far, we have described techniques that we adopted to implement the component-port model in C++. It should be noted that all these implementation details are transparent to users. Users do not need to have advanced knowledge of C++ templates to write simulations in COST.

35.3 Wireless Sensor Network Simulation

Building a wireless sensor network simulation in SENSE consists of the following steps:

- Designing a sensor node component
- Constructing a sensor network derived from *CostSimEng*
- Configuring the system and running the simulation

Here, we assume that all components needed by a sensor node component are available from the component repository. If this is not the case, the user must develop new components, as described in the COST website (http://www.cs.rpi.edu/~cheng3/cost). We should also mention that the first step of designing a sensor node component is not always necessary if a standard sensor node is to be used.

This first line of this source file demands that *HeapQueue* must be used as the priority queue for event management. For wireless network simulation, because of the inherent drawback of *CalendarQueue*, and also because of the particular channel component being used, *HeapQueue* is often faster.

```
#define queue_t HeapQueue
```

This header file is absolutely required.

```
#include "../../common/sense.h"
```

The following header files are necessary only if the corresponding components are needed by the sensor node component:

```
#include "../../app/cbr.h"
#include "../../mob/immobile.h"
#include "../../net/flooding.h"
#include "../../net/aodvi.h"
#include "../../net/dsri.h"
#include "../../mac/null_mac.h"
#include "../../mac/mac_80211.h"
#include "../../phy/transceiver.h"
#include "../../phy/simple_channel.h"
#include "../../energy/battery.h"
#include "../../energy/power.h"
#include "../../util/fifo_ack.h"
```

#cxxdef is similar to #define, except that it is only recognized by the CompC++ compiler. The following two lines state that the flooding component will be used for the network layer. These two macros can also be overridden by command line macros definitions (whose format is '−D=').

```
#cxxdef net_component Flooding
#cxxdef net_struct Flooding_Struct
```

For layer XXX, XXX_Struct is the accompanying class that defines data structures and types used in that layer. The reason we need a separate class for this purpose is that each XXX is a component, and that due to the particular way in which the CompC++ compiler was implemented, data structures and types defined inside any component is not accessible from outside. Therefore, for each layer XXX, we must define all those data structures and types in XXX_Struct, and then derive component XXX from XXX_Struct.

The following three lines state:

- The type of packets in the application layer is *CBR_Struct::packet_t*.
- The network layer passes application layer packets by reference (which may be faster than by pointer, for *CBR_Struct::packet_t* is small, so *app_packet_t* becomes the template parameter of *net_struct*; the type of packets in the network layer is then *net_packet_t*.
- Now that *net_packet_t* is more than a dozen bytes long, it is better to pass it by pointer, so *net_packet_t** instead of *net_packet_t* becomes the template parameter of the *MAC80211_Struct*; the type of packets in the mac layer is then *mac_packet_t*. Physical layers also use *mac_packet_t*, so there is no need to define more packet types.

```
typedef CBR_Struct::packet_t app_packet_t;
typedef net_struct<app_packet_t>::packet_t net_packet_t;
typedef MAC80211_Struct<net_packet_t*>::packet_t mac_packet_t;
```

Now we can begin to define the sensor node component. First, we instantiate every subcomponent used by the node component. We need to determine the template parameter type for each subcomponent, usually starting from the application layer. Normally the application layer component does not have any template parameter.

Sensor Network Component-Based Simulator

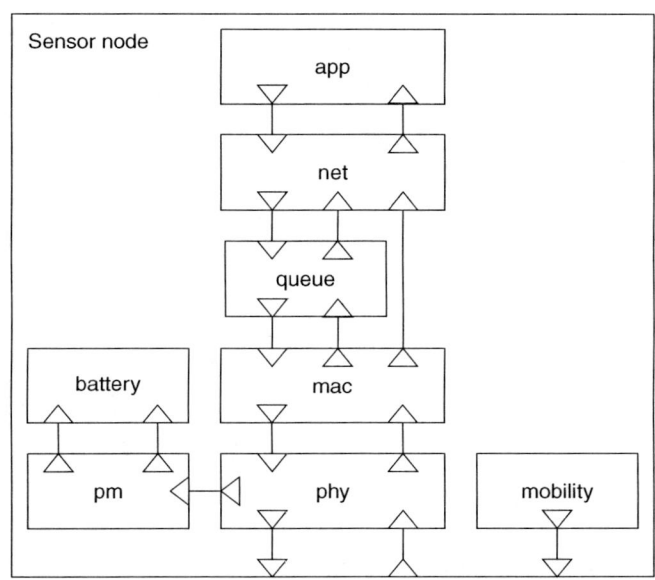

FIGURE 35.3 Sensor node components in SENSE.

Figure 35.3 shows the internal structure of a sensor node.

```
component SensorNode : public TypeII
{
public:

    CBR app;
    net_component <app_packet_t> net;
    MAC80211 <net_packet_t*> mac;
    // A transceiver that can transmit and receive at the same time (of course
    // a collision would occur in such cases)
    DuplexTransceiver < mac_packet_t > phy;
    // Linear battery
    SimpleBattery battery;
    // PowerManagers manage the battery
    PowerManager pm;
    // sensor nodes are immobile
    Immobile mob;
    // the queue used between network and mac
    FIFOACK3<net_packet_t*,ether_addr_t,unsigned int> queue;

    double MaxX, MaxY; // coordinate boundaries
    ether_addr_t MyEtherAddr; // the ethernet address of this node
    int ID; // the identifier

    virtual ~SensorNode();
    void Start();
    void Stop();
    void Setup();
```

The following lines define one inport and two outports to be connected to the channel components:

```
        outport void to_channel_packet(mac_packet_t* packet, double power, int id);
        inport void from_channel (mac_packet_t* packet, double power);
        outport void to_channel_pos(coordinate_t& pos, int id);
};
SensorNode::~SensorNode()
{
}

void SensorNode::Start()
{
}

void SensorNode::Stop()
{
}
```

This function must be called before running the simulation.

```
void SensorNode::Setup()
{
```

At the beginning, the amount of energy in each battery is 1,000,000 J.

```
    battery.InitialEnergy=1e6;
```

Each subcomponent must also know the ethernet address of the sensor node on which it resides. It must be remembered that the application layer is a CBR component, which would stop at FinishTime to give the whole network an opportunity to clean up any packets in transit. Assigning *false* to *app.DumpPackets* means that if COST_DEBUG is defined, *app* still will not print out anything.

```
    app.MyEtherAddr=MyEtherAddr;
    app.FinishTime=StopTime()*0.9;
    app.DumpPackets=false;
```

The coordinates of the sensor node must be set and the *mob* must be given *ID* since *ID* was used to identify the index of the sensor node when the position info is sent to the channel component.

```
    mob.InitX=Random(MaxX);
    mob.InitY=Random(MaxY);
    mob.ID=ID;
```

When a net component is about to retransmit a packet that it received, it cannot do so because all nodes that received the packet may attempt to retransmit the packet immediately, inevitably resulting in a collision. *ForwardDelay* gives the maximum delay time that the packet to be retransmitted may incur. The actual delay is randomly chosen between [0,*ForwardDelay*].

```
    net.MyEtherAddr=MyEtherAddr;
    net.ForwardDelay=0.1;
    net.DumpPackets=true;
```

If *Promiscuity* is true, then the mac component will forward every packet even if it is not destined to this sensor node, to the network layer. To debug the mac layer, set *mac.DumpPackets* to true.

    mac.MyEtherAddr=MyEtherAddr;
    mac.Promiscuity=true;
    mac.DumpPackets=true;

The PowerManager takes care of power consumption at different states. The following lines state that the power consumption is 1.6 W at transmission state, 1.2 at receive state, and 1.115 at idle state.

    pm.TXPower=1.6;
    pm.RXPower=1.2;
    pm.IdlePower=1.15;

phy.TxPower is the transmission power of the antenna. *phy.RXThresh* is the lower bound on the receive power of any packet that can be successfully received and *phy.CSThresh* the lower bound on the receive power of any packet that can be detected. *phy* also needs to know the *ID* because it needs to communicate with the channel component.

    phy.TXPower=0.0280;
    phy.TXGain=1.0;
    phy.RXGain=1.0;
    phy.Frequency=9.14e8;
    phy.RXThresh=3.652e-10;
    phy.CSThresh=1.559e-11;
    phy.ID=ID;

Now, we can establish the connections between components. The connections will become much clearer if we look at Figure 35.4 that represents the entire sensor network.

    connect app.to_transport, net.from_transport;
    connect net.to_transport, app.from_transport;

    connect net.to_mac, queue.in;
    connect queue.out, mac.from_network;
    connect mac.to_network_ack, queue.next;
    connect queue.ack, net.from_mac_ack;
    connect mac.to_network_data, net.from_mac_data ;

These three lines are commented out. They are used when the net component is directly connected to the mac component without going through the queue.

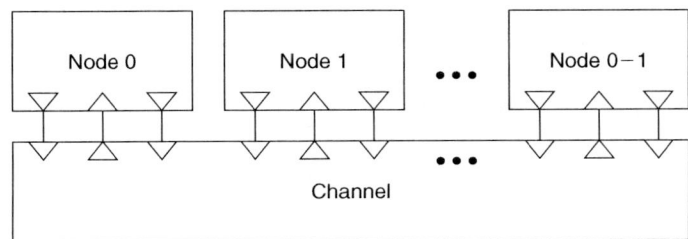

FIGURE 35.4 Sensor network components in SENSE where each node has a structure shown in Figure 35.3.

```
//connect mac.to_network_data, net.from_mac_data ;
//connect mac.to_network_ack, net.from_mac_ack;
//connect net.to_mac, mac.from_network;

connect mac.to_phy, phy.from_mac;
connect phy.to_mac, mac.from_phy;

connect phy.to_power_switch, pm.switch_state;
connect pm.to_battery_query, battery.query_in;
connect pm.to_battery_power, battery.power_in;
```

These three connect statements are different. All the above ones are between an outport of a subcomponent and an outport of another subcomponent, while these three are between a port of the sensor node and a port of a subcomponent. We can view these connections as mapping from the ports of subcomponents to its own ports, i.e., to expose the ports of internal components. Also remember that the connect statement is so designed that it can take only two ports, and then packets always flow through from the first port to the second port, so when connecting two inports, the inport of the subcomponent must be placed in the second place.

```
connect phy.to_channel, to_channel_packet;
connect mob.pos_out, to_channel_pos;
connect from_channel, phy.from_channel;
}
```

Once we have the sensor node component ready, we can start to build the entire simulation, which is named *RoutingSim*. It must be derived from the simulation engine class *CostSimEng*. This is the structure of the network.

```
component RoutingSim : public CostSimEng
{
public:
   void Start();
   void Stop();
```

These are simulation parameters. We do not want configurators of the simulation to access the parameters of those intercomponents.

```
   double MaxX, MaxY;
   int NumNodes;
   int NumSourceNodes;
   int NumConnections;
   int PacketSize;
   double Interval;
```

Here, we declare sense nodes as an array of *SensorNode*, and a channel component.

```
   SensorNode[] nodes;
   SimpleChannel < mac_packet_t > channel;

   void Setup();
};
void RoutingSim :: Start()
{

}
```

After the simulation is stopped, we will collect some statistics.

```
void RoutingSim :: Stop()
{
    int i,sent,recv;
    double delay;
    for(sent=recv=i=0,delay=0.0;i<NumNodes;i++)
    {
                    sent+=nodes[i].app.SentPackets;
                    recv+=nodes[i].app.RecvPackets;
                    delay+=nodes[i].app.TotalDelay;
    }
    printf("APP -- packets sent: %d, received: %d, success rate: %.3f, delay: %.3f\n",
                    sent,recv,(double)recv/sent,delay/recv);
    for(sent=recv=i=0;i<NumNodes;i++)
    {
                    sent+=nodes[i].net.SentPackets;
                    recv+=nodes[i].net.RecvPackets;
    }
    printf("NET -- packets sent: %d, received: %d\n",sent,recv);
    for(sent=recv=i=0;i<NumNodes;i++)
    {
                    sent+=nodes[i].mac.SentPackets;
                    recv+=nodes[i].mac.RecvPackets;
    }
    printf("MAC -- packets sent: %d, received: %d\n",sent,recv);
}
```

The simulation has a *Setup()* function that must be called before the simulation can be run. The reason we do not do this in the constructor is that we must assign values to its parameters after the simulation component has been instantiated. The *Setup()* function, which can be renamed to anything, first maps component parameters to corresponding simulation parameters (for instance, assign the value of the simulation parameter *interval* to the component parameter *source.interval*). It then connects pairs of inport and outports.

```
void RoutingSim :: Setup()
{
    int i,j;
```

The size of the sensor node array must be set using *SetSize()* before the array can ever be used.

```
    nodes.SetSize(NumNodes);
    for(i=0;i<NumNodes;i++)
    {
                    nodes[i].MaxX=MaxX;
                    nodes[i].MaxY=MaxY;
                    nodes[i].MyEtherAddr=i;
                    nodes[i].ID=i;
            nodes[i].Setup(); // don't forget to call this function for each sensor node
    }
```

The channel component needs to know the total number of sensor nodes. It also needs to know the value of *CSThresh* since it will not send packets to nodes that cannot detect them. *RXThresh* is also needed to produce the same receive power in those nodes that can just correctly receive packets when using different propagation models.

In this example *FreeSpace* is used.

```
channel.NumNodes=NumNodes;
channel.DumpPackets=false;
channel.CSThresh=nodes[0].phy.CSThresh;
channel.RXThresh=nodes[0].phy.RXThresh;
channel.PropagationModel=channel.FreeSpace;
```

The channel component also has a *Setup()* function that sets the size of its outport array.

```
channel.Setup();

for(i=0;i<NumNodes;i++)
{
            connect nodes[i].to_channel_packet,channel.from_phy;
            connect nodes[i].to_channel_pos,channel.pos_in;
            connect channel.to_phy[i],nodes[i].from_channel ;
}
```

This is to create communication pairs.

```
int src,dst;
for(i=0;i<NumSourceNodes;i++)
{
            for(j=0;j<NumConnections;j++)
            {
            do
            {
                        src=Random(NumNodes);
                        dst=Random(NumNodes);
            }while(src==dst);
            nodes[src].app.Connections.push_back(
    make_triple(ether_addr_t(dst),Random(PacketSize)+PacketSize/2,
                        Random(Interval)+Interval/2));
            }
            }
}
```

35.3.1 Running the Simulation

To run the simulation we first need to create a simulation object from the simulation component class. Several default simulation parameters must be determined. *StopTime* denotes the ending time of the simulation. *Seed* is the initial seed of the random number generator used by the simulation.

To compile the program, the following commands can be used:

```
../../bin/cxx sim_routing.cc
g++ -Wall -o sim_routing sim_routing.cxx
```

The following command line will start the simulation run:

sim_routing [StopTime] [NumNodes] [MaxX] [NumSourceNodes] [PacketSize] [Interval]

```
int main(int argc, char* argv[])
{
  RoutingSim sim;

  sim.StopTime = 1000;
  sim.Seed = 1234;

  sim.MaxX = 2000;
  sim.MaxY = 2000;

  sim.NumNodes = 110;
  sim.NumConnections = 2;
  sim.PacketSize = 2000;
  sim.Interval = 100.0;

  if(argc >= 2) sim.StopTime = atof(argv[1]);
  if(argc >= 3) sim.NumNodes = atoi(argv[2]);
  sim.NumSourceNodes = sim.NumNodes / 10;
  if(argc >= 4) sim.MaxX = sim.MaxY = atof(argv[3]);
  if(argc >= 5) sim.NumSourceNodes = atoi(argv[4]);
  if(argc >=6) sim.PacketSize = atoi(argv[5]);
  if(argc >= 7) sim.Interval = atof(argv[6]);

  printf("StopTime: %.0f, Number of Nodes: %d, Terrain: %.0f by %.0f\n",
         sim.StopTime, sim.NumNodes, sim.MaxX, sim.MaxY);
  printf("Number of Sources: %d, Packet Size: %d, Interval: %f\n",
         sim.NumSourceNodes, sim.PacketSize, sim.Interval);

  sim.Setup();
  sim.Run();

  return 0;
}
```

35.4 Conclusions

The example given in the previous section has been extended to simulate two innovative protocols for sensor networks: ESCORT (Branch et al., 2005) and SSR (Chen et al., 2005). In the first case, the focus was on identifying groups of sensor nodes that can share communication duties to save energy. The second protocol tested the efficiency of a self-selecting routing in which each hop decides its successor on the fly, using lecture hall algorithm for self-selection (Chen et al., 2006). Both systems used large sensor networks (several thousand nodes in each case) and many traffic scenarios requiring multiple runs for each combination of parameters. In all cases, sensor simulator performed reliably and efficiently. Moreover, introducing additional features in the simulation, or trying different variants of the implemented protocols required either small modifications in existing components or introduction of new components, greatly simplifying maintenance of the code.

More generally, COST has been used for other network simulations such as queuing networks, computer networks, and PCS simulations. These examples can be found at http://www.cs.rpi.edu/~cheng3/cost. COST is targeted at the simulation modelers who have a beginning or intermediate knowledge of the C++ language. Once they understand the basic component-port model and its support classes, it is fairly easy for them to write models with COST, and, more importantly, to take the component-based approach to model the system to be simulated. Once a component repository with a wide range of models is developed, the modeler will be able to construct a simulation just by connecting components obtained from the component.

COST is a discrete-event simulator written in C++ that embodies a component-oriented modeling style. At the heart of COST is a component-port model that is distinguished from many developed component models by the notion of output ports. Our simulation component classification allows us to extend such a component-port model to make it well suited for discrete-event simulation by introducing the implicit timestamp mechanism and timers.

The most distinct feature of COST is the component reusability. Components developed for one simulation can be effortlessly reused in other simulations. With an extensive set of library components, writing simulation in COST could be as simple as dragging a few components from the library and connecting them, as some commercial simulators do. The extra advantage of COST is that building components from scratch is simple.

The only inefficiency of COST simulations comes from the message exchange between components, which may involve several layers of function calls and a few virtual function table lookups. However, this is rather the deficiency of the C++ language, not of the underlying component-port model, because theoretically such overhead can be eliminated during the configuration phase. Had we had a truly component-oriented language, COST would have achieved perfect efficiency.

References

Austern, M. H., 1999. *Generic Programming and the STL*. Addison-Wesley, Reading, MA.

Branch, J., Chen, G., and Szymanski, B. K., 2005. ESCORT: Energy-Efficient Sensor Network Communal Routing Topology Using Signal Quality Metrics. *Proceedings of the International Conference on Networking—ICN 2005*, Vol. 3420, pp. 438–448. LNCS, Springer, Berlin.

Chen, G. G., Branch, J. W., and Szymanski, B. K., 2005. Self-Selective Routing for Wireless Ad Hoc Networks. *Proceedings of the IEEE International Conference on Wireless and Mobile Computing, Networking and Communications WiMob 2005*, Vol. 3, pp. 57–64.

Chen, G. G., Branch, J. W., and Szymanski, B. K., 2006. A Self-Selection Technique for Flooding and Routing in Wireless Ad-Hoc Networks. *Journal of Network and Systems Management*, 14 (3), 359–380, 2006.

Chen, G. and Szymanski, B. K., 2001. Component-Based Simulation. *Proceedings of the. 2001 European Simulation Multi-Conference*, pp. 68–75. SCS Press, Delft, Netherlands.

DEVS, 2004. http://acims.arizona.edu

Fishwick, P. A., 1992. SIMPACK: Getting Started with Simulation Programming in C and C++. *Proceedings of the 1992 Winter Simulation Conference*, eds. J. J. Swain, D. Goldsman, R. C. Crain, and J. R. Wilson, pp. 154–162.

GloMoSim, 2004. http://pcl.cs.ucla.edu/projects/glomosim/

Gomes, F., Franks, S., Unger, B., Xiao, Z., Cleary, J., and Covington, A., 1995. SIMKIT: A High Performance Logical Process Simulation Class Library in C++. *Proceedings of the 1995 Winter Simulation Conference*, eds. C. Alexopoulos, K. Kang, W. R. Lilegdon, and D. Goldsman, pp. 706–713.

Javasim, 2004. http://www.j-sim.org/

Law, A. M. and Kelton, W. D., 1982. *Simulation Modeling and Analysis*. McGraw-Hill, New York.

Musser, D. R. and Saini, A., 1996. *STL Tutorial and Reference Guide*. Addison-Wesley, Reading, MA. ns2, 1990. http://www.isi.edu/nsnam/ns/

Qualnet, 2004. http://www.scalable-networks.com/

ssfnet, 2006, http://www.ssfnet.org/

Zeigler, B. P., Kim, D., and Praehofer, H., 1997. DEVS formalism as a framework for advanced distributed simulation. *Proceedings of the 1st International Workshop on Distributed Interactive Simulation and Real Time Applications*.

V

Case Studies

36
Multidomain Modeling with Modelica

Martin Otter
DLR Institute of Robotics and Mechatronics

Hilding Elmqvist
Dynasim AB

Sven Erik Mattsson
Dynasim AB

36.1 Modelica Overview 36-1
36.2 Modelica Basics 36-3
   Component Coupling • Discontinuous Systems • Relation-Triggered Events • Variable Structure Systems • Other Language Elements
36.3 Modelica Libraries 36-17
36.4 Symbolic Processing of Modelica Models 36-19
   Sorting and Algebraic Loops • Reduction of Size and Complexity • Index Reduction • Example
36.5 Outlook 36-25

36.1 Modelica Overview

Modelica[1] is a freely available language to model the dynamic behavior of technical systems based on schematics. This chapter gives an overview of the language features, free and commercial Modelica libraries, as well as symbolic algorithms needed to transform the high-level description of Modelica into a form that is suited for a numerical integration algorithm.

Modelica is suited for multidomain modeling of large, complex, and heterogeneous technical systems, for example,

- mechatronic models in robotics, automotive, and aerospace applications involving mechanical, electrical, hydraulic, thermal, and control subsystems,
- process-oriented models with multiphase and multisubstance fluids in pipe networks such as air conditioning systems, batch processes or machine cooling, and
- generation, distribution, and consumption of electric power.

Modelica is designed such that it can be utilized in a similar way as an engineer builds a real system: first trying to find standard components like motors, pumps, and valves from manufacturers' catalogues with appropriate specifications and interfaces and only if there does not exist a particular subsystem, a component model would be newly constructed based on standardized interfaces.

A typical example of a Modelica model, the components of a system model of a vehicle, is sketched in Figure 36.1 below by screenshots of Modelica schematics and of three-dimensional animations of some of the vehicle components (component animation can also be specified in a Modelica model based on primitives and on CAD data).

---
[1] Modelica® is a registered trademark of the Modelica Association.

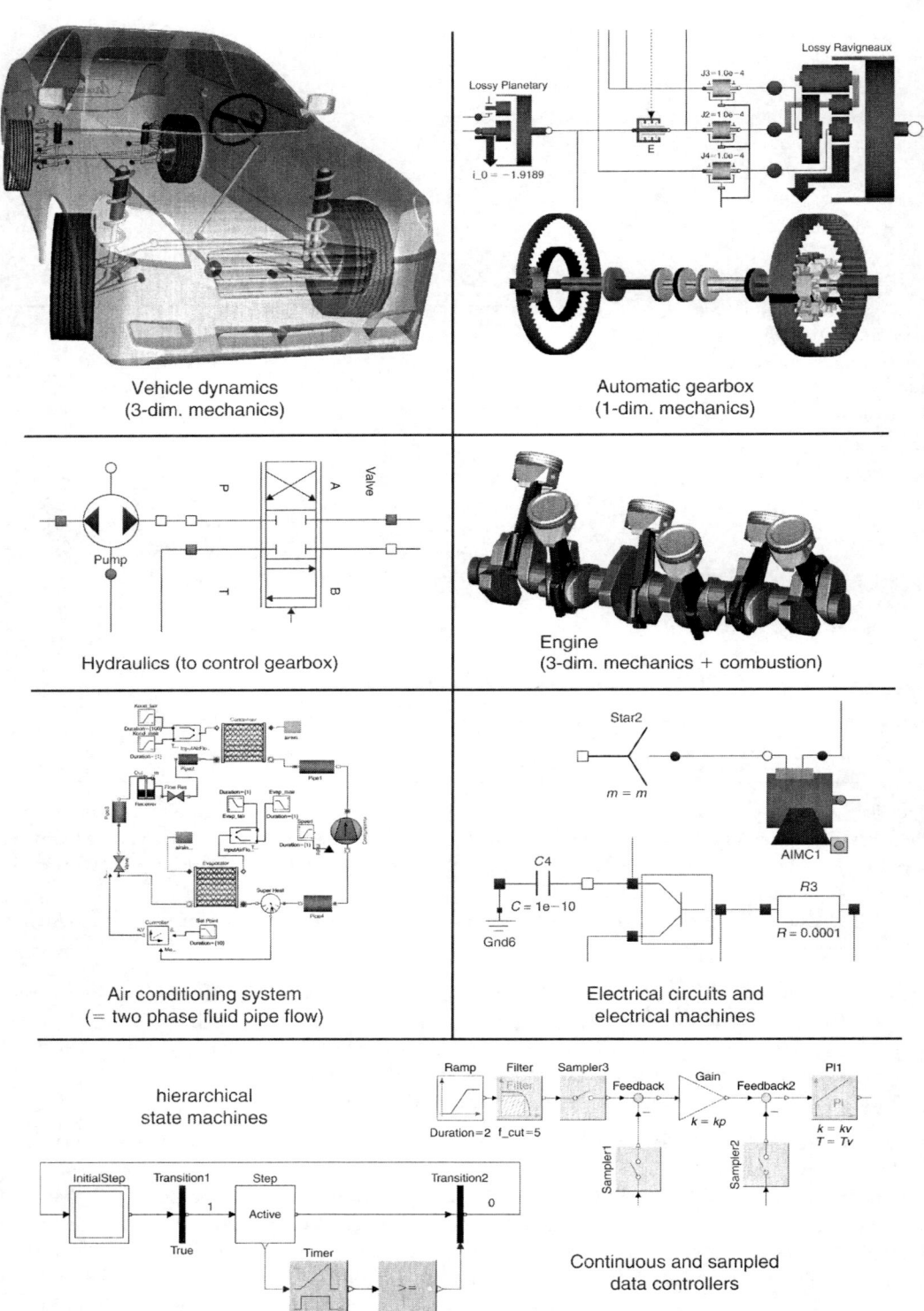

FIGURE 36.1 Schematics and three-dimensional animations of vehicle components modeled in Modelica. (From Hubertus Tummescheit Modelon AB, Sweden.)

Models in Modelica are mathematically described by *differential, algebraic,* and *discrete equations.*[2] Modelica is designed such that available, specialized algorithms can be utilized to enable efficient handling of large system models having more than hundred thousand equations. It is suited and used for hardware-in-the-loop simulations and for embedded control systems. Modelica is *not* designed for the direct description of partial differential equations, as, e.g., performed by finite element or computational fluid dynamics programs. However, results of finite element computations are utilized in Modelica, e.g., for the description of flexible bodies.

The Modelica language is a *textual specification* that describes details of a model on a high level. To be useful, a Modelica modeling and simulation environment is needed to graphically edit and browse a textually defined Modelica schematic, to transform the model in a form that is better suited for reliable integration, to simulate the model, to visualize the simulation results, and to import Modelica models in other simulation environments such as Simulink[3] (see Chapter 37). A growing number of such Modelica environments are available commercially and also in the "public domain." For an actual overview, see http://www.Modelica.org/tools.

Reuse is a key issue for handling complexity. There have been several attempts to define object-oriented languages for modeling of technical systems. However, the ability to reuse and exchange models relies on a standardized format. It was thus important to bring this expertise together to unify concepts and notations: the Modelica design effort was initiated and headed by Hilding Elmqvist and started in September 1996. The language has been designed by the developers of the object-oriented modeling languages Allan, Dymola, NMF, ObjectMath, Omola, SIDOPS+, Smile, by computer scientists, and by modeling specialists from the mechanical, electrical, electronic, hydraulic, pneumatic, fluid, and control domain. The nonprofit Modelica Association was formed in 2000 with Martin Otter as chairman, to manage the continually evolving Modelica language and the development of the free Modelica standard library. In the same year, Modelica was used the first time in actual applications. As of 2006, there are more than 1000 users of Modelica. More details, especially the actual language specification, free Modelica libraries, downloadable publications, links to Modelica modeling, and simulation environments as well as to Modelica consultants can be found at the Modelica homepage http://www.Modelica.org/. Also books about Modelica are available, such as Tiller (2001) and Fritzson (2003).

36.2 Modelica Basics

Modelica supports both high-level modeling by composition of components and low-level modeling by implementing basic components with equations. Models of standard components are typically available in model libraries. Using a graphical model editor from a Modelica environment, a model can be defined by drawing a composition diagram (also called schematics) by positioning icons that represent the models of the components, drawing connections, and giving parameter values in dialogue boxes. Constructs for including graphical annotations in the Modelica language make icons and composition diagrams portable between different tools. An example of a composition diagram of a simple drive train is shown in Figure 36.2 below.

The system can be decomposed into a set of connected components: an electrical motor, a gear, a load inertia, and a controller. Every component is represented by an illustrative icon. At the border of a component icon "connectors" are present. A line drawn between two connectors represents the actual physical coupling, such as electrical wire, rigid mechanical coupling, or fluid pipe flow. The textual representation of this Modelica model is (graphical annotations are not shown; from

---

[2]"Discrete" equations are either equations that are active at distinct points in time only (e.g. at a sampling instant) or equations that are used to compute Integer or Boolean variables.

[3]Simulink® is a registered trademark of The MathWorks Inc.

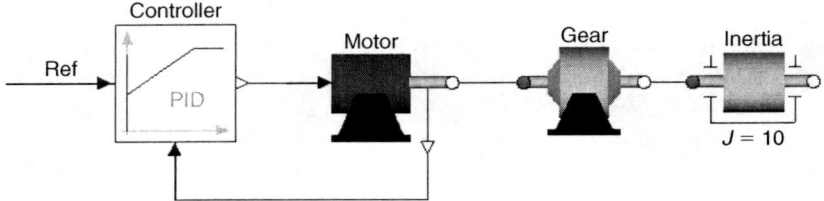

FIGURE 36.2 Modelica schematic of a simple drive train consisting of a controlled motor, a gear, and a load inertia.

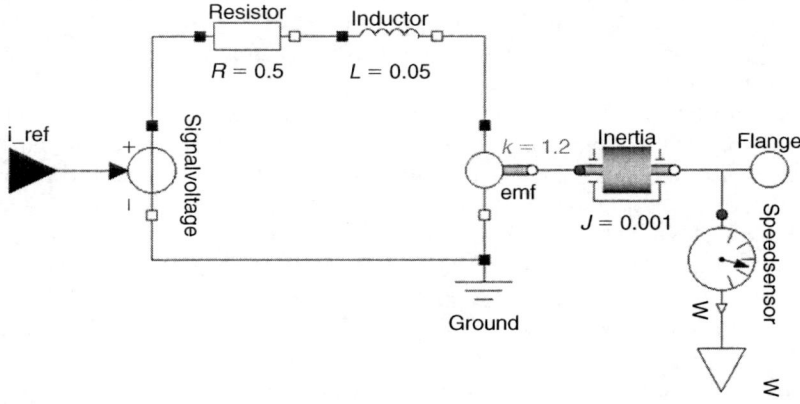

FIGURE 36.3 Modelica schematic of a direct current motor.

these graphical annotations the schematic in Figure 36.2 is constructed by a Modelica schematic editor):

```
model MotorDrive
    PID         controller;
    Motor       motor;
    Gearbox     gear      (ratio=100);
    Inertia     inertia(J=10);
equation
    connect(controller.y , motor.i_ref);
    connect(controller.u2, motor.w);
    connect(gear.flange_a, motor.flange);
    connect(gear.flange_b, inertia.flange_a);
end MotorDrive;
```

This is a composite model which specifies the topology of the system to be modeled in terms of components and connections between the components. The statement "`Gearbox gear(ratio=100);`" declares a component `gear` of model class `Gearbox` and sets the value of the gear ratio, `ratio`, to `100`. A component model may be a composite model to support hierarchical modeling. The composition diagram of the model class `Motor` is shown in Figure 36.3. On the left side, the reference current `i_ref` is an input signal, whereas on the right side the `flange` is a one-dimensional rotational mechanical connector.

The meaning of connections will be discussed below as well as the description of behavior on the lowest level using mathematical equations.

Physical modeling deals with the specification of relations between physical quantities. For the drive system, quantities such as angle and torque are of interest. Their types are declared in Modelica as

```
type Angle = Real(quantity="Angle" , unit="rad",
                 displayUnit="deg");
type Torque = Real(quantity="Torque", unit="N.m");
```

where Real is a predefined type, which has a set of attributes such as name of quantity, unit of measure, default display unit for input and output, minimum, maximum, nominal, and initial value. The Modelica Standard Library, which is a free library developed by the Modelica Association, includes about 450 of such type definitions based on ISO 31-1992 "General principles concerning quantities, units and symbols" and on ISO 1000-1992 "SI units and recommendations for the use of their multiples and of certain other units."

Connections specify interactions between components and are represented graphically as lines between connectors. A connector should contain all quantities needed to describe the interaction. For example, node potential v and current i are needed for electrical components. Angle phi and cut-torque tau are needed for drive train elements:

```
connector Pin                connector Flange
   Voltage      v;              Angle       phi;
   flow Current i;              flow Torque tau;
end Pin;                     end Flange;
```

Two connections,

```
connect(pin1, pin2);
connect(pin1, pin3);
```

with pin1, pin2, and pin3 of connector class Pin, connects the three pins together at one node. This implies three equations,

```
pin1.v = pin2.v
pin1.v = pin3.v
pin1.i + pin2.i + pin3.i = 0
```

The first two equations indicate that the electrical potentials on the connected branches are the same, and the third corresponds to Kirchhoff's current law stating that the current sums to zero at a node. Similar laws apply to mass flow rates in piping networks and to cut-forces and torques in mechanical systems. The sum-to-zero equations are generated when the prefix **flow** is used in the connector declarations. The Modelica Standard Library includes also connector definitions (details see below).

To build reusable descriptions, **partial** models can be defined and reused. For example, a common property of many electrical components is that they have two pins. This means that it is useful to define an interface model class OnePort, that has two pins, p and n, and a quantity, v, that defines the voltage drop across the component.

```
partial model OnePort
   Pin       p, n;
   Voltage v;
equation
   v = p.v - n.v;
   0 = p.i + n.i;
end OnePort;
```

The equations define common relations between quantities of a simple electrical component. The equations in an **equation** section are mathematical equations of the form "expression = expression." The alternative is an **algorithm** section that contains assignment statements of the form "variable := expression." During code generation, equations are sorted and further manipulated, whereas assignment statements in algorithm sections are not changed (details are given in Section 36.4 below).

The keyword **partial** indicates that the model is incomplete and cannot be instantiated. To be useful, a constitutive equation must be added. A model for an inductor extends OnePort by adding a parameter for the inductance and the appropriate equation:

```
model Inductor "Idealized inductor"
   extends OnePort;
   parameter Inductance L "Inductance";
equation
   L*der(p.i) = v;
end Inductor;
```

A string between the name of a class and its body as well at the end of each statement (before the semicolon) is treated as a description text. Tools may display this documentation in special ways, e.g., in parameter menus, or as automatic labels in plots. The keyword parameter specifies that the quantity is constant during a simulation experiment, but can change values between experiments. The built-in operator **der**(..) defines the time derivative of the referenced variable.

36.2.1 Component Coupling

An important and difficult design decision is the definition of component interfaces, called "connectors" in Modelica. One possibility would be to follow the bond graph approach (see, e.g., Karnopp et al., 2000; Cellier, 1991) that models primarily the energy flow between components. The energy flow, i.e., power P, is provided as the product of two variables, such as "$P =$ velocity $\times$ force" and the two variables are used in a connector (one of them has to have the "flow" prefix). This approach is only useful in special cases and is utilized in Modelica for electrical and magnetic connectors and in the free BondLib library (Cellier and Nebot, 2005). For other domains, a more general approach is needed that is based on the following requirements and that includes the bond graph connectors as a special case:

1. A connector should contain an *independent* set of variables that describes the desired physical effects in the interface. Usually, mean values over the interface area are used, e.g., a resultant cut-torque.[4]
2. When connecting components together, the *relevant balance equations* and the *relevant boundary conditions* for the infinitesimal small connection point must be generated by the Modelica connection semantics (variables without the flow prefix are equal and the sum of the variables with the flow prefix is zero).

Typical *balance equations* are energy balance, mass balance, momentum balance, equilibrium of cut-forces, and cut-torques. Typical *boundary conditions* are the equality of position, angle, temperature, or pressure of the connected components. This last rule guarantees the fundamental requirement that balance equations are not only fulfilled within a component, but also when components are connected together.

As an example for balance equations, assume that three rotational mechanical flanges f1, f2, and f3 are connected together. With the definition of connector "Flange" above, the following equations are generated:

```
f1.phi = f2.phi;
f2.phi = f3.phi;
0 = f1.tau + f2.tau + f3.tau;
```

---
[4]Variables c_i in a connector are *not* independent, if at least one variable c_j can be computed by algebraic equations, by differentiation and/or by integration from the other variables c_k in the connector. In Modelica it is possible to have a redundant set of variables in a connector. In such a case, there are unnecessary restrictions how components can be connected together, especially if the connection structure contains a loop.

The first two equations state the *boundary condition* that connected absolute flange angles at the connection point are identical. The last equation states *torque equilibrium* at the connection point. Other balance equations are not taken into account for one-dimensional rotational mechanics. From the above equations, it follows indirectly that also the energy balance is fulfilled, since

```
P = der(f1.phi)*f1.tau + der(f2.phi)*f2.tau + der(f3.phi)*f3.tau
  = der(f1.phi)*(f1.tau + f2.tau + f3.tau)
  = der(f1.phi)*0
  = 0
```

Note, in bond graph methodology, angular velocity and cut-torque are used as variables here. The disadvantage is that a connection with the bond graph connectors does no longer guarantee the equality of connected angles, as needed, e.g., for servo systems where angles are controlled.

A more complicated example is the connector for multiphase, thermo-fluid pipe flow of one substance (Elmqvist et al., 2003):

```
connector FlowPort "Thermo-fluid flow of multi-phase substance"
    AbsolutePressure      p       "Pressure in connection point";
    SpecificEnthalpy      h       "Specific mixing enthalpy";
    flow MassFlowRate     m_flow  "Mass flow in to component";
    flow EnthalpyFlowRate H_flow  "Enthalpy flow in to component";
end FlowPort;
```

When using this connector, it is required that the specific mixing enthalpy *h* in the port is only referenced in the following equation of a component that computes the enthalpy flow rate H_flow from the *upstream* specific enthalpy (= the specific enthalpy in the port, if the fluid flows from the port in to the component, and the specific enthalpy of a point inside the component, if the fluid flows from the component to the port):

```
port.H_flow = port.m_flow*(if port.m_flow > 0 then port.h
                                              else h_in_component);
```

This approach solves the difficult problem to handle reversing, splitting, and joining flow of connected ports automatically under the assumption of ideal mixing. More detailed models require special join/splitter elements where the losses are described. This means that mass and energy balance in a connection point are always fulfilled independently how components are connected together (the momentum balance is fulfilled only under some additional assumptions). As an example, let us assume that ports A.port, B.port, and C.port of the components A, B, and C are connected together. This results in the following equations:

Equations due to "`connect(A.port, B.port); connect(A.port,C.port)`":
```
    A.port.p = B.port.p = C.port.p;
    A.port.h = B.port.h = C.port.h;
    0 = A.port.m_flow + B.port.m_flow + C.port.m_flow;
    0 = A.port.H_flow + B.port.H_flow + C.port.H_flow;
```

Equations inside components A, B, C:

```
    A.port.H_flow = A.port.m_flow*(if A.port.m_flow > 0 then A.port.h
                                                        else A.h;
    B.port.H_flow = B.port.m_flow*(if B.port.m_flow > 0 then B.port.h
                                                        else B.h;
    C.port.H_flow = C.port.m_flow*(if C.port.m_flow > 0 then C.port.h
                                                        else C.h;
```

The first two lines state that pressure p and specific enthalpy h in the connection point are identical. Since all other intensive quantities of a multiphase fluid (e.g., density, or specific entropy) can be computed from p and h, the medium state is uniquely defined. The third and fourth equations state the mass and energy balance for an infinitesimal small control volume in the connection point. Together with the remaining equations for port.H_flow in the components, a linear system of equations is present to compute the mixing enthalpy A.port.h = B.port.h = C.port.h in the connection point. It has the solution:

```
A.port.h = -( (if A.port.m_flow > 0 then 0 else A.port.m_flow*A.h)+
              (if B.port.m_flow > 0 then 0 else B.port.m_flow*B.h)+
              (if C.port.m_flow > 0 then 0 else C.port.m_flow*C.h) )
         /( (if A.port.m_flow > 0 then A.port.m_flow else 0)+
            (if B.port.m_flow > 0 then B.port.m_flow else 0)+
            (if C.port.m_flow > 0 then C.port.m_flow else 0) )
```

Therefore, independently of the flow directions in the three ports, the mixing enthalpy is always uniquely computed, provided at least one mass flow rate does not vanish (see Elmqvist et al., 2003 for details how to handle the case if all mass flow rates vanish). From the mixing enthalpy and the port pressure, all other mixing quantities can be computed, such as mixing temperature.

In the Modelica standard library (see below), elementary connectors for all important technical domains are provided. The definition of these connectors is summarized in the Table 36.1.

Column "type" is the physical domain, column "potential" defines the variables in the connector that have no "flow" prefix, column "flow" defines the variables with a "flow" prefix, column "connector name" defines the connector class name where the definition is present and column "icon" is a screenshot of the connector icon.

For three-dimensional mechanical systems, the position vector r[3] from the origin of the world frame to the origin of the connector frame, the transformation matrix T[3,3] from the world to the connector frame, as well as the cut-force vector f[3] and the cut-torque vector t[3] are utilized. The transformation matrix T[3,3] contains a redundant set of variables. In Otter et al. (2003) it is described how the information about the constraint equations of T is defined in Modelica, and how a tool can automatically remove connection restrictions due to this redundancy based on this additional information. In Modelica, array dimensions are declared with square brackets, e.g., A[3,4] is a two-dimensional array where the first dimension has size 3 and the second size 4. A dimension size specified as ":", e.g., Xi[:], defines an unknown dimension that can be defined when using the array.

Type "thermal" refers to heat transfer. Note, in bond graph methodology temperature and entropy flow rate are used as connector variables, because the product of these two variables is the energy flow through the port. The definition in the Modelica Standard Library with temperature and heat flow rate also fulfills the energy balance in a connection point. It has the additional advantage that lumped elements, such as a thermal conductor, a thermal capacitor or the "fully isolated boundary condition" lead to linear equations in the connector variables, whereas the bond graph approach leads to nonlinear equations.

Type "thermo-fluid pipe flow" refers to one-dimensional thermo-fluid flow in pipe networks with incompressible or compressible media that may have one or multiple (homogenous) phases and/or one or multiple substances. The potential variables in the connector are pressure p, specific mixing enthalpy h, and the vector of independent mixture mass fractions Xi[:]. The flow variables are mass flow rate m_{flow}, enthalpy flow rate H_{flow} and the vector of the independent substance mass flow rates $mXi_{flow}[:]$. When only one substance is present, vectors Xi and mXi_{flow} have dimensions zero and are therefore not present. More detailed information of this connector is available in Elmqvist et al. (2003) and in the Modelica.Fluid.UsersGuide.

Besides physical connectors, also a set of signal connectors are provided, especially for block diagrams and for hierarchical state machines. Type "signal bus" characterizes an *empty* connector that has the additional prefix "**expandable.**" This connector type is used to define a hierarchical collection of named signals where the full connector definition (containing all signal definitions) is defined *implicitly* by the

TABLE 36.1 Connector Definitions in the Modelica Standard Library

Type	Potential	Flow Variables	Connector Name (Modelica Library)	Icon
Physical connectors				
Electric analog	Electric potential	Electric current	Modelica.Electrical.Analog.Interfaces.Pin	
Electric multi-phase	Vector of electrical pins		Modelica.Electrical.MultiPhase.Interfaces.Plug	
Magnetic	Magnetic potential	Magnetic flux	Modelica.Magnetic.Interfaces.MagneticPort	
Translational	Distance	Force	Modelica.Mechanics.Translational.Interfaces.Flange	
Rotational	Angle	Torque	Modelica.Mechanics.Rotational.Interfaces.Flange	
3D mechanics	$r[3]$, $T[3,3]$	$f[3]$, $t[3]$	Modelica.Mechanics.MultiBody.Interfaces.Frame	
Thermal	Temperature	Heat flow rate	Modelica.Thermal.HeatTransfer.Interfaces.HeatPort	
Thermo-fluid pipe flow	p, h, $Xi[:]$	m_{flow}, H_{flow}, $mXi_{flow}[:]$	Modelica.Fluid.Interfaces.FluidPort	
Signal connectors				
Signal	Real Integer Boolean	None	Modelica.Blocks.Interfaces.RealInput/Output Modelica.Blocks.Interfaces.IntegerInput/Output Modelica.Blocks.Interfaces.BooleanInput/Output	
Electric digital	Integer (1 … 9)	None	Modelica.Electrical.Digital.Interfaces.DigitalSignal	
State machine	Booleans	None	Modelica.StateGraph.Interfaces.Step_in/Transition_in	
Signal bus	Configurable	None	Modelica.Blocks.Interfaces.SignalBus	

set of variables occurring in all connections of this type. This connector is used to communicate a large amount of signals in a convenient way between components.

Modelica supports also *hierarchical connectors*, in a similar way as hierarchical models. As a result, it is, e.g., possible, to collect elementary connectors together. For example, an electrical plug consisting of two electrical pins can be defined as:

```
connector Plug
    Pin phase;
    Pin ground;
end Plug;
```

With one **connect**(..) equation, either two plugs can be connected (and therefore implicitly also the phase and ground pins) or a Pin connector can be directly connected to the phase or ground of a Plug connector, such as "**connect**(resistor.p, plug.phase)." A predefined connector of this type is the "electrical multiphase" connector (see table above), that contains a vector of pins, with a default dimension of 3, i.e., it allows a convenient description of multiphase networks.

Besides "one-to-one" connections of components on the same hierarchical level, also "one-to-many" connections over component hierarchies can be conveniently defined with the **inner/outer** language construct to model the coupling of components with *physical fields*. Typical examples are bodies in the gravity field of the earth where the gravity acceleration depends on the body position. A language construct to model conveniently connections where every object is connected to any other object of a general physical fields is currently under development.

36.2.2 Discontinuous Systems

After the presentation of the fundamental structuring mechanisms in Modelica and the means to describe continuous models, attention is now given to discrete modeling features, especially the difficult problem how to synchronize continuous and discrete components: In Modelica, the central property is the usage of *synchronous differential, algebraic*, and *discrete equations*. Synchronous languages (Halbwachs, 1993; Benveniste et al., 2003), such as SattLine (Elmqvist, 1992), Lustre (Halbwachs et al., 1991), or Signal (Gautier et al., 1994) have been used to model discrete controllers to yield safe implementations for real-time systems and to verify important properties of the discrete controller before executing it. The idea of generalizing the data flow principle of synchronous languages in the context of continuous/discrete systems was introduced in Elmqvist et al. (1993) and further improved in Otter et al. (1999) and Elmqvist et al. (2001).

A typical example of a (continuous/discrete) hybrid model is given in Figure 36.4 where a continuous plant

$$\frac{d\mathbf{x}_p}{dt} = f(\mathbf{x}_p, \mathbf{u})$$
$$\mathbf{y} = g(\mathbf{x}_p)$$

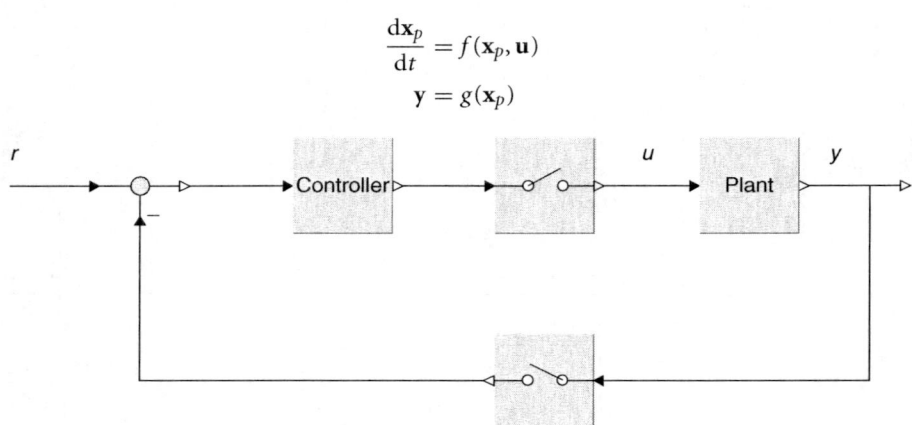

FIGURE 36.4 Continuous plant controlled by a discrete controller.

is controlled by a digital linear controller

$$\mathbf{x}_c(t_i) = \mathbf{A}\mathbf{x}_c(t_i - T_s) + \mathbf{B}(\mathbf{r}(t_i) - \mathbf{y}(t_i))$$
$$\mathbf{u}(t_i) = \mathbf{C}\mathbf{x}_c(t_i - T_s) + \mathbf{D}(\mathbf{r}(t_i) - \mathbf{y}(t_i))$$

using a zero-order hold to hold the control variable **u** between sample instants (i.e., $\mathbf{u}(t) = \mathbf{u}(t_i)$ for $t_i \leq t < t_i + T_s$), where T_s is the sample interval, $\mathbf{x}_p(t)$ is the state vector of the continuous plant, $\mathbf{y}(t)$ is the vector of measurement signals, $\mathbf{x}_c(t_i)$ is the state vector of the digital controller and $\mathbf{r}(t_i)$ is the reference input. In Modelica, the complete system can be easily described by connecting appropriate blocks. However, for simplicity of the discussion, here an overall description of the system is given in one model where the discrete equations of the controller are within the when clause below.

```
model SampledSystem
    parameter Real Ts=0.1 "sample period";
    parameter Real A[:, size(A,1)], B[size(A,1), :],
                   C[:, size(A,2)], D[size(C,1),size(B,2)];
    input   Real r[size(B,2)]   "reference";
    output  Real y[size(B,2)]   "measurement";
            Real u [size(C, 1)] "actuator";
            Real xc[size(A, 1)] "controller state";
            Real xp[:]          "plant state";
equation
    der(xp) = f(xp, u);         // plant
          y = g(xp);
    when sample(0,Ts) then      // digital controller
        xc = A*pre(xc) + B*(r-y);
        u  = C*pre(xc) + D*(r-y);
    end when;
end SampledSystem;
```

During continuous integration the equations within the when clause are deactivated. When the condition of the when clause *becomes* true, an event is triggered, the integration is halted and the equations within the when clause are active at this event instant. The operator sample(. ..) triggers events at sample instants with sample time T_s and returns **true** at these event instants. At other time instants it returns **false**. The values of variables are kept until they are explicitly changed. For example, **u** is computed only at sample instants. Still, **u** is available at all time instants and consists of the value calculated at the last event instant. The equations in a when clause are called "discrete equations" since they are only active at distinct points in time. Equations in a when clause have the restriction that on one side of the "=" sign only a variable reference and no expression is allowed to identify uniquely the variable of the equation that "holds" its value when the equation is not active.

At a sampling instant t_i, the controller needs the values of the controller state $\mathbf{x}_c$ for the time t_i and for the previous sample instant $t_i - T_s$, which is determined by using the **pre** operator. Formally, the left limit $x(t^-)$ of a variable x at a time instant t is characterized by **pre**(x), whereas x itself characterizes the right limit $x(t^+)$. Since x_c is only discontinuous at sample instants, the left limit $x_c(t_i^-)$ at sample instant t_i is identical to the right limit $x_c(t_i^+ - T_s)$ at the previous sample instant and therefore **pre**(xc) characterizes this value.

The *generalized synchronous principle* introduced in Elmqvist et al. (1993) states that at every time instant the *active* equations express *relations* between variables that have to be *fulfilled concurrently* at the current time instant. As a consequence, during continuous integration the equations of the plant have to be fulfilled, whereas at sample instants the equations of the plant and of the digital controller hold *concurrently*. To efficiently solve such types of models, all equations are *sorted* under the assumption that all equations are active (details see below). In other words, the order of the equations is determined by *data flow analysis* resulting in an *automatic synchronization* of continuous and discrete equations. For the example above, sorting results in an ordered set of assignment statements:

```
// "input" variables provided from environment: r, xp, pre(xc)
y := g(xp);
when sample(0,Ts) then
   xc := A*pre(xc) + B*(r-y);
   u  := C*pre(xc) + D*(r-y);
end when;
der(xp) := f(xp, u);
```

Note, that the evaluation order of the equations is correct (in the sense that all active equations are concurrently fulfilled) both when the controller equations are active (at sample instants) and when they are not active.

The (generalized) synchronous principle has several consequences. First, evaluation of discrete equations is performed in zero (simulated) time, i.e., Modelica has instantaneous communication (Benveniste et al., 2003). If needed, it is possible to model the computing time by explicitly delaying the assignment of variables. Second, unknown variables are uniquely computed from a system of equations,[5] i.e., Modelica has deterministic concurrency (Benveniste et al., 2003). This implies that the number of active equations and the number of unknown variables in the active equations at every time instant are identical. This requirement is, for example, violated in

```
equation // incorrect model fragment!
  when h1> 3 then
     close = true;
  end when;

  when h2 < 1 then
     close = false;
  end when;
```

If by accident or on purpose the relations $h1 > 3$ and $h2 < 1$ become **true** at the same time instant, we have two conflicting equations for `close` and it is not defined which equation should be used. In general, it is not possible to detect by source inspection whether conditions become **true** at the same event instant or not. Therefore, in Modelica the assumption is used that *all equations* in a model may potentially be active at the same time instant during simulation. Owing to this assumption, the total number of (continuous and discrete) equations shall be identical to the number of unknown variables. It is often possible to rewrite the model above by placing the when clauses in an **algorithm** section and changing the equations into assignments

```
algorithm                              algorithm
  when h1 > 3 then                       when h2 < 1 then
    close := true;                         close := false;
  end when;          or alternatively    elsewhen h1 > 3 then
  when h2 < 1 then                         close := true;
    close := false;                      end when;
  end when;
```

---

[5]Nonlinear equations may have multiple solutions. However, starting from user defined initial conditions, the desired solution is uniquely identified.

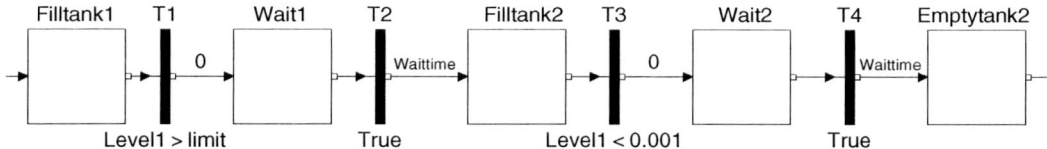

FIGURE 36.5 Part of a Modelica StateGraph to fill and empty a tank system.

The algorithm section groups the two **when** clauses so that they are evaluated sequentially in the order of appearance and the second one get higher priority. All assignment statements within the *same* **algorithm** section are treated as a set of n equations, where n is the number of different left-hand side variables (e.g., the model fragment above corresponds to one equation in the unknown "close"). An **algorithm** section is sorted as one entity together with the rest of the system. Another assignment to close somewhere else in the model would still yield an error.

Handling hybrid systems in this way has the advantage that the *synchronization* between the continuous time and discrete event parts is *automatic* and leads to a *deterministic* behavior. Problems are detected during translation, e.g., *nondeterministic concurrency* (a variable is defined by two or more equations) or *deadlock* (an algebraic loop between different when clauses).

The Modelica language elements for discrete systems are especially used to implement library components on a higher level to define discrete systems in a convenient way graphically, such as

- discrete input/output blocks (Modelica.Blocks.Discrete, Modelica.LinearSystems.Sampled),
- logical blocks (Modelica.Blocks.Logical), and
- hierarchical state machines (Modelica.StateGraph) (Otter et al., 2005).

The Modelica.StateGraph library is based on a subset of JGraphcharts (Årzen et al., 2002) that is an enhanced version of the industrial standard Grafcet/Sequential Function Charts (IEC 848 and IEC 61131-3). JGraphcharts and StateGraphs have a similar modeling power as Statecharts. A simple example is shown in Figure 36.5 containing part of the definition to fill and empty a tank system. For example, if the state machine is in step "fillTank1" and the Boolean condition "level1 > limit" of transition T1 becomes true, then the state machine switches to step "wait1." The basic ideas to implement state machines in Modelica with Boolean equations are described in Mosterman et al. (1998) and Elmqvist et al. (2001).

In Grafchart, Grafcet, Sequential Function Charts, and Statecharts, *actions* are usually formulated *within a step or a state*. Such actions are distinguished as *entry*, *normal*, *exit*, and *abort* actions. For example, a valve might be opened by an entry action of a step and might be closed by an exit action of the same step. In StateGraphs this is *not possible* due to the (generalized) synchronous principle explained above, which does not allow that one variable is defined by two equations. Instead, Boolean equations are added to a StateGraph to set the valve according to the StateGraph state, e.g.,

```
valve2 = fillTank2.active or emptyTanks.active and level1 > limit
```

This feature of a StateGraph is very useful, since it allows a Modelica translator to *guarantee* that a given StateGraph has *deterministic* behavior without conflicts. For example, if the valve is opened and closed in different steps, the Boolean equation for the valve defines uniquely the priority if several steps influencing the valve would be active at the same time instant, e.g., by parallel execution branches.

36.2.3 Relation-Triggered Events

During continuous integration it is advantageous that the model equations remain continuous and differentiable, since the numerical integration methods are based on this assumption. This requirement is often

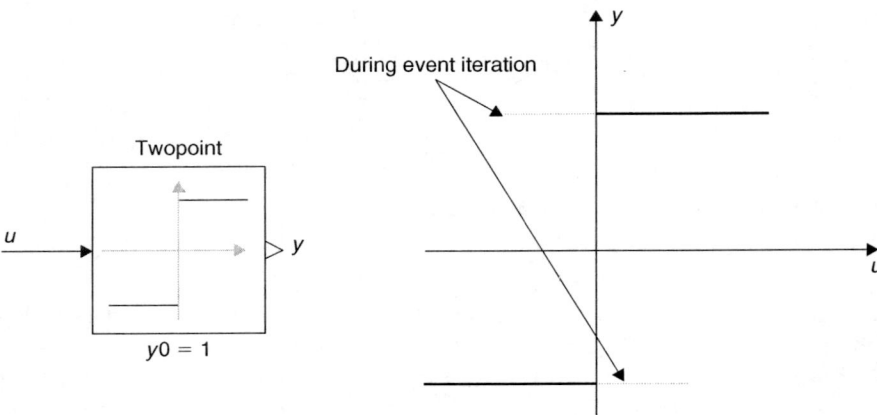

FIGURE 36.6 Two point switch block to demonstrate relation-triggered events.

violated by if-clauses. For example the two-point switch block in Figure 36.6 with input u and output y may be described by

```
model TwoPoint
  parameter Real y0=1;
  input Real u;
  output Real y;
equation
  y = if u > 0 then y0 else -y0;
end TwoPoint;
```

At point u=0 this equation is discontinuous, if the if-expression would be taken *literally*. A discontinuity or a nondifferentiable point can occur if a relation, such as $x_1 > x_2$ changes its value, because the branch of an if-statement may be changed. Such a situation can be handled in a numerical sound way by detecting the switching point within a prescribed accuracy, halting the integration, selecting the corresponding new branch, and restarting the integration, i.e., by triggering a *state event*.

In general, it is not possible to determine by source inspection whether a specific relation will lead to a discontinuity or not. Therefore, in Modelica it is by default assumed that every relation will potentially introduce a discontinuity or a nondifferentiable point in the model. Consequently, relations *automatically* trigger state events (or time events for relations depending only on time) at the time instants where their value is changed. This means, e.g., that model TwoPoint is treated in a numerical sound way: during the iteration to determine the time instant when u crosses zero, the if-expression remains on the branch that it had before the iteration started, i.e., during zero crossing detection the relation $u > 0$ is not taken literally.

Modelica has several operators for hybrid systems, such as **smooth**(…), **noEvent**(…) to switch off event handling and treat relations literally, **reinit(x, value)** to reinitialize a continuous state with a new value at an event instant, **initial**() to inquire the first and **terminal**() to inquire the last evaluation of the model during a simulation run.

36.2.4 Variable Structure Systems

If a physical component is modeled (macroscopically) precisely enough, there are no discontinuities in a system. When neglecting some "fast" dynamics, to reduce simulation time and identification effort,

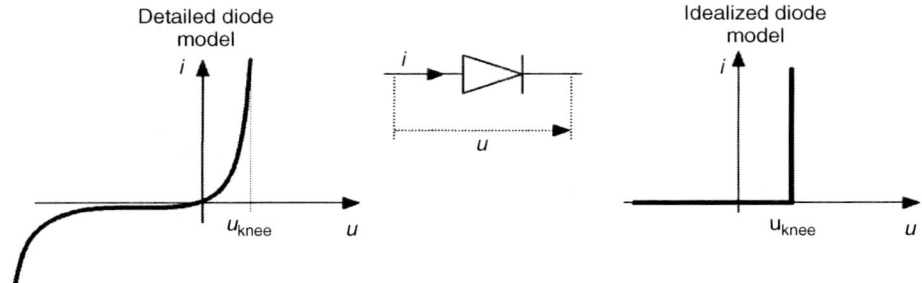

FIGURE 36.7 Detailed and idealized diode characteristic.

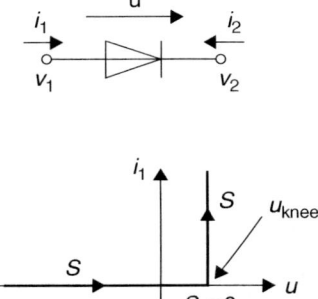

FIGURE 36.8 Declarative description of an ideal diode model.

discontinuities may appear. As a typical example, consider modeling of a diode, where i is the current through the diode and u is the voltage drop between its pins.

The diode characteristic is shown in the left part of Figure 36.7. If the detailed switching behavior can be neglected with regard to other modeling effects and the diode shall be used in its "operating range," it is advantageous to use the ideal diode curve shown in the right part of Figure 36.7. It typically gives a simulation speedup of 1–2 orders of magnitude.

It is straightforward to model the detailed diode curve, because the current i has just to be given as (analytic or tabulated) function of the voltage drop u. It is more difficult to model the ideal diode curve in the right part of Figure 36.8, because the current i at $u = u_{knee}$ is no longer a function of u, i.e., a mathematical description in the form $i = i(u)$ is no longer possible. This problem can be solved by introducing a curve parameter s and describing the curve as $i = i(s)$ and $u = u(s)$. This description form is more general and allows defining an ideal diode uniquely in a declarative way:

```
equation
    0  = i1 + i2;
    u  = v1 - v2;
    on = s > 0;
    u  = uknee + (if on then 0 else s);
    i1 =          if on then s else 0;
```

To understand the consequences of parameterized curve descriptions, the ideal diode is used in the simple rectifier circuit of Figure 36.9.

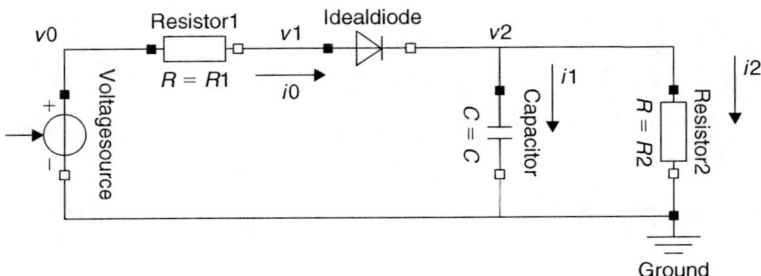

FIGURE 36.9 Simple electrical circuit to demonstrate the handling of an ideal diode model.

Collecting the equations of all components and connections, as well as sorting and simplifying the set of equations under the assumption that the input voltage $v_0(t)$ of the voltage source is a known time function and that the states (here: v_2) are assumed to be known and the derivatives should be computed, leads to

```
     on = s > 0;
      u = v1 - v2;
      u = uKnee + (if on then 0 else s);
     i0 = if on then s else 0;
  R1*i0 = v0 - v1;

  i2:= v2/R2;
  i1:= i0-i2;
  der(v2):= i1/C;
```

The first five equations build a system of equations in the five unknowns on, s, u, v1, and i0. The remaining assignment statements are used to compute the state derivative **der**(v2). During continuous integration the Boolean variables, i.e., on, are fixed and the Boolean equations are not evaluated. In this situation, the first equation is not touched and the next four equations form a *linear* system of equations in the four unknown variables s, u, v1, and i0, which can be solved by Gaussian elimination. An event occurs if one of the relations (here: s > 0) changes its value.

At an *event instant*, the first five equations are a mixed system of discrete and continuous equations that cannot be solved by, say, Gaussian elimination, since there are both Real and *Boolean* unknowns. However, appropriate algorithms can be constructed: (1) Make an *assumption* about the values of the relations in the system of equations. (2) Compute the discrete variables. (3) Compute the continuous variables by Gaussian elimination (discrete variables are fixed). (4) Compute the relations based on the solution of (2) and (3). If the relation values agree with the assumptions in (1), the iteration is finished and the mixed set of equations is solved. Otherwise, new assumptions on the relations are necessary, and the iteration continues. Useful assumptions on relation values are for example: (a) Use the relation values computed in the last iteration and perform a fixed-point iteration (the convergence can be enhanced by some algorithmic improvements). (b) Try all possible combinations of the values of the relations systematically (exhaustive search). In the above example, both approaches can be simply applied, because there are only two possible values (s > 0 is false or true). However, if n switches are coupled, there are n relations and therefore 2^n possible combinations which have to be checked in the worst case.

The technique of parameterized curve descriptions was introduced in (Claus et al., 1995) and a series of related papers. However, no proposal was given on how to implement such models in a numerically sound way. In Modelica, the solution method follows logically because the equation-based system naturally leads to a system of mixed continuous/discrete equations, which have to be solved at event instants (Otter et al., 1999).

Alternative approaches to treat ideal switching elements are (a) by using variable structure equations that are controlled by state machines to describe the switching behavior (see, e.g., Barton, 1992; Elmqvist et al., 1993; Mosterman and Biswas, 1996), or by (b) using a complementarity formulation (see, e.g., Lötstedt, 1982; Pfeiffer and Glocker, 1996; Schumacher and van der Schaft, 1998). The approach (a) has the disadvantage that the continuous part is described in a declarative way but not the part describing the switching behavior. As a result, algorithms with better convergence properties for the determination of consistent switching structure cannot be used. Furthermore, this involves a global iteration over all model equations whereas parameterized curve descriptions lead to local iterations over the equations of the involved elements. The approach (b) seems to be difficult to use in an object-oriented modeling language and seems to be applicable only in special cases (e.g., it does not seem possible to describe ideal thyristors).

Note, mixed systems of equations do not only occur if parameterized curve descriptions are used, but also in other cases, e.g., whenever an if-statement is part of an algebraic loop and the expression of the if-statement is a function of the unknown variables of the algebraic loop.

36.2.5 Other Language Elements

The most basic Modelica language elements have been presented. Modelica additionally supports enumerations as well as arrays and operations on arrays, for example,

```
parameter Real a[:]"Vector of coefficients";
final parameter Integer nx = size(a,1) - 1;
Real sigma[nx];
Real A[nx,nx] "System matrix";
equation
  A = [zeros(nx-1,1), identity(nx-1);
       -a[1:na-1]/a[na]              ];
  sigma = Modelica.Math.Matrices.singularValues(A);
```

The elements of arrays may be of the basic data types (Real, Integer, Boolean, and String) or in general component models, e.g., a resistor. This allows convenient description of simple discretized partial differential equations. Also mathematical functions without side effects can be defined. Functions may have optional input and output arguments. External C- and Fortran subroutines can be called via this function interface in a convenient way. A powerful package concept is available to structure large model libraries and to find a component or/and a library in the file system giving its hierarchical Modelica class name due to a name mapping of Modelica names to path names of the underlying operating system. Root level packages may have a version number and scripts can be defined to perform automatic conversion between versions. Instance and class definitions within a model may be replaceable and can be redeclared to another model class in a higher hierarchy. This allows the substitution of complete submodels from a higher level, e.g., to replace one controller or medium type by another one. Finally, a suite of annotations is standardized that defines the graphical appearance of Modelica components, as well as annotations to define the details of a parameter menu in a simple way. As an example, in Figure 36.10 the parameter menu of a revolute joint is shown as it is automatically constructed by Dymola (Dynasim, 2006) from the Modelica definition.

36.3 Modelica Libraries

In order that the Modelica language is useful for modelers, it is important that libraries of the most commonly used components are available, ready to use, and sharable between applications. For this reason, the Modelica Association develops and maintains a growing *Modelica Standard Library* called *package Modelica*. This is a free library that can be used in commercial Modelica simulation environments.

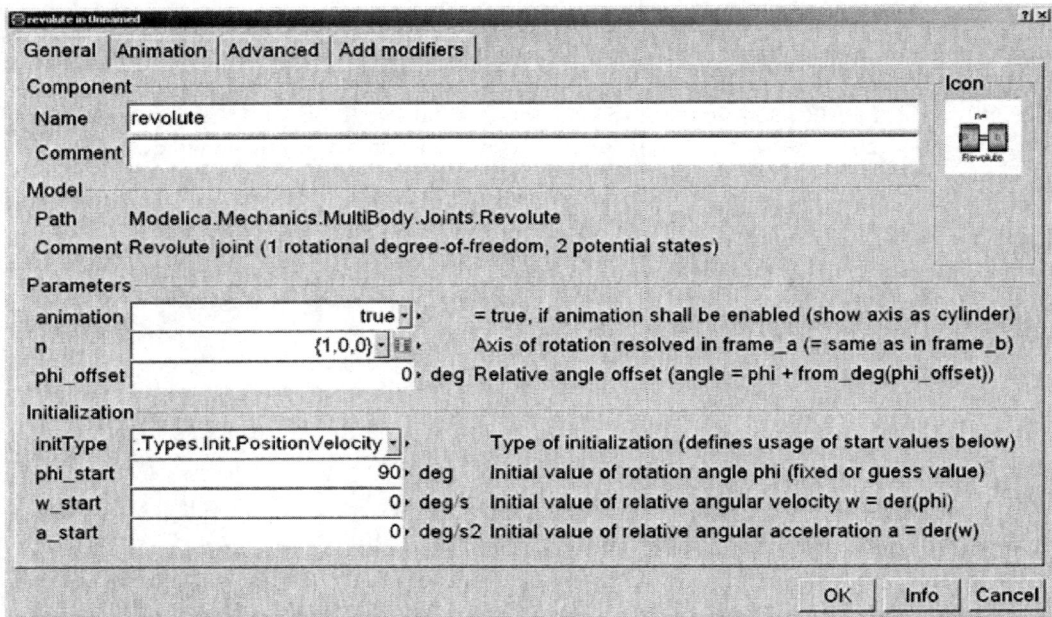

FIGURE 36.10 Automatically constructed parameter dialog.

Furthermore, other people and organizations are developing free and commercial Modelica libraries. For more information and especially for downloading the free libraries, see http://www.modelica.org/library.

Version 2.2.1 of the Modelica standard library from March 2006 contains about 650 model classes and 500 functions in the following sublibraries:

Sublibrary	Content
.Blocks	Continuous, discrete, logical, table input/output blocks
.Electrical.Analog	Analog electrical components (e.g., resistor, diode, and MOS transistors)
.Electrical.Digital	Digital electrical components based on the VHDL standard for 9-valued logic with conversions to 2-, 3-, 4- valued logic
.Electrical.Machines	Uncontrolled asynchronous, synchronous, DC-machines
.Electrical.MultiPhase	Analog electrical components for 2, 3, and more phases
.Math	Mathematical functions (e.g., sin, cos) and functions operating on matrices (e.g., norm, solve, eigenValues, and singularValues)
.Mechanics.MultiBody	Three-dimensional mechanical components (bodies, joints, sensors, forces, etc.)
.Mechanics.Rotational	One-dimensional rotational mechanical components. Includes convenient to use speed- and torque-dependent Coulomb friction elements
.Mechanics.Translational	One-dimensional translational mechanical components
.Media	Large library of fluids (1240 gases and mixtures between the gases, IF97 detailed water medium, simple liquid water medium, dry and moist air, table-based media)
.SIunits	450 type definitions based on ISO 31-1992 (e.g., Angle and Length)
.StateGraph	Hierarchical state machines with Modelica as action language
.Thermal.FluidHeatFlow	Simple components for one-dimensional incompressible thermo-fluid flow (e.g., to model cooling of machines with air or water).
.Thermal.HeatTransfer	One-dimensional heat transfer with lumped elements
.Utilities	Utility functions especially for scripting (e.g., operating on files, streams, strings, and system)

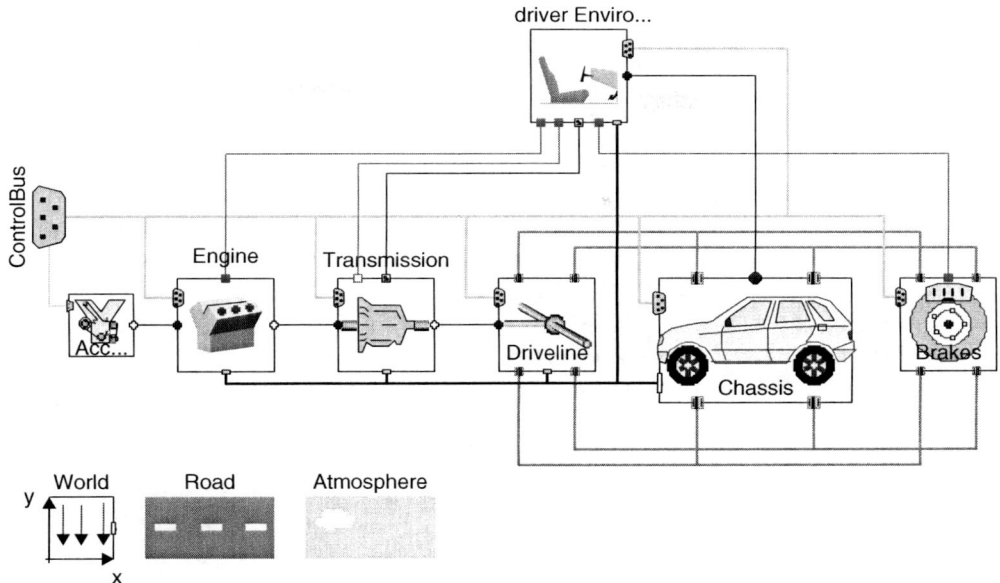

FIGURE 36.11 A VehiclesInterfaces architecture model consisting of accessory, engine, transmission, drive line, chassis, brakes, driver environment, road, and atmosphere components with standardized interfaces.

Component libraries in package Modelica are realized by specialists in the respective area. There is also a growing number of commercial Modelica libraries available, e.g.:

Library name	Content
Air Conditioning	Detailed models of vehicle air conditioning systems (from Modelon)
AlternativeVehicles	Alternative vehicle architectures, such as hybrid cars (from arsenal research and DLR)
FlexibleBodies	Modeling large motions of beams and of flexible bodies exported from finite element programs (DLR)
HyLib	Hydraulic components (from Modelon)
PneuLib	Pneumatic components (from Modelon)
PowerTrain	Vehicle power trains (from DLR)
SmartElectricDrives	Detailed models of controlled machines (from arsenal research)
Transmission	Detailed models of vehicle transmissions (from Ricardo)
VehicleDynamics	Three-dimensional mechanical models of complete vehicles (from Modelon)

The automotive related libraries are based on the free VehicleInterfaces package that provides standard component interfaces and vehicle architectures for complete vehicle system models, in order that automotive libraries can work seamlessly together (Dempsey et al., 2006). This library is based on ideas of Tiller et al. (2003) and of the PowerTrain library (PowerTrain, 2002). A screenshot of one of the VehicleInterfaces architecture models is shown in Figure 36.11.

36.4 Symbolic Processing of Modelica Models

The Modelica Language Specification (Modelica, 2005) defines how a Modelica model shall be mapped into a mathematical description as a mixed system of *differential-algebraic equations* (DAE) and *discrete equations* with Real, Integer, and Boolean variables as unknowns. There are no general-purpose solvers for such problems. There are numerical DAE solvers, which could be used to solve the continuous part.

However, if a DAE solver is used directly to solve the original model equations, the simulation will be very slow and initialization might be not possible for higher index systems (see below). It is therefore assumed that Modelica models are first symbolically transformed into a form that is better suited for numerical solvers. In this section, the transformation techniques are sketched that have been initially designed for the Dymola modeling language (Elmqvist, 1978), further developed in Omsim for the Omola language (Mattsson and Söderlind, 1993) and in the commercial Modelica simulation environment Dymola (Mattsson et al., 2000; Dynasim, 2006):

Dymola converts the differential-algebraic system of equations symbolically to ordinary differential equations in state-space form, i.e., solves for the derivatives. Efficient graph-theoretical algorithms are used to determine which variables to solve for in each equation and to find minimal systems of equations to be solved simultaneously (algebraic loops). The equations are then, if possible, solved symbolically or code for efficient numeric solution is generated. Discontinuous equations are properly handled by translation to state or time events as required by numerical integration routines.

36.4.1 Sorting and Algebraic Loops

The behavior of a Modelica model is defined in terms of genuine equations and a Modelica translator must assign an equation for each variable as part of the sorting procedure, which also identifies algebraic loops. To be able to process problems with hundred thousand unknowns, the idea is to focus on the structural properties, i.e., which variables that appear in each equation rather than how they appear. This information can be represented by a "structure" Jacobian, where for a system of equations, $\mathbf{h}(\mathbf{x}) = \mathbf{0}$, each element i, j, is zero if x_j does not appear in the expression h_i, otherwise it is one. The sorting procedure is to order unknowns and equations to make the structure Jacobian become Block Lower Triangular (BLT). A BLT partitioning reveals the structure of a problem. It decomposes a problem into subproblems, which can be solved in sequence. There are efficient algorithms (see, e.g., Duff et al., 1986), for constructing BLT partitions with diagonal blocks of minimum size (with respect to permutation of equations and variables). Each nonscalar block on the diagonal constitutes an algebraic loop. This sorting procedure identifies all algebraic loops in their minimal form that is unique. The sorting procedure is done in two steps. The first step is to assign each variable, x_j, to a unique equation, $h_i = 0$ such that x_j appears in this equation. It can be viewed as permuting the equations to make all diagonal elements of the structure Jacobian nonzero. If it is *impossible* to pair variables and equations in this way then the problem is structurally singular. The second step of the BLT partition procedure is to find the loops in a directed graph that has the variable/equation pairs of the first step as nodes. The basic algorithm was given by Tarjan (1972).

36.4.2 Reduction of Size and Complexity

A Modelica model has typically many simple equations, $v_1 = v_2$ or $v_1 = -v_2$ being the result of connections. These are easy to exploit for elimination.

From the BLT partition it is rather straightforward to find unknowns that actually are constant and can be calculated and substituted at translation. This may have considerable impact on the complexity of the problem that has to be solved numerically. For example, a multibody component is developed for free motion in a three-dimensional space. When using it we connect it to other components and set parameters implying restrictions on its motion. For example, it may be restricted to move in a plane. It means that coefficients in the equations become zero and terms disappear. This in turn may make algebraic loops to decompose into smaller loops or even disappear.

A linear small algebraic loop is solved symbolically. Otherwise code for efficient numeric solution is generated. To obtain efficient simulation, it is very important to reduce the size of the problem sent to a numerical solver. The work to solve a system of equations increases rapidly with the number of unknowns, because the number of operations is proportional to the cube of n, i.e., $O(n^3)$, where n is the number of unknowns. One approach to reduce size is called tearing (Elmqvist and Otter, 1994). Let $\mathbf{z}$ represent the unknowns to be solved from the system of equations. Let $\mathbf{z}$ be partitioned as $\mathbf{z}_1$ and $\mathbf{z}_2$ such that

$$\mathbf{L} \cdot \mathbf{z}_1 = \mathbf{f}_1(\mathbf{z}_2)$$
$$0 = \mathbf{f}_2(\mathbf{z}_1, \mathbf{z}_2)$$

where $\mathbf{L}$ is the lower triangular with nonzero diagonal elements. A numerical solver needs then only consider $\mathbf{z}_2$ as unknown. A numerical solver provides guesses for $\mathbf{z}_2$ and would like to have the $\mathbf{f}_2$ residuals calculated for these guesses. When having a value for $\mathbf{z}_2$, it is simple to calculate $\mathbf{z}_1$ from the first set of equations. Note, that it is very important to avoid divisions by zero. The assumption that the diagonal elements are nonzero guarantees this. It is then straightforward to calculate the $\mathbf{f}_2$ residuals. The $\mathbf{z}_1$ variables are in fact hidden from the numerical solver. The general idea of tearing is to decompose a problem into two sets, where it is easy to solve for the first set when the solution to the second set is known and to iterate over the second set. The aim is of course to make the number of components of $\mathbf{z}_2$ as small as possible. It is a hard (NP-complete) problem to find the minimum. However, there are fast heuristic approaches to find good partitions of $\mathbf{z}$. If the equations are linear, they can be written as

$$\mathbf{L}\mathbf{z}_1 = \mathbf{A}\mathbf{z}_2 + \mathbf{b}_1$$
$$0 = \mathbf{B}\mathbf{z}_1 + \mathbf{C}\mathbf{z}_2 + \mathbf{b}_2$$

and it is possible to eliminate $\mathbf{z}_1$ to get $\mathbf{J}\mathbf{z}_2 = \mathbf{b}$, where

$$\mathbf{J} = \mathbf{C} + \mathbf{B}\mathbf{L}^{-1}\mathbf{A}$$
$$\mathbf{b} = \mathbf{b}_2 + \mathbf{B}\mathbf{L}^{-1}\mathbf{b}_1$$

This may be interpreted as Gauss elimination of $\mathbf{z}_1$. The procedure may be iterated. Note, since $\mathbf{L}$ is a lower triangular matrix, the determination of $\mathbf{J}$ and $\mathbf{b}$ is at most $O(n^2)$.

When solving a linear equation system, a major effort is to calculate an LU or QR factorization of the Jacobian, $\mathbf{J}$. Back substitutions are much less computationally demanding. In some cases, the elements of the Jacobian does not vary continuously with time. The Jacobian may, for example, only change at events and it is then only necessary to calculate and factorize it during event iterations and not during continuous simulation. In other cases, it may depend only on parameters and constants and then it needs only to be calculated once, at the start of a simulation.

When using Newton methods for nonlinear equation systems, it is necessary to calculate the Jacobian. If this is made numerically from residuals, then n residual calculations are needed. Dymola provides analytic Jacobians. These are more accurate and much less computationally demanding, because there are many common subexpressions to exploit. Modelica provides facilities to provide derivatives also for external functions.

36.4.3 Index Reduction

When solving an ordinary differential equation (ODE) the problem is to integrate, i.e., to calculate the states when the derivatives are given. Solving a DAE may also include differentiation, i.e., to calculate the derivatives of given variables. Such a DAE is said to have high index. It means that the number of states needed for a model is less than the number of variables appearing differentiated. The number of states is equal to the number of independent initial conditions that can be imposed. Higher index DAEs are typically obtained because of constraints between models. To support reuse, model components are developed to be "general." Their behavior is restricted when they are used to build a model and connected to other components. Take as a very simple example two rotating bodies with inertia J_1 and J_2 connected rigidly to each other. The angles and the velocities of the two bodies should be equal. Not all four differentiated variables can be state variables with their own independent start values. The connection equation for the angles, $\varphi_1 = \varphi_2$, must be differentiated twice to get a relation for the accelerations to allow calculation of the reaction torque.

The reliability of a direct numerical solution is related to the number of differentiations needed to transform the system algebraically into ODE form. Modern numerical integration algorithms for DAEs, such as used by most simulators, can handle systems where equations needed to be at most differentiated once. However, reliable direct numerical solutions for nonlinear systems are not known if two or more differentiations are required. Furthermore, if mixed continuous and discrete systems are solved, the hybrid DAE must be initialized at every event instant. In this case, it is, in general, not sufficient to just fulfill the original DAE. Instead, also some differentiated equations have to be fulfilled, in order that the initialization is consistent. Direct numerical methods have problems at events to determine consistent restart conditions of higher index systems.

Higher index DAEs can be avoided by restricting how components may be connected together and/or include manually differentiated equations in the components for the most common connection structures. The drawback is (1) physically meaningful component connections may no longer be allowed in the model or (2) unnecessary "stiff" elements have to be introduced in order that a connection becomes possible. For example, if a stiff spring is introduced between the two rotating bodies discussed above, the problem has no longer a higher index.

Since most Modelica libraries are designed in a truly object-oriented way, i.e., every meaningful physical connection can also be performed with the corresponding Modelica components, this leads often to higher index systems, especially in the mechanical and thermo-fluid field. Also modern controllers based on nonlinear inverse plant models lead to higher index DAEs (Looye et al., 2005) and can be conveniently treated with Dymola.

Dymola transforms higher index problems by differentiating equations analytically. The standard algorithm by Pantelides (1988) is used to determine how many times each equation has to be differentiated. The algorithm by Pantelides is based on the structure of the equations. It means that there are examples where it does not give the optimal result (Reissig et al., 2000). However, the practical experience is very good. Moreover, for large problems a structural analysis is the only feasible approach. Selection of which variables to use as state variables is done statically during translation or in more complicated cases during simulation with the dummy derivative method (Mattsson and Söderlind, 1993; Mattsson et al., 2000). Let us make the example above a bit more realistic and put a gearbox with fixed gear ratio n between the two bodies. Dymola differentiates the position constraint twice to calculate the reaction torque in the coupling, and it is sufficient to select the angle and velocity of either body as state variables. The constraint leads to a linear system of simultaneous equations involving angular accelerations and torques. The symbolic solution contains a determinant of the form "$J_1 + n^2 J_2$". Dymola thus automatically deduces how inertia is transformed through a gearbox.

36.4.4 Example

To illustrate how Dymola's symbolic processing reduces the size and complexity, we will show the structure Jacobian at different stages when translating a mechanical model with a kinematic loop.

Figure 36.12 shows the structure Jacobian of the original model. There are about 1200 unknown variables and equations. Each row corresponds to an equation and each column corresponds to a variable. A blue marker indicates that the variable appears in the equation. There are 3995 markers. The upper half of the matrix has a banded structure. These equations are the equations appearing in the component models and such equations refer typically only to the local variables of the component. The equations in the lower part are equations deduced from the connections, which includes references to variables of two or more components.

Figure 36.13 shows the structure of the problem after exploitation of simple equations to eliminate alias variables and utilizing zero constants. The number of unknowns is reduced from about 1200 to 330.

Then equations are differentiated to reduce the DAE index and states are selected. After some further simplifications the number of unknowns is reduced to 250. A BLT partitioning reveals that there are three algebraic loops as indicated by Figure 36.14.

Figure 36.15 shows the structure after tearing. The first algebraic loop is a nonlinear loop with 12 unknowns. This loop includes the positional constraints of the kinematics loop. The tearing procedure reduces the number of iteration variables to 2. This is illustrated by turning the eliminated part from grey

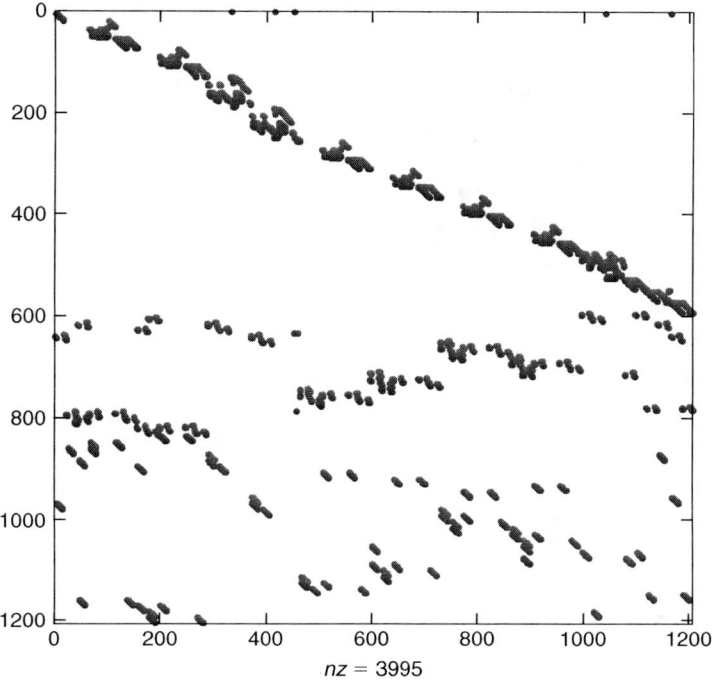

FIGURE 36.12 The structure Jacobian of the original model.

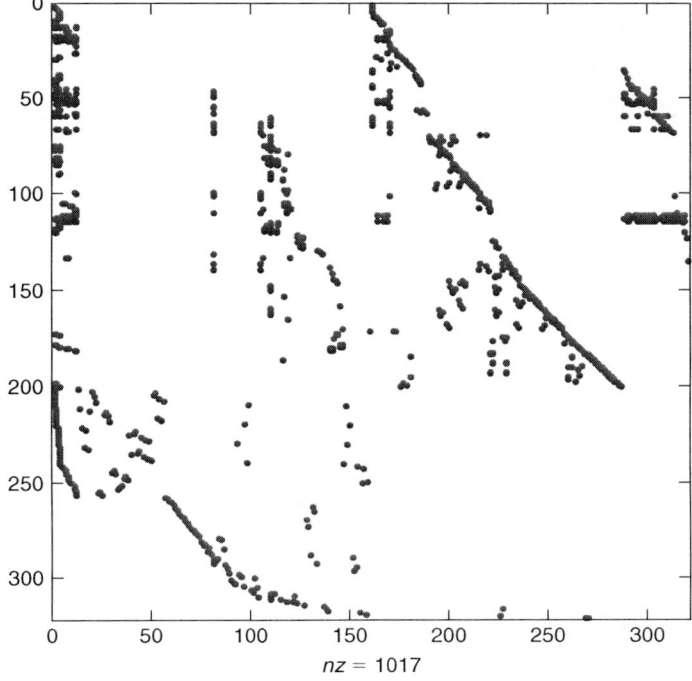

FIGURE 36.13 The structure Jacobian after elimination of alias variables.

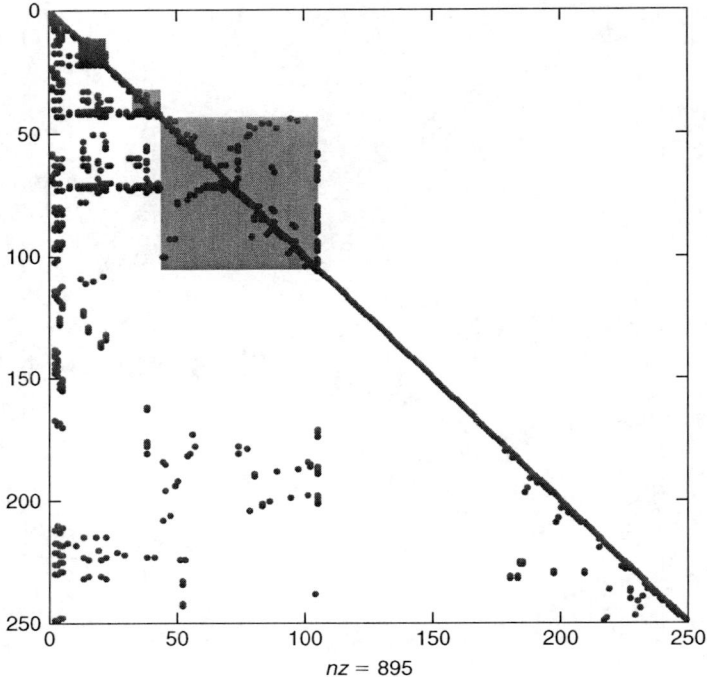

FIGURE 36.14 The BLT partitioning.

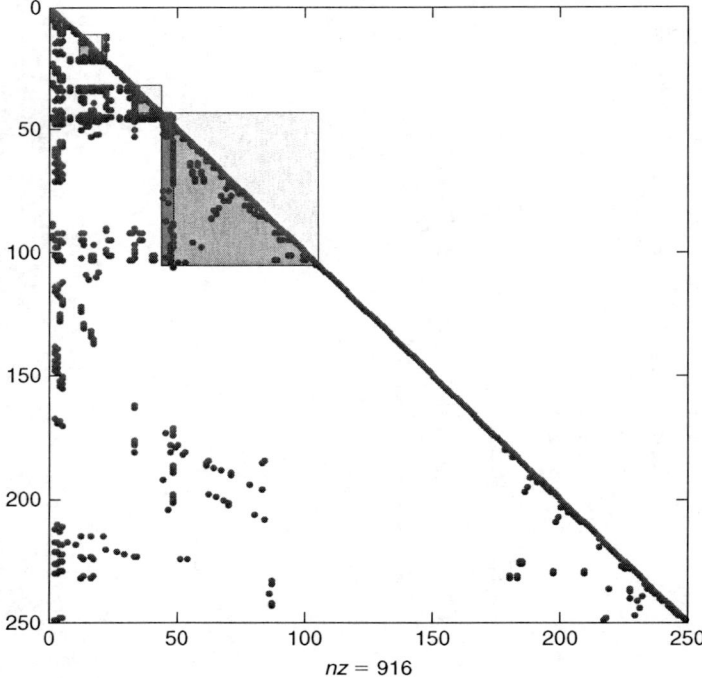

FIGURE 36.15 The structure after tearing.

to green. The second loop includes the velocity constraints due to the kinematic loop. It means that this loop includes the equations of the positional constraints differentiated. This loop has also 11 unknowns, but it is linear. The remaining two by two system can be solved symbolically or numerically. The third loop includes acceleration and force/torque as unknown variables. The loop is linear and has 62 unknowns and the tearing procedure eliminates 57 so a linear 5 by 5 system remains to be solved numerically.

36.5 Outlook

The Modelica Association will continue to further improve the Modelica language and new versions of the Modelica standard library will continue to occur every 6–12 months.

In June 2006, Dassault Systemes, the world leader in Computer Aided Design (CAD) and Product Lifecycle Management (PLM) announced its new product line CATIA Systems (see http://www.3ds.com/news-events/press-room). A central part will be behavioral modeling and simulation with Modelica and Dymola. This will add a new dimension to Modelica because system simulation with close integration of CAD and product databases will become possible. This means that the time consuming and error prone process to *parameterize* system models will be considerably improved. We expect that Dassault Systems' strategic decision will further accelerate the growth of the Modelica user community.

Acknowledgments

Up to 2006, about 50 people, mostly from the Modelica Association, have contributed to the Modelica language and the Modelica Standard Library. The contributors are listed at the end of the Modelica language specification (Modelica, 2005) and in the Modelica.UsersGuide of the Modelica standard library (http://www.modelica.org/library/Modelica/Modelica_2_2_1.zip). This chapter is based on their contributions.

References

Årzen, K.-E., R. Olsson, and J. Akesson 2002. Grafchart for Procedural Operator Support Tasks. *Proceedings of the 15th IFAC World Congress*, Barcelona, Spain.

Barton, P. 1992. *The Modelling and Simulation of Combined Discrete/Continuous Processes*. PhD thesis, University of London, Imperial College.

Benveniste, A., P. Caspi, S. Edwards, N. Halbwachs, P. le Guernic, and R. de Simone. 2003. The Synchronous Languages, 12 Years Later. *Proceedings of the IEEE*, 91 (1), 64–83.

Cellier, F. 1991. *Continuous System Modeling*. Springer, New York.

Cellier, F., and A. Nebot 2005. The Modelica Bond Graph Library. *Proceedings of the 4th International Modelica Conference*, Hamburg, Germany, ed. G. Schmitz, pp. 57–65, http://www.modelica.org/events/Conference2005/online_proceedings/Session1/Session1b1.pdf, http://www.modelica.org/library/BondLib.

Claus, C., J. Haase, G. Kurth, and P. Schwarz. 1995. Extended amittance description of nonlinear n-poles. *Archiv für Elektronik und Übertragungstechnik/International Journal of Electronics and Communications*, 40, 91–97.

Dempsey, M., H. Elmqvist, M. Gäfvert, P. Harman, C. Kral, M. Otter, and P. Treffinger. 2006. Coordinated automotive libraries for vehicle system modelling. *Proceedings of the 5th International Modelica Conference*, Sept. 4–5, Vienna.

Duff, I.S, A.M. Erisman, and J.K. Reid. 1986. *Direct Methods for Sparse Matrices*. Clarendon Press, Oxford.

Dynasim. 2006. Dymola Version 6.0. Dynasim AB, Lund, Sweden. Homepage: http://www.dynasim.se/.

Elmqvist, H. 1978. *A Structured Model Language for Large Continuous Systems.* Dissertation, Report CODEN:LUTFD2/(TFRT–1015), Department of Automatic Control, Lund Institute of Technology, Lund, Sweden.

Elmqvist, H. 1992. An object and data-flow based visual language for process control. *ISA/92-Canada Conference & Exhibit*, Toronto, Canada, Instrument Society of America.

Elmqvist, H., F. Cellier, and M. Otter. 1993. Object–oriented modeling of hybrid systems. *Proceedings ESS'93, European Simulation Symposium*, Delft, The Netherlands, pp. xxxi–xli.

Elmqvist, H. and M. Otter. 1994. Methods for tearing systems of equations in object-oriented modeling. *Proceedings ESM'94, European Simulation Multiconference*, Barcelona, Spain, June 1–3, pp.326–332.

Elmqvist, H., S. E. Mattsson, and M. Otter. 2001. Object-oriented and hybrid modeling in modelica. *Journal Europeen des Systemes Automatises*, 35, 1 a X.

Elmqvist, H., H. Tummescheit, and M. Otter. 2003. *Object-oriented modeling of thermo-fluid systems*. Proceedings of the 3rd International Modelica Conference, Linköping, Sweden, ed. P. Fritzson, pp. 269–286. http://www.modelica.org/Conference2003/papers/h40_Elmqvist_fluid.pdf.

Fritzson, P. 2003. *Principles of Object-Oriented Modeling and Simulation with Modelica 2.1.* IEE Press, Wiley-Interscience, Wiley, New York.

Gautier, T., P. le Guernic, and O. Maffeis. 1994. *For a New Real-Time Methodology*. Publication Interne No. 870, Institut de Recherche en Informatique et Systemes Aleatoires, Campus de Beaulieu, 35042 Rennes Cedex, France. ftp://ftp.inria.fr/INRIA/publication/publi-pdf/RR/RR-2364.pdf.

Halbwachs, N. 1993. *Synchronous Programming of Reactive Systems.* Springer. http://www.esterel-technologies.com/files/synchronous-programming-of-reactive-systems-tutorial-and-references.pdf.

Halbwachs, N., P. Caspi, P. Raymond, and D. Pilaud. 1991. The synchronous data flow programming language LUSTRE. *Proceedings of IEEE*, 79, 1305–1321, http://www.esterel-technologies.com/files/LUSTRE-synchronous-programming-language.pdf.

Karnopp, D. C., D. L. Margolis, and R. C. Rosenberg. 2000. *System Dynamics: Modeling and Simulation of Mechatronic Systems.* 3rd edition. McGraw-Hill, New York.

Looye, G., M. Thümmel, M. Kurze, M. Otter, and J. Bals. 2005. Nonlinear inverse models for control. *Proceedings of the 4th International Modelica Conference*, Hamburg, ed. G. Schmitz, http://www.modelica.org/events/Conference2005/online_proceedings/Session3/Session3c3.pdf.

Lötstedt, P. 1982. Mechanical systems of rigid bodies subject to unilateral constraints. *SIAM Journal of Applied Mathematics* 42, 281–296.

Mattsson, S. E. and G. Söderlind. 1993. Index reduction in differential-algebraic equations using dummy derivatives. *SIAM Journal of Scientific and Statistical Computing*, 14, 677–692.

Mattsson, S. E., H. Olsson, and H. Elmqvist. 2000. Dynamic selection of states in Dymola. *Proceedings of the Modelica Workshop 2000*, pp. 61–67. http://www.modelica.org/Workshop2000/papers/Mattsson.pdf.

Modelica. 2005. *Modelica® —A Unified Object-Oriented Language for Physical Systems Modeling— Language Specification, Version 2.2.* http://www.Modelica.org/Documents/ModelicaSpec22.pdf.

Mosterman, P., and G. Biswas. 1996. A formal hybrid modeling scheme for handling discontinuities in physical system models. *Proceedings of AAAI-96*, Portland, OR, USA, pp. 985–990.

Mosterman, P., M. Otter, and H. Elmqvist. 1998. Modeling petri nets as local constraint equations for hybrid systems using Modelica. *Proceedings of SCSC'98*, Reno, Nevada, USA, Society for Computer Simulation International, pp. 314–319.

Otter, M., H. Elmqvist, and S. E. Mattsson. 1999. Hybrid modeling in Modelica based on the synchronous data flow principle. *IEEE International Symposium on Computer Aided Control System Design (CACSD'99)*, Hawaii, USA.

Otter, M., H. Elmqvist, and S. E. Mattssson. 2003. The new multibody library. *Proceedings of the 3rd International Modelica Conference*, Linköping, Sweden, ed. P. Fritzson. http://www.modelica.org/Conference2003/papers/h37_Otter_multibody.pdf.

Otter, M., K.-E. Årzén, and I. Dressler. 2005. StateGraph—a modelica library for hierarchical state machines. *Proceedings of the 4th International Modelica Conference*, Hamburg, ed. G. Schmitz, pp. 569–578. http://www.modelica.org/events/Conference2005/online_proceedings/Session7/Session7b2.pdf.

Pantelides, C. 1988. The consistent initialization of differential-algebraic systems. *SIAM Journal of Scientific and Statistical Computing*, 9 (2), 213–231.

Pfeiffer, F. and C. Glocker. 1996. *Multibody Dynamics with Unilateral Contacts*. Wiley, New York.

PowerTrain. 2002. *Modelica Library PowerTrain, version 1.0*. DLR Institute of Robotics and Mechatronics.

Reissig, G., W. S. Martinsson, and P. I. Barton. 2000. Differential-algebraic equations of index 1 may have an arbitrarily high structural index. *SIAM Journal of Scientific Computing*, 21 (6), 1987–1990.

Schumacher, J.M. and A.J. van der Schaft. 1998. Complementarity modeling of hybrid systems. *IEEE Transactions on Automatic Control*, 43, 483–490.

Tarjan, R.E. 1972. Depth first search and linear graph algorithms, *SIAM Journal on Computing*, 1 (2), 146–160.

Tiller, M. 2001. *Introduction to Physical Modeling with Modelica*. Kluwer Academic Publishers, Dordrecht.

Tiller, M., P. Bowles, and M. Dempsey. 2003. Developing a vehicle modeling architecture in Modelica. *Proceedings of the 3rd International Modelica Conference*, Linköping, Sweden, ed. P. Fritzson. http://www.modelica.org/events/Conference2003/papers/h32_vehicle_Tiller.pdf.

37
On Simulation of Simulink® Models for Model-Based Design

Rohit Shenoy
The MathWorks Inc.

Brian McKay
The MathWorks Inc.

Pieter J. Mosterman
The MathWorks Inc.

37.1 Introduction 37-1
37.2 The Case Study Example 37-3
   Feedback Control
37.3 Designing with Simulation 37-4
37.4 Obtaining Computational Models 37-4
   Modeling from First Principles • Using Data and
   Simulation to Obtain or Tune Models
37.5 The Robotic Arm Model 37-7
   The Mechanical Model of Arm • Electromechanical
   Model of the Motor and Coupling • Tuning the Motor
   Parameters with Data • Generating the Motor Models
   from Data
37.6 Using Computational Models for Control Design 37-12
   Designing Controllers through Modeling and
   Simulation • Linearizing Models for Control
   System Design • Designing a Controller • Tuning
   Controller Designs Using Optimization Techniques
37.7 Testing with Model-Based Design 37-16
   Requirements-Based Testing through Simulation •
   Simulation with Hardware and Implemented Designs •
   Other Uses of Rapid Prototyping in the Design Process
37.8 Conclusions 37-19

37.1 Introduction

To remain competitive and reduce cost, industry increasingly relies on computational models (Mosterman, 2004). Designing computational models and using numerical simulation is an alternative to building hardware prototypes for testing purposes. Computational modeling (Aberg and Gage, 2004; Breunese et al., 1995; Cellier et al., 1996; Culley and Wallace, 1994) has a number of advantages over traditional engineering methods, such as:

- Computational models tend to be less expensive to produce and easier to modify.
- The ease of modification enables answering "what-if" questions by facilitating rapid exploration of design options, a task that is time consuming and expensive with physical prototypes. Additionally, some experiments require multiple simulations with different parameters which can be performed through Monte Carlo simulations. This is impossible with physical prototypes because these are prohibitively expensive.

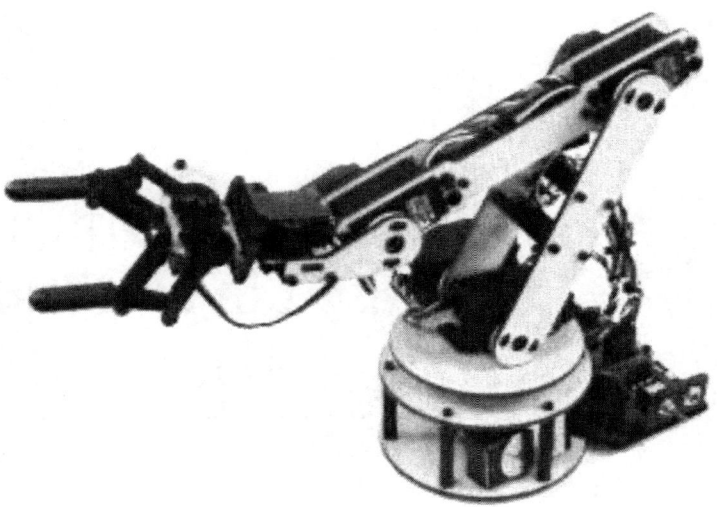

FIGURE 37.1 A robotic arm.

- In the case of complex systems such as aircraft, power grids, and communication networks, simulation enables unstable and emergency modes to be tested more safely, without risk to infrastructure or human life.
- Simulation of such unstable and emergency modes is also an order of magnitude less expensive. For example, testing an aircraft to failure would cost much more than simulating such an experiment, which, if the failure would lead to a crash and loss of life, would not be possible otherwise.
- Modern computational packages allow engineers to employ designs that have been used in simulation and use them in the eventual product. This is a major time-saving feature.

Building dynamic models in software is part of Model-Based Design (Barnard, 2004), which has all the merits listed above and is being adopted by engineers in many fields, in particular, the aerospace and automotive industry. It can be applied to applications such as feedback control, communications, signal and image processing, and financial modeling. Model-Based Design with Simulink® (2006) enables design, analysis, and simulation of processes in a safe software environment. Consider the example of a robotic arm designed for use in a production plant, shown in Figure 37.1. Creating a physical prototype would be expensive and time consuming. Testing it could be dangerous if the robot malfunctioned. The alternative is to model the mechanical, electrical, and electronic components of the robot in software, and then use the model to test the algorithms through simulation.

Model-Based Design enables continuous testing as algorithms and computational models are created and refined. The integration of testing into the design process can identify problems sooner than approaches that do not use modeling and simulation. Furthermore, tests can be designed earlier in the overall design process, which is particularly advantageous if carried out in parallel with other design tasks. Finally, Model-Based Design with Simulink provides the capability to generate computer code such as C or C++ from algorithms and computational system models, and use that code for implementation and testing. In particular, hardware-in-the-loop simulation which uses a real-time computational model of a physical system will be discussed as well as the systematic testing process of a designed system (Mosterman, 2004). These aspects help bridge the gap between the software and hardware design.

This chapter outlines workflow and software tools for computational modeling and numerical simulation of dynamic systems with MATLAB® (2006) and Simulink. It will not concentrate on the detailed methodological and technological aspects of numerical simulation of systems with combined continuous-time and discrete-event behavior, so-called *hybrid dynamic systems* (Mosterman and Biswas,

2002). Instead, one of the key goals of the chapter is to explain the selection criteria for modeling techniques based on simulation, design, and testing needs. Section 37.2 examines a robotic arm as a case study and several examples with the goal of understanding different modeling techniques. Section 37.3 discusses how simulation can be used in control system design. Section 37.4 presents the modeling, spectrum spanning first-principles modeling from equations, physics-based modeling, and data-driven modeling. Section 37.5 applies the different modeling approaches to the robotic arm of the case study. Section 37.6 discusses further usages of simulation in control system design. Section 37.7 gives an overview of the use of testing as a simulation tool in the design of control systems. Section 37.8 presents the conclusions of this work.

37.2 The Case Study Example

A robotic arm application, shown in Figure 37.1, is used as a case study because it is relatively generic and so will reduce the need for discussing overwhelming engineering theory. The robotic arm is a commercial off-the-shelf product.[1] It consists of a turntable base, an "upper arm," a "forearm," and a "hand." Two revolute joints connect the turntable to the upper arm, and a single revolute joint provides the connection to each other part. Revolute joints have one degree of rotational motion about a fixed axis, such as a wheel spinning on an axle. Each joint is connected to a DC motor, which serves as the joint actuator and a potentiometer, which serves as a position sensor. The motor of a joint moves the bodies connected at the joint relative to each other. The sensor measures and reports how far the bodies have moved. Each actuator/sensor pair provides an input/output point, which can be used to manipulate the device through the use of feedback control (Åström and Wittenmark, 1984; Dorf, 1987).

37.2.1 Feedback Control

Consider the task of moving the upper arm in the robotic arm to a specific position from its current position. The steps are as follows:

- The position sensor reports its current position.
- The position is compared with the desired position and it is decided in which direction the motor should turn.
- The motor starts turning in the desired direction.
- At a regular time interval, for example, every 0.01 s, the sensor value is read and compared with the desired position.
- The speed and direction of the motor are then adjusted depending on whether the required position is being approached or has been passed.
- These steps are repeated until the required position is reached plus or minus some acceptable threshold.

In this simple feedback-control algorithm, information is returned regarding the current state of the robotic arm position, which enables making a decision on what should happen in the next time step. Feedback control theory and strategies can be arbitrarily complex (Åström and Wittenmark, 1984; Franklin et al., 2002; Ogata, 2001; Zhou and Doyle, 1997), but the basic process outlined above provides the context to understand how any feedback-control application can be taken and the algorithm is designed, tested, and simulated. Figure 37.2 shows the basic feedback-control loop. The plant, the device to be controlled, is the upper arm with its motor. The sensor is the position sensor. The controller is the algorithm that adjusts the motor speed and direction on the basis of the difference between desired and actual position.

Now, consider the entire arm. By controlling the motion of each arm segment about the four joints, the hand can be moved. This is a multiple-input, multiple-output (MIMO) problem of great interest to

---

[1] It can be acquired through Lynxmotion, www.lynxmotion.com.

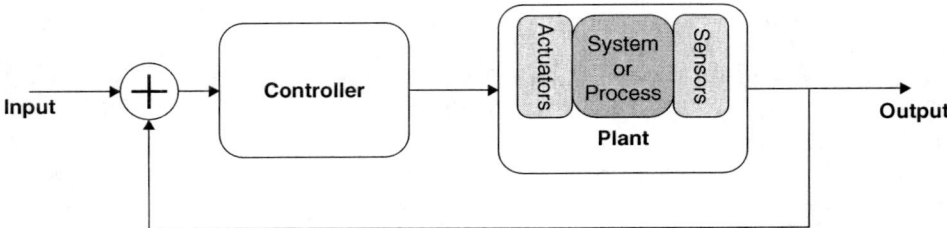

FIGURE 37.2 Elements of embedded control systems.

control engineers because many systems have multiple inputs and multiple outputs. Two examples of this are

- Washing machines must control water temperature and level, and motor speed based on the type of clothes and size of the load.
- Car engines must adjust intake and exhaust levels to meet performance and efficiency requirements.

In general, the use of computational modeling and simulation as described in this chapter has a prolific application in industry.

37.3 Designing with Simulation

A "design" as used hereafter is defined as a software component or algorithm that functions either alone or with other components or algorithms. To study the behavior of a new design, a model is needed that includes not only the design but also other components, systems, or algorithms as well with which the design interacts. Once a computational model of the entire system is available, simulation can be used to test and refine the design before going to a final implementation.

Consider the robotic arm. The design is the control algorithm that will be embedded on a microprocessor connected to the robotic arm. The entire system that must be modeled includes the algorithm, the actual robot mechanics, electrical circuitry, and any other devices and components that the robot will interact with. Once this has been modeled accurately, the feedback-control algorithm discussed in the previous section can be designed. As the algorithm is designed, testing of it in combination with the robot can be started in a simulation. Since simulated rather than actual hardware is used, it can be quickly experimented with algorithms as they are developed without costly damage. This would not be possible with a physical robot because hardware is likely to be damaged during experimentation with different designs. With actual hardware, a small fleet of robots would be needed to run the same tests.

37.4 Obtaining Computational Models

Here, focus is on the parts of the computational model other than the control algorithm. In the case of the robotic arm this is the plant, which consists of the actual mechanics, electrical circuitry, and dynamic effects such as friction. Computational models of physical devices and phenomena can be obtained by one or more of the following methods (this is not an exhaustive list):

1. Derive mathematical relationships from published or otherwise well-accepted theory, also called first-principles modeling. To an extent, this area pertains to phenomenological modeling (Broenink, 1990; Karnopp et al., 1990).
2. Derive mathematical relationships using data gathered from the devices being modeled. The data can then be used in a data-fitting approach (Bendat and Piersol, 2000; Ljung, 1998).

On Simulation of Simulink® Models for Model-Based Design

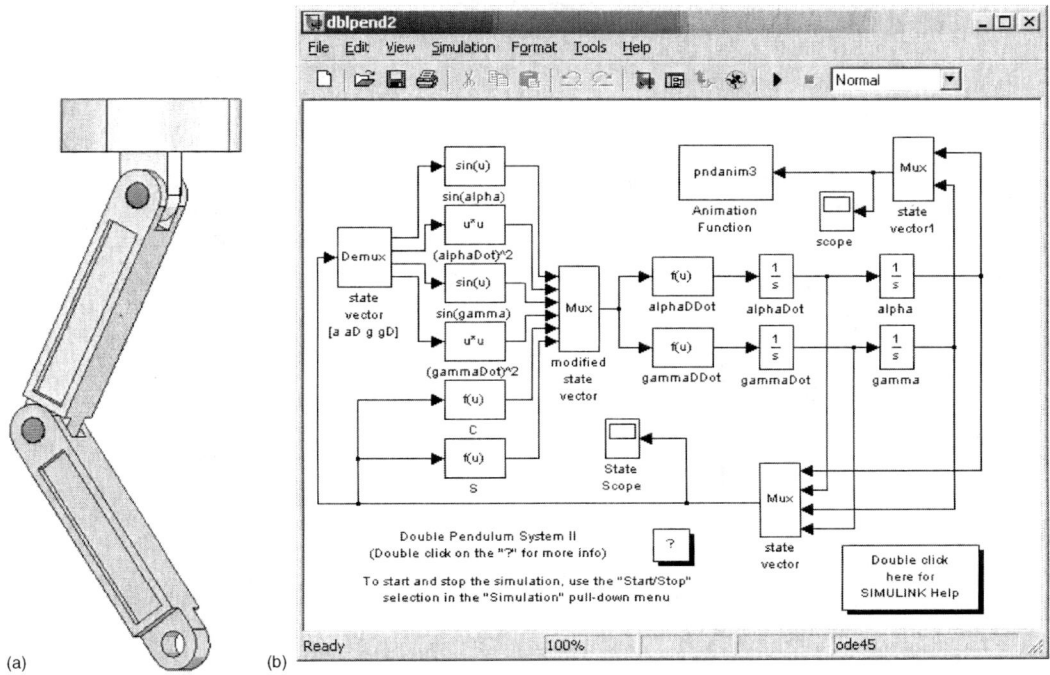

FIGURE 37.3 A pendulum. (a) Multibody model. (b) Simulink model.

3. Use modeling software that derives computational models from physical information such as mass, inertia, resistance, and capacitance (also considered as first principles). This is more closely affiliated with structural modeling approaches such as Modelica (Elmqvist et al., 1999; Tiller, 2001).

The first method is typically the most time consuming, but is widely accepted because the engineer has visual access to the equations that describe the system. The second method relies on techniques such as curve fitting, fuzzy logic, neural networks, and nonlinear optimizations to produce mathematical estimates of a systems behavior. While this method can be effective at obtaining models quickly, the engineer must consider several test scenarios to ensure that the computational models capture all the system dynamics. The third method is typically the most practical to implement, since the engineer needs to know only physical properties, not mathematical relationships.

37.4.1 Modeling from First Principles

First-principle modeling methods offer the capability of testing a design through simulation without committing to specific hardware. This enables development engineers to explore design options more thoroughly and discover problems early, saving time and money.

Before looking at obtaining the computational model of the robotic arm, consider the double pendulum in Figure 37.3(a) that is similar to the robotic arm. It contains one less linkage and no turntable. The kinematics equations could be derived based on first principles (Method 1). The dynamics, however, are much more difficult to model. The dynamics will require friction models that are difficult to come by, certainly by derivation from first principles (see, e.g., (Lötstedt, 1981). To do so would require knowledge much beyond that of multibody dynamics.

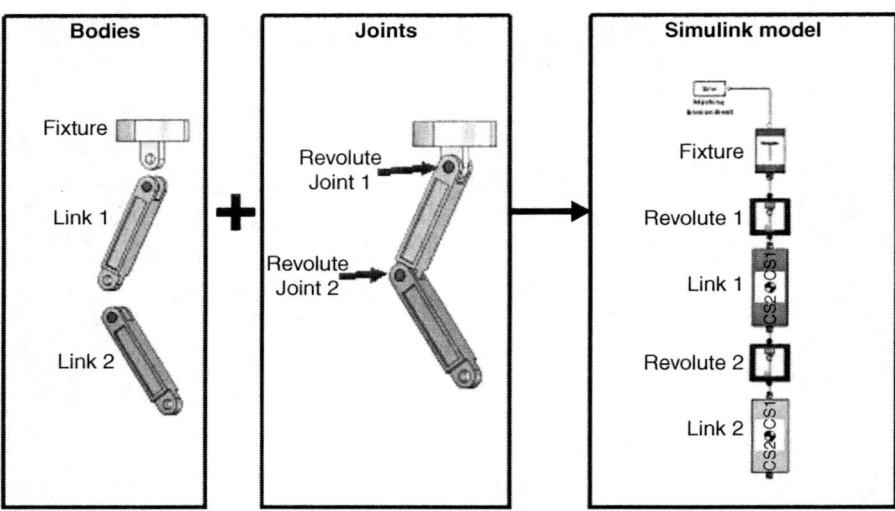

FIGURE 37.4 Elements of a multibody pendulum model.

The following equation represents the kinematics of the double pendulum in Figure 37.3(a):

$$\alpha = \int \frac{-L_2 \sin(\alpha) + mw_2(-\sin(\alpha-\gamma))\sin(\gamma) - ne(-\sin(\alpha-\gamma))\cos(\alpha-\gamma)\alpha^2 - n\cos(\alpha-\gamma)r^2}{1 - ne\sin^2(\alpha-\gamma)} \tag{37.1}$$

Note that, in spite of the apparent complexity, Eq. (37.1) does not yet include a friction model. The corresponding Simulink model for the generalized coordinates is given in Figure 37.3(b). Clearly, the equations for this relatively simple system can be complex. The more complicated robotic arm will have an even more complicated formulation.

To further compound matters, where Eq. (37.1) uses a generalized coordinate system with only two variables, an implementation with Cartesian coordinates would have 12 variables. It is obvious how this relatively simple device can become very complicated to model and analyze in detail.

The alternative is to use Method 3 for first-principles modeling. The pendulum can be considered a multirigid-body system that can be modeled in SimMechanics (2006), a modeling tool that works within Simulink and provides multibody mechanical modeling capabilities. The user only needs information regarding mass, inertia, dimension, and the position and orientation of joint axes of motion to create a SimMechanics model. With this information the model compiler can automatically derive the system of differential and algebraic equations (DAE) that capture the model behavior (Wood and Kennedy, 2003).

Users can assemble bodies and joints to model their systems, or model their systems in SolidWorks® (2002) and automatically extract SimMechanics models. Figure 37.4 shows the SimMechanics model with its one-to-one correspondence to the physical components. This model can then be integrated with standard Simulink blocks that can be used to model effects such as friction or a control algorithm in continuous-time, sampled time, or scheduled task form.

Computer Automated Multiparadigm Modeling is increasingly become important to negotiate the complexity of modern engineered systems (Mosterman and Vangheluwe, 2004). It relies heavily on domain-specific formalisms and tools. In this vein, tools to model physical systems other than Sim-Mechanics that generate the mathematical equations from higher-level component representations in different domains are available to the modeler as well. Simulink integrates tightly with four tools for component-level first-principles modeling of physical systems (Method 3):

- SimMechanics models three-dimensional mechanical systems.
- SimPowerSystems (2006) models electrical circuitry and power flow.

- SimDriveline (2006) models one-dimensional rotational motion.
- SimHydraulics™ (2006) models hydraulic systems.

These tools can be used in conjunction with equations derived in Simulink to model various types of systems for a variety of purposes.

37.4.2 Using Data and Simulation to Obtain or Tune Models

Often, one does not have the luxury of using Method 3 exclusively and must use Method 1 to model some components. Simulink provides an environment where all the tools for modeling physical systems that are referenced above can be mixed with equations represented with standard Simulink blocks. Methods 1 and 3 assume that the behavior of the system is well-understood. In an ideal world this is true, and one could model complex behaviors such as friction very accurately. In the real world, however, it must be dealt with manufacturing tolerances, lack of information about component behavior, and other uncertainties regarding the systems being modeled. This is where Method 2 is used, which relies on data collected from the components to either tune existing models built with Methods 1 and 3, or generate linear models for "black box systems." As mentioned earlier, there are a variety of data-fitting techniques but the basic principle is as follows:

- Apply some disturbance or other input to the system such as moving the double pendulum in Figure 37.3(b) in a measured way.
- Measure any change in the system with sensors, this is the system output (here, the pendulum angle versus time).
- Use the input/output data pairs to
  —tune parameters in existing models, using tools such as Simulink Parameter Estimation (2006)(considered "gray-box" modeling), or
  —generate models using an appropriate tool such as the System Identification Toolbox (2006) or the Fuzzy Logic Toolbox (2006) (considered black-box modeling).

The black-box techniques such as fuzzy logic are particularly useful in characterizing phenomena such as chemical processes for which equations might not be easily derived. These modeling methods can be applied to many systems and processes. Both types of data modeling (gray-box and black-box) are particularly interesting because they rely on simulation, to tune parameters and generate models, respectively. In the case of gray-box modeling, Simulink Parameter Estimation simulates a model hundreds or thousands of times while using optimization techniques to adapt parameter values until the actual system output matches the measured system output. This is referred to as *parameter tuning* as well. In the case of black-box modeling, the techniques vary slightly depending on the method. The System Identification Toolbox (2006) enables multiple simulations to generate a model that will give the desired measured output from the associated input. Once the parameters have been tuned so that the model reflects the observations of reality accurately, the model can be used in the larger system simulation. Models generated from data can be incorporated into larger system simulations as well.

Another approach to modeling with data is transfer function estimation. This is the process of taking experimental data and converting it using spectral estimation techniques to compute the frequency response of a system (Bendat and Piersol, 2000). The Signal Processing Toolbox (2006) has many functions that aid in the estimation of a transfer function. Additionally, a linear parametric model can be fit to the experimental frequency response function using the modeling tools in the System Identification Toolbox.

In some cases, linear models may not describe the model accurately and the underlying equations of motion may not be that well known. In this case, a nonlinear black-box neural network can be created. These types of models can be created using the Neural Network Toolbox (2006).

37.5 The Robotic Arm Model

A computational model of the robotic arm is obtained using the different approaches.

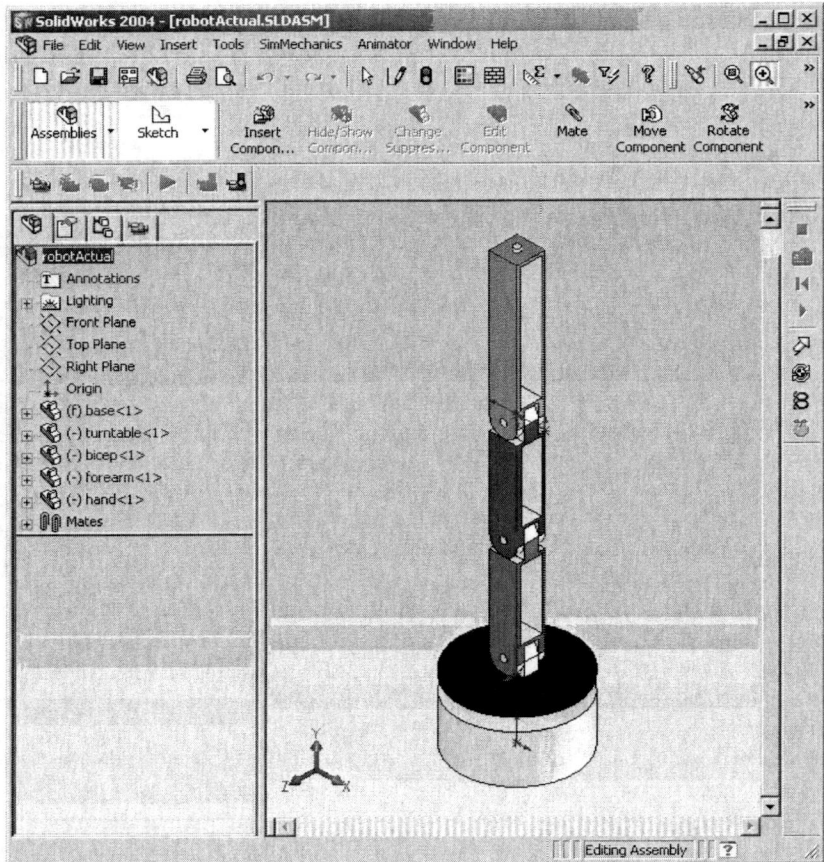

FIGURE 37.5 CAD model of a robotic arm.

37.5.1 The Mechanical Model of Arm

The computational model of the robotic arm consists of the mechanical model, the motor models, and the actual controller feedback loop. To model the mechanics, the robot is first modeled with a computer-aided design (CAD) tool, in this case SolidWorks (Figure 37.5) using the engineering drawings provided by the manufacturer (Figure 37.6). The SimMechanics model is then automatically extracted from the SolidWorks model.

SimMechanics models are composed of bodies connected to each other by joints that have user-specified ranges of motion. The dimensions of the bodies and the axes of motion for the joints are specified in a Cartesian coordinate space (x, y, z). The robot in Figure 37.5 has five bodies: three arm segments, the turntable, and the base. The base is fixed to a reference point, the other dimensions and axes are specified relative to this ground position. Each body is connected to the adjacent body by a joint as described in Section 37.2.

37.5.2 Electromechanical Model of the Motor and Coupling

Once the kinematics of the mechanical model are available, the dynamics for the motors and the couplings that connect the motors to the joints must be modeled. The joints are assumed to be frictionless (which is a coarse approximation of reality, but sufficient for this application). A pulse-width modulation scheme

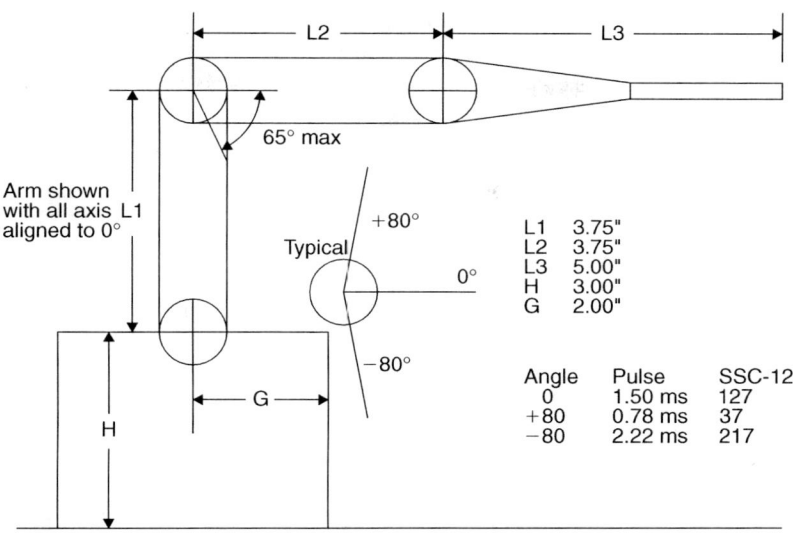

FIGURE 37.6 Engineering drawing of a robotic arm.

regulates the voltage to the electric motor and controls the motion of the linkage. The switching circuitry is modeled using an H-Bridge, a digital device that can switch the direction of current flow. The motor itself consists of an inductor, a resistor, and a model of back electromotive force (EMF). The component blocks for the motor and H-Bridge are from SimPowerSystems. The remaining components in the motor model are gearing, inertia, and other mechanical components, all modeled in SimDriveline. Since each joint has one motor, the model can be made a reference to a library component and a separate motor model connected at each joint. An individual motor model is shown in Figure 37.7. The motor models can then be coupled to the SimMechanics model of the robotic arm. Standard sensor and actuator interface blocks are necessary when moving from one modeling formalism to another, such as from SimMechanics to SimPowerSystems, or to standard Simulink. The blocks translate physical properties such as force and torque into the time-based signals of Simulink.

37.5.3 Tuning the Motor Parameters with Data

The DC motor model shows a relationship between current and torque. Torque causes the motor shaft to spin in accordance with a relationship to the back EMF. The remaining parameters include shaft inertia, viscous friction (damping), armature resistance, and armature inductance. Values for those parameters must be accurate for the motor model to behave similar to the actual motor. While manufacturers may provide the values, one should assume those to be averaged values with added manufacturing tolerances. It is necessary to estimate these parameters as precisely as possible for the model to ascertain whether it is an accurate representation of the actual DC servo motor system. Table 37.1 lists the model parameters and their initial values.

When a series of voltage pulses is input to the motor, the motor shaft turns in response. If there is a discrepancy between the model parameters and those of the physical system, however, the model response will not match that of the actual system. Figure 37.8 shows the response of the model using the initial parameter values listed in Table 37.1 together with the actual response of the motor. To obtain a more accurate response, the parameters must be reestimated. This is where Simulink Parameter Estimation plays a pivotal role.

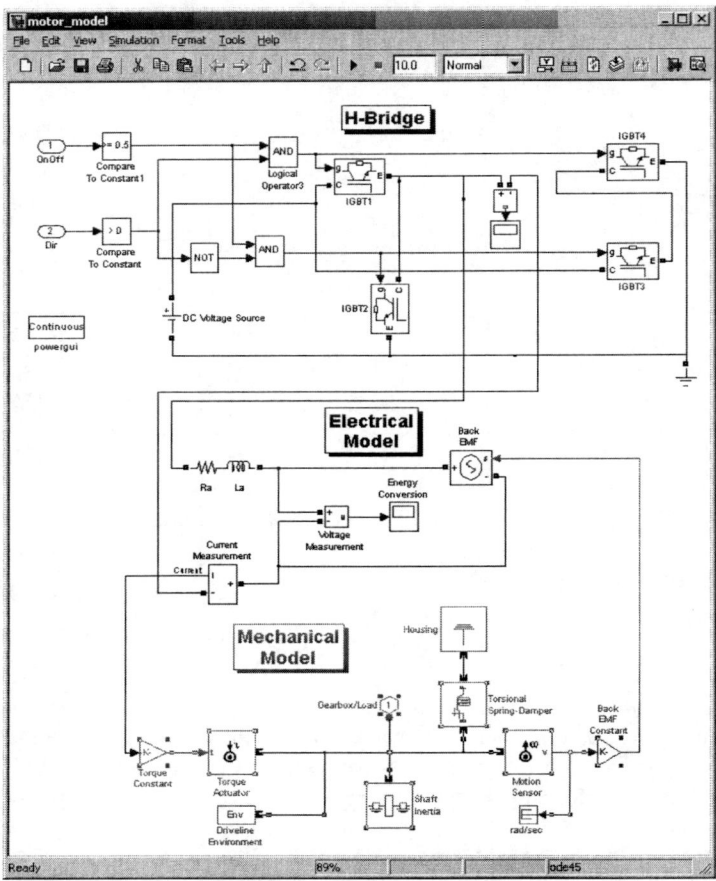

FIGURE 37.7 Model of a DC motor and the signal conditioning hardware.

TABLE 37.1 Estimated Parameter Set.

Name	Parameter	Value	Unit
Viscous friction	B	0.008	Nms/rad
Shaft inertia	J	$5.7e^{-7}$	kgm^2
Motor constant	Km	0.0134	Vs/rad
Armature inductance	La	$6.5e^{-5}$	H
Armature resistance	Ra	1.9	Ohm

Input/output data from the motors can be applied to tune parameters in the model until the computational model mimics the behavior of the real robotic arm with sufficient precision. The typical workflow is as follows:

- Connect a voltage source to a motor on the robotic arm.
- Input a voltage to the motor and read the resulting position sensor values, and save these as input/output datasets.
- Repeat this with different types of input such as steps, ramps, and frequency sweeps.
- Select the parameters to tune.

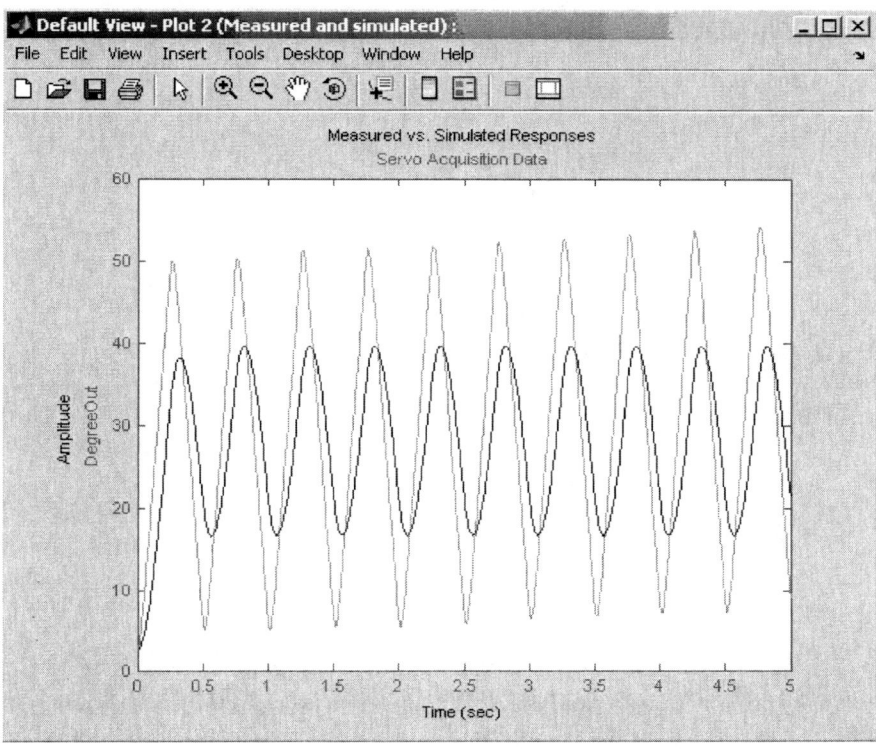

FIGURE 37.8 Actual response and model response *before* fine-tuning the parameter estimates.

- Some datasets will be used to tune the parameters while others can be used to validate the tuned values.
- Using the input/output datasets, tune the parameters until the input produces an output that matches the actual robotic arm with the desired precision.
- For validation, simulate the model with an input from one or more datasets that were not used for tuning.
- Compare the computational output to the output from the validation dataset. If they match within reasonable tolerances, the model is sufficiently tuned for design.

Figure 37.9 shows the comparison of the outputs from the computational model to the robotic arm outputs after tuning. Note that the results overlap closely. At this point, the model can be used to design the controller, for example, by employing classical, modern, and robust control design approaches (Franklin et al., 2002; MacIejowski, 1989; Ogata, 2001; Skogestad, 1996; Zhou and Doyle, 1997).

37.5.4 Generating the Motor Models from Data

In some cases, modeling the individual components of the overall system in detail may not be of great interest. Though it is of value when, for example, designing the DC motor configuration, for the design of the motor controller it may not be critical. In case of designing the controller, the entire motor model can often be treated as a black box. A linear model of the motor can be estimated using the same datasets used in Section 37.5.3. This approach works well when the system being modeled is relatively linear like the motor used here. Using the System Identification Toolbox, a model can be identified using a similar workflow as shown in Section 37.5.3.

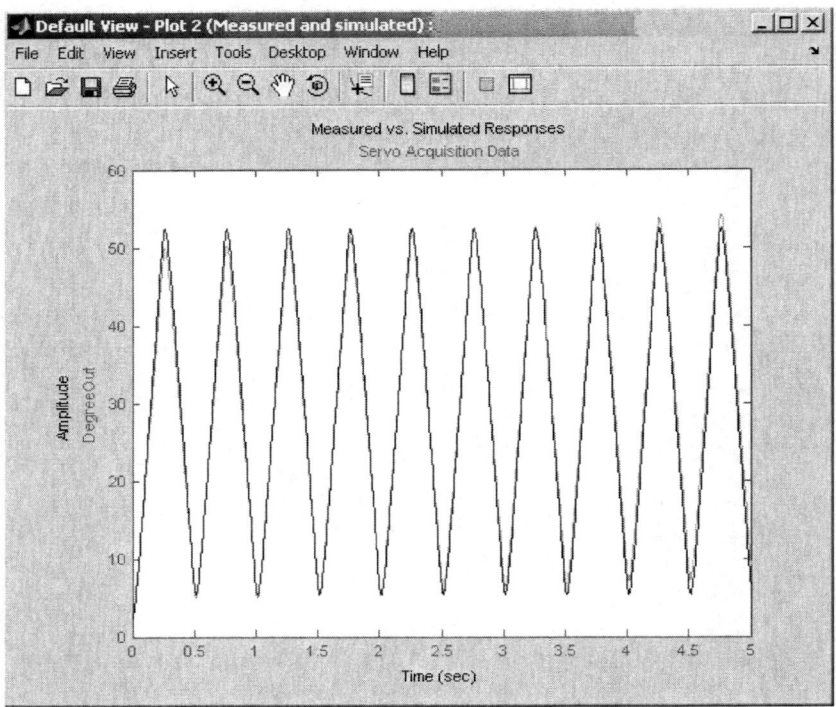

FIGURE 37.9 Actual response and model response *after* fine-tuning the parameter estimates.

- Connect a voltage source to one motor on the robotic arm.
- Input a voltage to the motor and read the resulting position sensor values, and save these as input/output datasets.
- Repeat this with other inputs such as steps, ramps, and frequency sweeps.
- Some of the inputs can be used to identify the model while others are used to validate the model.
- Using the input/output datasets, identify the model until the input produces an output that matches the robotic arm.
- For validation, use input from a dataset that was not used for identification.
- Compare the computational output to the output from the dataset. If they match within required tolerances, the model is sufficiently accurate for design.

37.6 Using Computational Models for Control Design

The preceding sections have investigated how to obtain accurate dynamic models of physical devices. As stated earlier, one important reason to develop accurate computational models of physical processes and devices is to facilitate the design of control laws for regulating or controlling their behavior.

37.6.1 Designing Controllers through Modeling and Simulation

The most basic approach to designing a control system need not rely on computational models, but rather can be developed through paper and pencil analysis, design, and iteration on the real systems that will be controlled. This approach normally requires a lot of experience, and is typically complex to implement. Additionally, it is more costly to implement because of the resources needed to mitigate the risks of damaging the system and actually making the physical system available for design work. Some benefits of using computational models and simulation to design controllers are

- Ability to use specialized simulation-based control algorithm development tools.
- Ability to evaluate hardware and system constraints such as actuator effort and response time in a safe, low-cost environment.
- Increased innovation through the ability to experiment with many possible solutions before implementation.

The actual process to design controllers in a modeling and simulation environment varies depending on the tools available. This section describes the approach used to design controllers in a simulation environment and also studies the use of some specialized tools that aim to simplify the design process. The typical process, which will be described in detail in the subsequent subsections, is

- Model the dynamics of the device or process being controlled (see Section 37.5 for the modeling of the robotic arm).
- Obtain a linear representation of this "plant" model about the relevant operating points.
- Use linear control design techniques to tune the controllers to meet performance requirements.
- Validate the linear control design on the nonlinear computational model.

An alternative to this process is to use tools that depend on simulation and optimization-based techniques. For the design of a controller for the robotic arm, a set of tools called Simulink Control Design (2006), Control System Toolbox (2006), and Simulink Response Optimization (2006) are used. A single workflow-based graphical user interface (GUI) serves as a task manager and portal to this set of tools.

In the case of the robotic arm, the control task is to move the hand to a desired point in space. This is done by manipulating the arm segments of the robot arm. The exact control design goals are as follows:[2]

- Robot arm position control:
    — Design joint angle controllers for the turntable, bicep, forearm, and hand joints.
    — Design prefilters to balance the bandwidths of the responses to reduce the impact of off-diagonal closed-loop responses.
- Joint angle loop control requirements:
    — The bandwidth is less than 50 Hz.
    — The gain margin is greater than 20 db.
- Closed-loop position control step-response requirements:
    — The overshoot is less than 10%.
    — The rise time is less than 1.5 s.
    — The cross-coupling is less than 10%.

The model developed in the previous sections is used as a starting point. It will need to be linearized to investigate how controllers can be designed to meet the specified requirements.

37.6.2 Linearizing Models for Control System Design

With the robotic arm, the first step is to obtain a linear representation of the nonlinear model. The linearized model is then used to compute pertinent open- and closed-loop response plots that are used directly in control design. To obtain the linearized model, the following steps need to be taken:

- Specify the control structure in Simulink, with feedback loops, compensators, and prefilters as shown in Figure 37.10.

---

[2] Detailed definitions and descriptions of these terms can be found in the online documentation of the Control System Toolbox (2006), http://www.mathworks.com.

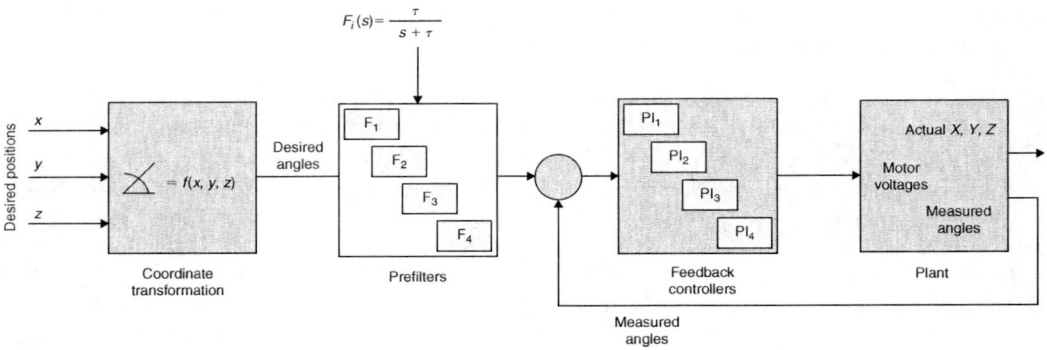

FIGURE 37.10 Feedback loop structure.

- Select the eight compensators to be tuned which include four prefilters and four feedback controllers.
- Select the closed-loop inputs and outputs which map the desired position to the actual measured position.
- Specify or compute the operating points for the linear analysis of the model.

Linear models are then automatically extracted using the information from these steps. These linear models are used to setup the control design task in the GUI.

The open- and closed-loop linearization results are highly dependent on the operating point—for example, the states of the integrators of the model at different operating points. Trim or equilibrium operating points are a special type of operating point that engineers find very useful. A basic description of equilibrium conditions is that, over time, the operating point remains steady and constant. In Simulink and other block diagram simulation tools there are two commonly used approaches to specify equilibrium conditions of a model of the physical system. The first method is that the users employ their intuitive knowledge about the system to pick an equilibrium condition. This can be a rather time-consuming and difficult process because of the large number of operating points that must be specified in a complicated model.

The second option is to employ an approach known as trim analysis. The approach uses optimization to solve for a set of operating points that satisfy the equilibrium conditions. Simulink Control Design provides trim analysis capabilities to obtain initial conditions for various operating points. Another alternative is to use "simulation snapshots" to specify operating points close to the region where the control effort is desired.

In the case of the robotic arm, the model was linearized at a number of operating points for different positions of the arm. A single operating point was selected for the design and the other operating points were used to verify the control system of the robot arm in different configurations.

37.6.3 Designing a Controller

Once a linearized open-loop model has been obtained, a typical next step is to select a control system structure and tune the individual compensators. In the case of designing a controller for the robotic arm, the control structure is specified in the earlier step to help the tool determine the linear representation. The robotic arm controller configuration consists of four feedback loops with prefilters as shown in Figure 37.10. For such a multiloop system, several input/output combinations have to be linearized to attempt to design multiple controllers such that the overall multiloop system meets controller performance requirements. Multiloop controller design can be approached in a number of ways:

- Sequential loop closure where the designer first tunes one loop with the others open, and then sequentially closes the other loops.

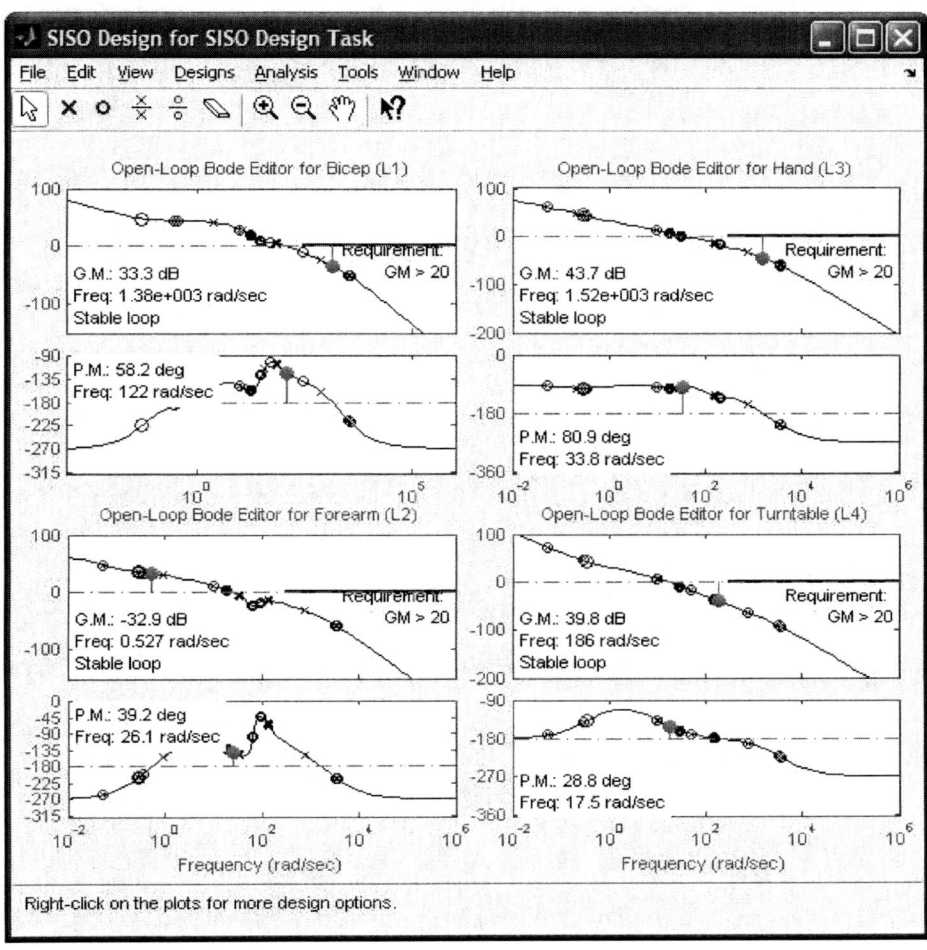

FIGURE 37.11 SISO design plots.

- Use traditional MIMO design techniques such as H-infinity to tune the loops simultaneously.
- Simultaneously tune the coupled single input single output (SISO) control loops.

For the robotic arm, the GUI is used with the linearized model to setup a control design task to design the controllers. The steps are

- Select design plots such as root locus, Bode and Nichols charts for each loop that needs to be designed.
- Select closed-loop analysis plots for viewing.
- Use design plots (Figure 37.11) to graphically shape all loops and edit the compensator structure, while viewing loop interactions and closed-loop responses in real time.

Because the tools can compute loop interactions while tuning all loops, the approach is to use visualization to see how changing one compensator affects all the other responses that are of interest. The graphical design tools can be employed in this way to design a set of compensators to try to best meet the performance requirements. Additionally, optimization techniques within these graphical tools can be exploited to help tune existing controllers, while trading off between multiple design requirements.

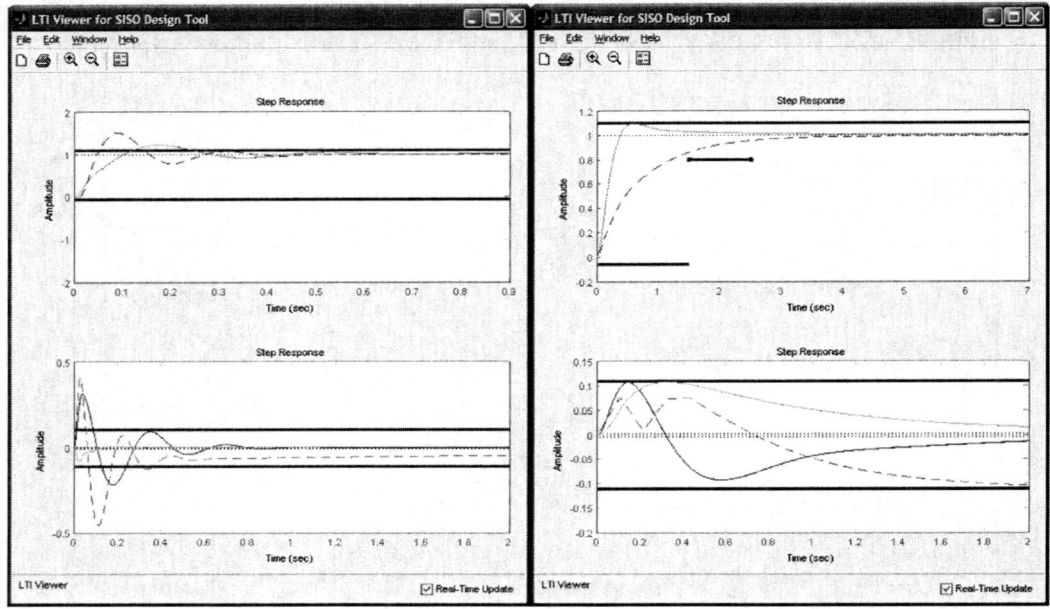

FIGURE 37.12 Closed-loop responses before (left) and after (right) optimization.

37.6.4 Tuning Controller Designs Using Optimization Techniques

The described graphical design tools simplify tuning an MIMO controller by enabling simultaneous tuning of loops. However, the process can still be quite complex; especially for systems with many tightly coupled loops. Because a computational model is available, this process can be simplified by leveraging optimization techniques to automate multiloop tuning. Using Simulink Response Optimization, performance bounds and requirements can be specified on the design and closed-loop response plots and then several compensators can be tuned simultaneously through optimization. This is done by graphically specifying requirements in either the time domain, as overshoot, settling time, etc. or in the frequency domain, for example, as gain/phase margin, bandwidth, or pole zero locations.

For the robotic arm, the performance requirements listed above are specified for the controller by using a combination of both time-domain and frequency-domain plots. The optimization is then run to obtain a valid solution. Figure 37.12 shows the final responses obtained and how these fit within the performance envelopes that are defined. Note that although the system being tuned is an MIMO system, the problem is configured as eight coupled SISO controllers that are tuned simultaneously. In the event that the optimization is not able to meet all requirements, the requirements can be relaxed, or, alternatively, a different control strategy can be evaluated using all the methods that have been discussed. Once satisfactory control performance is obtained, the design can be validated on a full nonlinear simulation by exporting the controller gains directly to the Simulink model.

37.7 Testing with Model-Based Design

One aspect of Model-Based Design is the testing of designs while they are under development. By testing early in the design phase, errors and deficiencies may be recognized and rectified early in the design phase, before the cost of correction becomes too high, or worse, the error makes it into the final implementation. Testing early and often is a good principle, and it is imperative to do so in a systematic manner.

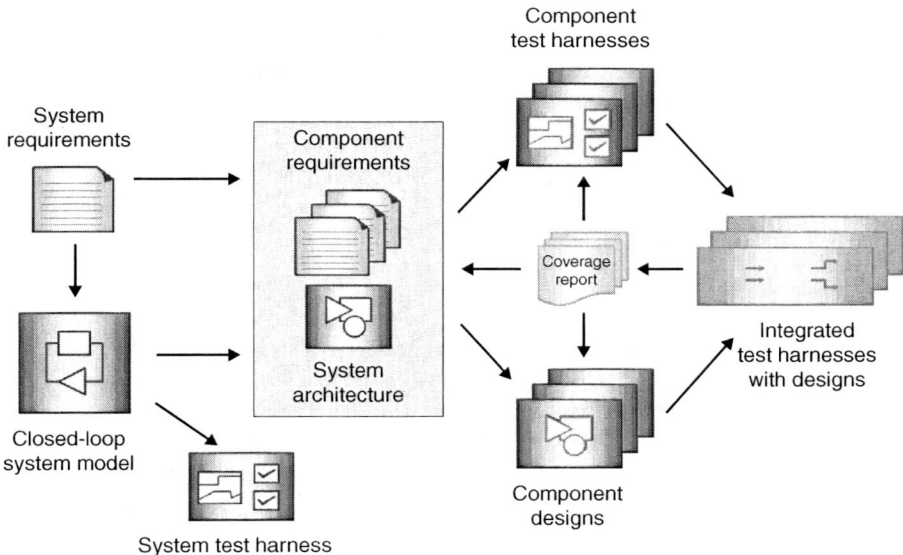

FIGURE 37.13 An overview of Model-Based Design.

37.7.1 Requirements-Based Testing through Simulation

Requirements-based testing, which can contribute to systematic testing, refers to defining test vectors for each requirement, and creating corresponding checks to verify that the requirements are met. Essentially, a test harness is created to verify that the design algorithm meets the requirements. This workflow is shown in Figure 37.13.

The test vectors typically consist of a series of inputs designed to exercise the computational model through a range of expected and unexpected behavior. In case of the latter, this enables testing of dangerous modes of operation, and the corresponding design of fail-safe elements. Often this kind of failure-mode testing in simulation is impossible to do with a physical prototype or implementation of the system. In addition to test vectors, verification blocks in Simulink can be used to check model output data ranges and ensure that requirements are met. Once a test harness with test vectors and verification blocks has been obtained, a coverage report can be generated to understand how well the algorithm was exercised. Using Simulink Verification and Validation (2006), coverage metrics are collected as the tests execute to quantify which elements of the design have been excited and which have not. Using coverage-based test verification, the following features and benefits are achieved:

- Measure how well the model has been exercised.
- Identify what additional test vectors are needed to exercise the model more thoroughly.
- Identify and remove unnecessary elements in the design.
- Ensure that the requirements, design, and tests are consistent and complete.

37.7.2 Simulation with Hardware and Implemented Designs

In addition to the use of computational models for simulation during the design stages, models can also be used in real-time simulations for verification and validation. In such a scenario, computer code is generated from the computational model and downloaded to dedicated computers to simulate the model in real time. This is often referred to as *rapid prototyping*, of which there are two variations. The first is *rapid controller prototyping*, which typically involves placing the controller algorithm on a processor such as a commercially available PC-based processor, and interfacing the processor with the plant, here the robotic arm. The entire system is then simulated in real time (Mosterman et al., 2005).

The second variation is *hardware-in-the-loop* (HIL), which refers to interfacing the processor on which the controller runs with a combination of real hardware and computational models, and running in real time on dedicated processors. HIL enables the integration of difficult-to-model hardware as part of the simulation environment. For example, actuators can be highly nonlinear, and so if a model of such an actuator is used, the analysis may not be sufficiently precise. HIL allows the model of the physical system minus the actuator to be connected to the real actuator and hence enables the full nonlinear behavior to be validated.

In the case of the robotic arm, rapid controller prototyping is used with xPC Target (2004). Code is generated from the controller algorithm using Real-Time Workshop® (2002). Next, using a commercially available PC running the real-time kernel of xPC Target, the controller algorithm is downloaded onto the PC and commercially available I/O boards are used to connect the PC to the robotic arm. The control algorithm running on the PC is then used to control the robotic arm and validate the algorithms. At this point, the controller algorithm can be tuned in real time, while connected to the robotic arm. Alternatively, the algorithm can be modified and code regenerated to test the new algorithm. This process is repeated till the algorithm operates satisfactorily.

Reasons to perform rapid controller prototyping include

- Quick algorithm testing and retesting using hardware that has been tested in simulation.
- Testing control algorithms with fixed-step solvers in real time, which is closer to real-world implementations.

To briefly study an example of HIL, consider a flight-control application with the flight controller implemented on the actual flight-control box, a dedicated computer that will go into the production aircraft. The flight-control box could then be integrated with a cockpit, a human pilot, and a flight-simulator package. The cockpit would enable the pilot to provide realistic inputs, while the flight simulator could be running the aircraft dynamics computational model (that have been modeled using a combination of the three modeling methods) to validate the behavior of the controller.

The flight simulator could be mounted on a motion-simulator platform to generate the appropriate forces and torques. Models of wind and turbulence and atmospheric effects can be included to test the behavior of the controller in "dangerous" situations without risk to the pilot or the aircraft. This can be implemented through the use of xPC Target and models generated with Simulink and the physical modeling products such as SimMechanics. This approach is very useful in safety-critical applications such as those found in aircraft and other vehicles, where testing with the real hardware is expensive, time-consuming, and often heavily regulated.

37.7.3 Other Uses of Rapid Prototyping in the Design Process

To design a control system, rapid prototyping tools can be used as shown in Figure 37.14 for other stages of the process including the actual computational modeling. At the start of the control design process, an engineer may have a rather inaccurate model or no model at all. So, at first a skeleton control system is developed to stabilize a system and to get the desired behavior to experiment with. Once this is achieved, experiments can be designed and performed to acquire responses of the system at various operating conditions. The acquired data can then be exploited to enhance the plant model, and to design a new control system using the more accurate plant model. Simulation of the combined control system and plant model then allows studying the performance of the system and the control system can be optimized using the full nonlinear plant simulation model. Finally, the control system can be implemented on a rapid prototyping system. If the system does not meet the performance of the control system as obtained in simulation, the model is further refined as well as the design of the control system to try to achieve improved performance.

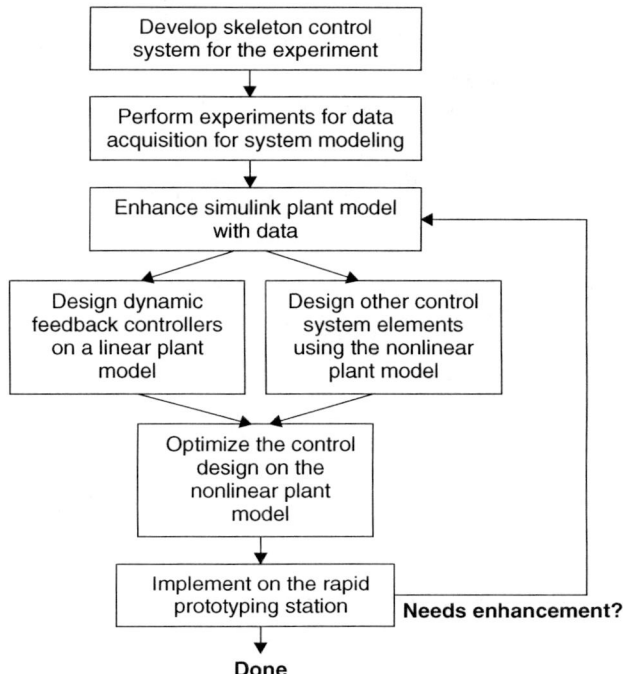

FIGURE 37.14 Rapid prototyping control design process.

37.8 Conclusions

Through the use of advanced computational modeling and numerical simulation capabilities available in Simulink, an accurate model of the robotic arm can be quickly obtained and fine-tuned with measured data. Control algorithms for the robot arm were then designed and tested rapidly and effectively. Using requirements-based testing and rapid controller prototyping each requirement was systematically tested and the entire set of requirements was formally verified. When the control algorithm was finally implemented its proper behavior to control the actual robotic arm was established with confidence. It was thoroughly analyzed to deliver the required performance, even under fault conditions.

MATLAB, Simulink, Stateflow, Handle Graphics, Real-Time Workshop, and xPC TargetBox are registered trademarks and SimBiology, SimEvents, and SimHydraulics are trademarks of The MathWorks Inc. Other product or brand names are trademarks or registered trademarks of their respective holders.

References

R. Aberg and S. Gage. Strategy for successful enterprise-wide modeling and simulation with COTS software. In *AIAA Modeling and Simulation Technologies Conference and Exhibit*. Providence, Rhode Island, August 2004. CD-ROM.

K.J. Åström and B. Wittenmark. *Computer Controlled Systems: Theory and Design*. Prentice-Hall, Englewood Cliffs, NJ, 1984.

P. Barnard. Graphical techniques for aircraft dynamic model development. In *AIAA Modeling and Simulation Technologies Conference and Exhibit*. Providence, Rhode Island, August 2004. CD-ROM.

J.S. Bendat and A.G. Piersol. *Random Data: Analysis & Measurement Procedures*. Wiley-InterScience, Hoboken, NJ, 2000.

A.P.J. Breunese, T.J.A. de Vries, J. van Amerongen, and P.C. Breedveld. Maximizing impact of automation on modeling and design. In *ASME Dynamic Systems & Control Div. '95*, pp. 421–430, San Francisco, CA, 1995.

J.F. Broenink. *Computer Aided Physical Modeling and Simulation: A Bond Graph Approach*. PhD dissertation, University of Twente, Enschede, The Netherlands, 1990.

F.E. Cellier, H. Elmqvist, and M. Otter. Modelling from physical principles. In W.S. Levine, editor, *The Control Handbook*, pp. 99–107. CRC Press, Boca Raton, FL, 1996.

Control System Toolbox. *Control System Toolbox User's Guide*. The MathWorks Inc., Natick, MA, March 2006.

S.J. Culley and A.P. Wallace. The modelling of engineering assemblies based on standard components. In J. Sharpe and V. Oh, editors, *Computer Aided Conceptual Design*, pp. 113–129, Lancaster, UK, 1994.

R.C. Dorf. *Modern Control Systems*. Addison-Wesley Reading, MA, 1987.

H. Elmqvist, S.E. Mattsson and M. Otter. Modelica™—a unified object-oriented language for physical systems modeling: Language specification, December 1999. Version 1.3, http://www.modelica.org/.

G.F. Franklin, J.D. Powell, and A. Emami-Naeini. *Feedback Control of Dynamic Systems*. Prentice-Hall, Englewood Cliffs, NJ, 2002.

Fuzzy Logic Toolbox. *Fuzzy Logic Toolbox User's Guide*. The MathWorks Inc., Natick, MA, March 2006.

Hydro-Queébec TransÉnergie Technologies. *SimPowerSystems User's Guide*. The MathWorks Inc., Natick, MA, March 2006.

D.C. Karnopp, D.L. Margolis, and R.C. Rosenberg. *Systems Dynamics: A Unified Approach*. Wiley, New York, 2nd edition, 1990.

L. Ljung. *System Identification: Theory for the User*. Prentice-Hall, Englewood Cliffs, NJ, 2nd edition, 1998.

P. Lötstedt. Coulomb friction in two-dimensional rigid body systems. *Z. angew. Math. u. Mech.*, 61: 605–615, 1981.

J.M. MacIejowski. *Multivariable Feedback Design*. Addison-Wesley, Reading, MA, 1989. Electronic Systems Engineering Series.

MATLAB. *The Language of Technical Computing*. The MathWorks Inc., Natick, MA, March 2006.

P.J. Mosterman and G. Biswas. A hybrid modeling and simulation methodology for dynamic physical systems. *SIMULATION: Transactions of The Society for Modeling and Simulation International*, 178(1): 5–17, January 2002.

P.J. Mosterman, S. Prabhu, A. Dowd, J. Glass, T. Erkkinen, J. Kluza, and R. Shenoy. Embedded real-time control via matlab, simulink, and xpc target. In D. Hristu-Varsakelis and W.S. Levine, editors, *Handbook on Networked and Embedded Systems*, pp. 419–446. Birkhäuser, Boston, MA, 2005.

P.J. Mosterman, S. Prabhu, and T. Erkkinen. An industrial embedded control system design process. In *Proceedings of The Inaugural CDEN Design Conference* Montreal, July 2004. CD-ROM.

P.J. Mosterman and H. Vangheluwe. Computer automated multi-paradigm modeling: An introduction. *SIMULATION: Transactions of The Society for Modeling and Simulation International*, 80(9): 433–450, September 2004.

Neural Network Toolbox. *Neural Network Toolbox User's Guide*. The MathWorks Inc., Natick, MA, March 2006.

K. Ogata. *Modern Control Engineering*. Prentice-Hall, Englewood Cliffs, NJ, 4th edition, 2001.

Real-Time Workshop. *Real-Time Workshop User's Guide*. The MathWorks Inc., Natick, MA, March 2002.

Signal Processing Toolbox. *Signal Processing Toolbox User's Guide*. The MathWorks Inc., Natick, MA, March 2006.

SimDriveline. *SimDriveline User's Guide*. The MathWorks Inc., Natick, MA, March 2006.

SimHydraulics. *SimHydraulics User's Guide*. The MathWorks Inc., Natick, MA, March 2006.

SimMechanics. *SimMechanics User's Guide*. The MathWorks Inc., Natick, MA, March 2006.

Simulink. *Using Simulink*. The MathWorks Inc., Natick, MA, March 2006.

Simulink Control Design. *Simulink Control Design User's Guide*. The MathWorks Inc., Natick, MA, March 2006.

Simulink Parameter Estimation. *Simulink Parameter Estimation User's Guide*. The MathWorks Inc., Natick, MA, March 2006.

Simulink Response Optimization. *Simulink Response Optimization User's Guide*. The MathWorks Inc., Natick, MA, March 2006.

Simulink Verification and Validation. *Simulink Verification and Validation User's Guide*. The MathWorks Inc., Natick, MA, March 2006.

I.P.S. Skogestad. *Multivariable Feedback Control: Analysis and Design*. Wiley, New York, 1996.

SolidWorks. *Introducing SolidWorks*. SolidWorks Corporation, Concord, MA, 2002.

System Identification Toolbox. *System Identification Toolbox User's Guide*. The MathWorks Inc., Natick, MA, March 2006.

M.M. Tiller. *Introduction to Physical Modeling with Modelica*. Kluwer, Boston, MA, 2001.

G.D. Wood and D.C. Kennedy. Simulating mechanical systems in simulink with simmechanics. Technical Report 91124v00, The MathWorks Inc., Natick, MA, 2003.

xPC Target. *xPC Target User's Guide*. The MathWorks Inc., Natick, MA, March 2004.

K. Zhou and J.C. Doyle. *Essentials of Robust Control*. Prentice-Hall, Englewood Cliffs, NJ, 1997.

Index

A

Abstract model, **3**-10
Abstraction and scaling, **5**-9 to **5**-17
   choice of, consequences, **5**-14 to **5**-15
   geometric scaling, **5**-11 to **5**-13
   lumped elements and, **5**-9 to **5**-11
   of data presentations, **5**-15 to **5**-17
   scale in equations, size and limits, **5**-13 to **5**-14
Adams' method, **17**-17
Agent-oriented modeling in simulation, **8**-1 to **8**-12
   agents for modeling, **8**-3 to **8**-6
      agent metaphor for, **8**-3 to **8**-6
   modeling and simulation for agents, **8**-6 to **8**-10
      for multiagent systems, **8**-7 to **8**-8
      multiagent systems, designing, **8**-8 to **8**-10
Algebraic reasoning and verification, **19**-11 to **19**-19
   bottle filling line example, **19**-12 to **19**-14
   normal form, syntax of, **19**-15 to **19**-16
   parallel composition, elimination of, **19**-17
   process instantiation, elimination of, **19**-15
   recursion scope operator, syntax and semantics of, **19**-14
   substitution of constants and additional elimination, **19**-17 to **19**-18
   tool-based verification, **19**-19
Agriculture and natural resources
   ontology-based simulation in, **30**-1 to **30**-13
      building, **30**-6 to **30**-10: collection of relevant documents, **30**-6; define equations, **30**-9 to **30**-10; generating program code for implementing the simulation, **30**-10; identifying classes, individuals, and properties, **30**-7 to **30**-9; initial values of state variables, constants, and database access, **30**-10; model defining in terms of elements, **30**-7; simulation execution, **30**-10
      connecting models with data sources, **30**-4 to **30**-5
      integrating documentation and training resources, **30**-5
      model base, **30**-3
      reasoning, **30**-6 to **30**-7
      representing equations and symbols, **30**-4
      system structure (logical and physical), **30**-3 to **30**-4
      tools for, **30**-10 to **30**-12: EquationEditor, **30**-11 to **30**-12; ontology editor, **30**-10 to **30**-11; SimulationEditor, **30**-12
Analogy-making
   architectural principles, **2**-8 to **2**-12, *See also separate entry*
   as a means of 'bootstrapping' cognition, **2**-3 to **2**-5
   as sameness, **2**-2 to **2**-3
   bottom-up/top-down interaction, **2**-6 to **2**-8
   computational modeling, dynamics of, **2**-1 to **2**-17
   context-dependent computational temperature, **2**-6
   definition, **2**-1
   mechanisms presented, impact of, **2**-16
   program, working, **2**-12 to **2**-13
      Copycat, **2**-12
      Tabletop, **2**-12
   proportional analogies, **2**-2
   representation-building in, dynamics of, **2**-5 to **2**-6
   scaling up issue, **2**-16
   top-down and bottom-up processes, interaction between, **2**-6 to **2**-8
Architectural principles,
   in analogy-making, **2**-8 to **2**-12
      'slipnet,' a semantic network, **2**-8 to **2**-9
      codelets, **2**-10
      coderack, **2**-10
      computational temperature, dynamic codelet selection via, **2**-10 to **2**-12
      local, stochastic processing, **2**-12
      representation-building and correspondence-finding, integration, **2**-12
      workspace, **2**-9: AAB, **2**-9
      worldview, **2**-10
Artificial Intelligence (AI), **2**-11
Asymptotic stability, **18**-9 to **18**-13

Asynchronous input property, **14**-16 to **14**-17
Automata theory, **19**-2
Automatic synchronization, **36**-12

B

Balloted product development process (BPDP), **9**-7
Blender interface, **14**-10 to **14**-11
Boiling water example, of multimodeling, **14**-20 to **14**-26
  2D representation, **14**-22
  code generation, **14**-23 to **14**-26
  FBM, multimodeling for, **14**-21
  first-level FSM for, **14**-22
  FSM, multimodeling for, **14**-21
  model creation, **14**-2 to **14**-23
    geometry and dynamic models, **14**-22 to **14**-23
    interaction model, **14**-23
  second-level FSM for, **14**-22
  third-level FBM, **14**-22
Bond graphs, **26**-5 to **26**-21
  appearance, **26**-6
  causality, **26**-14 to **26**-21
    arbitrary causality, **26**-16, **26**-18 to **26**-19
    causal analysis, feedback on modeling decisions, **26**-16 to **26**-17
    causal constraints, **26**-16
    causal port properties, **26**-14
  elementary behaviors, **26**-6 to **26**-14
    environment, sources, boundary conditions, constraints, **26**-7 to **26**-8
    fixed causality of second kind, **26**-15, **26**-18
    fixed causality of the first kind, **26**-14 to **26**-15, **26**-17
    power continuous structure, **26**-8 resistor, **26**-8 to **26**-9
    preferred causality, **26**-15 to **26**-16, **26**-17 to **26**-18
    storage, **26**-6 to **26**-7
  generalized junction structure (GJS), *See also separate entry*
Bottom-up/top-down interaction, **2**-6 to **2**-8
  computational models implementing, **2**-8
Boundedness, **20**-9
Buckingham Pi theorem, **5**-5, **5**-7 to **5**-9
Bulirsch–Stoer method, **17**-17
Burke's theorem, **25**-12, **25**-14

C

Cash–Karp formulas, **17**-12
Causality-based classification, **3**-9
Causal loop diagrams, **33**-17 to **33**-19
Cellular automata (CA)
  characteristics, **21**-3
  dynamic systems modeling with, **21**-1 to **21**-16
    history, **21**-2 to **21**-3
    lattice gas cellular automata (LGCA) models of fluid dynamics, **21**-5 to **21**-16, *See also separate entry*
    one-dimensional CAS, **21**-3 to **21**-5
    boundary conditions, **21**-3
Chemical reactions, dynamics of, **17**-2
Classification based on determinism, **3**-9
Colored Petri nets (CPNs), **24**-10 to **24**-12
Commercial off-the-shelf (COTS) simulation packages
  commercial off-the-shelf discrete-event simulation packages (CSPs), **9**-2
    CSP-based distributed simulation, **9**-5 to **9**-6:
      current progress, **9**-6; problem of, **9**-5 to **9**-6
  modeling with, **9**-2 to **9**-3
Component-oriented simulation toolkit (COST), **35**-3 to **35**-7
  functor, **35**-5
  implementation, **35**-4
  inport and outport class, **35**-6
  motivation, from object to component, **35**-3 to **35**-5
  simulation time and port index, **35**-6 to **35**-7
  timer, **35**-7
Concurrency theory, **19**-2
Connectionist networks (CNs), **22**-2 to **22**-4
  basic approach, **22**-2 to **22**-3
  function approximation, **22**-3
  learning, **22**-3 to **22**-4
Conservation laws, **11**-16 to **11**-18
Consistent-mass formulation, **13**-16
Context-dependent computational temperature, **2**-6
Continuous systems, discrete-event simulation of, **11**-1 to **11**-22
  conservation laws, **11**-16 to **11**-18
  coupled ordinary differential equations, simulating, **11**-6 to **11**-8
  DEVS representation of discrete-event integrators, **11**-8 to **11**-13, *See also separate entry*
  heat equation, **11**-13 to **11**-16
  single ordinary differential equation, simulating, **11**-2 to **11**-5
  two-point integration schemes, **11**-19 to **11**-21
Continuous-time models, **3**-7
Coordinate-wise monotonicity, **18**-10 to **18**-13
Copycat program, **2**-12
Critical damping, **13**-17

D

Daisy world model, **27**-6 to **27**-9
  background, **27**-6
  equilibrium diagram for, **27**-7

Index I-3

feedback loop control of Daisy world
  temperatures, **27**-8
management flight simulator, **27**-9 to **27**-13
Decision making, models to support, **33**-2 to **33**-4
Defense Advanced Research Project Agency
  (DARPA), **16**-2
Defense simulation
  interoperability and composability in, **16**-1 to **16**-9,
    See also under Simulation interconnection
Differential-algebraic equations (DAE), **36**-19
Determinism, **3**-9
DEVS representation of discrete-event integrators,
    11-8 to **11**-13
  atomic models, **11**-8
  coupled models, **11**-8
  DEVS simulation algorithm, **11**-9
    of two coupled ordinary differential equations
      simulation of, **11**-13
Differential equations, *See also* Ordinary differential
  equations
Dimensional analysis, **5**-4
  basic method of, **5**-5 to **5**-7
    Buckingham Pi theorem, **5**-5, **5**-7 to **5**-9
Discontinuity sticking, **15**-18
Discontinuous systems, **36**-10 to **36**-13
Discrete dynamical systems, difference equations as,
    18-1 to **18**-17
  basic concepts, **18**-2 to **18**-4
  first-order difference equations, **18**-4 to **18**-8
    asymptotic stability, **18**-4 to **18**-6
    chaos, **18**-7 to **18**-8
    cycles and limit cycles, **18**-6 to **18**-7
  higher order difference equations, **18**-8 to **18**-17,
    See also separate entry
Discrete-event execution, **10**-7 to **10**-13
  ATM multiplexer model, **10**-8
    cell arrival event handler, **10**-9
    cell departure event handler, **10**-9
    state, **10**-9
  conservative parallel execution, **10**-10 to **10**-11
  execution method, **10**-7
  optimistic parallel execution, **10**-11 to **10**-13
    mixed-mode parallel execution, **10**-13
    reverse computation in, **10**-12
  parallelizing, **10**-9 to **10**-13
  sequential, **10**-8
Discrete-event modeling ontology (DeMO), **3**-2, **3**-7
    to **3**-8
  DeModel class hierarchy, **3**-18
  model component class hierarchy, **3**-16
  model concept, **3**-14
  model mechanism, **3**-14
  overview, **3**-14 to **3**-18
Discrete-event models, **3**-7 to **3**-8
Discrete-event simulation (DES), **9**-2, **10**-2

conceptual basis for, **3**-4 to **3**-6
of continuous systems, **11**-1 to **11**-22, *See also*
  under Continuous systems
in management science, **33**-7 to **33**-16, *See also*
  under Management science
in manufacturing systems, **34**-7 to **34**-8, *See also*
  under Manufacturing systems
overview of, **3**-12 to **3**-14
Discrete-event systems modeling
  DEVS formalism for, **6**-1 to **6**-12
    for discrete-event systems modeling, **6**-1 to **6**-12
      atomic DEVS model, **6**-4 to **6**-5: composition
        of, **6**-8 to **6**-9; state equation form of, **6**-7
      composed DEVS model, system analysis by,
        6-9 to **6**-10
      coupled DEVS model, **6**-5
      DES analysis with, **6**-7 to **6**-10
      ping-pong protocol, DEVS modeling, **6**-2, **6**-5
        to **6**-7
      RECEIVER, **6**-2
      SENDER: external views, **6**-2; internal views,
        6-2
      simulation of, **6**-10 to **6**-12
      methodology and environment, **6**-10 to **6**-12
      simulation speedup and simulators
        interoperation, **6**-12
      system-theoretic DES modeling, **6**-3
Discrete-state transition
  basic forms, **15**-11
    combinational logic, **15**-11
    sequential logic, **15**-11 to **15**-12
Discrete-time models, **3**-8
Distributed interactive simulation (DIS), **16**-2
Distributed modeling, **9**-1 to **9**-16, *See also under*
  Commercial off-the-shelf
  case study, **9**-13 to **9**-16
    bicycle assembly SOM, **9**-15
    bicycle factor model, **9**-14
    bicycle manufacturing, system FOM, **9**-15
    frame production line SOM, **9**-14
    illustrative protocol, **9**-13 to **9**-16
    wheel production line SOM, **9**-14
  definition, **9**-1
  distributed simulation, **9**-3 to **9**-5
  standards-based approach, **9**-7 to **9**-13
    emerging standards and the CSPI-PDG, **9**-7 to
      9-8
    entity transfer specification, **9**-9 to **9**-13:
      architecture, **9**-10; interaction hierarchy,
      9-11; OMT tables used for, **9**-12
    simulation interoperability standards
      organization, **9**-7
    type I interoperability reference model, **9**-8 to
      9-9
Distributed simulation, **9**-3 to **9**-5, **10**-2

Domains, *See also under individual entries*;
    Mesh-based simulation
  continuum domains, **12**-7 to **12**-9
  definitions, transformations, and interactions,
    12-7 to **12**-9
  discrete domains, **12**-9
  discretized geometric domains, **12**-13
  geometric domain, **12**-13
  interactions of domains, **12**-9
Domain-specific languages (DSL), **7**-2
  language definition formalism, **7**-3 to **7**-4
    abstract syntax of, **7**-3
    concrete syntax of, **7**-3
Domain-specific modeling (DSM), **7**-1 to **7**-18
  case studies, **7**-6 to **7**-11
    customized Petri net modeling language, **7**-6 to
      7-11, *See also under* Petri nets
  components of, **7**-2 to **7**-6, *See also*
    Domain-specific languages
    application areas of, **7**-5 to **7**-6
    domain-specific modeling environments
      (DSME), **7**-4; characteristics, **7**-4
    model generators, **7**-4 to **7**-5
  conference registration application
    design of, **7**-14
    model, **7**-15 to **7**-16
    S60 simulator, **7**-16
  modeling language, defining, **7**-12 to **7**-13
  supporting tools, **7**-17 to **7**-18
    metamodeling tools, retrospective of, **7**-17
    modern metamodeling tools, **7**-17 to **7**-18
Dymola symbolic processing, **36**-22
Dynamic context-dependent
    representation-building, **2**-4
Dynamic exchange language (DXL), **14**-13 to **14**-20
  between heterogeneous models in DXL, **14**-20
  between homogeneous models, **14**-18 to **14**-20
  concepts, **14**-13
  FBM-to-DXL translation, **14**-18
  multimodel syntax, **14**-14
  programming semantics, **14**-14
  semantics of, **14**-15 to **14**-17
    asynchronous input property, **14**-16 to **14**-17
    augmented DXL syntax, **14**-15
    DXL block notation, **14**-15
    information stream mechanism, **14**-15 to **14**-16
    notation, **14**-15
    synchronous input property, **14**-16
  syntax of, **14**-14
Dynamic system modeling,
  categorizing models, ways of, **1**-2
    analysis, **1**-2
    synthesis, **1**-2
    theory, **1**-2
  examples, **1**-2 to **1**-3; types of, **1**-2

finite state machine, **1**-2
functional block model, **1**-2
ordinary differential equation, **1**-2
Petri net, **1**-2
language, **1**-3 to **1**-5, *See also separate entry*
languages of, **1**-1 to **1**-11
taxonomic approaches, **1**-3

E

Ecological modeling and simulation
  from historical development to individual-based
    modeling, **29**-1 to **29**-16
    applications, **29**-12 to **29**-15
    as scientific instruments, **29**-7 to **29**-8
    determinism or probability?, **29**-5
    individual-based models, **29**-9 to **29**-12
    modeling techniques, **29**-5 to **29**-7
    organization levels and methodological choices,
      29-8 to **29**-9
Effective process times (EPT), **34**-8 to **34**-10
Elman network, **22**-5
Energy systems language, dynamic simulation with,
    28-1 to **28**-32
  reading an, **28**-4 to **28**-8
    diagram layout: and its connection to energy
      theory, **28**-4 to **28**-6; sources arrangement,
      28-5 to **28**-6; transformity principle of,
      28-5, *See also under* Transformity principle
  Forrester's systems dynamics approach,
    comparison with, **28**-29 to **28**-31
  Marsh sector model, **28**-27 to **28**-29, *See also
    separate entry*
  model constants, calibration of, **28**-21 to **28**-22
  simulation, preparation for, **22**-27: simulation
    software and energy systems language,
    28-22 to **28**-26
  stylistic diagramming features, **28**-7 to **28**-8:
    converging and diverging flow lines, **28**-8;
    nested symbols, **28**-8; sector boundaries,
    28-7 to **28**-8; vertical and horizontal stacks
    of symbols, **28**-8
  symbols within the system boundary, **28**-6 to
    28-7, *See also under* Symbols
  timescales and numerical integration, **28**-26 to
    28-27
  translating a diagram to dynamic equations,
    28-8 to **28**-21: basic equation forms
    indicated in, **28**-16 to **28**-21; equation
    naming convention, **28**-10 to **28**-15;
    feedback effect of production processes on
    the environmental variables involved,
    28-20; flow-limited source, **28**-18 to **28**-19;
    multiple simultaneous interactions in,
    28-19 to **28**-20; of Marsh Sector, **28**-11 to

28-15; passive inputs and environmental conditions, 28-21; two-way interactions, 28-20 to 28-21
Environmental systems
  Daisy world model, 27-6 to 27-9, *See also separate entry*
  system dynamics modeling of, 27-1 to 27-13
    flowers and sales models, comparison of, 27-4 to 27-6: feedback loop structure of flowers model, 27-5; feedback loop structure of the sales model, 27-5
  Stella diagram, 27-2
  Vensim diagram, 27-2
Euler–Cauchy integration, 33-16
Euler's theorem, 26-22
Event relationship graphs (ERGs)
  enrichments to, 23-7 to 23-10
    building large and complex models, 23-8 to 23-9
    parametric event relationship graphs, 23-7 to 23-8
  process interaction flows, 23-13
  experimenting with, 23-17 to 23-20
  graph analysis, 23-16 to 23-17
  modeling causality with, 23-1 to 23-20
    background and definitions, 23-2 to 23-7
    discrete-event systems and models, 23-2 to 23-3, 23-10 to 23-16: attributes, 23-2; factors, 23-2; mapping Petri nets into event relationship graphs, 23-11 to 23-13; parameters, 23-2; stochastic timed Petri net (STPN), 23-10 to 23-11
    discrete-event system simulations, 23-3
    graph modeling element, ER in, 23-3 to 23-6
    verbal event graphs, 23-6
  reading, 23-6 to 23-7
  simulation of, 23-16
  variations of, 23-9 to 23-10
Execution, model, 10-1 to 10-13
  discrete-event execution, 10-7 to 10-13, *See also under* Discrete-event execution
  elements of, 10-2 to 10-3
  executable timelines, 5
    physical time, 10-5
    simulation time, 10-5
    wallclock time, 10-5
  execution platforms, 10-3 to 10-4
  generating executables from models, 10-4 to 10-5
  pacing the execution, 10-5
  simulating large-scale models/scenarios, approaches to, 10-3
  systems and models, 10-1 to 10-2
  time-stepped execution, 10-5 to 10-7
    for parallel execution, 10-7
    for sequential execution, 10-6
    parallelizing, 10-6 to 10-7

F

4D face-centered hyper cube (FCHC), 21-10
Faraday's law of induction, 26-13
Finite elements, 13-1 to 13-21
  dynamics, 13-16 to 13-21
  finite element theory (FEM), 13-1 to 13-9
    shape functions, 13-4
    simple FEM theory, 13-1 to 13-4
    tapered extensional example, 13-4 to 13-6; axial loaded tapered problem, 13-4; shape function models for, 13-5; axial displacement result, 13-5; axial stress for, 13-6
    available elements, 13-8 to 13-9
    mapping errors, 13-7 to 13-8
    numerical integration, 13-6 to 13-7
    shape function accuracy, 13-6
  membrane elements, 13-9 to 13-12, *See also separate entry*
  multiple degree of freedom (MDOF) dynamic analysis, 13-19 to 13-21
  single degree of freedom (SDOF) dynamic analysis, 13-17 to 13-19
  solid elements, 13-15 to 13-16
    behavior and DOF, 13-15 to 13-16
Flow models, 34-16 to 34-18
Forrester's systems dynamics approach, 28-29 to 28-31
Formalism, 1-4
'FSM synthesis', 1-9
  FSM Moore machine semantics, 1-9
Function
  set-theoretic concepts, 1-9

G

Gaia hypothesis, 27-6
Gauss–Legendre quadrature method, 13-6 to 13-7
Generalized junction structure (GJS)
  transduction and interconnection, 26-9 to 26-14
    junctions, 26-10
    decomposition, 26-12
Generalized semi-Markov Processes (GSMP), 23-13
Generic runtime infrastructure for distributed simulation (GRIDS), 9-6
Geometric scaling, 5-11 to 5-13
Giles et al network, 22-5

H

Heat equation, 11-13 to 11-16
Higher order difference equations, 18-8 to 18-17
  asymptotic stability
    coordinate-wise monotonicity, 18-10 to 18-13
    weak contractions, 18-9 to 18-10

Higher order difference equations, (*Contd.*)
  persistent oscillations and chaos, **18**-13 to **18**-14
  semiconjugacy, **18**-14 to **18**-17
Human behavior representation (HBR), **8**-5
Human interaction in organizational systems, modeling, **31**-1 to **31**-13
  systems and human interaction, **31**-2 to **31**-3
    case study at Ford Motor Company, **31**-8 to **31**-12
    designed abstract systems, **31**-2
    designed physical systems, **31**-2
    human activity systems, **31**-2
    human-to-human interaction, modeling, **31**-5
    human-to-system interaction, modeling, **31**-4 to **31**-5
    knowledge-based improvement (KBI) methodology, **31**-5 to **31**-8, *See also separate entry*
    natural systems, **31**-2
    need for, **31**-3 to **31**-4
    research and practice, **31**-4 to **31**-5
Hybrid dynamic systems, modeling and execution, **15**-1 to **15**-24
  advanced topics in, **15**-17 to **15**-22
    mode changes, **15**-19 to **15**-22: reinitialization, **15**-19 to **15**-20; sequences of, **15**-20 to **15**-22
    zero-crossing detection, **15**-17 to **15**-18: difficulty in, **15**-17
  behavior classes of, **15**-4, **15**-9 to **15**-12
    continuous-time behavior, **15**-9 to **15**-10
    handling mode transitions, **15**-10 to **15**-12: event detection and location, **15**-10 to **15**-11; mode transition inferencing, **15**-11 to **15**-12
    operational structure, **15**-9
    reinitialization of state variables, **15**-12
  description, **15**-6
  design, **15**-7 to **15**-9
    explicit models, **15**-8 to **15**-9
    implicit models, **15**-7 to **15**-8
  driver control in, **15**-4
  geometric representation of, **15**-6
  implementation, **15**-12 to **15**-16
    classes of events, **15**-12
    classes of temporal behavior, **15**-12 to **15**-13
    event-driven execution, **15**-14
    time-driven execution, **15**-13 to **15**-14
    types, combining, **15**-14 to **15**-16
  need for, **15**-3 to **15**-5
  pathological behavior classes, **15**-22 to **15**-23
Hybrid process algebra, **19**-3
Hypertext markup language (HTML), **3**-2 to **3**-3

I

Individual-based models, **29**-9 to **29**-12
Integrative multimodeling, **14**-2 to **14**-4
  general multimodeling, **14**-2 to **14**-4
    homogeneous and heterogeneous multimodels, **14**-3 to **14**-4
    intralevel and interlevel couplings, **14**-2 to **14**-3
    purpose of, **14**-2
Irreflexivity, **20**-7 to **20**-8

J

Jackson network, **25**-14

K

Kadanoff–Swift model, **21**-6
Kingman's equation, **34**-5
Kirchoff theory, **13**-13 to **13**-14, **36**-5
Knowledge-based improvement (KBI) methodology, **31**-5 to **31**-8
  data collection through simulation (stage 2), **31**-7 to **31**-8
  determining the consequences of the decision making strategies (stage 4), **31**-8
  determining the decision makers' decision making strategies (stage 3), **31**-8
  seeking improvements (stage 5), **31**-8
  understanding the decision making process (stage 1), **31**-7

L

Languages
  natural language, **1**-4
  of dynamic system modeling, **1**-3 to **1**-5
    pragmatics, **1**-4, *See also separate entry*
    semantics, **1**-4
    syntax, **1**-4, *See also separate entry*
Lattice Boltzmann method (LBM), **21**-14 to **21**-16
  Lattice-BGK (L-BGK) method, **21**-15
Lattice gas cellular automata (LGCA) models
  of fluid dynamics, **21**-5 to **21**-16
    FHP model, **21**-9
    HPP model, **21**-7
  research activity on, **21**-10
    applications, **21**-14
    fluid dynamics and, **21**-11 to **21**-12
    lattice Boltzmann method (LBM), **21**-14 to **21**-16, *See also separate entry*
    simulating an, **21**-12 to **21**-14
Learning algorithms, **22**-3
Lefkovitch models, **29**-3
Legendre transforms, **26**-22 to **22**-25
  causality and, **26**-24 to **26**-25

Index I-7

in electrical circuits, **26**-24
in mechanics, **26**-24
in thermodynamics, **26**-23 to **26**-24
Leslie matrix, **29**-3
Level-rate diagrams, **33**-19 to **33**-21
Lindley's equation, **25**-7
Linearity, **20**-8 to **20**-9
Linearity, **5**-19 to **5**-20
  geometric scaling and, **5**-19 to **5**-20
Little's Law, **25**-7 to **25**-8, **34**-6
Lookahead, **10**-10
Lotka–Volterra model, **17**-2
Lumped elements, **5**-9 to **5**-11
Lumped-mass formulation, **13**-16

M

M/M/1 queueing system, **25**-9 to **25**-11
  average queue length, **25**-10
  average system time, **25**-10
  departure process of, **25**-12 to **25**-13
  stability condition for, **25**-10
  utilization and throughput, **25**-10
Management science, dynamic modeling in, **33**-1 to **33**-22
  conceptual modeling skills, **33**-5 to **33**-6
  discrete event simulation, **33**-7 to **33**-16
    activity cycle diagrams in conceptual modeling, **33**-10 to **33**-13
    application areas, **33**-8 to **33**-9: business process re-engineering (BPR), **33**-8; in business process improvement, **33**-8; in health care, **33**-8; in transport and physical logistics, **33**-9
    DES model implementation in computer software, **33**-13
    discrete event simulation using a VIMS, **33**-14 to **33**-16
    suitable problems, **33**-7 to **33**-8: complicated characteristic, **33**-7; discrete events characteristic, **33**-8; dynamic characteristic, **33**-7; individual entities characteristic, **33**-8; interactive characteristic, **33**-7; stochastic behavior characteristic, **33**-8
    terminology, **33**-9 to **33**-10: simulation activities, **33**-10; simulation clock, **33**-10; simulation entities, **33**-9; simulation events, **33**-10; simulation processes, **33**-10; simulation resources, **33**-9
  model validation, **33**-21 to **33**-22
  models to support decision making, **33**-2 to **33**-4
    dynamic systems modeling in, **33**-4
  project management skills, **33**-4 to **33**-5
    initial negotiation and project definition, **33**-5
    project completion, **33**-5
    project management and control, **33**-5
  system dynamics, **33**-16 to **33**-21
    qualitative system and dynamics, **33**-17 to **33**-19
    quantitative system dynamics, **33**-19 to **33**-21
    system structure and system behavior, **33**-16 to **33**-17
  technical skills, **33**-6 to **33**-7
Manufacturing systems, modeling and analysis of, **34**-1 to **34**-19
  basic quantities for, **34**-2
  control of, **34**-10 to **34**-12
  discrete-event models, **34**-7 to **34**-8
  effective process times, **34**-8 to **34**-10
  flow models, **34**-16 to **34**-18
    traffic flow model, **34**-17 to **34**-18
  hybrid model, **34**-15 to **34**-16
  preliminaries, **34**-2 to **34**-3
  standard fluid model and extensions, **34**-12 to **34**-16
    common fluid model, **34**-12 to **34**-14
    extended fluid model, **34**-14: approximation to, **34**-14 to **34**-15
  steady-state analysis, analytical models for, **34**-3 to **34**-7
    mass conservation (throughput), **34**-4
    queueing relations (Wip, Flow Time), **34**-5 to **34**-7
MarkovChains, **29**-3
Markovian queueing models/networks, simple, **25**-8 to **25**-17
  closed queueing networks, **25**-15 to **25**-17
  M/M/1 queueing system, **25**-9 to **25**-11, *See also separate entry*
  mean value analysis (MVA), **25**-16 to **25**-17
  non-Markovian queueing systems, **25**-17 to **25**-18
  open queueing networks, **25**-13 to **25**-15
  product form networks, **25**-17
Marsh sector model
  dynamic output of, **28**-27 to **28**-29
    model output analysis, **28**-28 to **28**-29
    model validation, **28**-29
Mathematical modeling
  basic elements, **5**-1 to **5**-20
    abstraction and scale, **5**-9 to **5**-17, *See also separate entry*
    conservation and balance principles, **5**-17 to **5**-19
    dimensional analysis, **5**-5 to **5**-7, *See also separate entry*
    dimensional consistency and analysis, **5**-3 to **5**-9
    dimensional homogeneity, **5**-4 to **5**-5
    linearity, role of, **5**-19 to **5**-20
  types of, **3**-6 to **3**-8
    classification based on state, **3**-6 to **3**-7

Mathematical modeling (*Contd.*)
    time-based classification, **3**-7 to **3**-8:
        continuous-time models, **3**-7;
        discrete-event models, **3**-7 to **3**-8;
        discrete-time models, **3**-8; static models,
        3-8
Mathematical optimization programs,
    in ERGs, **23**-13 to **23**-16
MATLAB®, **37**-2
Maxwell symmetry, **26**-25
mean value analysis (MVA), **25**-16 to **25**-17
Meaning, **1**-8
    connotation concept, **1**-8
    defining, possibilities for, **1**-8
    denotation concept, **1**-8
Membrane elements, **13**-9 to **13**-12
    2-D shape functions, **13**-9 to **13**-10
    flat plate and shell elements, **13**-12 to **13**-15
        generalized stress, **13**-14 to **13**-15
        Kirchoff theory, **13**-13
        Mindlin theory, **13**-13 to **13**-14
        plate theory, **13**-13 to **13**-14
    mesh correctness and convergence, **13**-11
    stress difference to indicate mesh accuracy, **13**-11
    element meshing, **13**-11 to **13**-12
    membrane theory, **13**-9
    shear locking, **13**-10 to **13**-11
Mesh-based simulation
    classification, **12**-8
    reverse classification, **12**-8
Military modeling, **32**-1 to **32**-13
    applications, **32**-1 to **32**-2
    dynamics, **32**-4 to **32**-8
        dynamic environment, **32**-8
        engagement, **32**-7
        exchange, **32**-6 to **32**-7
        movement, **32**-5
        perception, **32**-5 to **32**-6
        reasoning, **32**-7 to **32**-8
    military simulation systems, **32**-11 to **32**-12
    modeling approach, **32**-8 to **32**-11
        artificial intelligence, **32**-10 to **32**-11
        logical process, **32**-10
        Mathematic, **32**-9 to **32**-10
        Physics, **32**-8 to **32**-9
        stochastic processes, **32**-10
    representation, **32**-2 to **32**-4
        constructive model, **32**-3
        engineering, **32**-2 to **32**-3
        environment, **32**-4
        live, **32**-4
        virtual model, **32**-3
Mindlin theory, **13**-13 to **13**-14
Modal methods, **13**-19
Model execution, *See under* Execution, model

Modelica, multidomain modeling with, **36**-1 to **36**-25
    basics, **36**-3 to **36**-17
    component coupling, **36**-6 to **36**-10
    connector definitions in, **36**-9
    discontinuous systems, **36**-10 to **36**-13
    Modelica libraries, **36**-17 to **36**-19
        automatically constructed parameter dialog, **36**-18
    overview, **36**-1
    relation-triggered events, **36**-13 to **36**-14
    symbolic processing of, **36**-19 to **36**-25
        index reduction, **36**-21 to **36**-22
        size and complexity, reduction of, **36**-20 to **36**-21
        sorting and algebraic loops, **36**-20
    variable structure systems, **36**-14 to **36**-17
    vehicle components modeled in, **36**-2
Molecular dynamics, **12**-5 to **12**-6
Multimodeling, **14**-1 to **14**-27
    boiling water example, **14**-20 to **14**-26, *See also under* Boiling water
    dynamic exchange language (DXL), **14**-13 to **14**-20, *See also separate entry*
    integrative multimodeling, **14**-2 to **14**-4, *See also separate entry*
    multimodeling exchange language (MXL), **14**-11 to **14**-13, *See also separate entry*
    RUBE framework, **14**-4 to **14**-5 scene construction, **14**-5 to **14**-11
        blender interface, **14**-10 to **14**-11
        interaction model creation, **14**-8 to **14**-10
        ontology, **14**-5 to **14**-8: classes and relationships, **14**-6 to **14**-8; for the boiling water, **14**-7; properties and relationships, **14**-8
Multimodeling exchange language (MXL), **14**-11 to **14**-13
    concepts, **14**-11 to **14**-12
    multimodeling in, **14**-12 to **14**-13
Multiple degree of freedom (MDOF) dynamic analysis, **13**-19 to **13**-21
    normal mode shapes for, **13**-20
Multiport generalizations, **26**-21 to **26**-28
    coenergy and Legendre transforms, **26**-22 to **22**-25, *See also under* Legendre transforms
    loudspeaker example, **26**-25 to **26**-28
Multiscale simulation, multimodel hierarchy for, **12**-1 to **12**-16
    composite material system, design, **12**-2
    for drug delivery system, **12**-3
    functional and information hierarchies in, **12**-4 to **12**-10
        domain definitions, transformations, and interactions, **12**-7 to **12**-9, *See also under* Domains

mathematical Physics description
  transformations and interactions, 12-4 to 12-7: molecular dynamics (MD), 12-5 to 12-6; partial differential equations (PDE), 12-4 to 12-5; PDEs and MD, interactions between, 12-6 to 12-7
  physical parameter definitions, transformations, and interactions, 12-9 to 12-10
functional components to support, 12-10 to 12-14
  equation parameters, 12-12 to 12-13
  interactions between components, 12-11
  problem definition, 12-12
  scale-linking operators, 12-14
  tensor fields, 12-13 to 12-14
multimodel simulation procedures, 12-14 to 12-15
  adaptive atomistic/continuum adaptive multiscale simulation, 12-15
  automated adaptive mesh-based simulation, 12-14 to 12-15
quasicontinuum method in, 12-3

N

Navier–Stokes equations, 21-12
Network traffic, 15-4

O

Object Model Template (OMT), 9-3
  tables, 9-4
Odum's systems philosophy and methodology, 28-2
Ontology
  for modeling and simulation, 3-12
  for scientific domains, 3-11
Open queueing networks, 25-13 to 25-15
Open Source Physics library (OSP), 17-3
Ordinary differential equations, 17-1 to 17-20
  adaptive step, 17-11 to 17-12
    adapting the step, 17-11 to 17-12
    embedded Runge–Kutta formulas, 17-11
    implementation, 17-12 to 17-15: interpolation, 17-15; multistepping, 17-15
    interval-halving, 17-11
  chemical reactions, dynamics of, 17-2
  implementation techniques, 17-5, 17-8 to 17-11, *See also* Runge–Kutta methods; Taylor methods
  numerical solution, 17-3 to 17-5
  OSP library, 17-20
  performance and other methods, 17-15 to 17-19
    extrapolation methods, 17-17 to 17-18
    implicit algorithms and stiff equations, 17-16 to 17-17
    multistep methods, 17-17
    symplectic integration methods, 17-18 to 17-19
  planetary motion, 17-2 to 17-3

predator–prey population dynamics, 17-2
simple pendulum, 17-1
state events, 17-19 to 17-20

P

PackageModelica, 36-17
Parallel simulation, 10-2
Parametric event relationship graphs, 23-7 to 23-8
Partial differential equations, 12-4 to 12-5
Pending events list, 23-3
Pendulum model, 37-5
Persistent oscillations and chaos, 18-13 to 18-14
Petri nets
  analysis of, 24-7 to 24-10
    incidence matrix and state equation, 24-8
    invariant analysis, 24-8 to 24-10: minimal-support invariant, 24-9; support of an invariant, 24-9
    reachability analysis, 24-7
    simulation, 24-10
  colored Petri nets (CPNs), 24-10 to 24-12
  definition, 24-1
  for dynamic event-driven system modeling, 24-1 to 24-16
    modeling power, 24-4 to 24-5: concurrency, 24-4; conflict, 24-4; mutually exclusive, 24-4; priorities, 24-4; resource constraint, 24-4; sequential execution, 24-4; synchronization, 24-4
    transition firing, 24-3 to 24-4, *See also separate entry*
  mapping, into ERGs, 23-11 to 23-13
  marking in, 24-2
  Petri net modeling language, 7-6 to 7-11
    constraint, 7-8
    defining the modeling language, 7-7 to 7-8
    dining philosophers in, 7-8 to 7-9
    Petri net language, dining philosophers in, 7-8 to 7-9
  places, 1-8
  properties, 24-5 to 24-7
    liveness, 24-6 to 24-7
    reachability, 24-6
    safeness, 24-6
  timed Petri nets, 24-12 to 12-16, *See also under* Petri nets
  tokens, 1-8
'PhysicalEntity' 3-13
Ping-pong protocol, DEVS modeling, 6-2, 6-5 to 6-7
Planetary motion, 17-2 to 17-3
Plate theory, 13-13 to 13-14
Poisson arrival process, 25-9

Port-based modeling
  bond graphs, **26**-5 to **26**-21, *See also separate entry*
  dynamic models, **26**-3 to **26**-4
  multiport generalizations, **26**-21 to **26**-28, *See also separate entry*
  of engineering systems in terms of bond graphs, **26**-1 to **26**-28
    power conjugation, **26**-4
  structured systems, physical components and interaction, **26**-4 to **26**-5
    dynamic conjugation, **26**-4
  versus traditional modeling, **26**-1 to **26**-3
Pragmatics, **1**-4 to **1**-5, **1**-10 affecting modeling, **1**-10
Predator–prey population dynamics, **17**-2
Process algebra, **19**-1 to **19**-20
  algebraic reasoning and verification, **19**-11 to **19**-19, *See also separate entry*
  calculation, **19**-2
  definition, **19**-1 to **19**-2
  history, **19**-3
  hybrid process algebra, **19**-3
  χ process algebra, **19**-4 to **19**-11, *See also separate entry*
χ Process algebra
  syntax and informal semantics of, **19**-4 to **19**-11
    assembly line example, **19**-6 to **19**-7: iconic model of, **19**-6
    atomic statements, semantics of, **19**-9 to **19**-10: skip and multiassignment, **19**-9; delay predicate, **19**-9; delay statement, **19**-10
    compound statements, semantics of, **19**-10 to **19**-11: alternative composition, **19**-11; guard operator, **19**-10; loop and while statement, **19**-11; parallelism, **19**-11; sequential composition, **19**-10; variable and channel scope operator, **19**-11
    controlled tank, **19**-4 to **19**-6: simulation of, **19**-4
  semantic framework, **19**-8 to **19**-9
  statement syntax, **19**-7 to **19**-8
Propositional logic, **20**-1 to **20**-3

Q

Queueing system models, **25**-1 to **25**-18
  Little's Law, **25**-7 to **25**-8
  non-Markovian queueing systems, **25**-17 to **25**-18
  performance of, **25**-4 to **25**-6
  queueing system dynamics, **25**-6 to **25**-7
  simple Markovian queueing models, **25**-8 to **25**-11, *See also under* Markovian Queueing models
  specification of, **25**-2 to **25**-4
    issues to consider, **25**-3
    notation, **25**-3
    operating policies, **25**-3
    stochastic models, **25**-2
    structural parameters, **25**-3

R

Rapid prototyping, **37**-17 to **37**-18
Rational equations, **5**-3 to **5**-4
Reactive systems, **19**-2
Recursion scope operator, **19**-14
Reflexivity, **20**-7
Relation
  set-theoretic concepts, **1**-9
Representation-building
  in analogy-making, dynamics of, **2**-5 to **2**-6
Robotic arm model, **37**-7 to **37**-12
  computational model of, **37**-7
    CAD model of, **37**-8
    electromechanical model of the motor and coupling, **37**-8 to **37**-9
    generating the motor models from data, **37**-11 to **37**-12
    mechanical model of arm, **37**-8
    tuning the motor parameters with data, **37**-9 to **37**-11
RUBE framework, **14**-4 to **14**-5
Runge–Kutta methods, **17**-7 to **17**-8, **17**-10 to **17**-18
  embedded Runge–Kutta formulas, **17**-11
  generic Runge–Kutta table of coefficients, **17**-7
Runtime Infrastructure (RTI), **9**-3
  services, **9**-4

S

Scaling up issue, **2**-16
Semantic Web impact, on modeling and simulation, **3**-1 to **3**-20
  adding semantics to simulation models, **3**-10 to **3**-12
    ontology for modeling and simulation, **3**-12
    ontology for scientific domains, **3**-11
  causality-based classification, **3**-9
  classification based on determinism, **3**-9
  discrete-event simulation (DeSO), conceptual basis for, **3**-4 to **3**-6
    entity, **3**-4
    event, **3**-4
    force case, **3**-6
    force, **3**-5
    no force case, **3**-6
    space, **3**-4
    state, **3**-5
    time, **3**-4
  mathematical models, types of, **3**-6 to **3**-9, *See also under* Mathematical models
  relevant issues, **3**-2 to **3**-4
  semantic Web languages, **3**-4

Index

Semantic Web rule language (SWRL), 3-3
Semantics, 1-4 to 1-5, 1-7 to 1-10
  functional semantic mappings, 1-7
  in temporal logic, 20-5 to 20-6
  meaning in, 1-8, *See also under* Meaning
  semantic Web, 1-10
Semiconjugacy, 18-14 to 18-17
Semiotics, 1-4
Sensor network component-based simulator, 35-1 to 35-16
  component-oriented simulation toolkit (COST), 35-3 to 35-7, *See also separate entry*
  currently available components and simulation engines, 35-2 to 35-3
  need for, 35-1 to 35-3
  SENSE, features of, 35-2
  wireless sensor network simulation, 35-7 to 35-15, *See also separate entry*
Sequential simulation, 10-2
Shrinking gradients problem, 22-8
Sign
  formation, triangular relationship in, 1-4
Simple pendulum, 17-1
Simulation interconnection, theory and practice for, 16-1 to 16-9
  composition issue, 16-2
  configuration issue, 16-2
  integration issue, 16-2
  interoperation issue, 16-2
  simulation composability, 16-6 to 16-9
    composability and complexity, 16-6 to 16-7:
    composability, formalisms for, 16-8;
    composability, restricting the scope of, 16-8 to 16-9
  history of, 16-6
  simulation interoperability, 16-2 to 16-5
    aggregate level simulation protocol, 16-4 to 16-5
    distributed interactive simulation, 16-4
    entity-level simulation, 16-4
    high level architecture, 16-5
    simulator networking, 16-3
Simulation models
  adding semantics to, 3-10 to 3-12
Simulink® models for model-based design, simulation, 37-1 to 37-19
  case study, 37-3 to 37-4
  computational models for control design, 37-12 to 37-16
    designing a controller, 37-14 to 37-16
    designing controllers through modeling and simulation, 37-12 to 37-13
    linearizing models for control system design, 37-13 to 37-14
    optimization technique, 37-16
    SISO design plots, 37-15

computational models, obtaining, 37-4 to 37-7
  modeling from first principles, 37-5 to 37-7
  using data and simulation to obtain or tune models, 37-7
designing with simulation, 37-4
model-based design, testing with, 37-16 to 37-19
  rapid prototyping control design process, 37-17 to 37-18
  requirements-based testing through simulation, 37-17
  simulation with hardware and implemented designs, 37-17 to 37-18
pendulum model, 37-5
robotic arm model, 37-7 to 37-12, *See also separate entry*
robotic arm, 37-2
SimMechanics models, 37-6
Single degree of freedom (SDOF) dynamic analysis, 13-17 to 13-19
S-invariant concept, 24-8
'Slipnet,' a semantic network, 2-8 to 2-9
Spatio-temporal connectionist networks (SCNs), 22-1 to 22-9
  applications, 22-8 to 22-9
  basic approach, 22-4 to 22-5
  connectionist networks (CNs), 22-2 to 22-4, *See also separate entry*
  learning, 22-6 to 22-8
  mathematical and numerical modeling approaches to, 21-1
  popular SCNs, 22-5
    Elman network, 22-5
    Giles et al network, 22-5
    Williams and Zipser network, 22-5
  representational power, 22-6
  unfolding an, 22-7
Static models, 3-8
Stella diagram, 27-2
Stochastic models, 25-2
Stochastic timed Petri net (STPN), 23-10 to 23-11
Stokes–Dirac structure, 26-14
Symbols, in energy systems language
  categories, 28-6
    circles, 28-6
    composite symbols, 28-6
    price-controlled flow symbol, 28-7
    rectangle, 28-7
    storage symbol, 28-6
    two-way interaction symbol, 28-7
  within the system boundary, 28-6 to 28-7
Symplectic gyrator (**SGY**), 26-12
Symplectic integration methods, 17-18 to 17-19
Synchronous input property, 14-16
Synchronous languages, 36-10

Syntax, 1-4 to 1-7
  in temporal logic, 20-5 to 20-6
  same state machine, interpretation, 1-6
    BrettBaskovich, 1-6
    Christina Sirois, 1-6
    EmilyWelles, 1-6
    TimWinfree, 1-6
Systems engineering, 4-1 to 4-9
  definition, 4-6
  levels for, 4-2
  life cycles, 4-3, 4-5
  phases, 4-5
  process-oriented view of, 4-5
  systems engineering life cycles (SELC), 4-7
    analysis and assessment of the alternatives, 4-7
    formulation of the issue, 4-7
    interpretation and selection, 4-7
    technical direction and systems management, importance, 4-4 to 4-8
      inactive, 4-4
      interactive, 4-4
      proactive, 4-4
      reactive, 4-4

T

Tabletop program, 2-12 to 2-13
  semantic memory used in, 2-14
Taylor methods, 17-4, 17-5 to 17-6
Temporal logic, 20-1 to 20-14
  introducing, 20-3 to 20-5
  models of time, 20-6 to 20-10
    boundedness, 20-9
    combinations of properties, 20-10
    density and discreteness, 20-9 to 20-10
    irreflexivity, 20-7 to 20-8
    linearity, 20-8 to 20-9
    reflexivity, 20-7
    transitivity, 20-7
  propositional logic, 20-1 to 20-3
  syntax and semantics, 20-5 to 20-6

Timed Petri nets, 24-12 to 12-16
  deterministictimed Petri nets (DTPNs), 24-12, 24-13 to 24-14
  stochastic timed Petri nets (STPNs), 24-12, 24-14 to 24-16
  topological structure of, 24-12
Time-stepped simulation method, 10-2
T-invariant concept, 24-8
Traffic flow model, 34-17 to 34-18
Transformity principle
  of diagram layout, 28-5
    overall effect of transformity on, 28-6
    sources arrangement, 28-5 to 28-6
Transition firing, 24-3 to 24-4
  enabling rule, 24-3
  firing rule, 24-3
  self-loop transition, 24-3
  sink transition, 24-3
  source transition, 24-3
Transitivity, 20-7
Two-point integration schemes, 11-19 to 11-21

V

Validation process, 1-2
Vensim diagram, 27-2 to 27-4
Verbal event graphs, 23-6
Verification process, 1-2

W

Web Ontology Language (OWL), 30-1
Williams and Zipser network, 22-5
Wireless sensor network simulation, 35-7 to 35-15
  running the simulation, 35-14 to 35-15

Z

Zero-crossing detection, 15-17 to 15-18